ORGANIC PHOSPHORUS COMPOUNDS

Volume 7

G. M. KOSOLAPOFF
Auburn University

and

L. MAIER
Monsanto Research S.A.

A Wiley-Interscience Publication

JOHN WILEY & SONS, New York • London • Sydney • Toronto

Library of Congress Cataloging in Publication Data:

Kosolapoff, Gennady M
Organic phosphorus compounds.

1950 ed. published under title: Organophosphorus compounds.
Includes bibliographies.
1. Organophosphorus compounds. I. Maier, L., joint author. II. Title.

QD412.P1K55 1975 547′.07 72-1359
ISBN 0-471-50446-7 (v. 7)

Printed in the United States of America

10 9 8 7 6 5 4 3 2 1

EDITORS NOTE

With the present Volume 7, the series on "Organic Phosphorus Compounds" comes to a conclusion. As demonstrated in these volumes, the chemistry of organic phosphorus compounds is a rather complex branch of organic chemistry showing great variety and richness. It is equally interesting for its theoretical implications, for the diversity of its synthetic methods, and for the physiological and industrial significance of organic phosphorus compounds.

More than 100,000 compounds are listed in the seven volumes and references for their preparation and to all their physical properties known are given. Furthermore, all the methods for the preparation of the various types of compounds are discussed, in detail, in the 21 chapters. Also, the general chemistry and the general physical properties of the various types of compounds are reported.

As already indicated in the preface in Volume 1, the unusually large number of organic phosphorus compounds is explained by the ability of phosphorus to occur in several valence and oxidation states and to bind up to six organic groups directly or through heteroatoms. The following classification may be useful:

Classification of Phosphorus Compounds

Number of attached atoms	1	2	3	4	5	6
σ-Bond hybridization for observed symmetry	sp	$sp^2 + \pi$	sp^3	sp^3	sp^3d	sp^3d^2
Directional characteristics	Linear	Planar	Trigonal pyramidal	Tetrahedral	Trigonal bipyramidal	Octahydral or tetragonal bipyramidal
Geometry	P—X	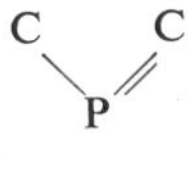	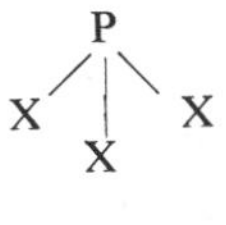	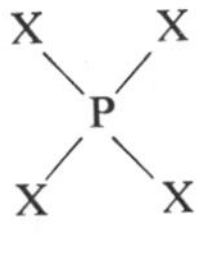	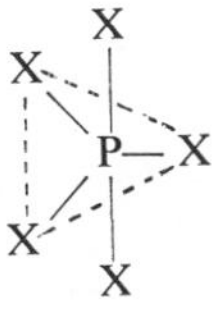	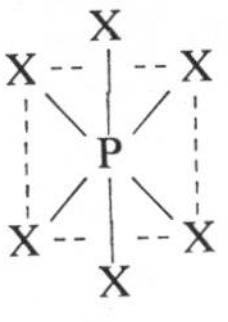
Number of known structures	~10	~50	~10,000	>1,000,000	~500	~100
Examples	PN, P_2 P≡CH	Phosphorins	PCl_3, PR_3 P $(OR)_3$	R_4P^+, PO_4^{3-} O=PCl_3	PCl_5, PPh_5	PF_6^-, $PhPF_5^-$
				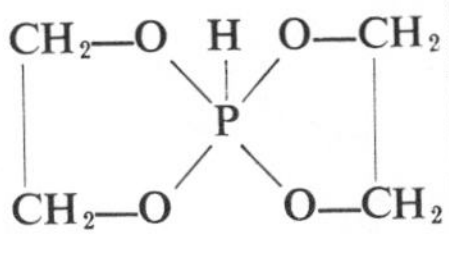	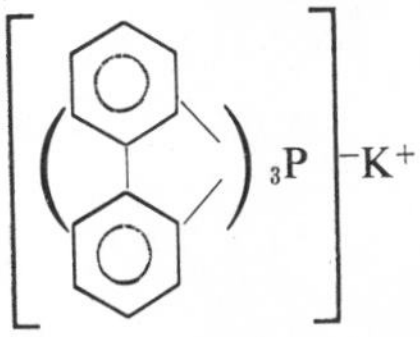	
Stability under normal conditions	Unstable	Stable	Stable	Most stable as a class	Very reactive	Very reactive
Discussed in Chapters	1	1	1, 2 3A, 3B 8, 10, 11 12, 13	3A, 3B, 4 5A, 6, 7, 9 10, 11, 12 13, 14, 15, 16 17, 18, 19	5B	5B

The chapter numbers of the present series, in which the various types of compound are discussed, are listed as the last entry in the above table.

Professor Karl Arnold August Michaelis and Professor Aleksandr Erminingel'dovich Arbuzov, whose pictures are shown below may be considered as the founders of organic phosphorus chemistry.

K. A. August Michaelis (1847–1916) Professor of chemistry in Rostock; founder of organic phosphorus chemistry. Courtesy Deutsches Museum, Munich.

Aleksandr E. Arbuzov (1877–1968) Professor of chemistry in Kazan; founder of the Russian school of phosphorus chemistry and codiscoverer of the Michaelis-Arbuzov reaction. Courtesy *Zhur. Obshch. Khim.*, 1969.

We dedicate these volumes on "Organic Phosphorus Compounds" to these two great scientists.

We believe that our coauthors have done an excellent job in exhaustively reviewing the literature and in making their chapters as concise and complete as possible. We again wish to thank them for their untiring and excellent cooperation.

A series of this size, although carefully checked, must suffer from some errors and omissions. We should be grateful if readers would notify us of any such faults that they may detect.

L. Maier
G. M. Kosolapoff

Zürich, Switzerland
Auburn, Alabama
March 1975

CONTENTS

VOLUME 2

VOLUME 3

VOLUME 4

VOLUME 6

Chapter 18. Phosphonic Acids and Derivatives

K.H. WORMS AND M. SCHMIDT-DUNKER

Henkel and Cie GmbH, Düsseldorf, Germany

Introduction

This chapter deals with phosphonic acids of general formula $RPO(OH)_2$ and their derivatives in the sequence given in the list of contents. The literature has been evaluated including 1971 as completely as possible, with a few selected references of 1972 and 1973 added.

In order to avoid misunderstandings, throughout this chapter H_3PO_3 is referred to as phosphorous acid and the derivatives have been named accordingly, irrespective whether they are derived from the ortho form $P(OH)_3$ or from the phosphonic form $HPO(OH)_2$. The same applies to the nomenclature used for phosphonous acids and their derivatives, which may be derived from the ortho form $RP(OH)_2$ or from the phosphinic form $RPO(OH)H$.

Phosphonic acid dihalides of formula $RPOX_2$ are not discussed here, since they were already listed in Chapter 9. However, monohalo acids, esters, amides, and other derivatives have been included.

A. METHODS OF PREPARATION

I. SYNTHESIS OF PHOSPHONIC ACIDS

1. HYDROLYSIS OF PHOSPHONIC ACID HALIDES. Phosphonic acid halides of general formula $RPOHal_2$ are hydrolyzed according to

$$RPOHal_2 + 2H_2O \rightarrow RPO(OH)_2 + 2HHal$$

The reaction may be accelerated by gentle heating. Phosphonic acids of low solubility, i.e., aryl phosphonic acids, crystallize upon cooling; short-chain aliphatic phosphonic acids are isolated by evaporation. Some phosphonic acids do not crystallize or only after several months standing in the air or a desiccator.[446, 1229, 1256, 1258, 1260, 1272, 2080]

Anhydrous phosphonic acids are easily obtained, when the reaction is carried out in a solvent, boiling under 100°C, with the theoretical amount of water.[1808] Short chain fatty acids are also suitable for the transformation of phosphonic acid halides to phosphonic acids:[1946]

$$RPOHal_2 + 2R'CO_2H \rightarrow RPO(OH)_2 + 2R'COHal$$

Phosphonic acid halides with several electronegative substituents on the α-C atom are more stable to hydrolysis. With CCl_3POCl_2, in cold $^n/2$ KOH, $CCl_3PO(OK)Cl$ is formed.[2088,2091] Upon heating, the P-C bond is split:

$$CCl_3PO(OK)Cl + 3KOH \rightarrow CCl_3H + K_3PO_4 + KCl + H_2O$$

Hydrolysis in acidic solution under pressure yields $CCl_3PO(OH)_2$, however.[2087]

Orthophosphonic acid tetrahalides, which are accessible by addition of Cl_2 to dihalophosphines, are also generally easily hydrolyzed to the corresponding phosphonic acids.[456, 729, 952, 1100, 1229, 1256, 1258, 1260, 1269, 1272, 1286, 1778, 2069, 2070]

$$RPHal_4 + 3H_2O \rightarrow RPO(OH)_2 + 4HHal$$

The adducts of PCl_5 and olefines ($R_2C{=}CHPCl_4 \cdot PCl_5$) yield upon hydrolysis phosphonic acids and monophosphoric acid.[238, 240, 241, 1019] The complexes of formula $[RP^{\oplus}Cl_3][Al^{\ominus}Cl_4]$, obtained by reaction of PCl_3 with alkyl halides and $AlCl_3$, are suitably reacted with SO_2 to yield phosphonic acid dichlorides (See chapter 9) before hydrolysis.

Certain carbinoles, i.e., triarylcarbinoles, react

with PCl_3 probably via unstable phosphorous acid ester chlorides to form triarylmethane phosphonyldichlorides, which are relatively stable to hydrolysis. Carboxylic acid hydroxymethylamides also react with PCl_3 to yield phosphonic acid dichlorides, which may be hydrolyzed without further purification to acylaminomethane phosphonic acid.[365, 506, 800, 1523]

$$RCONR'CH_2OH + PCl_3 \xrightarrow{-\ HCl} [RCONR'CH_2OPCl_2]$$
$$\rightarrow RCONR'CH_2POCl_2 \xrightarrow[-\ 2HCl]{+\ 2H_2O} RCONR'CH_2PO(OH)_2$$

Treatment with HCl conc. leads to dissociation of the acyl group and formation of aminomethane phosphonic acids.

$$RCONR'CH_2PO(OH)_2 \xrightarrow{H_2O} R'HNCH_2PO(OH)_2 + RCO_2H$$

The reaction of diazonium salts (tetrafluoroborates, fluorosilicates, or chlorozincates) with PCl_3 in presence of Cu(I) salts and subsequent hydrolysis also yields phosphonic acids.[479, 569]

$$[R\overset{\oplus}{N}\equiv N]BF_4^{\ominus} + PCl_3 \rightarrow [RN{=}N\text{-}\overset{\oplus}{P}Cl_3]BF_4^{\ominus} \xrightarrow{-N_2} [R\overset{\oplus}{P}Cl_3]\ BF_4^{\ominus}$$
$$\xrightarrow{3H_2O} RPO(OH)_2 + BF_3 + HF + 3HCl$$

The reactivity of the aromatic diazonium salts depends on the nature of substituents. Whereas most compounds react without difficulty, the 2-nitrophenyl and the o-tolyl compound did not react. With two diazonium groups on one aromatic ring, only one reacts according to the above equation, the other undergoes a Sandmeyer reaction.[187, 228, 290, 291, 479, 481, 570, 1018a, 1130, 1131a]

In the presence of O_2, PCl_3 (or $ROPCl_2$, R_2NPCl_2) reacts with alkanes to yield alkanephosphonyldihalides, (See Chapter 9), which may be hydrolyzed to the corresponding phosphonic acids.

$$RH + 2PCl_3 + O_2 \rightarrow RPOCl_2 + POCl_3 + HCl$$

Since the reaction of paraffins with PCl_3 yields mixtures of isomeric compounds, the reaction is of interest only for the preparation of cycloaliphatic phosphonic acids when defined compounds are required.[403, 657, 658]

In some cases, treatment of pyrazol derivatives with $POCl_3$ leads to substitution with the $-POCl_2$ group, which is subsequently hydrolyzed to the $-PO(OH)_2$ group:[659, 1288]

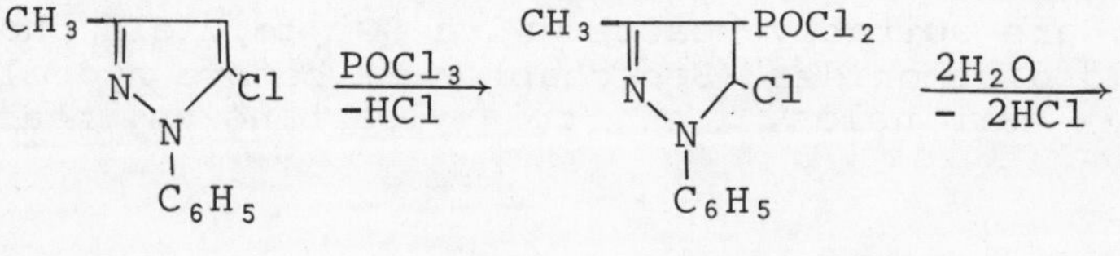

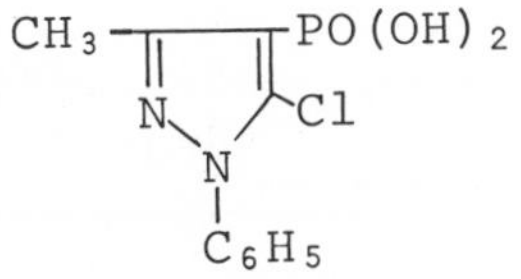

2. HYDROLYSIS OF ESTERS AND AMIDES. Hydrolysis of phosphonic acid esters (mono- or diesters) with mineral acids generally leads to phosphonic acids without cleavage of the P-C bond, even though the reaction normally requires severe conditions. The reaction rate increases in the order[1000, 1420]

$$HCl < HBr < HI$$

For example, α-aminomethane phosphonic acid esters, which are especially stable, are not hydrolyzed upon boiling with HCl conc., but do react with hot 48% HBr. With hydrohalide acids, occasionally alkyl halides are liberated instead of alcohols due to the acidolytic reaction

$$RPO(OR')_2 + 2HCl \rightarrow RPO(OH)_2 + 2R'Cl$$

Hydrolysis with bases leads to the corresponding phosphonic acid salts

$$RPO(OR')_2 + 2MOH \rightarrow RPO(OM)_2 + 2R'OH$$

With diluted aqueous bases under reflux, the $-PO(OR')_2$ group is often only partially hydrolyzed to the -PO(OR')(OH) group.[142, 276, 1418, 1419] In these cases, treatment with aqueous or alcoholic bases in an autoclave at temperatures of 130-150°C is required. If the compounds contain other groups sensitive to hydrolysis, these are generally also hydrolyzed.

Malonic acid derivatives with a $-PO(OR)_2$ group may be completely hydrolyzed with bases under pressure without loss of a carboxyl group.[2082] Upon acidic hydrolysis, one carboxyl group is eliminated.[1551]

In some cases, treatment with acids or bases leads to the cleavage of the P-C bond, especially with α-oxoalkane phosphonic acid esters. 2-Oxoalkane phosphonic acid esters are partially split, too; i.e., 2-oxopropane phosphonic acid esters yield acetone, H_3PO_4, and alcohol. On the other hand, 2-oxo-2-phenylethane phosphonic acid may be prepared from the corresponding alkyl esters by hydrolysis with 6N HCl under pressure.[1552]

In contrast to older reports, $CCl_3PO(OH)_2$ may be obtained through acidic hydrolysis of the esters.[231, 2088, 2091]

Phosphonoformic acid is obtained from the esters by hydrolysis with dilute aqueous base in low yields only, since the P-C bond is partially split.[1418]

Similar to phosphonic acid esters, the esteramides, mono- and diamides may also be hydrolyzed to the free acid. The reaction is mostly carried out with 12N HCl.[1339a, 1292]

3. HYDROLYSIS OF PHOSPHONIC ACID ANHYDRIDES, THIOPHOSPHONIC ACIDS, AND THEIR DERIVATIVES. Phosphonic acids are easily obtained through hydrolysis of phosphonic acid anhydrides $(RPO_2)_n$ or pyrophosphonic acids at elevated temperatures. However, since phosphonic acids are more easily accessible than their anhydrides, this reaction is of minor importance. Only a few aromatic hydrocarbons without polar groups react with P_4O_{10} directly to phosphonic acid anhydrides which upon hydrolysis yield the corresponding phosphonic acids.[583, 683, 1088]

The reaction of pyrophosphonic acids with alcohols or amines leads to 1:1 mixtures of phosphonic acids with phosphonic acid monoesters or monoamides, respectively.[379, 381]

$$RPO(OH)\text{-}O\text{-}PO(OH)R + R'OH \rightarrow RPO(OH)_2 + RPO(OR')(OH)$$

$$+ R_2'NH \rightarrow RPO(OH)_2 + RPO(NR_2')(OH)$$

Thiophosphonic acid derivatives easily yield phosphonic acids upon hydrolysis, since the intermediate monothiophosphonic acids are split to phosphonic acids and H_2S.[916, 1670] Since dithiophosphonic acid anhydrides are readily accessible from the reaction of P_4S_{10} with olefins or aromatic hydrocarbons, this reaction, in contrast to the hydrolysis of phosphonic acid anhydrides, is of considerable preparative interest.[583, 683, 1088]

4. HYDROGENOLYTIC CLEAVAGE OF BENZYL AND PHENYL ESTERS. Catalytic reduction of benzyl and phenyl esters with Pd/H_2 leads to the corresponding phosphonic acids. This method is preferable over hydrolysis, when the compound contains hydrolyzable groups that are to be retained.[204, 688] For example, by hydrogenation carbamidomethane phosphonic acid is obtained from its benzyl ester, whereas hydrolysis leads to phosphonoacetic acid:

$$RPO(OCH_2C_6H_5)_2 \xrightarrow{Pd/H_2} RPO(OH)_2 + 2CH_3C_6H_5$$

5. PYROLYSIS OF ESTERS. Phosphonic acid alkyl esters (with the exception of methyl esters) are cleaved upon heating into olefins and phosphonic acids.

$$Cl_2C\ [PO(OPr\text{-}i)_2]_2 \xrightarrow{180^0C}$$

$$Cl_2C\ [PO(OH)_2]_2 + 4\ CH_2{=}CHCH_3$$

The necessary temperatures depend on the nature of the

alkyl group. Ethyl esters require 250-280°C, i-propyl esters 120-180°C. Isobutyl esters and especially t-butyl esters are pyrolyzed at even lower temperatures. Since the reaction is catalyzed by traces of acids, the temperature may be reduced considerably after the reaction is started. In addition to olefins, small amounts of alcohols are formed in the reaction. Whereas hydrolysis of pure esters normally yields pure phosphonic acids, pyrolysis leads to formation of small amounts of side products, requiring purification by crystallization.[116, 1383, 2082]

6. OXIDATION OF PHOSPHORUS COMPOUNDS. Since the P-C-bond is normally resistent to oxidizing agents, compounds with a P-C bond in a lower oxidation state may be oxidized to phosphonic acids. Suitable starting compounds are primary phosphines, cyclopolyphosphines, and phosphonous acids. Useful oxidizing agents are HNO_3, halogens, H_2O_2 and $HgCl_2$.[212, 784, 792, 981, 1179, 1260, 1953]

Cyclopolyphosphines may also be oxidized by oxygen to phosphonic acid anhydrides, which are subsequently hydrolyzed in the usual manner.[1147]

The products resulting from catalytic oxidation of white phosphorus with oxygen in the presence of olefins may be transformed by hydrolysis and oxidation with HNO_3 to unsaturated phosphonic acids or hydroxyalkane phosphonic acids, respectively.[2060, 2078]

$$\text{(cyclohexane)} + 1/2\ P_4 + 2O_2 \rightarrow [\text{intermediate}] \xrightarrow{HNO_3}$$

$$\text{(cyclohexene)}PO(OH)_2 + H_3PO_4$$

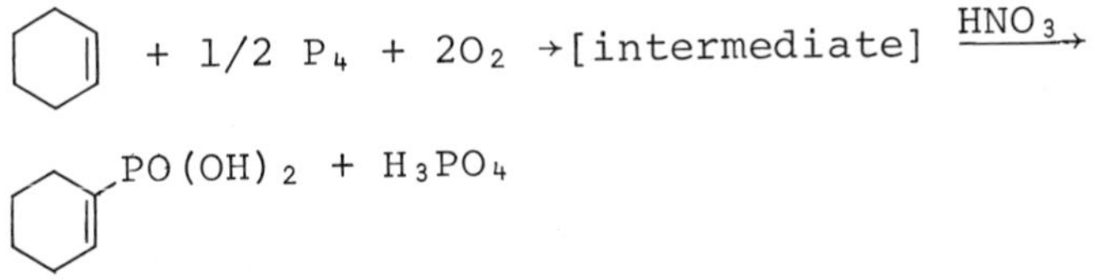

7. DISPROPORTIONATION OF PHOSPHONOUS ACIDS. Phosphonous acids slowly disproportionate at ambient temperature, fast at 100°C, to phosphonic acids and primary phosphines.[1277]

$$3RPO(OH)H \rightarrow RPH_2 + 2RPO(OH)_2$$

The reaction is of little preparative importance.

8. ADDITION OF H_3PO_3 AND PCl_3 TO DOUBLE BONDS. Whereas the addition of dialkyl phosphites to terminal double bonds in the presence of radical catalysts is carried out in many cases with good yields, the addition of H_3PO_3 often fails. Polymerization and double-bond isomerization of the olefin occur at much faster rates.[686, 696, 1677]

H_3PO_3 adds to activated double bonds without a catalyst, but the yields are again disappointing.[1644] Since, on the other hand, the base-catalyzed addition of dialkyl phosphites to activated double bonds results in yields of up to 90%, this synthesis with subsequent hydrolysis is generally preferred.

α,β-Unsaturated ketones react in glacial acetic acid with PCl_3 or PBr_3 to products which, upon hydrolysis, often produce γ-oxo-alkylphosphonic acids in good yields. In contrast to the above-mentioned reactions, in this case a different mechanism must be assumed.[409, 412, 493, 906]

$$3R_2C{=}CHCOR' + PCl_3 \rightarrow [P(OCR'{=}CHCR_2Cl)_3 \rightarrow$$

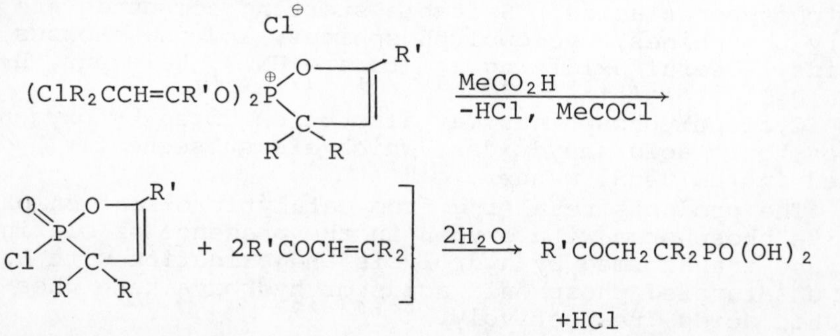

The proposed mechanism is supported by the cyclic structures that could be isolated from acetic anhydride solution.

9. REACTION OF CARBONYL COMPOUNDS WITH H_3PO_3 OR PCl_3. Phosphorous acid reacts with carbonyl compounds like aldehydes at elevated temperatures to α-hydroxyalkane phosphonic acids:[1181, 1182, 1186]

$$RCHO + HPO(OH)_2 \rightarrow RCH(OH)PO(OH)_2$$

Due to low yields, this reaction is of minor importance.

The reaction of PCl_3 with carbonyl compounds, however, is of preparative importance and has been investigated intensively. Fossek described the reaction of 3 moles aldehyde with PCl_3, which allows the synthesis of hydroxyalkane phosphonic acids.[553, 554, 555] The following mechanism has been proposed:[925, 927]

$$3RCHO + PCl_3 \rightarrow [RCHClO]_3P \rightarrow RCHClPO[OCHClR]_2$$

$$\xrightarrow{3H_2O} RCH(OH)PO(OH)_2 + 2RCHO + 3HCl$$

Paraformaldehyde reacts with PCl_3 at temperatures of about

$80^{0}C$ to give high yields of hydroxymethane phosphonic acid.[206, 1446]

Later on it was found by Conant, that aldehydes and PCl_3 may also be caused to react in a molar ratio of 1:1, if glacial acetic acid is present. Acetyl chloride is found in the reaction mixture. The reaction product upon hydrolysis yields hydroxyalkane phosphonic acids. With alcohols the monoalkyl esters of the corresponding acid are formed, but with dry HCl one obtains α-chloroalkane phosphonic acids.[410, 413-415, 417]

The following reaction mechanism has been proposed:

$$RCHO + PCl_3 + 2MeCO_2H \rightarrow [RC \overset{\overset{\displaystyle H}{|}}{\underset{\underset{\displaystyle O}{}}{-}}\overset{\overset{\displaystyle O}{\|}}{P} - OH] + HCl + 2MeCOCl$$

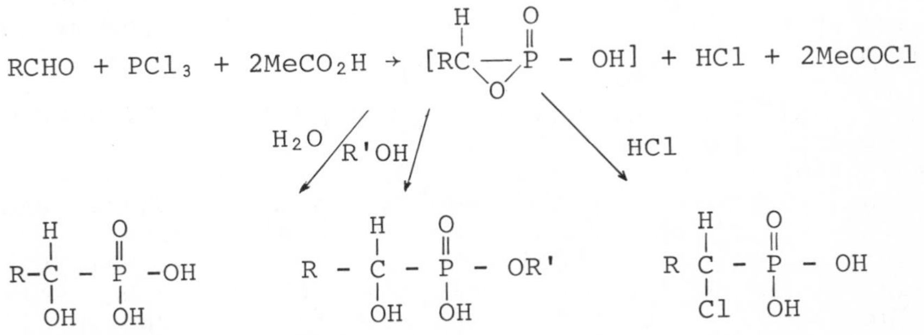

In the reaction of aldehydes with PCl_3 under pressure, α-chloroalkane phosphonic acid chlorides are formed.

Ketones also interact with PCl_3 similarly to aldehydes, but are less reactive. Benzoic acid instead of acetic acid is, therefore, used as the reaction medium. Due to partial loss of H_2O, mixtures of α-hydroxyalkane phosphonic acids with α,β-unsaturated acids are frequently obtained.

10. CONDENSATION OF H_3PO_3 WITH AMINES AND FORMALDEHYDE. Primary and secondary amines, as well as NH_3, react with formaldehyde and H_3PO_3 in mineral acid solution in the manner of a Mannich condensation to form aminomethylene phosphonic acids. In this reaction, products are generally obtained, in which all N-H bonds of the starting amine have been phosphonomethylated. The reaction is, therefore, well suited for preparation of amino-tris(methylene phosphonic acid) from NH_3, not however for amino-bis- or -mono (methylene phosphonic acid). Polyamines react in a similar manner; i.e., with ethylenediamine, ethylenediamine tetrakis (methylene phosphonic acid) is obtained.[1328]

The reaction of mono- or diethanolamine with H_3PO_3 and formaldehyde is more complex. As intermediate products, ethanolamine-bis(methylenephosphonic acid) or diethanolaminemethylene phosphonic acid are formed, respectively. Under the influence of acids, these undergo condensation and the internal esters are formed.

$$HOCH_2CH_2N[CH_2PO(OH)_2]_2 \underset{}{\overset{H^\oplus}{\rightleftharpoons}} (HO)_2OPCH_2\overset{\displaystyle O}{\underset{\displaystyle CH_2CH_2O}{\overset{\|}{\underset{|}{N-CH_2-P-OH}}}} + H_2O$$

$$(HOCH_2CH_2)_2NCH_2PO(OH)_2 \underset{}{\overset{H^\oplus}{\rightleftharpoons}} HOCH_2CH_2\overset{\displaystyle O}{\underset{\displaystyle CH_2CH_2O}{\overset{\|}{\underset{|}{N-CH_2-P-OH}}}} + H_2O$$

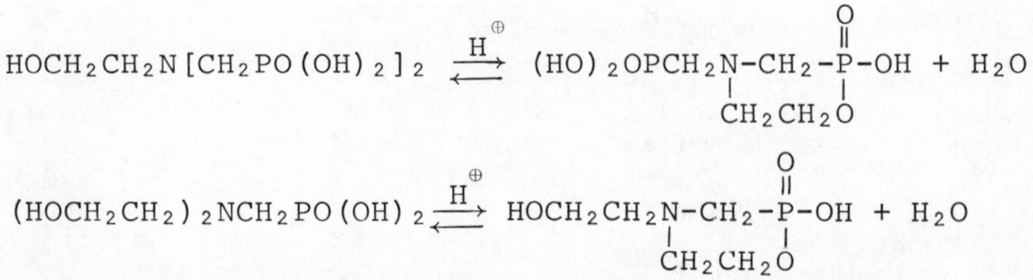

By precipitation of the cyclic esters with suitable solvents, the equilibrium is shifted to the right and the internal esters isolated from the mixture. With strong
bases above 100°C, they may be transformed to the salts of
the open-chain form. [2083]

11. DIRECT SYNTHESIS OF HYDROXYALKANE-1,1-DIPHOSPHONIC
ACIDS. The reaction of acetylating compounds with H_3PO_3 had
been described in old publications, but the isolation of
defined products from the complex reaction mixtures had
failed. If, however, one submits these mixtures to a steam
distillation until the distillate is no longer acidic,
hydroxyethane-1,1-diphosphonic acid (HEDP) may be isolated
in up to 97% yield. [629] In analogy to the synthesis of esters of this acid, the following mechanism is most probable:

$$CH_3COX + HPO(OH)_2 \xrightarrow[- HX]{} CH_3COPO(OH)_2 \xrightarrow{H_3PO_3}$$
$$CH_3C(OH)[PO(OH)_2]_2$$
$$[X = halide, CH_3COO^-]$$

The acetylating agents lead to further condensation
reactions, and in the complex reaction mixture acetylation
products of HEDP as well as an HEDP condensated dimer are
found. The latter can be isolated as a crystalline compound. The structure of the dimer has only recently been
elucidated.[265, 311a, 1503a, 1677d]
Similar to HEDP, other hydroxyalkane-1,1-diphosphonic
acids may be prepared. Several syntheses have been developed:
1. Acylchlorides + H_3PO_3.
2. Carboxylic acid anhydrides + H_3PO_3.
3. Carboxylic acids + defined amounts of H_2O + PCl_3.[262]
4. Carboxylic acids + P_4O_6. [1677a]
5. Carboxylic acids + H_3PO_3/P_2O_5 or acetylphosphates.[2082]

In all these reactions, primarily complex mixtures are obtained, which by subsequent steam distillation are converted
to hydroxyalkane-1,1-diphosphonic acids. A slightly different course of reaction is observed with benzoyl chloride
and H_3PO_3, which results in a mixture of phenylchlorometh-
ane diphosphonic acid and phenylhydroxymethane diphosphonic
acid. [275] The formed product may be isolated by crystalli-

zation. The most simple preparation of the hydroxyacid, however, is the reaction of phenylaminomethane diphosphonic acid with HNO_2 mentioned under Section I.13.

12. DIRECT SYNTHESIS OF AMINOALKANE-1,1-DIPHOSPHONIC ACIDS. For the direct synthesis of aminoalkane-1,1-diphosphonic acids a number of reactions are known, which probably all proceed via the following intermediate

$$\left[\begin{array}{c} R^2 \\ \diagdown \\ \overset{\oplus}{N} = C \diagup^{R'}_{\diagdown Y} \\ \diagup \\ R^3 \end{array} \right] X^{\ominus} \xrightarrow[-HX,\ HY]{2\ H_3PO_3} \begin{array}{c} R^2 \quad\quad PO(OH)_2 \\ \diagdown | \\ N - C - R' \\ \diagup | \\ R^3 \quad\quad PO(OH)_2 \end{array}$$

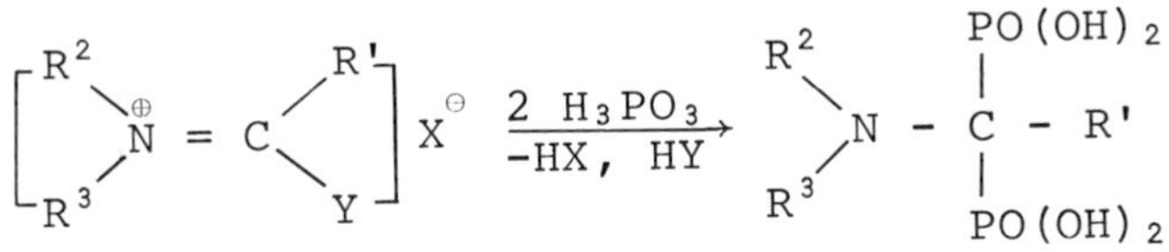

[X = Hal, Y = Hal or OH, R', R^2,R^3 = H, alkyl, aryl]

The following principal procedures have been described:
1. Nitril + $PHal_3$ and hydrolyzing agents. [1094]
2. N-unsubstituted carboxylic acid amide dihalogenides + H_3PO_3.
3. N,N-disubstituted carboxylic acid amide dihalogenides + H_3PO_3.
4. Carboxylic acid amide hydrohalogenides + H_3PO_3.
5. Carboxylic acid amides and $PHal_3$ or PCl_3/H_3PO_3 and subsequent hydrolysis. [1527]

To (1). Most nitriles are susceptible to this reaction. Benzonitrile reacts with special ease. PBr_3 results in much better yields than PCl_3.

To (2). The carboxylic acid amide dihalogenides do not have to be isolated as such. It is sufficient to conduct a stream of HCl or HBr gas into the molten mixture of nitrile and H_3PO_3.

To (3). The amiddihalogenides are produced by the reaction of the corresponding amide with PCl_5, oxalylchloride, etc. This method is especially suited for the preparation of aminodiphosphonic acids, which are N-di-substituted with short-chain alkyl groups.

To (4). Isolation of the carboxylic acid amide hydrohalides is not required. It suffices to react a mixture of carboxylic acid amide with hydrogen halides at 130-160°C. Suitable amides are those of formula

$R-CONH_2$; R $\geq$ C_2 alkyl, not H, C_1, aryl

To (5). Unsubstituted as well as mono- and disubstituted amides are suitable for this reaction. Good yields are obtained with dimethylformamide. The reaction of $HCONH_2$ with excess phosphorus trihalide yields a primary product, probably possessing the structure,

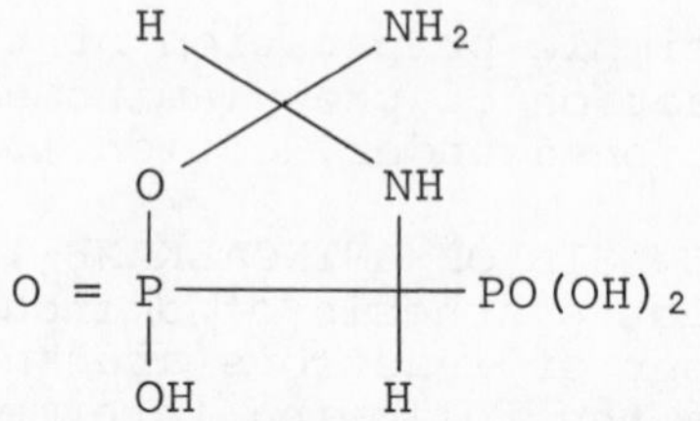

which upon alkaline hydrolysis is cleaved into aminomethane diphosphonic acid, ammonia, and formate.

13. VARIATIONS OF THE HYDROCARBON REST OF PHOSPHONIC ACIDS. The P-C-bond is extremely resistant to hydrolysis and the $-PO(OH)_2$ group is also quite inert. This explains why many reactions, i.e., dehalogenation, dehydrohalogenation, nitration, halogenation, dehydration, polymerization, hydrogenation, reduction, and oxidation, may be carried out with the organic part of phosphonic acids. To be mentioned is the reactivity of the α-Cl-atom in phosphonic acids. Thus chloromethane phosphonic acid reacts in alkaline aqueous solution with amines to aminomethane phosphonic acids. NH_2 groups may be changed with HNO_2 to OH groups. Phenylaminomethane diphosphonic acid, easily obtainable from benzonitrile and PBr_3, may thus be transformed to the corresponding hydroxy compound.[274]

Acylations, alkylations, diazotations, and azo-coupling reactions may also be carried out without cleavage of the P-C-bond. The synthesis of many phosphonic acids is simpler by this indirect method than by direct phosphonylation.

II. SYNTHESIS OF PHOSPHONIC ACID MONOHALIDES

1. HYDROLYSIS OF PHOSPHONIC ACID DIHALIDES. Phosphonic acid dihalides or orthophosphonic acid tetrahalides, with the exception of the fluorides, are so reactive towards hydrolyzing agents, that one rarely succeeds in isolating the stage of phosphonic acid monohalides. The preparation of phosphonic acid monohalides can be achieved, if the α-C-atom is substituted by strongly electronegative groups. Thus, $Cl_3CPO(OH)Cl$ is obtained by hydrolysis of the di- or tetrachloride. Also, triphenylmethane phosphonic acid monochloride is prepared by reaction of the dichloride with alcoholic bases in the form of the salt.[759, 2087, 2089, 2091]

Similar to fluorophosphates, the P-F bond of monofluorophosphonates is again more inert than a P-Cl bond. Thus, monofluorophosphonates are obtainable through hydrolysis (in the presence of weak bases, for example aniline).[870, 871]

2. PYROLYSIS OF PHOSPHONIC ACID ESTER HALIDES. Pyrolysis of ester halides generally does not succeed due to the tendency of P-Cl bonds to react with P-OH groups resulting in

formation of P-O-P bonds. Ester fluorides, however, may be pyrolyzed successfully. Thus, isopropylester fluorides yield the corresponding monofluorides.[786]

3. HALOGENATION OF PHOSPHONIC ACIDS. Due to the tendency of P-Cl bonds to undergo condensation reactions, careful chlorination succeeds only with the more inert phosphonic acid chlorides. The preparation of $Ph_3CPO(OH)Cl$ by treating the acid with PCl_5 under mild conditions has been described.[760]

III. SYNTHESIS OF PHOSPHONIC ACID MONOESTERS

1. ALCOHOLYSIS OF PHOSPHONIC ACID HALIDES. The alcoholysis of phosphonic acid dihalides or orthophosphonic acid tetrahalides with constant removal of the formed hydrogen halide from the equilibrium leads predominantly to formation of diesters. Without removal of hydrogen halide, especially when working under pressure, monoesters are produced.[416, 1157]

$$RPOCl_2 + 2HOR' \rightarrow RPO(OR')(OH) + R'Cl + HCl$$

$$RPCl_4 + 3HOR' \rightarrow RPO(OR')(OH) + 2R'Cl + 2HCl$$

The action of alcoholic bases on triaryl phosphonyldichlorides also leads to formation of monoesters.[299, 300, 760]

2. HYDROLYSIS OF PHOSPHONIC ACID ESTER HALIDES. Phosphonic acid ester halides are hydrolyzed by water or aqueous bases, if necessary at elevated temperatures, to the corresponding phosphonic acid monoesters or their salts, respectively.[657, 865, 1257, 1284]

The cleavage of cyclic enol ester chlorides of γ-oxoalkane phosphonic acids is a route to the preparation of γ-oxoalkane phosphonic acid monoesters. The reaction leads either to the desired product or to cyclic diesters, which are easily cleaved by acids or bases to the monoesters.[412, 493]

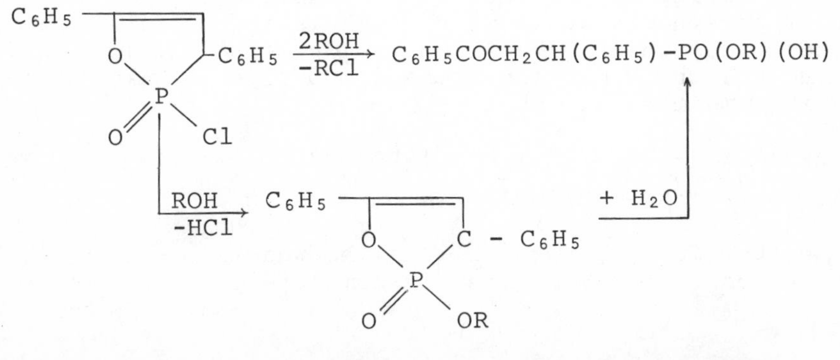

3. HYDROLYSIS OF PHOSPHONIC ACID DIESTERS. The hydrolysis of phosphonic acid diesters is a stepwise process. However, the difference in the reaction rates is generally not large enough to allow the isolation of the monoester intermediate. This applies especially when the hydrolysis is carried out with strong acids. Only in a few cases, for example in the acid-catalyzed hydrolysis of phosphonoacetic acid trialkyl esters, may the monoester be isolated.[1418]

In base-catalyzed hydrolysis, it is generally easier to select suitable reaction conditions for preparation of the monoesters. Aqueous or alcoholic bases, ammonia, amines, alkaliphenolates, or even neutral salts may be used.[412, 1484, 2017] Hydrolysis of alkylaryl diesters generally yields the monoalkyl esters.[592] Upon heating of tertiary amines with phosphonic acid diesters, ammonia salts of monoesters are formed

$$R''_3N + RPO(OR')_2 \rightarrow [RPO(OR')O^{\ominus}]\ [\overset{\oplus}{N}R'R''_3]$$

The cleavage of phosphonic acid diesters formed in the "Michaelis-Becker" reaction by neutral salts leads to appreciable losses in yield due to formation of monoesters.[13]

A further important method for the preparation of phosphonic acid monoesters is the hydrogenolysis of diesters. Phosphonic acid alkylbenzyl esters, for example, are easily reduced to the monoalkyl esters.[68]

4. CLEAVAGE OF P-O-P-BONDS BY ALCOHOLS. P-O-P bonds, for example of phosphonic acid anhydrides, are easily cleaved by alcohols.[372, 373]

$$(RPO_2)_n + nR'OH \rightarrow nRPO(OH)(OR')$$

Remarkably, monoethanolamine yields phosphonic acid mono-(2-aminoethyl) ester exclusively without formation of phosphonic acid amide.[375]

Pyrophosphonic acids react with alcohols at elevated temperatures to yield mixtures of phosphonic acid and the corresponding monoester, requiring complicated purification steps.[379, 381] These complications can be avoided by use of condensating agents such as carbodiimide. For the reaction of phosphonic acids with alcohols in the presence of carbodiimide, which gives high yields of phosphonic acid monoesters, the following mechanism implying pyrophosphonic acid intermediates is most probable:[326]

$$2RPO(OH)_2 + (C_6H_{11}N{=})_2C \rightarrow [RPO(OH)]_2O + (C_6H_{11}NH)_2CO$$

$$RPO(OH)_2 + RPO(OH)(OR') \xleftarrow{\quad R'OH \quad} [RPO(OH)]_2O$$

The phosphonic acid returns into the reaction cycle until consumption to the monoester is complete.

5. REACTION OF CARBONYL COMPOUNDS WITH $(RO)PCl_2$. In analogy to reactions of carbonyl compounds with PCl_3 (Section I. 9), alkylphosphite dichlorides yield α-hydroxyalkane phosphonic acid monoesters. The reaction of α,β-unsaturated ketones, however, leads to formation of γ-oxoalkane phosphonic acid monoesters.[418]

6. REACTION OF ACYLCHLORIDES AND DIALKYL PHOSPHITES Acylchlorides react with dialkyl phosphites to yield phosphonic acid monoesters and alkylchlorides.[262a]

$$RCOCl + HPO(OR')_2 \rightarrow RCOPO(OR')(OH) + R'Cl$$

By contrast, Cl_3CCOCl reacts with cleavage of HCl according to :[1934]

$$Cl_3CCOCl + HPO(OR)_2 \rightarrow Cl_3CCOPO(OR)_2 + HCl$$

7. OXIDATION OF PHOSPHONOUS ACID MONOESTERS. The reaction of phosphonous acid monoesters with suitable oxidizing agents, i.e., N_2O_4, yields phosphonic acid monoesters. The reaction is of minor importance.[1496]

IV. SYNTHESIS OF PHOSPHONIC ACID ESTER HALIDES AND PSEUDOHALIDES

1. HALOGENATION OF PHOSPHONIC ACID DIESTERS. Phosphonic acid diesters react with excess halogenating agents, such as PCl_5, to the phosphonyldihalides. (cf. Chapter 9) Under controlled conditions, i.e., molar ratio of 1:1 and inert solvents, phosphonic acid ester halides may be obtained. Other halogenating agents such as $COCl_2$ and $[COCl]_2$ may also be used. $SOCl_2$ is less well suited.[405, 1454, 1703, 1711, 1764]

$$RPO(OR')_2 + PCl_5 \rightarrow RPO(OR')Cl + R'Cl + POCl_3$$

Phosphonic acid alkylester chlorides decompose to phosphonic acid anhydrides and alkyl halides at higher temperatures. Only low-boiling compounds may, therefore, be purified by distillation.

It should be mentioned that monothiophosphonic acid O,O-diesters also yield phosphonic acid ester halides with $COCl_2$.[337]

$$COCl_2 + RPS(OR')_2 \rightarrow RPO(OR')Cl + R'Cl + COS$$

2. REACTION OF PHOSPHONIC ACID HALIDES AND ALCOHOLS. Phosphonic acid dihalides (chlorides and bromides) react with equimolar amounts of alcohols or phenols in the pres-

ence of tertiary amines to phosphonic acid ester halides.[677, 1254, 1336, 2003, 2005]

$$RPOCl_2 + R'OH + R''_3N \rightarrow RPO(OR')Cl + [R''_3NH]\,Cl$$

In the reaction of phenols, acid-binding agents are not required.

Orthophosphonic acid tetrachlorides as well as their aluminum halide complexes can also serve as starting materials.[787]

$$[R\overset{\oplus}{P}Cl_3]\,[AlCl_4]^{\ominus} + 2R'OH \rightarrow RPO(OR')Cl + R'Cl + 2HCl + AlCl_3$$

Phosphonic acid difluorides or -chloride fluorides react to yield phosphonic acid alkylester fluorides, which are dangerous poisons. These compounds are more stable than the other ester halides. This is also shown by the prevalent formation of the ester fluorides from mixtures of the dichloride and the difluoride.
Phosphonic acid ester fluorides may also be prepared from ester chlorides and alkalifluorides and also in a one-step reaction from phosphonic acid dihalides according to:[191, 871, 1982, 2098a]

$$RPOCl_2 + R'OH + 2MF \rightarrow RPO(OR')F + 2MCl + HF$$

The reaction of orthophosphonic acid monoester trihalides with acetic acid is of little preparative importance.

$$RP(OR')Cl_3 + CH_3CO_2H \rightarrow RPO(OR')Cl + CH_3COCl + HCl$$

3. ALKYLATION OF PHOSPHONIC ACID MONOHALIDES. The alkylation of phosphonic acid monohalides is of low importance due to the inaccessibility of this class of compounds. Only the reaction of triarylmethane phosphonic acid chloride with dialkylsulfate has been described.[760]

$$(C_6H_5)_3CPO(OH)Cl + (CH_3O)_2SO_2 \rightarrow (C_6H_5)_3CPO(OCH_3)Cl + CH_3O(OH)SO_2$$

4. HALOGENATION OF PHOSPHONOUS ACID MONOESTERS. Treatment of phosphonous acid monoesters with Cl_2, SO_2Cl_2 or CCl_3SCl leads to the formation of phosphonic acid alkylester chlorides.[68, 1704a]

$$RPO(OR')H + Cl_2 \rightarrow RPO(OR')Cl + HCl$$

The reaction of phosphonous acid diesters with CCl_3SCl has also been described. Analogously, with BrCN, phos-

The preparation of alkyl esters— with the exception of cyclic glycol esters— normally requires the presence of acid-binding agents, such as alkali, tertiary amines, or pyridine. Alternatively, the reaction may be carried out with alkali alcoholates. This procedure is preferred in the reaction of phosphonic acid ester halides. In order to avoid cleavage by neutral salts, these are separated before distillation of the product. The preparation of enol esters can be achieved by two routes, i.e., reaction of phosphonic acid halides with aldehydes in the presence of tertiary amines [614] or with ClHg-enol salts in the presence of pyridine: [1136]

$$RPOCl_2 + 2CH_3CHO + R_3'N \rightarrow RPO(OCH=CH_2)_2 + 2[R_3'\overset{\oplus}{N}H] \overset{\ominus}{Cl}$$

$$RPO(OR')Cl + ClHgOC(CH_3)=CH_2 \rightarrow RPO(OR')[OC(CH_3)=CH_2]$$
$$+ HgCl_2$$

However, the unsaturated esters may polymerize, thus resulting in low product yields.

The synthesis of β-chloroalkyl esters may be achieved by the reaction of phosphonic acid halides with epoxides in the presence of catalysts. [2023]

$$RPOCl_2 + 2\overline{CH_2CH_2O} \rightarrow RPO(OCH_2CH_2Cl)_2$$

3. FROM PHOSPHONIC ACID ANHYDRIDES. Phosphonic acid anhydrides react with epoxides with the formation of cyclic esters. [1464]

$$(RPO_2)_n + n\overline{OCH_2CHR} \rightarrow RP\overset{\overset{O}{\parallel}}{\underset{O-CH_2}{<}}\overset{O-CHR}{\underset{}{|}}$$

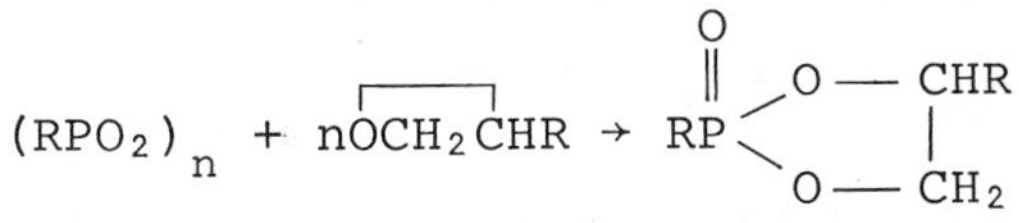

4. OXIDATION OF PHOSPHONOUS ACID ESTERS. The oxidation of phosphonous acid diesters with N_2O_4, MnO_2, or oxygen is of minor preparative importance. [614, 898, 1748]

$$2RP(OR')_2 + O_2 \rightarrow 2RPO(OR')_2$$

5. THE MICHAELIS-ARBUZOV REACTION. This reaction, which was found by Michaelis and Kaehne [1283] and later developed by Arbuzov, is one of the most important and best-investigated reactions for the preparation of phosphonic acid esters. Other compounds, for example phosphinic acid esters, may be prepared the same way.

Originally, it was observed that alkyl halides react with trialkyl phosphites with formation of phosphonic acid esters. Later it was found that other alkylating agents and especially acylating agents, i.e., acyl halides, may also be used in the reaction. According to numerous investigations, the following mechanism applies:

$$RHal + P(OR')_3 \rightarrow [R\overset{\oplus}{P}(OR')_3]\ Hal^{\ominus} \rightarrow RPO(OR')_2 + R'Hal$$

In the case of alkyl esters, the phosphonium intermediate cannot normally be isolated. In the reaction of carbonium salts such as triphenylmethyltetrafluoroborate, the intermediate may be obtained.[476, 477] The stable intermediate may subsequently be cleaved by alkali alcoholates or $NaHCO_3$ to the phosphonic acid diester. Triaryl phosphites and alkyl halides also form phosphonium salt-like compounds. By reaction with alcohols or bases or thermally > 200^0C, they are transformed to phosphonic acid esters.[117] Generally, the reactivity of organic halides follows the sequences:

$$RCOHal > RCH_2Hal > RR'CHHal \gg RR'R''CHal$$

$$RI > RBr > RCl$$

Aryl halides, vinyl halides and organic fluorides normally react only in the presence of a catalyst. Tertiary alkyl halides, upon heating with trialkyl phosphites, do not react until they decompose yielding olefins. Notable exceptions are triarylmethyl halides, which react smoothly to the corresponding phosphonic acid esters.[104, 119, 122]

In all the described reactions, complications arise from the self-alkylating ability of trialkyl phosphites. For instance, phosphonic acid esters are also formed upon heating of trialkyl phosphites, especially in the presence of catalytic amounts of alcohols, alkyl halides, aluminum alkyls, inorganic iodides, or thermally from the Cu(I) adducts. For the preparation of phosphonic acid esters with the same organic group bound to P and O, it suffices to treat a trialkyl phosphite with catalytic amounts of alkyl halide. If, however, the organic groups shall be different, the alkyl halide has to be applied in at least stoichiometric amounts, better in excess. Also, the alkyl halide formed in the reaction should constantly be removed from the reaction mixture. Tris(iso-alkylphosphites) are often used, since the secondary alkyl halides formed are of relatively low reactivity. Tris(isopropyl phosphite) is the most commonly used phosphite.

Tris(2-chloroethyl) phosphite, obtainable from PCl_3 and ethylene oxide, is easily converted to 2-chloroethylphosphonic acid diester upon heating.[920]

Some reactions, for example of iodobenzene with trialkyl phosphites, do not proceed under normal conditions, but may be photocatalyzed.[1523a] Certain metal halides also serve as catalysts. Thus, the reaction of $P(OEt)_3$ or

$P(OPr\text{-}i)_3$ with a number of α,β-unsaturated alkylhalides or aromatic halides is accelerated by Ni halides or Cu powder. An even better catalyst is $PdCl_2$.[1195, 1987-89]

In contrast to literature reports, the reaction of trans-dibromo-ethane with $P(OMe)_3$, catalyzed by $NiCl_2$, may become explosive.[1526a]

α-Allyl halides react with phosphites as expected, internal allyl halides may partially undergo allyl shift.[1646]

$$R'CHClCH{=}CH_2 \xrightarrow{P(OR)_3} R'CH{=}CHCH_2PO(OR)_2 + RCl$$

Propargyl halides react with $P(OEt)_3$ at 90^0C according to:[1563]

$$HC{\equiv}CCH_2Br + P(OEt)_3 \rightarrow HC{\equiv}CCH_2PO(OEt)_2$$
$$+ H_2C{=}C{=}CHPO(OEt)_2$$

With bases, $H_2C{=}C{=}CPO(OEt)_2$ is further isomerized to $CH_3C{\equiv}CPO(OEt)_2$.

Trialkyl phosphites react readily with acylchlorides to form α-ketophosphonic acid esters. By contrast with $COCl_2$ or $(COCl)_2$, $ClPO(OR)_2$ is obtained. Maleic acid dichloride only yields tarry products.

α-Halovinylketones and -esters react similarly to acetylenic halides to yield the corresponding phosphonic acid esters.[471] Saturated α-haloketones and -aldehydes, however, do not react by the normal Michaelis-Arbuzov route. Rather, by means of a Perkow reaction, dialkylvinylphosphates are formed.[1459] Phosphonous acid diesters also undergo the Perkow reaction, thus opening another route to phosphonic acid vinyl esters.

In addition to alkyl- and acyl halides, other alkylating or acylating agents also react with trialkyl phosphites according to the Michaelis-Arbuzov scheme. Lactames[421, 422] and lactones,[422, 1035] phenylisocyanatedichlorides,[1055a] sulfonic acid esters,[1366] sultones,[738] salts of Mannich bases,[1364] carboxylic acid anhydrides,[935] and imidchlorides[1036] may be used in the reaction. Chloroformic acid esters and trialkyl phosphites yield phosphonoformic acid esters.[112, 1418] Also, with carbamide- and thiocarbamide acid chlorides, phosphonoformamide- and phosphonothioformamide esters are obtained, respectively.[1173, 1723]

Of interest is also the reaction of chloroanil, which reacts with trialkyl phosphite esters of <u>secondary</u> alcohols according to :[1721]

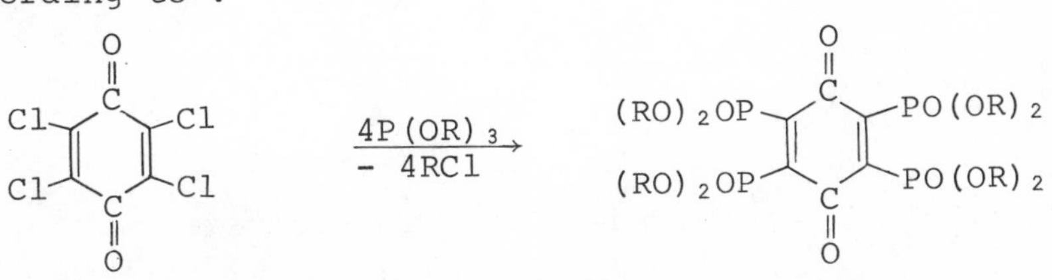

The tetraphosphonic acid ester may be reduced with Pd/H_2 to the corresponding hydroquinone derivative.

Halomalonic acid esters, haloacetic acid esters, and halogenated β-disulfones do not react with $P(OR)_3$ to phosphonic acid esters. While dihalomethanes, i.e., CH_2Br_2, give high yields of methane diphosphonic acid ester, only one chlorine atom of CCl_4 reacts to form $Cl_3CPO(OR)_2$. This reaction is accelerated by UV-irradiation or peroxides and probably proceeds via radical formation.[117] Trihalomethanes do not react according to the Michaelis-Arbuzov scheme. Interestingly, phosgene immoniumchlorides and $P(OR)_3$ form tris(alkyl phosphonic acid)esters:[1055b]

$$[R_2\overset{\oplus}{N}{=}CCl_2]\overset{\ominus}{Cl} + 3P(OR)_3 \rightarrow R_2N{-}C[PO(OR)_2]_3$$

Earlier claims regarding the synthesis of methane triphosphonic acid ester derivatives could not be verified.[715b,c]

Nearly all aliphatic phosphites are susceptible to the Michaelis-Arbuzov reaction. Only phosphites carrying electronegative groups such as $-OCH_2CN$, $-OCH_2CCl_3$ do not react in the desired manner. In the reaction with CH_3I, the reactivity of mixed trialkyl phosphites decreases with R in the following sequence:[1076]

$$Me > Et > i\text{-}Pr$$

Cyclic phosphites of the structures

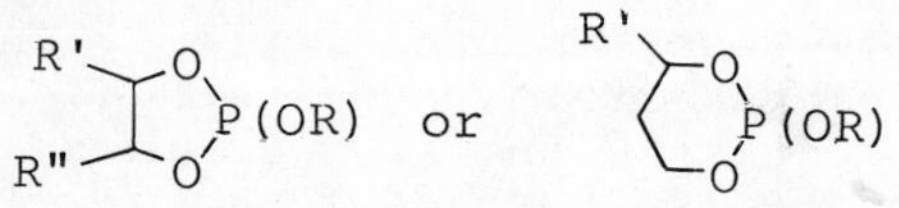

generally react with exchange of R.[122]
Compounds of structure

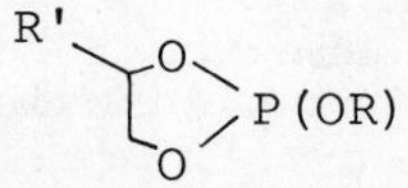

however, undergo ring opening according to

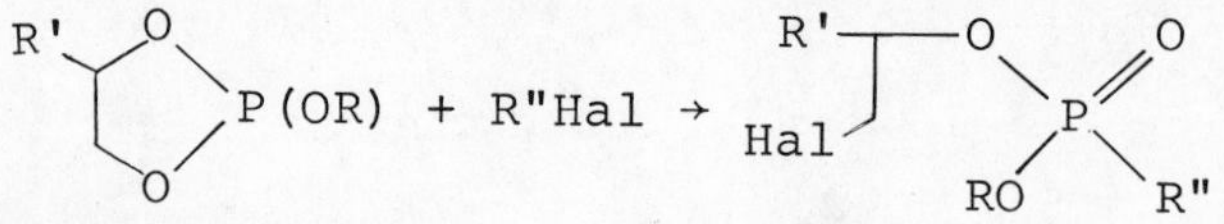

Upon heating, a new ring is formed by cleavage of RHal to yield[121, 122]

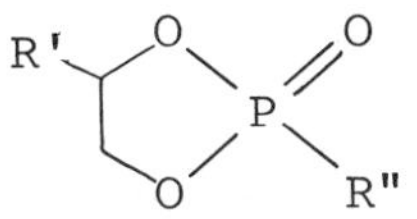

Tris(t-butylphosphite), in contrast to earlier reports, also yields phosphonic acid esters in some cases. However, due to side reactions, the yields are quite low.[754a]

6. THE MICHAELIS-BECKER REACTION. This reaction of dialkyl phosphite salts with alkyl halides was discovered by Michaelis and Becker[1278] and later investigated in detail by other researchers, notably by Nylén. Sodium salts of diethyl- and dibutylphosphite are preferably used, and instead of alkyl halides, other alkylating and acylating agents, as in the Michaelis-Arbuzov reaction, may be used. Alkali salts of dialkyl phosphites may be substituted by mixtures of dialkyl phosphite and tertiary amine.

Alkyl halides (except fluorides) react smoothly, splitting off alkali halides. With secondary alkyl halides it is difficult to obtain uniform products.[446] Alicyclic halides and phosphite salts yield the expected phosphonic acid esters, whereas the reaction with trialkyl phosphites often fails. Triarylmethane chlorides also react smoothly, other tertiary halides, however, are quite unreactive and eventually mixtures of products are formed.

The reactions of unsaturated alkyl halides are more complicated, since in some cases they are accompanied by addition of dialkyl phosphites to double bonds.[1871]

$$CH_2{=}CHCH_2Br + NaPO(OR)_2 \xrightarrow{-2NaBr} CH_2{=}CHCH_2PO(OR)_2$$

$$\xrightarrow{+\ HPO(OR)_2} (RO)_2OPCH(CH_3)CH_2PO(OR)_2$$

Halogen atoms in α-position to double bonds react similarly. With bromostyrene, $(RO)_2OPCH(C_6H_5)CH_2PO(OR)_2$ is obtained. Occasionally, double-bond isomerizations are observed, for example with allyl halides or in the reaction of bromocyclohexene-2 with $NaPO(OBu)_2$ yielding Δ^1-cyclohexene-phosphonic acid ester.[522]

While aromatic halides are quite inert, some heteroaromatic compounds, for example 2-chloroquinoline, do react with $NaPO(OR)_2$. α-Halocarboxylic acid esters also react smoothly. Aliphatic α,ω-dihaloalkanes yield the corresponding α,ω-diphosphonic acid esters, but with 2,3-dibromopropionitrile, only the α-Br atom reacts according to the Michaelis-Becker scheme. Simultaneously, HBr is split off.[142]

$$BrCH_2CHBrCN + 2NaPO(OR)_2 \rightarrow (RO)_2OPCH{=}CHCN + HPO(OR)_2 + 2\ NaBr$$

CHF_2Cl gives $CHF_2PO(OR)_2$ in smooth reaction, but the

yields of methylenediphosphonic acid ester obtained from CH_2J_2 are low due to side reactions. Starting with $ClCH_2PO(OR)_2$, much better yields are achieved.[1872]

The reaction of acyl halides and phosphite salts normally does not stop at the α-oxophosphonic acid ester stage These undergo further reaction according to:

$$R'COPO(OR)_2 + NaPO(OR)_2 \rightarrow R'C(ONa)[PO(OR_2]_2$$

$$\xrightarrow[-NaHal]{R'COHal} R'C[-OCOR'][PO(OR)_2]_2$$

Similar end products are supposedly obtained in the reaction of acid anhydrides with $NaPO(OR)_2$. $ClCO_2R$ and $NaPO(OR)_2$ yield phosphonoformic acid, if the reaction is carried out in aprotic solvents.[1418] Ethylene oxide yield 2-hydroxyethane phosphonic acid ester in an interesting reaction.[369] Carboxylic acid esters react according to:

$$R'CO_2R'' + NaPO(OR)_2 \rightarrow R''PO(OR)_2 + R'CO_2Na$$

Treatment of β- or γ-lactones or lactames with $NaPO(OR)_2$ leads to complex reaction mixtures, which are reported to contain small amounts of hydroxyalkane-1,1-diphosphonic aci esters, among other products.

Sulfonic acid esters react smoothly with $NaPO(OR)_2$, as they do with $P(OR)_3$. Interesting also is the reaction between $HPO(OR)_2$ and $NaPO(OR)_2$ in refluxing toluene, giving good yields of $RPO(OR)_2$.[1455]

To be mentioned is further the reaction of $COCl_2$ and $NaPO(OR)_2$, yielding hydroxymethane diphosphonic acid ester. The ester could not be purified by distillation, however.[1677b]

7. ADDITION OF PHOSPHITES TO C=C AND C≡C MULTIPLE BONDS. The addition of dialkyl and trialkyl phosphites to C-C multiple bonds without catalysts proceeds only at elevated temperatures with low rates. Nonactivated double bonds react only under radical conditions. Highest reactivity is found with terminal double bonds. Internal double bonds, for example of oleic acid, oleic alcohol, or cyclic olefins react at much lower rates. As a competing reaction, polymerization is often incurred.[1192, 1387, 1677, 1954]

Addition to polarized double bonds is difficult to achieve thermally, but proceeds smoothly with radical catalysis. With acetic acid vinylester, 2-acetoxyethane phosphonic acid ester is obtained.[427] The alkali-catalyzed addition, proceeding in the manner of a Michael addition, is often preferred. In the presence of alkali alcoholate, vinyl acetate and phosphite react to form 1-acetoxyethane phosphonic acid ester. With base catalysis, the $-PO(OR)_2$

group adds in β-position to the activating group. α,β-Unsaturated ketones yield γ-oxoalkane phosphonic acid esters, and with α,β-unsaturated carboxylic acids or nitriles, the corresponding 2-phosphonic acid esters are obtained.[1562]

It should be mentioned here that C,H-acidic compounds, for example phosphonoacetic acid esters or methane-diphosphonic acid esters may also add to activated double bonds (cf. Section V.14).[765, 1644]

$$H_2C\,[PO(OR)_2]_2 + R'O_2CCH{=}CHCO_2R' \rightarrow$$
$$[(RO)_2OP]_2CHCH(CO_2R')CH_2CO_2R'$$

While dialkyl phosphites add to the C=C double bond of α,β-unsaturated ketones exclusively, with α,β- unsaturated aldehydes only the C=O bond is attacked. With p-quinone, however, phosphoric acid dialkyl(4-hydroxyphenyl) ester is obtained.

Activated alkines may add 2 moles of dialkyl phosphite. With acetylenedicarboxylic acid ester, for example, 1,2-dicarbethoxy-ethane-1,2-bis(phosphonic acid ester) is obtained.[1544, 1548, 1670] Inamines, however, add only 1 mole of $HPO(OR)_2$ to yield enamine-1-phosphonic acid esters.[1809a]

In addition to carbonyl groups, sufficient activation of olefinic double bonds for addition of dialkyl phosphites also results from neighboring phosphono, nitro, or sulfonyl groups.

Trialkyl phosphites also add to activated double bonds and may behave as alkylating agents at the same time. Thus, acrylic acid and $P(OR)_3$ form β-phosphonopropionic acid trialkyl esters. Contrary to dialkyl phosphites, addition to α,β-unsaturated aldehydes takes place at the C=C double bond.[936a,b]

$$CH_2{=}CHCHO + P(OR)_3 \rightarrow (RO)_2OPCH_2CH{=}CHOR$$

The resulting vinyl ether is readily hydrolyzed to the phosphonoaldehyde.

With vinyl ethers carrying an activating group at the β-C atom, addition of the $-PO(OR)_2$ group is followed by elimination of an alkoxy group, resulting in the formation of α,β-unsaturated phosphonic acid esters.[1046a]

In the presence of catalysts such as BF_3, H_2SO_4 or pyridine, ketene also adds dialkyl phosphites. The α-oxophosphonic acid esters initially formed tend to add a second mole of ketene:[426]

$$H_2C{=}C{=}O + HPO(OR)_2 \rightarrow [H_3CCOPO(OR)_2 \rightleftarrows$$
$$H_2C{=}C(OH)PO(OR)_2] \xrightarrow{H_2C=C=O} H_3C\overset{\overset{\displaystyle O}{\|}}{C}{-}O{-}\overset{\overset{\displaystyle CH_2}{\|}}{C}{-}PO(OR)_2$$

The addition of a mixture of $NaPO(OR)_2$ and $HPO(OR)_2$ to 2-propinol represents a special case:[400]

$$HC{\equiv}CCH_2OH \xrightarrow{NaPO(OR)_2} HC{\equiv}CCH_2ONa + HPO(OR)_2$$

$$HC{\equiv}CCH_2ONa \xrightarrow{HPO(OR)_2} (RO)_2OPCH_2CH[PO(OR)_2]CH_2ONa$$

$$\xrightarrow[-NaOH]{} (RO)_2OPCH_2C[PO(OR)_2]{=}CH_2$$

$$\xrightarrow[(OH^-)]{+HPO(OR)_2} [(RO)_2OPCH_2-]_2CHPO(OR)_2$$

Analogously, one obtains a tetraphosphonic acid ester from butindiol.[400]

8. ADDITION OF PHOSPHITES TO C=O, C=S, AND ACETAL GROUPS. With the exception of α,β-unsaturated ketones, dialkyl phosphites add to the C=O double bond of ketones and aldehydes.

$$R'R''CO + HPO(OR)_2 \rightarrow R'R''C(OH)[PO(OR)_2]$$

The reaction may be reversed with aqueous base; the P-C bond of α-hydroxyphosphonic acid esters is cleaved. Dicarbonyl compounds yield only monoaddition products with $HPO(OR)_2$.

Dialkyl phosphites with β-chlorinated alkyl groups react smoothly at ambient temperature, but normally elevated temperatures are required. Radical or base catalysis, preferably with amines or alkali alcoholates, produces a fast, exothermic reaction.[530, 1340, 1611]

α-Halogen substitution increases the reactivity of the carbonyl component. Chloral, for example, reacts exothermically at ambient temperature even with diphenylphosphite.[1112] With α-haloketones, however, in the absence of catalysts reaction temperatures of 100-130°C are required.[28] In the reaction of diphenyl phosphite with aldehydes and ketones, acid catalysis is sometimes used.[2061a]

The base-catalyzed reaction of α-oxoalkane-phosphonic acid esters with dialkyl phosphites leads to formation of hydroxyalkane-1,1-diphosphonic acid esters.

$$R'COPO(OR)_2 + HPO(OR)_2 \rightarrow R'C(OH)[PO(OR_2]_2$$

Contrary to earlier reports,[333, 422] these diphosphonic acid esters cannot be distilled, because they undergo thermal rearrangements according to :[538]

$$R'C(OH)[PO(OR)_2]_2 \rightarrow R'CH[OPO(OR)_2][PO(OR)_2]$$

α-Hydroalkane monophosphonic acid diesters may also rearrange at elevated temperatures to phosphoric acid esters.[1619]

Trialkyl phosphites add to aldehydes only at elevated temperatures under pressure, generally with simultaneous alkylation of the O atom :[9, 178, 636]

$$R'CHO + P(OR)_3 \rightarrow [(RO)_3\overset{\oplus}{P}CHR'\overset{\ominus}{O}] \rightarrow R'CH(OR)PO(OR)_2$$

With aliphatic aldehydes, reaction of the intermediate with a second mole of aldehyde is also observed :

$$(RO)_3\overset{\oplus}{P}CHR'\overset{\ominus}{O} + R'CHO \rightarrow ROCHR'OCHR'PO(OR)_2$$

Instead of using $P(OR)_3$, the addition reactions may also be carried out with $P(OR)_2(O_2CR')$ or $P(OR)_2(NHR')$.

The reaction of dialkyl- or diarylchlorophosphites with carbonyl compounds proceeds in a similar manner. In glacial acetic acid, again α-hydroxyalkane phosphonic acid esters are formed.[418]

$$R'_2C{=}O + ClP(OR)_2 \xrightarrow[-\ CH_3COCl]{+\ CH_3CO_2H} R'_2C(OH)PO(OR)_2$$

In the absence of acetic acid (200°C, pressure), aldehydes and diaryl chlorophosphites react to form α-chloroalkane phosphonic acid esters.[926]

With α,β-unsaturated ketones, diaryl chlorophosphites add to the C=C double bond. Subsequently, with acetic acid, γ-oxoalkane phosphonic acid esters are formed.[493] In some cases, cyclic ester chlorides and α,β-unsaturated ketones form phostones.[1714]

Dialkyl phosphonites also add to carbonyl compounds. The primary products rearrange to phosphonic acid alkyl enolesters.

$$R'P(OR)_2 + R''COCH_3 \rightarrow RPO(OR)\left[-O-\overset{\overset{\displaystyle CH_2}{\|}}{C}-R''\right]$$

Alkalidialkyl phosphites may be added to C=S double bonds. Thus, the esters of phosphonomono- or dithioformic acid may be obtained, respectively, according to

$$(RO)_2OPNa + \underset{\underset{\displaystyle X}{\|}}{C}{=}S \rightarrow (RO)_2OP-\underset{\underset{\displaystyle X}{\|}}{C}-SNa$$

$$\xrightarrow[-NaJ]{+\ RJ} (RO)_2OP\underset{\underset{\displaystyle X}{\|}}{C}SR \quad [X = S, O]$$

9. MANNICH REACTION, ADDITION TO SCHIFF BASES, AND OTHER REACTIVE C-N BONDS. α-Aminoalkane phosphonic acid esters may be synthesized by various base-catalyzed addition and condensation reactions carried out with reactive C-N bonds. The most important reactions are:

a. Aldehydes and ketones react with primary or secondary amines RR'NH (R,R'=H, alkyl, aryl) and dialkyl phosphites to form α-aminoalkane phosphonic acid esters.[913, 1219, 1220] With primary amines and NH_3, better yields are obtained when the adduct with aldehydes or ketones is formed first and subsequently treated with dialkyl phosphite.[356] This indicates a mechanism similar to the Mannich reaction. (A mechanism via primary formation of α-hydroxyphosphonic acid esters is improbable, for these react only with NH_3 but not with primary amines under formation of the corresponding α-amino compounds.[356, 905]

In this context, the reaction of compounds of structure R_2NCH_2X ($X = -NR_2$, $-OR$) with $HPO(OR)_2$ should be mentioned:

$$R_2NCH_2X + HPO(OR)_2 \rightarrow R_2NCH_2PO(OR)_2 + HX$$

Secondary amines also react smoothly. With formaldehyde, the following mechanism is most probable:

$$R_2NH + HCHO \rightarrow R_2\overset{\oplus}{N}HCH_2\overset{\ominus}{O} \xrightarrow{H+} [R_2\overset{\oplus}{N}HCH_2OH]$$

$$\xrightarrow{-\ H_2O} [R_2\overset{\oplus}{N}{=}CH_2 \rightleftarrows R_2\overline{N}{-}\overset{\oplus}{C}H_2] \xrightarrow{\overset{\ominus}{P}O(OR)_2} R_2NCH_2PO(OR)_2$$

With secondary amines, only monophosphonic acid esters may be obtained in this reaction. The monoalkylaminoalkane phosphonic acid esters derived from primary amines, however, may again react due to their remaining N-H bond. With NH_3 suitable molar ratios of the reactants leads to formation of amino-tris(methylene- phosphonic acid esters).[1484a]

$$NH_3 + 3HCHO + 3HPO(OR)_2 \rightarrow N[CH_2PO(OR)_2]_3 + 3H_2O$$

Diamino compounds and their condensation products with aldehydes or ketones react in a similar fashion to monoamino compounds. With suitable molar ratios, the reaction proceeds at both N atoms.[914, 915]

Hydrobenzamide reacts in the presence of bases according to:

$$PhCH(N{=}CHPh)_2 + 2HPO(OR)_2 \rightarrow PhCH[NHCHPhPO(OR)_2]_2$$

The product is hydrolyzed with aqueous HCl to PhCHO and

$[PhCH(\overset{\oplus}{N}H_3)PO(OR)_2]Cl^-$.

b. In addition to Schiff bases, other compounds with a C=N double bond are also capable of adding dialkyl phosphites. Thus, base-catalyzed addition to carbodiimides yields phosphonoformamidines.[956]

$$(R'N=)_2C + HPO(OR)_2 \rightarrow R'N=C(NHR')PO(OR)_2$$

c. Isocyanates and isothiocyanates also react with dialkylphosphites either at elevated temperatures or with base catalysis at room temperature.[56, 559, 1489, 1639, 1720, 1723]

$$R'N=C=X + HPO(OR)_2 \rightarrow RNHC(X)PO(OR)_2$$

$$X = O,\ S$$

d. N-unsubstituted benzimido ethers react with 2 moles of $HPO(OR)_2$ to yield aminophenylmethane diphosphonic acid esters.[1036]

$$PhC(=NH)OEt + HPO(OR)_2 \rightarrow [PhC(NH_2)(OEt)PO(OR)_2]$$

$$\xrightarrow{-EtOH} PhC(=NH)PO(OR)_2 \xrightarrow{HPO(OR)_2}$$

$$PhC(NH_2)[PO(OR)_2]_2$$

In the addition of $HPO(OR)_2$ to phenyliminophenylmethane phosphonic acid ester, the $-PO(OR)_2$ group does not add to the carbon atom, as was originally reported,[1036] but rather to the nitrogen atom.[715d]

$$PhC(=NPh)PO(OR)_2 + HPO(OR)_2 \not\rightarrow PhC(NHPh)[PO(OR)_2]_2$$

$$\searrow PhN[PO(OR)_2]CHPhPO(OR)_2$$

e. Formamidines and "immoniumchlorides" readily add dialkyl phosphite according to:[715a]

$$PhCON=CHNMe_2 + HPO(OR)_2 \rightarrow PhCONHCH(NMe_2)PO(OR)_2$$

$$[(EtO)_2OPCH=\overset{\oplus}{N}Me_2]Cl^- + HPO(OR)_2 \rightarrow$$

$$[(EtO)_2OP]_2CHNMe_2 + HCl$$

10. REACTIONS WITH ACETALS AND ORTHOCARBOXYLIC ACID ESTERS. These reactions normally proceed by a Michaelis-Arbuzov analogue mechanism. Formaldehyde/alcohol adducts react with $(RO)_2PCl$, $ROPCl_2$ as well as PCl_3 in the presence of $ZnCl_2$ or $AlCl_3$ as catalysts to yield alkoxymethane phosphonic acid esters:[1038]

$$(RO)_2PCl + CH_2(OR)_2 \rightarrow [ClCH_2OR + P(OR)_3] \rightarrow$$

$$ROCH_2PO(OR)_2 + RCl$$

$$PCl_3 + 3CH_2(OR)_2 \rightarrow ROCH_2PO(OR)_2 + 2ClCH_2OR + RCl$$

The mechanism is similar to the reaction of PCl_3 or phosphorous acid ester chlorides with orthocarboxylic acid esters (especially orthoformic acid esters):[713]

$$(RO)_2PCl + HC(OR)_3 \rightarrow [ClCH(OR)_2 + P(OR)_3] \rightarrow$$
$$(RO)_2CHPO(OR)_2 + RCl$$

Again, in the reaction of PCl_3, 3 moles of $HC(OR)_3$ have to be used, otherwise the yields drop due to formation of $ClCH(OR)_2$ as a by-product. This can be avoided by addition of 2 moles $P(OR)_3$ to the 1:1 mixture of PCl_3 and $HC(OR)_3$.

O,N-acetals of formaldehyde react similarly to O,O-acetals. In this reaction, however, no catalyst is required:[1038]

$$(RO)_2PCl + H_2C(OR')NR''_2 \rightarrow R''_2NCH_2PO(OR)_2 + R'Cl$$

Dimethylformamidine diacetal does not react analogously with $(RO)_2PCl$ to form $(Me_2N)(OMe)CHPO(OR)_2$. Rather, $(RO)_2POR'$ is obtained. With dialkyl phosphite, however, at temperatures below 30°C smooth reaction is observed:

$$Me_2NCH(OMe)_2 + HPO(OR)_2 \rightarrow \underset{\mathbf{1}}{Me_2N(OMe)CHPO(OR)_2} + MeOH$$

The product **1** adds a second mole $HPO(OR)_2$ at elevated temperatures:

$$\mathbf{1} + HPO(OR)_2 \rightarrow \underset{\mathbf{2}}{Me_2NCH[PO(OR)_2]_2} + MeOH$$

If the reaction is carried out considerably above 30°C, **1** may not be isolated due to immediate reaction to **2**. With S-acetals, a similar reaction is observed. The initially formed $(Me_2N)(SR')CHPO(OR)_2$, however, is much less reactive than the O analogue **1** and only at considerably higher temperatures a second mole of $HPO(OR)_2$ is added to form **2**.[712,714]

11. REACTION OF PHOSPHORIC ACID ESTER HALIDES. The reaction of $PO(OR)_2Cl$ with organometallic compounds such as Grignard reagents or Li organyls is quite complicated, since the ester groups may also react with formation of phosphine oxides. However, with sterically hindered aryl Grignards, reasonable yields are obtained, if the ester chloride is added to a solution of the Grignard. Sufficient steric hindrance is generally provided by o-substit-

uents. With nonhindered Grignards, it is advantageous to add the Grignard solution to excess $PO(OR)_2Cl$.[328, 457]

Instead of Grignards or Li organyls, metalated CH-acidic compounds may also be used, for example Na-malonic esters or Na-acetylacetonates.[1034, 1643]

Fluoromethane phosphonic acid esters may be obtained by the reaction of monofluorophosphoric acid esters with diazomethane:[1790]

$$(RO)_2OPF + N_2CH_2 \rightarrow (RO)_2OPCH_2F + N_2$$

12. SYNTHESES WITH DIAZOACETIC ACID ESTERS. A reaction of limited preparative use is the treatment of diazoacetic acid esters with dialkyl phosphites:

$$(RO)_2OPH + N_2CH_2COOR' \rightarrow (RO)_2OPCH_2COOR' + N_2$$

13. TRANSESTERIFICATION. Similar to carboxylic acid esters, phosphonic acid esters may undergo ester interchange with alcohols or phenols. Preferably, the reaction is base-catalyzed and proceeds smoothly if a short-chain alcohol rest is to be replaced by an OR rest with a longer carbon chain. Transesterification may also be carried out with alkyl halides. With halogenated or alkoxylated alkyl rests of suitable chain length bound to the phosphorus atom, intramolecular ring closure leads to the formation of phostones. Heating in the absence of catalysts is normally sufficient to start this reaction:[173, 600]

$$HalCH_2CH_2CH_2PO(OR)_2 \rightarrow (RO)O\overline{PCH_2CH_2CH_2O} + RHal$$

Phosphonic acid amides and ester amides may also react with alcohols and phenols with interchange to yield phosphonic acid diesters.

14. MODIFICATIONS OF PHOSPHONIC ACID ESTERS. Phosphonic acid esters are relatively stable compounds and with few exceptions, the P-C bond is extremely unreactive. Cleavage of the P-O-C bond also requires drastic reaction conditions. It is, therefore, possible to carry out a number of reactions of the organic rest of phosphonic acid esters without interference from the $-PO(OR)_2$ group.

Most reactions, however, are influenced by the neighboring $-PO(OR)_2$ group. In aromatic systems, the phosphonic acid ester group behaves as a second-order substituent similar to the nitro group. Double bonds are activated by neighboring aliphatic $-PO(OR)_2$ groups and α-CH bonds are quite acidic. In the following incomplete list, a number of typical reactions of phosphonic acid esters are given:

$$(RO)_2OP\text{-}CH_2X + Na \xrightarrow{-\ 1/2\ H_2} (RO)_2OPCHNaX \xrightarrow{R'Hal}$$

$(RO)_2OP-CHR'X + NaCl$ [367, 1015, 1034]   (1)

$[X = -PO(OR)_2, -COOR, -CN]$

$R'CH=CHPO(OR)_2 + XH \xrightarrow{NaOR} R'CHXCH_2PO(OR)_2$

$[XH = -CH, -SH, -NH, -PH$ acidic compounds]   (2)

$R'CH=CHCH_2PO(OR)_2 + RCO(OOH) \xrightarrow{(H_2O)}$

$R'\overset{\frown{O}}{CH-CH}CH_2PO(OR)_2$ or $R'CH(OH)CH(OH)CH_2PO(OR)_2$ [133]   (3)

$R'CH=CHPO(OR)_2 + Br_2 \rightarrow RCHBrCHBrPO(OR)_2$ [888]   (4)

$H_2C=CHPO(OR)_2 + MeCH=CH-CH=CHMe \rightarrow$

$(RO)_2OP\overline{CHCHMeCH=CH-CHMeCH_2}$ [1592a]   (5)

Polymerization and copolymerization of unsaturated

phosphonic acid esters   (6)

$BrCH_2PO(OR)_2 \xrightarrow{H_2/Pd} CH_3PO(OR)_2$ [1143]   (7)

$HalCH_2CHHalPO(OR)_2 \xrightarrow{Zn} H_2C=CHPO(OR)_2$ [1935]   (8)

$HalCH_2CH_2PO(OR)_2 \xrightarrow{KOH \text{ or } NR_3} H_2C=CHPO(OR)_2$ [309, 552, 623]   (9)

$PhCCl=CHPO(OR)_2 \xrightarrow{KOH/ROH} PhC\equiv CPO(OR)_2$ [90]   (10)

$HalCH_2PO(OR)_2 + Na_2S \rightarrow S[CH_2PO(OR)_2]_2$ [128a]   (11)

$HOCH_2PO(OR)_2 + R'COX \rightarrow R'CO-O-CH_2PO(OR)_2$ [426]

$[X = -Hal, -OCOR']$ (12)

$PhCOPO(OR)_2 + 2Na/Hg + 2 H_2O \rightarrow PhCH(OH)PO(OR)_2$

$+ 2NaOH$ (13)

$HOCH_2PO(OR)_2 + NH_3 \rightarrow H_2NCH_2PO(OR)_2 + H_2O$ [905]   (14)

$H_2NC(Me)_2PO(OR)_2 + COCl_2 \rightarrow OCNC(Me)_2PO(OR)_2$

$+ 2HCl$ [1220a]   (15)

$O_2N-\langle\bigcirc\rangle-PO(OR)_2 + H_2/Pt \rightarrow H_2N-\langle\bigcirc\rangle-PO(OR)_2$ [932]   (16)

$H_2N-\langle\bigcirc\rangle-PO(OR)_2 + 2\overline{CH_2CH_2O} \rightarrow$

$(HOCH_2CH_2)_2N-\langle\bigcirc\rangle-PO(OR)_2$ [932]   (17)

$H_2N-\langle\bigcirc\rangle-PO(OR)_2 + EtONO \rightarrow HO-\langle\bigcirc\rangle-PO(OR)_2$ [576]   (18)

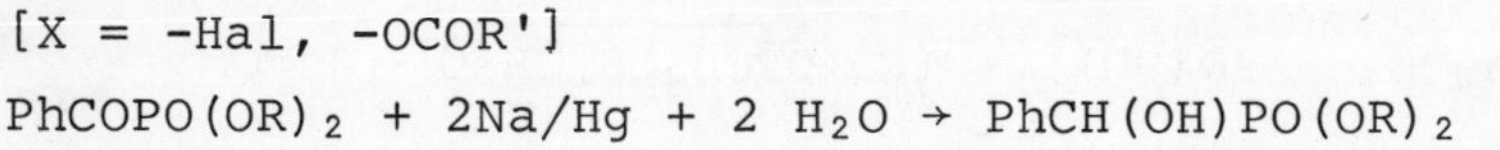

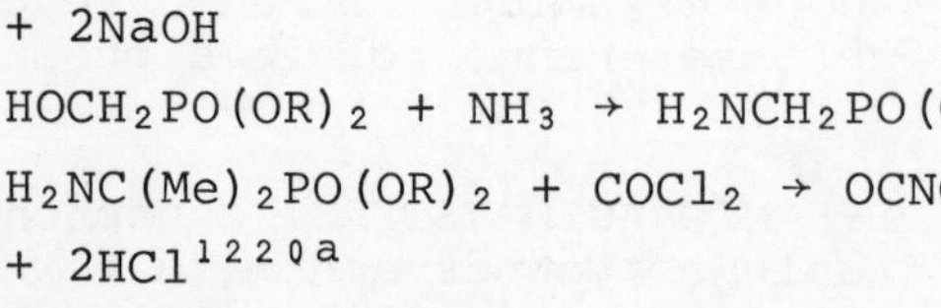

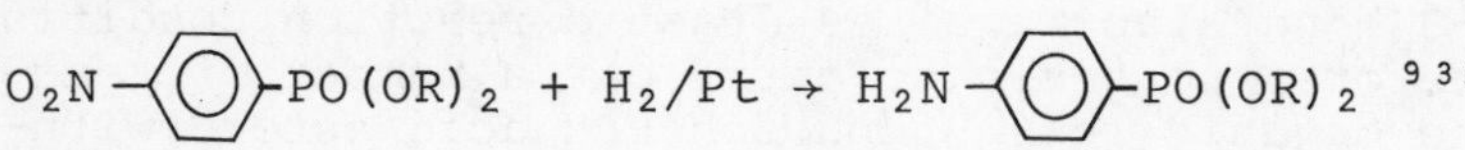

VI. VARIOUS PHOSPHONIC ACID DERIVATIVES

1. SILYL ESTERS. The preparation of silyl esters of phosphonic acids proceeds in a similar fashion to general silylation of acids by the following reactions:

$$RPO(OH)_2 + 2HalSiR'_3 \xrightarrow{\text{pyridine}} RPO(OSiR'_3)_2 + 2HHal \quad [951] \qquad (1)$$

$$RPO(OR')_2 \xrightarrow{HalSiR''_3} RPO(OR')(OSiR''_3) + R'Hal$$

$$RPO(OR')(OSiR''_3) \xrightarrow{HalSiR''_3} RPO(OSiR''_3)_2 + R'Hal \quad [1164] \qquad (2)$$

$$RPO(OH)_2 + 2R'_3SiOH \rightarrow RPO(OSiR'_3)_2 + 2H_2O \quad [1437] \qquad (3)$$

[removal of H_2O by azeotropic distillation with benzene]

$$HOCH_2PO(OH)_2 + 2R'_3SiOR'' \rightarrow HOCH_2PO(OSiR'_3)_2 + 2R''OH \quad [1434] \qquad (4)$$

[at lower temperatures, the alcoholic -OH group is normally not silylated]

$$RPO(OH)_2 + 2R'_3SiN{=}CMeOSiMe_3 \rightarrow RPO(OSiR'_3)_2 + 2MeCONHSiMe_3 \quad [974a] \qquad (5)$$

[under drastic reaction conditions, the second $-SiMe_3$ group may also be transferred]

$$RPO(OH)_2 + 2Me_3SiNHSiMe_3 \rightarrow RPO(OSiMe_3)_2 + 2H_2NSiMe_3 \qquad (6)$$

$$HOCH_2PO(OH)_2 + 3HSiR_3 \xrightarrow[80\text{-}135^\circ C]{Ni} R_3SiOCH_2PO(OSiR_3)_2 + 3H_2 \qquad (7)$$

[partial silylation is not possible] [1437]

A further reaction starting from trialkyl phosphites exists the mechanism of which is not completely understood : [1164]

$$2(RO)_3P \xrightarrow[-\ 3RHal]{3HalSiR'_3} RPO(OR)(OSiR'_3) + RPO(OSiR'_3)_2$$

The products are formed in varying ratios, probably via a Michaelis-Arbuzov mechanism.

2. HYDROXYLAMINE DERIVATIVES. Phosphonic acid ester oximides may be prepared by the reaction of oximides with phosphonic acid ester halides :[739]

$$R'CH{=}NOH + RPO(OR'')Cl \rightarrow RPO(OR'')ON{=}CHR' + HCl$$

N-Hydroxycarbamidic acid phenyl esters react with phosphonic acid ester halides according to : [1946a]

$$PhO_2CNHOH + RPO(OR')Cl \rightarrow RPO(OR')ONHCO_2Ph + HCl$$

Normally, hydroxamic acids would undergo a Lossen degradation under these conditions.

3. ORGANOMETALLIC DERIVATIVES. A number of organometallic compounds, especially Sn(IV) compounds, react with trialkyl phosphites under formation of phosphonic acid derivatives. Partially, polymeric compounds are formed, which received some interest as temperature-resistent plastics, whose structures, however, are not completely understood. Therefore, these polymers are not discussed here. The most important routes to low molecular weight compounds are the following:

$RPO(OR)_2 + R'_3SnJ \rightarrow RPO(OR)(OSnR'_3) + RJ$ [1168] (1)

$(RO)_3P + R'_3SnJ \rightarrow RPO(OR)(OSnR'_3) + RJ$ [152, 1168] (2)

$2(RO)_3P + R'_2SnJ_2 \rightarrow [RPO(OR)O-]_2SnR'_2 + 2RJ$ (3)

$3(RO)_3P + RSnJ_3 \rightarrow [RPO(OR)O-]_3SnR' + 3RJ$ [138, 1168] (4)

THIOPHOSPHONIC ACIDS AND ACID ESTERS (REMARKS)

There are three different types of thiophosphonic acid derivatives, which are named mono-, di-, and trithiophosphonic acid derivatives according to the number of S atoms bound to the phosphorus. Due to different types of bonding, isomers are possible as is seen from the following table:

1. Acids	Monothio	Dithio	Trithio compd.
	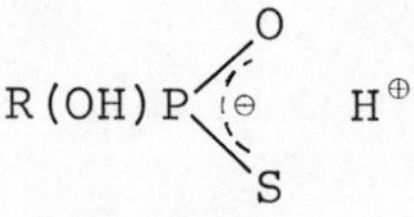	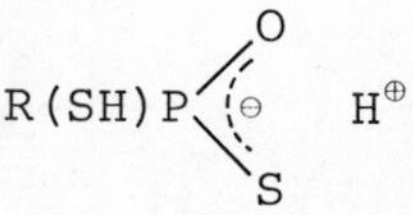	$RPS(SH)_2$
2. Mono-halides and pseudo-halides	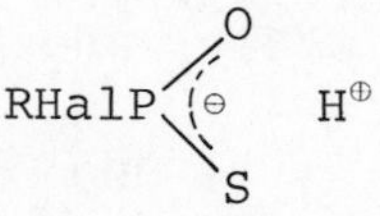	RPS(SH)Hal	
3. Mono-esters	R(R'S)PO(OH)	RPS(OR)(SH)	RPS(SR)(SH)
	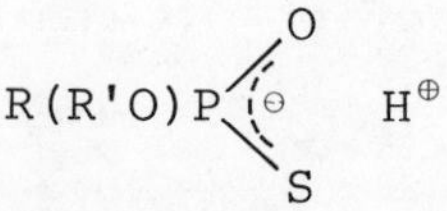	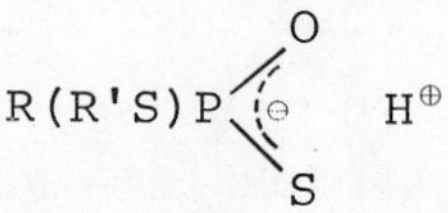	
4. Ester-halides	RPS(OR)Hal RPO(SR)Hal	RPS(SR)Hal	

5. Diesters $RPS(OR)_2$ $RPS(OR)(SR)$ $RPS(SR)_2$

$RPO(OR)(SR)$ $RPO(SR)_2$

Some of the above-listed compounds, especially most acid halides, are not yet known in pure form. Of the compounds mentioned under (5), only a few salts of mono- and dithiophosphonic acids are known. The mesomerism between P=S and P-S is only considered in the table, where distinctively different isomers may be obtained.

VII. SYNTHESIS OF THIOPHOSPHONIC ACIDS

1. BY HYDROLYSIS. Thiophosphonic acids, normally unstable oils, are difficultly isolable in a pure form. At the temperatures required for hydrolysis of esters or amides, simultaneous hydrolysis of the P-S bonds normally occurs, with formation of phosphonic acids. The hydrolysis of thiophosphonic acid dichlorides also proceeds only at elevated temperatures, phosphonic acids again being the major products. Hydrolysis with the calculated amount of base in some cases leads to formation of the pure salts of thiophosphonic acids:[907, 986]

$$RPSCl_2 \xrightarrow{4MOH} R(MO)P\genfrac{}{}{0pt}{}{\diagup O}{\diagdown S}\ominus \quad M^{\oplus} + 2MCl + 2H_2O$$

Dialkali salts of monothiophosphonic acids may also be obtained by partial hydrolysis of alkali salts of dithiophosphonic acids with H_2O under controlled conditions:[794]

$$R(NaS)P\genfrac{}{}{0pt}{}{\diagup O}{\diagdown S}\ominus \quad Na^{\oplus} + H_2O \rightarrow R(NaO)P\genfrac{}{}{0pt}{}{\diagup O}{\diagdown S}\ominus \quad Na^{\oplus} + H_2S$$

2. FROM P_4S_{10} AND GRIGNARD COMPOUNDS. The reaction of P_4S_{10} with Grignard compounds is another suitable method for the preparation of the salts of dithiophosphonic acids. As by-products, dithiophosphinic and also trithiophosphonic acid derivatives are formed. With the proper molar ratio of 1 mole P_4S_{10} and 3 moles of Grignard compound, the formation of dithiophosphonic acids prevails. The compounds may be purified by crystallization of their salts or by extraction of the Ni salts with benzene or ether:[794]

$$4RMgHal + P_4S_{10} \xrightarrow{8H_2O} 4R(SH)P\genfrac{}{}{0pt}{}{\diagup O}{\diagdown S}\ominus \quad H^{\oplus} + 2H_2S + 4Mg(OH)Hal$$

VIII. SYNTHESIS OF THIOPHOSPHONIC ACID MONOESTERS

1. FROM THIOPHOSPHONIC ACID ESTER HALIDES OR PHOSPHONIC ACID ESTER HALIDES. For the preparation of thiophosphonic

acid O-monoesters, an often-applied method consists in the hydrolysis of compounds of structure RPS(OR')Hal with H_2O/dioxane or diluted bases. Subsequent acidification yields the free acid monoester:[784a]

$$\text{RPS(OR')Hal} + 2\text{MOH} \rightarrow \text{R(OR')P}\langle^{O}_{S}\rangle^{\ominus}\ \text{M}^{\oplus} + \text{MHal} + \text{H}_2\text{O}$$

$$\xrightarrow{\text{H}^{\oplus}} \text{R(OR')P}\langle^{O}_{S}\rangle^{\ominus}\ \text{H}^{\oplus}$$

The free acids may be extracted with ether and after thorough drying can be purified by distillation.

Thiophosphonic acid S-monoesters are obtained in a similar manner by hydrolysis of thiophosphonic acid S-ester halides. The products are oils and tend to decompose upon distillation.

A further route to prepare O-monoesters is the reaction of RPO(OR)Cl with NaHS:[897]

$$\text{RPO(OR')Cl} + 2\text{NaHS} \rightarrow \text{R(OR')P}\langle^{O}_{S}\rangle^{\ominus}\ \text{Na}^{\oplus} + \text{NaCl} + \text{H}_2\text{S}$$

Similarly, the dithiophosphonic acid O-monoesters are obtained as sodium salts:[897]

$$\text{RPS(OR)Hal} + 2\text{NaHS} \rightarrow \text{RPS(OR)(SNa)} + \text{NaCl} + \text{H}_2\text{S}$$

Starting from thiophosphonic acid dichlorides, these salts may also be obtained by reaction with alcohols and KHS:[1171a]

$$\text{RPSCl}_2 + \text{R'OH} + 3\text{KHS} \rightarrow \text{RPS(OR')(SK)} + 2\text{KCl} + 2\text{H}_2\text{S}$$

2. BASE-CATALYZED HYDROLYSIS OF ESTERS. The cleavage of diesters is an often-used method, especially for the preparation of thiophosphonic acid O-monoesters. Aqueous or aqueous/alcoholic bases (MOH, tertiary amines) are used in this reaction. In the case of arylthiophosphonic acid O,O-di-alkylesters, even with an excess of base the reaction stops at the monoester stage:[1680]

$$\text{RPS(OR')}_2 + \text{NaOH} \rightarrow \text{R(OR')P}\langle^{O}_{S}\rangle^{\ominus}\ \text{Na}^{\oplus} + \text{R'OH}$$

With aliphatic compounds, preferably smaller amounts of alkali are used in order to avoid side reactions. This is especially important in the hydrolysis of unreactive O,O-diisoalkyl esters. The preparation of the monoester salts is suitably followed by acidification of the solution, extraction of the free acid monoester with ether and distillation.[784a, 897]

Dithiophosphonic acid monoesters may be prepared by reacting trithiophosphonic acid dialkyl esters with alkali alcoholate:[114]

$$PhPS(SR)_2 + NaOR \rightarrow Ph(SR)P\overset{O}{\underset{S}{\langle}}{}^{\ominus}\ Na^{\oplus} + R_2S$$

3. REACTION OF PHOSPHONOUS ACID MONOESTERS WITH SULFUR. Alkali salts of phosphonous acid monoesters react readily with sulfur to yield alkali salts of monothiophosphonic acid monoesters. Acidification and suitable workup leads to the free thiophosphonic acid monoesters.

The sulfur may also be added to the free phosphonous acid, preferably in presence of dioxane as a solvent:[288, 1471]

$$RPO(OR')X + S \rightarrow R(OR')P\overset{O}{\underset{S}{\langle}}{}^{\ominus}\quad X^{\oplus}$$

[X = H, alkali metal]

4. ALCOHOLYSIS OF THIOPHOSPHONIC ACID ANHYDRIDES. Thiophosphonic acid anhydrides react with alcohols in the expected manner:

$$(RPOS)_n + nR'OH \quad nR(OR')P\overset{O}{\underset{S}{\langle}}{}^{\ominus}\ H^{\oplus}$$ [for n = 3 [449a, 499a]]

The cleavage of S,S-anhydrides to yield dithiophosphonic acid monoesters with pentaerythritol has also been described:[707]

$$nC(CH_2OH)_4 + 4(RPS_2)_n \rightarrow nC[CH_2O(R)PS(SH)]_4$$

These compounds have been isolated as oils, but the Ni^{2+} and $Et_3\overset{\oplus}{N}H$ salts were obtained in crystalline form.

IX. SYNTHESIS OF THIOPHOSPHONIC ACID ESTER HALIDES AND PSEUDOHALIDES

1. FROM PHOSPHONIC OR THIOPHOSPHONIC ACID HALIDES. The reaction of thiophosphonic acid dihalides with alcohols or phenols in the presence of tertiary amines leads to the formation of thiophosphonic acid O-ester halides.[788, 870a]

$$RPSX_2 + R'OH + R''_3N \rightarrow RPS(OR')X + [R''_3\overset{\oplus}{N}H]X^{\ominus} \quad [X = F, Cl]$$

Analogously, phosphonic acid dihalides and mercaptans form thiophosphonic acid S-ester halides.

$$RPOCl_2 + R'SH + R''_3N \rightarrow RPO(SR')Cl + [R''_3\overset{\oplus}{N}H]Cl^{\ominus}$$

The cyclic enolester chlorides obtained from α,β-unsaturated ketones and PCl_3 may be converted to thiophosphonic acid O-enolester chlorides by reaction with H_2S:[1460]

$$CH_3COCH{=}C(CH_3)_2 + PCl_3 \rightarrow [Cl_2\overset{\oplus}{\overline{P}C(CH_3)_2CH{=}C(CH_3)\overline{O}}]Cl^{\ominus}$$

$$\xrightarrow[-2HCl]{H_2S} Cl(S)\overline{PC(CH_3)_2CH{=}C(CH_3)O}$$

Thiophosphonic acid ester fluorides may be prepared by halogen exchange of thiophosphonic acid ester chlorides with NaF or of thiophosphonic acid dichlorides with NaF in the presence of alcohols.[1834a]

$$RPSCl_2 + R'OH + 2NaF \rightarrow RPS(OR')F + 2NaCl + HF$$

In a similar manner, RPS(OR')Cl and KSCN react with the formation of RPS(OR')(NCS).

2. FROM THIOPHOSPHONIC ACID DIESTERS. In analogy to the preparation of phosphonic acid ester chlorides, thiophosphonic acid O,O-diesters may be carefully chlorinated with PCl_5:[1703]

$$RPS(OR')_2 + PCl_5 \rightarrow RPS(OR')Cl + R'Cl + POCl_3$$

Phosgene is not suited as a chlorinating agent in this reaction, since desulfurization to phosphonic acid esters occurs. However, thiophosphonic acid O,S-diesters are transformed by phosgene at ambient temperature to thiophosphonic acid S-ester chlorides.[337]

$$RPO(OR')(SR'') + COCl_2 \rightarrow RPO(SR'')Cl + R'Cl + CO_2$$

3. FROM PHOSPHONOUS AND THIOPHOSPHONOUS ACID DERIVATIVES. Phosphonous acid ester chlorides add elemental sulfur according to:

$$RP(OR')Cl + S \rightarrow RPS(OR')Cl$$

Instead of the ester chlorides, the crude reaction products obtained from $RPCl_2$ and R'OH are normally directly sulfurated according to:

$$RPCl_2 + R'OH + R''_3N \rightarrow RP(OR')Cl + [R''_3\overset{\oplus}{N}H]Cl^{\ominus}$$

$$\downarrow S$$

$$RPS(OR')Cl$$

Thiophosphonous acid monoesters are converted to thiophosphonic acid monoester halides by oxidizing chlorinating agents, as shown in the following equations:

$$RPS(OR')H + Cl_2 \rightarrow RPS(OR')Cl + HCl$$

$$RPS(OR')H + SO_2Cl_2 \rightarrow RPS(OR')Cl + SO_2 + HCl$$

$$RPS(OR')H + CCl_4 \rightarrow RPS(OR')Cl + CHCl_3$$ [1113a]

Both aliphatic and aromatic dichlorophosphines react with sulfenylchlorides in liquid SO_2 to yield thiophosphonic acid S-monoester chlorides:[1370]

$$RPCl_2 + R'SCl + SO_2 \rightarrow RPO(SR')Cl + SOCl_2$$

4. CLEAVAGE OF BIS(ALKOXYTHIOPHOSPHONYL)DISULFIDES OR THIOPHOSPHONIC ACID ANHYDRIDES. Bis(alkoxythiophosphonyl) disulfides are cleaved by alkalicyanides with formation of thiocyanates.

$$R(OR)(S)P\text{-}SS\text{-}P(S)(OR)R + MCN \rightarrow RPS(OR)(SCN) + RPS(OR)SM$$

Dithiophosphonic acid anhydrides add alkalicyanides or -fluorides. The resulting alkali salts of dithiophosphonic acid cyanides or fluorides, respectively, may be further treated with alkyl halides:[1844a, 1849a]

$$(RPS_2)_2 + KX \rightarrow [2RPS(SK)X] \xrightarrow{2R'Hal} RPS(SR')X + KHal$$

$$[X = \text{-CN, -F}]$$

5. CHLORINATION OF DITHIOPHOSPHONIC ACID O-MONOESTERS. The reactions of compounds of structure RPS(OR')SH with Cl_2 or SO_2Cl_2 produces RPS(OR')Cl in normally good yields.[395] Most probably, the reaction proceeds via disulfide intermediates:

$$2RPS(OR')SH + Cl_2 \rightarrow RP(S)(OR')\text{-}S\text{-}S\text{-}(S)P(OR')R + 2HCl$$

The disulfide intermediate is cleaved by additional Cl_2 to form the desired RPS(OR')Cl and some S and/or S_2Cl_2. The reaction with SO_2Cl_2 proceeds in a similar manner. As discussed later on, monothiophosphonic acid O-monoesters also react with SO_2Cl_2 with formation of monophosphonic acid sulfenylchlorides(cf. Section XI.1).

X. SYNTHESIS OF THIOPHOSPHONIC ACID DIESTERS

1. FROM PHOSPHONIC, THIOPHOSPHONIC, OR PHOSPHORIC ACID HALIDES. In analogy to the synthesis of phosphonic acid esters, thiophosphonic acid esters may also be prepared from

the corresponding dihalides. The following synthetic routes exist:

$$RPSCl_2 + 2R'OH \rightarrow RPS(OR')_2 + 2HCl$$

$$RPS(OR')Cl + R''OH \rightarrow RPS(OR')(OR'') + HCl$$

In the absence of acid-binding agents, the reaction of the acid halides with alcohols or phenols gives very low yields.[986] Good yields are normally obtained with alkali alcoholates or phenolates.[729,788] The reaction of RPS(OR')Cl with alcohols or phenols in the presence of tertiary amines leads to formation of the asymmetric diesters.[1812] Dihalides, however, are reported to react incompletely with alcohols even in the presence of tertiary amines.[1222,1710]

$$RPO(OR')Cl + NaSR'' \rightarrow RPO(OR')(SR'') + NaCl$$ [897]

This reaction proceeds smoothly with mercaptans as well as thiophenols. Starting from $RPOCl_2$, the O,S-diesters may be obtained without intermittent purification:[901]

$$RPOCl_2 + R'OH \xrightarrow[-HCl]{R_3N} RPO(OR')Cl \xrightarrow[-NaCl]{NaSR''} RPO(OR')(SR'')$$

RPO(SR')Cl and alcohols also react in the presence of tertiary amines according to:

$$RPO(SR')Cl + R''OH \xrightarrow[-HCl]{R_3N} RPO(SR')(OR'')$$

$$RPS(OR')Cl + NaSR'' \rightarrow RPS(OR')(SR'') + NaCl$$

This reaction is preferably carried out in inert solvents at slightly elevated temperatures.[1795,1827] Dithiophosphonic acid ester halides also react with formation of O,S-diesters:

$$RPS(SR')Cl + R''OH \xrightarrow[-HCl]{R_3N} RPS(OR'')(SR')$$

$$RPOCl_2 + 2HSR' \rightarrow RPO(SR')_2 + 2HCl$$

$$RPO(SR')Cl + HSR'' \rightarrow RPO(SR')(SR'') + HCl$$

These reactions are best carried out with alkali mercaptides or with mercaptanes and tertiary amines.[84] The symmetric diesters may also be prepared by cleavage of disulfides with phosphonic acid dihalides according to:

$$2(R'S)_2 + RPOCl_2 \rightarrow RPO(SR')_2 + 2R'SCl$$

$$RPSCl_2 + 2HSR' \xrightarrow[-2HCl]{} RPS(SR')_2$$

$$RPS(SR')Cl + HSR'' \xrightarrow[-HCl]{} RPS(SR')(SR'')$$

These reactions are performed with alkali mercaptides or thiophenolates in inert solvents or with mercaptanes or thiophenoles in the presence of tertiary amines. For the preparation of thiobenzyl esters, a different method has been described requiring only catalytic amounts of pyridine:

$$RPSCl_2 + 2ClH_2CPh + 2H_2S \xrightarrow[\text{pyridine}]{160\text{-}170\ ^0C} RPS(SH_2CPh)_2 + 4HCl$$

2. FROM PHOSPHONOUS AND THIOPHOSPHONOUS ACID ESTERS OR ESTER DERIVATIVES. For the conversion of phosphonous or thiophosphonous acid esters to thiophosphonic acid diesters essentially three types of reactions are known:

a. Addition of elemental sulfur

$$RP(OR')_2 + S \rightarrow RPS(OR')_2$$ [153, 1710]

$$RP(OR')(SR'') + S \rightarrow RPS(OR')(SR'')$$ [1787a]

$$RP(SR')_2 + S \rightarrow RPS(SR')_2$$

The products are diesters with P=S double bonds. Asymmetric thiophosphonic acid esters may be synthesized.

b. Oxidation. By oxidation of thiophosphonous acid esters with H_2O_2 or molecular oxygen the corresponding diesters of structure RP(O)< may be obtained. Again, symmetric or asymmetric products are formed depending on the starting thiophosphonous acid ester:

$$RP(OR')(SR'') \xrightarrow{O_2} RPO(OR')(SR'')$$

$$RP(SR')_2 \xrightarrow{H_2O_2} RPO(SR')_2$$ [1787a]

c. Reaction with S Compounds. Phosphonous acid diesters react with active sulfur compounds such as sulfenylchlorides, sulfenylamides, or alkylisothiocyanates with introduction of a mercapto group. The reactions probably proceed via formation of phosphonium-type intermediates similarly to the Michaelis-Arbuzov reaction.

$$RP(OR')_2 + R''SCl \rightarrow \{[R\overset{\oplus}{P}(OR')_2(SR'')]Cl^{\ominus}\}$$
$$\rightarrow RPO(OR')(SR'') + R'Cl$$ [1338]

$$RP(OR')_2 + R''SNR''_2 \rightarrow \{[R\overset{\oplus}{P}(OR')_2(SR'')]\ NR''^{\ominus}_2\}$$
$$\rightarrow RPO(OR')(SR'') + R'NR''_2$$ [1470]

$$RP(OR')_2 + R''SCN \rightarrow \{[R\overset{\oplus}{P}(OR')_2(SR'')]\ CN^{\ominus}\}$$
$$\rightarrow RPO(OR')(SR'') + R'CN$$ [1303]

In addition to diesters, monoesters of phosphonous or thiophosphonous acids or their alkali or ammonium salts react in a similar manner with reactive sulfur compounds such as thiosulfonic acid esters, salts of thiosulfuric acid monoesters, sulfenylamides, or organic disulfides:

$$RPS(OR')H + R'''SO_2SR'' \xrightarrow{NR_3}$$

$$RPS(OR')(SR'') + [R'''SO_2^{\ominus}][R_3\overset{\oplus}{N}H]$$ [1113c]

$$RPS(OR')Na + NaOSO_2SR'' \rightarrow$$

$$RPS(OR')(SR'') + Na_2SO_3$$ [1113b]

$$RPO(OR')H + R''SNR''_2 \rightarrow RPO(OR')(SR'') + HNR''_2$$ [1496]

$$RPO(OR')H + R''SSR'' \rightarrow RPO(OR')(SR'') + R''SH$$ [1787a]

3. ALKYLATION OF PHOSPHONIC OR THIOPHOSPHONIC ACID MONOESTERS.

a. Thiophosphonic acid monoesters may be alkylated with diazoalkanes:

$$(R'O)P\overset{O}{\underset{S}{\lessgtr}}{}^{\ominus}\ H^{\oplus} + CH_2N_2 \rightarrow \underset{(90\%)}{RPO(OR')(SCH_3)}$$

$$+ \underset{(10\%)}{RPS(OR')(OCH_3)} + N_2$$

Since both products have almost the same boiling points, isolation of the pure S-esters is very difficult.

b. More convenient is the alkylation of alkali or quaternary ammonium salts of thiophosphonic acid monoesters with alkyl halides:

$$(R'O)RP\overset{O}{\underset{S}{\lessgtr}}{}^{\ominus}\ M^{\oplus} + ClR'' \rightarrow RPO(OR')(SR'') + M^{\oplus}Cl^{\ominus}$$ [902]

$$(R'O)RP\overset{S}{\underset{S}{\lessgtr}}{}^{\ominus}\ M^{\oplus} + ClR'' \rightarrow RPS(OR')(SR'') + M^{\oplus}Cl^{\ominus}$$ [897]

c. The following reactions starting from dithiophosphonic acid O-monoesters are especially suited for the preparation of dithiophosphonic acid O,S-diesters. The O-monoester adds readily to olefins or acetylenes with terminal or activated internal multiple bonds:

$$RPS(OR')SH + H_2C{=}CHSR'' \rightarrow RPS(OR')(SCHMeSR'')$$ [1834]

Addition to nonactivated double bonds is facilitated by UV irradiation.

Monoesters also react with formaldehyde and mercaptans or thiophenols to yield dithiophosphonic acid esters with an $-SCH_2SR''$ group:[1787a]

$$RPS(OR')SH + HCHO + HSR'' \rightarrow RPS(OR')(SCH_2SR'') + H_2O$$

The addition of RPS(OR')SCl to terminal olefins has also been reported:[1092, 1338]

$$RPO(OR')SCl + H_2C{=}CHR'' \rightarrow RPO(OR')(SCH_2CHClR'')$$

4. CONVERSION OF PHOSPHONIC, THIOPHOSPHONIC, AND IMINOPHOSPHONIC ACID ESTERS Thiophosphonic acid O,O-diesters are converted upon heating, especially in the presence of alkyl halides (Pistshimuka reaction).[900, 901]

$$RPS(OR')_2 \xrightarrow{R''J} RPO(OR')(SR')$$

The conversion of cyclic thiophosphonic acid O,O-diesters by heating with PPh_3 has also been described:[1381]

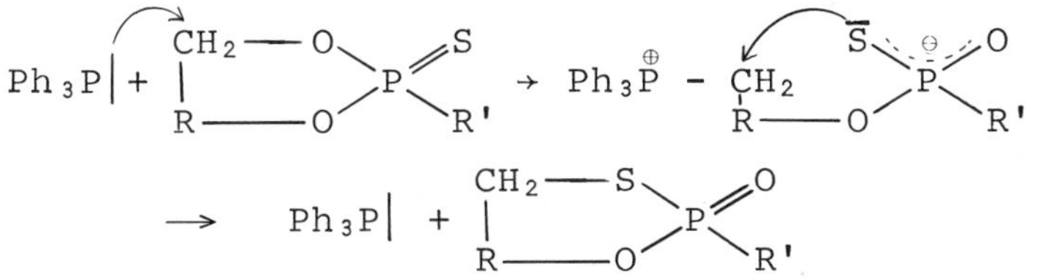

Also of minor preparative interest is the reaction of iminophosphonic acid esters with CS_2:[890]

$$RP(=NR'')(OR')_2 + CS_2 \rightarrow RPS(OR')_2 + R''NCS$$

Similarly, monothiophosphonic acid O,S-diesters upon treatment with P_4S_{10} are converted according to:[1428, 1735]

$$RPO(OR')(SR'') \xrightarrow{P_4S_{10}} RPS(OR')(SR'')$$

5. FROM DITHIOPHOSPHONIC ACID ANHYDROSULFIDES. Phosphonic acid anhydrides are cleaved by epoxides or alcohols with formation of cyclic phosphonic acid esters or phosphonic acid monoesters, respectively. Analogous reactions are observed with dithiophosphonic acid anhydrosulfides:

$$(RPS_2)_n + nH_2\overline{CCHRO} \rightarrow nR(S)\overline{POCHRCH_2S} \quad [1845]$$

$$(RPS_2)_n + nR'OH \rightarrow RPS(OR')(SH)$$

If the reaction is carried out with excess alcohol and at

elevated temperature, dithiophosphonic acid O,S-diesters are obtained:[522]

$$RPS(OR')SH + R'OH \rightarrow RPS(OR')(SR') + H_2O$$

6. THE MICHAELIS-ARBUZOV REACTION OF MONOTHIOPHOSPHITES. This reaction, which is most important for the synthesis of phosphonic acid diesters, may be modified for the preparation of thiophosphonic acid O,S-diesters.

$$P(SR)(OR')_2 + ClR'' \rightarrow \{[R''\overset{\oplus}{P}(SR)(OR')_2]Cl^{\ominus}\} \rightarrow R''PO(SR)(OR') + R'Cl$$

The importance of this modified synthesis is quite low.

7. THE MICHAELIS-BECKER REACTION. The Michaelis-Becker reaction may also be modified for the synthesis of thiophosphonic acid esters. As starting materials, sodium salts of thiophosphonous diesters are suitable.

$$(RO)_2SPNa + ClR' \rightarrow R'PS(OR)_2 + NaCl$$

Instead of the Na salts, mixtures of thiophosphonous acid diesters and Na alcoholates may be used. Since the reaction is normally carried out in inert organic solvents, for reasons of solubility the dibutyl ester is often preferred.[900, 901, 1633]

8. ADDITION OF PHOSPHONOUS ACID DERIVATIVES TO C-C MULTIPLE BONDS. Compounds of structure $HPS(OR)_2$ may be added to double and triple bonds in a manner similar to dialkylphosphites. With activated double bonds, base catalysis is indicated, whereas nonactivated double bonds best react with radical catalysis at elevated temperatures. UV-irradiation facilitates the reaction.

$$(RO)_2SPH + H_2C{=}CHR' \rightarrow (RO)_2SPCH_2CH_2R'$$ [1615]

$$(RO)_2SPH + H_2C{=}CHCO_2R' \rightarrow (RO)_2SPCH_2CH_2CO_2R'$$ [1670]

With activated triple bonds, addition of 2 moles $HPS(OR)_2$ may be achieved.[1670]

$$RO_2CC{\equiv}CCO_2R + 2HPS(OR)_2 \rightarrow RO_2C[PS(OR)_2]CHCH[PS(OR)_2]CO_2R$$

Similarly, $HPS(OR)_2$ also adds to ketene, with subsequent addition of a second mole of ketene to the primary product in the enol form:[1396]

$$HPS(OR)_2 + H_2C{=}C{=}O \rightarrow [CH_3COPS(OR)_2 \leftrightarrow CH_2{=}C(OH)PS(OR)_2] \xrightarrow{CH_2=C=O} CH_2{=}C(OCOCH_3)PS(OR)_2$$

Finally, α,β-unsaturated ketones (not, however, unsaturated aldehydes) also react with dialkylthiophosphites. Addition occurs exclusively to the C=C double bond and not to the carbonyl group. In the base catalyzed reaction, the thiophosphono group is attached to the β-C-atom. [1674]

$$HPS(OR)_2 + R'CH=CHCOR' \rightarrow (RO)_2SPCHR'CH_2COR'$$

In the addition of $\overline{SCH_2CH_2S}PCl$ to α,β-unsaturated ketones, a phosphonium salt intermediate is formed. Depending on the nature of the ketone, the intermediate undergoes rearrangement to dithiophosphonic acid S,S-diester or O,S-diester.

$$\overline{SCH_2CH_2S}\text{-}PCl + H_2C=CRCOCH_3 \rightarrow$$

$$\{[\overline{SCH_2CH_2S}\overset{\oplus}{P}CH_2CR=C(CH_3)O]Cl^{\ominus}\}$$

[R=CH₃]   [R=H]

$$(ClCH_2CH_2S)O\overline{P}CH_2C(CH_3)=C(CH_3)S \quad (ClCH_2CH_2S)\overline{S}PCH_2CH=C(CH_3)O$$

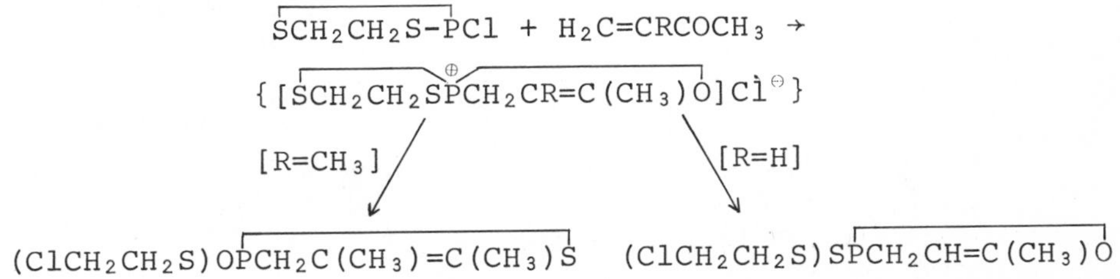

9. ADDITION OF THIODIALKYL OR THIODIARYL PHOSPHITES TO >C=O AND >C=S BONDS. In analogy to dialkylphosphites, $HPS(OR)_2$ also adds to >C=O or >C=S double bonds (with the exception of α,β-unsaturated ketones). [327, 1674]

$$(RO)_2SPH + R'R''C=O \rightarrow R'R''C(OH)PS(OR)_2$$

[R' = H, alkyl, aryl; R" = alkyl, aryl]

As catalysts, Na alcoholates are normally used. The addition to >C=S bonds, for example to CS_2, may be catalyzed by metallic sodium.

$$(RO)_2SPH + CS_2 \xrightarrow{Na} (RO)_2SPC(S)SNa$$

The product reacts further with alkyl halides according to:

$$(RO)_2SPC(S)SNa + R'Cl \rightarrow (RO)_2SPC(S)SR' + NaCl \quad [396a]$$

10. MANNICH REACTION, ADDITION TO SCHIFF BASES, AND OTHER REACTIVE C=N BONDS. The synthesis of phosphonic acid esters by addition of dialkylphosphites to compounds with C=N bonds may also be carried out with thiophosphonous acid diesters. Normally, the reaction of $HPS(OR)_2$ with C=N compounds leads to formation of thiophosphonic acid O,O-diesters. The reaction of aldehydes or ketones with RR'NH (R,R' = H, alkyl, aryl) and thiophosphonous acid diesters proceeds according to: [914, 916, 1221]

$$RR'NH + RR'C=O + HPS(OR)_2 \rightarrow RR'NCRR'PS(OR)_2$$

With NH_3, in a similar manner compounds of structure $H_2NCRR'PS(OR)_2$ are obtained. Subsequent reaction with isocyanates or thioisocyanates leads to formation of α-ureido- or α-thioureido-thiophosphonic acid esters, respectively. [906a]

The addition to azomethines is best carried out with alkaline catalysis: [1660]

$$R'N=CHR'' + HPS(OR)_2 \xrightarrow{NaOR} R''CH(NHR')PS(OR)_2$$

Addition of $HPS(OR)_2$ to aliphatic isocyanates proceeds without catalysis, whereas aromatic isocyanates require base catalysis.

$$R'N=C=O + HPS(OR)_2 \rightarrow R'NHCOPS(OR)_2$$

11. TRANSESTERIFICATIONS. While generally transesterifications are quite unimportant for the preparation of thiophosphonic acid esters, the following synthesis warrants some interest.

$$PCl_3 + 3RCHO \rightarrow [P(OCHClR)_3] \rightarrow RCHClPO(OCHClR)_2$$

While the product yields $RCHClPO(OH)_2$ upon hydrolysis, with mercaptanes the following subsequent reaction occurs:

$$RCHClPO(OCHClR)_2 \xrightarrow{3R'SH} RCH(SR')PO(SR')_2 + 2RCHO + 3HCl$$

12. MODIFICATION OF THE HYDROCARBON MOIETY OF THIO-PHOSPHONIC ACID DERIVATIVES. Similarly to phosphonic acid derivatives, reactions may also be carried out at the hydrocarbon moiety of thiophosphonic acid derivatives, especially if the organic rest is unsaturated or carries a reactive group (cf. Section V.14). The application of these reactions is somewhat limited, however, since both the P=S and P-S-C bonds are more easily hydrolyzed than the corresponding P-O bonds.

XI.   OTHER THIOPHOSPHONIC ACID ESTER DERIVATIVES

1. THIOPHOSPHONIC ACID ESTER SULFENYL CHLORIDES. These compounds may be prepared by the reaction of thiophosphonic acid O-monoesters with SO_2Cl_2. The synthesis proceeds with intermediate formation of bis(phosphono)disulfides: [288]

$$2(RO)RP\overset{O}{\underset{S}{\big\langle}} {}^{\ominus} \ H^{\oplus} + SO_2Cl_2 \rightarrow$$

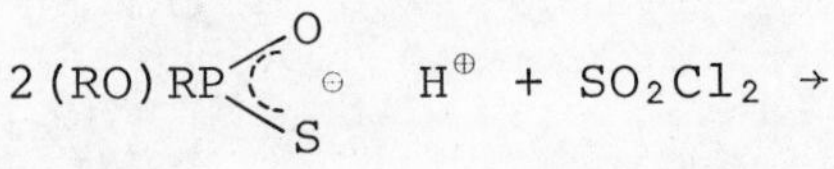

$$(RO)(R)OP\text{-}SS\text{-}PO(R)(OR) + SO_2 + 2HCl$$

$$\swarrow SO_2Cl_2$$

$$2RPO(OR)SCl + SO_2$$

Analogously, the sulfenyl chloride may be directly prepared from disulfide and SO_2Cl_2.

2. DISULFIDES OF PHOSPHONIC AND THIOPHOSPHONIC ACID ESTERS

a. If the reaction described in Section XI.1 is carried out with a stoichiometric ratio of ester: SO_2Cl_2 = 2:1, the intermediately formed disulfide may be isolated in satisfactory amounts.
b. By reaction of sulfenylchlorides or iodine with thiophosphonic acid monoesters, a number of disulfides may be obtained according to:

$$RPO(OR')SCl + (R'O)(R)P\begin{smallmatrix}O\\ S\end{smallmatrix}^{\ominus} \; H^{\oplus} \rightarrow$$

$$[RPO(OR')S\text{-}]_2 + HCl$$ [288]

$$RPS(OR')SH + ClSAr \rightarrow (R'O)(R)SPSSAr + HCl$$ [1859a]

$$2RPS(OR')(SK) + J_2 \rightarrow [(R'O)(R)SPS\text{-}]_2 + 2KJ$$ [1166]

XII. SELENOPHOSPHONIC ACID DERIVATIVES

1. SELENOPHOSPHONIC ACID MONOESTERS. Selenophosphonic acid esters are much less stable than the corresponding thio esters. The monoesters in form of their alkali salts have been prepared by addition of Se to alkali salts of phosphonous acid monoesters.[1190a]

$$(RO)OP(R)Na + Se \rightarrow (RO)RP\begin{smallmatrix}O\\ Se\end{smallmatrix}^{\ominus} \; Na^{\oplus}$$

2. SELENOPHOSPHONIC ACID DIESTERS. Selenophosphonic acid diesters may be prepared by the following methods:

a. Selenophosphonic acid dihalides react with alcohols in the presence of tertiary amines according to:[1710]

$$RPSeCl_2 + 2R'OH + 2NR_3 \rightarrow RPSe(OR')_2 + 2[R_3\overset{\oplus}{N}H]\overset{\ominus}{Cl}$$

b. Elemental selenium adds to dialkylphosphites according to:[1697, 1710]

$$RP(OR)_2 + Se \rightarrow RPSe(OR)_2$$

The exothermic reaction normally gives good yields. Instead of pure dialkylphosphites, the crude products from alkyldichlorophosphines and alcohols (in the presence of tertiary amines) may be directly caused to react with selenium. [723]

c. Similarly to dialkylphosphites, HPSe(OR)$_2$ also adds to activated double bonds with base catalysis: [1294]

$$NCCH=CH_2 + HPSe(OR)_2 \xrightarrow{NaOR'} NCCH_2CH_2PSe(OR)_2$$

XIII.   TELLUROPHOSPHONIC ACID ESTERS

Tellurophosphonic acid esters may be obtained by addition of tellurium to phosphonous acid dialkyl esters. The crude compounds decompose easily and may, therefore, not be purified by distillation. Adequate methods for purification of the crude compounds have not been developed. [723]

$$RP(OR')_2 + Te \rightarrow RPTe(OR')_2$$

XIV.   PHOSPHONIC ACID MONOAMIDES

1. HYDROLYSIS OF PHOSPHONIC ACID AMIDE HALIDES, IMIDES, OR DIAMIDES. Phosphonic acid monoamides are readily prepared by hydrolysis of phosphonic acid amide halides with diluted aqueous alkali bases or ammonia. With mineral acids, the monoamides are further hydrolyzed to free phosphonic acids. [1892]

$$RPO(NR_2')X + 2NaOH \longrightarrow RPO(NR_2')ONa + NaX + H_2O$$

The free monoamides are obtained by careful acidification of the solution. Some amide halides are already hydrolyzed by hot water. [851] Phosphonic acid amide azides behave similarly to the amide halides. [201]

Instead of the amide halides, phosphonic acid dihalides may be hydrolyzed in the presence of NH$_3$, primary, or secondary amines to yield the corresponding monoamides. [767, 1281]

The partial hydrolysis of aryl phosphonic acid imides also yields monoamides of aryl phosphonic acids: [1265, 1282]

$$Ph(O)P\underset{\underset{R}{N}}{\overset{\overset{R}{N}}{<}}P(O)Ph \xrightarrow{H_2O/OH^\ominus} 2PhPO(NHR)OH$$

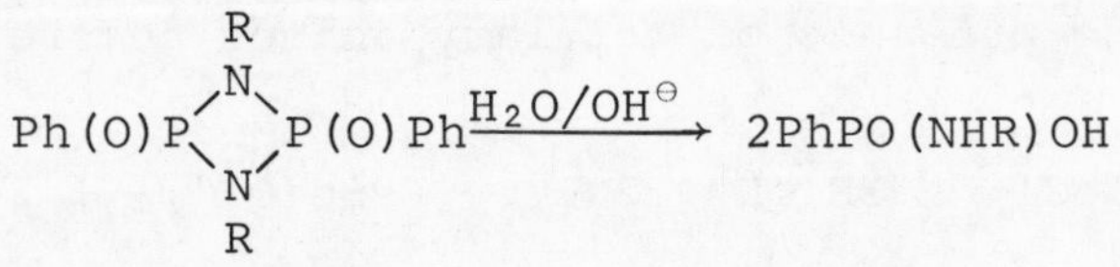

The hydrolysis of cyclic phosphonic acid diamides with aqueous bases proceeds in a similar fashion. [1947]

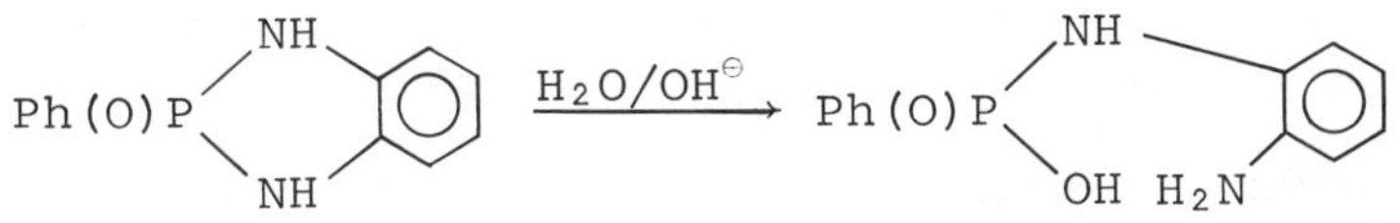

Phosphonic acid anhydrides react with secondary amines to yield monoamide salts according to:[375]

$$[PhPO_2]_n + 2nHNEt_2 \rightarrow [PhPO(NEt_2)O^{\ominus}][H_2\overset{\oplus}{N}Et_2]$$

2. HYDROLYSIS OF IMINOPHOSPHONIC ACID DERIVATIVES. N-Acyl- or aroyliminophosphonic acid dichlorides react with water under mild conditions according to:[465, 471, 1909]

$$RP(=NCOR')Cl_2 \xrightarrow{H_2O} RPO(NHCOR')OH + 2HCl$$

The reaction proceeds via intermediate formation of the corresponding phosphonic acid amide chlorides RPO(NHCOR')-Cl, which may be isolated if the dichloride is treated with anhydrous formic or acetic acid (cf. Section XV.2)

N-Arylsulfonyliminophosphonic acid diamides are quite stable towards water. By hot aqueous NaOH, however, they are hydrolyzed to yield arylsulfamides and sodium salts of phosphonic acid monoamides.[1895]

$$PhP(=NSO_2Ar)(NH_2)_2 \xrightarrow{H_2O/NaOH}$$

$$PhPO(NH_2)ONa + ArSO_2NH_2 + NH_3$$

XV. PHOSPHONIC ACID AMIDE HALIDES

1. REACTION OF PHOSPHONIC ACID DIHALIDES WITH AMINES. Phosphonic acid amide chlorides are most conveniently obtained from phosphonic acid dichlorides and 2 moles of NH_3, primary or secondary amines. Instead of a second mole of these amines, a tertiary amine may be used to bind the hydrochloric acid produced in the reaction. Normally, the reaction is carried out with cooling in inert solvents such as benzene, ether, or chlorinated hydrocarbons.

$$RPOCl_2 + NHR'_2 \xrightarrow{NR''_3} RPO(NR'_2)Cl + [R''_3\overset{\oplus}{N}H]Cl^{\ominus}$$

The ammoniumchloride may be separated from the product by filtration.

Phosphonic acid difluorides react simularly with exclusive formation of the amide fluoride.[1761, 1860] If the chloride fluorides are allowed to react instead of the difluorides, chloride is exchanged selectively to yield the amide fluoride.[851, 860]

Phosphonic acid amide fluorides may also be prepared by halide exchange from the corresponding chloride with alkali fluorides or alkali hydrogen fluorides. Similarly, amide chlorides exchange with alkali cyanates to form phosphonic acid amide isocyanates.[468]

Phosphonic acid diamides also react with KHF_2 to yield monoamide fluorides according to:[255a, 1704]

$$RPO(NR'_2)_2 + 2KHF_2 \rightarrow RPO(NR'_2)F + [R'_2\overset{\oplus}{N}H_2]F^{\ominus} + 2KF$$

2. ACIDOLYSIS OF IMINOPHOSPHONIC ACID DICHLORIDES N-Acyl- or aroyliminophosphonic acid dichlorides obtained from carboxylic acid amides and alkyl- or arylphosphorus tetrachloride are quite sensitive to water. They hydrolyze easily to yield the corresponding phosphonic acid monoamides (cf. Section XIV.2). By treatment with anhydrous formic acid, glacial acetic acid or the equimolar amount of H_2O in a benzene/ether mixture, phosphonic acid N-acyl or -aroyl amide chlorides may be obtained.

$$RCONH_2 + R'PCl_4 \xrightarrow[-2HCl]{} RCON{=}P(R')Cl_2 \xrightarrow{HCOOH} R'PO(NHCOR)Cl$$

With $R{=}CF_3$, the H_2O/benzene/ether hydrolysis gives better yields than the acidolysis.[1904]

The monomeric or dimeric phenyldichlorophosphazoaryles of composition $ArN{=}P(Ph)Cl_2$ obtained from $PhPCl_4$ and aromatic amines also react with anhydrous formic acid according to:[2100]

$$ArN{=}P(Ph)Cl_2 \xrightarrow{HCOOH} PhPO(NHAr)Cl$$

The raw product of $PhPCl_4$ and amine may be used without purification.

3. OXIDATION OF PHOSPHONOUS ACID DERIVATIVES. The oxidative addition of phosphonous acid dialkylamide dichlorides to olefines proceeds according to :[2108]

$$2(R_2N)PCl_2 + ClCH{=}CH_2 + O_2 \rightarrow$$
$$ClCH_2CHClPO(NR_2)Cl + Cl_2PO(NR_2)$$

The adduct of alkylphosphonous acid amide fluorides with chlorine may be treated with alcohols to yield phosphonic acid amide fluorides :[496]

$$CH_3P(NEt_2)FCl_2 \xrightarrow{2EtOH} CH_3PO(NEt_2)F + 2EtCl + H_2O$$

Arylphosphonous acid ester amides react with BrCN with formation of arylphosphonic acid amide cyanides in a Michaelis-Arbuzov-type reaction :[1826]

$$ArP(OR)(NR_2') + BrCN \rightarrow ArPO(NR_2')CN + RBr$$

4. REACTION OF PHOSPHONIC ACID ISOCYANATE AND THIOISOCYANATE HALIDES WITH AMINES, ALCOHOLS, AND C,H-ACTIVE COMPOUNDS. Phosphonic acid isocyanate or thioisocyanate halides are easily prepared by reaction of the dichloride or chloride fluoride with NaOCN or NaSCN, respectively. The very reactive monoisocyanates react readily in ethereal solution with equimolar amounts of alcohols or aromatic amines according to :[1891, 1892]

$$RPO(NCO)X \xrightarrow{R'OH} RPO(NHCO_2R')X$$

$$RPO(NCO)X \xrightarrow{NH_2R'} RPO(NHCONHR')X$$

The addition is exothermic. Thioisocyanate halides react in a similar manner.

Compounds with reactive C-H bonds such as ketene O,N- or N,N-acetals also add to equimolar amounts of isocyanate halides in exothermic reaction.[853, 855]

$$RPO(NCO)F + H_2C{=}C(Ph)(NC_5H_{10}) \rightarrow RPO[NHCOCH{=}CPh(NC_5H_{10})]\ F$$

5. FURTHER REACTIONS Phosphonic acid azide halides react with phosphines or phosphites with liberation of N_2 to yield compounds of composition $RPO(N{=}PR_3)X$.[1910]

From a mixture of $CH_3PO(NHCOCCl_3)Cl$ and PCl_5, upon heating to 120^0, $CH_3P(N{=}CClCCl_3)Cl$ is obtained as a product[1906].

XVI. PHOSPHONIC ACID ESTER AMIDES

1. REACTION OF PHOSPHONIC ACID ESTER HALIDES AND AMINES. Phosphonic acid ester halides react readily with 2 moles of NH_3, primary, or secondary amines in nonaqueous solvents. One mole of the amine for binding of HX may be substituted by a tertiary amine.[861]

$$RPO(OR')X + 2HNR''_2 \rightarrow RPO(OR')(NR''_2) + [R''_2\overset{\oplus}{N}H_2]X^{\ominus}$$

$$RPO(OR')X + H\overline{NCH_2CH_2} \xrightarrow{NEt_3} RPO(OR')(\overline{NCH_2CH_2}) + [Et_3\overset{\oplus}{N}H]X^{\ominus}$$

The reaction proceeds stereospecifically with inversion of configuration.[2]

Phosphonic acid ester amides may undergo exchange reactions with other amines.[2097]

$$CH_3PO(OPh)(NHPh) + HOCH_2CH_2NH_2 \rightarrow$$
$$CH_3PO(OPh)(NHCH_2CH_2OH) + PhNH_2$$

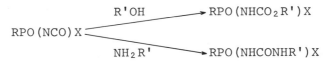

Pyrophosphonic acid diesters are cleaved by amines to yield phosphonic acid ester amides.[326]

2. REACTION OF PHOSPHONIC ACID AMIDE HALIDES AND ALCOHOLS. Phosphonic acid amide halides react with alcohols in the presence of tertiary amines to yield ester amides. Complications in this reaction may arise from the formation of anhydrides as by-products.[1703]

$$RPO(NR'_2)Cl + R''OH \xrightarrow[-HCl]{NR'''_3} RPO(OR'')(NR'_2)$$

With phenols, it is often preferred to use the alkali phenolates.[1893, 1906]

Without isolation of the intermittently formed amide halides, phosphonic acid ester amides may be directly obtained from the reaction of the dichloride with an alcohol and an amine in a molar ratio of 1:1:3.[517, 1722] Phosphonic acid diamides may also react with alcohols to yield ester amides.[1931]

$$PhPO(NH_2)_2 + BuOH \rightarrow PhPO(OBu)(NH_2) + NH_3$$

3. MICHAELIS-ARBUZOV REACTION OF PHOSPHOROUS ACID DIESTER AMIDES. Phosphorous acid diester amides react with alkyl halides,[891] acyl halides,[23] or aroyl halides[61] with formation of Michaelis-Arbuzov-type intermediates. Reaction with CCl_4 is also observed.[27] As a complication, quaternization of the amide nitrogen atom may occur. The phosphonium intermediates may rearrange at the P atom to yield phosphonic acid ester amides, or a Perkow rearrangement may lead to phosphoric acid diester amides. The Perkow reaction is favored, when the halogen atom is in conjugation with a carbonyl group.[19, 20]

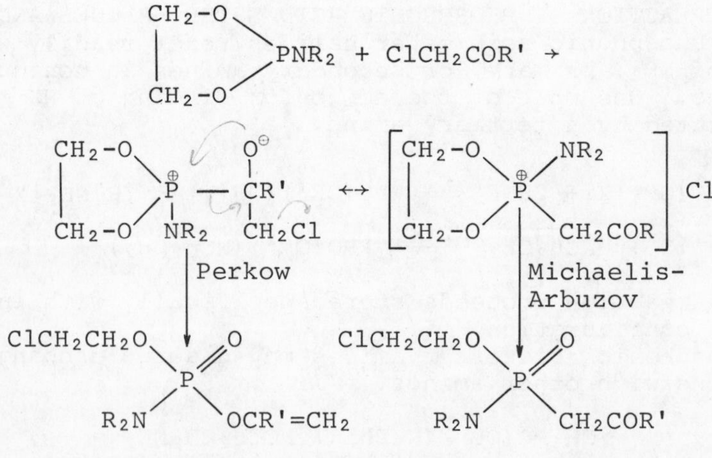

Normally, the reaction with alkyl halides is run in a sealed tube. With α-halo-ketones, -aldehydes, or esters, one may carry out the reaction in refluxing toluene or without a solvent at 140-160°. [1238, 1658]

Phosphorous acid diester amides also undergo thermal rearrangement with formation of phosphonic acid ester amides. [1637] In the case of propargyl esters, spontaneous exothermic rearrangement leads to allene phosphonic acid ester amides. [24]

$$HC\equiv CCH_2OP(OR)(NR'_2) \rightarrow H_2C=C=CHPO(OR)(NR'_2)$$

Epichlorohydrine also reacts with $(RO)_2P(NR_2)$ partially in a Michaelis-Arbuzov-type reaction to yield methylphosphonic acid ester amides. [39]

4. ADDITION OF PHOSPHOROUS ACID ESTER AMIDES TO C=C, C=O BONDS AND LACTONES. Phosphorous acid ester amides $(RO)(R'_2N)POH$ react with activated C=C double bonds, for example, those of acrylic acid esters or acrylonitrile. [1636] The addition is best catalyzed by NaOR.

$$(RO)(R'_2N)P(O)H + CH_2=CHCN \xrightarrow{NaOEt} NCCH_2CH_2PO(OR)(NR'_2)$$

Similar products are obtained from the addition of phosphorous acid diester amides to acrylic acid. [1635] No catalysts are required in this reaction.

$$(Et_2N)P(OEt)_2 + CH_2=CHCOOH \rightarrow EtO_2CCH_2CH_2PO(OEt)(NEt_2)$$

With lactones, the diester amides react upon heating according to : [425]

$$(Et_2N)P(OEt)_2 + \begin{array}{c} CH_2-C \diagup^O \\ | \qquad\quad \diagdown O \\ CH_2-CH_2 \end{array} \rightarrow EtO_2CCH_2CH_2CH_2PO(OEt)(NEt_2)$$

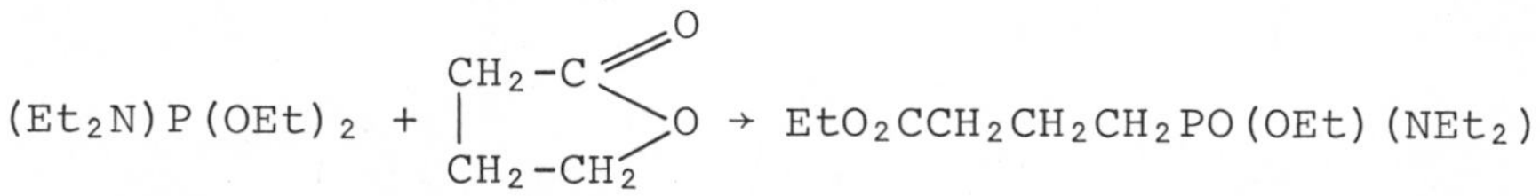

Phosphorous acid ester amides also add to C=O double bonds of α-keto acid esters, [1591] ketones, [28] and aldehydes. [28, 724]

The exothermic reactions require no catalysts. At higher temperatures, the products tend to rearrange to phosphate esters.

$$(MeO)(NEt_2)P(O)H + Me_2CO \rightarrow Me_2C(OH)PO(OMe)(NEt_2)$$

In a similar manner, phosphorous acid diester amides add to aldehydes [25, 724] and ketones. [25]

$$(NHPh)P(OPh)_2 + O=C \overset{}{\bigcirc} \rightarrow \bigcirc\!\!-\!\!\underset{OPh}{\overset{}{\text{—}}}\!PO(OPh)(NHPh)$$

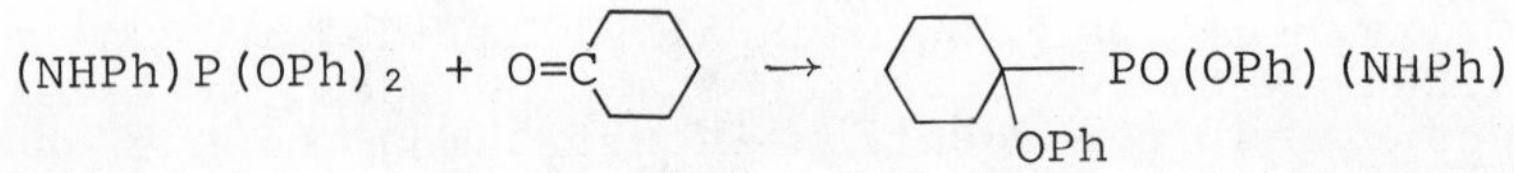

These addition reactions probably proceed via onium-type adducts.

With α-oxophosphonic acid diesters, diphosphonic acid ester amides are formed. [1591]

$$(EtO)(NEt_2)P(O)H + MeCOPO(OEt)_2 \rightarrow$$

$$\underset{Me}{\overset{OH}{(EtO)_2P(O)\overset{|}{\underset{|}{C}}\text{-}PO(OEt)(NEt_2)}}$$

5. FROM IMINOPHOSPHONIC ACID DERIVATIVES. Iminophosphonic acid dihalides react exothermically with anhydrous alcohols or phenols to yield the corresponding phosphonic acid ester amides. [464, 1902]

$$ArP(=NCOAr')Cl_2 \xrightarrow{2ROH} ArPO(OR)(NHCOAr') + RCl + HCl$$

Iminophosphonic acid diesters undergo a facile thermal rearrangement to yield phosphonic acid ester amides according to: [1914]

$$MeP(=NCO_2Me)(OMe)_2 \xrightarrow{120\text{-}140^0} MePO(OMe)(NMeCO_2Me)$$

The imide-amide rearrangement may be catalyzed by HF or alkyl halides. [627, 1138]

N-Sulfonyliminophosphonic acid dichlorides react in benzene solution with an excess sodium alcoholate and subsequent acidification to yield N-sulfonyl phosphonic acid ester amides. [1889, 1896]

$$PhP(=NSO_2R)Cl_2 \xrightarrow[2.\ H^{\oplus}]{1.\ NaOR'} PhPO(OR')(NHSO_2R)$$

Diesters of N-sulfonyliminophosphonic acids isomerize thermally at temperatures above 200^0. [1898] They may also be hydrolyzed with aqueous/alcoholic NaOH. [1897]

$$PhP(=NSO_2Ar)(OR)_2 \xrightarrow{H_2O/OH^{\ominus}} PhPO(OR)(NHSO_2Ar) + ROH$$

6. REACTION OF PHOSPHONIC ACID ESTER ISOCYANATES WITH AMINES, ALCOHOLS, OR C,H-ACTIVE COMPOUNDS. Phosphonic acid ester isocyanates are generally prepared by the reaction of the corresponding ester chlorides with NaOCN. They react vigorously with aromatic amines and alcohols to yield the corresponding ureides and urethanes, respectively.

$$RPO(OR')NCO \xrightarrow{ArNH_2} RPO(OR')(NHCONHAr)$$

$$RPO(OR')NCO \xrightarrow{R''OH} RPO(OR')(NHCOOR'')$$

The corresponding reactions may be carried out with ester thioisocyanates. With moisture, the ester isocyanates decompose with formation of the N-unsubstituted amides.[280]

$$RPO(OR')NCO + H_2O \rightarrow RPO(OR')NH_2 + CO_2$$

The isocyanates also combine readily with equimolar amounts of methylene bases such as nitrogen heterocycles, ketone O,N-, and N,N-acetals.[594, 853]

$$RPO(OR')NCO + H_2C{=}CXY \rightarrow RPO(OR')(NHCOCH{=}CXY)$$

7. REACTION OF PHOSPHONOUS ACID MONOESTERS WITH CCl_4 AND AMINES. Phosphonous acid monoesters react with primary and secondary amines in CCl_4 solution to form phosphonic acid ester amides in high yields (See also Chapter 10).[68]

$$PhP(OCH_2Ph)(O)H + CCl_4 + 2C_6H_{11}NH_2 \rightarrow$$

$$PhPO(OCH_2Ph)(NHC_6H_{11}) + CHCl_3 + [C_6H_{11}\overset{\oplus}{N}H_3]Cl^{\ominus}$$

In a similar manner, chloromethylphosphonous acid esters react with secondary amines through a series of elimination-addition reactions and a hydrogen shift to yield methylphosphonic acid ester amides.[648]

$$ClCH_2P(OMe)(OH) + 2Me_2NH \xrightarrow{Et_2O}$$

$$CH_3PO(OMe)(NMe_2) + [Me_2\overset{\oplus}{N}H_2]Cl^{\ominus}$$

8. FURTHER METHODS. Phosphonic acid ester azides react with phosphines or phosphites by loss of nitrogen according to:[202, 1913]

$$RPO(OR')N_3 + PR''_3 \rightarrow RPO(OR')(N{=}PR''_3) + N_2$$

In the case of phosphite adducts, thermal rearrangement of the products may occur:

$$CH_3PO(OEt)[N{=}P(OMe)_3] \xrightarrow{175^0} CH_3PO(OEt)[N(Me)PO(OMe)_2]$$

Phosphonic acid ester isothiocyanates react with chlorine according to:

$$CH_3PO(OR)NCS + Cl_2 \rightarrow CH_3PO(OR)N{=}CCl_2 + SCl_2$$

The methylene Cl atoms may be further reacted with amines.[845]

Phosphonous acid ester amides react with chloral via phosphonium salt intermediates to yield vinyl ester amides:[1746]

$$C_2H_5P(OR)(NMe_2) + CCl_3CHO \rightarrow C_2H_5PO(NMe_2)(OCH{=}CCl_2) + RCl$$

9. MODIFICATION OF PHOSPHONIC ACID ESTER AMIDES. Modifications of phosphonic acid ester amides may be carried out at the organic group attached to the P atom or with the amide group. An unsaturated P-bonded organic group may be isomerized[24] or brominated,[1897] and subsequent dehydrobrominations may be carried out.[1897]

An unsubstituted amide group may react with organic isocyanates to yield urethanes;[517] a secondary amide group may react with formaldehyde[663] or be metallated and further treated with acyl halides.[1914] Unsaturated amide groups may be hydrogenated.[1722]

XVII. PHOSPHONIC ACID DIAMIDES

1. FROM PHOSPHONIC ACID DIHALIDES AND AMINES. Phosphonic acid diamides are generally prepared by the reaction of the dichlorides with 4 moles of NH_3, primary or secondary amines. The amount of amine necessary to bind HCl may be substituted by a tertiary base. Normally, the reaction is carried out in inert solvents such as benzene, ether, or chlorinated hydrocarbons. In aqueous solutions, hydrolysis to monoamides or free phosphonic acids may occur. With primary amines, the reaction is smooth and heating should be avoided to prevent formation of phosphonic acid imides. Secondary amines are less reactive and require heating. Unreactive amines may be heated with the dichloride without solvents.

$$RPOCl_2 + 4HNR'_2 \rightarrow RPO(NR'_2)_2 + 2[R'_2\overset{\oplus}{N}H_2]Cl^{\ominus}$$

In the case of N-heterocycles, the metallated hetero-cycles may be used.[1447]

With diamines or oxalimidic esters, formation of five- and six-membered heterocycles is observed.[466, 1792, 2042] Aromatic o-diamines react in a 2-step fashion and require elevated temperatures. With aromatic 1,4-diamines, polymeric phosphonic acid diamides are formed. Instead of ethylene diamine, silicon or germanium imidazolidines may be used.[2096]

A halogenorganic group bound to a P atom normally remains unchanged. However, with 1,2-dihaloethane- and 4-chlorobutenephosphonic acid dihalides, hydrohalides are cleaved to yield 1-haloethene- and butadiene phosphonic acid diamides, respectively.[1223, 1916]

Upon treatment with NaOH/MeOH, γ- and δ-chloroalkane phosphonic acid diamides of primary amines are transformed to phostamic acid amides.[767, 768]

$$Cl(CH_2)_3PO(NHPh)_2 \rightarrow Ph\overline{N(CH_2)_3P}O(NHPh) + HX$$

The preparation of asymmetric phosphonic acid diamides may be achieved by the stepwise action of two different amines on phosphonic acid dichlorides. The intermediately formed phosphonic acid amide chlorides may be isolated, but this is normally not necessary. Phosphonic acid diamides react at elevated temperatures (>200°) with other amines by exchange of the amide group.[681, 731, 2007, 2097]

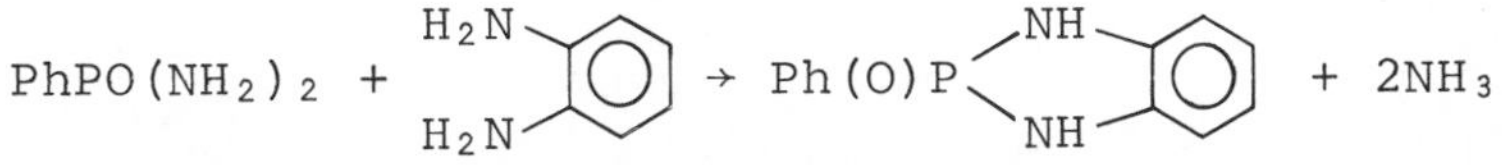

2. CLEAVAGE OF PHOSPHONIC IMIDES BY AMINES. Phosphonic acid imides are cleaved by amines at elevated temperatures with formation of symmetric or asymmetric phosphonic acid diamides.[253, 1282]

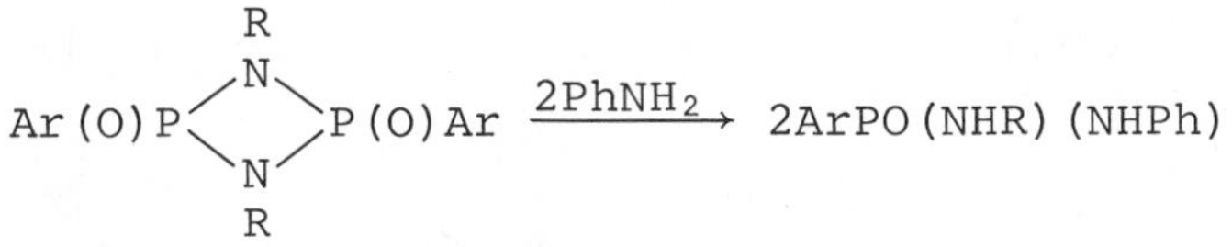

This method is of negligible preparative importance.

3. MICHAELIS-ARBUZOV REACTION OF PHOSPHOROUS ACID ESTER DIAMIDES. Phosphorous acid ester diamides react with organic halides to form phosphonium halide compounds. The reaction is not smooth, since the presence of tertiary nitrogen atoms leads to quaternization and by-product formation. Upon heating, alkyl halide is cleaved from the phosphonium halide intermediates, and via a Michaelis-Arbuzov rearrangement phosphonic acid diamides are formed.[1275]

As a competing reaction, a Perkow rearrangement leads to unsaturated esters of diamidophosphoric acid. The relative rates of these reactions, for example, with keto-halide compounds, are largely dependant on the reaction temperature and the nature and the position of the halide atom.[61, 516, 1570]

$$(Et_2N)_2P(OR) + ClCH_2COCH_3 \xrightarrow{\text{Michaelis-Arbuzov}} (Et_2N)_2P(O)CH_2COCH_3 + RCl$$

$$(Et_2N)_2P(OR) + ClCH_2COCH_3 \xrightarrow{\text{Perkow}} (Et_2N)_2P(O)OC(Me){=}CH_2 + RCl$$

The reaction is normally carried out in refluxing toluene or xylene or without solvents.[1238]

With chloroform and $(R_2N)_2P(OR')$, both dichloromethane phosphonic acid diamides and diamidophosphoric acid chlorides are obtained. CCl_4 only forms the latter products.[27]

4. ADDITION OF PHOSPHOROUS ACID DERIVATIVES TO C=C C=N, C=O DOUBLE BONDS, AND β-LACTONES. The addition of phosphorous acid diamides to activated olefines, azomethines, or arylsulfonylimines and azodicarboxylic acid bismorpholide leads to phosphonic acid diamide derivatives.[1248]

$$(R_2N)_2P(O)H + R'CH{=}CHY \rightarrow (R_2N)_2P(O)CR'HCH_2Y$$

The addition proceeds smoothly even at ambient temperature, if NaH, NaOR, or NEt_3 are used as catalysts.[1248]

Phosphorous acid ester diamides also add to acrylic acid or acrylic acid derivatives.[1657, 1665]

$$2(Et_2N)_2P(OEt) + 2CH_2{=}CHCOOH \rightarrow$$

$$EtO_2CCH_2CH_2PO(NEt_2)_2 + Et_2NCOCH_2CH_2PO(OEt)(NEt_2)$$

With β-lactones, phosphorous acid ester diamides react according to :[425]

$$(R_2N)_2P(OEt) + \begin{array}{l} CH_2{-}O \\ | \quad\;\; | \\ CH_2{-}C{=}O \end{array} \rightarrow EtO_2CCH_2CH_2PO(NR_2)_2$$

Phosphorous acid triamides add to aldehydes with catalysis by amine hydrochlorides.[1390]

$$P(NMe_2)_3 + C_6H_5CHO \rightarrow C_6H_5CH(NMe_2)PO(NMe_2)_2$$

Phosphorous acid diamides may also be aminoalkylated directly.[1390]

$$(Me_2N)_2P(O)H + C_6H_5CH(NMe_2)_2 \rightarrow$$

$$C_6H_5CH(NMe_2)PO(NMe_2)_2 + Me_2NH$$

5. HYDROLYSIS OF IMINOPHOSPHONIC ACID DERIVATIVES Iminophosphonic acid dichlorides, obtained from carboxylic acid amides and $RPCl_4$, react with amines. Subsequent hydrolysis yields phosphonic acid diamides.[1890, 1906]

$$CCl_3CON{=}P(Me)Cl_2 \xrightarrow[2.\ H_2O]{1.\ PhNH_2} MePO(NHPh)(NHCOCCl_3)$$

Iminophosphonic acid amide chlorides obtained from $RP(NR_2)C$ and aromatic azides are hydrolyzed to yield phosphonic acid diamides.[790]

N-Arylsulfonyliminophosphonic acid amide halides are also readily hydrolyzed to phosphonic acid diamides.[1894]

$$PhP(=NSO_2Ar)(NHPh)Cl \xrightarrow{H_2O} PhPO(NHPh)(NHSO_2Ar) + HCl$$

Iminophosphonic acid diamides, upon treatment with aqueous alcoholic solutions of NaOH, yield sodium salts according to:

$$PhP(=NSO_2Ar)(NHPh)_2 \xrightarrow{NaOH/H_2O} PhPO(NHPh)(\underset{\underset{Na}{|}}{NSO_2Ar})$$

Acidification leads to the free acid.[1894]

6. REACTION OF PHOSPHONIC ACID DIISOCYANATES WITH AMINES, ALCOHOLS, AND C,H-ACIDIC COMPOUNDS. Phosphonic acid dichlorides react readily with NaOCN in benzene solution with formation of the corresponding diisocyanates. These combine vigorously with amines and alcohols or phenols to yield diureides and diurethanes, respectively.[469, 679, 762]

$$RPO(NCO)_2 \underset{2R'OH}{\overset{2R'NH_2}{<}} \begin{array}{l} RPO(NHCONHR')_2 \\ RPO(NHCOOR')_2 \end{array}$$

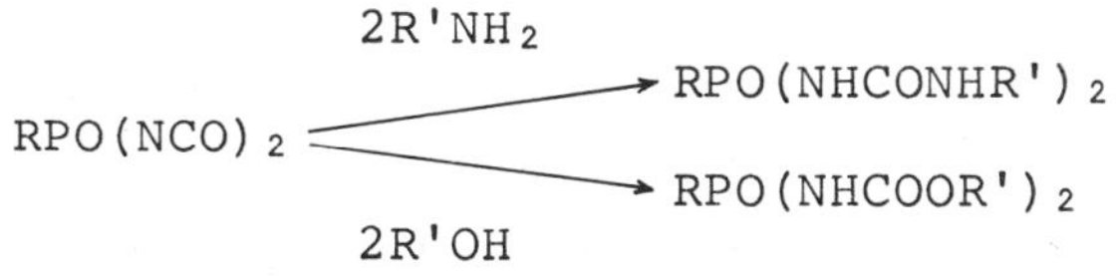

Analogously, from amide isocyanates the corresponding amide ureides and amide urethanes may be obtained.[468]

Methylene compounds with activated C-H bonds add to diisocyantes according to:[853, 855]

$$RPO(NCO)_2 + H_2C=C(NC_5H_{10})(Ph) \longrightarrow$$

$$R(O)P\underset{NHCO}{\overset{NHCO}{<}}C=C(NC_5H_{10})(Ph)$$

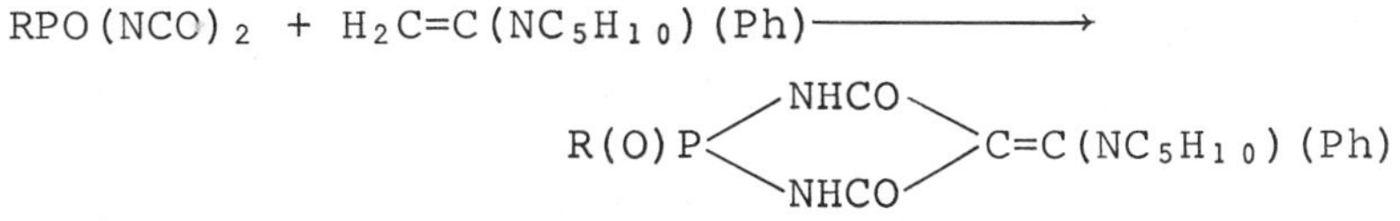

The diurethanes react with phosgene or simply by heating to yield cyclic derivatives.[2010, 2011]

$$RPO(NHCONHR')_2 \xrightarrow{-NH_2R'} R(O)P\underset{NHCO}{\overset{NHCO}{<}}NR'$$

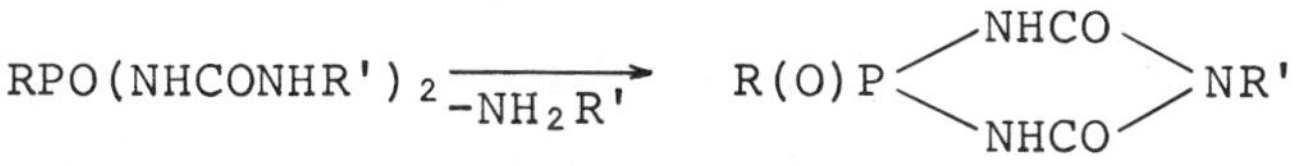

These compounds may also be obtained from the reaction of equimolar amounts of diisocyanates and amine.[2011]

Hydrazines and oximes add to phosphonic acid diisocyanates to yield the corresponding derivatives.[854]

$$RPO(NCO)_2 + 2H_2NNHPh \longrightarrow RPO(NHCONHNHPh)_2$$

$$RPO(NCO)_2 + 2Me_2C=NOH \longrightarrow RPO(NHCOON=CMe_2)_2$$

7. SYNTHESIS VIA METALLATION OF PHOSPHORIC ACID AMIDES. Phosphoric acid diamide chlorides react with Na metal to

yield $NaPO(NR_2)_2$. With alkyl halides, the corresponding phosphonic acid diamides are obtained.[1406]

$$ClPO(NR_2)_2 \xrightarrow[-\ NaCl]{2Na} Na^{\oplus}[\overset{\ominus}{P}O(NR_2)_2] \xrightarrow[-\ NaX]{RX} RPO(NR_2)_2$$

Phosphoric acid diamide chlorides also react directly with Grignard compounds. This reaction is important for the synthesis of aromatic phosphonic acid diamides.[1339a]

$$PhMgBr + ClPO(NHPh)_2 \longrightarrow PhPO(NHPh)_2 + MgBrCl$$

Hexamethylphosphoric triamide $PO(NMe_2)_3$ reacts with Na metal by cleavage of the anion $\overset{\ominus}{N}Me_2$ and formation of the salt $NaPO(NMe_2)_2$.[1409]

With α,β-unsaturated ketones, this salt reacts according to:[1407]

$$PhCH{=}CHCOPh \xrightarrow[-NaOH]{1.\ NaPO(NMe_2)_2 \quad 2.\ H_2O} PhCOCH_2CHPhPO(NMe_2)_2$$

8. OXIDATION OF PHOSPHONOUS ACID DIAMIDES. Phosphonous acid diamides add chlorine, and hydrolysis of the adducts yields phosphonic acid diamides.[1291]

$$RP(NR'_2)_2 + Cl_2 \longrightarrow RPCl_2(NR'_2)_2 \xrightarrow[-2HCl]{H_2O} RPO(NR'_2)_2$$

The oxidation may also be carried out directly with air,[361] H_2O_2,[809] or N_2O_4.[466] In the case of N_2O_4, careful dosage must be applied in order to avoid further oxidation and polymerization.

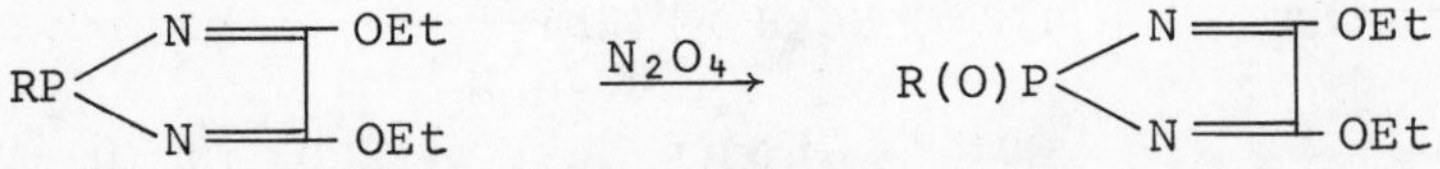

9. FURTHER METHODS. Phosphonic acid diazides or amide azides react readily at ambient temperature with phosphines or phosphites to form phosphonic acid phosphoranes.[202, 191]

$$RPO(N_3)_2 + 2PR''_3 \xrightarrow[-\ 2N_2]{} RPO(N{=}PR''_3)_2$$

$$RPO(NR'_2)N_3 + PR''_3 \xrightarrow[-\ N_2]{} RPO(NR'_2)(N{=}PR''_3)$$

Phosphorous acid diamides add to aromatic isocyanates according to:[1391]

$$(Et_2N)_2P(O)H + ArNCO \rightarrow ArNHCOPO(NEt_2)_2$$

$(Me_2N)_2PCl$ reacts with propargylalcohol to yield vinylphosphonic acid diamide.[1189]

Butadiene and PCl_5 react in acetic acid solution to yield 4-chlorobutenephosphonic acid dichloride, which with secondary amines yields butadienephosphonic acid diamides.[1916]

10. MODIFICATION OF PHOSPHONIC ACID DIAMIDES. The most important method for the modification of alkylphosphonic acid diamides is the lithiation of the acidic methylene group in α-position to the P atom. The organometallic derivative thus obtained reacts with alkyl halides simply by formation of secondary alkylphosphonic acid diamides.[439, 1081]

With aldehydes or ketones, however, β-hydroxyalkyl phosphonic acid diamides are obtained, which upon thermolysis by refluxing in benzene or toluene yield substituted olefines that may only be obtained with great difficulty by other routes.[438, 439]

$$R^1R^2CHPO(NR_2)_2 \xrightarrow[2.\ R^3R^4C=O]{1.\ BuLi} R^1R^2C(OH){-}CR^3R^4PO(NR_2)_2$$

$$\xrightarrow{\Delta} R^1R^2C{=}CR^3R^4 + (R_2N)_2PO_2H$$

The reaction is stereospecific; i.e., from a pair of diastereomers, two isomeric olefines may be obtained.[438] With allylphosphonic diamides, the method can be extended to the synthesis of substituted dienes.[438]

Vinylphosphonic acid diamides add Me_2NH to yield β-dimethylaminoethylphosphonic acid diamides.[908] Nitroaromatic phosphonic acid diamides may be reduced to the corresponding amino derivatives.[572] Further methods to modify diamides include the reaction of an NH bond of the amide group with acyl or aroyl halides[768] or with potassium and subsequent alkylation with alkyl halides.[735]

XVIII. PHOSPHONIC ACID HYDRAZIDES AND AZIDES

1. PHOSPHONIC ACID MONO- AND DIHYDRAZIDES. Phosphonic acid dichlorides react with aqueous solutions of substituted hydrazines to yield monohydrazides.[1240]

$$MePOCl_2 + H_2NNHCOPh \xrightarrow[-\ 2HCl]{NEt_3} MePO(NHNHCOPh)OH$$

In inert solvents, dichlorides and excess hydrazine or 2 moles of hydrazine and tertiary amine yield phosphonic acid dihydrazides.[1260, 1281, 1930]

$$RPOCl_2 + 2H_2NNR'_2 \xrightarrow[-\ 2HCl]{NEt_3} RPO(NHNR'_2)_2$$

N-Unsubstituted dihydrazides form dihydrazones with carbonyl compounds.[1930]

Phosphonic acid dichlorides react with benzoylhydrazide at elevated temperatures by formation of phosphaoxotetraazaring compounds.[662, 1240]

$$\text{RPOCl}_2 + 2\text{H}_2\text{NNHCOPh} \xrightarrow{4\text{NEt}_3} R(O)P\begin{array}{c}\text{NH}-\text{N}=\text{C}-\text{Ph}\\ \diagdown O\\ \text{NH}-\text{N}=\text{C}-\text{Ph}\end{array}$$

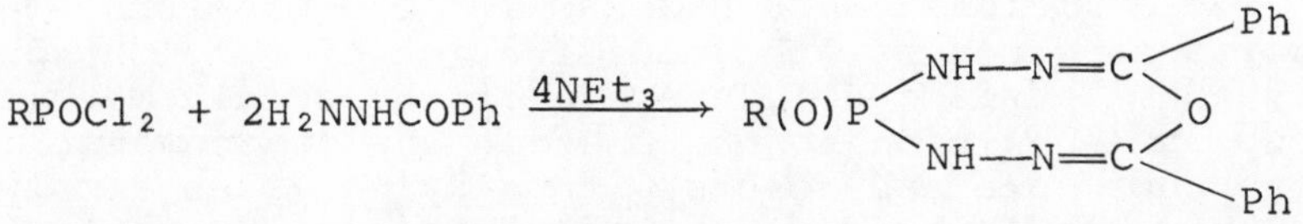

2. PHOSPHONIC ACID ESTER HYDRAZIDES. Phosphonic acid ester halides and 2 moles of hydrazine or 1 mole of hydrazine and a tertiary amine react to form phosphonic acid ester hydrazides.

$$\text{RPO(OR')Cl} + \text{H}_2\text{NNR}_2'' \xrightarrow{\text{NEt}_3} \text{RPO(OR')(NHNR}_2'')$$

Unsubstituted hydrazide esters react with carbonyl compounds by formation of ester hydrazones.[661]

$$\text{MePO(OPh)(NHNH}_2) + O=\text{CMe}_2 \rightarrow \text{MePO(OPh)NHN=CMe}_2$$

3. PHOSPHONIC ACID AMIDE HYDRAZIDES. Phosphonic acid am halides react with hydrazines according to:

$$\text{RPO(NR}_2')\text{X} + \text{H}_2\text{NNR}_2'' \xrightarrow[-\text{HX}]{\text{NEt}_3} \text{RPO(NR}_2')(\text{NHNR}_2'')$$

HX may be bound either by a tertiary amine or by excess hydrazine. Phosphonic acid diamides also react with hydrazines by exchange of an amide vs. a hydrazide group.[1931]

$$\text{RPO(NH}_2)_2 + \text{H}_2\text{NNH}_2 \rightarrow \text{RPO(NH}_2)(\text{NHNH}_2) + \text{NH}_3$$

4. PHOSPHONIC ACID AZIDE DERIVATIVES The halide groups of phosphonic acid dihalides, amide halides, or ester halides are readily exchanged with NaN_3 to yield the corresponding azide derivatives. With dihalides, the reaction proceeds stepwise.[202, 1910]

$$\text{RPOCl}_2 + \text{NaN}_3 \rightarrow \text{RPO(Cl)N}_3 + \text{NaCl}$$

$$\text{RPO(Cl)N}_3 + \text{NaN}_3 \rightarrow \text{RPO(N}_3)_2 + \text{NaCl}$$

$$\text{RPO(NR}_2')\text{Cl} + \text{NaN}_3 \rightarrow \text{RPO(NR}_2')\text{N}_3 + \text{NaCl}$$

Amide azides may also be obtained by the reaction of azide halides with amines. With NaOCN, phosphonic acid azide halides yield the corresponding azide isocyanates, which react further with amines or alcohols to form azide ureides and azide urethanes, respectively.[1910]

With triarylphosphines, phosphonic acid diazides react in a stepwise process with loss of N_2 to yield phosphoranes according to:[202]

$$\text{RPO(N}_3)_2 + \text{PPh}_3 \rightarrow \text{RPO(N=PPh}_3)(\text{N}_3) + \text{N}_2$$

$$\text{RPO(N=PPh}_3)(\text{N}_3) + \text{PPh}_3 \rightarrow \text{RPO(N=PPh}_3)_2 + \text{N}_2$$

XIX. THIOPHOSPHONIC ACID MONOAMIDES

The preparation of thiophosphonic acid amides has not been investigated in any detail. Possibly, phenyl thiophosphonic acid monoanilide is formed in the reaction of phenyl phosphine with thionylaniline.[98]

XX. THIOPHOSPHONIC ACID AMIDE HALIDES

1. THIOPHOSPHONIC ACID DIHALIDES AND AMINES. Thiophosphonic acid dichlorides react with 2 moles of NH_3, primary or secondary amines, or 1 mole of an amine in the presence of tertiary bases to yield thiophosphonic acid amide chlorides.[1254]
Dibromides react in a similar manner.[1151]

$$RPSCl_2 + 2HNR'_2 \rightarrow RPS(NR'_2)Cl + HNR'_2 \cdot HCl$$

The difluorides react even with an excess of amine below 40° to yield the amide fluorides selectively.[1843]

Unsubstituted thiophosphonic acid amide fluorides react with PCl_5 or Me_3SiCl in the presence of NEt_3 according to:[1759]

$$RPS(NH_2)F + PCl_5 \xrightarrow{CCl_4} RPS(N{=}PCl_3)F + 2HCl$$

$$RPS(NH_2)F + Me_3SiCl \xrightarrow[-\ HCl]{Et_3N} RPS(NHSiMe_3)F$$

2. FROM PHOSPHONOUS ACID AMIDE HALIDES. Phosphonous acid amide halides readily add elemental sulfur upon heating in the presence of catalytic amounts of $AlCl_3$.[154, 458]

$$EtP(NEt_2)Cl + S \rightarrow EtPS(NEt_2)Cl$$

3. REACTION OF THIOPHOSPHONIC ACID ISOCYANATE HALIDES WITH AMINES OR C,H-ACTIVE COMPOUNDS. Phosphonous acid isocyanate fluorides react with $PSCl_3$ to yield thiophosphonic acid isocyanate fluorides. These isocyanates add aromatic amines and C,H-acidic compounds according to:

$$RPS(NCO)F + ArNH_2 \rightarrow RPS(NHCONHAr)F$$

$$RPS(NCO)F + H_2C{=}C(Ph)(NC_5H_{10}) \rightarrow RPS[NHCOCH{=}C(Ph)(NC_5H_{10})]F$$

4. FROM PHOSPHONIC ACID AMIDE HALIDES AND P_4S_{10} Phosphonic acid amide halides react with P_4S_{10} upon prolonged heating by replacement of oxygen with sulfur.[646]

$$RPO(NR'_2)Cl \xrightarrow{P_4S_{10}} RPS(NR'_2)Cl$$

XXI. THIOPHOSPHONIC ACID O-ESTER AMIDES

1. FROM THIOPHOSPHONIC ACID O-ESTER HALIDES AND AMINES
Thiophosphonic acid O-ester halides react with NH_3, primary
or secondary amines at ambient temperature according to:[173]

$$RPS(OR')Cl + 2HNR_2'' \rightarrow RPS(OR')(NR_2'') + [H_2\overset{\oplus}{N}R_2'']Cl^{\ominus}$$

The reaction with amidines proceeds in a similar manner.[1175]

$$RPS(OR')Cl + HN=CR_2'' \xrightarrow[-HCl]{NEt_3} RPS(OR')(N=CR_2'')$$

2. FROM THIOPHOSPHONIC ACID AMIDE HALIDES AND ALCOHOLS
Thiophosphonic acid amide halides react with alcohols or
phenols under mild conditions in the presence of hydrogen
halide acceptors.[1853a, 665]

$$MePS(NHR)Cl + R'OH \xrightarrow[-HCl]{NEt_3} MePS(NHR)(OR')$$

3. FROM PHOSPHONOUS ACID ESTER AMIDES. Phosphonous acid
ester amides readily add elemental sulfur. The reaction
requires no catalyst.[154]

$$RP(OR')(NR_2'') \xrightarrow{S} RPS(OR')(NR_2'')$$

4. FROM THIOPHOSPHONIC ACID ESTER THIOISOCYANATES AND
AMINES. Thiophosphonic acid ester thioisocyanates readily
add both aliphatic and aromatic amines to yield the corresponding ester ureides.[1856]

$$RPS(OR')(NCS) + R''NH_2 \rightarrow RPS(OR')(NHCSNHR'')$$

5. SYNTHESIS BY GRIGNARD COMPOUNDS. A convenient method
for the preparation of thiophosphonic acid ester amides
exists in the reaction of thiophosphoric acid ester amide
chlorides with Grignard compounds.

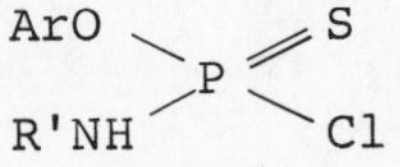

$$\begin{array}{c} ArO \\ R'NH \end{array}\!\!>\!P\!\!<\!\!\begin{array}{c} S \\ Cl \end{array} \xrightarrow{RMgBr} RPS(OAr)(NHR') + MgBrCl$$

By attack at the N atom, thiophosphoric acid imides are
obtained as by-products.

XXII. THIOPHOSPHONIC ACID S-ESTER AMIDES

1. FROM THIOPHOSPHONIC ACID S-ESTER CHLORIDES. The reaction of thiophosphonic acid S-ester chlorides and 2 moles of
NH_3, primary or secondary amines, leads to the formation
of thiophosphonic acid S-ester amides.[337]

$$CH_3PO(SEt)Cl + 2H_2NPh \rightarrow CH_3PO(SEt)(NHPh) + [Ph\overset{\oplus}{N}H_3]Cl^{\ominus}$$

2. FROM THIOPHOSPHONIC ACID AMIDE CHLORIDES. More important from a preparative point of view is the reaction of phosphonic acid amide chlorides and mercaptanes in the presence of tertiary amines.[1165, 1853a]

$$C_2H_5PO(NMe_2)Cl + HSCH_2CH_2NEt_2 \xrightarrow[- HCl]{NEt_3}$$

$$C_2H_5PO(NMe_2)(SCH_2CH_2NEt_2)$$

3. OXIDATION OF THIOPHOSPHONOUS ACID S-ESTER AMIDES. Thiophosphonous acid S-ester amides may be oxidized with N_2O_4 in inert solvents to yield thiophosphonic acid S-ester amides.[50, 1747]

$$EtP(SPr)(NEt_2) \xrightarrow{N_2O_4} EtPO(SPr)(NEt_2)$$

4. FURTHER METHODS. Equimolar amounts of dichlorophosphines and phosphonous acid diesters react with sulfenyl amides to yield thiophosphonic acid S-ester amides.[1470]

$$RPCl_2 + RP(OR_2') \xrightarrow[- 2R'Cl]{2R''SNR_2} 2RPO(SR'')(NR_2)$$

Similarly, phosphonous acid diesters react with aminosulfenyl chlorides.[1299]

$$EtP(OEt)_2 + Et_2NSCl \rightarrow EtPO(OEt)(SNEt_2) + EtCl$$

N-γ-Halopropyl thiophosphonic ester amides are quite unstable and undergo cyclization according to:[351]

$$PhPS(OEt)[NH(CH_2)_3Br] \rightarrow Ph(O)P \underset{S-CH_2}{\overset{NH-CH_2}{<}} {>} CH_2 + EtBr$$

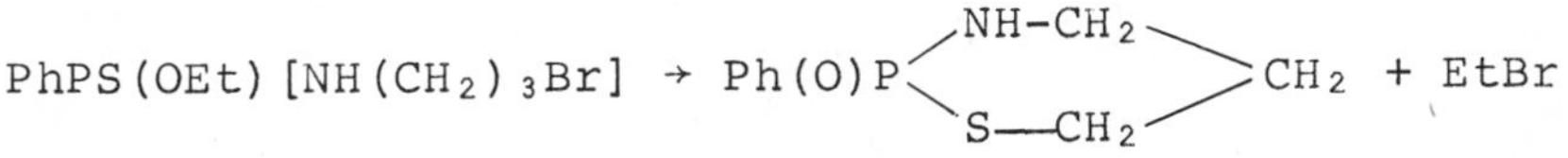

XXIII. THIOPHOSPHONIC ACID DIAMIDES

1. FROM THIOPHOSPHONIC ACID HALIDES. Thiophosphonic acid dihalides react with NH_3, primary or secondary amines to yield thiophosphonic acid diamides. The reaction is less vigorous than with phosphonic acid dihalides.[1254]

$$RPSCl_2 + 4HNR_2' \rightarrow RPS(NR_2')_2 + 2[H_2\overset{\oplus}{N}R_2']Cl^{\ominus}$$

The reaction with NH_3 may be carried out in liquid ammonia; otherwise, inert solvents are preferred.[1931]
With aromatic o-diamines, five-membered phosphadiazaheterocycles are formed.[452]
Asymmetric diamides may be obtained by the subsequent

reactions of 2 moles of different amines, with or without isolation of the intermediately formed amide chloride.[1232] These reactions proceed stereospecifically with inversion of configuration at the thiophosphonyl center.[1876]

4-Chlorobutene thiophosphonic acid dichloride reacts with excess amines by HCl abstraction to yield butadiene thiophosphonic acid diamides.[1629]

Thiophosphonic acid diamides of primary amines condense upon heating to yield a mixture of cis- and trans-thiophosphonic acid imides.[541, 2029]

$$4PhPS(NHMe)_2 \xrightarrow[-4MeNH_2]{250^\circ} \underset{S}{\overset{Ph}{\diagdown}}P\underset{\underset{Me}{N}}{\overset{\overset{Me}{N}}{\diagup\diagdown}}P\overset{Ph}{\underset{S}{\diagup}} + \underset{S}{\overset{Ph}{\diagdown}}P\underset{\underset{Me}{N}}{\overset{\overset{Me}{N}}{\diagup\diagdown}}P\overset{S}{\underset{Ph}{\diagup}}$$

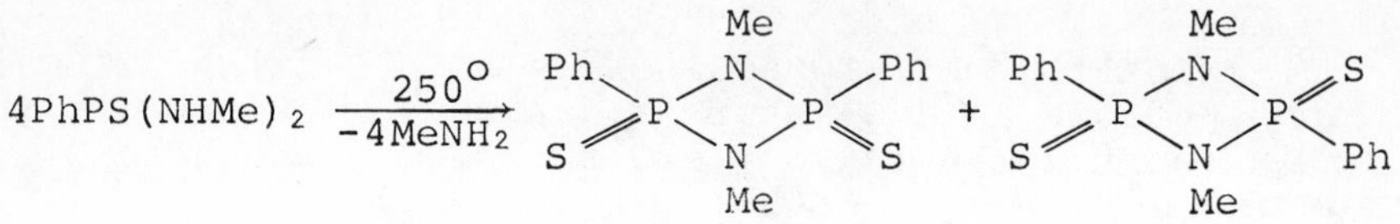

2. CLEAVAGE OF THIOPHOSPHONIC ACID IMIDES. Thiophosphon acid imides react with amines, upon heating to about 150° in a sealed tube, by forming thiophosphonic acid diamides. The cleavage proceeds stepwise.[668]

$$Me(S)P\underset{\underset{Ph}{N}}{\overset{\overset{Ph}{N}}{\diagup\diagdown}}P(S)Me + 2PhNH_2 \rightarrow 2MePS(NHPh)_2$$

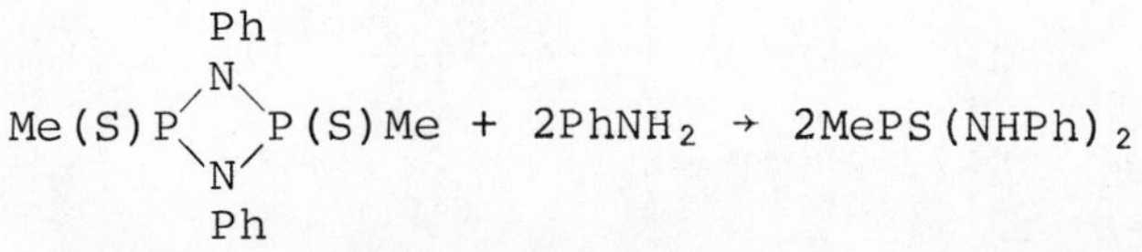

3. REACTION OF THIOPHOSPHONIC ACID DIISOCYANATES WITH AMINES OR C,H-ACTIVE COMPOUNDS. Thiophosphonic acid diisocyanates react vigorously with amines and C,H-active compounds.[855]

$$RPS(NCO)_2 + 2PhNH_2 \rightarrow RPS(NHCONHPh)_2$$

$$RPS(NCO)_2 + H_2C=C(Ph)(NC_5H_{10}) \rightarrow$$

$$R(S)P\underset{NHCO}{\overset{NHCO}{\diagup\diagdown}}C=C(Ph)(NC_5H_{10})$$

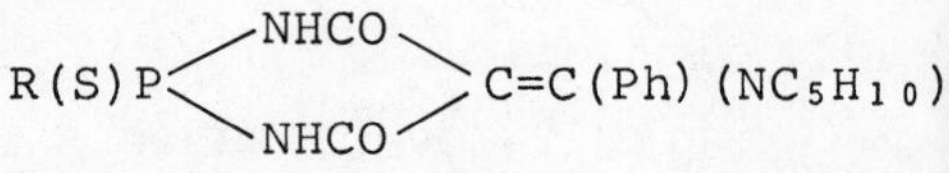

4. FROM PHOSPHONOUS ACID DIAMIDES AND SULFUR. Upon heating with elemental sulfur to about 100-150°, phosphonous acid diamides are readily converted to the corresponding thiophosphonic acid diamides.[1291]

$$RP(NR_2')_2 + S \rightarrow RPS(NR_2')_2$$

In the absence of acid catalysts, phosphonous acid diamides abstract sulfur from alkylmercaptanes.[1389]

5. ALKYLATION OF THIOPHOSPHORIC DIAMIDE CHLORIDES. Thio phosphoric acid diamide chlorides react with aluminium trialkyls according to:[1150]

$$3(R_2N)_2P(S)Cl + Et_3Al \rightarrow 3EtPS(NR_2)_2 + AlCl_3$$

XXIV. THIOPHOSPHONIC ACID HYDRAZIDE AND AZIDE DERIVATIVES

1. THIOPHOSPHONIC ACID HYDRAZIDE DERIVATIVES. Thiophosphonic acid dichlorides and 4 moles of a hydrazine react to yield dihydrazides.[729, 1930] With a molar ratio of 1 : 3, cyclic dihydrazides are obtained by reaction at both nitrogen atoms.[2006]

$$RPSCl_2 + 4H_2NNHR' \rightarrow RPS(NHNHR')_2 + 2[H_2N\overset{\oplus}{N}H_2R']Cl^{\ominus}$$

$$2RPSCl_2 + 6H_2NNH_2 \rightarrow R(S)P \overset{\displaystyle NHNH}{\underset{\displaystyle NHNH}{\diagup\diagdown}} P(S)R + 4[H_2N\overset{\oplus}{N}H_3]Cl^{\ominus}$$

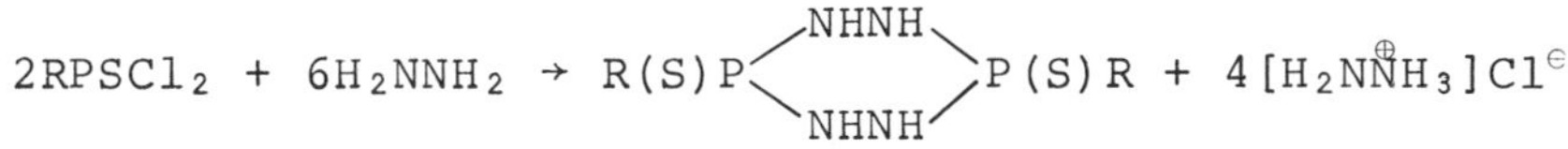

With $RPSF_2$, exchange of only one F atom leads to the hydrazide fluoride.[1762]

Ester chlorides and amide chlorides react similarly to yield ester hydrazides and amide hydrazides, respectively. With ester chlorides and hydrazine in a 2 : 3 molar ratio, hydrazine reacts with both NH_2 groups.[507, 1500]

$$MePS(OEt)Cl + 2H_2NNH_2 \rightarrow Me(S)P \overset{\displaystyle OEt\ EtO}{\underset{\displaystyle NH-NH}{\diagup\diagdown}} P(S)Me + [H_2N\overset{\oplus}{N}H_3]Cl^{\ominus}$$

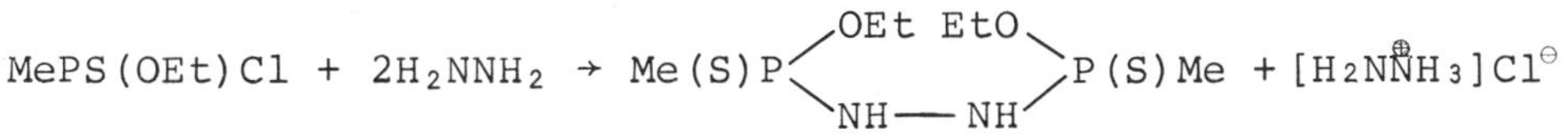

N-Unsubstituted hydrazides may be acetylated[1930] or caused to react with carbonyl compounds to give the corresponding hydrazones.[661]

Thiophosphonic acid hydrazide esters may also be obtained by the addition of sulfur to phosphonous acid hydrazide esters.[371]

2. THIOPHOSPHONIC ACID AZIDE DERIVATIVES. Thiophosphonic acid dichlorides exchange with sodium azide to yield the corresponding diazides.[202] Similarly, with ester chlorides the ester azides may be obtained.[1837a] Thiophosphonic acid anhydrides are cleaved by NaN_3 to yield sodium salts of dithiophosphonic acid monoazides.[1758, 1838]

$$(RPS_2)_2 + 2NaN_3 \rightarrow 2RPS(N_3)SNa$$

By reaction of the salts with alkyl halides the corresponding ester azides may be obtained.[1838]

XXV. DITHIOPHOSPHONIC ACID MONOAMIDES AND ESTER AMIDES

1. FROM THIOPHOSPHONIC ACID HALIDE DERIVATIVES. The alkali salts of dithiophosphonic acid monoamides may be prepared

by the reaction of thiophosphonic acid amide halides with KSH.[1846a]

$$RPS(NR_2')Cl + 2KSH \rightarrow RPS(NR_2')SK + KCl + H_2S$$

Thiophosphonic acid amide halides also react with mercaptanes in the presence of tertiary amines or bases (NaOEt) to yield dithiophosphonic acid ester amides.[666, 1855]

$$EtPS(NMe_2)Cl + PhSH \xrightarrow[- HCl]{NR_3} EtPS(NMe_2)SPh$$

Thiophosphonic acid dichlorides and β-mercaptoalkylamines react to yield cyclic dithiophosphonic acid ester amides.

$$RPSCl_2 + HSCH_2CH_2NH_2 \xrightarrow[- 2HCl]{2NEt_3} R(S)P\begin{array}{c} \diagup NHCH_2 \\ | \\ \diagdown S-CH_2 \end{array}$$

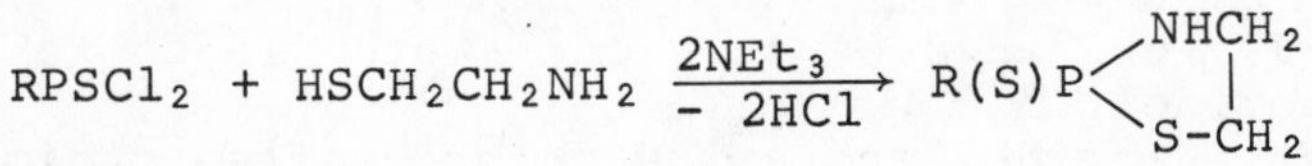

2. BY CLEAVAGE OF DITHIOPHOSPHONIC ACID ANHYDRIDES. Ammonium salts of dithiophosphonic acid monoamides are conveniently obtained by cleavage of dithiophosphonic acid anhydrides with primary or secondary amines.[798a, 1427]

$$(RPS_2)_2 + 2HNR_2' \rightarrow 2[RPS(NR_2')S^{\ominus}][H_2\overset{\oplus}{N}R_2']$$

If the salts are treated with alkyl halides or chloroacetic acid esters, dithiophosphonic acid ester amides are formed.[703, 706, 1846a]

$$[RPS(NR_2')\overset{\ominus}{S}] H_2\overset{\oplus}{N}R_2' + R''X \rightarrow RPS(NR_2')SR'' + [H_2\overset{\oplus}{N}R_2'] X^{\ominus}$$

With tertiary amines and dithiophosphonic acid anhydrides, one obtains internal salts according to:[704]

$$(RPS_2)_2 + 2NR_3' \rightarrow 2RPS(\overset{\oplus}{N}R_3')S^{\ominus}$$

3. ADDITION OF SULFUR TO THIOPHOSPHONOUS ACID ESTER AMIDES. Thiophosphonous acid ester amides readily add sulfur to form dithiophosphonic acid ester amides.[50, 207, 1747]

$$PhP(SR)(NR_2') + S \rightarrow PhPS(SR)(NR_2')$$

XXVI. PYROPHOSPHONIC ACIDS

1. FROM PHOSPHONIC ACID DERIVATIVES. The most general method for the preparation of pyrophosphonic acids is the thermal dehydration of phosphonic acids.[379, 904]

$$2PhPO(OH)_2 \xrightarrow{200^\circ, 48\text{-}72\ hr} PhP(O)(OH)\text{-}O\text{-}P(O)(OH)Ph$$

The reaction proceeds much more readily in the presence of dehydrating agents such as isocyanates,[556, 558] benzimide chlorides,[189] or carbodiimides.[326]

Pyrophosphonic acids are also prepared by partial hydrolysis of phosphonic acid anhydrides.[97, 776]

$$2(RPO_2)_n + nH_2O \rightarrow n\ RP(O)(OH)\text{-}O\text{-}P(O)(OH)R$$

The hydration may be carried out with anhydrous formic, acetic, or benzoic acid according to:[514, 720]

$$2(MePO_2)_n + 2nRCOOH \rightarrow n\ MeP(O)(OH)\text{-}O\text{-}P(O)(OH)Me + n(RCO)_2O$$

Incomplete hydrolysis of phosphonic acid dichlorides with less than 2 moles of water leads to mixtures of phosphonic acid and pyrophosphonic acid.[97]

$$2RPOCl_2 + 3H_2O \rightarrow RP(O)(OH)\text{-}O\text{-}P(O)(OH)R + 4HCl$$

2. FROM PHOSPHONOUS ACIDS. Phenylphosphonous acid and phenylisocyanate in an exothermic reaction yield the dianilinium salt of diphenylpyrophosphonic acid among other products.[558] As a by-product, diphenylpyrophosphonic acid is also obtained from the reaction of phenylphosphonous acid with pentaphenylcyclopentaphosphine.[597-599]

3. FURTHER METHODS. Pyrophosphonic acids may be liberated from their esters by reaction with HI.[760]

$$Ph_3CP(O)(OR)\text{-}O\text{-}P(O)(OR)CPh_3 + 2HI \rightarrow Ph_3CP(O)(OH)\text{-}O\text{-}P(O)(OH)CPh_3 + 2RI$$

Pyrolysis of dimethyl phosphite at about 250°C gives good yields of dimethylpyrophosphonic acid.[254, 1455]

$$2(MeO)_2P(O)H \xrightarrow{\Delta} [2MePO(OMe)(OH)] \rightarrow MeP(O)(OH)\text{-}O\text{-}P(O)(OH)Me + Me_2O$$

Triethyl phosphite reacts with tetramethylformamidinium chloride according to:[1784a]

$$P(OEt)_3 + [Me_2NCH=\overset{\oplus}{N}Me_2]Cl^{\ominus} \rightarrow$$

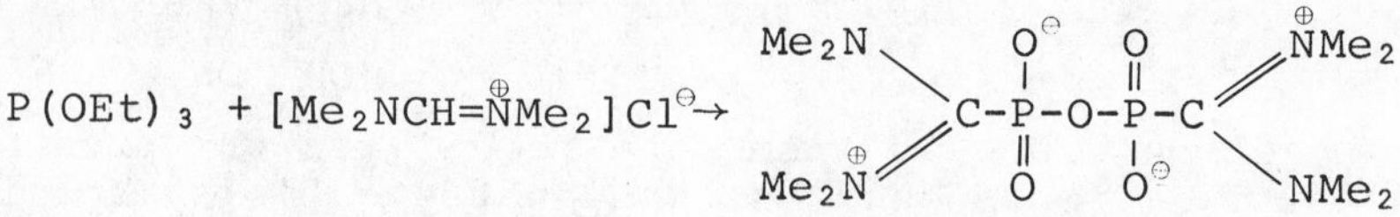

XXVII. PYROPHOSPHONIC ACID DIESTERS

1. FROM PHOSPHONIC ACID ESTER DERIVATIVES. In analogy to free phosphonic acids, phosphonic acid monoesters also react with dehydrating agents, such as carbodiimides, to yield pyrophosphonic acid esters.[268, 453, 647]

Phosphonic acid ester amides react with acid chlorides to yield pyrophosphonic acid diesters.[1703, 1716] The treatment of phosphonic acid diesters with inorganic acid halides e.g., SOCl$_2$, at 80-120° leads to products containing pyrophosphonic acid diesters.[434, 743]

A convenient method for the preparation of pyrophosphonic acid esters is the pyrolysis of mixed anhydrides of phosphonic acid monoesters and carboxylic acids. Disproportionation proceeds according to:[1490, 1596, 1647]

$$2RPO(OR')O_2CR'' \longrightarrow \underset{OR'\quad OR'}{RP-O-PR} + (R''CO)_2O$$

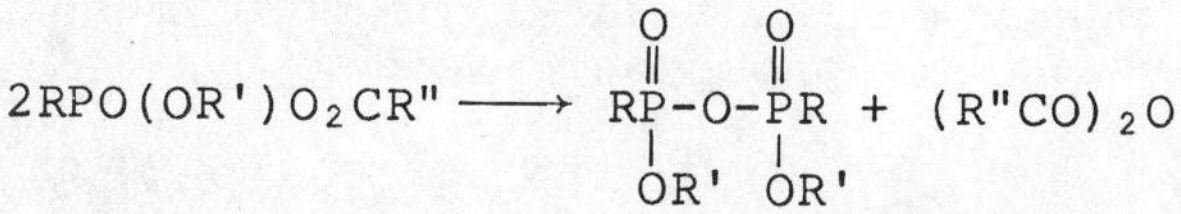

2. ALKYLATION OF PYROPHOSPHONIC ACIDS. Triphenylmethane pyrophosphonic acid may be alkylated directly by dialkylsulfate.[766]

XXVIII. PYROPHOSPHONIC ACID AMIDES

1. FROM PHOSPHONIC ACID ANHYDRIDES AND AMINES. Secondary amines react with phosphonic acid anhydrides to yield pyrophosphonic acid monoamide ammonium salts. Cleavage of the remaining P-O-P bond occurs only above 60°C and results in phosphonic acid monoamides [cf. Section XIV.1].[375]

$$2(PhPO_2)_n + 2nHNEt_2 \rightarrow n[PhP-O-PPh][H_2\overset{\oplus}{N}Et_2]$$

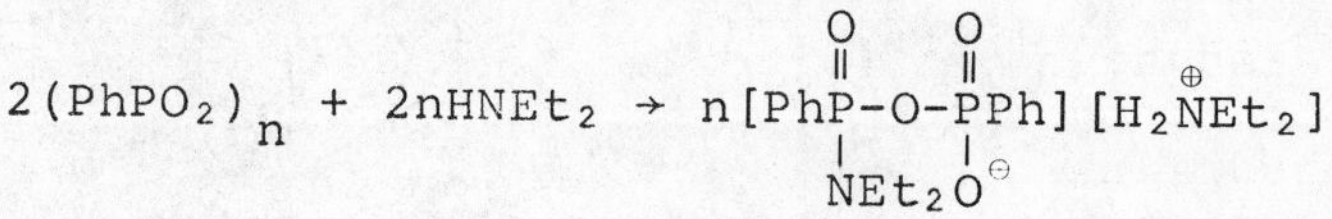

2. FROM PHOSPHONIC ACID HALIDE DERIVATIVES. Similar to phosphonic acid ester chlorides, phosphonic acid amide chlorides may be hydrolyzed with the calculated amount of water in the presence of acid-binding agents to yield pyrophosphonic acid diamides.[875, 1703, 1711a]

$$2EtPO(NEt_2)Cl + H_2O \xrightarrow[- HCl]{NR_3} \underset{\underset{NEt_2}{|}}{EtP}\overset{\overset{O}{\|}}{} - O - \underset{\underset{NEt_2}{|}}{PEt}\overset{\overset{O}{\|}}{}$$

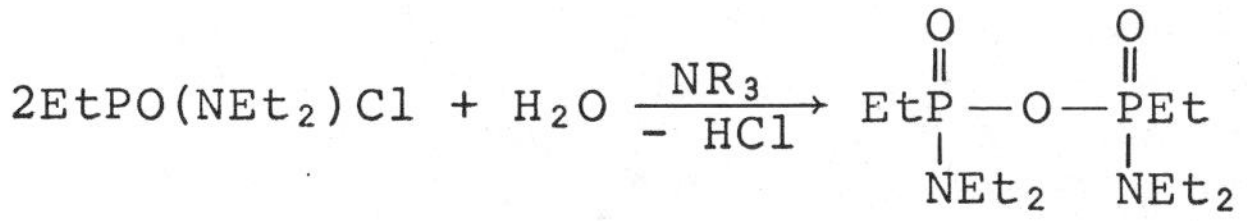

In the synthesis of the diamides, it is not necessary to isolate the phosphonic acid amide chlorides. Rather, the crude product from phosphonic acid dichloride and secondary amine (molar ratio 1 : 1) in the presence of tertiary amine may be hydrolyzed.[572, 767]

Asymmetric pyrophosphonic diamides may be obtained from salts of phosphonic acid monoamides or phosphonic acid ester amides and phosphonic acid amide chlorides.[255]

$$RPO(NR_2')ONa + RPO(NR_2')Cl \rightarrow \underset{\underset{NR_2'}{|}}{RP}\overset{\overset{O}{\|}}{} - O - \underset{\underset{NR_2'}{|}}{PR}\overset{\overset{O}{\|}}{} + NaCl$$

$$RPO(NR_2')(OR'') + RPO(NR_2')Cl \xrightarrow{180^{\circ}C} \underset{\underset{NR_2'}{|}}{RP}\overset{\overset{O}{\|}}{} - O - \underset{\underset{NR_2'}{|}}{PR}\overset{\overset{O}{\|}}{} + R''Cl$$

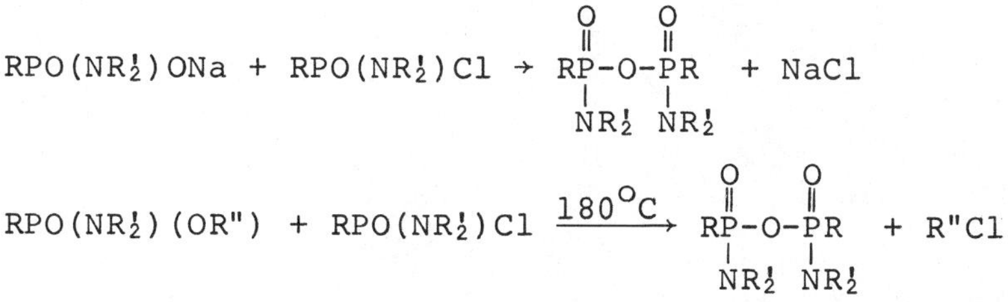

XXIX. PHOSPHONIC ACID ANHYDRIDES

1. FROM PHOSPHONIC ACID HALIDES. Phosphonic acid anhydrides may be prepared by heating equimolar amounts of phosphonic acid and the dichloride in refluxing benzene.[384, 729, 1281, 1289]

$$nRPOCl_2 + nRPO(OH)_2 \rightarrow (RPO_2)_n + 2nHCl$$

Alternatively, the phosphonic acid dichlorides may be reacted with the stoichiometric amount of H_2O,[1481] paraformaldehyde,[1326] or anhydrous formic acid.[383, 509, 510]

$$nRPOCl_2 \xrightarrow{HCOOH} (RPO_2)_n + 2nHCl + nCO_2$$

Phosphonic acid alkylester chlorides are thermally unstable and split off alkyl halides upon heating to yield phosphonic acid anhydrides.[776] Pyrophosphonic acid dialkylesters also have been reported to undergo thermal disproportionation to phosphonic acid dialkylesters and phosphonic acid anhydrides.[1490]

The anhydrides obtained are oligomers or polymers, depending on the synthetic conditions.

2. PHOSPHONYLATION OF HYDROCARBONS. The direct interaction of hydrocarbons with P_4O_{10} proceeds only under drastic conditions (autoclave, 275-325°C). The reaction yields phosphonic acid anhydrides only as by-products, the main

products being mixed anhydrides of phosphonic and phosphoric acid. With naphthalene, the phosphono group is introduced in 2-position, and with chlorobenzene and toluene, mixtures of o- and p-isomers are obtained.[1084, 1088]

XXX. THIOPYROPHOSPHONIC ACID ESTERS

1. FROM PHOSPHONIC ACID DERIVATIVES. Phosphonic acid ester chlorides react with H_2S in the presence of pyridine to yield thiopyrophosphonic acid O,O-diesters. In these compounds, the sulfur atom is bound as a thiono group.[405, 1114, 1293, 1297]

$$2RPO(OR')Cl \xrightarrow[-\ 2HCl]{H_2S} \begin{array}{cc} O & S \\ \| & \| \\ RP\!-\!O\!-\!PR \\ | & | \\ OR' & OR' \end{array}$$

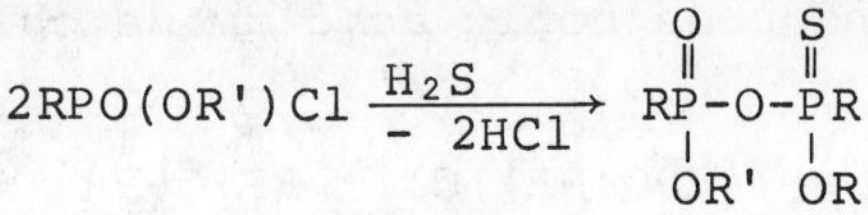

Asymmetric O,O-diesters may be obtained by the reaction of phosphonic acid ester chlorides and thiophosphonic acid O-monoesters in the presence of acid-binding agents.[1113]

2. FROM THIOPHOSPHONIC ACID DERIVATIVES. Alkali salts of thiophosphonic acid O-alkylmonoesters react with phosgene to yield thiopyrophosphonic acid O,O-diesters by liberation of COS.[337]

$$2RPO(OR')SK + COCl_2 \rightarrow \begin{array}{cc} O & S \\ \| & \| \\ RP\!-\!O\!-\!PR \\ | & | \\ OR' & OR' \end{array} + 2KCl + COS$$

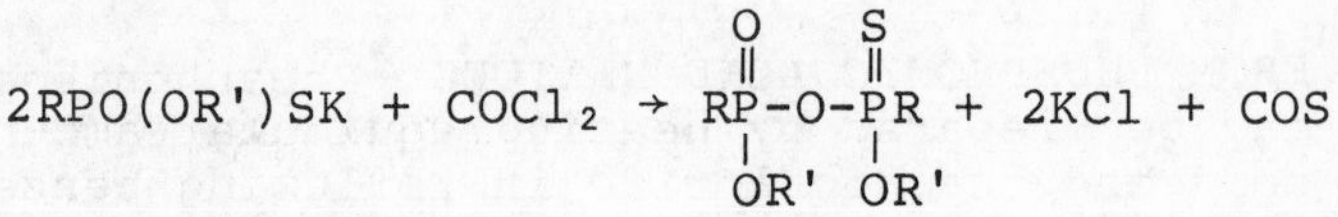

Condensation of thiophosphonic acid monoesters may be achieved with carbodiimides according to:[1308]

$$2RPS(OR')OH \xrightarrow{C_6H_{11}N=C=NC_6H_{11}} \begin{array}{cc} O & S \\ \| & \| \\ RP\!-\!O\!-\!PR \\ | & | \\ OR' & OR' \end{array} + (C_6H_{11}NH)_2CS$$

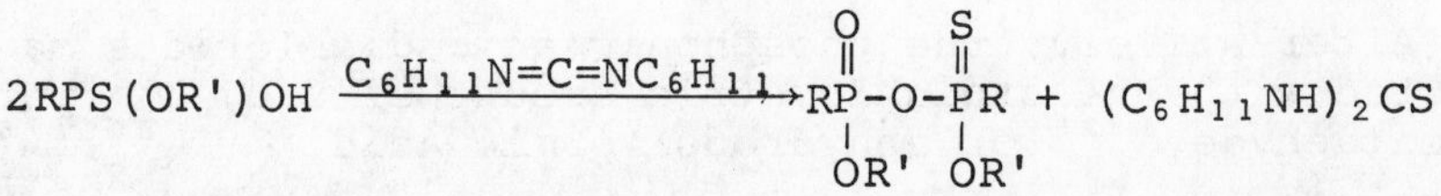

The reaction proceeds by attack of the thiophosphonyl ester anion on the O atom of the adduct

$$\begin{array}{cc} R'O & O^{\ominus} \\ \diagdown & | \\ & P\!-\!S\!-\!C\!=\!\overset{\oplus}{N}HC_6H_{11} \\ \diagup & | \\ R & NHC_6H_{11} \end{array}$$

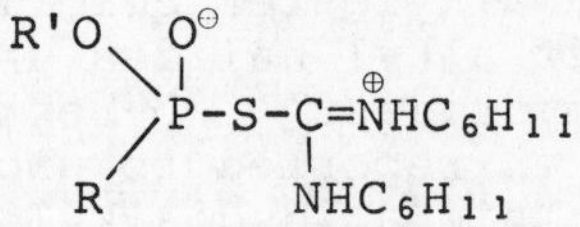

In the absence of bases, the reaction is stereospecific. S-Chlorothiophosphonic acid monoesters react with phosphonous acid diesters in benzene solution to yield thiopyrophosphonic acid diesters among other products.[1297]

$$EtPO(OEt)SCl + EtP(OEt)_2 \rightarrow \left[\begin{array}{cc} O & OEt \\ \| & |^{\oplus} \\ EtP\!-\!S\!-\!PEt \\ | & | \\ OEt & OEt \end{array} \right] Cl^{\ominus}$$

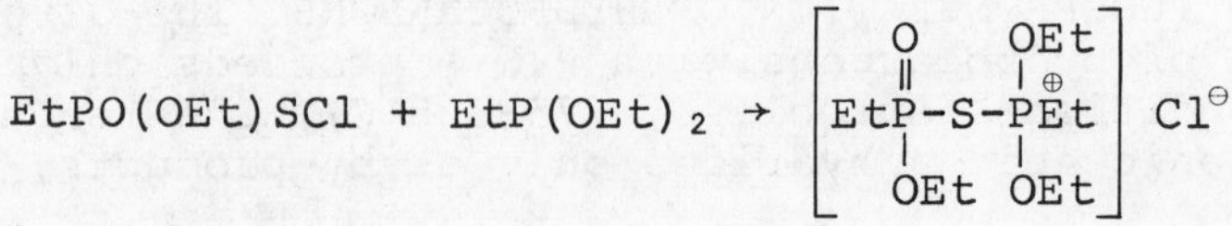

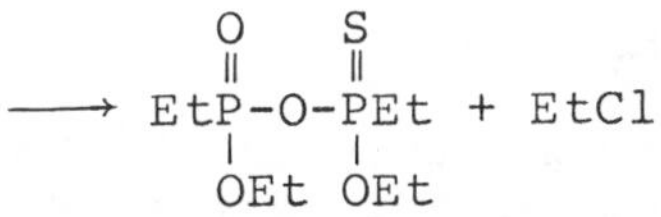

$$\longrightarrow \underset{\underset{OEt}{|}}{\overset{\overset{O}{\parallel}}{Et P}}-O-\underset{\underset{OEt}{|}}{\overset{\overset{S}{\parallel}}{P Et}} + EtCl$$

Thiophosphonic acid O,S-diesters react with acetylbromide to yield dithiopyrophosphonic acid S,S-diesters.[640]

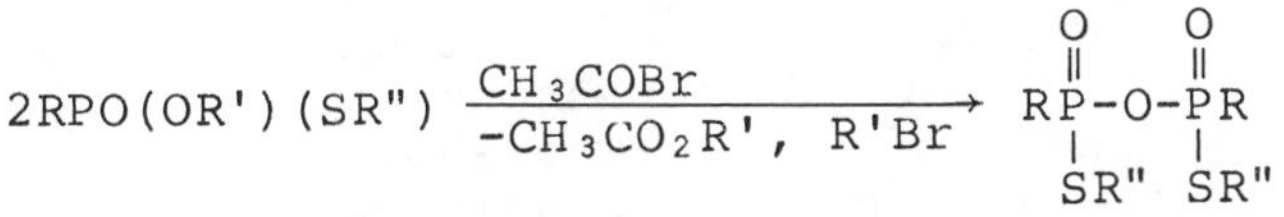

$$2RPO(OR')(SR'') \xrightarrow[-CH_3CO_2R',\ R'Br]{CH_3COBr} \underset{\underset{SR''}{|}}{\overset{\overset{O}{\parallel}}{RP}}-O-\underset{\underset{SR''}{|}}{\overset{\overset{O}{\parallel}}{PR}}$$

Thiophosphonic acid O-ester chlorides yield upon hydrolysis with the stoichiometric amount of water in the presence of pyridine dithiopyrophosphonic acid O,O-diesters.[1841]

$$2RPS(OR')Cl \xrightarrow[-2HCl]{H_2O} \underset{\underset{OR'}{|}}{\overset{\overset{S}{\parallel}}{RP}}-O-\underset{\underset{OR'}{|}}{\overset{\overset{S}{\parallel}}{PR}}$$

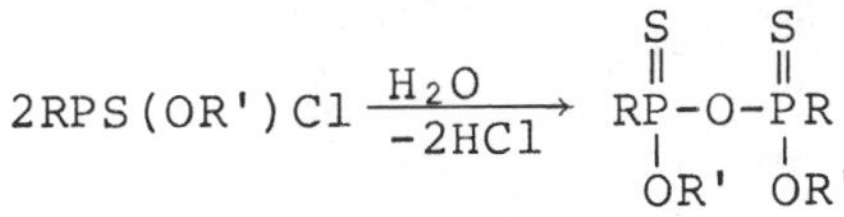

3. FROM PHOSPHONOUS ACID MONOESTERS. Phosphonous acid monoesters react with sulfur dichloride with subsequent addition of pyridine to yield thiopyrophosphonic acid diesters.[1153]

$$2RPO(OR')H \xrightarrow[-2HCl]{\substack{1.\ SCl_2 \\ 2.\ pyridine}} \underset{\underset{OR'}{|}}{\overset{\overset{O}{\parallel}}{RP}}-O-\underset{\underset{OR'}{|}}{\overset{\overset{S}{\parallel}}{PR}}$$

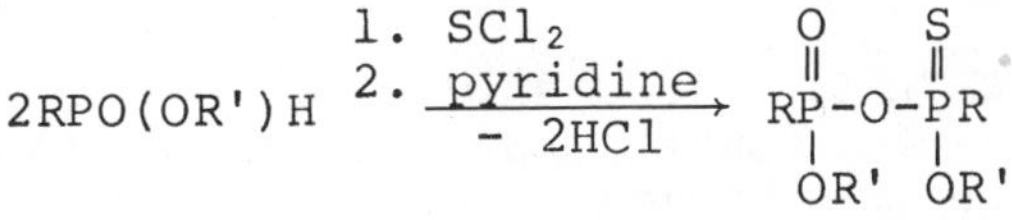

XXXI. THIOPYROPHOSPHONIC ACID AMIDES

The reaction of mixed phosphonic monoamide-phosphonous acid anhydrides with thiophosphonic acid monoesters is exothermic and leads to thiopyrophosphonic acid ester monoamides.[1323]

$$\underset{\underset{NEt_2}{|}}{\overset{\overset{O}{\parallel}}{MeP}}-OP(OPr)_2 + MePS(OBu)OH \rightarrow \underset{\underset{NEt_2}{|}}{\overset{\overset{O}{\parallel}}{MeP}}-\!\!\!-O-\!\!\!-\underset{\underset{OBu}{|}}{\overset{\overset{S}{\parallel}}{PMe}} + (PrO)_2P(O)H$$

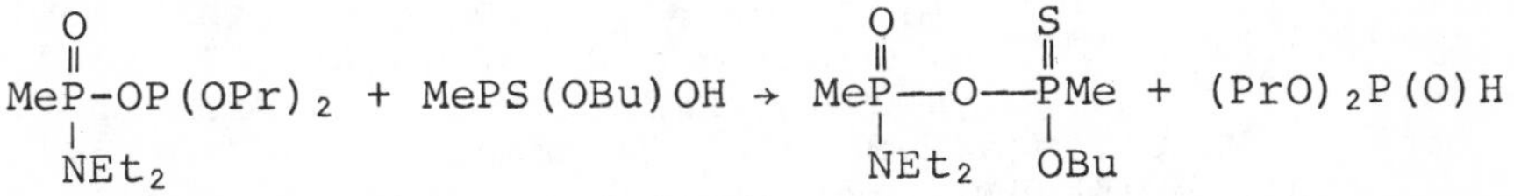

Hydrolysis of thiophosphonic acid amide chlorides with the calculated amount of water leads to dithiopyrophosphonic acid diamides.[1842]

XXXII. DITHIOPHOSPHONIC ACID ANHYDRIDES AND MONOTHIOPHOSPHONIC ACID ANHYDRIDES

1. FROM THIOPHOSPHONIC ACID DICHLORIDES. Thiophosphonic acid dichlorides react with H_2S at elevated temperatures to yield dithiophosphonic acid anhydrides. Instead of the dichlorides, alkylphosphonic tetrachlorides may be used.[408, 965]

$$nRPSCl_2 \xrightarrow[-2nHCl]{nH_2S} (RPS_2)n$$

The reaction proceeds much more smoothly in the presence of tertiary amines.[1810] The anhydrides are generally dimeric and exist predominantly in the trans-configuration.[1379]

2. FROM HYDROCARBONS AND P_4S_{10}. Unsaturated aliphatic and alicyclic hydrocarbons react with P_4S_{10} at elevated temperatures to yield dithiophosphonic acid anhydrides. With cyclohexene, the P atom was shown to add in α-position to the double bond. The resulting anhydride was a dimer.[522]

$2C_6H_{10} + 1/2\ P_4S_{10} \rightarrow$ 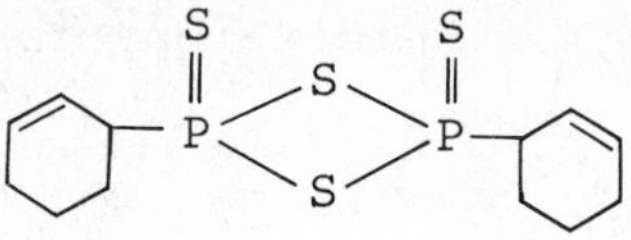 $+ H_2S$

Aromatic hydrocarbons without polar groups react with P_4S_{10} only slowly to yield aryldithiophosphonic acid anhydrides.[360] The reaction proceeds smoothly in specific temperature ranges with phenol ethers. The products are difficult to obtain in a pure form.[1086]

3. FURTHER METHODS. Phenyl phosphine reacts with elemental sulfur in a 1:3 molar ratio to yield the corresponding dithiophosphonic acid anhydride, probably via a $(PhPS)_4$ intermediate.

$2PhPH_2 + 6S \rightarrow$ 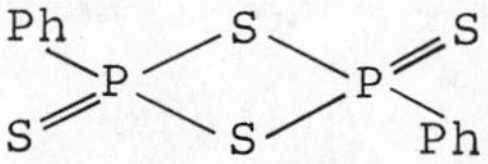 $+ 2H_2S$

Dithiophosphonic acid anhydrides may also be obtained by the reaction of alkyl- and aryldichlorophosphines with H_2S_2[217] or K_2S_2[646a] or by treatment of $[RPCl_3]^+[AlCl_4]^-$ with H_2S.[965] The polymeric $(PhPS)_5$ resulting from the reaction of $PhPCl_2$ and H_2S adds elemental sulfur to yield the anhydride $[PhPS_2]_2$.[1321] With equimolar amounts of S_2Cl_2, primary arylphosphines were reported to form trimeric dithiophosphonic acid anhydrides.[1426] However, dithiophosphonic acid anhydrides were later shown to exist as dimers only.[1148, 1154]

Phosphonic acid dichlorides react with H_2S, preferably in the presence of a tertiary amine, to yield monothiophosphonic acid anhydrides.[449a, 1758] These anhydrides are also formed in the reaction of primary aromatic phosphines with thionyl chloride,[98] by controlled hydrolysis of thiophosphonic acid dichlorides in the presence of tertiary amines[509a] and by the reaction of thiophosphonic acid ester chlorides with H_2S.[499a]

$$3RPOCl_2 + 3H_2S \xrightarrow{R_3'N} \underset{\substack{S \diagup \\ O \diagdown_{P \diagdown} O}}{\overset{S \diagdown_{P} - O - P \diagup S}{R \mid \qquad \mid R}} R + 6HCl$$

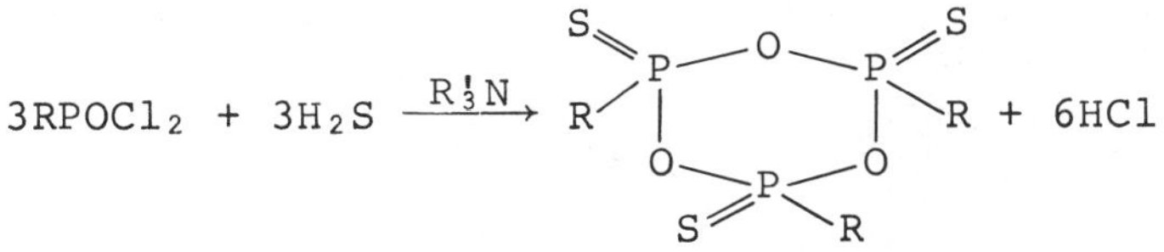

Monothiophosphonic acid anhydrides are also obtained by hydrolysis of dithiophosphonic anhydrides.[965] The anhydrides were shown to be cyclic trimers.[449a]

XXXIII. MIXED ANHYDRIDES OF PHOSPHONIC AND CARBOXYLIC ACIDS.

1. FROM PHOSPHONIC ACIDS AND MONOESTERS. Phosphonic acids in form of their mono-t-ammonium salts react with iso-cyanates to yield O-carbamoyl phosphonic acids. With diluted mineral acids, the free phosphonic acid anhydride can be isolated.[557]

$$[PhPO(OH)O^{\ominus}][H\overset{\oplus}{N}R_3] + OCNPh \rightarrow [PhPO(OCONHPh)O^{\ominus}][H\overset{\oplus}{N}R_3]$$

Phosphonic acid monoesters also add to isocyanates. Formally, the mixed anhydrides of phosphonic acid monoesters and the oximes of carboxylic acids are obtained in this reaction.[338]

$$RPO(OR')(OH) + OCNAr \rightarrow RPO(OR')(O\underset{\substack{\| \\ NOH}}{C}Ar)$$

Phosphonic acid monoesters react at low temperatures with carboxylic acid halides in the presence of tertiary amines to yield O-acylphosphonic acid esters.[865]

$$RPO(OR')(OH) + ClCOR'' \xrightarrow[- HCl]{NR_3} RPO(OR')O_2CR''$$

2. FROM PHOSPHONIC ACID ESTER CHLORIDES. Phosphonic acid ester chlorides react with the Ag salts of carboxylic acids to yield O-acyl phosphonic acid esters.[1490] The potassium salts of carboxylic acids may also be used in this reaction.[1596]

$$RPO(OR')Cl + AgO_2CR'' \rightarrow RPO(OR')O_2CR'' + AgCl$$

3. FROM PHOSPHOROUS ACID ESTERS. Acyl phosphorous acid dialkyl esters react with alkyl halides in a Michaelis-Arbuzov-type reaction to yield O-acylphosphonic acid alkyl esters.[103a]

$$(RO)_2PO_2CR' + R''Hal \rightarrow R''PO(OR)O_2CR' + RHal$$

Trialkyl phosphites react with carboxylic acid anhydrides to give mixed anhydrides according to :[936]

$$(RO)_3P + (R'CO)_2O \xrightarrow[-R'CO_2R]{} R'COPO(OR)_2 \xrightarrow[-\ R'CO_2R]{(R'CO)_2O}$$

$$RCOPO(OR)O_2CR'$$

These mixed anhydrides are thermally unstable. Upon heating they decompose to yield pyrophosphonic acid esters (cf. Section XXVII.1) and carboxylic acid anhydrides.

4. FORMATION OF CYCLIC ANHYDRIDES Phosphonic acid dichlorides react with salicylic acid upon heating to yield cyclic anhydrides according to:[1375, 1376]

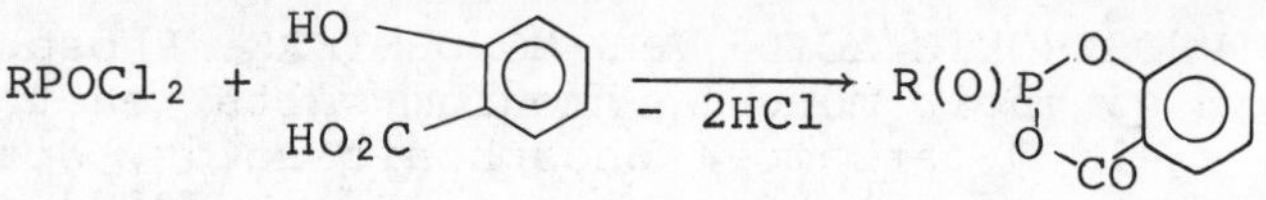

Cyclic O-acylphosphonic acid alkylesters are formed upon heating of 2-carbalkoxypropane phosphonic acid alkylester chlorides[1492] or 2-chlorocarboxyethane phosphonic acid diesters.[2032]

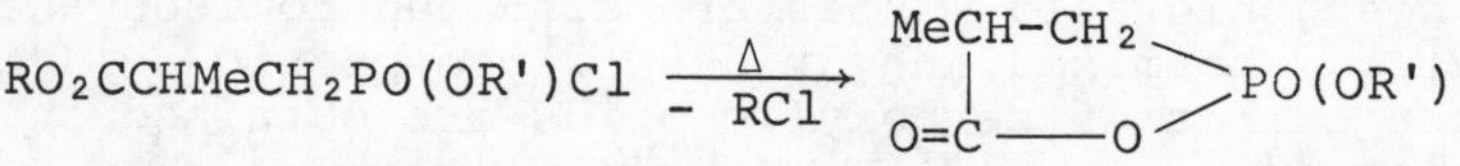

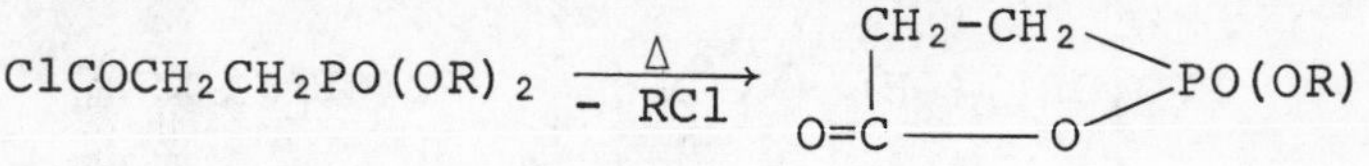

5-ring mixed anhydrides may also be obtained from phosphorous acid diester chlorides and acrylic acid.[504]

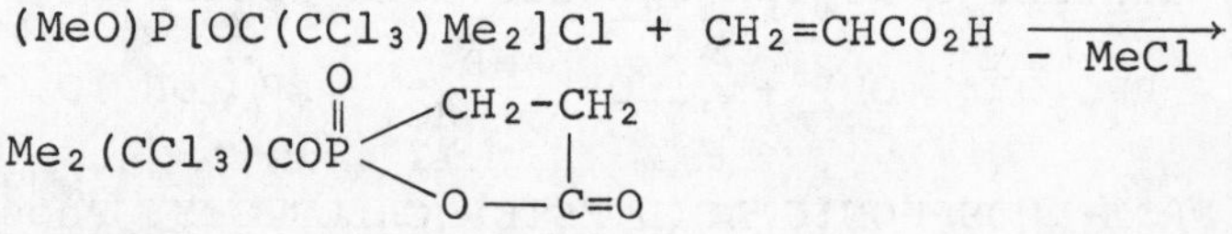

Phosphonic acid anhydrides react with α-hydroxynitriles to yield compounds that are formally anhydrides of a phosphonic acid and a carboxylic acid imide.[380]

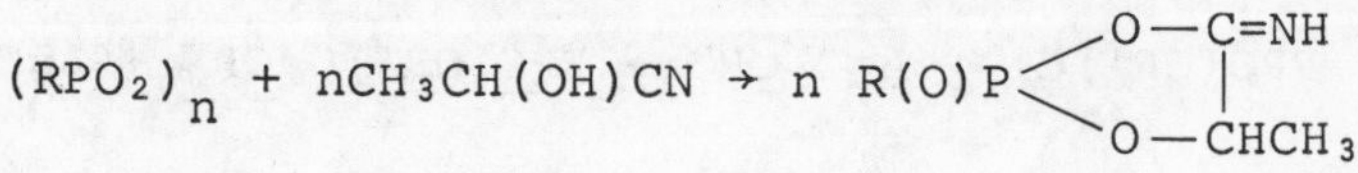

XXXIV. DERIVATIVES OF MIXED ANHYDRIDES OF PHOSPHONIC ACIDS WITH OTHER P ACIDS

1. WITH PHOSPHORIC ACID DERIVATIVES. Triesters of the mixed anhydrides are most readily prepared by the reaction

of trialkyl phosphites with phosphoric acid diester chlorides. $ZnCl_2$ may be used as a catalyst.[214, 215, 1304]

$$P(OR)_3 + ClPO(OR')_2 \rightarrow R\underset{\substack{| \\ OR}}{\overset{\substack{O \\ \|}}{P}}\text{-}O\text{-}\underset{\substack{| \\ OR'}}{\overset{\substack{O \\ \|}}{P}}\text{-}OR' + RCl$$

Instead of trialkyl phosphites, phosphonic acid diester may be use.[214]

$$RPO(OR')_2 + ClPO(OR'')_2 \rightarrow R\underset{\substack{| \\ OR'}}{\overset{\substack{O \\ \|}}{P}}\text{-}O\text{-}\underset{\substack{| \\ OR''}}{\overset{\substack{O \\ \|}}{P}}\text{-}OR'' + R'Cl$$

The triesters may also be prepared from mixed anhydrides of phosphorous and phosphoric acid with alkyl halides in a Michaelis-Arbuzov-type reaction.[105] As a further method, the reaction of phosphonic acid dialkylesters with metaphosphoric acid alkylesters has been reported.[1010]

Phosphonic acid dialkylesters react with phosphoric acid bis(dialkylamide) chlorides in a 1:1 molar ratio according to:[2004]

$$RPO(OR')_2 + ClPO(NR''_2)_2 \rightarrow R\underset{\substack{| \\ OR'}}{\overset{\substack{O \\ \|}}{P}}\text{-}O\text{-}\underset{\substack{| \\ NR''_2}}{\overset{\substack{O \\ \|}}{P}}\text{-}NR''_2 + R'Cl$$

With a molar ratio of 1:2, the reaction proceeds further to yield:[2002]

$$RPO(OR')_2 + 2ClPO(NR''_2)_2 \rightarrow \overset{\substack{O \\ \|}}{RP}(\text{-}O\underset{\substack{| \\ NR''_2}}{\overset{\substack{O \\ \|}}{P}}\text{-}NR''_2)_2 + 2R'Cl$$

Anhydrides made up from 1 phosphonic acid unit and 2 phosphoric acid units are also obtained from phosphonic acid dichlorides and 2 moles of phosphoric acid monoester diamides.[2002]

2. WITH PHOSPHOROUS ACID DERIVATIVES. Phosphonic acid anhydrides react with dialkylphosphite amides and phosphorous triamides according to:[512, 1323]

$$(RPO_2)_n + n(R'O)_2PNR''_2 \rightarrow nR\underset{\substack{| \\ NR''_2}}{\overset{\substack{O \\ \|}}{P}}\text{-}O\text{-}P(OR')_2$$

$$(RPO_2)_n + nP(NR'_2)_3 \rightarrow nR\underset{\substack{| \\ NR'_2}}{\overset{\substack{O \\ \|}}{P}}\text{-}O\text{-}P(NR'_2)_2$$

3. WITH THIOPHOSPHORIC ACID DERIVATIVES. Phosphonic acid dichlorides react with monothiophosphonic acid O,O-diesters in a 1:2 molar ratio according to:[103b]

$$RPOCl_2 + 2HOPS(OR')_2 \rightarrow R\overset{O}{\overset{\|}{P}}(-O\underset{OR'}{\underset{|}{\overset{S}{\overset{\|}{P}}}}OR')_2 + 2HCl$$

Phosphonous acid diesters and S-chlorothiophosphoric acid diesters react with phosphonous acid diesters by formation of mixed anhydride esters in about 70% yield.[1301]

$$(RO)_2P(O)SCl + R'P(OR'')_2 \rightarrow (RO)_2\overset{S}{\overset{\|}{P}}\text{-}O\text{-}\underset{OR''}{\underset{|}{\overset{O}{\overset{\|}{P}}}}R' + R''Cl$$

The reaction proceeds by nucleophilic attack at the S atom with intermediary formation of a quasi-phosphonium complex.

XXXV. SELENO- AND TELLUROPHOSPHONIC ACID AMIDE DERIVATIVES

Selenophosphonic acid diamides are readily obtained from selenophosphonic acid dichlorides and amines.[723]

$$RPSeCl_2 + 4PhNH_2 \rightarrow RPSe(NHPh)_2 + 2[Ph\overset{\oplus}{N}H_3]Cl^{\ominus}$$

Another method is the addition of elemental selenium to phosphonous acid diamides.[1054, 1701]

$$RP(NR'_2)_2 + Se \rightarrow RPSe(NR'_2)_2$$

The addition of selenium to thiophosphonous acid S-ester amides leads to the corresponding thioselenophosphonic acid S-ester amides.[1747, 2075]

$$RP(SR')(NR''_2) + Se \rightarrow RPSe(SR')(NR''_2)$$

The corresponding tellurophosphonic acid amide derivatives have not been prepared.

XXXVI. IMINOPHOSPHONIC ACID DERIVATIVES

1. IMINOPHOSPHONIC ACID DIHALIDES Alkyl- and arylphosphonic acid tetrachlorides react with N-unsubstituted carboxylic acid amides by loss of HCl to yield the corresponding N-substituted iminophosphonic acid dichlorides.[1889, 1904, 1906, 1909]

$$RPCl_4 + H_2NCOR' \rightarrow RP(=NCOR')Cl_2 + 2HCl$$

Similarly, from arylsulfonamides and $RPCl_4$ or from the sodium salts of N-chlorosulfonamides and $RPCl_2$ one obtains N-arylsulfonyliminophosphonic acid dichlorides.[1889]

$$RPCl_4 + H_2NSO_2R' \xrightarrow{-\,2HCl} RP(=NSO_2R')Cl_2$$

$$RPCl_2 + NaNClSO_2R' \xrightarrow{-\,NaCl} RP(=NSO_2R')Cl_2$$

The N-carboxy iminophosphonic dichlorides exchange halide with SbF_3 to form the corresponding difluorides.[1138]

$$RP(=NCOR')Cl_2 \xrightarrow{SbF_3} RP(=NCOR')F_2$$

Carboxylic acid ester imides also react with arylphosphonic tetrachlorides.[741]

$$RC(=NH)OAlk + ArPCl_4 \xrightarrow[-\,HCl,-AlkCl]{} ArP(=NCOR)Cl_2$$

Activated aromatic amines, e.g., O-aminobenzonitrile, react with alkylphosphonic tetrachlorides upon melting according to:[1821]

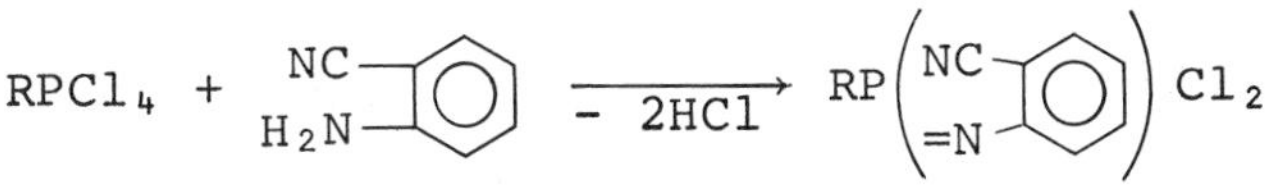

2. IMINOPHOSPHONIC ACID ESTERS. N-Carboxy- and N-sulfonyl iminophosphonic acid dichlorides react with alcohols and phenols in the presence of acid-binding agents to yield in a stepwise reaction iminophosphonic acid ester chlorides and diesters.[1138] Mostly, Na alcoholates or phenolates are used.[1896, 1897]

$$RP(=NCOR')Cl_2 + R''OH \xrightarrow{NR_3} RP(=NCOR')(OR'')Cl + HCl$$

$$RP(=NSO_2Ar)Cl_2 + 2NaOR' \rightarrow RP(=NSO_2Ar)(OR')_2 + 2NaCl$$

Phosphonous acid diesters react with aryl azides[627, 890] or alkyl azidoformates[1914] by loss of N_2 to yield the corresponding N-substituted iminophosphonic acid diesters.

$$RP(OR')_2 + ArN_3 \rightarrow RP(=NAr)(OR')_2 + N_2$$

$$RP(OR')_2 + R''O_2CN_3 \rightarrow RP(=NCO_2R'')(OR')_2 + N_2$$

Dialkylphosphorous isocyanates react with aldehydes to give cyclic iminophosphonic acid diesters.[1984]

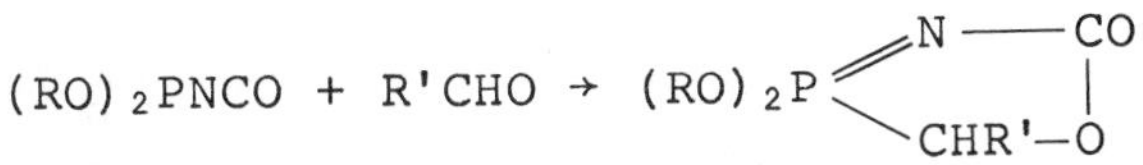

Dialkylphosphorous monoaryl amides add to acrylonitrile

or acrylic acid esters to form aryliminophosphonic acid diesters.[1567, 1635]

$$(RO)_2PNHAr + CH_2=CHX \quad XCH_2CH_2P(=NAr)(OR)_2$$

$$[X = CN, CO_2R]$$

A similar addition reaction is observed with Schiff bases.[1568]

$$(RO)_2PNHAr + R''CH=NR' \rightarrow R'NHCHR''P(=NAr)(OR)_2$$

3. IMINOPHOSPHONIC ACID AMIDES. N-Carboxyalkyl and N-arylsulfonyl iminophosphonic acid dichlorides react with NH_3 or primary amines in the presence of acid-binding agents to yield the corresponding amide chlorides and diamides, depending on the molar ratio of the reactants.[279, 1894, 1895]

$$RP(=NCOR')Cl_2 + 2NH_3 \rightarrow RP(=NCOR')(NH_2)Cl + NH_4Cl$$

$$RP(=NSO_2Ar)Cl_2 + 4R'NH_2 \rightarrow RP(=NSO_2Ar)(NHR)_2 + 2[R'\overset{\oplus}{N}H_3]C$$

4. HYDRAZINOPHOSPHONIC ACID ESTERS. Phosphorous acid dialkylester hydrazides react with Schiff bases to yield hydrazinophosphonic acid diesters.[17]

$$(RO)_2PNHNHPh + R'CH=NR'' \longrightarrow R''NHCHR'P(=NNHPh)(OR)_2$$

XXXVII. PHOSPHONIC ACID IMIDES

Dimeric phosphonic acid imides may be obtained by the reaction of phosphonic acid dichlorides with primary amines in a 1:3 molar ratio.[768, 1241, 1282]

$$2RPOCl_2 + 6R'NH_2 \rightarrow R(O)P\overset{\displaystyle NR'}{\underset{\displaystyle NR'}{\diagup\!\!\!\diagdown}}P(O)R + 4[R'\overset{\oplus}{N}H_3]Cl^{\ominus}$$

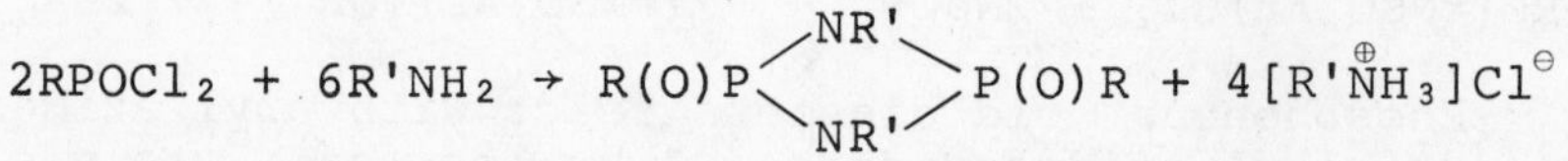

Another method of preparation is the thermal decomposition of phosphonic acid diamides of primary amines.[1261, 1282, 2098]

$$2RPO(NHR')_2 \xrightarrow[-2R'NH_2]{\Delta} R(O)P\overset{\displaystyle NR'}{\underset{\displaystyle NR'}{\diagup\!\!\!\diagdown}}P(O)R$$

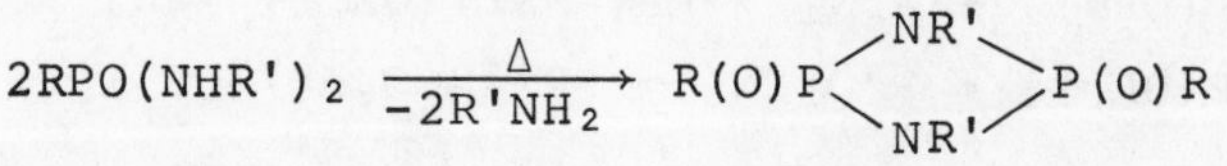

By the reaction of $RPO(NHR')_2$ with p-phenylene diamine, polymeric phosphonic acid imides have been prepared.[1448]

$$2nPhPO(NHR)_2 + nH_2NC_6H_4NH_2 \xrightarrow{-4nRNH_2} \left[\begin{array}{c} Ph \diagdown\!\!\!\diagup\!P\!\diagup\!\!\!\diagdown O \\ -N \diagup\!\!\!\diagdown\!\!\!\diagup\!\!\!\diagdown N-C_6H_4- \\ O \diagdown\!\!\!\diagup\!P\!\diagdown\!\!\!\diagup Ph \end{array} \right]_n$$

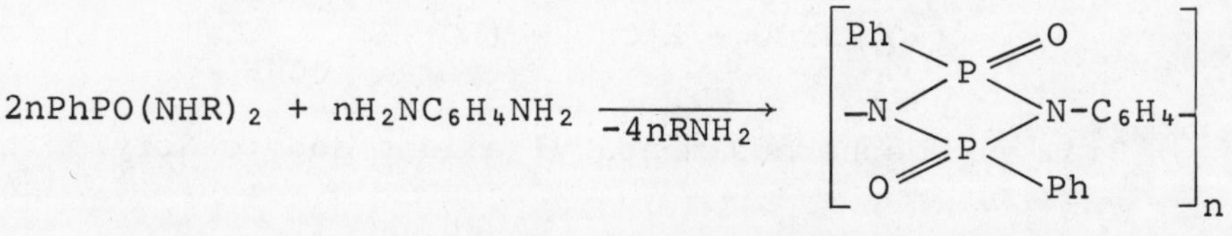

B. PROPERTIES OF PHOSPHONIC ACIDS, THIOPHOSPHONIC ACIDS, AND THEIR DERIVATIVES

B.1. Physical and Chemical Properties

Phosphonic acids are dibasic and generally less acidic than monophosphoric acid. There are two series of alkali salts: primary salts, which react slightly acidic in water, and secondary salts, which react alkaline. Both series are readily soluble in water. The alkaline earth salts are much less soluble, occasionally (e.g., with hydroxymethane phosphonic acid) a negative temperature coefficient of solubility is observed. The majority of Mg salts of phosphonic acids substituted in 2-position are readily soluble.[571] For the preparation of the free acids, the earlier used methods of treating Pb or Ba salts with H_2SO_4 have been replaced by ion exchange.

Primary as well as secondary Ag salts have been isolated.[792, 1683] The secondary Ag salt of trichloromethane phosphonic acid detonates upon heating.[2088] For the separation of aromatic phosphonic acids from aqueous solutions, crystallization of the sparely soluble hemi-salts of Na, K, or $NH_4^{\oplus}$ [$RPO(OH)(OM) \cdot RPO(OH)_2$] is a suitable method.[291, 479, 1084, 1088, 1286]

A number of di- and polyphosphonic acids have gained technical importance. Especially diphosphonic acids with geminal phosphono groups, e.g., 1-hydroxyethane-1,1-diphosphonic acid (HEDP),[264] aminopolymethylene phosphonic acids such as amino-tris(methylene phosphonic acid) (ATMP)[1328] and phosphonocarboxylic acids (e.g., 2-phosphono-butane-1,2,4-tricarboxylic acid)[612] should be mentioned. These complexing agents exhibit a strong threshold-effect; i.e., in understoichiometric amounts they retard the precipitation of various sparely soluble alkaline earth salts, especially those imparting hardness to industrial and household water supplies.[264] Therefore they are finding increasing use for the prevention of deposits in boiler systems. As compared to the previously used polyphosphates they offer the advantage of being stable vs. hydrolysis. Cleavage proceeds only > 150°C with measurable rates. Further technically important uses are in the liquefaction of mineral suspensions.[261]

HEDP and complexing aminomethylene phosphonic acids have been reported to stabilize peroxide in detergents. HEDP is also commercially used in soaps.[263] A number of diphosphonic acids, e.g., HEDP, methylenediphosphonic acid (MDP) and dichloromethane diphosphonic acid (Cl_2MDP) have recently stirred interest in the pharmaceutical field for the regulation of calcium metabolism.[540a]

β-Aminoethanephosphonic acid has been observed in nature in lower organisms.[1677d] Cis-1,2-epoxypropane phosphonic acid has interesting antibiotic properties.[769a] Phosphonic acids are quite stable towards hydrolysis, even at elevated temperatures. Electronegative substituents, however, increase the reactivity. Thus, trichloromethane phosphonic acid is cleaved in boiling aqueous bases to yield $HCCl_3$ and H_3PO_4. The P-C bond of aromatic phosphonic acids carrying electronegative substituents in 2- or 4-position (e.g., OH or RO substituted) is also cleaved upon boiling with dilute acids with formation of H_3PO_4.[228, 581, 958, 2053] In some cases, treatment with Br_2 also splits the P-C bond.[480, 1009]

Phosphonic acid diesters are also relatively stable compounds. Boiling with alkali bases normally hydrolyzes the diesters to the monoester stage. Complete hydrolysis is achieved only above 130°C under pressure. Diaryl esters are more easily hydrolized than dialkyl esters. Boiling with mineral acids results in hydrolysis only after several hours. Normally, the P-C bond is not cleaved under these conditions, with the exception of α-oxoalkane phosphonic acid esters, which are cleaved to yield carboxylic acids and phosphorous acid esters. The preparation of phosphono formic acid by hydrolysis of the ester with diluted bases is accompanied by partial decomposition. Hydrolysis of α-hydroxyalkane- and trichloromethane phosphonic acid ester may only be achieved with mineral acids without cleavage of the P-C bond.[230, 2087, 2089] α-Hydroxyalkane phosphonic acid esters with halogen substituents in β-position are converted to β-haloalcohols and phosphoric acid ester in acidic solution. In alkaline solutions, conversion to phosphoric acid dialkylvinylester is observed.[1112]

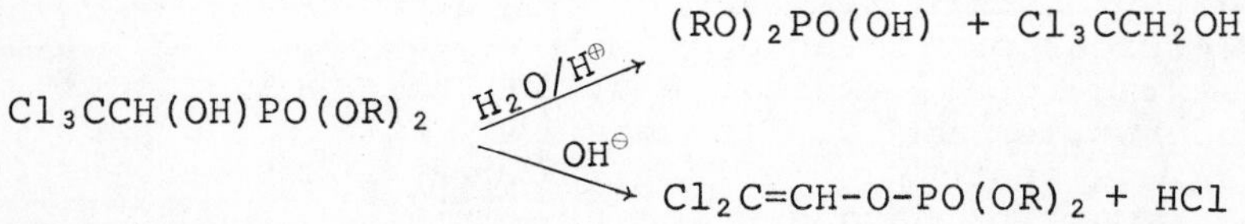

Phosphonic acid esters with strongly activated methylene groups in α-position react with carbonyl compounds in the presence of alkali hydrides, -amides, or -alcoholates with formation of olefins and dialkylphosphates.[799a] This reaction is analogous to the preparation of olefins, which are difficult to prepare by other means, from phosphine alkylenes and carbonyl compounds. (cf. Chapter 5A)

Many phosphonic acid esters with acidic groups have insecticidal properties often paralleled by high mammalian toxicity (cholinesterase inhibitors). Thiophosphoric and thiophosphonic acid esters are generally preferred, since they offer a better ratio of insecticidal properties to mammalian toxicity as compared to their oxygen analogues.

Monoester halides of phosphonic acids (especially chlorides) are important chemical intermediates for the preparation of monoesters, mixed diesters, monoester derivatives, and pyrophosphonic acid esters.

The esterfluorides are extremely toxic. Methylphosphonic acid i-propylester fluoride (Sarin) was developed as a nerve gas during World War II.

Thiophosphonic acids are unstable and have not been isolated as free acids. The salts of thiophosphonic acids, however, obtained by hydrolysis of the diesters, are relatively stable. In some cases, the salts of monothiophosphonic acids have been isolated as analytically pure compounds.

The hydrolysis of thiophosphonic acid esters with diluted mineral acids proceeds with simultaneous cleavage of the P-S bonds and leads to formation of phosphonic acids.

Monothiophosphonic acid monoesters may be prepared by various methods, preferably by controlled alkaline hydrolysis of the corresponding diesters. The resulting monoacids are stable enough to be purified by vacuum distillation.[784a, 1680] Dithiophosphonic acid O-monoesters may also be purified by distillation.[1864]

Thiophosphonic acid esters have found limited interest as insecticides.

B.2. Spectral Properties

The structural elucidation and characterisation of phosphonic acids and derivatives has been greatly accelerated by spectroscopic methods, especially by infrared and nuclear magnetic resonance spectroscopy. References to spectral data for individual substances are given in the list of compounds.

B.2.1. Infrared Spectra

IR spectra have been recorded for a large part of the compounds in this chapter, but few are discussed in any detail. Apart from the normal absorptions of the organic rest, the spectra contain the bonds attributable to the tetrahedral RP(O)XY group (X,Y = OH, OR, NRR', Hal, etc). The P=O stretching frequency is normally found in the region of 1320-1200 cm^{-1}. Due to inter- or intramolecular hydrogen bonding effects, it may be shifted by 50-100 cm^{-1} to lower wave numbers.

In compounds containing a pyramidal $-PO_3$ group (C_{3v} symmetry), both the symmetrical and antisymmetrical stretching modes are IR active, resulting in strong absorptions in the 1170-1030 (ν_{as}) and 1025-940 cm^{-1} (ν_s) region. The P=S stretching vibration of thiophosphonic acid derivatives is normally found from 750-550 cm^{-1}

Phosphorus-carbon stretching frequencies normally occur in the 800-600 cm^{-1} region, but have been found very difficult to assign.

B.2.2. Nuclear Magnetic Resonance Spectra

^{31}P NMR spectra have been recorded for a considerable number of phosphonic and thiophosphonic acid derivatives. Mainly the ^{31}P chemical shift data, but also the spin-spin coupling constants have been employed in structural proofs.

The chemical shifts for the quadruply connected compounds vary from +10 to - 100 ppm. Positive chemical shifts are recorded for alkinylphosphonic acid derivatives, negative shifts in the region -50 to -100 ppm are found for thiophosphonic acid derivatives, due to changes of the π-bonding around the phosphorus atom when sulfur is the electron-pair acceptor.

The coupling constants for 2-bond splittings (J_{PCH}, J_{PNH}) and 3-bond splittings (J_{PCCH}, J_{POCH}, J_{PSCH}, J_{PNCH} are usually in the order of 5 to 20 Hz. Phosphorus-fluorine coupling constants are somewhat larger, the range being 50 to 150 Hz for J_{PCF}.

C. LIST OF COMPOUNDS

REMARKS

The ordering principle used in this section is by the carbon number and composition of the organic group directly attached to the phosphorus atom by a P-C bond. The sequence is identical with that in Chemical Abstracts. This allows to treat as a group different derivatives of a particular phosphonic acid in the sequence given in the list of contents. A chemical variation of the P-bonded organic rest results in listing in different groups; for instance, $HO_2CCH_2PO(OMe)_2$ is found under $C_2H_3O_2$, whereas $CH_3O_2CCH_2PO(OMe)_2$ appears under $C_3H_5O_2$.

For practical reasons, all amide derivatives and anhydrides appear in separate groups, but registered by the same ordering principle as above.

Polyphosphonic and polythiophosphonic acid derivatives, where more than one phosphonate or thiophosphonate group is attached to the same organic rest by P-C bonds, are also treated in separate groups. The ester $(RO)_2OP(CH_2)_xPO(OH)_2$, for example, is listed as a polyphosphonic acid derivative. However, compounds of composition $(RO)_2OPO(CH_2)_xOP(CH_3)O(OR)$ or $(RO)O(CH_3)PO(CH_2)_xPO(OR')_2$ are found in the group of the corresponding monophosphonic acid, i.e., in both cases under CH_3.

The following examples may clarify the ordering principle:

Compound searched	Listed under
$CH_3PO(OH)_2$, its monoester, esterhalides, diesters	Monophosphonic acids, CH_3
$H_2NCOCH_2PO(OR)_2$	C_2H_4NO

$\underline{(CH_3)_2NCOCH_2}PO(OR)_2$	C_4H_8NO
$O(CH_2)_4PO(OR)$	C_4H_8
$(CH_3O)_2OPO(CH_2)_2PO(OR)_2$	$C_4H_{10}O_4P$
$(C_6H_5)_2P(CH_2)_2PO(OR)_2$	$C_{14}H_{14}P$
$CH_3PO(NR_2)(OH)$ $CH_3PO(NR_2)(RO')$ $CH_3PO(NR_2)_2$	Monophosphonic acid amide derivatives, CH_3
$(RO)_2OP(CH_2)_2PO(OR)_2$ $(RO)_2SP(CH_2)_2PS(OR)_2$	Polyphosphonic acids, C_2H_4 Polythiophosphonic acids, C_2H_4
$(CH_3PO_2)_2$	Phosphonic acid anhydride derivatives, CH_3

C.1. Phosphonic Acids

$\underline{C_1}$

$BrH_2CPO(OR)_2$.

R = Et. V.2. b_1 99°.[2086]

R = i-Pr. From $(i\text{-}PrO)_2OPCH_2CO_2Ag$ and Br_2, $b_{0.5}$ 80-92°, n_D^{25} 1.4652.[1143]

$ClH_2CPO(OH)_2$. I.9. m. 88°,[2080] ^{31}P -17.8 ppm.[1189a]
$ClH_2CPO(OH)O^{-}\,{}^{\oplus}H_3NC_6H_5$. m. 187-188°.[2038]
$ClH_2CPO_3^{2-}(H_3\overset{\oplus}{N}C_6H_{11}\text{-cycl.})_2$. m. 199-200°.[2038]
$ClH_2CPO(OH)F$. II.1. As aniline salt, m. 91-92°, ^1H[1717]
$ClH_2CPO(OR)OH$.

R = Me. III.4. $b_{0.006}$ 130-135°.[1464]

R = $C_{10}H_{21}$. Const. for extraction from U (VI), dissociation const. in 75% alcohol.[809]

R = Et. As $H_3\overset{\oplus}{N}C(=NH)SEt$ salt, m. 125-126° (acetone/EtOH). ^1H, ^{31}P -12.6 ± 0.5 ppm, UV, IR, synthesis from $ClH_2CPO(OEt)_2$ and $(H_2N)_2CS$.[1324]

R = Et. III.4. As $4\text{-}MeC_6H_4\overset{\oplus}{N}H_3$ salt, m. 195-197° (alcohol).[1464]

R = i-Pr. III.4. As $4\text{-}MeC_6H_4\overset{\oplus}{N}H_3$ salt, m. 190-193° (alcohol).[1464]

$ClH_2CPO(OR)Cl$.

R = Ph. IV.2. b_1 123-124°, n_D^{20} 1.5354, d_4^{20} 1.3989.[615]

R = 4-ClC_6H_4. IV.2. $b_{0.14}$ 119-121°, n_D^{25} 1.5515, d_4^{25} 1.4985.[1234]

$ClH_2CPO(OR)NCO$. From $ClHCPO(OR)NH_2$ and $COCl_2$.

R = Ph.[1911]

R = 4-MeC_6H_4-. $b_{0.25}$ 126-128°, n_D^{20} 1.5294, d_4^{20} 1.3273.[1781]

$ClH_2CPO(OR)_2$.

R = Me. V.2. b_4 79.5-80°,[2038] ^{31}P -18.5 ppm.[1189a]

R = Et. V.2. $b_{2.5}$ 78-79°.[2038] $b_{2.5}$ 86-87°,[2090] ^{31}P -18.1 ppm.[1189a]

R = -CH_2CH_2Cl. V.2. $b_{0.075}$ 98-100°.[943] V.2. $b_{3.2}$ 161-165°.[419] V.2. b_1 158-160°.[994]

R = -CH_2CCl_3. V.2. $b_{4.5}$ 177-181°.[419]

R = i-Pr. V.2. b_2 72-73°.[1485]

R = -$CH_2C{\equiv}CH$. V.2. $b_{0.5}$ 122-124°, IR.[377]

R = -$CH_2CBr_2CHBr_2$. From $ClH_2CPO(OCH_2C{\equiv}CH)_2$ and Br_2, m. 70-72°.[377]

R = -$CH_2CJ{=}CHJ$. From $ClH_2CPO(OCH_2C{\equiv}CH)_2$ + J_2, m. 67-69°.[377]

R = Bu. V.2. b_2 110-111°.[1485]

R = -$CH_2CH_2O_2CCH{=}CH_2$. V.2.[1901]

R = -$CH_2CH_2O_2CCMe{=}CH_2$. V.2.[1901]

R = Ph. V.2. b_2 170-171°, m. 42-42.5° (ether/petroleum ether).[2038]

R = 4-ClC_6H_4-. V.2. m. 58-59°, n_D^{25} 1.5718.[2024]

R = 2-ClC_6H_4-. V.2.[341]

R = 4-MeC_6H_4-. V.2. m. 48-48.5°.[2038, 2026]

R = 4-t-BuC_6H_4-. V.2. b_{27} 171-172°.[341]

R = -$SiEt_3$ VI.1. $b_{1.5}$ 130-131°.[1439]

R = -$SiEt_2Me$. VI.1. b_3 112°.[1439]

R = -$SiEtMe_2$. VI.1. b_3 100-107°.[1439]

$ClH_2CPO(OEt)(OR)$.

R = 4-$O_2NC_6H_4$-. V.2. n_D^{25} 1.5412, herbicide.[2024]

R = Ph. V.2. b_1 140°, n_D^{20} 1.5082, d_4^{20} 1.2557.[615]

$(ClH_2C-)OP-O-R-O$. (ring: P bonded to terminal O)

R = $-CH_2CH_2-$. V.3. $b_{0.007}$ 105-115°.[1464]

R = $-CHMeCH_2-$. V.3. $b_{0.007}$ 100-110°.[1464]

R = $-CH_2CMe_2CH_2-$. V.2. m. 115-116°.[424]

R = 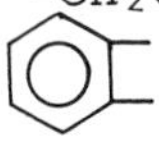 V.8.[926]

R = 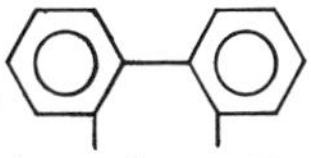 V.2. m. 161-163°.[1923]

$Cl_2HCPO(OEt)_2$. From $P(OEt)_3$ + $Cl_3CPO(OEt)_2$ in hexanol, b_{10} 133°,[190] ^{31}P -9.3 ppm.[1189a]

$Cl_3CPO(OH)_2 \cdot 2\ H_2O$. I.2. m. 81-82°.[231] ^{31}P -5.2 ppm.[1189a]

$Cl_3CPO(OH))Cl$. II.1. m. 79°.[2089]

$Cl_3CPO(OEt)OH$. From $Cl_3CPO(OEt)_2$ + alcohol or PhOH under reflux.[560]

$Cl_3CPO(OR)_2$.

R = Me. V.5. b_9 110-112°,[933, 934] d_0^{15} 1.4594.[934]

R = Et. V.5,[933, 934, 1004] b_{16} 135-137°,[1004] b_{16} 129-130.5°,[934] b_{13} 127-128°,[1004] b_{7-10} 122-123°, d_0^{14} 1.3664,[934] ^{31}P -6.5 ppm.[1189a] n_D^{25} 1,4582, n_D^{14} 1.4585.[934] V.3. b_{14} 130-131°.[955] V.5.[330] ^1H.[248]

R = Pr. V.5.[933, 934] b_{12} 144-145°,[934] b_{7-10} 145°,[933] d_0^{15} 1.2459, n_D^{15} 1.4582.[934] ^1H.[248]

R = i-Pr. V.5.[933,934] b_{12} 127-130°,[934] b_{7-10} 127-130°,[933] d_0^{22} 1.2206, n_D^{20} 1.4478.[934] ^1H.[248]

R = $-CH_2CH{=}CH_2$. V.5. $b_{0.5}$ 118°.[957] V.5. b_{7-10} 136-138°,[933] V.5. b_{10} 136-138°, d_0^{20} 1.500, n_D^{20} 1.4552.[934]

R = Bu. V.5. b_{7-10} 145-146°.[933] V.5. b_7 145-146°, n_D^{18} 1.4521.[934] V.5. b_5 150-155°, n_D^{25} 1.4490.[1004]

R = i-Bu. V.5. b_{7-10} 144-145°,[933] V.5. b_9 144-145°, d_0^{17} 1.1942, n_D^{17} 1.4487.[934]

R = Ph. m. 66°, ^{31}P +4.54 ppm.[248]

R = $2-MeC_6H_4-$. $b_{0.0005}$ 165°.[248]

R = $3-MeC_6H_4-$. $b_{0.002}$ 170°.[248]

R = $4-MeC_6H_4-$. m. 57°, ^{31}P +3.70 ppm.[248]

R = $4-ClC_6H_4-$. m. 48°, ^{31}P +3.82.[248]

R = $4-BrC_6H_4-$. m. 48°.[248]

R = $2-O_2NC_6H_4-$. m. 116°, ^{31}P +4.89 ppm.[248]

R = 3,4-$Cl_2C_6H_3$-. m. 92.5°.[248]

R = 3,5-$Cl_2C_6H_3$-. m. 89°.[248]

R = 4-$MeOC_6H_4$-. m. 48.5°, ^{31}P +2.96 ppm.[248]

R = 4-$EtOC_6H_4$-. m. 53.5°.[248]

R = 3,4,5-$Me_3C_6H_2$-. m. 115°.[248]

R = 4-$MeSC_6H_4$-. m. 72°, ^{31}P +3.62.[248]

R = 3-Me-4-MeS-C_6H_3-. m. 78°.[248]

$FH_2CPO(OPr\text{-}i)Cl$. IV.1. b_{11} 75-76°, n_D^{20} 1.4158.[721]

$FH_2CPO(OR)_2$.

R = Pr. V.6. b_{13} 97-98°, n_D^{20} 1.4080, d_4^{20} 1.0532.[721]

R = sec.-Bu. V.6. b_3 96-100°.[1790] V.6.$b_{0.1}$ 71-72°, n_D^{20} 1.4174, d_4^{20} 1.0333.[721]

$F_2HCPO(OR)_2$.

R = Et. V.6.b_{12} 85.6-86.5°.[1933]

R = i-Pr. V.6. b_{12} 89-90°.[1933]

R = Bu. V.6. b_{12} 124-125°.[1933]

$F_3CPO(OH)_2$. I.1/I.6. m. 81-82°.[234]

$N_2CHPO(OEt)_2$. V.14. $b_{0.35}$ 69-70°, n_D^{20} 1,4534, IR.[1732]

$HO_2CPO(OH)_2$. The acid is unstable. Rate const. of the acid-catalyzed decarboxylation.[2064]

$NaO_2CPO(ONa)_2 \cdot 6H_2O$. I.2.[2064]

$H_2NC(O)PO(OR)_2$.

R = Et. From $EtO_2CPO(OEt)_2$ + NH_3/H_2O at room temperature, m. 134-135° (benzene).[1418, 1420] By aminolysis of $MeSC(O)PO(OEt)_2$, m. 138-139°.[709]

R = i-Pr. By aminolysis of $MeSC(O)PO(OPr\text{-}i)_2$, m. 95-97°.[709]

$IH_2CPO(OH)_2$. From $HOH_2CPO(OH)_2$ + P_{red} + HJ (57%), m. 79-81°.[1524]

$IH_2CPO(OR)_2$.

R = Et, V.5. b_6 133°, d_0^{19} 1.6662,[124] $b_{0.7}$ 101°, n_D^{17} 1.4975,[552] $b_{0.2}$ 90-100°, n_D^{20} 1.4816,[334] ^{31}P -20.9 ppm.[1189a]

R = i-Pr. V.5. $b_{0.5}$ 80-83° n_D^{25} 1.4802.[1143] V.14. $b_{0.5}$ 84-90°, n_D^{25} 1.4718.[1143]

R = $-CH_2CH=CH_2$. V.5. $b_{0.6}$ 95-106°, n_D^{20} 1.4995.[334]

(MeO)$_2$OPCH$_2$OPMeO(OMe). V.8. b$_{0.6}$ 145-145.5^O, ^{31}P δ_1
-36, δ_2 -22.6 ppm.[755]
(R$_3$SiO)$_2$OPCH$_2$O-PO(CH$_2$OSiR$_3$)OSiR$_3$.

R$_3$Si = -SiEt$_3$. VI.1. b$_3$ 238-240°,.[1433a]

R$_3$Si = -SiEt$_2$Me. VI.1. b$_3$ 216-218°.[1433a]

R$_3$Si = -SiPr$_2$Me. VI.1. b$_3$ 244-246°.[1433a]

(RO)OP   OCH$_2$ PO(OR).
         CH$_2$O

R = Me. V.8. m. 151.5-152° (benzene).[755]

R = Et. By therm. decomp. of (EtO)$_2$OPCH$_2$OSiEt$_2$(OEt),
    m. 143.5-144° (benzene).[1436]

CH$_3$PO(OH)$_2$. I.1,[1283],[1420] I.2,[446] I.6.[792] m. 104°,
[1420] m. 104-106°,[446] m. 105°,[792] hygroscopic, pK$_1$
2.35, pK$_2$ 7.1,[1776a] ^{31}P -31.0 ± 0.1, ^{1}H,[528]
synthesis of salts from MePO(OPr-i)$_2$ and MCl,[1311]
IR (salts).[319]
CH$_3$PO(OH)F. II.1. As aniline salt, m. 135°, ^{1}H,
^{19}F.[1717] II.1. As cyclohexylamine salt, m. 164-166°.
[292]

[CH$_3$PO(F)O$^\ominus$]$_2$ (Me$_2$N$\overset{\oplus}{}$⟨ ⟩$\overset{\oplus}{}$NMe$_2$). By decomp. of

MePO(OCH$_2$CH$_2$NMe$_2$)F, m. >300°.[236]
CH$_3$PO(OR)OH.

R = Me. III.4. b$_1$ 121-123°.[1466] III.1,[948] III.2.[948]
    UV, IR.[1312] Complexes with Ti(III), V(III),
    Cr(III), UV,IR, X-rays.[1312]

R = Et. III.4. b$_{0.1}$ 108-110°.[1466] III.1.,[948] III.2,
    [948] III.3.[949] III.4. b$_{0.3}$ 117-118°.[510] Salt with
    H$_2$NC(=NH)SEt, m. 145-145.5^O (acetone/EtOH),
    ^{1}H, ^{31}P -24.7 ± 1 ppm, UV, IR.[1324] III.3.[678]

R = -CH$_2$CH$_2$Cl. III.4. b$_{0.03}$ 100°.[510]

R = Pr. III.4. n$_D^{20}$ 1.4425, d$_4^{20}$ 1.005.[509] Tetramethyl
    ammonium salt, III.3. m. 174°.[345],[353]

R = i-Pr. III.4. b$_{0.1}$ 102-104°.[1466] III.1,[948] III.2.
    [948] Complexes with Ti(III), V(III), Cr(III), UV,
    IR, X-rays.[1312] Salts with Eu, Al, Ga, In, Sc,
    V, Cr, Y from the m. chlorides and MePO(OPr-i)$_2$

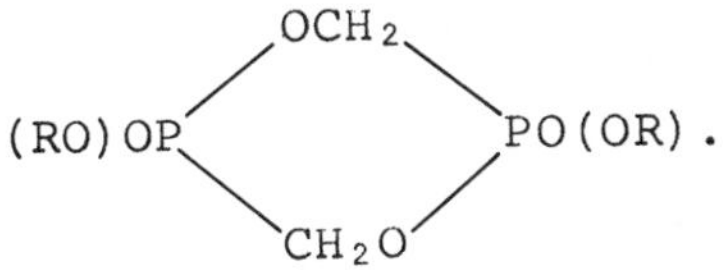

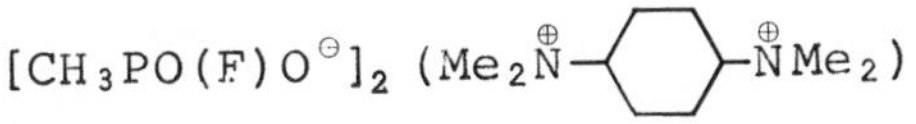

at 50-200°.[1311]

R = Bu. III.4. $b_{0.1}$ 116-117°.[1466]

R = i-Bu. III.4. $b_{0.5}$ 142-143°.[1466] III.4. $b_{2.5}$ 148-150°.[510]

R = $EtSCH_2CH_2$-. III.4. m. 138-140°.[510]

R = i-Am. III.4. $b_{0.3}$ 138-140°.[1466]

R = $-CH_2CMe_3$. III.1,[948] III.2.[948]

R = cycl.-C_6H_{11}-. III.4. m. 45-48°.[1466]

R = Ph. III.3. As Li salt, UV.[226] III.1. As dicyclohexylamine salt, m. 105-106°.[510] III.3. Reactions with nucleophilic reagents.[225] III.2.[1498]

R = 3-$O_2NC_6H_4$-. III.4. m. 103-104°.[1466]

R = 4-$O_2NC_6H_4$-. III.4. m. 90-90.5°.[1466] III.4. Reactions with nucleophilic reagents.[225] III.4. m. 113-114°, UV, alkaline hydrolysis.[226]

R = 2,4-$Cl_2C_6H_3$-. III.1. m. 72.5-73.5°.[1234]

R = 2,4,5-$Cl_3C_6H_2$-. III.4. m. 118-120°.[1466]

R = 4-MeC_6H_4-. III.2. m. 78-79°.[1498]

R = 4-Cl-3-MeC_6H_3-. III.2. m. 93-95°.[1498]

R = $-CH_2CHEtBu$. III.4. n_D^{20} 1.4423, d_4^{20} 1.020.[1465]

R = $-C_9H_{19}$. III.4. m. 31-32°.[1466]

R = $-C_{10}H_{21}$. III.4. m. 38-39°.[1466] III.2. m. 69-70°.[270]

R = -Bornyl. III.4. m. 120-122°.[1466]

R = $-C_{12}H_{25}$. III.4. m. 41-43°.[1466]

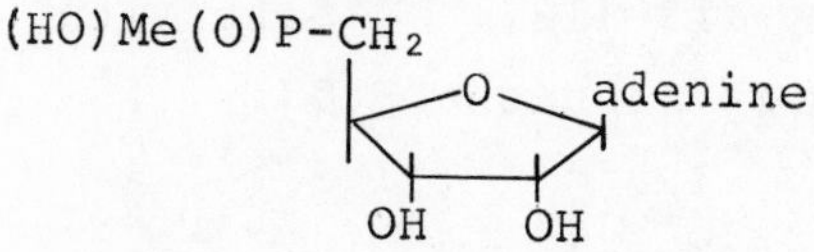

. III.4. m. 180-181° (dec.).[1255]

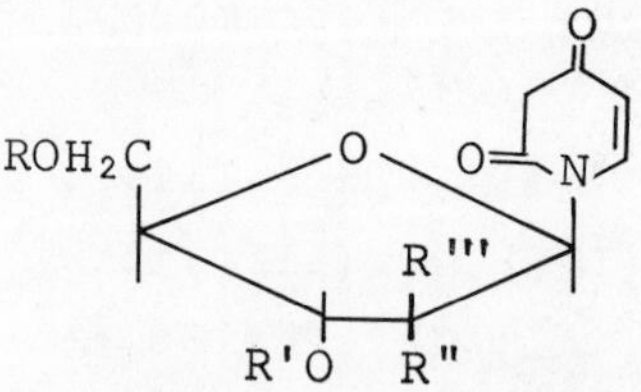

R = H, $CH_3C(O)$-, 4-$O_2NC_6H_4$-C(O)-;

R' = -P(O)Me(OH); R" = -OH; R"' = H. III.4. UV.[2050]

R = H, CH_3CO_2-,4-$O_2NC_6H_3$-C(O)-; R' = H; R" = -OP(O)Me(OH) III. 4, 4. UV.[2050]

R = H; R' = H; R" = H; R''' = OP(O)Me(OH). III.4. UV.[2050]

R = -P(O)Me(OH); R' = H; R" = OH; R''' = H. III.4.[795]

R = H, CH_3C(O)-; R' = -P(O)Me(OH);

R" = H; R = -OH. III.4. UV.[2050]

CH_3PO(OR)CN.

R = Et. IV.5. b_3 73-74°.[1474]

R = Pr. IV.5. b_6 95°, $n_D^{24.5}$ 1.4125, $d_4^{24.5}$ 1.0385.[1474]

CH_3PO(OR)Cl.

R = Me. IV.1. b_{22} 73°.[1454]

R = Et. IV.1. b_{21} 83°.[1454] IV.2. b_{10} 66°, n_D^{20} 1.4370 d_4^{20} 1.2413.[1596] IV.2.[1169] IV.1. $b_{0.7}$ 37°.[405]

R = $-CH_2CH_2Cl$. IV.1. b_5 103-105°.[622]

R = Pr. IV.1. b_{21} 94°.[1454] IV.2. b_{12} 79°, n_D^{20} 1.4351, d_4^{20} 1.1821.[1596] IV.2.[1169]

R = i-Pr. IV.1. b_{22} 83°.[1454] IV.4. b_3 47-48°.[639] IV.4. S(+) from R(-) MePO(OPr-i)H + N-Cl-succinimide.[1734] IV.1. $b_{0.5}$ 36-37°.[1711] From MePS(OPr-i)ONa and $COCl_2$, $b_{1.2}$ 37°, n_D^{25} 1.4281.[2] IV.4. b_3 47-48°.[639]

R = Bu. IV.1. b_{19} 105°.[1454]

R = sec.-Bu. IV.2.[4]

R = $-CH_2CMe_3$. IV.2.[4]

R = $-CHMeCMe_3$. IV.2.[4]

R = Ph. IV.2. b_2 114°, n_D^{20} 1.5223, d_4^{20} 1.2829.[615] From MePO(OPh)NH_2 and $COCl_2$, $b_{0.05}$ 95-98°, n_D^{20} 1.5263.[1911] IV.2.[70] IV.2. b_{21} 153-155°.[1498]

R = 2-MeC_6H_4-. IV.2. b_{18} 160-163°.[1498]

R = 3-MeC_6H_4-. IV.2.[77] IV.2. b_{24} 163-165°.[1498]

R = 4-MeC_6H_4-. IV.2. b_{20} 162-165°.[1498]

R = 4-Cl-3-MeC_6H_3-. IV.2. b_{20} 185-187°.[1498]

R = -CHMe-C_6H_{13}. IV.2.[4]

CH_3PO(OR)F.

R = Me. From WF_6 and P(OMe)$_3$.[1405] IV.2. b_{55} 71-72°,

IR, ^{1}H.[1717]

R = Et. IV.2. b_{11} 49°.[191] From 2.4.6-$(O_2N)_3C_6H_2F$ and $Me(EtO)P\langle^{O}_{S}$ ⊖ $H_2N^{\oplus}(C_6H_{11}$-cycl.$)_2$ in acetone, n_D^{25} 1.3773.[292] From 2.4-$(O_2N)_2C_6H_3F$ and MePO(OEt)SR b_{20} 49°, n_D^{25} 1.3778.[220]

R = i-Pr. From 2,4,6-$(O_2N)_3C_6H_2F$ and $Me(i\text{-}PrO)P\langle^{O}_{S}$⊖ salt in acetone, n_D^{25} 1.3811.[292, 293] From MePO(OPr-i)[$-OCF_2CH(CF_3)_2$] and NaO_2CCH_3.[978]

R = Bu. From 2,4,6-$(O_2N)_3C_6H_2F$ and $Me(BuO)P\langle^{O}_{S}$⊖ $H_2\overset{\oplus}{N}(C_6H_{11}$-cycl.$)_2$ in acetone, n_D^{25} 1.3930.[292]

R = t-Bu. From 2,4,6-$(O_2N)_3C_6H_2F$ and $Me(t\text{-}BuO)P\langle^{O}_{S}$⊖ $H_2\overset{\oplus}{N}(C_6H_{11}$-cycl.$)_2$, m. -18°, n_D^{25} 1.3902.[292]

R = $Me_2NCH_2CH_2$-. IV.2. $b_{0.4}$ 43°, n_D^{25} 1.4131, ^{1}H.[236]

R = $Me_2N(CH_2)_3$-. IV.2. $b_{0.6}$ 65°, ^{1}H.[236]

$CH_3PO(F)(OCH_2CH_2\overset{\oplus}{N}Me_3)$ $J^{\ominus}$. Half-life time at pH 8.5 and 25° 3.7 min, inhibition of ACHE at pH 7.6, $1.3\cdot10^{9}$ mole^{-1} min^{-1}.[882]

$CH_3PO(OR)NCO$.

R = Et. IV.5. b_{13-14} 82-83°.[726]

R = Ph. IV.2. $b_{0.07}$ 105-108°.[468]

R = 4-ClC_6H_4-. IV.2. $b_{0.06}$ 110-112°.[468]

$CH_3PO(OR)NCS$.

R = Me. IV.2. $b_{0.3}$ 70°, n_D^{20} 1.5192, d_4^{20} 1.2651.[852]

R = Et. IV.2. $b_{0.5}$ 57-60°, n_D^{20} 1.5065, d_4^{20} 1.1965.[852]

R = Pr. IV.2. $b_{0.5}$ 64°, n_D^{20} 1.4794, d_4^{20} 1.1583.[852]

R = i-Pr. IV.2. $b_{0.5}$ 81°, n_D^{20} 1.4978, d_4^{20} 1.1435.[852]

$CH_3PO(OR)_2$.

R = Me. V.5. b_{76} 82°, n_D^{25} 1.4118.[1079] V.5. b_{12} 67°, n_D^{20} 1.4105.[1371] V.5. b_{12} 66°.[1076] V.5.[346, 1638, 14 ^{31}P -32.4 ppm.[1189a] V.2.[1363] By reaction of $ClH_2CPO(OMe)H$ with NaOMe/MeOH.[648]

R = Et. V.5. b 192-194°.[102] V.5. b_4 64°, n_D^{16} 1.4120,[552] b_{11} 80.5-81°, d_0^{0} 1.0725, n_D^{14} 1.4062.[115] V.1.[1283, 46 787, 2092] V.5. b_{10} 90°.[1420] V.6,[1924] ^{31}P -30 ppm.[1189a]

R = $-CH_2CCl_3$, V.2. m. 94-94.5°.[1363]

R = $-CH_2CH_2F$. V.2. $b_{1.5}$ 101°.[897]
R = $-CH_2CF_3$. V.5. b_4 51-52°, n_D^{20} 1.3387.[222a]
R = $-CH_2CH_2NH_2 \cdot HCl$. V.2.[1936]
R = Pr. V.13. b_9 93-96°.[1483] V.5. b_{12} 105-106°, d_0^0 1.0683, n_D^{18} 1.4082.[115]
R = i-Pr. V.5. b_3 66°, $n_D^{16.5}$ 1.4120.[552] V.6. $b_{3.5}$ 58-60°,[1764] ^{31}P -27.4 ppm.[1189a] V.5. b_1 51°.[551] IR and ^1H investigations of Sn(II), Sn(IV), and Ti(IV) halogenids.[1445] Complexes with M salts.[939]
R = $HC{\equiv}CCH_2-$. V.2. b_2 139-140°.[1142]
R = $(ClH_2C)_2CH-$. V.2. b_1 127-129°.[1751]
R = $MeOCH_2CH_2-$. V.5. b_{12} 152-153°, n_D^{20} 1.4345, d_4^{20} 1.1407.[1373]
R = t-Bu. V.5, V.6. ^1H, ^{31}P -21.2 ppm, IR.[1190] V.13.[1495] V.2. $b_{1.5}$ 86-91°.[446] V.5. b_5 105.5-106°.[160] V.2.[1363]
R = $-CH_2CF_2CF_2CF_3$. V.2. b_9 92-93°, n_D^{20} 1.3220, d_4^{20} 1.6242.[630]
R = $-CH_2CH_2OEt$. V.5. b_{19} 147-148°, n_D^{20} 1.4348, d_4^{20} 1.0912.[1373]
R = $-CH_2CO_2Et$. V.13. b_2 148-148.5°.[1646]
R = $-CH_2CH_2SEt$. V.6. $b_{0.0003}$ 108-109°.[1222]
R = i-Am. V.2. $b_{0.5}$ 102-105°, n_D^{20} 1.4265, d_4^{20} 0.9499.[270] IR. n_D^{20} 1.4292, d_4^{20} 0.9553, extraction of $Ce(NO_3)_3$.[1431]
R = $-(CH_2)_4CCl_3$. V.2. b_1 177-178°.[1363]
R = $-(CH_2)_2O_2CCH{=}CH_2$. V.2. Polymerizable.[1901]
R = $-CH_2CMe_3$. V.1. $b_{0.02}$ 110°, n_D^{25} 1.4250.[338]
R = $C_6H_{13}-$. V.13.[1495] V.2. b_1 125-128°, n_D^{20} 1.4359, d_4^{20} 0.9500.[270]
R = $BuOCH_2CH_2-$. V.2. b_1 148-150°, n_D^{20}1.4370, d_4^{20} 1.0271.[270]
R = $-(CH_2)_2O_2CCMe{=}CH_2$. V.2. Polymerizable.[1901]
R = $-(CH_2)_6NH_2 \cdot HCl$. V.2.[1936]
R = Ph. V.5. m. 36-37°, b_{11} 190-195°.[1283] V.5. b_{11} 201-202°, d_0^{20} 1.2051.[102] V.5. m. 35-36°.[1076] V.5.

[461] From $[PhO)_3\overset{\oplus}{P}Me]J^{\ominus}$ and Ag salts.[802] V.13. m. 33-35°,[1483] ^{31}P -23.5 ppm[1189a]

R = 4-$O_2NC_6H_4$-. V.2. m. 119-120° .[1234] V.1. Rate const. of hydrolysis.[1099]

R = 4-$ClO_2SC_6H_4$-. V.14. m. 84.5 - 87°, IR.[777]

R = 4-ClC_6H_4-. V.5. b_{20} 245°.[1283]

R = -C_7H_{15}. V.2. $b_{2,5}$ 160-162°, n_D^{20} 1.4373, d_4^{20} 0.9355.[270]

R = 3-MeC_6H_4-. V.5. b_7 200-205°.[1283]

R = 4-MeC_6H_4-. V.5. b_{12} 220-225°.[1283]

R = -C_8H_{17}. V.5. $b_{5.5}$ 200-200.5°.[160] V.13.[1495]

R = sec. -C_8H_{17}. V.13. b_1 78-83°.[1495]

R = -CH_2CHEtBu. V.2. b_3 158-161°, n_D^{20} 1.4412, d_4^{20} 0.9207.[1465]

R = -$(CH_2)_2O(CH_2)$OBu. V.2. $b_{0.5}$ 202-205°, n_D^{20} 1.4436, d_4^{20} 1.0422.[270]

R = 2-Cl-4-(CH_2=CH)C_6H_3-. V.5, V.2. $b_{0.13}$ 155-156°.[621]

R = 2-(CH_2=$CHCH_2$)C_6H_4-. V.2. b_1 172°, n_D^{20} 1.5502, d_4^{20} 1.1130.[1981]

R = 2-($\overline{OCH_2-C}HCH_2$)C_6H_4-. V.14. n_D^{20} 1.5512, d_4^{20} 1.2405.[1981]

R = -$C_{12}H_{25}$. V.13. $b_{0.04}$∿ 157°.[1079]

R = -CHMeCH$_2$- 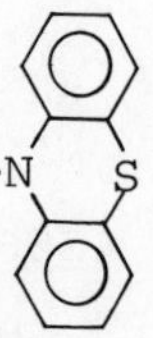 . V.2. m. 46-48°.[1480]

$CH_3PO(OMe)OR$.

R = Pr. V.1. b_{13} 84°, n_D^{20} 1.4175, ^1H.[1994] V.1. b_{13} 84°.[353]

R = i-Pr. V.1. b_{13} 69°, n_D^{20} 1.4175, ^1H.[1994] V.1. b_{13} 69°.[353]

R = -CH_2CH=CH_2. V.1. b_{13} 89-90°, n_D^{20} 1.4350, ^1H.[1994] V.1. b_{13} 89°.[353] V.13. b_{13} 88-90°,[975]

R = Bu. V.13.[975] V.1. b_{13} 97°.[353] V.1. b_{13} $96\text{-}97^\circ$, n_D^{20} 1.4205, ^1H.[1994]

R = $-CH_2CH{=}CH\text{-}CH_3$. V.13. n_D^{20} 1.4402.[975]

R = $-CH_2CO_2Et$. V.13. $b_{1.5}$ $97\text{-}98^\circ$.[1646]

R = i-Am. V.13. b_{11} $103\text{-}103.5^\circ$.[975]

R = $-CHMeCO_2Et$. V.1. $b_{0.004}$ 69°, n_D^{20} 1.4280, ^1H.[1994]

R = Ph. V.2.[1498]

R = $-CH_2Ph$. V.1. b_{13} 159°, n_D^{20} 1.5095, ^1H.[1994] V.1. b_{13} 159°.[353] V.13. b_{12} $156\text{-}157^\circ$.[975]

R = $2\text{-}ClC_6H_4CH_2-$. V.1. $b_{0.08}$ $105\text{-}110^\circ$, n_D^{20} 1.5202, ^1H.[1994]

R = $4\text{-}ClC_6H_4CH_2-$. V.1. $b_{0.05}$ $90\text{-}94^\circ$.[353] V.1. $b_{0.05}$ $90\text{-}94^\circ$, n_D^{20} 1.5191. ^1H.[1994]

R = $2,4\text{-}Cl_2C_6H_3CH_2-$. V.1. $b_{0.06}$ 119°, n_D^{20} 1.5335, ^1H.[1994]

R = $3\text{-Me-}4\text{-MeS-}C_6H_3-$. V.2, $b_{0.01}$ 108°, insecticide.[1862]

R = $-C_{12}H_{25}$. V.13. $b_{0.03}$ $\sim 115^\circ$, n_D^{25} 1.4406, IR.[1079] V.13. $b_{1.5}$ $153\text{-}157^\circ$, n_D^{20} 1.4410.[975]

$CH_3PO(OEt)OR$.

R = $-CH_2CH_2F$. V.5. $b_{0.5}$ $61\text{-}62^\circ$.[897]

R = i-Pr. V.13. $b_{3.5}$ 60°, n_D^{25} 1.4095.[2]

R = $-CH_2CO_2Et$. V.13. b_2 $106\text{-}107^\circ$.[1646]

R = $-CH_2CH_2SEt$. V.2. $b_{0.03}$ $93\text{-}94^\circ$, b_2 $101\text{-}102^\circ$.[895]

R = Ph. V.13.[160] b_2 136°.[225a]

R = $2\text{-}ClC_6H_4-$. b_1 120°.[225a]

R = $2\text{-}O_2NC_6H_4-$. $b_{1.5}$ 157°.[225a]

R = $4\text{-}O_2NC_6H_4-$. V.2. Cholinesterase inhibition.[1169] $b_{1.5}$ 178° [225a]

R = $4\text{-}MeSC_6H_4-$. V.2. $b_{0.01}$ 90°.[1796]

R = $-CHMeC_6H_{13}$. V.2. $b_{0.2}$ $73\text{-}74^\circ$.[949]

R = 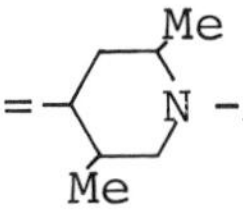 -Me. V.2. $b_{2.5}$ $110\text{-}113^\circ$.[1472]

R = 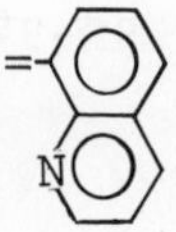 V.2. $b_{0.03}^{0}$.[527]

R = 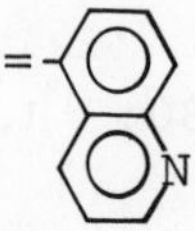 V.2. $b_{0.05}$ 130°.[527]

$CH_3PO(OCH_2CH_2Cl)OPh$. V.2. b_1 138-139°, n_D^{20} 1.5113, d_4^{20} 1.2570.[615]

$CH_3PO(OPr)OR$.

R = 4-ClC_6H_4-. b_3 162°.[255a]

R = 4-$O_2NC_6H_4$. V.2. Cholinesterase inhibition.[1169]

R = 2,4-$(O_2N)_2C_6H_3$-. b_1 160°.[225a]

R = -CH_2Ph. V.1. $b_{0.02}$ 95-100°.[353]

R = -C_8H_{17}. V.13 b_2 158-160°.[1483]

R = -$CH_2C(O)Ph$. V.1. $b_{0.02}$ 134°.[345]

R = 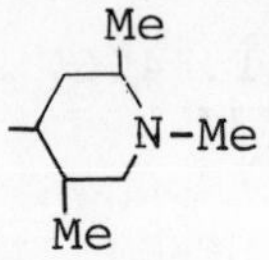. V.2. b_3 125-126°.[1472]

$CH_3PO(OPr$-$i)OCH_2CH_2NMe_2$. V.2. $b_{0.2}$ 83°, n_D^{25} 1.4291. ^1H.[236]

$CH_3PO(OCH_2CH_2OMe)OAm$. V.5. b_{13} 147-150°, n_D^{20} 1.4327, d_4^{20} 1.0191.[1373]

$CH_3PO(OBu)O$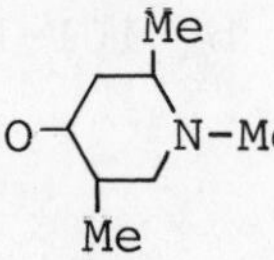. V.2. b_2 130°.[1472]

$CH_3PO(OCMe_3)OPh$. V.5.[960]

$CH_3PO(OCH_2CH_2OEt)OAm$. V.5. b_{19} 158-164°, n_D^{20} 1.4302, d_4^{20} 1.0010.[1373]

$CH_3PO(OPh)OR$.

R = 2-MeC_6H_4-. V.2. $b_{1.5-2}$ 188-190°, n_D^{20} 1.5455, d_4^{20} 1.1723.[615]

R = 4-MeC_6H_4-. V.2. $b_{1.5-2}$ 186-188°, n_D^{20} 1.5454, d_4^{20} 1.1727.[615]

R = 2-MeO_2C-C_6H_4-. V.2. $b_{1.5-2}$ 207-210°, n_D^{20} 1.5663.[615]

$CH_3(OPh)OPO$ 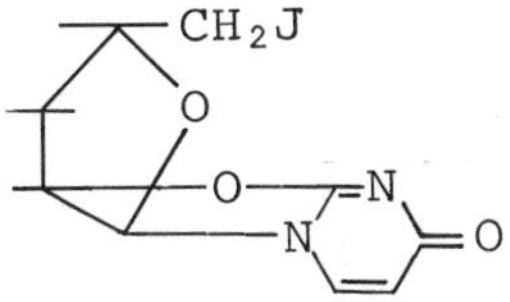

Diastereomers. (a) m. 222-224°, (b) 196-197°, UV, ^1H.[2049]

$CH_3PO(OC_6H_4NO_2-4)[OCH_2C(O)Ph]$. V.2. Rate const. of hydrolysis.[1099]
$CH_3PO(OC_6H_4NO_2-4)[OCH_2C(O)C_6H_4Cl-4]$. V.1. Rate const. of hydrolysis.[1099]
$CH_3PO(OC_6H_4NO_2-4)[-OCH_2C(O)C_6H_4OMe-4]$. V.1. Rate const. of hydrolysis.[1099]
$CH_3(O)\dot{P}O-R-\dot{O}$.

R = $-CH_2CH_2-$. V.2. b_3 104-105°.[995] V.3. $b_{0.2}$ 63-64°.[1464] V.13.[1649]

R = $-(CH_2)_3-$. V.2.[1363] V.2. b_3 97-99°.[995]

R = $-CHMeCH_2-$. V.13.[120] V.3. $b_{0.2}$ 78-80°.[1464] V.2. $b_{0.5}$ 86-88°.[994] V.5. b_{13} 140.5-142°, n_D^{20} 1.4415.[120]

R = $-CHMeCH_2CHMe-$. V.2. $b_{0.85}$ 99°.[943]

R = $-CH_2(CF_2)_3CH_2-$. V.2. m. 88-89°.[1881]

R = $-CH_2-CMe(CH_2J)-CH_2-$. V.5. m. 120°.[2059]

R = $-CH_2-C(CH_2OH)(CH_2J)-CH_2-$. V.5. m. 134°.[2059]

R = 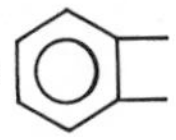 V.2. b_2 120°, m. 84°, ^1H.[2075]

R = $-CH_2CMe[-CHMe(OMe)]-CH_2-$. V.5. b_6 155°, n_D^{20} 1.4639, d_4^{20} 1.1769.[284a] V.5. b_6 155°.[285]

R = $-CH_2C[-CHMe(OMe)](CHMe_2)-CH_2-$. V.5. b_5 180°.[285]

R = $-CH_2C[-CHMe(OCHMe_2)](-CHMe_2)-CH_2-$. V.5. b_1 135°.[285]

$(EtO)Me(O)P-OCH_2O-P(O)Me(OEt)$. V.13. b_8 149-150°.[1648]
$(EtO)Me(O)P-O(CH_2)_4O-P(O)Me(OEt)$. V.2. $b_{0.03}$ 141-145°.[1957]
$(PhO)Me(O)P-O(CH_2)_2O-P(O)Me(OPh)$. V.2. $b_{0.85}$ 123-125°.[626]
$CH_3PO(OR)_2$, R = SiR'R"R'".

R = $-SiMe_3$. VI.1. b_{27} 105-107.5°.[1682]

R = $-SiMeEt_2$. VI.1. b_3 122°.[1442]

R = $-SiEt_3$. VI.1. b_4 145°.[1442]

R = $-SiMePr_2$. VI.1. b_4 160°.[1442]

R = $-SiMePh_2$. VI.1. b_3 270°.[1442]

R = $-SiEtPh_2$. VI.1. b_6 300°.[1442]

R = $-SiPh_3$. VI.1. m. 197°.[1442]

Me(O)P-OSiMe$_2$OSiMe$_2$O (cyclic). VI.1. $b_{1.2}$ 141-143°.[74]

Me(O)P-O(-SiMe$_2$O-)$_2$SiMe$_2$O (cyclic). VI.1. $b_{1.2}$ 134-136°.[74]

Me(O)P-O(-SiMe$_2$O-)$_3$SiMe$_2$O (cyclic). VI.1.[74]

Me(O)P-OSiMe$_2$-O-P(O)Me-OSiMe$_2$O (cyclic). VI.1. $b_{1.2}$ 173-175°.[74]

$CH_3PO(OCH_2CH_2Cl)-ON=CClF$. From Cl_2FC-NO + $MePCl_2$ + $(CH_2OH)_2$ + NEt_3. $b_{0.1}$ 63°.[1199]

$CH_3PO(OPr-i)-ON=CH-$ 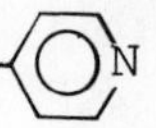 . VI.2.[739]

$CH_3PO(OPr-i)-ON=CMe-$ 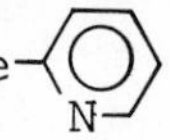 VI.2. m. 40°.[260]

$[CH_3PO(OPr-i)-ON=CH-$ 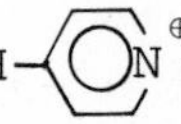 $-Me]J^{\ominus}$. V.14. m. 97° (dec.).[739]

$[CH_3PO(OPr-i)-ON=CMe-$ 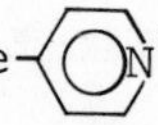 $-Me]$ $J^{\ominus}$. VI.2. m. 124.[260]

$[CH_3PO(OPr-i)-ON=CMe-$ 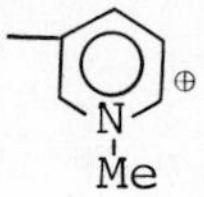 $]J^{\ominus}$. VI.2. m. 163°.[260]

$HOCH_2PO(OH)_2$. I.2. m. 84-85°,[41] m. 84.5-86°,[124] m. 85° [1446] pyridine salt, m. 145°.[1446] I.2. m. 87-88°, aniline salt, m. 167-168° (ethanol).[963] I.9. m. 88-90° (EtOH/$MeCO_2Et$).[206] I.9. Salts from sodium and org. amines,[835] ^{31}P -23.5 ppm.[1189a]

$HOCH_2PO(OEt)_2$. V.8. b_3 124-126°.[12] V.8. $b_{0.2}$ 103-105°, n_D^{20} 1.4342.[963] V.7. b_5 72° (? !), d_0^0 1.0726 [124]. From $(EtO)_2OPNa$ + HCHO + CH_3CO_2M, $b_{1.5}$ 112-115°, n_D^{25} 1.4310, n_D^{20} 1.4322, IR.[1444]

$NaOH_2CPO(OEt)_2$. From polyoxymethylene and $NaPO(OEt)_2$ [127]

$(HOH_2C-)(HO)OPO-CH_2$ 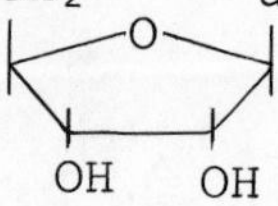 uracil (or cytosine). III. 4.[797]

4.[797]

$HOCH_2PO(OR)_2$.

R = $-SiMeEt_2$. VI.1. n_D^{20} 1.4423, d_4^{20} 1.0067.[1434]

R = $-SiEt_3$. VI.1. n_D^{20} 1.4472, d_4^{20} 0.9956.[1434]

R = $-SiMePr_2$. VI.1. n_D^{20} 1.4465, d_4^{20} 0.9804.[1434]

R = $-SiMePh_2$. VI.1. n_D^{20} 1.5755.[1434]

R = $-SnBu_3$. From $HOCH_2PO(OH)_2$ and $(Bu_3Sn)_2O$, cryst. n_D^{20} 1.5122.[1948]

$HSCH_2PO(OH)_2$. I.2.[859]

$H_2NCH_2PO(OH)_2$. I.2. m. 286.5 (dec.).[357] IR.[602,473] Crystal structure.[602] Stability const. of complexes with Mg, Ca, Zn, Co, Cu.[2084] I.2.[365,366] m. >300 .[365,366, 506,1523] pK_1 1.85, pK_2 5.35, pk_3 10.0.[365]

NaS(O)CPO(OEt)$_2$. From $(EtO)_2OPNa$ and COS.[711]

$\underline{C_2}$

BrC≡CPO(OPr-i)$_2$. V.14. $b_{0.001}$ 76-78°.[1788]

H_2C=CBrPO(OH)$_2$. I.2. Monoaniline salt, m. 187-188°.[1223]

H_2C=CBrPO(OEt)$_2$. V.2. b_2 78-78.5°.[1223] V.14 b_3 88-90°.[904]

$BrH_2CCH_2PO(OH)_2$. I.2. I.6. m. 83-85°. I.1. m. 86-87°.[917a]

$BrH_2CCH_2PO(OH)OCH_2$-CH[$-O_2C-C_{15}H_{31}$]-CH_2[$-O_2C-C_{15}H_{31}$]. III. 1. m. 54-57°[193]

$BrH_2CCH_2PO(OR)_2$.

R = Et. V.5.[552,1007] $b_{0.8}$ 101°, b_2 86-87°.[552]

R = Br_2HCCH_2-. V.5. m. 48-49° (ligroin), $b_{2.5}$ 190-191° .[917a]

R = Pr. V.5. b_1 115°.[309]

R = i-Pr. V.5. b_1 115°.[309]

BrHC=CBr-PO(OPr-i)$_2$. V.14. $b_{0.1}$ 87-98°, n_D^{21} 1.4939.[1788]

BrH_2C-CHBr-PO(OH)$_2$. I.1. m. 115-116°.[1223]

BrH_2C-CHBr-PO(OEt)$_2$. V.14. b_4 129-131.5°, d_4^{20} 1.6596, n_D^{20} 1.4939.[888] V.14. b_3 123-125°.[904]

ClC≡CPO(OEt)$_2$. V.5. $b_{2.5}$ 88.5-89.5°, n_D^{20} 1.4473, d_4^{20} 1.1539.[827]

CHF=CClPO(OEt)$_2$. V.14. b_3 80-82°.[2109]

ClFCHC(S)-PO(OEt)$_2$. V.5. n_D^{20} 1.3939, d_4^{20} 1.0988[548]

ClH_2CCHF-P(O)(OEt)$_2$. V.14. b_5 118-121°, n_D^{20} 1.4270, d_4^{20} 1.2444.[2109]

$ClFCHCF_2$-PO(OH)$_2$. I.1. I.2. m. 63-63.5°.[324]

ClHC=CHPO(OR)$_2$.

R = Me. V.14. b_4 76-78°.[1935]

R = Et. V.2. cis- and trans-isomers, IR, ^1H.[450] V.5. $b_{0.25}$ 64-65°, n_D^{25} 1.4505, IR, ^1H.[1989] V.14. b_3 80-81°.[1935]

R = i-Pr. V.14. b_3 90-93°.[1935]

R = Bu. V.14. b_4 119-120°.[1935]

H_2C=CCl-PO(OH)$_2$. I.1.[1223]

H_2C=CCl-PO(OEt)$_2$. V.2.[450, 1223] $b_{0.2}$ 49-50°.[450]

Cl(O)CCH$_2$PO(OR)$_2$.

R = Et. From (t-BuO)C(O)CH$_2$PO(OEt)$_2$ + PCl_5. Decomp. by distillation, n_D^{20} 1.4540, d_4^{20} 1.2680.[282]

R = Pr. From (t-BuO)C(O)$CH_2PO(OEt)_2$ + PCl_5. Decomp. by distillation, n_D^{20} 1.4490, d_4^{20} 1.1790.[282]

$HO_2CCHCl-PO(OH)_2$. I.2. m. 176-178°.[1888]
$O_2NCH_2CHClPO(OEt)_2$. V.14.[864]
$O_2NOCH_2CHClPO(OEt)_2$. V.14. $b_{0.1}$ 114-115°.[537]
$ClCH_2CH_2PO(OH)_2$. I.1, I.2. m. 74-75° (benzene).[920] I.2.[1218] The compound is cleaved to yield C_2H_4,$Cl^{\ominus}$ and H_3PO_4 at pH >4, plant growth regulator.[430a, 500a, 1776b, 2063a]
$ClCH_2CH_2PO(OH)OC_6H_4$-OH-2. III.2. m. 100-102°.[923]
$ClCH_2CH_2PO(OPh)Cl$. IV.2. b_1 130-132°, n_D^{20} 1.5329, d_4^{20} 1.3528.[615]
$ClH_2CCH_2PO(OR)_2$.

R = Me. V.1. b_1 65-67°, n_D^{20} 1.4490, d_4^{20} 1.2666.[923]

R = Et. V.2. $b_{2.5}$ 92-93°, n_D^{20} 1.44.2, d_4^{20} 1.1570.[270] V.e.[919] V.2. b_7 103-110°, b_4 92-94°, n_D^{20} 1.4390, d_4^{20} 1.1558.[923]

R = ClH_2CCH_2-. V.5. m. 37°, b_5 170-172°, d_4^{26} 1.3892, n_D^{26} 1.4828.[920] V.5.[619,623]

R = CH_2=$CHCH_2$-. V.2.[617]

R = Bu. V.2. b_2 128-130°, n_D^{20} 1.4445, d_4^{20} 1.0824.[270]

R = Ph. V.2. b_2 189-190°, b_1 176-178.5°. d_4^{15} 1.2663, n_D^{15} 1.5577.[923]

$ClH_2CCH_2PO(OPh)(OEt)$. V.2. $b_{1.5-2}$ 161°, n_D^{20} 1.5048, d_4^{20} 1.2176.[615]
(ClH_2CCH_2-)OPO-R-O.

R = -CHMeCH -. V.5. b_1 123-125°.[1575]

R = -$CH_2CH(CH_2OMe)$-. V.5. b_3 143-144°.[1575]

R = -CHMeCHMe-. V.5. $b_{1.5}$ 143-144°.[1575]

R = 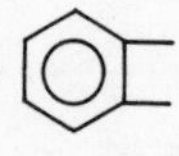 V.2. $b_{4.5}$ 167-170°, d_4^{20} 1.4015, n_D^{20} 1.5502.[923]

R = -$SiEt_3$. VI.1. b_2 145-147°.[1439]

$CH_3CHClPO(OEt)_2$. V.2. b_5 93°.[2086]
$CH_3OCHClPO(OR)_2$.

R = Me. V.5, V.6, V.4, $b_{0.01}$ 78°, n_D^{23} 1.4375, ^1H.[719]

R = Et. V.5, V.6, V.4, $b_{0.01}$ 87°, n_D^{23} 1.4348, ^1H.[719]

$ClH_2CCH(OH)-PO(OEt)_2$. V.8. $b_{0.06}$ 116-118°, n_D^{20} 1.4574 .[47]
$CH_3SCHCl-PO(OR)_2$.

R = Me. V.5, V.6, V.4, $b_{0.01}$ 81°, n_D^{20} 1.4984, ^1H .[719]

R = Et. V.5, V.6, V.4, $b_{0.01}$ 88°, n_D^{20} 1.4849, ^1H .[719]

$Cl(O)_2SOCH_2CH_2PO(OR)_2$.

R = Et, V.5, V.6, b_4 120-121°.[1174]

R = Pr. V.5, V.6, b_2 133-134°.[1174]

R = iPr. V.5, V.6, b_1 63-64°.[1174]

$O=C=CClPO(OEt)_2$. V.14. n_D^{20} 1.4720, d_4^{20} 1.3160.[282]
$ClFHCCHCl-PO(OEt)_2$. V.14. b_3 103-106°.[2109]
$Cl_2HCCF_2PO(OR)_2$.

R = Me. V.7. b_3 73-75°.[822]

R = Et. V.7. b_3 87-90°.[822]

R = Pr. V.7. b_3 98-100°.[822]

$Cl_2FCCH(OH)PO(OMe)_2$. V.8.[974]
$ClFHCCFCl-PO(OR)_2$.

R = Me. V.7. b_6 91-92°.[822]

R = Et. V.7. b_6 100-102°.[822]

R = Pr. V.7. b_6 110-112°.[822]

$ClHC=CClPO(OEt)_2$. V.14. b_7 108°.[2109]
$Cl(O)CCHCl-PO(OEt)_2$. V.14. $b_{0.07}$ 95-96°, n_D^{20} 1.4575, d_4^{20} 1.3130.[282]
$Cl_2C=C(OH)-(O)POCH_2CMe_2CH_2O$ (ring). V.14. Insecticide.[436]
$HO_2CCCl_2PO(OH)_2.2H_2O$. I.1. m. 85-125° (? !).[1887]
$HO_2CCCl_2PO(OR)_2$.

R = Et. V.14. n_D^{20} 1.4700, d_4^{20} 1.4710.[282]

R = Pr. V.14. n_D^{20} 1.4680, d_4^{20} 1.3290.[282]

$H_2N(O)CCCl_2PO(OR)_2$.

R = Et. V.14. m. 74-76°.[281]

R = Pr. V.14 m. 55-57°.[281]

R = i-Pr. V.14. m. 86-88°.[281]

R = Bu. V.14. m. 46-48°.[281]

$Cl_2HCCH_2PO(OEt)_2$. V.2. $b_{0.1}$ 45°, ^1H, IR.[450]
$ClH_2CCHClPO(OR)_2$.

R = Et. V.2. $b_{0.1}$ 93-97°, IR.[450] V.14. b_3 110-111° .[1935]

R = i-Pr. V.14. b_3 115-118°.[1935]

R = Bu. V.14. b_3 143-145°.[1935]

$ClSCH_2CHClPO(OCH_2CH_2Cl)_2$. V.14.[863]

$CH_3SCCl_2PO(OEt)_2$. V.5, V.14, $b_{0.01}$ 106°, n_D^{20} 1.4810 (1.4814).[719]

$Cl_2HCCFClPO(OR)_2$.

R = Me. V.7. b_4 93-94°.[822]

R = Et. V.7. b_4 100-104°.[822]

R = Pr. V.7. b_4 105-108°.[822]

$Cl_2HCCHClPO(OR)_2$.

R = Et. V.2. b_4 134-135°.[2109]

R = Pr. V.2. b_3 145-148°.[2109]

$Cl_3CCH(OH)PO(OR)_2$.

R = Me. V.8. m. 80-82°.[273] V.8.[491] Metabolism.[756] V.8. Insecticide.[1978] V.8. m. 81°.[1112] V.8.[966] V.8. m. 78.4-79.7°.[1772] V.8. m. 78-80°.[209]

R = Et. V.8. m. 55-56°.[209]

R = $ClCH_2CH_2$-. V.8. m. 88-89°.[1400]

R = Cl_3CCH_2-. V.8. m. 157-158°.[137]

R = Pr. V.8. m. 68-70.5°.[209]

R = i-Pr. V.8. m. 105-106.5°.[209]

R = $(ClH_2C)_2CH$-. V.8. m. 110-112°.[961]

R = Bu. V.8. n_D^{25} 1.4718[209]

R = CCl_3CMe_2-. V.8. m. 181-182°.[137]

R = $-CMe_2(CN)$. V.8. m. 150-151°.[137]

R = $-\overline{C\ (CCl_3)(CH_2)_3\ C}H_2$. V.8. m. 163-164°.[33]

$[Cl_3CCH(OH)-](O)\overline{P-O-CH_2CMe_2CH_2O}$. V.8. m. 209-210°.[424] V.5. m. 209-210°, insecticide.[436]

$[Cl_3CCH(OH)-](O)P<(O(CH_2)_2O)_2>P(O)[-(HO)HCCCl_3]$. V.8.[1493]

$Cl(O)CCCl_2PO(OR)_2$.

R = Et. V.14. $b_{0.07}$ 82-83°, n_D^{20} 1.4690, d_4^{20} 1.4108.[282]

R = Pr. V.14. $b_{0.07}$ 80-82°, n_D^{20} 1.4650, d_4^{20} 1.3390.[282]

$Cl_3CC(O)PO(OR)_2$.

R = Me. From $(MeO)_2OPH + Cl_3CCHO$, $b_{0.05}$ 85-86°, n_D^{20}

1.4740, d_4^{20} 1.5300.[1934]

R = Et. From $(EtO)_2OPH + Cl_3CCHO$, $b_{0.05}$ 95-97°, n_D^{20} 1.4655, d_4^{20} 1.4100.[1934]

$FHC{=}CHPO(OEt)_2$. V.14 $b_{3.5}$ 76-77°, n_D^{20} 1.4270, d_4^{20} 1.1706.[2109]

$FCH_2CH_2PO(OEt)_2$. V.5, V.6. b_{11} 74-75°, b_{18} 86.5-88°.[1790]

$F(O)_2SCH_2CH_2PO(OR)_2$.

R = Me. v.7. $b_{0.4}$ 127-128°.[1491]

R = Et. V.7. $b_{0.5}$ 132-134°.[1491]

$F_2HCCH_2PO(OH)_2$. I.1. As monoaniline salt, m. 224-226° (dec.).[324] I.b. As monoaniline salt, m. 224° (dec.).[676]

$F_2HCCH\ PO(OEt)_2$. V.7. b_5 73-76°.[2041]

$F_2C{=}CFPO(OEt)_2$. V.5. b_7 81°, n_D^{20} 1.377.[980]

$F_3CCH_2PO(OBu)_2$. From $P(OCH_2CF_3)_3$ + LiBu, b_1 80-81°, n_D^{20} 1.4081.[1212]

$F_2HCCF_2PO(OH)_2$. I.1. m. 50-51°. As monoaniline salt, m. 225-228° (dec.).[324]

$F_2HCCF_2PO(OR)_2$.

R = Me. V.7. b_{10} 72°, n_D^{25} 1.4041.[302]

R = Et. V.7. b_{10} 87°, b_{20} 100°, n_D^{25} 1.3700.[302]

R = Bu. V.7. $b_{0.45}$ 71-74°, n_D^{25} 1.3970.[302]

$HC{\equiv}C{-}PO(OR)_2$.

R = Et. V.14. $b_{0.5}$ 94°.[1963] VI. $b_{0.1}$ 75-77°.[1788]

R = i-Pr. V.14. $b_{0.1}$ 50-60°, ^1H.[1788]

$O{=}C{=}CH{-}PO(OEt)_2$. V.14. Identification by transformation to $Me(O)CCH_2PO(OEt)_2$.[282]

$NC{-}CH_2PO(OBu)_2$. V.5. b_3 155-156°.[1672]

$O{=}C{=}N{-}CH_2PO(OR)_2$.

R = Et. V.5. $b_{0.02}$ 69-70°, n_D^{20} 1.4361, d_4^{20} 1.1695.[1903]

R = Pr. V.5. $b_{0.02}$ 75-76°, n_D^{20} 1.4400, d_4^{20} 1.0999.[1903]

R = i-Pr. V.5. $b_{0.03}$ 65-66°, n_D^{20} 1.4315, d_4^{20} 1.0889.[1903]

$H_2C{=}CHPO(OH)_2$. I.1. n_D^{20} 1.4710, d_4^{20} 1.398.[1808,904]

$H_2C{=}CHPO(OR)OH$.

R = Et. n_D^{20} 1.4516, d_4^{20} 1.4521.[509]

R = Am. d_4^{20} 1.087.[509]

$H_2C{=}CHPO(OEt)Cl$. IV.2, b_2 138°, n_D^{20} 1.4509, d_4^{20} 1.1663.[903]

$H_2C=CHPO(OCH_2CH_2Cl)Cl$. IV.1. b_1 94-100°.[2068] V.14.[623, 920]

$H_2C=CHPO(OBu)Cl$. IV.4. $b_{1.5}$ 57°.[898]

$H_2C=CHPO(OR)_2$.

R = Me. V.2. $b_{0.4}$ 37-40°.[309]

R = Et. V.14. b_1 50°, n_D^{25} 1.4260.[1007] V.5. b_{13} 80-82°.[1985] V.14, b_3 65-70°.[355] V.14. $b_{2.3}$ 69°.[451] V.14. b_3 68-70°, n_D^{20} 1.4300.[888] V.14. b_1 50°, n_D^{25} 1.4260.[1007] n_D^{15} 1.4320.[552] V.5. b_9 85.5°, n_D^{20} 1.4252, IR, ^{1}H.[1989] V.2.[904]

R = ClH_2CCH_2-. V.14. b_4 137-139°, n_D^{20} 1.4772, d_4^{20} 1.3182,[888] ^{31}P - 22ppm.[1189a] V.14. $b_{2.1}$135-136°.[3...] V.14.[623, 920]

R = Pr. V.14. $b_{0.8}$ 63-65°.[309]

R = i-Pr. V.14. $b_{0.8}$ 65-66°.[309] V.5. b_{13} 88°, n_D^{20} 1.4221, IR, ^{1}H.[1989]

R = $CH_2=CHCH_2$-. V.14.[617]

R = $ClH_2CCH(OH)CH_2$-. V.14. $b_{0.8}$ 130-150°.[1752]

R = Bu. V.14. $b_{1.5}$ 101-102°.[1935]

$H_2C=CHPO(OMe)OCH_2CH_2Cl$. V.2. b_1 82-85°, IR.[2068]

$H_2C=CHPO(OEt)OCH_2CH_2SEt$. V.2. b_2 115-117°.[895]

$(H_2C=CH-)OPOCH_2CH_2O$. V.14. $b_{1.5}$ 118-119°.[1575]

$(H_2C=CH-)OPOCHMe-CHMeO$. V.14. $b_{1.5}$ 106-107°.[1575]

$(H_2C=CH-)OPOCH_2(CF_2)_3CH_2O$. V.2. m. 79-80°.[1881]

$(H_2C=CH-)(EtO)OP-OCH_2CH_2O-PO(OEt)(-CH=CH_2)$. V.2. b_1 182-184°.[626]

$H_2C=CHPO(OSiMe_3)_2$. VI.1. b_{14} 102.3-104°.[1682]

$CH_3C(N_2)PO(OMe)_2$. V.14. $b_{0.2}$ 50-52°, IR, ^{1}H.[1191]

$CH_3C(O)PO(OR)_2$.

R = Me. V.5. b_6 72°, IR.[1424] V.5. $b_{6.5}$ 82-86°, n_D^{25} 1.4229, IR, ^{1}H.[1191] V.14. $b_{0.2}$ 50-52°, n_D^{25} 1.4583.[1191] V.5. $b_{6.5}$ 83-85°.[918] V.5. b_5 76-78°, n_D^{20} 1.4210, d_4^{20} 1.2109.[922] V.8. b_{16} 93-95°.[1396]

R = Et. V.5. $b_{1.5}$ 62-65°.[422] V.5. b_3 75-80°, $b_{1.5}$ 60-61°, n_D^{20} 1.4200, d_4^{20} 1.0991, p-nitrophenylhydrazone m. 131-132° (EtOH).[922] V.5. b_2 70-73°.[918] V.8. b_{13} 103-105°.[1396] From $(EtO)_2POP(OEt)_2$ + MeC(O)Cl, b_{12} 100-103°, n_D^{20} 1.4203, d_4^{20} 1.1002.[604] V.5. b_4 78-80°, IR.[1424] V.14. b_2 63-64°, n_D^{20}

1.4210.[1414] V.5. b_{3-4} 181-184° (? !), IR, ^1H.[244] V.6. b_{10} 98-100°.[1617]

R = Pr. V.5. b_{16} 127-128°.[607] V.5. b_{11} 117-118°.[1584]

R = i-Pr. V.5. b_4 73-74°, IR.[1424]

R = $NCCH_2CH_2$-. V.5. $b_{4-4.5}$ 113-113.5°, b_3 95°, n_D^{20} 1.4092, d_4^{20} 1.1965, p-nitrophenylhydrazone m. 189.5-190° (MeOH).[922]

R = Bu. V.5. $b_{1.5}$ 87-88°, n_D^{20} 1.4301, d_4^{20} 1.0199. p-nitrophenylhydrazone m. 104-104.5° (EtOH), adduct with $NaHSO_3$ m. 135-136°.[918] V.5. b_{10} 135°.[607] V.8. b_{13} 128-130°.[1396]

R = i-Bu. V.8. b_{14} 119-120°.[1396] V.5. b_{11} 128-130°.[1584]

R = sec.-Bu. V.5. b_5 101-102°.[1424]

$CH_3C(O)PO(OEt)(-OCH_2CHBrCH_2OMe)$. V.5. b_2 131-132°.[122]
$[(H_3CC(O)-]\overline{OPOCH(-CH_2OMe)CH_2O}$. V.5. b_2 141-142.5°.[122]
$[H_3CC(O)-]\overline{OPOCHMeCMe_2O}$. V.5. b_1 144-145°.[1917]
$OHCCH_2PO(OR)_2$.

R = Me. V.14. b_1 87-88°, n_D^{20} 1.4410, d_4^{20} 1.2605.[1135] V.14.[722]

R = Et. V.14. $b_{0.13}$ 82-83°, n_D^{20} 1.4749, d_4^{20} 1.1182.[1347] V.14. $b_{0.17}$ 78-80°.[1705] V.14.[1132] V.14. $b_{0.35}$ 89-90°, IR, ^1H.[691] V.14. $b_{0.2}$ 72-73°, n_D^{20} 1.4335, IR $\nu_{C=O}$ 1732, $\nu_{P=O}$ 1240, ν_{P-O-C} 1050, 1023, 952 cm^{-1}, 2,4-dinitrophenylhydrazone m. 108°.[1986]

R = i-Pr. V.14. 2,4-Dinitrophenylhydrazone m. 134-135°.[1788]

R = $-C_8H_{17}$. V.14. $b_{0.03}$ 165-167°.[1502]

$\overline{OCH_2CH}PO(OR)_2$.

R = Me. V.14. $b_{0.4}$ 58°, IR, ^1H.[691]

R = Et. V.14. $b_{0.2}$ 66-68°, ^1H.[691] V.14. b_2 86-89°. IR.[47]

R = ClH_2CCH_2-. V.14. $b_{0.06}$ 72-82°.[47]

R = Bu. V.14. $b_{0.1}$ 81-87°.[47]

R = cycl-C_6H_{11}-. V.14. $b_{0.06}$ 105-110°.[47]

$CH_3SC(O)-PO(OEt)ONa$. III.3. m. 205-207°, IR, ^{31}P+3.2 ppm.

[709,710]

$CH_3SC(O)PO(OR)_2$.

R = Et. V.14. $b_{0.2}$ 76.2-78.5°, n_D^{27} 1.4692, herbicide.[711] V.6. $b_{0.1}$ 80-81°, $b_{1.7}$ 116-118°, n_D^{25} 1.4705, IR, ^{31}P+4.6 ppm.[709]

R = i-Pr. V.6. $b_{0.1}$ 76-77°, $b_{0.5}$ 100-101°, n_D^{25}1.4634, ^{31}P+6.5 ppm.[709]

$CH_3O_2CPO(OEt)_2$. V.5. b_1 57-59.5°.[1723]

$HO_2CCH_2PO(OH)_2$. I.2. m. 139.5 ($MeCO_2H$).[112b] I.2.[114] I.2. I.2.[1418] m. 142-143°.[1419,1420] NMe_4 salt ^{31}P - 13.3 ppm.[1189a]

$HO_2CCH_2PO(OR)_2$.

R = Me. V.14. m. 35-36° (butanone/ether), K salt, m. 122-123° (MeOH).[1172]

R = Et. V.14. m. 27-28° (ether).[1172] V.14. n_D^{20} 1.4450, d_4^{20} 1.2470.[282] K salt, m. 106-108°.[1172] V.6. Cyclohexylamine salt m. 107-108° (acetone/ether).[1172] V.14. Ba salt, m. 204-206° (dec.).[1707]

R = Pr. V.14. n_D^{20} 1.4420, d_4^{20} 1.1540.[282]

R = Ph. V.14. m. 143-144° (butanone).[1172]

$CH_3SC(S)PO(OPr\text{-}i)_2$. V.6. $b_{0.4}$ 116-117°, n_D^{25} 1.5168, ^{31}P+4.[illegible] ppm.[709]

$H_2NC(O)CH_2PO(OR)_2$.

R = Et. V.5. m. 80-82°.[1941]

R = Pr. V.5. m. 78-80°.[281]

R = i-Pr. V.5. m. 98-99°.[281]

R = Bu. V.5. m. 94-96°.[281]

$CH_3NHC(O)$-$PO(OEt)_2$. V.9. $b_{0.5}$ 107°.[1720] V.9. b_3 142°.[1639]

$Cl^{\ominus}$ $[H_3\overset{\oplus}{N}CH_2C(O)$-$PO(OEt)_2]$ V.5. m. 129°.[174]

CH_3NH-$C(S)PO(OR)_2$.

R = Et. V.9. $b_{0.2}$ 133-134°.[1489]

R = i-Pr. V.9. $b_{0.08}$ 113-113.5°.[1489]

R = Bu. V.9. $b_{0.05}$ 140.5°.[1489]

$(EtO)_2OPCHMeOPS(OEt)C_2H_5$. V.14. $b_{0.01}$ 105°.[1861]

$C_2H_5PO(OH)_2$. I.6,[792,729,1259a] I.1.[729] I.2,[1000,1002,1278,1420] I.6.[1167] m. 30-35°[1167] m. 44°,[792] m. 61-62°,[1000,1002] m. 61.5-62.5°[1420] b_8 330-340°,[1167] pK_1 2.45, pk_2 7.85.[1776a] I.5.[1638] NMe_4 salt, ^{31}P - 24.1 ppm .[1189a]

$C_2H_5PO(OEt)OH$. From $C_2H_5PS(OEt)OH$ or $C_2H_5PSe(OEt)OH$ and

Me_2SO.[1307,1309] Complexes with Ti(III), V(III), Cr (III).[1312] UV, IR, X-rays.[1312] Sodium salt, m. 420-426°.[1468] $4\text{-}ClC_6H_4CH_2\overset{\oplus}{S}{=}C(NH_2)_2$ salt, m. 166-167° (acetone).[753]

$C_2H_5PO(OBu)OH$. III.4. b_1 140-142°.[1464]

$C_2H_5PO[OCH_2CH_2SP(S)(OMe)_2]OH$. From $(C_2H_5)O\overline{POCH_2CH_2O}$ and $PS(OMe)_2SH$.[1052]

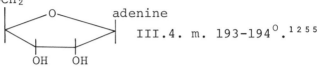

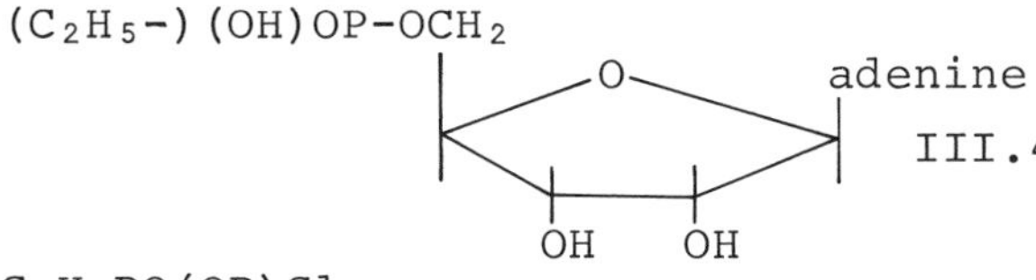

III.4. m. 193-194°.[1255]

$C_2H_5PO(OR)Cl$.

R = Et. IV.1. b_{10} 73-74°.[1711] IV.1.[68,1704a] IV.b.[1169]

R = Pr. IV.1. b_{11-12} 85-87°.[1711] IV.b.[1169]

R = i-Pr. IV.1. $b_{9.5}$ 76.5-78.5°.[1711]

R = Bu. IV.1. b_{10-11} 100-102°.[1711]

R = i-Bu. IV.1. $b_{13.5}$ 98-99.5°.[1711]

$C_2H_5PO(OR)F$.

R = Me. Electronic structure.[1075]

R = Et. Electronic structure.[1075] From $C_2H_5PO(OEt)SR$ and $2,4\text{-}(O_2N)_2C_6H_3F$.[220]

R = Pr. Electronic structure.[1075]

R = i-Pr. Electronic structure.[1075]

$C_2H_5PO(OEt)NCO$. IV.5. $b_{0.35-0.4}$ 55-56°.[726]

$C_2H_5PO(OR)_2$.

R = Et. V.5.[102] V.5. b_2 62°, n_D^{18} 1.4172.[552] V.5. b_{16} 90-92°.[1790] V.6, V.2, b_{760} 198°, b_{20} 90-95°, d_0^{21} 1.025.[1278] V.6. b_9 86-88°.[1420] V.2.[787] V.7. b_{27-29} 98-99°.[747] V.5. b_{15} 87°.[1553] V.5.[331] By pyrolysis of $C_2H_5PO(OEt)ONa$, b_{750} 203°.[1667] Dipol moment, magnetic moment.[1216] V.5. b_{11} 80-83°.[404] V. 5. $b_{1.5}$ 52-54°,[1366] ^{31}P - 32.5 ppm.[1189a] V.5.[1924] Rotation of the bonds.[1069] V.5.[980] Addition compound $C_2H_5PO(OEt)_2 \cdot Al(Bu\text{-}i)_3$.[1784]

R = F_3CCH_2-. V.5. b_1 41°, n_D^{20} 1.3440.[1211]

R = $H_2C{=}CMe$-. V.5. b_3 71.5-72°.[1136]

R = $\overline{OCH_2CH}CH_2$-. V.4.[1748] V.13. b_8 125-126°, n_D^{20} 1.4312,

d_4^{20} 0.9674.[38]

R = Bu. V.6. b_{17} 137-139°, n_D^{25} 1.4258, d_4^{25} 0.9623,[1000,1002] ^{31}P - 31 ppm.[1189a]

R = EtO_2CCH_2-. V.13. b_1 133-134°.[1648]

R = Ph. V.5. b_{13} 202°.[117]

R = C_6H_{13}-. V.13. b_7 160-162°.[1646]

R = $PhCH_2$-. V.13. $b_{0.6}$ 168°.[771]

R = 2-Cl-4-H_2C=CHC_6H_3-. V.5. b_1 205-207°.[621]

$C_2H_5PO(OMe)OR$.

R = Cl_2C=CH-. V.5.[1397] V.8. b_{13} 119-120°.[1713]

R = Cl_3CCHCl-. V.14. $b_{0.28}$ 101-102°.[1713]

R = C_6H_{13}-. V.13. b_8 115-117°.[1646]

$C_2H_5PO(OEt)OR$.

R = BrH_2CCH_2-. V.5. b_{11} 129-130°, n_D^{25} 1.4258, d_4^{25} 0.9623,[1000,1002] V.5. b_{11} 131-132°.[1553]

R = Cl_2C=CH-. V.5.[1397]. V.8. $b_{0.2}$ 95-96°.[1713]

R = Cl_3CCHCl-. V.14. $b_{0.2}$ 103.5-104°.[1713]

R = FH_2CCH_2-. V.5. b_2 70-71°.[897]

R = H_2C=CMe-. V.2. $b_{0.5}$ 56-57°.[1136,168]

R = BrH_2CCHMe-. V.5. b_{14} 133.5-134.5°, n_D^{20} 1.4558.[120]

R = Bu. V.13. b_{13} 111-113°, n_D^{20} 1.4175°.[975]

R = (NC)Me_2C-. V.5. $b_{0.017}$ 68-69°.[182]

R = EtO_2CCH_2-. V.13. b_1 88-89°.[1648]

R = EtO_2CCHMe-. From MeC(O)CO_2Et and EtPO(OEt)H, b_7 134°.[1626]

R = C_6H_{13}-. V.13, b_1 125-126°.[1646]

R = $CH_2(CH_2)_3CH$=C-. V.5. $b_{0.5}$ 85-86°.[169]

R = O=C-O-$(CH_2)_2$C=CMe-. V.5. $b_{0.01}$ 150-155°, n_D^{20} 1.4843.[635]

R = 4-$O_2NC_6H_4$-. V.2. Cholinesterase inhibitor.[1169]

R = $PhCH_2$-. V.13. $b_{18.5}$ 174°.[771]

R = $(EtO_2C)_2CH$-. V.8. $b_{2.5}$ 133-134°, n_D^{20} 1.4350, d_4^{20} 1.1459.[1623]

R = PhMeCH-. From MeC(O)Ph + EtPO(OEt)H + NaOEt, b_1

123°.[1626]

R = 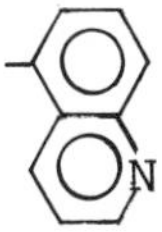 V.2. $b_{0.07}$ 131-132°, insecticide.[527]

R = 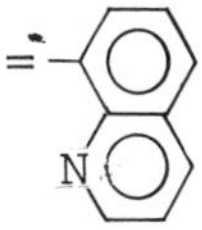 V.2. $b_{0.03}$ 130 , insecticide.[527]

R = 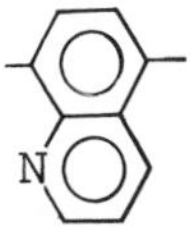 Cl. V.2. $b_{0.05}$ 145°, insecticide.[527]

R = $C_{12}H_{25}$-. V.13. $b_{0.14}$ 123°, n_D^{25} 1.4383.[1079]

$C_2H_5PO(OCH{=}CH_2)OBu$. In addition to other products by reaction of $EtPO(OBu)_2$ with OCH_2CH_2Cl, b_8 100-103°.[38]

$C_2H_5PO(OCH{=}CCl_2)OR$.

R = Pr. V.5.[1397] V.8. $b_{0.5}$ 99.5-100.5°.[1713]

R = i-Pr. V.5.[1397] V.8. $b_{0.3}$ 87-88°.[1713]

R = Bu. V.5.[1397] V.8. $b_{0.4}$ 108-109°.[1713]

R = i-Bu. V.5.[1397] V.8. $b_{0.3}$ 98-101°.[1713]

$C_2H_5PO(OCHClCCl_3)OR$.

R = Pr. V.14. $b_{0.3}$ 115-117°.[1713]

R = i-Pr. V.14. $b_{0.25}$ 99-102°.[1713]

R = Bu. V.14. $b_{0.3}$ 120-122°.[1713]

R = i-Bu. V.14. $b_{0.35}$ 116.5-120°.[1713]

$C_2H_5PO(OPr)OR$.

R = $H_2C{=}CHOCH_2CH_2$-. V.5. b_1 87-89°.[1917]

R = EtO_2CCHMe-. From $MeC(O)CO_2Et$ + EtPO(OPr)H + PrONa, b_7 139°.[1626]

R = C_6H_{13}-. V.13. b_9 137-147°.[1646]

R = $4\text{-}O_2NC_6H_4$-. Cholinesterase inhibitor.[1196]

R = PhMeCH-. From MeC(O)Ph + EtPO(OPr)H + PrONa, b_1 114°.[1626]

$C_2H_5PO(OPr\text{-}i)OR$.

R = $(EtO_2C)CHMe$-. From $MeC(O)CO_2Et$ + EtPO(OPr-i)H +

i-PrONa, b_7 136°.[1626]

R = PhCHMe-. From MeC(O)Ph + EtPO(OPr-i)H + i-PrONa, b_4 149°.[1626]

$C_2H_5PO(OCMe{=}CH_2)(OBu)$. V.2.[1136]

$C_2H_5PO(OBu)(OC_6H_{13})$. V.13. b_9 146-147°.[1646]

$(C_2H_5-)(MeO)OP-OC(=CCl_2)-PEt(O)OMe$. From $EtPCl_2$ and $Cl_3CC(O)Cl$ at 120-130° and reaction with MeOH/ NEt_3, m. 83°.[1417]

$(C_2H_5-)OPO-R-O$ (cyclic)

R = $-(CH_2)_2-$. V.13.[1649] V.5.[1557] V.5. b_1 94-95°.[1575] V.3. $b_{0.2}$ 83-84°.[1464]

R = $-CHMeCH_2-$. V.3. $b_{0.2}$ 87-88°.[1464]

R = $-(CH_2)_4-$. V.13. b_1 98-99°.[1649]

R = $-CH(-CH_2OMe)CH_2-$. V.5. b_1 126-128°.[1575] V.5. b_3 140-141°.[122]

R = $-CMe_2CHMe-$. V.5. b_1 100°.[1917]

R = $-CMe_2CMe_2-$. V.5. b_1 90-93°.[1917]

R = 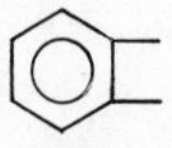 . V.5. m. 69-70°.[120a]

R = 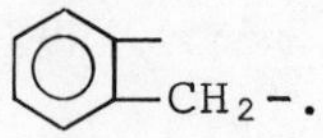 V.2. Insecticide.[508]

R = $-CH_2C[-CHMe(OMe)]Me-CH_2-$. V.5. b_3 143°.[285]

R = $-CHCH_2OCH_2O-CH-CHOCH_2OCHCH-$ (cyclic). V.5. ^{31}P -32 ppm.[2051]

R = $-CH_2C[CHMe(OMe)](CHMe_2)CH_2-$. V.5. b_3 160°.[285]

R = $-CH_2C[CHMe(OCHMe_2)](CHMe_2)CH_2-$. V.5. b_1 150°.[285]

R = 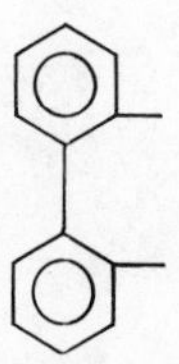 . V.5. ^{31}P -45 ppm.[2051]

R = 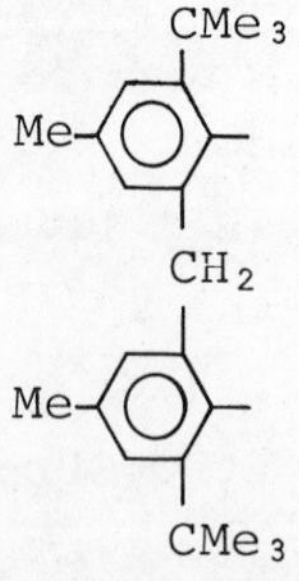 . V.5. ^{31}P -26 ppm.[2051]

$(RO)O(C_2H_5)P\text{-}OCH_2O\text{-}P(C_2H_5)O(OR)$.

R = Et. V.13. b_9 153-154°.[1648]

R = Pr. V.13 b_{12} 172-173°.[1648]

$(C_2H_5\text{-})O(EtO)POC(=CCl_2)PO(OEt)\text{-}C_2H_5$. From $EtPCl_2$ and $Cl_3C(O)Cl$ at 120-130°, m. 68°.[1417]

$(EtO)O(C_2H_5\text{-})POCH(Me)\text{-}PO(OEt)_2$. V.2. $b_{0.01}$ 98°, insecticide.[1861]

$(EtO)O(C_2H_5\text{-})POCH_2P(C_2H_5)S(OEt)$. V.2. $b_{0.01}$ 102°, insecticide.[1861]

$(EtO)O(C_2H_5\text{-})P\text{-}O(CH_2)_4O\text{-}P(\text{-}C_2H_5)O(OEt)$. V.2.[1957] V.2. $b_{3.5}$ 173-174°.[1561]

$(EtO)O(C_2H_5\text{-})P\text{-}O(CH_2)_2O(CH_2)_2O\text{-}P(C_2H_5)O(OEt)$. V.2. b_4 189-190°.[1561]

$(EtO)O(C_2H_5\text{-})P\text{-}O$-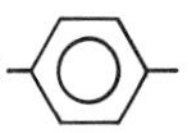-$O\text{-}P(\text{-}C_2H_5)O(OEt)$. V.5. $b_{0.053}$ 167-169°.[1554]

$(EtO)O(C_2H_5\text{-})P\text{-}OCHMe\text{-}P(\text{-}C_2H_5)S(OEt)$. V.2. $b_{0.01}$ 98°, insecticide.[1861]

$C_2H_5PO(OEt)(OSiEt_3)$. From $P(OEt)_3$ and $BrSiEt_3$ at 70° or from $(EtO)_2POSiEt_3$ and EtBr.[322] Pyrolysis of $Et_3SiOP(OEt)_2$ at 160°.[1440]

$C_2H_5PO(OSiMe_3)_2$. From Me_3SiOEt and PBr_3, b_{25} 107-113°.[2056]

$C_2H_5PO(OSiEt_3)_2$. From Et_3SiOEt and PBr_3, $b_{8.5}$ 163.5-170°.[2056] Pyrolysis of $Et_3SiOP(OEt)_2$ or $(Et_3SiO)_2$ POEt at 160°.[1440] VI.1.[1164]

$C_2H_5PO(OPr\text{-}i)ON{=}CMe$-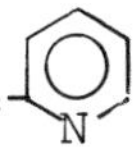. VI.2. m. 31°.[260]

$C_2H_5PO(OPr\text{-}i)ON{=}CMe$-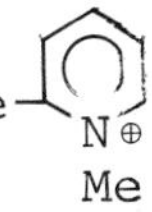 $J^{\ominus}$. VI.2. m. 142°.[260]

$C_2H_5PO(OPr\text{-}i)ON{=}CMe$-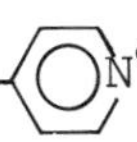 $J^{\ominus}$. VI.2. m. 118°.[260]

$C_2H_5PO(OEt)(OSnEt_3)$. From $C_2H_5PO(OEt)_2 + Et_3SnJ$. n_D^{14} 1.4902[1168]

$H_2NC(O)NHCH_2PO(OEt)_2$. V.14. m. 127-128°.[1903]

$H_2NC(=NH)SCH_2PO(OH)_2$. I.13. m. 208-209°.[859]

$H_2NC(=NH)SCH_2PO(OEt)OH$. From $[ClH_2CPO(OEt)O^{\ominus}]H_3\overset{\oplus}{N}C(=NH)SEt$ and $(H_2N)_2CS$ at 125-130°, m. 228-230°, ^1H, ^{31}P -12.1± 0.8 ppm, UV, IR.[1324]

$CH_3CH(OH)PO(OH)_2$. I.9. m. 74-78°.[555] I.6.[1182,1186] I.9. m. 74-78°.[206] I.2. aniline salt, m. 168-169°.[963]

$CH_3CH(OH)PO(OR)_2$.

R = Me. V.10. n_D^{20} 1.4287.[1038] V.8, b_{18} 138°[40].

R = Et. V.8. $b_{1.5}$ 116-119°, b_6 139-140°, n_D^{25} 1.4298,

n_D^{20} 1.4308, IR[1444]. V.8. $b_{0.5}$ 108-110°, n_D^{20} 1.4348 [963] V.8. b_6 139-140°.[12]

R = Bu. V.8. b_9 162-163°.[18]

R = $(ClH_2C-)Me_2C-$. V.8. m. 118.5-120° (ligroine).[34]

R = $\overline{CH_2(CH_2)_3C}(CCl_3)-$. V.8. m. 143-145°.[33]

$[CH_3CH(OH)-]$ $O\overline{PO-R-O}$.

R = $-(CH_2)_3-$. V.8. m. 120-122°.[1588]

R = -CHMeCHMe-. b_2 134-135°.[1588]

$HOCH_2CH_2PO(OH)_2$. I.6. Yellow Ag salt.[981]

$HOCH_2CH_2PO(OEt)_2$. From $\overline{CH_2CH_2O}$ and $HPO(OEt)_2$, b_9 120-130°.[369] V.6. $b_{0.35}$ 105-109°, n_D^{20} 1.4372, $IR_1\nu_{P=O}$ 1247, ν_{OH} 3430 cm^{-1}.[1938]

$CH_3OCH_2PO(OEt)_2$. V.6. b_9 102-103°.[103] V.5. $b_{0.1}$ 65-67°, n_D^{20} 1.421.[675] V.14.[849]

$(CH_3OCH_2-)O\overline{POCH_2CEt(-CH_2Cl)CH_2O}$. V.5. m. 85°.[2059]

$HOCH_2CH(OH)PO(OR)_2$.

R = Et. V.14. $b_{0.3}$ 140°, ^1H.[691] V.14. n_D^{20} 1.4540 d_4^{20} 1.232.[501]

R = Bu. V.14. n_D^{20} 1.4550.[501]

$CH_3SCH_2PO(OEt)_2$. V.14.[632]

$CH_3CH(NH_2)PO(OH)_2$. I.2. m. 283-285° (dec.).[357] I.13. m. 272-274°.[2040]

$CH_3CH(NH_2)PO(OEt)_2$. V.14. b_3 70-73°.[356]

$H_2NCH_2CH_2PO(OH)_2$. I.2. m. 285°.[1005] I.2. Dec. at 250°.[366] I.2. m. 281-282°.[537a] pK_1 2.45, pK_2 7.0, pK_3 10.8.[1776] I.2. m. 271-278°.[210] Reviews of properties.[626a,1677e] Isolation from Rumen protozoa.[799b]

$H_3\overset{\oplus}{N}CH_2CH_2PO(O^{\ominus})OCH_2CH[-O_2CC_{15}H_{31}]CH_2[-O_2CC_{15}H_{31}]$ (as L-α). I.2. m. 180-181° ($CHCl_3$).[1771]

$H_3\overset{\oplus}{N}CH_2CH_2PO(O^{\ominus})OCH_2CH[-O_2CC_{17}H_{35}]CH_2[-O_2CC_{17}H_{35}]$ (as DL). I.2. m. 181-181.5° ($CHCl_3$).[1771]

$H_2NCH_2CH_2PO(OBu)_2$. V.6. b_{25} 134-135°.[366]

$CH_3NHCH_2PO(OH)_2$. I.13. m.>240°.[1523]

$H_2NCH_2CH(OH)PO(OEt)_2$. V.14. IR.[691]

$\underline{C_3}$

$(CF_3)_2BrCPO(OMe)_2$. From $(CF_3)_2C=P(OMe)_3$ and Br_2, $b_{0.5}$ 53-56°, m. 52-54°.[1305]

$H_2C=CBrCH_2-PO(OEt)_2$. V.5. b_1 98°, n_D^{19} 1.4730.[1961]

$BrHC=CHCH_2-PO(OEt)_2$. V.5. b_1 98°, n_D^{19} 1.4730.[1961,1082]

$CH_3O_2CCHBrPO(OEt)_2$. V.14. b_3 142-147°, n_D^{20} 1.4610, d_4^{20} 1.3985.[697]

$BrH_2C(CH_2)_2PO(OH)_2$. I.2. m. 113-113.5°.[767] I.2. m. 107-108 [999] ^{31}P -30 ppm.[1189a]

$BrH_2C(CH_2)_2PO(OEt)_2$. V.5.[767,999]

$BrH_2CCH(OMe)PO(OR)_2$.

R = Et. V.5. b_4 129-131°.[10]

R = i-Pr. V.5. b_4 128-129°.[10]

R = Bu. V.5. b_3 143-144°.[10]

BrHC=CBrCH$_2$PO(OEt)$_2$. V.5. $b_{0.7}$ 125°, n_D^{21} 1.5090.[1961]
BrH$_2$CCHBrCH$_2$PO(OCH$_2$CHBrCH$_2$Br)$_2$. V.14.[1745]
F$_2$C=CClCF$_2$PO(OH)$_2$. I.1.[259]
F$_2$C=CClCF$_2$PO(OMe)$_2$. V.5. $b_{0.1}$ 45-48°, IR, ^{1}H, ^{19}F.[259]
F$_3$CCCl=CFPO(OMe)$_2$. V.14. $b_{0.01}$ 35-45°.[649]
F$_2$ClCC(O)CF$_2$PO(OMe)$_2$. V.5. b_{18} 85-90°, n_D^{25} 1.3854.[1968]
H$_2$C=C=CClPO(OR)$_2$.

R = Et. From HC≡CCH$_2$OH and ClP(OEt)$_2$, $b_{2.5}$ 102°.[1450]

R = Pr. From HC≡CCH$_2$OH and ClP(OPr)$_2$, $b_{1.5}$ 110°.[1450]

R = Bu. From HC≡CCH$_2$OH and ClP(OBu)$_2$, $b_{0.04}$ 112°.[1450]

ClCH$_2$C(O)NHC(O)PO(OPr-i)$_2$. V.9. m. 64-67°.[463]
CH$_3$O$_2$CN=CClPO(OR)$_2$.

R = Me. V.2. $b_{0.05}$ 105-107°, n_D^{20} 1.4560, d_4^{20} 1.3886.[490]

R = Et. V.2. $b_{0.05}$ 95-96°, n_D^{20} 1.4530, d_4^{20} 1.2544.[490]

ClHC=CHCH$_2$PO(OEt)$_2$. V.5. $b_{0.15}$ 79°; n_D^{20} 1.4562.[1082,1960]
H$_2$C=CClCH$_2$PO(OR)$_2$.

R = Me. V.5. b_2 60-62°.[431]

R = Bu. V. 6. b_2 98-99°.[431]

CH$_3$OCH=CClPO(OR)$_2$.

R = Me. V.2. $b_{0.06}$ 86-87°, n_D^{20} 1.4663, d_4^{20} 1.3012.[1348]

R = Et. V.2. $b_{0.2}$ 96-97°, n_D^{20} 1.4666, d_4^{20} 1.2243.[1348]

ClC(O)CHMePO(OEt)$_2$. V.14. n_D^{20} 1.4550, d_4^{20} 1.2290.[282]
CH$_3$C(O)CHClPO(OEt)$_2$. V.14. $b_{0.05}$ 96-97°, n_D^{20} 1.4520 d_4^{20} 1.2630.[282]
ClC(O)CH$_2$CH$_2$PO(OPh)[OCMe$_2$(CCl$_3$)]. From (PhO)[(Cl$_3$C)CMe$_2$O]PCl and H$_2$C=CHCO$_2$H.[504]
H$_2$NC(O)CClMePO(OEt)$_2$. V.14. m. 86-87°.[280]
Cl(CH$_2$)$_3$PO(OH)$_2$. I.2. m. 101-103°.[767] I.3.[767]
Cl(CH$_2$)$_3$PO(OR)$_2$.

R = Et. V.10.[767]

R = Bu. V.6. $b_{0.01}$ 115-117°.[767]

R = EtSCH$_2$CH$_2$-. V.2. $b_{0.0006}$ 87-89°.[895]

H$_3$CCH(-CH$_2$Cl)PO(OEt)$_2$. V.2. b_{10} 126-128°.[2109]
H$_3$CCHClCH$_2$PO(OR)$_2$.

R = Me. V.2. b_6 100-102°.[2109]

R = Et. V.2. b_6 105-107°.[2109]

(CH$_3$CHClCH$_2$-)OPO-R-O. (ring: P–O–R–O)

R = -CH$_2$CH$_2$-. V.5. b_4 164-166°.[1575]

R = -CHMeCH$_2$-. V.5. b$_1$ 136°.[1575]

R = -CHMeCHMe-. V.5. b$_{0.5}$ 131-134°.[1575]

R = -CH$_2$CH(CH$_2$OMe)-. V.5. b$_3$ 151-153°.[1575]

ClH$_2$CCMe(OH)PO(OCMe$_2$CCl$_3$)$_2$. V.8. m. 151°.[961]
ClH$_2$CMe(OH)PO(OPh)$_2$. I.9. m. 119°.[418]
CH$_3$C(O)CCl$_2$PO(OEt)$_2$. V.14. b$_{0.05}$ 82-83°.[282]
CH$_3$O$_2$CCCl$_2$PO(OEt)$_2$. V.14. b$_{0.05}$ 81-82°.[281]
ClH$_2$CCH$_2$CHClPO(OH)$_2$. I.1. m. 67-68°.[1935]
ClH$_2$CCH$_2$CHClPO(OR)$_2$.

R = Me. V.2. b$_3$ 111°.[1935]

R = Et. V.2. b$_3$ 116-118°.[1935]

R = H$_2$C=CHCH$_2$-. V.2. b$_2$ 137-140°.[1935]

ClH$_2$CCHClCH$_2$PO(OR)$_2$.

R = Et. V.2. b$_3$ 116-118°.[1935]

(ClH$_2$CCHClCH$_2$-)OPOCH$_2$CH$_2$O. V.5. b$_1$ 134-137°.[1575]
(ClH$_2$C-)$_2$C(OH)PO(OR)$_2$.

R = Me. V.8. m. 57-58°, n$_D^{20}$ 1.4790, d$_4^{20}$ 1.4295.[30]

R = Et. V.8. m. 45-46°.[30]

R = Pr. V.8. n$_D^{20}$ 1.4670, d$_4^{20}$ 1.2235.[30]

R = i-Pr. V.8. m. 79-80°.[30]

R = Bu. V.8. n$_D^{20}$ 1.4600, d$_4^{20}$ 1.1711.[30]

R = i-Bu. V.8. m. 55-56°.[30]

R = -CMe$_2$(CCl$_3$). V.8. m. 176-177°.[961]

(OCN-)C(O)CCl$_2$PO(OR)$_2$.

R = Et. V.14. b$_{0.07}$ 89-90°, n$_D^{20}$ 1.4715, d$_4^{20}$ 1.3933.[281]

R = Pr. V.14. b$_{0.08}$ 108-109°, n$_D^{20}$ 1.4690, d$_4^{20}$ 1.3269.[2]

R = i-Pr. V.14. b$_{0.4}$ 105-106°, n$_D^{20}$ 1.4660, d$_4^{20}$ 1.3025.
[281]

R = Bu. V.14. b$_{0.1}$ 120-121°, n$_D^{20}$ 1.4685, d$_4^{20}$ 1.2677.[28]
Cl$_3$CC(O)NHC(O)PO(OPr-i)$_2$. V.9. m. 64-67°.[463]
Cl$_2$C=CClCH(OH)-PO(OMe)$_2$. V.8. m. 90°.[1755]
(Cl$_2$HC)(CH$_2$Cl)C(OH)PO(OR)$_2$.

R = Et. V.8. n$_D^{20}$ 1.4830, d$_4^{20}$ 1.397.[37]

R = ClCH$_2$CH$_2$-. V.8. n$_D^{20}$ 1.5062, d$_4^{20}$ 1.530.[37]

R = Pr. V.8. n$_D^{20}$ 1.4840. d$_4^{20}$ 1.336.[37]

R = Bu. V.8. n$_D^{20}$ 1.4792, d$_4^{20}$ 1.259.[37]

R = i-Bu. V.8. n$_D^{20}$ 1.4750, d$_4^{20}$ 1.259.[37]

R = EtOCH$_2$CH$_2$-. V.8. n$_D^{20}$ 1.4827, d$_4^{20}$ 1.338.[37]

R = C_6H_{13}-. V.8. n_D^{20} 1.4735, d_4^{20} 1.180.[37]

$Cl_3CCH(OH)-O-CH_2PO(OR)_2$.

R = Et. V.14.[100]

R = i-Pr. V.14.[100]

$Cl_3CCH(Cl)-O-CH_2PO(OPr\text{-}i)_2$. V.14.[100]

$Cl_3CCCl_2CH(OH)-PO(OMe)_2$. V.8. m. 130°.[1755]

$MeO_2CCHFPO(OMe)_2$. V.6. b_1 112-115°, n_D^{20} 1.4138, d_4^{20} 1.1960.[549]

$F(O)_2SNHCHEtPO(OEt)_2$. V.8. m. 51°.[789]

$(F_3CCH_2CH_2-)OPOCH_2(CF_2)_2CH_2O$. V.2. $b_{4.5}$ 107-109°, m. 55-56.5°.[1881]

$F_3CCMe(OH)-PO(O^{\ominus})_2(H_3\overset{\oplus}{N}Ph)_2$. I.6. m. 190-195° (dec.).[698]

$F_3CC(O)CHF-PO(OH)_2$. I.2. As aniline salt m. 146-147°.[979]

$(F_2HC-)_2C(OH)-PO(OMe)_2$. V.8. Insecticide.[631]

$F_3CCF=CFPO(OEt)F$. V.5. b_{18} 43°.[983]

$F_3CCF=CFPO(OEt)_2$. V.5. b_8 83°.[980] V.5.[983]

$(F_3C)_2CHPO(OMe)_2$. V.14. $b_{0.4}$ 44°, ^{19}F.[1305]

$F_3CCHFCF_2PO(OEt)_2$. b_{10} 87°, n_D^{25} 1.3606.[301]

$NC-CH=CHPO(OR)_2$.

R = Et. V.5. $b_{1.5}$ 106°[1874] V.5. $b_{0.4}$ 94°, V.7. $b_{0.15}$ 88°.[1040] V.6. b_1 124-127°.[142]

R = Bu. V.5. b_3 119-120°[1874] V.5.[567]

$H_2C=C(CN)PO(OR)_2$.

R = Et. V.7. b_1 82-83°.[1672]

R = Bu. V.7. b_1 108-109°.[1672]

$OHCCH(CN)-PO(OR)_2$.

R = Et. V.14. b_1 94-97°, n_D^{20} 1.4725, d_4^{20} 1.1836.[1047]

R = Pr. V.14. b_1 106-109°, n_D^{20} 1.4745, d_4^{20} 1.1232.[1047]

R = Bu. V.14. b_1 116-119°, n_D^{20} 1.4665, d_4^{20} 1.0855.[1047]

$(OCN-)C(O)CH_2PO(OR)_2$.

R = Et. V.14. $b_{0.05}$ 120-122°, n_D^{20} 1.4550, d_4^{20} 1.2482.[281]

R = Pr. V.14. $b_{0.1}$ 110-111°, n_D^{20} 1.4450, d_4^{20} 1.1754.[281]

R = Bu. V.14. $b_{0.7}$ 145-146°, n_D^{20} 1.4530, d_4^{20} 1.1159.[281]

$H_2C=C=CHPO(OR)_2$.

R = Me. V.2. $b_{0.4}$ 75-76°, ^1H.[2000]

R = Et. V.5. $b_{0.4}$ 89°, n_D^{25} 1.4544. ^{31}P -14.4 ppm.[1188] V.5. $b_{0.4}$ 90°.[1189] V.5. $b_{0.025}$ 60-62°.[1559] V.14.

$b_{0.9}$ 86°.[1961]

R = Pr. V.5. $b_{0.06}$ 86-87°.[1559] V.5.[1556]

R = i-Pr. V.5. b_2 83-84°.[1559]

R = HC≡CCH$_2$-. V.5. $b_{0.3}$ 117°, n_D^{25} 1.4842, ^{31}P -17.9 ppm.[1188] From HC≡CCH$_2$OH and PCl$_3$/NR$_3$, $b_{0.25}$ 115-116°.[382]

R = t-Bu. V.2. $b_{0.1}$ 54-56°, IR.[641]

HC≡CCH$_2$PO(OEt)$_2$. V.6. $b_{0.4}$ 89-91°, n_D^{22} 1.4475.[1961] V.5. $b_{0.8}$ 78-80°.[942]

CH$_3$C≡C-PO(OR).

R = Me. V.5. V.6. $b_{2.0-2.1}$ 100-101°, IR, ^{1}H.[653]

R = Et. V.5. V.6. $b_{0.9}$ 101-106°, IR, ^{1}H.[653] V.6. $b_{0.3}$ 84-87°.[1963] V.14. b_1 91-91.5°, n_D^{20} 1.4472, d_4^{20} 1.0717.[824] V.14. $b_{0.5}$ 95°, n_D^{21} 1.4471, d_4^{20} 1.079 [1961] V.5. b_4 115-116°, n_D^{20} 1.4460, d_4^{20} 1.0734. V.6. $b_{0.5}$ 95°.[1962] V.14 b_4 115-116°.[1559]

R = Pr. V.14. b_4 127-128°.[1559] V.5. b_4 127-128°, n_D^{20} 1.4478, d_4^{20} 1.0279.[1555] V.14.[1556]

R = i-Pr. V.14. b_3 105-107°.[1559]

R = Bu. V.5. b_6 152-153°, n_D^{20} 1.4498, d_4^{20} 1.0029.[1555] IR, μ = 3.93 D, n_D^{20} 1.4488, d_4^{20} 1.0023.[1431]

R = Ph. V.14. $b_{0.45}$ 171-173°.[1559]

H$_2$NCH=C(CN)PO(OEt)$_2$. V.14. m. 122-123°.[1033]

HC≡CH(OH)PO(OR)$_2$.

R = Me. V.8. m. 76.5°.[1572]

R = Et. V.8. m. 57°.[1572]

R = Pr. V.8. b_1 132-132.5°.[1572]

R = Bu. V.8. $b_{0.5}$ 132°.[1572]

HOH$_2$CC≡CPO(OEt)$_2$. V.6, V.14. $b_{0.3}$ 160°.[1963]

HO$_2$CCH=CHPO(OH)$_2$. I.13. Mono-p-toluidine salt, m. 191°.[10]

OCH$_2$-O-CH=CPO(OR)$_2$.

R = Et. V.6. $b_{0.2-0.5}$ 128-134°, n_D^{20} 1.4485.[502]

R = Bu. V.6. $b_{0.25}$ 127-132°, n_D^{20} 1.4495.[502]

O(O)COCH$_2$CHPO(OEt)$_2$. V.5. $b_{0.02}$ 115-116°, n_D^{19} 1.4460, IR.[717]

NCCH$_2$CH$_2$PO(OR)$_2$.

R = Me. V.7. b_{11} 158°.[1562]

R = Et. V.7. b_7 150-152°.[142] V.14. b_1 130°, n_D^{20} 1.4382.[1414] V.7. $b_{0.6-0.7}$ 107-111°.[1073,1074] V.7. b_2 128°, n_D^{20} 1.4370.[638] V.7. $b_{0.1}$ 110°.[751]

R = cycl.-C_6H_{11}. V.7. $b_{0.01}$ 120-124°.[1634]

R = $PhCH_2$-. V.7. $b_{0.001}$ 160-162°.[1634]

$NCCH_2CH_2PO(OBu)(OC_9H_{19})$. V.7. $b_{0.0097}$ 80-85°.[1162]

$(NCCH_2CH_2-)\overline{OPOCHMeCH_2O}$. V.7. b_2 182°, n_D^{20} 1.4660.[1588]

$O_2NCH=CMe-PO(OEt)_2$. V.14 $b_{0.04}$ 102-104°, n_D^{20} 1.4578.[1214]

$[H_2C=CHCH_2PO(OMe)O^{\ominus}]\ \overset{\oplus}{N}Me_4$. III.3.[345] III.3. m. 17°.[353]

$H_2C=CHCH_2PO(OR)_2$.

R = Me. V.2. b_{12} 85-86°.[309]

R = Et. V.6. b_{10} 92-93.5.[1616] V.5. b_{16} 97-98°, n_D^{20} 1.4313. d_4^{20} 1.0342.[1558] V.2. b_{18} 99-102°.[309] V.6. b_2 78-79°.[1486] V.5. b_2 78-81°, n_D^{21} 1.4320.[552]

R = Pr. V.2. b_{30} 140°.[309]

R = i-Pr. V.2. b_{19} 106-108°.[309]

R = $H_2C=CHCH_2$-. V.5. $b_{0.5}$ 82°.[957]

R = Bu. V.5, V.4. $b_{1.5}$ 93-94.5°,[898] n_D^{20} 1.4390, d_4^{20} 0.9811, μ = 2.98 D, IR.[1431] V.2. $b_{1.5}$ 88-88.5°.[898] V.5.[1697]

$H_2C=CHCH_2PO(OMe)OR$.

R = 2,4,5-$Cl_3C_6H_2CH_2$-. V.2. $b_{0.05}$ 148-150°.[1994]

R = $PhC(O)CH_2$-. V.1. $b_{0.02}$ 140°.[345]

$H_2C=CHCH_2PO(OC_6H_{13})OCH=CCl_2$. V.8. $b_{0.4}$ 133-135°.[1698]

$(H_2C=CHCH_2-)\overline{OPO-R-O}$.

R = $-CHMeCMe_2$-. V.5. $b_{1.5}$ 120-124°.[1917]

R = $-CMe_2CMe_2$-. V.5. b_1 120-122°.[1917]

$CH_3CH=CHPO(OH)_2$ (cis). I.2. m. 55-57°, IR.[641]

$CH_3CH=CHPO(OR)_2$ (cis).

R = Me. V.14. b_7 78-78.5°, ^1H.[2000]

R = Et. V.14. b_{13} 142°, n_D^{20} 1.4359, d_4^{20} 1.0381, ^1H.[1479]

R = t-Bu. V.14. $b_{0.1}$ 45-46°, IR.[641]

$H_3CCH=CHPO(OR)_2$ (trans or probably trans).

R = Me. V.2. b_{10} 95-96°, ^1H.[2000]

R = Et. V.14. b_{13} 111-113°, n_D^{20} 1.4407, d_4^{20} 1.0435. V.5. $b_{2.5}$ 78-80°.[1616]

$(H_3CCH=CH-)\overline{OPO-R-O}$

R = $-CH_2CH_2-$. V.14 b_1 149-150°.[1575]

R = $-CHMe-CH_2-$. V.14 $b_{1.5}$ 122-124°.[1575]

R = $-CHMe-CHMe-$. V.14. b_1 119-120°.[1575]

$H_2C{=}CMePO(OH)_2$. I.13.[165]

$H_2C{=}CMePO(OR)_2$.

R = Me. V.14. b_8 66.5-67°, n_D^{20} 1.4318, d_4^{20} 1.1054.[1589] V.2. b_{1-2} 44-46°, n_D^{20} 1.4340.[744]

R = Et. V.14.[1416] V.14. b_{12} 86°.[757]

R = Bu. V.2. $b_{0.25}$ 86-87°, n_D^{20} 1.4376.[744] V.14. $b_{0.6}$ 103°.[757]

$H_2CCH_2NH-N{=}C-PO(OR)_2$.

R = Et. V.14. $b_{0.05}$ 124-126°, n_D^{20} 1.4784, d_4^{20} 1.1556, IR, ^1H.[1580] V.14. $b_{0.5}$ 125-127°, n_D^{20} 1.4779, d_4^{20} 1.1338.[1578]

R = Pr. V.14. $b_{0.0001}$ 132-133°, n_D^{20} 1.4720, d_4^{20} 1.1028 IR, ^1H.[1580]

R = Bu. V.14. $b_{0.005}$ 142-143°, n_D^{20} 1.4700, d_4^{20} 1.0591, IR, ^1H.[1580]

$CH_3OCH_2C(N_2)PO(OMe)_2$. V.14. $b_{0.1}$ 56-58°, n_D^{25} 1.4610, IR, ^1H.[1191]

$C_2H_5C(O)PO(OR)_2$.

R = Et. V.5. b_7 105°. As 2,4-dinitrophenylhydrazone, m 93-94°, IR, ^1H.[244]

R = Pr. V.5. b_{13} 135-137°.[607]

R = Bu. V.5. b_{14} 157-158°.[607]

$CH_3C(O)CH_2PO(OR)_2$.

R = Me. V.5. b_{10} 123-124°.[1566] V.5. $b_{0.03}$ 53-54°, n_D^{22} 1.4335.[390]

R = Et. V.5. $b_{1.5}$ 97-98°.[1039] V.5.[866] V.14. b_1 98°.[195] V.5. b_9 126°.[1552] V.14.[165] V.14. $b_{0.1}$ 88°, n_D^{21} 1.4350, d_4^{21} 1.111.[1961] V.14. $b_{0.2}$ 92°.[1963] V.14.[14] V.14.[1140] ^1H. ^{31}P - 20.0 ppm.[1093]

R = Bu. V.14. b_1 126-127°. n_D^{20} 1.4400, d_4^{20} 1.078.[1133] V.5. b_3 137-138°.[1566]

$[CH_3C(O)CH_2-]OPOCH_2C(CH_2Cl)Et-CH_2O$. V.5. m. 98°.[2059]

$CH_3C(OMg/_2){=}CH-PO(OEt)_2$. From $CH_3C(O)CH_2PO(OEt)_2$ and Mg in liq. NH_3, m. 186-187°.[971]

$CH_3OCH=CHPO(OMe)(OEt)$. V.14. $b_{0.04}$ 52°, n_D^{25} 1.4506, IR. ^1H.[1191]

$H_2C=CHCH(OH)PO(OR)_2$.

R = Me. V.8. b_{10} 154-155°.[1611]

R = Et. V.8. b_{13} 154-155°.[1542]

$OHCCH_2CH_2PO(OR)_2$.

R = Me. From $(MeO)_2\overline{PCH_2CH=CH-O}$ and H_2O or $MeCO_2H$ in Et_2O.[166]

R = Et. V.14. $b_{0.2}$ 93°, n_D^{22} 1.4391.[1082] V.14. $b_{0.23}$ 92-93°.[1705]

$OHCCHMe-PO(OR)_2$.

R = Et. V.14. b_{10} 108-115°, As 2,4-dinitrophenylhydrazone, m. 131-132°.[1733] V.14. $b_{0.8}$ 85-87°.[1502]

R = i-Pr. V.14. $b_{0.7}$ 102-103°, n_D^{20} 1.4398, d_4^{20} 1.0482, As 2,4-dinitrophenylhydrazone, m. 93-95°.[1350]

$[CH_3\overline{CHOCH}-PO(OH)O]^{\ominus}\ H_3\overset{\oplus}{N}CHMePh$. V.14. m. 132-134°.[641]

$H_2\overline{COC}Me-PO(OEt)_2$. V.14. $b_{1.5}$ 75.5-77°.[165] V.6. $b_{0.6}$ 75°, n_D^{20} 1.4320, d_4^{20} 1.122, IR.[1958]

$H_2\overline{COCH}CH_2PO(OEt)_2$. V.5. $b_{0.6}$ 98°, IR, ^1H.[691] V.5. $b_{0.1}$ 68-72°.[1536]

$CH_3CO_2CH_2PO(OR)_2$.

R = Me. V.14. $b_{0.08}$ 63°.[1878a]

R = Et. V.14 b_4 106.5-107°.[1589]

R = Me_2EtSi-. VI.1. $b_{2.5}$ 112°, n_D^{20} 1.4332, d_4^{20} 1.0237.[1435]

R = $MeEt_2Si-$. VI.1. $b_{5.5}$ 138-139°, n_D^{20} 1.4402, d_4^{20} 1.0125.[1435]

R = Et_3Si-. VI.1. $b_{5.5}$ 155°, n_D^{20} 1.4470, d_4^{20} 1.0103.[1435]

R = $MePr_2Si-$. VI.1. $b_{4.5}$ 166°, n_D^{20} 1.4431, d_4^{20} 0.9752.[1435]

$CH_3O_2CCH_2PO(OR)Cl$.

R = Et. IV.1. $b_{0.07}$ 82-83°, n_D^{20} 1.4530, d_4^{20} 1.3040.[280]

R = Ph. IV.1. $b_{0.07}$ 140-141°, n_D^{20} 1.5180, d_4^{20} 1.3390.[280]

$CH_3O_2CCH_2PO(OR)NCO$.

R = Et. IV.2. $b_{0.1}$ 103-104°, n_D^{20} 1.4500, d_4^{20} 1.2810.[280]

R = Ph. IV.2. $b_{0.07}$ 145-146°, n_D^{20} 1.5110, d_4^{20} 1.3010.[280]

$CH_3O_2CCH_2PO(OR)_2$.

R = Me. V.2. b_5 117-118°, n_D^{20} 1.4373, d_4^{20} 1.2644.[1172]

R = Et. V.6. b_9 131.5-132°.[1418]

$C_2H_5O_2CPO(OR)_2$.

R = Et. V.5. $b_{12.5}$ 133°, IR.[1424] V.14.[1042] V.5. b_{12} 135.3°.[112] V.5. $b_{12.5}$ 138.25°, d_0^0 1.1422.[112a] V.5. b_{10} 130-131°.[115] V.5. b_{8-10} 122.5-123°.[1418,1419,1420] V.5.[1980]

R = i-Pr. V.5. b_9 117-118°, IR.[1424]

$C_2H_5O_2CPO(OEt)(OCH_2CHClCH_2OMe)$. V.5. $b_{1.5}$ 144.5-146°, n_D^{20} 1.4520.[122]

$HO_2CCH_2CH_2PO(OH)_2$. I.13. m. 166-167°.[1143] I.13. as p-toluidine salt, m. 158°.[1041] I.2. m. 167°.[638] I.6.[1953] I.2. m. 167-168°.[112] I.2. m. 170°.[116] I.2. m. 178-180°.[1419,1420]

$HO_2CCH_2CH_2PO(OR)OH$.

R = Me. III.3. m. 100-101°.[2032]

R = Et. III.3. m. 155-157°.[2032]

R = Pr. III. 3. m. 159-160°.[2032]

R = Cl_3CCMe_2-. III.3. m. 133-134.5°.[504]

$HO_2CCHMe\text{-}PO(OH)_2$. I.2. m. 119-132°.[112]

$HO_2CCHMe\text{-}PO(OEt)_2$. V.14. n_D^{20} 1.4480, d_4^{20} 1.2070.[282]

$OCH_2CH_2OCHPO(OEt)ONa$. III.3. m. 177-178°, ^1H.[715]

$OCH_2CH_2OCHPO(OEt)_2$. V.5. n_D^{23} 1.4320.[715]

$H_2CSCMePO(OR)_2$.

R = Me. V.14.[1416]

R = Et. V.14.[1416]

$OCH_2CH_2CH_2PO(OR)$.

R = Et. V.14. $b_{0.55}$ 74°, IR.[499] V.5. $b_{9.6}$ 72-73°.[1649]

R = Pr. V.5. b_7 76-78°.[1649]

$CH_3C(O)NHCH_2PO(OH)_2$. I.13. m. 185°.[912]

$CH_3C(O)NHCH_2PO(OR)_2$.

R = Me. V.9.[837] V.9. V.8.[834]

R = Et. V.9.[837] V.9. V.8.[834,838]

R = Pr. V.9.[837]

R = Bu. V.9.[837] V.9, V.8.[834,838]

$C_2H_5NHC(O)\text{-}PO(OEt)_2$. V.5. b_1 104-105°.[1723] V.9. b_{15} 167-168°.[1631] V.9. $b_{0.5}$ 104-105°.[1720]

$H_2NC(O)\text{-}CH_2CH_2PO(OEt)_2$. V.7. m. 74-76°.[210] V.7. m. 74.5-76.2°, ^1H.[751]

$(CH_3)_2NC(O)PO(OEt)_2$. Complexes with Ni(II), Co(II), Cu(II).[652]
$[H_3\overset{\oplus}{N}CH_2CH_2C(O)PO(OEt)_2]Cl^{\ominus}$. V.5. m. 340°.[184]
$CH_3OCH_2CH(-CN)PO(OEt)_2$. V.14. b_1 114-116°.[1651]
$HO_2CCH_2NHCH_2PO(OH)_2$. I.13. m. 230° (dec.).[1330]
$HO_2CCH(NH_2)-CH_2PO(OH)_2$. V.7. and I.2. m. 228° (dec.).[357]
$O_2NCH_2CMe(OH)-PO(OMe)_2$. V.14. m. 68-69°.[1214]
$(CH_3)_2C(N_2)-PO(OMe)_2$. V.14. $b_{0.11}$ 50°, ^1H.[1191]
$(CH_3)_2NC(S)-PO(OR)_2$.

R = Et. V.5. n_D^{20} 1.5062, d_4^{20} 1.1551.[1173]

R = Pr. V.5. n_D^{20} 1.5063, d_4^{20} 1.0982.[1173]

R = i-Pr. V.5. n_D^{20} 1.4971, d_4^{20} 1.0899.[1173]

R = Bu. V.5. n_D^{20} 1.4791, d_4^{20} 1.0539.[1173]

R = i-Bu. V.5. n_D^{20} 1.4929, d_4^{20} 1.0585.[1173]

R = Am. V.5. n_D^{20} 1.4813, d_4^{20} 1.0211.[1173]

R = i-Am. V.5. n_D^{20} 1.4905, d_4^{20} 1.0281.[1173]

$OCH_2CH_2OCH_2PO(OEt)$. V.14. $b_{0.5}$ 97-97.5°.[173]
$C_3H_7PO(OH)_2$. I.5.[1638] I.2. m. 73°.[446] I.1, I.6. m. 66°.[729] I.2. m. 72.5-74.5°.[1420]
$C_3H_7PO(OC_6H_{11}$-cycl.)Cl. IV.2. $b_{0.05}$ 79°, n_D^{25} 1.4688.[740]
$C_3H_7PO(OC_6H_{11}$-cycl.)F. IV.2. $b_{0.03}$ 43°, n_D^{25} 1.4378.[740]
$C_3H_7PO(OPr)NCO$. IV.5. $b_{0.35}$ 64-66°.[726]
$C_3H_7PO(OR)_2$.

R = Et. V.5. $b_{8.5}$ 86-88°.[102] V.6. b_9 92-93°.[1420]

R = F_3CCH_2-. V.5. b_1 50-53°, n_D^{20} 1.3526.[1211]

R = Pr. Magn. rotation of the bonds.[1069] μ = 2.98 D.[1216] As by-product by pyrolysis of PrPO(OPr)ONa.[1667] V.5.[102] V.5. b_{18} 126°.[1315]

R = Bu. V.6. b_{10} 133-137°, n_D^{20} 1.4272, d_4^{20} 0.9598.[1431]

R = EtO_2CCH_2-. V.13. b_2 154-156°.[1646]

$C_3H_7PO(OEt)(OCH_2CO_2Et)$. V.13. b_2 122-123°.[1646]
$C_3H_7PO(OPr)OR$.

R = $(ClH_2C)_2CH-$. V.5. b_3 132-134°.[1574]

R = $(EtOCH_2-)(ClH_2C-)CH-$. V.5. $b_{1.5}$ 125-126°.[1574]

R = $(EtO_2C)_2CH-$. V.8. $b_{2.5}$ 137-138°, n_D^{20} 1.4370, d_4^{20} 1.1242.[1623]

$(C_3H_7-)OPO-(CH_2)_2O$. V.13. b_1 119-120°.[1649]
$(C_3H_7-)OPO-(CH_2)_4O$. V.13. $b_{0.008}$ 80-81°.[1649]
$(C_3H_7-)OPO(CH_2)_2O(CH_2)_2O$. V.13. $b_{0.03}$ 93-95°.[1649]
$(C_3H_7-)(OPr)OPO(CH_2)_2OPO(OPr)(-C_3H_7)$. V.5. b_2 201-202°, n_D^{20} 1.4438, d_4^{20} 1.0760.[1553]
$(CH_3)_2CHPO(OH)_2$. I.6. Aniline salt, m. 175-177°.[315] I.1. m.

74-75°.[446] I.6. I.1. m. 71°.[729] I.6.[791] I.6.[1167]
$(CH_3)_2CHPO(OEt)OH$. III.4. b_1 116-117°, n_D^{20} 1.4306, d_4^{20} 1.0902.[509] III.2. $b_{0.05}$ 80-82°, n_D^{25} 1.4199.[338]
$(CH_3)_2CHPO(OPr\text{-}i)F$. Calc. of the electronic structure.[1075]
$(CH_3)_2CHPO(OR)_2$.
R = Et. V.5.[1366] V.2. $b_{1.8}$ 56-58°.[787]
R = $4\text{-}HOC_6H_4\text{-}$. V.2. m. 141°.[1461]
$(CH_3)_2CHPO(OEt)OR$.

R = 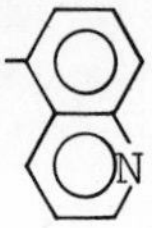 V.2. $b_{0.3}$ 145-150°, insecticide.[527]

R = 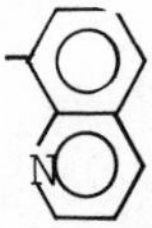 V.2. $b_{0.04}$ 126°, insecticide.[527]

$(CH_3)_2CHPO(OPr\text{-}i)OCH(CO_2Et)_2$. V.8. $b_{2.5}$ 124-125°, n_D^{20} 1.4353, d_4^{20} 1.1200.[1623]
$[(EtO)_2OP\text{-}CH{=}\overset{\oplus}{N}Me_2]Cl^{\ominus}$. V.5. From $(EtO)_2OPCH(OMe)NMe_2$ and $SOCl_2$.[714]
$(MeO)_2OPNHC(O)\text{-}PO(OR)_2$.
R = Me. V.8. n_D^{20} 1.4510, d_4^{20} 1.3614.[972]
R = Et. V.8. n_D^{20} 1.4440, d_4^{20} 1.2689.[972]
R = Pr. V.8. n_D^{20} 1.4478, d_4^{20} 1.2061.[972]
R = i-Pr. V.8. n_D^{20} 1.4460, d_4^{20} 1.2040.[972]
R = Bu. V.8. n_D^{20} 1.4400, d_4^{20} 1.1386.[972]
R = i-Bu. V.8. n_D^{20} 1.4430, d_4^{20} 1.1472.[972]
R = Am. V.8. n_D^{20} 1.4434, d_4^{20} 1.1259.[972]
R = i-Am. V.8. n_D^{20} 1.4428, d_4^{20} 1.1121.[972]
R = Ph. V.8. n_D^{20} 1.5510, d_4^{20} 1.321.[972]
R = $4\text{-}ClC_6H_4\text{-}$. V.8. n_D^{20} 1.5420, d_4^{20} 1.4209.[972]
R = $4\text{-}MeC_6H_4\text{-}$. V.8. n_D^{20} 1.5473, d_4^{20} 1.3140.[972]
$CH_3OCH_2CH_2PO(OEt)_2$. V.14. b_3 81°, n_D^{20} 1.4259, d_4^{20} 1.0709.[1640]
$C_2H_5OCH_2PO(OEt)_2$. V.5. b_{10} 105-107°.[836]
$HOCH_2CH_2CH_2PO(OH)_2$. I.6.[1953]
$(CH_3)_2C(OH)\text{-}PO(OH)_2$. I.5.[165] I.2.[1777a,1178,1178a,1179] m. 175°.[1179] m. 167-169°.[415]
$(CH_3)_2C(OH)\text{-}PO(OR)_2$.
R = Me. V.8. m. 72-73°.[40] V.1. m. 76°.[1178,1186]

R = Et. V.8. n_D^{20} 1.4320.[963] V.1. m. 14-15°, b_{20} 145°.[1178]

R = CH_2=$CHCH_2$-. V.8. b_{12} 132°.[18]

R = $(ClH_2C)_2CH$-. V.8. m. 79-80°.[961]

R = Bu. V.8. b_7 154-155°.[18] V.8. b_{10} 124-126°.[1340]

R = sec.-Bu. V.8. b_{13} 153-154°.[35]

R = Cl_3CCMe_2-. V.8. m. 147°.[34]

R = $EtOCH_2CH_2$-. V.8. b_2 144-145°.[16]

R = Ph. V.8. m. 113-114°.[418]

$C_2H_5CH(OH)$-$PO(OH)_2$. I. 9. m. 162°.[555]

$C_2H_5CH(OH)$-$PO(OR)_2$.

R = Me. By hydrolysis of $(MeO)_3$P-CHEt-OCHEt-O. $b_{0.5}$ 100-103°, ^{31}P -31.1 ppm.[1691]

R = Et. V.8. $b_{1.5}$ 120-121°.[21]

R = H_2C=CHCH -. V.8. b_1 129-130°.[21]

R = $(ClH_2C)_2CH$-. V.8. m. 79-80°.[961]

R = sec.-Bu. V.8. b_2 139-140°.[21]

R = Cl_3CCMe_2-. V.8. m. 126-127°.[34]

R = $CH_2(CH_2)_3C(CCl_3)$-. V.8. m. 156.5-157.5°.[33]

$HOCH_2CH(OH)CH_2PO(OEt)_2$. V.14. $b_{0.25}$ 142°, ^1H.[691]

$CH_3OCH_2CH(OH)PO(OEt)_2$. V.14. $b_{0.25}$ 108°, ^1H.[691]

$(CH_3O)_2CH$-$PO(OR)_2$.

R = Me. b_{19} 113-114°, n_D^{20} 1.4280, d_4^{20} 1.2042.[1706]

R = Et. b_8 112-114°, n_D^{20} 1.4246, d_4^{20} 1.1043.[1706]

$CH_3OCH_2OCH_2PO(OMe)_2$. V.8. b_{12} 130°.[755]

$CH_3CH[$-$OS(O)OMe]$-$PO(OR)_2$.

R = Me. V.14. b_1 107°, n_D^{20} 1.4477, d_4^{20} 1.3113.[1589]

R = Et. V.14. b_3 119-120°, n_D^{20} 1.4427, d_4^{20} 1.2122.[1589]

CH_3SCHMe-$PO(OEt)_2$. V.14. $b_{0.2}$ 68-70°, n_D^{23} 1.4620, IR, ^1H.[442]

$C_2H_5SCH_2PO(OEt)_2$. V.6. b_{10} 131°.[1044] V.14. b_1 127-127.5°.[128a]

$C_2H_5S(O)_2CH_2PO(OEt)_2$. V.14. $b_{11.5}$ 194°.[128a]

$C_2H_5CH(NH_2)$-$PO(OH)_2$. I.13. m. 264-266°.[2040]

$C_2H_5CH(NH_2)$-$PO(OEt)_2$. V.14. $b_{2.6}$ 45-47°.[356] V.14. $b_{0.05}$ 75°.[789]

$(CH_3)_2C(NH_2)$-$PO(OR)_2$.

R = Et. V.9. b_3 65-66°.[1220]

R = $EtSCH_2CH_2$-. V.9. As picrate, m. 99-100°.[1222]

$H_2NCH_2CHMePO(OH)_2$. I.3. m. 282-286°.[210]

$H_2NCHMeCH_2PO(OH)_2$. I.3. m. 278-284°.[210]
$(CH_3)_2NCH_2PO(OR)_2$.

R = Me. V.14. b_2 88°.[529]

R = Et. V.9. $b_{0.5}$ 83-85°, n_D^{25} 1.4281.[329] V.9. b_1 85°, n_D^{20} 1.4288.[284] V.9.[834]

R = Bu. V.9.[834]

$H_3\overset{\oplus}{N}CMe_2PO(OH)O^{\ominus}$. I.2. m. 260°.[2016]
$H_3\overset{\oplus}{N}CMe_2PO(OEt)O^{\ominus}$. III.3. m. 274-275°.[2016]
$H_2NCH_2CH(OH)CH_2PO(OEt)_2$. V.14. IR, ^1H.[691]
$H_2NCH_2CH_2\overset{\oplus}{N}H_2CH_2PO(OH)O^{\ominus}$. I.13. m. 251-253°.[2041a]

$\underline{C_4}$

$\overline{OCH_2CH_2OAs}CHMePO(OEt)_2$. V.14. $b_{0.2}$ 112-113°, ^{31}P -24.8 ppm[1742]

$BrCH=CH-S(O)_2-CH=CHPO(OEt)_2$. V.5. $b_{0.02}$ 60-65°.[941]
$CH_3CH=CBrCH_2PO(OEt)_2$. V.5. $b_{0.6}$ 102°, n_D^{23} 1.4743.[1961]
$BrCH=CMe-CH_2PO(OEt)_2$. V.5. $b_{0.3}$ 96°, n_D^{24} 1.4749.[1082]
$C_2H_5OCH=CBrPO(OEt)_2$. V.2. $b_{0.05}$ 108-109°, n_D^{20} 1.4765, d_4^{20} 1.3553.[1348]
$C_2H_5O_2CCHBrPO(OEt)_2$. V.14. b_1 128-131°, n_D^{20} 1.4632, d_4^{20} 1.3742.[697]
$BrH_2CCH(OEt)-PO(OMe)_2$. V.5. b_{13} 137-138°.[31]
$Br(CH_2)_4PO(OH)_2$. I.2. m. 126.5-127.8°, IR.[499] I.2.[767]
$Br(CH_2)_4PO(OEt)_2$. V.5.[767]
$\overline{OCBr(-CH_2Br)-CHBrCHBrP}(O)Cl$. From Br_2 and $Cl(O)\overline{P-CH=CH-C(=CH_2)O}$ in CCl_4, m. 147-148°.[2033]

Cl-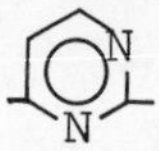-$PO(OH)_2$. I.2. Dec. 157-160°.[1022]

Cl-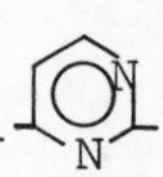-$PO(OPr\text{-}i)_2$. V.5. $b_{0.15}$ 125.5-126.5°, n_D^{25} 1.4817. d_4^{25} 1.1860.[1022]

$CH_3CHCl-C\equiv CPO(OEt)_2$. V.14. b_1 123°.[1963]
$H_2C=CCl-CH=CHPO(OR)Cl$.

R = Me. IV.2. $b_{2.5}$ 98°.[2034]

R = Et. IV.2. $b_{1.5}$ 95.5°.[2034]

R = Pr. IV.2. b_2 102°.[2034]

R = i-Pr. IV.2. b_3 100.5°.[2034]

R = $CH_2=CHCH_2-$. IV.2. b_2 108.5°.[2034]

R = Bu. IV.2. b_3 123°.[2034]

R = i-Bu. IV.2. $b_{2.5}$ 116°.[2034]

R = Am. IV.2. b_1 124°.[2034]

R = i-Am. IV.2. $b_{2.5}$ 123°.[2034]

$C_2H_5O_2CN=CCl-PO(OEt)_2$. V.2. $b_{0.05}$ 110-111°, n_D^{20} 1.4510 d_4^{20} 1.2135.[490]

$\overline{OCH_2CH_2CHCl-CHPO}(OEt)_2$. V.5. b_{12} 61-67°, n_D^{22} 1.4187, IR.[717]

$ClH_2CCH=CHCH_2PO(OEt)_2$. V.2. V.5. $b_{1.5}$ 121-122°, n_D^{20} 1.4670, d_4^{20} 1.1483.[1616] V.5. $b_{0.2}$ 92-94°.[63]

$ClH_2CCMe=CHPO(OR)_2$.

R = Et. V.2. b_4 127-128°, n_D^{20} 1.4622, d_4^{20} 1.1948.[992]

R = Pr. V.2. b_4 140-141°, n_D^{20} 1.4611, d_4^{20} 1.1289.[992]

R = Bu. V.2. b_1 140-142°, n_D^{20} 1.4629, d_4^{20} 1.0916.[992]

R = Am. V.2. b_4 162-163°, n_D^{20} 1.4607, d_4^{20} 1.0706.[992]

$CH_3CCl=CHCH_2PO(OEt)_2$. V.5. $b_{0.4}$ 92-93°, n_D^{21} 1.4568.[825,1082] V.5. b_3 102.5-103.5, n_D^{20} 1.4582, d_4^{20} 1.1330.[825]

$H_2C=CH-CHClCH_2PO(OBu)_2$. V.2. b_2 147°.[79]

$C_2H_5OCH=CClPO(OEt)Cl$. IV.2. $b_{0.04}$ 84°, n_D^{20} 1.4920, d_4^{20} 1.3196.[1348]

$C_2H_5OCH=CClPO(OR)_2$.

R = Me. V.2. $b_{0.1}$ 88-89°, n_D^{20} 1.4711, d_4^{20} 1.2692.[1348]

R = Et. V.2. $b_{0.03}$ 95-96°, n_D^{20} 1.4648, d_4^{20} 1.1903.[1348]

R = Pr. V.2. $b_{0.06}$ 110-111°, n_D^{20} 1.4596, d_4^{20} 1.1338.[1348]

R = i-Pr. V.2. $b_{0.06}$ 104-105°, n_D^{20} 1.4542, d_4^{20} 1.1146.[1348]

R = Bu. V.2. $b_{0.06}$ 125-126°, n_D^{20} 1.4608, d_4^{20} 1.0963.[1348]

R = i-Bu. V.2. $b_{0.06}$ 113-114°, n_D^{20} 1.4538, d_4^{20} 1.0776.[1348]

$CH_3CHClOCH=CHPO(OR)_2$.

R = Et. V.2. b_5 109-110°, n_D^{20} 1.4575, d_4^{20} 1.1170.[585]
V.2. b_3 114-115°.[2034]

R = Pr. V.2. b_3 126°.[2034]

R = Bu. V.2. b_4 136.5°, n_D^{20} 1.4462, d_4^{20} 1.0140.[585]

R = i-Bu. V.2. b_4 122°, n_D^{20} 1.4410, d_4^{20} 1.004.[585]

R = i-Am. V.2. b_9 135-136°, n_D^{20} 1.4458, d_4^{20} 0.9831.[585]

$ClH_2CCH_2OCH=CHPO(OR)_2$.

R = Me. V.2. b_4 147-148°, n_D^{20} 1.4771, d_4^{20} 1.282.[2035]

R = Et. V.2. b_4 143°, n_D^{20} 1.4580, d_4^{20} 1.171.[2035]

R = Pr. V.2. b_3 160°, n_D^{20} 1.4648, d_4^{20} 1.131.[2035]

R = i-Pr. V.2. b_3 136°, n_D^{20} 1.4621, d_4^{20} 1.140.[2035]

R = $H_2C=CHCH-$. V.2. b_3 154°, n_D^{20} 1.4941, d_4^{20} 1.180.[2035]

R = Bu. V.2. b_4 192°, n_D^{20} 1.4639, d_4^{20} 1.087.[2035]

R = i-Bu. V.2. b_2 151°, n_D^{20} 1.4625; d_4^{20} 1.091.[2035]

R = i-Am. V.2. b_2 185°, n_D^{20} 1.4555, d_4^{20} 1.040.[2035]

$Cl(CH_2)_3C(O)PO(OMe)_2$. V.5. $b_{0.13-0.15}$ 110-120°, IR, n_D^{25} 1.4560.[1191]

$CH_3CO_3CHClCH_2PO(OEt)_2$. V.2. m. 64-65°.[1132]

$CH_3O_2CCClMePO(OEt)_2$. V.14. $b_{0.05}$ 95-98°, n_D^{20} 1.4455, d_4^{20} 1.2130.[280]

$CH_3CO_2CHClCH_2PO(OMe)_2$. V.2. m. 29-30°.[1134]

$CH_3CO_2CH(CH_2Cl)$-$PO(OR)_2$.

R = Et. V.8. $b_{0.19}$ 104.5-105°, ^{31}P -17.5 ppm.[1587]

R = Pr. V.8. $b_{0.035}$ 98-100°, ^{31}P -17.8 ppm.[1587]

R = Bu. V.8. $b_{0.19}$ 128.5 ^{31}P -18.1 ppm.[1587]

R = i-Bu. V.8. $b_{0.65}$ 127.5-128.5°, ^{31}P -21.8 ppm.[1587]

$C_2H_5O_2C$-$CHClPO(OEt)_2$. From $HO_2CCH_2PO(OEt)_2$ and SO_2Cl_2, b_2 121-123°, n_D^{20} 1.4402, d_4^{20} 1.205.[1888]

$\overline{O(CH_2)_2OC}(CH_2Cl)$-$PO(OMe)_2$. V.5. b_{10} 153-154°, n_D^{23} 1.4608, IR.[718]

$Cl(CH_2)_4PO(OH)_2$. I.2. m. 110-120°.[767]

$Cl(CH_2)_4PO(OEt)Cl$. IV.1. $b_{0.4}$ 112°.[767]

$Cl(CH_2)_4PO(OR)_2$.

R = Et. V.6. b_{12} 150-152°, $b_{0.05}$ 106°.[767]

R = Bu. V.6. $b_{0.05}$ 120°.[767]

$ClH_2CCH_2CHMePO(OBu)_2$. V.2. b_3 148-153°.[2109]

$(CH_3)_2CClCH_2PO(OH)_2$. I.1. m. 95-97°.[241]

$ClH_2CCH(-SEt)PO(OEt)_2$. V.14.[793] V.14. b_{3-4} 124-126°.[1807]

$ClH_2CCHClCH{=}CHPO(OEt)_2$. V.14. $b_{0.04}$ 144°, n_D^{20} 1.4782, d_4^{20} 1.2410.[1629]

CH_3O_2C-$NHC(O)CCl_2PO(OEt)_2$. V.9. $b_{0.1}$ 148-149°.[281]

$Cl_2C{=}CH$-O-$(CH_2)_2PO(OBu)_2$. V.5.[188]

$C_2H_5O_2C$-$CCl_2PO(OEt)$. V.14. $b_{0.07}$ 98-99°, n_D^{20} 1.4560, d_4^{20} 1.3020.[281]

$CH_3CO_2CH(CHCl_2)PO(OR)_2$.

R = Et. V.8. $b_{0.032}$ 97-98°, ^{31}P -13.2 ppm.[1587]

R = Pr. V.8. $b_{0.08}$ 114-115°, ^{31}P -13.75 ppm.[1587]

R = Bu. V.8. $b_{0.08}$ 130-131°, ^{31}P -13.5 ppm.[1587]

R = i-Bu. V.8. $b_{0.09}$ 126-127°, ^{31}P -14.1 ppm.[1587]

$\overline{O(CH_2)_2OC}(CHCl_2)$-$PO(OMe)_2$. V.5. $b_{0.02}$ 110°, m. 46°, IR.[718]

$ClH_2CCH_2SCH_2CHCl$-$PO(OCH_2CH_2Cl)_2$. V.14.[862]

$Cl_3CCH[-O_2CNHMe]PO(OMe)_2$. V.14. m. 112-114°.[987]

$CH_3CO_2CH(CCl_3)$-$PO(OR)_2$.

R = Me. V.8. $b_{0.027}$ 100-101°.[1586] V.14.[347] V.8.[99]

R = Et. V.14. $b_{0.03}$ 101-103°.[1398] V.8. $b_{0.15}$ 127-127.5°.[1586]

R = Pr. V.14. $b_{0.026}$ 107-108°.[1398] V.8. $b_{0.033}$ 113-115°.[1586]

R = i-Pr. V.14. $b_{0.03}$ 95-98°.[1398]

R = Bu. V.14. $b_{0.03}$ 118-120°.[1398]

R = i-Bu. V.14. $b_{0.03}$ 115-117°.[1398] V.8. $b_{0.035}$ 118-120°.[1586]

$Cl_3CC(O)-C(OH)MePO(OR)_2$.

R = Me. V.14. $b_{1.5}$ 122-124°.[1650]

R = Et. V.14. $b_{1.5}$ 130-131°.[1650]

$O(CH_2)_2OC(CCl_3)-PO(OMe)_2$. V.5. b_{12} 190°, m. 117°, IR.[718]

$CH_3C(O)NHCH(CCl_3)-PO(OR)_2$.

R = Et. m. 128-130°.[492]

R = i-Pr. m. 134-135°.[492]

$Cl_3CCH(OH)-O-CHMePO(OR)_2$.

R = Et. V.14.[100]

R = Pr. V.14.[100]

$ClH_2CO_2-CH(-CCl_3)PO(OR)_2$.

R = $ClCH_2CH_2$-. V.14. m. 52-53°.[1400]

R = Bu. V.14. $b_{0.03}$ 134-136°.[1399]

R = i-Bu. V.14. $b_{0.04}$ 128°.[1399]

$Cl_3C-CH(Cl)-O-CHMePO(OR)_2$.

R = Et. V.14. b_1 119-120°.[100]

R = i-Pr. V.14. b_1 128°.[100]

$Cl_2HCCO_2CH(-CCl_3)-PO(OR)_2$.

R = Me. V.14. $b_{0.032}$ 119-120°.[1399]

R = Et. V.14. $b_{0.044}$ 126-127°.[1399]

R = Pr. V.14. $b_{0.03}$ 130-132°.[1399]

R = i-Pr. V.14 $b_{0.028}$ 121-123°.[1399]

R = i-Bu. V.14. $b_{0.03}$ 132°.[1399]

$Cl_3CCO_2CH(-CCl_3)-PO(OR)_2$.

R = $ClCH_2CH_2$-. V.14. m. 61-63°.[1400]

R = i-Pr. V.14. m. 44-45°.[1399]

R = Bu. V.14. $b_{0.03}$ 147-149°.[1399]

R = i-Bu. V.14. m. 31-32°.[1399]

$C_2H_5O_2C-CHFPO(OEt)_2$. V.5. $b_{0.01}$ 75°.[1141] V.6. $b_{0.7}$ 125°, n_D^{20} 1.4196, d_4^{20} 1.1890.[549]

$(CH_3)_2CHCH(-NHSO_2F)-PO(OEt)_2$. V.9. m. 88°.[789]

$F_2CCH_2CF_2CH_2PO(OEt)_2$. V.7. b_1 100-103°, ^1H.[2041]

$F_2C-CF_2-CF=CPO(OR)_2$.

R = Me. V.7. b_5 80-81°, n_D^{20} 1.3773, d_4^{20} 1.476.[2041]

R = Et. V.7. b_1 73-74°, n_D^{20} 1.3829, d_4^{20} 1.359.[2041] V.5[983]

R = Pr. V.7. b_1 82-83°, n_D^{20} 1.3910, d_4^{20} 1.298.[2041]

$(CF_3)_2CMe-PO(OMe)_2$. From $(CF_3)_2C=S$ and $P(OMe)_3$, $b_{0.8}$ 40-41°, n_D^{25} 1.3604, ^{19}F.[1305]

$(CF_3)_2C=CFPO(OEt)F$. From $(CF_3)_2C=CF_2$ and $P(OEt)_3$. b_{20} 50-52°.[983]

$(CF_3)_2C=CFPO(OEt)_2$. From $(CF_3)_2C=CF_2$ and $P(OEt)_3$, b_5 80°.[983,980]

$HCF_2(CF_2)_3PO(OH)_2$. I.2. m. 40.5-43.5°.[302]

$HCF_2(CF_2)_3PO(OR)_2$.

R = Me. V.7. $b_{1.1}$ 58°, n_D^{25} 1.3521.[302]

R = Et. V.7. $b_{0.2}$ 54°, $b_{5.5}$ 95-96°, n_D^{25} 1.3577.[302]

R = Bu. V.7. $b_{0.5}$ 88°, n_D^{25} 1.3800.[302]

$H_2C=CH-C\equiv C-PO(OEt)_2$. V.5. $b_{2.5}$ 107-108°, n_D^{20} 1.4696, d_4^{20} 1.0630, μ = 3.88 D.[826]

$NCCH_2CH(-CN)PO(OEt)_2$. V.7. $b_{0.005}$ 135°, n_D^{28} 1.4432.[751]

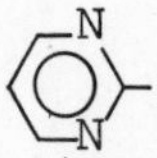
$-PO(OH)_2$. I.2. m. 212-214°.[1022]

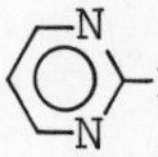
$-PO(OPr-i)_2$. V.5. $b_{0.03}$ 125-126°, m. 59.5-61°.[1022]

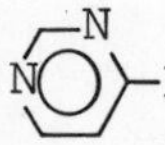
$-PO(OPr-i)_2$. V.14. $b_{0.015}$ 91-92°, m. 30.5-31°.[1022]

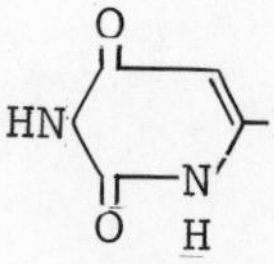
$-PO(OH)_2$. V.14. m. (as 1-hydrate) 270° (dec.).[1145]

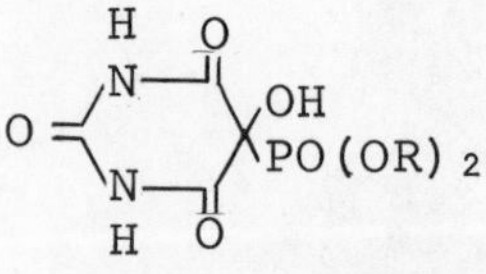

R = Me. V.8. m. 189-190°.[1120]

R = Et. V.8. m. 197-198°.[1120]

R = Pr. V.8. m. 193°.[1120]

R = i-Pr. V.8. m. 193-194°.[1120]

R = Bu. V.8. m. 184°.[1120]

R = i-Bu. V.8. m. 178-179°.[1120]

O-CH=CH-CH=C-PO(OMe)$_2$. V.5. $b_{0.4}$ 111-114°.[1422] ^{1}H.[954]

O-(O)C-CH$_2$-CH=C-PO(OEt)$_2$. V.5. $b_{0.02}$ 133-135°, n_D^{20} 1.4620.[1042]

H$_2$CCO$_2$C(O)-CHPO(OR)$_2$.

R = Et. V.7. n_D^{20} 1.4540.[1929]

R = ClH$_2$CCH$_2$-. V.7. n_D^{20} 1.4904.[1929]

R = Ph. V.7. n_D^{20} 1.5532.[1929]

S-CH=CH-CH=C-PO(OH)$_2$. I.1. m. 159° (water).[1778]

S-CH=CH-CH=C-PO(OR)$_2$.

R = Me. V.5. $b_{0.3}$ 101-103°.[1422] ^{1}H.[954]

R = Et. V.5. $b_{0.1}$ 101-104°.[1985] V.2. $b_{0.1}$ 103-104°, n_D^{20} 1.5001.[1987]

HN-CH=CH-CH=C-PO(OEt)$_2$. V.2. b_{14} 117-119°, $b_{0.6}$ 77°, IR. UV.[694]

O-C(=CH$_2$)-CH=CHPO(OR).

R = Me. V.2. m. 71-72°, ^{1}H.[2033]

R = Et. V.2. b_5 120°, ^{1}H.[2033]

R = Pr. V.2. b_3 134°, ^{1}H.[2033]

R = i-Pr. V.2. b_2 129°, ^{1}H.[2033]

R = H$_2$C=CHCH$_2$-. V.2. b_5 150°, ^{1}H.[2033]

R = Bu. V.2. b_4 152 , ^{1}H.[2033]

R = i-Bu. V.2. b_3 136.5°, ^{1}H.[2033]

R = Am. V.2. $b_{1.5}$ 150°, ^{1}H.[2033]

R = i-Am . V.2. b_3 158°, ^{1}H.[2033]

OCN-C(O)CHMe-PO(OEt)$_2$. V.14. $b_{0.07}$ 78-79°, n_D^{20} 1.4400, d_4^{20} 1.1681.[280]

CH$_3$C(O)-CH(-CN)-PO(OR)$_2$.

R = Pr. b_2 117-118°, d_4^{20} 1.1088, IR.[1087]

R = Bu. b_2 139-140 , d_4^{20} 1.0499, IR.[1087]

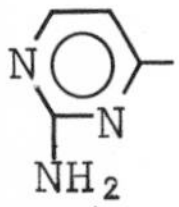 PO(OH)$_2$. I.2. Subl. 250°.[1022]

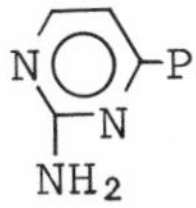PO(OPr-i)$_2$. V.14. m. 163-165°.[1022]

H$_2$C=CH-CH=CH-PO(OR)$_2$.

R = Me. V.2. b_1 74-75°.[1207]

R = Et. V.14 b_{11} 117-124°, n_D^{25} 1.4400[997]. V.2. b_{13}
122-123°.[1616]

$H_2C=CH-CH=CH-PO(OMe)(OEt)$. V.2. b_1 79-80°.[1208]
$H_3CCH=C=CH-PO(OEt)_2$. V.14. b_1 105°, n_D^{25} 1.4497, ^{31}P -14.8
ppm.[1188] V.5. b_{12} 123-125°.[1559]
$H_2C=C=CMe-PO(OEt)_2$. V.14.[1556] V.14. $b_{0.1}$ 73°, n_D^{25} 1.4587,
^{31}P -17.3 ppm.[1188,1189] V.5. b_8 112-118°.[1559]
$C_2H_5C\equiv C-PO(OEt)_2$. V.14. b_3 115-117°, n_D^{20} 1.4470, d_4^{20}
1.0325.[827] V.6. $b_{0.4}$ 100°.[1963]
$CH_3C\equiv C-CH_2PO(OEt)_2$. V.5. b_3 98°, n_D^{20} 1.4540, d_4^{20} 1.0489.
[826] V.14. $b_{0.7}$ 98°, n_D^{24} 1.4530.[1961]
$\overline{H_2CCH_2CH=C}-PO(OMe)_2$. V.14. $b_{2.8}$ 84°, n_D^{25} 1.4596, IR,
^1H.[1191]

$\overline{O-C(=CH_2)-CH=CH}-P(O)Cl$. IV. m.   62-63°, b_4 124-125°.[2033]
$H_3C-C=CH-NH-N=C-PO(OR)_2$.

R = Me. V.14. $b_{0.01}$ 138-140°.[1599a]

R = Et. V.14.[1599a]

R = Bu. V.14. $b_{0.008}$ 151°.[1599a]

$\overline{H_3C-N-N=CH-CH=C}-PO(OPr-i)_2$. V.14. $b_{0.0007}$ 85-86°, n_D^{19}
1.4622.[1788]
$HOCHMeC\equiv C-PO(OEt)_2$. V.6. V.14. $b_{0.4}$ 162°.[1963]
$HC\equiv C-CMe(OH)-PO(OMe)_2$. V.7. m. 92°.[1572]
$CH_3C(O)CH=CH-PO(OBu)_2$. V.5. $b_{0.01}$ 116-118°, n_D^{20} 1.4522.
[1046]

$CH_3CH=CHC(O)-PO(OMe)_2$. V.5. $b_{0.23-0.46}$ 86-100°, n_D^{25} 1.4721,
IR, ^1H.[1191]
$\overline{H_2CCH_2CH}-C(O)PO(OR)_2$.

R = Me. V.5. $b_{0.02}$ 66-67°, n_D^{25} 1.4543, IR, ^1H.[1191]

R = Et. V.5. b_5 119-120°, IR, ^1H, 2,4-dinitrophenylhy-
drazone m. 114-115°.[244]

$H_2C=CH-O-CH=CHPO(OR)_2$.

R = Me. V.14. b_5 114-115°, n_D^{20} 1.4446, d_4^{20} 1.154.[2035]

R = Et. V.14. b_2 105°, n_D^{20} 1.4410, d_4^{20} 1.069.[2035]

OHCCMe=CH-PO(OEt)$_2$. V.14. n_D^{20} 1.4456, d_4^{20} 1.0765.[993,991]
$\overline{OCH=CH-CH=C}-CH(OH)-PO(OMe)_2$. V.8. m. 62-63°.[1776]
$CH_3CO_2-CH=CHPO(OR)_2$.

R = Me. V.14. $b_{2.5}$ 114-116°, n_D^{20} 1.4530, d_4^{20} 1.2344.
[1135]

R = Et. V.14 b_2 123-124°, n_D^{20} 1.4511, d_4^{20} 1.1570.[1135]
$CH_3O_2C-CH=CHPO(OR)_2$.

R = Me. V.7. b_{15} 149.5-151°.[968]

R = Et. V.5. b_1 109-110°.[432]

$H_2C=CHO_2CCH_2PO(OEt)_2$. V.5. $b_{0.25}$ 98-100°.[1785]
$H_2C=CHCO_2CH_2PO(OEt)_2$. From $(EtO)_2OPNa + CH_2O +$ $H_2C=CHC(O)Cl$, $b_{0.6}$ 108-110°, n_D^{25} 1.4340, IR.[1444] V.14 b_3 114° [1598]
$H_2C=C[-O_2CMe]PO(OR)_2$.

R = Me. V.8, V.7. b_{11} 129.5-130°.[1396]

R = Et. V.8. V.7. b_{13} 135-135.5°.[1396] V.8.[426] V.14. b_2 110-112°.[1642]

R = Bu. V.8, V.7. b_{10} 161-162°.[1396]

R = i-Bu. V.8, V.7. b_{12} 152-153°.[1396]

$HO_2CCH=CMePO(OEt)_2$. V.14. As p-toluidine salt, m. 209°.[1040]
$MeO_2C-CH=CH-PO(OR)_2$.

R = Me. V.7. b_{15} 148-151°.[967]

R = Et. V.5. b_1 109-110°.[432] V.7. $b_{0.4}$ 98-99°.[1313]

$C_2H_5O_2CC(O)PO(OEt)_2$. V.5. n_D^{20} 1.4432.[1042]
$HO_2CCH_2CH(-CO_2H)-PO(OH)_2$. I.2. Complexing agent.[611]
$NC(CH_2)_3PO(OEt)_2$. V.6. b_8 163-164°, d_4^{17} 1.0885.[1419]
$NCCH_2CHMePO(OMe)_2$. V.7. b_{10} 144-145°.[277,1652]
$NCCHMeCH_2PO(OMe)_2$. V.7. b_{11} 141-142°.[277]
$NCCH_2CH_2OCH_2PO(OR)_2$.

R = Me. V.7. b_3 157-159°, n_D^{20} 1.4460, d_4^{20} 1.2260.[1675]

R = Et. V.7. b_2 155-157°, n_D^{20} 1.4440, d_4^{20} 1.1401.[1675]

$CH_3OCH_2CH(-CN)PO(OEt)_2$. V.14. b_1 114-116°.[1651]
$H_2C=CHC(O)NHCH_2PO(OR)_2$.

R = Me. V.9.[837]

R = Et. V.9.[837]

R = Pr. V.9.[837]

R = Bu. V.9.[837]

$H_3CC(O)NHCH_2C(O)-PO(OEt)_2$. V.5. m. 155°.[174]
$H_3CO_2CN=C(OMe)-PO(OEt)_2$. V.14. $b_{0.04}$ 55-57°, n_D^{20} 1.4300, d_4^{20} 1.1851.[490]
$C_2H_5O_2C-NHC(O)-PO(OPr\text{-}i)_2$. V.9. m. 73-76°.[463]
$H_2C=CH-CH_2-NHC(S)PO(OR)_2$.

R = Et. V.9. $b_{0.07}$ 125-126°.[1489]

R = i-Pr. V.9. $b_{0.05}$ 123-124°.[1489]

$NC-\overset{\ominus}{C}(-\overset{\oplus}{S}Me_2)-PO(OEt)_2$. V.14. m. 72-73°, IR, ¹H.[430]

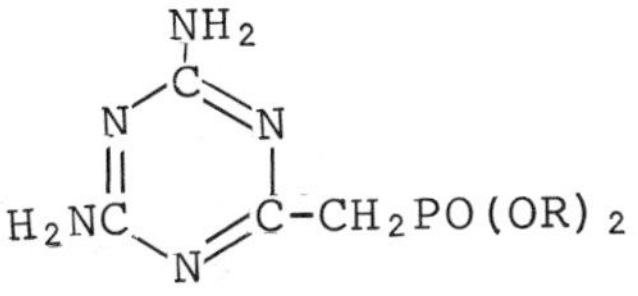

R = Me. V.5. m. 220°.[1991]

R = Et. V.5. m. 223°.[1991]

$(CH_3)_2$C=CH-PO(OR)$_2$.

R = Me. V.14. $b_{2.6}$ 75-76°, n_D^{25} 1.4502, IR, ^{1}H.[1191] V.14. $b_{0.08}$ 86-87°.[1346]

R = H_2C=CH-CH_2-. V.2. b_{3-4} 110-115°, n_D^{25} 1.4666.[2021a]

R = H_2C=CMe-CH_2-. V.2. b_{1-2} 110-120°, n_D^{26} 1.4667.[2021a]

$(CH_3)_2$C=CHPO(OEt)OCH=CCl_2. V.13 b_{10} 131-139°, n_D^{20} 1.4780, d_4^{20} 1.2260.[937]

CH_3CH=CHCH_2PO(OR)$_2$.

R = Et. V.5. b_7 100.5°, n_D^{20} 1.4379, d_4^{20} 1.0214.[1558] V.6. b_{12} 112°.[1616] V.5.[1556]

CH_3CH=CHCH_2PO(OCH=CCl_2)OR.

R = i-Pr. V.13. $b_{0.04}$ 86-88°.[1102]

R = Bu. V.13. $b_{0.06}$ 100-102°.[1102]

R = i-Bu. V.13. $b_{0.07}$ 93-95°.[1102]

C_2H_5CH=CHPO(OEt)OH. III.3. Dicyclohexylammonium salt, m. 185-186°.[770]

C_2H_5CH=CHPO(OEt)$_2$ (cis). V.14. b_1 69.5°, n_D^{20} 1.4400, d_4^{20} 1.0132, ^{1}H.[1479]

H_2C=CH-CHMePO(OEt)$_2$. V.5. b_{14} 104-105°, n_D^{20} 1.4323, d_4^{20} 1.0139, IR.[1558] V.5.[1091]

CH_3CH=CMePO(OEt)$_2$ (cis). V.14. b_7 84-85°, n_D^{20} 1.4381, d_4^{20} 1.0210, ^{1}H.[1479]

CH_3CH=CMePO(OEt)$_2$. V.5.[1556] V.14. b_{12} 109°, n_D^{20} 1.4405.[757]

H_2CCH_2-NH-N=C-CH_2PO(OEt)$_2$. V.14. $b_{0.8}$ 130-132°, n_D^{20} 1.4769, d_4^{20} 1.1590.[1578] V.14. $b_{0.005}$ 129-130°, n_D^{20} 1.4758, d_4^{20} 1.1306, IR, ^{1}H.[1580]

CH_3CHCH_2-NH-N=C-PO(OEt)$_2$. V.14. $b_{0.01}$ 130°, n_D^{20} 1.4765, d_4^{20} 1.3335, IR, ^{1}H.[1580]

CH_3C(O)CH_2CH_2PO(OR)$_2$.

R = Et. V.13.[1364] V.7. $b_{0.15}$ 96°, n_D^{28} 1.4345.[751] V.14. $b_{0.3}$ 104°, n_D^{21} 1.4382.[1082] ^{1}H, ^{31}P -30.8 ppm.[1093] V.6. $b_{0.3}$ 104°, n_D^{21} 1.4382, d_4^{21} 1.090.[1958] V.14. $b_{0.5}$ 103°, IR.[1959] V.14. b_5 119-123°, n_D^{20} 1.4378, d_4^{20} 1.0833.[825] V.7. $b_{0.05}$ 81-83°.[1750]

R = Pr. V.7. $b_{0.05}$ 95-97°.[1750]

R = Bu. V.6. $b_{1.5}$ 134-136°, n_D^{20} 1.4390, d_4^{20} 1.0476, IR.[1431] V.8.[840] V.7. b_8 180-182°.[1750]

R = i-Bu. V.7. $b_{0.01}$ 104-105°.[1750]

R = Ph. V.1. $b_{0.0003}$ 95-112°, n_D^{20} 1.5575, d_4^{20} 1.222.[493] V.7. $b_{0.01}$ 170-172°.[1750]

R = $C_{10}H_{21}$-. V.2. $b_{0.0001}$ 120-170° (!?), n_D^{20} 1.4528, d_4^{20} 0.9287.[493]

$C_2H_5C(O)CH_2PO(OEt)_2$. V.1, $b_{0.3}$ 94°, n_D^{20} 1.4370.[1082] V.14. $b_{0.5}$ 100°, IR.[1959] ^1H, ^{31}P -20,1 ppm.[1094] V.14. b_1 103°, n_D^{21} 1.4378, d_4^{21} 1.081.[1961]

$(CH_3)_2CHC(O)PO(OEt)_2$. V.5, b_4 99-100°, IR, ^1H; as 2,4-dinitrophenylhydrazone, m. 126-127°.[244]

$CH_3C(O)CHMe$-$PO(OEt)_2$. V.14.[148] V.14. b_6 115°, IR.[1959] V.14. $b_{0.5}$ 88°, $n_D^{20.5}$ 1.4369, $d_4^{20.5}$ 1.089.[1961]

CH_3OCMe=$CHPO(OEt)_2$. V.14. b_9 125°.[1599]

CH_3OCH=CMe-$PO(OR)_2$.

R = Et. V.2. $b_{0.035}$ 52°, n_D^{20} 1.4480, d_4^{20} 1.0704.[1350]

R = Pr. V.2. $b_{0.055}$ 74°, n_D^{20} 1.4500, d_4^{20} 1.0331.[1350]

C_2H_5OCH=$CHPO(OR)O^{\ominus}\overset{\oplus}{N}H_4$.

R = Et. III.4. m. 130-132°.[1481]

R = i-Pr. III.4. m. 114-115°.[1481]

R = i-C_8H_{17}-. III.4. m. 138-140°.[1481]

C_2H_5OCH=$CHPO(OPh)OH$. III.4. m. 82-84°.[1482]

C_2H_5OCH=$CHPO(OC_6H_{13})O^{\ominus}]H_3\overset{\oplus}{N}C_6H_4Me$. III.4. m. 67-70°.[1481]

C_2H_5OCH=$CHPO(OEt)Cl$. IV.1. $b_{0.3}$ 92°.[1347]

C_2H_5OCH=$CHPO(OR)_2$.

R = Me. V.2. b_2 87°.[81] V.5, V.2.[722] V.2. b_1 86°.[83]

R = Et. V.5. $b_{0.05}$ 76-77°, n_D^{20} 1.4815, d_4^{20} 1.0814.[1347] ^1H, ^{31}P -21.4 ppm.[1093] V.2. b_1 100°.[83]

R = i-Pr. V.14. $b_{0.001}$ 70-72°, n_D^{22} 1.4335.[1788]

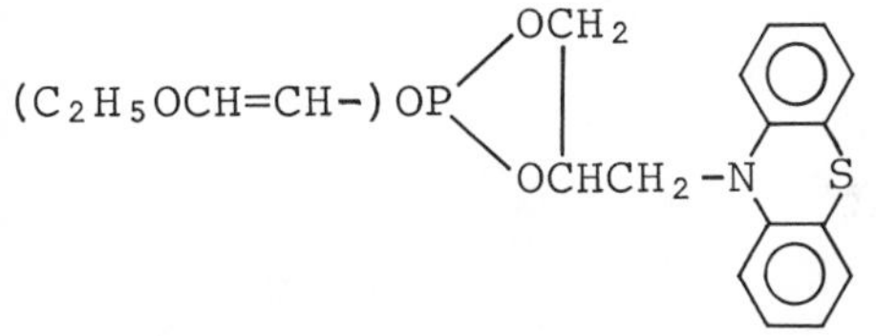
V.3. m. 188-190°.[227]

H_2C=CH-O-$CH_2CH_2PO(OR)_2$.

R = Me. V.5. b_{25} 135-136°.[1679]

R = Et. V.5. b_{17} 138°, n_D^{27} 1.4351.[1679] V.5.[1681]

R = i-Pr. V.5. b_6 121-122°.[1679] V.5.[1681]

CH_3CH=$CHCH(OH)PO(OEt)_2$. V.8. b_9 163-164°.[1542]

H_2C=$CHMe(OH)$-$PO(OEt)_2$. V.6. $b_{0.5}$ 105°.[1958]

OCH_2-CEt-$PO(OEt)_2$. V.14. $b_{0.5}$ 84°, n_D^{20} 1.4357, d_4^{20} 1.096.[1958]

$OHCCH_2CHMe$-$PO(OEt)_2$. V.7. $b_{0.05}$ 99°, n_D^{30} 1.4320.[751]

$OHCCHMeCH_2PO(OEt)_2$. V.14. $b_{0.4}$ 95°, n_D^{20} 1.4394.[1082]

$OHCCHEtPO(OR)_2$.

R = Et. V.14. $b_{0.5}$ 83-84°; 2,4-dinitrophenylhydrazone, m. 122-123°.[1502] V.14. $b_{0.065}$ 59°, n_D^{20} 1.4370, d_4^{20} 1.0780.[1350]

R = Pr. V.14. $b_{0.8}$ 107-108°; 2,4-dinitrophenylhydrazone, m. 130°.[1502]

R = Bu. V.14. $b_{0.18}$ 104-105°, n_D^{20} 1.4436, d_4^{20} 1.0073; 2,4-dinitrophenylhydrazone m. 90-92°.[1350]

$C_2H_5SCH_2C(O)PO(OEt)_2$. V.5. b_3 94°, n_D^{20} 1.4462, d_4^{20} 1.0960.[2112]

$C_2H_5SC(O)CH_2PO(OR)_2$.

R = Et. V.5. $b_{0.05}$ 104-106°, n_D^{20} 1.4705, d_4^{20} 1.1509.[650] V.5. $b_{0.2}$ 115-120°.[1239] V.5. n_D^{20} 1.4701, d_4^{20} 1.1568.[1109]

R = Pr. V.5. $b_{0.05}$ 100°.[651]

R = Bu. V.5. b_2 155°.[651]

$HO_2C(CH_2)_3PO(OH)_2$. I.2. m. 127-128.5° (water).[1419,1420]
$CH_3CO_2CH_2CH_2PO(OEt)_2$. V.5. b_2 162°.[45] V.7. $b_{0.7}$ 96-97°.[427]
$CH_3O_2CCH_2CH_2PO(OMe)Cl$. IV.2. $b_{0.3}$ 84°.[1712]
$CH_3O_2CCH_2CH_2PO(OR)_2$.

R = Me. V.7. b_1 102-103°.[2032] V.7. b_{10} 137-138°.[1546]

R = Et. V.14. b_{15} 156°, n_D^{20} 1.4340.[1414] V.7. $b_{0.4}$ 103-103.2°.[1074]

R = cycl.-C_6H_{11}-. V.7. b_{7-8} 174-176°.[1643]

R = $PhCH_2$-. V.7. $b_{0.001}$ 143-145°.[1643]

$CH_3O_2CCH_2CH_2PO(OBu)(-OC_9H_{19})$. V.7. $b_{0.009}$ 50-55°.[1162]
$(CH_3O_2CCH_2CH_2-)OPO-R-O$ (cyclic).

R = $-CHMeCH_2-$. V.7. b_2 169°, n_D^{20} 1.4590, d_4^{20} 1.268.[1588]

R = $-(CH_2)_3-$. V.7. $b_{1.5}$ 158-162°, n_D^{20} 1.4420, d_4^{20} 1.2190.[1588]

R = -CHMeCHMe-. V.7. b_4 181-182°, n_D^{20} 1.4528, d_4^{20} 1.2182.[1588]

R = $-CHMeCH_2CH_2-$. V.7. b_2 188°, n_D^{20} 1.4575, d_4^{20} 1.2140.[1588]

$[C_2H_5O_2CCH_2PO(OMe)O^{\ominus}]\overset{\oplus}{N}Me_4$. III.3. m. 62°.[353]
$[C_2H_5O_2CCH_2PO(OEt)O^{\ominus}]H_3\overset{\oplus}{N}C_6H_{11}$-cycl. III.2. m. 53-54°.[1172]
$C_2H_5O_2CCH_2PO(OEt)Cl$. IV.1. m. 13-15°, n_D^{20} 1.4430, d_4^{20} 1.2232.[1172]
$C_2H_5O_2CCH_2PO(OR)_2$.

R = Me. V.12. $b_{0.4}$ 91-92°.[1578]

R = Et. V.12. $b_{0.4}$ 100-103°.[1578] V.5.[102] V.5. b_{12} 149-150°.[112a] V.5. V.6.[1418] V.6. b_{14-15} 147-149°.[1420] V.6.[1003] V.6. b_{10} 138.5°.[112b] V.6. b_{9-10} 140-141°.[114] V.5. n_D^{20} 1.4320, $b_{0.05}$ 72-80°.[112] V.14. $b_{0.05}$ 82-83°.[1346] V.14. $b_{0.1}$ 102-105°.[1707] V.2. $b_{0.09}$ 84-85°.[1348] V.6. b_3 124-125°.[1172]

R = Pr. V.12. $b_{0.1}$ 104-106°.[1578]

R = i-Pr. V.5. $b_{0.07}$ 96-97°, n_D^{20} 1.4285, d_4^{20} 1.0700.[281]

R = i-Bu. V.5. b_{10} 170-171°, d_0^{17} 1.0212.[119a]

R = Am. V.5. b_2 157-160°, n_D^{20} 1.4383, d_4^{20} 1.0168.[819]

R = C_6H_{13}-. V.5. b_2 164-167°, n_D^{20} 1.4401, d_4^{20} 0.9973.[819]

R = $C_{10}H_{21}$-. V.5. m. 83.5-84°.[819]

$C_2H_5O_2CCH_2PO(OMe)(-OCH_2C_6H_3Cl_2$-2.4). V.1. $b_{0.05}$ 145-148°, ^1H.[1994]

$(C_2H_5O_2CCH_2-)OPOCH_2C(CH_2Br)Et-CH_2O$ (ring). V.e. m. 50°.[2059]

$CH_3O_2CCHMePO(OEt)Cl$. IV.1. $b_{0.07}$ 95-96°, n_D^{20} 1.4480, d_4^{20} 1.2480.[280]

$CH_3O_2CCHMePO(OEt)(-NCO)$. IV.2. $b_{0.06}$ 82-84°.[280]

$CH_3O_2CCHMePO(OEt)_2$. V.14. $b_{0.05}$ 74-76°, n_D^{20} 1.4330, d_4^{20} 1.1274.[280]

$CH_3O_2CCHMePO(OEt)(OPh)$. V.14. $b_{0.07}$ 124-125°, n_D^{20} 1.4981 d_4^{20} 1.2240.[280]

$CH_3CH(-CO_2Me)-PO(OEt)_2$. V.5. b_1 95°.[432]

$HO_2CCMe_2PO(OH)_2$. By oxidation of $MeC(O)CH_2CMe_2PO(OH)$ with fuming HNO_3, cryst.[95,1260a]

$CH_3CH[-O_2CMe]PO(OR)_2$.

R = Me. V.14. b_2 85-86°, n_D^{20} 1.4302, d_4^{20} 1.1908.[1589]

R = Et. V.14. b_8 124.5-125°, n_D^{20} 1.4260, d_4^{20} 1.1030.[1589] V.6. b_{12} 130-132°.[1617] V.14.[1617] V.7.[1622]

R = Pr. V.14. b_8 140-143°, n_D^{20} 1.4290, d_4^{20} 1.0543.[1589]

R = i-Pr. V.14. b_2 88.5°, n_D^{20} 1.4251, d_4^{20} 1.0530.[1589]

R = Bu. V.14. b_2 121.5-122°, n_D^{20} 1.4340, d_4^{20} 1.0335.[1589]

R = i-Bu. V.14. b_{10} 148-149°, n_D^{20} 1.4300, d_4^{20} 1.0340.[1589]

$CH_3C(O)CH(-OMe)PO(OEt)_2$. V.5. $b_{0.005}$ 65-70°, IR, ^1H.[684]

$C_2H_5OCH_2C(O)PO(OEt)_2$. V.5. b_{10} 102-104°, n_D^{20} 1.4158, d_4^{20} 1.0966.[2112]

$CH_3C(O)CMe(OH)PO(OR)_2$.

R = Me. V.8. b_4 116°.[1056]

R = Et. V.8.[15] V.8. b_{10} 138-140°.[1056]

R = Bu. V.8. b_4 142°.[1056]

$(CH_3O)_2C{=}CHPO(OMe)_2$. V.14. $b_{0.08}$ 86-87°, n_D^{20} 1.4489, d_4^{20} 1.2081.[1346]

$\overline{OCH_2CH_2O}CMePO(OMe)_2$. V.5. b_{12} 117°, IR.[718]

$\overline{OCH_2CH}CH_2OCH_2PO(OR)_2$.

R = Et. V.14.[1441]

R = Pr. V.14.[1441]

R = i-Pr. V.14.[1441]

R = Bu. V.14.[1441]

$CH_3O_2CCH(OMe)PO(OR)_2$.

R = Me. V.5. $b_{0.02}$ 83-85°, n_D^{21} 1.4427, IR.[717]

R = Et. V.5. b_8 154-155°, $n_D^{21.5}$ 1.4382, IR, ^{31}P -14.9 ppm.[717]

$C_2H_5OC(O)OCH_2PO(OEt)_2$. V.14. $b_{11.5}$ 152-153°.[127]

$C_2H_5SCH{=}CHPO(OR)_2$.

R = Pr. V.2. b_2 147°.[86]

R = i-Pr. V.14. n_D^{15} 1.4830.[1788]

$HSCH_2CH{=}CHCH_2PO(OEt)_2$. V.14. $b_{0.1}$ 98-100°.[63]

$\overline{O(CH_2)_4P}O(OH)$. I.13. m. 104.5-106°.[499]

$\overline{O(CH_2)_4P}O(OR)$.

R = Et. V.14. $b_{0.5}$ 88-90°.[173] V.14. $b_{0.08}$ 56°, IR.[499] V.5. $b_{0.008}$ 75-77°.[1649] V.4. $b_{0.8}$ 89-90°.[1938]

R = Pr. V.5. $b_{0.007}$ 70-71°.[1649]

R = Bu. V.5. $b_{0.008}$ 75-77°.[1649]

$\overline{H_2CCH_2}N-(CH_2)_2PO(OEt)_2$. V.14.[1177] V.14. $b_{0.01}$ 52-56°, n_D^{20} 1.44505.[1176]

$(CH_3)_2NC(O)CH_2PO(OEt)_2$. V.5. $b_{0.15}$ 139-140°, n_D^{20} 1.4581, d_4^{20} 1.1335.[1108]

$C_2H_5NHC(O)CH_2PO(OEt)_2$. V.5. $b_{0.2}$ 141°, n_D^{20} 1.4549, d_4^{20} 1.1304.[1108]

$H_2NC(O)CH_2CHMePO(OEt)_2$. V.7. m. 72-76°.[210]

$H_2NC(O)CHMeCH_2PO(OEt)_2$. V.7. b_{13} 146-147°.[1671] V.14. m. 75-77°.[210]

$C_2H_5O_2CNHCH_2PO(OR)_2$.

R = Et. V.14. $b_{0.15}$ 126-127°, n_D^{20} 1.4460, d_4^{20} 1.1531.[1903]

R = Pr. V.14. $b_{0.02}$ 122-124°, n_D^{20} 1.4490, d_4^{20} 1.1138.[1903]

$H_2NCO_2(CH_2)_3PO(OMe)_2$. V.7. m. 50-52°.[1463]
$CH_3C(O)NHCH_2CH(OH)PO(OEt)_2$. V.14. $b_{0.25}$ 160°, IR, ^1H.[691]
$H_2NC(O)CH(-OEt)PO(OEt)_2$. V.5. m. 90-94°, IR, ^1H.[684]
$HO_2CCH(NH_2)-(CH_2)_2PO(OH)_2$. I.13. m. 226° (dec.).[357]
$HO_2CCH_2NMeCH_2PO(OH)_2$. I.13. m. 223-224°.[2113]
$O_2NCHMe-CMe(OH)PO(OEt)_2$. V.14. $b_{0.0035}$ 140-142°.[1214]
$CH_3CH(-OEt)PO(OEt)Cl$. From MeCHCl-OEt + $(EtO)_2PCl$, $b_{0.02}$ 57°.[608]
$H_2C=CH-C(OH)MePO(OEt)_2$. V.6. $b_{0.5}$ 105°, n_D^{20} 1.4485, d_4^{20} 1.089.[1958]
$C_4H_9PO(OH)_2$. I.2.[1011] I.2. m. 101-103°.[108] I.2. m. 103.5-104°.[1000,1002] I.2. m. 102°,[737] ^{31}P -32 ppm.[1189a]
$C_4H_9PO(OR)OH$.

R = Et. III.4. b_1 147-149°.[326]

R = Bu. Complexes MeL_3 with Ti(III), V(III), Cr(III).[1312]

$C_4H_9PO(OC_6H_{11}$-cycl.)Cl. IV.2. $b_{0.1}$ 90°, n_D^{25} 1.4680.[740]
$C_4H_9PO(OR)F$.

R = i-Pr. Calculation of the electronic structure.[1075]

R = cycl.-C_6H_{11}-. IV.2. $b_{0.15}$ 77°, n_D^{25} 1.4391.[740]

$C_4H_9PO(OEt)NCO$. IV.5. $b_{0.5}$ 122-124°.[726]
$C_4H_9PO(OR)_2$.

R = Et. V.7. b_{28} 121°.[747] V.5. $b_{0.1}$ 53-55°.[1366] V.5. b_1 74°, n_D^{17} 1.4244,[552] ^{31}P -32.3 ppm.[1189a]

R = FH_2CCH_2-. From $P(OCH_2CF_3)_3$ + LiBu or BuMgBr, b_2 53-53.3°, n_D^{20} 1.3310.[1211]

R = Bu. V.6.[2079] V.6. b_{10} 150-151°. V.6. b_{20} 160-162°.[1000,1002] In 6% yield by autoxidation of Bu_3P.[314] μ = 2.89 D.[1216] Magn. rotation of bonds.[1069] V.2. b_2 84°, n_D^{20} 1.4320, d_4^{20} 0.9242.[286] ^{31}P -30.9 ppm.[1189a]

R = i-Bu. Complexes with 3 d-metals.[938]

R = EtO_2CCH_2-. V.13. b_1 148-149°.[1646]

R = Am. V.2. b_4 127-130°, n_D^{20} 1.4371, d_4^{20} 0.9131.[286]

R = i-Am. V.2. b_4 130°, n_D^{20} 1.4340, d_4^{20} 0.9025.[286]

$C_4H_9PO(OMe)(OCMe=\overline{C-CH_2CH_2OC}=O)$. V.13. $b_{0.007}$ 148°, n_D^{20} 1.4878.[635]
$C_4H_9PO(OEt)(OR)$.

R = EtO_2CCH_2-. V.13. b_1 122-123°.[1646]

R = $O=\overline{C-OCH_2CH_2C}=CMe$-. V.13. $b_{0.001}$ 147°, n_D^{20} 1.4818.[635]

R = $O{=}COCH_2CH_2C{=}C(-CO_2Et)-$. V.13. $b_{0.005}$ 160°, n_D^{20} 1.4803.[635]

R = $O{=}COCHMeCH_2C{=}CMe-$. V.13. $b_{0.001}$ 140°, n_D^{20} 1.4767.[635]

R = $O{=}COCHMeCH_2C{=}C(-CO_2Et)-$. V.13. $b_{0.005}$ 163°, n_D^{20} 1.4765.[635]

R = $O{=}COCHMeCH_2CH_2C{=}CMe-$. V.13. $b_{0.005}$ 164°, n_D^{20} 1.4803.[635]

R = $O{=}COCHMeCH_2CH_2C{=}C(-CO_2Et)$. V.13. $b_{0.008}$ 164° n_D^{20} 1.4887.[635]

R = $H_2CCH_2O_2C-C{=}CMe-$. V.5. $b_{0.001}$ 147°, n_D^{20} 1.4818.[318]

$C_4H_9PO(OBu)(-OCBu{=}CCH_2CH_2OC{=}O)$. V.13. $b_{0.005}$ 150°, n_D^{20} 1.4750.[635]

$C_4H_9PO(OAm\text{-}i)OR$.

R = $C_6H_{13}-$. V.6. $b_{0.4}$ 132-134°.[1161]

R = $C_7H_{15}-$. V.6. $b_{0.3}$ 137-138°.[1161]

R = sec. $C_8H_{17}-$. V.6. $b_{0.1}$ 128-130°.[1161]

$(C_4H_9-)OP-O-R-O$.

R = $-CH_2CH_2-$. V.13. b_1 130-132°.[1649]

R = $-CHMeCH_2-$. V.5. $b_{0.2}$ 121.5-123°. n_D^{20} 1.4482.[120]

R = $-CH_2C(-CH_2J)Me-CH_2-$. V.5. m. 55°.[2059]

$(C_4H_9-)(EtO)OP-O(CH_2)_4O-PO(OEt)(-C_4H_9)$. V.2. $b_{0.005}$ 167°.[19]

$(BuO)(O)C_4H_9-POCH_2PO(OBu)_2$. V.8. $b_{1.1}$ 161-171°.[755]

$C_4H_9PO(OSiMe_3)_2$. VI.1.[951]

$(CH_3)_2CHCH_2PO(OH)_2 \cdot 1/2\ H_2O$. I.2. m. 124°.[112b] I.6.[729,791]

$(CH_3)_2CHCH_2PO(OR)_2$.

R = F_3CCH_2-. V.5. b_2 63-64°, n_D^{20} 1.3585.[1211]

R = i-Bu. V.5. b_{10} 133.5-134°, d_4^{20} 0.9459.[112b]

$(C_2H_5)(CH_3)CH-PO(OH)_2$. I.2. m. 118.5-119°.[446]

$(C_2H_5)(CH_3)CH-PO(OR)_2$.

R = Bu. V.7. b_{13} 143-146°.[1954]

R = Am. V.2. b_3 142°, n_D^{20} 1.4361, d_4^{20} 0.9423.[286]

R = i-Am. V.2. b_3 132°, n_D^{20} 1.4375, d_4^{20} 0.9528.[286]

$(CH_3)_3CPO(OH)_2$. I.1. m. 191.5-192° (ligroine/acetic acid).[446] ^{31}P -37 ppm.[1189a]

$(CH_3)_3CPO(OEt)OH$. III.2 $b_{0.03}$ 72-76°, n_D^{25} 1.4253.[338]

$(CH_3)_3CPO(OR)F$.

R = Me. IV.2. $b_{0.1}$ 49°.[1860]

R = Et. IV.2. $b_{0.1}$ 55°.[1860]

R = i-Pr. IV.2. $b_{0.1}$ 60°.[1860]

$[(CH_3)_3C-]$OPOCMe=CMeO. By heating of $SCH_2CH_2OP(CMe_3)OCMe=CMeO$, m.79-80°.[1951]

$CH_3NHC(O)-NMe-CH_2PO(OH)_2$. I.1/I.2. m. 154°.[1462]

$C_2H_5OCH_2CH_2PO(OEt)_2$. V.14. b_3 87°, n_D^{20} 1.4258, d_4^{20} 1.0425.[1640]

$C_2H_5OCHMe-PO(OEt)_2$. V.8. b_1 69-70°, n_D^{20} 1.4218, d_4^{20} 1.1328.[830]

$(C_2H_5OCHMe-)O$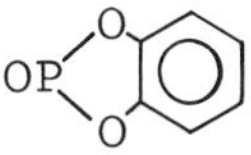 V.13. $b_{0.03}$ 112-113°, n_D^{20} 1.5060, d_4^{20} 1.2145.[609]

$C_3H_7CH(OH)PO(OH)_2$. I.2. m. 154.5-155°.[963]

$C_3H_7CH(OH)PO(OR)_2$.

R = Me. V.8. b_{15} 143-145°.[40]

R = Et. V.8. $b_{0.3}$ 111-112°, n_D^{20} 1.4378.[963] V.13. b_1 126°.[530]

R = Bu. V.8. b_6 168-170°.[18]

R = $H_2C(CH_2)_3C(-CCl_3)-$. V.13. m. 136-137°.[33]

$(C_2H_5)(CH_3)C(OH)-PO(OH)_2$. I.1. I.9. m. 158-159° (MeOH/Me_2CO).[415,1183,1184,1185,1186]

$(C_2H_5)(CH_3)C(OH)-PO(OR)_2$.

R = Me. V.8. b_{13} 141-142°.[40]

R = Et. V.8. n_D^{20} 1.4331, d_4^{20} 1.0540.[1589]

R = i-Pr. V.8. b_{13} 138-140°.[35]

R = Ph. V.1. m. 128.5 (EtOH).[418]

$(CH_3)_2CHCH(OH)-PO(OH)_2$. I.2. m. 165-166°.[963] I.9. m. 168-169°.[555]

$(CH_3)_2CHCH(OH)-PO(OEt)_2$. V.8. $b_{0.3}$ 96-97°, n_D^{20} 1.4392.[963] V.8. b_3 112-113°, n_D^{20} 1.4390, d_4^{20} 1.0758.[1589]

$CH_3OCH_2CH(OMe)PO(OR)_2$.

R = Et. V.14. b_8 130-131°.[1651]

R = Bu. V.14. b_9 147-148°.[1651]

$CH_3OCH_2CH(OH)CH_2PO(OEt)_2$. V.14. $b_{0.16}$ 105°, ^1H.[691]

$C_2H_5OCH_2CH(OH)-PO(OR)_2$.

R = Me. V.14. $b_{0.3}$ 105°.[691]

R = Et. V.14. $b_{0.25}$ 110°, ^1H.[691]

$C_2H_5-S(O)_2-CH_2CH_2PO(OR)_2$.

R = Me. V.7. b_3 180°.[1662]

R = Et. V.7. $b_{1.5}$ 184°.[1662]

R = Pr. V.7. b_2 190°.[1662]

R = Bu. V.7. b_3 208°.[1662]

$CH_3OS(O)OCMe_2PO(OMe)_2$. V.14. b_2 109-110°, n_D^{20} 1.4497, d_4^{20} 1.2778.[1589]

$H_2N(CH_2)_4PO(OH)_2$. I.2. m. 133-134°, pK_1 2.55, pK_2 7.55, pK_3 10.9.[366,1776a]

$C_3H_7CH(-NH_2)-PO(OH)_2$. I.13. m. 262-264°.[2040]

$(CH_3)_2NCH_2CH_2PO(OEt)_2$. V.14. b_1 100°, n_D^{25} 1.4329[329]. V.2. $b_{0.0007}$ 88-90°.[895]

$(CH_3)_2NCH_2CH_2PO(OEt)OCH_2CH_2SEt$. V.2. $b_{0.0007}$ 88-90°.[895]

$(CH_3)_2CHCH(-NH_2)PO(OH)_2$. I.2. m. 271°.[789]

$(CH_3)_2CHCH(-NH_2PO(OEt)_2$. V.14. $b_{0.05}$ 74°.[789]

$CH_3CH(-NHEt)PO(OEt)_2$. V.8. b_1 84°.[529]

$(CH_3O-)(Me_2N-)CHPO(OR)_2$.

R = Et. V.14. $b_{0.01}$ 65-66°, n_D^{20} 1.4370, ^1H.[715a] V.8. $b_{0.02}$ 65°.[714]

R = i-Pr. $b_{0.01}$ 62°.[714]

R = Ph. m. 69-76° (not pure).[714]

$(CH_3)_2NS(O)_2CH_2CH_2PO(OMe)_2$. V.7. b_1 184-185°, n_D^{20} 1.4690.[491]

$(MeO)_2OPOCHMe-PO(OMe)_2$. V.14. $b_{0.07}$ 126-128°, n_D^{25} 1.4300, d_4^{25} 1.269.[333] The old structure as diphosphonic ester is not correct. By distillation the diphosphonic ester is transformed to the above cited structure.[537a]

$[Me_2\overset{\oplus}{S}CH_2CH_2PO(OH)_2]Br^{\ominus}$. I.13. m. 133-136°.[628]

$[Me_3\overset{\oplus}{N}CH_2PO(OEt)_2]Cl^{\ominus}$. V.14. m. 178°.[478]

$(CH_3)_3SiOCH_2PO(OEt)_2$. V.14. b_9 113°, n_D^{20} 1.4205, d_4^{20} 1.0116.[1436]

$(CH_3)_3SiCH_2PO(OR)_2$.

R = Et. V.5. b_3 79-80°.[637] V.6. b_{22} 118.5-121°.[344]

R = Bu. V.6. b_{17} 154-158°.[344] V.6. b_{17} 154-158°.[509]

R = Am. V.6. b_{18} 186-188°, n_D^{20} 1.4418.[344]

$\underline{C_5}$

$(CH_3)_2AsOCHEtPO(OEt)_2$. V.8. $b_{0.1}$ 89-90°.[1742]

$C_3H_7O_2CCHBrPO(OEt)_2$. V.14. b_2 136-142°, n_D^{20} 1.4640, d_4^{20} 1.3330.[697]

$(CH_3)_3SiCHBrCH_2PO(OEt)_2$. V.5. $b_{1.5}$ 95-98°, n_D^{20} 1.4562, d_4^{20} 1.1433.[920]

$F_2\overline{CCF_2CF_2CCl=C}-PO(OH)_2$. I.1. m. 107-109°.[566] I.1, I.2.[562]

$F_2\overline{CCF_2CF_2CCl=C}-PO(OR)_2$.

R = Me. V.5. $b_{4.5}$ 64-66°, n_D^{25} 1.3817, d_4^{20} 1.6669.[563,566]

R = Et. V.5. $b_{1.8}$ 87-88°, n_D^{25} 1.3998.[563,566] I.2. $b_{1.5}$

90-92° n_D^{24} 1.4010 , IR.[562]

R = Bu. V.5. $b_{0.0001}$ 48-51°(bath temp.), n_D^{25} 1.4091.[563,566]

$CH_3C(O)C(=CMeCl)PO(OBu)_2$. V.2. b_3 159-160°.[1133]
$CH_3OCH=C(-CCl=CH_2)-PO(OMe)_2$. V.2. $b_{0.01}$ 90-91°.[1920]
$C_2H_5C(O)C(=CHCl)-PO(OEt)_2$. V.2. $b_{1.5}$ 117°, n_D^{20} 1.4670 , d_4^{20} 1.1604.[1133]
$CH_3SCH=C(-CCl=CH_2)-PO(OR)_2$.

R = Me. V.2. $b_{0.007}$ 78°, n_D^{20} 1.5791, d_4^{20} 1.3011.[1918]

R = Et. V.2. $b_{0.005}$ 80°, n_D^{20} 1.5498, d_4^{20} 1.2121.[1918]

$(CH_3)_2CHO_2CN=CClPO(OMe)_2$. V.2. $b_{0.06}$ 98-100°, n_D^{20} 1.4521, d_4^{20} 1.2631.[490]
$ClH_2CCHMe-OCH=CHPO(OR)_2$.

R = Pr. V.2. b_2 138°, n_D^{20} 1.4638°, d_4^{20} 1.1024.[586]

R = $H_2C=CHCH_2$-. V.2. b_2 153°, n_D^{20} 1.4870, d_4^{20} 1.1542.[586]

R = Bu. V.2. b_3 173°, n_D^{20} 1.4598 , d_4^{20} 1.0520.[586]

R = i-Bu. V.2. b_3 157°, n_D^{20} 1.4612 , d_4^{20} 1.0708.[586]

R = i-Am. V.2. b_2 164 , n_D^{20} 1.4657 , d_4^{20} 1.0495.[586]

$C_2H_5CO_2CHClCH_2PO(OR)_2$.

R = Me. V.2. m. 13-14°.[1134]

R = Et. V.2. m. 25-26°.[1134]

$H_2C=C(-CH_2Cl)CCl=CHPO(OR)_2$.

R = Me. V.2. b_3 136°.[89]

R = Et. V.2. b_2 142.5°.[89]

R = Pr. V.2. b_2 141°.[89]

$C_2H_5O_2C-NHC(O)CCl_2PO(OEt)_2$. V.9. $b_{0.4}$ 140-141°.[281]
$C_3H_7O_2C-CCl_2PO(OEt)_2$. V.14. $b_{0.07}$ 97-98°, n_D^{20} 1.4485 , d_4^{20} 1.2667.[281]
$CH_3O_2CCCl_2CMe(OH)-PO(OMe)_2$. V.8. m. 79-82°.[1995]
$(ClCH_2CH_2O)_2CHPO(OCH_2CH_2Cl)_2$. From $HC(OCH_2CH_2Cl)_2$ and PCl_3, $b_{0.06}$ 180-181°.[713]
$H_2C=CH-CO_2CH(-CCl_3)PO(OEt)_2$. V.14. $b_{0.065}$ 120-121°, n_D^{20} 1.4750, d_4^{20} 1.3516.[604]
$C_2H_5-CO_2CH(-CCl_3)PO(OCH_2CH_2Cl)_2$. V.14. $b_{0.013}$ 143-145°.[1400,347]
$C_2H_5OC(O)OCH(-CCl_3)PO(OCH_2CH_2Cl)_2$. V.14. $b_{0.012}$ 166-168°.[1400]
$CH_3-CO_2CH(-CCl_3)OCH_2PO(OEt)_2$. V.14. b_1 139-140°.[99]
$Cl_3CCH(OH)NH-C(O)CH_2CH_2PO(OMe)_2$. V.14. m. 140-141°.[399]
$C_2H_5NH-CO_2CH(-CCl_3)PO(OMe)_2$. V.14. m. 97-98°.[987]
$(CH_3)_2NS(O)_2NH-CO_2CH(-CCl_3)PO(OMe)_2$. V.14. m. 57°.[1999]
$(MeO)_2OPNH-CO_2CH(-CCl_3)PO(OMe)_2$. V.14. m. 38-40°.[467]

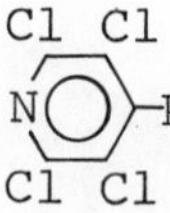
PO(OR)$_2$.

R = Et. V.5. b_1 159-160°.[858]

R = Pr. V.5. $b_{0.35}$ 161-163°.[858]

R = Bu. V.5. $b_{0.08}$ 154-156°.[858]

Cl_2C-CCl=CCl-C(O)-CCl-PO(OEt)$_2$. V.5. $b_{0.77}$ 152-160°.[1743]

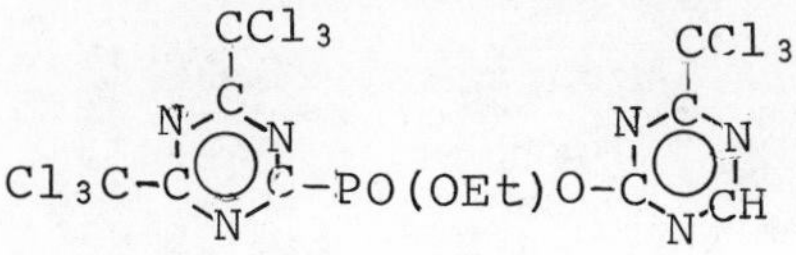
. V.5/V.13. b_1 118°, n_D^{29} 1.5230.[1867]

F_2C-CF_2C(-OMe)=C-PO(OMe)$_2$. V.14.[2041]
F_3C-C(-OEt)=CFPO(OEt)$_2$. V.14. b_4 110-111°.[979]
$F_3CCF_2CF_2CH_2CH_2PO(OCH_2CF_3)_2$. V.5. b_2 72°.[1211]
CH_3-C≡C-C≡CPO(OEt)$_2$. V.5. $b_{1.5}$ 134.5-136.5°.[823]
OCH=CH-CH=C-C(O)PO(OEt)$_2$. V.5. b_1 117-118°.[1540]

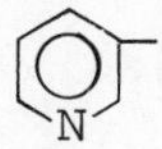
PO(OH)$_2$. I.1. m. 258-260°.[233]

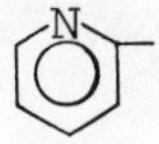
PO(OH)$_2$. I.2. m. 224-227°.[1718] I.2.[1884]

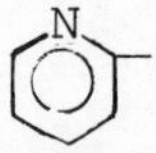
PO(OEt)$_2$. V.6. $b_{0.03}$ 96-97°, IR, ^{1}H.[1718]

HC≡CCH$_2$CH(-CN)PO(OR)$_2$.

R = Et. V.14. b_3 163-165°.[1604]

R = Pr. V.14. b_8 170-171°.[1604]

R = Bu. V.14. b_1 138-140°.[1604]

CH=CHO-CH(CN)-CH-PO(OEt)$_2$. V.7. b_6 130-133°.[277]
CH_3CH=CH-C≡CPO(OEt)$_2$. V.5. b_2 112-115°, n_D^{20} 1.4790, d_4^{20} 1.0610, μ = 4.06 D.[826]
H_2C=CH-C≡CCH$_2$PO(OEt)$_2$. V.2. b_{20} 49-52°, n_D^{20} 1.4064, d_4^{20} 1.5460.[824]
H_2C=CMe-C≡C-PO(OEt)$_2$. V.5. b_3 110-111°, n_D^{20} 1.4696, d_4^{20} 1.0549, μ = 3.84 D.[826] V.6. $b_{0.4}$ 90°.[1962]

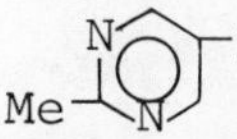
PO(OBu)$_2$. V.6. $b_{0.05}$ 134°, n_D^{25} 1.4540.[1022]

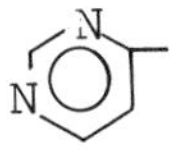
$CH(OH)PO(OEt)_2$. V.8. m. 183-183.5°.[327]

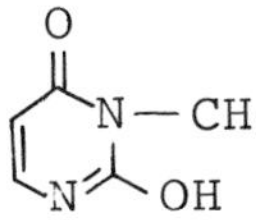
$N-CH_2PO(OH)_2$. I.13. Isolation as Ba salt.[1739]

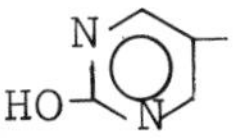
$CH(OH)PO(OH)_2$. I.2. m. 300-303°(dec.).[783]

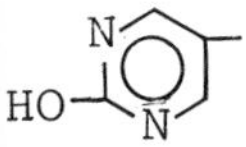
$CH(OH)PO(OEt)_2$. V.14. m. 88-91°.[783]

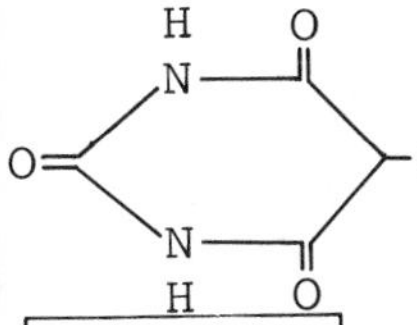
$CH_2PO(OEt)_2$. V.14. 197-198°.[1043]

$OCH=CH-CH=C-CH(OH)PO(OR)_2$.

R = Me. V.8. m. 62-63°.[1776]

R = Et. V.8. b_3 179-180°.[29]

R = ClH_2CCH_2-. V.8. m. 46-48°.[886]

$OCH=CH-CH=C-CH(OH)-P(O)-OCHMeCH_2CH_2O$. V.8. m. 81°.[1588]
$H_2C=CHCO_2-C(=CH_2)PO(OEt)_2$. V.8. $b_{0.04}$ 84-85°.[605]
$HC\equiv C-CH[-O_2CMe]PO(OEt)_2$. V.8. $b_{0.07}$ 100°.[1572]
$SCH=CH-CH=C-CH_2PO(OH)_2$. I.2. m. 108-109°.[1006]
$SCH=CH-CH=C-CH_2PO(OR)_2$.

R = Et. V.5. b_5 133-138°.[211]

R = Bu. V.6. b_3 147-150°.[1006]

$CH_3-NCH=CH-CH-C-PO(OEt)OH$. III.3. m. 131-132°.[694]
$CH_3-NCH=CH-CH=C-PO(OEt)_2$. V.9. b_3 122.5-126°, IR, UV.[694]
$NCCH=CMeCH_2PO(OEt)_2$. V.5.[591]
$C_2H_5C(O)CH(-CN)PO(OR)_2$.

R = Pr. V.14. b_2 127-129°, n_D^{20} 1.4540, d_4^{20} 1.0746, IR.[1047]

R = Bu. V.14. b_3 155-157°, n_D^{20} 1.4484, d_4^{20} 1.0334, IR.[1047]

$CH_3C(O)SCH_2CH(-CN)PO(OEt)_2$. V.7. $b_{0.04}$ 133-135°.[1672]
$O=C=N-C(O)CMe_2PO(OEt)_2$. V.14. $b_{0.05}$ 74-75°, n_D^{20} 1.4380, d_4^{20} 1.1520.[280]
$(HO_2CCH_2-)_2NCH_2PO(OH)_2$. I.13. dec. 208°.[914] I.13.

Complexing agent.[1870a]

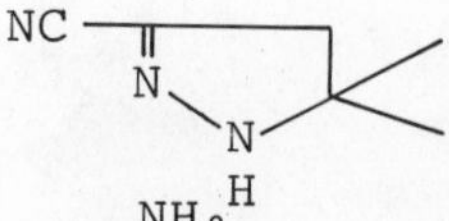
$PO(OMe)_2$. V.14. m. 89-92°, IR, ^1H.[1878a]

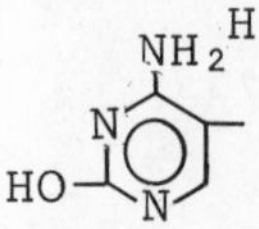
$CH(OH)PO(OH)_2$. I.2. m. 300-303° (dec.).[783]

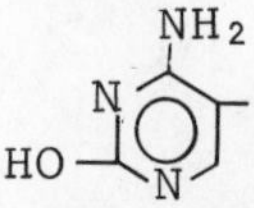
$CH(OH)PO(OEt)_2$. V.14. m. 88-91°.[783]

$C_3H_7C{\equiv}CPO(OR)_2$.

R = Me. V.2. b_1 83°.[363]

R = Et. V.11. $b_{0.2}$ 115°.[363] V.6. $b_{0.4}$ 110°.[1963]

$(CH_3)_2C{=}C{=}CHPO(OR)_2$.

R = Et. V.5. $b_{0.7}$ 95°, n_D^{25} 1.4588.[1188,1189] V.6. b_9 119-121°.[1601] V.5.[1556] V.5. b_{10} 122-123°.[1559]

R = Pr. V.5.[1556] V.5. b_{10} 136°.[1559]

R = Bu. V.5. b_1 123-124°.[1559]

R = $HC{\equiv}CCMe_2$-. V.2. n_D^{20} 1.4859, IR.[287]

$[(CH_3)_2C{=}C{=}CH\text{-}]\overline{OPOCH_2CHMeO}$. V.2. n_D^{20} 1.4917, IR.[287]

$CH_3CH{=}CH\text{-}CH{=}CHPO(OR)Cl$.

R = Et. IV.2. b_2 108-110°.[1208]

R = i-Pr. IV.2. b_{4-5} 135-137°.[1208]

$CH_3CH{=}CH\text{-}CH{=}CHPO(OR)_2$.

R = Me. V.2. b_2 101-103°.[1207]

R = Bu. V.2.[620]

$CH_3CH{=}CH\text{-}CH{=}CHPO(OMe)(OPr\text{-}i)$. V.2. b_3 84-86°.[1208]

$H_2C{=}CHCMe{=}CHPO(OEt)Cl$. IV.2. $b_{1.5}$ 88°.[1208]

$H_2C{=}CHCMe{=}CHPO(OR)_2$.

R = Me. V.2. b_1 86-87°.[1207]

R = Et. V.2. b_1 92-93°.[1616]

$H_2C{=}CHCMe{=}CHPO(OMe)OR$.

R = Et. V.2. b_1 85.5-86°.[1208]

R = i-Pr. V.2. b_2 91-93°.[1208]

$\overline{H_2CCH_2CH_2CH{=}C}PO(OR)_2$.

R = Me. V.14. $b_{1.7}$ 85-87°, n_D^{25} 1.4669, IR, ^1H.[1191]

R = Et. V.14. b_{12} 129 , n_D^{20} 1.4609.[757]

$\overline{H_2CCH_2CH{=}CH\text{-}C}HPO(OR)_2$.

R = Et. V.6. b_3 100-100.5°, n_D^{20} 1.4579, d_4^{20} 1.0726.[133]

R = Pr. V.6. $b_{1.5}$ 110-110.5°, n_D^{20} 1.4553, d_4^{20} 1.0235.[133]

R = i-Pr. V.6. b_2 96-96.5°, n_D^{20} 1.4483, d_4^{20} 1.0176.[133]

R = Bu. V.6. b_1 122-123°, n_D^{20} 1.4544, d_4^{20} 1.0053.[133]

R = i-Bu. V.6. b_4 126.5-127°, n_D^{20} 1.4531, d_4^{20} 0.9990.[133]

OCH_2CH_2OP-OCMe(CN)PO(OEt)$_2$. From OCH_2CH_2OPCl, MeC(O)CN, and P(OEt)$_3$, $b_{0.05}$ 139-140°, ^{31}P δ_1 -141.8, δ_2 -13.5 ppm, IR.[257]

HN-N=CMe-CMe=C-PO(OMe)$_2$. V.14. m. 141-142°, IR, ^1H.[1191]

H_2CCH_2N[-C(O)Me]-N=CPO(OR)$_2$.

R = Et. V.14. $b_{0.001}$ 135-136°, n_D^{20} 1.4900, d_4^{20} 1.1906.[1580]

R = Pr. V.14. $b_{0.0001}$ 143°, n_D^{20} 1.4830, d_4^{20} 1.1331.[1580]

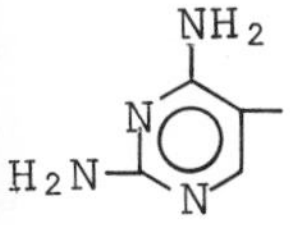
CH(OH)PO(OEt)$_2$. V.14. m. >220°(dec.).[783]

CH_3C(O)CH=CMePO(OEt)$_2$. V.14.[126]

CH_3CH=CMeC(O)PO(OMe)$_2$. V.5. $b_{0.02}$ 79-84°, IR, ^1H.[1191]

H_2C-CH_2CH_2CH-C(O)PO(OR)$_2$.

R = Me. V.5. $b_{0.07}$ 82.5°, n_D^{25} 1.4532, IR, ^1H.[1191]

R = Et. V.5. b_5 120-124°; as 2,4-dinitrophenylhydrazone, m. 136-136.5°.[244]

$H_2CCH_2CH_2C$(O)CHPO(OH)$_2$. I.13. m. 93-95°.[773]

$H_2CCH_2CH_2C$(O)CHPO(OEt)OH. III.3. As dicyclohexylammonium salt, m. 144-144.5°.[773]

$H_2CCH_2CH_2C$(O)CHPO(OEt)$_2$. V.14.[162] V.14. $b_{0.8}$ 116-118°.[773]

H_2CC(O)CH_2CH_2CHPO(OEt)$_2$. V.7. $b_{0.15}$ 104°.[751]

$HOCMe_2$-C≡C-PO(OEt)$_2$. V.6. V.14. $b_{0.4}$ 133°.[1963]

$(CH_3)_2C$(OH)-C≡C-PO(OEt)$_2$. V.6. $b_{0.4}$ 133°.[1962]

H_2CCH_2CH(OH)CH=CPO(OR)$_2$.

R = Et. V.14. $b_{0.025}$ 127-128°.[1687]

R = Bu. V.14. $b_{0.022}$ 153-154°.[1687]

HC≡C-$CH_2OCHMePO$(OR)$_2$.

R = Et. V.14. b_2 98-100°.[1610]

R = Pr. V.14. b_1 107-110°.[1610]

R = Bu. V.14. b_1 113-115°.[1610]

$H_2CCH_2CH_2CH$-O-C-PO(OR)$_2$.

R = Et. V.14. b_1 115-115.5°, n_D^{20} 1.4566, d_4^{20} 1.1528.[133] V.6. b_1 104-105°.[163]

R = Pr. V.14. b_2 129-130°, n_D^{20} 1.4555, d_4^{20} 1.0922.[133]

R = i-Pr. V.14. b_2 126-126.5°, n_D^{20} 1.4494, d_4^{20} 1.0861.[133]

R = Bu. V.14. b_2 145.5-146°, n_D^{20} 1.4562, d_4^{20} 1.0660.[13]

R = i-Bu. V.14. $b_{1.5}$ 140-141°, n_D^{20} 1.4535, d_4^{20} 1.0595.[133]

$H_2\overline{CCH_2CH_2SC(O)C}HPO(OEt)_2$. V.14. $b_{0.05}$ 95-110°, n_D^{20} 1.5020, UV λ_{max} 242 mμ.[996]

$[CH_3C(O)-]_2CHPO(OEt)_2$. V.14. $b_{0.5}$ 80°.[1045]

$CH_3-CO_2CH=CHCH_2PO(OR)_2$.

R = Me. V.14. $b_{0.02}$ 104-105°.[166]

R = Bu. V.14. $b_{0.25}$ 119-119.5°.[1322]

$CH_3-CO_2CMe=CHPO(OEt)_2$. V.14. b_1 99-100°, n_D^{20} 1.4468, d_4^{20} 1.1207.[971,1135]

$C_2H_5-CO_2CH=CHPO(OEt)_2$. V.14. $b_{1.5}$ 119-121°, n_D^{20} 1.4502, d_4^{20} 1.1190.[1135]

$H_2C=CH-CO_2CHMePO(OR)_2$.

R = Me. V.14. b_1 86-87°.[1598]

R = Et. V.14. b_6 129-130°.[1598] From $(EtO)_2P(O)Na$ + CH_3CHO and $CH_2=CHC(O)Cl$, $b_{0.05}$ 90-93°, n_D^{25} 1.4333, IR.[1444]

R = Pr. V.14. b_3 136-138°.[1598]

R = Bu. V.14. b_3 140-141°.[1598]

$H_2C=CMeCO_2CH_2PO(OEt)_2$. V.14. b_2 104.5-105°.[1598] From $(EtO)_2P(O)Na$ + CH_2O + $H_2C=CMeC(O)Cl$, $b_{0.3}$ 92-94°, n_D^{23} 1.4398, IR.[1444]

$C_2H_5O_2C-CH=CHPO(OEt)_2$. V.7. b_5 138-142°.[968,967]

$CH_3O_2C-CH=CMePO(OMe)_2$. V.7. b_{10} 134°.[967]

$O=\overline{CCH_2CH_2CMe}PO(OR)_2$ (ring through O).

R = Me. V.8. $b_{0.35}$ 118-120°.[798]

R = Et. V.8. $b_{0.2}$ 100-104°.[335]

R = i-Bu. V.8. $b_{0.3-0.4}$ 135-141°.[798]

$H_2\overline{CCH_2CH_2OC}(-CHO)PO(OEt)_2$. V.6. b_2 115-120°.[143] V.5. b_2 110-112°.[143]

$C_2H_5O_2C-CH_2C(O)PO(OEt)_2$. V.5. $b_{0.9}$ 115°. As 2,4-dinitrophenylhydrazone, m. 77-78°.[1145]

$HO_2CCH_2CMe(-CO_2H)-PO(OH)_2$. I.2. Complexing agent.[613]

$H_2C=CMe-C(O)NHCH_2-PO(OR)_2$.

R = Me. V.8.[837]

R = Et. V.8.[837,838]

R = Pr. V.8.[837]

R = Bu. V.8.[837,838]

$H_2C{=}CH{-}CH_2NHC(O)CH_2PO(OEt)_2$. V.5. $b_{0.22}$ 147-147.5°, n_D^{20} 1.4579, d_4^{20} 1.1240.[1108]

$NCCH_2CH_2OCHMePO(OR)_2$.

R = Me. V.7. b_3 151-152°, n_D^{20} 1.4440, d_4^{20} 1.1775.[1675]

R = Et. V.7. b_2 153-155°, n_D^{20} 1.4410, d_4^{20} 1.1070.[1675]

R = i-Pr. V.7. b_2 141-142°, n_D^{20} 1.4372, d_4^{20} 1.0540.[1675]

$CH_3O_2C{-}N{=}C(-OEt)PO(OMe)_2$. V.14. $b_{0.04}$ 70-71°, n_D^{20} 1.4452, d_4^{20} 1.2225.[490]

$(HO_2CCH_2-)_2N{-}CH_2PO(OH)_2$. I.10. m. 210°, ^{31}P.[1328]

$O_2NCH_2CMe[-O_2CMe]PO(OR)_2$.

R = Et. V.14. $b_{0.05}$ 131-132°.[1214]

R = i-Pr. V.14. $b_{0.02}$ 110-111°.[1214]

$C_2H_5SCH_2CH(-CN)PO(OEt)_2$. V.7. $b_{0.04}$ 118-119°.[1672]

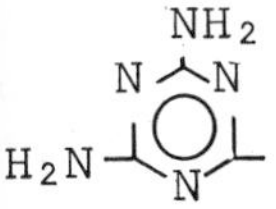
$CH_2CH_2PO(OEt)_2$. V.14. m. 171-174°.[1991]

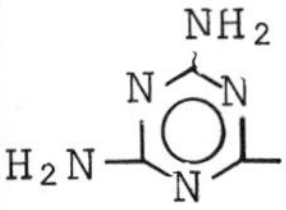
$SCH_2CH_2PO(OEt)_2$. V.14, m. 145-147°.[1991]

$(CH_3)_2C{=}CH{-}CH_2PO(OMe)_2$. V.14. $b_{0.2}$ 51-52°, IR, ^1H.[1196]

$CH_3CH{=}CEtPO(OMe)_2$. V.14. Chromatographic cis-trans separation, IR, ^1H.[229]

$(CH_3)_2CHCH{=}CHPO(OEt)_2$. V.14. $b_{0.5}$ 55-57°, n_D^{20} 1.4392, d_4^{20} 0.9900, ^1H.[1479]

$H_2C{=}CMeCH_2CH_2PO(OEt)_2$. V.14. b_{15} 70°.[89]

$H_2C(CH_2)_3CHPO(OH)_2$. I.6. As aniline salt, m. 181-183°.[315]

$(CH_3)_2CCH_2CHPO(OMe)_2$. V.14. $b_{0.9}$ 52°, IR, ^1H.[1878a]

$SCMe{-}NH{-}CH_2CH{-}CH_2PO(OEt)_2$. V.5. b_3 147-150°.[141]

$(CH_3)_3CCH(N_2)PO(OMe)_2$. V.5. $b_{0.1}$ 41.5°, n_D^{25} 1.4562, IR, ^1H.[1191]

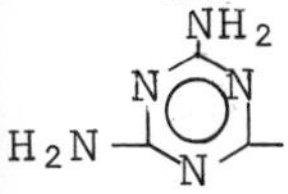
$NHCH_2CH_2PO(OEt)_2$. V.14. m. 232°.[1991]

$CH_3C(O)(CH_2)_3PO(OEt)_2$. V.14. $b_{0.5}$ 115°, IR.[1959]

$CH_3C(O)CHEtPO(OEt)_2$. V.14. b_3 100-103°, n_D^{20} 1.4430, d_4^{20} 1.0971.[971]

$CH_3C(O)CH_2CHMePO(OR)_2$.

R = Me. V.7. b_{10} 134-135°.[1545]

R = Et. V.7. b_{10} 139-140°.[1545]

R = Bu. V.7. b_{10} 156-158°.[1545]

R = i-Bu. V.7. b_{10} 172-173°.[1545]

$C_2H_5C(O)CH_2CH_2PO(OR)_2$.

R = Me. V.7. $b_{0.15}$ 96-97°, n_D^{30} 1.4316, d_4^{30} 1.1621.[1787]

R = Et. V.6. $b_{0.4}$ 116°, n_D^{21} 1.4401, d_4^{21} 1.070.[1958]
V.7. $b_{0.06}$ 103-104°, n_D^{30} 1.4291, d_4^{30} 1.0936.[1787]

R = Bu. V.7. $b_{0.01}$ 126-128°, n_D^{30} 1.4340, d_4^{30} 1.0232.[1787]

R = $C_4H_9CHEtCH_2$-. V.7. n_D^{30} 1.4437, d_4^{30} 0.9692.[1787]

$(CH_3)_3CC(O)PO(OR)_2$.

R = Me. V.5. $b_{2.0-2.1}$ 76-78°, n_D^{25} 1.4280, IR, ^1H.[1191]

R = Et. V.5. b_5 97-100°, As 2,4-dinitrophenylhydrazone m. 140-141°, IR, ^1H.[244]

$C_2H_5OCH{=}CH{-}CH_2PO(OEt)_2$. V.14. b_3 132-133°.[1082] ^{31}P -27 ppm ^1H.[1093] V.8.[636] V.14.[175]

$C_2H_5OCH{=}CMePO(OR)_2$.

R = Et. V.5. b_{11} 136-137.5°, n_D^{23} 1.4443, IR.[1733] V.2. $b_{0.085}$ 69°, n_D^{20} 1.4492, d_4^{20} 1.0452.[1350]

R = Pr. V.2. $b_{0.035}$ 76-77°, n_D^{20} 1.4486, d_4^{20} 1.0133.[135

$C_2H_5OCMe{=}CHPO(OEt)_2$. V.14. b_{12} 137°.[1599] V.14. $b_{0.9}$ 98°, n_D^{21} 1.4507, d_4^{21} 1.058.[1961] V.6. $b_{0.1}$ 98°, IR.[1959] ^1H, ^{31}P -21.6 ppm.[1093]

$C_2H_5OC(=CH_2){-}CH_2PO(OEt)_2$. V.6. $b_{0.1}$ 98°, n_D^{19} 1.4510, d_4^{19} 1.056.[1410]

$H_2\overline{CCH_2CH_2O}CMe{-}PO(OEt)_2$. V.6. b_1 89°, n_D^{20} 1.4441, d_4^{20} 1.097[1958]

$CH_3C(O)CH_2CH_2OCH_2PO(OEt)_2$. V.7. $b_{1.5}$ 137°, n_D^{20} 1.4430, d_4^{20} 1.1271.[1675]

$C_2H_5OCH_2C(O)CH_2PO(OEt)_2$. V.14. $b_{0.2}$ 105°, n_D^{21} 1.4408, d_4^{21} 1.115.[1961] ^1H, ^{31}P -19.5 ppm.[1093]

$CH_3C(O)CH_2CMe(OH)PO(OEt)_2$. V.8.[126] V.5. b_{12} 138°, n_D^{20} 1.4570.[1040] V.8.[15]

$HOCMe_2C(O)CH_2PO(OPr\text{-}i)_2$. V.14. $b_{0,1}$ ∿86°, n_D^{19} 1.4366.[1788]

$CH_3OCH_2OCMe{=}CHPO(OEt)_2$. V.14. b_1 108-110°.[1651]

$CH_3{-}CO_2(CH_2)_3PO(OEt)_2$. V.5. $b_{0.1}$ 96°, IR.[499]

$CH_3{-}CO_2CMe_2PO(OR)_2$.

R = Me. V.14. $b_{15.5}$ 126°, n_D^{20} 1.4360, d_4^{20} 1.1660.[1589]

R = Et. V.14. b_{20} 137.5°, n_D^{20} 1.4315, d_4^{20} 1.0907.[1589]

$CH_3{-}CO_2CHEtPO(OEt)_2$. V.14. b_3 104.5°, n_D^{20} 1.4302, d_4^{20} 1.0865.[1589]

$CH_3{-}CO_2CHMeCH_2PO(OEt)_2$. V.7. $b_{0.5}$ 89-93°.[1536]

$CH_3O_2C-(CH_2)_3PO(OEt)_2$. V.5. $b_{0.01}$ 104-105°, n_D^{20} 1.4376.[516]

$CH_3O_2C-CHMeCH_2PO(OR)_2$.

R = Me. V.7. b_{10} 137-138°.[1565]

R = Et. V.5. b_1 95°.[432]

R = $EtOCH_2CH_2-$. V.7. b_{12} 185-186°, n_D^{20} 1.4432, d_4^{20} 1.106.[1344]

R = cycl.-$C_6H_{11}-$. V.7. b_3 190-193°.[1634]

R = $BuOCH_2CH_2-$. V.7. b_{12} 223-226°, n_D^{20} 1.4470, d_4^{20} 1.049.[1344]

R = $PhCH_2-$. V.7. $b_{0.001}$ 150-153°.[1634]

R = $PhOCH_2CH_2-$. V.7. $b_{0.005}$ 155°, n_D^{20} 1.4440, d_4^{20} 1.019.[1344]

$CH_3O_2C-CHMeCH_2PO(OBu)OAm$. V.7. $b_{0.015}$ 75°.[1162]

$(CH_3O_2C-CHMeCH_2-)OPO-R-O$.

R = $-CHMeCH_2-$. V.7. b_2 152°, n_D^{20} 1.4435, d_4^{20} 1.1920.[1588]

R = $-(CH_2)_3-$. V.7. $b_{1.5}$ 177-180°, n_D^{20} 1.4580, d_4^{20} 1.2310.[1588]

R = $-CHMeCHMe-$. V.7. b_3 170-171°. n_D^{20} 1.4547, d_4^{20} 1.1912.[1588]

R = $-CH_2CH_2CHMe-$. V.7. b_3 167-169°, n_D^{20} 1.4580, d_4^{20} 1.183.[1588]

$CH_3O_2C-CMe_2PO(OEt)_2$. V.14. $b_{0.05}$ 67-68°, n_D^{20} 1.4310. d_4^{20} 1.0989.[280]

$C_2H_5O_2C-CH_2CH_2PO(OEt)Cl$. IV.2. $b_{0.4}$ 105-107°.[1712]

$C_2H_5O_2C-CH_2CH_2PO(OEt)_2$. V.7. b_6 130-131°.[2032] V.7. $b_{0.25}$ 98°, n_D^{30} 1.4302, ^1H.[751] V.14. $b_{1.5}$ 109-110°.[605] V.14. $b_{0.15}$ 109-110°, n_D^{20} 1.4325, d_4^{20} 1.0995.[610] V.5, V.6. b. 287-288°, b_{10} 151°, d_0^{17} 1.1015.[116] V.6. b_{10} 149.5-150°.[1419] V.7. b_{12} 156-158°.[1043] V.7. b_3 137-142°, n_D^{20} 1.4330.[638]

$EtO_2C-CH_2CH_2PO(OEt)(OCMe_2CCl_3)$. V.2. $b_{0.001}$ 129-131°[504]

$EtO_2C-CHMePO(OEt)_2$. V.5. b_{10} 138.5-138.75°.[112a] V.14. b_{12} 143-144°.[112] Mass spect.[1403]

$C_3H_7O_2C-CH_2PO(OEt)_2$. V.5. $b_{0.07}$ 88-89°, n_D^{20} 1.4310, d_4^{20} 1.1154.[281]

$HO_2C(CH_2)_4PO(OH)_2$. I.2. isolated as Pb salt.[1041]

$HOCH-CH(OH)CH_2CH_2CHPO(OR)_2$.

R = Pr. V.14. $b_{1.5}$ 167-169°, n_D^{20} 1.4691, d_0^{20} 1.1494.[133]

R = i-Pr. V.14. $b_{1.5}$ 148-150°, n_D^{20} 1.4640, d_0^{20} 1.1427.[133]

R = Bu. V.14. b_2 177-178°, n_D^{20} 1.4627, d_0^{20} 1.0889.[133]

R = i-Bu. V.14. b_1 182-184°, n_D^{20} 1.4654, d_0^{20} 1.1092.[13]

CH_3-$CO_2CH_2CH_2OCH_2PO(OEt)_2$. V.5. $b_{0.5}$ 123-123,5°.[173]
CH_3O_2C-$CH_2CH_2OCH_2PO(OR)_2$.

R = Me. V.7. b_2 125-126°.[1675]

R = Et. V.7. b_{12} 180°.[1675]

$C_2H_5O_2C$-$CH_2OCH_2PO(OEt)_2$. V.14. $b_{1.5}$ 147-149°.[127]
CH_3O_2C-$CH_2CMe(OH)PO(OMe)_2$. V.8. m. 55-56°.[1995]
$C_2H_5O_2C$-$CMe(OH)PO(OMe)_2$. V.8. m. 57-58°.[14]
$C_2H_5SCH{=}CHCH_2PO(OEt)_2$. V.14. $b_{0.4}$ 123°, n_D^{22} 1.4840, d_4^{22} 1.089.[1082]
$C_2H_5SCMe{=}CHPO(OR)_2$.

R = Et. V.14. b_2 123-125°, n_D^{20} 1.4788, d_4^{20} 1.0892.[1607]

V.14. $b_{0.4}$ 113-114°, n_D^{23} 1.4891.[1082]

R = Pr. V.14. b_3 130°.[1602]

R = Bu. V.14. b_3 148°.[1602]

$S(CH_2)_3SCMePO(OMe)_2$. From $S(CH_2)_3SC{=}P(OMe)_3$ at 70°-80°, $b_{0.1}$ 90°.[441]
$O(CH_2)_5PO(OEt)$. V.6. $b_{0.7}$ 93-95°.[1938]
$(CH_3)_2N$-$CMe{=}CHPO(OEt)_2$. ^1H, ^{31}P-26,6 ppm.[1093]
$H_2C(CH_2)_3N$-$CH_2PO(OEt)_2$. V.9. $b_{0.1}$ 110°, n_D^{25} 1.4471.[329]
$(CH_3)_2NCH_2C(O)CH_2PO(OEt)_2$. V.14. $b_{0.1}$ 99°.[1961]
$(CH_3)_2NC(O)CH_2CH_2PO(OR)_2$.

R = i-Pr. V.8. IR, ^1H.[988]

R = sec.-Bu. V.8. IR, ^1H.[988]

$CH_3C(O)NHCMe_2PO(OEt)_2$. V.14. m. 100-101°.[1589]
$(CH_3)_2CHNHC(O)CH_2PO(OEt)_2$. V.5. $b_{0.18}$ 137°, n_D^{20} 1.4513, d_4^{20} 1.0906.[1108]
$C_4H_9NHC(O)PO(OEt)_2$. V.9.[1720]
$H_2CCH_2OCH_2CH_2NCH_2PO(OEt)_2$. V.14. b_3 137°.[529] V.9. $b_{0.1}$ 124°, n_D^{25} 1.4549.[329]
CH_3O_2C-$CH(NMe_2)PO(OEt)_2$. V.5. $b_{0.01}$ 80-81°, n_D^{19} 1.4450, IR.[717]
$C_2H_5O_2C$-$NHCH_2CH_2PO(OEt)_2$. V.5. $b_{0.08}$ 126-128°, n_D^{20} 1.4470, d_4^{20} 1.1353.[1903]
$CH_3C(O)NHCH_2CH(OH)CH_2PO(OEt)_2$. V.14. $b_{0.25}$ 162°, IR, ^1H.[691]

$(CH_3)_2CHCH(CH_2NO_2)PO(OR)_2$.

R = Me. V.7.[1662]

R = Et. V.7. b_4 144°.[1662]

$C_5H_{11}PO(OH)_2$. I.2. m. 120.5-121° (ligroin),[1000, 1002] ^{31}P -33 ppm.[1189a]
$C_5H_{11}PO(OC_6H_{11}$-cycl.)Cl. IV.2. $b_{0.1}$ 99°, n_D^{25} 1.4688.[740]
$C_5H_{11}PO(OC_6H_{11}$-cycl.)F. IV.2. $b_{0.005}$ 60°, n_D^{25} 1.4409.[740]

$C_5H_{11}PO(OR)_2$.

R = Et. V.5. $b_{1.5}$ 86°, $n_D^{16.5}$ 1.4282.[552]

R = Bu. V.6. b_{17} 167-169°, n_D^{25} 1.4318, d_4^{25} 0.9428.[1000, 1002] V.2. b_{12} 159°, n_D^{20} 1.4360, d_4^{20} 0.9452.[286]

R = Am. V.2. b_2 150-151°, n_D^{20} 1.4378, d_4^{20} 0.9408.[286] Magn. rotation of the bonds.[1069]

R = i-Am. V.2. b_3 152°, n_D^{20} 1.4360, d_4^{20} 0.9344.[286]

R = $PhCH_2CH_2$-. V.6. $b_{0.003}$ 165°, n_D^{20} 1.5239, d_4^{20} 1.0665.[1499]

i-$C_5H_{11}PO(OH)_2$. I.1, I.7. m. 166°.[729] I.6. m. 139°.[190a] I.6. m. 160°.[791]

i-$C_5H_{11}PO(OEt)_2$. V.5. $b_{0.8}$ 75°, $n_D^{16.5}$ 1.4266.[552]

$(C_2H_5)_2CHPO(OH)_2$. I.6. As aniline salt, m. 141-143°.[315]

$C_3H_7CHMePO(OH)_2$. I.6. As aniline salt, m. 145-146°.[315]

$C_2H_5CMe_2PO(OR)_2$.

R = Bu. V.2. m. 110°.[286]

R = Am. V.2. m. 127°.[286]

R = i-Am. V.2. m. 132°.[286]

$CH_3O(CH_2)_4PO(OMe)_2$. V.6, V.14. $b_{0.001}$ 67-68°, n_D^{20} 1.4415, d_4^{20} 1.0312.[730]

$C_3H_7OCH_2CH_2PO(OEt)_2$. V.14. b_2 100°, n_D^{20} 1.4283, d_4^{20} 1.0206.[1640]

$C_4H_9OCH_2PO(OBu)_2$. V.5, V.10. $b_{0.1}$ 112-114°, b_1 127°, n_D^{20} 1.4320, (1.4328), (1.4340).[1038]

$(CH_3)_2CHOCH_2CH_2PO(OEt)_2$. V.14. b_9 119-120°, n_D^{20} 1.4250, d_4^{20} 1.0145.[1640]

$C_3H_7CH(OMe)PO(OEt)(-OC_6H_4NO_2-4)$. V.5. n_D^{25} 1.516.[741]

$(CH_3)_2CHCH_2CH(OH)PO(OH)_2$. I.6, I.9. m. 191° (rapid heating), dec. 183-184° (slow heating).[1182, 1186] I.9. m. 188°.[1446] I.9. m. 184-185°.[555]

$(CH_3)_2CHCH_2CH(OH)PO[-O-\overline{C(CCl_3)CH_2CH_2CH_2CH_2}]_2$. V.8. m. 155-157°.[33]

$C_3H_7CMe(-OH)PO(OH)_2$. I.6, I.9. m. 139-140° (acetone/ether).[1184, 1186]

$(C_2H_5)_2C(OH)PO(OH)_2$. I.6, I.9. m. 108°.[1183, 1186]

$(C_2H_5)_2C(OH)PO(OMe)_2$. V.1. m. 64-65°.[229]

$C_2H_5CHMeCH(OH)PO[OC(CH_2Cl)_3]_2$. V.8. m. 134-135°.[34]

$H_2C=C(OSiMe_3)PO(OR)_2$.

R = Me. V.8. b_2 79°, ^1H.[1791]

R = Et. V.5. b_{10} 109-110°, b_1 85-86°, n_D^{20} 1.4380, d_4^{20} 1.0114.[1412, 1414]

$(CH_3O)_2CHCH_2CH_2PO(OMe)_2$. By heating of $(MeO)_3\overline{PCH_2CH=CHO}$ at

120°.[166, 175]

$(C_2H_5O)_2CHPO(OEt)ONa$. III.3. m. 240-244°, ^1H.[715]

$(C_2H_5O)_2CHPO(OR)_2$.

R = Me. From $HC(OEt)_3$ and $P(OMe)_3$.[1706]

R = Et. From $HC(OEt)_3$ and $HPO(OEt)_2$ or $P(OEt)_3$.[1706, 1709] From $HC(OEt)_3$ + PCl_3 + $P(OEt)_3$, b_{14} 132-134°, n_D^{19} 1.4261, IR.[718] From $HC(OEt)_3$ + $(EtO)_2PCl$ in the presence of $ZnCl_2/AlCl_3$, $b_{0.02}$ 83°, n_D^{20} 1.4250 IR.[475]

R = ClH_2CCH_2-. From $HC(OEt)_3$ and $P(OCH_2CH_2Cl)_2Cl$, $b_{0.3}$ 148-151°, n_D^{23} 1.4606.[713]

R = Ph. From $HC(OEt)_3$ and $P(OPh)_2Cl$, $b_{0.08}$ 138-140°, n_D^{23} 1.5580.[713]

$(C_2H_5O)_2CHPO(OMe)OEt$. V.1. b_{12} 122-124°, n_D^{24} 1.4240.[715] From $(EtO)_2CHPO(OEt)ONa$ and MeHal in DMF or dioxane, b_{14} 128°.[715]

$(C_2H_5O)_2CHPO(OEt)OR$.

R = ClH_2CCH_2-. From $P(OEt)(OCH_2CH_2Cl)Cl$ and $HC(OEt)_3$, $b_{0.03}$ 102-105°, n_D^{24} 1.4430.[713]

R = $MeOCH_2$-. From $(EtO)_2CHPO(OEt)ONa$ and $MeOCH_2Hal$ in DMF or dioxane, $b_{0.01}$ 90-93°.[715]

R = Pr. From $(EtO)_2CHPO(OEt)ONa$ and PrHal in DMF or dioxane, $b_{0.01}$ 84-85°.[715]

R = $CH_2{=}CHCH_2$-. From $(EtO)_2CHPO(OEt)ONa$ and $CH_2{=}CH{-}CH_2$-Hal in DMF or dioxane, $b_{0.01}$ 79°.[715]

R = Ph. From $P(OEt)(OPh)Cl$ and $HC(OEt)_3$, $b_{0.02}$ 104-105°, n_D^{26} 1.4852.[713, 715]

R = $4\text{-}ClC_6H_4OCH_2$-. From $(EtO)_2CHPO(OEt)ONa$ and 4-ClC_6-H_4OCH_2Cl in DMF or dioxane, $b_{0.01}$ 150-153°.[715]

R = Me_3Si-. From $(EtO)_2CHPO(OEt)ONa$ and Me_3SiCl in DMF or dioxane, $b_{0.02}$ 74-76°.[715]

$[(C_2H_5O)_2CH-]\overline{OPOCH_2CMe_2CH_2O}$. From $HC(OEt)_3$ and $\overline{ClPOCH_2CMe_2\text{-}CH_2O}$ in the presence of $ZnCl_2/AlCl_3$, $b_{0.02}$ 147-150°, n_D^{20} 1.4533, IR.[475]

$C_2H_5OCH_2CH(OH)CH_2PO(OEt)_2$. V.14. $b_{0.2}$ 105°, ^1H.[691]

$(CH_3)_2C(-OH)-CH(OH)CH_2PO(OPr\text{-}i)_2$. V.14. IR.[1788]

$C_2H_5OS(O)_2CH_2CH_2CH_2PO(OEt)_2$. V.5. $b_{0.4}$ 165-171°.[595] V.5.[73]

$(CH_3)_2N(CH_2)_3PO(OEt)(OCH_2CH_2SEt)$. V.2. $b_{0.0001}$ 72-74°.[895]

$(C_2H_5)_2NCH_2PO(OH)_2$. I.2.[529]

$(C_2H_5)_2NCH_2PO(OR)OH$.

R = Et. III.3. m. 150°.[868]

R = Bu. III.3. m. 117°.[868]

$(C_2H_5)_2NCH_2PO(OR)_2$.

R = Et. V.14. b_3 96°.[529] V.14. b_3 95°.[531] V.9. $b_{0.4}$ 82–84°, n_D^{21} 1.4334.[868] V.5.[836] From Et_2NCH_2OEt and PCl_3, b_3 96–98°.[1038]

R = Pr. From $(Et_2N)_2CH_2$ and $HPO(OPr)_2$.[1323] V.9.[513]

R = Bu. V.9.[513] V.9. $b_{0.3}$ 104°, n_D^{20} 1.4370.[868] From Et_2NCH_2OBu and PCl_3, b_1 132°.[1038]

$(CH_3)_2NCMe_2PO(OR)OH$.

R = Me. V.9. m. 168–169°.[813]

R = Et. V.9. m. 161–162°.[813]

R = i-Pr. V.9. m. 171–172°.[813]

$(C_2H_5)_2NCH_2PO(OEt)OH$. III.8. m. 153°.[1484]

$[(C_2H_5)_2NCH_2-]OP\underset{O(CH_2)_2O}{\overset{O(CH_2)_2O}{\diamondsuit}}PO[-CH_2N(C_2H_5)_2]$. From

$(Et_2N)_2CH_2$ and $H(O)P\underset{O(CH_2)_2O}{\overset{O(CH_2)_2O}{\diamondsuit}}P(O)H$ at 130°.[1493]

$C_4H_9NHCH_2PO(OEt)_2$. V.8. $b_{2.5}$ 114°.[529]

$C_2H_5NHCHEtPO(OEt)_2$. V.8. $b_{1.5}$ 93°.[529]

$C_2H_5NHCMe_2PO(OEt)_2$. V.8. $b_{2.3}$ 91°.[529]

$H_2N(CH_2)_5PO(OH)_2$. I.2.[366] pK_1 2.6 , pK_2 7.6, pK_3 11.0.[1776a]

$C_4H_9CH(-NH_2)PO(OH)_2$. I.13. m. 267–269°.[2040]

$HOCH_2CH_2NHCMe_2PO(OMe)_2$. V.9. Picrate m. 128,5–129° (dec.).[782]

$(HOCH_2CH_2-)_2NCH_2PO(OH)_2$. I.13. Isolated as disodium salt – 5 hydrate.[1693] I.2. Cryst.[2083]

$C_2H_5SCH(-NMe_2)PO(OEt)_2$. V.14. $b_{0.01}$ 86–88°, n_D^{21} 1.4756.[715a]

$(EtO)O(C_2H_5-)PCH_2PO(OEt)_2$. V.6. b_1 140–141°.[772]

$(EtO)_2OPOCH_2PO(OEt)_2$. V.14. $b_{1.5}$ 162.5–163.5^O.[127] V.14. $b_{0.01}$ 104°.[1861]

$(EtO)_2SP-S-CH_2PO(OEt)_2$. V.14. $b_{0.06}$ 128–130°, n_D^{20} 1.4945, d_4^{20} 1.2331.[2055]

$(CH_3)_3\overset{\oplus}{N}CH_2CH_2PO(OH)O^{\ominus}$. I.2. m. 227.5–228.5^O.[1769]

$[(CH_3)_3\overset{\oplus}{N}CH_2CH_2-]OP[OCH_2CH(-O_2CC_{15}H_{31})-CH_2(-O_2CC_{15}H_{31})]O^{\ominus}$. From (RO-)(OH)OPCH$_2CH_2$Br and NMe$_3$ in DMF.[193]

$[(CH_3)_3\overset{\oplus}{N}CH_2CH_2-]OP[OCH_2CH(-O_2CC_{17}H_{35})-CH_2(-O_2CC_{17}H_{35})]O^{\ominus}$. From (RO-)(OH)OPCH$_2CH_2$Br and NMe$_3$ in DMF.[192]

$(C_2H_5)_2\overset{\oplus}{N}H-CH_2-PO(OEt)O^{\ominus}$. By heating from $[Et_2\overset{\oplus}{N}HCH_2PO(OEt)_2]$ $Cl^{\ominus}$, m. 153°.[1484]

$[Me_3\overset{\oplus}{N}CH(OMe)PO(OR)_2]J^{\ominus}$.

R = Et. V.14. m. 98–101°.[714]

R = i-Pr. V.14. m. 116–118°.[714]

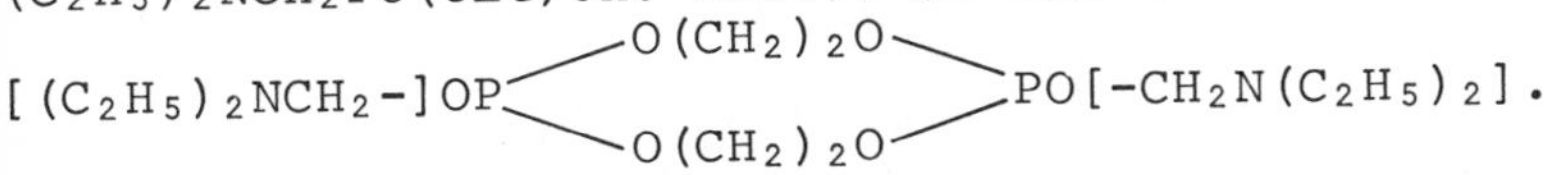

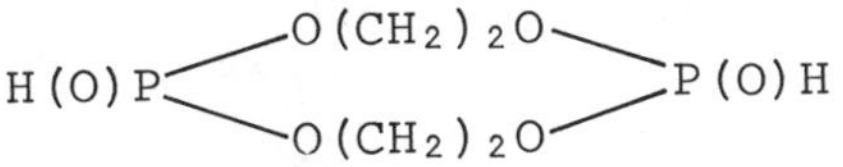

$(CH_3)_3SiOCHMePO(OR)_2$.

R = Me. V.14. b_{10} 99-100°, n_D^{20} 1.4279, d_4^{20} 1.0393, ^{1}H. [1374]

R = Et. V.14. b_8 109-110°, n_D^{20} 1.4210, d_4^{20} 0.9980, ^{1}H. [1374] V.5. $b_{0.1-0.2}$ 55-56°.[258] V.14. $b_{0.1-0.2}$ 55-56°, n_D^{25} 1.4204.[258] V.14. b_1 82°, n_D^{20} 1.4252, d_4^{20} 0.9723.[1414]

$(C_2H_5-)SiMe_2OCH_2PO(OEt)_2$. V.14. b_5 104°, n_D^{20} 1.4292, d_4^{20} 1.0145.[1436]
$(C_2H_5O)Me_2SiCH_2PO(OEt)_2$. V.5. b_3 79-81°.[321]
$(CH_3)_3SiCH_2CH_2PO(OH)_2$. I.2. m. 147°.[208]
$(CH_3)_3SiCH_2PO(OBu)_2$. V.7. b_1 128°, n_D^{25} 1.4353, d_4^{25} 0.936. [208]

EtMe$_2$SiCH$_2$PO(OEt)$_2$. V.5. b_{13} 128.5-131.5^O.[321]

C_6
2-Br-3-O$_2$NC$_6$H$_3$PO(OH)$_2$. I.1. m. 224^O(dec.).[460]
2-Br-4-O$_2$NC$_6$H$_3$PO(OH)$_2$. I.1. m. 217-220°.[228]
2-Br-5-O$_2$NC$_6$H$_3$PO(OH)$_2$. I.1. m. >230^O(dec.).[582]
4-Br-3-O$_2$NC$_6$H$_3$PO(OH)$_2$. I.13. m. 185°.[1260, 1269]
5-Br-2-O$_2$NC$_6$H$_3$PO(OH)$_2$. I.1. m. 214-217^O(dec.).[579]
2-BrC$_6$H$_4$PO(OMe)$_2$. V.5. $b_{0.2}$ 113-115°.[1421]
3-BrC$_6$H$_4$PO(OH)$_2$. I.13. m. 152-153°.[1008]
3-BrC$_6$H$_4$PO(OMe)$_2$. V.5. $b_{0.2}$ 104-105°.[1421]
4-BrC$_6$H$_4$PO(OH)$_2$. I.1. m. 202°.[1260, 1269] I.2. m. 198-199. [1008] I.1. m. 201-202°.[481]
4-BrC$_6$H$_4$PO(OR)$_2$.

R = Me. V. 5. $b_{0.1}$ 95-96°.[1421]

R = Et. V.2. $b_{0.5}$ 126-128°, n_D^{25} 1.5188.[1008]

4-BrC$_6$H$_4$PO(OCH=CCl$_2$)OR.

R = Et. V.8. b_5 170-172°.[964]

R = Bu. V.8. b_4 190-191°.[964]

R = Am. V.8. b_4 202.5-203.5.[964]

O=C-CBr=CMe-N=C(OH)-N-CH$_2$PO(OH)$_2$. I.13. IR.[1739]
O=C-CBr=CMe-N=C(OH)-N-CH$_2$PO(OMe)$_2$. V.2.[1738]
O=C-CBr=CMe-N=C(-NH$_2$)-N-CH$_2$PO(OH)$_2$. I.13. IR.[1739]
C$_4$H$_9$O$_2$C-CHBrPO(OEt)$_2$. V.14. b_1 131-136°, n_D^{20} 1.4618, d_4^{20} 1.3244.[697]
2,5-Br$_2$C$_6$H$_3$PO(OH)$_2$. I.1. m. 204-208°.[460]
3-H$_2$N-2.4.6-Br$_3$C$_6$HPO(OH)$_2$. I.13. m. 222°.[1395]
2-Cl-4-O$_2$NC$_6$H$_3$PO(OH)$_2$. I.1. m. 221.5-224°.[582]
4-Cl-3-O$_2$NC$_6$H$_3$PO(OH)$_2$. I.13. m. 166-168°.[1260, 1269] I.13. m. 166°.[186, 1395]
2-ClC$_6$H$_4$PO(OH)$_2$. I.1. m. 182-184°.[479]
2-ClC$_6$H$_4$PO(OR)$_2$.

R = Me. V.5. $b_{0.3}$ 106-107°.[1421]

R = Et. V.5. $b_{0.3}$ 113-115°.[1987]

3-$ClC_6H_4PO(OH)_2$. I.13. m. 136-137°.[1008] I.1. m. 137-137.5°.[479]

3-$ClC_6H_4PO(OMe)_2$. V.14. $b_{0.5}$ 109-110°.[1421]

4-$ClC_6H_4PO(OH)_2$. I.2. m. 187-188°.[1020] I.1. m. 184-185°.[1260, 1269] I.1. m. 184°.[1395] I.1.[213] I.1. m. 187-188.5°.[479]

4-$ClC_6H_4PO(OEt)CN$. IV.4. $b_{0.5}$ 140°.[1826]

4-$ClC_6H_4PO(OR)_2$.

R = Me. V.5. $b_{0.1}$ 95-96°.[1421]

R = Et. V.5. $b_{0.15}$ 105-108°.[1987] V.5. $b_{0.3}$ 124°, n_D^{20} 1.5038.[1988] V.2. b_4 144-146°, $b_{1.5}$ 117-119°, n_D^{25} 1.5068.[1020] Reactivity with PhMgBr.[763] V.5. $b_{0.05}$ 105-108°.[1985]

R = $Cl^{\ominus}[H_3\overset{\oplus}{N}CH_2CH_2$-. V.2.[1936]

R = $H_2C{=}CHCH_2$-. V.2. b_2 136-139°, n_D^{25} 1.5208.[2019]

R = $H_2C{=}CMeCH_2$-. V.2. b_1 137-140°, n_D^{25} 1.5162.[2019]

R = $Cl^{\ominus}[H_3\overset{\oplus}{N}(CH_2)_6$-. V.2.[1936]

2-HO-5-$ClC_6H_3PO(OH)_2$. I.6. cryst.[1131]

6-H_2N-3-$ClC_6H_3PO(OH)_2$. I.1.[1130]

3-H_2N-4-$ClC_6H_3PO(OH)_2$. I.13. Dec. 270°.[1260, 1269]

$HC{\equiv}C-CH_2CH{=}CCl-CH_2PO(OH)_2$. I.1. IR, ^{1}H.[1228]

$HC{\equiv}C-CH_2CH{=}CCl-CH_2PO(OEt)_2$. V.2. IR, ^{1}H.[1228]

$H_2C-CH{=}CH-CH_2CH_2CClPO(OEt)_2$. V.14. $b_{0.3}$ 103-105°.[451]

$C_2H_5C(O)C({=}CClMe)-PO(OEt)_2$. V.2. b_2 123-124°, n_D^{20} 1.4720, d_4^{20} 1.1525.[1133]

$C_3H_7C(O)C({=}CHCl)-PO(OEt)_2$. V.2. b_1 122-123°, n_D^{20} 1.4656, d_4^{20} 1.1355.[1133]

$C_2H_5SCH{=}C(-CCl{=}CH_2)-PO(OR)_2$.

R = Me. V.2. $b_{0.0085}$ 80-81°, n_D^{20} 1.5700, d_4^{20} 1.2531.[1918]

R = Et. V.2. $b_{0.004}$ 84°, n_D^{20} 1.5460, d_4^{20} 1.1861.[1918]

R = Pr. V.2. $b_{0.005}$ 90°. n_D^{20} 1.5402, d_4^{20} 1.1574.[1918]

R = Bu. V.2. $b_{0.0015}$ 82°, n_D^{20} 1.5294, d_4^{20} 1.1014.[1918]

$C_3H_7CH{=}CCl-CH_2PO(OEt)_2$. V.2. IR, ^{1}H.[1228]

$C_3H_7-CO_2CHClCH_2PO(OEt)_2$. V.2. m. 21-22°.[1134]

$C_4H_9CHClCH_2PO(OH)_2$. I.1. m. 81-82°.[524]

$C_4H_9CHClCH_2PO(OR)_2$.

R = Me. V.2. b_5 124-125°, n_D^{20} 1.4509, d_4^{20} 1.1196.[524]

R = Et. V.2. b_2 119-120°, n_D^{20} 1.4459, d_4^{20} 1.0632.[524]

R = Pr. V.2. b_2 143-144°, n_D^{20} 1.4479, d_4^{20} 1.0316.[524]

R = $H_2C{=}CHCH_2$-. V.2. b_5 147-148°, n_D^{20} 1.4649, d_4^{20} 1.0672.[524]

R = Bu. V.2. b_4 164-166°, n_D^{20} 1.4483, d_4^{20} 1.0131.[524]

$ClH_2CSiMe_2OSiMe_2$-$CH_2PO(OEt)_2$. V.5.[321]
2,3-$Cl_2C_6H_3PO(OH)_2$. I.1. m. 200-202°.[460]
2,5-$Cl_2C_6H_3PO(OH)_2$. I.2. m. 194-197°.[1020] I.1. m. 193-196°[582]

2,5-$Cl_2C_6H_3PO(OEt)_2$. V.2. b_3 160-164°, n_D^{25} 1.5105.[1020]
3,5-$Cl_2C_6H_3PO(OH)_2$. I.1. m. 188-190°.[460]
$Cl_2HCSiMe_2$-OCHEtPO(OEt)$_2$. VI.1. $b_{0.15-0.2}$ 114-118°.[258]
$(CH_3)_2CHO_2C$-NHC(O)$CCl_2PO(OR)_2$.

R = Et. V.9. $b_{0.01}$ 150-151°.[281]

R = Bu. V.9. $b_{0.1}$ 156-157°.[281]

2,4,5-$Cl_3C_6H_2PO(OH)_2$. I.1. m. 252-254°.[460]
CH_3-$CO_2CMe[-C(O)CCl_3]PO(OR)_2$.

R = Me. V.14. b_1 109-111°.[1650]

R = Et. V.14. b_1 115-116°.[1650]

$C_3H_7CO_2CH(-CCl_3)PO(OMe)_2$. From $(MeO)_2OPH$ and $Cl_3CCHClO_2$-CC_3H_7, insecticide.[347] V.14.[348] V.14. $b_{0.013}$ 154-155°.[1400]

$(CH_3)_2CH$-$CO_2CH(-CCl_3)PO(OCH_2CH_2Cl)_2$ V.14. $b_{0.015}$ 137-139°.[1400] V.14.[347]
$MeCO_2CH(-CCl_3)OCHMePO(OEt)_2$. V.14. b_1 139-140°.[100]
$C_2H_5OC(O)OCH(-CCl_3)$-$OCH_2PO(OEt)_2$. V.14. b_1 148-149°.[100]
2-$FC_6H_4PO(OH)_2$. I.1. m. 146-149°.[582]
4-$FC_6H_4PO(OH)_2$. I.1. m. 125-127°.[291]
4-$FC_6H_4PO(OR)OBa_{1/2}$.

R = Me. III.4.[384]

R = ClH_2CCH_2-. III.4.[384]

R = $H_2NCH_2CH_2$-. III.4.[384]

R = i-Pr. III.4.[384]

R = $HOCH_2CH_2CH_2$-. III.4.[384]

R = $NCCH_2CH_2$-. III.4.[384]

R = $H_2NC(O)CH_2CH_2$-. III.4.[384]

R = $Me_2NCH_2CH_2$-. III.4.[384]

R = Me_2EtC-. III.4.[384]

R = $C_{10}H_{21}$-. III.4.[384]

4-$FC_6H_4PO(OR)_2$.

R = Me. V.5. $b_{0.2}$ 80-81°.[1421]

R = Et. V.2. b_8 136-137°, n_D^{20} 1.4788, ^1H, ^{19}F.[1809]

$(C_2H_5)_2NC(O)CHFPO(OEt)_2$. V.6. $b_{0.5}$ 131-132°, n_D^{20} 1.4319,

d_4^{20} 1.1210.[549]
$F_2\overline{CCF_2-C(OEt)=C}PO(OEt)_2$. V.14. b_1 92-95°.[2041]
$(F_3C)_2C=C(OEt)-PO(OEt)_2$. V.5.[979] V.14. b_7 111-112°.[979]
$F_2HC(CF_2)_5PO(OR)_2$.

R = Me. V.7. $b_{0.6}$ 55°, n_D^{25} 1.3486.[302]

R = Et. V.7. $b_{0.22}$ 80°, n_D^{25} 1.3507.[302]

2-$IC_6H_4PO(OH)_2$. I.1. m. 219-222°.[582]
3-$IC_6H_4PO(OH)_2$. I.1. m. 183-184°.[460] I.13. m. 182-183°.[1008]
4-$IC_6H_4PO(OH)_2$. I.1. m. 221-223° (cryst. as 1-hydrate).[579] I.13. m. 228-229°.[1008]
2-$O_2NC_6H_4PO(OEt)_2$. V.5. $b_{0.01}$ 120°.[1922]
3-$O_2NC_6H_4PO(OH)_2$. I.13. m. 140°.[1279, 1280, 1395] I.13.[1008] Ba salt is soluble in water.[1279, 1280] I.1. m. 155-156°.[479]
3-$O_2NC_6H_4PO(OR)_2$.

R = Me. V.2. b_1 148-150°.[350]

R = Pr. V.2. b_1 161-164°.[350]

R = i-Pr. V.2. $b_1$156-160°.[350]

R = Bu. V.2. $b_{0.6}$ 162-167°.[350]

R = i-Bu. V.2. $b_{0.6}$ 161-163°.[350]

4-$O_2NC_6H_4PO(OH)_2$. I.1. m. 195-197° (EtOH/hexane).[1824]
4-$O_2NC_6H_4PO(OR)OH$.

R = Et. III.4. m. 111-112.5°.[326]

R = Ph. III.4. m. 134-141°.[326]

4-HO-3-$O_2NC_6H_3PO(OH)_2$. V.14.[228] V.14. m. 214-216° (water).[186]

$C_6H_5PO(OH)_2$. I.1. m. 162.5-163°.[479] I.1. m. 161-162°.[2019] I.1. m. 158°.[1256, 1257] I.2. m. 158-159°.[1020] I.1. m. 158-160°.[1395] I.1.[1285, 1285a] I.6. m. 158°.[1257] I.6.[985] I.2/I.6.[1869, 1870] I.2. m. 157-158°.[1292] I.2. m. 161-163°.[328] I.6/I.2.[67] Sn salt.[754] I.2. m. 162.5-164°,[1789] ^{31}P -18.5 ppm.[1189a]
$C_6H_5PO(OH)F$. II.1. As aniline salt, m. 155-158°.[1717]
$C_6H_5PO(OR)OH$.

R = Me. III.4. Isolated as Ba salt.[373, 378]

R = Et. III.4. Isolated as Ba salt.[373 378] III.3. Isolated as dicyclohexylammonium salt, m. 140.7-141.8°.[1680] III.4. n_D^{20} 1.5258, d_4^{20} 1.2331.[509]

R = Br_3CCH_2-. III.4. Isolated as Na salt.[376]

R = ClH_2CCH_2-. III.4. Isolated as Ba salt.[372]

R = $H_2NC(O)CH_2$-. III.4. Isolated as Ba salt.[380]

R = Pr. III.4. Isolated as Ba salt.[373, 378]

R = i-Pr. III.4. Isolated as Ba salt.[373, 378]

R = HC≡C-CH_2-. III.4. Isolated as Ba salt.[376]

R = $ClH_2CCH_2CH_2$-. III.4. Isolated as Ba salt.[372]

R = $FH_2CCH_2CH_2$-. III.4. Isolated as Ba salt.[372] III.4. Isolated as Ca salt.[381]

R = $(FH_2C)_2CH$-. III.4. Isolated as Ba salt.[372]

R = $F_2HCCHF_2CH_2$-. III.4. Isolated as Ca salt.[381]

R = $NCCH_2CH_2$-. III.4. Isolated as Ba salt.[380]

R = $H_2NC(O)CHMe$-. II.2. Isolated as Ba salt.[380] III.3.[380]

R = $(HOCH_2)_2CH$-. III.4. Isolated as Ba salt.[376]

R = $HOCH_2CH(OH)CH_2$-. III.4. Isolated as Ba salt.[376]

R = Bu. III.4. $b_{0.009}$ 145-148°, n_D^{20} 1.5119, d_4^{20} 1.1399.[509] III.4. IR.[1458] III.4. Isolated as Ba salt.[373, 378] III.2. n_D^{25} 1.5092.[865] III.7.[1496]

R = t-Bu. III.1. Isolated as Na salt, m. 237°, free acid m. 88.5°.[448]

R = $ClH_2C(CH_2)_3$-. III.4. Isolated as Ba salt.[372]

R = $NC(CH_2)_3$-. III.4. Isolated as Ba salt.[380]

R = $NCCHMeCH_2$-. III.4. Isolated as Ba salt.[380]

R = $H_2NC(O)CMe_2$-. III.4. Isolated as Ba salt.[380]

R = Am. III.4. Isolated as Ba salt.[373, 378]

R = $(CH_3)_2C{=}CH{-}CH_2$-. III.4. Isolated as Ba salt.[385]

R = $NC(CH_2)_4$-. III.4. Isolated as Ba salt.[380]

R = Ph. III.4. m. 70-72°.[326] III.4. $b_{0.001}$ 170-180°.[509] III.2. m. 57°.[1257, 1284]

R = $HCF_2(CF_2)_3CHMe$-. III.4. $b_{0.001}$ 130-140°, n_D^{20} 1.4370, d_4^{20} 1.5229.[509]

R = $(CH_3)_3\overset{\oplus}{N}(CH_2)_3$-. III.3.[354]

R = 2-HOC_6H_4-. III.1. m. 123-124°.[247]

R = $PhCH_2$-. III.3. IR.[1458]

R = $C_4H_9CHEtCH_2$-. III.3. IR.[1458]

R = $(CH_3)_2C{=}CHCH_2CMe{=}CHCH_2$-. III.4. Isolated as Ba salt.[385]

R = $PhNHC(S)NHCH_2CH_2$-. III.8. Isolated as Ca salt.[374]

R = 4-$BrC_6H_4NHC(S)NHCH_2CH_2$-. III.8. Isolated as Ca salt.[374]

R = $(CH_3)_2C$=$CHCH_2CHCMe$=$CHCH_2$-. III.4. Isolated as Ba salt.[385]

R = $H_2CCHMeCH_2CH_2CH(-CHMe_2)CH$-. III.4. Isolated as Ba salt.[376]

R = 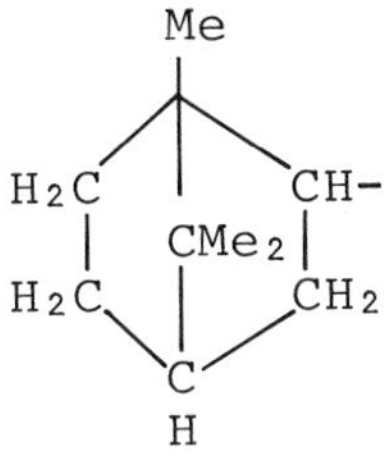 III.4. m. 104-105°.[376]

R = $PhNHC(S)NMeCH_2CH_2$-. III.8. Isolated as Ca salt.[374]

R = 4-$BrC_6H_4NHC(S)NMeCH_2CH_2$-. III.8. Isolated as Ca salt.[374]

R = $C_{13}H_{27}$-. III.3. IR.[1458]

$C_6H_5PO(OR)Cl$.

R = ClH_2CCH_2-. IV.1. b_2 155-158°.[625]

R = Bu. IV.1. $b_{0.75}$ 116-117.5°.[865]

R = Ph. IV.2. $b_{0.025}$ 121-125°.[776, 1193]

R = cycl.-C_6H_{11}-. IV.2. n_D^{25} 1.5328.[740]

$C_6H_5PO(OR)F$.

R = Me. IV.2. b_4 90-91°, n_D^{23} 1.4863, IR, ^1H.[1717]

R = Bu. IV.2. $b_{0.1}$ 81-82°, n_D^{22} 1.4770, IR.[1717]

R = cycl.-C_6H_{11}-. IV.2. $b_{0.05}$ 94°, n_D^{25} 1.5021.[740]

$C_6H_5PO(OR)_2$.

R = Me. V.2.[1824] V.5. $b_{0.9}$ 115°, n_D^{20} 1.5060.[1988] V.5.[1528, 1529] V.9. b. 247°.[1257, 1279, 1280]

R = Et. V.5.[1529] V.5. $b_{0.1}$ 97°, n_D^{20} 1.4921.[1988] V.14. $b_{0.1}$ 95-100°.[451] V.5. $b_{0.2}$ 96-98°, n_D^{20} 1.4926.[1987] V.6.[689] V.2.[1537] V.5. $b_{0.2}$ 94-98°.[1985] V.5.[1195] Reactivity with PhMgBr.[763] V.13.[848] ^{31}P -16.9 ppm.[1189a]

R = ClH_2CCH_2-. V.4.[625] V.2. b_1 155-156°.[614]

R = $Cl^{\ominus}[H_3\overset{\oplus}{N}CH_2CH_2]$-. V.2.[1936]

R = i-Pr. V.5. $b_{0.1}$ 96-97°.[1985] V.5. $b_{0.1}$ 94°, n_D^{20} 1.4810.[1988] V.5. $b_{0.1}$ 96-97°, n_D^{20} 1.4810.[1987] V.5.[1529] ^{31}P -15.7 ppm.[1189a]

R = H_2C=$CHCH_2$-. V.2. b_1 128°, n_D^{25} 1.5128.[2019]

R = HC≡CCH_2-. V.2. $b_{0.15}$ 129-133°.[377] V.2.[616]

R = $Br_2HCCBr_2CH_2$-. V.14. m. 132-133°.[377]

R = JHC=CJ-CH_2-. V.14. m. 68-70°.[377]

R = Bu. V.2. b_4 166°.[2021] V.5.[1529] Extraction of $FeCl_3$.[1353]

R = C_2H_5CHMe-. Extraction of $FeCl_3$.[1353]

R = H_2C=$CMeCH_2$-. V.2. b_{2-3} 140-143°.[2019]

R = $C_2H_5OCH_2CH_2$-. V.2. b_{17} 220°.[2021]

R = Am. V.2. b_5 170°.[2021]

R = H_2C=CH-$CO_2CH_2CH_2$-. V.2.[1901]

R = Ph. V.2. m. 63.5°.[1257, 1284] Extraction of $FeCl_3$.[1353] From $[Ph\overset{\oplus}{P}(OPh)_3]J^{\ominus}$ and dil. NaOH (reflux). m. 72.5-73.5°.[1529]

R = cycl.-C_6H_{11}-. V.2.[2021]

R = $C_4H_9OCH_2CH_2$-. V.2. b_4 207-210°,[2021] ^{31}P -19 ppm.[1189a]

R = H_2C=$CMeCO_2CH_2CH_2$-. V.2.[1901]

R = 2,3,4,6-Cl_4C_6H-. V.2. m. 170°.[831]

R = C_4H_9-$CHEtCH_2$-. V.2. b_4 204-207°.[2021]

$C_6H_5PO(OMe)OR$.

R = Cl_2C=CH-. V.5.[1978] V.5. b_1 125-137°.[62]

R = H_2C=CMe-. V.5. b_8 137-138°.[1566]

R = $H_2NC(O)CHMe$-. V.13.[380]

R = Bu. V.13. n_D^{20} 1.4898.[975]

R = 2,4-$Cl_2C_6H_3$-. V.2. $b_{0.5}$ 130°.[1394]

$C_6H_5PO(OEt)OCH$=CCl_2. V.5. b_{2-3} 131-144°.[62]

$C_6H_5PO(OBu)OCH_2CH_2Cl$. V.2. $b_{1.5}$ 165-168°.[625]

$(C_6H_5-)\overline{OPO-R-O}$.

R = -CH_2CH_2-. V.2. b_{6-7} 210°.[2020]

R = -$(CH_2)_3$-. V.2. $b_{7.5}$ 212-214°.[2020] Conformation.[1159] V.2. $b_{0.05}$ 175°.[354]

R = -CHMeCHMe-. V.2. b_{15} 210-215°.[2020]

R = $-CH_2CMe_2CH_2-$. V.5. $b_{0.1}$ 150-160°, m. 109-111°.[1985, 1987] V.2. m. 103-105°.[424]

R = $-CH_2(CF_2)_3CH_2-$. V.2. m. 80-82°.[1881]

R = 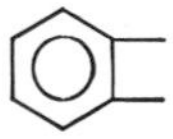 V.2. m. 124-125°, b_9 206°.[96]

R = 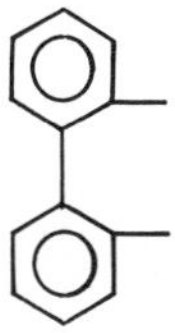 V.13. m. 116-117°.[1923]

R = 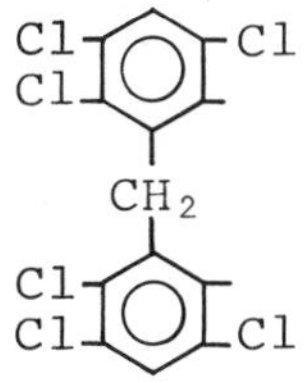 V.2. m. 195°.[831]

$(C_6H_5-)(EtO)OPO(CH_2)_4OPO(OEt)(-C_6H_5)$. V.2. $b_{0.05}$ 200-205°.[1957]
$C_6H_5PO(OSiMe_3)_2$. V.1. b_8 96.5-98°.[1682]

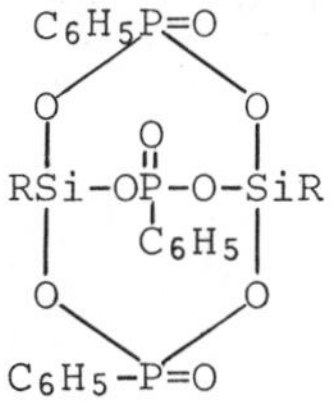

R = Me. VI.1. $b_{0.09}$ 282-283°.[73]

R = Et. VI.1. $b_{0.043}$ 290-295°.[73]

R = Bu. VI.1. $b_{0.035}$ 269-270°.[73]

R = Ph. VI.1. $b_{0.028}$ 270-275°.[73]

$C_6H_5PO(OSnBu_3)_2$. From $PhPO(OH)_2$ + $(Bu_2Sn)_2O$.[1948]
$C_2H_5-C{\equiv}C-C{\equiv}C-PO(OEt)_2$. V.5. $b_{0.5}$ 130-131°.[823]
$4-H_2N-3-O_2NC_6H_3PO(OH)_2$. I.13. Dec. 231°.[1395]
$2-HOC_6H_4PO(OH)_2$. I.13. m. 124-127°.[581] I.1, I.13.[1131]
$2-HOC_6H_4PO(OR)_2$.

R = Me. V.5. m. 97-98°, IR.[1422]

R = $(ClH_2C-)_3C-$. V.8. m. 143-144°.[34]

$3-HOC_6H_4PO(OMe)_2$. V.5. m. 91-92°, IR.[1422]

$4\text{-}HOC_6H_4PO(OH)_2$. From $Si(OPh)_4$ + P_4S_{10} and hydrolysis of the reaction product. m. 173-173.5°.[1065] I.13.[480]

$4\text{-}HOC_6H_4PO(OR)_2$.

R = Me. V.5. m. 77-78°, IR.[1422]

R = Et. V.5. m. 93-94° (benzene).[1987]

SCH=CH-CH=C-CH=CHPO(OEt)OH. I.2. As dicyclohexylammonium salt, m. 170-170.5°.[770]

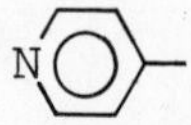 $CH_2PO(OEt)_2$. V.6. $b_{0.05}$ 89°[1206]

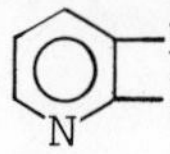 Me, $PO(OH)_2$. I.2. m. 279-280°.[1718]

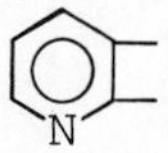 Me, $PO(OEt)_2$. V.6. $b_{0.07}$ 109-110°.[1718]

Me 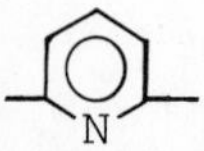 $PO(OH)_2$. I.2. m. 277-280°.[1718]

Me 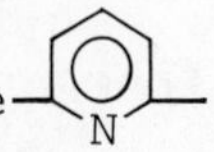 $PO(OEt)_2$. V.6. $b_{1.5}$ 140°, IR, ^1H.[1718]

Me 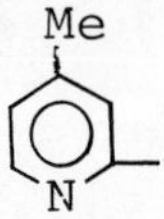 $PO(OH)_2$. I.2. m. 272-276°.[1718]

Me 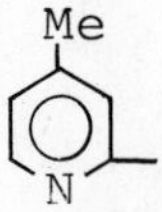 $PO(OEt)_2$. V.6. $b_{0.05}$ 109-112°, IR, ^1H.[1718]

$2\text{-}H_2NC_6H_4PO(OH)_2$. I.13. m. 199-200°.[480, 577]

$2\text{-}H_2NC_6H_4PO(OMe)_2$. V.5. $b_{0.25}$ 117-118°, IR.[1422]

$3\text{-}H_2NC_6H_4PO(OH)_2$. I.13. Dec. 280°.[1279, 1280] I.13. Dec. 290°.[1008] I.13. Dec. 275-280°.[1395]

$4\text{-}H_2NC_6H_4PO(OH)_2$. I.13. m. 245°.[213] I.13.[480]

$4\text{-}H_2NC_6H_4PO(OMe)_2$. V.5. m. 107-108°, IR.[1422] V.14, m. 105-106.5°.[932]

$C_2H_5O_2CC(-CN)=CHPO(OEt)_2$. From $EtO_2CC(-CN)=CHOEt$ and $NaPO(OEt)_2$, $b_{0.1}$ 128°, n_D^{20} 1.4624.[1046]

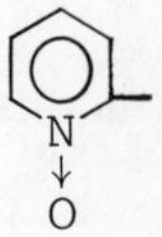 $CH_2PO(OH)_2$. I.13.[278]

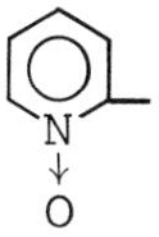
$CH_2PO(OEt)_2$. V.6.[278]

$CH_2PO(OEt)_2$
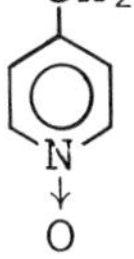
V.14.[278]

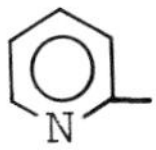
$CH(OH)PO(OEt)_2$. V.8. m. 142.5-143°.[327]

$3\text{-}H_2N\text{-}4\text{-}HOC_6H_3PO(OH)_2$. I.13. m. 200°.[186]
$4\text{-}H_2N(O)_2SC_6H_4PO(OH)_2$. I.1. m. 224-225°.[479]

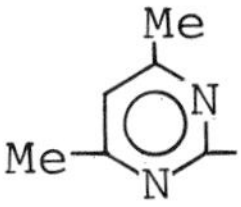
$PO(OH)_2$. I.2. m. 153-154 .[1022]

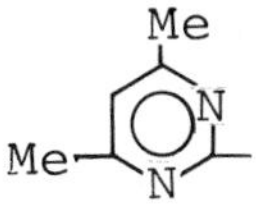
$PO(OEt)_2$. V.6. $b_{0.07}$ 121-122°, n_D^{30} 1.4847, d_4^{30} 1.1298.[1022]

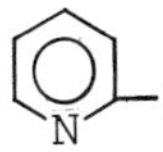
$CH(NH_2)PO(OH)_2$. I.2. m. 165-166°.[1203]

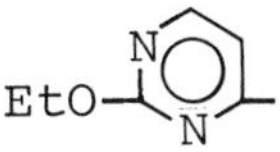
$PO(OPr\text{-}i)_2$. V.14. $b_{0.05}$ 120-122°, n_D^{29} 1.4753, d_4^{29} 1.1090.[1022]

$O{=}C\text{-}CH{=}CMe\text{-}N{=}C(OH)\text{-}N\text{-}CH_2PO(OH)_2$ (ring). I.13.[1739]
$O{=}C\text{-}CH{=}CMe\text{-}N{=}C(OH)\text{-}N\text{-}CH_2PO(OR)_2$ (ring).

R = Me. V.2.[1738]

R = Bu. V.2.[1739]

$O{=}C\text{-}C(\text{-}NO_2){=}CMe\text{-}N{=}C(\text{-}NH_2)\text{-}NCH_2PO(OH)_2$ (ring). V.14.[1739]

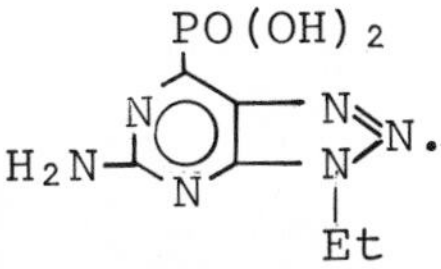
I.2. m. 284-285°.[1883]

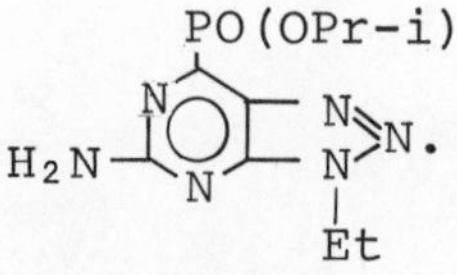
V.5. m. 119°.[1883]

$HC{\equiv}C\text{-}CH_2CH[\text{-}C(O)Me]\text{-}PO(OR)_2$.

R = Et. V.14, V.7. b_8 129-130°.[1605]

R = Pr. V.14, V.7. b_4 150-152°.[1605]

R = Bu. V.14, V.7. b_1 138-140°.[1605]

$[CH_3C(O)]_2C{=}CH\text{-}PO(OR)_2$. From $[MeC(O)]_2C{=}CHOR$ and $NaPO(OR)_2$.

R = Et. $b_{0.01}$ 127°, n_D^{20} 1.4622.[1046]

R = Bu. $b_{0.01}$ 157°, n_D^{20} 1.4648.[1046]

$H_2C{=}CHCH_2O_2CCH{=}CHPO(OCH_2\text{-}CH{=}CH_2)_2$. V.7. b_5 154-156°.[967]

$O{=}COCH_2CH_2C[\text{-}C(O)Me]\text{-}PO(OEt)_2$. V.6. $b_{0.05}$ 130°.[317]

$MeO_2CCH{=}C[\text{-}O_2CMe]\text{-}PO(OMe)_2$ (cis or trans). V.7. $b_{0.02}$ 112-115°, n_D^{25} 1.4586, ^{31}P-12.0 ppm.[1382]

$HO_2CCH_2CH(\text{-}CO_2H)CH(CO_2H)\text{-}PO(OH)_2{\cdot}H_2O$. I.2. m. 123.5°.[2082, 765]

$HC{=}CMe\text{-}NH\text{-}CMe{=}C\text{-}PO(OEt)_2$. V.11. $b_{1.5}$ 53°, IR, ^1H.[694]

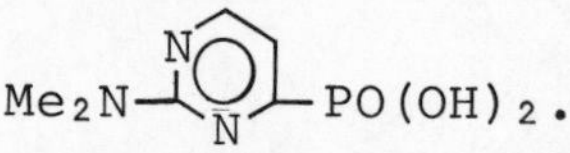
. I.2. m. 196-198°.[1022]

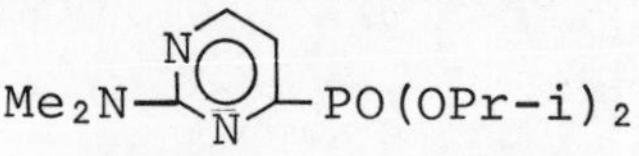
. V.14. $b_{0.05}$ 110-112°, m. 28-29°, n_D^{25} 1.5005, d_4^{25} 1.1006.[1022]

$O{=}C\text{-}CH{=}CMe\text{-}N{=}C(\text{-}NH_2)\text{-}N\text{-}CH_2PO(OH)_2$. I.13.[1739]

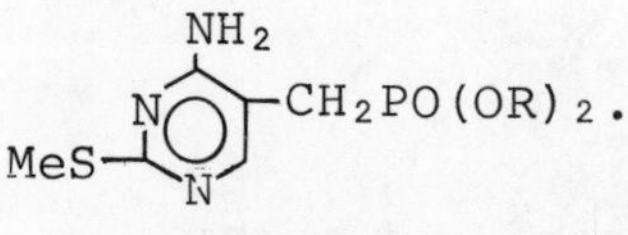

R = Me. V.5. m. 281-284° (dec.).[783]

R = Et. V.5. m. 268-270° (dec.).[783]

R = i-Pr. V.5. m. 286-289° (dec.).[783]

R = Ph. V.5. m. 197-200° (dec.).[783]

R = C_6H_{13}-. V.5. m. 121-123°.[783]

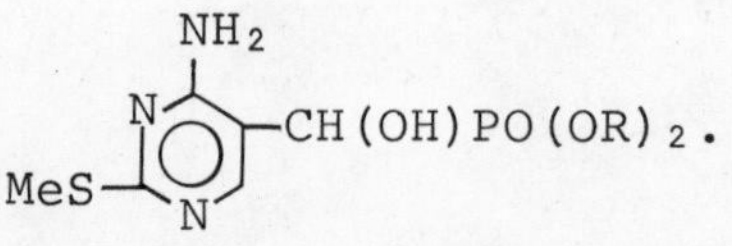

R = Me. V.8. m. $109-112^\circ$.[783]

R = Et. V.8. m. $120-122^\circ$.[783]

R = Bu. V.8. m. $171-173^\circ$(dec.).[783]

R = Ph. V.8. m. $134-136^\circ$.[783]

$MeS(O)_2$ — [pyrimidine ring with NH_2] — $CH(OH)PO(OEt)_2$. V.14. m. $181-183^\circ$.[783]

[structure: ring with CO_2H and $PO(OEt)$] — $C(O)Me$. V.14. As p-toluidide; m. $149-165^\circ$.[318]

$C_4H_9C\equiv CPO(OR)_2$.

R = Et. V.11. $b_{0.1}$ 96°.[363] V.6. $b_{0.6}$ 115°.[1962, 1963]

R = Bu. V.6. $b_{0.5}$ 143°.[1963]

$(C_3H_7CH=C=CH-)\overline{OP-O-R-O}$.

R = $-CH_2CHMe-$. V.2. n_D^{20} 1.4910, IR.[287]

R = $-CH_2CH_2CHMe-$. V.2. n_D^{20} 1.4950, IR.[287]

$H_2\overline{C(CH_2)_3CH=C}PO(OH)_2$. I.13.[522] I.2. m. $129-131^\circ$.[959]

$H_2\overline{C(CH_2)_3CH=C}PO(OEt)_2$. V.14. b_5 118°.[757]

$H_2\overline{C(CH_2)_2CH=CH-C}HPO(OH)_2$. I.3. m. $104-106^\circ$.[522]

$H_2\overline{C(CH_2)_2CH=CH-C}HPO(OR)_2$.

R = Me. V.6. $b_{1.5}$ $92.5-93.5^\circ$.[132]

R = Et. V.6. b_1 $93.5-94^\circ$.[132]

R = Pr. V.6. b_1 $119-120^\circ$.[132]

R = i-Pr. V.6. b_2 $107.5-108^\circ$.[132]

R = Bu. V.6. b_2 $129.5-130^\circ$.[132]

R = i-Bu. V.6. b_2 $133-134^\circ$.[132]

$H_2\overline{C-CH=CHCH_2CH_2CH}-PO(OEt)_2$. V.14. $b_{1.5}$ $114-115^\circ$.[451]

$H_2\overline{C(CH_2)_3C}=CHPO(OMe)_2$. V.14. $b_{0.08}$ 67°, n_D^{25} 1.4750, IR, ^1H.[1191]

$Me\overline{C=N-NMe-CMe=C}-PO(OR)_2$.

R = Et. V.2. b_{13} $173-176^\circ$.[659]

R = Pr. V.2. b_{18} $198-200^\circ$.[659]

R = Bu. V.2. b_{18} $212-214^\circ$.[659]

$H_2\overline{CCH_2N[-C(O)Me]-N=C}-CH_2PO(OEt)_2$. V.14. $b_{0.002}$ $142-143^\circ$.

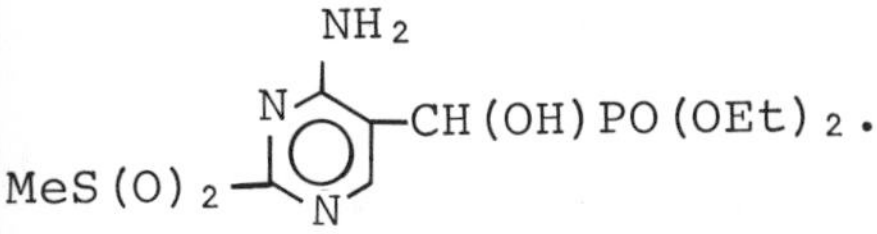

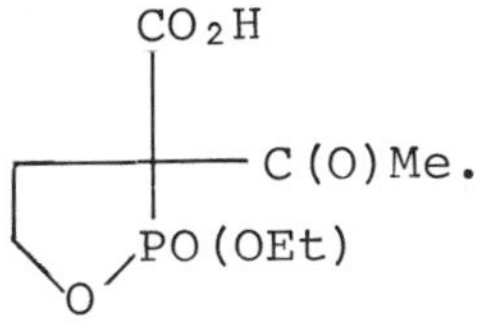

n_D^{20} 1.4873, d_4^{20} 1.1703, IR, ^1H.[1580]

MeCH-CH$_2$N[-C(O)Me]-N=C-PO(OEt)$_2$. V.14. $b_{0.002}$ 139-140°, n_D^{20} 1.4875, d_4^{20} 1.1628, IR, ^1H.[1580]

H_2C-C[-C(O)Me]=N-NH-CMe-PO(OMe)$_2$. V.14. m. 111-112°, IR, ^1H.[1878a]

H_2C(CH$_2$)$_3$CH-C(N$_2$)-PO(OMe)$_2$. V.14. $b_{0.2}$ 80°, n_D^{25} 1.4800, IR, ^1H.[1191]

(CH$_3$)$_2$C=CMe-C(O)PO(OMe)$_2$. V.5. $b_{0.22}$ 80-86°, n_D^{25} 1.4678, IR, ^1H.[1191]

HC≡CCMe$_2$OCH$_2$PO(OEt)$_2$. V.14. b_6 138-140°.[1610]

O=C(CH$_2$)$_4$CHPO(OEt)$_2$. V.14. b_1 111-113°.[161, 148]

H_2CC(O)(CH$_2$)$_3$CHPO(OEt)$_2$. V.7. b_{15} 170°.[1612]

H_2C(CH$_2$)$_2$CH-O-CHCHPO(OR)$_2$.

R = Et. V.6.[163] V.14. b_1 118.5-119°.[132]

R = Pr. V.14. b_1 131-132°.[132]

R = i-Pr. V.14. b_2 129-130°.[132]

R = Bu. V.14. $b_{0.5}$ 165-166°.[132]

H_2C(CH$_2$)$_3$CHC(O)PO(OR)$_2$.

R = Me. V.5. $b_{0.05}$ 72-75°, n_D^{25} 1.4588, IR, ^1H.[1191]

R = Et. V.5. b_3 121-126°. As 2,4-dinitrophenylhydrazone, m. 123-125°, IR, ^1H.[244]

CH$_3$C(O)SCH$_2$CH=CH-CH$_2$PO(OEt)$_2$. V.14. b_8 180-181°.[1630]

[CH$_3$C(O)]$_2$CHCH$_2$PO(OBu)$_2$. V.7.[840]

C$_3$H$_7$CO$_2$CH=CHPO(OR)$_2$.

R = Me. V.14. b_1 114-116°, n_D^{20} 1.4590, d_4^{20} 1.1727.[1135]

R = Et. V.14. $b_{1.5}$ 125-126°, n_D^{20} 1.4510, d_4^{20} 1.1023.[113]

H_2C=CMeCO$_2$CHMePO(OR)$_2$.

R = Et. V.14. $b_{0.6}$ 91-92°.[1598]

R = Pr. V.14. b_2 121-123°.[1598]

R = i-Pr. V.14. b_2 117°.[1598]

R = Bu. V.14. b_3 136-138°.[1598]

R = i-Bu. V.14. b_3 119-120°.[1598]

C$_2$H$_5$O$_2$C-CH=CMePO(OEt)$_2$. V.7. b_9 150-151.5°.[968] Cis-comp. V.5.[1062] Trans-comp. V.7.[1062] V.7. b_3 119-120°.[967] From MeCH=C(-NO$_2$)CO$_2$Et and P(OEt)$_3$, cis:trans - 7:3, b_2 118-121°, IR, ^1H.[392] V.5. $b_{0.8}$ 112-114°, n_D^{20} 1.4482.[104] From MeC(O)CH$_2$CO$_2$Et and HPO(OEt)$_2$.[840]

C$_3$H$_7$O$_2$C-CH=CHPO(Pr)$_2$. V.7. b_{15} 184-186°.[967, 968]

C$_2$H$_5$O$_2$CCH$_2$CH$_2$C(O)PO(OEt)$_2$. V.5. $b_{0.25}$ 133-135°, n_D^{20} 1.4412.[1042]

C$_2$H$_5$O$_2$CCH[-C(O)Me]PO(OEt)$_2$. V.11. b_1 129-130°.[1034]

(HO)HC=C(-CO$_2$Et)CH$_2$PO(OEt)$_2$. V.14. $b_{0.5}$ 134-135°.[1043]

CH$_3$C(O)SCH$_2$CH(-CO$_2$Me)-PO(OEt)$_2$. V.14. b_4 178°, n_D^{20} 1.4650, d_4^{20} 1.2183.[1642]

$CH_3O_2CCH_2CH(CO_2Me)PO(OR)_2$.

R = Me. V.7. b_2 142°.[1548] V.7. $b_{0.2-0.25}$ 116-122°.[781]

R = H_2C=$CHCH_2$-. V.7. $b_{0.18}$ 144-148°, n_D^{20} 1.4048, IR.[781]

$(C_2H_5)_2N$-C≡C-PO(OEt)$_2$. B.14. b_1 145-147°, n_D^{20} 1.4680, d_4^{20} 1.0538.[828]

HC≡$CCH_2NHCMe_2PO(OR)_2$.

R = Me. V.14. $b_{0.8}$ 87-89°.[1610]

R = Et. V.14. b_6 126-127°.[1610]

R = Pr. V.14. $b_{0.3}$ 84-85°.[1610]

R = Bu. V.14. $b_{0.8}$ 102-106°.[1610]

$C_4H_9CH(-CN)PO(OEt)_2$. V.5.[1358]

H_2C=CH-C(O)NH-$CMe_2PO(OEt)_2$. V.14. m. 66°.[1597]

$NCCH_2CH_2OCHEtPO(OEt)_2$. V.14. 147-148°, n_D^{20} 1.4425, d_4^{20} 1.0915.[1675]

$H_2\overline{CCH_2OCH_2CH_2N}$-C(O)$CH_2PO(OEt)_2$. V.5. $b_{1.4}$ 167-170°.[1940]

$CH_3C(O)CH_2NHC(O)CHMePO(OEt)_2$. V.5. $b_{0.1}$ ~146°.[1526]

O=$\overline{CCH_2NH-C(O)N}$-CH(-NMe_2)-PO(OEt)$_2$. V.14. m. 113-117°, ^1H.[715a]

C_4H_9CH=$CHPO(OR)_2$.

R = Me. V.14. $b_{0.03}$ 57°, n_D^{25} 1.4450, IR, ^1H.[1191] V.14. b_4 103-105°.[524]

R = Et. V.14. b_2 114-116°.[524]

R = Pr. V.14. b_2 126-128°.[524]

R = H_2C=$CHCH_2$-. V.14. b_2 132-134°.[524]

R = Bu. V.14. b_5 155-156°.[524]

$(C_2H_5)_2C$=$CHPO(OR)_2$.

R = Me. V.14. $b_{0.03}$ 92-93°.[1346]

R = Et. V.14. $b_{0.05}$ 91-92°.[1346]

$(CH_3)_2CHCH$=CMe-PO(OEt)$_2$. V.14. $b_{0.6}$ 91°.[757]

cycl.-$C_6H_{11}PO(OH)_2$. I.13.[580] I.1. m. 160-161°.[2108] I.1. m. 166-167°.[403] I.8. m. 166-167°.[696] I.2. As mono NH_4^+ salt, m. 149°.[1503] NMe_4 -salt ^{31}P - 24.8 ppm.[1189a]

cycl.-$C_6H_{11}PO(OR)OH$.

R = Et. III.4. n_D^{20} 1.4777, d_4^{20} 1.1393.[509]

R = Bu. III.3. As dicyclohexylamine salt, m. 160-163°.[1680]

R = i-Bu. III.4. n_D^{20} 1.4679, d_4^{20} 1.0792.[509]

R = Am. III.4. n_D^{20} 1.4701, d_4^{20} 1.0732.[509]

R = Ph. III.2. m. 80-105° (as 1-hydrate).[657]

R = $C_4H_9CHEtCH_2$-. III.4. n_D^{20} 1.4693, d_4^{20} 1.0382.[509]

cycl.-$C_6H_{11}PO(OR)Cl$.

    R = Me. IV.6. $b_{0.03}$ 69°.[973]

    R = Et. IV.6. b_3 160°.[973] IV.6.[2108]

cycl.-$C_6H_{11}PO(OR)_2$.

    R = Me. V.2. b_{10} 124°.[500] V.7. b_2 96°, n_D^{18} 1.4720, d_4^{18}
    1.090, IR.[1503]

    R = Et. V.7. b_{13} 140-144°, n_D^{18} 1.4615, d_4^{20} 1.032, IR.
    [1503] V.14. $b_{0.3}$ 82-84°.[451] V.5.[1366] From cyclohex-
    ane, Cl_2 and $HPO(OEt)_2$ (h·ν), $b_{3.5}$ 125-127°.[1354]

    R = Pr. V.7. b_5 164-165°, n_D^{18} 1.4680, d_4^{20} 0.997, IR.
    [1503]

    R = i-Pr. V.7. b_{11} 141-145°, n_D^{18} 1.4620, d_4^{20} 1.012, IR
    [1503]

    R = $HC{\equiv}CCH_2$-. V.2. $b_{1.2}$ 155-157°, IR.[377]

    R = Br_2HC-CBr_2CH_2-. V.14. m. 95-98°.[377]

    R = $JHC{=}CJ$-CH_2-. V.14. m. 102-105°.[377]

    R = Bu. V.7. b_1 134-140°.[1952]

    R = Ph. From cyclohexane, Cl_2, and $HPO(OPh)_2$, m. 62°.
    [1354]

    R = $4-HOC_6H_4$-. V.2. m. 182°.[1461]

    R = $(CH_3)_3CO$-. V.2. m. 58.5-60°.[448]

    R = Me_3Si-. VI.1.[1341]

$C_5H_{11}C(N_2)PO(OMe)_2$. V.14. $b_{0.44}$ 82°, IR, ^1H.[1191]

$(CH_3)_2C\underset{N=N}{\overset{CH_2}{<}}CMePO(OMe)_2$. V.14. $b_{0.007}$ 94-95°, n_D^{20}
    1.4571, d_4^{20} 1.1304, IR, ^1H.[1581]

$CH_3C(O)(CH_2)_4PO(OEt)_2$. V.14. IR.[1959]

$CH_3C(O)CH_2CHMeCH_2PO(OEt)_2$. V.7.[642]

$CH_3C(O)CHMeCHMePO(OR)_2$.

    R = Me. V.7. b_{20} 158°.[1550]

    R = Bu. V.7. $b_{2.8}$ 158°.[877]

$CH_3C(O)CH_2CMe_2PO(OH)_2$. I.2.[1550] I.9. m. 63-64^O(1-hydrate).
    I.9. m. 62-63^O(1-hydrate).[493],[1260a]

$CH_3C(O)CH_2CMe_2PO(OR)OH$.

    R = Me. III.3.[170]

    R = Bu. III.3.[170] III.2. $b_{0.0002}$ 82-100°.[493]

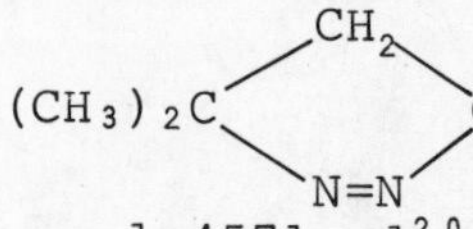

$CH_3C(O)CH_2CMe_2PO(OR)_2$.

R = Et. V.13.[171]

R = Ph. V.13. $b_{0.0008}$ 136-150°.[493]

$CH_3C(O)CH_2CHEtPO(OEt)_2$. V.8.[840]

$C_2H_5C(O)(CH_2)_3PO(OEt)_2$. V.6. $b_{0.4}$ 123°, n_D^{20} 1.4415, d_4^{20} 1.049, IR.[1958]

$C_3H_7C(O)CH_2CH_2PO(OEt)_2$. V.6. $b_{0.9}$ 128°, n_D^{23} 1.4390, d_4^{21} 1.050[1958]

$C_4H_9C(O)CH_2PO(OR)_2$.

R = Et. V.14. $b_{0.3}$ 100°.[1963]

R = Bu. V.14. $b_{0.5}$ 138°.[1963]

$C_5H_{11}C(O)PO(OMe)_2$. V.5. $b_{0.07-0.12}$ 73.5-75.5°, n_D^{25} 1.4339, IR, ^1H.[1191]

$(CH_3)_2CHC(-OMe)=CHPO(OEt)_2$. V.14. b_{10} 137°, n_D^{20} 1.4505.[1606]

$(CH_3)_2C=C(-OMe)CH_2PO(OR)_2$.

R = Pr. V.14. $b_{1.5}$ 120-122°.[1603]

R = Bu. V.14. b_1 132-135°.[1603]

$C_2H_5OCH=CMeCH_2PO(OEt)_2$. V.14. $b_{0.4}$ 94.5°, n_D^{24} 1.4462, d_4^{24} 1.043.[1082]

$C_2H_5OCMe=CHCH_2PO(OEt)_2$. V.14. b_3 105.5-106°, n_D^{20} 1.4504, d_4^{20} 1.0363.[825] ^{31}P -27.7 ppm, ^1H.[1093] V.14. $b_{0.2}$ 100°, n_D^{22} 1.4470, d_4^{22} 1.040.[1082]

$C_2H_5OCH_2CMe=CHPO(OEt)_2$. V.2. b_2 122-123°, n_D^{20} 1.4553, d_4^{20} 1.0894.[992]

$C_2H_5OCEt=CHPO(OEt)_2$. V.14. $b_{0.8}$ 104°, n_D^{22} 1.4455.[1961] ^{31}P -21.7 ppm, ^1H.[1093] V.14. $b_{0.2}$ 94°, IR,[1082]

$C_2H_5OCH=CEt-PO(OR)_2$.

R = Me. V.2. $b_{0.08}$ 63.5°, n_D^{20} 1.4532, d_4^{20} 1.0825.[1350]

R = Et. V.2. $b_{0.095}$ 75.5°, n_D^{20} 1.4476, d_4^{20} 1.0275.[1350]

R = Pr. V.2. $b_{0.035}$ 98°, n_D^{20} 1.4532, d_4^{20} 1.0100.[1350]

R = Bu. V.2. $b_{0.2}$ 115°, n_D^{20} 1.4512, d_4^{20} 0.9822.[1350]

$C_2H_5OC(=CH_2)-CHMePO(OEt)_2$. V.6. $b_{0.1}$ 102°, n_D^{18} 1.4570, d_4^{18} 1.051.[1410]

$C_2H_5OC(=CHMe)-CH_2PO(OEt)_2$. V.6. $b_{0.1}$ 100°, IR.[1959] V.6. $b_{0.1}$ 100°, n_D^{18} 1.4481, d_4^{18} 1.045.[1410]

$C_2H_5OCMe=CMePO(OEt)_2$. V.6. $b_{0.1}$ 102°, IR.[1959]

$(CH_3)_2CHOCMe=CHPO(OEt)_2$. V.14. $b_{0.2}$ 86°, n_D^{24} 1.4436.[1961]

$(C_3H_7OCH=CMe-)$ 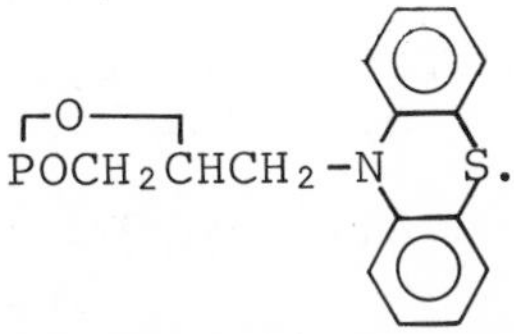

. V.3. m. 140-142°.[227]

$H_2C(CH_2)_2OCEtPO(OEt)_2$. V.6. $b_{0.5}$ 88°, n_D^{20} 1.4463, d_4^{20} 1.079.[1958]
$H_2C(CH_2)_4C(OH)PO(OH)_2$. I.9. m. 191-192°.[522] I.9. m. 187.5-188.5°.[959]
$H_2C(CH_2)_4C(OH)PO(OR)_2$.

R = $(ClH_2C)_2CH$-. V.8. m. 89-91°.[961]

R = Bu. V.8. b_7 178-180° [18] (Cf.[1619]).

R = Cl_3C-CMe_2-. V.8. m. 168.5-169°.[34]

R = $H_2C(CH_2)_3C(-CCl_3)$-. V.8. m. 179-180°.[33]

$C_4H_9SC(O)CH_2PO(OEt)_2$. V.14. n_D^{20} 1.4680, d_4^{20} 1.1012.[1109]
$C_4H_9SCH_2C(O)PO(OEt)_2$. V.5. b_3 100°, n_D^{20} 1.4431, d_4^{20} 1.0700 [2112]
$C_2H_5SCH_2CH_2SC(O)CH_2PO(OEt)_2$. V.5. $b_{0.5}$ 150-153°.[1109]
$CH_3C(O)CH_2CH_2OCHMePO(OMe)_2$. V.7. $b_{1.5}$ 133-135°, n_D^{20} 1.4470 d_4^{20} 1.1711.[1675]
$CH_3C(O)CMe(-OEt)$-$PO(OEt)_2$. V.8. b_{10} 127-130°, n_D^{20} 1.4250. [1060]

$(C_2H_5O)_2C{=}CHPO(OR)_2$.

R = Me. V.14. $b_{0.03}$ 92-93°, n_D^{20} 1.4510, d_4^{20} 1.1301. [1346]

R = Et. V.14. $b_{0.05}$ 91-92°, n_D^{20} 1.4460, d_4^{20} 1.0926. [1346]

R = Pr. V.14. $b_{0.06}$ 111-112°, n_D^{20} 1.4403, d_4^{20} 1.0278. [1346]

R = Bu. V.14. $b_{0.08}$ 128-130°, n_D^{20} 1.4498, d_4^{20} 1.0206. [1346]

$(C_2H_5O$-$)(CH_3O$-$)CH$-$CH{=}CHPO(OMe)_2$. V.2. b_2 105-106°, n_D^{20} 1.4504, d_4^{20} 1.1319.[2037]
$C_2H_5OCHMeOCH{=}CHPO(OEt)_2$. V.2. $b_{0.09}$ 108-109°, n_D^{20} 1.4410, d_4^{20} 1.0760.[585]
OCH_2CH_2O-CMe-$(CH_2)_2PO(OEt)_2$. V.6. $b_{0.1}$ 127°.[1959] V.6. $b_{0.1}$ 128°.[1410]
$C_3H_7CO_2CH_2CH_2PO(OR)_2$.

R = Me. V.7. $b_{0.01}$ 93-94°.[1787]

R = Et. V.7. $b_{0.01}$ 98-99°.[1787]

R = Bu. V.7. $b_{0.13}$ 142-143°.[1787]

$CH_3CO_2(CH_2)_4PO(OEt)_2$. V.5.[173]
$(CH_3)_2CH$-$CH[-O_2CMe]PO(OEt)_2$. V.14. $b_{1.5}$ 100°.[1589]
$C_2H_5CMe[-O_2CMe]PO(OEt)_2$. V.14. $b_{2.5}$ 100-100.5°, n_D^{20} 1.4382, d_4^{20} 1.0991.[1589]
$C_2H_5O_2C(CH_2)_3PO(OH)_2$. From the acid and EtOH/HCl. m. 76-77°.[1419,1420]
$C_2H_5O_2C(CH_2)_3PO(OEt)_2$. V.5. $b_{0.02}$ 120-121°.[516] Mass spect. [1403]

$C_2H_5O_2CCH_2CHMePO(OR)_2$.

R = Et. V.7. b_{11} 146-147°.[1549]

R = Bu. V.7. $b_{0.02}$ 83-84°.[276]

$C_2H_5O_2CCHMeCH_2PO(OEt)_2$. V.7. b_4 119-119.5°.[1671] V.7. b_1 108-109°.[2032] V.7. b_2 138-140°.[638]

$C_2H_5O_2CCHEtPO(OEt)_2$. V.5. $b_{10.5}$ 147.5-148°.[112a] Mass spect.[1403] V.5. b_{10} 65° (?!).[42]

$C_3H_7O_2CCH_2CH_2PO(OPr)_2$. V.7. b_1 136°, n_D^{20} 1.4315.[638]

$(CH_3)_3CO_2CCH_2PO(OEt)_2$. V.6. $b_{0.5}$ 82-82.5°.[1143]

$C_4H_9O_2CCH_2PO(OBu)_2$. V.6. b_8 182-183°, n_D^{20} 1.4383, d_4^{20} 1.0184.[1172]

(HO)HC-CH(OH)-$(CH_2)_3$CHPO(OR)$_2$. (ring: HC and CH joined)

R = Et. V.14. b_1 155-156°.[132]

R = Pr. V.14. $b_{1.5}$ 176-178°.[132]

R = i-Pr. V.14. b_1 151-153°.[132]

$C_2H_5SCH_2CH_2O_2CCH_2PO(OR)_2$.

R = Et. V.5. $b_{0.75}$ 145-148°.[1109]

R = Pr. V.5. $b_{0.5}$ 130-135°.[1109]

R = Bu. V.5. $b_{0.4}$ 138-142°.[1109]

$CH_3O_2CCH_2CH_2OCHMePO(OEt)_2$. V.7. b_3 139°.[1675]

$CH_3OCH_2CH(-CO_2Et)PO(OEt)_2$. V.14. b_{10} 155-156°.[1651]

$C_2H_5O_2CCH(-OEt)-PO(OEt)_2$. V.5, V.6. $b_{0.01}$ 92-95°, IR, ^{1}H.[684]

$CH_2PO(OH)O^{\ominus}NH_4^{\oplus}$

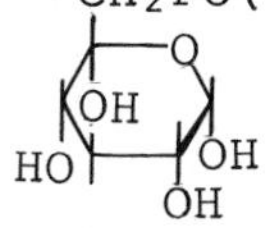

I.6.[1387]

$C_2H_5SCH_2CH_2CH{=}CHPO(OR)_2$.

R = Et. V.14. b_{1-2} 149-150°[990]

R = Bu. V.14. b_2 178-180°[990]

$C_2H_5SCH_2CH{=}CHCH_2PO(OEt)_2$. V.14. b_1 144-145°.[1630]

$C_2H_5SCMe{=}CHCH_2PO(OEt)_2$. V.14. $b_{0.2}$ 110-113°.[1082]

$C_2H_5SCH{=}CMe-CH_2PO(OEt)_2$. V.14. $b_{0.3}$ 116-117°.[1082]

$CH_3CH{=}C(-SEt)CH_2PO(OEt)_2$. V.14. $b_{0.4}$ 107°.[1082]

$C_3H_7SCMe{=}CHPO(OR)_2$.

R = Et. V.14. b_1 113°.[1602]

R = Pr. V.14. b_1 123°.[1602]

$(C_2H_5)_2NCH{=}CHPO(OR)_2$.

R = Et. V.14. $b_{0.0001}$ 95-96°.[1700]

R = i-Pr. V.14. $b_{0.05}$ 114°.[1788]

$H_2C(CH_2)_3NCH_2CH_2PO(OEt)_2$ (ring). Kinetics of the reaction of $H_2C(CH_2)_3NH$ (ring) and $H_2C{=}CHPO(OEt)_2$.[1885] V.14. b_2 120°.[329]

$H_2\overline{C(CH_2)_4N}-CH_2PO(OR)_2$.

R = Et. V.5. $b_{0.4}$ 98°, n_D^{20} 1.4563.[283] V.14. b_3 124°.[52] V.10.[1038] V.9. $b_{0.1}$ 128-131°.[329] From $[H_2\overline{C(CH_2)_4N}-CH_2$ and $PO(Et)_3$, b_2 125°.[284] V.9. $b_{0.01}$ 74-75°.[142]

R = Bu. From PCl_3 and $H_2\overline{C(CH_2)_4N}CH_2OBu$, $b_{0.2}$ 137°, n_D^{20} 1.4562.[283]

$H_2\overline{CCH_2O-CH_2CH_2N}-CH_2CH_2PO(OEt)_2$. V.14. b_1 137°.[329] V.14.[188]
$C_4H_9NHC(O)CH_2PO(OEt)_2$. V.5. $b_{0.18-0.2}$ 152-154°.[1108]
$(CH_3)_2NC(O)(CH_2)_3PO(OR)_2$.

R = Et. V.5. $b_{0.003}$ 112-114°.[515]

R = Pr. V.5. $b_{0.005}$ 127°, n_D^{20} 1.4642.[515]

R = i-Pr. V.5. $b_{0.005}$ 110-112°, n_D^{20} 1.4569.[515]

R = C_6H_{13}-. V.5. n_D^{20} 1.4495.[515]

$C_2H_5NHC(O)CHMeCH_2PO(OEt)_2$. V.7. $b_{0.05}$ 143-145°, n_D^{20} 1.4600 d_4^{20} 1.0954.[1986]
$(C_2H_5-)(CH_3-)C[-NHC(O)Me]PO(OEt)_2$. V.14. m. 78-79°, (petroleum ether).[1589]
$(C_2H_5)_2NC(O)CH_2PO(OR)_2$.

R = Et. V.5. $b_{1.2}$ 135°, n_D^{25} 1.4560.[1940] V.5. $b_{0.1}$ 123-124°.[1239] V.14. b_1 138.5-139.5.[828] V.5. $b_{1.2}$ 135°.[1941]

R = i-Pr. Complexes with rare earth metals.[1950]

$H_2NCO_2CH_2CH_2CHMeCH_2PO(OMe)_2$. V.7. m. 71°.[1463]
$C_2H_5O_2CCH_2NMeCH_2PO(OEt)_2$. V.1. $b_{0.001}$ 92°.[2113]
$Me_3\overset{\oplus}{N}-\overset{\ominus}{C}(-CO_2Me)-PO(OEt)_2$. V.14. m. 61-62°, ^1H, ^{31}P-17ppm.[71]
$NCCH(-SiMe_3)CH_2PO(OEt)_2$. V.7. b_1 132°, n_D^{20} 1.4520, d_4^{20} 1.0364.[1414]
$\overline{O-CMe_2-CH_2CMe(-OH)-P}O(OR)$.

R = Me. V.8. m. 127-128°.[1799]

R = Et. V.8. m. 100-101°.[1799]

R = Pr. V.8. m. 69-70°.[1799]

R = Bu. V.8. m. 60-61°.[1799]

$CH_3CH{=}CH-CH[-OPO(OMe)_2]-PO(OMe)_2$. V.14. $b_{0.1}$ 140-142°;[333] the structure as diphosphonic acid ester is incorrect.[538]

$C_6H_{13}PO(OH)_2$. I.8. m. 105-106°.[686] I.5. m. 104-105°; the temperature of the pyrolysis falls in the sequence Et - Pr - i-Pr - sec.-Bu.[343] I.2. m. 104.5-106° (ligroin).[1000,1002] pK_1 2.6, pK_2 7.9.[1776a] I.1. m. 98°.[737]
$C_6H_{13}PO(OMe)OH$. From $(MeO-)OPH(OH)$ and $H_2C{=}CHC_4H_9$ ($h\nu$).[1678]
$C_6H_{13}PO(OR)_2$.

R = Et. V.5. b_{17} 138-140°.[343] V.5. b_{17} 140-145°.[1000, 1002] V.5. b_2 103°, n_D^{17} 1.4311.[552] V.7. b_{10} 126°.[1954] V.7. $b_{8.5}$ 125-126°.[1614]

R = Pr. V.5. $b_{0.5}$ 109-110°.[343]

R = i-Pr. V.5. b_2 111.5-113°.[343]

R = Bu. V.5. b_{20} 183-185°.[343] V.6. b_{20} 182-184°, n_D^{25} 1.4332, d_4^{25} 0.9366.[1000,1002]

R = sec.-Bu. V.5. b_1 125-128°.[343]

R = C_6H_{13}-. μ=2.88 D.[1216] Magn. rotation of the bonds.[1069]

R = Et_3Si-. VI.1. $b_{1.5}$ 168-172°, n_D^{20} 1.4485, d_4^{20} 0.9372.[487]

$(CH_3)_2CH-CHMeCH_2PO(OH)_2$. I.2. m. 95-96°.[1252]

$(CH_3)_2CH-CHMeCH_2PO(OEt)_2$. V.6. $b_{5.5}$ 108-109°, n_D^{20} 1.4303, d_4^{20} 0.9705.[1552] V.5, V.7. b_3 88.5-91.5°, n_D^{25} 1.4302.[1552]

$(CH_3)_2CHCMe_2PO(OH)_2$. I.2. m. 163.5-164°.[1551]

$(CH_3)_2CHCMe_2PO(OEt)_2$. V.7. $b_{1.3}$ 67-68°, n_D^{25} 1.4335, d_4^{25} 0.9774.[1551]

$Me_3\overset{\oplus}{N}CH(-CO_2Me)PO(OEt)O^{\ominus}$. By heating from $[(EtO)_2OPCH(\overset{\oplus}{N}Me_3)CO_2Me]\overset{\ominus}{X}$, m. 202-204°.[716]

$[(EtO)_2OPCH(\overset{\oplus}{N}Me_3)CO_2Me]X^{\ominus}$.

X = $MeOSO_3^{\ominus}$. From $(EtO)_2OPCH(-NMe_2)CO_2Me$ and Me_2SO_4.[716]

X = $I^{\ominus}$. From $(EtO)_2OPCH(-NMe_2)CO_2Me$ and MeI.[716]

X = $Br^{\ominus}$. From $(EtO)_2OPCH(-NMe_2)CO_2Me$ and MeBr.[716]

$H_2\overline{CCH_2NMe-CH_2CH_2N}CH_2PO(OH)_2$. I.2. As HBr adduct, m. 243-244°.[1351]

$H_2\overline{CCH_2NMe-CH_2CH_2N}CH_2PO(OEt)_2$. V.9. $b_{0.07}$ 87-89°, n_D^{25} 1.4586. V.9. $b_{0.5}$ 110°.[329]

$(C_2H_5)_2NC(O)NHCH_2PO(OEt)_2$. V.14. $b_{0.2}$ 148-150°, n_D^{20} 1.4648, d_4^{20} 1.1061.[1903]

$C_2H_5NHC(O)NEt-CH_2PO(OH)_2$. I.1, I.7.[1462]

$C_2H_5O(CH_2)_4PO(OEt)_2$. V.6. $b_{0.02}$ 73-74°, n_D^{20} 1.4328, d_4^{20} 1.0153.[2048]

$C_2H_5OCH_2CHEtPO(OEt)_2$. From $EtOCH_2CH(OH)Et$ and PBr_3, $b_{0.02}$ 71-72°, n_D^{20} 1.4332, d_4^{20} 1.0175.[2048]

$C_4H_9OCH_2CH_2PO(OR)_2$.

R = Et. V.7. b_9 129-130°, n_D^{20} 1.4295, d_4^{20} 1.0045.[1640] V.5. b_{12} 145.5-146°.[1628] V.7. $b_{0.5}$ 95-97°, ^1H. IR.[1402]

R = Pr. V.7. $b_{1.5}$ 122-125°, ^1H, IR.[1402]

R = i-Pr. V.7. $b_{0.2}$ 80-84°, ^1H, IR.[1402]

R = Bu. V.7. $b_{0.2}$ 110-113°, ^1H, IR.[1402]

$(CH_3)_2CHCH_2OCH_2CH_2PO(OR)_2$.

R = Me. V.7. $b_{0.5}$ 71-75°, n_D^{25} 1.4198, d_4^{25} 1.0460.[1401]

R = Et. V.7. $b_{0.2}$ 78-81°, n_D^{25} 1.4229, IR, ^1H.[1402]

R = Pr. V.7. $b_{0.5}$ 103-107°, n_D^{25} 1.4283, d_4^{25} 0.9676.[140]

R = Bu. V.7. b_2 121-125°, n_D^{25} 1.4310, d_4^{25} 0.9567.[1401]

R = $C_4H_9CHEtCH_2$-. V.7. $b_{0.2}$ 166-170°, n_D^{25} 1.4414, d_4^{25} 0.9205.[1401]

$(CH_3)_3COCH_2CH_2PO(OR)_2$.

R = Me. V.7. $b_{0.5}$ 85°, IR, ^{31}P.[691]

R = Et. V.7. $b_{0.2}$ 82°, IR, ^{31}P.[691]

$C_4H_9CH(OMe)PO(OEt)(OC_6H_4NO_2$-4). V.5. n_D^{25} 1.514.[741]

$CH_3CH(OBu)PO(OR)Cl$.

R = Et. From MeCHClOBu and $(EtO)_2PCl$, $b_{0.08}$ 73°.[608]

R = Bu. From MeCHClOBu and $(BuO)_2PCl$, $b_{0.08}$ 96°.[608]

R = Am. From MeCHClOBu and $(AmO)_2PCl$, $b_{0.04}$ 92°.[608]

$(C_2H_5O)_2CHCH_2PO(OR)_2$.

R = Et. V.5. $b_{0.12}$ 97-98°.[1705] V.5. $b_{0.3}$ 83-88°, n_D^{20} 1.4338.[1986]

R = i-Pr. $b_{0.001}$ 88-90°.[1788]

$(C_2H_5O)_2CMePO(OR)_2$.

R = Et. From $MeC(OEt)_3$ and $HPO(OEt)_2$, b_{10} 120-122°.[134]

R = Ph. From $MeC(OEt)_3$ and $HPO(OPh)_2$, $b_{0.063}$ 123-126°.[1349]

$C_4H_9OCH_2OCH_2PO(OBu)_2$. V.8. $b_{2.5}$ 153-157°.[755]

$C_2H_5OCH_2CH(OMe)$-$CH_2PO(OH)_2$. I.2. As monoacridinium salt, m. 70-72°.[1770]

$C_2H_5OCH_2CH(OMe)CH_2PO(OPr$-i$)_2$. V.5.[1770]

$(CH_3)_2CH_2CH[-O_2CMe]$-$PO(OEt)_2$. V.8. $b_{0.1}$ 93-94°, n_D^{20} 1.4312 d_4^{20} 1.0560.[1377]

$(CH_3OCH_2-)_2C(OMe)PO(OBu)_2$. V.14. b_{10} 160-161°.[1651]

$(CH_3)_3SiCH_2CH{=}CHPO(OH)_2$. I.1.[1530]

$(CH_3)_3SiCH_2CH{=}CHPO(OEt)_2$. V.2.[1530]

$H_2C{=}CHCH_2SiMe_2CH_2PO(OEt)_2$. V.5. $b_{1.5}$ 73-75°, n_D^{20} 1.4490, d_4^{20} 0.9854.[320]

$(C_2H_5)_2NCH_2CH_2PO(OEt)_2$. V.14. b_3 106-107°.[1007] V.6. b_{12} 125-130°.[336]

$(C_2H_5)_2NCHMePO(OEt)_2$. V.14. b_1 94°.[529]

$(CH_3)_2CHCH(-NMe_2)$-$PO(OMe)_2$. V.7. $b_{0.3-0.4}$ 64-66°, n_D^{20} 1.4452, IR, ^1H.[332]

$(HOCH_2CH_2)_2NCHMePO(OCH_2CH_2Cl)(OCH_2CH_2OH)$. V.9. n_D^{25} 1.4985,

^{31}P -21.9 ppm.[256]

$(C_2H_5)_2NS(O)_2CH_2CH_2PO(OR)_2$.

R = Me. V.7. b_1 177-179°.[1491]

R = Et. V.7. $b_{0.25}$ 171-173°.[1491]

$(EtO)_2SP\text{-}SCH_2CH_2PO(OEt)_2$. V.14. b_1 164-165°.[1600]
$(EtO)_2SPOCHMe\text{-}PO(OEt)_2$. V.14. $b_{0.01}$ 106°.[1861]
$CH_3CH[\text{-}O\text{-}PO(OEt)_2]\text{-}PO(OEt)_2$. V.6. b_1 150°.[1618] V.14. $b_{0.01}$ 108°.[1861] V.14. $b_{0.06}$ 125°.[333] The structure as diphosphonic acid ester is incorrect.[538] V.14, V.6.[1617] V.14, V.7.[1622]
$C_2H_5O_2C\text{-}N{=}C(\text{-}NEt_2)PO(OEt)_2$. V.14. $b_{0.05}$ 120-122°, n_D^{20} 1.4737, d_4^{20} 1.1223.[490]
$MeEt_2SiOCH_2PO(OSiEt_2Me)_2$. VI.1. b_3 133°, n_D^{20} 1.4436, d_4^{20} 0.9645.[1437] VI.1. b_3 135°, n_D^{20} 1.4423, d_4^{20} 0.9647.[1438]

$(EtO)Me_2SiCH_2CH_2PO(OEt)_2$. V.7. b_1 115°.[208]
$Me_3SiOCMe_2PO(OMe)_2$. VI.1. b_{11} 96.5-97.5°, ^1H.[1374]
$(EtO)_2MeSiCH_2PO(OEt)_2$. V.5. b_1 99-99.5°.[320]
$Me_3SiOSiMe_2CH_2PO(OR)_2$.

R = Et. V.6. b_{36} 154-156°.[344]

R = Bu. V.6. b_{10} 150-152°.[344]

C_7

$\overline{O\text{-}CH{=}CH\text{-}CH{=}C}\text{-}CH(\text{-}OAsMe_2)PO(OEt)_2$. From Me_2AsBr, $\overline{OCH{=}CH\text{-}CH{=}C}CHO$; and $P(OEt)_3$, $b_{0.15}$ 132-136°.[1742]
$4\text{-}BrC_6H_4C(N_2)\text{-}PO(OEt)_2$. V.14. m. 31-32°, IR.[883]
$C_6H_5CHBrPO(OEt)_2$. V.14. b_1 164-166°.[697]
$2\text{-}Br\text{-}4\text{-}MeC_6H_3PO(OH)_2$. I.1. m. 199-203°.[582]
$3\text{-}Me\text{-}6\text{-}BrC_6H_3PO(OH)_2$. I.13. m. 198°.[1260,1269]
$3\text{-}Me\text{-}6\text{-}BrC_6H_4PO(OH)_2$. I.13. m. 198°.[697]
$4\text{-}BrC_6H_4CHBrPO(OEt)_2$. V.14. b_1 159-164°.[697]
$4\text{-}ClC_6H_4C(N_2)PO(OEt)_2$. V.14. m. 39-40°, IR.[883]
$2\text{-}ClC_6H_4C(O)PO(OMe)_2$. V.5. $b_{0.1}$ 102-109°, ^1H. As 2,4-dinitrophenylhydrazone, m. 112-113°.[249]
$4\text{-}ClC_6H_4C(O)PO(OMe)_2$. V.5. $b_{1.5}$ 136°, ^1H. as 2,4-dinitrophenylhydrazone, m. 164-167°.[249]
$3\text{-}Cl\text{-}4\text{-}HO_2CC_6H_3PO(OH)_2$. I.13. m. 254°.[1229]

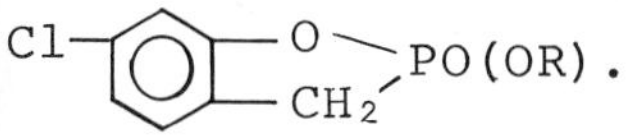

R = Et. V.9. $b_{0.2}$ 138-140°.[847]

R = Pr. V.9. $b_{0.2}$ 145-148°, m. 55-57°.[847]

R = Bu. V.9. $b_{0.2}$ 158°, m. 40-42°.[847]

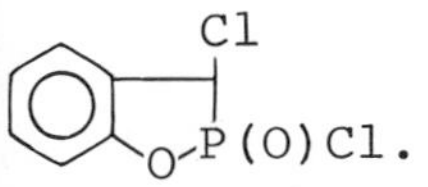
From $2\text{-}HOC_6H_4CHO$ and PCl_3.[926]

$4\text{-}ClC_6H_4NHC(O)PO(OR)_2$.

R = Et. V.14.[1023]

R = i-Pr. V.9. m. 122-127°.[559]

$2\text{-}O_2N\text{-}4\text{-}Me\text{-}5\text{-}ClC_6H_2PO(OH)_2$. I.13. m. 223-227° (dec.).[578]
$2\text{-}ClC_6H_4CH_2PO(OH)_2$. I.2. m. 183° (water).[1037]
$2\text{-}ClC_6H_4CH_2PO(OMe)OH$. III.2. As NMe_4 salt; m. 196°.[353]
$3\text{-}Cl\text{-}4\text{-}MeC_6H_3PO(OH)_2$. I.1. m. 190° (water).[1229] I.1. m. 160-162°.[578]
$3\text{-}Me\text{-}4\text{-}ClC_6H_3PO(OH)_2$. I.1. m. 158-161°.[578]
$2\text{-}Me\text{-}5\text{-}ClC_6H_3PO(OH)_2$. I.13. m. 205°.[1260,1269]
$2\text{-}Cl\text{-}5\text{-}MeC_6H_3PO(OH)_2$. I.13. m. 176°.[1260,1269]
$C_6H_5CHClPO(OR)_2$.

R = Me. V.14. $b_{0.07}$ 105°, ^1H, IR.[1878a]

R = Et. V.2. $b_{1.25}$ 128-129°.[924]

R = $SiEtMe_2$. VI.1. b_4 160-161°.[1439]

R = $SiEt_2Me$. VI.1. b_4 170°.[1439]

R = $SiEt_3$. VI.1. b_2 175-176°.[1439]

$C_6H_5\text{-}Hg\text{-}CHClPO(OR)_2$.

R = Me. V.5.[1878]

R = Et. V.5.[1878]

$C_6H_5CH(\text{-}MgCl)PO(OEt)_2$ V.14.[267]
$2\text{-}MeO\text{-}5\text{-}ClC_6H_3PO(OH)_2$. I.1. m. 222.5-225°.[574]
$4\text{-}ClC_6H_4CH(OH)\text{-}PO(OCH_2CH_2Cl)_2$. V.8. m. 107-108°.[885]
$3\text{-}ClO_2S\text{-}4\text{-}MeOC_6H_3PO(OH)_2$. I.13. m. 138-140° (dec.).[749]
$4\text{-}ClC_6H_4S(O)_2OCH_2PO(OEt)_2$. V.14. m. 42-43°.[2054]
$2\text{-}H_2N\text{-}4\text{-}Me\text{-}5\text{-}ClC_6H_2PO(OH)_2$. I.13.[2074]
$CH_3CCl{=}C[\text{-}C(O)Pr]PO(OEt)_2$. V.2. b_4 150-152°, n_D^{20} 1.4725, d_4^{20} 1.1345.[1133]
$C_3H_7SCH{=}C[\text{-}CCl{=}CH_2]PO(OR)_2$.

R = Me. V.2. $b_{0.005}$ 84-85°, n_D^{20} 1.5623, d_4^{20} 1.2248.[1918]

R = Et. V.2. $b_{0.004}$ 84-86°, n_D^{20} 1.5436, d_4^{20} 1.1535.[1918]

$C_5H_{11}CCl{=}CHPO(OEt)_2$. V.1. b_{17} 152°.[240]
$C_2H_5O(CH_2)_3\text{-}CCl{=}CHPO(OH)_2$. I.1.[87]
$C_2H_5O(CH_2)_3\text{-}CCl{=}CHPO(OEt)_2$. V.2. b_3 158-160°.[87]
$C_5H_{11}CHClCH_2PO(OH)_2$. I.1. m. 91-92°.[524]
$C_5H_{11}CHClCH_2PO(OR)_2$.

R = Me. V.2. b_4 128-130°, n_D^{20} 1.4511, d_4^{20} 1.1068.[524]

R = Et. V.2. b_5 137-139°, n_D^{20} 1.4480, d_4^{20} 1.0553.[524]

R = Pr. V.2. b_4 151-152°, n_D^{20} 1.4480, d_4^{20} 1.0301.[524]

R = $H_2C{=}CH\text{-}CH$ -. V.2. b_5 160-162°, n_D^{20} 1.4640, d_4^{20}

1.0654.[524]

R = Bu. V.2. b_3 170-171°, n_D^{20} 1.4491, d_4^{20} 1.0153.[524]

$CHPO(OH)_2$. I.2. m. 204-206°, ^1H.[715]

CHPO(OEt)ONa. III.3 m. 314-316°, ^1H.[715]

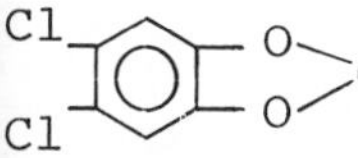
CHPO(OEt)Cl. IV.1. ^1H, mass spect.[715]

$CHPO(OEt)_2$. V.5. m. 65-66°, IR.[717]

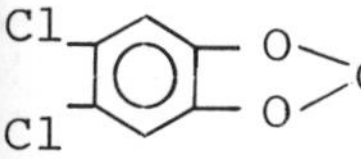
$CHPO(OEt)OC_6H_2Cl_3$-2.4.6. V.2. m. 137-141°.[715]

$(2,4\text{-}Cl_2C_6H_3CH_2)\overline{OPOCH_2CEt(-CH_2Cl)CH_2O}$. V.5. m. 125°.[1059]
$C_6H_5HgCCl_2PO(OR)_2$.

R = Me. V.5.[1878]

R = Et. V.5.[1878]

$C_6H_5CCl(-HgCl)PO(OMe)_2$. V.14. m. 151°.[1878a]
$C_6H_5OCCl_2PO(OEt)_2$. V.5. $b_{0.02}$ 116°, n_D^{23} 1.4603.[719]

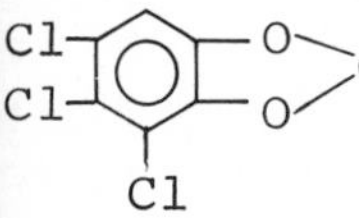
$CHPO(OEt)_2$. V.5. m. 118-120°, IR.[717]

2,3,6-$Cl_3C_6H_2CH_2PO(OH)_2$. I.1. m. 158-159.2°.[656]
2,4,5-$Cl_3C_6H_2CH_2PO(OEt)_2$. ^1H.[1921]
3-Me-2,5,6-$Cl_3C_6HPO(OH)_2$. I.13. m. 220°.[1260,1269]
2,3,6-$Cl_3C_6H_2CH(OH)PO(OR)_2$.

R = Me. V.8. m. 139.5-140.5°.[2066]

R = Ph. V.8. m. 53-56°.[2066]

$(EtO)_2(O)POC(=CCl_2)\text{-}CCl_2PO(OEt)_2$. V.8. $b_{0.1}$ 136°, n_D^{25} 1.4764.[1970]

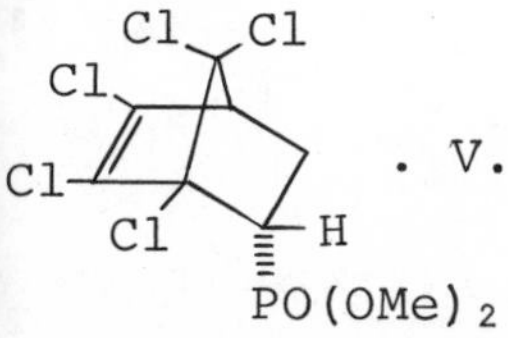
. V.14. ^1H.[342]

$FO_2SNHCHPh\text{-}PO(OEt)_2$. V.8. m. 117°.[789]
$C_2H_5O_2CCH{=}C(\text{-}CH_2F)CH_2PO(OEt)_2$. cis:trans = 2:1. V.5. $b_{0.01}$ 85°, IR.[1139]
$C_2H_5O_2CCH{=}CMe\text{-}CHFPO(OEt)_2$. cis:trans : 1:1. $b_{0.1}$ 91°, IR.[1139]
$FH_2CCH(OH)CH_2CH[\text{-}CO_2Et]PO(OEt)_2$. V.14. b_1 139-141°.[1774]
$C_2H_5O_2CCH{=}C(\text{-}CH_2F)\text{-}CHFPO(OEt)_2$. cis/trans. V.5. $b_{0.1}$ 93°, IR.[1139]
$4\text{-}F_3CC_6H_4PO(OH)_2$. I.1. m. 177-179°[871]
$4\text{-}F_3CC_6H_4PO(OH)F$. II.1. m. as aniline salt 173-174°[871]
$4\text{-}F_3CC_6H_4PO(OPr\text{-}i)F$. IV.2. b_8 94-95°[871]
$(C_2H_5)_2NC(\text{-}CF_3){=}CFPO(OEt)_2$. V.14. b_6 108-109°[979]
$F_2CCF_2C(OPr){=}CPO(OPr)_2$. V.14, V.7. b_1 106-107°[2041]
$F_2CCF_2CF_2C(OEt){=}CPO(OEt)_2$. V.14. $b_{3.5}$ 122-124°, IR.[564]
$F_3C(CF_2)_4CHFCF_2PO(OH)_2$. I.2.[301]
$F_3C(CF_2)_4CHFCF_2PO(OEt)_2$. V.7. b_{10} 105-113°(some dec.) n_D^{25} 1.3816.[301]
$4\text{-}NCC_6H_4PO(OH)_2$. I.1. I.13. m. 142-142°.[484]
$4\text{-}NCC_6H_4PO(OEt)_2$. V.5. $b_{0.01}$ 132°, m. 31-35°.[1987]

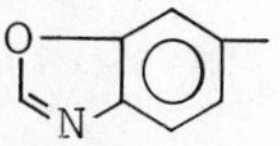
$PO(OH)_2$. I.13. m. 204-205°[290]

$2\text{-}HO_2C\text{-}3\text{-}O_2NC_6H_3PO(OH)_2$. I.1. m. 204-215°(dec.).[576]
$2\text{-}HO_2C\text{-}4\text{-}O_2NC_6H_3PO(OH)_2$. I.13. m. 192-194°.[576]
$2\text{-}HO_2C\text{-}5\text{-}O_2NC_6H_3PO(OH)_2\cdot H_2O$. I.13. m. 192-194°.[576]
$3\text{-}HO_2C\text{-}4\text{-}O_2NC_6H_3PO(OH)_2$. I.13. m. 190-206°(dec.).[576]
$3\text{-}HO_2C\text{-}5\text{-}O_2NC_6H_3PO(OH)_2$. I.13. m. 230.5-233°.[576]
$4\text{-}HO_2C\text{-}2\text{-}O_2NC_6H_3PO(OH)_2\cdot H_2O$. I.13. m. 228-238°(dec.).[576]
$4\text{-}HO_2C\text{-}3\text{-}O_2NC_6H_3PO(OH)_2$. I.13. m. 213-217.5°.[576]
$5\text{-}HO_2C\text{-}2\text{-}O_2NC_6H_3PO(OH)_2$. I.13. m. 201-218°(dec.).[576]
$C_6H_5C(N_2)PO(OR)_2$.

R = Me. V.14. m. 44-44.5°.[1878a]

R = Et. V.14.[883]

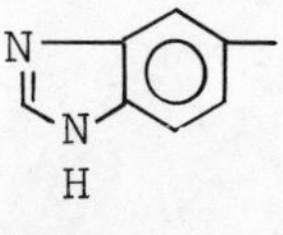
$PO(OH)_2$. I.13. m. 250°[290]

$4\text{-}O_2NC_6H_4NHC(O)PO(OEt)_2$. V.9. m. 144-145°.[559]
$2,4\text{-}(O_2N)_2C_6H_3CH_2PO(OH)_2$. I.13.[304]
$2,4\text{-}(O_2N)_2C_6H_3S\text{-}O\text{-}CH_2PO(OEt)_2$. V.14. m. 50-51°.[815]
$C_6H_5C(O)PO(OR)_2$.

R = Me. V.5. $b_{0.25}$ 146°, IR, ^1H. As 2,4-dinitrophenylhydrazone, m. 198-199°.[249] V.5. $b_{3.5}$ 144.5-146°.[92] As 4-nitrophenylhydrazone, m. 128-129°.[918] As $NaHSO_3$ adduct (2-hydrate), m. 84°.[918]

R = Et. V.5.[1005] V.5. $b_{2.5}$ 141°.[922] As 2,4-dinitrophenylhydrazone, m. 171-172°.[1005]

R = Pr. V.5. $b_{0.08}$ 123-124°.[607]

R = i-Pr. V.5. b_3 147-149°, IR.[1424]

2-$OHCC_6H_4PO(OMe)_2$. V.5. $b_{0.45}$ 126-127°, IR.[1422]
2-$HO_2CC_6H_4PO(OH)_2$. I.2.[654] I.1. m. 175.5-179°.[582] I.13. m. 172 .[1260,1269]
3-$HO_2CC_6H_4PO(OH)_2$. I.13. m. 245-256°.[1260,1269]
4-$HO_2CC_6H_4PO(OH)_2$. From dibenzoylperoxide + HPO(OEt) and acidic hydrolysis.[534] From $PhCO_2Me$ + bis-t-butylperoxide + $HPO(OEt)_2$ and acidic hydrolysis, m. 377-379°.[535,1763] I.13. m. > 300°.[1260,1269] I.1. m. > 300°.[479]

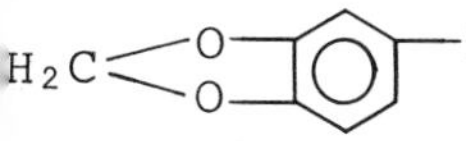
$PO(OH)_2$. I.1. As toluidine salt, m. 160-161°.[291]

$C_6H_5C(S)PO(OMe)_2$. V.5. $b_{0.01}$ 103°, ^1H, ^{31}P -16.5 ppm, UV, IR.[2085]

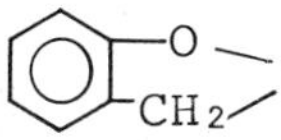
PO(OEt). V.9. $b_{0.03}$ 112-114°.[847]

CH=CHPO(OEt)OH. III.2. As picrate, M. 63.5-65°.[770]

$C_6H_5NHCOPO(OMe)_2$. V.9. m. 103°.[1639]
2-$O_2NC_6H_4CH_2PO(OH)_2$. I.2.[304]
2-$O_2NC_6H_4CH_2PO(OMe)_2$. V.5.[304]
4-$O_2NC_6H_4CH_2PO(OH)_2$. I.13. Dec. 217°.[1106] I.13. m. 148-153°,[932] ^1H.[1921] Split by dil. NaOH at 72° to yield 4-$O_2NC_6H_4Me$ and H_3PO_4; The monoester is stable under this condition.[1226]
4-$O_2NC_6H_4CH_2PO(OEt)_2$. V.14 $b_{0.1}$ 148-153°, n_D^{25} 1.5220,[932] ^1H.[1921]
2-Me-4-$O_2NC_6H_3PO(OH)_2$. I.1. m. 225-228°.[575]
2-Me-5-$O_2NC_6H_3PO(OH)_2$. I.1. m. 222-225°.[575]
3-Me-4-$O_2NC_6H_3PO(OH)_2$. I.1. m. 173-176°.[576]
4-Me-2-$O_2NC_6H_3PO(OH)_2$. I.13. m. 128-146°.[576]
4-Me-3-$O_2NC_6H_3PO(OH)_2$. I.13. m. 191°.[1260,1269] I.13. m. 156-158°.[576]
2-$O_2NC_6H_4CH(OH)PO(OEt)_2$. V.8. m. 109-109.5°.[11]
3-$O_2NC_6H_4CH(OH)PO(OR)_2$.

R = Et. V.8. m. 92-93°.[11]

R = $(ClH_2C)_2CH$-. V.8. m. 120-121°.[961]

R = $H_2C(CH_2)_3C(-CCl_3)$-. V.8. m. 156-157°.[33]

4-$O_2NC_6H_4CH(OH)PO(OH)_2$. I.9.[1095]
2-MeO-4-$O_2NC_6H_3PO(OH)_2$. I.1. m. 226.5-227.5°.[582]
2-HO-5-$O_2NC_6H_3CH_2PO(OH)_2$. I.2. m. 224-229°.[1127]
2-HO-5-$O_2NC_6H_3CH_2PO(OEt)_2$. V.5. m. 137°.[1127]
3-$O_2NC_6H_4S(O)_2OCH_2PO(OR)_2$.

R = Et. V.14. m. 70-71°.[2054]

R = i-Pr. V.14. m. 63-64°.[2054]

$C_6H_5NHC(S)PO(OR)_2$.

R = Et. V.8. m. 57.3-58°, n_D^{20} 1.5813, d_4^{20} 1.1758.[56]

R = Pr. V.8. n_D^{20} 1.5562, d_4^{20} 1.1336.[56]

R = i-Pr. V.8. m. 52-55°.[56]

R = Bu. V.8. n_D^{20} 1.5521, d_4^{20} 1.1032.[56]

R = i-Bu. V.8. m. 41-45°.[56]

R = Am. V.8. n_D^{20} 1.5510, d_4^{20} 1.0524.[56]

R = i-Am. V.8. n_D^{20} 1.5448, d_4^{20} 1.0750.[56]

$C_6H_5CH_2PO(OH)_2$. I.2. m. 167-169°.[1680] I.9. m. 166°.[1106] I.1. m. 169°.[1094a] I.2. m. 169-170°.[120a] I.2. m. 166°.[1001] I.9. m. 168-169°.[1037] I.2. m. 173-175°.[932]

$C_6H_5CH_2PO(OMe)O^{\ominus}\overset{\oplus}{N}Me_4$. III.3. m. 200°.[353]

$C_6H_5CH_2PO(OEt)OH$. III.3. m. 63-64°.[1680]

$C_6H_5CH_2PO(OC_6H_{11}$-cycl.)Cl. IV.2. n_D^{25} 1.5309.[740]

$C_6H_5CH_2PO(OC_6H_{11}$-cycl.)F. IV.2. $b_{0.03}$ 96°.[740]

$C_6H_5CH_2PO(OR)_2$.

R = Et. V.2. $b_{0.2}$ 115°.[1983] V.5. b_{25} 169-171°, V.6. b_{14} 155°.[1790] V.5. b_{15} 160-164°.[932]

R = $H_2C{=}CHCH_2$-. V.2. $b_{1.5}$ 141-143°.[2022]

R = Bu. V.6. b_4 140-143°.[1001]

$C_6H_5CH_2PO(OMe)OCH_2CO_2Et$. V.1. $b_{0.05}$ 132-135°.[1994]

$C_6H_5CH_2PO(OEt)OCH_2CH_2Cl$. V.5. $b_{7.5}$ 182-183°.[122]

$(C_6H_5CH_2-)\overline{OPO-R-O}$.

R = $-CH_2CH_2-$. V.5. m. 123°.[121]

R = $-CHMeCH_2-$. V.5. m. 122-122.5°.[120]

R = 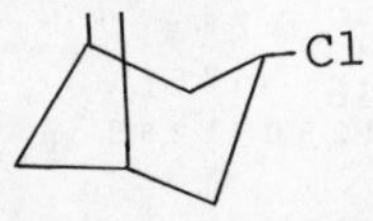

. V.5. IR. white plates from methanol[245]

R = $-CH_2CEt(-CH_2Cl)CH_2-$. V.5. m. 118°.[2059]

R = $-CH_2C(-CH_2Cl)_2-CH_2-$. V.5. m. 123°.[2059]

R = $-CH_2CMe[-CH_2OS(O)_2C_6H_4Me-_4]-CH_2-$. V.13. m. 171°.[2052]

R = 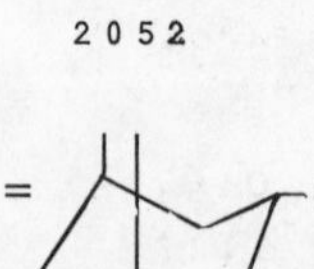
$-OS(O)_2C_6H_4Me$-4. V.13. m. 168°.[2052]

2-$MeC_6H_4PO(OH)_2$. I.1. m. 141°.[1286] I.6. m. 141°.[1260,1269]
2-$MeC_6H_4PO(OR)_2$.

R = Me. V.5. $b_{0.1}$ 107-110.5°, IR, ^{1}H.[690] V.5. $b_{0.25}$ 96-97°.[1422]

R = Et. V.5. $b_{0.3}$ 105-107°, IR.[1422] V.5. $b_{0.01}$ 117-118°.[1987] V.5. $b_{0.01}$ 117-119°.[1985] V.11. b_{14} 148-150°.[328]

3-$MeC_6H_4PO(OH)_2$. I.1. m. 121°.[187] I.1. m. 116-117°.[1260,1269]

3-$MeC_6H_4PO(OR)_2$.

R = Me. V.5. b_4 141°, IR, ^{1}H.[690] V.5. $b_{0.45}$ 106-107°, IR.[1422]

R = Et. V.5. $b_{0.4}$ 104-105°, IR.[1422] V.5. $b_{0.02}$ 119-120°.[1987] V.5. $b_{0.02}$ 118-121°.[1985]

4-$MeC_6H_4PO(OH)_2$. I.1. m. 198-199°.[187,479] I.1.[1100,1256a,1260,1269,1286] I.2. m. 187.5-189°.[1020]
4-$MeC_6H_4PO(OR)_2$.

R = Me. V.5. $b_{0.05}$ 93-94°.[1422] V.5. $b_{0.1}$ 100°.[1988] V.5. b_1 107-110.5°, IR, ^{1}H.[690]

R = Et. V.5. b_4 148-149°.[1422] V.5. $b_{0.05}$ 118-119°.[1987] V.5. $b_{0.01}$ 117-120°.[1985] V.2. b_1 122°.[1020] V.11. b_1 122°.[328] V.5. $b_{0.1}$ 110°.[1988] Reactivity with PhMgBr.[763]

R = H_2C=CH-CH_2-. V.2. b_1 134-136°.[2019]

R = H_2C=CMe-CH_2-. V.2. b_1 146-149°.[2019]

(4-MeC_6H_4-)OP 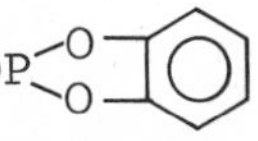 . V.1. m. 81°.[1260,1269]

NCCH$(CH_2)_2$CH=CH-CH-PO$(OEt)_2$. V.14. $b_{0.045}$ 135°, n_D^{20} 1.4850, d_4^{20} 1.1401.[1629]

HN=C(-NH_2)  -PO$(OH)_2$. I.13. m. 300°.[484]

2-$MeOCH_6H_4PO(OH)_2$. I.1. m. 179°.[228]
2-$MeOC_6H_4PO(OR)_2$.

R = Me. V.5. $b_{0.1}$ 110-112°.[1422]

R = Et. V.5. b_2 120-121°, IR.[1422] V.14.[451]

3-$MeOC_6H_4PO(OMe)_2$. V.5. $b_{0.1}$ 116-117°, IR.[1422]

4-MeOC$_6$H$_4$PO(OH)$_2$. I.2. m. 179-179.5°.[360] I.1. m. 158°.[126
1269]

4-MeOC$_6$H$_4$PO(OR)$_2$.

R = Me. V.5. b$_{0.1}$ 113-114°.[1422]

R = Et. V.5. b$_{0.4}$ 177-178°.[1422] V.5. b$_{1.5}$ 168-169°.[198
V.5. b$_{1.5}$ 168-169°.[1987]

C$_6$H$_5$OCH$_2$PO(OH)$_2$. I.13. m. 141-142°.[2062]
4-HOC$_6$H$_4$CH$_2$PO(OEt)$_2$. V.5. m. 89-91°, ^{1}H.[1452]
C$_6$H$_5$CH(OH)-PO(OH)$_2$. I.2. m. 167-168°(dec.).[963] I.9. m. 170-172°.[414] I.9. As aniline salt, m. 201°.[415] I.9. m. 177-178°.[922] I.9.[555] I.6.[1181] I.6. m. 195°.[1186] I.9.[1446] I.2.[1095] I.2. m. 173°(benzene/MeCO$_2$H).[1037]
C$_6$H$_5$CH(OH)PO(OR)$_2$.

R = Me. V.1. m. 99°.[1181] V.14. m. 101-102°.[922]

R = Et. V.8. m. 83-83.2°.[963] V.14. m. 83-84°.[1414] V.8.
83°.[12] V.14. m. 83-84°.[922]

R = ClH$_2$CCH$_2$-. V.8. m. 64-66°.[886]

R = Pr. V.8. b$_9$ 127-128° [40](? cf. Ref.[1619])

R = (ClH$_2$C)$_2$CH-. V.8. m. 80.5-81.5°.[961]

R = H$_2$C(CH$_2$)$_3$C(-CCl$_3$)-. V.8. m. 143-144°.[33]

[C$_6$H$_5$CH(OH)-]OPO-R-O.

R = -(CH$_2$)$_3$-. V.8. m. 161-163°.[1588]

R = -CHMe-CHMe-. V.8. b$_3$ 124-126°.[1588]

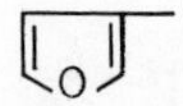-CH=CH-CH(OH)PO(OEt)$_2$. V.8. m. 101-102°.[1611]

2-HOC$_6$H$_4$CH(OH)PO[OC(-CCl$_3$)(CH$_2$)$_3$CH$_2$]$_2$. V.8. m. 149-150°.[3
C$_6$H$_5$S(O)$_2$CH$_2$PO(OH)$_2$. I.13. m. 261°.[1041]
C$_6$H$_5$S(O)$_2$OCH$_2$PO(OR)$_2$.

R = Et. V.14. b$_{0.02}$ 130°.[2054]

R = Pr. V.14. b$_{0.02}$ 140°.[2054]

R = i-Pr. V.14. m. 54-55°.[2054]

R = Bu. V.14. b$_{0.02}$ 146°.[2054]

4-MeSC$_6$H$_4$PO(OH)$_2$. I.1.[660] I.13. m. 242°.[1041]
4-MeSC$_6$H$_4$PO(OEt)$_2$. V.5. b$_{0.4}$ 126-128°.[675]
C$_6$H$_5$SCH$_2$PO(OH)$_2$. I.13. m. 242°.[1041]
C$_6$H$_5$SCH$_2$PO(OEt)$_2$. V.14. b$_7$ 177-179°.[128a] V.14. b$_{0.4}$ 126-128°.[675]
4-(H$_2$NCH$_2$)C$_6$H$_4$PO(OH)$_2$. I.13. m. >300°.[484] I.2. m. 280°.[2016]
4-(H$_3\overset{\oplus}{N}$CH$_2$)C$_6$H$_4$PO(OEt)O$^{\ominus}$. III.3. m. 247° [2016]
4-H$_2$NC$_6$H$_4$CH$_2$PO(OH)$_2$. I.13. Dec. 323-325°.[1005]

$4-H_2NC_6H_4CH_2PO(OEt)_2$. V.14. m. 91-92°.[932]
$C_6H_5CH(-NH_2)_2PO(OH)_2$. I.2. m. 272-273°.[1003a,1005] I.2. m. 280° (as hydrochloride).[789] I.13.[2040] I.2.[1036]
$C_6H_5CH(-NH_2)PO(OEt)OH$. III.3. m. 247°. As hydrochloride; m. 161° (dec.).[1036]
$[C_6H_5CH(-\overset{\oplus}{N}H_3)PO(OEt)_2]Cl^{\ominus}$. V.14. m. 158°.[789] V.14. m. 158-159°.[913]
$C_6H_5NHCH_2PO(OH)_2$. I.13. m. 221°.[1041]
$4-MeNHC_6H_4PO(OH)_2$. I.13. m. 177.5-179.5°.[481]
$2-H_2N-4-MeC_6H_3PO(OH)_2$. I.13. m. 213-215°.[577]
$2-H_2N-5-MeC_6H_3PO(OH)_2$. I.13. m. 219-222°.[577]
$3-H_2N-4-MeC_6H_3PO(OH)_2$. I.13. m. 290° (dec.)[1260,1269]

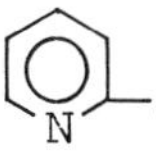
$-CH_2CH_2PO(OH)_2$. I.2. m. 149-150° (as HCl adduct).[1204]

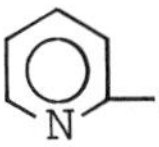
$-CH_2CH_2PO(OR)_2$.

R = Et. V.7. $b_{0.05}$ 101°.[1204] V.7. $b_{0.2}$ 112-113°.[1203]
R = i-Pr. V.7. $b_{0.05}$ 99°.[1204]

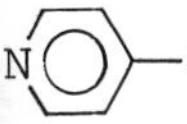
$-CH_2CH_2PO(OH)_2$. I.2. m. 139-141° (as HCl adduct).[1204]

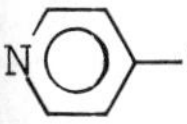
$-CH_2CH_2PO(OEt)_2$. V.7. $b_{0.2}$ 117°.[1203,1204]

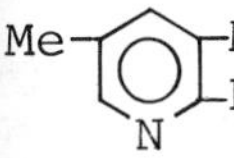
$PO(OH)_2$. I.2. m. 300°.[1718]

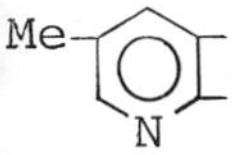
$PO(OEt)_2$. V.6. $b_{0.03}$ 107°, IR, ^1H.[1718]

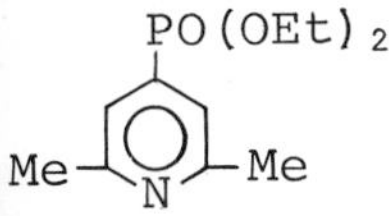
. V.6. $b_{0.2}$ 105°, IR, ^1H.[1718]

$\overline{N-CMeCH_2CH=CH-CH=C}-PO(OR)_2$.

R = Me. V.14. $b_{0.04}$ 105°.[340]
R = Et. V.14. $b_{0.02}$ 115-120°.[340]

$\overline{N=CHCH_2CMe=CH-CH=C}-PO(OMe)_2$. V.14. $b_{0.05}$ 110°.[340]
$2-HOC_6H_4CH(-NH_2)PO(OEt)_2$. V.9. m. 310°.[356]

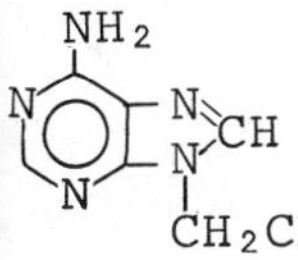
$CH_2CH_2PO(OCH_2CH_2Cl)_2$. V.14. m. 180.5-181.5°.[1103]

$H_2C(CH_2)_3C$=C=CHPO(OEt)$_2$. V.14.[1188] V.5. $b_{0.1}$ 135°.[1189]

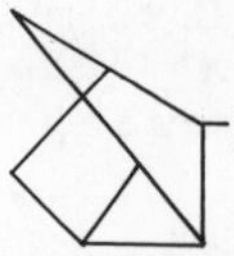
-PO(OEt)$_2$. V.7. $b_{0.5}$ 95-96°, IR, ^1H.[2028]

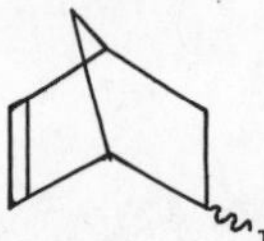
PO(OEt)$_2$. V.14. $b_{0.3}$ 88-91°.[451]

2-$HOC_6H_4CH(-NH_2)PO(OEt)_2$. V.9. m. 310°.[356]

OCH_2C(-CN)[-C(O)OEt]CH_2-PO(OEt). V.14.[841,844]

O=C-NMe-CMe=CH-C(O)-N-$CH_2PO(OH)_2$. V.14.[1739]

O=C-NH-C(O)NHC(O)CH-$(CH_2)_3PO(OEt)_2$. V.14. m. 229°.[1041]

O=C-$(CH_2)_4$C=CHPO(OEt)$_2$. V.8. b_{14} 140°.[1040]

$[CH_3C(O)-]_2C=CHCH_2PO(OEt)_2$. V.7. $b_{0.1}$ 129°.[1046]

$MeCHCH_2OC(O)C$[-C(O)Me]-PO(OEt)$_2$. V.6. $b_{0.05}$ 113.5°.[318]

HC≡$CCH_2CH(-CO_2Et)PO(OR)_2$.

R = Et. V.14. b_6 154-155°.[1604]

R = Pr. V.14. $b_{0.3}$ 120°.[1604]

R = Bu. V.14. $b_{0.8}$ 143°.[1604]

O=C-O-CMe$(CH_2)_2$CH(-CHO)-PO(OEt)$_2$. V.6. $b_{0.05}$ 125-128°.[318]

O=C-O$(CH_2)_2$-C[-C(O)Et]-PO(OEt)$_2$. V.6. $b_{0.1}$ 125-126°.[318]

$(CH_3O)(HO)C=C(-CO_2Me)$-CH=CH-PO(OPr-i)$_2$. V.14. $b_{0.002}$ 83°.[1788]

$HO_2CCH_2-C(-CO_2H)[PO(OH)_2]-CH_2CH_2CO_2H$. I.2. Complexing agen[612]

$HO_2CCH_2C(CO_2H)_2CH_2CH_2PO(OH)_2$. I.2. Isolated as oily K salt[2082]

$HO_2CCH_2CH(CO_2H)CH(CO_2H)CH_2PO(OH)_2$. I.2. m. 154-155°, complexing agent.[765]

$HO_2CCH_2CH(CO_2H)CMe(CO_2H)[PO(OH)_2]$. I.2. Isolated as Na_5 salt, complexing agent.[765]

$H_2C=C(-CN)$-CHPrPO(OMe)$_2$. V.5. $b_{0.001}$ 103°.[2018]

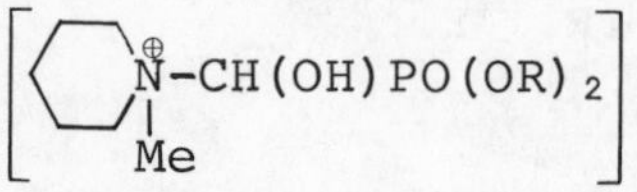

R = Et. V.8/V.14. m. 111°.[1121]

R = i-Pr. V.8/V.14. m. 146°.[1121]

R = i-Bu. V.8/V.14. m. 151°.[1121]

O-CH=CH-CH=C-CH(-NHEt)PO(OEt)$_2$. V.8. $b_{0.75}$ 127°.[529]

$O(CH_2)_2C(-CO_2H)$[-C(O)Et]PO(OEt). V.14. As p-toluidide m. 130°.[318]

$O(CH_2)_2C(-CO_2Me)$[-C(O)Me]PO(OEt). V.14. $b_{0.1}$ 126-128°.[318]

$OCHMeCH_2C(-CO_2H)$[-C(O)Me]PO(OEt). V.14. as p-toluidide, m. 187°.[318]

$C_5H_{11}C$≡CPO(OEt)$_2$. V.6. $b_{0.4}$ 125°.[1962,1963]

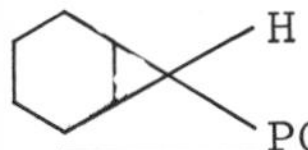
PO(OMe)$_2$. V.14. $b_{0.02}$ 61-63°, IR, ^{1}H.[1878a]

$H_2C(CH_2)_4C$=CHPO(OMe)$_2$. V.14. $b_{0.07}$ 76°, IR, ^{1}H.[1191]

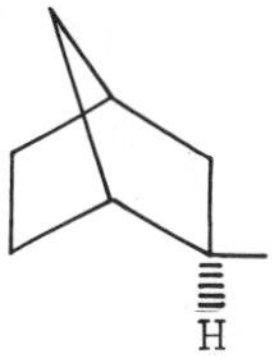
PO(OMe)$_2$. V.14. b_{10} 127-132°, ^{1}H.[342]

cycl.-$C_6H_{11}C(N_2)$PO(OMe)$_2$. V.14. $b_{0.1-0.15}$ 85-89°, IR, ^{1}H.[1191]

HN-N=C(-CO_2Et)CH_2CMePO(OMe)$_2$. V.13. m. 120-123°, IR, ^{1}H.[1878a]

$(CH_3)_2C$=CHC(O)CH_2CH_2PO(OR)$_2$.

R = Me. V.7. b_{13} 169-171°.[1562,1564]

R = Et. V.7. b_2 149°.[1562,1564]

cycl.-C_6H_{11}C(O)PO(OR)$_2$.

R = Me. V.5. $b_{0.05}$ 80-82°, IR, ^{1}H.[1191] V.5. $b_{0.5}$ 98-99°. As 2,4-dinitrophenylhydrazone, m. 125-126°, IR, ^{1}H.[244]

R = Et. V.5. $b_{0.2}$ 89-90°. As 2,4-dinitrophenylhydrazone, m. 113.5-115°, IR, ^{1}H.[244]

R = i-Pr. V.5. $b_{0.4}$ 100-102°, IR, ^{1}H.[244]

MeCHC(O)CH_2CHMeCHPO(OH)$_2$. I.8. As p-toluidine salt m. 153-154°.[906]

MeCHC(O)CH_2CHMeCHPO(OEt)$_2$. V.7. b_3 119-120°.[906]

H_2CC(O)$(CH_2)_3$CMePO(OH)$_2$. I.8. As aniline salt, m. 152°.[906]

H_2CC(O)$(CH_2)_3$CMePO(OEt)$_2$. V.7. b_2 120-124°.[906]

MeOCHCH=CH$(CH_2)_2$CHPO(OEt)$_2$. V.14. $b_{0.3}$ 102-105°.[451]

$H_2C(CH_2)_4$C—O—CHPO(OEt)$_2$. V.14. b_1 117-118°.[1201]

[MeC(O)]$_2$CHCHMePO(OR)$_2$.

R = Me. V.7. n_D^{20} 1.4472, d_4^{20} 1.1679 (not pure)[150]

R = Et. V.7. n_D^{20} 1.4510, d_4^{20} 1.1238.[150]

$C_2H_5O_2$C-CH=CEtPO(OEt)$_2$. cis : trans 6 : 4. V.7. $b_{1.5}$ 121-123°, IR, ^{1}H.[392]

$C_2H_5O_2$CCH=CMeCH_2PO(OEt)$_2$. V.5.[591]

$C_3H_7O_2$CCH=CMePO(OPr)$_2$. V.7. b_6 170-172°.[967,968]

$C_4H_9O_2$CCH=CHPO(OBu)$_2$. V.7.[968] V.7. b_3 171-173°.[967]

$CH_3O_2CCH_2CH_2$C(=CHMe)PO(OEt)$_2$. V.14. b_1 135-136°.[1621]

$C_2H_5O_2$CCHCH_2CHCH_2PO(OR)$_2$.

R = Me. V.14. $b_{1.7}$ 132-135°.[1577,1578]

R = Et. V.14. b_3 143-146°.[1577,1578]

$H_2C{=}CHCH_2CH(-CO_2Et)PO(OEt)_2$. V.14. $b_{0.05}$ 95-110°.[996]
$C_3H_7CO_2CMe{=}CHPO(OEt)_2$. V.14. b_1 116-118°.[1135]
$C_2H_5O_2CCH[-C(O)Me]-CH_2PO(OEt)_2$. V.8/V.14.[840]
$C_2H_5O_2CCH_2CH[-SC(O)Me]-PO(OEt)_2$. V.14. $b_{0.5}$ 149-150°.[1642]
$(C_2H_5O_2C-)_2CHPO(OR)_2$.

R = Me. V.5. b_{3-4} 153-154°.[115]

R = Et. V.5. b_{3-4} 154-156°.[115] V.5, V.14. $b_{0.2}$ 109-110°.[1992] V.5. b_3 153-156°.[1003]

R = Pr. V.5. b_{3-4} 169-170°.[115]

R = Bu. V.5.[115,1003] V.6.[1003]

$H_2C(-CO_2Me)C(-CO_2Me)MePO(OMe)_2$. V.14. b_1 127-130°.[613]
$H_2C{=}CMeC(O)NHCMe_2PO(OMe)_2$. V.14.[1597]
$H_2C(CH_2)_4NC(O)CH_2PO(OEt)_2$. V.5. b_2 156°.[1940]
$C_5H_{11}CH{=}CHPO(OR)_2$.

R = Me. V.7. b_{20} 154-155°, n_D^{20} 1.4725, d_4^{20} 1.0612.[1627] V.14. b_2 113-115°, n_D^{20} 1.4489, d_4^{20} 1.0114.[524]

R = Et. V.7. b_{12} 149-150°, n_D^{20} 1.4630, d_4^{20} 1.0201.[1627] V.14. b_3 114-116°, n_D^{20} 1.4468, d_4^{20} 0.9845.[524]

R = Pr. V.14. b_3 136-137°, n_D^{20} 1.4451, d_4^{20} 0.9698.[524]

R = $H_2C{=}CHCH_2-$. V.14. b_4 150-152°, n_D^{20} 1.4632, d_4^{20} 0.9969.[524]

R = Bu. V.14. b_3 148-150°, n_D^{20} 1.4480, d_4^{20} 0.9450.[524]

$C_4H_9CMe{=}CHPO(OPr)_2$. V.2. b_1 138-140°, n_D^{20} 1.4549, d_4^{20} 1.0391.[992]
$H_2CCH_2CHMeCH_2CH_2CHPO(OEt)Cl$. IV.2. $b_{0.1}$ 80-82°.[1899]
$H_2CCH_2CHMeCH_2CH_2CHPO(OR)_2$.

R = Me. V.2. $b_{0.2}$ 70-72°, n_D^{20} 1.4643, d_4^{20} 1.1052.[1899]

R = Et. V.2. $b_{0.2}$ 82-84°, n_D^{20} 1.4602, d_4^{20} 1.0580.[1899]

R = Pr. V.2. $b_{0.35}$ 115-117 , n_D^{20} 1.4597, d_4^{20} 1.0301.[1899]

R = Bu. V.2. $b_{0.35}$ 125-127°, n_D^{20} 1.4562, d_4^{20} 0.9940.[1899]

$CH_3O_2CN{=}C(-NEt_2)PO(OMe)_2$. V.2. $b_{0.05}$ 114-116°, n_D^{20} 1.4868, d_4^{20} 1.1916.[490]
$(C_2H_5-)(EtO)OPCH_2CH(-CN)PO(OEt)_2$. V.7. $b_{0.08}$ 168-170°.[1672]
$CH_3C(O)(CH_2)_5PO(Et)_2$. V.14. IR.[1959]
$C_2H_5C(O)(CH_2)_4PO(OEt)_2$. V.6. $b_{0.5}$ 138°.[1958]
$C_5H_{11}C(O)CH_2PO(OEt)_2$. V.14. $b_{0.2}$ 112°.[1963]
$CH_3C(O)CH_2CHPr-PO(OEt)_2$. V.14.[840] V.7. b_8 139-140°.[1659]
$(CH_3)_2CHOCH{=}CEtPO(OR)_2$.

R = Me. V.2. $b_{0.078}$ 69-70°, n_D^{20} 1.4544, d_4^{20} 1.0578.[135]

R = Et. V.2. $b_{0.07}$ 84-85°, n_D^{20} 1.4528, d_4^{20} 1.0250.[1350]

$C_4H_9OCMe{=}CHPO(OBu)_2$. V.14. $b_{2.5}$ 148-150°.[1133]
$(CH_3)_3COCMe{=}CHPO(OEt)_2$. V.14. $b_{0.2}$ 95°.[1961]
$C_2H_5OCEt{=}CHCH_2PO(OEt)_2$. V.14. $b_{0.3}$ 105°.[1082]
$(CH_3)_2C{=}C(-OEt)CH_2PO(OR)_2$.

R = Pr. V.14. b_5 135-138°.[1603]

R = Bu. V.14. $b_{0.8}$ 129-130°.[1603]

$(CH_3)_2CH{-}C(-OEt){=}CHPO(OEt)_2$. V.14. b_8 133-135°.[1606]
$EtOCH(CH_2)_3CHPO(OEt)_2$. V.2. $b_{1.5}$ 125-126°.[162]
$O(CH_4)_4CEt{-}PO(OEt)_2$. V.6. $b_{0.3}$ 91°, n_D^{21} 1.4525, d_4^{21} 1.073.[1958]

$CH_3C(O)CMe(-OPr)PO(OPr)_2$. V.8. b_{10} 140-142°.[1060]
$(C_2H_5O)_2CHCH{=}CHPO(OEt)_2$. V.2. b_1 118-119°, n_D^{20} 1.4429, d_4^{20} 1.0536.[2037]
$(C_3H_7O-)(CH_3O)CHCH{=}CHPO(OMe)_2$. V.2. $b_{0.045}$ 90-92°, n_D^{20} 1.4542, d_4^{20} 1.1169.[2037]
$C_3H_7OCH_2OCMe{=}CHPO(OEt)_2$. V.14. b_1 109-110°.[1651]
$C_2H_5OCH_2C(-OEt){=}CHPO(OEt)_2$. V.14. $b_{0.2}$ 115°.[1961] ^1H, ^{31}P -25ppm.[1093]
$OCH_2CH_2OCMe(CH_2)_3PO(OEt)_2$. V.6. $b_{0.05}$ 128°, IR.[1959,1410]
$CH_2CH_2OCMe_2CH_2C(-OH)PO(OEt)_2$. V.8. m. 75-76°.[1221]
$C_2H_5O_2C(CH_2)_4PO(OEt)_2$. V.5. $b_{0.05}$ 125-130°.[516]
$C_3H_7O_2C(CH_2)_3PO(OEt)_2$. V.5. $b_{0.09}$ 112°.[516]
$(CH_3)_2CHO_2C(CH_2)_3PO(OR)_2$.

R = Et. V.5. $b_{0.02}$ 128-135°.[516]

R = i-Pr. V.5. $b_{0.08}$ 130°.[516]

$(CH_3)_2CHO_2C(CH_2)_3PO(OEt)OC_8H_{17}$. V.5. $b_{0.03}$ 158°.[516]
$C_4H_9O_2C(CH_2)_2PO(OR)_2$.

R = Et. V.7. b_3 165°.[638]

R = Bu. V.7. b_1 140-145°.[638]

$(CH_3)_3CO_2CCHMePO(OEt)_2$. V.14. $b_{0.05}$ 89-90°, n_D^{20} 1.4315, d_4^{20} 1.0600.[280]
$C_2H_5O_2CCH_2CH(-SEt)PO(OEt)_2$. V.14. $b_{0.5}$ 121-122°, n_D^{20} 1.4592, d_4^{20} 1.1150.[1642]
$HS(CH_2)_3CH(-CO_2Et)PO(OEt)_2$. V.14. $b_{0.05}$ 115-120°, UV λ_{max} 210 mμ, IR.[996]
$CH_3O_2CCHMeCH_2OCHMePO(OMe)_2$. V.7. b_3 133°, n_D^{20} 1.4430, d_4^{20} 1.1220.[1675]

$C_2H_5O_2CCH_2CH(-OEt)PO(OEt)_2$. V.8. $b_{0.25}$ 132°.[751]
$C_2H_5SCH_2CMe{=}CHCH_2PO(OEt)_2$. V.14. $b_{1.5}$ 132°.[1630]
$C_2H_5SCH_2CHMe{-}CH{=}CHPO(OR)_2$.

R = Et. V.14. b_1 148.5°.[990]

R = Bu. V.14. b_{1-2} 170-171°.[990]

$(C_2H_5)_2NCH_2CH{=}CHPO(OEt)_2$. V.14. $b_{0.5}$ 114-116°.[1082]

$(C_2H_5)_2NCMe=CHPO(OEt)_2$. V.14. b_2 129°.[1602]
$CH_3CH=C(-NEt_2)PO(OR)_2$.

R = Et. V.7. $b_{0.1}$ $60-65^{\circ}$.[1809a]

R = i-Pr. V.7. $b_{0.005}$ $63-70^{\circ}$, $b_{0.1}$ $100-109^{\circ}$.[1809a]

R = Ph. V.7. $b_{0.1}$ $158-162^{\circ}$.[1809a]

$H_2C=CHCH(-NEt_2)PO(OEt)_2$. V.14. b_4 116°.[529]
$MeCH(CH_2)_2NCMe_2PO(OPr)_2$. V.9. b_5 $119-120^{\circ}$.[962]
$H_2C(CH_2)_4NCH_2CH_2PO(OR)_2$.

R = Et. V.6. b_2 $123-125^{\circ}$.[144] V.14. b_2 120°.[329]

R = Bu. V.4, V.6. b_1 $141.5-142.5^{\circ}$, n_D^{20} 1.4601, d_4^{20} 0.9966.[931]

$H_2C(CH_2)_4NCHMePO(OBu)_2$. V.9. b_1 $118-118.5^{\circ}$.[931]
$H_2C(CH_2)_4CHCH(-NH_2)$ $PO(OEt)_2$. V.9. As picrate, m. 172°.[1220]

$(C_2H_5)_2NCH_2C(O)CH_2PO(OEt)_2$. V.14. $b_{0.4}$ 118°.[1961]
$C_5H_{11}NHC(O)CH_2PO(OEt)_2$. V.5. $b_{1.2}$ 158°.[1940]
$(CH_3)_2CH(CH_2)_2NHC(O)CH_2PO(OEt)_2$. V.5. $b_{0.25}$ $136-139^{\circ}$.[1108]
$(C_2H_5)_2NC(O)CH_2CH_2PO(OR)_2$.

R = Et. V.5, V.7. b_{11} 191°.[1658]

R = Pr. V.5, V.7. b_{10} $187-189^{\circ}$.[1658]

$(C_2H_5)_2NC(O)CH_2CH_2PO(OEt)OCH_2CH_2Br$. V.7. $b_{0.05}$ $170-171^{\circ}$.[16
$[(C_2H_5)_2NC(O)CH_2CH_2-]P(O)-O-R-O$.

R = $-CH_2CH_2-$. V.7. $b_{0.005}$ $171-173^{\circ}$.[1656]

R = $-CHMe-CH_2-$. V.7. $b_{0.002}$ $169-171^{\circ}$.[1656]

$H_2C(CH_2)_4NCH(-OMe)PO(OMe)_2$. V.5. $b_{0.04}$ $115-120^{\circ}$, ^{31}P -18.2 ppm, IR.[717]
$H_2CCH_2OCMe_2CH_2C(-NH_2)PO(OEt)_2$. V.9. $b_{0.01}$ $79-81^{\circ}$.[1221]
$C_3H_7NHC(O)CH(-OEt)PO(OEt)_2$. V.5. $b_{0.1}$ $115-120^{\circ}$, IR, ^1H.[684]
$C_2H_5O_2CCH_2NHCMe_2PO(OEt)_2$. V.14, V.9. $b_{0.003}$ $77-81^{\circ}$[914]
$(EtO)_2SP-S-CMe=CHPO(OR)_2$.

R = Et. X.3. b_2 $171-172^{\circ}$.[1600]

R = Bu. X.3. $b_{0.05}$ $168-169^{\circ}$.[1600]

$C_7H_{15}PO(OH)_2$ I.13. m. 106°.[555] I.13. m. $103-103.5^{\circ}$.[1000,100
$C_7H_{15}PO(OR)_2$.

R = Me. V.7. $b_{15.5}$ 144°.[1614]

R = Et. V.5. $b_{1.9}$ 113°.[552]

R = Bu. V.6. b_{17} $188-190^{\circ}$.[1000,1002]

$C_5H_{11}CHMePO(OH)_2$. I.6. As aniline salt, m. $140-141^{\circ}$.[315]
$(C_3H_7)_2CHPO(OH)_2$. I.6. As aniline salt, m. $152-153^{\circ}$.[315]
$(CH_3)_2NCH_2C(-NMe_2)=CHPO(OEt)_2$. V.14. ^1H, ^{31}P -24.7 ppm.[1093]

$C_5H_{11}CH(-OMe)PO(OEt)OC_6H_4NO_2-4$. V.5. n_D^{25} 1.512.[741]

$C_6H_{13}CH(OH)PO(OH)_2$. I.9. m. 185°.[555] I.9. m. 165-173 ° (water).[415]
$(C_2H_5O)_2CHCH_2CH_2PO(OEt)_2$. V.5. $b_{0.07}$ 98-100°.[1705] V.14.[175]
$[(CH_3)_2CHO-]_2CHPO(OPr-i)_2$. From $HC(OPr-i)_3$ and PCl_3, b_{12} 140°, n_D^{21} 1.4225, ^1H.[713]
$(C_2H_5O)_2CEtPO(OEt)_2$. From $HC(OEt)_3$ and PCl_3, $b_{0.04}$ 75-77°, n_D^{23} 1.4332.[713]
$(C_2H_5S-)_2CMeCH_2PO(OR)_2$.

R = Pr. V.14. $b_{3.5}$ 145°.[1602] V.14. $b_{0.8}$ 118-120°.[1607]

R = Bu. V.14. b_2 155°.[1602]

$(CH_3)_3SiCH_2\overline{CHCH_2C}HPO(OMe)_2$. V.14. $b_{0.02}$ 55-57°, IR, ^1H.[1878a]

$(C_2H_5)_2N(CH_2)_3PO(OR)_2$.

R = Et. V.14. $b_{0.004}$ 116-117°[54]

R = Bu. V.14. $b_{0.04}$ 152-153°[54]

$(C_2H_5)_2NCHMeCH_2PO(OEt)_2$. V.14. b_2 137°.[1602]
$(C_2H_5)_2NCHEtPO(OEt)_2$. V.14. $b_{0.8}$ 95°.[529]
$(C_2H_5)_2NCMe_2PO(OR)OH$.

R = Me. From $(MeO)(Et_2N)OPH$ and Me_2CO, m. 141-142°.[813]

R = Et. From $(EtO)(Et_2N)OPH$ and Me_2CO, m. 138-139°.[813]

R = i-Pr. From $(i-PrO)(Et_2N)OPH$ and Me_2CO, m. 149-150°.[813]

$(C_2H_5)_2NCMe_2PO(OEt)_2$. V.9. b_{2-5} 103-105°.[529]
$(MeO-)O(C_4H_9-)PCMe(OH)-PO(OEt)_2$. V.14. $b_{0.28}$ 131-132°[14] (?).
$C_2H_5CH[-O-PO(OEt_2]PO(OEt)_2$. V.6. b_1 153-154°.[1618]
$[Et_3\overset{\oplus}{N}CH_2PO(OEt)_2]Cl^{\ominus}$. V.14. m. 197°.[478]
$^{\ominus}O_3SCH_2CH_2\overset{\oplus}{N}H_2CH_2CH_2\overset{\oplus}{N}H(CH_2CH_2SO_3^{\ominus})CH_2PO(OH)_2$. I.13. Dissociation const. pK values for complexes, IR.[818]
$(C_2H_5)_3SiOCH_2PO(OR)_2$.

R = Et. V.14. $b_{2.5}$ 115°, n_D^{20} 1.4360, d_4^{20} 1.0086.[1436]

R = i-Pr. V.14. b_2 121°, n_D^{20} 1.4340, d_4^{20} 0.9712.[1436]

R = Et_3Si-. V.14/VI.1. $b_{2.5}$ 162°, n_D^{20} 1.4512, d_4^{20} 0.9653.[1438] V.14/VI.1 b_2 160-162°.[1435] V.14/VI.1. b_3 163°.[1437]

$Et_2MeSiCH_2OCH_2PO(OH)_2$. I.2.[1433]
$Et_2MeSiCH_2OCH_2PO(OR)_2$.

R = Et V.14. b_6 140°[1433]

R = i-Pr. V.2. b_5 135°.[1433]

$Me_3SiOCMeEt-PO(OEt)_2$. V.8. b_1 92°.[1414]
$(EtO)_2MeSiCH_2CH_2PO(OR)_2$.

R = Me. V.7. b_{1-2} 146-154°.[1101]

R = Et. V.7. b_{3-4} 176°.[1101]

$(EtO)Et_2SiOCH_2PO(OEt)_2$. V.14. b_3 124-125°, n_D^{20} 1.4309, d_4^{20} 1.0283.[1436]

$(EtO)Me_2SiOCMe_2PO(OEt)_2$. V.8/V.14. $b_{0.4}$ 80°.[258]

$(EtO)Me_2SiOCHEtPO(OEt)_2$. V.8/V.14. $b_{0.4}$ 92-93°.[258]

$(EtO)_3SiCH_2PO(OEt)_2$. V.5. b_2 96.5-97°, n_D^{20} 1.4234, d_4^{20} 1.0547.[320] V.5. b_3 133-135°.[637]

$(EtO)_2MeSiCH_2OCH_2PO(OEt)_2$. V.14. b_2 120°.[1432,1433]

$(C_2H_5)_3SiCH_2PO(OEt)_2$. V.5. $b_{2.5}$ 93-95°.[387]

C_8

BrHC=CPhPO(OH)$_2$, cis-trans isomers. I.13. m. 143-145°.[959] I.13. m. 133-135°.[413]

$C_6H_5C(O)CHBrPO(OEt)_2$. V.14. m. 75°.[697]

$BrH_2CCBrPhPO(OH)_2$. I.13. m. 186-188°.[413]

2-$ClC_6H_4C{\equiv}CPO(OH)_2$. I.13. m. 134°.[240]

3-$ClC_6H_4C{\equiv}CPO(OH)_2$. I.1. m. 168°.[239]

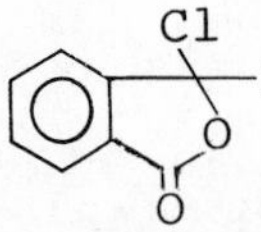
$-PO(OMe)_2$. V.5. m. 77.5-78.5°, ^1H.[243]

$C_6H_5CCl{=}CHPO(OEt)_2$. V.2. b_2 159°.[90]

$ClHC{=}CPhPO(OH)_2$. I.13. m. 145-146.5°, IR, ^1H, ^{31}P-12.3ppm.

4-$ClC_6H_4SC(O)CH_2PO(OEt)_2$. V.5. $b_{0.3}$ 143-146°.[1239]

4-$ClC_6H_4O_2CCH_2PO(OEt)_2$. V.5. $b_{0.3}$ 152-154°.[1239]

4-$ClC_6H_4CH_2SC(S)$-PO(OPr-i)$_2$. V.8. ^{31}P + 4.2ppm.[709]

2-$ClC_6H_4C(O)NHCH_2PO(OH)_2$. I.13. m. 176-177°.[1198]

4-$ClC_6H_4C(O)NHCH_2PO(OH)_2$. I.13. m. 179-180°.[1198]

$C_6H_5C(OH)(-CH_2Cl)PO(OMe)_2$. V.8. m. 161.5-163.5°.[1995]

$C_6H_5CMeClPO(OH)_2$. I.9. m. 174-175°.[413]

4-$ClC_6H_4NHC(O)NHCH_2PO(OR)_2$.

R = Et. V.14. m. 103-104°.[1903]

R = i-Pr. V.14. m. 82-83°.[1903]

2-MeO-5-$ClC_6H_3CH_2PO(OR)_2$.

R = Et. V.5. V.6. $b_{0.25}$ 140-141°, m. 25-27°, n_D^{20} 1.51 d_4^{20} 1.2176.[1538]

R = i-Pr. V.5. V.6. $b_{0.18}$ 125-127°, n_D^{20} 1.4932, d_4^{20} 1.1528.[1538]

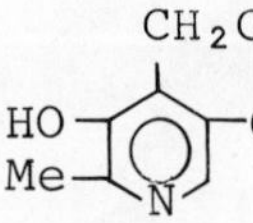
$-CH_2PO(OPr-i)_2$. V.14. m. as hydrochloride 172-173° (dec.)[232]

$H_2C(CH_2)_3$-CH=C-CCl=CHPO(OR)$_2$.

R = Et. V.2. b_1 132°.[88]

R = i-Pr. V.2. b_1 131°.[88]

H_2C-CH(-CH=CH$_2$)(CH$_2$)$_2$CHClCHPO[OCHCHClCH$_2$CH(-CH=CH$_2$)CH$_2$CH$_2$]$_2$.
V.5. n_D^{20} 1.5140, d_4^{20} 1.1781.[621]

C_4H_9OCH=C(-CCl=CH$_2$)PO(OR)$_2$.

R = Me. V.2. $b_{0.002}$ 79.5-81°.[1920]

R = Et. V.2. $b_{0.006}$ 90-91°.[1920]

C_4H_9SCH=C(-CCl=CH$_2$)PO(OR)$_2$.

R = Me. V.2. $b_{0.0002}$ 83-84°.[1918]

R = Et. V.2. $b_{0.004}$ 94-95°.[1918]

R = Pr. V.2. $b_{0.0002}$ 93°.[1918]

R = Bu. V.2. $b_{0.008}$ 92°.[1918]

C_5H_{11}CH=CCl-CH$_2$PO(OH)$_2$. I.1. m. 87-89°, ^{1}H.[1228]
C_5H_{11}CH=CCl-CH$_2$PO(OEt)$_2$. V.2. $b_{0.25}$ 107-108°, ^{1}H.[1228]
C_6H_{13}CHClCH$_2$PO(OH)$_2$. I.1. m. 89-91°.[1218]
2-ClC$_6$H$_4$CCl=CHPO(OH)$_2$. I.1. m. 187°.[240]
C_6H_5C(O)CCl$_2$PO(OH)$_2$. I.1. m. 152-153°.[223]
3,4-Cl$_2$C$_6$H$_3$CONHCH$_2$PO(OH)$_2$. I.13. m. 179-180°.[1198]
ClH$_2$CCClPhPO(OH)$_2$. I.13. m. 175-178°.[413]
2,4,5-Cl$_3$C$_6$H$_2$SC(O)CH$_2$PO(OEt)$_2$. V.5. $b_{0.2}$ 164-166°.[1239]
2,4,5-Cl$_3$C$_6$H$_2$O$_2$C-CH$_2$PO(OEt)$_2$. V.5. $b_{0.1}$ 139-141°.[1239]
4-ClC$_6$H$_4$NHC(O)CCl$_2$PO(OEt)$_2$. V.14. m. 92-93°.[282]
3-Cl$_3$CSS(O)$_2$-4-MeOC$_6$H$_3$PO(OH)$_2$. I.13. m. 166-168° (dec.).[749]
H_2C(CH$_2$)$_4$NS(O)$_2$NHCO$_2$CH(-CCl$_3$)-PO(OMe)$_2$. V.14. m. 121°.[1999]
H_2C(CH$_2$)$_4$N-C(-CF$_3$)=CFPO(OEt)$_2$. V.14. b_8 140°.[979]
C_6H_5C≡CPO(OH)$_2$. I.13. m. 142°.[240]
C_6H_5C≡CPO(OEt)$_2$. V.5. $b_{0.1}$ 132°, V.6. $b_{0.2}$ 139°.[1962,1963]
V.5. b_4 181-184°.[827] V.5. Kinetics of the reaction.[590]
V.5. $b_{0.09}$ 120-120.3°.[590a]
4-O$_2$NC$_6$H$_4$-N=N-CH(-CN)PO(OEt)$_2$. V.14. m. 121-122°[1343]

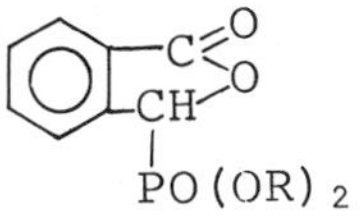

R = Me. V.2. m. 96-97°.[1692]

R = Et. V.5.[1692]

R = i-Pr. V.5. m. 98-99°.[1692]

(HC≡C-CH$_2$-)$_2$C(-CN)PO(OR)$_2$.

R = Et. V.14. b_1 136-137°.[1604]

R = Pr. V.14. b_1 149-150°.[1604]

R = Bu. V.14. $b_{0.8}$ 154°.[1604]

4-NCC$_6$H$_4$CH$_2$PO(OEt)$_2$. V.5. $b_{0.03}$ 146-150°.[1956]
4-OCNC$_6$H$_4$CH$_2$PO(OEt)$_2$. V.5. $b_{0.02}$ 133-135°.[1903]
C_6H_5C(OH)(-CN)PO(OMe)$_2$. V.14. b_1 143-143.5°[918] (cf. Ref. [1619])

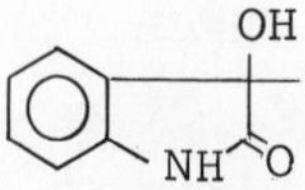

R = Me. V.8. m. 182-184°.[1362]

R = Et. V.8. m. 161-162°.[1362]

R = i-Pr. V.8. m. 158-160°.[1362]

4-$O_2NC_6H_4C(O)CH_2PO(OEt)_2$. V.5.[185]

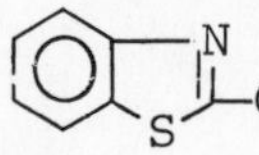 $-CH_2PO(OEt)_2$. V.5. $b_{0.2}$ 170-176°.[977]

C_6H_5-N=NCH(-CN)$PO(OEt)_2$. V.14. m. 84-85°.[1343]

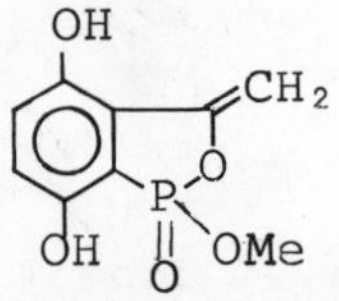 . From 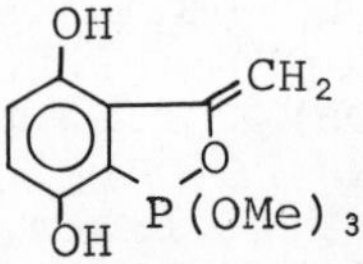 + $MeCO_2H$.[14

$C_6H_5CH=CHPO(OH)_2$. I.1. m. 146°.[238] I.1.[1991a] I.1. m. 154.5
155° (water); monosodium salt; m. 224-227°.[1019] I.2.
142-142.5°.[692]

$C_6H_5CH=CHPO(OR)OH$.

R = Me. III.3, III.4. m. 83-86°.[1217]

R = Et. III.3. m. 80-81.5°. As dicyclohexylammonium salt, m. 179-180°[770].

$C_6H_5CH=CHPO(OR)_2$.

R = Me. V.2. b_2 129°, m. 41-42° [91]. V.7. b_1 138.5-139° IR, ^1H.[692]

R = Et. V.2. b_2 138°.[91] V.7. b_2 138°, IR, ^1H.[692] V.5. $b_{0.15}$ 127°, n_D^{20} 1.5279, IR, ^1H.[1989]

R = H_2=$CHCH_2$-. V.2. b_1 152-158°.[2021a]

R = H_2C=$CMeCH_2$-. V.2. b_2 158-164°.[2021a]

R = 3-$H_2NC_6H_4$-. V.2.[1423]

R = 3-$OCNC_6H_4$-. V.14.[1423]

R = 3-H_2N-4-MeC_6H_3-. V.2. m. 152-154°.[1423]

$C_6H_5CH=CHPO(OMe)OBu$. V.13. b_1 149-150°.[975]

$H_2C=CPhPO(OH)_2$. I.9. m. 112-113°. As aniline salt, m. 180-
181°.[415]

$H_2C=CPhO(OPh)Cl$. From $PhPCl_2$ and MeC(O)PH at 150°, $b_{0.09}$
165-170°, n_D^{20} 1.6011, d_4^{20} 1.3070.[945]

$H_2C=CPhPO(OEt)_2$. V.13. $b_{0.3}$ 127°.[267] V.5. $b_{0.1}$ 101-103°,

n_D^{20} 1.5158, IR, ^1H.[1989]
4-$MeC_6H_4C(N_2)$-$PO(OEt)_2$. V.14. m. 24-25°, IR.[883]
4-$O_2NC_6H_4C(O)NHCH_2PO(OH)_2$. V.14. m. 206-207°.[1198]
$C_6H_5C(O)CH_2PO(OH)_2$. I.1. m. 139-141°.[525] I.2.[1552] I.2. m. 135°.[821]
$C_6H_5C(O)CH_2PO(OR)_2$.

R = Me. V.5. b_9 189°.[820]

R = Et. V.5. b_2 181-182°.[820] V.5, V.6. $b_{2.5}$ 174-176°.[119a] V.14.[165] V.6.[167] V.5. b_{11} 192-193°.[1552] V.14. $b_{0.2}$ 148°.[1963]

R = Pr. V.5. b_2 181-182°.[820]

R = Bu. V.5. b_9 209-211°.[820]

R = Am. V.5. b_1 204-205°.[820]

R = C_6H_{13}-. V.5. $b_{0.7}$ 202-203°.[820]

R = C_7H_{15}-. V.5. $b_{0.4}$ 206-207°.[820]

R = C_8H_{17}-. V.5. $b_{0.3}$ 227-229°.[820]

$C_6H_5C(O)CH_2PO(OPh)OR$.

R = Pr. V.5. n_D^{20} 1.5402, d_4^{20} 1.1911, IR.[821]

R = Bu. V.5. n_D^{20} 1.5388, d_4^{20} 1.1669, IR.[821]

4-[MeC(O)-]$C_6H_4PO(OH)_2$. I.1. m. 163-167°.[579]
4-[MeC(O)-]$C_6H_4PO(OEt)_2$. V.5. $b_{0.06}$ 148-150°.[1987] V.5. $b_{0.06}$ 145-150°.[1985]
$C_6H_5CH(-CHO)PO(OEt)_2$. V.14. m. 58°, n_D^{25} 1.5320.[1983]
C_6H_5CH-O-$CHPO(OR)_2$.

R = Me. V.14. b_2 146°, n_D^{20} 1.5184.[1202]

R = Et. V.14. b_2 161-162°, n_D^{20} 1.5147.[1202]

4-$MeOC_6H_4C(O)PO(OR)_2$.

R = Me. V.5. $b_{2.5}$ 170°, As 2,4-dinitrophenylhydrazone, m. 203-204°, IR, ^1H.[249]

R = Et. V.5. $b_{0.4}$ 158°. As 2,4-dinitrophenylhydrazone, m. 177-178°, IR, ^1H.[249]

R = i-Pr. V.5. ^1H. As 2,4-dinitrophenylhydrazone, m. 144-146°.[249]

R = H_2C=$CHCH_2$-. V.5. ^1H. As 2,4-dinitrophenylhydrazone, m. 123-125°.[249]

R = Bu. V.5. b_2 175°, ^1H. As 2,4-dinitrophenylhydrazone, m. 127-129°.[249]

$C_6H_5C(O)CH(OH)$-$PO(OR)_2$.

R = Me. V.8. n_D^{20} 1.5235, d_4^{20} 1.2914.[1624]

R = Et. V.8. n_D^{20} 1.5165, d_4^{20} 1.2392.[1624]

4-[CH_3CO_2-]$C_6H_4PO(OEt)_2$. V.5. $b_{0.1}$ 144-148°.[1985,1987]
$C_6H_5O_2CCH_2PO(OEt)_2$. V.5, V.6. b_9 153.5-157°.[119a] V.5. $b_{0.1}$ 115-120°.[1239]
$HO_2CCHPhPO(OH)_2$. I.2. As p-toluidine salt, m. 202°.[1037]
$C_6H_5S(O)_2CH$=$CHPO(OEt)_2$. V.5. $b_{0.01}$ 122-124°, n_D^{20} 1.5190, IR.[941]

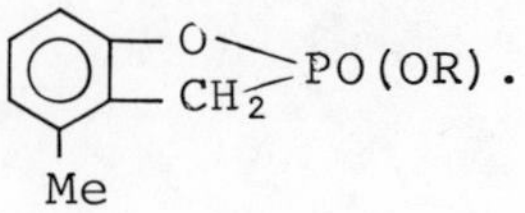

R = Et. V.9. $b_{0.04}$ 101-102.5°, m. 86-87°.[847]

R = Pr. V.9. $b_{0.2}$ 130-132°.[847]

R = Bu. V.9. $b_{0.1}$ 129-130°.[847]

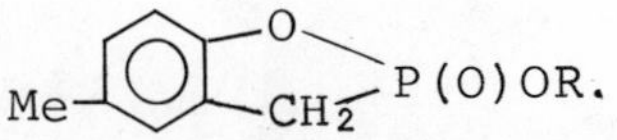

R = Et. V.9. $b_{0.1}$ 125-127°.[847]

R = Pr. V.9. $b_{0.1}$ 130-132°.[847]

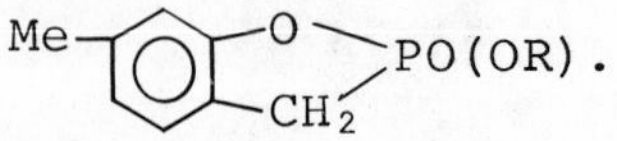

R = Et. V.9. $b_{0.08}$ 117-117.5°.[847]

R = Pr. V.9. $b_{0.08}$ 125-127°.[847]

R = Bu. V.9. $b_{0.2}$ 134-136°.[847]

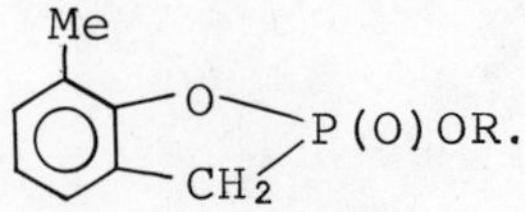

R = Et. From 2-HO-3-$MeC_6H_3NEt_2$, and $(EtO)_2PNEt_2$, $b_{0.35}$ 120°.[833]

R = 2-Et_2NCH_2-6-MeC_6H_3-. V.9.[833]

C_6H_5C(=NMe)$PO(OR)_2$.

R = Et. V.8. $b_{0.02}$ 101.5°.[2103]

R = Pr. V.8. $b_{0.001}$ 102°.[2103]

$CH_3C(O)NHC_6H_4PO(OH)_2$. I.13. m. 229°.[213]
$C_6H_5C(O)NHCH_2PO(OH)_2$. I.13. m. 182°.[506]

$C_6H_5C(O)NHCH_2PO(OR)_2$.

R = Me. V.9.[837]

R = Et. V.9.[837,838]

R = Pr. V.9.[837]

R = Bu. V.9.[837,838]

$C_6H_5NHC(O)CH_2PO(OEt)_2$. V.5. $b_{0.25}$ 174-177°.[1239] V.14. $b_{0.07}$ 167-168°.[282]

$C_6H_5NHC(S)OCH_2PO(OR)_2$.

R = Et. V.14. m. 75-75.7°.[55]

R = Pr. V.14. m. 50.8-52.3°.[55]

R = i-Pr. V.14. m. 97.2-97.9°.[55]

R = i-Bu. V.14. m. 70-71.5°.[55]

4-$O_2NC_6H_4CH_2CH_2PO(OH)_2$. I.2. m. 174-175°.[932]
4-$O_2NC_6H_4CH_2CH_2PO(OEt)_2$. V.14 $b_{0.05}$ 160-165°.[932]
4-$O_2NC_6H_4CHMePO(OEt)_2$. V.14. $b_{0.08}$ 153-158°.[932]

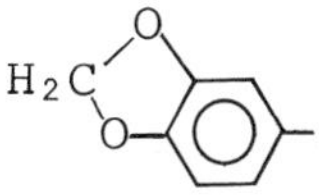

$-CH(-NH_2)PO(OEt)_2$. V.9. As picrate, m. 145°.[913]

2-$O_2NC_6H_4CH(-OMe)PO(OMe)_2$. V.8. m. 113-115°.[1059]
4-$O_2NC_6H_4CH(-OMe)PO(OMe)_2$. V.8. m. 132-134°.[1059]
$C_6H_5CH_2CH_2PO(OH)_2$. I.13. m. 137-138°.[238] I.2. m. 84-85°.[1499] I.1. m. ~90°.[737] I.2. m. 138.5-140°.[932]
$C_6H_5CH_2CH_2PO(OC_6H_{11}$-cycl.)Cl. IV.2. n_D^{25} 1.5240.[740]
$C_6H_5CH_2CH_2PO(OC_6H_{11}$-cycl.)F. IV.2. $b_{0.03}$ 113°, n^{25} 1.5001.[740]

$C_6H_5CH_2CH_2PO(OR)_2$.

R = Et. V.5. b_{14} 173-177°, n_D^{27} 1.4910.[932]

R = Bu. V.5. b_2 152°, n_D^{20} 1.4854, d_4^{20} 1.0113.[1499]

R = i-C_8H_{17}-. V.5. b_{1-2} 218-219°, n_D^{20} 1.4791, d_4^{20} 0.9278.[1499]

R = $PhCH_2CH_2$-. V.5. $b_{0.018}$ 185°, n_D^{20} 1.5594, d_4^{20} 1.1282.[1499]

$C_6H_5CHMePO(OH)_2$. I.6. As aniline salt, m. 114-116°.[315] I.2. m. 153-154.5°.[932]
$C_6H_5CHMePO(OEt)_2$. V.5. b_{15} 158-160°.[932]
3-$MeC_6H_4CH_2PO(OEt)_2$. V.5. $b_{0.7}$ 126-127°, n_D^{20} 1.4450, d_4^{20} 1.0775.[1237]
4-$MeC_6H_4CH_2PO(OH)_2$. I.2. m. 185-186°.[1128] I.1. m. 189°.[1685] I.2. m. 185°.[1001] I.2. m. 185-186°.[1127]
4-$MeC_6H_4CH_2PO(OR)_2$.

R = Et. V.5. b_2 160-163°, n_D^{20} 1.4958.[1127] V.5. $b_{0.2}$ 132-

135°.[1237]

R = Bu. V.6. b_{20} 213-215°.[1001]

2-$EtC_6H_4PO(OH)_2$. I.1. m. 145.5-147°.[570]
2-$EtC_6H_4PO(OEt)_2$. V.5. b_3 130-132°.[1422]
4-$EtC_6H_4PO(OH)_2$. I.1. m. 176-177.5°.[570] I.1. m. 164° (water).[1260,1269]
4-$EtC_6H_4PO(OEt)_2$. From 4-$EtC_6H_4NO_2$ and $P(OEt)_3$ in 5-7% yield.[340]
2,4-$Me_2C_6H_3PO(OH)_2$. I.1. m. 194°.[2069,2070]
2,5-$Me_2C_6H_3PO(OH)_2$. I.1. m. 179-180°.[2069,2070]
2,6-$Me_2C_6H_3PO(OMe)_2$. V.5. $b_{0.3}$ 92-93°.[1422]
3,5-$Me_2C_6H_3PO(OH)_2$. I.1. m. 161°.[2069,2070]
$C_6H_5NHC(O)NHCH_2PO(OR)_2$.

R = Me. V.14. m. 109-111° (MeOH).[1903]

R = Et. V.14. m. 78-79° (MeOH).[1903]

R = Pr. V.14. m. 68-69°.[1903]

$C_6H_5OCH_2CH_2PO(OH)_2$. I.2. m. 130-131°.[1306]
$C_6H_5OCH_2CH_2PO(OEt)_2$. V.5. b_{10} 185-187°.[1306] V.5. $b_{0.1}$ 135-140°, n_D^{25} 1.4947.[932]
$C_6H_5CH_2OCH_2PO(OR)_2$.

R = Et. V.5, V.6. b_4 169-170°, b_{12} 180-183°.[41]

R = Bu. V.6. b_3 168-170°.[41]

$CH_3OCHPhPO(OEt)_2$. V.5. $b_{0.4}$ 128°.[741]
$CH_3OCHPhPO(OEt)OC_6H_4NO_2$-4. V.5. m. 201-202°.[741]
4-$MeOC_6H_4CH_2PO(OR)_2$.

R = Et. V.5.[846] V.6. $b_{0.25}$ 123-125°, n_D^{20} 1.5040, d_4^{20} 1.1349.[1538] V.5.[839]

R = i-Pr. V.6. $b_{0.3}$ 132-135°, n_D^{20} 1.4920, d_4^{20} 1.0920.[1538]

4-$EtOC_6H_4PO(OH)_2$. I.1. m. 165°.[1260,1269] I.1.[291]
2-$MeC_6H_4OCH_2PO(OH)_2$. I.13. m. 157-158°.[2062]
3-$MeC_6H_4OCH_2PO(OH)_2$. I.13. m. 115-117°.[2062]
4-$MeC_6H_4CH(OH)PO(OR)_2$.

R = Me. V.8. m. 92-93°.[1222]

R = $H_2\overline{C(CH_2)_3C}(-CCl_3)$-. V.8. m. 162°.[33]

$C_6H_5CMe(-OH)PO(OH)_2$. I.9. I.6. m. 170°.[413,1183] V.14.[1220]
$C_6H_5CMe(-OH)PO(OR)_2$.

R = Et. V.8. m. 72-73°.[1619]

R = Pr. V.8. m. 59-59.5°.[1619]

R = i-Pr. V.8. m. 58-58.5°.[1619]

R = Ph. V.8. m. 143.5° (EtOH).[418]

$HOCH_2CHPhPO(OEt)_2$. V.14. $b_{0.1}$ 130°, n_D^{25} 1.5040.[1983]
$\overline{2\text{-}HO\text{-}3\text{-}MeC_6}H_3CH_2PO(OEt)_2$. V.5. ^1H.[1452]
$\overline{OCH=CH\text{-}CH=C}\text{-}CH_2CH[\text{-}C(O)Me]\text{-}PO(OMe)_2$. V.7. b_6 163-164°.[1545]
$2,4\text{-}(MeO)_2C_6H_3PO(OH)_2$. I.1.[1541]
$3,5\text{-}(MeO)_2C_6H_3PO(OMe)_2$. V.5. $b_{0.2}$ 120-122°, IR.[1422]
$\overline{OCH=CH\text{-}CH=C}\text{-}CH=CHCMe(OH)\text{-}PO(OR)_2$.

R = Me. V.8. b_8 108-110° [1319] (? cf. Ref.[1619]).

R = Et. V.8. b_8 112-114° [1319] (? cf. Ref.[1619]).

R = i-Pr. V.8. b_{11} 129-133° [1319] (? cf. Ref.[1619]).

R = Bu. V.8. b_{11} 170-171° [1319] (? cf. Ref. [1619]).

R = i-Bu. V.8. b_{11} 160-162° [1319] (? Cf. Ref.[1619]).

R = i-Am. V.8. b_{11} 189-190° [1319] (? cf. Ref.[1619]).

$HOCH_2CPh(\text{-}OH)PO(OH)_2$. I.13. m. 143-145° (Me_2CO).[413]
$H_2\overline{CCH_2O_2C\text{-}C}[\text{-}C(O)\overline{CHCH_2CH_2}]PO(OEt)_2$. V.6. $b_{0.2}$ 116-124°.[318]
$4\text{-}MeC_6H_4S(O)_2OCH_2PO(OR)_2$.

R = Et. V.14. $b_{0.02}$ 137°.[2054]

R = Pr. V.14. $b_{0.02}$ 147°.[2054]

R = i-Pr. V.14. m. 31-32°.[2054]

R = Bu. V.14. m. 35-36°.[2054]

$H_2\overline{CCH_2O_2C\text{-}C}[\text{-}C(O)CO_2Et]PO(OR)_2$.

R = Me. V.5. $b_{0.01}$ 122-125°.[317,318]

R = Et. V.5. $b_{0.05}$ 160°.[317,318]

$C_6H_5CH_2SCH_2PO(OEt)_2$. V.5. $b_{0.2}$ 70-72°, ^1H.[675]
$4\text{-}EtSC_6H_4PO(OH)_2$. I.6. m. 159-160°.[291]
$4\text{-}Me_2NC_6H_4PO(OH)_2$. I.6. m. 133° (EtOH).[1290]
$4\text{-}Me_2NC_6H_4PO(OEt)_2$. V.5. $b_{0.01}$ 155-158°.[1985]
$CH_3NHCHPhPO(OEt)_2$. V.7. m. 38-39°, $b_{0.11}$ 106°.[60]
$4\text{-}(H_2NCH_2)C_6H_4CH_2PO(OEt)_2$. V.14. $b_{0.05}$ 130-139°.[1956]
$4\text{-}H_2NC_6H_4CHMePO(OEt)_2$. V.14. m. 116-117°.[932]
$C_6H_5CH_2CH(\text{-}NH_2)PO(OH)_2$. I.2. m. 281° (dec.).[357]
$4\text{-}MeC_6H_4CH(\text{-}NH_2)PO(OH)_2$. I.2. m. 276°.[905]
$H_2NCH_2CHPhPO(OH)_2$. I.13. m. 274-276°.[210]
$C_6H_5CMe(\text{-}NH_2)PO(OH)_2$. I.13. m. 214-215°.[1005]
$C_6H_5CMe(\text{-}NH_2)PO(OEt)_2$. V.9. As picrate, m. 174-175°.[1220]
$\overline{N=CEtCH_2CH=CH\text{-}CH=C}PO(OEt)_2$. V.5. $b_{0.02}$ 103°, n_D^{20} 1.5066.[340]

$\text{-}(CH_2)_3PO(OH)_2$. I.13. m. 123-124°.[1205]

$CH_3C(O)CH=CHC(\text{-}CN)Et\text{-}PO(OEt)_2$. V.14. b_1 104-105°.[1651]
$4\text{-}MeOC_6H_4CH(\text{-}NH_2)PO(OH)_2$. I.2. m. 279° (dec.).[1036]
$[4\text{-}MeOC_6H_4CH(\text{-}\overset{\oplus}{N}H_3)\text{-}PO(OEt)_2]Cl^{\ominus}$. V.9. m. 160°.[356,905]
$C_6H_5NHCH_2CH(\text{-}OH)PO(OEt)_2$. V.14. m. 95-97°, IR, ^1H.[691]
$(EtO_2C\text{-})(NC\text{-})CH\text{-}CMe=CHPO(OR)_2$.

R = Et. V.14. $b_{0.8}$ 143°.[1609]

R = Pr. V.14. b_1 158-159°.[1609]

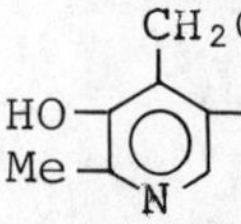

$CH_2PO(OPr\text{-}i)_2$. V.14. As hydrochloride m. 187.5-188° (dec.)[232]

$H_2C(CH_2)_4C{=}C{=}CHPO[-O(HC{\equiv}C-)C(CH_2)_4CH_2]_2$. V.2. m. 78-79°, IR.[287] From PCl_3 and $H_2C(CH_2)_4C(-C{\equiv}CH)OH$ in the presenc of NR_3, m. 78-79°.[382]

$[H_2C(CH_2)_4C{=}C{=}CH-]OPOCHMeCH_2O$. V.2. m. 89-91°, IR.[287]

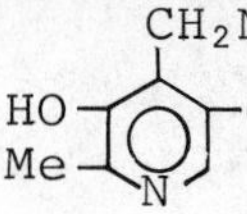

$CH_2PO(OPr\text{-}i)_2$. V.14. As hydrochloride m. 186° (dec.)[232]

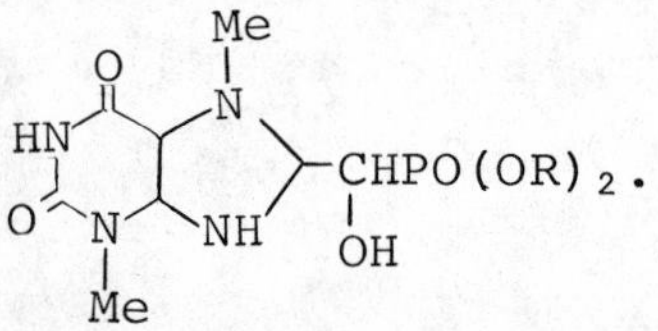

R = Me. V.8. m. 223-224°.[1119]

R = Et. V.8. m. 211-212°.[1119]

R = i-Pr. V.8. m. 216°.[1119]

R = i-Bu. V.8. m. 214-215°.[1119]

$OHCCHCH_2CMe{=}CHCH_2CH\text{-}PO(OR)_2$.

R = Me. V.14. $b_{2.5}$ 123-125°.[2037]

R = Et. V.14. b_2 129-130°. As 2,4-dinitrophenylhydrazone, m. 164-164.5°.[2037]

R = Pr. V.14. b_2 152-153°.[2037]

R = Bu. V.14. $b_{1.5}$ 156-157°. As 2,4-dinitrophenylhydrazone, m. 130-131°.[2037]

R = Am. V.14. $b_{0.029}$ 156-157°. As 2,4-dinitrophenylhydrazone, m. 124-125°.[2037]

$CH_3C(O)CH{=}CH\text{-}CMe[-C(O)Me]PO(OMe)_2$. V.14. b_1 124-125°.[1651]

$CH_3C(O)CH{=}CH\text{-}CH(-CO_2Et)PO(OEt)_2$. V.14. b_1 130-131°.[1651]

$O{=}C(CH_2)_3C(-CO_2Et)\text{-}PO(OEt)_2$. V.14. $b_{0.5}$ 145°.[773]

$H_2CCH_2CHMeO_2C\text{-}C[-C(O)Me]\text{-}PO(OEt)_2$. V.6. $b_{0.05}$ 122-126°.[318]

$(H_2C{=}CHCH_2-)_2NC(O)CH_2PO(OEt)_2$. V.5. $b_{0.15}$ 137-138°, n_D^{20} 1.4732, d_4^{20} 1.0881.[1108]

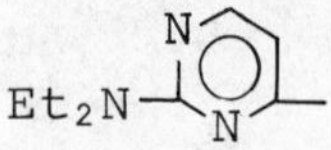

$PO(OPr\text{-}i)_2$. V.14. $b_{0.05}$ 113-114°.[1022]

$OCH_2(EtO_2C-)[MeC(O)-]CCH_2PO(OEt)$. From $EtO_2CCH_2C(O)CH_3$, $P(OEt)_3$ and paraformaldehyde.[844] V.14.[841]
$OCH_2CHMe-C(-CO_2Me)[-C(O)Me]PO(OEt)$. V.14. $b_{0.01}$ 95°.[318]
$OCHMe(CH_2)_2C(-CO_2H)[-C(O)Me]PO(OEt)$. V.14.[318]
$OCHMe(CH_2)_2C(-CO_2Me)(-CHO)PO(OEt)$. V.14. $b_{0.02}$ 96-98°.[318]
$C_6H_{13}C{\equiv}CPO(OEt)_2$. V.11. $b_{0.57}$ 133°.[363] V.6. $b_{0.4}$ 128°.[1962,1963]
$H_2CCMe{=}CMeCH_2CH_2CHPO(OEt)_2$. V.14. $b_{0.3}$ 103-107°.[451]

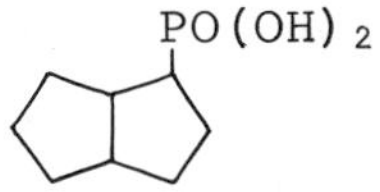

I.2. m. 101.5-102.5°.[587]

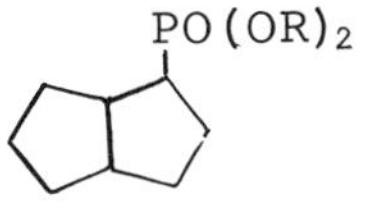

R = Et. V.7. $b_{0.5}$ 92-94°.[587]

R = Bu. V.7. $b_{0.3}$ 126°.[587]

$(CH_3)_3C-CMe{=}CH-C(O)PO(OMe)_2$. V.5. $b_{0.13-0.18}$ 99-102°. IR, ^1H.[1191]
$H_2C(CH_2)_2CH{=}C(-OEt)CHPO(OEt)_2$. V.6. b_1 125°, IR, n_D^{18} 1.4728, d_4^{18} 1.087[1410,1959]
$H_2C(CH_2)_3C(-OEt){=}CPO(OEt)_2$. V.2.[161]
cycl.-$C_6H_{11}CHOCHPO(OEt)_2$. V.14. b_1 117-118°.[1202]
$(CH_3)_2C{=}C{=}C(CMe_2OH)PO(OR)_2$.

R = Et. V.7. b_2 109°.[1560]

R = Pr. V.7. b_3 141°.[1560]

R = Bu. V.7. b_1 127-129°.[1560]

$Me_2COCMe_2OC{=}CHPO(OPr\text{-}i)_2$. V.14. $b_{0.2}$ 115-117°, n_D^{20} 1.4469.[1788]

$C_4H_9O_2CCH{=}CMePO(OBu)_2$. V.7. b_2 169-170.5°.[967,968]
$C_2H_5O_2CCH{=}CPrPO(OEt)_2$. cis:trans = 7:3. V.7. b_1 117-112°, ^1H, IR.[392]
$C_2H_5O_2CCH{=}C(-CHMe_2)PO(OEt)_2$. cis:trans = 7:3. V.7. $b_{1.5}$ 119-122°, ^1H, IR.[392]
$(CH_3)_2C{=}C(-CO_2Et)-CH_2PO(OEt)OH$. III.8. ^1H.[1196]
$C_2H_5O_2CCH[-C(O)Me]CHMePO(OR)_2$.

R = Me. V.7. n_D^{20} 1.4510, d_4^{20} 1.1914.[150]

R = Et. V.7. n_D^{20} 1.4497, d_4^{20} 1.2238, b_1 122-123°.[150]

$EtO_2CCH_2CH(-CO_2Et)PO(OH)_2$. I.8. b. 256-268°.[1644]
$EtO_2CCH_2CH(-CO_2Et)PO(OR)_2$.

R = Et. V.7. $b_{0.05}$ 140°, ^1H.[751]

R = Bu. V.7. n_D^{20} 1.4428, d_4^{20} 1.0840, extraction of $Ce(NO_3)_3$, IR.[1431]

R = i-Am. V.7. b_6 198°.[1644]

$(EtO_2C)_2CHCH_2PO(OEt)OH$. III.3.[1041]
$(EtO_2C)_2CHCH_2PO(OEt)_2$. From $(EtO_2C)_2CHCH_2OH$ and $P(OEt)_3$.[842] V.14. $b_{0.2}$ 141°, n_D^{20} 1.4420.[1041,1043] V.7. b_{10} 179-180°.[1643] $b_{0.05}$ 131-133°.[1043]
$H_2CCMe_2O_2CC[-C(O)CO_2Et]PO(OMe)_2$. V.5. $b_{0.05}$ 150°.[318]
$HC{\equiv}C-CMe_2NHCMe_2PO(OR)_2$.

R = Me. V.14. b_2 106-108°.[1610]

R = Et. V.14. $b_{0.3}$ 84°.[1610]

$H_2C(CH_2)_4NCH_2CH{=}CHPO(OEt)_2$. V.14. $b_{0.4}$ 131-133°.[1082]
$H_2C(CH_2)_4NCMe{=}CHPO(OEt)_2$. V.14. b_2 156°.[1602]
$H_2C(CH_2)_4NCH{=}CMePO(OBu)_2$. V.14. b_1 120-120.5°, n_D^{20} 1.4740, d_4^{20} 1.0010.[931]

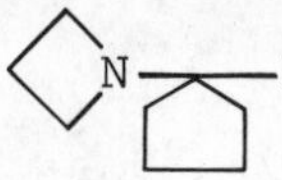—$PO(OPr)_2$. V.9. $b_{1.5}$ 125-126°.[962]

$H_2C(CH_2)_4NCH(-CO_2Me)PO(OR)_2$.

R = Me. V.5. $b_{0.05}$ 96-100°, IR, ^{31}P -20.7 ppm, n_D^{20} 1.4732.[717]

R = Et. V.5. $b_{0.05}$ 110-112°, IR.[717]

$H_2CCH_2OCH_2CH_2NCO(CH_2)_3PO(OR)_2$.

R = Pr. V.5. $b_{0.01}$ 154-156°.[515]

R = i-Pr. V.5. $b_{0.01}$ 161-162°.[515]

$Et_2NC(O)CH_2C(O)CH_2PO(OEt)_2$. V.6. $b_{0.2}$ 132°.[242]
$H_2CCH_2OCH_2CH_2NC(O)CH(-OEt)PO(OEt)_2$. V.5. $b_{0.01}$ 120-126°, ^1H.[684]
$OCH_2CMe_2CH_2O_2CCH_2CH_2PO(OPh)$. V.7. m. 97°, IR.[403a]
$(CH_2{=}CH-CH_2O-)_2OP-O-CHMePO(OCH_2CH{=}CH_2)_2$. V.14. $b_{0.15}$ 137°; the structure cited in the publication as diphosphonic ester is incorrect.[333,538]
$C_6H_{13}CH{=}CHPO(OEt)_2$. V.7. b_9 151-152°, n_D^{20} 1.4688, d_4^{20} 1.0122.[1627]
$C_5H_{11}CMe{=}CHPO(OBu)_2$. b_1 164-165°, n_D^{20} 1.4525, d_4^{20} 1.0001.[992]

$Me_3CCH_2CMe{=}CHPO(OH)_2$. I.1. m. 104-105°.[1019]
$Me_3CCH_2CMe{=}CHPO(OR)_2$.

R = $H_2C{=}CH-CH_2-$. V.2. b_{3-4} 135-140°.[2021a]

R = $H_2C{=}CMeCH_2-$. V.2. b_1 110-122°.[2021a]

$C_5H_{11}CHCH_2CHPO(OMe)_2$. V.14. $b_{0.03}$ 63-65°, ^1H, IR.[1878a]
$C_6H_{13}C(O)CH_2PO(OEt)_2$. V.14. $b_{0.4}$ 123°.[1963]
cycl.-$C_6H_{11}OCH_2CH_2PO(OEt)_2$. V.14. b_4 130-131°, n_D^{20} 1.4545, d_4^{20} 1.0511.[1640]
$C_4H_9OCH{=}CEtPO(OBu)Cl$. IV.2. $b_{0.5}$ 97°.[1501]

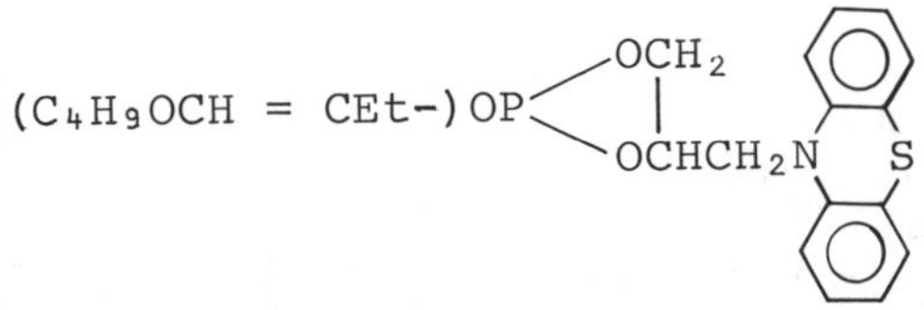

V.14. m. 106-107°.[227]

$Me_2C=C(-OPr)CH_2PO(OEt)_2$. V.14. b_9 145°.[1603]
$Me_2CHCH_2CH_2CMeOCHPO(OMe)_2$. V.14. b_4 112-113°.[1202]
$CH_3C(O)CMe(-OBu)PO(OBu)_2$. V.8. b_{10} 148-150°, n_D^{20} 1.4240, d_4^{20} 1.0192.[1060]
$(EtO)_2CHCMe=CHPO(OEt)_2$. V.14. b_8 146°.[991]
$(EtO)(PrO)CHCH=CHPO(OPr)_2$. V.14. b_2 146-147°.[2037]
$OCH_2CH_2OCMe(CH_2)_4PO(OEt)_2$. V.6. $b_{0.05}$ 138°, IR.[1410,1959]
$C_2H_5O_2C(CH_2)_5PO(OEt)_2$. V.5. $b_{0.03}$ 114-118°. Mass spect.[1403]
$C_2H_5O_2CCHBuPO(OEt)_2$. V.5. b_4 141°.[44]
$Me_2CHO_2C(CH_2)_4PO(OPr\text{-}i)(OC_6H_{13})$. V.5. $b_{0.01}$ 133-139°.[516]
$C_4H_9O_2C(CH_2)_3PO(OR)_2$.

R = Et. V.5. $b_{0.03}$ 124°.[516]

R = Pr. V.5. $b_{0.04}$ 122-123°.[516]

R = i-Pr. V.5. $b_{0.03}$ 135°.[516]

$(CH_3)_3CO_2C(CH_2)_3PO(OPr\text{-}i)_2$. V.5. $b_{0.02}$ 114-123°.[516]
$C_4H_9O_2CCHMeCH_2PO(OR)_2$.

R = $C_2H_5OCH_2CH_2$-. V.7. b_{10} 205-207°.[1344]

R = $C_4H_9OCH_2CH_2$-. V.7. b_8 235-236°.[1344]

R = cycl.-C_6H_{11}-. V.7. b_1 193-196°.[1634]

R = $PhCH_2$-. V.7. $b_{0.001}$ 169-172°.[1634]

R = $PhOCH_2CH_2$-. V.7. $b_{0.005}$ 168°.[1344]

$C_4H_9O_2CCHMeCH_2PO(OBu)OC_7H_{15}$. V.7. $b_{0.008}$ 100-105°.[1162]
$C_4H_9O_2CCHEtPO(OEt)_2$. V.5. Mass spect.[1403]
$C_4H_9SCH_2CH=CHCH_2PO(OR)_2$.

R = Et. V.14. b_1 158-160°.[990]

R = Bu. V.14. $b_{0.5}$ 178-190°.[990]

$(C_3H_7)_2NCH=CHPO(OMe)_2$. V.14. $b_{0.03}$ 122°, n_D^{20} 1.4840, d_4^{20} 1.0542.[1700]
$H_2C(CH_2)_4C(-NMe_2)PO(OPr\text{-}i)_2$. V.8. m. 140-141°.[813]
$(C_2H_5)_2NCH_2\text{-}CH=CHCH_2PO(OEt)_2$. V.14. $b_{1.5}$ 122-122.5°.[1630]
$H_2C(CH_2)_3NCH(-CHMe_2)PO(OEt)_2$. V.7. $b_{0.01}$ 88-89°, IR.[1429]
$H_2C(CH_2)_4NCHEtPO(OMe)_2$. V.7. $b_{0.01}$ 88-89°.[1429]
$Et_2NC(O)(CH_2)_3PO(OR)_2$.

R = Et. V.5. $b_{0.01}$ 128-131°.[515]

R = Pr. V.5. $b_{0.001}$ 148-150°.[515]

R = i-Pr. V.5. $b_{0.08}$ 143-145°.[515]

$Et_2NC(O)CHMeCH_2PO(OR)_2$.

R = Et. V.7. $b_{0.09}$ 120-121°.[1658]

R = Pr. V.7. b_8 186-187.5°.[1658]

$[Et_2NC(O)CHMeCH_2-]\overline{OPOCH_2CH_2O}$. V.7. $b_{0.004}$ 170-171°.[1656]
$Pr_2NC(O)CH_2PO(OEt)_2$. V.5. $b_{0.23}$ 143-147°.[1108]
$[(CH_3)_2CH]_2NC(O)CH_2PO(OEt)_2$. V.5. $b_{0.25}$ 133-135°.[1108]
$MeCHCH_2NMe-CHMeCH_2C(-OH)PO(OEt)_2$. V.8. m. 82-83°.[1221]
$C_8H_{17}PO(OH)_2$. I.2. m. 99.5-100.5°.[1000,1002] I.8. m. 99-100°.[696] I.2. m. 101.8-102.1°. As monoaniline salt, m. 143-145°. As monocyclohexylamine salt, m. 187-188°.[963] I.1. m. 98° (heptane).[737] As mono-4-$ClC_6H_4CH_2\overset{\oplus}{S}=C(NH_2)_2$ salt, m. 192-193°.[753]
$C_8H_{17}PO(OEt)NCO$. IV.5. $b_{0.06}$ 90-92°.[726]
$C_8H_{17}PO(OR)_2$.

R = Me. V.7. b_{14} 156°.[1614]

R = Et. V.7. b_{16} 167°.[1614] V.5. $b_{1.2}$ 119°.[552] V.7. $b_{0.}$ 91.8°, n_D^{20} 1.4361.[963]

R = Pr. V.7. b_{15} 184°.[1614]

R = Bu. V.7. b_{15} 201°.[1614] V.6. b_2 147-148°, n_D^{25} 1.437 d_4^{25} 0.9262.[1000,1002] V.7.[746]

R = Ph. V.5. $b_{0.006}$ 148°, n_D^{25} 1.5221.[1080]

R = C_8H_{17}-. V.6. b_3 236.5-237°.[160]

$C_4H_9CHEtCH_2PO(OH)OCH_2CHEtC_4H_9$. III.3. IR.[1458]
C_6H_{13}*$CHMePO(OEt)_2$. V.6. $[\alpha]_{546}^{27}$ + 6.83 (23% $CHCl_3$).
$MeCHCH_2NMeCHMeCH_2C(-NH_2)PO(OEt)_2$. V.9. $b_{0.1}$ 81-83°.[1221]
$C_6H_{13}CH(-OMe)PO(OEt)OC_6H_4NO_2$-4. V.5. n_D^{25} 1.503.[741]
$(EtO)_2CMeCH_2CH_2PO(OEt)_2$. V.14. $b_{0.4}$ 111°, n_D^{20} 1.4340, d_4^{21} 1.028.[1082]
$(EtO)_2MeSiCH=CHCH_2PO(OEt)_2$. V.14.[1875]
$C_4H_9NHCH(-CHMe_2)PO(OMe)_2$. V.7. $b_{0.25}$ 96-98°, ^1H.[332]
$Et_2NCHPrPO(OPr)_2$. V.9. $b_{0.03}$ 58-59°.[513]
$(EtO)_2OP-O-CMeEtPO(OEt)_2$. V.14/V.8. b_3 156-157°, n_D^{20} 1.4345 d_4^{20} 1.1420.[1620]
$(i-PrO)_2OP-O-CHMePO(OPr-i)_2$. V.14. $b_{0.06}$ 115°.[333] The structure as diphosphonic acid ester cited in the publication is incorrect.[538]
$Et_3SiCH_2OCH_2PO(OH)_2$. I.2. As monoaniline salt, m. 145-146°.[1433]
$Et_3SiCH_2OCH_2PO(OEt)_2$. V.14. b_3 121°.[1433]
$Et_3SiOCHMePO(OMe)_2$. V.6. b_{11} 138-140°, ^1H.[1374]
$Pr_2MeSiOCH_2PO(OR)_2$.

R = Et_2MeSi-. V.14. b_3 155-156°.[1434]

R = Pr_2MeSi-. V.14. $b_{2.5}$ 170°.[1437]

$(BuO)Me_2SiCH_2CH_2PO(OBu)_2$. V.7. b_1 157°.[208]
$(EtO)_2MeSi(CH_2)_3PO(OEt)_2$. V.5, V.2. b_2 125-126°[386]

$(EtO)_3SiCH_2CH_2PO(OR)_2$.

R = Me. V.7.[1192]

R = Et. V.5. b_2 135-140°, n_D^{20} 1.4253, d_4^{20} 1.0326.[320] V.7. b_2 141°.[208]

R = Bu. V.7. b_{1-2} 149-150°.[1101]

$(EtO)_2MeSiOCHEtPO(OEt)_2$. V.8. $b_{0.33-0.4}$ 102-104°.[258]
$Et_3SiCH_2CH_2PO(OEt)_2$. V.5, V.7. b_2 116-118°.[387]

$\underline{C_9}$
4-$MeC_6H_4C(O)CHBrPO(OEt)_2$. V.14. b_1 180-185°.[697]
$C_6H_5CH_2CCl=CHPO(OH)_2$. I.1. m. 179°.[240]
$H_2NC(O)CH-NH-C(-C_6H_4Cl-4)PO(OR)_2$.

R = Me. V.7. m. 137-138° (THF/hexane).[1404]

R = Et. V.7. m. 149-150° ($MeCO_2Et$/hexane).[1404]

2-$MeOC_6H_4CCl=CHPO(OH)_2$. I.1. m. 125-127°. As 1-hydrate, m. 64-67°.[240]
4-$MeOC_6H_4CCl=CHPO(OH)_2$. I.1. m. 105°.[240]

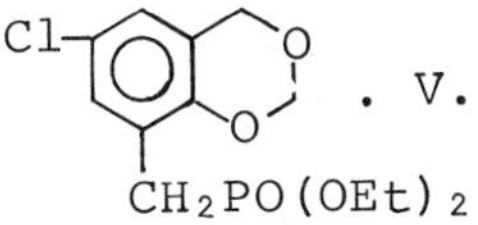
. V.5. b_3 195-198°.[1127]

4-$ClC_6H_4CH_2C(O)NHCH_2PO(OH)_2$. I.13. m. 172-173°.[1198]
2,4,5-Me_3-6-$ClC_6HPO(OH)_2$. I.13. m. 235° ($MeCO_2H$). As phenylhydrazine salt, m. 197.5° (EtOH).[1260,1269]
2,4,5-Me_3-3-O_2N-6-$ClC_6PO(OH)_2$. I.13. m. 227-228°.[1260, 1269]

2-EtO-5-$ClC_6H_3CH_2PO(OR)_2$.

R = Et. V.6. $b_{0.35}$ 135-140°, n_D^{20} 1.5084, d_4^{20} 1.1813.[1538] V.9.[848]

R = i-Pr. V.6. $b_{0.17}$ 135-137°, n_D^{20} 1.4940, d_4^{20} 1.1281.[1538]

2-EtS-5-$ClC_6H_3PO(OR)_2$.

R = Et. V.6. $b_{0.12}$ 134-135°, n_D^{20} 1.5440, d_4^{20} 1.2300.[1538]

R = i-Pr. V.6. $b_{0.14}$ 142-145°, n_D^{20} 1.5050, d_4^{20} 1.1373.[1538]

4-$ClC_6H_4SiMe_2CH_2PO(OBu)_2$. V.6. b_2 186.5-188°.[344]
$C_6H_5NHC(O)NHC(O)CCl_2PO(OR)_2$.

R = Et. V.14. m. 86-87°.[281]

R = Pr. V.14. m. 73-75°.[281]

4-$ClC_6H_4NHC(O)NHC(O)CCl_2PO(OR)_2$.

R = Et. V.14. m. 102-104°.[281]

R = Pr. V.14. m. 92-93°.[281]

R = Bu. V.14 m. 68-69°.[281]

4-$O_2NC_6H_4OS(O)_2NHCO_2CH(-CCl_3)PO(OMe)_2$. V.14. m. 173°.[1999]
$C_6H_5NHCO_2CH(-CCl_3)PO(OMe)_2$. V.14. m. 157-159°.[987]
$C_6H_5OS(O)_2NHCO_2CH(-CCl_3)PO(OMe)_2$. V.14. m. 155°.[1999]
$C_6H_5NHS(O)_2NHCO_2CH(-CCl_3)PO(OR)_2$.

R = Me. V.14. m. 181°.[1999]

R = Et. V.14. m. 185°.[1999]

4-$ClC_6H_4OS(O)_2NHCO_2CH(-CCl_3)PO(OMe)_2$. V.14. m. 176°.[1999]
4-$ClC_6H_4NHS(O)_2NHCO_2CH(-CCl_3)PO(OMe)_2$. V.14. m. 168°.[1999]
2,4-$Cl_2C_6H_3NHS(O)_2NHC(O)OCH(-CCl_3)PO(OMe)_2$. V.14. m. 163°.[1999]

3,4-$Cl_2C_6H_3NHS(O)_2NHC(O)OCH(-CCl_3)PO(OMe)_2$. V.14. m. 174°.[1999]

$C_2F_5CPh(OH)PO(OMe)_2$. V.8. Insecticide and acaricide.[631]
$H_2C(CH_2)_4NCH[-CH(CF_3)_2]PO(OEt)_2$. V.14. $b_{0.1}$ 60-62°.[979]
$C_7F_{15}CH_2CH_2PO(OH)_2$. I.2. I.1. m. 155-158°[1335]
$C_7F_{15}CH_2CHJPO(OEt)_2$. V.14. $b_{0.03}$ 108-114°.[1390]
$Et_3GeCH(-CN)CH_2PO(OEt)_2$. V.7. $b_{0.01}$ 122°, n_D^{20} 1.4708, d_4^{20} 1.1562.[1413]
$(EtO)_3GeOCHEtPO(OEt)_2$. V.8. $b_{0.1}$ 98-101°.[1741]
$C_6H_5CH=C(-CN)PO(OEt)_2$. V.14. $b_{0.5}$ 165°.[2058]
$C_6H_5C(O)CH(-CN)PO(OEt)_2$. V.14. $b_{0.1-0.2}$ 130-135°, IR.[1197]

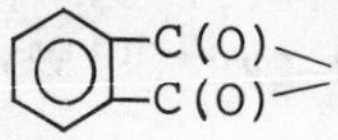
$NCH_2PO(OH)_2$. I.13. m. 280-285°[912,838]

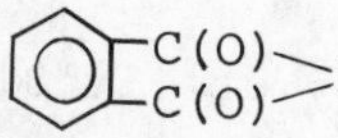
$NCH_2PO(OR)_2$.

R = Me. V.9.[837]

R = Et. V.9.[837,838]

R = Pr. V.9.[837]

R = Bu. V.9.[837]

$H_2C=C=CPhPO(OEt)_2$. V.5. $b_{0.6}$ 157-159°.[1559] V.14.[1556]

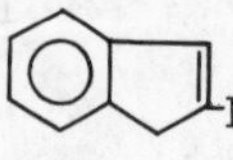
$PO(OH)_2$. I.1. m. 184°.[238,1991a]

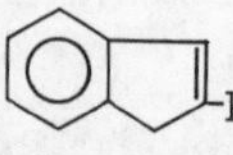
$PO(OEt)_2$. V.2. b_1 150°.[92]

$HC=N-NPh-CH=CPO(OBu)_2$. V.2. b_{20} 263-265°.[659]
$HN-NPh-C(O)CH=CPO(OEt)_2$. V.2. m. 89-91°.[1145]

$C_6H_5CO_2CH{=}CH{-}PO(OR)_2$.

R = Me. V.14. $b_{1.5}$ 161-162°.[1135]

R = Et. V.14. $b_{1.5}$ 181-182°.[1135]

$C_6H_5O_2CCH{=}CHPO(OPh)_2$. V.7. m. 142-143°.[967]

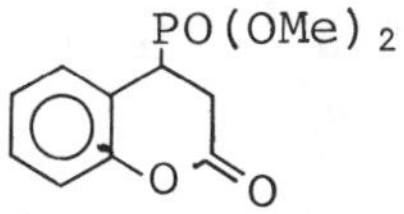

V.7. b_7 190-195°.[178]

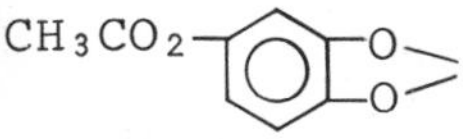

$CHPO(OEt)_2$. V.5. $b_{0.04}$ 142-145°, IR.[717]

2,5-HO_2C-4-$MeC_6H_2PO(OH)_2$. I.13. m. 185-190°. The positions of the carboxyl groups are not certain.[1260,1269]

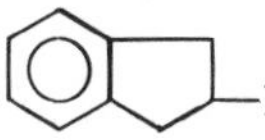

$-PO(OH)_2$. I.8. m. 195-196°.[696] I.13. m. 196°.[238]

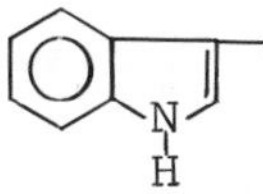

$CH_2PO(OH)_2$. I.2. m. 218.4°, plant growth antagonist[53]

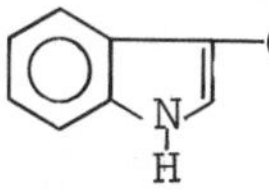

$CH_2PO(OEt)OH$. III.3. m. 136-137°.[2017,53]

$C_6H_5C(O)NHCH_2C(O)PO(OEt)_2$. V.5. b_1 107-109°.[174]

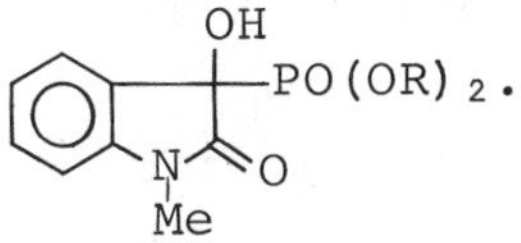

R = Me. V.8. m. 184-186°.[1362]

R = Et. V.8. m. 125-127°.[1362]

R = i-Pr. V.8. m. 132-133°.[1362]

4-$HO_2CC_6H_4NHC(O)CH_2PO(OR)_2$.

R = Am. V.14.[819]

R = C_6H_{13}-. V.14.[819]

4-$O_2NC_6H_4N{=}N{-}CH[-C(O)Me]PO(OEt)_2$. V.13. m. 117-118°.[1343]

$C_6H_5CH{=}CHCH_2PO(OEt)_2$. V.6. $b_{0.3}$ 130°.[267] V.5. b_4 169-171°, n_D^{20} 1.5268, d_4^{20} 1.0915, IR.[825]

$C_6H_5CMe{=}CHPO(OH)_2$. I.1. m. 95°.[238]

$C_6H_5CMe{=}CHPO(OEt)_2$. V.5. b_1 149-150°[142]

$C_6H_5CH{=}CMePO(OEt)_2$. V.7. $b_{0.5}$ 128°, b_1 144-145°.[692]

$H_2C{=}CPhCH_2PO(OEt)_2$. V.5. b_3 147-149°.[1451]

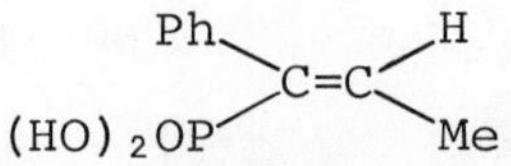

Ph$\diagdown$ $\diagup$H
$\quad$C=C
(HO)$_2$OP$\diagup$ $\diagdown$Me   .I.9. As 5-methylthiuronium hydrogen
$\qquad\qquad\qquad\qquad$ phosphonate, m. 184-186.5°, Ir.[959]

Ph$\diagdown$ $\diagup$Me
$\quad$C=C
(HO)$_2$OP$\diagup$ $\diagdown$H   .I.9. As 5-methylthiuronium hydrogen
$\qquad\qquad\qquad\qquad$ phosphonate, m. 177-179°(dec.),
$\qquad\qquad\qquad\qquad$ IR.[959]

(H$_2$C=CH)C$_6$H$_4$CH$_2$PO(OEt)$_2$. V.5. misture of o- and p- isomers
$\quad$b$_{0.5}$ 112-126°.[1213]

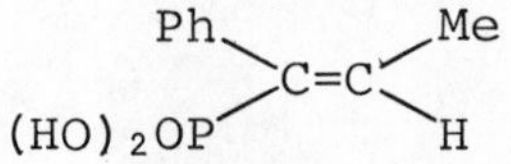

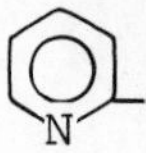

—CH(-CN)CH$_2$CH$_2$PO(OEt)$_2$. V.7.[1205]

H$_2$NC(O)$\overline{\text{CHNHCPhPO}}(OR)_2$.

$\quad$R = Me. Vg, m. 155-156°(benzene/hexane), IR.[1404]

$\quad$R = Et. Vg, m. 183-184°(benzene/hexane), IR.[1404]

C$_6$H$_5$NHC(O)NHC(O)CH$_2$PO(OEt)$_2$. V.14. m. 91-93°.[281]
C$_6$H$_5$C(O)CH$_2$CH$_2$PO(OEt)$_2$. V.5. b$_{0.008}$ 163-167°.[1364]
C$_6$H$_5$C(O)CHMePO(OEt)$_2$. V.5.[167] V.5. b$_{1.5}$ 156.5-158°.[159]
(HC≡CCH$_2$-)$_2$[MeC(O)-]CPO(OR)$_2$.

$\quad$R = Et. V.7 , V.14. b$_8$ 151-153°.[1605]

$\quad$R = Pr. V.7 , V.14. b$_4$ 161-163°.[1605]

$\quad$R = Bu. V.7 , V.14. b$_1$ 148-150°.[1605]

OHCCH$_2$CHPhPO(OEt)$_2$. V.7. b$_{0.05}$ 148°, ^{1}H.[751]
C$_6$H$_5$OCH=CHCH$_2$PO(OEt)$_2$. V.14. b$_{0.05}$ 134-135°, n$_D^{22}$ 1.5070,
$\quad$d$_4^{22}$ 1.129.[1082]
CH$_3$OCH=CPhPO(OR)$_2$.

$\quad$R = Me. V.8. b$_3$ 163-165°, m. 67-68°(ether).[1882]

$\quad$R = Et. V.8. b$_{0.5}$ 137-138°.[1882]

$\quad$R = Pr. V.8. b$_2$ 167-168°.[1882]

$\quad$R = Bu. V.8. b$_1$ 170-172°.[1882]

4-MeC$_6$H$_4$$\overline{\text{CH-O-CHPO}}(OMe)_2$. V.14. b$_2$ 151-152°.[1202]
C$_6$H$_5$$\overline{\text{CMe-O-CHPO}}(OMe)_2$. V.14. b$_2$ 140-141°, m. 84-85°.[1202]

$\qquad\qquad$PO(OEt)$_2$. V.5. b$_{0.1}$ 125°, IR.[717]
C$_6$H$_5$CH=CH-CH(OH)PO(OR)$_2$.

$\quad$R = Me. V.8. m. 101-102°.[1542]

$\quad$R = Et. V.8. m. 106.5-107°.[963]

$\quad$R = ClH$_2$CCH$_2$-. V.8. m. 68-70°.[886]

[C$_6$H$_5$CH=$\overline{\text{CHCH(OH)}}$-]OPO(CH$_2$)$_3$O. V.8. m. 151°.[1588]
4-MeOC$_6$H$_4$CH-O-CHPO(OMe)$_2$. V.14. m. 129-131°.[1202]

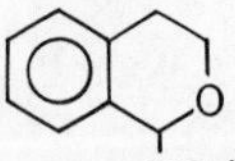

2-MeO-5-(OHC-)$C_6H_3CH_2PO(OEt)_2$. V.5. b_2 184-186°. As phenylhydrazone, m. 117°.[1127]

$CH_3CO_2CHPhPO(OR)_2$.

R = Me. V.14. $b_{0.03}$ 74°, IR, ^1H.[1878a] V.8. $b_{0.015}$ 116°, n_D^{20} 1.4866, d_4^{20} 1.1571.[1377]

R = Et. V.14. $b_{1.5}$ 146-148°, n_D^{20} 1.4920, d_4^{20} 1.1572.[1589]

4-MeO_2C-$C_6H_4CH_2PO(OEt)_2$. V.5. $b_{0.15}$ 137-138.5.[1237]

2-$EtO_2CC_6H_4PO(OEt)_2$. V.5. $b_{0.05}$ 132-134°.[1987]

3-$EtO_2CC_6H_4PO(OEt)_2$. V.5. $b_{0.25}$ 149-151°.[1987]

4-$EtO_2CC_6H_4PO(OH)_2$. I.13. m. 78°.[1260,1269]

4-$EtO_2CC_6H_4PO(OEt)_2$. V.13. $b_{0.05}$ 147-148°.[1987] V.5. $b_{0.1}$ 148-153°.[1985] V.5. $b_{0.4}$ 170°.[1988]

$C_6H_5O_2CCHMePO(OEt)_2$. V.5. $b_{7.5}$ 165-168.5°.[119a]

2,4-Me_2-5-$HO_2CC_6H_2PO(OH)_2$. I.13. m. 258°.[1260,1269]

$C_6H_5CH_2O_2CCH_2PO(OPr\text{-}i)_2$. V.5. $b_{0.7}$ 143-145°.[1143]

4-$MeO_2CC_6H_4CH_2PO(OEt)_2$. V.5. $b_{0.15}$ 137-138.5°.[1237]

$CH_3O_2CCHPhPO(OEt)_2$. V.14. b_1 125-130°.[267]

$HO_2CCH_2CHPhPO(OH)_2$. I.2. m. 202°.[1037]

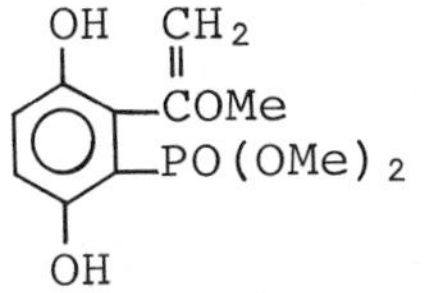
. V.8.[147]

4-$MeC_6H_4S(O)_2OCH$=$CHPO(OMe)_2$. V.5. m. 108-110°.[1225]

($HC\equiv CCH_2$-)($NCCH_2CH_2$-)[MeC(O)]$CPO(OEt)_2$. V.7, V.14. b_2 163°.[1605]

$C_6H_5CH_2C(O)NHCH_2PO(OH)_2$. I.13. m. 166-167°.[1198]

4-$MeC_6H_4C(O)NHCH_2PO(OH)_2$. I.13. m. 202.5-203°.[1198]

$C_6H_5NHC(O)CH_2CH_2PO(OR)OH$.

R = Et. III.3. m. 78-80°.[2032]

R = Pr. III.3. m. 90-91°.[2032]

$C_6H_5NHC(O)CH_2CH_2PO(OEt)_2$. V.7.[1664] V.14. $b_{0.08}$ 210-213°, n_D^{20} 1.5308.[1569]

$C_2H_5OC(=NH)$-C_6H_4-$PO(OH)_2$. I.13. m. 220° (as hydrochloride).[484]

$H_2NC(O)CH_2CHPhPO(OEt)_2$. V.7. m. 135-136°.[210]

4-$O_2NC_6H_4(CH_2)_3PO(OEt)_2$. V.14. $b_{0.1}$ 163-168°, n_D^{25} 1.5123.[932]

3-$EtO_2CNHC_6H_4PO(OH)_2$. I.13. m. 140°.[1395]

4-$MeOC_6H_4C(O)NHCH_2PO(OH)_2$. I.13. m. 200-202°.[1198]

$C_6H_5(CH_2)_3PO(OH)_2$. I.2. m. 123-125°.[932]

$C_6H_5(CH_2)_3PO(OC_6H_{11}\text{-cycl.})Cl$. IV.2. n_D^{25} 1.5244.[740]

$C_6H_5(CH_2)_3PO(OC_6H_{11}\text{-cycl.})F$. IV.2. $b_{0.005}$ 132°, n_D^{25} 1.4988.[740]

$C_6H_5(CH_2)_3PO(OEt)_2$. V.5. $b_{0.1}$ 125-128°.[932]

2,4,5-$Me_3C_6H_2PO(OH)_2$. I.1. m. 212°, bromine water splits off bromopseudocumene.[1260,1269]

2,4,5-$Me_3C_6H_2PO(OPh)_2$. V.2. m. 62.5°.[1260,1269]

2,4,6-$Me_3C_6H_2PO(OH)_2$. I.1. m. 167°.[1260,1269]
2,4,6-$Me_3C_6H_2PO(OEt)_2$. V.5. $b_{0.05}$ 105-107°.[1985] V.5. $b_{0.05}$ 111-112°.[1987]
4-$EtC_6H_4CH_2PO(OH)_2$. I.2. m. 178-178.5°(water).[1001]
4-$EtC_6H_4CH_2PO(OR)_2$.

R = Et. V.5. b_{14} 176-178°.[1001]

R = Bu. V.6. b_3 147-150°.[1001]

2,4-$Me_2C_6H_3CH_2PO(OH)_2$. I.2. m. 184-186°.[1127]
2,4-$Me_2C_6H_3CH_2PO(OEt)_2$. V.5. b_2 130-132°.[1127]
4-PrNH-3-$O_2NC_6H_3PO(OH)_2$. I.13. Dec. at 178-179°.[186]
$[H_3\overset{\oplus}{N}CH_2C(O)NHCHPhPO(OEt)_2]Br^{\ominus}$. V.14. dec. 220°.[1534]
4-$MeC_6H_4S(O)_2NHN{=}CMePO(OMe)_2$. V.14. m. 183°(dec.).[1191]

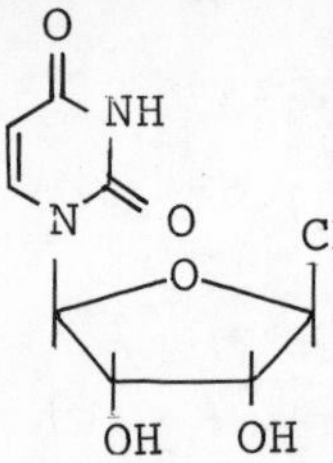
$CH_2PO(OH)_2$. I.2. UV λ_{max} 260nm[796]

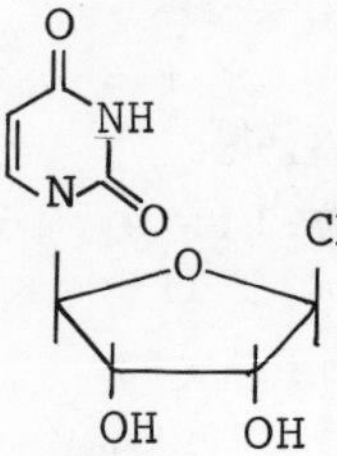
$CH_2PO(OEt)_2$. V.5. m. 224-225[207]

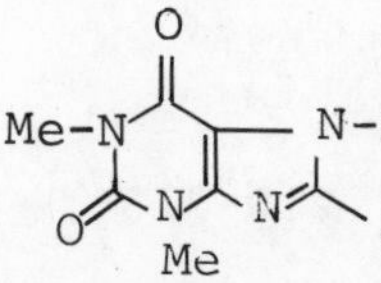
$CH(OH)PO(OR)_2$.

R = Me. V.8. m. 132-132.5°.[1118]

R = Et. V.8. m. 80-80.5°.[1118]

R = Pr. V.8. m. 77-77.5°.[1118]

R = i-Pr. V.8. m. 144-145°.[1118]

R = Bu. V.8. m. 54-55°.[1118]

R = i-Bu. V.8. m. 106-107°.[1118]

$CH_3OCH_2CHPhPO(OEt)_2$. V.14. b_1 122-124°.[1651]
$C_6H_5O(CH_2)_3PO(OEt)_2$. V.5. $b_{0.1}$ 140-145°.[932]
$C_6H_5CH_2OCH_2CH_2PO(OEt)_2$. V.6. $b_{1.2}$ 161-162°, IR.[1770]
2-$EtOC_6H_4CH_2PO(OEt)_2$. V.9. $b_{0.018}$110-113°.[848]
4-$EtOC_6H_4CH_2PO(OR)_2$.

R = Et. V.5. $b_{0.15}$ 124-127°, n_D^{20} 1.4920, d_4^{20} 1.0838.[1237] V.5, V.6. $b_{0.4}$ 138-139°, n_D^{20} 1.5005, d_4^{20} 1.1182.[1538]

R = i-Pr. V.5, V.6. $b_{0.2}$ 123-125°, n_D^{20} 1.4900, d_4^{20} 1.0526.[1538]

2-MeO-5-$MeC_6H_3CH_2PO(OEt)_2$. V.5. b_1 140°.[846]
2-Me-5-$MeOC_6H_3CH_2PO(OEt)_2$. V.5. $b_{0.3}$ 138°.[799]
$C_6H_5CH_2CH(-OMe)PO(OEt)OC_6H_4NO_2$-4. V.2. n_D^{25} 1.553.[741]
$C_2H_5OCHPhPO(OEt)_2$. V.5. $b_{0.01}$ 110-112°, IR, ^1H.[684] V.8.[636]

$C_6H_5CH_2CH_2CH(OH)PO(OH)_2$. I.2. m. 173-173.5°.[963]
$C_6H_5CEt(OH)PO(OR)_2$.

R = Me. V.8. m. 139.5-140°.[1619]

R = Et. V.8. m. 65-67°.[1619]

$HOCH_2CH_2CHPhPO(OEt)_2$. V.14. $b_{0.5}$ 135°.[267]
$Me_2P(O)OCPhPO(OMe)_2$. V.14. $b_{0.2}$ 179-180°.[455]
$CH_3OCH_2CPh(OH)PO(OMe)_2$. V.8. m. 102-103° (hexane/benzene).[1882]

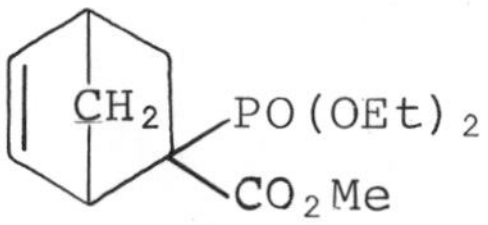
. V.14. b_2 139-141°, n_D^{20} 1.4700, d_4^{20} 1.1642.[1641]

$H_2CCHMeO_2CC[-C(O)CO_2Et]PO(OR)_2$.

R = Me. V.5. $b_{0.02}$ 145°.[317,318]

R = Et. V.5. $b_{0.02}$ 138-140°.[317,318]

$MeCHCH_2O_2CC[-C(O)CO_2Et]PO(OR)_2$.

R = Me. V.5. $b_{0.05}$ 145-150°.[317,318]

R = Et. V.5. $b_{0.03}$ 135°.[317,318]

$H_2CCH_2CH_2O_2C-C[-C(O)CO_2Et]PO(OR)_2$.

R = Me. V.5. $b_{0.02}$ 142°.[317,318]

R = Et. V.5. $b_{0.03}$ 143-147°.[317,318]

4-$EtSC_6H_4CH_2PO(OR)_2$.

R = Et. V.5. $b_{0.4-0.6}$ 174-180°, n_D^{20} 1.5270, d_4^{20} 1.1222.[1237] V.5. $b_{0.2}$ 132-135°, n_D^{20} 1.5320, d_4^{20} 1.1618.[1538]

R = i-Pr. V.5. $b_{0.4}$ 149-153°, n_D^{20} 1.5166, d_4^{20} 1.0844.[1538]

$C_6H_5NH(CH_2)_3PO(OH)_2$. I.13. m. 129-130° (water).[999,1000,1002] pK_1 2.1, pK_2 4.25, pK_3 7.15.[1776a] Monosodium salt, m. 187-189.5° (water).[999]

$C_6H_5CH_2CH_2NHCH_2PO(OEt)_2$. V.8. $b_{0.4}$ 109°.[529]
4-$Me_2NC_6H_4CH_2PO(OEt)_2$. V.8.[839]
$C_6H_5CH_2NHCHMePO(OH)_2$. I.2. m. 236-238°.[2040]
$C_6H_5NHCMe_2PO(OEt)_2$. V.8. m. 119-120°.[1632]
$C_2H_5NHCHPhPO(OEt)OH$. III.3. m. 233°.[868]
$C_2H_5NHCHPhPO(OEt)_2$. V.9. $b_{0.15}$ 112-115°.[868] V.9. $b_{0.9}$ 132°
60

$Me_2NCHPhPO(OR)OH$.

R = i-Pr. From (i-PrO)(NMe_2)P(O)H + PhCHO, m. 202-203°[813]

R = Bu. From (BuO)(NMe_2)P(O)H + PhCHO, m. 196-198°.[813]

$C_6H_5NHCH_2CH(OH)CH_2PO(OEt)_2$. V.14. m. 91-92°, IR, ^1H.[691]

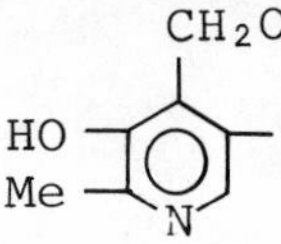

$-CH_2PO(OPr\text{-}i)_2$. V.14. m. 137-138°.[232]

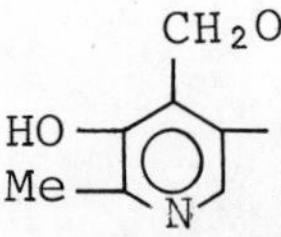

$-CH_2CH_2PO(OEt)_2$. V.14. m. 108-109°.[807]

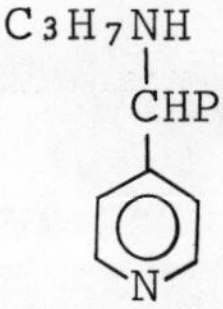

V.9. m. 102°.[2104]

$CH_3C(O)CH{=}CHCEt[-C(O)Me]PO(OR)_2$.

R = Me. V.14. b_1 128-130°.[1651]

R = Et. V.14. b_2 145-147°.[1651]

$HC{\equiv}CCH_2CEt(-CO_2Et)PO(OEt)_2$. V.14. b_6 160°.[1604]
$(EtO_2C-)[MeC(O)-]CH-CMe{=}CHPO(OR)_2$.

R = Et. V.14. $b_{1.5}$ 148°.[1609]

R = Pr. V.14. $b_{0.8}$ 151°.[1609]

$CH_3C(O)CH{=}CH-CMe(-CO_2Et)PO(OEt)_2$. V.14. b_2 160-161°.[1651]
$H_2C(CH_2)_4C{=}C[-O_2CMe]PO(OEt)_2$. V.14. b_2 123-123.5°, n_D^{20} 1.4600, d_4^{20} 1.1185.[1589]
$EtO_2CCHCH_2CH{=}CHCH_2CHPO(OEt)_2$. V.14. b_1 126-128°, n_D^{20} 1.464 d_4^{20} 1.1200.[1641]

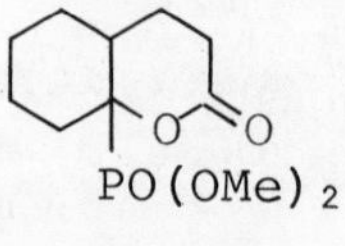

V.8. IR, m. 160°.[940]

$(EtO_2C-)[EtO(OH)C=]C-CH=CHPO(OPr-i)_2$. V.14. Sinters at 124°.[1788]

$EtO_2CCH_2C(-CO_2Et)=CHPO(OEt)_2$. V.5. $b_{0.3}$ 152-165°, n_D^{20} 1.4552.[1040] V.7. $b_{0.25}$ 154-158°, n_D^{20} 1.4538.[1040]

$MeO_2C-CH_2C(-CO_2Me)(-CH_2C{\equiv}CH)PO(OMe)_2$. V.14. b_1 133-140° (92% pure).[613]

$EtO_2CC(O)CH(-CO_2Et)CH_2PO(OEt)_2$. V.14.[1043]

$PhMe_2SiCH_2PO(OR)_2$.

R = Et. V.5. b_2 123-124°, n_D^{20} 1.4966, d_4^{20} 1.0571.[320]

R = Bu. V.6. b_{10} 200-202°, n_D^{20} 1.4864.[344]

$OCH=CH-CH=C-CH(NEt_2)PO(OEt)_2$. V.9. b_2 140°.[529]

$OCMe=C(-CO_2Et)CHEtPO(OEt)$. V.8.[840]

$OCHMe(CH_2)_2C(-CO_2Me)[-C(O)Me]PO(OEt)$. V.14. $b_{0.1}$ 104-106°.[318]

$OCH_2C(-CO_2Et)_2-CH_2PO(OEt)$. From $(EtO_2C)_2C(CH_2OH)_2$ and $P(OEt)_3$ in tetraline.[842] V.8/V.14. $b_{0.001}$ 120-121°.[841,844]

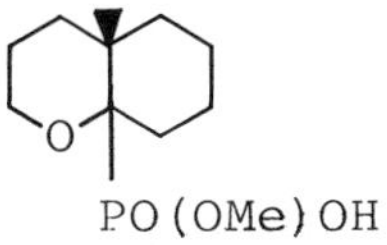

III.3. m. 214°, IR, opt. activity.[940]

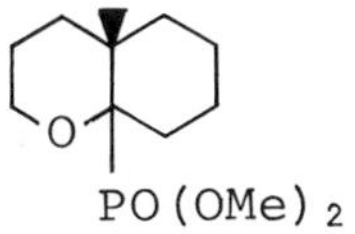

V.8. m. 116-117° (hexane), IR, opt. activity.[940]

$H_2CC(O)CH_2CMe_2CH_2CMePO(OBu)_2$. V.7. $b_{2.2}$ 190°.[877]

$EtO_2CCH_2C(O)CH_2CMe_2PO(OEt)_2$. V.14. $b_{0.1}$ 141-142°.[1043]

$CH_3C(O)S(CH_2)_3CH(-CO_2Et)PO(OEt)_2$. V.14. $b_{0.05}$ 149-152°, n_D^{20} 1.4715, IR, UV λ_{max} 232 mμ.[996]

$(EtO_2C-)_2CHCHMePO(OEt)_2$. V.7. b_{11} 180°.[1551]

$EtO_2CCH_2CMe(-CO_2Et)PO(OEt)_2$. V.7. b_9 182°.[1631]

$H_2C(CH_2)_3CH=CCH(-NMe_2)PO(OPr-i)OH$. From $(i-PrO)(NMe_2)P(O)H$ and $H_2C(CH_2)_3CH=CCHO$, m. 144-146°.[813]

$H_2C(CH_2)_4NCH_2CMe=CHPO(OEt)_2$. V.14. $b_{0.3}$ 122-124°.[1082]

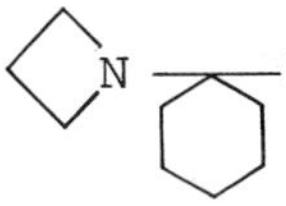

$PO(OPr)_2$. V.9. $b_{1.5}$ 125-126°.[962]

$H_2C(CH_2)_4NC(O)(CH_2)_3PO(OPr-i)_2$. V.5. $b_{0.01}$ 146-150°, n_D^{20} 1.4768.[515]

$Et_2NCH_2C{\equiv}CCH_2OCH_2PO(OR)_2$.

R = Et. V.14. b_2 148°.[1610]

R = Bu. V.14. $b_{0.8}$ 148-150°.[1610]

$Et_2NC(O)CMe[-C(O)Me]PO(OEt)_2$. V.6. $b_{0.15}$ 141-148°.[242]

$C_5H_{11}OCH_2CMe{=}CHPO(OAm)_2$. V.2. b_1 179-181°, n_D^{20} 1.4532, d_4^{20} 0.9774.[992]

$(CH_3)_2CHO_2CN{=}C(NEt_2)PO(OMe)_2$. V.14. $b_{0.06}$ 120-122°, n_D^{20} 1.4775, d_4^{20} 1.1482.[490]

$EtO(PrO)CH{-}CMe{=}CHPO(OPr)_2$. V.2. b_1 148-149 , n_D^{20} 1.4518, d_4^{20} 1.0229.[991]

$EtO(BuO)CH{-}CH{=}CHPO(OBu)_2$. V.2. $b_{1.5}$ 163-165°, n_D^{20} 1.4469, d_4^{20} 0.9817.[2037]

$OCH_2CH_2OCMe(CH_2)_5PO(OEt)_2$ (ring). V.6. $b_{0.05}$ 155°, n_D^{22} 1.4491, d_4^{2} 1.064.[1410] V.6. $b_{0.05}$ 155°, IR.[1959]

$BuO_2C{-}CH_2CHMeCH_2PO(OEt)_2$. V.5. $b_{0.04}$ 110°.[516]

$C_4H_9SCH_2CMe{=}CH{-}CH_2PO(OR)_2$.

R = Et. V.14. b_1 158-160°.[990]

R = Bu. V.14. $b_{0.5}$ 178-180°.[990]

$Et_2NCH_2CMe{=}CHCH_2PO(OEt)_2$. V.14. b_1 126-127°.[1630]

$(CH_3)_2C{=}C(-NEt_2)CH_2PO(OR)_2$.

R = Et. V.14. b_2 118-120°.[1603]

R = Pr. V.14. $b_{3.5}$ 140-143°.[1603]

R = Bu. V.14. b_1 135-137°.[1603]

$H_2C(CH_2)_3C(-NEt_2)PO(OPr)_2$ (ring). V.8. $b_{0.3}$ 80°, n_D^{20} 1.4488, d_4^{20} 0.9696.[1783]

$EtO_2CCH_2CH(-NEt_2)PO(OEt)_2$. V.14. b_2 135-137°.[1642]

$EtO_2CCH_2CH[-PO(OEt)Et]PO(OEt)_2$. V.14. b_5 193-195°, n_D^{20} 1.4520, d_4^{20} 1.1450.[1642]

$C_9H_{19}PO(OH)_2$. I.2. m. 99-100°.[1000,1002]

$C_9H_{19}PO(OR)_2$.

R = Me. V.7. b_{16} 171°.[1614]

R = Et. V.7. b_{14} 175°.[1614] V.5. b_{17} 177-186°.[1000,1002]

R = Bu. V.6. b_2 159-161°.[1000,1002]

$(C_2H_5O)_2CEt{-}CH_2CH_2PO(OEt)_2$. V.14. $b_{0.3}$ 113°, n_D^{21} 1.4357, d_4^{21} 1.025.[1082]

$C_7H_{15}CH(OMe)PO(OEt)OC_6H_4NO_2{-}4$. V.5. n_D^{25} 1.503.[741]

$H_2C(CH_2)_4C(-OSiMe_3)PO(OEt)_2$ (ring). V.8. b_1 135°.[1414]

$EtO_2CCH_2CH_2CH(-OSiMe_3)PO(OEt)_2$. V.8. $b_{0.45}$ 125-126°.[258]

$EtO_2CCH_2CMe(-OSiMe_3)PO(OEt)_2$. V.8. b_1 128-129°.[1414]

$(C_3H_7S)_2CMeCH_2PO(OR)_2$.

R = Et. V.14. b_4 151°.[1602]

R = Pr. V.14. $b_{0.85}$ 137°.[1602]

$(CH_3)_2CHCH_2CH[-O{-}PO(OEt)_2]{-}PO(OEt)_2$. V.6. b_3 170°, IR.[1618]

$(C_2H_5S)_3SiOCHEtPO(OEt)_2$. V.8. $b_{0.6-0.7}$ 172-173°.[258]

$(C_2H_5O)_2EtSi(CH_2)_3PO(OEt)_2$. V.5, V.2. b_3 136-139°.[386]

$(C_2H_5O)_3Si(CH_2)_3PO(OEt)_2$. V.5, V.14. b_3 143-146°.[386]

$(C_2H_5)_3Si(CH_2)_3PO(OEt)_2$. V.5, V.8, V.14. $b_{1.5}$ 118-120°.[387,]

C_{10}

$2{-}Br{-}4{-}MeC_6H_3NHC(O)NHC(O)CCl_2PO(OR)_2$.

R = Et. V.14. m. 93-95°.[281]

R = Pr. V.14. m. 59-60°.[281]

$(NC-)_2CBr-CHPhPO(OMe)_2$. V.14.[1877]
2-Br-4-$MeC_6H_3NHC(O)NHC(O)CH_2PO(OEt)_2$. V.14. m. 144-145°[281]
$MeC{=}N{-}NPhCCl{=}CPO(OH)_2$. I.1. m. 191°.[1288]
$CH_3C(O)CH_2CH(-C_6H_4Cl-4)PO(OEt)_2$. V.8.[840]
$Me_3SiOCH(-C_6H_4Cl-2)PO(OEt)_2$. V.8. $b_{0.6}$ 132-134°.[258]
$C_7H_{15}-CH{=}CCl-CH_2PO(OH)_2$. I.1. m. 90-92°, ^1H.[1228]
$C_7H_{15}-CH{=}CCl-CH_2PO(OEt)_2$. V.2. $b_{0.5}$ 129-130°, ^1H.[1228]
$C_8H_{17}CHClCH_2PO(OH)_2$. I.1. m. 96-99°.[1218]
$C_8H_{17}CHClCH_2PO(OMe)_2$. V.1. $b_{0.005}$ 104°.[1217]
4-$NCSC_6H_4NHC(O)NHC(O)CCl_2PO(OR)_2$.

R = Et. V.14. m. 95-96°.[281]

R = Bu. V.14. m. 62-64°.[281]

$(BuO)_2OP-O-CH(-CCl_3)PO(OBu)_2$. From Cl_3CCHO and $P(OBu)_2Cl$. $b_{1.5}$ 202-203°.[99]

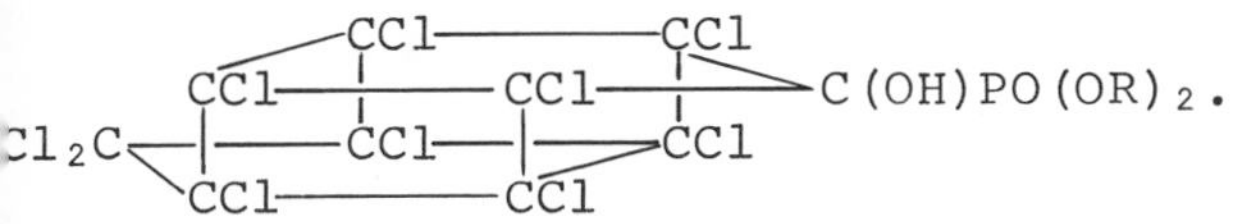

R = Me. V.8. m. 210°.[2067]

R = $H_2C{=}CHCH_2-$. V.8. m. 210°.[2067]

R = Bu. V.8. m. 150°.[2067]

R = Ph. V.8. m. 239°.[2067]

$(NC)_2C{=}CPh-PO(OMe)_2$. V.8. m. 88-89°.[1877]
4-$O_2NC_{10}H_6PO(OH)_2$-1. I.1. m. ~220° (dec.).[1021]
1-$C_{10}H_7PO(OH)_2$. I.1. m. 190°.[952] I.1. 189°.[1100]
1-$C_{10}H_7PO(OR)_2$.

R = Me. V.5. $b_{0.15}$ 129-131°.[1422]

R = Et. V.5. $b_{0.05}$ 147-149°.[1985]

2-$C_{10}H_7PO(OEt)_2$. V.5. $b_{0.05}$ 153-155°.[1987] V.5. $b_{0.1}$ 151°.[1988]

$HC{=}CPh-O-CH{=}CPO(OMe)_2$. V.14. m. 72-74°, IR, ^1H.[1689]
5-HO-$C_{10}H_6-PO(OH)_2$-2. I.13. m. 195-196°.[2073]
5-$HOS(O)_2-C_{10}H_6-PO(OH)_2$-2. I.13. As bis-p-toluidine salt m. 246-250°.[2073]
7-$HOS(O)_2-C_{10}H_6-PO(OH)_2$-2. I.13. Isolated as tri Na salt.[2073]

5-$H_2N-C_{10}H_6-PO(OH)_2$-2. I.13.[2073]
7-$H_2N-C_{10}H_6PO(OH)_2$-2. I.13. m. 295° (dec.).[2073]
$CH(OH)PO(OMe)_2$. V.8. As picrate, m. 142.5-143°.
$CH(OH)PO(OMe)_2$. V.8. As picrate, m. 142.5-143°.[327]

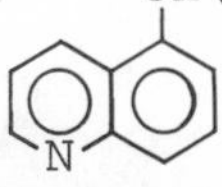

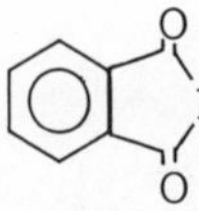
$NCH_2CH_2PO(OH)_2$. I.2. m. 198.5-199.5°.[1771]

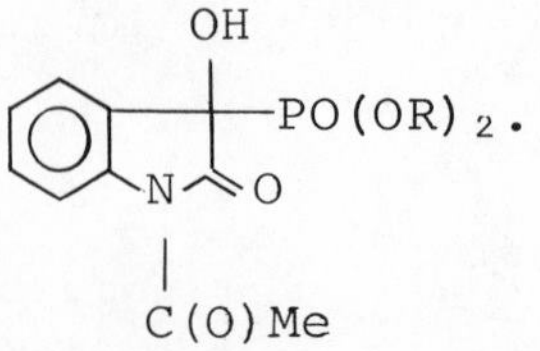

R = Me. V.8. m. 173-175°.[1362]

R = Et. V.8. m. 145-147°.[1362]

$H_2CC(-CN)=N-NHCPhPO(OMe)_2$. V.14. m. 137.5-138.5°, IR, ^{1}H.[1878a]

$C_6H_5CH=CH-CH=CHPO(OH)_2$. I.1. m. 192°.[239,1019]

$C_6H_5CH=CH-CH=CHPO(OEt)OH$. III.3. m. 128.5-129° ($CCl_4$); as dicyclohexylammonium salt, m. 190.5-191.5° (Me_2CO/H_2O).[770]

$MeC=CH-NPh-N=C-PO(OR)_2$.

R = Et. V.14. $b_{0.009}$ 151-153°, n_D^{20} 1.5420, d_4^{20} 1.1710.[1608]

R = Pr. V.14. $b_{0.006}$ 163-165°, n_D^{20} 1.5330, d_4^{20} 1.1316.[1608]

R = Bu. V.14. $b_{0.009}$ 170-172°, n_D^{20} 1.5263, d_4^{20} 1.0987.[1608]

R = Am. V.14. $b_{0.009}$ 172-173°, n_D^{20} 1.5200, d_4^{20} 1.0720.[1608]

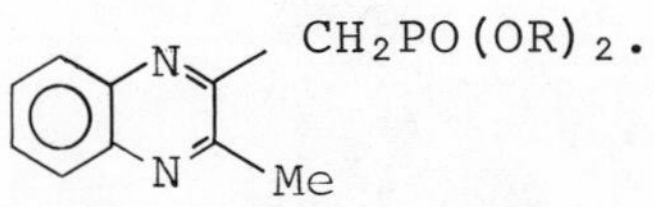

R = Et. V.5. m. 84-85°.[180]

R = i-Pr. V.5. m. 59°.[180]

R = Bu. V.5. $b_{0.015}$ 170-171°.[180]

$O=C-NPh-N=CMe-CHPO(OEt)_2$. V.5.[1998]

$C_6H_5CO_2CMe=CHPO(OEt)_2$. V.14. b_1 158-159°, n_D^{20} 1.5100, d_4^{20} 1.1586.[1135] V.2. $b_{1.5}$ 164-165°.[1133]

$CH_3O_2C-CH=CPhPO(OMe)_2$. V.7. b_9 189-192°.[967]

$C_6H_5O_2C-CH=CMePO(OPh)_2$. V.7. b_3 235-236°, m. 58-60°.[967]

$HO_2CC(O)CH_2CHPhPO(OH)_2$. I.13. dec. 183°.[411]

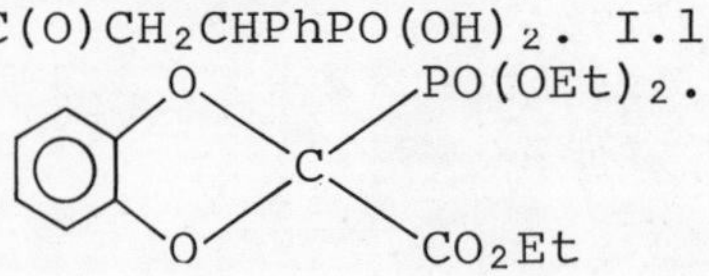
. V.5. $b_{0.04}$ 125-128°, IR.[717]

OCMe=CHCHPhPO(OR).

R = ClH_2CCH_2-. V.8. $b_{0.08}$ 114-115°.[1714]

R = ClH_2CCHMe-. V.8. $b_{0.07}$ 113-114°.[1714]

R = ClH_2CCH_2CHMe-. V.8. $b_{0.5}$ 140°.[1714]

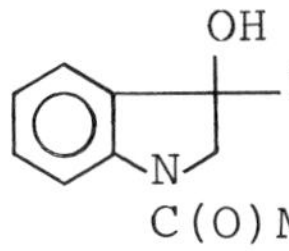

$PO(OPr\text{-}i)_2$. V.8. m. 140-142°.[1362]

4-$O_2NC_6H_4N$=NCH(-CO_2Et)-PO$(OEt)_2$. V.14. m. 104°.[1343]
4-EtC_6H_4CH=CHPO$(OH)_2$. I.1. m. 138-140°.[1019]
2,4-$Me_2C_6H_3$CH=CHPO$(OH)_2$. I.1. m. 142-143°.[1019]
H_2NC(O)CHNHC(-C_6H_4Me-4)PO$(OR)_2$.

R = Me. From 4-MeC_6H_4C=N-O-C(NH_2)=CH and $P(OMe)_3$, m. 159°, IR.[1404]

R = Et. From 4-MeC_6H_4C=N-O-C(NH_2)=CH and $P(OEt)_3$, m. 177-178°, IR.[1404]

$CH_3C(O)CH_2$CHPhPO$(OR)_2$.

R = Me. V.7. b_{10} 189-190°.[1542]

R = Et. V.8.[840] V.7. $b_{0.04}$ 150°, n_D^{25} 1.4985, ^{1}H.[751] V.7. b_{16} 203°.[1545]

R = Bu. V.7. $b_{0.02}$ 117°.[276]

$C_6H_5CO_2CMe_2PO(OH)_2$. I.13. m. 102°.[1178]
EtO_2CCHPhPO$(OEt)_2$. V.5. $b_{0.01}$ 127-129°.[1037]
HO_2CCHMeCHPhPO$(OH)_2$. I.2.[149]
(HC≡C-CH_2-$)_2$(EtO_2C-)CPO$(OR)_2$.

R = Et. V.14. $b_{0.2}$ 122°.[1604]

R = Pr. V.14. b_1 147°.[1604]

R = Bu. V.14. $b_{0.3}$ 151°.[1604]

C_6H_5SCH=CMe-CH_2PO$(OEt)_2$. V.14. $b_{0.4}$ 162-163°, n_D^{24} 1.5425, d_4^{24} 1.132.[1082]
C_6H_5CH=C(-NMe_2)PO$(OEt)_2$. V.7. $b_{0.01}$ 110-114°.[1809a]
4-[C_3H_7C(O)NH-]C_6H_4PO$(OH)_2$. I.13. m. 284°.[213]
3,4-$Me_2C_6H_3$C(O)NHCH_2PO$(OH)_2$. I.13. m. 206-207.5°.[1198]
($NCCH_2CH_2$-)(HC≡C-CH_2-)(EtO_2C-)CPO$(OEt)_2$. V.7. $b_{1.5}$ 170°.[1604]

$C_6H_5(CH_2)_4PO(OH)_2$. I.13. m. 95°.[239]
$C_6H_5(CH_2)_4$PO(OC_6H_{11}-cycl.)Cl. IV.2. n_D^{25} 1.5192.[740]
$C_6H_5(CH_2)_4$PO(OC_6H_{11}-cycl.)F. IV.2. $b_{0.03}$ 159°, n_D^{25} 1.4972.[740]

4-t-BuC_6H_4PO$(OH)_2$. I.1. m. 199.5-200°.[1016]
C_6H_5C(O)NHCH(-NMe_2)PO$(OEt)_2$. V.14. m. 130-133°, ^{1}H.[715a]

3-O_2N-4-$BuNHC_6H_3PO(OH)_2$. I.13. Dec. 176-178°.[186]
3-O_2N-4-i-$BuNHC_6H_3PO(OH)_2$. I.13. Dec. 176-180°.[186]
$(NCCH_2CH_2-)_2(EtO_2C-)CPO(OEt)OCH_2CH_2Cl$. V.7. $b_{0.005}$ 150°.[54]
3-O_2N-4-$H_2CCH_2OCH_2CH_2N-C_6H_3PO(OH)_2$. I.13. Dec. 176°.[186]
4-$MeC_6H_4S(O)_2NHN=C(-CH_2OMe)PO(OMe)_2$. V.14. m. 157-157.5° (dec.).[1191]
$C_6H_5CH_2CH_2CH(-OMe)PO(OEt)OC_6H_4NO_2$-4. V.5. n_D^{25} 1.548.[741]
2-EtO-3$MeC_6H_3CH_2PO(OEt)_2$. V.9. $b_{0.02}$ 86-88°.[848]
2-EtO-5-$MeC_6H_3CH_2PO(OEt)_2$. V.9. $b_{0.012}$ 102-103°.[848]
2-EtO-4-$MeC_6H_3CH_2PO(OEt)_2$. V.9. $b_{0.04}$ 127-128°.[848]

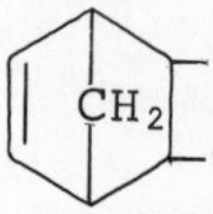
-PO(OEt)$_2$. V.14. b_1 134-136°.[1641]
-CO_2Et

O=$\overline{COCMe_2CH_2C}$[-C(O)CO_2Et]$PO(OMe)_2$. V.5. $b_{0.05}$ 150°.[317]
O=$\overline{COCHMeCH_2CH_2C}$[-C(O)$CO_2Et$]$PO(OR)_2$.

R = Me. V.5. $b_{0.02}$ 135-138°.[317,318]

R = Et. V.5. $b_{0.01}$ 135-137°.[317,318]

4-$Et_2NC_6H_4PO(OEt)_2$. V.5. $b_{0.15}$ 178-184°.[1985] V.5. $b_{0.03}$ 155-158°, n_D^{20} 1.5341.[1987]
$C_3H_7NHCHPhPO(OEt)_2$. V.9. $b_{0.05}$ 118°, n_D^{25} 1.4866.[60]
$C_6H_5CH_2NHCHEtPO(OH)_2$. I.2. m. 222-224°.[2040]
Me_2N-CMePh-PO(OR)OH.

R = Me. From (MeO)(Me_2N-)PHO and PhMeC=O , m. 174-175°[813]

R = Et. From (EtO)(Me_2N-)PHO and PhMeC=O , m. 153-154°[813]

R = i-Pr. From (i-PrO)(MeN-)PHO and PhMeC=O , m. 176-177°.[813]

$(EtO_2C-)(NC-)EtC-CMe=CHPO(OEt)_2$. V.14. b_1 153°.[1609]
$C_2H_5O_2CCH_2CH[-CH(-CN)(-CO_2Et)]PO(OEt)_2$. V.14. b_3 167-170°.[1642]

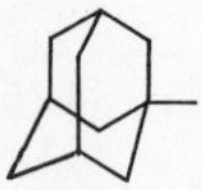
-$PO(OH)_2$. I.1. m. 297-305° (dec.). I.2. m. 308-310° (dec.).[1949]

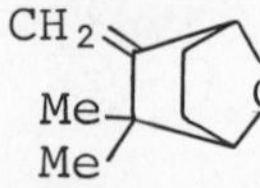
CHPO(OH)$_2$ I.1. 2 stereoisomers (?), m. 184°. (sesquihydrate, insol. in ether) dec. 167° (sol. in ether).[1194]

$Me_3SiOCHPhPO(OR)_2$.

R = Me. V.8. $b_{0.4}$ 100°, n_D^{20} 1.4865, d_4^{20} 1.0986.[1374]

R = Et. V.8. $b_{0.3}$ 105°, n_D^{20} 1.4767. d_4^{20} 1.0557.[1374] V.8. $b_{1-1.5}$ 124-125°, n_D^{20} 1.4842, d_4^{20} 1.0473.[1412,1414]

$H_2C(CH_2)_3C{=}C(-CO_2Et)CH_2PO(OEt)OH$. From cyclopentanone and $EtO_2C{-}CH_2CH_2PO(OEt)OH$, IR, ^1H.[1196]

$EtO_2C{-}CHCHMeCH{=}CHCH_2CHPO(OEt)_2$. V.14. b_1 131-132°, n_D^{20} 1.4640, d_4^{20} 1.1010.[1641]

$CH_3C(O)CH(-CO_2Et)CH_2CH{=}CHCH_2PO(OEt)_2$, V.14. b_8 180-180.5°.[1630]

$EtO_2C{-}C(O)CH_2C(O)CH_2CMe_2PO(OEt)_2$. V.14.[1043]

$(EtO_2C-)_2CH{-}CMe{=}CHPO(OR)_2$.

R = Et. V.14. $b_{1.5}$ 160-162°.[1609]

R = Pr. V.14. $b_{0.8}$ 158°.[1609]

R = Bu. V.14. b_1 178°.[1609]

$MeO_2CCH_2CH(-CO_2Me)CH(-CO_2Me)CH_2PO(OMe)_2$. V.7. $b_{0.1}$ 180-190°.[765]

$4{-}MeC_6H_4SiMe_2CH_2PO(OBu)_2$. V.6. $b_{2.6}$ 187-189°, n_D^{20} 1.4879.[344]

$(H_2C{=}CHCH_2-)_2NC(O)(CH_2)_3PO(OR)_2$.

R = Et. V.5. $b_{0.002}$ 127-128°, n_D^{20} 1.4774.[515]

R = Pr. V.5. $b_{0.01}$ 147-148°, n_D^{20} 1.4718.[515]

R = i-Pr. V.5. $b_{0.01}$ 138-139°, n_D^{20} 1.4672.[515]

$H_2C(CH_2)_4NCH_2C{\equiv}CCH_2OCH_2PO(OR)_2$.

R = Et. V.14. b_2 162°.[1610]

R = Bu. V.14. $b_{1.5}$ 183-184°.[1610]

$OCMe{=}C(-CO_2Et)CHPrPO(OEt)$. V.8.[840]

$OCMe_2(CH_2)_2C(-CO_2Me)[-C(O)Me]PO(OEt)$. V.14. $b_{0.1}$ 96°.[318]

$CH_2PO(OEt)_2$. V.7. b_{17} 186° (l-form).[1677]

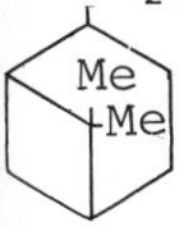

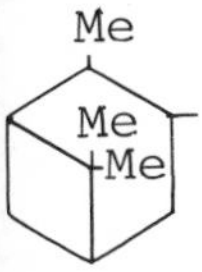

$-PO(OH)_2$. I.1., I.8.[1677]

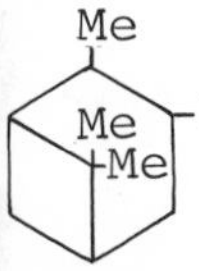

$-PO(OEt)_2$. V.7. b_1 117° (l-form), b_1 116-117° (d-form).[1677]

cycl.$-C_6H_{11}O_2C(CH_2)_3PO(OR)_2$.

R = Et. V.5. $b_{0.1}$ 168-170°.[516]

R = Pr. V.5. $b_{0.05}$ 135-138°.[516]

R = i-Pr. V.5. $b_{0.05}$ 130°.[516]

$(EtO_2C-)_2CH(CH_2)_3PO(OEt)_2$. V.14. $b_{0.25}$ 162-164°.[1041]

$H_2C(CH_2)_4NC(=CMe_2)CH_2PO(OR)_2$.

R = Et. V.14. b_1 128-130°.[1603]

R = Pr. V.14. $b_{2.5}$ 156°.[1603]

$H_2C(CH_2)_4N-CH(-CH_2CO_2Et)PO(OEt)_2$. V.14. b_4 163-164°, n_D^{20} 1.4583, d_4^{20} 1.0870.[1642]

$C_8H_{17}CH=CHPO(OH)_2$. I.13.[1218]

$C_8H_{17}CH=CHPO(OR)OH$

R = Me. III.4.[1217]

R = Ph. III.4.[1217]

$C_8H_{17}CH=CHPO(OMe)_2$. V.1.[1217]

(EtO)(AmO)CH-CH=CH-PO(OAm)$_2$. V.2. $b_{0.0025}$ 151-152°, n_D^{20} 1.4483, d_4^{20} 0.9658.[2037]

(EtO)(BuO)CH-CMe=CHPO(OBu)$_2$. V.14. b_1 161-162°, n_D^{20} 1.4504 d_4^{20} 0.9942.[991]

$BuO_2C(CH_2)_5PO(OBu)_2$. V.8. $b_{4.8-5.3}$ 137.5-140°, n_D^{20} 1.4319.[533]

$C_6H_{13}O_2C(CH_2)_3PO(OPr-i)_2$. V.5. $b_{0.01}$ 155-157°.[516]

$C_8H_{17}CH(-CO_2H)PO(OH)_2$. I.2. Glassy at 131.2°, m. 162-164°.[1

$H_2C(CH_2)_4C(-NEt_2)PO(OPr)_2$. V.9. $b_{0.05}$ 83-84°, n_D^{20} 1.4702, d_4^{20} 1.0071.[1783]

$Pr_2NC(O)(CH_2)_3PO(OR)_2$.

R = Et. V.5. $b_{0.004}$ 126-127°, n_D^{20} 1.4602.[515]

R = i-Pr. V.5. $b_{0.001}$ 117-119°, n_D^{20} 1.4571.[515]

$(Me_2CH-)_2NC(O)(CH_2)_3PO(OR)_2$.

R = Et. V.5. $b_{0.01}$ 133-135°, n_D^{20} 1.4593.[515]

R = Pr. V.5. $b_{0.005}$ 136-137°, n_D^{20} 1.4576.[515]

$NCCH_2CH_2CMe_2CH(-OSiMe_3)PO(OEt)_2$. V.8/V.14. $b_{0.7}$ 143-145°.[258]

$C_{10}H_{21}PO(OH)_2$. I.1. m. 99-100°.[737] I.8. m. 101.5-103°.[686] I.2. m. 102-102.5°.[1000,1002]

$C_{10}H_{21}PO(OR)_2$.

R = Me. V.7. b_{16} 182°.[1614]

R = Et. V.5. b_{17} 186-193°.[1000,1002] V.7. b_{12} 184°.[1614]

R = Bu. V.6. b_1 161°, n_D^{25} 1.4402, d_4^{20} 0.9232.[1000,1002] V.7. b_1 157°.[1954]

R = Ph. V.5. $b_{0.1}$ 178°, n_D^{25} 1.5202.[1080]

$C_4H_9CHEtCH_2OCH_2CH_2PO(OR)_2$.

R = Et. V.7. $b_{0.2}$ 116-118°, IR, ^1H.[1402]

R = Pr. V.7. $b_{0.2}$ 119-123°, IR, ^1H.[1402]

R = Bu. V.7. $b_{0.2}$ 128-131°, IR, ^1H.[1402]

$H_2\overline{C(CH_2)_4C}[-OSiMe_2(OEt)]PO(OEt)_2$. V.8/V.14. $b_{0.33}$ 125°.[258]
$C_8H_{17}CH(-SMe)PO(OEt)_2$. V.14. $b_{0.12}$ 132-133°, IR, ^1H.[442]
$H_2N(CH_2)_{10}PO(OH)_2$. I.13. m. 35-36° (EtOH).[366] pK_2 8.0; pK_3 11.25.[1776a]
$Bu_2NCH_2CH_2PO(OEt)_2$. V.14. b_3 140-142°.[1007]
$(BuO)_2OP-O-CHMePO(OBu)_2$. V.14. $b_{0.12}$ 159°, n_D^{25} 1.4364, d_4^{25} 1.039;[333] the structure as diphosphonic acid ester cited in the literature is not correct.[538]
$(EtO-)_3SiOCHPrPO(OEt)_2$. V.8/V.14. $b_{0.23-0.25}$ 114.5-117°.[258]
$Et_3Si(CH_2)_4PO(OEt)_2$. V.5, V.7. b_2 142-143°.[387]

<u>C_{11}</u>
$CH_3O_2CC(-CN)Br-CHPhPO(OR)_2$.

R = Me. V.14. m. 130°.[1877]

R = Et. V.14. m. 108°.[1877]

$4-BrC_6H_4CH[-OPO(OEt)_2]PO(OEt)_2$. V.8. b_3 210°.[1592]
$C_6H_5O(CH_2)_3-CCl=CHPO(OH)_2$. I.1.[87]
$C_6H_5O(CH_2)_3-CCl=CHPO(OR)_2$.

R = Me. V.2. b_2 197-198°.[87]

R = Et. V.2. b_2 199-200°.[87]

R = Bu. V.2. b_1 220-221°.[87]

$Cl(CH_2)_3C[=NNHS(O)_2C_6H_4Me-4]PO(OMe)_2$. V.14. m. 140-141°.[1191]

$2-ClC_6H_4CH[-O-PO(OEt)_2]PO(OEt)_2$. V.8. b_4 220-221°.[1592]
$4-ClC_6H_4CH[-O-PO(OEt)_2]PO(OEt)_2$. V.14.[1023]
$HO_2C(CH_2)_8CHClCH_2PO(OH)_2$. I.1. m. 96-99°.[1218]
$4-EtO_2CC_6H_4NHC(O)CCl_2PO(OEt)_2$. V.14. m. 101-102° (MeOH).[282]
$4-[(ClH_2CCH_2)_2N-]C_6H_4CH_2PO(OEt)_2$. V.14. m. 58-59°.[932] V.14. m. 58-59° (petroleum ether).[2044]
$1-C_{10}H_7CH(-NHSO_2F)PO(OEt)_2$. V.8. m. 168°.[789]

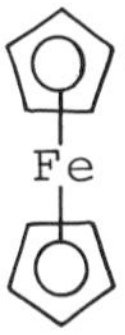
$CH_2PO(OEt)_2$. V.14.[2009]

$C_6H_5CH=CH-CH=C(-CN)PO(OEt)_2$. V.14.[2058]
$1-C_{10}H_7NHC(O)PO(OR)_2$.

R = Me. V.9. m. 139.5-140.5°.[559]

R = Et. V.9. m. 93-94° (EtOH/hexane).[1073,1074]

1-$C_{10}H_7CH_2PO(OH)_2$. I.2. m. 212-212.5° (water).[1001]
1-$C_{10}H_7CH_2PO(OEt)_2$. V.5. b_5 205-206°.[1001]
2-$C_{10}H_7CH_2PO(OH)_2$. I.2. m. 229-230°.[139]
2-$C_{10}H_7CH_2PO(OEt)_2$. V.5. b_1 170-172°.[139]
$(NC-)_2CMeCHPhPO(OMe)_2$. V.7.[149]
$O=C-NPhNHC(-CO_2H)=CCH_2PO(OH)_2$. I.2. As mono-p-toluidine salt, m. 215-216°.[1043]
$O=CNPhNHC(-CO_2H)=CCH_2PO(OEt)OH$. III.3. Isolated as Ba salt[1043]
$O=COCH_2CH_2C[-C(O)Ph]PO(OEt)_2$. V.6. $b_{0.01}$ 162°.[318]
1-$C_{10}H_7CH(-NH_2)PO(OH)_2$. I.2. m. 270°.[789]
$O=CONHCMe=C-CHPhPO(OMe)_2$. V.7. m. 165-166°, IR.[136]
$HN-N=C[-C(O)Me]-CH_2CPhPO(OMe)_2$. V.14. m. 70°.[1878a]
$MeC=N-NPh-CMe=CPO(OBu)_2$. V.2. b_{17} 253-254°.[659]
$CH_3C(O)CH=CHCHPhPO(OEt)_2$. V.14. b_2 134-136°.[1651]
$C_6H_5O(CH_2)_3C{\equiv}CPO(OR)_2$.

R = Me. V.14. b_4 206°.[87]

R = Et. V.14. b_3 190°.[87]

$EtO_2C-CH=CPhPO(OEt)_2$. V.7. b_4 185-186°.[967] From PhCH=C$(-NO_2)-CO_2Et$ and $P(OEt)_3$, cis:trans ratio 7:3, $b_{0.5}$ 150-152°, IR, ^1H.[392]
$C_6H_5CH=CHCH(-CO_2Me)PO(OEt)_2$. V.14. $b_{0.5}$ 150-155°.[267]
2,4,6-$Me_3C_6H_2CH=CHPO(OH)_2$. I.1. m. 176-178°.[1019]

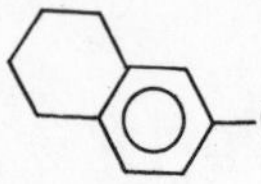
$-CH_2PO(OH)_2$. I.2. m. 165°.[1127]

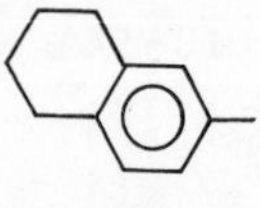
$-CH_2PO(OEt)_2$. V.5. b_3 165-167°.[1127]

4-$MeC_6H_4S(O)_2NHN=C(-CHCH_2CH_2)PO(OMe)_2$. V.14. m. 201-202° (dec.).[1191]

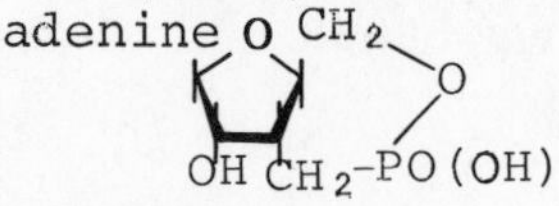

III.3. m. 220° (dec.).[879]

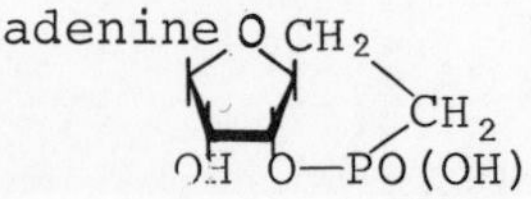

III.3. m. 220°, UV.[879]

2,4,6-$Me_3C_6H_2C(O)CH_2PO(OR)_2$.

R = Me. V.5. $b_{0.1}$ 80-90°[803]

R = Et. V.5. $b_{0.5}$ 170°[803]

4-t-$BuC_6H_4C(O)PO(OMe)_2$. V.5. $b_{1.2}$ 171-173°, ^1H; as 2,4-dinitrophenylhydrazone, m. 155-156°.[249]
$C_6H_5CH_2O_2C(CH_2)_3PO(OR)_2$.

R = Et. V.5. $b_{0.005}$ 136-140°.[516]

R = i-Pr. V.5. $b_{0.04}$ 153°.[516]

$EtO_2CCH_2CHPhPO(OR)_2$.

R = Me. V.7. b_{10} 192°.[1549]

R = Et. V.7. $b_{0.15}$ 140°, n_D^{25} 1.4872, ^1H.[751]

$EtO_2CCH[-C(O)Me]CH(-\overline{C=CH-C=CH-O})PO(OEt)_2$. From $MeC(O)CH_2CO_2Et$ + furfurol + $P(OEt)_3$.[840]

4-$[(ICH_2CH_2-)_2N]C_6H_4CH_2PO(OEt)_2$. V.14. m. 62-64°.[932] V.14.[2044]

$PhMeNC(O)(CH_2)_3PO(OR)_2$.

R = Et. V.5. $b_{0.001}$ 160-161°.[515]

R = Pr. V.5. $b_{0.001}$ 162-164°.[515]

$[CH_3C(O)-]EtNCHPhPO(OH)_2$. I.2. m. 165-166°.[60] I.13. Dec. 176°.[868]
$PhEtNC(O)CH_2CH_2PO(OEt)_2$. V.7. $b_{0.02}$ 140°[1664]
$(EtO_2C-)MeNCHPhPO(OEt)_2$. V.7. $b_{2.7}$ 180°.[60]
$NCCH_2CHMe-(HC{\equiv}CCH_2-)(EtO_2C-)CPO(OEt)_2$. V.7. b_1 137°.[1604]

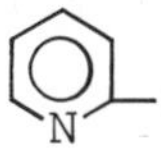$-CH(-CO_2Et)CH_2CH_2PO(OEt)_2$. V.7. $b_{0.01}$ 122-123°[1205]
$HO_2CCH(-CH_2Ph)NHCH_2CH_2PO(OCH_2CH_2Cl)_2$. V.14. IR. m. 154°[588]
2-Me-5-i-$PrC_6H_3CH_2PO(OH)_2$. I.2. m. 175-177°.[1127]
2-Me-5-i-$PrC_6H_3CH_2PO(OEt)_2$. V.5. b_2 146-148°.[1127]
4-$BuC_6H_4CH_2PO(OH)_2$. I.2. m. 162-163°.[1001]
4-$BuC_6H_4CH_2PO(OBu)_2$. V.6. b_2 175-178°.[1001]
$C_6H_5CH_2C(O)NHCH(-NMe_2)PO(OEt)_2$. V.14. m. 63-66°, ^1H.[715a]
3-O_2N-4-$AmNHC_6H_3PO(OH)_2$. I.13. Dec. 132-134°.[186]
3-O_2N-4-i-$AmNHC_6H_3PO(OH)_2$. I.13. Dec. 171-173°.[186]
4-$MeC_6H_4S(O)_2NHN{=}C(-CHMe_2)PO(OMe)_2$. V.14. m. 181-182°(dec.).[1191]

$PhMeC(O)(CH_2)_3PO(OPr-i)_2$. V.5. $b_{0.001}$ 162-164°.[515]

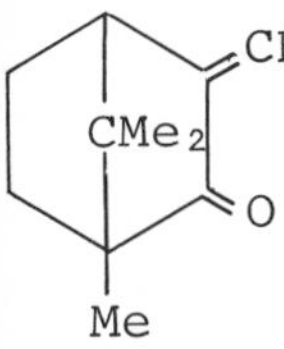

$=CHPO(OH)_2$. I.2. m. 106-107°.[1040]

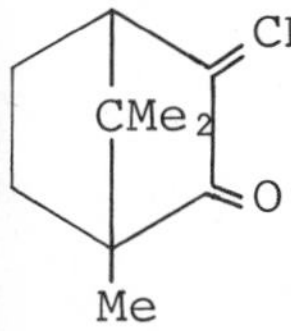

$=CHPO(OEt)_2$. V.8. $b_{0.05}$ 124-126°.[1040] V.5. $b_{0.2}$ 134°, n_D^{20} 1.4860.[1040]

$C_3H_7OCH_2CHPhPO(OEt)_2$. V.14. b_1 127-129°.[1651]
3-t-Bu-4-$HOC_6H_3CH_2PO(OC_{18}H_{37})OH$. III.4. m. 98-100° (petroleum ether/hexane).[1942]

$C_6H_5CH_2NHCHPrPO(OH)_2$. I.2. m. 214-216°.[2040]
$Me_2NCH_2CH_2CHPhPO(OMe)_2$. V.14. $b_{1.5}$ 142°[1170]
$Et_2NCHPhPO(OPr)_2$. V.9. b_2 75-79°, n_D^{20} 1.4255.[1390] V.9. $b_{0.04}$ 90-91°.[513]
$(EtO)_2OPO-CHPhPO(OEt)_2$. V.8/V.14. $b_{0.07}$ 149°;[333] the structure as diphosphonic acid ester cited in the literature is not correct.[538]
$Me_3SiOCH(-CH_2Ph)PO(OEt)_2$. V.8/V.14. $b_{0.4-0.5}$ 129-130°.[258]
$Me_3SiOCH(-C_6H_4Me-4)PO(OMe)_2$. V.8/V.14. $b_{0.44}$ 126°.[258]
$Me_3SiOCMePhPO(OMe)_2$. V.8. $b_{0.05}$ 90.5°, n_D^{20} 1.4933, d_4^{20} 1.0903, ^1H.[1374]
4-$MeOC_6H_4CH(-OSiMe_3)PO(OEt)_2$. V.8/V.14. $b_{0.6}$ 144-145°.[258]
$H_2C(CH_2)_4C=C(-CO_2Et)CH_2PO(OEt)OH$. From cyclohexanone and $EtO_2CCH_2CH_2PO(OEt)_2$, ^1H.[1196]
$CH_3C(O)CH(-CO_2Et)CH_2CMe=CHCH_2PO(OEt)_2$. V.14. b_5 178-179°.[1630]
$(EtO_2C-)_2CHCH_2CH=CHCH_2PO(OEt)_2$. V.14. b_2 182°.[1630]
$EtO_2C-C(-OEt)=C(-CO_2Et)CH_2PO(OEt)_2$. V.14. $b_{0.05}$ 147-149°.[1043]

$C_8H_{17}C(O)CH_2CH_2PO(OR)_2$.

R = Me. V.7. $b_{0.02}$ 125-126°, n_D^{30} 1.4407, d_4^{30} 0.9580.[1787]

R = Et. V.7. n_D^{30} 1.4373, d_4^{30} 1.0102.[1787]

R = Bu. V.7. n_D^{30} 1.4400, d_4^{30} 0.9738.[1787]

R = $C_4H_9CHEtCH_2$-. V.7. n_D^{30} 1.4472, d_4^{30} 0.9429.[1787]

$C_4H_9CHEt-CH[-CH_2C(O)Me]PO(OH)_2$. I.9. m. 66-69°.[493]
$AmO(EtO-)CHCMe=CHPO(OAm)_2$. V.2. b_7 170-171°, n_D^{20} 1.4501, d_4^{20} 0.9672.[991]
$C_6H_{13}O_2C(CH_2)_4PO(OPr-i)_2$. V.5. $b_{0.01}$ 158-159°.[516]
$C_6H_{13}O_2CCH_2CHMeCH_2PO(OPr-i)_2$. V.5. $b_{0.001}$ 140-152°.[516]
$C_7H_{15}CH(-CO_2Et)PO(OEt)_2$. V.5. b_4 168°.[43,44]
$C_{11}H_{23}PO(OMe)_2$. V.7. b_{20} 192°.[1614]
$(AmO)_2CHPO(OAm)_2$. From $HC(OAm)_3$ + PCl_3, $b_{0.06}$ 142-144°, ^1H.[713]

$C_6H_{13}CH(-NEt_2)PO(OR)_2$.

R = Pr. V.9. $b_{0.05}$ 83-84°.[513]

R = Bu. V.9. $b_{0.05}$ 93-95°.[513]

$Et_3SiOCHBuPO(OMe)_2$. V.8. b_3 103-106°, n_D^{20} 1.4480, d_4^{20} 1.0049.[1374]

C_{12}

4-$BrC_6H_4OC_6H_4PO(OH)_2$. I.13. m. 209°.[456] I.1.[660]
2-$BrC_6H_4SC_6H_4PO(OH)_2$. I.1.[660]
4-(2-ClC_6H_4O-)-3-$O_2NC_6H_3PO(OH)_2$. I.13. m. >200°.[186]
4-(4-ClC_6H_4O-)-3-$O_2NC_6H_3PO(OH)_2$. I.13. m. >200°.[186]
4-(4-ClC_6H_4O-)$C_6H_4PO(OH)_2$. I.1.[660]
4-(4-ClC_6H_4S-)$C_6H_4PO(OH)_2$. I.1.[660]
4-(2-ClC_6H_4O-)-3-$H_2NC_6H_3PO(OH)_2$. I.13. m. >200°.[186]

4-(4-ClC$_6$H$_4$O-)-3-H$_2$NC$_6$H$_3$PO(OH)$_2$. I.13. m. >200$^{\rm O}$.[186]
4-[(ClCH$_2$CH-)$_2$N-]C$_6$H$_4$CHMePO(OEt)$_2$. V.14. m. 61.5-62.5$^{\rm O}$.
   [932] V.14. m. 61-62$^{\rm O}$.[2044]
4-(4-FC$_6$H$_4$S-)C$_6$H$_4$PO(OH)$_2$. I.1.[660]
4-PhO-3-O$_2$NC$_6$H$_3$PO(OH)$_2$. I.13. m. >200$^{\rm O}$.[186]
2-C$_{10}$H$_7$CH=CHPO(OH)$_2$. I.1. m. 181.5-182$^{\rm O}$.[1019]
4-PhC$_6$H$_4$PO(OH)$_2$. I.1. m. 218-220$^{\rm O}$.[867]
2-PhOC$_6$H$_4$PO(OH)$_2$. I.13. m. 200-202$^{\rm O}$[574]
3-PhOC$_6$H$_4$PO(OH)$_2$. I.1. m. 92.5-93.5$^{\rm O}$.[579]
4-PhOC$_6$H$_4$PO(OH)$_2$. I.1. m. 185$^{\rm O}$(as 1-hydrate).[456]
4-PhO-3-H$_2$NC$_6$H$_3$PO(OH)$_2$. I.13. m. >200$^{\rm O}$.[186]

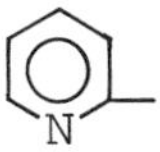

—CHPhPO(OR)$_2$.

   R = Me. V.8. m. 100$^{\rm O}$, as picrate m. 172$^{\rm O}$.[1121]

   R = Et. V.8. m. 106$^{\rm O}$, as picrate m. 175$^{\rm O}$.[1121]

   R = i-Pr. V.8. m. 102$^{\rm O}$, as picrate m. 178$^{\rm O}$.[1121]

—CH(-HNPh)PO(OPh)$_2$. V.8. m. 116$^{\rm O}$.[2104]

—CH(-HNPh)PO(OPh)$_2$. V.8. m. 149$^{\rm O}$.[2104]

CH(-HNPh)PO(OPh)$_2$. V.8. m. 158$^{\rm O}$.[2104]

— CH$_2$ — CH—   V.14. b$_{0.01}$ 131-133$^{\rm O}$.[1206]
                 |
               PO(OEt)$_2$
C$_2$H$_5$O$_2$CCH(-CN)CHPhPO(OEt)$_2$. V.14, V.8. n$_D^{20}$ 1.4999.[1037]
O=CONMeCMe=CCHPhPO(OMe)$_2$. V.8, V.14. m. 112-114$^{\rm O}$[136]
—C(-CH$_2$CH$_2$CN)$_2$PO(OEt)$_2$. V.14. b$_{0.05}$ 185-190$^{\rm O}$.[1206]

[MeC(O)-]$_2$CHCHPhPO(OMe)$_2$. V.7. b$_5$ 198-200$^{\rm O}$.[1643] By hydrolysis of O-CMe=C[-C(O)Me]CHPhP(OMe)$_3$.[1690]
PrO$_2$CCH=CPhPO(OPr)$_2$. V.7. b$_3$ 184-185$^{\rm O}$.[967]
PhCH=C(-CO$_2$Et)-CH$_2$PO(OEt)OH (cis and trans). From PhCHO
   and EtO$_2$CCH$_2$CH$_2$PO(OEt)$_2$, ^{1}H.[1196]

—CH=CHPO(OEt)$_2$. V.14. m. 67-79$^{\rm O}$.[807]

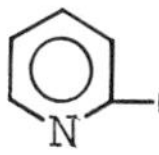

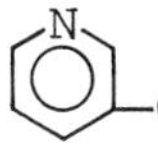

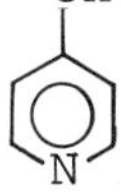

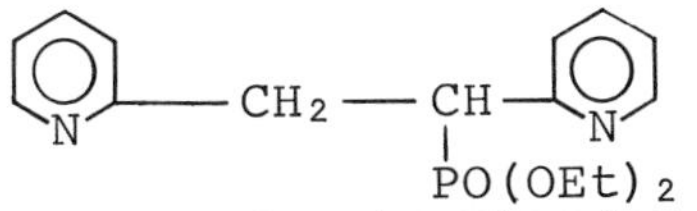

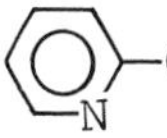

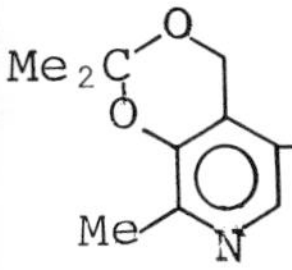

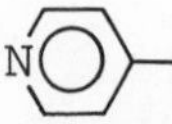$-C(-CH_2CH_2CO_2H)_2PO(OH)_2$. I.2. m. 220°.[1206]

2-t-$BuC_6H_4CH{=}CHPO(OH)_2$. I.1. m. 188-189°.[1019]
4-t-$BuC_6H_4CH{=}CHPO(OH)_2$. I.1. m. 150.5-151.5°.[1019]
$CHMe{=}CMe{-}C[{=}NNHS(O)_2C_6H_4Me{-}4]PO(OMe)_2$. V.14. m. 155-156° (dec.).[1191]
$Me_2C{=}CHC[{=}NNHS(O)_2C_6H_4Me{-}4]PO(OMe)_2$. V.14. m. 115-115.5°.[1191]
$H_2C(CH_2)_2CHC[{=}NNHS(O)_2C_6H_4Me{-}4]PO(OMe)_2$. V.14. m. 169.5-171° (dec.).[1191]
uracil $CH_2{-}PO(OR)(OR')$.

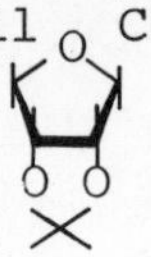

R = R' = H. I.2. R^f values.[796]

R = ICH_2CH_2-, R' = H. III.8.[796]

R = $HOCH_2CH_2-$, R'=H. III.3.[796]

R = R' = $PhCH_2OCH_2CH_2-$. V.5.[796]

4-$[(IH_2CCH_2)_2N-]C_6H_4CHMePO(OEt)_2$. V.14. m. 80-82°.[932, 2044]
$H_2C(CH_2)_4C(-HNPh)PO(OEt)_2$. V.9. m. 110-112°.[805]
4-$[C_5H_{11}C(O)NH-]C_6H_4PO(OH)_2$. I.13. m. 204°.[213]
$(EtO_2C-)EtNCHPhPO(OEt)_2$. V.9. $b_{0.4}$ 156°.[60]
$Me_3C{-}C[{=}NNHS(O)_2C_6H_4Me{-}4]PO(OMe)_2$. V.14. m. 88.5-89.5°.[1191]
3-t-Bu-4-HO-5-$MeC_6H_2CH_2PO(OC_{18}H_{37})_2$. V.5. m. 69-71° (acetone).[1943]
$C_6H_5CH_2NHCHBuPO(OH)_2$. I.2. m. 205-207°.[2040]
i=$PrCH(CH_2)_2CHMeCH_2CHO_2C{-}CH_2PO(OEt)_2$. V.5, V.6. m. 125-130°.[2001]
$(EtO_2C-)_2CHCH_2CMe{=}CHCH_2PO(OEt)_2$. V.14. b_5 212-213°.[1630]
$(EtO_2C-)_2CMeCH_2CH{=}CHCH_2PO(OEt)_2$. V.14. $b_{3.5}$ 192°.[1630]
$(EtO_2C-)_2CEtCMe{=}CHPO(OR)_2$.

R = Et. V.14. b_1 156-158°.[1609]

R = Pr. V.14. b_1 162-163°.[1609]

$EtO_2CCH_2CH[CH(-CO_2Et)_2]PO(OEt)_2$. V.14. b_2 175-177°, n_D^{20} 1.4455, d_4^{20} 1.1417.[1642]
$EtO_2CCH_2CH(-CO_2Et)CMe(-CO_2Me)PO(OEt)_2$. V.7. $b_{0.05}$ 164-170°[765]
$H_2C(CH_2)_4CH{-}CH(CH_2)_4CHPO(OH)_2$. I.8. m. 98-99.5°.[686]
$H_2C(CH_2)_4NC(O)CH[-N(CH_2)_4CH_2]PO(OR)_2$.

R = Me. V.5. m. 105-106°, IR, ^{31}P -25.2 ppm.[717]

R = Et. V.5. $b_{0.03}$ 160-163°, m. 56-58°, IR, ^{31}P -21.5 ppm.[717]

i-Pr-$CH(CH_2)_2CHMeCH_2CHO_2CCH_2PO(OEt)_2$. V.5, V.6. m. 125-130[2001]

$BuO_2CCH_2CH(-CO_2Bu)PO(OEt)_2$. V.7. n_D^{20} 1.4527.[1073]
$H_2C(CH_2)_{10}C(-OH)PO(OMe)_2$. V.8. m. 27°.[229] Crystal and mol. structure.[1782]
$MeO_2C(CH_2)_{10}PO(OMe)_2$. V.5. $b_{0.08}$ 153-154°, IR, n_D^{30} 1.4456, d_4^{30} 1.0385.[1786] V.7. $b_{0.06}$ 154-155°.[1787]
$C_7H_{15}O_2C(CH_2)_4PO(OEt)_2$. V.5. $b_{0.03}$ 140°.[516]
$C_8H_{17}O_2C(CH_2)_3PO(OEt)_2$. V.5. $b_{0.05}$ 142°.[516]
$C_4H_9O_2CCH(-C_6H_{13})PO(OC_6H_{13})_2$. V.5. $b_{0.3}$ 163-171°, n_D^{30} 1.4427, d_4^{30} 0.9438.[44]
$i\text{-}Bu_2NC(O)(CH_2)_3PO(OR)_2$.

R = Et. V.5. $b_{0.01}$ 133-136°.[515]

R = Pr. V.5. $b_{0.005}$ 148-150°.[515]

R = i-Pr. V.5. $b_{0.01}$ 134-135°.[515]

$C_{12}H_{25}PO(OH)_2$. I.1. m. 92°.[737] I.2. m. 99-100°.[1079] I.2. m. 100.5-101.5o.[1000,1002] pK_1 - , pK_2 8.25.[1776a]
$C_{12}H_{25}PO(OR)_2$.

R = Me. V.5. $b_{0.02}$ 100°, n_D^{25} 1.4345, IR.[1079] V.2. $b_{0.03}$ 110-116°.[1079]

R = Et. V.5. b_1 160°.[552] V.5. b_3 165-175°.[1000,1002] V.5. b_1 160°.[1117] V.5. $b_{0.02}$ 106°, IR.[1079] V.5. $b_{0.12}$ 116-120°.[1079]

R = Bu. V.6. b_3 196-199°, n_D^{25} 1.4432, d_4^{25} 0.9153.[1000, 1002] V.7.[746]

R = Ph. V.5. $b_{0.01}$ 165°.[1080]

$C_6H_{13}CHBuCH_2PO(OH)_2$. I.8. m. 99-100°.[686]

$\underline{C_{13}}$
$(4\text{-}BrC_6H_4\text{-})[2,4\text{-}(O_2N)_2C_6H_3NHN{=}]CPO(OR)_2$.

R = Me. V.14. m. 198.5-200°.[780,2103]

R = Et. V.14. m. 201-203°.[780,2103]

R = Pr. V.14. m. 131.5-133.5°.[780,2103]

R = i-Pr. V.14. m. 148-150°.[2103]

R = Bu. V.14. m. 116-118°.[780,2103]

$4\text{-}BrC_6H_4CH(-HNPh)PO(OR)_2$.

R = Et. V.14. m. 99.5-100.5°, kinetics of the reaction.[883]

R = Ph. V.14. m. 143-144°.[1027]

$(4\text{-}BrC_6H_4\text{-})[(EtO)_2OP\text{-}O\text{-}]EtCPO(OEt)_2$. V.8/V.14. $b_{0.05}$ 149-150°, n_D^{20} 1.5090, d_4^{20} 1.3450.[1590]
$(4\text{-}ClC_6H_4\text{-})[2,4\text{-}(O_2N)_2C_6H_3NHN{=}]CPO(OR)_2$.

R = Me. V.14. m. 190-192°.[780,2103]

R = Et. V.14. m. 203.5-206.5°.[780,2103]

R = Pr. V.14. m. 154-156°.[780,2103]

R = Bu. V.14. m. 137.5-139°.[780,2103]

$4\text{-ClC}_6\text{H}_4\text{C(=NPh)PO(OEt)}_2$. V.5. $b_{0.005}$ 155°, n_D^{20} 1.5742.[780,2103]

$(4\text{-ClC}_6\text{H}_4\text{NH-})(4\text{-O}_2\text{NC}_6\text{H}_4\text{-})\text{CHPO(OPh)}_2$. V.9. m. 165-167°.[2105]
$4\text{-ClC}_6\text{H}_4\text{CH(-HNPh)PO(OEt)}_2$. V.14. m. 75°.[883]
$4\text{-ClC}_6\text{H}_4\text{NHCHPhPO(OEt)}_2$. V.9. m. 111-113°.[1632] V.9. m. 112-114°.[1655]
$(4\text{-ClC}_6\text{H}_4\text{NHCHPh-})\text{OPO-R-O}$.

R = $-(\text{CH}_2)_3-$. V.9. m. 188-189°.[1588]

R = $-\text{CHMe-}(\text{CH}_2)_2-$. V.9. m. 207°.[1588]

$4\text{-ClC}_6\text{H}_4\text{CHC(-CO}_2\text{Et)=CMeOPO(OEt)}$. V.8/V.14.[840]
$[(\text{EtO})_2\text{OP-O-}](2\text{-ClC}_6\text{H}_4\text{-})\text{EtCPO(OEt)}_2$. V.8/V.14. $b_{0.04}$ 145-147°, n_D^{20} 1.5050, d_4^{20} 1.2464.[1590]
$\text{F}_2\text{CCF}_2\text{CF}_2\text{C(-NBu}_2\text{)=CPO(OEt)}_2$. V.14. $b_{0.2}$ 136-138°, IR.[564]

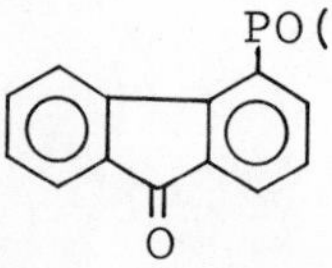
PO(OH)_2. I.13. m. 306-307.5°.[1137]

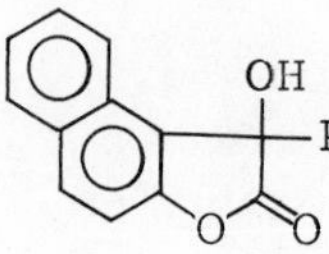
PO(OEt)_2. V.8. m. 154-156°.[1362]

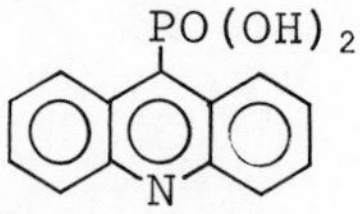
I.2. m. 248-250°.[488] I.2.[1884] I.2. m. 247-249° (dec.).[1004]

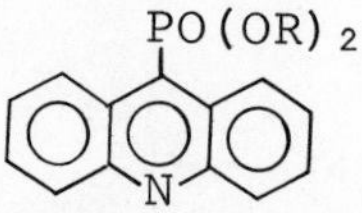

R = Et. V.14.[1884] V.2. m. 165-167°.[1004]

R = Pr. V.14.[1884]

$(4\text{-O}_2\text{NC}_6\text{H}_4\text{-})[2,4\text{-(O}_2\text{N)}_2\text{C}_6\text{H}_3\text{NHN=]CPO(OR)}_2$.

R = Me. V.14. m. 221-224° (dec.)[780,2103]

R = Et. V.14. m. 176-177.5°.[780,2103]

R = Pr. V.14. m. 177-179°.[780,2103]

R = i-Pr. V.14. m. 174° (dec.).[780,2103]

R = Bu. V.14. m. 127.5-129°.[780,2103]

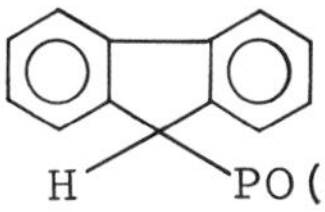
PO(OH)$_2$. I.2. Dec. 258°[325]. I.5.[1676]

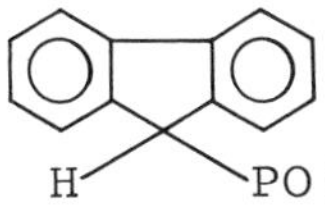
PO(OCH$_2$CH$_2$CH$_2$NEt$_2$)$_2$. V.1. m. 153-154°.[325]

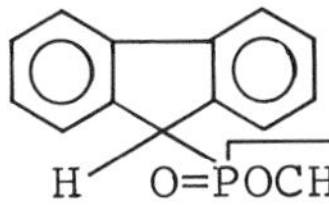
O=POCH$_2$CEt(-CH$_2$Br)CH$_2$O. V.5. m. 235°.[2059]

2,4-(O$_2$N)$_2$C$_6$H$_3$-S-O-CHPhPO(OEt)$_2$. V.5. m. 121.5-122.5°.[815]

2,4-(O$_2$N)$_2$C$_6$H$_3$NHN=CPhPO(OR)$_2$.

R = Me. V.14. m. 189.5-191°.[780,2103]

R = Et. V.14. m. 171-172°.[780,2103]

R = Pr. V.14. m. 142.5-145.5°.[780,2103]

R = i-Pr. V.14. m. 150-152°.[780,2103]

R = Bu. V.14. m. 113-114.5°.[2103]

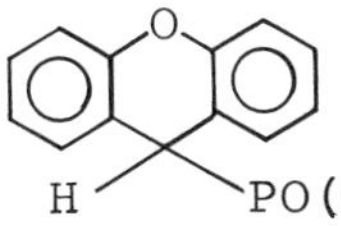
PO(OMe)$_2$. V.5. m. 128-129°[1361]

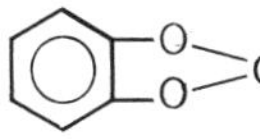
CPhPO(OEt)$_2$. V.5. $b_{0.02}$ 146-148°, m. 57-58°. IR.[717]

2-(2-HO$_2$CC$_6$H$_4$-)C$_6$H$_4$PO(OH)$_2$. I.13. m. 234-237°.[1137]

C$_6$H$_5$N=CPhPO(OR)$_2$.

R = Et. V.9. $b_{0.05}$ 163°.[1036] V.5.[1036] V.5. $b_{0.1}$ 151-152°.[2103]

R = Pr. V.5. $b_{0.03}$ 157.5-158°.[780]

R = Bu. V.5. $b_{0.04}$ 170-171°.[780]

4-[PhC(O)NH-]C$_6$H$_4$PO(OH)$_2$. I.13. m. 287°.[213]

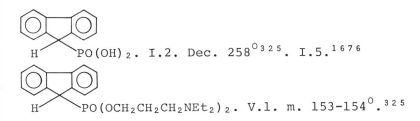

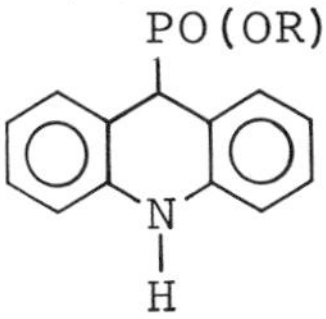

R = Et. From acridine hydrochloride and P(OEt)$_3$.[1884]

R = Pr. From acridine hydrochloride and P(OPr)$_3$.[1884]

4-O$_2$NC$_6$H$_4$CH(-OPh)PO(OPh)$_2$. V.8. m. 116-118°.[1059]

2-$O_2NC_6H_4$-S-OCHPhPO(OEt)$_2$. V.14. m. 104-105°.[815]
4-$O_2NC_6H_4$NHCH(-$C_6H_4NO_2$-4)PO(OPh)$_2$. V.8. m. 195-197°.[2105]
Ph_2CHPO(OH)$_2$. I.2. m. 235-238°.[325] I.6.[1676]
Ph_2CHPO(OEt)OH. III.3. m. 120-122°.[325]
Ph_2CHPO(OR)$_2$.

R = Et. V.5. b_2 180-181°.[128]

R = i-Pr. V.8. m. 83-84°.[1676]

Ph_2CHPO(OEt)OR.

R = $Me_2NCH_2CH_2$-. V.1. m. 125-126°.[325]

R = $Et_2NCH_2CH_2$-. V.1. m. 127-128°.[325]

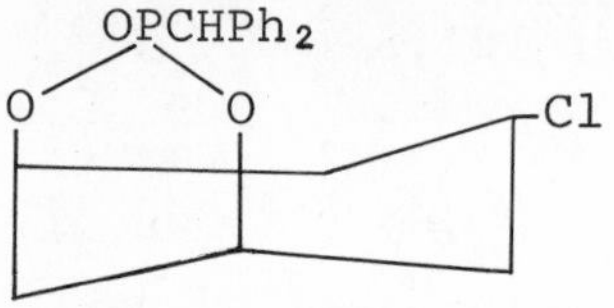
V.5. m. 219-220°.[245]

2-Ph$C_6H_4CH_2$PO(OH)$_2$. I.2. m. 167-169°.[1137]
2-Ph$C_6H_4CH_2$PO(OEt)$_2$. V.5. $b_{0.3}$ 148-152°.[1137]
4-Ph$C_6H_4CH_2$PO(OH)$_2$. I.2. Dec. 250°.[1001]
4-Ph$C_6H_4CH_2$PO(OBu)$_2$. V.6.[1001]
(PhO)$_2$OPNHC(O)PO(OR)$_2$.

R = Me. V.9. n_D^{20} 1.5400, d_4^{20} 1.3318.[972]

R = Et. V.9. n_D^{20} 1.5280, d_4^{20} 1.2760.[972]

R = Pr. V.9. n_D^{20} 1.5130, d_4^{20} 1.2160.[972]

R = i-Pr. V.9. n_D^{20} 1.5170, d_4^{20} 1.2179.[972]

R = Bu. V.9. n_D^{20} 1.5085, d_4^{20} 1.1860.[972]

R = i-Bu. V.9. n_D^{20} 1.5056, d_4^{20} 1.1889.[972]

R = Am. V.9. n_D^{20} 1.4985, d_4^{20} 1.1502.[972]

R = i-Am. V.9. n_D^{20} 1.4990, d_4^{20} 1.1487.[972]

R = Ph. V.9. n_D^{20} 1.5577, d_4^{20} 1.2633.[972]

R = 4-ClC_6H_4-. V.9. n_D^{20} 1.5565, d_4^{20} 1.3400.[972]

R = 4-MeC_6H_4-. V.9. n_D^{20} 1.5500. d_4^{20} 1.2351.[972]

PhN=C(-NHPh)PO(OEt)$_2$. From (EtO)$_2$PNHPh + PhNCO.[804]
[2,4-($O_2N)_2C_6H_3$NHN=](4-MeOC_6H_4-)CPO(OR)$_2$.

R = Me. V.14. m. 198-200°.[780]

R = Et. V.14. m. 177-178°.[780]

R = Pr. V.14. m. 113.5-115°.[780]

R = i-Pr. V.14. m. 140.5-141.5°.[780]

R = Bu. V.14. m. 139.5-140.5°.[780]

2-$O_2NC_6H_4$CH(-HNPh)PO(OEt)$_2$. V.9. m. 155°.[1632]
4-$O_2NC_6H_4$CH(-HNPh)PO(OPh)$_2$. V.9. m. 150-151°.[1027] V.9. m.

155.5°.[2104] V.9. m. 154-155°.[2105]

$3\text{-}O_2NC_6H_4NHCHPhPO(OPh)_2$. V.9. m. 144-145°.[1027]

$4\text{-}O_2NC_6H_4NHCHPhPO(OR)_2$.

R = Ph. V.9. m. 149-150°.[2105] V.9. m. 161°.[1027]

R = cycl.-C_6H_{11}-. V.9. m. 141-143°.[1634]

$4\text{-}Me_2NC_6H_4\text{-}C(\text{-}CN)[\text{-}CH(CN)_2]PO(OEt)_2$. V.7. m. 140-145°.[1868]

$2\text{-}(4\text{-}MeC_6H_4O\text{-})C_6H_4PO(OH)_2$. I.1. m. 203-204°.[579]

$4\text{-}(4\text{-}MeC_6H_4O\text{-})C_6H_4PO(OH)_2$. I.1.[660]

$Ph_2C(OH)PO(OH)_2$. I.9. m. 184-185° (water).[1184,1186] I.9.[415]

$Ph_2C(OH)PO(OR)_2$.

R = Me. V.8. m. 103-104°.[40]

R = Et. V.8. $b_{0.00025}$ 150°, n_D^{21} 1.5281[1340] (? cf. Ref. [1619]).

$(PhO)_2CHPO(OEt)OH$. III.3. m. 104-105°.[715]

$(PhO)_2CHPO(OEt)Cl$. IV.1. ^1H, mass spect.[715]

$(PhO)_2CHPO(OEt)_2$. V.5. $b_{0.01}$ 158-159°, m. 38-40°, ^1H.[715]

$2\text{-}[PhCH_2O\text{-}]C_6H_4PO(OH)_2$. I.13. From O-benzyloxibenzenediazonium fluoroborate and PCl_3, m. 156-157°.[581]

$C_6H_5NHCHPhPO(OR)OH$.

R = Et. III.3. m. 113°; as Na salt, m. 232°.[868]

R = C_8H_{17}-. III.3. m. 97-100°; as Na salt, m. 224°.[868]

$C_6H_5NHCHPhPO(OR)_2$.

R = Me. V.5. m. 88-89°.[1921]

R = Et. V.5. m. 91°.[1921] V.14. m. 91-92°.[883] V.9. m. 90-91°.[1073,1074] V.9. m. 92°.[529,1547] V.9. m. 91°.[1655] V.9. m. 93-94°.[805] V.9. 92-93°.[868]

R = Pr. V.9. m. 75-77°.[1655]

R = Ph. V.9. m. 158-159°.[2105] V.9. m. 154-155°.[805] V.9. m. 154-155°.[1027] V.9. m. 154.5°.[2104] V.9. m. 150-152°.[1634]

R = cycl.-C_6H_{11}-. V.9. m. 89-90°.[1634]

R = $PhCH_2$-. V.9. m. 109-110°.[1634]

R = C_8H_{17}-. V.9. cryst.[868]

$(C_6H_5NHCHPh\text{-})\overline{OPO\text{-}R\text{-}O}$.

R = $-(CH_2)_3-$. V.9. m. 169-170°.[1588]

R = $-CHMeCH_2CH_2-$. V.9. m. 181°.[1588]

R = $-CHMeCHMe-$. V.9. m. 145°.[1588]

$[Ph_2C(\text{-}\overset{\oplus}{N}H_3)PO(OEt)_2]Cl^{\ominus}$. V.9. m. 144°.[1219]

$O{=}\underline{C\text{-}NPh\text{-}NHC(\text{-}CO_2Et){=}C}CH_2PO(OEt)_2$. V.14.[1043]

$2\text{-}C_{10}H_7(CH_2)_3PO(OH)_2$. I.2. m. 181-182°, IR, ^1H.[370]

$2\text{-}C_{10}H_7(CH_2)_3PO(OEt)_2$. V.5. $b_{0.0003-0.0004}$ 155-160°, IR.[370]

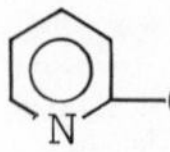—$CH(\text{-}HNC_6H_4Me\text{-}2)PO(OR)_2$.

R = Me. V.9. m. 68°, as picrate m. 156°.[1121]

R = Et. V.9. m. 56°, as picrate m. 137°.[1121]

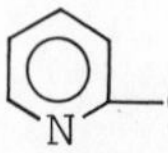—$CH(\text{-}HNC_6H_4Me\text{-}3)PO(OR)_2$.

R = Me. V.9. m. 102°, as picrate m. 169°.[1121]

R = Et. V.9. m. 116°, as picrate m. 187°.[1121]

R = i-Pr. V.9. m. 110°, as picrate m. 191°.[1121]

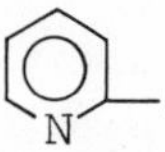—$CH(\text{-}HNC_6H_4Me\text{-}4)PO(OR)_2$.

R = Me. V.9. m. 112°, as picrate m. 164°.[1121]

R = Et. V.9. m. 125°, as picrate m. 172°.[1121]

R = i-Pr. V.9. m. 111°, as picrate m. 184°.[1121]

$BuO_2C\text{-}CH{=}CPhPO(OBu)_2$. V.7. b_2 205-210°.[967]

$OCMe{=}C(\text{-}CO_2Et)\text{-}CHPhPO(OMe)$. V.14.[131] V.8/V.14 $b_{0.015}$ 154-155°.[840]

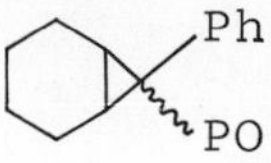

$PO(OH)_2$. I.2. m. 274-275°, ^1H, IR.[1878a]

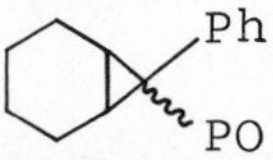

$PO(OMe)_2$. V.14. m. 187.5-188°, IR, ^1H.[1878a]

$PhMeC{=}C(\text{-}CO_2Et)CH_2PO(OEt)OH$(cis and trans). From PhC(O)Me and $EtO_2CCH_2CH_2PO(OEt)_2$, ^1H.[1196]

$(EtO_2C\text{-})[MeC(O)\text{-}]CH\text{-}CHPhPO(OR)_2$.

R = Me. By hydrolysis of $OCMe{=}C(\text{-}CO_2Et)CHPhP(OMe)_3$.[131]

R = Et. V.7. b_7 204-206°.[1551]

$CH_3C(\text{-}NHOH){=}C(\text{-}CO_2Et)\text{-}CHPhPO(OMe)_2$. V.14. m. 169-170°.[135]

$CH_3C({=}NOH)\text{-}CH(\text{-}CO_2Et)\text{-}CHPhPO(OMe)_2$. V.14. m. 119-121°.[135]

uracil 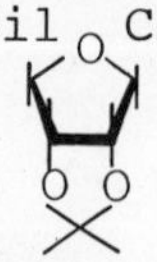 $CH{=}CHPO(OPh)_2$. V.14. m. 146-147, ^1H.[880]

$Me_2C{=}CMe\text{-}C[{=}NNHS(O)_2C_6H_4Me\text{-}4]PO(OMe)_2$. V.14. m. 84-85°.[1191]

$H_2C(CH_2)_3CHC[{=}NNHS(O)_2C_6H_4Me\text{-}4]$. V.14. As syn-form, m. 67-68°; anti-form, m. 145-146°.[1191]

$C_5H_{11}C[=NNHS(O)_2C_6H_4Me\text{-}4]PO(OMe)_2$. V.14. m. 106-107°.[1191]

Me_3SiCH_2  Ph, $PO(OMe)_2$. V.14. $b_{0.02}$ 109-111°, ^1H, IR.[1878a]

$Me_2\overline{C(CH_2)_3CMe=C}CH=CHCHMePO(OEt)_2$. From β- jonol and $P(OEt)_3$.[584]

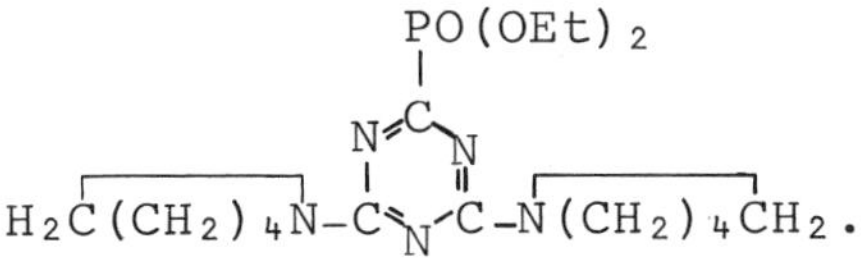

$H_2C(CH_2)_4N\text{-}C_3N_3(PO(OEt)_2)\text{-}N(CH_2)_4CH_2$. V.5. m. 104-106°.[778]

$(EtO)_2OP\text{-}O\text{-}CPhEtPO(OEt)_2$. V.8. b_1 168-169°, n_D^{20} 1.4870, d_4^{20} 1.1749.[1590]
$(EtO_2C\text{-})_2CEt\text{-}CH_2CH=CHCH_2PO(OEt)_2$. V.14. b_4 200°.[1630]
$(EtO_2C\text{-})_2CPr\text{-}CMe=CHPO(OEt)_2$. V.14. $b_{0.8}$ 151°.[1609]
$EtO_2CCH_2C(\text{-}CO_2Et)_2CH_2CH_2PO(OEt)_2$. V.14. $b_{0.04}$ 185-188°.[2082]

$(cycl.\text{-}C_6H_{11})_2NCH_2PO(OEt)_2$. V.14. b_2 149°.[529]
$[H_2C(CH_2)_4N\text{-}]_2CMeCH_2PO(OEt)_2$. V.14. b_2 165°.[1602]
$H_2C(CH_2)_{11}C(\text{-}OH)PO(OMe)_2$. V.8. m. 114-115°.[229]
$EtO_2C(CH_2)_{10}PO(OR)_2$.

R = Et. V.5. $b_{0.15}$ 163-164°, IR, n_D^{30} 1.4410, d_4^{30} 0.9910.[1786,1787]

R = Bu. V.5. $b_{0.1}$ 184-186°, IR.[1786]

R = $C_4H_9CHEtCH_2$-. V.5. $b_{0.001}$ 160°, IR, n_D^{30} 1.4500, d_4^{30} 0.9401.[1786,1787]

$(EtO)_2MeSiCH_2CH_2$- 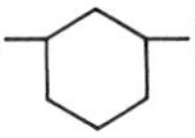 -$PO(OEt)_2$. V.7. b_2 198°, n_D^{25} 1.4515, d_4^{25} 1.071.[208]

C_{14}

3-$MeC_6H_4NHCH(\text{-}C_6H_4Br\text{-}4)PO(OPh)_2$. V.9. m. 127-128° (alcohol/benzene).[1027]
4-$MeC_6H_4S(O)_2NHN=C(\text{-}C_6H_4Br\text{-}4)PO(OEt)_2$. V.14. syn-form, m. 70-71°; anti-form, m. 145-146°.[883]
4-$ClC_6H_4CPh=CHPO(OH)_2$. I.1. m. 181°.[239]
4-$MeC_6H_4NHCH(\text{-}C_6H_4Cl\text{-}4)PO(OPh)_2$. V.9. m. 129-130° (alcohol/benzene).[1027]
4-$MeC_6H_4S(O)_2NHN=C(\text{-}C_6H_4Cl\text{-}4)PO(OEt)_2$. V.14. syn-form, m. 76-77°; anti-form, m. 147-148°.[883]
$(4\text{-}ClC_6H_4\text{-})_2C=CHPO(OH)_2$. I.1. m. 158-159°.[239]
$(3\text{-}Cl\text{-}4\text{-}HOC_6H_3\text{-})_2CMePO(OR)_2$.

R = Me. V.14. m. 213°.[817]

R = Et. V.14. m. 143.5°.[817]

$2\text{-}FC_6H_4CPh{=}CHPO(OH)_2$. I.1. m. 180°.[239]

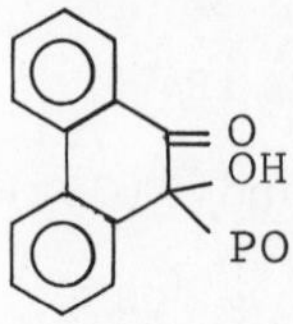

$PO(OH)_2$. I.9. m. 125-128°.[555]

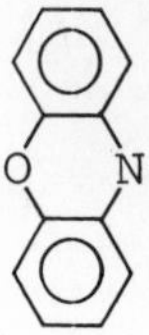

$N\text{-}C(O)CH_2PO(OR)_2$.

R = Et. V.5. m. 124-126°.[2094]

R = Pr. V.4. m. 73-75°.[2094]

R = i-Pr. V.5. m. 119-121°.[2094]

R = Bu. V.5. m. 51-53°.[2094]

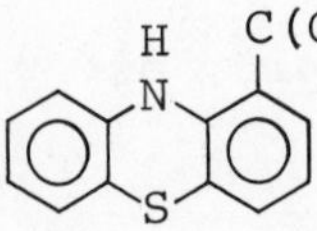

$C(O)CH_2PO(OEt)OCH_2CH_2Cl$. From $R\text{-}C(O)CH_2Cl$ and $H(O)P(OEt)(OCH_2CH_2OH)$, m. 133-134° (benzene).[2093]

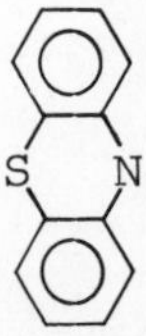

$N\text{-}C(O)CH_2PO(OR)_2$.

R = Et. V.5. m. 106-107°.[172]

R = Pr. V.5. m. 89-91°.[172]

R = i-Pr. V.5. m. 102-103°.[172]

R = Bu. V.5. m. 72-73°.[172]

R = i-Bu. V.5. m. 139°.[172]

$Ph_2C{=}CHPO(OH)_2$. I.1. m. 167°.[238]
$PhCH{=}CPhPO(OH)_2$. I.9. m. 208.5-210°, IR ^{31}P-11.0 ppm.[959]
$2\text{-}PhC_6H_4CH{=}CHPO(OH)_2$. I.1. m. 186-188° (water).[1019]
$3\text{-}PhC_6H_4CH{=}CHPO(OH)_2$. I.1. m. 156-157.5°.[1019]
$4\text{-}PhC_6H_4CH{=}CHPO(OH)_2$. I.1. m. 193-193.5° (butanol).[1019]
$2\text{-}O_2NC_6H_4CH(OH)\text{-}O\text{-}CH(\text{-}C_6H_4NO_2\text{-}2)PO(OR)_2$.

By hydrolysis of

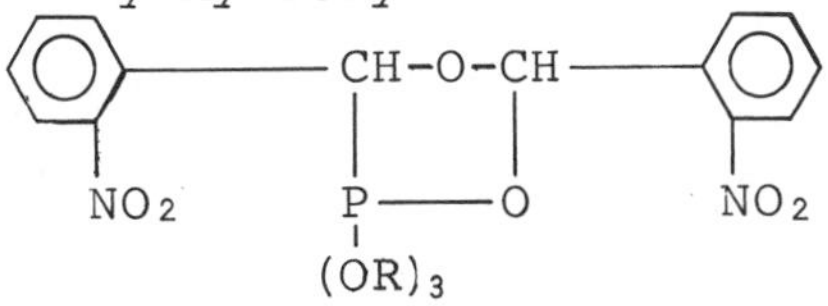

R = Et. m. 140-142°.[1059]

R = Pr. m. 128-130°.[1059]

2,4-$(O_2N)_2C_6H_3NHN=C(-C_6H_4Me-4)PO(OR)_2$.

R = Me. V.14. m. 190-192°.[780,2103]

R = Et. V.14. m. 180-181°.[780,2103]

R = Pr. V.14. m. 132-133°.[780,2103]

R = Bu. V.14. m. 117-118°.[780,2103]

2,4-$(O_2N)_2C_6H_3NHN=C(-C_6H_4OMe-4)PO(OR)_2$.

R = Me. V.14. m. 198-200°.[2103]

R = Et. V.14. m. 177-178°.[2103]

R = Pr. V.14. m. 113.5-115°.[2103]

R = i-Pr. V.14. m. 140.5-141.5°.[2103]

R = Bu. V.14. m. 139.5-140.5°.[2103]

$Ph_2C=C(OH)PO(OEt)_2$. V.5. m. 99-103°, UV, IR, ^1H. ^{31}P -9ppm, mass spect.[203]

4-$MeC_6H_4C(=NPh)PO(OEt)_2$. V.5. $b_{0.03}$ 159°, n_D^{20} 1.5661.[780,2103]

$Ph_2NC(O)CH_2PO(OEt)_2$. V.5. m. 67-68°(ligroine).[1696]

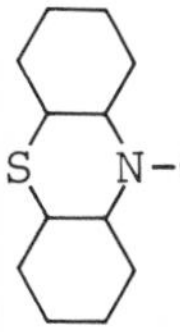$N-CH_2CH_2PO(OEt)_2$. V.14. $b_{0.002}$ 160-180°.[1480]

4-$(PhCH_2CH_2-)C_6H_4PO(OH)_2$. I.1, I.6. m. 256°.[1260,1269,1272]

$Ph_2CHCH_2PO(OH)_2$. I.13. m. 213°.[238]

2-$MeC_6H_4NHCH(-C_6H_4NO_2$-4)$PO(OPh)_2$. V.9. m. 101-103°.[2105]

4-$MeC_6H_4NHCH(-C_6H_4NO_2$-3)$PO(OPh)_2$. V.9. m. 147-148°(alcohol/benzene).[1027]

4-$MeC_6H_4NHCH(-C_6H_4NO_2$-4)$PO(OPh)_2$. V.9. m. 150-151°(alcohol/benzene).[1027]

4-$O_2NC_6H_4NHCH(-C_6H_4OMe$-4)$PO(OPh)_2$. V.9. m. 184-186°.[2105]

4-$MeOC_6H_4NHCH(-C_6H_4NO_2$-4)$PO(OPh)_2$. V.9. m. 106-107°.[2105]

$Ph_2C(OH)CH_2PO(OEt)_2$. V.14.[449]

$(4-HOC_6H_4-)_2CMePO(OH)_2$. I.2. m. 254°(dec.).[814]

$(4-HOC_6H_4-)_2CMePO(OR)_2$.

R = Me. V.8/V.14. m. 217°.[814,817]

R = Et. V.8/V.14. m. 160°.[814]

$4\text{-MeC}_6\text{H}_4\text{S(O)}_2\text{OCHPhPO(OMe)}_2$. V.14. m. 86-87°. IR, ^{1}H.[1878a]
$\text{C}_6\text{H}_5\text{CH}_2\text{NHCHPhPO(OH)}_2$. I.2. m. 236-237°.[2040]
$2\text{-MeC}_6\text{H}_4\text{NHCHPhPO(OPh)}_2$. V.9. m. 89-90°.[2105]
$3\text{-MeC}_6\text{H}_4\text{NHCHPhPO(OR)}_2$.

R = Et. V.9. m. 134-135°.[1655]

R = Ph. V.9. m. 146-146.5°(alcohol/benzene).[1027]

$4\text{-MeC}_6\text{H}_4\text{NHCHPhPO(OR)}_2$.

R = Ph. V.9. m. 170-171°(alcohol/benzene).[1027] V.9. m. 170-171°.[1634]

R = cycl.-C_6H_{11}-. V.9. m. 83-85°.[1634]

$(4\text{-MeC}_6\text{H}_4\text{NHCHPh-})\text{OPO-R-O}$.

R = $-(\text{CH}_2)_3-$. V.9. m. 201°.[1588]

R = $-\text{MeCHCH}_2\text{CH}_2-$. V.9. m. 169-170°.[1588]

$4\text{-MeC}_6\text{H}_4\text{CH(-HNPh)PO(OR)}_2$.

R = Me. V.9. m. 121-122°.[1632]

R = Et. V.14. m. 63.5-64.5°, kinetics of the reaction.[883]

$4\text{-MeOC}_6\text{H}_4\text{CH(-HNPh)PO(OEt)}_2$. V.14. m. 101.5-102.5°(ligroine), kinetics of the reaction.[883]
$4\text{=MeOC}_6\text{H}_4\text{CH(-HNPh)PO(OPh)}_2$. V.9. m. 142-143°.[2105]
$4\text{-[4-Me}_2\text{NC}_6\text{H}_4\text{-N=N-]C}_6\text{H}_4\text{PO(OH)}_2$. I.13. Dark red microcryst. powder.[1021]
$\text{MePh}_2\text{SiOCH}_2\text{PO(OSiPh}_2\text{Me)}_2$. V.14. $b_{1.5}$ 306°, n_D^{40} 1.5808, d_4^{40} 1.1520.[1437]
$4\text{-MeOC}_6\text{H}_4\text{CHC(-CO}_2\text{Et)=CMe-O-PO(OEt)}$. V.8/V.14.[840]
adenine $\text{CH}_2\text{CH}_2\text{PO(OPh)}_2$. V.14. m. 135-136°, ^{1}H.[880]

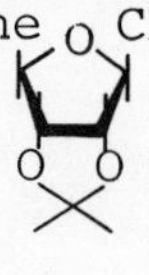

$(\text{EtO}_2\text{C-})_2\text{CHCHPhPO(OEt)}_2$. V.7. b_{11} 212-214°.[1551]
$\text{C}_5\text{H}_{11}\text{CHCH}_2\text{CPhPO(OMe)}_2$. V.14. $b_{0.04}$ 108°, IR, ^{1}H.[1878a]
$\text{H}_2\text{C(CH}_2)_4\text{CH-C[=NNHS(O)}_2\text{C}_6\text{H}_4\text{Me-4]PO(OMe)}_2$. V.14. syn-form, m. 82.5-83° anti-form, m. 167-168°(dec.).[1191]
$3\text{-(Et}_2\text{NCH}_2\text{-)-2-EtOC}_6\text{H}_3\text{CH}_2\text{PO(OEt)}_2$. V.9. $b_{0.02}$ 112-115°.[848]
$3,5\text{-(Me}_3\text{C-)}_2\text{-4-HOC}_6\text{H}_2\text{PO(OEt)}_2$. V.5. m. 115-116°.[1985] V.5 $b_{0.05}$ 172-176°, m. 105-106°.[1987]
$\text{Et}_3\text{SiOCMePhPO(OMe)}_3$. V.8. $b_{0.07}$ 119-121°, n_D^{20} 1.4972, d_4^{20} 1.0750, ^{1}H.[1374]
$(\text{EtO}_2\text{C-})_2\text{CEt-CH}_2\text{CMe=CHCH}_2\text{PO(OEt)}_2$. V.14. b_1 187°.[1630]
$(\text{i-AmO}_2\text{C-})\text{CH}_2\text{CH(-CO}_2\text{Am-i)PO(OAm-i)}_2$. V.7. b_4 214-216°.[1644]
$\text{C}_{11}\text{H}_{23}\text{CO}_2\text{CH}_2\text{CH}_2\text{PO(OR)}_2$.

R = Et. V.5. $b_{0.2}$ 162-164°.[46]

R = Bu. V.5. $b_{0.1}$ 164-172°.[46]

R = C_6H_{13}. V.5. $b_{0.1}$ 193-197°.[46]

$C_{10}H_{21}CH(-CO_2Et)PO(OR)_2$.

R = Et. V.5. $b_{0.1}$ 153-156°, n_D^{30} 1.4398, d_4^{30} 0.9782.[44]

R = Bu. V.5. $b_{0.18}$ 210°, n_D^{30} 1.4413, d_4^{30} 0.9547.[44]

R = C_6H_{13}-. V.5. $b_{0.25}$ 173°, n_D^{30} 1.4440, d_4^{30} 0.9394.[44]

$C_{14}H_{29}PO(OH)_2$. I.1. m. 92-93°.[737] I.2. m. 97-98° (ligroine).[1000,1002]

$C_{14}H_{29}PO(OR)_2$.

R = Et. V.5. b_3 ~200°.[1000,1002]

R = Bu. V.6. b_3 217-219°, n_D^{25} 1.4460, d_4^{25} 0.9114.[1000,1002]

R = Ph. V.5. m. 41.5-42° (pentane).[1080]

$C_{12}H_{25}OCH_2CH_2PO(OEt)_2$. V.7. $b_{0.2}$ 143-147°, IR, ^1H.[1402]

<u>C_{15}</u>

4-$ClC_6H_4C(O)CHBrCHPhPO(OH)_2$. I.13. m. 204°.[416]
4-$ClC_6H_4C(O)CHBrCHPhPO(OPh)_2$. V.14. m. 127-129°.[416]
$PhC(O)CHBrCHPhPO(OH)_2$. I.13. m. 196°.[412]

2,4,5-$Br_3C_6H_2O$-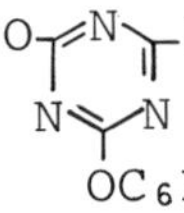-$PO(OEt)_2$. V.5. m. 80°.[832]

$OC_6H_2Br_3$-2,4,5

4-$ClC_6H_4C(O)CH_2CHPhPO(OH)_2$. I.1, I.9. m. 112-114° (as 1-hydrate).[416]
4-$ClC_6H_4C(O)CH_2CHPhPO(OR)OH$.

R = Me. I.2. m. 152-153°.[416]

R = Ph. I.2. m. 180°.[416]

4-$ClC_6H_4C(O)CH_2CHPhPO(OR)_2$.

R = Me. V.1. m. 123-124°.[416]

R = Ph. V.1. m. 109°.[416]

(4-ClC_6H_4-)(4-$MeOC_6H_4$-)$C{=}CHPO(OH)_2$. I.1. m. 132-133°.[239]
(4-ClC_6H_4-)(4-$MeOC_6H_4$-)$CHCH_2PO(OH)_2$. I.13. m. 126-127°.[239]
[4-$(ClCH_2CH_2$-$)_2NC_6H_4$-][$(NC)_2CH$-](NC-)$CPO(OR)_2$.

R = Me. V.7. m. 186-188°.[1868]

R = Et. V.7. m. 173-174°.[1868]

2,4,5-$Cl_3C_6H_2O$-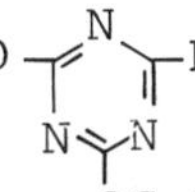-$PO(OEt)_2$. V.5. m. 78°.[832]

$OC_6H_2Cl_3$-2,4,5

2,3,4,5-Cl_4C_6HO– 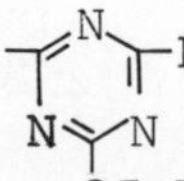 –$PO(OCH_2CH_2Cl)$. V.5. m. 112-114°.[832]
OC_6HCl_4-2,3,4,5

Cl_5C_6O– 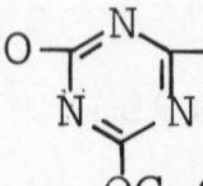 –$PO(OEt)_2$. V.5. m. 157-158°.[832]
OC_6Cl_5

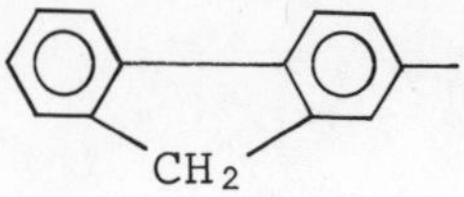 –$CH=CHPO(OH)_2$. I.1. Dec. 200-205°.[1019]

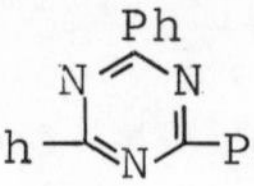 –$PO(OR)_2$

R = Me. V.5. m. 125-126°.[778]

R = Et. V.5. m. 98-99.5°.[778]

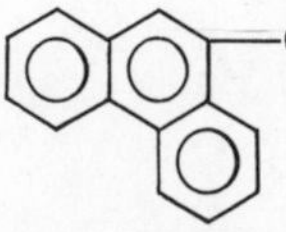 –$CH_2PO(OH)_2$. I.2. Dec. 212°.[1001]

[S-C(S)NMeC(O)-CH-]$(1-C_{10}H_7-)CHPO(OEt)_2$. V.7. m. 117-118°.[1125a]

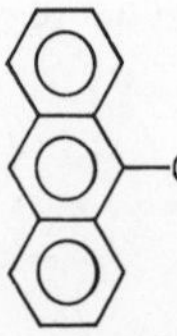 –$CH(OH)PO(OR)_2$.

R = Me. V.7. m. 199-201°.[183]

R = Et. V.7. m. 149-151°.[183]

R = i-Pr. V.7. m. 168.5-170°.[183]

R = i-Bu. V.7. m. 125-126.5°.[183]

2-$MeC_6H_4CPh=CHPO(OH)_2$. I.1. m. 154°.[241]

$Ph_2CCH_2CHPO(OR)_2$.

R = Me. V.14. m. 74-75°.[1582] V.14. IR, ^1H.[1583]

R = Et. V.14. m. 58.5°.[1582]

R = Pr. V.14. $b_{0.002}$ 163-164°.[1582]

R = Bu. V.14. $b_{0.005}$ 180-181°.[1582]

$H_2CCPh_2NH-N=CPO(OR)_2$.

R = Me. V.14. m. 109-110°, IR.[1583]

R = Et. V.14. m. 101-102°.[1582]

$H_2CCPh_2-N=N-CHPO(OMe)_2$. V.14. Dec. 88-89°.[1583]

$PhC(O)CH_2CHPhPO(OH)_2$. I.9. m. 116° (as 1-hydrate),[409] m. 117-118° (as 1-hydrate), m. 165-176° (as O-hydrate).[493]

$[PhC(O)CH_2CHPhPO(OR)OH]$.

R = Ph. III.2. m. 146°.[412]

R = $C_{10}H_{21}$-. III.2. m. 107-108°.[493]

R = $C_{12}H_{25}$-. III.2. m. 110-113°.[493]

R = $C_{14}H_{29}$-. III.2. m. 112-114°.[493]

R = $C_{16}H_{33}$-. III.2. m. 108-110°.[493]

R = $C_{18}H_{37}$-. III.2. m. 105-109°.[493]

R = $C_8H_{17}CH=CH-C_8H_{17}$-. III.2. m. 89-90°.[493]

$PhC(O)CH_2CHPhPO(OPh)_2$. V.1. m. 125°.[412] V.1. m. 116-117°.[418]

$CH_3C(O)CPh_2PO(OMe)_2$. V.14. m. 130° (benzene/hexane).[1581a]

$Ph_2C=C(-OMe)PO(OEt)_2$. V.14. ^1H, UV λ_{max} (hexane), 260 mμ, mass spect.[203]

$4-MeOC_6H_4-CPh=CHPO(OH)_2$. I.1. m. 145°.[239]

$OCPh_2CMePO(OMe)_2$. V.14. m. 87-88°.[1581a]

$(PhCH_2-)_2CHPO(OH)_2$. I.9. By heating of red P, $(PhCH_2)_2CO$ and conc. HJ at 180°, m. 142° (water), as aniline salt, m. 126°; as phenylhydrazine salt, m. 148-149°.[1281]

$(PhCH_2-)_2CHPO(OR)_2$.

R = Et. V.2. b_{20} 240°.[1281]

R = Ph. V.2. m. 120°.[1281]

$2-MeC_6H_4CHPhCH_2PO(OH)_2$. I.13. m. 160-161°.[241]

$PhCH_2CH_2CHPhPO(OH)_2$. I.13. m. 168-171° (cyclohexane).[493]

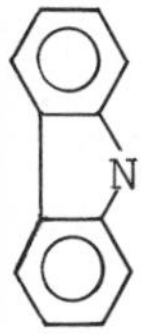
N-CH(-NMe_2)$PO(OEt)_2$. V.14. m. 98-101°, ^1H.[715a]

$(4-MeC_6H_4-)(3-MeC_6H_4NH-)CHPO(OEt)_2$. V.9. m. 102-103°.[1632]

$PhCH_2O_2C-C(-CH_2CH_2CN)_2PO(OEt)_2$. V.7. $b_{0.01}$ 160-162°.[540]

$(4-MeC_6H_4-)(4-MeC_6H_4S(O)_2NHN=)CPO(OEt)_2$. V.14. syn-form, m. 65-66°; anti-form, m. 100-101°.[883]

$(4-MeOC_6H_4-)(4-MeC_6H_4S(O)_2NHN=)CPO(OEt)_2$. V.14. anti-form, m. 132-133°.[883]

$4-MeOC_6H_4CHPhCH_2PO(OH)_2$. I.13. m. 102-103°.[239]

$(PhCH_2)_2C(OH)PO(OH)_2$. I.9. m. 181-182° (benzene).[415]

$(PhCH_2-)_2C(OH)PO(OMe)_2$. V.8. m. 126°.[40]
$PhCH_2CH_2CPh(-OH)PO(OH)_2$. I.9. m. 165-168° (benzene).[415]

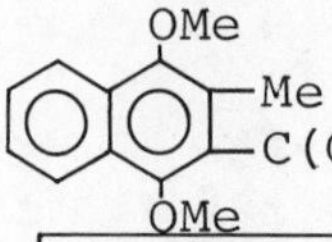
$-C(O)CH_2PO(OMe)_2$. V.14. ^{1}H[1932]

$EtO_2C-CHC(-CO_2Et)=N-NHCPhPO(OMe)_2$. V.14. m. 148.5-149°, I
^{1}H.[1878a]

$(4-Me_2NC_6H_4-)(4-O_2NC_6H_4NH-)CHPO(OPh)_2$. V.9. m. 183-184°.[2105]

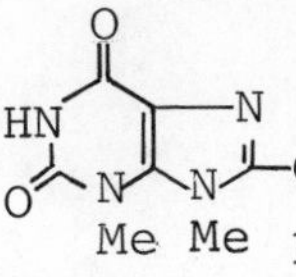
$CH-NHC_6H_4Me-3$. V.9. m. 159-160°.[1123]
$PO(OMe)_2$

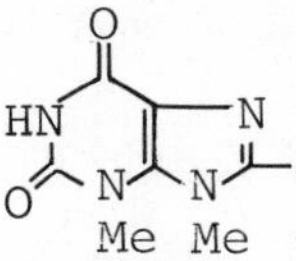
$CH-NHC_6H_4Me-4$.
$PO(OR)_2$

R = Me. V.9. m. 220-221°.[1123]
R = Et. V.9. m. 207-208°.[1123]

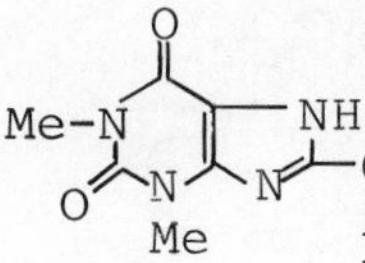
$CHNHC_6H_4Me-2$. V.9. m. 189-190°.[1123]
$PO(OMe)_2$

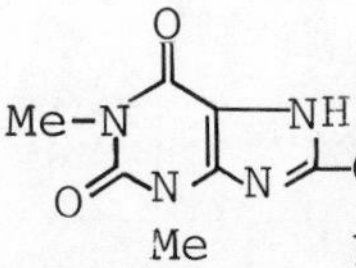
$CHNHC_6H_4Me-3$. V.9. m. 160-161°.[1123]
$PO(OMe)_2$

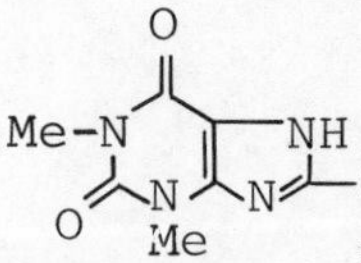
$CHNHC_6H_4Me-4$.
$PO(OR)_2$

R = Me. V.9. m. 200-201°.[1123]
R = Et. V.9. m. 194-195°.[1123]

$Ph_2P(O)CH(-NMe_2)PO(OEt)_2$. V.14. m. 153-154°. ^{1}H.[712] V.14. m. 150-153 .[714]

4-$Me_2NC_6H_4CH(-HNPh)PO(OPh)_2$. V.9. m. 125-126°.[2105]

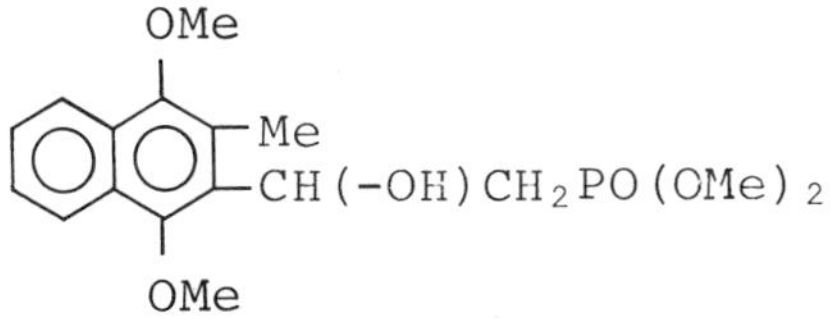

. V.14. m. 117°, ^{1}H.[1932]

$MeCO_2CMe=C(-CO_2Et)-CHPhPO(OMe)_2$. From $OCMe=C(-CO_2Et)CHPhP(OMe)_3$ and $[MeC(O)]_2O$. $b_{0.015}$ 172-173°[131]

$Me_3C-CMe=CHC[=NNHS(O)_2C_6H_4Me-4]PO(OMe)_2$. V.14. m. 118-118.5°.[1191]

3,5-$(Me_3C-)_2$-4-$HOC_6H_2CH_2PO(OR)_2$.

R = Me. V.5. m. 155-157°.[643]

R = Et. V.5. m. 119-120°.[643]

R = $ClCH_2CH_2$-. V.5. m. 113-117°.[1942]

R = i-Pr. V.5. m. 104-105°.[643]

R = sec.-Bu. V.5. m. 78-80°.[643]

R = C_6H_{13}-. V.5. $b_{0.7}$ 222-225°.[643]

R = $C_{12}H_{25}$-. V.5. m. 45-48°(acetone).[1943]

R = $C_{18}H_{37}$. V.5. m. 55-57°(acetone/hexane).[1943]

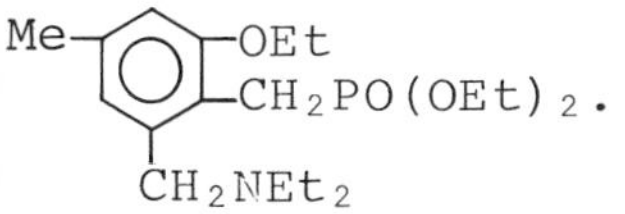

. V.9. $b_{0.07}$ 144-145°.[848]

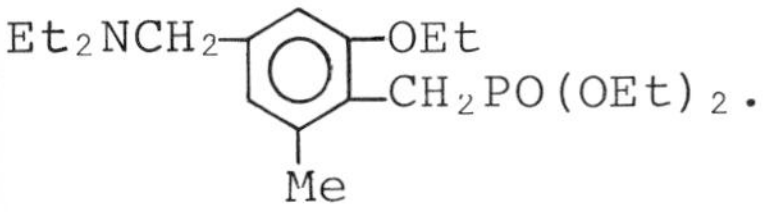

. V.9. $b_{0.02}$ 140-142°.[848]

$(BuO)_2OP-O-CHPhPO(OBu)_2$. V.8, V.14. $b_{0.1}$ 180-184°,[333] the structure as diphosphonic acid ester cited in the literature is not correct.[538]

$H_2C(CH_2)_{13}C(OH)PO(OMe)_2$. V.8. m. 106-107°.[229]

$C_{11}H_{23}O_2C-(CH_2)_3PO(OBu)_2$. V.5. $b_{0.25}$ 178-185°.[46]

$BuO_2C(CH_2)_{10}PO(OR)_2$.

R = Bu. V.7. n_D^{30} 1.4458.[1787] V.5. $b_{0.01}$ 193-194°, IR, n_D^{30} 1.4458, d_4^{30} 0.9573.[1786]

R = $C_4H_9CHEtCH_2$-. V.5. $b_{0.003}$ 170°, IR, n_D^{30} 1.4503, d_4^{30} 0.9316.[1786]

C_{16}

$[(PrO)_2OP-O-](4-BrC_6H_4-)CPrPO(OEt)_2$. V.14. $b_{0.035}$ 164-165°, n_D^{20} 1.5070, d_4^{20} 1.2944.[1590]

$[(PrO)_2OP-O-](2-ClC_6H_4-)CPrPO(OEt)_2$. V.14. $b_{0.04}$ 154-155°, n_D^{20} 1.5000, d_4^{20} 1.1980.[1590]

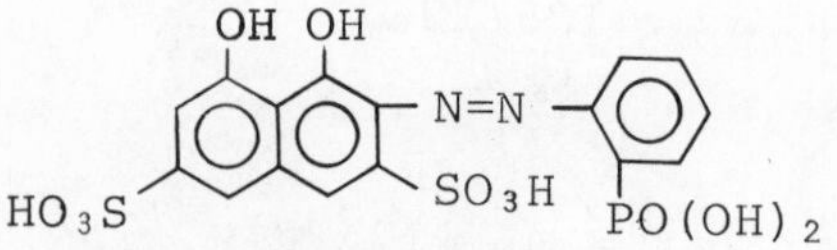

V.14. Reddish brown needles[1320]

HC=CPh-O-CPh=CPO(OH)$_2$. I.2. m. 196-197°.[1047] I.9.[687]

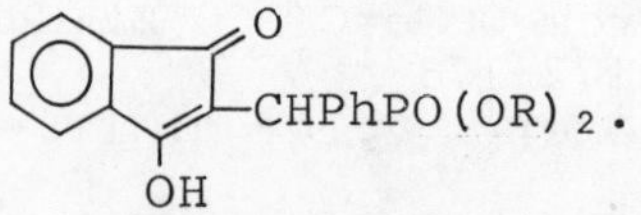

R = Me. V.7, V.14. Mono enol form m. 88-89°, diketo fo m. 110-112°.[157]

R = Et. V.7, V.14.[157] V.7. m. 90-92°, IR, ^{1}H.[1360]

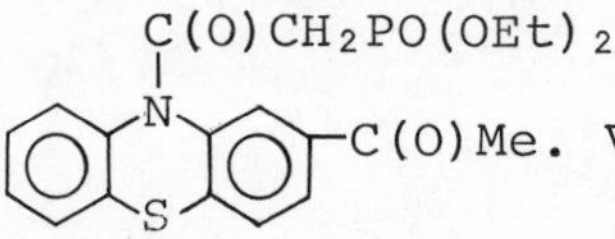

C(O)Me. V.5. m. 160-161°[1055]

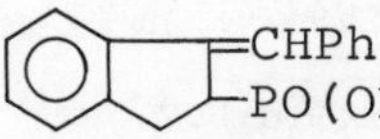

I.13. m. 188-189°.[238,1991a]

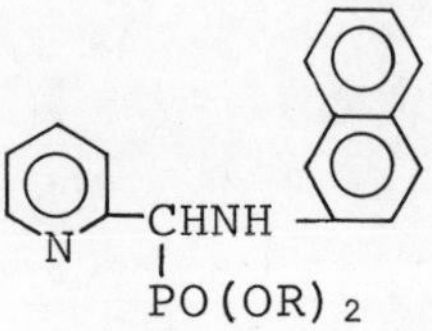

R = Me. V.9. m. 123°, as picrate m. 146°.[1121]

R = Et. V.9. m. 110°, as picrate m. 179°.[1121]

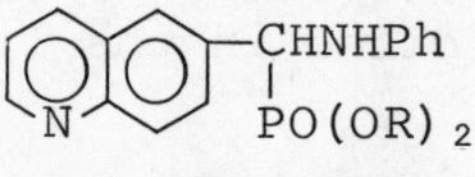

R = Me. V.9. m. 157-158°.[1122]

R = Et. V.9. m. 122-123°.[1122]

R = i-Pr. V.9. m. 154-155°[1122]

[PhC(O)-][PhC(O)CH$_2$-]CHPO(OH)$_2$. I.1, I.9. Dec. 183-185°.[49]

[PhC(O)-][PhC(O)CH$_2$-]CHPO(OR)$_2$.

R = Me. V.7. m. 121-122°, IR, ^{1}H.[1689]

R = Et. V.7. $b_{0.0001}$ 146-148°.[1042]

Ph$_2$C=C(-O$_2$CMe)PO(OEt)$_2$. V.14. IR, ^{1}H, mass spect.[203]

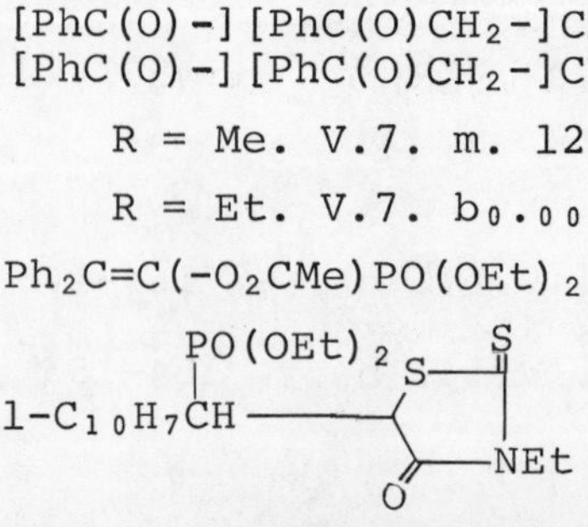

V.7. m. 124-125°.[1125a]

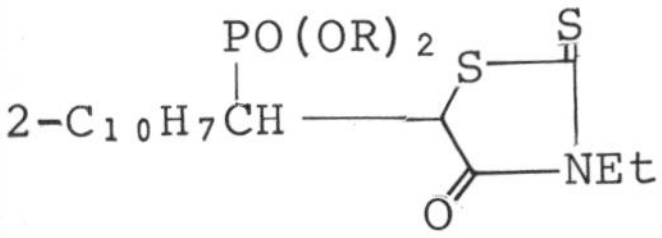

R = Me. V.7. m. 144-145°.[1125a]

R = Et. V.7. m. 99-100°.[1125a]

$H_2C-CPh_2-CMePO(OR)_2$.

R = Me. V.14. m. 71°.[1579]

R = Et. V.14. $b_{0.002}$ 158-159°.[1579]

R = Ph. V.2, V.14. m. 104-106°.[1579]

$H_2C-CPh_2CHCH_2PO(OEt)_2$. V.14. $b_{0.0007}$ 157-158°.[1582]

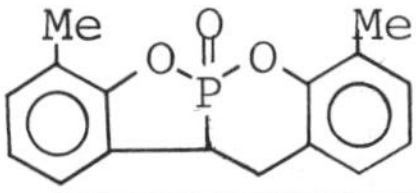

V.9/V.14. m. 225°.[833]

$H_2NCH-CPh=C(-NH_2)-O-CPhPO(OR)_2$.

R = Me. V.7. m. 167-168° (dec.).[1404]

R = Et. V.7. m. 188-189°.[1404]

$H_2C-CPh_2-N=N-CMePO(OPh)_2$. V.14. m. 107-108°.[1579]

$[2-O_2NC_6H_4CH(-OEt)-O-](2-O_2NC_6H_4-)CHPO(OEt)_2$. V.8. m. 144-145°.[1059]

$[4-O_2NC_6H_4CH(-OEt)-O-](4-O_2NC_6H_4-)CHPO(OEt)_2$. V.8. m. 139-140°.[1059]

$(4-MeOC_6H_4-)[PhC(O)CH_2-]CHPO(OH)_2$. I.1, I.9. m. 189°; as oxime, m. 156°.[409]

$PhC(O)NEtCHPhPO(OH)_2$. I.2. m. 163-165°.[60]

$PhC(O)NEtCHPhPO(OEt)_2$. V.9. n_D^{25} 1.5415.[60]

$[3,4-(MeO-)_2C_6H_3-](4-MeC_6H_4NH-)CHPO(OPh)_2$. V.9. m. 142-143° (alcohol/benzene).[1027]

$MeC(O)CH_2-[CH(-O_2CMe)]_4-CH(OH)PO(OMe)_2$. V.8. m. 172-173°.[59]

$3,5-(t-Bu-)_2-4-HOC_6H_2CHMePO(OR)_2$.

R = $C_{14}H_{29}-$. V.5. m. 40-42° (acetone).[1943]

R = $C_{16}H_{33}-$. V.5. m. 44-46.5° (acetone).[1943]

R = $C_{18}H_{37}-$. V.5. m. 72-74° (acetone).[1943]

$(PrO)_2OP-O-CPhPrPO(OPr)_2$. V.14. b_2 210-212°, n_D^{20} 1.4870. d_4^{20} 1.1095.[1590]

$BuO_2CCH_2CH(-CO_2Bu)CH(-CO_2Et)PO(OEt)_2$. V.7. $b_{2.5}$ 195-196°.[1644]

$i-BuO_2CCH_2CH(-CO_2Bu-i)CH(-CO_2Et)PO(OEt)_2$. V.7. $b_{2.5}$ 190°.[1644]

$(cycl.-C_6H_{11})_2NC(O)(CH_2)_3PO(OPr)_2$. V.5. $b_{0.01}$ 182-184°.[515]

$cycl.-C_6H_{11}CH_2CH_2CH(-C_6H_{11}-cycl.)CH_2PO(OH)_2$. I.8. m. 100.5-101.5°.[686]

$C_{13}H_{27}C(O)CH_2CH_2PO(OR)_2$.

R = Me. V.7. n_D^{30} 1.4477, d_4^{30} 1.0020.[1787]

R = Et. V.7. n_D^{30} 1.4450, d_4^{30} 0.9773.[1787]

R = Bu. V.7. n_D^{30} 1.4462, d_4^{30} 0.9603.[1787]

R = $C_4H_9CHEtCH_2$-. V.7. n_D^{30} 1.4517, d_4^{30} 0.9331.[1787]

$C_{13}H_{27}CO_2CH_2CH_2PO(OEt)_2$. V.5. $b_{0.1}$ 157-159°, n_D^{30} 1.4440. d_4^{30} 0.9703.[46]

$C_{12}H_{25}CH(-CO_2Et)PO(OEt)_2$. V.5. $b_{0.1}$ 173-176°, n_D^{30} 1.4420, d_4^{30} 0.9658.[44]

$C_{10}H_{21}CH(-CO_2Bu)PO(OR)_2$.

R = Et. V.5. $b_{0.3}$ 186°.[44]

R = Bu. V.5. $b_{0.6}$ 182°.[44]

$(C_6H_{13}-)_2NC(O)(CH_2)_3PO(OPr\text{-}i)_2$. V.5. $b_{0.01}$ 170-176°.[515]

$C_{16}H_{33}PO(OH)_2$. I.2. m. 94.5-95.5°.[1000,1002] I.1. m. 93°.[737]

$C_{16}H_{33}PO(OR)_2$.

R = Bu. V.6. b_2 226-228°, n_D^{25} 1.4481, d_4^{25} 0.9090.[1000,1002]

R = Ph. V.5. m. 49.5-51°(pentane).[1080]

$C_{14}H_{29}OCH_2CH_2PO(OEt)_2$. V.7. $b_{0.2}$ 168-172°, IR, ^{1}H.[1402]

C_{17}

PhCH=CBr-C(O)CH$_2$CHPhPO(OH)$_2$. I.13. m. 130-132°.[411]

HC=C(-C_6H_4Br-4)-O-C(-C_6H_4Br-4)=CHCHPO(OH)$_2$. I.2. m. 181°.[2057]

PhCHBrCHBrC(O)CH$_2$CHPhPO(OH)$_2$. I.13. m. 180-182°(as 2,5-hydrate).[411]

MeC=N-N(-C_6H_4Cl-4)-C(OH)=CCHPhPO(OMe)$_2$. V.7. m. 149-150°.[1124]

HC=CPh-O-CPh=CHCHPO(OH)$_2$. I.2.[1049]

HC=CPh-O-CPh=CHCHPO(OEt)$_2$. V.5.[1049]

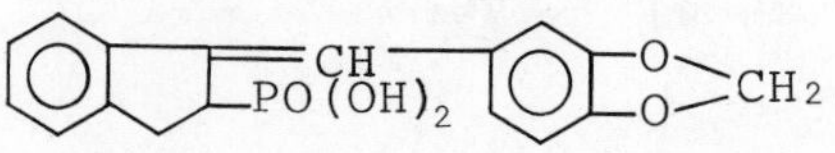

. I.13. m. 194°.[238,1991a]

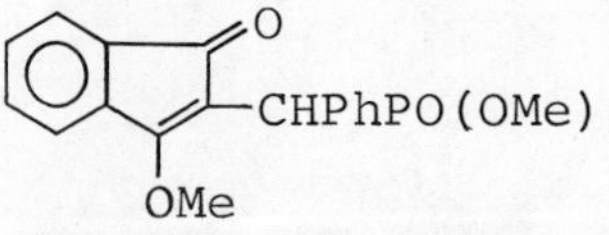

. V.7. V.14.[156]

1-$C_{10}H_7$NHCHPhPO(OR)$_2$.

R = Me. V.9. m. 138-139°.[1125]

R = Et. V.9. m. 105-106°.[1125]

2-$C_{10}H_7$NHCHPhPO(OEt)$_2$. V.9. m. 107-108°.[1547]

PhC(O)CH$_2$CHPhCH(-CN)PO(OEt)$_2$. V.7.[540]

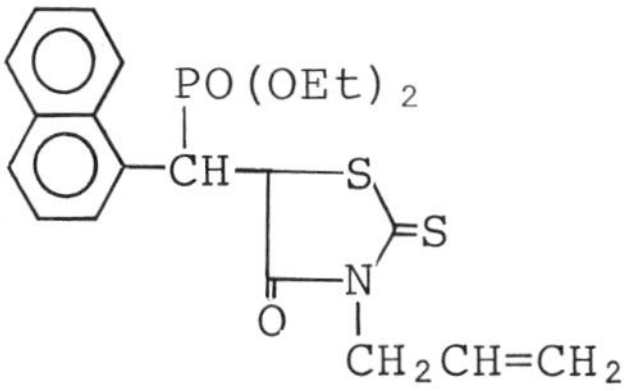

V.7. m. 96-97°.[1125a]

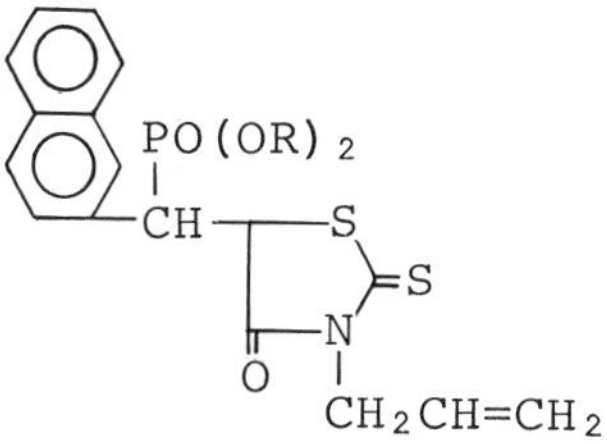

R = Me. V.7. m. 125-126°.[1125a]

R = Et. V.7. m. 97-98°.[1125a]

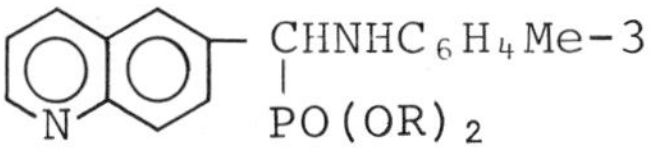

R = Me. V.9. m. 142-143°.[1122]

R = Et. V.9. m. 117-118°.[1122]

R = i-Pr. V.9. m. 125-126°.[1122]

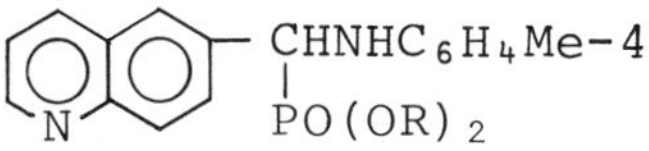

R = Me. V.9. m. 167-168°.[1122]

R = Et. V.9. m. 153-154°.[1122]

R = i-Pr. V.9. m. 154-155°.[1122]

O=C-NPh-NHCMe=C-CHPhPO(OMe)$_2$. V.7. $b_{0.002}$ 247-248°[134]

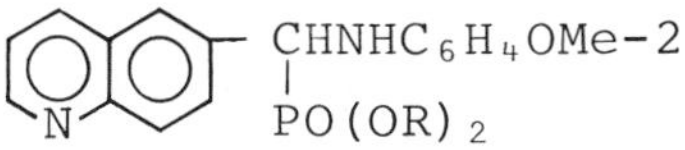

R = Me. V.9. m. 150-151°.[1122]

R = i-Pr. V.9. m. 149-150°.[1122]

R = Me. V.9. m. 176-177°.[1122]

R = Et. V.9. m. 156-157°.[1122]

R = i-Pr. V.9. m. 165-166°.[1122]

MeC=N-NPh-C(OH)=C-CHPhPO(OMe)$_2$. V.7. m. 115-116°.[1124]
[PhC(O)CH$_2$-](PhCH=CH-)CHPO(OH)$_2$. I.9. m. 159-161°.[411]
PhCH=CH-C(O)CH$_2$CHPhPO(OH)$_2$. I.9. m. 108° (as 1.25-hydrate).[409]

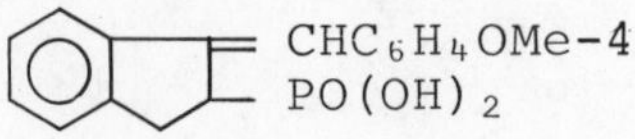
I.13. m. 192°.[238,1991a]

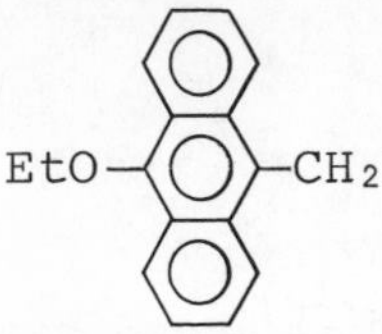
V.7. m. 130-131.5°.[176]

[4-(MeO$_2$C-)C$_6$H$_4$-]$_2$C(OH)PO(OMe)$_2$. V.8. m. 109° (heptane).[532]
PhCH$_2$O$_2$CNHCH$_2$C(O)NHCHPhPO(OEt)$_2$. V.14. m. 81-82°.[1534]
(PhCH$_2$CH$_2$-)$_2$C(OH)PO(OH)$_2$. I.9. m. 173-174°.[415]
4-MeC$_6$H$_4$NHCH[-C$_6$H$_4$(-CHMe$_2$)-4]PO(OPh)$_2$. V.9. m. 151-152° (alcohol/benzene).[1027]
PhO$_2$C(CH$_2$)$_{10}$PO(OBu)$_2$. V.5. n_D^{30} 1.4734, d_4^{30} 1.0080. IR.[1786]
H$_2$C(CH$_2$)$_{15}$C(OH)PO(OMe)$_2$. V.8. m. 85-86°.[229]
C$_6$H$_{13}$O$_2$C(CH$_2$)$_{10}$PO(OC$_6$H$_{13}$)$_2$. V.8. n_D^{30} 1.4492, d_4^{30} 0.9364.[1786]

C$_{18}$

MeC=N-N(-C$_6$H$_4$Cl-4)-C(OH)=CCH(-C$_6$H$_4$OMe-4)PO(OMe)$_2$. V.7. m. 151-152°.[1124]
1-C$_{10}$H$_7$-CPh=CHPO(OH)$_2$. I.1. m. 188°.[241]
2-C$_{10}$H$_7$-CPh=CHPO(OH)$_2$. I.1. m. 220°.[241]

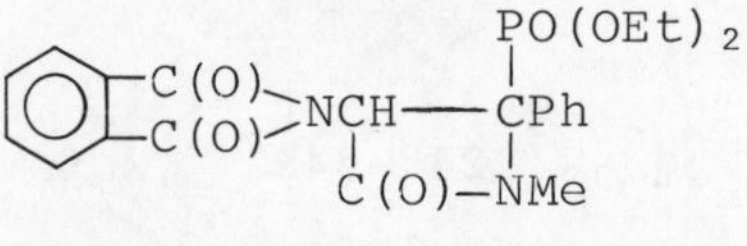
V.14. m. 174-174.5°.[145]

1,5-(PhSO$_2$NH-)$_2$C$_6$H$_3$PO(OMe)$_2$. V.7/V.14. m. 80°, IR, ^{1}H.[1095]

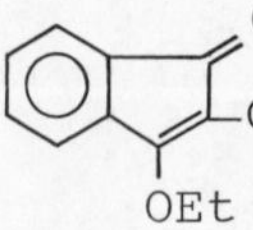
CHPhPO(OEt)$_2$. V.7. b$_{0.02}$ 220-222°, m. 86-89°[156]

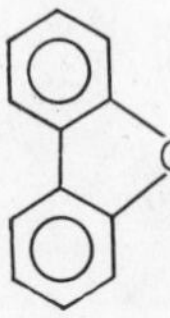
C=C(-CO$_2$Et)CH$_2$PO(OEt)OH. From 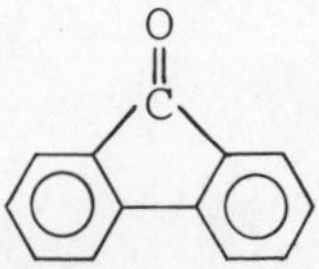 and EtO$_2$CCH$_2$CH$_2$PO(OEt)$_2$, IR, ^{1}H.[1196]

1-$C_{10}H_7NH-CH(-C_6H_4OMe-4)PO(OR)_2$.

R = Me. V.9. m. 117-118°.[1125]

R = Et. V.9. m. 110-111°.[1125]

2-$C_{10}H_7NH-CH(-C_6H_4OMe-4)PO(OR)_2$.

R = Me. V.9. m. 111-112°.[1125]

R = Et. V.9. m. 112-113°.[1125]

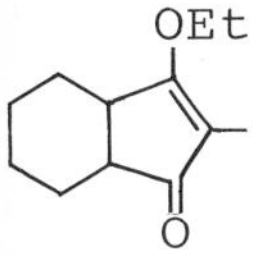
$-CHPhPO(OEt)_2$. V.14. m. 89-90°, IR.[1360]

$-CH(-HNC_6H_4OEt-4)PO(OR)_2$.

R = Me. V.9. m. 141-142°[1122]

R = Et. V.9. m. 145-146°[1122]

O=C-NPh-NMe-CMe=C-CHPhPO(OMe)$_2$. V.7. m. 133-134°.[134]

MeC=N-NPh-C(-OMe)=C-CHPhPO(OMe)$_2$. V.7. m. 58-64° (not pure).[134]

MeC=N-N(-C_6H_4Me-4)-C(OH)=C-CHPhPO(OMe)$_2$. V.7. m. 147-148°.[1124]

MeC=N-NPh-C(-OH)=CCH(-C_6H_4OMe-4)PO(OMe)$_2$. V.7. m. 139-140°.[1125]

(EtO_2C-)[PhC(O)-]CHCHPhPO(OMe)$_2$. By hydrolysis of OCPh=C(-CO_2Et)-CHPhP(OMe)$_3$.[130]

$Ph_2C(-CO_2Et)CH_2PO(OH)$. I.2. m. 179-181°, ^{1}H.[1196]

$Ph_2C=C(-CO_2Et)CH_2PO(OEt)OH$. From $Ph_2C=O$ and $EtO_2CCH_2CH_2PO(OEt)_2$, m. 141-142°, IR, ^{1}H.[1196]

[2-$O_2NC_6H_4$-CH(-OBu)-O-](2-$O_2NC_6H_4$-)CHPO(OBu)$_2$. V.8. m. 108°.[1059]

4-$Et_2NC_6H_4CH(-HNC_6H_4Me-4)PO(OPh)_2$. V.9. m. 113-114° (alcohol/benzene).[1027]

$BuO_2C-CH_2CH(-CO_2Bu)-CH(-CO_2Bu)PO(OEt)_2$. V.7, V.14. $b_{5.5}$ 222°.[1644]

$BuO_2C-CH_2CH(-CO_2Bu)-CH(-CO_2Bu-i)PO(OEt)_2$. V.7, V.14. b_3 199-200°.[1644]

i-$BuO_2C-CH_2CH(-CO_2Bu-i)-CH(-CO_2Bu)PO(OEt)_2$. V.7, V.14. b_5 220-222°.[1644]

i-$BuO_2C-CH_2-CH(-CO_2Bu-i)-CH(-CO_2Bu-i)PO(OEt)_2$. V.7, V.14. b_8 220°.[1644]

$C_{15}H_{31}CO_2CH_2CH_2PO(OEt)_2$. V.5. $b_{0.1}$ 169-172°, n_D^{30} 1.4452, d_4^{30} 0.9607.[46]

$MeO_2C(CH_2)_7CH(-C_8H_{17})PO(OMe)_2$. V.7. n_D^{30} 1.4531, d_4^{30} 0.9889.[1787]

$EtO_2CCH(-C_{14}H_{29})PO(OEt)_2$. V.5. $b_{0.2}$ 185-187°, n_D^{30} 1.4432, d_4^{30} 0.9652.[44]

$C_{18}H_{37}PO(OH)_2$. I.2. m. 98.5-99° (ligroine).[1000,1002]

$C_{18}H_{37}PO(OBu)OH$. III.3. m. 58-58.5°.[1770]
$C_{18}H_{37}PO(OBu)_2$. V.6. b_2 248-250°, n_D^{25} 1.4499, d_4^{25} 0.9037.[1000,1002]
$\overline{OCHMeCH_2CMe_2O}SiMeOCH(-C_{10}H_{21})PO(OEt)_2$. V.8/V.14. $b_{0.35-0.4}$ 178-179°.[258]

C19

$4-BrC_6H_4CPh_2PO(OH)_2$. I.2. m. 297°.[300]
$[(BuO)_2OP-O-](4-BrC_6H_4-)CBu-PO(OEt)_2$. V.14. $b_{0.04}$ 179-180°, n_D^{20} 1.5070, d_4^{20} 1.2675.[1560]
$4-ClC_6H_4CPh_2PO(OH)_2$. I.2. m. 273°.[300]
$(3-Cl-4-HOC_6H_3-)_2CPhPO(OR)_2$.

R = Me. V.14. m. 259-260°.[817]

R = Et. V.14. m. 209°(dec.).[814]

$[(BuO)_2OP-O-](2-ClC_6H_4-)CBu-PO(OEt)_2$. V.14. $b_{0.035}$ 161-163°, n_D^{20} 1.4930, d_4^{20} 1.1522.[1590]
$Ph_3CPO(OH)_2$. I.1, I.2. m. 279°.[104] I.2. m. 279°.[102a] I.1. m. 237°.[298] I.6. m. 283-283.5°.[759,760] I.1.[300]
$Ph_3CPO(OEt)OH$. III.3. m. 259°(benzene).[759,760]
$Ph_3CPO(OR)_2$.

R = Me. V.5, V.6. m. 157-158°.[102a,104] V.1. m. 154.5-155.5°.[759,760] V.5. m. 157°.[477]

R = Et. V.5. m. 121-122°(acetone).[477] V.5, V.6. m. 120-121°.[102a,104] V.1, V.2. m. 121-122°.[759,760] V.[111]

R = Pr. V.5. m. 109-110°.[102a,104]

R = $H_2C{=}CH-CH_2-$. V.5. m. 85°.[477]

R = i-Bu. V.5, V.6. m. 96-96.5°.[102a,104]

R = $2-EtC_6H_4-$. V.5. m. 119.5°.[477]

$Ph_3CPO(OR)(OCH_2CH_2Br)$.

R = Me. V.5. m. 153-155°.[122]

R = Et. V.5. m. 99-101°.[122]

$(Ph_3C-)O\overline{PO-R-O}$.

R = $-(CH_2)_3-$. V.5. m. 228°.[121]

R = $-CHMeCH_2CH_2-$. V.5. m. 192-193°.[122]

R = $-CH_2CMe(-CH_2Cl)CH_2-$. V.5. m. 201°.[2052]

R = 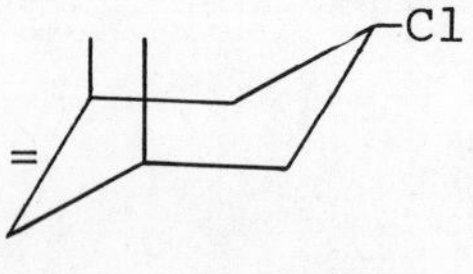

V.5.[2052]

$3-HOC_6H_4CPh_2PO(OH)_2$. I.1, I.2. m. 248°(cryst. as 2-hydrate).[300]

$(4\text{-}HOC_6H_4\text{-})_2CPhPO(OR)_2$.

R = Me. V.14. m. 270°.[817]

R = Et. V.14. m. 226°(dec.).[814]

$Ph_3P{=}N\text{-}N{=}CHPO(OR)_2$.

R = Me. V.14. m. 132-133°, IR.[1878a]

R = Et. V.14. Dec. 114-115°.[1732]

$Ph_2P(O)\text{-}CHPhPO(OEt)_2$. V.14. m. 161-162°.[267]
$Ph_2P(O)CPh(OH)PO(OMe)_2$. V.14. m. 120-121°.[455]
$Ph_2P(O)OCHPhPO(OMe)_2$. V.14. $b_{0.3}$ 192-194°.[455]
$Ph_3SiOCH_2PO(OSiPh_3)_2$. V.14. m. 241-242°.[1438]
$MeC{=}N\text{-}N(\text{-}C_6H_4Me\text{-}4)\text{-}C(\text{-}OH){=}C\text{-}CH(\text{-}C_6H_4OMe\text{-}4)PO(OMe)_2$. V.7. m. 129-130°.[1124]

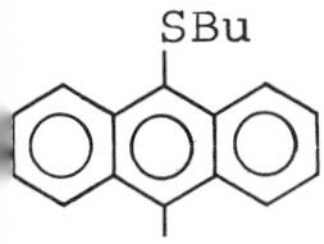

$CH(OH)PO(OEt)_2$. V.8. m. 80-82°.[183]
$PhC(O)CH_2CHPhCH(\text{-}CO_2Et)PO(OEt)_2$. V.14. m. 52-54°, IR.[308] V.7.[540]
$PhCH_2O_2CNHCH_2C(O)\text{-}NHCH_2C(O)\text{-}NHCHPhPO(OEt)_2$. V.14. m. 73°.[1534]

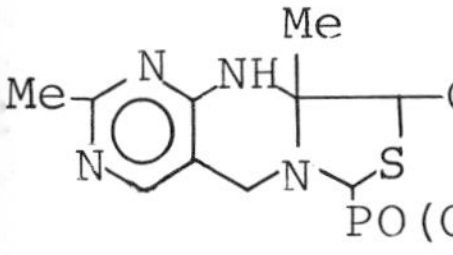

$\text{-}CH_2CH_2O_2CPh$. V.14. m. 107°, IR.[1979]

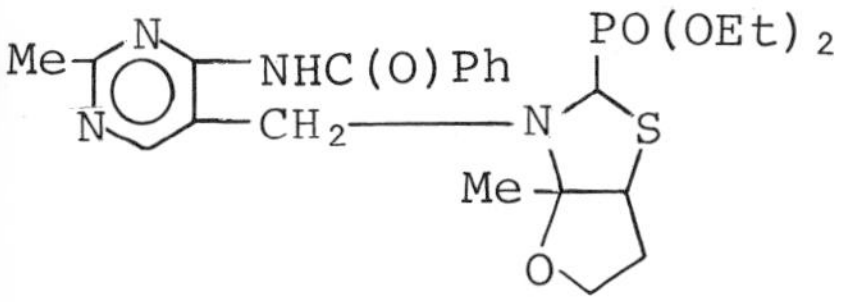

. V.14. m. 114°, IR.[1979]

$(EtO_2CCH_2CH_2\text{-})_2(PhCH_2O_2C\text{-})CPO(OEt)_2$. V.7. $b_{0.01}$ 212-214°[540]
$(BuO)_2OP\text{-}O\text{-}CBuPhPO(OBu)_2$. V.14. $b_{0.05}$ 196-197°, n_D^{20} 1.4870, d_4^{20} 1.0734.[1590]
$C_4H_9CHEtCH_2O_2C(CH_2)_{10}PO(OR)_2$.

R = Et. V.5, V.7. n_D^{30} 1.4472, d_4^{30} 0.9611.[1786,1787]

R = Bu. V.5. n_D^{30} 1.4490, d_4^{30} 0.9426.[1786]

R = $C_4H_9CHEtCH_2\text{-}$. V.5, V.7. n_D^{30} 1.4530, d_4^{30} 0.9257.[1786,1787]

$EtO_2C(CH_2)_7CH(\text{-}C_8H_{17})PO(OR)_2$.

R = Et. V.7. n_D^{30} 1.4492, d_4^{30} 0.9596.[1787]

R = Bu. V.7. n_D^{30} 1.4504, d_4^{30} 0.9435.[1787]

$C_{17}H_{35}C(O)NHCH_2PO(OH)_2$. I.2. Sintering at 108°.[506]

C_{20}

$OCPh_2C(-C_6H_4Br-4)PO(OR)_2$.

R = Me. V.14. m. 156-157°.[1581a]

R = Et. V.14. m. 162-163°.[1581a]

$MeC{=}N{-}N(-C_6H_4Cl{-}4){-}C(OH){=}C{-}CH[-C_6H_4(-CHMe_2){-}4]PO(OMe)_2$. V.7. m. 150-151°.[1124]

$OCPh_2C(-C_6H_4Cl-4)PO(OR)_2$.

R = Me. V.14. m. 159-160°.[1581a]

R = Et. V.14. m. 160 161°.[1581a]

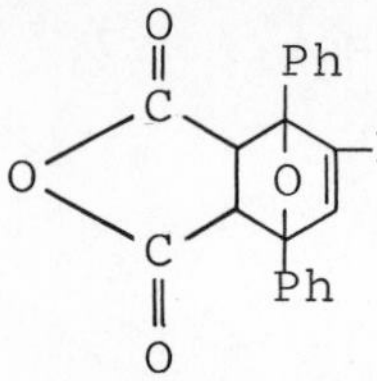

$-PO(OH)_2$. V.14. m. 132-134°.[687]

$4{-}PhC_6H_4CPh{=}CHPO(OH)_2$. I.1. m. 201°.[239]

$Ph_2C{=}CPhPO(OEt)_2$. V.14. m. 183-185°.[267]

$OCPh_2CPhPO(OR)_2$.

R = Me. V.14. m. 129-130°.[1581a]

R = Et. V.14. m. 97-98°.[1581a]

$4{-}PhC_6H_4CHPhCH_2PO(OH)_2$. I.13. m. 236°.[239]

$4{-}MeC_6H_4CPh_2PO(OH)_2$. I.2. m. 254° (benzene).[300]

$Ph_2CHCHPhPO(OH)_2$. I.2. m. 186-188°.[267]

$Ph_2CHCHPhPO(OEt)_2$. V.14. m. 181-183°.[267]

$3{-}MeOC_6H_4CPh_2PO(OH)_2$. I.13. m. 197° ($MeCO_2H$).[300]

$4{-}MeOC_6H_4CPh_2PO(OH)_2$. I.2. m. 210°.[300] I.2. m. 212°.[110]

$4{-}MeOC_6H_4CPh_2PO(OEt)_2$. V.6. m. 118.5-119°.[110]

$O{=}C{-}C(=CHPh)(CH_2)_3CHCHPhPO(OR)_2$. By hydrolysis of

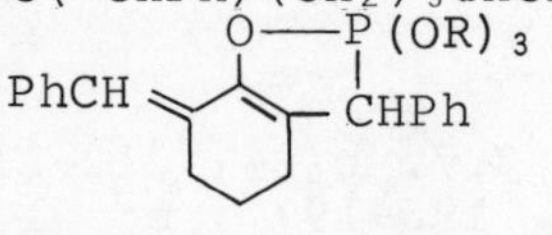

R = Me.[177]

R = Et.[177]

$(1{-}C_{10}H_7NH{-})[4{-}(Me_2CH{-})C_6H_4{-}]CHPO(OR)_2$.

R = Me. V.9. m. 114-115°.[1125]

R = Et. V.9. m. 101-102°.[1125]

$(2{-}C_{10}H_7NH{-})[4{-}(Me_2CH{-})C_6H_4{-}]CHPO(OR)_2$.

R = Me. V.9. m. 118-119°.[1125]

R = Et. V.9. m. 110-111°.[1125]

$MeC{=}N{-}NPh{-}C(OH){=}C{-}C[-C_6H_4(-CHMe_2){-}4]PO(OMe)_2$. V.7. m. 141-

142°.[1124]
MeO_2CCH-NH-CPh-N-CH(-CO_2Me)CPhPO(OEt)$_2$. V.7/V.14. m. 124-125°.[1404]
$MeCO_2CPh$=C(-CO_2Et)-CHPhPO(OMe)$_2$. V.14.[130]
i-AmO_2CCH_2CH(-CO_2Am-i)-CH(-CO_2Bu)PO(OEt)$_2$. V.7, V.14. b_1 232°.[1644]
i-AmO_2CCH_2CH(-CO_2Am-i)-CH(-CO_2Bu-i)PO(OEt)$_2$. V.7, V.14. b_4 204-205°.[1644]
$C_8H_{17}CH$=CH$(CH_2)_7CO_2CH_2CH_2PO(OEt)_2$. V.5. $b_{0.1}$ 181°, n_D^{20} 1.4542, d_4^{20} 0.9680.[46]
$C_8H_{17}O_2CCH_2CH(-CO_2C_8H_{17})PO(OEt)_2$. V.7. b_2 209°.[1644]
$C_{17}H_{35}C(O)CH_2CH_2PO(OR)_2$.

R = Me. V.7. n_D^{50} 1.4434, d_4^{50} 0.9585.[1787]

R = Et. V.7. m. 45°, n_D^{50} 1.4410, d_4^{50} 0.9405.[1787]

R = Bu. V.7. m. 47°.[1787]

R = $C_4H_9CHEtCH_2$-. V.7. n_D^{30} 1.4530, d_4^{30} 0.9215.[1787]

$C_{17}H_{35}CO_2CH_2CH_2PO(OEt)_2$. V.5. $b_{0.1}$ 185-191°, n_D^{50} 1.4411, d_4^{50} 0.9391.[46]
$EtO_2CCH(-C_{16}H_{33})PO(OEt)_2$. V.5. $b_{0.2}$ 185-187°, n_D^{30} 1.4452, d_4^{30} 0.9497.[44]
$(C_{10}H_{21}-)(C_8H_{17}-)CHCH_2PO(OH)_2$. I.8. m. 94-95°.[686]

$\underline{C_{21}}$

MeC=N-N(-C_6H_4Cl-4)C(OH)=CCH(-$C_{10}H_7$-1)PO(OR)$_2$.

R = Me. V.7. m. 149-150°.[1126]

R = Et. V.7. m. 135-136°.[1126]

MeC=N-N(-C_6H_4Cl-4)C(OH)=CCH(-$C_{10}H_7$-2)PO(OMe)$_2$. V.7. m. 134-135°.[1126]
4-ClC_6H_4NHCHPh-CH(-C_6H_4Me-2)PO(OEt)$_2$. V.14. m. 132-133°. IR.[1048]
4-ClC_6H_4NHCHPhCH(-C_6H_4Me-4)PO(OR)$_2$.

R = Et. V.14. m. 176-177°(172-174°).[1048]

R = i-Pr. V.14. m. 159-161°.[1048]

PhNHCH(-C_6H_4Cl-4)CH(-C_6H_4Me-4)PO(OEt)$_2$. V.14. m. 171-173°.[1048]

(4-PhC_6H_4-)(4-MeC_6H_4-)C=CHPO(OH)$_2$. I.1. m. 201°(cyclohexanone).[239]
MeC=N-NPh-C(OH)=CCH(-$C_{10}H_7$-1)PO(OR)$_2$.

R = Me. V.7. m. 127-128°.[1126]

R = Et. V.7. m. 123-124°.[1126]

MeC=N-NPh-C(OH)=CCH(-$C_{10}H_7$-2)PO(OR)$_2$.

R = Me. V.7. m. 127-128°.[1126]

R = Et. V.7. m. 105-106°.[1126]

OCPh_2C(-C_6H_4Me-4)PO(OR)$_2$.

R = Me. V.14. m. 139-140°.[1581a]

R = Et. V.14. m. 161°.[1581a]

$(4-MeOC_6H_4-)_2CPhPO(OH)_2$. I.2. m. 214°.[110]
$(4-MeOC_6H_4-)_2CPhPO(OEt)_2$. V.6.[110]
$PhNHCHPhCH(-C_6H_4Me-2)PO(OEt)_2$. V.14. m. 154-156°, IR.[1048]
$PhNHCHPhCH(-C_6H_4Me-4)PO(OR)_2$.

R = Et. V.14. m. 138-140°(135-137°).[1048]

R = i-Pr. V.14. m. 176-178°.[1048]

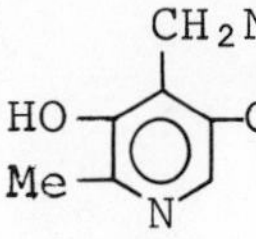
$-CH_2PO(OPr-i)_2$. V.14. m. 174-175.5°.[232]

$Me\overline{C=N-N(-C_6H_4Me-4)-C(OH)=C}CH[-C_6H_4(-CHMe_2)-4]PO(OMe)_2$. V.7
m. 130-131°.[1124]
$C_{17}H_{35}CHMeNHC(O)CH_2PO(OEt)_2$. V.5. m. 67-69°.[1108]

$\underline{C_{22}}$

$Me\overline{C=N-N(-C_6H_4Me-4)-C(OH)=C}-CH(-C_{10}H_7-1)PO(OR)_2$.

R = Me. V.7. m. 152-153°.[1126]

R = Et. V.7. m. 155-156°.[1126]

$Me\overline{C=N-N(-C_6H_4Me-4)-C(OH)=C}-CH(-C_{10}H_7-2)PO(OR)_2$.

R = Me. V.7. m. 138-139°.[1126]

R = Et. V.7. m. 114-115°.[1126]

$H_2\overline{C(CH_2)_2C(=CHPh)-C(-O_2CMe)=C}CHPhPO(OR)_2$. By hydrolysis of

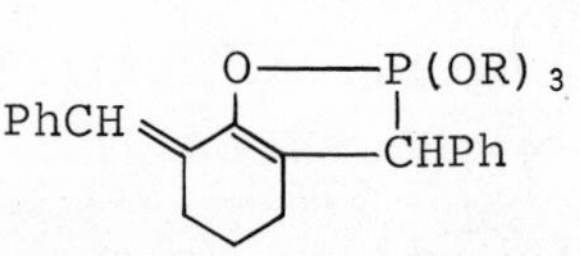

R = Me.[177]

R = Et.[177]

$(4-MeOC_6H_4-)_3CPO(OEt)_2$. V.6. m. 139-140.5°.[110]
$3-MeC_6H_4NHCHPhCH(-C_6H_4Me-2)PO(OEt)_2$. V.14. m. 135-137°.[104
$2-MeC_6H_4NHCHPhCH(-C_6H_4Me-4)PO(OEt)_2$. V.14. m. 155-157°.[104
$3-MeC_6H_4NHCHPhCH(-C_6H_4Me-4)PO(OEt)_2$. V.14. m. 142-144°.[104
$4-MeC_6H_4NHCHPhCH(-C_6H_4Me-4)PO(OEt)_2$. V.14. m. 148-150°, IR
[1048]

$CH_3(CH_2)_{7(8)}CH(CH_2)_{8(7)}CO_2Bu$
$\quad\quad|$
$\quad PO(OR)_2$ (Mixture of the isomers.)

R = Et. V.7. n_D^{30} 1.4481, d_4^{30} 0.9473.[1787]

R = Bu. V.7. n_D^{30} 1.4302, d_4^{30} 0.9372.[1787]

$C_{18}H_{37}OCH_2CH(OMe)CH_2PO(OH)_2$. I.2. m. 59.5-60.5°.[1770]

C_{23}

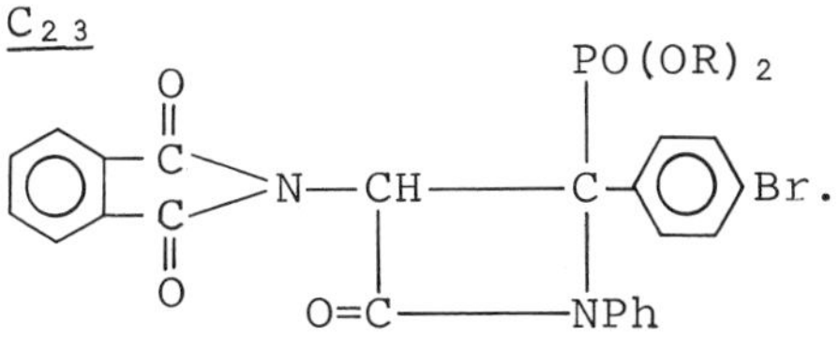

R = Me. V.14. m. 222.5-224°.[1453]

R = Et. V.14. m. 223-224°.[1453]

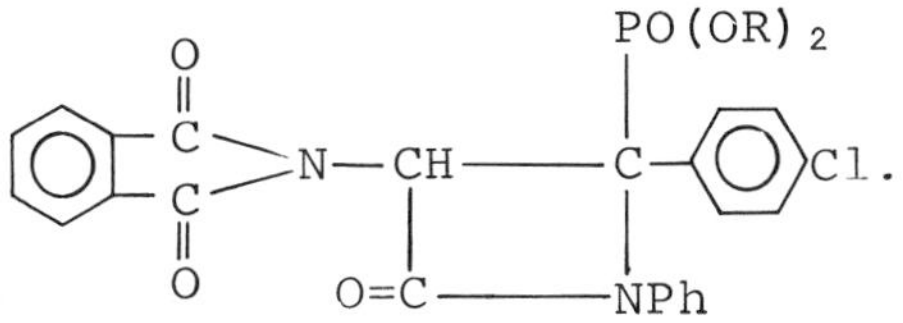

R = Me. V.14. m. 225.5-229°.[1453]

R = Et. V.14. m. 223-225°.[1453]

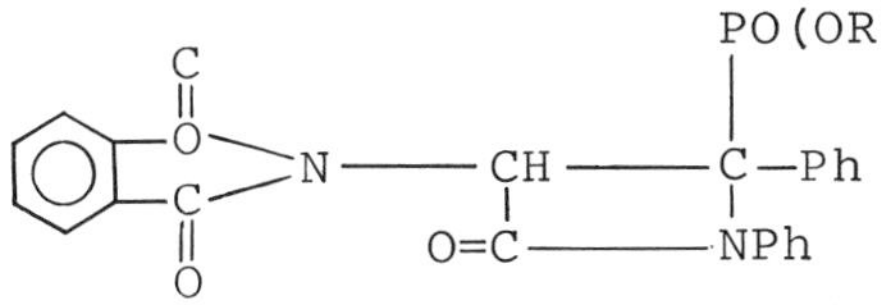

R = Me. V.14. m. 219.5-221°.[1453]

R = Et. V.14. m. 203-205°.[1453]

1-$C_{10}H_7CPh_2PO(OH)_2$. I.2. m. 256°.[300]
2-$C_{10}H_7CPh_2PO(OH)_2$. I.2. m. 247.5°.[300]
$PhCH_2O$-CPh=C[-$CH_2C(O)Ph$]$PO(OMe)_2$. V.7. m. 122-123°, IR. ^1H.[1689]

$C_{12}H_{25}O_2C(CH_2)_{10}PO(OR)_2$.

R = Bu. V.5. n_D^{30} 1.4500, d_4^{30} 0.9298, IR.[1786]

R = -$C_{12}H_{25}$-. V.5. n_D^{30} 1.4559, d_4^{30} 0.9040, IR.[1786]

C_{24}

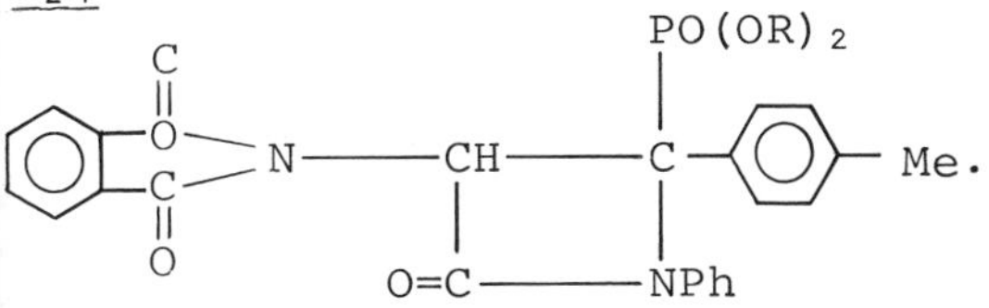

R = Me. V.14. m. 213-213.5°.[1453]

R = Et. V.14. m. 235-238°.[1453]

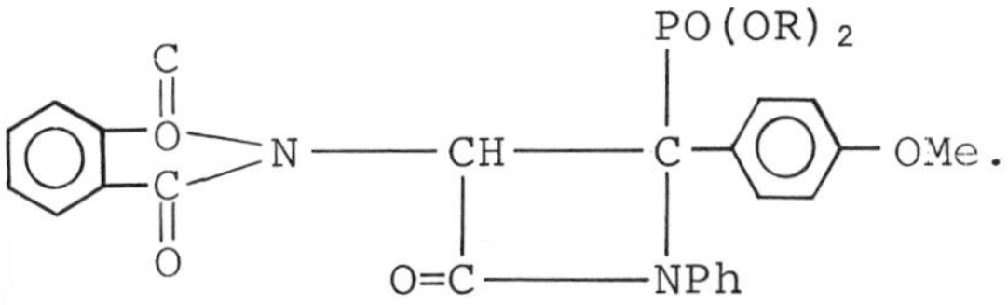

R = Me. V.14. m. 206-207°.[1453]

R = Et. V.14. m. 241.5-243°.[1453]

O=C-CPh=CPh-CHPh-CMePO(OMe)$_2$. V.7. m. 169-171°.[1314]

C_{25}

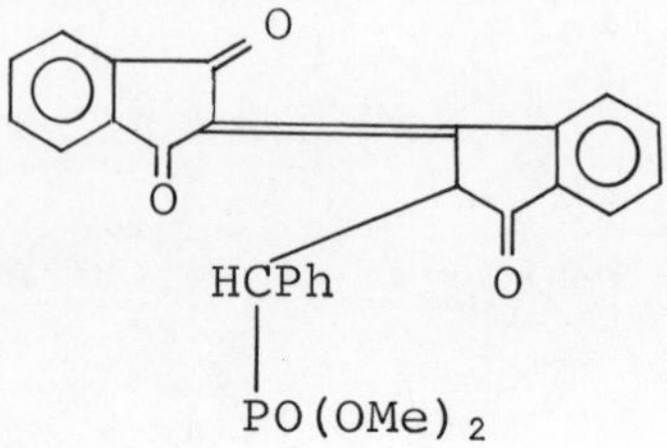

V.7. m. 85-87°.[158]

4-PhC$_6$H$_4$-CPh$_2$PO(OH)$_2$. I.1. m. 270-272°.[119]

4-PhC$_6$H$_4$-CPh$_2$PO(OR)$_2$.

R = Me. V.5. m. 124-125°.[119]

R = Et. V.5. m. 147-148°.[119] V.2. m. 145-146°.[118]

R = Pr. V.5. m. 131-133°.[119] V.2. m. 131-132°.[118]

R = i-Pr. V.5. m. 196.5-199.5°.[119]

R = Bu. V.5. V.2. m. 90-91°.[118,119]

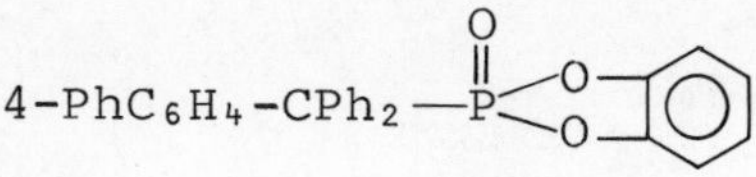

V.2. m. 163-164°.[119]

Ph$_3$P=N-N=CPhPO(OMe)$_2$. V.14. m. 132-133°.[1878a]

C_{26}

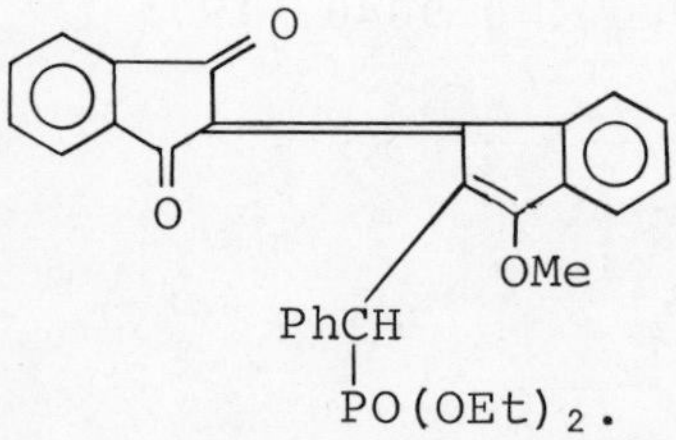

V.7. m. 76-78°.[158]

CH$_3$(CH$_2$)$_{8(7)}$CH(CH$_2$)$_{7(8)}$CO$_2$CH$_2$CHEtC$_4$H$_9$. (Mixture of the isomers)
(CH bears PO(OR)$_2$)

R = Et. V.7. n_D^{30} 1.4512, d_4^{30} 0.9362.[1787]

R = C$_4$H$_9$CHEtCH$_2$-. V.7. n_D^{30} 1.4555, d_4^{30} 0.9148.[1787]

C$_{27}$

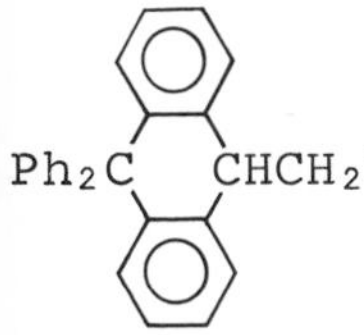

Ph$_2$C⟨⟩CHCH$_2$PO(OH)$_2$. I.1. m. 205°.[241]

C$_{28}$
Ph$_2$CHCO$_2$C(=CPh$_2$)PO(OEt)$_2$. V.14. UV, IR, ^{31}P-7ppm, ^{1}H, mass spect.[203]

C$_{29}$
O=CCPh=CPh-CHPhCPhPO(OMe)$_2$. V.7. (Reflux.) m. 210-212°, UV, IR, ^{1}H.[1314]
O=C-CHPhCPh=CPh-CPhPO(OMe)$_2$. V.7. (At room temperature in the dark.) m. 184-186°.[1314]
PhC=CPh-CPh=CPhC(OH)PO(OMe)$_2$. V.7. (At room temperature in the dark.) m. 168-171°, UV, IR.[1314]

C$_{31}$
(4-PhC$_6$H$_4$-)$_2$CPhPO(OH)$_2$. I.1. m. 164-165°.[119]
(4-PhC$_6$H$_4$-)$_2$CPhPO(OR)$_2$.

    R = Me. V.5. m. 159°.[119]

    R = Et. V.5. m. 140-141°.[119] V.2. m. 139-141°.[118]

    R = Pr. V.5. m. 110-111°.[119]

    R = i-Pr. V.5. m. 159-160°.[119]

    R = Bu. V.5. m. 88-89°.[119]

    R = i-Bu. V.5, V.2. m. 118-120°.[118,119]

(4-PhC$_6$H$_4$-)$_2$CPhP(=O)⟨O-/O-⟩⟨⟩   V.2. m. 74-88°.[119]

C$_{34}$
PhCHC[-C(O)Ph]=CPhCH$_2$CHPhCH(-CO$_2$Et)PO(OEt)$_2$. V.14. m. 207-209°, IR.[308]
[PhC(O)CH$_2$CHPh-]$_2$C(-CO$_2$Et)PO(OEt)$_2$. V.7. m. 261.262°.[540]

C$_{35}$
C$_{16}$H$_{33}$CH=C(-C$_{17}$H$_{35}$)PO(OH)$_2$. I.13. m. 35-40°.[744]

C$_{37}$
[4-PhC$_6$H$_4$-]$_3$CPO(OH)$_2$. I.1, I.2. m. 292-294°.[119]
[4-PhC$_6$H$_4$-]$_3$CPO(OR)$_2$.

    R = Me. V.5. m. 200-201°.[119]

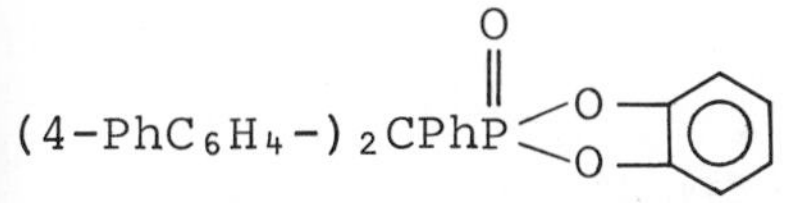

R = Et. V.5. m. 144-145°.[119] V.2. m. 143-145°.[118]

R = Pr. V.5. m. 148-149°.[119]

R = i-Pr. V.5. m. 177-178°.[119]

R = Bu. V.5. m. 132-133°.[119] V.2. m. 130-132°.[118]

R = i-Bu. V.5. m. 120-121°.[119]

[4-PhC$_6$H$_4$-]$_3$CP(=O)(-O-)$_2$C$_6$H$_4$ V.2. m. 212-214°.[119]

$C_{18}H_{37}OCH_2CH(-OC_{16}H_{33})CH_2PO(OH)_2$. I.2. m. 74.5-75.5°.[1770]

$C_{18}H_{37}OCH_2CH(-OC_{16}H_{33})CH_2PO(OMe)OH$. III.3. m. 50-51°, IR.[1770]

$C_{18}H_{37}OCH_2CH(-OC_{16}H_{33})CH_2P(O)(OCH_2CH_2\overset{\oplus}{N}H_3)O^\ominus$. III.3. m. 169-170°.[1771]

$C_{18}H_{37}OCH_2CH(-OC_{16}H_{33})CH_2PO(OEt)_2$. V.5. m. 42-43°.[1770]

$[PhC(O)CH_2CHPh-]_2CPhPO(OEt)_2$. V.14. m. 249-250°, IR.[241a]

C_{40}

$C_{18}H_{37}OCH_2CH(-OC_{18}H_{37})CH_2CH_2PO(OH)_2$. I.2. m. 77.5-78.5°.[17...]

C.2. Polyphosphonic Acids

C_1

$BrHC[PO(OR)_2]_2$.

R = Et. V.14. b$_{0.08}$ 127-128°, ^{31}P-13.0 ppm.[1384a]

R = i-Pr. V.14. b$_{0.03}$ 140°, ^{31}P-11.5 ppm.[1384a]

$Br_2C[PO(OR)_2]_2$.

R = Et. V.14. b$_{0.08}$ 117-120°, n$_D^{20}$ 1.4750, ^{31}P-8.5 ppm.[1677c]

R = i-Pr. V.14. n$_D^{20}$ 1.4910, ^{31}P-6.5 ppm.[1677c]

$ClHC[PO(OPr-i)_2]_2$. V.14. b$_{0.05}$ 105-108°, ^{31}P-11.5 ppm.[1384]

$Cl_2C[PO(OH)_2]_2$. I.5. Cryst. from ether-chloroform, ^{31}P--.8 ppm[1677b]. As disodium salt-5 hydrate.[2082]

$Cl_2CPO(OR)_2$.

R = Et. V.14. b$_{0.05}$ 119-120°, n$_D^{20}$ 1.4619, ^{31}P-8.5 ppm.[1677c]

R = i-Pr. V.14. m. 50°, ^{31}P -6.5 ppm.[1677c]

$H_2C[PO(OH)_2]_2$. I.2. m. 201°(t-BuOH).[334] I.1/I.6.[1937] I.2.[1872] I.2. m. 200-201°(MeCO$_2$H).[1420] I.2.[1740] Complexing agent,[1442a,899a] ^{31}P -16.7 ppm,[1189a] -17.8 ppm.[655a]

$H_2C[PO(OR)ONa]_2$.

R = Et. III.3.[1420]

R = H$_2$C=CHCH$_2$-. III.3 m. 209-211°.[334]

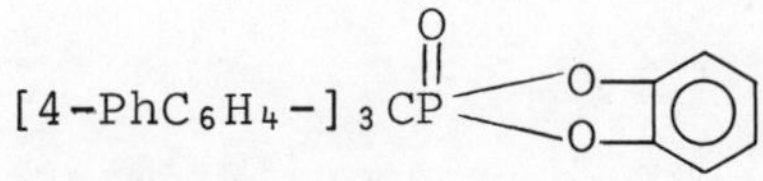

$(HO)_2OPCH_2PO(OH)OH_2C$  adenine III.3. m. 203-205° (dec.)[1255]

$H_2C[PO(OR)_2]_2$.

R = Me. V.6.[1740] V.1. $b_{0.05}$ 87-90°, ^{31}P -23.0 ppm.[1383]

R = Et. V.5. $b_{1.5}$ 143°.[552] V.5. $b_{0.9}$ 128-129°.[539] V.1. $b_{0.1}$ 90-94°, ^{31}P -19.0 ppm.[1383] V.5. $b_{0.5}$ 122-128°, n_D^{20} 1.4300.[334] V.5.[1017] V.6. b_{11} 171-174°.[1872] IR, ^{31}P -19.0 ppm.[1329]

R = i-Pr. V.5. $b_{0.1}$ 114°.[401]

R = $H_2C{=}CHCH_2$-. V.5. $b_{0.6}$ 138-140°.[334] V.1. Caution! the product may explode by distillation even in vacuum.[1383]

R = Bu. V.6. b_2 193-194°.[1485] V.14.[1740]

R = sec.-Bu. V.5.[401]

R = $C_{18}H_{37}$-. V.1. m. 60-62° (hexane).[1383]

$(EtO)_2OPCH_2PO(OBu)_2$. V.6. b_2 179-180°.[1485] V.6. b_2 178-181°.[2054]

$NaHC[PO(OR)_2]_2$.

R = Me. V.14. ^{31}P -45.5 ppm.[1677c]

R = Et. V.14. ^{31}P -41.5 ppm.[1677c]

R = i-Pr. V.14. ^{31}P -40.5 ppm.[1677c]

$(HO)HC[PO(ONa)_2]_2$. I.2./I.13. ^{31}P-15.0 ppm (as disodium salt).[1677b] ^{31}P-14.6 ppm [as $(NMe_4)_2$ salt].[655a]

$(HO)_2C[PO(OH)_2]_2$. I.2., isolated as disodium salt.[1677b] ^{31}P-13 ppm (acid); + 3.4 ppm [as $(NMe_4)_2$ salt];[655a] -15.0 ppm (as disodium salt).[1677b]

$H_2NCH[PO(OH)_2]_2$. I.13., I.12.[1527]

$OC[PO(ONa)_2]_2 \cdot xH_2O$. I.2.[1677b] ^{31}P +0.91 ppm [as $(NMe_4)_4$ salt].[655a]

C_2

$(EtO)_2OPC{\equiv}CPO(OEt)_2$. V.5. $b_{2.5}$ 181.5-184°, n_D^{20} 1.4475, d_4^{20} 1.1196.[827] ^{31}P +11 ppm, ^1H.[1158]

$(HO)_2OPCF_2CF_2PO(OH)_2$. I.1/I.6. m. 89-92°.[324]

$(RO)_2OPCH{=}CHPO(OR)_2$ (trans).

R = Me. V.5. m. 27-30°, IR, ^1H.[1989] Caution! The reaction mixture may explode.[1526a]

R = Et. V.14. $b_{0.3}$ 144-146°, n_D^{20} 1.4488, UV, IR.[1986] V.5. $b_{0.8}$ 156°.[1985,1989] ^{31}P -13 ppm. ^1H.[1158]

$[(i\text{-PrO})_2OP]_2\text{-CHSSCH-}[PO(OPr\text{-}i)_2]_2$. V.14. m. 99.5-101.5°, ^{31}P -15.2 ppm.[1384a]

$(i\text{-PrO})_2OPCH_2CH[PO(OPr\text{-}i)_2]_2$. V.7. $b_{0.0006}$ 142°, n_D^{20} 1.4430.[1788]

$(HO)_2OPCH_2CH_2PO(OH)_2$. I.2. m. 220-223°, IR, ^{31}P -27.4 ppm.[1329] I.2. m. 220-221°.[917b] I.1/I.6.[1937] I.2. m. 223-224°.[1938] I.2. m. 220-221°.[1900]

$[^{\ominus}O(EtO)OPCH_2CH_2PO(OEt)O^{\ominus}][H_3\overset{\oplus}{N}C(=NH)SEt]_2$. III.3. m. 205-206°, ^1H, ^{31}P -19.7 ± 1 ppm, UV, IR.[1324]

$(RO)_2OPCH_2CH_2PO(OR)_2$.

R = Et. V.6. b_2 160°, IR, ^{31}P -26.8 ppm.[1329] V.5. b_2 160°.[539] V.5. b_1 167°.[552] V.5, V.6. b_{14} 200-202°.[1790] V.6. $b_{0.3}$ 133-134°.[1938] V.14.[1986] V.6. $b_{0.05}$ 125-134°.[1327] V.6. b_1 164-165°.[1486]

R = ClH_2CCH_2-. V.7. m. 100.7-101.7°.[1900]

R = i-Pr. V.6. $b_{0.4}$ 144-145°.[1486]

R = Bu. V.6. b_3 202-204°.[1486]

R = Ph. V.5. m. 155-155.5°.[917b]

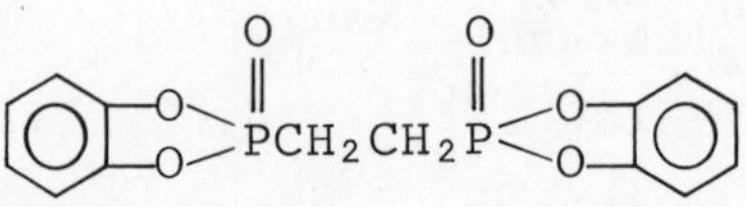

V.5. b_5 265°.[917b]

$CH_3CH[PO(OH)_2]_2$. I.2. m. 179-181°, ^{31}P -23.0 ppm,[1677c] -22 ppm.[665a]

$(HO)_2OPCH_2OCH_2PO(OH)_2$. I.2. m. 96-98°.[13]

$(RO)_2OPCH_2OCH_2PO(OR)_2$.

R = Et. V.5, V.6. $b_{7.5}$ 193-194°, n_D^{20} 1.4470.[13,41]

R = i-Pr. V.6. $b_{1.75}$ 171-172°.[1487]

R = Bu. V.6. $b_{0.4}$ 200-202°.[1487]

$(EtO)_2PCH_2CH(OH)PO(OEt)_2$. V.8. IR.[1986]

$MeC(OH)[PO(OH)_2]_2$. I.2.[265,629,262,1677a] I.13.[274] m. 105° (as 1-hydrate),[2082] complexing agent,[264, 899a] stab. of peroxides,[266] deflocculating agent,[261] stab. of soaps against deterioration,[263] Ca salts.[2077] ^{31}P -19.8 ppm.[655a]

$MeC(OH)[PO(OR)_2]_2$.

R = Me. V.1. m. 67-70° (acetone/hexane), ^{31}P -22.0 ppm.[1383,1384]

R = Et. V.8.[1622] ^{31}P -20.8 ppm.[538]

Condensation product of two molecules $CH_3C(OH)[PO(OH)_2]_2$ with the structure $[(HO)_2OP\text{-}]MeC\text{-}O\text{-}PO(OH)$ / $(HO)OP\text{-}O\text{-}CMe[\text{-}PO(OH)_2]$ (ring-linked). I.11.[265,311a,1677d] cryst. as 4-hydrate (water); X-ray investigation.[1503]

Hexasodium salt-14-hydrate.[265] The structure

(HO)OP - O - PO(OH)
[(HO)$_2$OP-]MeC - O - CMe[-PO(OH)$_2$] is wrong.[1503,1677d]

[(MeO)$_2$OP-]MeC - O - PO(OMe)
(MeO)OP - O - CMe[-PO(OMe)$_2$] . V.1. m. 141-144°. (acetone-hexane).[1383]

(RO)$_2$OPCH$_2$SCH$_2$PO(OR)$_2$.

R = Et. V.6. b_2 179-180°.[1487] V.14. $b_{2.5}$ 191-192°.[128a]

R = i-Pr. V.6. $b_{0.4}$ 164-165°.[1487]

R = Bu. V.6. b_2 230-232°.[1487]

CH$_3$SCH[PO(OEt)$_2$]$_2$. V.5, V.6. $b_{0.02}$ 134°.[719]

(EtO)$_2$OPCH$_2$S(O)$_2$CH$_2$PO(OEt)$_2$. V.14. m. 76-78° (petroleum ether).[128a]

MeHNCH[PO(OH)$_2$]$_2$. I.12. cryst., complexing agent.[1527]

MeC(-NH$_2$)[PO(OH)$_2$]$_2$. I.12. m. 277°.[1527]

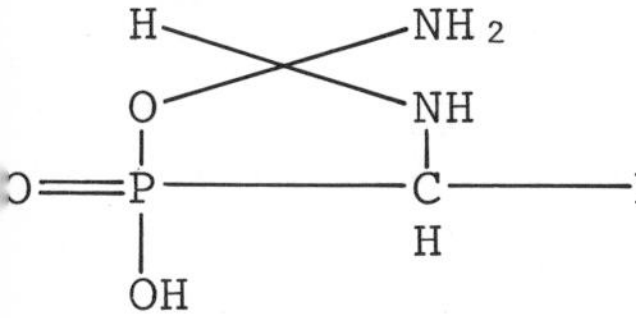
PO(OH)$_2$. From HC(O)NH and PCl$_3$/ H$_2$O.[1527]

HN[CH$_2$PO(OH)$_2$]$_2$. I.2. Ag salt.[1484a] ^{31}P -9.8 ppm[1329a]

HN[CH$_2$PO(OR)$_2$]$_2$.

R = Et. V.9. $b_{0.3}$ 150-151°[1484a]

R = Bu. V.9. $b_{0.6}$ 195-196°[1484a]

C$_3$

O=P[CCl$_2$PO(OEt$_2$]$_3$. V.14. m. 130°, ^{1}H.[1333]

O=C[CF$_2$PO(OEt)$_2$]$_2$. V.5. $b_{1.5}$ 154°.[1968]

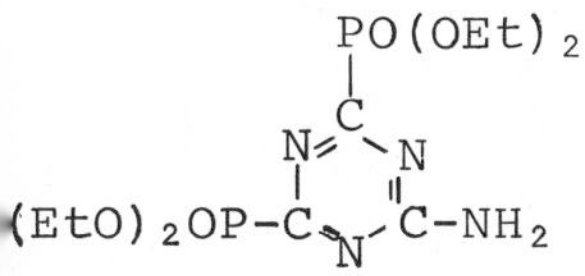
V.14. m. 104-105°, IR, ^{1}H.[1334]

(HO)$_2$OPCH(-CN)-CH$_2$PO(OH)$_2$. I.2.[843]

(EtO)$_2$OPCH(-CN)-CH$_2$PO(OEt)$_2$. From (EtO)$_2$OPCH$_2$CN, CH$_2$O, and P(OEt)$_3$.[843] V.7. b_2 173-174°.[1672]

(EtO)$_2$OPCH(-CN)-CH$_2$PO(OR)$_2$.

R = Pr. V.7. $b_{0.07}$ 175-176°.[1672]

R = Bu. V.7. $b_{0.06}$ 165-168°.[1672]

(EtO)$_2$OPCH=CMePO(OEt)$_2$. V.7. $b_{0.55}$ 132-134°, n_D^{20} 1.4531, d_4^{20} 1.1385.[1607] V.5. $b_{0.4}$ 150°.[1985]

(EtO)$_2$OPCH=CMePO(OMe)$_2$. V.7. $b_{0.8}$ 145°.[1602]

(EtO)$_2$OPCH$_2$C(O)CH$_2$PO(OEt)$_2$. V.14. $b_{0.6}$ 175°, n_D^{24} 1.4490, d_4^{24} 1.194.[1961] ^{31}P -19.4 ppm, ^{1}H.[1093]

(EtO)$_2$OPCH$_2$CH$_2$C(O)PO(OEt)$_2$. From (EtO)$_2$PCl + H$_2$C=CHCO$_2$H and $P(OEt)_3$. V.5. $b_{0.05}$ 165-168°, n_D^{20} 1.4612, d_4^{20} 1.2184.[610]

$HO_2CCH_2CH[PO(OH)_2]_2$. I.2. m. 192°, complexing agent.[2082]

$(EtO)_2OPCH_2OC(O)OCH_2PO(OEt)_2$. V.14. $b_{0.01}$ 173°.[816]

$(HO)_2OPCH[CH_2PO(OH)_2]_2$. I.2. cryst., diss. constants[400]

$(EtO)_2OPCH[CH_2PO(OEt)_2]_2$. V.7. $b_{0.1}$ 170°.[400] V.1. ^{31}P -28. ppm.[1383]

$(EtO)_2OPCH_2CH_2CH[PO(OEt)_2]_2$. V.7. $b_{0.08}$ 183-185°.[1673]

$(EtO)_2OPCH_2CMe[PO(OR)_2]_2$.

R = Me. V.7. $b_{0.6}$ 195°.[1602]

R = Et. V.7. $b_{0.4}$ 175-177°, n_D^{20} 1.4539, d_4^{20} 1.1795.[160

R = Pr. V.7. $b_{0.5}$ 194°.[1602]

$(EtO)_2OPCH_2CMe[PS(OEt)_2]_2$. V.14. $b_{0.3}$ 150°.[1602]

$HO_2CCH_2C(-NH_2)[PO(OH)_2]_2$. I.12.[1527]

$H_2C[-CH_2PO(OH)_2]_2$. I.1/I.6.[1937] I.2. m. 178°, IR, ^{31}P -28. ppm.[1329] I.2. m. 168-169°.[999] I.2. m. 170.5-172°.[1419, 1420]

$H_2C[-CH_2PO(OEt)_2]_2$. V.5. b_2 170-172°.[999] V.5. $b_{0.8}$ 175°.[55 V.6. b_8 198-199°.[1419,1420] V.6. b_1 178-180°, IR, ^{31}P -29.3 ppm.[1329] V.6. $b_{0.65}$ 80-81°, n_D^{20} 1.4508, IR.[1938]

$(CH_3)_2C[PO(OH)_2]_2$. I.1. m. 228.5-229.5°, ^{31}P -27.5 ppm.[167

$(EtO)_2OPCH_2CHMePO(OEt)_2$. V.7. b_3 164-165°.[1486]

$N[CH_2PO(OH)_2]_3$. I.10. m. 210-215°, ^{31}P -9.5 ppm.[1328] From NTA and PCl_3.[1051] From NH_3, CH_2O, and $Cl\overline{POCH_2CH_2O}$.[1050] Stab. of peroxides.[829] Threshold substance.[1688]

$N[CH_2PO(OEt)_2]_3$. V.9. $b_{0.8}$ 202-204°.[1484a] ^{31}P -22.6 ppm.[1329a]

$C_2H_5C(-OH)[PO(OH)_2]$. I.11.[265,629,262] I.13.[274] m. 108-110°[2082] also isolated as sodium salt, complexing agent,[26 stab. of peroxides,[266] deflocculating agent,[261] stab. of soaps against deterioration.[263]

$(EtO)_2OPCHMeCH(ONa)PO(OC_6H_{11}\text{-cycl.})_2$. V.8. m. >300°.[1502]

$O{=}P_\alpha[-CH_2P_\beta O(OEt)OH]_3$. III.4. $^{31}P_\alpha$ -36.5 ppm; P_β -16.4 ppm.[1155]

$(CH_3)_2NCH[PO(OH)_2]_2$. I.6. I.12. m. 246°(after drying in vacuum at 50°) complexing agent.[1527]

$(CH_3)_2NCH[PO(OR)_2]_2$.

R = Me. V.9. $b_{0.03}$ 95-97°, n_D^{23} 1.4611, ^1H.[712]

R = Et. V.9. $b_{0.03}$ 114-115°, n_D^{21} 1.4508, ^1H.[712,714]

V.9. $b_{0.01}$ 105-110°.[715a]

R = Ph. V.9. m. 210-212°.[714]

$CH_3C(-NHMe)[PO(OH)_2]_2$. I.12. Cryst. complexing agent[1527]

$C_2H_5C(-NH_2)[PO(OH)_2]_2$. I.12. Cryst. complexing agent[1527]

$CH_3N[-CH_2PO(OH)_2]_2$. I.10. m. 210-212°, ^{31}P.[1328] From MeN $[-CH_2PO(OH)_2]CH_2CO_2H$ and PCl_3.[1051]

$CH_3N[-CH_2PO(OR)_2]_2$

R = Et. V.9. $b_{0.5}$ 149-150°.[1484a]

R = i-Pr. V.9. $b_{0.3}$ 146°.[1484a]

R = Bu. V.9. b_1 195-197°.[1484a]

$(MeO)OP[-CH_2PO(OEt)_2]_2$. V.5. $b_{0.15}$ 186-188°.[1332] V.5. $b_{0.15}$ 188°.[976]

$O{=}P_\alpha[CH_2P(O)(OH)_2]_3$. I.1. Viscous oil, P_α -37.8 ppm, P_β -15.3 ppm.[1154a] For esters see Chapter 6.

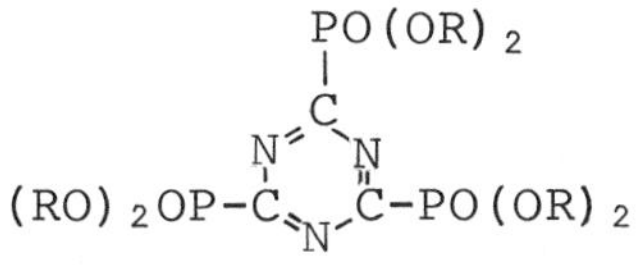

R = Me. V.5. m. 124-125°.[778]

R = Et. V.5. m. 95-96°, m. 82-84°.[778] V.5. m. 94-95°.[1337]

R = i-Pr. V.5. m. 67-70°.[778]

R = Ph. V.5. m. 94-95°.[778]

C_4

$(EtO)OP[CBr_2PO(OEt)_2]_2$. V.14. m. 67-80°, ^1H.[1333]

$(EtO)_2OPCHCF_2CF{=}CPO(OEt)_2$. V.14. b_1 122-125°.[2041]

$(EtO)_2OP-C{=}C-PO(OEt)_2$ (ring closed by $(CF_2)_2$). V.5. n_D^{25} 1.4208.[561]

$(EtO)_2OPCH(CF_2)_2CHPO(OEt)_2$. V.7/V.14. b_1 122-125°.[2041]

$(EtO)_2OPCH{=}CH{-}CH{=}CHPO(OEt)_2$. V.14. $b_{0.045}$ 162°.[1629]

$(MeO)_2OPCMe{=}C{=}CHPO(OEt)_2$. V.7. $b_{0.025}$ 141-142°.[1572]

$(MeO)_2OPC(O)(CH_2)_2C(O)PO(OMe)_2$. V.5. b_1 161-163°.[1352]

$(HO)_2OPCH(-CO_2H)CH(-CO_2H)PO(OH)_2$. I.2. m. 209-210°, ^{31}P -15.0 ppm.[1382]

$(RO)_2OPCH_2CH[-PO(OEt)_2]-C[-PO(OEt)_2]{=}CH_2$.

R = Me. V.7. b_1 210 , n_D^{20} 1.4700, d_4^{20} 1.2250.[1594]

R = Et. V.7. b_1 215-217°, n_D^{20} 1.4620, d_4^{20} 1.1800.[1594]

$(RO)_2OPCH_2CH{=}CHCH_2PO(OR)_2$.

R = Me. V.7. b_2 195°.[989]

R = Et. V.5. b_2 181-182°.[1616]

R = Pr. V.7. $b_{1.5}$ 205-207.5°.[989]

R = Bu. V.7. b_1 211.5-212.5°.[989]

$(RO)_2OPCH_2CH{=}CHCH_2PS(OMe)_2$.

R = Pr. V.7. $b_{1.5}$ 169-170°, n_D^{20} 1.4915, d_4^{20} 1.1531.[1593]

R = Bu. V.7. b_3 198-200°, n_D^{20} 1.4750, d_4^{20} 1.1150.[1593]

$(RO)_2OPCH_2CH{=}CHCH_2PS(OEt)_2$.

R = Me. V.7. $b_{1.5}$ 163°, n_D^{20} 1.4900, d_4^{20} 1.1803.[1593]

R = Et. V.7. b_3 197°, n_D^{20} 1.4812, d_4^{20} 1.1323.[1593]

R = i-Bu. V.7. b_3 187°, n_D^{20} 1.4860, d_4^{20} 1.1003.[1593]

$(MeO)_2OPCMe{=}CHCH_2PO(OMe)_2$. V.14. $b_{0.03}$ 111-113°, IR, ^1H.[1191]

$(HO)_2OPCH_2CH[-PO(OH)_2]CH[-PO(OH)_2]CH_2PO(OH)_2$. I.2. cryst.[400]

$(EtO)_2OPCH_2CH[-PO(OEt)_2]CH[-PO(OEt)_2]CH_2PO(OEt)_2$. V.1. ^{31}P -30.3 ppm.[1383]

$CH_3C(O)CH_2CH[PO(OBu)_2]_2$. V.7. $b_{0.01}$ 128°, n_D^{25} 1.4412.[1046]

$(EtO)_2OP\overline{CH[-C(O)Me]}CH_2PO(OEt)_2$. V.8/V.14. $b_{0.015}$ 138-139°.

$(EtO)_2OPCH\text{-}O(CH_2)_2OCHPO(OEt)_2$. V.7. $b_{0.1}$ 100°, n_D^{20} 1.4600, IR.[717]

$CH_3CO_2CMe[PO(OR)_2]_2$.

R = Me. V.8/V.14. $b_{0.02}$ 126-128°.[389]

R = Et. V.14. $b_{1.5}$ 147-148°, IR.[1618] V.14. $b_{0.1}$ 168-171°, IR.[1986]

$MeO_2CCH_2CH[-PO(OR)_2]_2$.

R = Me. V.7. $b_{3.5}$ 176-178°, n_D^{20} 1.4573, d_4^{20} 1.3159.[164]

R = Et. V.7. $b_{0.5}$ 166-167°.[1313]

$(EtO)_2OPCH(-CO_2Me)CH_2PO(OR)_2$.

R = Me. V.7. b_2 161-162°.[1642]

R = Et. V.7. b_2 166°.[1642]

R = Pr. V.7. b_2 174-175°.[1642]

R = i-Pr. V.7. b_4 170°.[1642]

R = Bu. V.7. b_2 179°.[1642]

R = i-Bu. V.7. b_3 172°.[1642]

$HO_2CCH_2CMe[-PO(OH)_2]_2$. I.2. m. 198-201°, complexing agent.[2082]

$HO_2CCH_2N[-CH_2PO(OH)_2]_3$. I.10 m. 203°, ^{31}P.[1328]

$OCH_2CH_2OP\text{-}OCMe[-PO(OEt)_2]_2$. V.8/V.14. n_D^{25} 1.4628, ^{31}P δ_1 -124.8, δ_2 -17.0 ppm, $\delta_1:\delta_2$ =1:2.[257]

$(HO)_2OP(CH_2)_4PO(OH)_2$. I.2. m. 217-220°, IR ^{31}P -31.6 ppm.[1329] I.6.[1937]

$(RO)_2OP(CH_2)_4PO(OR)_2$.

R = Et. V.6. $b_{0.1}$ 171°, IR, ^{31}P -31.5 ppm.[1329]

R = i-Am. V.6. n_D^{20} 1.4500, d_4^{20} 0.9986.[1160]

$H_2C\text{-}P(O)(OH)\text{-}O(CH_2)_2NCH_2PO(OH)_2$. I.10. Cryst.[2083]

$(RO)_2OPCH_2CH_2OCH_2CH_2PO(OR)_2$.

R = i-Pr. V.6. $b_{0.7}$ 154-155°.[1487]

R = Bu. V.6. $b_{0.6}$ 210-211°.[1487]

R = i-Am. V.6. m. 14°, n_D^{20} 1.4503, d_4^{20} 1.0000.[1160]

R = C_6H_{13}-. V.6. m. 15.2°, n_D^{20} 1.4520, d_4^{20} 0.9828.[1160]

R = C_7H_{15}-. V.6. m. 16.0°, n_D^{20} 1.4530, d_4^{20} 0.9644.[1160]

R = C_8H_{17}-. V.6. m. 17.5°, n_D^{20} 1.4550, d_4^{20} 0.9554.[1160]

R = C_9H_{19}-. V.6. m. 22.5°, n_D^{20} 1.4565, d_4^{20} 0.9475.[1160]

$C_3H_7C(-OH)[PO(OH)_2]_2$. I.11.[265,629,262] I.13.[274] m. 108-110°,[2082] also isolated as Na_4-salt-17-hydrate, complexing agent.[264]

$(EtO)_2OPCH_2-CH(OEt)PO(OEt)_2$. ^1H, ^{31}P δ_1-27.7 ppm, δ_2-30.6 ppm.[1093]

$(RO)_2OPCHEtCH(ONa)PO(OC_6H_{11}$-cycl$)_2$.

R = Et. V.8. m. >300°.[1502]

R = Pr. V.8. m. >300°.[1502]

$S[CH_2CH_2PO(OR)_2]_2$.

R = i-Pr. V.6. $b_{0.8}$ 150-151°.[1487]

R = Bu. V.6. $b_{0.6}$ 202-204°.[1487]

$O_2S[-CH_2CH_2PO(OPr\text{-}i)_2]_2$. V.7.[1663]

$[(EtO)_2OP-]_2\overset{\ominus}{C}-\overset{\oplus}{N}Me_3$. V.14. m. 38-40°, ^{31}P -28 ppm.[716]

$C_2H_5N[-CH_2PO(OH)_2]_2$. I.10. m. 205°, ^{31}P.[1328]

$CH_3C(-NMe_2)[-PO(OH)_2]_2$. I.12. Cryst., complexing agent.[1527]

$CH_3C(-NHEt)[-PO(OH)_2]_2$. I.12. Cryst., complexing agent.[1527]

$C_3H_7C(-NH_2)[-PO(OH)_2]_2$. I.12. Cryst., complexing agent.[1527]

$HOCH_2CH_2N[-CH_2PO(OH)_2]_2$. I.13. Isolated as Na salt.[2083]

$(EtO)OP_\alpha[-CH_2P_\beta O(OEt)_2]_2$. V.5. $b_{0.001}$ 146-153°.[1154b,1332] V.5. $b_{0.15}$ 180-182°,[976] n_D^{20} 1.4610, $^{31}P_\alpha$ -37.3 ppm, $^{31}P_\beta$ -20.0 ppm.[1154b]

$Me_3\overset{\oplus}{N}CH[PO(OR)_2]PO(OR)O^{\ominus}$.

R = Me. V.14. m. 136-139°.[716]

R = Et. V.14. m. 164-167°, ^{31}P δ_1 -11, δ_2 +4 ppm.[716]

$[(EtO)_2OP-]_2CH\overset{\oplus}{N}Me_3]X^{\ominus}$.

X = $J^{\ominus}$. V.14. m. 105-106°, ^{31}P -11 ppm.[716]

X = $MeSO_4^{\ominus}$. V.14. m. 116-118, ^{31}P -11 ppm.[716]

X = $ClO_4^{\ominus}$. V.14. m. 134-135°.[716]

$^{\ominus}O(HO)OPCH_2\overset{\oplus}{N}H_2CH_2CH_2\overset{\oplus}{N}H_2CH_2PO(OH)O^{\ominus}$. I.13. m. 282-286°(dec.) pK values.[2041a] I.2.[914]

$(EtO)_2POCH_2NHCH_2CH_2NHCH_2PO(OEt)_2$. V.9. As dipikrate dec. 191-192°.[914]

$Me_2Si[-OCH_2PO(OEt)_2]_2$. V.14. $b_{1.5}$ 164-166°, n_D^{20} 1.4387, d_4^{20} 1.1371.[1436]

C_5

ClC=CCl-CCl=CCl-C[-PO(OEt)$_2$]$_2$ (ring closed between the first and last C). V.5. b_{13} 82-85°.[312]

$(HO)_2OPC{=}CPO(OH)_2$ (ring closed by $-(CF_2)_3-$ between the two C). I.2. m. 194-199°(dec.).[562]

$(RO)_2OPC{=\!=}CPO(OR)_2$.
   └$(CF_2)_3$┘

R = Me. V.5. $b_{0.4}$ 132-134°.[561]

R = Et. V.5. $b_{0.1}$ 111-112°.[561,565,566]

R = F_3CCH_2-. V.2. $b_{0.1}$ 84-86°.[562]

R = Bu. V.5. $b_{0.004}$ 90-105°.[561]

R = Ph. V.5. m. 176-177°.[561]

$(BuO)_2OPC{=\!=}CPO(OEt)_2$. V.5. $b_{0.35}$ 150-152°.[562]
   └$(CF_2)_3$┘

$EtO(PhO)OPC{=\!=}CPO(OPh)OEt$. V.5. $b_{0.45}$ 174-176°.[561]
   └$(CF_2)_3$┘

$HO_2CCH_2CH(-CO_2H)CH[-PO(OH)_2]_2$. I.2. m. 149.5° (as 1-hydrate), complexing agent.[765]

$(RO)_2OPCH_2CMe{=}CHCH_2PO(OR)_2$.

R = Pr. V.7. $b_{0.5}$ 201-202°.[989]

R = Bu. V.7. b_1 220-222°.[989]

$(EtO)_2OPCH_2C(=CMe_2)PO(OR)_2$.

R = Me. V.7. b_1 147°.[1603]

R = Et. V.6/V.14. b_1 153-155°.[1601]

R = Pr. V.7. $b_{1.5}$ 156°.[1603]

R = Bu. V.7. $b_{0.3}$ 157°.[1603]

$(PrO)_2OPCH_2C(=CMe_2)PO(OR)_2$.

R = Me. V.7. $b_{1.5}$ 154-155°.[1603]

R = Et. V.7. $b_{1.5}$ 157-158°.[1603]

R = Pr. V.7. $b_{1.5}$ 165°.[1603]

R = Bu. V.7. b_4 203°.[1603]

$(BuO)_2OPCH_2C(=CMe_2)PO(OR)_2$.

R = Me. V.7. b_1 163-165°.[1603]

R = Et. V.7. $b_{0.8}$ 167°.[1603]

R = Pr. V.7. $b_{0.8}$ 172-174°.[1603]

$(EtO)_2OPCH{=}C(-OEt)CH_2PO(OEt)_2$. V.5. $b_{0.3}$ 166°.[1961]
$MeO_2CCH_2CH_2CH[-PO(OEt)_2]_2$. V.7. $b_{0.02}$ 155-156°.[1673]
$EtO_2CCH_2CH[-PO(OR)_2]_2$.

R = Me. V.7. $b_{1.5}$ 163-165°.[1642]

R = Et. V.7. $b_{0.5}$ 155-156°.[1642]

$EtO_2CCH_2CH[-PO(OR)_2]PO(OEt)_2$.

R = Me. V.7. b_1 164-166°.[1642]

R = Pr. V.7. b_4 185-186°.[1642]

R = i-Pr. V.7. b_3 172-174°.[1642]

R = Bu. V.7. b_3 197-199°.[1642]

R = i-Bu. V.7. $b_{0.5}$ 171-173°.[1642]

$EtO_2CCH[-PO(OEt)_2]CH_2PO(OEt)_2$. V.8/V.14. $b_{0.001}$ 147-148°.[843]
$(EtO)_2OPCH_2C(-SEt)=CHPO(OEt)_2$. V.14. $b_{0.025}$ 158-158.5°.[1572]
$H_2C=CHCH_2N[-CH_2PO(OH)_2]_2$. I.10. m. 190°, ^{31}P.[1328]
$H_2CCH_2OCH_2CH_2NCH[-PO(OH)_2]_2$. I.12. Cryst., complexing agent.[1527]
$(HO)_2OP(CH_2)_5PO(OH)_2$. I.2. m. 155°, IR, ^{31}P -31.9 ppm.[1329] I.2. m. 160°.[309]
$(EtO)_2OP(CH_2)_5PO(OEt)_2$. V.6. $b_{0.1}$ 180-181°, IR, ^{31}P -30.8 ppm.[1329] V.6. $b_{0.25}$ 190°.[309]
$C_4H_9CH[PO(OH)_2]_2$. I.1/I.6. m. 150-152°, ^{31}P -23.5 ppm.[764] I.2. m. 151-152°, ^{31}P -22.5 ppm.[1677c] I.2. m. 163-165°.[1015]
$C_4H_9CH[-PO(OEt)_2]_2$. I.14. $b_{0.3}$ 147-149°.[1015]
$C_4H_9C(-OH)[PO(OH)_2]_2$. I.11.[265,629,262] I.13.[274] Cryst., very hygrosc.,[2082] complexing agent.[264]
$Me_3CC(-OH)[PO(OMe)_2]_2$. V.8. m. 108-110° (benzene/hexane), ^{31}P -22.0 ppm.[1384]
$(HO)CMe-CH_2CMe[-PO(OEt)_2]OPO(OEt)$. V.8.[126]
$(EtO)_2OP_{\gamma}CH_2P_{\alpha}(O)[-CH_2CH_2P_{\beta}O(OEt)_2]_2$. V.5. $^{31}P_{\alpha}$ -45.1 ppm, $^{31}P_{\beta}$ -32.2 and -2.90 ppm, $^{31}P_{\gamma}$ -21.2 ppm.[1156]
$C_3H_7N[-CH_2PO(OH)_2]_2$. I.10. m. 183°, ^{31}P.[1328]
$Me_2CHN[-CH_2PO(OH)_2]_2$. I.10 m. 235°, ^{31}P.[1328]
$Et_2NCH[-PO(OH)_2]_2$. I.12. Cryst., complexing agent.[1527]
$C_4H_9NHCH[-PO(OH)_2]_2$. I.12. Cryst., complexing agent.[1527]

C_6

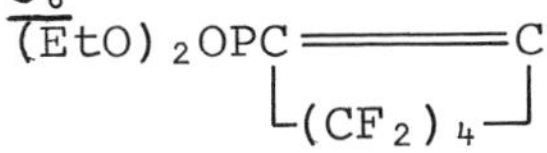

. V.5. $b_{0.5}$ 120-135°.[561]

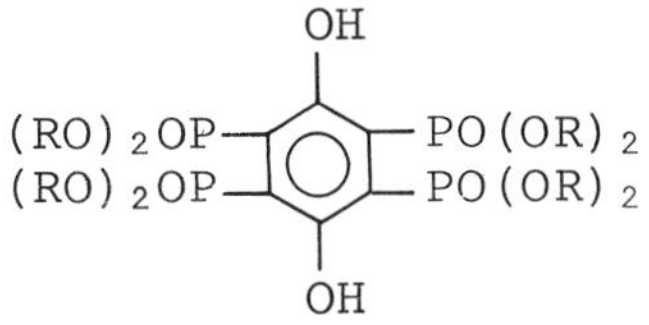

R = i-Pr. V.14. Cryst.[1721]

R = sec.-Bu. V.14. m. 120-121°.[1721]

1,2-$[(MeO)_2OP-]_2C_6H_4$. V.5. m. 80-81°.[1421]
1,3-$[(RO)_2OP-]_2C_6H_4$.

R = Me. V.5.[1421]

R = Et. V.5. $b_{0.04}$ 191-194°.[1987]

1,4-$[(RO)_2OP-]_2C_6H_4$

R = Me. V.5. m. 100-101°.[1421]

R = Et. V.5.[1987] V.5. m. 71-72°.[1985]

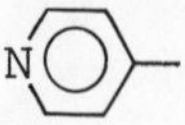—NHCH[-PO(OH)$_2$]$_2$. I.12. Cryst., complexing agent.[1527]

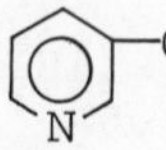—C(-NH$_2$)[-PO(OH)$_2$]$_2$ I.12. m. 228-232°.[1094]

(RO)$_2$OP-CH$_2$CH=C——CH-CH$_2$PO(OR)$_2$.
O——C(O)

R = Et. V.7/V.14. b_3 208-209°.[1596]

R = Pr. V.7/V.14. $b_{2.5}$ 222-223°.[1596]

(EtO)$_2$OP-CH$_2$-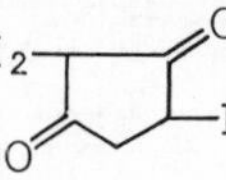-PO(OEt)$_2$. V.14. $b_{0.0001}$ 135-140°, IR.[1043]

(EtO)$_2$OP-C(-CO$_2$Me)=C(-CO$_2$Me)PO(OEt)$_2$. V.14. $b_{0.2}$ 175-180°, ^{31}P -5.5 ppm.[1382]

(EtO)$_2$OPCHCH$_2$CH=CHCH$_2$CHPO(OEt)$_2$. V.14. IR.[1986]

(RO)$_2$OP-C(O)(CH$_2$)$_4$C(O)-PO(OR)$_2$.

R = Me. V.5. n_D^{25} 1.4642.[1352]

R = Et. V.5. n_D^{25} 1.4552.[1352]

(EtO)$_2$OPCH$_2$CH$_2$C(O)O(O)CCH$_2$CH$_2$PO(OEt)$_2$.

1. From H$_2$C=CHCO$_2$H and (EtO)$_2$PO$_2$CMe, or
2. from (EtO)$_2$PO$_2$CCH=CH$_2$ and MeCO$_2$H, or
3. from (EtO)$_2$P(O)H and H$_2$C=CHC(O)OC(O)Me; $b_{0.055}$ 176-177°.[606]

(RO)$_2$OPCH(-CO$_2$Me)-CH(-CO$_2$Me)-PO(OR)$_2$.

R = Me. V.7. m. 89.5 - 91.5° (toluene).[1382] V.7. b_4 208-210°.[967]

R = Et. V.7. $b_{0.1}$ 172-175°.[1382] V.7. $b_{0.075-0.15}$ 175-180°, ^{31}P -18.6 ppm.[1382a]

HO$_2$CCH$_2$C(-CO$_2$H)[-PO(OH)$_2$]CH$_2$CH$_2$PO(OH)$_2$. I.2. Complexing agent.[2082]

(EtO)$_2$OPC(=CHMe)CHMeCH$_2$PO(OEt)$_2$. V.14. b_7 194-195°, n_D^{20} 1.4562, d_4^{20} 1.1043.[1621]

(HO)$_2$OPCH(CH$_2$)$_4$CHPO(OH)$_2$. I.2. m. 217-220°.[500] I.2. m. 216-219°.[1988]

(EtO)$_2$OPCH(CH$_2$)$_4$CHPO(OEt)$_2$. V.14. $b_{0.3}$ 164-167°.[1986]

(EtO)$_2$OPCH(-CH$_2$SEt)-C(=CH$_2$)PO(OEt)$_2$. V.14. b_2 186°, n_D^{20} 1.4787, d_4^{20} 1.1455.[1594]

H$_2$C(CH$_2$)$_4$N-CH[PO(OH)$_2$]$_2$. I.12. Cryst., complexing agent.[1527]

$(HO)_2OP(CH_2)_6PO(OH)_2$. I.2. m. 206-208°, ^{31}P. -30.6 ppm.[1329]

$(EtO)_2OP(CH_2)_6PO(OEt)_2$. V.6. $b_{0.1}$ 185°, ^{31}P. -29.6 ppm.[1329]

$N[-CH_2CH_2PO(OH)_2]_3$. I.2. m. 224-227°, ^1H, ^{31}P -20.8 ppm.[1158]

$N[-CH_2CH_2PO(OEt)_2]_3$. V.6, V.5. n_D^{20} 1.4699, ^1H, ^{31}P -29.8 ppm.[1158]

$N[-CHMePO(OCH_2CH_2Cl)(OCH_2CH_2OH)]_3$. V.9. ^{31}P -28.2 ppm.[256]

$(HO)_2OPCH_2N<[(CH_2)_2]_2>NCH_2PO(OH)_2$. I.2. m. 278-280° (dec.) as HBr adduct.

$(EtO)_2OPCH_2N<[(CH_2)_2]_2>NCH_2PO(OEt)_2$. V.9.[1351]

$[(HO)_2OPCH_2-]_2\ N(CH_2)_2N[-CH_2PO(OH)_2]_2$. I.10. m. 214°, ^{31}P.[1328] I.2.[1050] From EDTA and PCl_3.[1051] Threshold effect.[1688] Stab. of H_2O_2.[547] pK values of Y- and Ce-complexes.[1064] Complexing agent in galvanic baths.[1083] pK values of complexes.[1997]

$C_5H_{11}C(OH)[-PO(OH)_2]_2$. I.11.[265,629,262] I.13.[274] m. 110-112°.[2082] complexing agent.[264]

$O{=}P_\alpha[-CH_2CH_2P_\beta O(OEt)_2]_3$. V.5. $^{31}P_\alpha$ -48.9 and -46.3 ppm; P_β -31.3 and -29.4 ppm. n_D^{20} 1.4742.[1156]

$(EtO)_2OPCMe(OH)-(CH_2)_2CMe(OH)-PO(OEt)_2$. V.8. Cryst. (dioxane).[1964]

$C_4H_9N[-CH_2PO(OH)_2]_2$. I.10. m. 196°, ^{31}P.[1328]

$Me_2CHCH_2N[-CH_2PO(OH)_2]_2$. I.10. m. 224°, ^{31}P.[1328]

$MeEtCHN-[CH_2PO(OH)_2]_2$. I.10. m. 189°, ^{31}P.[1328]

$Me_2CH(CH_2)_2C(-NH_2)[-PO(OH)_2]_2$. I.12. m. 228° (dec.)[1094]

$(BuO)OP[-CH_2PO(OBu)_2]_2$. V.5. $b_{0.12}$ 179-182°.[976]

$(EtO)OP[-CHMePO(OEt)_2]_2$. V.14. ^1H.[1155]

$Me_2Si[OCHMePO(OEt)_2]_2$. V.8/V.14. $b_{0.2-0.4}$ 148-154°, n_D^{25} 1.4355.[258]

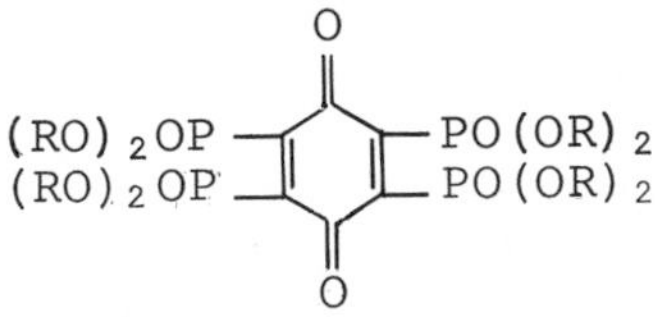

R = i-Pr. V.5. m. 154.5-155°.[1721]

R = sec.-Bu. V.5. m. 120-121°.[1721]

C7

$3-BrC_6H_4C(OH)[-PO(OMe)_2]_2$. V.8. m. 120-122° (ether), ^{31}P -18.0 ppm.[1384]

$4-ClC_6H_4CH[PO(OH)_2]_2$. I.11. m. 287-288°, complexing agent.[2082]

$PhOCCl[-PO(OEt)_2]_2$. V.5, V.14. $b_{0.01}$ 164°, n_D^{23} 1.4770.[719]

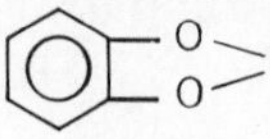$C[-PO(OEt)_2]_2$. V.5. n_D^{22} 1.4886, IR.[717]

$PhS(O)_2N{=}C[PO(OR)_2]_2$.

R = Me. V.5. $b_{0.001}$ 110-112°.[1369]

R = Et. V.5. $b_{0.001}$ 120-122°.[1369]

$C_6H_5C(OH)[-PO(OH)_2]_2$. I.11.[2082,265,262] I.13.[274] m. 191° (as 2-hydrate).[2082] complexing agent.[264]

$C_6H_5C(OH)[-PO(OMe)_2]_2$. V.8. m. 130-131° (ether), ^{31}P -18.0 ppm.[1384]

$C_6H_5C(-NH_2)[PO(OH)_2]_2$. I.12.[1094,1527] m. 225°(dec.).[1094]

$C_6H_5C(-NH_2)[-PO(OEt)_2]_2$. V.9. n_D^{18} 1.4931.[1036]

$(NCCH_2CH_2)_2C[-PO(OEt)_2]_2$. V.7. m. 78-79°.[1673]

$(MeO)_2OPC(O)(CH_2)_5C(O)PO(OMe)_2$. V.5. n_D^{25} 1.4647.[1352]

$(EtO)_2OP{-}CH(-CO_2Et){-}CMe{=}CHPO(OR)_2$.

R = Et. V.7, V.14. b_3 198-200°.[1609]

R = Pr. V.7, V.14. $b_{0.8}$ 162°.[1609]

$H_2\overline{C(CH_2)_4C}HNHCH[-PO(OH)_2]_2$. I.12. Cryst., complexing agent.[1527]

$(EtO)_2OPCHAmCH_2PS(OEt)_2$. V.7. b_2 169-170°, n_D^{20} 1.4640, d_4^{20} 1.0810.[1627]

$C_6H_{13}NHCH[-PO(OH)_2]_2$. I.12. Cryst., complexing agent.[1527]

$C_5H_{11}N[-CH_2PO(OH)_2]_2$. I.10. m. 186°, ^{31}P.[1328]

$Me_2CHCH_2CH_2N[-CH_2PO(OH)_2]_2$. I.10. m. 212°, ^{31}P.[1328]

$(EtO)_2OPCH(-NMe_2)NHC(O)NHCH(-NMe_2)PO(OEt)_2$. V.14. m. 126-129°, ^1H.[715a]

$(EtO)_2OPCH(-NMe_2)$ $NHC(S)NHCH(-NMe_2)PO(OEt)_2$. V.14. m. 116-121°, ^1H.[715a]

$H_2C[-CH_2CH_2PO(OH)CH_2PO(OH)_2]_2$. I.2. ^1H ^{31}P δ_1-17.1, δ_2-48.9 ppm.[1157]

$\underline{C_8}$

$[(EtO)_2OP]_2C{=}CHPh$. V.14. $b_{0.5}$ 186-188°.[1673]

4-$[(HO)_2OPCH_2-]C_6H_4CH_2PO(OH)_2$. I.2. m. 268-271°.[1127]

4-$[(EtO)_2OPCH_2-]C_6H_4CH_2PO(OEt)_2$. V.5. $b_{0.1}$ 203-206°, m. 67°.[1237] V.6.[359] V.5. m. 72-73°, b_2 204°.[1127]

$(EtO)_2OPCHPh{-}CH_2PO(OEt)_2$. V.7. b_1 184-186°, IR ^1H.[692]

$C_6H_5CH_2CH[-PO(OH)_2]_2$. I.2. m. 210-212°, ^{31}P -21.0 ppm.[1677c]

4-$MeOC_6H_4C(OH)[-PO(OMe)_2]_2$. V.8. m. 103-104° ^{31}P -18.0 ppm.[1384]

4-$[(RO)_2OPCH(OH)-]C_6H_4CH(OH)PO(OR)_2$.

R = Me. V.8. m. 196-197° (pyridine).[1573]

R = Et. V.8. m. 205-206° (pyridine).[1573]

R = Pr. V.8. m. 146-147° (pyridine).[1573]

R = Bu. V.8. m. 116-117° (pyridine).[1573]

(EtO)OPCH=CHCHCH$_2$CH$_2$CH=CHCHPO(OEt)$_2$. V.14. b$_{0.04}$ 182°, n$_D^{20}$
1.4845, d$_4^{20}$ 1.1480.[1629]

C$_6$H$_5$CH$_2$NHCH[PO(OH)$_2$]$_2$. I.12. Cryst., complexing agent.[1527]

C$_6$H$_5$CH$_2$C(-NH$_2$)[-PO(OH)$_2$]$_2$. I.12. Cryst.,[1094,1527] complexing agent.[1527]

(PhO)OP[-CH$_2$PO(OEt)$_2$]$_2$. V.5. b$_{0.12}$ 212-218°.[976]

4-[(RO)$_2$OPCH(-NH$_2$)-]C$_6$H$_4$CH(-NH$_2$)PO(OR)$_2$.

R = Me. V.9.[1573]

R = Et. V.9.[1573]

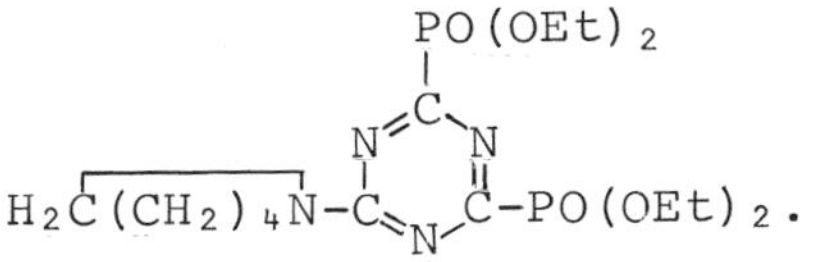

PO(OEt)$_2$

H$_2$C(CH$_2$)$_4$N-C$\underset{N}{\overset{\displaystyle N}{\diagdown}}$C-PO(OEt)$_2$. V.5. m. 68-70°.[778]

(RO)$_2$OPCH$_2$CMe=C —— CMe-CH$_2$PO(OR)$_2$
                    | |
                    O —— C(O)

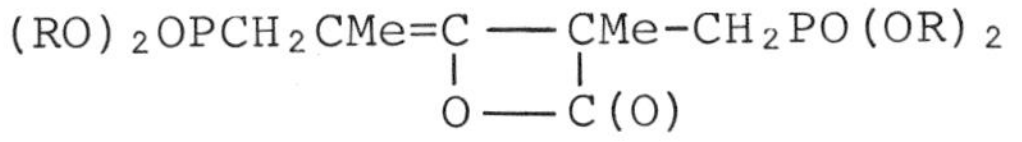

R = Et. V.7/V.14. b$_2$ 195°.[1596]

R = Pr. V.7/V.14. b$_{1.5}$ 197°.[1596]

(RO)$_2$OPC(=CMe$_2$)-C(=CMe$_2$)PO(OR)$_2$. From P(OR)$_2$Cl and Me$_2$C(OH)
-C≡C-C(OH)Me$_2$.

R = Et. b$_2$ 153-154°.[1560]

R = Pr. b$_2$ 179-181°.[1560]

R = Bu. b$_{0.025}$ 167-169°.[1560]

(EtO)$_2$OPCHCH$_2$CMe=CMeCH$_2$CHPO(OEt)$_2$. V.14. b$_{0.01}$ 145-146°,
n$_D^{20}$ 1.4685.[1986]

(MeO)$_2$OP-CMe(-CO$_2$Me)CMe(-CO$_2$Me)PO(OMe)$_2$. V.7. b$_{0.5}$ 190°,
m. 164.5-165.5°.[693]

(EtO)$_2$OPCH(-CO$_2$Et)CH(CO$_2$Et)PO(OEt)$_2$. V.7. b$_5$ 213-214°.[1548]
V.7. b$_3$ 208-212°.[967]

(RO)$_2$OP(CH$_2$)$_3$O$_2$C-CO$_2$(CH$_2$)$_3$PO(OR)$_2$.

R = Me. V.7. n$_D^{25}$ 1.4720, d$_4^{25}$ 1.3149.[1966]

R = Et. V.7. b$_{0.003}$ 150°, n$_D^{25}$ 1.4542, d$_4^{25}$ 1.1975.[1966]

R = Bu. V.7. b$_{0.006}$ 160°, n$_D^{25}$ 1.4524, d$_4^{25}$ 1.0942.[1966]

(EtO)$_2$OPCH$_2$CH=CH-CH$_2$S-]$_2$. V.14.[63]

[(HO)$_2$OPCH$_2$-](HO$_2$CCH$_2$-)N(CH$_2$)$_2$N(-CH$_2$CO$_2$H)[-CH$_2$PO(OH)$_2$]. I.
13. Isolated as Na salt.[914] Diss. and complex-forming
constants.[1686]

(HO$_2$CCH$_2$-)$_2$N(CH$_2$)$_2$N[-CH$_2$PO(OH)$_2$]$_2$. pK values of Y(III) and
Ce(III) complexes.[1996]

(RO)$_2$OPC(O)NH(CH$_2$)$_6$NHC(O)PO(OR)$_2$.

R = Me. V.9. m. 88°.[1066]

R = Et. V.9. m. 54°.[1066]

R = Pr. V.9. m. 38°.[1066]

R = i-Pr. V.9. m. 102°.[1066]

cycl.$-C_6H_{11}N[-CH_2PO(OH)_2]_2$. I.10. m. 237°, [31]P.[1328]

$(EtO)_2OPCH(-CH_2NEt_2)-C(=CH_2)PO(OEt)_2$. V.14. b_1 164°, n_D^{20} 1.6200, d_4^{20} 1.0970.[1594]

$(MeO)_2OPCH(-C_6H_{13})CH_2PO(OMe)_2$. V.1. $b_{0.03}$ $147-148^O$, n_D^{20} 1.4561.[500]

$[(HO)_2OPCH_2-]_2N(CH_2)_4N[-CH_2PO(OH)_2]_2$. I.10. m. $237-238^O$, [31]P.[1328]

$(HO)_2OPCH_2N$⟨CH₂-CHMe / CHMe-CH₂⟩$NCH_2PO(OH)_2$. I.2. m. $278-281^O$ (dec. (as HBr adduct).[1351]

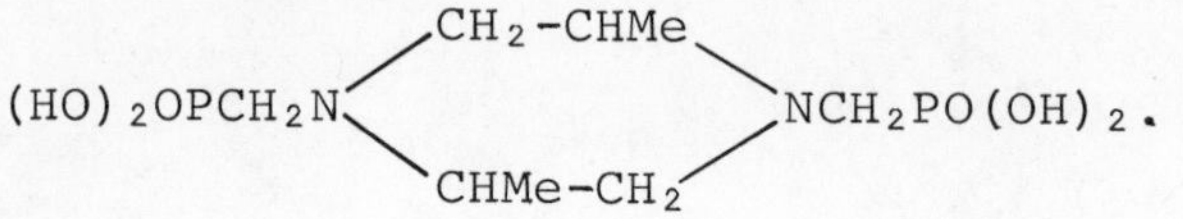

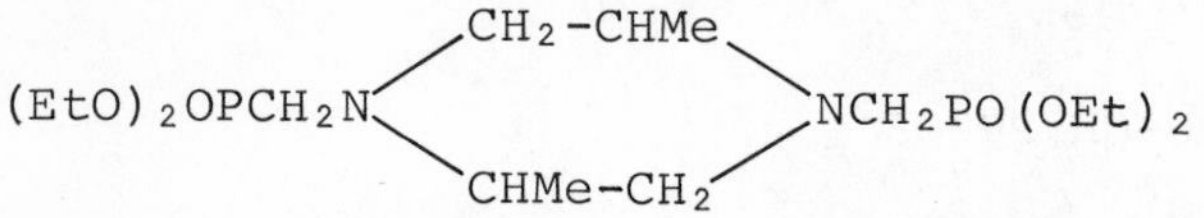

$(EtO)_2OPCH_2N$⟨CH₂-CHMe / CHMe-CH₂⟩$NCH_2PO(OEt)_2$ (trans). V.9.[1351]

$C_6H_{13}N[-CH_2PO(OH)_2]_2$. I.10. m. $205-207^O$, [31]P.[1328]
$C_7H_{15}C(-NH_2)[-PO(OH)_2]_2$. I.12. Cryst.[1527]
$C_6H_{13}NMe-CH[-PO(OH)_2]_2$. I.12. Cryst.[1527]
$(EtO)_2OPCH(-NHCHMe_2)-CH(-NHCHMe_2)PO(OEt)_2$. V.9. m. 151° (as picrate).[915]
$(HO)_2OPCMe_2HN(CH_2)_2NHCMe_2PO(OH)_2$. I.2.[914] pK values of Y and Ce-complexes.[1064]

C_9

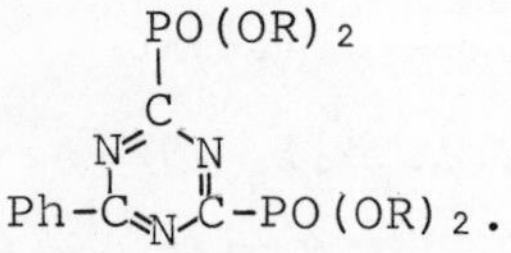

$Ph-C$⟨N-C(-PO(OR)₂)=N⟩$C-PO(OR)_2$.

R = Me. V.5. m. $114-116^O$.[778]

R = Et. V.5. m. $61-63^O$.[778]

$2,4-[(RO)_2OPC(O)NH-]_2C_6H_3Me$.

R = Et. V.9. m. $71-72^O$.[1068]

R = Pr. V.9. m. 82°.[1068]

R = i-Pr. V.9. m. $120-121^O$.[1068]

R = Bu. V.9. Cryst. after 10-12 months.[1068]

R = i-Bu. V.9. m. 109°.[1068]

$(EtO)_2OPCH[-C(O)Ph]-CH_2PO(OEt)_2$. V.8, V.14. $b_{0.015}$ $138-139^O$.
$(MeO)_2OPCHPh-CH(-CO_2H)PO(OMe)_2$. V.5.[472]
$3,5-[(EtO)_2OPCH_2-]_2-2-HOC_6H_2CO_2H$. V.5. m. $70-72^O$.[846]
$2,5-[(EtO)_2OPCH_2-]_2-4-MeC_6H_2OH$. V.5. b_1 $215-216^O$.[846]
$C_6H_5CH_2N[-CH_2PO(OH)_2]_2$. I.10. m. 248° [31]P.[1328]

$C_6H_5C(-NMe_2)[-PO(OH)_2]_2$. I.12. Cryst., complexing agent.[1527]

$C_6H_5CH_2CH_2C(-NH_2)[-PO(OH)_2]_2$. I.12. Cryst., complexing agent.[1527]

$(RO)_2OP(CH_2)_3O_2CCH_2CO_2(CH_2)_3PO(OR)_2$.

  R = Me. V.7. n_D^{25} 1.4700, d_4^{25} 1.2915.[1966]

  R = Et. V.7. n_D^{25} 1.4517, d_4^{25} 1.1833.[1966]

  R = Bu. V.7. n_D^{25} 1.4536, d_4^{25} 1.0917.[1966]

  R = BuCHEtCH_2-. V.7. n_D^{25} 1.4615, d_4^{25} 1.0267.[1966]

$EtO_2CCH_2CH(-CO_2Et)CH[-PO(OEt)_2]_2$. V.7. $b_{0.05}$ 197-213°.[765]
$[(EtO)_2OPCMe_2-CH_2]_2C=O$. V.7. n_D^{20} 1.4538.[1073]
$Bu_2C[PO(OH)_2]_2$. I.2. m. 239-241°, ^{31}P -26.0 ppm.[1677c]
$(BuO)_2OPCH(-C_7H_{15})-CH_2PO(OBu)_2$. V.7. $b_{0.5}$ 175-176°, IR, ^1H.[692]
$C_7H_{15}N[-CH_2PO(OH)_2]_2$. I.10. m. 215°, ^{31}P.[1328]
$C_8H_{17}NHCH[PO(OH)_2]_2$. I.10. Cryst.[1527]
$Bu_2NCH[PO(OH)_2]_2$. I.10. Cryst.[1527]

$\underline{C_{10}}$

$HO(EtO)OP-CH=CH-\langle\bigcirc\rangle-CH=CHPO(OEt)OH$. III.3. m. 205.5-207°.[770]

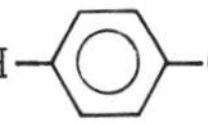

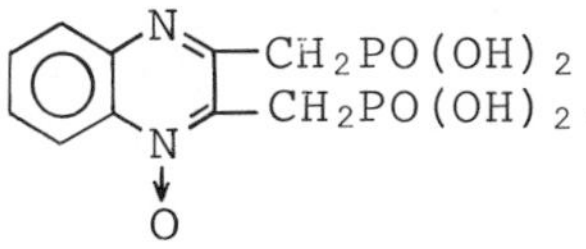

$\begin{array}{c}-CH_2PO(OH)_2\\-CH_2PO(OH)_2\end{array}$. I.2. Dec. 210-215°.[181]

$\begin{array}{c}-CH_2PO(OR)_2\\-CH_2PO(OR)_2\end{array}$

  R = Me. V.5. m. 111-112°.[181]

  R = Et. V.5. m. 110-112°.[181]

  R = i-Pr. V.5. m. 104-105.5°.[181]

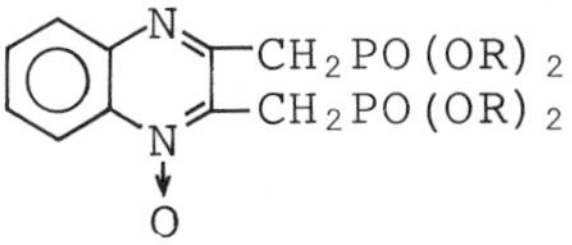

$\begin{array}{c}-CH_2PO(OH)_2\\-CH_2PO(OH)_2\end{array}$. I.2. Dec. 210-212°.[181]

$\begin{array}{c}-CH_2PO(OR)_2\\-CH_2PO(OR)_2\end{array}$

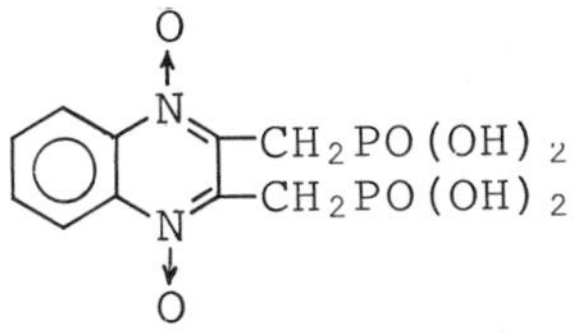

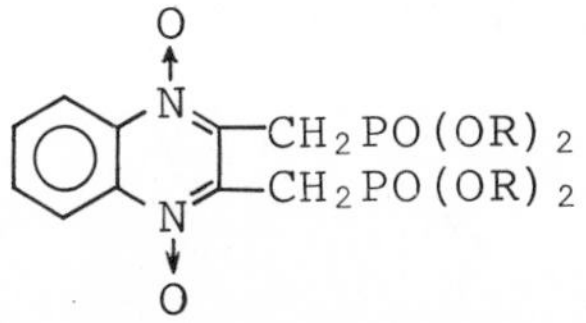

R = Me. V.5. m. 148-150°.[181]

R = Et. V.5. m. 136-138°.[181]

R = i-Pr. V.9. m. 127-129°.[181]

$(EtO)_2OPCHPhCH(-CO_2Me)PO(OEt)_2$. V.7. b_2 208°.[1670]
1,3-[$(HO)_2OPCH_2$-]$_2$-4,6-$Me_2C_6H_2$. I.2. m. 264°.[1127]
1,3-[$(EtO)_2OPCH_2$-]$_2$-4,6-$Me_2C_6H_2$. V.5. b_2 192-194°.[1127]
1,4-[$(HO)_2OPCH_2$-]$_2$-3,5-$Me_2C_6H_2$. I.2. m. 340-350°.[1127]
1,4-[$(EtO)_2OPCH_2$-]$_2$-3,5-$Me_2C_6H_2$. V.5. m. 58-59°.[1127]
$C_6H_5CH_2CH_2N[-CH_2PO(OH)_2]_2$. I.10. m. 237°, ^{31}P.[1328]
2,4,6-$Me_3C_6H_2NHCH[PO(OH)_2]_2$. I.10. Cryst., complexing agent.[1527]
$(MeO)_2POCH_2CHMeCO_2(CH_2)_2O_2CCHMeCH_2PO(OMe)_2$. V.7. b_7 127-128°.[1653]
$(RO)_2OP(CH_2)_3O_2C-(CH_2)_2CO_2(CH_2)_3PO(OR)_2$.

R = Me. V.7. n_D^{25} 1.4670, d_4^{25} 1.2685.[1966]

R = Et. V.7. n_D^{25} 1.4548, d_4^{25} 1.1802.[1966]

R = Bu. V.7. n_D^{25} 1.4526, d_4^{25} 1.0831.[1966]

R = $BuCHEtCH_2$-. V.7. n_D^{25} 1.4605, d_4^{25} 1.0186.[1966]

$(PrO)_2OPCH(-CO_2Pr)-CH(-CO_2Pr)PO(OPr)_2$. V.7. b_2 203-208°.[96]
$EtO_2CCH_2C(-CO_2Et)[-PO(OEt)_2]CH_2CH_2PO(OEt)_2$. V.14. $b_{0.15}$ 212-220°.[2082]
$(HO)_2OP(CH_2)_{10}PO(OH)_2$. I.2. m. 208-209°. ^{31}P -28.6 ppm.[132]
$(EtO)_2OP(CH_2)_{10}PO(OEt)_2$. V.6. $b_{0.8}$ 225°, ^{31}P -30.9 ppm.[132]
$[(HO)_2OPCH_2-]_2N(CH_2)_6N[-CH_2PO(OH)_2]_2$. I.10. m. 249°, ^{31}P.[1328]
$H_2C=CH-CH_2SiMe[-OCHEtPO(OEt)_2]_2$. V.8, V.14. $b_{0.3-0.4}$ 168-175°.[258]
$C_8H_{17}N[-CH_2PO(OH)_2]_2$. I.10. m. 213-215°, ^{31}P.[1328]
$C_8H_{17}NMe-CH[-PO(OH)_2]_2$. I.12. Cryst.[1527]
$BuEtCHCH_2N[-CH_2PO(OH)_2]_2$. I.10. m. 191°, ^{31}P.[1328]
$C_9H_{19}C(-NH_2)[PO(OH)_2]_2$. I.12. Cryst.[1527]
$O[-(CH_2)_2NHCMe_2PO(OH)_2]$. I.2. m. 210-212°.[1224]
$O[-(CH_2)_2NHCMe_2PO(OEt)_2]_2$. V.9. n_D^{20} 1.4580, d_4^{20} 1.0855.[122]

$\underline{C_{11}}$
$C_7F_{15}CH_2CH[-PO(OH)_2]CH_2CHJPO(OH)_2$. I.2.[1335]
$C_7F_{15}CH_2CH[-PO(OEt)_2]CH_2CHJPO(OEt)_2$. From $C_7F_{15}J$ and $H_2C=CHPO(OEt)_2$.[1335]
1,3-[$(HO)_2OPCH_2$-]$_2$-2,4,6-Me_3C_6H. I.2. m. >300°.[1001]
1,3-[$(BuO)_2OPCH_2$-]$_2$-2,4,6-Me_3C_6H. V.6. Cryst.[1001]
$C_9H_{19}N[-CH_2PO(OH)_2]_2$. I.10. m. 215°, ^{31}P.[1328]
$C_{10}H_{21}NHCH[-PO(OH)_2]_2$. I.12. Cryst.[1527]
$H_2C[-CH_2CH_2-P(O)(-OEt)CH_2PO(OEt)_2]_2$. V.14. $b_{0.001}$ 220-230° n_D^{20} 1.4598.[1157]

$\underline{C_{12}}$
2-[$(HO)_2OP$-]-5-$ClC_6H_3N=NC_6H_3Cl$-5-[$-PO(OH)_2$]-2. From 2-O_2N-$C_6H_4N_2Cl-BF_3$ and PCl_3/Cu_2Cl_2.

4-[(HO)$_2$OP-]C$_6$H$_4$-C$_6$H$_4$[-PO(OH)$_2$]-4. V.14. m. >300°, UV.[568]
4-[(EtO)$_2$OP-]C$_6$H$_4$-C$_6$H$_4$[-PO(OEt)$_2$]-4. V.5. m. 68-69°.[1985]
4-[(HO)$_2$OP-]C$_6$H$_4$N=N(→O)-C$_6$H$_4$[-PO(OH)$_2$]. V.14. m. 270-275° (dec.).[579]
2-[(HO)$_2$OP-]C$_6$H$_4$OC$_6$H$_4$[-PO(OH)$_2$]-2. V.14. m. 233-235°.[581]
(RO)$_2$OP(CH$_2$)$_3$O$_2$C(CH$_2$)$_4$CO$_2$(CH$_2$)$_3$PO(OR)$_2$.

R = Me. V.7. n_D^{25} 1.4692, d_4^{25} 1.2330.[1966]

R = Et. V.7. n_D^{25} 1.4530, d_4^{25} 1.1475.[1966]

R = Bu. V.7. n_D^{25} 1.4544, d_4^{25} 1.0698.[1966]

R = BuEtCHCH$_2$-. n_D^{25} 1.4635, d_4^{25} 1.0221.[1966]

(BuO)$_2$OPCH(-CO$_2$Bu)-CH(-CO$_2$Bu)PO(OBu)$_2$. V.7. b$_3$ 215-222°.[967]
[(HO)$_2$OPCH$_2$-]$_2$NCH$_2$ 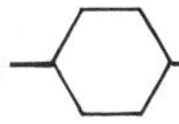 CH$_2$N[-CH$_2$PO(OH)$_2$]$_2$. I.10. m. 262°, ^{31}P.[1328]
[(HO)$_2$OPCH$_2$-]$_2$N(CH$_2$)$_8$N[-CH$_2$PO(OH)$_2$]$_2$. I.10. m. 241°, ^{31}P.[1328]
C$_{10}$H$_{21}$N[-CH$_2$PO(OH)$_2$]$_2$. I.10. m. 247°, ^{31}P.[1328]
C$_{11}$H$_{23}$C(-NH$_2$)[-PO(OH)$_2$]$_2$. I.12. Cryst.[1527]

C$_{13}$

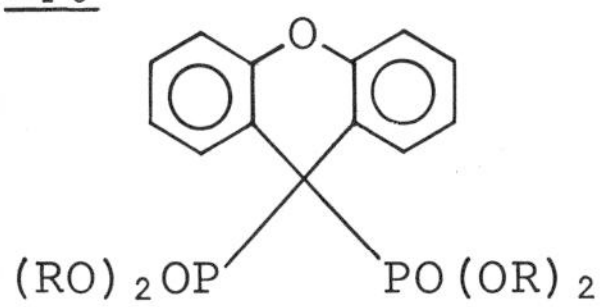

R = Et. V.5. m. 88-90°.[1361]

R = i-Pr. V.5. m. 136-138°.[1361]

[(HO)$_2$OPCH$_2$-]$_2$N(CH$_2$)$_9$N[-CH$_2$PO(OH)$_2$]$_2$. I.10. m. 217°, ^{31}P.[1328]
(C$_6$H$_{13}$)$_2$NCH[-PO(OH)$_2$]$_2$. I.12. Cryst.[1527]
H$_2$C[-CH$_2$NBuCH$_2$PO(OH)$_2$]$_2$. I.13.[251]

C$_{14}$
(RO)$_2$OPCHPh-CHPhPO(OR)$_2$.

R = Et. V.7. b$_1$ 195-196°, ^{1}H.[692]

R = Bu. V.7. b$_{0.2}$ 152-154°, IR, ^{1}H.[692]

(EtO)$_2$OPCHPhNH-NH-CHPhPO(OEt)$_2$. V.9. m. 102-103°.[1653]
[(HO)$_2$OPCH$_2$-]$_2$N(CH$_2$)$_{10}$N[-CH$_2$PO(OH)$_2$]$_2$. I.10. m. 225°, ^{31}P.[1328]
C$_{12}$H$_{25}$N[-CH$_2$PO(OH)$_2$]$_2$. I.10. m. 215°, ^{31}P.[1328]

C$_{15}$

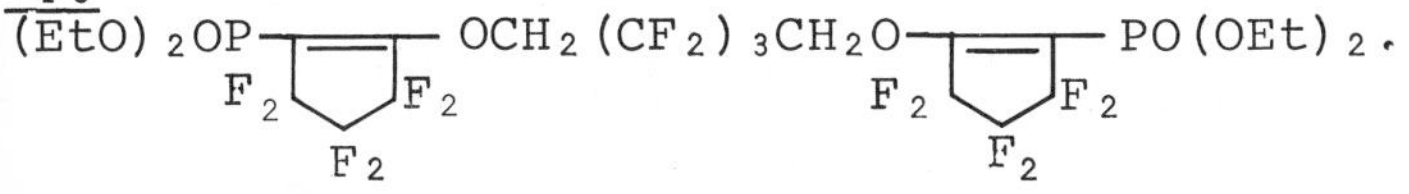

V.14. n_D^{25} 1.4039, IR.[564]

$H_2C[-\langle\bigcirc\rangle-NHC(O)PO(OR)_2]_2$.

R = Me. V.9. m. 155-156°.[1067]

R = Et. V.9. m. 115-117°.[1067]

R = ClH$_2$CCH$_2$-. V.9. m. 111-113°.[1067]

R = Pr. V.9. m. 88-89°.[1067]

R = i-Pr. V.9. Dec. 178-179°.[1067]

R = Bu. V.9. m. 68-70°.[1067]

R = i-Bu. V.9. m. 102-103°.[1067]

$(MeO)_2OP-CH[-C(O)Ph]-CHPhPO(OMe)_2$. V.5.[472]
$O=C[-OCHPhPO(OEt)_2]_2$. V.14. m. 94.5^O.[816]
$H_2C[-(CH_2)_3CO_2(CH_2)_3PO(OR)_2]_2$.

R = Me. V.7. n_D^{25} 1.4685, d_4^{25} 1.1776.[1966]

R = Et. V.7. n_D^{25} 1.4546, d_4^{25} 1.1171.[1966]

R = Bu. V.7. n_D^{25} 1.4552, d_4^{25} 1.0515.[1966]

R = BuEtCHCH$_2$-. V.7. n_D^{25} 1.4625, d_4^{25} 1.0052.[1966]

$C_{14}H_{29}CH[-PO(OH)_2]_2$. I.2. m. 156-162°.[1677c]
$[(HO)_2OPCH_2-]_2N(CH_2)_{11}N[-CH_2PO(OH)_2]_2$. I.10. m. 205°, ^{31}P.[1328]

$\underline{C_{16}}$

CH$_2$PO(OH)$_2$

$\langle\bigcirc\bigcirc\bigcirc\rangle$ I.2. Dec. 215°.[1001]

CH$_2$PO(OH)$_2$

$(MeO)_2OPCH[-C(O)Ph]CH[-C(O)Ph]-PO(OMe)_2$. V.5. m. 134-135°.[4]
$(PhO)_2OPCH(-CO_2Ph)-CH(-CO_2Ph)-PO(PhO)_2$. V.7. m. 175-176°.[967]

$\langle\bigcirc\rangle$-CH-NH(CH$_2$)$_2$NHCH-$\langle\bigcirc\rangle$
        | |
        PO(OEt)$_2$ PO(OEt)$_2$
OH   OH

V.9. m. 125-135° (rapid heating), m. >290° (slow heating).[589]

$\langle\bigcirc\rangle$ PO(OEt)$_2$
                     |
$(O)_2S$ N-CH$_2$C-CH$_2$PO(OEt)$_2$. V.5. m. 149°.[1926]
$\langle\bigcirc\rangle$ |
                     Me

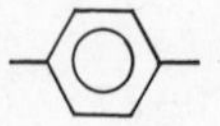

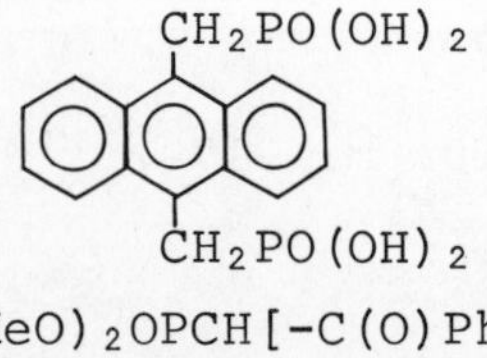

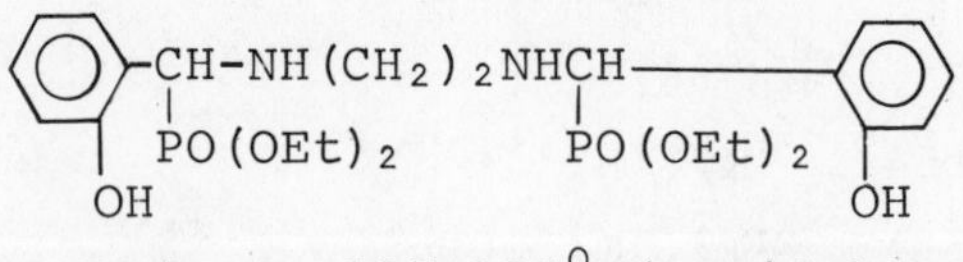

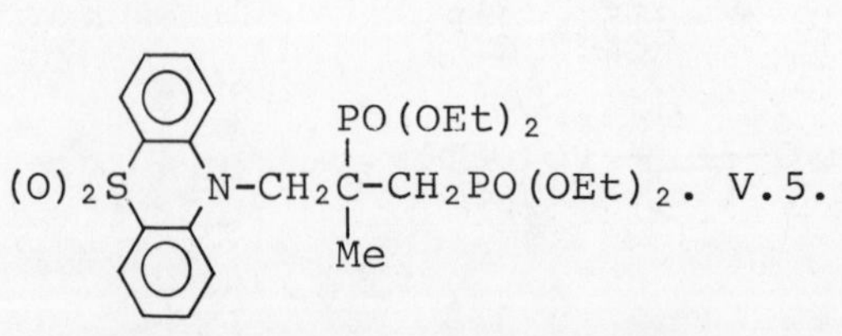

$(EtO)_2OPCHPh-NH(CH_2)_2NH-CHPhPO(OEt)_2$. V.9. m. 193-194°.[915]
$(EtO)_2OPCPh(-NHMe)-CPh(-NHMe)PO(OEt)_2$. V.9. m. 95°.[529]
$(RO)_2OP(CH_2)_3O_2C(CH_2)_8CO_2(CH_2)_3PO(OR)_2$.

R = Me. V.7. n_D^{25} 1.4681, d_4^{25} 1.1686.[1966]

R = Et. V.7. n_D^{25} 1.4538, d_4^{25} 1.1047.[1966]

R = Bu. V.7. n_D^{25} 1.4545, d_4^{25} 1.0448.[1966]

R = BuEtCHCH₂-. V.7. n_D^{25} 1.4600, d_4^{25} 0.9923.[1966]

$[(HO)_2OPCH_2-]_2N(CH_2)_{12}N[-CH_2PO(OH)_2]_2$. I.10. m. 212°, ^{31}P.[1328]

$C_{14}H_{29}N[-CH_2PO(OH)_2]$. I.10. m. 216°, ^{31}P.[1328]
$C_{15}H_{31}C(-NH_2)[PO(OH)_2]_2$. I.12. Cryst.[1527]

$\underline{C_{17}}$
2-$HOC_6H_4-CH[-PO(OH)_2]NH(CH_2)_3NHCH[-PO(OH)_2]C_6H_4OH$-2. I.2. m. 277-279° (dec.).[589]

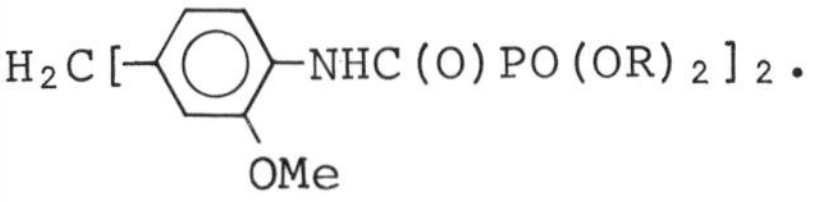

R = Me. V.9. m. 102-104°.[1067]

R = Et. V.9. m. 148-150°.[1067]

R = ClH_2CCH_2-. V.9. m. 127-128°.[1067]

R = i-Pr. V.9. m. 146°.[1067]

R = i-Bu. V.9. m. 125-126°.[1067]

$(C_8H_{17}-)_2NCH[-PO(OH)_2]_2$. I.12. Cryst.[1527]

$\underline{C_{18}}$

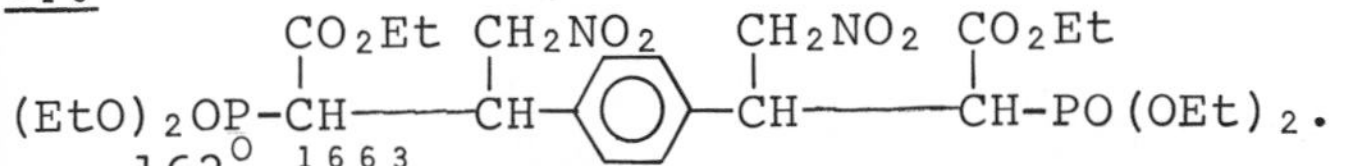
V.7. m. 161-162°.[1663]

$C_{16}H_{33}N[-CH_2PO(OH)_2]_2$. I.10. m. 215°, ^{31}P.[1328]

$\underline{C_{19}}$

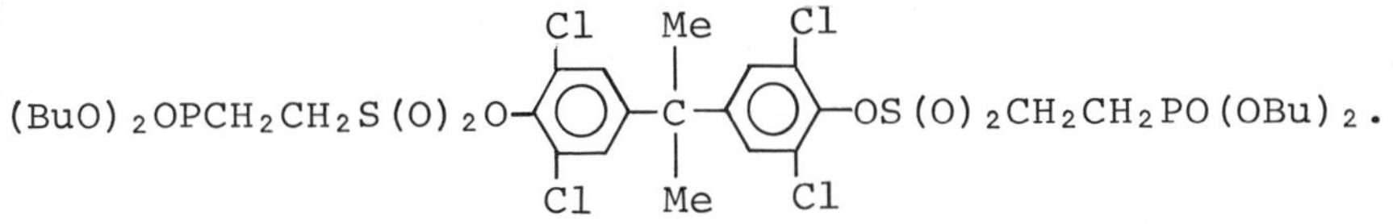

V.7. m. 133-135°.[1513]

$\underline{C_{20}}$

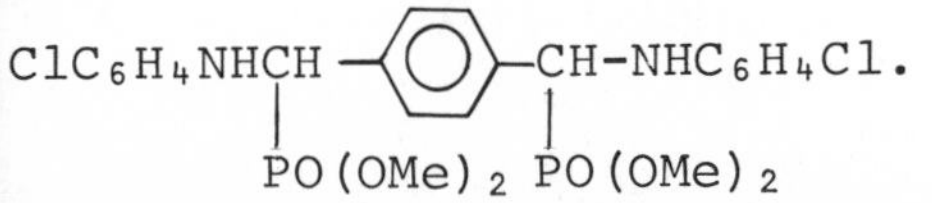
V.9. m. 263-264°.[1654]

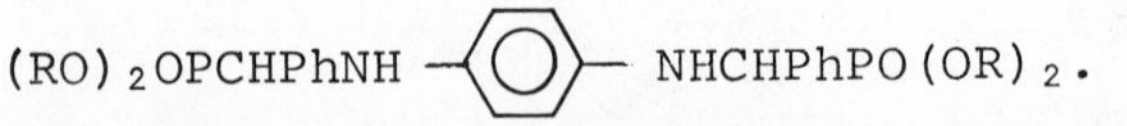

(RO)$_2$OPCHPhNH —⟨O⟩— NHCHPhPO(OR)$_2$.

R = Me. V.9. m. 187-188°.[1654] V.9. m. 180-181°.[1653]

R = Et. V.9. m. 193-194°.[1654] V.9. m. 199-200°.[1653]

R = Pr. V.9. m. 206-207°.[1654]

R = Bu. V.9. m. 211-212°.[1654]

(RO)$_2$OPCH(-NHPh) —⟨O⟩— CH(-NHPh)PO(OR)$_2$.

R = Me. V.9. m. 164-165°.[1654]

R = Et. V.9. m. 184-185°.[1654]

R = Pr. V.9. m. 189-190°.[1654]

R = i-Pr. V.9. m. 206-207°.[1654]

$C_{18}H_{37}N[-CH_2PO(OH)_2]_2$. I.10. m. 219°, ^{31}P.[1328]

$\underline{C_{21}}$
$C_6H_5CH[-HNCHPhPO(OEt)_2]_2$. V.9.[1036]
$(C_{10}H_{21})_2NCH[-PO(OH)_2]_2$. I.12. Cryst.[1527]

$\underline{C_{22}}$

Cl—⟨O⟩— N = N —[naphthalene with OH OH, HO$_3$S, SO$_3$H, PO(OH)$_2$]— N = N —⟨O⟩— Cl   V.14.[1129]

⟨O⟩— N = N —[naphthalene with OH OH, HO$_3$S, SO$_3$H, PO(OH)$_2$]— N = N —⟨O⟩   V.14.[1129]

(HO)$_2$OPCH=CPh—⟨O⟩— CPh=CHPO(OH)$_2$. I.1. m. 210°.[239]

4-MeC$_6$H$_4$CHNH—⟨O⟩— NHCHC$_6$H$_4$Me-4. V.9. m. 172-173°.[1653]
            | |
        PO(OMe)$_2$ PO(OMe)$_2$

$\underline{C_{24}}$

(EtO)$_2$OPCHPhN[-C(O)Me]—⟨O⟩— N[-C(O)Me]CHPhPO(OEt)$_2$. V.14.m

222-225°.[1653]

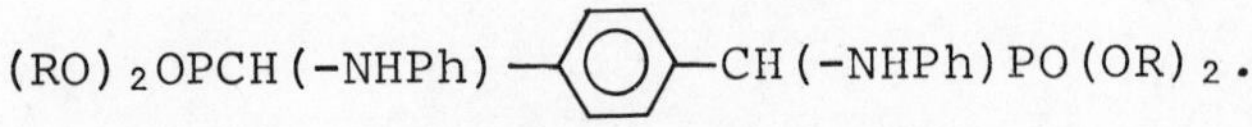

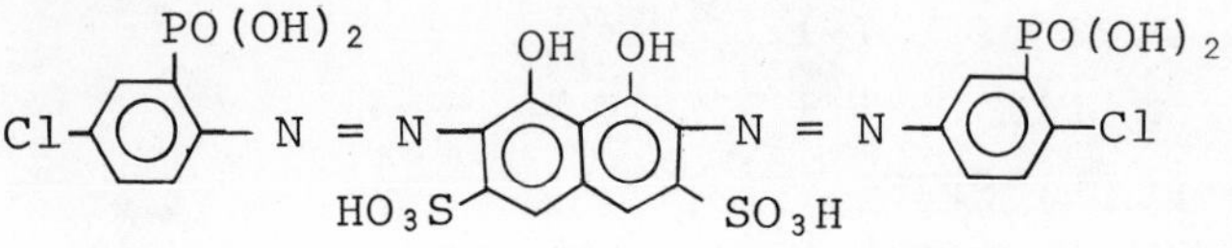

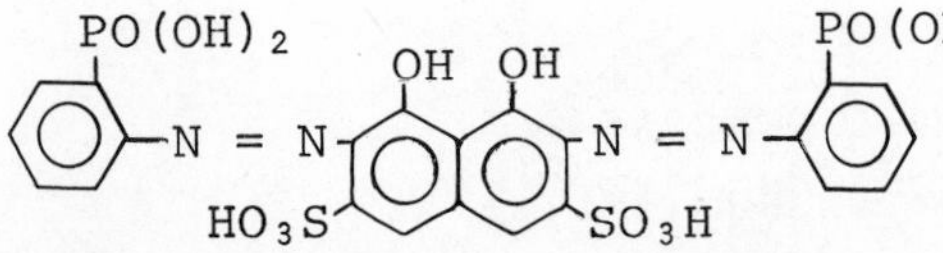

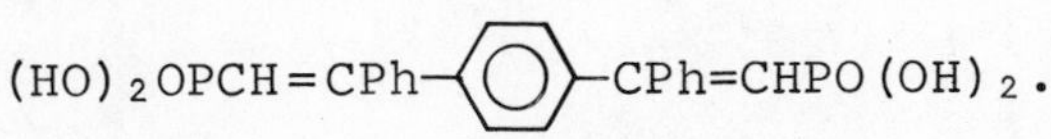

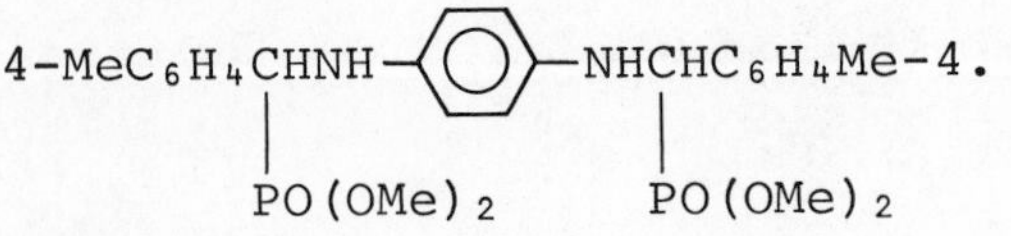

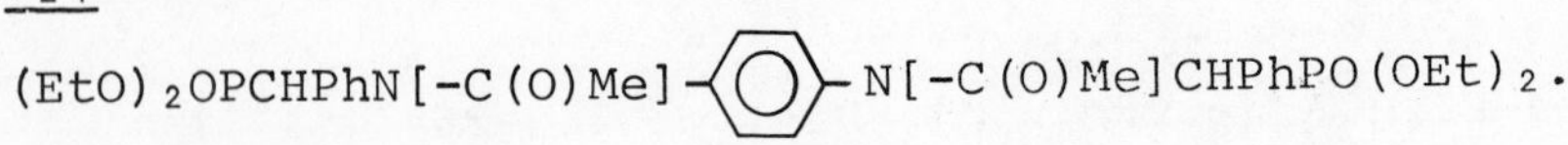

C_{26}

$(RO)_2OPCHPhNH$ —⟨O⟩-⟨O⟩— $NHCHPhPO(OR)_2$.

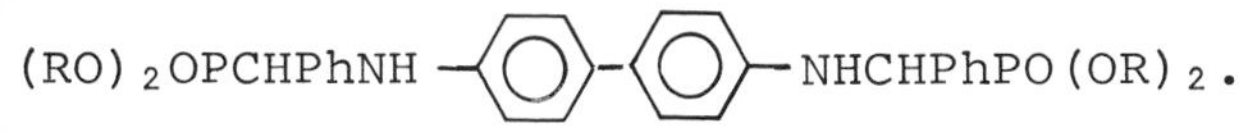

R = Me. V.9. m. 186.5-188°.[1653]
R = Et. V.9. m. 220-221°.[1653]

C_{28}

Et_2N-⟨O⟩-CHNH —⟨O⟩—NHCH-⟨O⟩-NEt_2.
          | |
       $PO(OEt)_2$ $PO(OEt)_2$

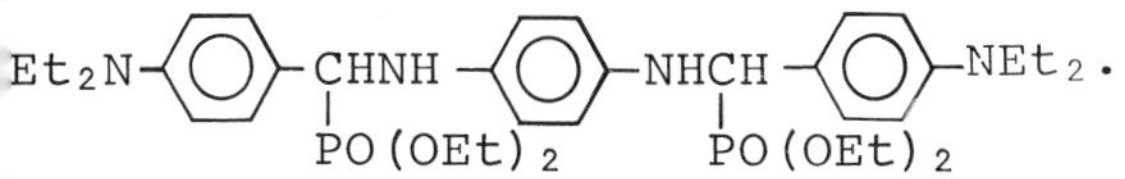

V.9. m. 157-158°.[1653]

C_{29}
$(C_{14}H_{29})_2NCH[PO(OH)_2]_2$. I.12. Cryst.[1527]

C_{37}
$(C_{18}H_{37})_2NCH[PO(OH)_2]_2$. I.12. Cryst.[1527]

C.3. Thiophosphonic Acids

C_1
$ClH_2CPS(OR)Cl$.

R = Et. IX.1.[523]
R = i-Pr. IX.1.[523]
R = $4-ClC_6H_4-$. IX.1. n_D^{25} 1.5956, d_4^{25} 1.479.[1230]
R = $2,4-Cl_2C_6H_3-$. IX.1. $b_{0.1}$ 122-123°.[1230]
R = $2,4,5-Cl_3C_6H_2-$. IX.1. $b_{0.17}$ 140.5-143°.[1230]

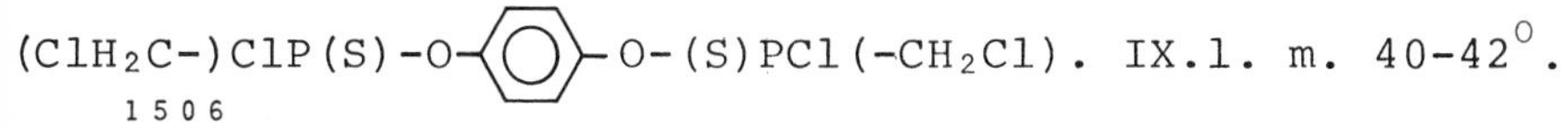

$(ClH_2C-)ClP(S)-O$-⟨O⟩-$O-(S)PCl(-CH_2Cl)$. IX.1. m. 40-42°.
[1506]

$ClH_2CP(S)(OR)NCS$.
R = Me. IX.1. $b_{0.01}$ 58°.[1858]
R = Et. IX.1. $b_{0.01}$ 63°.[1858]
$ClH_2CPS(OEt)OC_6H_4NO_2-4$. X.1.[1811]
$ClH_2CPS(OEt)SR$.
R = $4-FC_6H_4-$. X.1. m. 47°.[523]
R = $4-MeC_6H_4-$. X.1. m. 36°.[523]
$ClH_2CPS(OPr-i)SR$.
R = $4-ClC_6H_4-$. X.1. m. 30°.[523]
R = $4-FC_6H_4-$. X.1. m. 29°.[523]
R = $4-MeC_6H_4-$. X1. m. 20°.[523]

R = 4-MeOC$_6$H$_4$-. X.1. n_D^{25} 1.587[523]

ClH$_2$CPO(SR)$_2$.

R = Et. X.1. b$_1$ 130°, n_D^{25} 1.5559.[2027]

R = Bu. X.1. b$_1$ 150-151°, n_D^{25} 1.5287.[2027] X.1. b$_3$ 153-154.[1886]

R = t-Bu. X.1. n_D^{25} 1.5317.[2027]

R = Ph. X.1. n_D^{25} 1.6454.[2027]

R = 4-ClC$_6$H$_4$-. X.1. m. 110° (ether).[2027]

R = C$_{12}$H$_{25}$-. X.1. n_D^{25} 1.4925.[2027]

ClH$_2$CPS(SBu)$_2$. X.1. b$_{1.5}$ 165°.[2027]

Cl$_3$CPS(OR)$_2$.

R = Me. X.1.[1515]

R = Et. X.7. b$_1$ 88°, n_D^{20} 1.5053, d$_4^{20}$ 1.3687.[66]

H$_3$CPS(SH)F. VII.3. b$_{12-14}$ 49-50°, n_D^{20} 1.5557, d$_4^{20}$ 1.3520, IR,[1756] complexes with Ti and Sn.[1760]

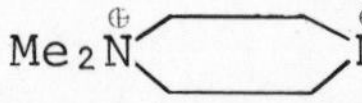

Me$_2$N$\overset{\oplus}{}$⟨   ⟩$\overset{\oplus}{}$NMe$_2$− salt.[236]

(RO)CH$_3$P⟨$\overset{O}{\underset{S}{}}$ H.

R = Me. VIII.1. b$_1$ 63-64°.[897] Me$_4$N$^\oplus$ salt. VIII.2. m. 105°. Rate of isomerisation to (MeS)CH$_3$P(O)-O$^\ominus$$\overset{\oplus}{N}Me_4$.[352] VIII.1. b$_{0.01}$ 68°, ^{1}H.[391] Isolation of opt. isomers.[294] VIII.3. Or from CH$_3$POCl$_2$, MeOH, a NaHS.[1456]

R = Et. VIII.1. b$_{0.01}$ 74°, ^{1}H.[391] VIIIc, n_D^{20} 1.4951, d$_4^{20}$ 1.1760, m. 216-218° (as Na-salt).[1471]

R = Pr. VIII.1. b$_{0.05}$ 87°, ^{1}H.[391]

R = i-Pr. VIII.1. b$_{0.3}$ 73-74°.[1457] (S)(+) from (R)(-) (i-PrO)MeP(O)H and S.[1734]

R = Bu. VIII.1. b$_{0.05}$ 107°, ^{1}H.[391] VIII.3. n_D^{20} 1.4925, d$_4^{20}$ 1.1067, m. 193-194° (as Na salt).[1471]

R = sec.-Bu. VIII.1. b$_{0.6-0.7}$ 99-100°.[1457]

R = C$_6$H$_{13}$-. VIII.3. n_D^{20} 1.4918, d$_4^{20}$ 1.0560, m. 161-163 (as Na salt).[1471]

CH$_3$P-O(CH$_2$)$_2$O-PCH$_3$. VIII.2. m. 101-103.5^O.[1497]

S$\overset{\ominus}{}$O   S$\overset{\ominus}{}$O

Na$^\oplus$ Na$^\oplus$

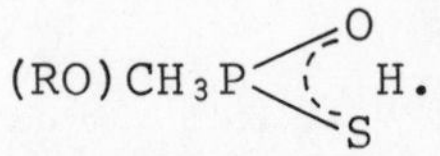

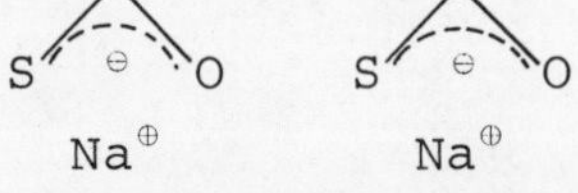

$(CH_3-)(O)\overset{\ominus}{O}PSCH_2CH_2\overset{\oplus}{N}H_3$. VIII.2. m. 190°.[351]
$(CH_3-)(O)\overset{\ominus}{O}PSCH_2CH_2\overset{\oplus}{N}HPr_2$. VIII.2. m. 106°, ^1H.[236]
$CH_3PS(OR)SH$.

R = Me. VIII.4. $b_{0.01}$ 28°.[395] $Br_3Sn^{\oplus}$ and $Cl_3Sn^{\oplus}$ salts.[1357] VIII.1. $b_{1.2}$ 66-67°.[1793]

R = Et. VIII.4. $b_{0.2}$ 45-48°.[395] VIII.4. $b_{0.05}$ 32°.[1864] VIII.1. $b_{1.2}$ 66-67°[1793]

R = ClH_2CCH_2-. VIII.4. $b_{0.007}$ 58°.[395]

R = Cl_2CHCH_2-. VIII.4. $b_{0.06}$ 66.5°.[395]

R = Pr. VIII.4. $b_{0.02}$ 50°.[395]

R = i-Pr. VIII.4. $b_{0.01}$ 38°.[1864]

R = $MeOCH_2CH_2-$. VIII.4. $b_{0.01}$ 58-60°.[395]

R = Bu. VIII.4. $b_{0.005}$ 47°.[395]

R = i-Bu. VIII.4. $b_{0.003}$ 37°.[395]

R = $MeO(CH_2)_3-$. VIII.4. $b_{0.002}$ 57°.[395]

R = $EtOCH_2CH_2-$. VIII.4. $b_{0.005}$ 92°.[395]

R = Am. VIII.4. $b_{0.04}$ 61-64°.[395]

R = Ph. VIII.4. m. 200-202°(as Ni salt).[395]

R = $Me_3CCHMe-$. VIII.4. $b_{0.01}$ 56°.[1864]

$C[CH_2OPS(-CH_3)SH]_4$. VIII.4. m. 129-130°, IR.[707]
$CH_3PS(OR)Br$.

R = Et. IX.1. b_{20} 87-88°.[1151]

R = i-Pr. IX.1. b_5 73-74°.[1151]

R = Bu. IX.1. b_{20} 135-138°.[1151]

$CH_3PS(OR)Cl$.

R = Me. IX.1.[1247] IX.1. b_{21} 54-55°.[896] IX.1. b_{20} 59°.[391]

R = Et. IX.1. b_{15} 65°.[391] IX.1. b_8 55-55.5°.[896] IX.1.[1247] IX.1. b_4 41°.[788] IX.3. b_5 47°.[788] IX.1. b_{10} 55-56°, ^1H, ^{31}P -9.6 ppm.[1372]

R = BrH_2CCH_2-. IX.5. $b_{0.3}$ 74-75°.[395]

R = ClH_2CCH_2-. IX.5. $b_{0.3}$ 72-73°.[395]

R = Cl_2HCCH_2-. IX.5. $b_{0.1}$ 66°.[395]

R = Cl_3CCH_2-. IX.5. $b_{0.4}$ 69-72°.[395]

R = Pr. IX.5. b_2 54-55°.[395] IX.1. $b_{0.5}$ 45°.[788] IX.1.[1247] IX.1. b_7 66-67°.[896] IX.1. b_{20} 86°.[391] IX.5. b_2 50-51°.[1467] IX.3. $b_{0.8}$ 47°.[788] IX.1. b_{19} 84-85°.[6]

R = i-Pr. IX.3. b_2 45-46°.[788] IX.1. $b_{0.2}$ 45-46°.[788] IX.2. $b_{0.05}$ 22°.[1428]

R = $ClH_2C(CH_2)_2$-. IX.5. $b_{0.4}$ 73-75°.[395]

R = $ClHC{=}CHCH_2$-. IX.5. $b_{0.2}$ 69-79°.[395]

R = $H_2C{=}CClCH_2$-. IX.5. $b_{0.2}$ 54-59°.[395]

R = $Cl_2HC(CH_2)_2$-. IX.5. $b_{1.2}$ 108-110°.[395]

R = $ClH_2CCHClCH_2$-. IX.5. $b_{0.2}$ 84-86°.[395]

R = $(ClH_2C)_2CH$-. IX.5. $b_{0.3}$ 77-86°.[395]

R = $ClHC{=}CClCH_2$-. IX.5. $b_{0.6}$ 87-90°.[395]

R = $Cl_2C{=}CHCH_2$-. IX.5. b_2 98-102°.[395]

R = $Cl_2C{=}CClCH_2$-. IX.5. $b_{0.2}$ 85-88°.[395]

R = $CH_3OCH_2CH_2$-. IX.5. $b_{2.8}$ 82-83°.[395]

R = Bu. IX.1. b_{17} 98°.[391] IX.3. $b_{1.3}$ 48°.[788] IX.1. $b_{0.4}$ 41°.[788] IX.5. $b_{1.5}$ 58°.[395] IX.1. b_{20} 104-105°.[6]

R = i-Bu. IX.5. b_5 68°.[395]

R = $ClH_2C(CH_2)_3$-. IX.5. $b_{0.4}$ 92-94°.[395]

R = $CH_3CCl{=}CHCH_2$-. IX.5. $b_{0.3}$ 76°.[395]

R = $MeO(CH_2)_3$-. IX.5. $b_{0.4}$ 77-90°.[395]

R = $EtO(CH_2)_2$-. IX.5. $b_{0.2}$ 71-72°.[395]

R = Am. IX.5. $b_{1.8}$ 78-82°.[395] IX.1. b_2 67°.[6]

R = Me_3CCH_2-. IX.5. m. 64-66°.[395]

R = C_6H_{13}-. IX.1. b_2 86-87°.[6]

R = $ClH_2C(CH_2)_5$-. IX.5. $b_{0.1}$ 96-100°.[395]

R = Ph. IX.5. $b_{0.3}$ 92°.[395] IX.1. b_{10} 121-123°.[1372] IX.1. $b_{0.16}$ 97°.[1380] IX.1. $b_{0.09}$ 81-82°.[1230]

R = 2,4-$Br_2C_6H_3$-. IX.1. b_1 159-162°.[1506]

R = 4-ClC_6H_4-. IX.1. $b_{0.03}$ 96°.[1380] IX.1. b_1 120-122°.[1506]

R = 2-Cl-4-FC_6H_3-. IX.1. b_1 114-116°.[1506]

R = 2,4-$Cl_2C_6H_3$-. IX.1. b_1 140-142°.[1506] IX.1. $b_{0.16}$ 105-111°.[1230]

R = 2,3-$Cl_2C_6H_3$-. IX.1. $b_{0.5}$ 130.5-131.5°.[1380]

R = 2,4,5-$Cl_3C_6H_2$-. IX.1. $b_{0.17}$ 138-140°, m. 56.5-58°.[1230]

R = 2,4,6-$Cl_3C_6H_2$-. IX.1. b_1 142-144°.[1506]

R = 4-$O_2NC_6H_4$-. IX.5. m. 64-66°.[395] IX.1. m. 67-68.5°.[1380] IX.1. m. 56-58°.[1506]

R = 2-MeC_6H_4-. IX.5. $b_{0.2}$ 94°.[395]

R = 3-MeC_6H_4-. IX.5. $b_{0.1}$ 87°.[395]

R = 4-MeC_6H_4-. IX.5. $b_{0.4}$ 111°.[395] IX.1. $b_{0.18}$ 100-101°.[1230]

R = 4-$CF_3C_6H_4$-. IX.1. $b_{0.3}$ 70-72°.[1380]

R = 2-$MeOC_6H_4$-. IX.1. $b_{1.5}$ 139-141°.[1506]

R = 4-$MeOC_6H_4$-. IX.1. $b_{0.002}$ 88-90°.[1380] IX.5. $b_{0.1}$ 109°.[395]

R = 4-$MeSC_6H_4$-. IX.1. $b_{0.008}$ 97-98°.[1380] IX.1. b_1 141-143°.[1506]

R = $C_4H_9CHEtCH_2$-. IX.5. $b_{0.4}$ 94-95°.[395]

R = $PhCH_2CH_2$-. IX.5. $b_{0.1}$ 110°.[395]

R = $C_{10}H_{21}$-. IX.5. $b_{0.4}$ 133-135°.[395]

R = 4-t-BuC_6H_4-. IX.1. m. 55-57°.[1380]

R = $C_{10}H_7$-. IX.1. m. 71-72°.[1506, 1511]

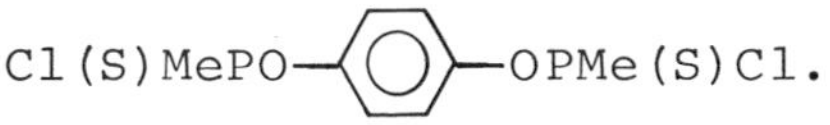

. IX.1. m. 160-161°.[1506]

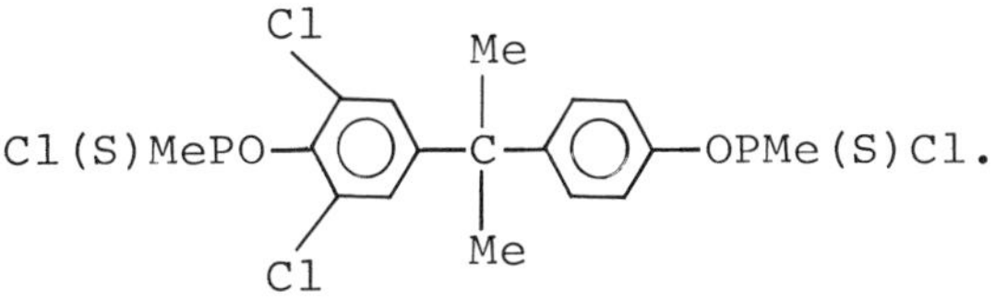

. IX.1. m. 142-143°.[1506]

$CH_3PS(OPr\text{-}i)F$. From MePS(OPr-i)OH and $F_2C{=}C(-CF_3)_2$ in the presence of NEt_3, or from $MePS[OCF_2CH(CF_3)_2]_2$ and NEt_3.[978]

$CH_3PS(OCH_2CH_2NMe_2)F$. IX.1. $b_{0.5}$ 52°.[236]

$CH_3PS(F)(OCH_2CH_2\overset{\oplus}{N}Me_3)I^{\ominus}$. IX.5. m. 109°, ^1H.[126] Kinetics of hydrolysis, AChE inhibition.[882]

$CH_3PS(OR)NCS$.

R = Me. IX.1. b_1 66°.[1858]

R = Et. IX.1. $b_{0.01}$ 39°.[1858]

$CH_3PS(OEt)SCN$. IX.4. $b_{0.01}$ 40°.[1863]

$CH_3PO(SR)Cl$.

R = Me. IX.3. b_2 70-71°.[1370]

R = Et. IX.3. b_3 75-76°.[1370] IX.1. b_{10} 93-95°, ^{31}P

-60 ppm.[1372]

R = Pr. IX.3. b_5 97°.[1370]

R = i-Pr. IX.3. b_5 78-79°.[1370]

R = Ph. IX.1. $b_{0.04}$ 116-118°, ^1H, ^{31}P -52 ppm.[1372]

$CH_3PO(SEt)F$. MO calculation

$CH_3PS(SR)Cl$.

R = Me. By heating of MePS(OMe)Cl, b_{23} 106.5-107°.[720]

R = Ph. IX.1. $b_{0.04}$ 106-107°, ^1H, ^{31}P -90 ppm.[1372]

$CH_3PS[-SCH_2CH_2NHC(O)Me]F$. From MePS(SH)F, $\overline{HNCH_2CH_2}$ and MeC(O)Cl, $b_{0.5}$ 120-125°.[1519]

$CH_3PS(SEt)SCN$. IX.4. $b_{0.01}$ 46°.[1850]

$CH_3PS(OR)_2$.

R = Me. X.1. b_4 40°.[788] X.1. $b_{2-2.5}$ 40-42 .[1457]

R = Et. X.1. $b_{2.3}$ 46°.[788] X.1. b_{25} 89-92°.[1457] X.1. b_{14} 77-79°, ^{31}P -96 ppm.[1372] X.7. b_{13} 76-78°.[901] X.1. b_5 57-59°.[1243] X.1. $b_{1.5}$ 58.5-60°, IR, ^{31}P -94.3 ppm.[1151]

R = ClH_2CCH_2-. X.2. $b_{0.01}$ 70°.[1853]

R = FH_2CCH_2-. X.1. $b_{0.5}$ 64-65°.[897]

R = Pr. X.1. n_D^{24} 1.4627, d_4^{20} 1.019.[1243] X.1. $b_{0.18}$ 39°[788] X.1. $b_{0.45-0.5}$ 71-73°.[1457]

R = i-Pr. X.1. b_1 42°.[788] X.1. $b_{0.25}$ 42-44°.[1457]

R = $ClHCMeCH_2$-. X.2. $b_{0.01}$ 76°.[1853]

R = $(ClH_2C)_2CH$-. X.2. $b_{0.01}$ 118°.[1853]

R = Bu. X.1. $b_{0.7}$ 93-94°.[1457] X.1. $b_{0.09}$ 51°.[788] X.7. b_1 72-74°.[901] X.2. b_1 69-70°.[928]

R = sec.-Bu. X.1. $b_{0.6}$ 58-60°.[1457]

R = Am. X.1. $b_{0.08}$ 74°.[788]

R = Ph. X.1. $b_{0.04}$ 131-131.5°, m. 35-36°, ^1H, ^{31}P -90 ppm.[1372]

$CH_3PS(OMe)OR$.

R = $EtSCH_2CH_2$-. X.1. n_D^{20} 1.5204, d_4^{20} 1.1861.[896]

R = 3-Me-4-$O_2NC_6H_3$-. X.1.[1811]

R = 2-$MeSC_6H_4$-. X.1. $b_{0.01}$ 104°.[1797]

R = 4-$MeSC_6H_4$-. X.1. $b_{0.01}$ 108°.[1797] X.1.[1737]

R = 4-Cl-2-$MeSC_6H_3$-. X.1. $b_{0.01}$ 111°.[1797]

R = 4-Me(O)S-C_6H_4-. X.1. $b_{0.01}$ 110°.[1797]

R = 4-$Me(O)_2S-C_6H_4$-. X.1. m. 67° (EtOH).[1797]

R = 4-$EtSC_6H_4$-. X.1. $b_{0.01}$ 110°.[1797]

R = 2-MeS-4-MeC_6H_3-. X.1. $b_{0.01}$ 117°.[1797]

R = 4-MeS-3-MeC_6H_3-. X.1. $b_{0.01}$ 96°.[1797]

R = 4-Me_3C-2-ClC_6H_3-. X.1. $b_{0.01}$ 111°.[1814]

(MeO)(S)MePOOPMe(S)OMe. X.1. m. 94-98°.[1506]

$CH_3PS(OEt)OR$.

R = FH_2CCH_2-. X.1. $b_{0.5}$ 43-44°.[897]

R = $EtSCH_2CH_2$-. X.1. $b_{0.01}$ 63°.[1812] X.2. $b_{0.6}$ 88°.[785] X.1. b_1 78°, n_D^{20} 1.5050, d_4^{20} 1.1124.[896] X.1. $b_{0.1}$ 101-105°.[1151]

R = Ph. X.1.[1512]

R = 4-ClC_6H_4-. X.1. $b_{0.01}$ 94°.[1813]

R = 2,4-$Cl_2C_6H_4$-. X.1. $b_{0.5}$ 142-143°.[1506]

R = 2-$MeOC_6H_4$-. X.1. b_1 139-141°.[1506]

R = 2-$MeSC_6H_4$-. X.1. $b_{0.01}$ 108°.[1797]

R = 4-$MeSC_6H_4$-. X.1. $b_{0.01}$ 102°.[1797]

R = 4-MeC(O)-C_6H_4-. X.1. n_D^{23} 1.5418.[1820]

R = 4-$EtSC_6H_4$-. X.1. $b_{0.01}$ 125°.[1797]

R = 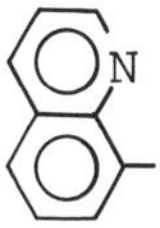 X.1. $b_{0.07}$ 132°.[527]

R = 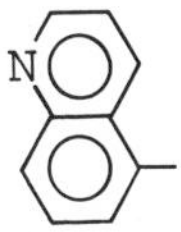 X.1. $b_{0.01}$ 125°.[527]

R =  X.1. m. 63-65°.[527]

R = 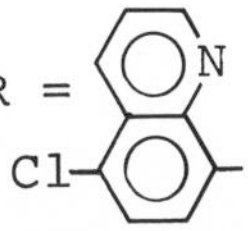 X.1. m. 64°.[527]

R = 4-Me_3C-2-ClC_6H_3. X.1. $b_{0.01}$ 112°.[1814]

R = 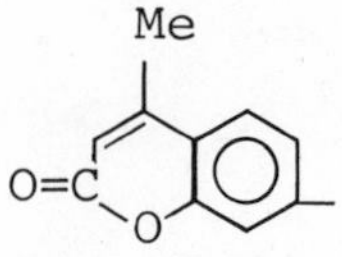 X.l. m. 73-74°.[1151] X.l. m. 72-73°.[896]

R = 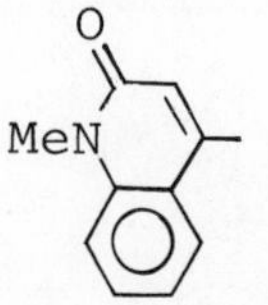 X.l. m. 120-122°.[1111]

R = 2,4-Me_3C-6-$MeSC_6H_2$-. X.l. $b_{0.01}$ 122°.[1797]

$CH_3PS(OCH_2CH_2F)OCH_2CH_2SEt$. X.l. $b_{0.02}$ 70-72°.[897]

$CH_3PS(OPh)OR$.

R = 4-NCC_6H_4-. X.l. m. 73-74°.[1380]

R = 3-F_3C-4-NCC_6H_3-. X.l. m. 44-47°.[393]

$CH_3PS(OC_6H_4NO_2$-4)OR.

R = Me. X.l. m. 52-53°.[896]

R = Et. X.l. m. 35-36°.[896] X.l. m. 37-39°, $b_{0.005}$ 160-165°, ^{31}P -94.3 ppm, IR.[1151] X.l.[1811]

R = FH_2CCH_2-. X.l. $b_{0.01}$ 102-103°.[897]

R = Pr. X.l. $b_{0.003}$ 118-121°.[1116]

R = H_2C=$CClCH_2$-. X.l. $b_{0.003}$ 140°.[1378]

R = Cl_2C=$CClCH_2$-. X.l. m. 68.5-69.5° (hexane).[1378]

R = Ph. X.l. m. 80-81°.[1380]

R = 2-BrC_6H_4-. X.l. m. 71-73°.[394, 1380]

R = 3-BrC_6H_4-. X.l . Solid.[394]

R = 2-ClC_6H_4-. X.l. m. 84-86°.[394, 1380]

R = 3-ClC_6H_4-. X.l. m. 75-77°.[1380]

R = 4-ClC_6H_4-. X.l. m. 111-113°.[394, 1380]

R = 2,4-$Cl_2C_6H_3$. X.l. m. 84.5-85°.[394, 1380]

R = 3-$O_2NC_6H_4$-. X.l. m. 148-149°.[1380]

R = 2-MeC_6H_4-. X.l. m. 71-71.5°.[1380]

R = 3-MeC_6H_4-. X.l. m. 57-59°.[1380]

R = 4-MeC_6H_4-. X.l. m. 107-108°.[1380]

R = 4-$F_3CC_6H_4$-. X.l. m. 75-77°.[1380]

R = 4-$MeOC_6H_4$-. X.l. m. 105.5-106°.[1380]

R = 2-EtC_6H_4-. X.1. m. 64-66°.[1380]

R = 4-$Me_3CC_6H_4$-. X.1. m. 83°.[1380]

R = 1-$C_{10}H_7$-. X.1. m. 104-106°.[1380]

R = 2-$C_{10}H_7$-. X.1. m. 93-95°.[1380]

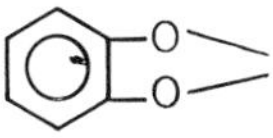
P(S)Me. X.2. b_2 88°, m. 50°, ^1H.[2075]

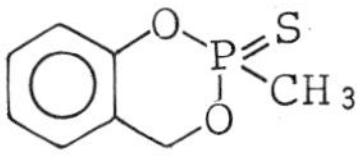
X.1. $b_{0.6}$ 130°.[508]

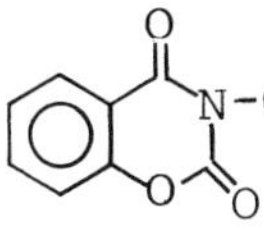
-OPMe(S)OMe. X.1. m. 141°.[519]

$CH_3PS(OPr$-i)ON=C(-CN)-C_6H_4Cl-2. VI.2. m. 66-68°.[224]

$CH_3PO(OMe)SR$.

R = FH_2CCH_2-. X.3. b_1 59-60°.[897]

R = Bu. X.3. b_1 64-65°.[6]

R = $EtSCH_2CH_2$-. X.3. $b_{0.01}$ 93°.[1847]

R = Ph. X.6. $b_{0.04}$ 101-102°.[1317]

$CH_3PO(OEt)SR$.

R = Et. X.3.[644] X.3. b_{10} 96-98°, ^{31}P -48 ppm.[1372] X.2. b_{10} 95-97°.[1469] X.3. b_9 92-93°.[1471] X.3. b_7 87-88°, n_D^{20} 1.4768.[902]

R = BrH_2CCH_2-. X.3. $b_{0.1}$ 80°.[1844]

R = FH_2CCH_2-. X.3. b_1 82°.[897]

R = $MeSCH_2$-. X.3. b_2 117.5-120°.[902]

R = Pr. X.3. $b_{2.5}$ 85-86°.[644]

R = $HO_2CCH_2CH_2$-. X.8. $b_{0.02}$ 81-82°.[1533]

R = $EtSCH_2$-. X.3. $b_{0.01}$ 73°.[1827] X.12. $b_{0.01}$ 70°.[1849]

R = Bu. X.3. b_9 112-114°.[1471] X.3. b_2 100-102°.[644]

R = $Me_2NC(S)SCH_2$-. X.12. m. 141°.[1849]

R = $EtSCH_2CH_2$-. X.1. $b_{0.01}$ 83°.[1827] X.1. b_1 98-99°.[895]

R = i-Am. X.3. b_1 73-74°.[644]

R = $EtS(CH_2)_3$-. X.1. b_2 112-114°.[895]

R = C_6H_{13}-. X.3. b_1 105-106°.[644]

R = Ph. X.6. $b_{0.03}$ 96-97°.[895]

R = $EtS(CH_2)_4$-. X.1. b_1 126-126.5°.[895]

R = $BuSCH_2CH_2$-. X.3. b_1 132-133°.[894]

R = $Me_3C-(CH_2)_3$-. X.3. b_1 76-77°.[644]

R = $PhOCH_2CH_2$-. X.3. $b_{0.01}$ 106°, m. 36.5-37°.[645]

R = $4-BrC_6H_4OCH_2CH_2$-. X.3. $b_{0.01}$ 140°, m. 46-46.5°.[645]

R = $3-ClC_6H_4OCH_2CH_2$-. X.3. $b_{0.01}$ 137°.[645]

R = $4-ClC_6H_4OCH_2CH_2$-. X.3. $b_{0.01}$ 132°.[645]

R = $PhSCH_2CH_2$-. X.3. b_1 155-156°.[894]

R = $C_6H_{13}SCH_2CH_2$-. X.3. b_1 137-138°.[894]

R = $3-MeC_6H_4OCH_2CH_2$-. X.3. b_1 132°.[645]

R = $4-MeC_6H_4OCH_2CH_2$-. X.3. $b_{0.01}$ 119°.[645]

R = $3-MeOC_6H_4OCH_2CH_2$-. X.3. $b_{0.01}$ 125°.[645]

R = $4-MeOC_6H_4OCH_2CH_2$. X.3. $b_{0.01}$ 126°.[645]

R = $C_8H_{17}SCH_2CH_2$-. X.3. b_1 170-171°.[894]

R = $4-Me_3CC_6H_4OCH_2CH_2$-. X.3. $b_{0.01}$ 140°.[645]

R = $C_{10}H_{21}SCH_2CH_2$-. X.3. b_1 192-195°.[894]

$CH_3PO(OPr)SR$.

R = Et. X.2. b_{12} 109-110°.[1469]

R = Bu. X.3. b_1 85-86°.[6]

$CH_3PO(OPr-i)SR$.

R = Me. X.3. $b_{0.3}$ 45°(+), n_D^{25} 1.4697.[235] X.4. $b_{4.5}$ 76° n_D^{25} 1.4705.[2]

R = Et. X.2. b_9 94-95°.[1469] X.11. $b_{0.35}$ 46° (+), n_D^{25} 1.4661.[235] X.3. b_1 54° (-), n_D^{20} 1.4668.[1428]

R = Pr. X.4. b_2 80°, n_D^{25} 1.4661.[2]

R = $Me_2NCH_2CH_2$-. AChE inhibition.[294]

R = Ph. X.6. $b_{0.04}$ 113-114°[1317] X.2. $b_{0.1}$ 95-97°, IR, ^1H.[1342] X.2, X.1. (S) (+) aus (R) (-).[1734]

$CH_3PO(OBu)SR$.

R = Et. X.2. b_9 115-117°.[1469]

R = Bu. X.2. b_{13} 143-146°.[1470] X.3. b_1 92-93°.[6]

$CH_3PO(OBu-i)SR$.

R = Et. X.2. b_1 111-112°.[1469]

R = Bu. X.2. b_8 127-128°.[1469]

R = i-Bu. X.2. b_{12} 137-141°.[1470]

$CH_3PO(OBu\text{-}sec.)SBu$. X.1. AChE inhibition[4]

$CH_3PO(OCH_2CH_2NMe_2)SR$.

R = Me. X.1. $b_{0.3}$ 119°.[1171]

R = Et. X.1. $b_{0.6}$ 124°, AChE inhibition is very high.[1171]

R = MeCH(OH)-. From MePO(OR)SH and $MeCHO/NEt_3$.[195]

$CH_3PO(OAm)SBu$. X.3. $b_{1.5}$ 87°.[6]

$CH_3PO(OCH_2CMe_3)SBu$. X.1. AChE inhibition[4]

$CH_3PO(OC_6H_{13})SBu$. X.2. b_9 163-165°.[1470] X.3. b_2 116°.[6]

$CH_3PO(OCHMeCMe_3)SBu$. X.1. AChE inhibition[4]

$CH_3PO(OPh)SR$.

R = Ph. X.3. $b_{0.03}$ 137-138°, ^1H, ^{31}P -46 ppm.[1372]

R = $PhOCH_2CH_2$-. Antienzymatic activity.[306]

R = $4\text{-}BrC_6H_4OCH_2CH_2$-. Antienzymatic activity.[306]

R = $3\text{-}ClC_6H_4OCH_2CH_2$-. Antienzymatic activity.[306]

R = $4\text{-}ClC_6H_4OCH_2CH_2$-. Antienzymatic activity.[306]

R = $3\text{-}O_2NC_6H_4OCH_2CH_2$-. Antienzymatic activity.[306]

R = $4\text{-}O_2NC_6H_4OCH_2CH_2$-. Antienzymatic activity.[306]

R = $3\text{-}HOC_6H_4OCH_2CH_2$-. Antienzymatic activity.[306]

R = $4\text{-}HOC_6H_4OCH_2CH_2$-. Antienzymatic activity.[306]

R = $3\text{-}MeC_6H_4OCH_2CH_2$-. Antienzymatic activity.[306]

R = $4\text{-}MeC_6H_4OCH_2CH_2$-. Antienzymatic activity.[306]

R = $3\text{-}MeOC_6H_4OCH_2CH_2$-. Antienzymatic activity.[306]

R = $4\text{-}MeOC_6H_4OCH_2CH_2$-. Antienzymatic activity.[306]

R = $4\text{-}Me_3C\text{-}C_6H_4OCH_2CH_2$-. Antienzymatic activity.[306]

$CH_3PO(OC_7H_{15})SBu$. X.3. $b_{0.02}$ 70°.[6]

$CH_3PO(OC_8H_{17})SBu$. X.3. $b_{0.03}$ 95°.[6]

$CH_3PO(OC_9H_{19})SBu$. X.3. $b_{0.02}$ 110°.[6]

$CH_3PO(OC_{10}H_{21})SBu$. X.3. $b_{0.02}$ 120°.[6]

$CH_3PO(SMe)OCHCH_2CHMe(CH_2)_2CH(\text{-}CHMe_2)$. X.2, X.3. m. 45-47°, ^1H.[237]

(RS-)(O)$MePO(CH_2)_2OPMe(O)$(-SR).

R = Me. X.2. $b_{0.0001}$ 84-85°.[1497]

R = Et. X.2. $b_{0.0001}$ 90-93°.[1494]

R = Bu. X.2. $b_{0.0001}$ 104-105 .[1497]

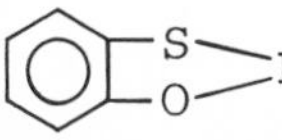P(O)Me. X.2. ^1H.[2075]

$CH_3PS(OMe)SR$.

R = FH_2CCH_2-. X.3. b_3 84-85°.[897]

R = $EtSCH_2CH_2$-. X.1. $b_{0.01}$ 74°, DL_{50} 2.5 mg/kg (rat).[1827]

R = $HC\equiv CCH_2NMe$-C(O)CH_2-. X.3. n_D^{25} 1.566.[474]

R = $MeO_2CCH_2CH(-CO_2Me)$-. X.1. $b_{3.5}$ 155-156°, n_D^{20} 1.5269. d_4^{20} 1.2765.[896]

R = EtO_2CCH(-SMe)-CHMe-. X.3. $b_{0.5}$ 134°.[1804]

R = EtO_2CCH(-SEt)-CHMe-. X.3. $b_{0.3}$ 130-132°.[1804]

R = 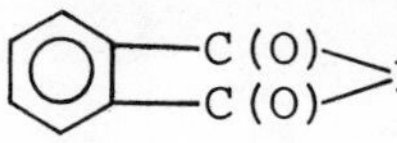

NCH_2-. X.3. m. 89°, insecticide[1973]

$CH_3PS(OEt)SR$.

R = FH_2CCH_2-. X.3. $b_{0.5}$ 68°.[897]

R = $H_2NC(S)CH_2$-. X.12. m. 67-68°.[1805]

R = $MeSCH_2CH_2$-. X.1. $b_{0.01}$ 60°.[1827]

R = MeC(O)NHC(O)CH_2-. X.3. n_D^{22} 1.5538.[1815]

R = EtSCHMe-. X.3. $b_{0.01}$ 78°.[1834]

R = $EtSCHMeCH_2$-. X.12. $b_{0.18}$ 100-104°.[1443]

R = $Et_2NCH_2CH_2$-. X.1. $b_{0.001}$ 67°.[1795]

R = $MeO_2CCMe(-SMe)CH_2$-. X.3. $b_{0.05}$ 128°.[1803, 1804]

R = MeO_2CCH(-SMe)CHMe-. X.3. $b_{0.05}$ 109°.[1804]

R = $MeO_2CCH_2CH(-CO_2Me)$-. X.1. b_2 166-167°, n_D^{20} 1.5065, d_4^{20} 1.1818.[896]

R = EtO_2CCH(-SMe)CHMe-. X.3. $b_{0.3}$ 126°.[1804]

R = MeO_2CCH(-SEt)CHMe-. X.3. $b_{0.01}$ 108°.[1804]

R = 4-MeC_6H_4-. X.1. $b_{0.01}$ 147°.[1798]

R = EtO_2CCH(-SEt)CHMe-. X.3. $b_{0.1}$ 124°.[1804]

(EtO)(S)MeP-SCH_2SPMe(S)(OEt). X.3. insecticide.[1840]

(EtO)(S)MeP-SCH(-CN)-$(CH_2)_2$CH(-CN)-SPMe(S)(OEt). X.3.[1852]

(EtO)(S)MePSCH[-C(O)NH_2]-$(CH_2)_2$-CH[-C(O)NH_2]-SPMe(S)(OEt). X.3. m. 116°.[1852]

(EtO)(S)MePSCH(-CO_2Et)-S-CH(-CO_2Et)-SPMe(S)(OEt). X.3. n_D^{24} 1.5402.[1802]

$CH_3PS(OCH_2CH_2Cl)SC_6H_4Cl$-4. X.1. m. 93-95°.[1331]

$CH_3PS(OCH_2CHCl_2)SC_6H_4Cl$-4. X.1. m. 77-78°.[1331]

$CH_3PS(OCH_2CH_2F)SR$.

R = $MeO_2CCH_2CH(-CO_2Me)$-. X.1. $b_{0.008}$ 96-97°.[897]

R = $EtO_2CCH_2CH(-CO_2Et)-$. X.1. $b_{0.005}$ 109-110°.[897]

R = $i-PrO_2CCH_2CH(-CO_2Pr-i)-$. X.1. $b_{0.015}$ 100-101°.[897]

$CH_3PS(OPr-i)SR$.

R = Et. X.1. (+), $b_{0.7}$ 57°, n_D^{20} 1.5158.[1428] X.4. (-).[1428]

R = i-Pr. X.4.[1735]

R = $MeSCH_2CH_2-$. X.3. $b_{0.01}$ 74°.[407]

R = Ph. X.4. b_{10} 71-73°.[1735] Synthesis of the opt. isomers, X.1. $b_{0.04}$ 105°.[1974]

$CH_3PS(OCH_2CHClCH_2Cl)SC_6H_4Cl-4$. X.1. m. 77-78°.[1331]

$CH_3PS(OBu)SBu$. X.2. b_3 115-116°, n_D^{20} 1.5120, d_4^{20} 1.0222.[1473]

$CH_3PS(OBu-i)SCH_2N$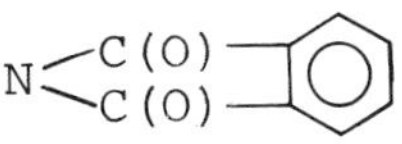 X.3. m. 49°, insecticide.[1973]

$CH_3PS(SPr)O-$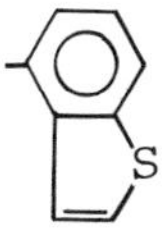 X.1. $b_{0.23}$ 153-158°.[1245]

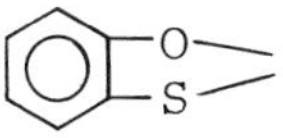P(S)Me. X.2. ^1H.[2075]

$CH_3PO(SR)_2$.

R = Et. X.1.[1495] X.6. $b_{0.045}$ 63.5-64°.[1316] X.1. b_{12} 120-123°, ^{31}P -58 ppm.[1372]

R = Pr. X.6. $b_{0.04}$ 60-61°.[1316]

R = Bu. X.6. $b_{0.04}$ 82.5-83.5°.[1316]

R = Ph. X.1. m. 82-83°, ^{31}P -56 ppm.[1372]

$CH_3PO(SEt)SPh$. X.1. $b_{0.07}$ 135-136°. ^{31}P -58 ppm.[1372]

$CH_3PO(SPr)(-SCH_2CHMeSEt)$. X.11. $b_{0.4}$ 132-137°.[1443]

$CH_3PO(SCH_2CH_2NMe_2)SR$.

R = Me. X.1. $b_{0.5}$ 128°.[1171]

R = Et. X.1. $b_{2.5}$ 140°.[1171]

R = $Me_2NCH_2CH_2-$. X.1. $b_{0.5}$ 145°.[1171]

R = $Et_2NCH_2CH_2-$. X.1. b_1 158°.[1171]

$CH_3PS(SR)_2$.

R = Me. X.1. b_{14} 131-133°, n_D^{20} 1.6386, d_4^{20} 1.2548.[720]

R = Et. X.1. b_{10} 131-133°, ^1H, ^{31}P -77 ppm.[1372]

R = ClH_2CCH_2-. X.1.[1510]

R = Ph. X.1. $b_{0.03}$ 169-171°, m. 49-50°, ^1H, ^{31}P -79 ppm.[1372]

R = 4-ClC_6H_4-. X.1. m. 94°.[1857]

R = 3,4-$Cl_2C_6H_3$-. X.1. m. 99°.[1857]

R = $PhCH_2$-. X.1. m. 47-48°.[1521]

R = C_8H_{17}-. X.1.[1504]

$CH_3PS(SEt)SR$.

R = ClH_2CCH_2-. X.1.[1507]

R = $EtSCH_2CH_2$-. X.3. $b_{0.01}$ 89°.[1859]

R = Ph. X.3. $b_{0.01}$ 108°.[1859]

$CH_3PS(SPr)S$- 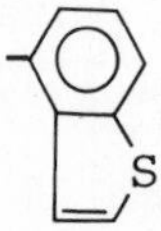 X.1. b_{22-38} 168-178°, n_D^{22} 1.6915, d_4^{20} 1.294.[1245, 1325]

$CH_3PS(SBu)SCH_2CH_2NMe_2$. X.1.[1509]
$CH_3PS(SPh)SCH_2CH_2SEt$. X.3.[1859]

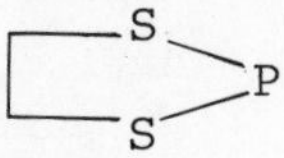 P(S)Me. X.2. b_2 132-134°, m. 74°.[2076]

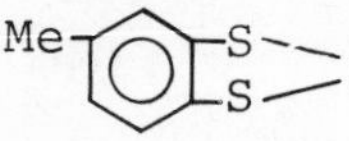 P(S)Me. X.2.[2076]

<u>C_2</u>
H_2C=$CClPO(SEt)_2$. X.1. $b_{0.1}$ 110-112°.[1729]
$ClH_2CCH_2PS(OC_6H_3Cl_2$-2,4)Cl. IX.1. $b_{0.17}$ 140-141°.[665]
$ClH_2CCH_2PO(SR)_2$.

R = Et. X.1. $b_{0.05}$ 82-85°.[596]

R = Bu. X.1. b_2 168-170°.[1886]

$ClH_2CCH_2PS(SPh)_2$. X.1. m. 64-66.5°.[1695]
$Cl_3CCH(OH)PS(OR)_2$.

R = Et. X.9. m. 124-126°.[32]

R = Cl_3CMe_2-. X.9. m. 181-182°.[32]

[$Cl_3CCH(OH)$-]SP 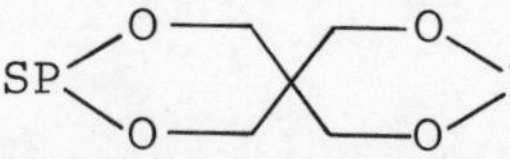 PS[-(HO)$HCCCl_3$]. X.1.[1684]

H_2C=$CHPS(OH)_2$. VII.1. As aniline salt, dec. 105-106°.[907]
H_2C=$CHPS(OEt)Cl$. IX.1. b_{12} 71-73°.[907]

$H_2C{=}CHPS(OEt)NCS$. IX.1. b_1 73°.[1858]

$H_2C{=}CHPS(OR)_2$.

R = Me. X.1. b_{20} 82-83°.[907]

R = Et. X.1. b_{10} 75-76°.[907]

R = Bu. X.2. $b_{2.5}$ 96.2-96.8°.[898, 930]

$H_2C{=}CHPS(OEt)OCH_2CH_2SEt$. X.1. b_4 135-136°.[907]

$H_2C{=}CHPO(OEt)SCH_2CH_2SEt$. X.4. $b_{0.0002}$ 84-85°.[907]

$H_2C{=}CHPO(SEt)_2$. X.1. $b_{0.03}$ 150°, n_D^{25} 1.5721.[1729]

$CH_3C(O)PS(OEt)_2$. X.7. b_2 92-93°, n_D^{20} 1.4500, d_4^{20} 1.0926.[1668]

$OHCCH_2PS(OR)_2$.

R = Me. X.12. $b_{0.08}$ 67-68°; as 2,4-dinitrophenylhydrazone, m. 111.4-114.5°.[1345]

R = Et. X.12. $b_{0.13}$ 82-83°; as 2,4-dinitrophenylhydrazone, m. 86-87°.[1345]

R = Pr. X.12. $b_{0.06}$ 82-84°; as 2,4-dinitrophenylhydrazone, m. 74-74.5°.[1345]

$OHCCH_2PO(OEt)SEt$. X.12. b_{13} 122°; as 2,4-dinitrophenylhydrazone, m. 107-108°.[1347]

$[CH_3NHC(O)-]SP$  $PS[-(O)CNHCH_3]$. X.10. m. 250-251°.[269, 1684]

$[CH_3NHC(S)-]SP$  $PS[-(S)CNHCH_3]$. X.10.[1684]

$(C_2H_5-)(NaO)P$ 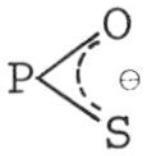 $Na^{\oplus}$. VII.2.[794]

$C_2H_5PS(SH)F$. VII.3. b_{12-13} 64-67°, n_D^{20} 1.5433, IR.[1756] Complexes with As^{3+}, Sb^{3+}, Cl_2Ti^{2+}, Cl_2Sn^{2+}, Br_2Sn^{2+}.[1760]

$(RO)C_2H_5P$ 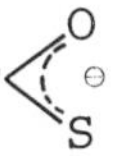 $H^{\oplus}$.

R = Et. VIII.1. $b_{0.01}$ 49-51°, n_D^{20} 1.4905, as cyclohexylammonium salt, m. 126-127°.[1310] VIII.1, VIII.3. Isolated as Na salt, separ. of opt. isomers by chinine salt.[1300] VIII.4. $b_{0.4}$ 56°.[395] VIII.1. $b_{0.01}$ 38°.[1864]

R = i-Pr. VIII.4. $b_{0.01}$ 37°.[1864]

R = Bu. VIII.4. $b_{0.006}$ 50°.[395] VIII.3. Isolated as Na salt.[288]

$C_2H_5PS(OR)SH$.

R = Me. VIII.1. $b_{0.6}$ 60°.[1793]

R = Et. VIII.1. b_1 70°.[1793]

R = Pr. VIII.1. b_1 78-79°.[1793]

R = i-Pr. VIII.1. $b_{0.4}$ 80-90°.[1793]

R = i-Bu. VIII.1. $b_{1.4}$ 108°.[1793]

$C_2H_5PS(OR)Cl$.

R = Me. IX.1.[1247]

R = Et. IX.1.[1247] IX.2.[1296] IX.3.[1298] IX.2.(-) form.[12 IX.1.[1300] IX.3. b_5 60°.[788] IX.1. $b_{2.3}$ 45°.[788] IX.5 $b_{2.8}$ 55-55.5°.[395]

R = BrH_2CCH_2-. IX.5. $b_{0.2}$ 79°.[395]

R = ClH_2CCH_2-. IX.5. $b_{0.13}$ 68-72°.[395]

R = Pr. IX.1. $b_{1.5}$ 54°.[788] IX.1.[1247]

R = i-Pr. IX.1. $b_{3.5}$ 60°.[788]

R = $Cl(CH_2)_3$-. IX.5. $b_{0.15}$ 78°.[395]

R = Bu. IX.5. $b_{0.2}$ 67°.[395]

R = Ph. IX.5. $b_{0.24}$ 99°.[395]

R = $C_{10}H_{21}$-. From $EtPO(OC_{10}H_{21})Cl$ and P_4S_{10}, $b_{0.2}$ 74°, IR.[973]

$C_2H_5PS(OEt)F$. IX.1. b_{15} 50°.[1717]

EtPS(OR)NCS.

R = Me. IX.1. b_1 66°.[1858]

R = Et. IX.1. $b_{0.01}$ 46°.[1858]

EtPS(OEt)SCN. IX.4. $b_{0.01}$ 47°.[1863]

$C_2H_5PO(SR)Cl$.

R = Me. IX.3. $b_{2.5}$ 81°.[1370]

R = Et. IX.3. $b_{2.5}$ 84-85°.[1370]

R = Pr. IX.3. b_2 98°.[1370]

R = i-Pr. IX.3. b_3 98°.[1370]

$C_2H_5PS(SR)Cl$. By reaction of $EtP(S)Cl_2$ with $EtP(SR)_2$.

R = Et. b_{14} 64°.[1029]

R = Bu. b_9 96-97°.[1029]

$C_2H_5PS(OR)_2$.

R = Me. X.1. $b_{2.2}$ 40°.[788]

R = Et. X.1. b_1 41-43°.[788] X.7. $b_{13.5}$ 82-83.5°.[900] X.2.

b_{10} 80-82°.[153]

R = ClH_2CCH_2-. X.2. $b_{0.01}$ 71°.[1853]

R = Pr. X.1. $b_{0.13}$ 42°.[788]

R = i-Pr. X.1. $b_{0.5}$ 48°.[788]

R = $ClCHMeCH_2$-. X.2. $b_{0.01}$ 75°.[1853]

R = $\overline{OCH_2C}HCH_2$-. X.2. b_1 129-130°[1748]

R = Bu. X.1. $b_{0.09}$ 58°.[788] X.2. b_1 76-77°.[928] X.7. b_2 97-98°.[899]

R = Am. X.1. $b_{0.06}$ 80-82°.[788]

R = Ph. X.2. b_1 144-145°.[153]

$C_2H_5PS(OMe)O$- 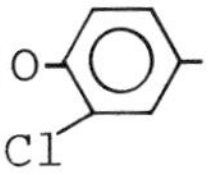 -CMe_3. X.1. $b_{0.01}$ 115°.[1814]

$C_2H_5PS(OEt)OR$.

R = $NCCMe_2$-. X.2. $b_{0.017}$ 71-74°[182]

R = (EtO)(S)$EtPCH_2$-. X.1. $b_{0.01}$ 92°.[1861]

R = 2,4-$Cl_2C_6H_3$-. X.1.[1811]

R = 2,4,5-$Cl_3C_6H_2$-. X.1.[1811]

R = $PhCH_2$-. X.1. $b_{0.01}$ 74-76°[1298]

R = 2-MeO-4-NCC_6H_3-. X.1. m. 74.5-76.5°.[1945]

R = 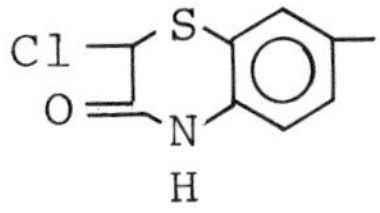 X.12.[1801]

R = 4-[MeC(O)-]C_6H_4-. X.1. n_D^{23} 1.5496.[1820]

R = 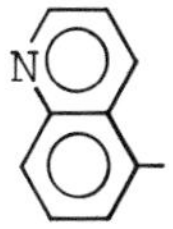 X.1. $b_{0.05}$ 115°.[527]

R = 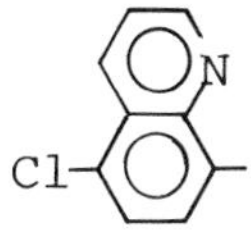 X.1. $b_{0.05}$ 138°.[527]

R = 4-Me_3C-2-ClC_6H_3-. X.1. $b_{0.01}$ 119°.[1814]

R = 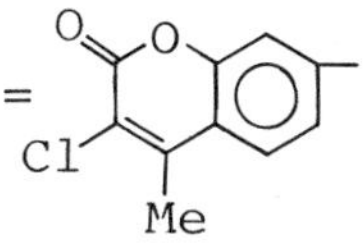 X.1.[1825]

R = 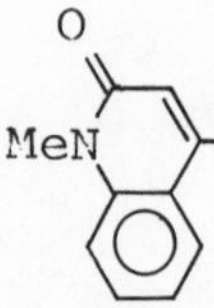 X.1. m. 58°.[1111]

R = $PhCH_2O_2CCMe$=CH-. X.1.[397]

$C_2H_5PS(OPr\text{-}i)OR$.

R = 4-O_2N-2-ClC_6H_3-. X.11. m. 48°.[1967]

R = 4-$O_2NC_6H_4$-. X.11. n_D^{25} 1.4555.[1967]

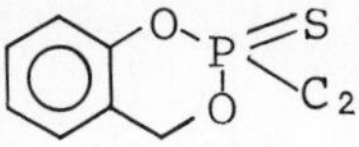 C_2H_5. X.1. $b_{0.6}$ 120°.[508]

$C_2H_5PO(OMe)SCH_2$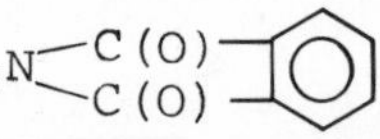 X.3. m. 76°.[1973]

$C_2H_5PO(OEt)SR$.

R = Et. X.2. b_{12} 103-104°.[1303] X.1.[901] X.4.[900] X.6. (-), $b_{0.6}$ 55°, $[\alpha]_D$ - 70.90°.[1428] X.2.[1303]

R = BrH_2CCH_2-. X.3. $b_{0.1}$ 85°.[1844]

R = ClH_2CCH_2-. X.3.[1092] X.3. Reaction proceeds without racemization.[1300]

R = $NCCH_2$-. X.3. $b_{0.01}$ 90°.[1847]

R = $Me_2NC(S)CH_2$-. X.12.[1849]

R = $Et_2NCH_2CH_2$-. X.1. $b_{0.5}$ 99°. X.3. $b_{0.01}$ 94°.[1847]

R = $PhCH_2CH_2$-. X.3. $b_{0.01}$ 118°.[1865]

R = 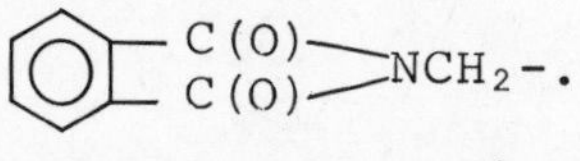 . X.3. m. 71°.[1973]

$C_2H_5PO(OPr)SCH_2$ X.3. m. 71°.[1973]

$C_2H_5PS(OMe)SR$.

R = $EtSCHMeCH_2$-. X.12. $b_{0.06}$ 99-105°.[1443]

R = HC≡CCH_2NMe-C(O)CH_2-. X.3. n_D^{25} 1.553.[474]

R = SCH=CH-N=CNHC(O)CH_2-. X.3. m. 90-93°.[526]

R = SCPh=CMe-N=CNHC(O)CH_2-. X.3. m. 110-111°.[526]

R = S-C(-$C_6H_3Cl_2$-2.5.)=CMeN=CNHC(O)CH_2-. X.3. m. 138-141°.[526]

R = $\overline{SCHC}(-2-C_{10}H_7)-N=\overline{C}NHC(O)CH_2-$. X.3. m. 140-144°.[526]

$C_2H_5PS(OEt)SR$.

R = Et. X.4. (-), $b_{0.5}$ 49°, $[\alpha]_D$ - 72.00.[1428] X.1. (+),
$\qquad b_{0.2}$ 45-47°, $[\alpha]_D$ + 62.40.[1428]

R = MeHC=CH-. X.3.[1443]

R = (MeS-)(NC-)CH-. X.3. $b_{0.01}$ 98°.[1866]

R = MeC(O)NHC(O)CH$_2$-. X.3. n_D^{22} 1.5422.[1815]

R = MeSCH(-CN)-CH$_2$-. X.3. $b_{0.01}$ 100°.[1804]

R = (MeSO$_2$)$_2$NCH$_2$CH$_2$-. X.3. m. 70°.[1971]

R = MeSCHMeCH$_2$-. X.12. $b_{0.2}$ 103-106°.[1443]

R = $\overline{S-CH=CH-N=C}$-NHC(O)CH$_2$-. X.3. m. 93-95°.[526]

R = $\overline{S-CBr=CH-N=C}NHC(O)CH_2-$. X.3. m. 77-80°.[526]

R = $\overline{S-C(-NO_2)=CH-N=C}NHC(O)CH_2-$. X.3. m. 83-86°.[526]

R = Ph. X.1. $b_{0.1}$ 130-132°.[1969]

R = $\overline{SCH=CMe-N=C}$-NHC(O)CH$_2$-. X.3. m. 116-123°.[526]

R = MeO$_2$CCMe(-SMe)CH$_2$-. X.3. $b_{0.1}$ 126°.[1803, 1804]

R = 2-MeC$_6$H$_4$-. X.1. $b_{0.1}$ 135°.[1969]

R = EtO$_2$CCH(-SMe)-CHMe-. X.3. $b_{0.15}$ 126°.[1804]

R = MeO$_2$CCH(-SEt)-CHMe-. X.3. $b_{0.15}$ 120°.[1804]

R = MeO$_2$CCMe(-SEt)-CH$_2$-. X.3. $b_{0.07}$ 128-130°.[1804]

R = EtO$_2$CCH(-SEt)-CHMe-. X.3. $b_{0.1}$ 127°.[1804]

R = 4-ClC$_6$H$_4$C(O)NHC(O)CH$_2$-. X.3. m. 141-142°.[1815]

R = 4-MeC$_6$H$_4$SCH(-CN)-. X.3. n_D^{25} 1.5865.[1806]

R = 4-t-BuC$_6$H$_4$-. X.1. $b_{1.5}$ 165°, m. 56°.[1969]

R = $\overline{SCHCPh-N=C}NHC(O)CH_2-$. X.3. m. 70-74°.[526]

R = $\overline{SCH}=C(-C_6H_4NO_2-4)-N=\overline{C}NHC(O)CH_2-$. X.3. m. 135-138°.[526]

R = $\overline{SCPh=CMe-N=C}NHC(O)CH_2-$. X.3. m. 132-135°.[526]

(EtO)(S)EtPSCH$_2$SPEt(S)(OEt). X.3.[1840]

(EtO)(S)EtPS(CH$_2$)$_2$O(CH$_2$)$_2$SPEt(S)(OEt). X.3. n_D^{25} 1.5458.[1972]

(EtO)(S)EtPSCH————S————CHSPEt(S)(OEt). X.3. n_D^{24} 1.5433.
$\qquad\qquad$ | $\qquad\qquad\qquad$ |
$\qquad\qquad$ CO$_2$Et $\qquad\qquad\quad$ CO$_2$Et
[1802]

$C_2H_5PS(OCH_2CH_2Cl)SR$.

R = Et. X.2. b_7 100-102°.[49]

R = i-Pr. X.2. $b_{0.005}$ 72-73°.[49]

$C_2H_5PS(OPr)SCH_2CHMeSEt$. X.12. $b_{0.09}$ 112-113°.[1443]
$C_2H_5PS(OPr\text{-}i)SR$.

R = $\overline{SCH=CH-N=C}$-NHC(O)CH_2-. X.3. m. 93-95°.[526]

R = 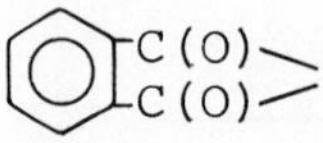 NCH_2-. X.3. m. 39°.[1973]

R = 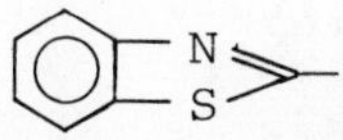 -NHC(O)CH_2- . X.3. n_D^{30} 1.6135.[1945]

R = $\overline{SCH=C(-C_6H_4NO_2\text{-}4)-N=C}$NHC(O)$CH_2$-. X.3. m. 65-68°.[52]

$C_2H_5PS(OBu\text{-}i)SR$.

R = Ph. X.1. $b_{0.05}$ 129°.[310]

R = 4-ClC_6H_4-. X.1. $b_{0.07}$ 154-156°.[310]

R = 2-Me-5-ClC_6H_3-. X.1. $b_{0.06}$ 138°.[310]

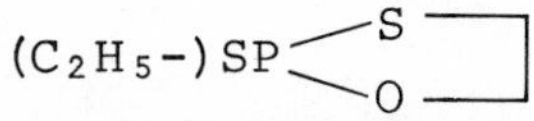 X.5.[1845]

$C_2H_5PO(SEt)_2$. X.1. $b_{0.04}$ 69-70°.[50]
$C_2H_5PO(SPr)SCH_2CH=CH_2$. X.11. $b_{0.08}$ 104-105°.[1443]
$C_2H_5PS(SR)_2$.

R = Et. X.2. b_1 91-93°.[155] By reaction of $Et_2PS(SEt)$ with EtP(SEt)Cl in 10% yield, b_9 128-130°.[1030] By reaction of MeEtPS(SEt) with EtP(SEt)Cl.[1031] By reaction of EtP(S)Cl_2 with $EtP(SEt)_2$, b_{14} 152°.[1029] As by-product by reaction of $EtP(SEt)_2$ with EtHal, $b_{0.28}$ 89-90°.[1028]

R = Bu. By reaction of EtP(S)Cl_2 with $EtP(SBu)_2$, $b_{0.04}$ 106°.[1029] As by-product by reaction of $EtP(SBu)_2$ with BuHal, $b_{0.02}$ 106-108°.[1028] X.1. $b_{0.04}$ 103-104.5°.[50]

R = Am. As by-product by reaction of $EtP(SAm)_2$ with AmHal, $b_{0.6}$ 152°.[1028]

R = i-Am. As by-product by reaction of $EtP(SAm\text{-}i)_2$ with i-AmHal, b_7 159-161°.[1028]

R = Ph. X.1. $b_{0.01}$ 144°.[1816]

R = 4-ClC_6H_4-. X.1. m. 76°.[1857]

R = 3,4-$Cl_2C_6H_3$-. X.1. m. 68°.[1857]

$C_2H_5PS(SEt)SPh$. X.3. $b_{0.01}$ 108°.[1859]

$C_2H_5PO(OR)SCl$.

R = Et. XI.1. $b_{0.002}$ 33-34°, n_D^{20} 1.4805, d_4^{20} 1.2283.[649a] XI.1. The reaction proceeds without racemization.[1300] XI.1. $b_{0.5}$ 33-34°.[288]

R = Bu. XI.1. $b_{0.4}$ 74°, n_D^{25} 1.4528.[288]

$CH_3CH(OH)PS(OEt)_2$. X.9. b_{1-2} 75-77°, n_D^{20} 1.4715, d_4^{20} 1.1034.[1411] X.9.[603] X.9.[1674]

C_3

$(CH_3)_2CBrPO(SEt)Br$. IX.1. $b_{0.005}$ 100°.[1727]

$(CH_3)_2CBrPO(SEt)OEt$. X.1. $b_{0.02}$ 79°.[1727]

$Cl(CH_2)_3PS(OEt)_2$. X.8. b_3 95°, n_D^{20} 1.4687, d_4^{20} 1.1212.[1628]

$(CH_3)_2CClPO(SR)Cl$.

R = Me. IX.1. $b_{0.01}$ 63°.[1727]

R = Et. IX.1. $b_{0.05}$ 55-60°.[1727]

R = i-Pr. IX.1. $b_{0.02}$ 70°.[1727]

R = $Cl(CH_3)_2$-. IX.1. $b_{0.03}$ 100°.[1727]

R = Bu. IX.1. $b_{0.04}$ 80-85°.[1727]

R = i-Bu. IX.1. $b_{0.04}$ 75°.[1727]

R = sec.-Bu. IX.1. $b_{0.02}$ 75°.[1727]

$(CH_3)_2CClPO(SMe)OMe$. X.1. $b_{0.005}$ 49-53°.[1727]

$(CH_3)_2CClPO(SEt)OR$.

R = Me. X.1. $b_{0.008}$ 58-62°.[1727]

R = Et. X.1. $b_{0.02}$ 80°.[1727]

R = $CH_3CH{=}CH$-. X.1. $b_{0.007}$ 79°.[1727]

R = Ph. X.1. $b_{0.005}$ 125°.[1727]

R = 2-ClC_6H_4-. X.1. $b_{0.005}$ 130°.[1727]

R = 3-ClC_6H_4-. X.1. $b_{0.01}$ 135°.[1727]

R = 4-ClC_6H_4-. X.1. $b_{0.01}$ 128-132°.[1727]

R = 2,4-$Cl_2C_6H_3$-. X.1. $b_{0.005}$ 140°.[1727]

$(CH_3)_2CClPO(SBu)OR$.

R = Me. X.1. $b_{0.01}$ 72-74°.[1727]

R = Et. X.1. $b_{0.01}$ 96°.[1727]

R = Ph. X.1. $b_{0.002}$ 133°.[1727]

R = 4-ClC_6H_4-. X.1. $b_{0.002}$ 170°.[1727]

$(CH_3)_2CClPO(SBu\text{-sec.})OMe$. X.1. $b_{0.05}$ 59-62°.[1727]

$CH_3C{\equiv}CPS(OEt)_2$. X.1. $b_{0.3}$ 82-83°, IR.[363]

$H_2C{=}C{=}CHPS(OR)_2$. From $(RO)_2PCl$ and $HSCH_2C{\equiv}CH$.

R = i-Pr. $b_{0.02}$ 56°.[1449]

R = i-Bu. b_2 105-108°.[1449]

$H_2C{=}C{=}CHPO(SMe)_2$. X.8. $b_{0.3}$ 100-105°.[1189]
$HC{\equiv}CCH(OH)PS(OEt)_2$. X.9. b_1 85.5°.[1572]
$NCCH_2CH_2PS(OEt)_2$. X.4. $b_{0.06}$ 130-132°, n_D^{20} 1.4960, d_4^{20} 1.1010.[1567]
$(NCCH_2CH_2-)\overline{SPOCHMeCHMeO}$. X.8. b_1 99-100°, n_D^{20} 1.4948, d_4^{20} 1.2360.[1588]
$H_2C{=}CHCH_2PS(OR)_2$.

R = Et. X.7. b_9 88°, n_D^{20} 1.4733, d_4^{20} 1.0390.[1669] X.2. b_{11} 91-91.5°.[1702]

R = Pr. X.2. b_{11} 117-117.5°.[1702]

R = i-Pr. X.2. b_7 89-90°.[1702]

R = Bu. X.2. b_{10} 133-135°.[1697]

R = Am. X.2. $b_{0.15}$ 104-106°.[1702]

R = C_6H_{13}-. X.2. $b_{0.22}$ 124-126°.[1698]

$H_2C{=}CHCH_2PO(SR)_2$.

R = Me. X.1. $b_{0.1}$ 84°.[1730]

R = Et. X.1. $b_{0.02}$ 92°.[1730]

R = Pr. X.1. $b_{0.03}$ 112°.[1730]

R = i-Pr. X.1. $b_{0.03}$ 92°.[1730]

R = Bu. X.1. $b_{0.02}$ 118°.[1730]

R = Am. X.1. $b_{0.005}$ 133°.[1730]

$C_2H_5C(O)PS(OEt)_2$. X.7. b_3 70-72°, n_D^{20} 1.4690, d_4^{20} 1.0886.[1668]

$CH_3C(O)CH_2PS(OR)_2$.

R = Et. X.12. b_3 111°, IR.[1959]

R = $H_2C{=}CMe$-. From PCl_3 and $Et_3SnCH_2C(O)CH_3$ and addition of S.[1415]

$CH_3OCH{=}CHPS(OR)_2$.

R = Me. X.1. b_{12} 122°.[1345]

R = Et. X.1. $b_{0.04}$ 89°.[1345]

R = Pr. X.1. $b_{0.03}$ 105°.[1345]

$CH_3OCH{=}CHPO(OEt)SCH_2CH_2SEt$. X.3. $b_{0.01}$ 90°.[1846]
$CH_3OCH{=}CHPO(SR)_2$.

R = Et. X.1. $b_{0.05}$ 115°.[1728]

R = i-Pr. X.1. $b_{0.08}$ 117°.[1728]

$H_2C{=}CHCH(OH)PS(OEt)_2$. X.9. b_{10} 120-122°.[1674] X.9.[603]

$CH_3S(O)CH{=}CHPO(SR)_2$.

R = Me. X.12. n_D^{23} 1.5964.[1729, 1731a]

R = Et. X.12. n_D^{25} 1.5853.[1729, 1731a]

R = Pr. X.12. n_D^{20} 1.5301.[1729, 1731a]

R = Bu. X.12. n_D^{23} 1.5400.[1729, 1731a]

OHCCHMePS(OPr-i). X.12. $b_{0.025}$ 64°.[1350]

$CH_3SCH{=}CHPO(SR)_2$.

R = Me. X.1. $b_{0.07}$ 137°, m. 34°.[1731a]

R = Et. X.1. $b_{0.04}$ 135°.[1729, 1731a]

R = Pr. X.1. $b_{0.03}$ 150°.[1729, 1731a]

R = i-Pr. X.1. $b_{0.25}$ 145°.[1729, 1731a]

R = Bu. X.1. $b_{0.04}$ 155°.[1729]

$C_2H_5S(S)CPS(OEt)_2$. X.4/X.12. $b_{0.12}$ 103°[65]

$C_2H_5NH(S)C$ 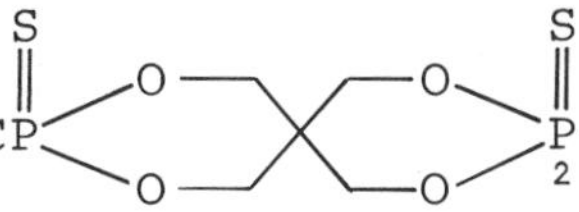 $-C(S)NHC_2H_5$. X.10. m. $186-187^{\circ}$.[269]

$CH_3SCH_2NHC(S)$  $C(S)NHCH_2SCH_3$. X.10. m. $169-170^{\circ}$.[269]

$C_3H_7PS(OR)Cl$.

R = Me. IX.1.[1247]

R = Et. IX.1.[1247]

R = ClH_2CCH_2-. IX.5. $b_{0.3}$ $78-80^{\circ}$, n_D^{25} 1.5116.[395]

R = Pr. IX.1.[1247]

$C_3H_7PS(OEt)SCN$. IX.1. $b_{0.01}$ 56°.[1858]

$C_3H_7PS(SEt)SCN$. IX.4. $b_{0.01}$ 56°.[1858]

$C_3H_7PS(OEt)_2$. X.7. b_2 $63.5-65.5^{\circ}$.[901]

$C_3H_7PO(OEt)SCH_2CH_2NEt_2$. X.1. $b_{0.015}$ $100-102^{\circ}$.[811]

$C_3H_7PO(SPh)_2$. X.1. m. $66-67^{\circ}$.[1816]

$C_3H_7PO(SPr)SCH_2CH{=}CH_2$. X.11. $b_{0.1}$ $100-107^{\circ}$.[1443]

$(CH_3)_2CHPS(OR)SH$.

R = ClH_2CCH_2-. VIII.4. $b_{0.002}$ 57°.[395]

R = $HC{\equiv}CCH_2$-. VIII.4. Ni salt, m. $84-86^{\circ}$.[395]

$(CH_3)_2CHPS(OR)Cl$.

R = ClH_2CCH_2-. IX.5. $b_{0.7}$ $78-82^{\circ}$.[395]

R = Pr. IX.5. $b_{0.7}$ $58-59^{\circ}$.[395]

$(CH_3)_2CHPS(OEt)SCN$. IX.1. $b_{0.01}$ 44°.[1858]

$(CH_3)_2CHPS(SEt)SCN$. IX.4. $b_{0.01}$ 51°.[1850]

$(CH_3)_2CHPS(OMe)OC_6H_3SMe$-4-Me-3. X.3.[1797]
$(CH_3)_2CHPO(OEt)SCH_2CH_2NEt_2$. X.1. $b_{0.0007}$ 82-84°.[811]
$(CH_3)_2CHPS(OEt)SCHMeCH(-SEt)CO_2Et$. X.3. $b_{0.1}$ 130°.[1804]
$(CH_3)_2CHPO(SPh)_2$. X.1. b_2 165°.[1816]
$(CH_3)_2C(OH)PS(OR)_2$.

R = Et. X.9. b_{10} 107-109°.[1674]

R = Cl_3CMe_2-. X.9. m. 153-155°.[32]

R = $H_2C(CH_2)_3C(-CCl_3)$-. X.9. m. 162-163°.[32]

$CH_3SCHMePO(SMe)_2$. X.11. $b_{0.04}$ 110°.[296]
$C_2H_5SCH_2PO(SEt)_2$. X.11. $b_{0.1}$ 120°.[296]
$Me_2(S)POCH_2PS(OR)_2$.

R = Me. X.12. $b_{0.01}$ 98°.[1861]

R = Et. X.12. $b_{0.01}$ 106°.[1861]

<u>C_4</u>

EtOCH=CBrPS$(OR)_2$.

R = Me. X.1. $b_{0.06}$ 94-96°, n_D^{20} 1.5287, d_4^{20} 1.4483.[1348]

R = Et. X.1. $b_{0.053}$ 106-108°, n_D^{20} 1.5220, d_4^{20} 1.3573.[1348]

EtO_2C-CHBrPS$(OEt)_2$. X.12. $b_{0.3}$ 116-118°, n_D^{20} 1.4983, d_4^{20} 1.3623.[1346]
ClH_2C-CMe=CHPS$(OR)_2$.

R = Et. X.1. $b_{0.5}$ 114-116°.[992]

R = Pr. X.1. $b_{0.5}$ 124-126°.[992]

R = Bu. X.1. $b_{0.3}$ 136-137°.[992]

R = Am. X.1. $b_{0.5}$ 157-159°.[992]

$Cl(CH_2)_4PS(OEt)_2$. X.2. $b_{0.01}$ 87-88°.[766]
$ClH_2CCHEtPS(OEt)_2$. X.8. b_3 110°, n_D^{20} 1.4821, d_4^{20} 1.0842.[1628]

$C_2H_5CMeClPO(SR)Cl$.

R = Me. IX.1. $b_{0.15}$ 75°.[1727]

R = Et. IX.1. $b_{0.15}$ 80°.[1727]

R = $Cl(CH_2)_3$-. IX.1. $b_{0.04}$ 125°.[1727]

R = Bu. IX.1. $b_{0.1}$ 100°.[1727]

$C_2H_5CMeClPO(SMe)OC_6H_4NO_2$-4. X.1. $b_{0.07}$ 180°.[1727]
$C_2H_5CMeClPO(SEt)OR$.

R = Me. X.1. $b_{0.06}$ 81°.[1727]

R = Et. X.1. $b_{0.02}$ 72°.[1727]

R = 2,4,5-$Cl_3C_6H_2$-. X.1. $b_{0.3}$ 168°.[1727]

R = EtO_2CCH=CMe-. X.1. $b_{0.04}$ 125-130°.[1727]

R = 3-MeC$_6$H$_4$-. X.1. b$_{0.01}$ 128°.[1727]

C$_2$H$_5$CMeClPO(SBu)OPh. X.1. b$_{0.005}$ 150°.[1727]

H$_2$C=CH-CH=CHPS(OR)$_2$.

R = Me. X.1. b$_2$ 80°, n$_D^{20}$ 1.5300, d$_4^{20}$ 1.1109.[1593]

R = Et. X.1. b$_2$ 95°, n$_D^{20}$ 1.5026, d$_4^{20}$ 1.0481.[1629]

R = Pr. X.1. b$_2$ 93-93.5^O, n$_D^{20}$ 1.4971, d$_4^{20}$ 1.0234.[1593]

R = Bu. X.1. b$_3$ 115°, n$_D^{20}$ 1.4984, d$_4^{20}$ 0.9968.[1593]

CH$_3$CO$_2$C(=CH$_2$)PS(OR)$_2$.

R = Me. X.8/X.9. b$_{13}$ 121-122°.[1396]

R = Et. X.8/X.9. b$_{11}$ 126-128°.[1396]

R = i-Bu. X.8/X.9. b$_{11}$ 153-154°.[1396]

$\overline{\text{S-CMe}}$=CH-CH$_2$$\overline{\text{P}}$(O)SCH$_2CH_2$Cl. X.8. b$_{0.5}$ 130°.[1026]

[H$_2$C=CHCH$_2$NHC(S)-]SP$\diagdown$O—⟨ ⟩—O$\diagup$PS[-(S)CNHCH$_2$CH=CH$_2$].

X.8. m. 148-149°.[269]

CH$_3$CH=CHCH$_2$PS(OR)$_2$.

R = Et. X.7. b$_9$ 103°, n$_D^{20}$ 1.4772, d$_4^{20}$ 1.0230, IR, Raman spect.[1669] X.8. b$_3$ 88-90°, n$_D^{20}$ 1.4791, d$_4^{20}$ 1.0340.[1628]

R = Pr. X.2. b$_{0.1}$ 76-78°.[1102]

R = Bu. X.2. b$_{0.07}$ 86-88°.[1102]

R = i-Bu. X.2. b$_{0.08}$ 78-80°.[1102]

Me$_2$C=CHPS(OEt)NCS. IX.1. b$_{0.01}$ 69°.[1858]

Me$_2$C=CHPS(OR)$_2$.

R = Et. X.2. b$_{11}$ 117-119°, n$_D^{20}$ 1.4830, d$_4^{20}$ 1.0310.[937]

R = 4-O$_2$NC$_6$H$_4$-. X.1. m. 103-105°.[2063]

Me$_2$C=CHPS(OEt)OR.

R = EtSCH$_2$CH$_2$-. X.1. b$_{0.01}$ 103°.[1795]

R = EtSCHMe-. X.3. b$_{0.01}$ 97°.[1834]

R = Et$_2$NCH$_2$CH$_2$-. X.1. b$_{0.01}$ 89°.[1851] X.3. b$_{0.01}$ 96°.[1846]

(EtO)(Me$_2$C=CH-)SPSCH$_2$SPS(-CH=CMe$_2$)OEt. X.3.[1840]

H$_2$C=CH-CHMePS(OEt)$_2$. X.7. b$_9$ 90-95°, IR, Raman spect.[1669]

OHCCHEtPS(OEt)$_2$. X.12. b$_{0.55}$ 62-64°, n$_D^{20}$ 1.4750, d$_4^{20}$ 1.0759; as 2,4-dinitrophenylhydrazone, m. 92.5-93°.[1350]

C$_2$H$_5$OCH=CHPS(OMe)OR.

R = Me. X.1. b$_1$ 86°.[83]

R = $Et_2NCH_2CH_2$-. X.1. $b_{0.01}$ 80°.[1851]

$C_2H_5OCH{=}CHPS(OEt)OR$.

R = $Me_2NCH_2CH_2$-. X.1.[1851]

R = $Et_2NCH_2CH_2$-. X.1. $b_{0.01}$ 114°.[1851, 1854]

R = 4-$O_2NC_6H_4$-. X.1. $b_{0.01}$ 85°.[1836]

R = 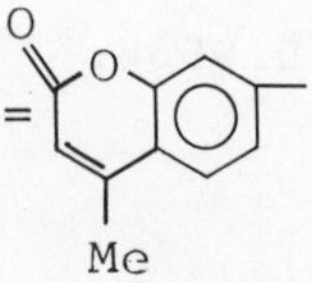 X.1.[1825]

$C_2H_5OCH{=}CH$-PO(OEt)SR.

R = Et. X.1. $b_{0.1}$ 90°, n_D^{20} 1.4963, d_4^{20} 1.1160.[1347]

R = $EtSCH_2CH_2$-. X.1, X.3. $b_{0.01}$ 110°.[1829]

R = C_6H_{13}-. X.3. $b_{0.01}$ 110°.[1829]

R = $C_{12}H_{25}$-. X.3. $b_{0.01}$ 160°.[1829]

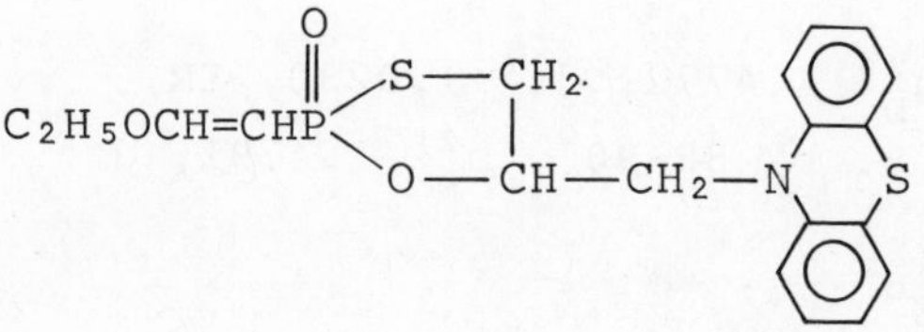 X.5. m. 145-146°.[227]

$C_2H_5OCH{=}CHPS(OEt)SR$.

R = Et. X.1. $b_{0.01}$ 90°.[1828]

R = Ph. X.1. $b_{0.01}$ 130°.[1828]

R = cycl.-C_6H_{11}-. X.1. $b_{0.01}$ 120°.[1828]

$C_2H_5OCH{=}CHPO(SR)_2$.

R = Me. X.1. $b_{0.05}$ 125°.[1728, 1731]

R = Et. X.1. $b_{0.1}$ 126°.[1728, 1731] X.1. b_2 135°.[84]

R = Pr. X.1. $b_{0.05}$ 136°.[1728, 1731]

R = i-Pr. X.1. $b_{0.02}$ 115°.[1728, 1731]

R = Bu. X.1. $b_{0.05}$ 147°.[1728, 1731] X.1. b_2 154°.[84]

R = Ph. X.1. m. 82°.[1728, 1731]

$C_2H_5OCH{=}CHPS(SEt)_2$. X.1. b_1 134°.[84]

$CH_3CH{=}CH$-CH(OH)$PS(OEt)_2$. X.9.[603, 1674]

$C_2H_5S(O)CH{=}CHPO(SR)_2$.

R = Me. X.1. n_D^{20} 1.5990.[1729]

R = Et. X.1. n_D^{20} 1.5810.[1729]

R = Pr. X.1. n_D^{25} 1.5626.[1729]

R = Bu. X.1. n_D^{24} 1.5502.[1729]

EtO_2CCH_2**PS(OEt)**$_2$. X.12. $b_{0.03}$ 80-82°, $b_{0.2}$ 89-90°.[1346] X.7. b_5 105-106°.[900]

$MeO_2CCH_2CH_2PO(SR)Cl$.

R = i-Pr. IX.1. $b_{0.01}$ 86-88°.[1475]

R = i-Bu. IX.1. $b_{0.01}$ 90-92°.[1475]

$MeO_2CCH_2CH_2PS(OEt)_2$. X.8. $b_{1.5}$ 110°.[1670]

$CH_3C(O)CMe(OH)PS(OEt)_2$. X.9. b_{10} 124°.[1625]

$C_2H_5SCH{=}CHPS(OEt)OCH_2CH_2NEt_2$. X.1. $b_{0.01}$ 118°.[1839]

$C_2H_5SCH{=}CHPO(OEt)SR$.

R = MeO_2CCH_2-. X.3. $b_{0.01}$ 108°.[1833]

R = $EtSCH_2CH_2$-. X.3. $b_{0.01}$ 114°.[1832]

$C_2H_5SCH{=}CHPS(OEt)SC_6H_{13}$. X.1. $b_{0.01}$ 104°.[1831]

$C_2H_5SCH{=}CHPO(SR)_2$.

R = Me. X.1. $b_{0.05}$ 140°.[1729, 1731a]

R = Et. X.1. $b_{0.01}$ 145°.[1729, 1731a]

R = Pr. X.1. $b_{0.02}$ 154°.[1729, 1731a]

R = i-Pr. X.1. $b_{0.1}$ 150°.[1729, 1731a]

R = Bu. X.1. $b_{0.03}$ 170°.[1729, 1731a]

R = Am. X.1. $b_{0.2}$ 194°.[1729, 1731a]

R = Ph. X.1. $b_{0.05}$ 195°.[1729, 1731a]

$H_2C{=}C(\text{-}SEt)PO(SEt)_2$. X.12. $b_{0.03}$ 150°.[1729, 1731a]

$C_2H_5SCH_2NHC(S)$- 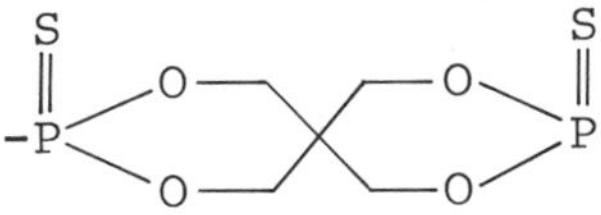 $-C(S)NHCH_2SC_2H_5$. X.10. m. 145-146°.[269]

$C_4H_9PS(OR)SH$.

R = Me. VIII.4. $b_{0.02}$ 61.5°.[702]

R = Et. VIII.4. $b_{0.18}$ 66-67°.[701] VIII.4. $b_{0.01}$ 82-85°.[708] VIII.1. $b_{0.6-1.4}$ 97-100°.[1793]

R = ClH_2CCH_2-. VIII.1. $b_{0.024}$ 89°.[395]

R = Pr. VIII.1. $b_{0.003}$ 60°.[395]

R = i-Pr. VIII.4. $b_{0.2}$ 74.5-75°.[702] VIII.4. $b_{0.01}$ 77-79°.[708]

R = $EtOCH_2CH_2$-. VIII.4. $b_{0.15}$ 98°.[702]

R = cycl.-C_6H_{11}-. VIII.4. $b_{0.23}$ 111-112°.[701] VIII.4.

$b_{0.06}$ 121-122°.[708]

R = C_6H_{13}-. VIII.4. $b_{0.22}$ 104°.[701]

R = i-C_6H_{13}-. VIII.4. $b_{0.04}$ 116-117°.[708]

$C_4H_9PS(OR)Cl$.

R = ClH_2CCH_2-. IX.5. $b_{0.3}$ 87-90°.[395]

R = Pr. IX.5. $b_{0.7}$ 70°.[395]

$C_4H_9PS(OR)_2$.

R = Et. X.7. $b_{2.5}$ 74.5-77.5°.[901]

R = Bu. X.7. b_2 108-109°.[1633] Magn. rotation of the bonds.[1069]

$C_4H_9PO(OEt)SR$.

R = Et. X.3. b_1 82.5-84°.[902]

R = $Et_2NCH_2CH_2$-. X.1. $b_{0.0025}$ 93-96°.[811]

$C_4H_9PS(OEt)SR$.

R = Bu. X.3. $b_{0.2}$ 99-100°.[701]

R = $MeO_2CCH_2CH_2$-. X.3. $b_{0.3}$ 116-117°.[701]

R = $PhCH_2$-. X.3. $b_{0.01}$ 130°.[701]

$C_4H_9PS(-SCH_2CH_2CN)OR$.

R = Et. X.3. $b_{0.2}$ 142.5°.[701]

R = cycl.-C_6H_{11}-. X.3. $b_{0.25}$ 158°.[701]

R = C_6H_{13}-. X.3. $b_{0.2}$ 155.5.[701]

$C_4H_9PS(-SCH_2CH_2SEt)OR$.

R = Me. X.3. $b_{0.04}$ 109-112°.[706]

R = i-Pr. X.3. $b_{0.02}$ 110-112°.[706]

R = Ph. X.3. $b_{0.01}$ 162-165°.[706]

R = C_8H_{17}-. X.3. $b_{0.02}$ 160-162°.[706]

$C_4H_9PO(SR)_2$.

R = Bu. X.1. $b_{0.02}$ 125°.[1725]

R = Ph. X.1. m. 50-53°.[1816]

$C_4H_9PS(SBu)_2$. X.5. b_1 164-166°.[1520] From $BuPH_2$ and BuSSBu, $b_{0.17}$ 138-144°, IR.[669]

$Me_2CHCH_2PS(OCH_2CHMe_2)_2$. X.2. b_{12} 136-136.5°.[1710]

$Me_2CHCH_2PS(OC_6H_4SMe\text{-}4)OR$.

R = Me. X.1. $b_{0.01}$ 114-117°.[1797]

R = Et. X.1. $b_{0.02}$ 120-125°.[1797]

$Me_2CHCH_2PS(OEt)SR$.

R = EtO$_2$CCH(-SMe)-CHMe-. X.3. b$_{0.15}$ 134°.[1804]

R = EtO$_2$CCH(-SEt)CHMe-. X.3. b$_{0.8}$ 136°.[1804]

Me$_2$CHCH$_2$PS(OR)SCH$_2$CO$_2$Et.

R = Ph. X.5. b$_{0.02}$ 150-154°.[699]

R = cycl.-C$_6$H$_{11}$-. X.5. b$_{0.1}$ 147-149°.[699]

Me$_2$CHCH$_2$PS(OR)SCHMe-CO$_2$Et.

R = Me. X.5. b$_{0.01}$ 112-113°.[699]

R = Et. X.5. b$_{0.01}$ 109-110°.[699]

R = i-Pr. X.5. b$_{0.01}$ 108-110°.[699]

R = MeOCH$_2$CH$_2$-. X.5. b$_{0.01}$ 132-134°.[699]

R = EtOCH$_2$CH$_2$-. X.5. b$_{0.02}$ 124-126°.[699]

Me$_2$CHCH$_2$PS(SEt)SR.

R = i-Pr. X.3. b$_{0.03}$ 104-106°.[705]

R = Bu. X.3. b$_{0.01}$ 114-115°.[705]

R = Ph. X.3. b$_{0.01}$ 147-148°.[705]

Me$_2$CHCH$_2$PS(SPr)SCHMeCO$_2$Et. X.3. b$_{0.03}$ 134-136°.[705]

Me$_2$CHCH$_2$PS(SPr-i)SR.

R = H$_2$C=CH-CH$_2$-. X.3. b$_{0.04}$ 111-112°.[705]

R = EtO$_2$CCH$_2$-. X.3. b$_{0.02}$ 138.5-140°.[705]

Me$_2$CHCH$_2$PS(SBu)SCH$_2$SPr. X.3. b$_{0.03}$ 156-158°.[705]

Me$_2$CHCH$_2$PS(SAm-i)SCH$_2$Ph. X.3. b$_{0.03}$ 173-174°.[705]

C$_2$H$_5$CHMePS(OR)SH.

R = Et. From i-BuPS($\overset{\oplus}{N}$R$_3$)S$^{\ominus}$ and EtOH; as $\overset{\oplus}{N}$Et$_2$H$_2$ salt, m. 105-107°; as $\overset{\oplus}{N}$Et$_3$H salt, m. 29-30°.[704]

R = i-Pr. VIII.4. b$_{0.15}$ 62°.[702]

R = H$_2$C=CHCH$_2$-. VIII.4. b$_{0.13}$ 77.5-78°.[702]

R = EtOCH$_2$CH$_2$-. VIII.4. b$_{0.19}$ 104-104.5^O.[702]

C$_2$H$_5$CHMePS(OEt)SCN. IX.1. b$_{0.01}$ 52°.[1858]

(MeO)Me$_3$C-P$\diagdown\!\!\!\diagdown$ $\overset{O}{\underset{S}{}}$ H. Synthesis and opt. active derivatives.

[1032]

C$_2$H$_5$OCH$_2$CH$_2$PS(OEt)OCH$_2$CH$_2$SEt. X.12. b$_{0.0001}$ 85-86.5^O.[907]

C$_2$H$_5$SCH$_2$CH$_2$PS(OEt)OCH$_2$CH$_2$NEt$_2$. X.1. b$_{0.01}$ 118°.[1839]

C$_2$H$_5$OCH$_2$CH$_2$PO(OEt)SCH$_2$CH$_2$SEt. X.12. b$_{0.0001}$ 110-112°.[907]

C$_2$H$_5$SCHMePO(SEt)$_2$. X.11. b$_{0.007}$ 119°.[296]

CH$_3$SCHEtPO(SMe)$_2$. X.11. b$_{0.06}$ 120°.[296]

Me$_2$NCH$_2$CH$_2$PS(OEt)OCH$_2$CH$_2$SEt. X.12. b$_{0.00006}$ 80-82°.[907]

Me$_2$P(S)OCHMePS(OEt)$_2$. X.12. b$_{0.01}$ 96°.[1861]

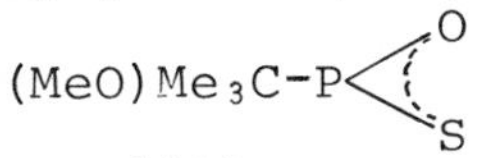

C_5

$H_2C(CH_2)_3CClPO(SR)Cl$.

R = Me. IX.1. $b_{0.1}$ 89°.[1727]

R = Et. IX.1. $b_{0.06}$ 82°.[1727]

R = Pr. IX.1. $b_{0.03}$ 90°.[1727]

R = Bu. IX.1. $b_{0.15}$ 118°.[1727]

$H_2C(CH_2)_3CClPO(OMe)SMe$. X.1. $b_{0.08}$ 83°.[1727]

$H_2C(CH_2)_3CClPO(OEt)SR$.

R = Et. X.1. $b_{0.02}$ 82°.[1727]

R = Pr. X.1. $b_{0.01}$ 97°.[1727]

$Et_2CClPO(SR)Cl$.

R = Me. IX.1. $b_{0.15}$ 94°.[1727]

R = Et. IX.1. $b_{0.3}$ 93°.[1727]

$Et_2CClPO(OEt)SEt$. X.1. $b_{0.03}$ 77°.[1727]

$C_3H_7CMeClPO(SMe)Cl$. IX.1. $b_{0.02}$ 94-98°.[1727]

HC=CH-CH=CH-CHPS(OBu)$_2$. X.2. b_1 114-115°.[929]

$C_3H_7C{\equiv}CPS(OEt)_2$. X.1. $b_{0.1}$ 95-96°, IR.[363]

O-CMe=CMe-CH_2P(S)SCH_2CH_2Cl. X.8. $b_{0.5}$ 114-115°, IR.[1026]

cycl.-$C_5H_9PS(OBu)_2$. X.2. b_1 108-108.5°.[929]

$CH_3C(O)-(CH_2)_3PS(OEt)_2$. X.12. $b_{0.7}$ 118°.[1959] X.8. b_{12} 142-144°.[1674]

$CH_3C(O)CHMeCH_2PS(OMe)_2$. X.11. $b_{0.5}$ 75°, IR.[1026]

$C_2H_5OCMe{=}CHPS(OEt)_2$. X.7. $b_{0.5}$ 104°, IR.[1959]

$C_2H_5OCH{=}CMePS(OPr\text{-}i)$. X.1. $b_{0.05}$ 79.5°, n_D^{20} 1.4820, d_4^{20} 1.0238.[1350]

$H_2C(CH_2)_3C(OH)PS(OCMe_2CCl_3)_2$. X.9. m. 172-174°.[32]

$EtSC(O)CH_2CH_2PO(SEt)_2$. X.1. $b_{0.2}$ 134-135°.[1712]

(EtO)(MeO)C=CHPS(OMe)$_2$. X.12. $b_{0.07}$ 87-88°, n_D^{20} 1.4946, d_4^{20} 1.1901.[1346]

$CH_3OCH_2CH_2OCH{=}CHPS(OEt)_2$. X.1. b_1 120°.[80]

$MeO_2CCHMeCH_2PS(OEt)_2$. X.1. $b_{0.5}$ 98-98.5°.[1670]

$(MeO_2CCHMeCH_2-)$SPOCHMeCHMeO. X.8. b_1 107-112°, n_D^{20} 1.4820, d_4^{20} 1.1823.[1588]

$EtO_2CCMe(OH)PS(OEt)_2$. X.9. b_5 121°.[1625]

$C_3H_7SCH{=}CHPO(SR)_2$.

R = Et. X.1. $b_{0.2}$ 158°.[1729, 1731a]

R = i-Pr. X.1. $b_{0.15}$ 152°.[1729, 1731a]

$C_4H_9NHC(O)$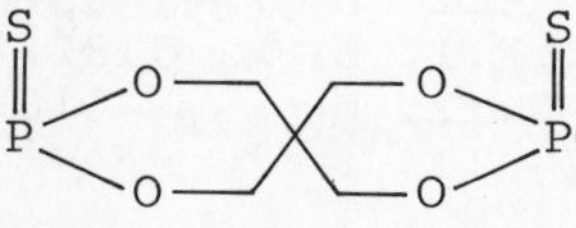$PC(O)NHC_4H_9$. X.10. m. 209-210°.[269]

$O_2NCH_2CH(-CHMe_2)PS(OMe)_2$. X.8. b_2 131.5°.[1662]

$C_5H_{11}PS(OR)SH$.

R = Me. VIII.4. $b_{0.04}$ 78°.[702] VIII.4. $b_{0.01}$ $56\text{-}63^\circ$.[708]

R = Et. VIII.4. $b_{0.08}$ $81\text{-}84^\circ$.[708]

R = Pr. VIII.4. $b_{0.04}$ $95\text{-}96^\circ$.[702]

R = i-Pr. VIII.4. $b_{0.02}$ $72\text{-}78^\circ$.[708]

R = cycl.-C_6H_{11}-. VIII.4. $b_{0.03}$ $112\text{-}115^\circ$.[708]

R = C_6H_{13}-. VIII.4. $b_{0.01}$ $111\text{-}114^\circ$.[708]

$Me_2CHCH_2CH_2PS(OEt)_2$. X.1. b. $250\text{-}255^\circ$.[729]
$C_3H_7CHMePS(OR)SH$.

R = Me. VIII.4. $b_{0.2}$ 68.5°.[702]

R = Pr. VIII.4. $b_{0.2}$ $92\text{-}92.5^\circ$.[702]

R = $MeOCH_2CH_2$-. VIII.4. $b_{0.15}$ $97\text{-}98^\circ$.[702]

R = $EtOCH_2CH_2$-. VIII.4. $b_{0.14}$ 104°.[702]

$Me_3SiOC(=CH_2)$-$PS(OMe)_2$. X.9. b_2 83°, ^1H.[1791]
$(EtO)_2CHPS(OEt)_2$. From $(EtO)_2P(S)H$ and $HC(OEt)_3$, b_{12} $127\text{-}130^\circ$, n_D^{20} 1.4550, d_4^{20} 1.0624.[1708]
$(EtO)_2SPOCH_2PS(OEt)_2$. X.12. $b_{0.01}$ 104°.[1861]
$C_3H_7SCHMePO(SPr)_2$. X.11. $b_{0.002}$ 132°.[296]
$C_2H_5SCHEtPO(SEt)_2$. X.11. $b_{0.07}$ 120°.[296]
$CH_3SCHPrPO(SMe)_2$. X.11. $b_{0.03}$ 120°.[296]

C_6

o/p-$ClC_6H_4PS(OEt)Cl$. IX.1. $b_{0.005}$ $106\text{-}110^\circ$, ^{31}P -84.3 ppm, IR.[1151]
4-Cl-$C_6H_4PS(OBu)Cl$. IX.1. $b_{0.005}$ 140°.[1726]
4-Cl-$C_6H_4PS(OR)F$.

R = Me. IX.1. b_4 $109\text{-}110^\circ$.[850]

R = Et. IX.1. b_{10} $133\text{-}134^\circ$.[850]

R = i-Pr. IX.1. b_4 $125\text{-}126^\circ$.[850]

4-$ClC_6H_4PS(OCH_2CH_2Cl)_2$. X.2. b_2 $178\text{-}180^\circ$.[624]
o/p-$ClC_6H_4PS(OEt)OC_6H_4NO_2$-4. X.2. $b_{0.005}$ $195\text{-}205^\circ$, ^{31}P -83.4 ppm, IR.[1151]

(EtO)S(4-ClC_6H_4-)P-O—  —OP(-C_6H_4Cl-4)(S)OEt. X.1. m. $96\text{-}104^\circ$.[1506]

(EtO)S(4-ClC_6H_4-)P-S—  —SP(-C_6H_4Cl-4)(S)OEt. X.3.[1840]

4-$ClC_6H_4PO(SBu)_2$. X.1. $b_{0.3}$ $170\text{-}180^\circ$.[1752]
$C_2H_5SCH=C(-CCl=CH_2)PO(SEt)_2$. X.1. $b_{0.015}$ $122\text{-}123^\circ$.[2047]
$H_2C(CH_2)_4CClPO(SR)Cl$.

R = Et. IX.1. $b_{0.2}$ 110°.[1727]

R = Bu. IX.1. $b_{0.1}$ 130°.[1727]

$H_2\overline{C(CH_2)_4CC}lPO(OEt)SR.$

 R = Et. X.1. $b_{0.03}$ 99°.[1727]

 R = Bu. X.1. $b_{0.01}$ 130°.[1727]

$Me_2CHCH_2CMeClPO(SR)Cl.$

 R = Me. IX.1. $b_{0.1}$ 82°.[1727]

 R = Et. IX.1. $b_{0.12}$ 88°.[1727]

$Me_2CHCH_2CMeClPO(OMe)SEt.$ X.1. $b_{0.08}$ 75°.[1727]
$CH_3CHClCH_2CH(-SEt)-PO(SEt)_2.$ X.11. $b_{0.01}$ 145°.[296]
$o/p-FC_6H_4PS(OEt)Cl.$ IX.1. $b_{0.04}$ $101-105^{\circ}$.[1151]
$o/p-FC_6H_4PS(OEt)OC_6H_4NO_2-4.$ X.1. ^{31}P -82.6 ppm, IR.[1151]
$C_6F_5PS(OR)_2.$

 R = Me. X.2. $b_{0.1}$ $68-70^{\circ}$, ^{31}P -72.6 ppm, ^{19}F.[536]

 R = Et. X.2. $b_{0.2}$ $80-82^{\circ}$, ^{31}P -69.2 ppm, ^{19}F.[536]

 R = Ph. X.2. m. 68°, ^{31}P -58.2 ppm, ^{19}F.[536]

$4-O_2NC_6H_4PO(SR)_2.$

 R = Me. X.1.[1525]

 R = Et. X.1.[1525]

 R = Pr. X.1.[1525]

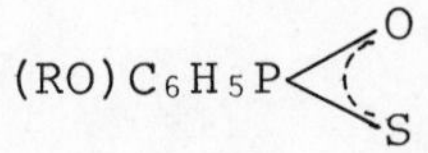

$(RO)C_6H_5P{\Large\langle}{\substack{O \\ \\ S}}H.$

 R = Me. As $\overset{\oplus}{N}Me_4$ salt, m. 135°; kinetics of the isomerization to $(MeS)C_6H_5P(O)O^{\ominus}\overset{\oplus}{N}Me_4.$[352] VIII.2.[1993]

 R = Et. VIII.2. As dicyclohexylamine salt, m. 151.5-
  154°.[1630] VIII.4. $b_{0.04}$ $93-94^{\circ}$, n_D^{20} 1.5643, d_4^{20}
  1.2202.[1321]

 R = Pr. VIII.4. $b_{0.04}$ $94-95^{\circ}$, n_D^{20} 1.5636, d_4^{20} 1.1900.
  [1321]

 R = Bu. VIII.4. $b_{0.03}$ $83-85^{\circ}$, n_D^{20} 1.5460, d_4^{20} 1.1754.
  [1321]

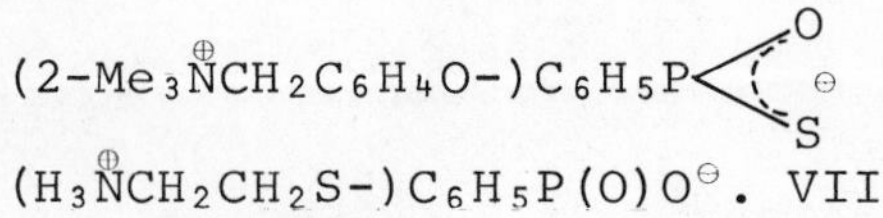

$(2-Me_3\overset{\oplus}{N}CH_2C_6H_4O-)C_6H_5P{\Large\langle}{\substack{O \\ \ominus \\ S}}$. VIII.2. m. 183°.[1381]

$(H_3\overset{\oplus}{N}CH_2CH_2S-)C_6H_5P(O)O^{\ominus}.$ VIII.2. m. 260°.[351]
$C_6H_5PS(OR)SH.$

 R = Et. VIII.4. As Ni salt, m. $184-185^{\circ}$.[395] VIII.4.
  $b_{0.003}$ $83-85^{\circ}$, n_D^{20} 1.6020, d_4^{20} 1.2310.[1321]

 R = Bu. VIII.4. $b_{0.0014}$ $86-88^{\circ}$, n_D^{20} 1.5683, d_4^{20}
  1.1620.[1321]

$C_6H_5PS(SMe)Br$. From $(C_6H_5PS_2)_2$ and MeBr, $b_{0.01}$ 175°, ^{31}P
   -71.0 ppm.[542]
$C_6H_5PS(OR)Cl$.

   R = Me. IX.1. $b_{0.005}$ 90°.[1726]

   R = Et. IX.1. $b_{0.001}$ 90°.[1726] IX.1. $b_{0.32}$ 90°.[874]

   R = ClH_2CCH_2-. IX.1. $b_{0.05}$ 120°.[1726]

   R = $ClH_2CCHClCH_2$-. IX.1. $b_{0.05}$ 160°.[1726]

   R = $(ClH_2C)_2CH$-. IX.1. $b_{0.05}$ 160°.[1726]

   R = Bu. IX.1. $b_{0.002}$ 95°.[1726]

   R = Ph. IX.1. $b_{0.45}$ $131-135^O$.[776] IX.1. $b_{0.8}$ $143-145^O$.
      [1380] IX.1. b_1 $165-168^O$.[1506] IX.1. $b_{0.001}$ 144°.[1726]

   R = $4-ClC_6H_4$-. IX.1. $b_{0.008}$ 140°.[1726]

   R = $2,4-Cl_2C_6H_4$-. IX.1. $b_{0.001}$ 190°.[1726]

   R = $2-MeC_6H_4$-. IX.1. b_1 $173-176^O$.[1506] IX.1.[1726]

   R = $3-MeC_6H_4$-. IX.1. $b_{0.01}$ 145°.[1726]

   R = $4-MeC_6H_4$-. IX.1. $b_{0.05}$ 160°.[1726]

   R = $2,4-Me_2C_6H_3$-. IX.1. $b_{0.01}$ 160°.[1726]

   R = $3,4-Me_2C_6H_3$-. IX.1. $b_{0.025}$ 180°.[1726]

   R = C_8H_{17}-. IX.1. $b_{0.001}$ $134-135^O$.[1726]

   R = $2-EtOC_6H_4$-. IX.1. $b_{0.004}$ 174°.[1726]

   R = $2,6-(MeO)_2C_6H_3$-. IX.1. $b_{0.025}$ 180°.[1726]

$(C_6H_5-)(S)ClP-O-\langle\bigcirc\rangle-OPCl(S)-C_6H_5$. X.1. m. $127-130^O$.[1506]

$(C_6H_5-)(S)ClPO-\langle\bigcirc\rangle-\overset{\overset{Me}{|}}{\underset{\underset{Me}{|}}{C}}-\langle\bigcirc\rangle-O-PCl(S)-C_6H_5$. X.1.[1506]

$C_6H_5PO(SR)Cl$.

   R = Me. IX.3. $b_{0.28}$ $104.5-105.5^O$.[1370]

   R = Et. IX.3. $b_{0.1}$ $107-109^O$.[1370]

   R = Pr. IX.3. $b_{0.25}$ $112-114^O$.[1370]

   R = i-Pr. IX.3. $b_{0.5}$ $115-117^O$.[1370]

$C_6H_5PS(OR)NCS$.

   R = Me. IX.1. $b_{0.03}$ 105°.[1726]

   R = Et. IX.1.[1858] IX.1. $b_{0.03}$ 108°.[1726]

   R = ClH_2CCH_2-. IX.1. $b_{0.03}$ 137°.[1726]

   R = Pr. IX.1. $b_{0.02}$ 125°.[1726]

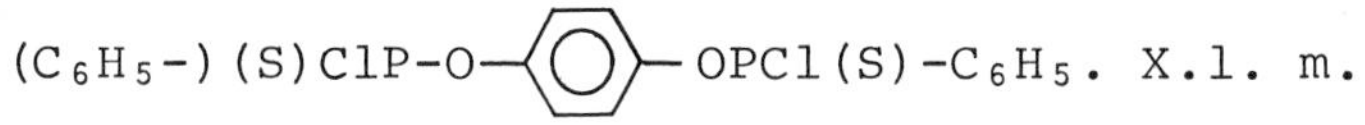

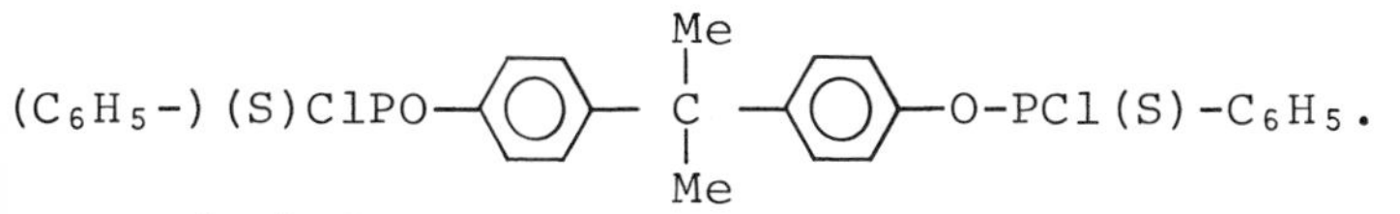

R = i-Pr. IX.1. $b_{0.02}$ 110°.[1726]

R = Bu. IX.1. $b_{0.001}$ 135°.[1726]

R = Ph. IX.1. $b_{0.02}$ 162°.[1726]

R = 3-MeC_6H_4-. IX.1. $b_{0.04}$ 164°.[1726]

R = 4-MeC_6H_4-. IX.1. $b_{0.02}$ 160°.[1726]

R = 3,4-$Me_2C_6H_3$-. IX.1. $b_{0.03}$ 175°.[1726]

$C_6H_5PS(OR)_2$.

R = Et. X.1. $b_{0.3}$ 94-96°.[1321]

R = OCH_2CHCH_2-. X.2. $b_{0.007}$ 116-120°.[1748]

R = Cl_5C_6-. X.1. m. 221°.[831]

R = Et_3Si-. X.2. b_2 169-172°.[1430]

$C_6H_5PS(OMe)OR$.

R = 2,5-Cl_2-4-BrC_6H_2-. X.1. n_D^{22} 1.6385.[1744]

R = $ClHCCH_2CHJCHClCH_2CH$-. X.1. m. 99-101°.[398]

$C_2H_5PS(OEt)OR$.

R = H_2CSCH_2CH-. X.2. n_D^{20} 1.2211, d_4^{20} 1.5791[146]

R = $Me_2NCH_2CH_2$-. X.1. $b_{0.01}$ 110°.[1851]

R = 2,5-Cl_2-4-BrC_6H_2-. X.1. n_D^{26} 1.6203.[1744]

R = 4-$O_2NC_6H_4$-. X.1.[221]

R = $Et_2NCH_2CH_2$-. X.2. $b_{0.01}$ 110-115°.[1832] X.1. $b_{0.01}$ 120°.[1851, 1854]

R = 4-ClH_2CS-C_6H_4-. X.12. n_D^{24} 1.6115.[1800]

R = 4-$MeC(O)C_6H_4$-. X.1. n_D^{23} 1.5917.[1820]

$C_6H_5PS(OPr)OC_6H_2Br$-4-Cl_2-2,5. X.1. n_D^{22} 1.5992.[1744]
$C_6H_5PS(OPr\text{-}i)OPh$. X.11.[1967]

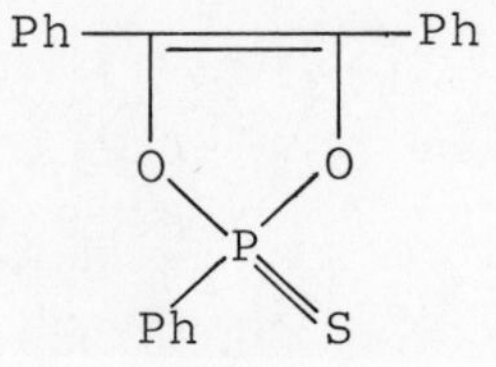

X.1. m. 126.5-127°, ^{31}P -104 ppm, mass spect.[1368]

$C_6H_5PO(OMe)SEt$. X.11. $b_{2.2}$ 112°.[752]
$C_6H_5PO(OEt)SR$.

R = Et. X.3. b_8 100-101.5°.[902] X.2.[1338]

R = EtO_2CCH_2-. X.3. $b_{0.01}$ 110°.[1835]

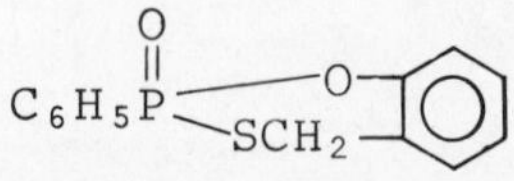

X.4. m. 87°.[1381]

$C_6H_5PS(OMe)SR$.

R = EtO_2CCH_2-. X.5. $b_{0.03}$ 134-135°.[699]

R = EtO_2CCHMe-. X.5. $b_{0.01}$ 113-115°.[699]

R = C_6H_{13}-. X.1. $b_{0.01}$ 103°.[1837]

$C_6H_5PS(OEt)SR$.

R = Et. X.1. $b_{0.01}$ 94°.[1837]

R = CH_3CH=CH-. X.3. $b_{0.15}$ 131-132°.[1443]

R = CH_3CH=CH-CH_2-. X.3. $b_{0.2}$ 143-147°.[1443]

R = $CH_3C(O)NHC(O)CH_2$-. X.3. m. 73-75°.[1815]

R = $C_2H_5SCH_2CH_2$-. X.1. $b_{0.01}$ 126°.[1827]

R = S-CH=CMe-N=C-SCH_2-. X.1.[1110]

R = cycl.-C_6H_{11}-. X.1. $b_{0.01}$ 125°.[1837]

$C_6H_5PS(SMe)OCHCH(-Pr-i)CH_2CH_2CHMeCH_2$. X-ray investigation.[489]

$C_6H_5PO(SR)_2$.

R = Et. X.1. n_D^{20} 1.5955.[1725]

R = Bu. X.1. $b_{0.08}$ 148-154°.[1725]

$(C_6H_5-)OPSCH_2CH_2S$. From $PhP(O)Cl_2$ and $Me_2SiSCH_2CH_2S$ or $MePrSiSCH_2CH_2S$.

$C_6H_5PS(SR)_2$.

R = Et. X.2. $b_{3.5}$ 191-192°.[114] X.1. $b_{0.05}$ 109-117°, ^{31}P -80.5 ppm.[1368]

R = ClH_2CCH_2-. X.1.[1510]

$(C_6H_5-)SPSCH_2CH_2S$. From $PhPSCl_2$, $BrCH_2CH_2Br$, and H_2S, m. 66-68°.[271] X-ray investigation, ^1H, ^{31}P -96.8 ppm.[1089]

$C_4H_9C{\equiv}CPS(OEt)_2$. X.1. $b_{0.07}$ 94°, IR.[363]

$H_2CC(O)(CH_2)_3CHPS(OEt)_2$. X.8. b_{15} 182°.[1612]

Me_2CCH=CMeCOP(S)Cl. IX.1. b_{12} 105-110°.[1460]

cycl.-$C_6H_{11}PS(OR)SH$.

R = Et. VIII.4. As Ni salt, m. 167-169°;[395] as $\overset{\oplus}{N}BuH_3$ salt, m. 104-105°; as $\overset{\oplus}{N}Et_2H_2$ salt, m. 135-136°; as $\overset{\oplus}{N}(i\text{-}Bu)_2H_2$ salt, m. 134-136°.[704] VIII.4. $b_{0.015}$ 100-103°.[702]

R = i-Bu. VIII.1. As Ni salt, m. 147°.[973]

R = C_8H_{17}-. VIII.1. As Ni salt, m. 143°.[973]

R = $C_{10}H_{21}$-. VIII.1. As Ni salt, m. 157°.[973]

$C[CH_2OPS(-C_6H_{11}\text{-cycl.})SH]_4$. VIII.4. m. 126-127°.[707]

cycl.-$C_6H_{11}PS(OR)Cl$. From cycl.-$C_6H_{11}PO(OR)Cl$ and P_4S_{10}.[973]

R = i-Bu. $b_{0.1}$ 66°.[973]

R = C_8H_{17}-. $b_{0.1}$ 63-65°.[973]

R = $C_{10}H_{21}$-. $b_{0.1}$ 72-74°.[973]

cycl.-$C_6H_{11}PS(OR)_2$.

R = Et. X.8. b_2 116.5-117°.[1615]

R = Pr. X.8. b_2 129°.[1615]

R = Bu. X.8. b_2 141-142.5°.[1615]

cycl.-$C_6H_{11}PS(OEt)OCH_2CH_2NEt_2$. X.1. $b_{0.01}$ 122°.[1851]

cycl.-$C_6H_{11}PO(OEt)SR$.

R = $EtSCH_2$-. X.1. $b_{0.01}$ 110°.[1848]

R = $EtSCH_2CH_2$-. X.1. $b_{0.01}$ 118°.[1848]

R = $Et_2NCH_2CH_2$-. X.1. $b_{0.01}$ 118°.[1848]

cycl.-$C_6H_{11}PS(OEt)SR$.

R = Et. X.5. $b_{0.01}$ 99-100°.[699]

R = sec.-Bu. X.3.[1834]

R = EtO_2CCH_2-. X.5. $b_{0.01}$ 133-135°.[699]

R = $EtSCH_2CH_2$-. X.3. $b_{0.02}$ 151-153°.[1834]

cycl.-$C_6H_{11}PS(OPr$-i)SR.

R = EtO_2CCH_2-. X.5. $b_{0.01}$ 158-159°.[699]

R = $EtSCH_2CH_2$-. X.3. $b_{0.01}$ 135-138°.[706]

cycl.-$C_6H_{11}PS(OPh)SR$.

R = Et. X.5. $b_{0.01}$ 213-215°.[699]

R = EtO_2CCH_2-. X.5. $b_{0.01}$ 185-186°.[699]

R = $EtSCH_2CH_2$-. X.3. $b_{0.01}$ 187-188°.[706]

cycl.-$C_6H_{11}PS(OC_6H_{11}$-cycl.)SR

R = Et. X.5. $b_{0.02}$ 153-154°.[699]

R = EtO_2CCH_2-. X.5. $b_{0.05}$ 165-167°.[699]

(EtO)(cycl.-C_6H_{11}-)$SPSCH_2SPS$(-C_6H_{11}-cycl.)OEt. X.3.[1840]

cycl.-$C_6H_{11}PO(SR)_2$.

R = Et. X.1. $b_{0.8}$ 126-128°.[973]

R = Bu. X.1. $b_{0.05}$ 135°.[1725]

R = C_8H_{17}-. X.1. $b_{0.01}$ 224-225°.[973]

cycl.-$C_6H_{11}PS(SPr$-i)SR.

R = H_2C=$CHCH_2$-. X.3. $b_{0.02}$ 140-143°.[705]

R = $PrSCH_2$-. X.3. $b_{0.02}$ 172-173°.[705]

cycl.-$C_6H_{11}PS(SBu)SCHMeCO_2Et$. X.3. $b_{0.01}$ 171-172°.[705]

cycl.-$C_6H_{11}PS(SAm\text{-}i)SCH_2CO_2Et$. X.3. $b_{0.02}$ 174-175°.[705]
$CH_3C(O)CH_2CMe_2PS(OEt)_2$. X.8. b_7 128-130°.[1674]
$Me_2CHCH_2OCH{=}CHPO(SPh)_2$. X.1. m. 72-73°.[84]

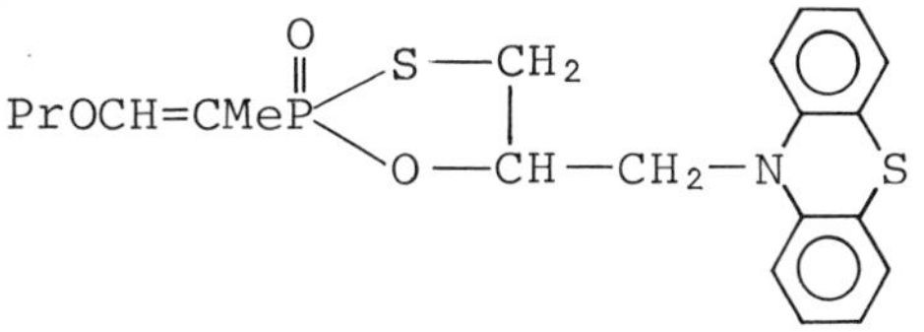

. X.5. m. 128-130°.[227]

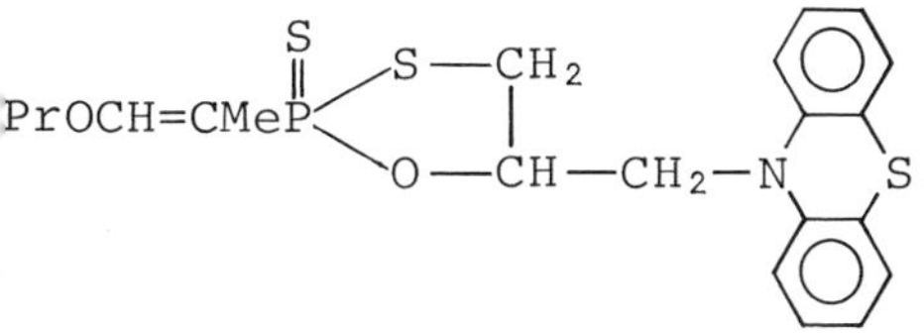

. X.5. m. 138-140°.[227]

$EtMeCHOCH{=}CHPO(SR)_2$.

R = Et. X.1. n_D^{20} 1.5291.[1731]

R = i-Pr. X.1. $b_{0.6}$ 145°, n_D^{20} 1.5226.[1728, 1731]

$EtOCH{=}CEtPS(OR)_2$.

R = Me. X.1. $b_{0.08}$ 62°, n_D^{20} 1.4950, d_4^{20} 1.0882.[1350]

R = Et. X.1. $b_{0.065}$ 68.5°, n_D^{20} 1.4820, d_4^{20} 1.0372.[1350]

$(EtO)_2C{=}CHPS(OEt)_2$. X.12. $b_{0.03}$ 95-96°, n_D^{20} 1.4839, d_4^{20} 1.0889.[1346]

$(EtO)(MeO)CHCH{=}CHPO(OMe)SMe$. X.1. $b_{0.5}$ 102-104°. n_D^{20} 1.4773, d_4^{20} 1.1445.[2037]

$H_2C(CH_2)_4C(OH)PS(OR)_2$.

R = Cl_3CCMe_2-. X.9. m. 181-183°.[32]

R = $H_2C(CH_2)_3C(-CCl_3)$-. X.9. m. 158-160°.[32]

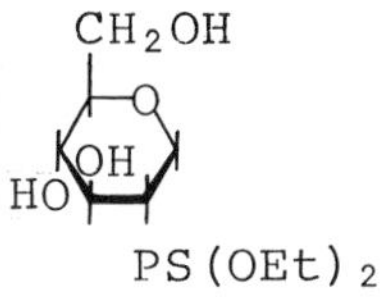

X.8. m. 93-95°.[946]

$C_2H_5SCH_2CH{=}CHCH_2PS(OMe)_2$. X.12. $b_{1.5}$ 120-122°, n_D^{20} 1.5320, d_4^{20} 1.1500.[1593]

$C_6H_{13}PS(OR)SH$.

R = Me. VIII.4. $b_{0.02}$ 68-71.5°.[708]

R = Et. VIII.4. $b_{0.03}$ 87-91°.[708]

R = i-Am. VIII.4. $b_{0.02}$ 113-114°.[708]

R = cycl.-C_6H_{11}-. VIII.4. $b_{0.01}$ 152-156°.[708]

$C_6H_{13}PO(OEt)SCH_2CH_2NEt_2$. X.1. $b_{0.003}$ 118°.[811]
$Me_2CH(CH_2)_3$-PS(OR)SH.
R = Et. VIII.4. $b_{0.04}$ 87-93°.[708]
R = i-Pr. VIII.4. $b_{0.2}$ 105-107°.[708]
R = $Me_2CH(CH_2)_3$-. VIII.4. $b_{0.04}$ 121-124°.[708]
R = C_8H_{17}-. VIII.4. $b_{0.02}$ 153-156°.[708]
(EtO)(EtS)$CHCH_2PO$(OEt)SEt. X.12. $b_{0.038}$ 99°, n_D^{20} 1.4886, d_4^{20} 1.0818.[1347]
$C_2H_5SCHPrPO(SR)_2$.
R = Et. X.11. $b_{0.03}$ 130°.[296]
R = Pr. X.11. $b_{0.03}$ 140°.[296]
$C_4H_9SCHMePO(SBu)_2$. X.11. $b_{0.18}$ 155°.[296]
$Et_2NCH_2CH_2PS(OEt)OCH_2CH_2SEt$. X.1. $b_{0.00002}$ 89-91°.[907]
$(EtO)_2OPOCHMePS(OEt)_2$. X.12. $b_{0.01}$ 101°.[1861] X.9/X.12. b_2 146°.[1625]
$(EtO)_2SPOCHMePS(OEt)_2$. X.12. $b_{0.01}$ 102°.[1861]
$(EtO)_2SPOCMe(OH)$-$PS(OEt)_2$. X.8/X.12. b_2 140-142°.[1668]

$\underline{C_7}$

4-$ClC_6H_4NHC(O)$ 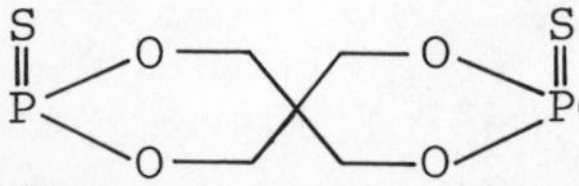 $C(O)NHC_6H_4Cl$-4. X.10. m. 239-240°.[269]
$ClC_6H_4CH_2PS(OMe)OC_6H_2Cl_3$-2,4,6. X.1. m. 103-105°.[1506]
$C_3H_7SCH=C(-CCl=CH_2)PO(SEt)_2$. X.1. $b_{0.015}$ 128-129°.[2047]
$C_5H_{11}CMeClPO(SEt)Cl$. IX.1. $b_{0.1}$ 113°.[1727]
$C_4H_9CEtClPO(SEt)Cl$. IX.1. $b_{0.02}$ 100°.[1727]
$C_4H_9CEtClPO(OMe)SEt$. X.1. $b_{0.04}$ 103°.[1727]

2,5-$Cl_2C_6H_3NHC(O)$ 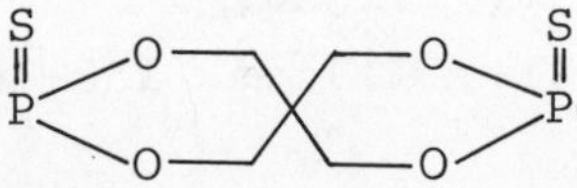 $C(O)NHC_6H_3Cl_2$-2,5. X.10 [269]

3,4-$Cl_2C_6H_3NHC(O)$ 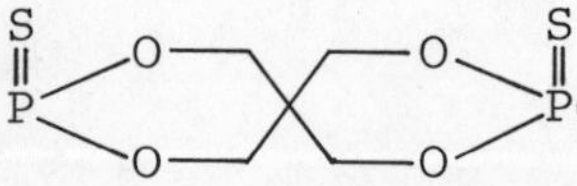 $C(O)NHC_6H_3Cl_2$-3,4. X.10 [269]

$C_6H_5NHC(O)PS(OMe)_2$. X.10. m. 107-108°.[1639]

$C_6H_5NHC(O)$ 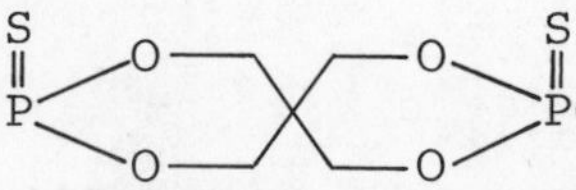 $C(O)NHC_6H_5$. X.10. m. 227-228°.[269]

3-$O_2NC_6H_4CH(OH)PS[O-(Cl_3C-)C(CH_2)_3CH_2]_2$. X.9. m. 157-158°.[32]

$C_6H_5CH_2PS(OR)Cl$.

R = 2,4,5-$Cl_3C_6H_2$-. X.1. n_D^{20} 1.6250.[1506]

R = 1-$C_{10}H_7$-. X.1. m. 108-113°.[1506]

R = 2-$C_{10}H_7$-. X.1. n_D^{20} 1.6495.[1506]

$C_6H_5CH_2PS(OEt)_2$. X.7. $b_{3.5}$ 124-125°.[900]

$C_6H_5CH_2Ps(OMe)OR$.

R = 2,4,6-$Cl_3C_6H_2$-. X.1. m. 66-68°.[1506]

R = 2-$C_{10}H_7$-. X.1. n_D^{20} 1.5996.[1506]

$C_6H_5CH_2PS(OEt)OC_6H_4Cl_2$-2,4. X.1. b_1 219-222°.[1506]

$(MeO)(C_6H_5CH_2-)SP-O-$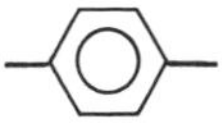$-O-PS(-CH_2C_6H_5)(OMe)$. X.1. m. 105-112°.[1506]

$C_6H_5CH_2PO(SR)_2$.

R = Et. X.6. $b_{0.065}$ 117-118°.[1316]

R = Pr. X.6. $b_{0.04}$ 132-134°.[1316]

4-$MeC_6H_4PS(OBu)Cl$. IX.1. $b_{0.02}$ 150°.[1726]

4-$MeOC_6H_4PS(OCH_2CH_2Cl)_2$. X.2. n_D^{20} 1.5623.[618]

(4-$MeOC_6H_4$-)$SPSCH_2CH_2O$. X.5.[1845]

$C_6H_5CH(OH)PS(OR)_2$.

R = Cl_3CCMe_2-. X.9. m. 167-168°.[32]

R = $H_2C(CH_2)_3C(-CCl_3)$-. X.9. m. 142-143°.[32]

cycl.-$C_5H_9C{\equiv}CPS(OEt)_2$. From cycl.-$C_5H_9C{\equiv}CLi$ and $ClPS(OEt)_2$, $b_{0.12}$ 110°.[363]

$C_5H_{11}C{\equiv}CPS(OEt)_2$. From $C_5H_{11}C{\equiv}CLi$ and $ClPS(OEt)_2$, $b_{0.07}$ 101°, IR.[363]

$H_2C{=}CMeC(O)NHCMe_2PS(OMe)_2$. X.12. b_5 127-128°.[1597]

$C_5H_{11}CH{=}CHPS(OEt)_2$. X.8. b_1 119°, n_D^{20} 1.4800, d_4^{20} 0.9972.[1627]

$CH_3C(O)CH_2CHPrPS(OEt)_2$. X.8. b_4 133-134°.[1659]

$(CH_3)_2CHOCH{=}CEtPS(OEt)_2$. X.1. $b_{0.058}$ 76.5°, n_D^{20} 1.0251, d_4^{20} 1.4828.[1350]

$(EtO)_2CH-CH{=}CHPO(OEt)SEt$. X.1. b_1 123-124°, n_D^{20} 1.4665, d_4^{20} 1.0739.[2037]

$OCH_2CH_2OCMe(CH_2)_3PS(OEt)_2$. X.7. $b_{0.01}$ 134°, IR.[1959]

$H_2C(CH_2)_4\overset{\oplus}{N}H-CH_2CH_2PS(OBu)O^{\ominus}$. VIII.3. m. 181-181.5°.[931]

$H_2C(CH_2)_4NCH_2CH_2PS(OBu)_2$. X.2. $b_{1.5}$ 140-143°, n_D^{20} 1.4870, d_4^{20} 0.9977.[931]

$C_7H_{15}PS(OR)_2$.

R = Et. X.8. b_{13} 146-148°.[1615]

R = Pr. X.8. b_2 146-147°.[1615]

R = Bu. X.8. b_3 161°.[1615]

R = C_7H_{15}-. Magn. rotation of the bonds.[1069]

$(EtO)_2SPOCEt(OH)PS(OEt)_2$. X.8/X.12. b_2 147°, n_D^{20} 1.4810, d_4^{20} 1.1465.[1668]

$\underline{C_8}$

$4\text{-}ClC_6H_4SCH=CHPO(SR)_2$.

R = Et. X.1. n_D^{25} 1.6358.[1729, 1731a]

R = Pr. X.1. $b_{0.04}$ 212°, n_D^{20} 1.6158.[1729, 1731a]

$2\text{-}MeO\text{-}5\text{-}ClC_6H_3CH_2PS(OEt)_2$. X.7. $b_{0.22}$ 136-138°, n_D^{20} 1.530(d_4^{20} 1.1812.[1538]

$C_4H_9SCH\text{-}C(\text{-}CCl=CH_2)\text{-}PO(SR)_2$.

R = Et. X.1. $b_{0.018}$ 158-160°.[2047]

R = Pr. X.1. $b_{0.012}$ 148°.[2047]

$C_6H_5C\equiv CPS(OR)_2$.

R = Me. X.1. $b_{0.2}$ 119°, IR.[363]

R = Et. X.1. $b_{0.15}$ 134-135°, IR.[363]

$C_6H_5CH=CHPS(OR)_2$.

R = Me. X.1. b_1 124°.[93]

R = Et. X.8. b_3 146°, n_D^{20} 1.5450, d_4^{20} 1.1190.[1627]

$C_6H_5CH=CHPS(OEt)SCH_2CH_2SEt$. X.3. $b_{0.01}$ 134°.[1846]

$C_6H_5CH=CHPS(OEt)SCH_2SEt$. X.1. $b_{0.01}$ 130°.[1830]

$C_6H_5CH=CHPO(SR)_2$.

R = Et. X.1. b_1 168°.[93]

R = Ph. X.1. m. 117°.[93]

$C_6H_5OCH=CHPS(OEt)_2$. X.1. b_2 139°.[83]

$C_6H_5OCH=CHPO(SEt)_2$. X.1. b_1 162°.[84]

$4\text{-}EtC_6H_4PS(OBu)Cl$. IX.1. $b_{0.03}$ 150°.[1726]

$OCH=CH\text{-}CH=C\text{-}CH[\text{-}CH_2C(O)Me]PS(OEt)_2$. X.8. b_7 164-166°.[1674]

$CH_3SCHPhPO(SMe)_2$. X.11. m. 90°.[296]

cycl.-$C_6H_{11}C\equiv CPS(OEt)_2$. X.1. $b_{0.45}$ 134°, IR.[363]

$C_6H_{13}C\equiv CPS(OEt)_2$. X.1. $b_{0.2}$ 130°, IR.[363]

$CH_3C(O)CH(\text{-}CO_2Et)CHMePS(OEt)_2$. X.8. b_{13} 151-152°.[1661]

$(EtO_2C\text{-})_2CHCH_2PS(OEt)_2$. X.8. b_{11} 180-182°.[1643]

$EtO_2CCH_2CH(\text{-}CO_2Et)PS(OEt)_2$. X.8. b_1 146.5°.[1670]

$C_6H_{13}CH=CHPS(OEt)_2$. X.8. b_9 149-150°.[1627]

$Me_2C=C(\text{-}CMe_3)PS(OEt)OCH_2CH_2NEt_2$. X.1. $b_{0.01}$ 107°.[1851]

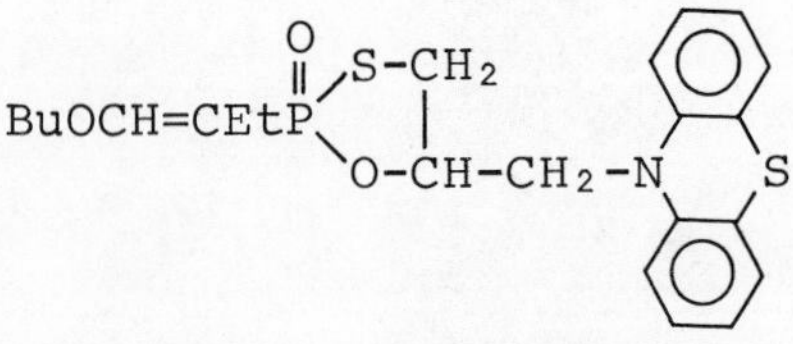

. X.5. m. 138-140°.[227]

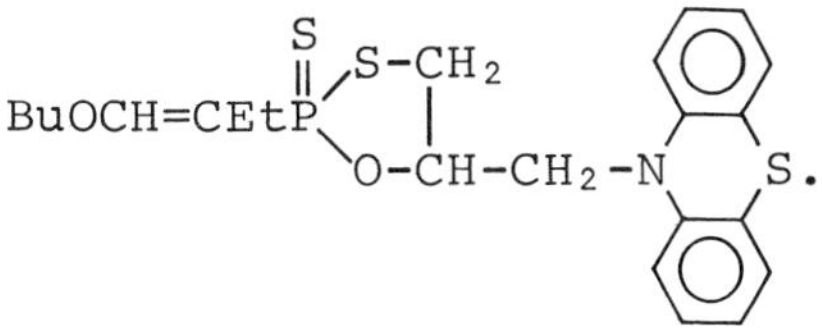
X.5. m. 105-106°.[227]

(PrO)(EtO)CH-CH=CHPO(OPr)SPr. X.1. $b_{1.5}$ 131-133°.[2037]
$CH_3CH(-SEt)CH_2CH(-SEt)PO(SEt)_2$. X.11. $b_{0.08}$ 169°.[296]

<u>C_9</u>
$4\text{-}ClC_6H_4CH(-SEt)PO(SEt)_2$. X.11. $b_{0.02}$ 170°.[296]
$(i\text{-}Bu)_2CClPO(SEt)Cl$. IX.1. $b_{0.2}$ 110°.[1727]
$Cl_5C_6SCH_2CMe(-OH)PS(OR)_2$.

R = Me. X.9. m. 188-189°.[2095]

R = Et. X.9 m. 200-201°.[2095]

R = Pr. X.9 m. 167-168°.[2095]

$C_2H_5SCHPhPO(SEt)_2$. X.11. $b_{0.01}$ 170°.[296]
$C_7H_{15}C{\equiv}CPS(OEt)_2$. X.11. $b_{0.07}$ 121°, IR.[363]
$(EtO_2C-)_2CHCHMePS(OEt)_2$. X.8. b_3 180-181°.[1631]
$EtO_2CCH_2CMe(-CO_2Et)PS(OEt)_2$. X.8. b_9 180-181°.[1631]
$CH_2PS(OEt)_2$. X.8. b_2 149°.[946]
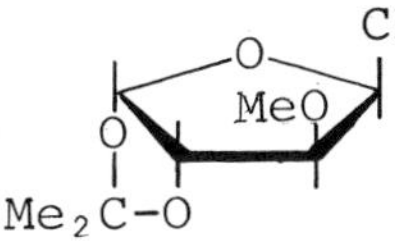

$C_4H_9CHEtCH(-SMe)PO(SMe)_2$. X.11. $b_{0.01}$ 128°.[269]

<u>C_{10}</u>
$C_6H_5CH_2CH_2C{\equiv}CPS(OEt)_2$. X.1. $b_{0.08}$ 135°, IR.[363]
MeC=CH-NPh-N=CPS(OEt)$_2$. X.12. $b_{0.009}$ 147-149°, n_D^{20} 1.5700, d_4^{20} 1.1810.[1608]
$CH_3C(C)CH_2CHPhPS(OEt)_2$. X.8. b_7 177-178°.[1674]
$EtO_2CCHPhPS(OEt)_2$. X.8. b_1 158.5°.[1670]
$(EtO_2C-)_2CHCMe_2PS(OBu)_2$. X.8. b_4 200-202°.[1661]
$(i\text{-}Pr-)_2NC(O)(CH_2)_3PS(OPr\text{-}i)_2$. X.7. $b_{0.004}$ 135-139°.[515]

<u>C_{11}</u>
$4\text{-}BrC_6H_4CH[-OP(O)(OEt)_2]PS(OEt)_2$. X.9/X.12. b_1 179-180°.[1592]

$2\text{-}ClC_6H_4CH[-OP(O)(OEt)_2]PS(OPr)_2$. X.9/X.12. b_1 179°.[1592]
$EtO_2CCH=CPhPS(OEt)_2$. X.8. b_1 168-169°.[1670]
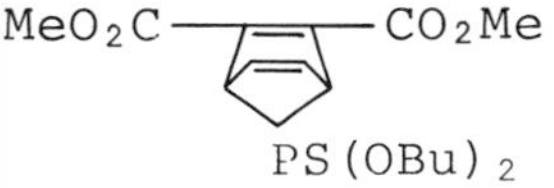
X.12. b_2 194-195°.[929]

$C_{11}H_{23}PS(OEt)_2$. X.8. b_{14} 197-198°.[1615]

$\underline{C_{13}}$
PhNHCHPhPS(OBu)$_2$. X.10. m. 53°[1660]
(PhNHCHPh-)SPOCHMe-CHMeO. X.10. m. 148-150°.[1588]

$\underline{C_{14}}$
4-MeOC$_6$H$_4$NHCHPhPS(OEt)$_2$. X.10. m. 48°.[1660]
(EtO$_2$C)$_2$CHCHPhPS(OEt)$_2$. X.10. b$_4$ 210-212°.[1661]

$\underline{C_{15}}$
4-(Me$_2$N-)C$_6$H$_4$CH(-NHPh)PS(OR)$_2$.

  R = Et, X.10. m. 100-100.5°.[1660]

  R = Bu. X.10. m. 80°.[1660]

4-HO-3,5-(BuS-)$_2$C$_6$H$_2$CH$_2$PO(SC$_{18}$H$_{37}$)$_2$. X.6. m. 68-70°
  (hexane).[1942]

$\underline{C_{17}}$
1-C$_{10}$H$_7$-NHCHPhPS(OEt)$_2$. X.10. m. 136.5°.[1660]

$\underline{C_{19}}$
Ph$_3$CPS(OEt)$_2$. From (EtO)$_2$SPH and ClCPh$_3$, m. 128°.[1666]

C.4. Polythiophosphonic Acids

$\underline{C_3}$
(EtO)$_2$SPCH$_2$CHMePS(OEt)$_2$. X.8. b$_8$ 182-184°, n_D^{20} 1.4924.
  d_4^{20} 1.1229.[1669]

$\underline{C_4}$
(EtO)$_2$SPCH$_2$CH=CHCH$_2$PS(OMe)$_2$. X.8. b$_{1.5}$ 161-162.5°, n_D^{20}
  1.5115, d_4^{20} 1.1774.[1593]
CH$_3$CO$_2$CMe[-PS(OEt)$_2$]$_2$. X.12. b$_2$ 152°, n_D^{20} 1.4928, d_4^{20}
  1.1833.[1668]
(EtO)$_2$SPCH$_2$CHEtPS(OEt)$_2$. X.8. b$_{2.5}$ 165-172°, n_D^{20} 1.4928,
  d_4^{20} 1.1176.[1669]

$\underline{C_6}$
(PhCH$_2$S)$_2$SP—⟨O⟩—PS(SCH$_2$Ph)$_2$. X.1. m. 114-115°.[1517]

$\underline{C_7}$
(EtO)$_2$SPCH$_2$CHAmPS(OEt)$_2$. X.8. b$_8$ 206-207°, n_D^{20} 1.4854,
  d_4^{20} 1.0765.[1627]

$\underline{C_8}$

(EtO)$_2$SPCH—⟨O⟩—CHPS(OEt)$_2$. X.10. m. 134-136°.[1573]
      | |
      OH   OH

(EtO)$_2$SP-CH—⟨O⟩—CHPS(OEt)$_2$. X.9.[1573]
      | |
      NH$_2$ NH$_2$

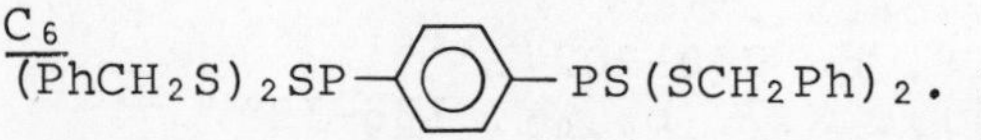

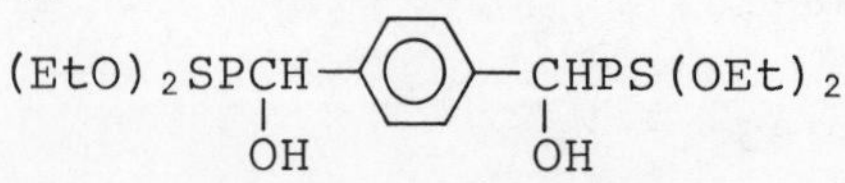

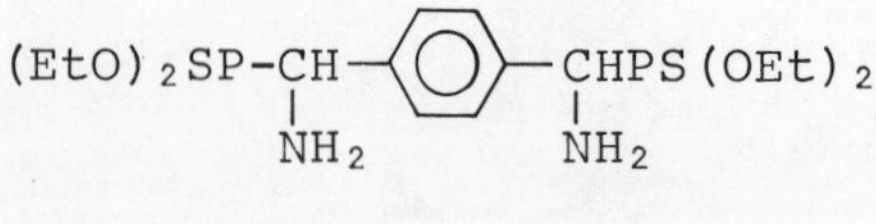

$(EtO)_2SPCH(-CO_2Et)-CH(-CO_2Et)-PS(OEt)_2$. X.8. b_1 185°.[1670]

$\underline{C_{10}}$
$(EtO)_2SPCH_2CHMeCO_2CH_2CH_2O_2CCHMePS(OEt)_2$. X.8. b_6 135-136°.[1653]

$\underline{C_{11}}$
$(EtO)_2SPCHPhCH(-CO_2Et)-PS(OEt)_2$. X.8. b_1 212°.[1670]

$\underline{C_{17}}$
$O=C[CH_2CHPhPS(OEt)_2]_2$. X.8. m. 113.5-115°.[1653]

$\underline{C_{20}}$

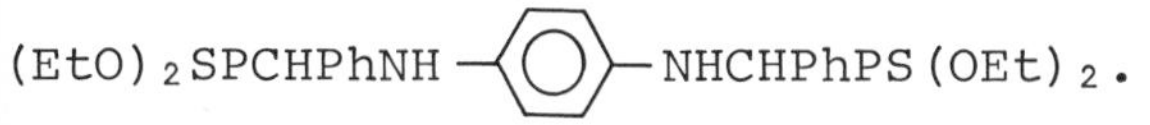

$(EtO)_2SPCHPhNH$—⟨O⟩—$NHCHPhPS(OEt)_2$. X.10. m. 166-167°.[1653]

$\underline{C_{26}}$

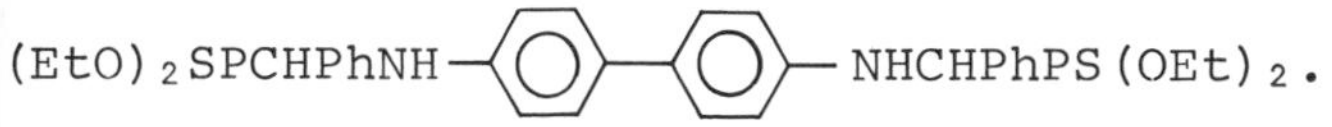

$(EtO)_2SPCHPhNH$—⟨O⟩—⟨O⟩—$NHCHPhPS(OEt)_2$. X.10. m. 149-150°.[1653]

$\underline{C_{28}}$

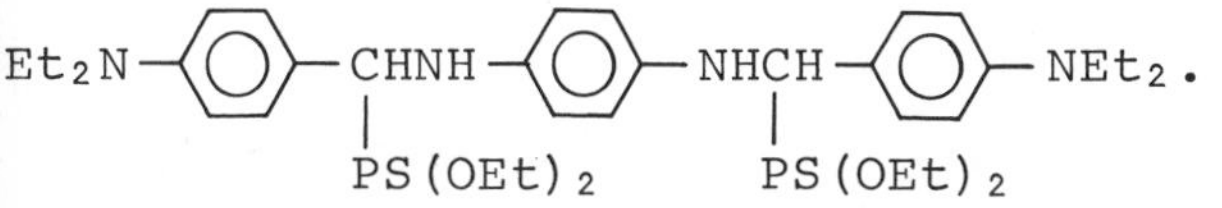

Et_2N—⟨O⟩—$CHNH$—⟨O⟩—$NHCH$—⟨O⟩—NEt_2. X.10. m. 82-83°.[1653]
　　　　　　　　|　　　　　　　|
　　　　　　$PS(OEt)_2$　　$PS(OEt)_2$

C.5. Selenophosphonic Acids

$\underline{C_1}$
$(CH_3-)Se\overline{PSCH_2CH_2}S$. XII.1. b_2 148°, m. 68°.[2076]

$(CH_3-)SeP{<}^{O}_{O}{-}$⟨O⟩ XII.1. b_2 90°, m. 46°, ^1H.[2075]

$\underline{C_2}$
$C_2H_5PO(OEt)SeNa$. XI.1.[52]
$C_2H_5PO(OEt)SeCH_2CH_2NEt_2$. From $EtPO(OEt)SeNa$ and $ClCH_2CH_2-NEt_2$, very toxic. LD_{50} 0.021 mg/kg in mice, n_D^{25} 1.4950.[52]

$\underline{C_3}$
$NCCH_2CH_2PSe(OEt)_2$. XII.1. $b_{0.15}$ 82-83°.[1294]
$H_2C=CHCH_2PSe(OR)_2$.

　　R = Et. XII.1. b_{10} 101-102°.[1702]
　　R = Pr. XII.1. b_{11} 126-127°.[1702]
　　R = i-Pr. XII.1. b_7 101-102°.[1702]
　　R = Bu. XII.1. b_{10} 145-147°.[1697]
　　R = Am. XII.1. $b_{0.47}$ 122°.[1702]

R = C_6H_{13}-. XII.1. $b_{0.2}$ 124-126°.[1698]

$\underline{C_4}$

$CH_3CH{=}CHCH_2PSe(OR)_2$.

R = Me. XII.3. b_3 98-100°, n_D^{20} 1.5164, d_4^{20} 1.3343.[1593]

R = Pr. XII.1. $b_{0.04}$ 77-80°.[1102]

R = Bu. XII.1. $b_{0.06}$ 90-91°.[1102]

R = i-Bu. XII.1. b_9 144-146°.[1102]

Me_2CHCH_2PSe(OBu-sec.)$_2$. XII.1. b_{14} 144-146°.[1710]

C.6. Tellurophosphonic Acids

$\underline{C_3}$

$H_2C{=}CHCH_2PTe(OEt)_2$. XIII.1. $b_{0.32}$ 65-66°.[1702]

C.7. Phosphonic Acid Amide Derivatives

$\underline{C_1}$

$CH_2ClPO(NHCOR)OH$.

R = $-CHCl_2$. XIV.2. m. 113-116°, IR.[1909]

R = $-CF_3$. XIV.2. m. 137-139°, IR.[1909]

$CH_2ClPO(NR_2)Cl$.

R = Me. XV.1. $b_{0.2}$ 79-81°, n_D^{20} 1.4953, d_4^{20} 1.3666.[1242]

R = Et. XV.1. $b_{0.3}$ 96°, n_D^{20} 1.4858, d_4^{20} 1.3122.[1242]

$CH_2ClPO(NHCOR)Cl$.

R = $-CHCl_2$. XV.2. m. 78-80°, IR.[1909]

R = $-CCl_3$. XV.2. m. 112-114°, IR.[1909]

$CH_2ClPO(NHC_6H_4Cl\text{-}4)F$. XV.1. m. 122-124°.[860]
$CH_2ClPO(NHCOOMe)Cl$. XV.4. m. 102-104°, IR.[1908]
$CH_2ClPO(NHCONHAr)Cl$

Ar = Ph. XV.4. m. 112-114°, IR.[1908]

Ar = 4-ClC_6H_4-. XV.4. m. 141-142°, IR.[1908]

Ar = 4-$O_2NC_6H_4$-. XV.4. m. 153-155°, IR.[1908]

$CH_2ClPO(NHCOOR)F$.

R = Me. XV.4. m. 99-101°, IR.[1908]

R = Ph. XV.4. m. 98-100°, IR.[1908]

$CH_2ClPO(NHCONHAr)F$.

Ar = Ph. XV.4. m. 122-124°, IR.[1908]

Ar = 4-ClC_6H_4-. XV.4. m. 143-145°, IR.[1908]

$CH_2ClPO(OMe)(NHCOR)$.

R = $-CHCl_2$. XVI.5. m. 103-104°.[1909]

R = $-CCl_3$. XVI.5. m. 125-126°.[1909]

$CH_2ClPO(OMe)(NHCONH_2)$. XVI.2. m. 163°.[1462]

$CH_2ClPO(OPh)(NRR')$.

R = H, R' = Et. XVI.2. m. 41-42.5°.[1234]

R = R' = Et. XVI.2. m. 113-114°.[1234]

$CH_2ClPO(OC_6H_4Cl\text{-}4)(NHR)$.

R = Me. XVI.2. m. 46.5-49°.[1234]

R = Et. XVI.2. m. 51-53°.[1234]

R = i-Pr. XVI.2. m. 113.5-114.5°.[1234]

R = Bu. XVI.2. m. 44.5-47°.[1234]

R = sec.-Bu. XVI.2. $b_{0.15}$ 89.5-91°.[1234]

$CH_2ClPO(OC_6H_4Cl\text{-}4)[N(Pr\text{-}i)(CH_2OH)]$. XVI.9. By reaction of CH_2O with $ClH_2CPO(OC_6H_4Cl\text{-}4)(NHPr\text{-}i)$, m. 112-113.5°.[663]

$CH_2ClPO(OC_6H_4CH_3\text{-}4)(NHCONHPh)$. XVI.6. m. 142-143°.[1781]

$CH_2ClPO(OC_6H_{11}\text{-cycl.})[N(C_3H_5)_2]$. XVI.2. $b_{0.05}$ 122-124°.[1722]

$CH_2ClPO(OC_6H_{11}\text{-cycl.})[N(Pr)_2]$. XVI.9. By hydrogenation of the diallyl analogue, $b_{0.05}$ 128°.[1722]

$CH_2ClPO(OEt)(N{=}PR_3)$.

R = Et. XVI.8. $b_{0.04}$ 132-133°, IR, ^{31}P.[1913]

R = i-Pr. XVI.8. $b_{0.06}$ 125-126°, IR, ^{31}P δ_1 -8.3, δ_2 + 6.2 ppm.[1913]

$CH_2ClPO(OPh)(N{=}PR_3)$.

R = Et. XVI.8. $b_{0.05}$ 157-158°, IR, ^{31}P.[1913]

R = i-Pr. XVI.8. $b_{0.06}$ 125-133°, IR, ^{31}P δ_1 -8.3, δ_2 + 6.2 ppm.[1913]

$CH_2ClPO(OPh)NHCOCONHPO(OPh)CH_2Cl$. XVI.8. m. 187-189° (dec.).[1911]

$CH_2ClPO(OEt)NHCOCONHPO(OEt)CH_2Cl$. XVI.8. m. 204-205°.[1911]

$CH_2ClPO(NH_2)_2$. XVII.1. m. 120-122°.[349]

$CH_2ClPO(NR_2)_2$.

R = Me. XVII.1. b_4 117-118°, m. 48.5-49.5°.[2038]

R = Et. XVII.1. b_3 125-126°.[2038]

$CH_2ClPO(\overline{NCH_2CH_2})_2$. XVII.1. $b_{0.001}$ 85-90°.[1476]

$CH_2ClPO(NMe_2)(NHPh)$. XVII.1. m. 134-136°.[1242]

$CH_2ClPO(NEt_2)(NHAr)$.

Ar = Ph. XVII.1. m. 82-83°.[1242]

Ar = $3\text{-}ClC_6H_4$-. XVII.1. m. 66.5-68.5°.[1242]

$CH_2Cl(O)\overline{PN(Me)CH_2CH_2N}(Me)$. XVII.1. m. 75-77°, ^1H.[2043]

$CH_2ClPO(NHCOOR)_2$.

R = Me. XVII.6. m. 172-174°, m. 181-183°.[762]

R = Et. XVII.6. m. 157-158°.[469]

R = i-Pr. XVII.6. m. 142-144°.[469]

$CH_2ClPO(NHCONHPh)_2$. XVII.6. m. 186-188°(dec.).[469] m. 199-200°.[762]

$CH_2Cl(O)P$⟨$NHCO$⟩$C=C(Ph)NC_5H_{10}$. XVII.6. m. 153-155°.[855]
         $NHCO$

$CH_2Cl(O)P$⟨$NHCO$⟩$C=C(OEt)_2$. XVII.6. m. 125-127°.[853]
         $NHCO$

$CH_2Cl(O)P$⟨$NHCO$⟩$CHCOOEt$. XVII.6. m. 180-183°.[853]
         $NHCO$

$CH_2ClPO(NH_2)[NHCOCH(CO_2)CO_2Et]$. XVII.6. (By hydrolysis of the previous compound.) m. 155-157°.[853]

$CH_2Cl(O)P$ ... XVII.6. m. 155-156°.[853]

$CHCl_2PO(NHC_6H_4Cl-4)F$. XV.1. m. 167-171°.[860]
$CHCl_2PO(NHCOR)Cl$.

R = $-CCl_3$. XV.2. m. 138-140°, IR.[1909]

R = $-CF_3$. XV.2. m. 93-95°, IR.[1909]

$CHCl_2PO(NHCOOMe)Cl$. XV.4. m. 118-119°.[1907]
$CHCl_2PO(NHCONHAr)Cl$.

Ar = Ph. XV.4. m. 118-120°, IR.[1907]

Ar = $4-ClC_6H_4-$. XV.4. m. 130-132°(dec.), IR.[1907]

Ar = $4-O_2NC_6H_4-$. XV.4. m. 150-151°(dec.), IR.[1907]

$CHCl_2PO(NHCOOMe)F$. XV.4. m. 95-97°.[1907]
$CHCl_2PO(NHCONHAr)F$.

Ar = Ph. XV.4. m. 120-122°, IR.[1907]

Ar = $4-ClC_6H_4-$. XV.4. m. 149-151°, IR.[1907]

Ar = $4-O_2NC_6H_4-$. XV.4. m. 155-157°, IR.[1907]

$CHCl_2PO(OMe)(NHCOR)$.

R = $-CHCl_2$. XVI.5. m. 108-110°, IR.[1909]

R = $-CCl_3$. XVI.5. m. 154-155°, IR.[1909]

$CHCl_2PO(OC_6H_4Cl-2)(NHPr-i)$. XVI.1. m. 99.5-100.5°.[664]
$CHCl_2PO(NEt_2)_2$. XVII.3. b_8 157-160°, m. 124-125°.[27]
$CHCl_2PO(NHCOOR)_2$.

R = Me. XVII.6. m. 175-176°.[469]

R = Et. XVII.6. m. 135-137°.[469]

R = i-Pr. XVII.6. m. 168-170°.[469]

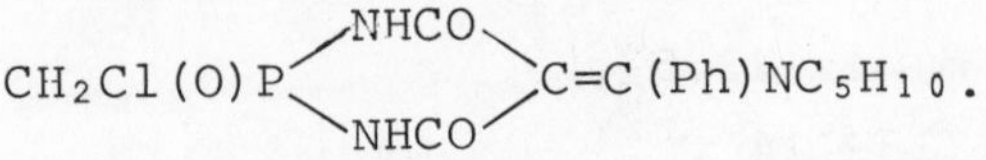

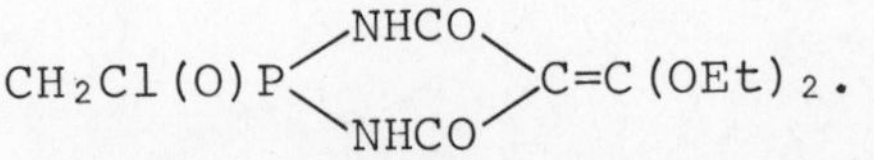

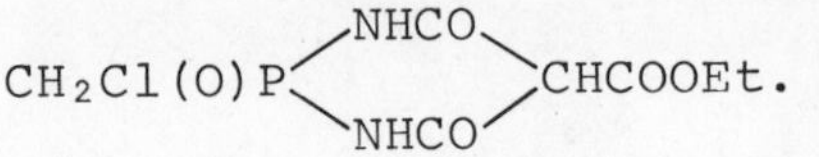

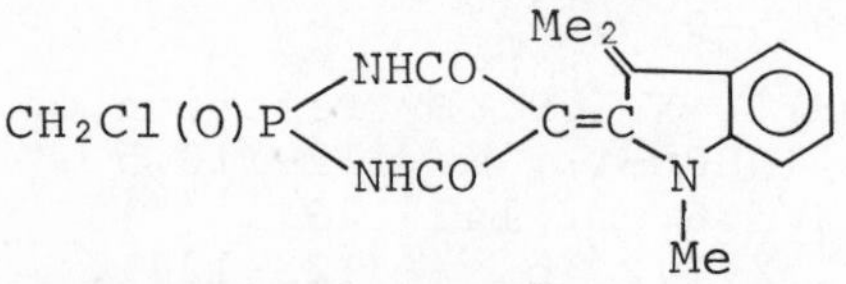

$CHCl_2PO(NHCONHPh)_2$. XVII.6. m. $192-194^{\circ}$(dec.).[469]

$CHCl_2(O)P\langle(NHCO)_2\rangle C=C(Ph)NC_5H_{10}$. XVII.6. m. 184°.[855]

$CCl_3PO(NHR)OH$.

R = Ph. XIV.1. m. $190-191^{\circ}$.[955]

R = $4-CH_3OC_6H_4-$. XIV.1. m. $169-170^{\circ}$.[955]

$CCl_3PO(OH)NHPO(OH)CCl_3$. XIV.1. m. $286-288^{\circ}$.[955]

$CCl_3PO(NRR')Cl$.

R = H, R' = Ph. XV.1. m. $112-113^{\circ}$.[955] m. $113-113.5^{\circ}$.[2088]

R = H, R' = cycl.-$C_6H_{11}-$. XV.1. m. $118-119^{\circ}$.[955]

R = Me, R' = Ph. XV.1. m. $57-58^{\circ}$.[955]

R = Me, R' = cycl.-$C_6H_{11}-$. XV.1. $b_{0.5}$ $144-145^{\circ}$, $b_{1.5}$ 164°.[955]

$CCl_3PO(NHCOR)Cl$.

R = $-CHCl_2$. XV.2. m. $147-148^{\circ}$.[1904]

R = $-CCl_3$. XV.2. m. $147-148^{\circ}$.[1904]

R = $-CF_3$. XV.2. m. $108-110^{\circ}$,[1904] 105°.[1138]

R = $4-O_2NC_6H_4-$. XV.2. m. $133-134^{\circ}$.[1904]

$CCl_3PO(NHCOOR)F$.

R = Me. XV.4. m. $73-75^{\circ}$.[1905]

R = i-Pr. XV.4. m. $76-77^{\circ}$.[1905]

R = Ph. XV.4. m. $84-86^{\circ}$.[1905]

R = $4-ClC_6H_4-$. XV.4. m. $94-96^{\circ}$.[1905]

$CCl_3PO(NHCONHAr)F$.

Ar = Ph. XV.4. m. $137-138^{\circ}$.[1905]

Ar = $4-CH_3OC_6H_4-$. XV.4. m. $125-127^{\circ}$.[1905]

Ar = $4-ClC_6H_4-$. XV.4. m. $141-143^{\circ}$.[1905]

Ar = $4-BrC_6H_4-$. XV.4. m. $140-142^{\circ}$.[1905]

Ar = $4-O_2NC_6H_4-$. XV.4. m. $143-145^{\circ}$.[1905]

$CCl_3PO(N=R)Cl$.

R = $-CH(CH=CH)_2NMePh$. XV.1. m. $155.5-157^{\circ}$(HCl adduct).[955]

R = $-CH(CH=CH)_2NMeC_6H_4Cl-4$. XV.1. m. $184-185^{\circ}$ (HCl adduct).[955]

$CCl_3PO(OMe)(NRR')$.

R = H, R' = Ph. XVI.1. m. $146-147^{\circ}$.[955]

R = H, R' = $-COCCl_3$. XVI.5. m. 143-145°, IR.[1904]

R = H, R' = $4-ClCOC_6H_4-$. XVI.5. m. 145-146°, IR.[1904]

R = H, R' = $4-O_2NCOC_6H_4-$. XVI.5. m. 152-153°, IR.[1904]

$CCl_3PO(OEt)(NRR')$.

R = R' = Et. XVI.3. b_9 144-146°, n_D^{20} 1.4792, d_4^{20} 1.2864.[27]

R = H, R' = Ph. XVI.1. m. 135.5-136.5°.[955]

R = H, R' = $-COCHCl_2$. XVI.5. m. 154-156°, IR.[1904]

R = H, R' = $4-ClCOC_6H_4-$. XVI.5. m. 139-140°, IR.[1904]

R = H, R' = $4-O_2NCOC_6H_4-$. XVI.5. m. 146-147°, IR.[1904]

$CCl_3PO(OPr-i)(NRR')$.

R = R' = Et. XVI.3. b_3 130-132°, n_D^{20} 1.4792, d_4^{20} 1.2512.[27]

R = H, R' = $-COCHCl_2$. XVI.5. m. 100-102°, IR.[1904]

$CCl_3PO(OBu)NR_2$.

R = Et. XVI.3. b_8 161-163°, n_D^{20} 1.4755, d_4^{20} 1.2157.[27]

$CCl_3PO(OC_6H_4Cl-2)(NHPr-i)$. XVI.1. m. 144-145°.[664]

$CCl_3PO(NHR)_2$.

R = Me. XVII.1. m. 133-135°.[184]

R = Ph. XVII.1. m. 172°.[872a, 2088]

$CCl_3PO(NHC_6H_{11}-cycl.)_2$. XVII.1. m. 191-192°.[955]

$CCl_3PO(NC_5H_{10})_2$. XVII.1. m. 238-240°.[955]

$CCl_3PO(NHPh)(NHC_6H_{11}-cycl.)$. XVII.1. m. 198-199°.[955]

$CCl_3PO(NHCONHPh)_2$. XVII.6. m. 186-188°.[469]

$CCl_3PO(NHCOOR)_2$.

R = Me. XVII.6. m. 165-166°.[469]

R = Et. XVII.6. m. 142-144°.[469]

$CCl_3(O)\overline{PN(Me)CH_2CH_2N}(Me)$. XVII.1. m. 88°, ^1H.[2043]

$CCl_3PO(NHPh)(NHNH_2)$. XVIII.3. m. 191-193°.[955]

$CCl_3PO(NHPh)(NHN{=}CMe_2)$. XVIII.3. m. 147-148°.[955]

$CCl_3PO(NHPh)(NHN{=}CHPh)$. XVIII.3. m. 195.5-197°.[955]

$NH_2COPO(NEt_2)_2$. XVII.10. From $EtO_2CPO(NEt_2)_2$ and NH_4OH, m. 119-120°.[61]

$CH_3PO(NEt_2)Cl$. XV.1. b_9 115-115.5°, n_D^{20} 1.4648, d_4^{20} 1.1274,[1703, 1704] b_2 80°, n_D^{17} 1.4678.[677]

$CH_3PO(NHCOR)Cl$.

R = $-CCl_3$. XV.2. m. 162-164°, IR.[1906]

R = $-CF_3$. XV.2. m. 98-100°, IR.[1906]

R = Ph. XV.2. m. 112-114°, IR.[1906]

R = 4-$O_2NC_6H_4$-. XV.2. m. 164-166°, IR.[1906]

$CH_3PO(N=CClCCl_3)Cl$. XV.5. m. 92-94°.[1906]

$CH_3PO(NH_2)F$. XV.1. m. 42°, IR, ^1H, ^{19}F, mass spect.[1761]

$CH_3PO(NR_2)F$.

R = Me. XV.1, XV.3. b_2 52°.[496, 497, 1787a] MO calc., ^{31}P.[1075]

R = Et. XV.1. b_{19} 101°, n_D^{20} 1.4195, d_4^{20} 1.0735.[1704] XV.3. b_5 75-77°, n_D^{20} 1.4210, d_4^{20} 1.0747, IR, ^1H, ^{19}F.[496, 497]

$CH_3PO(NHR)F$.

R = Me. XV.1. b_4 103°.[1787a]

R = 4-$CH_3C_6H_4$-. XV.1. m. 81-83°.[860]

R = 4-ClC_6H_4-. XV.1. m. 128-131°.[860]

$CH_3PO[NHCOCH=C(Ph)NC_5H_{10}]F$. XV.4. m. 144-145°.[855]

$CH_3PO(NHCSNHAr)F$.

Ar = Ph. XV.4. m. 105-107°.[860]

Ar = 4-ClC_6H_4-. XV.4. m. 106-108°.[860]

$CH_3PO(N=PPh_3)Cl$. XV.5. m. 139-141°, IR.[1910]

$CH_3PO(NR_2)NCO$.

R = Me. XV.5. $b_{0.05}$ 49-51°, n_D^{20} 1.4690, d_4^{20} 1.1687, IR.[468]

R = Et. XV.5. $b_{0.04}$ 76-78°, n_D^{20} 1.4555, d_4^{20} 1.1057, IR.[468]

$CH_3PO(OMe)(NRR')$.

R = R' = Me. XVI.7. $b_{0.5}$ 50-51°, ^1H.[648]

R = R' = Et. XVI.1. $b_{0.5}$ 71-74°, n_D^{20} 1.4355, d_4^{20} 1.0264.[1388, 1703] XVI.3. b_{32} 118-120°, n_D^{20} 1.4382, d_4^{20} 1.0222.[1388]

R = H, R' = Me XVI.1. $b_{0.02}$ 72-73°,[1912] ^1H.[648]

R = H, R' = Et. XVI.1. $b_{0.02}$ 78-79°.[1912]

R = H, R' = i-Pr. XVI.1. $b_{0.03}$ 81-83°.[1912]

R = H, R' Bu. XVI.1. $b_{0.1}$ 95-96°,[1912] b_{1-2} 130-132°, n_D^{20} 1.4420.[2015]

R = H, R' = sec.-Bu. XVI.1. b_5 146°, n_D^{20} 1.4414.[2015]

R = H, R' = Ph. XVI.1. m. 76.5-77.5°. XVI.3. m. 77.5-78°.[890]

$CH_3PO(OMe)(NHCOR)$.

R = $-CCl_3$. XVI.5. m. 118-120°.[1906]

R = Ph. XVI.5. m. 170-172°.[1906]

R = $4-O_2NC_6H_4-$. XVI.5. m. 164-166°.[1906]

$CH_3PO(OMe)[N(Me)COOR]$.

R = Me. XVI.5. $b_{0.02}$ 55-56°, n_D^{20} 1.4452, d_4^{20} 1.2106, IR.[1914]

R = Et. XVI.5. $b_{0.02}$ 58-60°, n_D^{20} 1.4392, d_4^{20} 1.1598, IR.[1914]

R = i-Pr. XVI.5. $b_{0.02}$ 60-62°, n_D^{20} 1.4360, d_4^{20} 1.1139, IR.[1914]

$CH_3PO(OEt)NHR$.

R = Me. XVI.1. $b_{0.5}$ 86-88°.[1912]

R = Et. XVI.1. $b_{0.4}$ 91-93°.[1912]

R = i-Pr. XVI.1. $b_{0.03}$ 66-67°,[1912] b_3 118°, n_D^{20} 1.4350.[2015]

R = Bu. XVI.1. $b_{0.1}$ 100-101°,[1912] b_{10} 152°, n_D^{20} 1.4403,[2015] b_{20} 110-112°.[950]

R = sec.-Bu. XVI.1. b_5 138°, n_D^{20} 1.4382,[2015] b_{30} 121-125°.[950]

R = t-Bu. XVI.1. b_{25} 134-140°.[950]

R = Ph. XVI.1. b_7 192°, n_D^{20} 1.5188.[2015]

R = $-CH_2CH_2SEt$. XVI.1. $b_{1.5}$ 140-150°.[861]

R = $-CH_2CH_2NHPh$. XVI.1. $b_{1.5}$ 160-170°.[861]

$CH_3PO(OEt)NR_2$.

R = Me. XVI.1, XVI.3. b_1 55°, n_D^{20} 1.4272, d_4^{20} 1.0328.[406, 1664a] XVI.3. b_{30} 98-100°, n_D^{20} 1.4272, d_4^{20} 1.0328.[1388]

R = Et. XVI.1. b_{22} 110-114°,[950] b_1 54°.[406] XVI.3. b_7 94.5°, n_D^{20} 1.4380, d_4^{20} 1.0053.[39] XVI.1. $b_{1.5}$ 77-79°, n_D^{20} 1.4350, d_4^{20} 0.9944.[1703]

R = i-Pr. XVI.1. $b_{0.1}$ 80-81°, n_D^{20} 1.4395, d_4^{20} 1.0763.[1703]

R = Bu. XVI.1. b_{30} 108-110°.[950] XVI.3. b_2 117-119°, n_D^{20} 1.4432, d_4^{20} 0.9533.[39]

R = sec.-Bu. XVI.1. b_{30} 115-120°.[950]

$CH_3PO(OEt)(\overline{NCH_2CH_2})$. XVI.1. b_3 76-78°.[861] XVI.7. b_6 84-85°, n_D^{20} 1.4514, d_4^{20} 1.1071.[1392]

$CH_3PO(OEt)(\overline{NCOCH_2CH_2CO})$. XVI.7. m. 82°.[2107]

$CH_3PO(OEt)(NEtPh)$. XVI.5. $b_{0.5}$ 110-112°, n_D^{20} 1.5111, d_4^{20} 1.0907.[627]

$CH_3PO(OEt)(NHCOR)$.

R = Me. XVI.3. $b_{0.0011}$ 82.5-83°, n_D^{20} 1.4515, d_4^{20} 1.1632, m. 42-43°, IR.[89]

R = $-CCl_3$. XVI.5. m. 78-80°.[1906]

R = $4-O_2NC_6H_4-$. XVI.5. m. 126-128°.[1906]

$CH_3PO(OEt)[N(Me)COOR]$.

R = Me. XVI.9. $b_{0.02}$ 69-71°, n_D^{20} 1.4433, d_4^{20} 1.1557.[1914]

R = Et. XVI.9. $b_{0.02}$ 75-76°, n_D^{20} 1.4396, d_4^{20} 1.1179.[1914] (By reaction of $CH_3PO(OEt)NMeNa$ and ROCOCl.)

$CH_3PO(OEt)(NRSO_2R')$.

R = Pr, R' = Me. XVI.2. b_{3-4} 190-200°, n_D^{20} 1.4590, d_4^{20} 1.1951.[725]

R = R' = Ph. XVI.2. m. 91-93°.[725]

$CH_3PO(OEt)[NRPO(OR)_2]$.

R = Me. XVI.8. $b_{0.04}$ 101-103°.[1913]

R = Et. XVI.8. $b_{0.05}$ 106-107°.[1913]

$CH_3PO(OEt)$(NHCONH 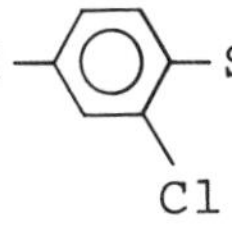 SCN). XVI.6. m. 154-156°.[726]

$CH_3PO(OEt)NHCY$ 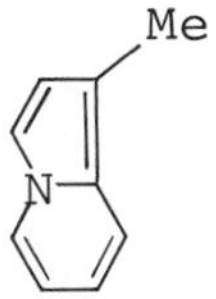

Y = O. XVI.6. m. 114-115°.[594]

Y = S. XVI.6. m. 125-126°.[594]

$CH_3PO(OEt)NHCO$ 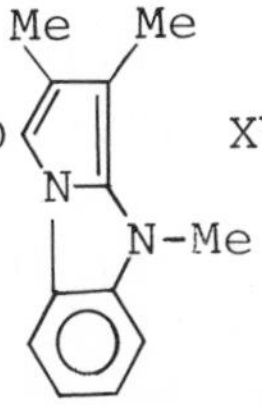 XVI.6. m. 185-186°.[594]

$CH_3PO(OCH{=}CH_2)NR_2$.

R = Me. XVI.1. $b_{0.1}$ 58-60°, n_D^{20} 1.4493, d_4^{20} 1.0722.[131]

R = Et. XVI.1. $b_{0.1}$ 69-70°, n_D^{20} 1.4478, d_4^{20} 1.0195.[131]

$CH_3PO(OEt)(N{=}PRR'_2)$.

R = R' = OMe. XVI.8. $b_{0.03}$ 96-97°, IR, ^{31}P δ_1 -17.8, δ_2 -0.6 ppm.[1913]

R = R' = OEt. XVI.8. $b_{0.05}$ 100-102°, IR, ^{31}P δ_1 -18.1, δ_2 +3.6 ppm.[1913]

R = R' = OPr-i. XVI.8. $b_{0.07}$ 103-105°, IR, ^{31}P δ_1 -17.9, δ_2 +7.6 ppm.[1913]

R = Me, R' = OMe. XVI.8. $b_{0.03}$ 104°, IR.[1913]

R = Me, R' = OEt. XVI.8. $b_{0.03}$ 102-103°, IR, ^{31}P δ_1 -21.0, δ_2 -32.8 ppm.[1913]

R = Me, R' = NMe_2. XVI.8. $b_{0.04}$ 112-113°, IR.[1913]

R = R' = Ph. XVI.8. m. 99-100°, IR.[1913]

$CH_3PO(OC_3H_7)NMe_2$. XVI.1. b_1 62°.[406]

$CH_3PO(OC_3H_7)N{=}CCl_2$. XVI.8. b_2 77°, n_D^{20} 1.4794, d_4^{20} 1.2898.[845]

$CH_3PO(OC_3H_7\text{-}i)(NHR)$.

R = Me. XVI.1. $b_{0.06}$ 73-75°.[1912]

R = Et. XVI.1. $b_{0.03}$ 69-71°.[1912]

R = i-Pr. XVI.1. $b_{0.07}$ 85-87°,[1912] b_3 120°, n_D^{20} 1.4319.[2015]

R = Bu. XVI.1. b_{11} 138-139°,[1912] b_7 148°, n_D^{20} 1.4373.[2015]

R = sec.-Bu. XVI.1. b_{10} 144-146°, n_D^{20} 1.4358.[2015]

R = Ph. XVI.1. Separation of opt. isomers : (-)phosphonamidate $[\alpha]_D^{30}$ -116°, m. 122-124°, and (+)phosphonamidate, $[\alpha]_D^{26}$ +115°, m. 120-122°.[2]

$CH_3PO(OC_3H_7\text{-}i)NR_2$.

R = Me. XVI.2. b_1 48°.[1722] XVI.1. $b_{0.04}$ 43°.[406]

R = Et. XVI.1. $b_{0.1}$ 80-81°, n_D^{20} 1.4395, d_4^{20} 1.0763.[1703]

$CH_3PO(OC_3H_7\text{-}i)(NC_5H_{10})$. XVI.1. b_1 90°.[406]

$CH_3PO(OC_3H_7\text{-}i)(\overline{NCH_2CH_2OCH_2CH_2})$. XVI.1. b_1 86°.[406]

$CH_3PO(OC_3H_7\text{-}i)(N{=}CCl_2)$. XVI.8. $b_{0.4}$ 67-68°, n_D^{20} 1.4758, d_4^{20} 1.2874.[845]

$CH_3PO(OC_3H_7\text{-}i)$ 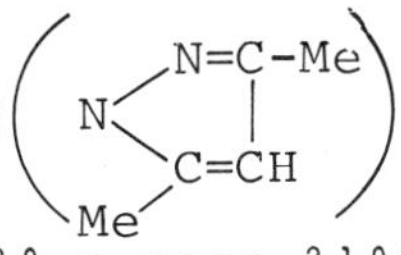. XVI.7. b_1 80.5-81°, n_D^{20} 1.4740, d_4^{20} 1.0806.[2107]

$CH_3PO(OC_4H_9)NR_2$.

R = Me. XVI.1. $b_{0.5}$ 64°.[406]

R = Et. XVI.3. b_1 99°, n_D^{20} 1.4374, d_4^{20} 0.9681.[39]

$CH_3PO(OC_4H_9\text{-}i)(\overline{NCOCH_2CH_2CO})$. XVI.7. m. 62°, $b_{0.003}$ 140°.[2107]

$CH_3PO(OC_4H_9\text{-}i)$ 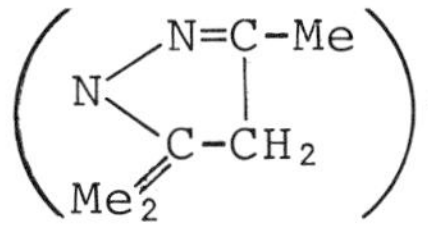. XVI.7. $b_{0.001}$ 76-78°, n_D^{20} 1.4645, d_4^{20} 1.0175.[2107]

$CH_3PO(OC_5H_{11})(NMe_2)$. XVI.1. b_1 91°.[406]

$CH_3PO(OC_6H_4Cl\text{-}4)(NHR)$.

R = Et. XVI.1. $b_{0.15}$ 133-133.5°.[664]

R = i-Pr. XVI.1. m. 74-74.5°.[664]

R = sec.-Bu. XVI.1. m. 49.5-51°.[664]

$CH_3PO(OC_6H_4Cl\text{-}3)(NHR)$.

R = Me. XVI.1. $b_{0.17}$ 142-142.5°.[1235]

R = Et. XVI.1. $b_{0.15}$ 133-135°.[1235]

R = i-Pr. XVI.1. $b_{0.1}$ 128-130°.[1235]

R = Bu. XVI.1. $b_{0.29}$ 162°.[1235]

R = sec.-Bu. XVI.1. $b_{0.13}$ 139-142°.[1235]

R = t-Bu. XVI.1. $b_{0.28}$ 137-138°.[1235]

$CH_3PO(OC_6H_4Cl\text{-}3)(NEt_2)$.XVI.1. $b_{0.34}$ 138°.[1235]

$CH_3PO(OC_6H_4Cl\text{-}4)(NHR)$.

R = Me. XVI.1. m. 60-61°.[1235]

R = Et XVI.1. m. 51-53.5°.[1235]

R = i-Pr. XVI.1. m. 92-93°.[1235]

R = Bu. XVI.1. $b_{0.15}$ 142-143°.[1235]

R = Ph. XVI.1. m. 146-147°.[1235]

R = 2-$CH_3C_6H_4$-. XVI.1. m. 97-97.5°.[1235]

R = 4-$CH_3C_6H_4$-. XVI.1. m. 100-100.5°.[1235]

R = 2-ClC_6H_4-. XVI.1. m. 75.5-76.5°.[1235]

R = 3-ClC_6H_4-. XVI.1. m. 92-92.5°.[1235]

R = 4-ClC_6H_4-. XVI.1. m. 73.5-74.5°.[1235]

R = 4-$EtOC_6H_4$-. XVI.1. m. 73-74°.[1235]

$CH_3PO(OC_6H_4Cl\text{-}4)(NR_2)$.

R = Me. XVI.1. $b_{0.17}$ 114°.[1235] XVI.2. Herbicide.[200]

R = Et. XVI.1. $b_{0.2}$ 122-123°.[1235]

$CH_3PO(OC_6H_4Cl\text{-}4)(NHCOR)$.

R = Me. XVI.6. m. 127-129°, IR.[468]

R = Et. XVI.6. m. 124-126°, IR.[468]

R = i-Pr. XVI.6. m. 134-136°, IR.[468]

$CH_3PO(OC_6H_4Cl\text{-}4)(NHCONHPh)$. XVI.6. m. 165-167°, IR.[468]

$CH_3PO(OC_6H_3Cl_2\text{-}2,4)(NEt_2)$. XVI.2. $b_{0.3}$ 136-137°.[1234]

$CH_3PO(OC_6H_2Cl_3\text{-}2,4,5)(NR_2)$.

R = Me. XVI.2. m. 106-108°.[1234]

R = Et. XVI.2. m. 47-48°.[1234]

$CH_3PO(OC_6Cl_5)(NR_2)$.

R = Me. XVI.2. m. 149-151.5°.[1234]

R = Et. XVI.2. m. 95-96°.[1234]

$CH_3PO(OC_6H_4NO_2\text{-}4)(NR_2)$.

R = Me. XVI.2. m. 119-120°.[1234]

R = Et. XVI.2. m. 55-58°.[1234]

$CH_3PO(OC_6H_4NO_2\text{-}4)(NHCOCCl_3)$. XVI.5. m. 179-181°.[1906]

$CH_3PO(OC_6H_4NO_2\text{-}4)[N{=}C(CCl_3)OC_6H_4NO_2\text{-}4]$. XVI.2. m. 150-151°.[1906]

$CH_3PO(OPh)(NHR)$.

R = H. XVI.1. m. 98°.[1488]

R = Me. XVI.1. $b_{0.2}$ 132°, n_D^{20} 1.5252, d_4^{20} 1.1600.[1488]

R = Et. XVI.1. $b_{0.2}$ 142°, n_D^{20} 1.5169, d_4^{20} 1.1330.[1488]

R = $-CH_2CH_2OH$. XVI.1. m. 54°, IR.[2097]

R = $-C_4H_9$. XVI.1. b_{10} 206°, n_D^{20} 1.5125.[2015]

$CH_3PO(OPh)(NHAr)$. XVI.1. Kinetics of thermal decomp.[2014, 2015]

Ar = Ph. XVI.1. m. 94°.

Ar = $2\text{-}CH_3C_6H_4$-. XVI.1. m. 80°.

Ar = $3\text{-}CH_3C_6H_4$-. XVI.1. m. 108-109°.

Ar = $4\text{-}CH_3C_6H_4$-. XVI.1. m. 110°.

Ar = $2\text{-}ClC_6H_4$-. XVI.1. m. 99°.

Ar = $3\text{-}ClC_6H_4$-. XVI.1. 88-89°.

Ar = $4\text{-}ClC_6H_4$-. XVI.1. m. 104.5°.

$CH_3PO(OPh)(NHCOOR)$.

R = Me. XVI.6. m. 122-124°, IR.[468]

R = Et. XVI.6. m. 136-137°, IR.[468]

R = i-Pr. XVI.6. m. 62-64°, IR.[468]

$CH_3PO(OPh)(NHCOCCl_3)$. XVI.5. m. 95-97°.[1906]
$CH_3PO(OPh)[NHCOCONHPO(OPh)CH_3]$. XVI.8. m. 179-180°.[1911]
$CH_3PO(OPh)(NHCONHPh)$. XVI.6. m. 138-140°, IR.[468]

$CH_3PO(OPh)NHCS$— 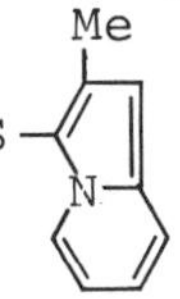 XVI.6. m. 125-126°.[594]

$CH_3PO(OPh)(N{=}PRR'_2)$.

R = R' = OMe. XVI.8. n_D^{20} 1.5050, d_4^{20} 1.2629, IR, ^{31}P δ_1 -16.7, δ_2 -1.25 ppm.[1913]

R = R' = OEt. XVI.8. $b_{0.05}$ 148-150°, IR.[1913]

R = R' = OPr-i. XVI.8. $b_{0.04}$ 137-138°, IR.[1913]

R = Me, R' = OMe. XVI.8. $b_{0.03}$ 150-151°, IR.[1913]

R = Me, R' = OEt. XVI.8. $b_{0.04}$ 143-144°, IR.[1913]

R = Me, R' NMe_2. XVI.8. $b_{0.035}$ 157-158°, IR.[1913]

R = R' = NMe_2. XVI.8. $b_{0.045}$ 161-162°, IR.[1913]

$CH_3PO(OPh)[N(Me)PO(OMe)_2]$. XVI.8. $b_{0.07}$ 143°.[1913]
$CH_3PO(OC_6H_{11}$-cycl.)(NRR').

R = H, R' = Me. XVI.2. $b_{0.5}$ 125-126°.[1722]

R = R' = Me. XVI.2. $b_{0.1}$ 82-84°.[1722]

$CH_3PO(OC_8H_{17})NMe_2$. XVI.2. $b_{0.02}$ 88-89°.[1722]
$CH_3PO[OC(Ph)CH_2CH(Me)N(Me)CH_2CH(Me)](NEt_2)$. XVI.8. As picrate, m. 76-76.5°.[1472] (By oxidation of the phosphonous acid derivative with O_2/UV.)
$CH_3PO(NH_2)_2$. XVII.1. m. 128-129°.[1683]
$CH_3PO(NHR)_2$.

R = Me. XVII.1. m. 63-65°.[768]

R = Ph. XVII.1. m. 156.5-158.5°.[1231]

R = $3\text{-}CH_3C_6H_4$-.XVII.1. m. 164-165°.[1231]

R = $3\text{-}ClC_6H_4$-. XVII.1. m. 174-175°.[1231]

R = $4\text{-}ClC_6H_4$-. XVII.1. m. 167-168°,[1231] 217-220°.[768]

$CH_3PO(NHPh)(NHR)$.

R = $-CH_2CH_2OH$. XVII.1. m. 55-56°, IR.[2097]

R = Bu. XVII.1. m. 76-77°.[2098]

R = $-CH_2C_6H_5$. XVII.1. m. 59-60°.[2098]

R = 4-ClC_6H_4-. XVII.1. m. 56°.[2098]

$CH_3PO(NR_2)_2$.

R = Me. XVII.1. $b_{1.5}$ 74-75°, ^{1}H.[439] b_{32} 138°, n_D^{30} 1.4539, d_4^{30} 1.0157.[1018] XVII.7. $b_{0.4}$ 55°, n_D^{21} 1.4572, d_4^{24} 1.022.[1406] XVII.1. b_3 62-63°, m. -3.2° n_D^{20} 1.4598, d_4^{20} 1.022, dipole moment μ_{25} = 4.8 D, useful as a solvent, dissolves Na, K.[1695a] m. -3.2°, b_{760} 240°, b_{138} 132°, b_2 64°, n_D^{20} 1.4598, d_4^{20} 1.022, dielect. const. ε_{25} = 31.6, dipole moment μ_{25} = 4.8 D.[1098]

R = Et. XVII.1. b_{33} 166°, n_D^{30} 1.4565, d_4^{30} 0.9562.[1018] XVII.3. b_3 123°, n_D^{20} 1.4598, d_4^{20} 0.9679.[39] b_{22} 145-148°.[1275]

R = Pr. XVII.1. $b_{3.5}$ 137°, n_D^{30} 1.4603, d_4^{30} 0.9395.[1018] XVII.3. b_{25} 176-180°.[1275]

R = Bu. XVII.3. b_1 172-174°, n_D^{20} 1.4612, d_4^{20} 1.9173.[39]

$CH_3PO(\overline{NCH_2CH_2})_2$. XVII.1. b_2 82-83°, IR.[1416a]

$CH_3PO(\overline{NCH_2CHEt})_2$. XVII.1. $b_{0.5}$ 77-78°, n_D^{20} 1.4695, d_4^{20} 1.0169, IR.[1416a]

$CH_3PO(\overline{NCH_2CMe_2})_2$. XVII.1. $b_{0.5}$ 93-94°, n_D^{20} 1.4668, d_4^{20} 1.0200, IR.[1416a]

$CH_3PO(NMe_2)(NHR)$.

R = i-Pr. XVII.1. $b_{0.15}$ 89-90°.[667]

R = Ph. XVII.1. m. 74-75°, ^{1}H.[1231]

R = 3-$CH_3C_6H_4$-. XVII.1. m. 86-88°, ^{1}H.[1231]

R = 4-$CH_3C_6H_4$-. XVII.1. m. 139-141°, ^{1}H.[1231]

R = 3-ClC_6H_4-. XVII.1. m. 124-125°, ^{1}H.[1231]

R = 4-ClC_6H_4-. XVII.1. m. 158-160°, ^{1}H.[1231]

$CH_3PO(NMe_2)(NR_2)$.

R = Et. XVII.1. $b_{0.4}$ 86-86.5°.[667]

R = Pr. XVII.1. $b_{0.2}$ 89.5-90.5°.[667]

$CH_3PO(NEt_2)(NHR)$.

R = i-Pr. XVII.1. $b_{0.2}$ 108-110°.[667]

R = i-Bu. XVII.1. $b_{0.2}$ 126-128°.[667]

$CH_3PO(NEt_2)(NR_2)$.

R = Et. XVII.1. $b_{0.15}$ 78-80°.[667]

R = Pr. XVII.1. $b_{0.4}$ 103-105°.[667]

$CH_3(O)P\langle^{NH}_{NH}\rangle$⬡   XVII.1. m. 150-173°.[1947]

$CH_3PO(NMe_2)(NHCOOR)$.

    R = Me. XVII.6. m. 108-110°.[468]

    R =˙Et. XVII.6. m. 103-105°.[468]

    R = i-Pr. XVII.6. m. 90-93°.[468]

$CH_3PO(NEt_2)(NHCOOR)$.

    R = Me. XVII.6. m. 79-81°.[468]

    R = Et. XVII.6. m. 63-65°.[468]

    R = i-Pr. XVII.6. m. 65-67°.[468]

$CH_3PO(NR_2)(NHCONHPh)$.

    R = Me. XVII.6. m. 132-134°.[468]

    R = Et. XVII.6. m. 123-124°.[468]

$CH_3PO(NHPh)(NHCOCCl_3)$. XVII.5. m. 180-182°, IR.[1906]
$CH_3PO(NHC_6H_4Cl-4)(NHCOCCl_3)$. XVII.5. m. 200-202°, IR.[1906]
$CH_3PO(NHCONHPh)_2$. XVII.6. m. 194-195°.[470]
$CH_3PO(NHCOOR)_2$.

    R = $-C_6H_{13}$. XVII.6. m. 84-86°.[469]

    R = $-C_7H_{15}$. XVII.6. m. 87-89°.[469]

    R = $-C_8H_{17}$. XVII.6. m. 91-94°.[469]

    R = $-C_9H_{19}$. XVII.6. m. 78-81°.[469]

    R = $-C_{10}H_{12}$. XVII.6. m. 90-92°.[469]

$CH_3(O)P\langle^{NHCO}_{NHCO}\rangle C=C(Ph)NC_5H_{10}$. XVII.6. m. 180-181°.[855]

$CH_3(O)P\langle^{NHCO}_{NHCO}\rangle C=C(OEt)_2$. XVII.6. m. 148-150°.[853]

$CH_3(O)P\langle^{NHCO}_{NHCO}\rangle CHCO_2Et$. XVII.6. m. 149-150°[853] (by hydro-

           lysis of the previous compound).
$CH_3(O)P(NH_2)[NHCOCH(CO_2H)CO_2Et]$. XVII.6. m. 135-136°.[853]
    (by hydrolysis of the previous compound).

$CH_3(O)P\langle^{NHCO}_{NHCO}\rangle C=C\langle^{Me_2}$⬡ XVII.6. m. 148°.[853]
                     Me

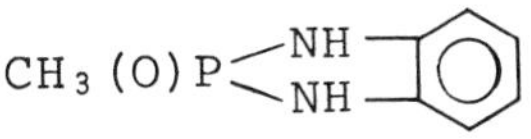

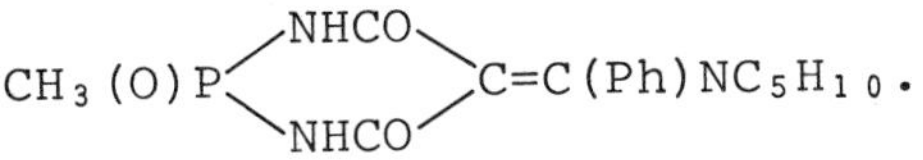

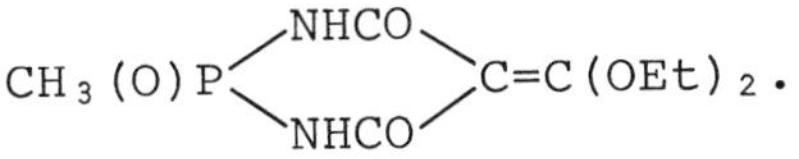

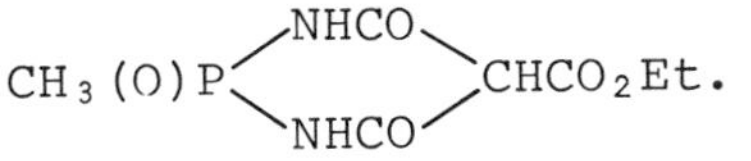

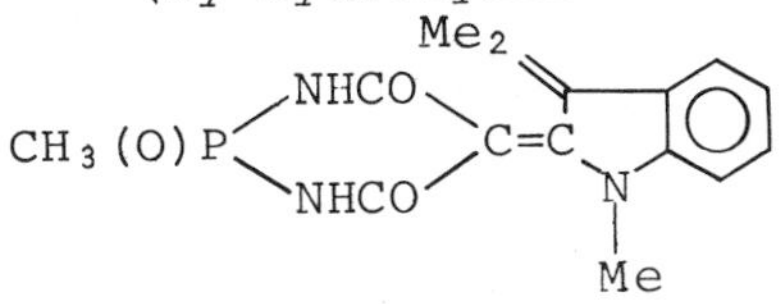

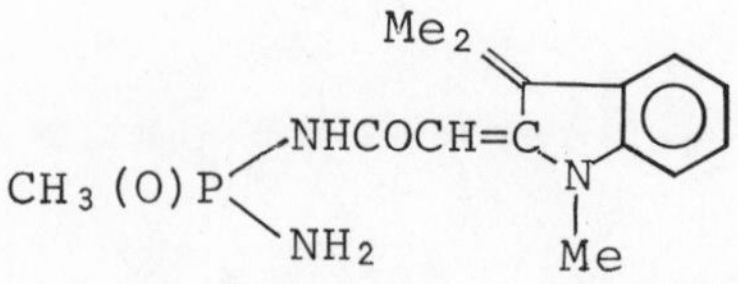

XVII.6. m. 162-164°.[853]

(by hydrolysis of the previous compound)

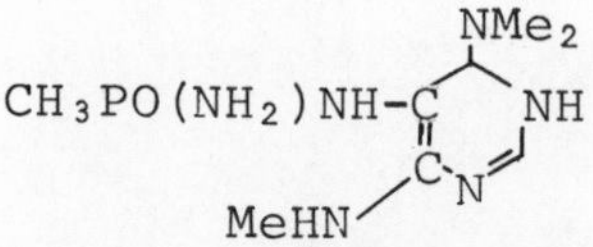

. m. (dihydrate) >150°.[1105]

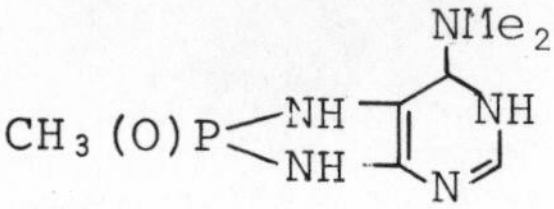

. m. (hemihydrate) >250°.[1105]

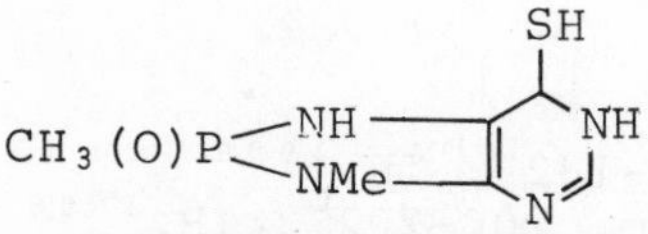

. m. (1½ hydrate) >200°.[1105]

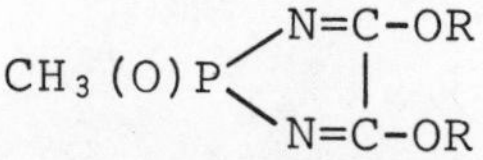

R = Me. XVII.8. m. 137-138° (dec.).[466]

R = Et. XVII.8. m. 118-120°.[466]

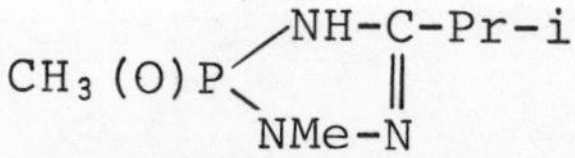

XVII.8. m. 131°, IR, ^{1}H, ^{31}P -36.6 ppm.[361]

$CH_3PO(NHPh)(N=PPh_3)$. XVII.9. m. 76-78°.[1910]
$CH_3PO(NHCONHPh)(N=PPh_3)$. XVII.9. m. 128-130°.[1910]
$CH_3PO(NHCOOR)(N=PPh_3)$.

R = Me. XVII.9. m. 118-120° (dec.).[1910]

R = Et. XVII.9. m. 154-156°.[1910]

R = i-Pr. XVII.9. m. 73-75° (dec.).[1910]

R = Ph. XVII.9. m. 85-87° (dec.).[1910]

$CH_3PO(NHCOOEt)(N=PR_3)$.

R = -OEt. XVII.9. m. 95-97°.[1910]

R = -OPr-i. XVII.9. m. 124-125°.[1910]

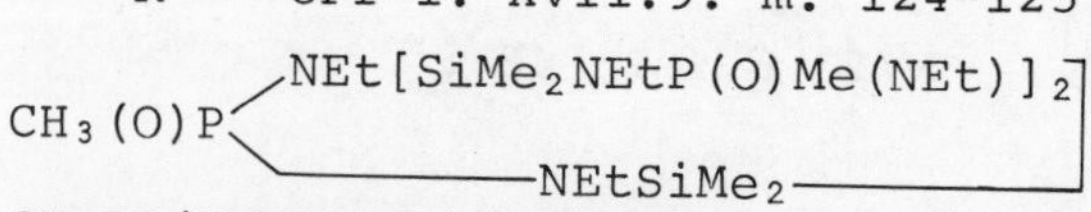

. XVII.10. b_1 170°.[76]

$CH_3PO(NHNHCOPh)(OH)$. XVIII.1. m. 195-197°, IR.[1240]
$CH_3PO(OCH=CH_2)(NRNMe_2)$.

R = H. XVIII.2. b_3 115-117°, n_D^{20} 1.4529, d_4^{20} 1.0771.[1318]

R = Pr. XVIII.2. $b_{0.001}$ 75-78°, n_D^{20} 1.4547, d_4^{20} 1.0312.[1318]

$CH_3PO(OEt)(NHNH_2)$. XVIII.2. Viscous oil.[507]
$CH_3PO(OPr\text{-}i)(NHNH_2)$. XVIII.2. m. 62°.[507]
$CH_3PO(OBu\text{-}i)(NRNR'_2)$.

R = R' = H. XVIII.2. b_3 140°.[507]

R = H, R' = Me. XVIII.2. $b_{0.5}$ 103°, n_D^{20} 1.4378, d_4^{20} 1.0086.[2107]

R = Pr, R' = Me. XVIII.2. $b_{0.001}$ 77-78°, n_D^{20} 1.4438, d_4^{20} 0.9718.[2107]

$CH_3PO(OC_6H_4Cl\text{-}2)(NHNMe_2)$. XVIII.2. m. 66°.[1500]
$CH_3PO(OC_6H_4Cl\text{-}4)[NHN(Me)R]$.

R = Me. XVIII.2. m. 86-87°.[661]

R = Ph. XVIII.2. m. 119-120°.[661]

$CH_3PO(OC_6H_4Cl\text{-}4)[NHNH(4\text{-}ClC_6H_4O)P(O)CH_3]$. XVIII.2. m. 138-140°.[661]
$CH_3PO(OC_6H_3Cl_2\text{-}2,4)(NHNHPh)$. XVIII.2. m. 128-129°.[661]
$CH_3PO(OPh)(NHNPhR)$.

R = H. XVIII.2. m. 118-119°.[661]

R = Me. XVIII.2. 84-86°.[661]

$CH_3PO(OPh)[NHN{=}C(Me)_2]$. XVIII.2. m. 99-100°.[661]
$CH_3PO(OC_6H_{11}\text{-cycl.})(NHNH_2)$. XVIII.2. m. 77°.[507]
$CH_3PO(NMe_2)(NHNMe_2)$. XVIII.3. $b_{0.25}$ 95-96°, n_D^{25} 1.4634, d_4^{25} 1.0452.[661]
$CH_3PO(NMe_2)(NHNHPh)$. XVIII.3. m. 168-170°.[667]
$CH_3PO(NEt_2)(NHNHPh)$. XVIII.3. m. 103-104.5°.[667]

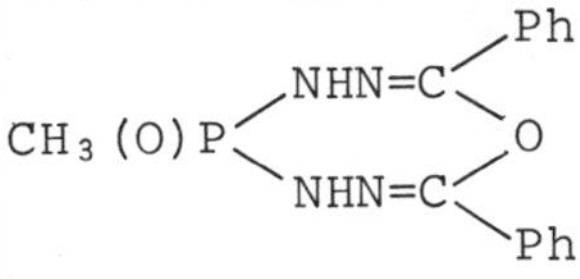

XVIII.1. m. 212-213°.[1240] m. 215-216.5°, ^1H, IR.[662]

$CH_3PO(N_3)Cl$. XVIII.4. $b_{0.03}$ 46-47°, n_D^{20} 1.4850, d_4^{20} 1.4125.[1910]
$CH_3PO(N_3)(NCO)$. XVIII.4. $b_{0.04}$ 49-50°, n_D^{20} 1.4835, d_4^{20} 1.3975.[1910]
$CH_3PO(N_3)_2$. XVIII.4. $b_{0.03}$ 49-50°, n_D^{20} 1.4980, d_4^{20} 1.3826.[1910]
$CH_3PO(NEt_2)(N_3)$. XVIII.4. $b_{0.03}$ 76-77°, n_D^{20} 1.4680, d_4^{20} 1.0891.[1910]
$CH_3PO(NHCOOR)(N_3)$.

R = Me. XVIII.4. m. 82-83°.[1910]

R = Et. XVIII.4. m. 127-128°.[1910]

R = i-Pr. XVIII.4. m. 114-115°.[1910]

R = Ph. XVIII.4. m. 96-98°.[1910]

$CH_3PO(NHCONHPh)(N_3)$. XVIII.4. m. 158-160° (dec.).[1910]

$\underline{C_2}$

$CH_2{=}CBrPO(NHPh)_2$. XVII.1. m. 156-157°[1223] (by simultaneous HBr cleavage from $CH_2BrCHBrPOCl_2$).

$CH_2{=}CClPO(NHPh)_2$. XVII.1. m. 167-169°[1223] (by simultaneous HCl cleavage from $CH_2ClCHClPOCl_2$).

$ClCH_2CH_2PO(NHC_6H_4Cl\text{-}4)F$. XV.1. m. 84-88°.[860]

$ClCH_2CH_2PO(NHCONHAr)F$.

Ar = Ph. XV.4. m. 123-125°.[860]

Ar = $4\text{-}ClC_6H_4\text{-}$. XV.4. m. 156-157°.[860]

$ClCH_2CH_2PO(NR_2)_2$.

R = Me. XVII.1. b_2 102-103°, n_D^{20} 1.4771, d_4^{20} 1.1315.[90]

R = Et. XVII.1. b_3 128-130°, n_D^{20} 1.4763, d_4^{20} 1.0667.[90]

$ClCH_2CH_2PO(NHPh)_2$. XVII.1. m. 169-170°.[918]

$ClCH_2CH_2PO(NHCOOR)_2$.

R = Et. XVII.6. m. 160-163°.[854]

R = i-Pr. XVII.6. m. 171-172°.[854]

$ClCH_2CHClPO(NMe_2)Cl$. XV.3. $b_{1.5}$ 114-115°, n_D^{20} 1.4780, d_4^{20} 1.4036.[2108]

$Cl_3CCH(OH)PO(OEt)(NRSO_2R')$.

R = Pr, R' = Me XVI.4. m. 123-124°.[724]

R = R' = Ph. XVI.4. m. 109-110°.[724]

$Cl_3CCH(OH)PO(OBu)(NRSO_2R')$.

R = Me, R' = Ph. XVI.4. b_2 150-155°, n_D^{20} 1.5160, d_4^{20} 1.3602.[724]

R = R' = Ph. XVI.4. m. 101-102°.[724]

$CH_2{=}CHPO(OCH_2CH_2Cl)NMe_2$. XVI.1. $b_{1.5}$ 118-119.5°.[2068]

$CH_2{=}CHPO(NR_2)_2$.

R = Me. XVII.1. b_3 82°, n_D^{20} 1.4732, d_4^{20} 1.0257.[904]

R = Et. XVII.1. b_2 89-91°, n_D^{20} 1.4690, d_4^{20} 0.9736.[908]

$CH_2{=}CHPO(\overline{NCH_2CH_2})_2$. XVII.1. $b_{0.1-0.15}$ 75-78°.[1476]

$CH_2{=}CHPO(NHCOOEt)_2$. XVII.6. m. 118-120°.[854]

$CH_3COPO(OPr)[N(Me)SO_2Bu]$. XVI.3. b_1 112-114°, n_D^{20} 1.4642, d_4^{20} 1.1535.[23]

$CH_3COPO(OBu)[N(Me)SO_2Bu]$. XVI.3. b_1 113-114°, n_D^{20} 1.4662, d_4^{20} 1.1425.[23]

$CH_3COPO(NEt_2)_2$. XVII.3. b_1 98-99°, n_D^{20} 1.4790, d_4^{20} 1.0348.

$CH_3O_2CPO(OEt)(\overline{NCH_2CH_2})$. XVI.1. $b_{1.5}$ 108-110°, n_D^{20} 1.4552, d_4^{20} 1.1790.[1475]

$C_2H_5PO(NR_2)Cl$.

R = Me. XV.1. b_{10} 102°.[446a] b_{16} 51-52°, n_D^{20} 1.4855, d_4^{20} 1.0271.[1746]

R = Et. XV.1. $b_{3.5}$ 101.5°, n_D^{20} 1.4643, d_0^{20} 1.0965.[1703, 1704] b_{27} 140°.[2002a]

$C_2H_5PO(NHCOCF_3)Cl$. XV.2. m. 45-46°, IR.[1138]
$C_2H_5PO(NH_2)F$. XV.1. m. 43°, IR, ^1H, ^{19}F, mass spect.[1761]
$C_2H_5PO(NEt_2)F$. XV.1. b_{15} 103°, n_D^{20} 1.4130, d_4^{20} 1.0036.[1704]
$C_2H_5PO(NHC_6H_4Cl\text{-}4)F$. XV.1. m. 67-73°.[860]
$C_2H_5PO(NHCONHAr)F$.

Ar = Ph. XV.4. m. 127-128°.[860]

Ar = 4-ClC_6H_4-. XV.4. m. 149-150°.[860]

$C_2H_5PO(OMe)(NEt_2)$. XVI.1. b_4 87-87.5°, n_D^{20} 1.4381, d_0^{20} 1.0043.[1703]
$C_2H_5PO(OCH{=}CCl_2)(NR_2)$.

R = Me. XVI.8. b_2 80-82°, n_D^{20} 1.4714, d_4^{20} 1.2482.[1746]

R = Et. XVI.8. b_1 92-93°, n_D^{20} 1.4758, d_4^{20} 1.1974.[1746]

$C_2H_5PO(OEt)(NHR)$.

R = Et. XVI.1. $b_{2.5}$ 111.5-112°, n_D^{20} 1.4380, d_0^{20} 1.0232.[1703]

R = $-CH_2CH_2Cl$. XVI.1. $b_{0.75}$ 144-145°, n_D^{20} 1.4565, d_0^{20} 1.1515.[1703]

R = $-CH(Me)(CH_2)_3NEt_2$. XVI.1. $b_{0.25}$ 153-154°, n_D^{20} 1.4530, d_0^{20} 0.9594.[1703]

R = $-CH_2CO_2Et$. XVI.3. $b_{0.001}$ 127°, m. 45-47°.[671]

$C_2H_5PO(OEt)(NRR')$.

R = R' = Me. XVI.1. $b_{0.5}$ 63-64°, n_D^{20} 1.4305, d_0^{20} 1.0100.[1703]

R = R' = Et. XVI.1. $b_{3.5}$ 98-98.5°, n_D^{20} 1.4355, d_0^{20} 0.9866.[1703] XVI.3. b_6 80-81°, n_D^{20} 1.4315, d_4^{20} 0.9945.[1637] XVI.3. b_8 58-60°, n_D^{20} 1.4275, d_4^{20} 0.9365.[1636]

R = Et, R' = Ph. XVI.5. b_{3-4} 129-131°, n_D^{20} 1.5098, d_4^{20} 1.0697, kinetics.[627]

R = R' = Ph. XVI.3. $b_{0.03}$ 100-101°, n_D^{20} 1.5698, d_4^{20} 1.1068.[1637]

$C_2H_5PO(OEt)(\overline{NCH_2CH_2})$. XVI.1. $b_{1.5}$ 89-92°, n_D^{20} 1.4305, d_0^{20} 1.0100.[1703]

$C_2H_5PO(OEt)(NHCH_2COOEt)$. XVI.3. $b_{0.001}$ 127°, m. 45-47°.[671]

$C_2H_5PO(OEt)(NHCOR)$.

R = Me. XVI.3. $b_{0.0011}$ 84°, n_D^{20} 1.4509, d_4^{20} 1.1265, IR.[891]

R = $-CF_3$. XVI.5. $b_{0.5}$ 105°, n_D^{20} 1.4023, d_4^{20} 1.3250.[113]

R = $-CH{=}C(NEt_2)_2$. XVI.6. m. 114-116°.[853]

R = $-C[CONHPO(OEt)Et]{=}C(NEt_2)_2$. XVI. 6. m. 137-139°.[85]

$C_2H_5PO(OEt)(NHCONHAr)$.

Ar = Ph. XVI.6. m. 100-102°.[726]

Ar = $4-ClC_6H_4-$. XVI.6. m. 143-144°.[726]

Ar = 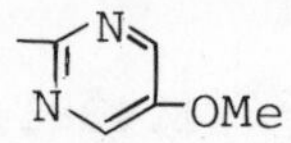

XVI.6. Tumor growth inhibitor.[727]

$C_2H_5PO(OEt)(NHCSNHAr)$.

Ar = Ph. XVI.6. m. 82°.[1856]

Ar $4-ClC_6H_4-$. XVI.6. m. 88°.[1856]

$C_2H_5PO(OEt)NHCO-$ 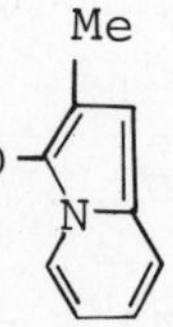

XVI.6. m. 117-118°.[594]

$C_2H_5PO(OEt)NHCO-$ 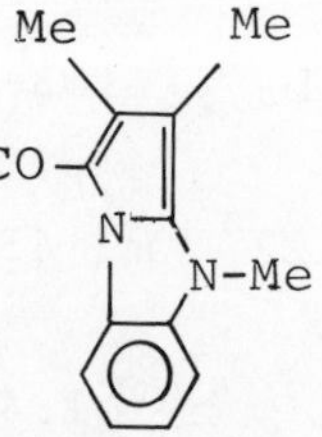

XVI.6. m. 181-182°.[594]

$C_2H_5PO(OEt)NHCO-$ 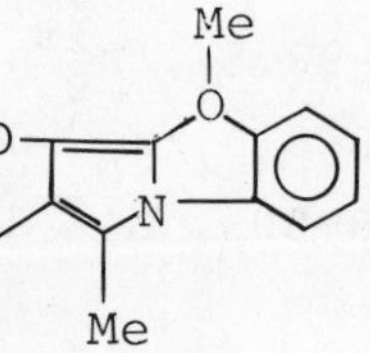

XVI.6. m. 164-165°.[594]

$C_2H_5PO(OPr)(NEt_2)$. XVI.1. b_2 98-99°, n_D^{20} 1.4368, d_4^{20} 0.9783.[1703]

$C_2H_5PO(OPr-i)(NEt_2)$. XVI.1. $b_{1.5}$ 81-82°, n_D^{20} 1.4335, d_4^{20} 0.9749.[1703]

$C_2H_5PO(OBu)(NEt_2)$. XVI.1. b_1 98.5-99.5°, n_D^{20} 1.4390, d_0^{20} 0.9562.[1703]

$C_2H_5PO(OBu-sec.)(NRR')$.

R = H, R' = Et. XVI.1. b_1 129-129.5°, n_D^{20} 1.4358, d_0^{20} 0.9749.[1703]

R = R' = Et. XVI.1. b_1 95.5-97°, n_D^{20} 1.4357, d_0^{20} 0.9553.[1703]

$C_2H_5PO(OBu)[N(i\text{-}Bu)SO_2Me]$. XVI.3. b_2 146-148°, n_D^{20} 1.4620, d_4^{20} 1.1088.[23]

$C_2H_5PO(OBu)[N(Me)SO_2Ph]$. XVI.2. b_2 150-155°, n_D^{20} 1.5200, d_4^{20} 1.2285.[725]

$C_2H_5PO(OBu)[N(Ph)SO_2Ph]$. XVI.2. m. 99-101°.[725]

$C_2H_5PO(OCH_2CH_2NEt_2)(NR_2)$.

R = Me. XVI.1. $b_{0.6}$ 82°.[1165]

R = Et. XVI.1. $b_{0.4}$ 110°.[1165]

$C_2H_5PO(OCH_2CH_2NEt_2)(\overline{NCH_2CH_2OCH_2CH_2})$. XVI.1. $b_{0.5}$ 135°.[1165]

$C_2H_5PO(OC_6H_4NO_2\text{-}4)(NEt_2)$. XVI.1. $b_{0.8}$ 172.5°, n_D^{20} 1.5309, d_0^{20} 1.1861.[1703]

$C_2H_5PO(OC_6H_{11}\text{-cycl.})(NRR')$.

R = H, R' = $-C_6H_{11}$-cycl. XVI.2. m. 117°.[253]

R = R' = Et. XVI.1. $b_{0.5}$ 118-120.5°, n_D^{20} 1.4630, d_0^{20} 1.0066.[1703]

$C_2H_5PO(OCH_2Ph)(NRR')$.

R = H, R' = $-C_6H_{11}$-cycl. XVI.2. m. 194-195°.[253]

R = R' = Et. XVI.1. b_2 162-163°, n_D^{20} 1.4975, d_0^{20} 1.0554.[1703]

$C_2H_5PO(OC_6H_{13})(NEt_2)$. XVI.1. $b_{0.5}$ 116.5-118°, n_D^{20} 1.4417, d_0^{20} 0.9425.[1703]

$C_2H_5PO(NR_2)_2$.

R = Me. XVII.1. b_{31} 142°, n_D^{30} 1.4550, d_4^{30} 0.9971.[1018]

XVII.7. $b_{0.5}$ 63.5°, n_D^{23} 1.4567, d^{26} 1.002.[1406]

R = Et. XVII.1. b_{26} 183°, n_D^{30} 1.4591, d_4^{30} 0.9689.[1018]

R = Pr. XVII.1. $b_{1.2}$ 137°, n_D^{30} 1.4600, d_4^{30} 0.9346.[1018]

$C_2H_5PO(NHC_6H_{11}\text{-cycl.})_2$. XVII.2. m. 161-162°.[253]

$C_2H_5PO(NHPh)(NHC_6H_{11}\text{-cycl.})$. XVII.2. m. 153-155°.[253]

$C_2H_5PO(NMe_2)(NHC_6H_4Cl\text{-}3)$. XVII.1. m. 109-110°.[661]

$C_2H_5PO(NEt_2)(NHAr)$.

Ar = Ph. XVII.1. m. 59-61°.[661]

Ar = $4\text{-}CH_3C_6H_4$-. XVII.1. m. 127-128°.[661]

Ar = $3\text{-}ClC_6H_4$-. XVII.1. m. 79-80°.[661]

$C_2H_5(O)\overline{PN(Me)CH_2CH_2N}(Me)$. XVII.1. $b_{0.5}$ 93°.[2042] $b_{0.8}$ 90-92°, ^1H.[2043]

$C_2H_5(O)\overline{PN(Me)CH_2CH_2CH_2N}(Me)$. XVII.1. $b_{0.3}$ 98°, ^1H.[2043]

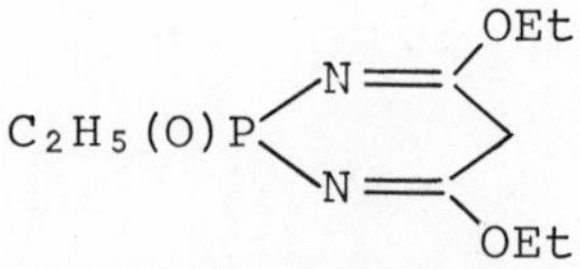 XVII.1. m. 147-148°, ^1H.[2043]

$C_2H_5PO(NHCONHPh)_2$. XVII.6. m. 180-182°.[470]

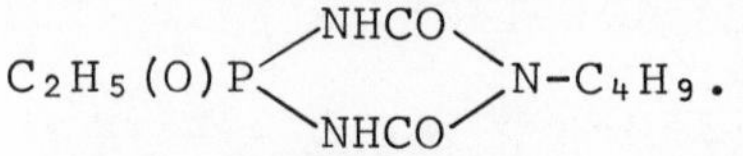 XVII.6. m. 165-175°.[2010]

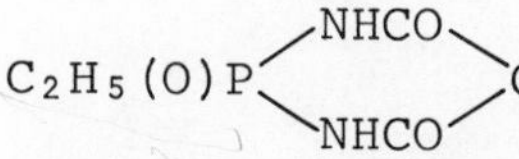 $C{=}C(Ph)NC_5H_{10}$. XVII.6. m. 180-182°.[855]

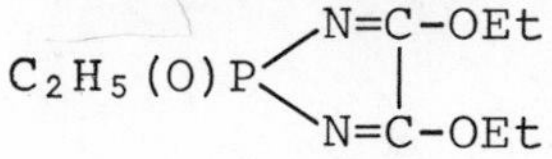 XVII.8. m. 61-64°.[466] Dichloride and oxalimidic ester, subsequent oxidation.

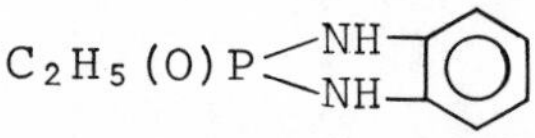 XVII.1. m. 193-194°.[452]

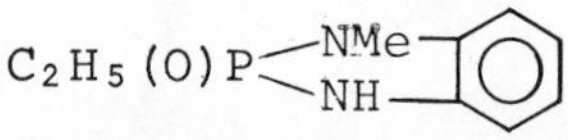 · 2HCl. XVII.1. m. 107.5-108°.[452]

$CH_3OCH_2PO(NMe_2)_2$. XVII.7. $b_{0.6}$ 77°, n_D^{23} 1.4567, d^{25} 1.053.[1406]

$\underline{C_3}$

$CF_2ClCOCF_2PO(OMe)(NEt_2)$. XVI.3. $b_{0.7}$ 75-76°, n_D^{20} 1.4045.[19]

$CH_2ClCH_2CH_2PO(NEt_2)Cl$. XV.1. $b_{0.05}$ 114-115°, n_D^{20} 1.4878, d_4^{20} 1.1870, IR.[57]

$CH_2ClCH_2CH_2PO(NEt_2)_2$. XVII.1. $b_{0.04}$ 146-147°, n_D^{20} 1.4810, d_4^{20} 1.0590, IR.[57]

$CH_2ClCH_2CH_2PO(\overline{NCH_2CH_2})_2$. XVII.1. $b_{0.005}$ 92-93°, n_D^{20} 1.5085 d_4^{20} 1.2334, IR.[57]

$CH_2ClCH_2CH_2PO(NHPh)_2$. XVII.1. m. 116-118 , IR.[57] m. 161-162°.[767]

$CH_2ClCH_2CH_2PO(NHCH_2CH{=}CH_2)_2$. XVII.1. $b_{0.04}$ 134-135°, n_D^{20} 1.5233, d_4^{20} 1.1343, IR.[57]

$CH_3C{\equiv}CPO(OEt)NEt_2$. XVI.9. m. 111-113°.[24]

$CH_3C{\equiv}CPO(OBu)NEt_2$. XVI.9. m. 135-137°.[24]

$CH_2{=}C{=}CHPO(OEt)NEt_2$. XVI.3. m. 105-107°.[24]

$CH_2{=}C{=}CHPO(OBu)NEt_2$. XVI.3. m. 122-124°.[24]

$CH_2{=}C{=}CHPO(NMe_2)_2$. XVII.4. $b_{0.17}$ 85-88°, n_D^{25} 1.5046.[1189]

$CH_2{=}C{=}CHPO(NHCOOR)_2$.

R = Me. XVII.6. m. 143-145°.[854]

R = Et. XVII.6. m. 116-117°.[854]

R = i-Pr. XVII.6. m. 143-146°.[854]

R = $-N{=}CMe_2$. XVII.6. m. 106-107°.[854]

$NCCH_2CH_2PO(OEt)(NEt_2)$. XVI.4. $b_{0.5}$ 102-104°, n_D^{20} 1.4450,

d_4^{20} 1.0861.[1636]
$C_3H_5PO(OPr)[N(Bu\text{-}i)SO_2Me]$. XVI.3. b_2 138-139°, n_D^{20} 1.4660, d_4^{20} 1.1061, IR.[23]
$CH_2{=}CH\text{-}CH_2PO(NMe_2)_2$. XVII.1. b_{2-3} 76-80°.[744] $b_{0.03}$ 61°.[438]
$CH_2{=}CH\text{-}CH_2PO(NCH_2CH_2)_2$. XVII.1. $b_{0.035}$ 86-87°, n_D^{20} 1.5020,[673] $b_{0.0001}$ 70-80°, n_D^{20} 1.5082, d_4^{20} 1.1300.[1478]
$CH_2{=}CHCH_2(O)PN(Me)CH_2CH_2N(Me)$. XVII.1. $b_{0.4-0.45}$ 98.5-100.5°, ^1H, IR, mass spect.[438]
$CH_3COCH_2PO[OCH(Me)CH_2Cl](NR_2)$.

R = Et. XVI.3. b_3 120-121°, IR.[20]

R = Pr. XVI.3. b_2 123.5-124.5°, IR.[20]

R = Bu. XVI.3. b_2 143-144.5°, IR.[20]

$CH_3CH_2COPO(OPr)[N(Me)SO_2Bu]$. XVI.3. b_1 115-117°, n_D^{20} 1.4644, d_4^{20} 1.1288.[23]
$CH_3CH_2COPO(OBu)[N(Me)SO_2Bu]$. XVI.3. b_1 116-117°, n_D^{20} 1.4660, d_4^{20} 1.1202.[23]
$CH_3COCH_2PO(OPr)(NRSO_2R')$.

R = Bu, R' = Me. XVI.3. b_1 138-139°, n_D^{20} 1.4585, d_4^{20} 1.1608.[22]

R = Me, R' = Bu. XVI.3. b_1 133-134°, n_D^{20} 1.4590, d_4^{20} 1.1407.[22]

$CH_3COCH_2PO(OBu)[N(Bu)SO_2Bu]$. XVI.3. b_1 134-136°, n_D^{20} 1.4610, d_4^{20} 1.1278.[22]
$CH_3COCH_2PO(NEt_2)_2$. XVII.3. b_8 149-150°, n_D^{20} 1.4709, d_4^{20} 1.0413.[1570]
$CH_3O_2CCH_2PO(OEt)(NH_2)$. XVI.6. m. 76-77°.[280]
$CH_3O_2CCH_2PO(OPh)(NH_2)$. XVI.6. m. 75-76°.[280]
$CH_3O_2CCH_2PO(OEt)(NHCONHAr)$.

Ar = $4\text{-}ClC_6H_4$-. XVI.6. m. 148-149°.[280]

Ar = $-C_6H_3Me\text{-}4\text{-}Br\text{-}2$. XVI.6. m. 111-112°.[280]

$CH_3CH_2O_2CPO(OMe)(NEt_2)$. XVI.3. $b_{1.5}$ 111-112°, n_D^{20} 1.4527, d_4^{20} 1.0922.[61]
$CH_3O_2CCH_2PO(OCH_2CH_2Cl)NEt_2$. XVI.3. b_2 124.5-126°, n_D^{20} 1.4650, d_4^{20} 1.1936.[19]
$CH_3O_2CCH_2PO[OCH(Me)CH_2Cl]NEt_2$. XVI.3. b_2 120-121.5°, n_D^{20} 1.4656, d_4^{20} 1.1656.[20]
$C_2H_5O_2CPO(NEt_2)_2$. XVII.3. $b_{0.5}$ 114-117°, n_D^{20} 1.4650, d_4^{20} 1.0524.[61]
$CH_2CH_2CH_2PO(NPh)(NHPh)$. XVII.10. m. 193-196°.[767]
$C_3H_7PO(NEt_2)Cl$. XV.1. $b_{3.5}$ 112.5-113.5°, n_D^{20} 1.4642, d_0^{20} 1.0707.[1703, 1704]
$C_3H_7PO(NEt_2)F$. XV.1. b_{12} 110.5°, n_D^{20} 1.4238, d_4^{20} 1.0301.[1704]
$C_3H_7PO(OEt)(NHCONHC_6H_4Cl\text{-}4)$. XVI.6. m. 134-135°.[726]
$C_3H_7PO(OEt)(NHCONHPh)$. XVI.6. m. 82°.[1856]

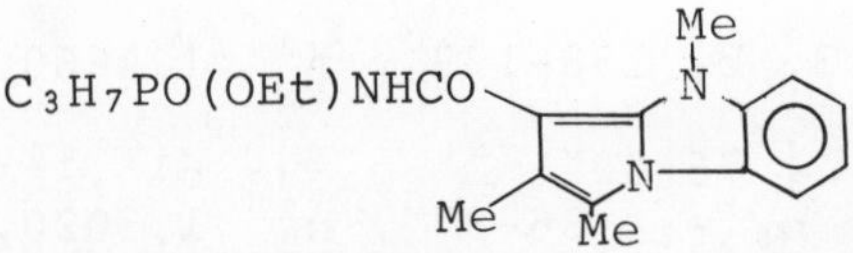

$C_3H_7PO(OEt)NHCO$—

XVI.6. m. $151-152^O$.[594]

$C_3H_7PO(OPr)[N(Me)SO_2Bu]$. XVI.3. b_2 $147-148^O$, n_D^{20} 1.4640,
d_4^{20} 1.1112, IR.[23]

$C_3H_7PO(NR_2)_2$.

    R = Me. XVII.1. b_{33} 149°, n_D^{30} 1.4564, d_4^{30} 0.9758.[1018]

        XVII.7. $b_{0.4}$ 70°. n_D^{22} 1.4553, d_4^{24} 0.983.[1406]

    R = Et. XVII.1. b_{28} 189°, n_D^{30} 1.4580, d_4^{30} 0.9542.[1018]

    R = Pr. XVII.1. $b_{1.2}$ 141°, n_D^{30} 1.4620, d_4^{30} 0.9289.[1018]

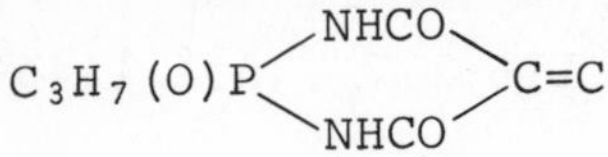

$C_3H_7(O)P$⟨$\genfrac{}{}{0pt}{}{NHCO}{NHCO}$⟩$C=C(Ph)NC_5H_{10}$. XVII.6. m. $192-193^O$.[855]

$C_3H_7(O)P$⟨$\genfrac{}{}{0pt}{}{N=C-OEt}{N=C-OEt}$⟩ XVII.8. m. $64-66^O$.[466]

$i-C_3H_7PO(NR_2)Cl$.

    R = Me. XV.1. b_{12} 104°, n_D^{25} 1.4628.[446a]

    R = Et. XV.1. b_1 90-91 , n_D^{20} 1.4641, d_0^{20} 1.0527.[1703,] [1704]

$i-C_3H_7PO(NEt_2)F$. XV.1. b_{17} 82°, n_D^{20} 1.4231, d_4^{20} 1.0263.[170]
$i-C_3H_7PO(NHPh)F$. XV.1. m. $83-84.5^O$.[860]
$i-C_3H_7PO(NHCONHC_6H_5)F$. XV.4. m. $140-141^O$.[855]
$i-C_3H_7PO(NHCONHC_6H_4Cl-4)F$. XV.4. m. 163°.[855]
$i-C_3H_7PO[NHCOCH=CHPh(NC_5H_{10})]F$. XV.4. m. $136-138^O$.[855]
$[i-C_3H_7P(O)F(NHCO)-]_2C=C(OEt)(NEt_2)$. XV.4. m. $96-97^O$.[853]
$i-C_3H_7PO(NHCSNHC_6H_4Cl-4)F$. XV.4. m. $119-120^O$.[855]
$i-C_3H_7PO(OEt)(NEt_2)$. XVI.4. $b_{0.5}$ $74.5-75^O$, n_D^{20} 1.4360,
d_0^{20} 0.9709.[1703]
$i-C_3H_7PO(OEt)[NHCOCH=C(NEt_2)_2]$. XVI.6. m. $138-140^O$.[853]
$i-C_3H_7PO(OEt)[NHCOCH=C(OEt)(NEt_2)]$. XVI.6. m. 102°.[853]

$i-C_3H_7PO(OEt)NHCOCH=C$

XVI.6. m. $151-153^O$.[853]

$i-C_3H_7PO(OEt)NHCOCH=C$

XVI.6. m. $202-204^O$.[853]

$i-C_3H_7PO(OEt)NHCOC=C(OEt)NEt_2$ XVI.6. m. $133-134^O$.[853]
                |
             $CONHPO(OEt)Pr-i$

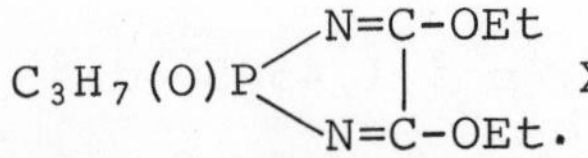

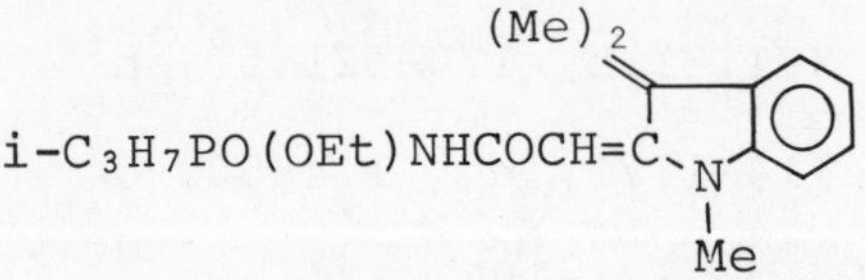

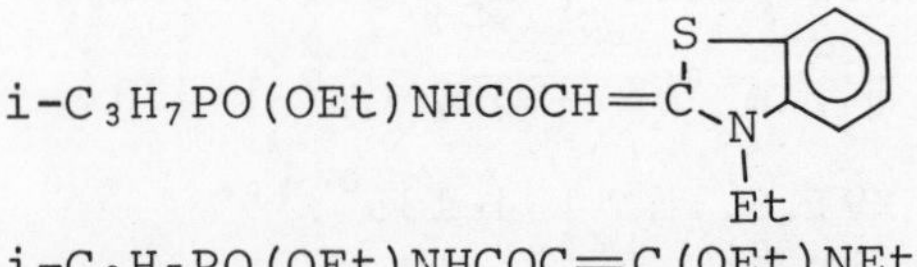

$i\text{-}C_3H_7PO(NMe_2)(NR_2)$.

$\quad$ R = Me. XVII.1. $b_{0.15}$ 74-76°, d_4^{25} 0.9835.[1715]

$\quad$ R = Et. XVII.1. $b_{0.15}$ 83-85°, d_4^{25} 0.9665.[1715]

$\quad$ R = Pr. XVII.1. $b_{0.2}$ 104-106°, d_4^{25} 0.9753.[1715]

$i\text{-}C_3H_7PO(NMe_2)(NHAr)$.

$\quad$ Ar = Ph. XVII.1. m. 159-160°.[1242]

$\quad$ Ar = 4-ClC$_6$H$_4$-. XVII.1. m. 171-173°.[1242]

$i\text{-}C_3H_7PO(NEt_2)(NHAr)$.

$\quad$ Ar = Ph. XVII.1. m. 126-127°.[1242]

$\quad$ Ar = 4-ClC$_6$H$_4$-. XVII.1. m. 138-139°.[1242]

$i\text{-}C_3H_7PO(NHCONHPh)_2$. XVII.6. m. 192-193°.[762]

$i\text{-}C_3H_7PO(NHCONHAr)_2$.

$\quad$ Ar = 4-NCSC$_6$H$_4$-. XVII.6. m. 168-172°.[854]

$\quad$ Ar = -C$_6$H$_3$-Me-2-J-5. XVII.6. m. 183-185°.[854]

$\quad$ Ar = 4-EtO$_2$CC$_6$H$_4$-. XVII.6. m. 197-198°.[854]

$\quad$ Ar = 2-C$_{10}$H$_7$-. XVII.6. m. 205-206°.[854]

$i\text{-}C_3H_7PO(NHCO_2Me)_2$. XVII.6. m. 163-166°.[762]

$i\text{-}C_3H_7PO(NHCONHNHPh)_2$. XVII.6. m. 147°.[854]

$i\text{-}C_3H_7(O)P\langle$ NHCO, NHCO $\rangle$C=C(OEt)(NEt$_2$). XVII.6. m. 183°.[853]

$i\text{-}C_3H_7(O)P\langle$ NHCO, NHCO $\rangle$C=C(NEt$_2$)$_2$. XVII.6. m. 220-222°.[853]

$i\text{-}C_3H_7(O)P(NH_2)(NHCOCH_2CONEt_2)$. By hydrolysis of the 1,3,2-diazaphosphorinane, m. 147-148°.[853]

$i\text{-}C_3H_7(O)P\langle$ NHCO, NHCO $\rangle$C=C$\langle$Me$_2$, N(Me)... $\rangle$ XVII.6. m. 171-172°.[853]

$i\text{-}C_3H_7PO(NHCOON\text{-}CMe_2)_2$. XVII.6. m. 115-117°.[854]

$i\text{-}C_3H_7(O)P\langle$ N=C—OEt, N=C—OEt $\rangle$ XVII.8. m. 68-71°.[466]

$i\text{-}C_3H_7(O)P\langle$ NHN=C$\langle$Ph, O, Ph $\rangle$, NHN=C $\rangle$ XVIII.1. m. 186°, IR.[1240]

$(CH_3)_2C(OH)PO(OMe)(NEt_2)$. XVI.4. m. 141-142°.[28]

$C_2H_5OCH_2PO(NMe_2)_2$. XVII.7. $b_{0.5-0.6}$ 80-81°, n_D^{23} 1.4542, d^{23} 1.022.[1406]

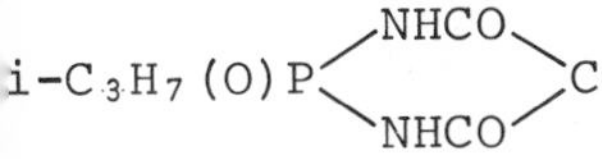

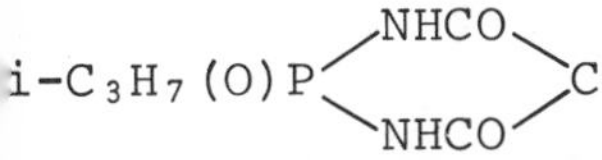

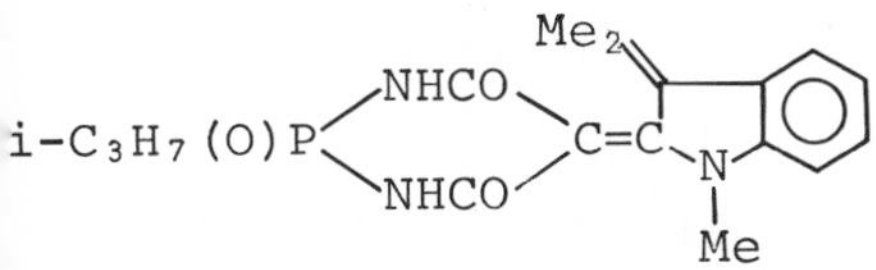

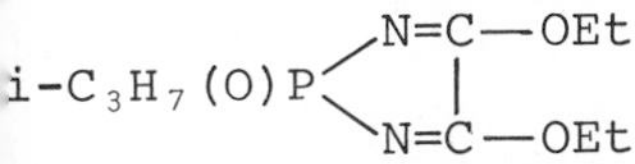

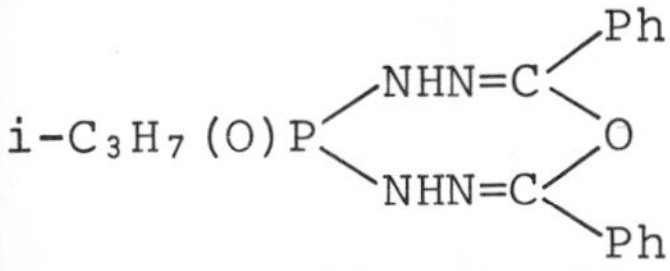

$(CH_3)_2NCH_2PO(OEt)(NMe_2)$. XVI.8. b_{14} 110-112°, n_D^{20} 1.4822[28]
(from $(EtO)_2P$-O-COPh and $Me_2NCH_2NMe_2$).

C4

CH_2=CClCH=CHPO(OEt)(NR_2).

R = Et. XVI.1. b_4 165°.[2034]

R = Pr. XVI.1. b_3 178°.[2034]

CH_2=CClCH=CHPO(OPr-i)(NEt_2). XVI.1. b_1 149°.[2034]
CH_3CH_2OCH=CClPO(OEt)(NEt_2). XVI.1. $b_{0.04}$ 102-103°, n_D^{20} 1.4812, d_4^{20} 1.1443.[1348]
CH_3CH_2OCH=CClPO$(NEt_2)_2$. XVII.1. $b_{0.06}$ 123-124°, n_D^{20} 1.5283 d_4^{20} 1.1500.[1348]

$Cl(CH_2)_4PO$(NH- 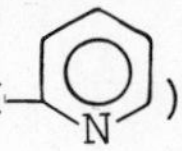)OH. XIV.1. m. 134-138°.[768]

$Cl(CH_2)_4PO$(NHAr)OH.

Ar = Ph. XIV.1. m. 135-140°.[767]

Ar = 4-$CH_3C_6H_4$-. XIV.1. m. 125-130°.[767]

Ar = 4-$CH_3OC_6H_4$-. XIV.1. m. 101-105° (hemihydrate).[767]

$Cl(CH_2)_4PO(NH_2)_2$. XVII.1. m. 95-100°.[768]
$Cl(CH_2)_4PO(NHR)_2$.

R = -CH_2COOEt. XVII.1. m. 140-150°.[768]

R = -CH_2Ph. XVII.1. m. 64-68°.[768]

R = 3-ClC_6H_4-. XVII.1. m. 122-125°.[768]

R = 4-ClC_6H_4-. XVII.1. m. 175-177°.[768]

R = 4-$CH_3SC_6H_4$-. XVII.1. m. 161-162°.[768]

R = 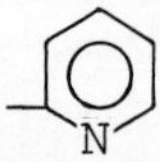 XVII.1. m. 175-176°.[768]

$Cl(CH_2)_4PO(NHAr)_2$.

Ar = Ph. XVII.1. m. 120-121°.[767]

Ar = 4-$CH_3C_6H_4$-. XVII.1. m. 143-144°.[767]

Ar = 4-$CH_3OC_6H_4$-. XVII.1. m. 103-104°.[767]

Ar = 4-$EtO_2CC_6H_4$-. XVII.1. m. 221°.[767]

$[Cl(CH_2)_4PO(NEt_2)NHC_6H_4$-4-$]_2$. XVII.1. m. 210-215°.[768]
CH_2=CHCH=CHPO(OEt)(NEt_2). XVI.1. $b_{3.5}$ 109-111°, n_D^{20} 1.4852 d_4^{20} 1.0214, IR, ^1H.[1208]
CH_2=CHCH=CHPO$(NMe_2)_2$. XVII.9. $b_{0.1}$ 102-120°.[1916]
CH_2=CHCH=CHPO$[N(CH_2CH=CH_2)_2]_2$. XVII.9. b_1 157-170°.[1916]
CH_3O_2CCH=CHPO(NR_2)(NR'_2).

R = R' = Me. XVII.4. $b_{0.5}$ 109-112°.[437]

R = Me, R' = Et. XVII.4. $b_{0.5}$ 131-134°.[437]

R = R' = Et. XVII.4. $b_{0.5}$ 152-155°.[437]

CH_3CH_2OCH=CHPO$(NR_2)_2$.

R = Me. XVII.1. b_1 108° [82], $b_{0.5}$ 119°, n_D^{20} 1.4765, d_4^{20} 1.0462.[1501]

R = Et. XVII.1. b_2 127°, n_D^{20} 1.4745, d_4^{20} 0.9958.[1501]

$CH_3CH_2OCH{=}CHPO(\overline{NCH_2CH_2})_2$. XVII.1. $b_{0.08}$ 101°, m. 40-41°.[673]

$CH_3CH_2OCH{=}CHPO(NHCONHAr)_2$.

Ar = Ph. XVII.6. m. 102-104°.[854]

Ar = 4-ClC_6H_4-. XVII.6. m. 106-107°.[854]

$CH_3CH_2OCH{=}CHPO(NHCOOR)_2$.

R = Me. XVII.6. m. 164° (dec.).[854]

R = Et. XVII.6. m. 163-164° (dec.).[854]

R = -N=CMe_2. XVII.6. m. 96-99° (dec.).[854]

$CH_3CH_2OCH{=}CH(O)P\langle(NHCO)(NHCO)\rangle C{=}C(OEt)_2$. XVII.6. m. 143-144°.[853]

$CH_3CH_2OCH{=}CH(O)P\langle(NHCO)(NHCO)\rangle C{=}C(Ph)(NC_5H_{10})$. XVII.6. m. 200-201°.[855]

$CH_3CH_2OCH{=}CHPO(NHCONHNHPh)_2$. XVIII.1. m. 152° (dec.).[854]

$CH_3CH_2O_2CCH_2PO(OEt)(NEt_2)$. XVI.3. $b_{0.4}$ 103-108°, n_D^{20} 1.4475, d_4^{20} 1.0599.[1238]

$CH_3O_2CCH_2CH_2PO(OEt)(NEt_2)$. XVI.4. $b_{0.4}$ 104-105°, n_D^{20} 1.4410, d_4^{20} 1.1191.[1636]

$CH_3CH_2O_2CCH_2PO(OEt)(\overline{NCH_2CH_2})$. XVI.1. b_3 120-123°, n_D^{20} 1.4593, d_4^{20} 1.1610.[1475]

$CH_3CH_2O_2CCH_2PO(OCH_2CH_2Cl)(NEt_2)$. XVI.3. b_2 135-136°, n_D^{20} 1.4660, d_4^{20} 1.1650, IR.[19]

$CH_3CH_2O_2CCH_2PO[OCH(Me)CH_2Cl](NEt_2)$. XVI.3. b_2 129-130.5°, n_D^{20} 1.4640, d_4^{20} 1.1425, IR.[20]

$CH_3O_2CCH_2CH_2PO(OEt)(\overline{NCH_2CH_2})$. XVI.1. $b_{1.5}$ 128-130°, n_D^{20} 1.4600, d_4^{20} 1.1480.[1475]

$CH_3O_2CCH(CH_3)PO(OEt)(NHCONHC_6H_4Cl$-4). XVI.6. m. 138-140°.[280]

$CH_3CH_2O_2CCH_2PO(NEt_2)_2$. XVII.3. $b_{0.45-0.7}$ 142-147°, n_D^{20} 1.4668, d_4^{20} 1.0712.[1238]

$CH_3O_2CCH_2CH_2PO(\overline{NCH_2CH_2})_2$. XVII.1. $b_{0.0001}$ 95-100°, n_D^{20} 1.4900, d_4^{20} 1.1930.[1475]

$CH_3CH_2O_2CCH_2PO(\overline{NCH_2CH_2})(NHCH_2CH{=}CH_2)$. XVII.1. $b_{0.001}$ 120-125°, n_D^{20} 1.4975, d_4^{20} 1.2300.[1475]

$CH_3O_2CC(Me)(OH)PO(OEt)(NEt_2)$. XVI.4. b_5 130°, n_D^{20} 1.4461, d_4^{20} 1.1673.[1591]

$CH_3CH_2SCOCH_2PO(OEt)(NR_2)$.

R = Me. XVI.3. $b_{0.15}$ 75-77°, n_D^{20} 1.4721, d_4^{20} 1.0894.[1238]

R = Et. XVI.3. $b_{1.3-1.4}$ 102-112°, n_D^{20} 1.4760, d_4^{20} 1.0967.[1238]

$CH_3CH_2SCOCH_2PO(NEt_2)_2$. XVII.3. $b_{0.4}$ 108-113°, n_D^{20} 1.4980, d_4^{20} 1.0473.[1238]

$\overline{CH_2CH_2CH_2CH_2PO(N}R)(NHR)$.

R = $-CH_2Ph$. XVII.9. m. 55-57°.[767, 768]

R = Ph. XVII.9. m. 156-159°.[767, 768]

R = $4-CH_3C_6H_4-$. XVII.9. m. 160-165°.[767, 768]

R = $4-ClC_6H_4-$. XVII.9. m. 207-210 .[767, 768]

R = $4-CH_3OC_6H_4-$. XVII.9. m. 177.5-178°.[767, 768]

R = $4-CH_3SC_6H_4-$. XVII.9. m. 187-188°.[767, 768]

R = $4-EtO_2CC_6H_4-$. XVII.9. dec. >180°.[767, 768]

R = 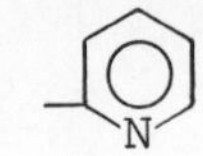 XVII.9. m. 168-170°.[767, 768]

$\overline{CH_2CH_2CH_2CH_2PO[N}(C_6H_4Cl-4)][N(C_6H_4Cl-4)(COPh)]$. XVII.9. m. 186-190°.[768]

$\overline{CH_2CH_2CH_2CH_2PO[N}($ 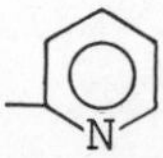 $)][N($ 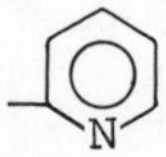 $)(CONHPh)]$. XVII.9. m. 160-165°.[768]

$C_4H_9PO(NEt_2)Cl$.[1704] XV.1. b_3 113-115°, n_D^{20} 1.4641, d_4^{20} 1.0526.

$C_4H_9PO(NEt_2)F$.[1704] XV.1. b_{16} 128-130°, n_D^{20} 1.4358, d_4^{20} 1.0380.

$C_4H_9PO(OEt)(NHCONHAr)$.

Ar = $4-ClC_6H_4-$. XVI.6. m. 98-100°.[726]

Ar = $4-O_2NC_6H_4-$. XVI.6. m. 167-168°.[726]

$C_4H_9PO(NH_2)_2$. XVII.1. m. 90-110°.[768]

$C_4H_9PO(NHR)_2$.

R = $-CH_2C_6H_5$. XVII.1. m. 91-93°.[768]

R = $4-ClC_6H_4-$. XVII.1. m. 163-166°.[768]

$C_4H_9PO(NR_2)_2$.

R = Me. XVII.7. b_{15} 135-140°, n_D^{21} 1.4577, d_4^{21} 0.982,[1409] $b_{0.4}$ 80°, n_D^{22} 1.4564, d^{24} 0.972.[1406] XVII.1. b_{33} 157°, n_D^{30} 1.4542, d_4^{30} 0.9618.[1018]

R = Et. XVII.1. $b_{3.5}$ 137°, n_D^{30} 1.4585, d_4^{30} 0.9314.[1018]

$C_4H_9PO(\overline{NCH_2CH_2})_2$. XVII.1. $b_{0.5}$ 105-106°.[670]

$C_4H_9(O)P$ 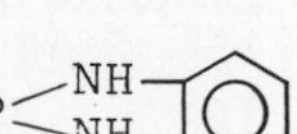 XVII.1. m. 177°.[454]

$C_4H_9PO(NHCH_2Ph)_2$. XVII.1. m. 91-93°.[768]

$C_4H_9PO(NHCH_2Ph)[N(COPh)CH_2Ph]$. XVII.9. m. 88-90° [768] (from previous compound with PhCOCl).

sec.-$C_4H_9PO(NR_2)Cl$

R = Me. XV.1. $b_{0.2}$ 73-75°, d_4^{25} 1.0984.[1715]

R = Et. XV.1. $b_{0.15}$ 85-86°, d_4^{25} 1.0600.[1715]

sec.-$C_4H_9PO(NMe_2)(NRR')$

R = R' = Et. XVII.1. $b_{0.15}$ 87-89°, d_4^{25} 0.9748.[1715]

R = R' = Pr. XVII.1. $b_{0.25}$ 132-136°, d_4^{25} 0.9672.[1715]

R = H,R' = Ph. XVII.1. m. 170-171°.[1715]

t-$C_4H_9PO(NRR')F$

R = H, R' = Me. XV.1. m. 72°.[1860]

R = R' = Me. XV.1. m. 43°.[1860]

t-$C_4H_9PO[N(CH_2)_3CH_2]F$. XV.1. b_1 78°.[1860]

t-$C_4H_9PO(OEt)(NHCSNHC_6H_4Cl$-4). XVI.6. m. 161°.[1856]

$CH_3CH_2OCH_2CH_2PO(OEt)(NEt_2)$. XVI.7. $b_{0.4}$ 113-114°, n_D^{20} 1.4390, d_4^{20} 1.0962.[1636]

$CH_3CH_2CH_2OCH_2PO(NMe_2)_2$. XVII.7. $b_{0.5}$ 78-82°, n_D^{23} 1.4494, d^{25} 0.993.[1406]

$HOCH(CH_3)CH(CH_3)PO(NMe_2)_2$. Diastereomers. XVII.1.[438]

R-form, oil, ^1H, IR, mass spect. of Me_3Si ether, yields cis-2-butene on thermolysis. S-form, m. 76-80°, ^1H, IR, sublimes >45°/10^{-3} mm, yields trans-2-butene on thermolysis.[438]

$(CH_3)_2NCH_2CH_2PO(NEt_2)_2$. XVII.9. b_2 127-130°, n_D^{20} 1.4697, d_4^{20} 0.9725.[908]

C_5

$CH_3SCH=C(CCl=CH_2)PO(NEt_2)_2$. XVII.1. $b_{0.009}$ 101°, n_D^{20} 1.5606, d_4^{20} 1.1107.[1918a]

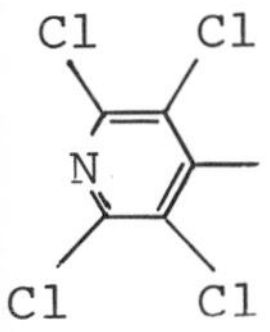

$PO(NHPh)_2$. XVII.1. m. 227-229°.[858]

$CH_2=CHC(CH_3)=CHPO(OMe)(NEt_2)$. XVI.1. b_1 93-94°, n_D^{20} 1.4928, d_4^{20} 1.0225, IR, ^1H.[1208]

$C_3H_7OCH=CHPO(NR_2)_2$.

R = Me. XVII.1. $b_{2.5}$ 146°, n_D^{20} 1.4756, d_4^{20} 1.0183.[1501]

R = Et. XVII.1. $b_{2.0}$ 141°, n_D^{20} 1.4762, d_4^{20} 1.0085.[1501]

$CH_3CH_2O_2CCH_2CH_2PO(OEt)(NRR')$.

R = H, R' = Ph. XVI.4. b_8 214-215°, n_D^{20} 1.5300, d_4^{20} 1.2080, IR.[1635]

R = R' = Me. XVI.4. $b_{0.7}$ 110-116°.[425]

R = R' = Et. XVI.4. b_{10} 154-156°, n_D^{20} 1.4370, d_4^{20} 1.0400, IR.[1635] XVI.3. $b_{0.01}$ 105-107°, n_D^{20} 1.4472, d_4^{20} 1.0538.[1658] XVI.4. $b_{0.4}$ 113-114°, n_D^{20} 1.4390, d_4^{20} 1.0962.[1636]

R = $-CH_2CH_2CN$, R' = Ph. XVI.4. n_D^{20} 1.5158, IR.[1635]

$CH_3CH_2O_2CCH_2CH_2PO(OCH_2CH_2Cl)(NEt_2)$. XVI.3. b_2 154.5-155.5° n_D^{20} 1.4640, d_4^{20} 1.1456.[19]

$CH_3CH_2O_2CCH_2CH_2PO[OCH(Me)CH_2Cl](NEt_2)$. XVI.3. b_3 153-154°, n_D^{20} 1.4630, d_4^{20} 1.1244.[20]

$C_3H_7O_2CCH_2PO(OCH_2CH_2Cl)(NEt_2)$. XVI.3. b_2 144-144.5°, n_D^{20} 1.4662, d_4^{20} 1.1494, IR.[19]

$C_3H_7O_2CCH_2PO[OCH(Me)CH_2Cl](NEt_2)$. XVI.3. b_2 140-141°, n_D^{20} 1.4641, d_4^{20} 1.1281, IR.[20]

i-$C_3H_7O_2CCH_2PO(OCH_2CH_2Cl)(NEt_2)$. XVI.3. b_2 140-141°, n_D^{20} 1.4629, d_4^{20} 1.1427, IR.[19]

i-$C_3H_7O_2CCH_2PO[OCH(Me)CH_2Cl](NEt_2)$. XVI.3. b_2 136-137.5°, n_D^{20} 1.4647, d_4^{20} 1.1250, IR.[20]

$CH_3CH_2O_2CCH_2CH_2PO(NR_2)_2$.

R = Me. XVII.4. $b_{0.9}$ 116-120°.[425]

R = Et. XVII.4. $b_{0.08}$ 134-135°, n_D^{20} 1.4643, d_4^{20} 1.0279 [1657]

$C_2H_5O_2CC(CH_3)(OH)PO(OEt)(NEt_2)$. XVI.4. b_5 130-132°, n_D^{20} 1.4375, d_4^{20} 1.1226.[1591]

$(C_2H_5)_2NCOPO(NEt_2)_2$. XVII.3. $b_{0.008}$ 95.5-98°, defoliant, IR.[1531, 1532]

$(C_2H_5)_2NCSPO(NEt_2)_2$. XVII.3. $b_{0.025-0.075}$ 120-126°, defoliant, IR. [1531, 1532]

$C_4H_9OCH_2PO(NMe_2)_2$. XVII.7. $b_{0.5}$ 96-98°, $n_D^{24.5}$ 1.4533, d^{27} 0.993.[1406]

$C_4H_9SCH_2PO(NMe_2)_2$. XVII.7. $b_{0.5-0.6}$ 118-120°, n_D^{22} 1.4927, d^{23} 1.050.[1406]

<u>C_6</u>

$4-ClC_6H_4PO(NHCOPh)OH$. XIV.2. m. 117-118°.[471]

$4-ClC_6H_4PO(NHCOAr)Cl$.

Ar = Ph. XV.2. m. 117-118°.[471]

Ar = $4-O_2NC_6H_4-$. XV.2. m. 169-171°.[471]

$4-ClC_6H_4PO(OMe)NHCOAr$.

Ar = Ph. XVI.5. m. 147-148°.[464]

Ar = $4-O_2NC_6H_4-$. XVI.5. m. 178-180°.[464]

$4\text{-}ClC_6H_4PO(NHCONHR)_2$.

R = Bu. XVII.6. m. 181-182°.[2013]

R = Ph. XVII.6. m. 228-229°.[2013]

$C_2H_5SCH{=}C(CCl{=}CH_2)PO(NEt_2)_2$. XVII.1. $b_{0.007}$ 107°, n_D^{20} 1.5564, d_4^{20} 1.0992.[1918a]

$C_2H_5SCH{=}C(CCl{=}CH_2)PO(\overline{NCH_2CH_2})_2$. XVII.1. $b_{0.15}$ 133°, n_D^{20} 1.6023, d_4^{20} 1.2706.[2047]

$4\text{-}O_2NC_6H_4PO(OMe)(NHR)$.

R = 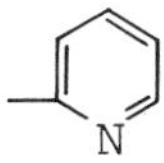 XVI.1. m. 113-115°.[326]

R = 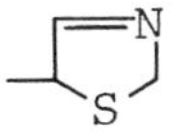 XVI.1. m. 135-136.5°.[326]

$4\text{-}O_2NC_6H_4PO(OEt)(NHR)$.

R = $-CH_2Ph$. XVI.1. m. 81.5-82.5°.[326]

R = 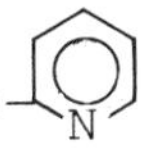 XVI.1. m. 92-94°.[326]

R = 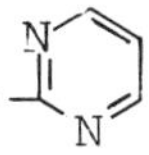 XVI.1. m. 192-195°.[326]

R = 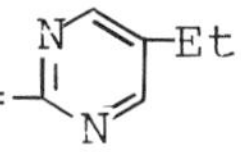 XVI.1. m. 142-144°.[326]

R = 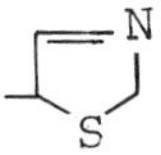 XVI.1. m. 138-139°.[326]

$4\text{-}O_2NC_6H_4PO(OCH_2CF_3)$(NH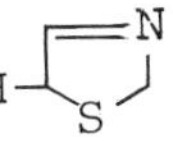). XVI.1. m. 143-145°.[326]

$4\text{-}O_2NC_6H_4PO(OPh)$(NH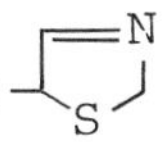). XVI.1. m. 172-174°.[326]

$4\text{-}O_2NC_6H_4PO(NHR)_2$.

R = Ph. XVII.1. m. 195-197.5°.[483]

R = $2\text{-}ClC_6H_4$-. XVII.1. m. 187-190°.[483]

R = $3\text{-}ClC_6H_4$-. XVII.1. m. 187-188°.[483]

R = $4\text{-}ClC_6H_4$-. XVII.1. m. 187-190°.[483]

R = $3\text{-}BrC_6H_4$-. XVII.1. m. 190.5-193.5°.[483]

R = $4\text{-}MeC_6H_4$-. XVII.1. m. 185-187°.[483]

R = $4\text{-}EtO_2SC_6H_4$-. XVII.1. m. 213-215°.[483]

R = $3\text{-}F_3CC_6H_4$-. XVII.1. m. 178-180°.[483]

R = $3\text{-}EtOC_6H_4$-. XVII.1. m. 131.5-132°.[483]

R = $4\text{-}EtO_2CC_6H_4$-. XVII.1. m. 198-200°.[483]

R = $4\text{-}NCC_6H_4$-. XVII.1. m. 153-156°.[483]

R = $2\text{-}C_{10}H_7$-. XVII.1. m. 215-217°.[438]

R = 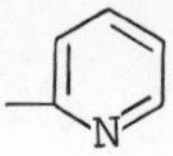

. XVII.1. m. 200-202°.[483]

R = 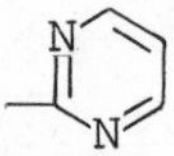

.XVII.1. m. 182-184°.[483]

$4\text{-}O_2NC_6H_4PO(\overline{NCH_2CH_2})_2$. XVII.1. m. 107-110°.[572]
$4\text{-}O_2NC_6H_4PO(NEt_2)_2$. XVII.1. m. 88-92°.[572]
$4\text{-}O_2NC_6H_4PO[\overline{N(CH_2)_3CH_2}]_2$. XVII.1. m. 67.5-69.5°.[572]
$4\text{-}O_2NC_6H_4PO[\overline{N(CH_2)_4CH_2}]_2$. XVII.1. m. 139-142°.[572]
$4\text{-}O_2NC_6H_4PO[\overline{N(CH_2)_2O(CH_2)_2}]_2$. XVII.1. m. 204-206°.[572]
$4\text{-}O_2NC_6H_4PO(NHCONHR)_2$.

R = H. XVII.6. m. 195-198°.[2013]

R = Bu. XVII.6. m. 186-187°.[2013]

R = cycl.-C_6H_{11}-. XVII.6. m. 195-196°.[2013]

R = Ph. XVII.6. m. 186-187°.[2013]

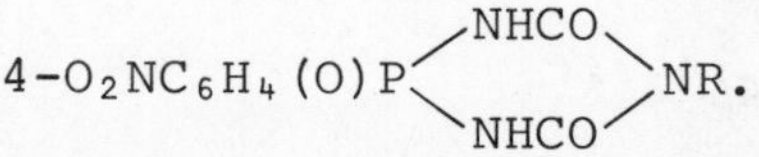

R = H. XVII.6. m. 194-196°.[2011]

R = Me. XVII.6. m. 220-221°.[2011]

$C_6H_5PO(NH_2)OH$. XIV.2. Isolated as Na salt, m. 164-165°.[189]
$C_6H_5PO(NHR)OH$.

R = $-CH_2CH_2CH_2NH_2$. XIV.1. m. 250°.[1385]

R = $-CH_2CH(OH)CH_2NH_2$. XIV.1. m. 204-207°.[1385]

R = Ph. XIV.1. m. 125°,[1260, 1282] 210-212°.[851]

R = $2\text{-}H_2NC_6H_4$-. XIV.1. m. 246-248°.[1947]

$C_6H_5PO(NEt_2)OH$. XIV.1. $Et_2NH_2^{\oplus}$ salt, m. 65-69°.[375]
$C_6H_5PO(NHCOAr)OH$.

Ar = Ph. XIV.2. m. 116-118°.[465]

Ar = 4-ClC_6H_4-. XIV.2. m. 127-128°.[471]

Ar = 4-BrC_6H_4-. XIV.2. m. 138-140°.[465]

Ar = 3-$O_2NC_6H_4$-. XIV.2. m. 170-171°.[471]

$C_6H_5PO(NHCOOR)OH$.

R = Me. XIV.2. m. 128-129°.[1919]

R = Et. XIV.2. m. 125-126°.[1919]

$C_6H_5PO(NHCONHAr)OH$.

Ar = Ph. XIV.1. m. 138-139°.[1892]

Ar = 4-$CH_3C_6H_4$-. XIV.1. m. 146-147°.[1892]

Ar = 4-ClC_6H_4-. XIV.1. m. 142-143°.[1892]

Ar = $C_6H_3Cl_2$-2.4. XIV.1. m. 120-121°.[1892]

Ar = 4-BrC_6H_4-. XIV.1. m. 146-147°.[1892]

Ar = 2-$O_2NC_6H_4$-. XIV.1. m. 135-136°.[1892]

Ar = 3-$O_2NC_6H_4$-. XIV.1. m. 164-166°.[1892]

Ar = 4-$O_2NC_6H_4$-. XIV.1. m. 145-147°.[1892]

Ar = 1-$C_{10}H_7$-. XIV.1. m. 135-137°.[1892]

Ar = 2-$C_{10}H_7$-. XIV.1. m. 144-145°.[1892]

$C_6H_5PO(N{=}PPh_3)OH$. XIV.1. m. 205-206°,[202] pKa (THF - H_2O = 60:40) = 7.34, (EtOH - H_2O = 77:23) = 7.74, IR.[201]

$C_6H_5PO(OH)N{=}P(Ph)_2$-C_6H_4-4-$P(Ph)_2{=}NPO(Ph)OH$. XIV.1. m. 248-251°, IR.[201]

$C_6H_5PO(OH)N{=}PPh_2$-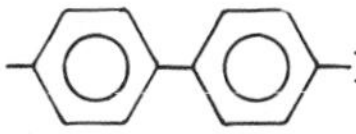-$P(Ph)_2{=}NPO(Ph)OH$. XIV.1. m. 230-234°, pK_a (THF - H_2O = 60:40) = 7.40, IR.[201]

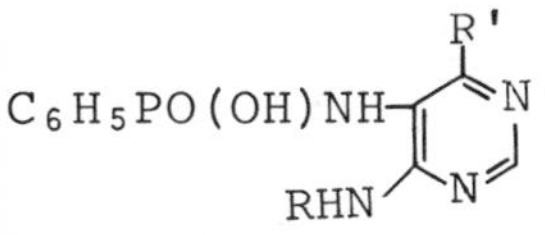

R = H, R' = Cl. XIV.1. m. 160-161° (hemihydrate).[1105]

R = H, R' = -NMe_2. XIV.1. m. 245-247°.[1105]

R = Me, R' = -NMe_2. XIV.1. m. 199-200°.[1105]

R = 4-$CH_3C_6H_4$-, R' = -NMe_2. XIV.1. m. 180-182°.[1105]

$C_6H_5PO(NEt_2)Cl$. XV.1. $b_{0.5}$ 132°, n_D^{25} 1.5243, d_4^{25} 1.217.[2007]

$C_6H_5PO(NHAr)Cl$.

Ar = Ph. XV.2. m. 139-140°.[2100]

Ar = 2-MeC_6H_4-. XV.2. m. 107-108°.[2100]

Ar = 3-MeC_6H_4-. XV.2. m. 125-127°.[2100]

Ar = 4-MeC_6H_4-. XV.2. m. 137-138°.[2100]

Ar = 2-ClC_6H_4-. XV.2. m. 102-104°.[2100]

Ar = 3-ClC_6H_4-. XV.2. m. 131-133°.[2100]

Ar = 4-ClC_6H_4-. XV.2. m. 140-142°.[2100]

Ar = 4-$MeOC_6H_4$-. XV.2. m. 165-167°.[2100]

Ar = 4-$EtOC_6H_4$-. XV.2. m. 176-177°.[2100]

Ar = 2-BrC_6H_4-. XV.2. m. 101-103°.[2100]

Ar = 3-BrC_6H_4-. XV.2. m. 136-138°.[2100]

Ar = 4-BrC_6H_4-. XV.2. m. 160-162°.[2100]

Ar = 2-$O_2NC_6H_4$-. XV.2. m. 114-116°.[2100]

Ar = 3-$O_2NC_6H_4$-. XV.2. m. 144-145°.[2100]

Ar = 4-$O_2NC_6H_4$-. XV.2. m. 189-191°.[2100]

Ar = -$C_6H_3Cl_2$-2,4. XV.2. m. 100-102°.[2100]

Ar = -$C_6H_3Cl_3$-2,4,6. XV.2. m. 184-185°.[2100]

Ar = -$C_6H_3Br_2$-2,4. XV.2. m. 125-127°.[2100]

Ar = -$C_6H_2Br_3$-2,4,6. XV.2. m. 210-212°.[2100]

Ar = -$C_6H_3(NO_2)_2$-2,4. XV.2. m. 141-143°.[2100]

Ar = -$C_6H_2NO_2$-4-Cl_2-2,6. XV.2. m. 179-180°.[2100]

$C_6H_5PO(NHCOR)Cl$.

R = -CCl_2H. XV.2. m. 143-144°.[1902]

R = -CCl_3. XV.2. m. 135-137°.[1902]

R = Me. XV.2. m. 68-70°.[465]

R = Ph. XV.2. m. 128-129°.[465, 471]

R = 3-$O_2NC_6H_4$-. XV.2. m. 151-152°.[471]

R = 4-$O_2NC_6H_4$-. XV.2. m. 161-162°.[471]

$C_6H_5PO[NHCOC(CN)Alk_2]Cl$.

Alk = Me. XVI.5. m. 90-91°.[1890]

Alk = Et. XVI.5. m. 161-162°.[1890]

Alk = Pr. XVI.5. m. 138-139°.[1890]

Alk = Bu. XVI.5. m. 131-132°.[1890]

Alk = -C_5H_{11}. XVI.5. m. 105-106°.[1890]

$C_6H_5PO(NHCOOMe)Cl$. XV.2. m. 87-89°.[1919]

$C_6H_5PO(NHCONHAr)Cl$.

Ar = Ph. XV.4. m. 129-130°.[1891]

Ar = 4-$CH_3C_6H_4$-. XV.4. m. 107-108°.[1892]

Ar = 4-BrC_6H_4-. XV.4. m. 137-139°.[1892]

Ar = $4\text{-}ClC_6H_4$-. XV.4. m. 137-138°.[1892]

Ar = $-C_6H_3Cl_2$-2,4. XV.4. m. 141-142°.[1892]

Ar = $2\text{-}O_2NC_6H_4$-. XV.4. m. 148-150°.[1892]

Ar = $3\text{-}O_2NC_6H_4$-. XV.4. m. 146-148°.[1892]

Ar = $4\text{-}O_2NC_6H_4$-. XV.4. m. 157-158°.[1892]

Ar = $1\text{-}C_{10}H_7$-. XV.4. m. 81-84°.[1892]

Ar = $2\text{-}C_{10}H_7$-. XV.4. m. 132-134°.[1892]

$C_6H_5PO(NH_2)F$. XV.4. m. 86°, IR, ^1H, ^{19}F, mass spect.[1761]

$C_6H_5PO(NEt_2)F$. XV.1. b_{15} 153-156°, n_D^{21} 1.4970, d_4^{21} 1.3100.[851]

$C_6H_5PO(NHAr)F$.

Ar = Ph. XV.1. m. 109-112°.[851]

Ar = $4\text{-}CH_3C_6H_4$-. XV.1. m. 81-82°.[851]

Ar = $4\text{-}ClC_6H_4$-. XV.1. m. 122-123°.[851]

$C_6H_5PO(NMe_2)CN$. XV.3. b_1 155°.[1826]

$C_6H_5PO(OMe)(NHAr)$.

Ar = Ph. XVI.2. m. 123-125°.[2099]

Ar = $2\text{-}CH_3C_6H_4$-. XVI.2. m. 120-122°.[2099]

Ar = $3\text{-}CH_3C_6H_4$-. XVI.2. m. 113-115°.[2099]

Ar = $4\text{-}CH_3C_6H_4$-. XVI.2. m. 124-125°.[2099]

Ar = $3\text{-}BrC_6H_4$-. XVI.2. m. 105-107°.[2099]

Ar = $4\text{-}BrC_6H_4$-. XVI.2. m. 127-129°.[2099]

Ar = $3\text{-}ClC_6H_4$-. XVI.2. m. 97-99°.[2099]

Ar = $4\text{-}ClC_6H_4$-. XVI.2. m. 130-131°.[2099]

Ar = $-C_6H_3Cl_2$-2,6. XVI.2. m. 178-179°.[2099]

Ar = $-C_6H_2Cl_3$-2,4,6. XVI.2. m. 136-138°.[2099]

Ar = $-C_6H_2Br_3$-2,4,6. XVI.2. m. 168-170°.[2099]

Ar = $2\text{-}O_2NC_6H_4$-. XVI.2. m. 66-68°.[2099]

Ar = $3\text{-}O_2NC_6H_4$-. XVI.2. m. 169-171°.[2099]

Ar = $4\text{-}O_2NC_6H_4$-. XVI.2. m. 118-119°.[2099]

Ar = $-C_6H_3(NO_2)_2$-2,4. XVI.2. m. 100-102°.[2099]

Ar = $4\text{-}CH_3OC_6H_4$-. XVI.2. m. 125-127°.[2099]

Ar = $4\text{-}EtOC_6H_4$-. XVI.2. m. 96-97°.[2099]

$C_6H_5PO(OMe)(NHCOR)$.

R = $-CCl_2H$. XVI.5. m. 123-124°.[1902]

R = $-CCl_3$. XVI.5. m. 130-132°.[1902]

R = Ph. XVI.5. m. 142-143°.[464]

R = 3-$O_2NC_6H_4$-. XVI.5. m. 143-144°.[464]

R = 4-$O_2NC_6H_4$-. XVI.5. m. 155-156°.[464]

$C_6H_5PO(OMe)(NHCOOMe)$. XVI.2. m. 117-119°.[1919]
$C_6H_5PO(OMe)(NHCONHAr)$.

Ar = Ph. XVI.2. m. 154-156°.[1893]

Ar = 4-$CH_3C_6H_4$-. XVI.2. m. 185-186°.[1893]

Ar = 4-BrC_6H_4-. XVI.2. m. 177-178°.[1893]

Ar = 4-$O_2NC_6H_4$-. XVI.2. m. 183-185°.[1893]

Ar = 1-$C_{10}H_7$-. XVI.2. m. 98-101°.[1893]

Ar = 2-$C_{10}H_7$-. XVI.2. m. 168-170°.[1893]

$C_6H_5PO(OMe)(NHSO_2Ar)$.

Ar = Ph. XVI.5. m. 172-173°.[1896]

Ar = 2-$CH_3C_6H_4$-. XVI.5. m. 166-168°.[1896]

Ar = 4-$CH_3C_6H_4$-. XVI.5. m. 156-157°.[1896]

Ar = 4-ClC_6H_4-. XVI.5. m. 165-167°.[1896]

Ar = 2-$O_2NC_6H_4$-. XVI.5. m. 161-162°.[1896]

Ar = 4-$O_2NC_6H_4$-. XVI.5. m. 175-176°.[1896]

Ar = 1-$C_{10}H_7$-. XVI.5. m. 165-167°.[1869]

Ar = 2-$C_{10}H_7$-. XVI.5. m. 165-166°.[1869]

$C_6H_5PO(OMe)N(Me)SO_2C_6H_4Me$-4. XVI.5. $b_{0.05}$ 175-180°.[1898]
$C_6H_5PO(OMe)(N{=}PPh_3)$. XVI.8. m. 133-135°.[202]
$C_6H_5PO(OEt)(NH_2)$. XVI.2. m. 127°.[1931]
$C_6H_5PO(OEt)(NHPh)$. XVI.2. m. 105°.[1282]
$C_6H_5(O)POCH_2CH_2NMe$. XVI.8. ^1H, ^{31}P -35.7 ppm.[679] By reaction of $PhPOCH_2NMe$ with PhNCO.
$C_6H_5PO(OEt)(NEtPh)$. XVI.5. b_1 147°, n_D^{20} 1.5588, d_4^{20} 1.1265.[627]
$C_6H_5PO(OEt)(NHCOR)$.

R = -CCl_2H. XVI.5. m. 90-92°.[1902]

R = -CCl_3. XVI.5. m. 108-109°.[1902]

$C_6H_5PO(OEt)(NHCONHAr)$.

Ar = Ph. XVI.2. m. 72-74°.[1893]

Ar = 4-MeC_6H_4-. XVI.2. m. 120-121°.[1893]

Ar = 4-BrC_6H_4-. XVI.2. m. 65-66°.[1893]

Ar = 4-$O_2NC_6H_4$-. XVI.2. m. 170-172°.[1893]

Ar = 1-$C_{10}H_7$-. XVI.2. m. 97-98°.[1893]

Ar = 2-$C_{10}H_7$-. XVI.2. m. 98-101°.[1893]

$C_6H_5PO(OEt)(NHSO_2Ar)$.

Ar = Ph. XVI.5. m. 141-142°.[1896]

Ar = 2-MeC_6H_4-. XVI.5. m. 156-158°.[1896]

Ar = 4-MeC_6H_4-. XVI.5. m. 165-167°.[1896]

Ar = 4-ClC_6H_4-. XVI.5. m. 158-160°.[1896]

Ar = 3-$O_2NC_6H_4$-. XVI.5. m. 140-143°.[1896]

Ar = 1-$C_{10}H_7$-. XVI.5. m. 148-150°.[1896]

Ar = 2-$C_{10}H_7$-. XVI.5. m. 195-197°.[1896]

$C_6H_5PO(OCH_2CH=CH_2)(NHSO_2Ar)$.

Ar = Ph. XVI.5. m. 136-138°.[1897]

Ar = 4-MeC_6H_4-. XVI.5. m. 145-147°.[1897]

Ar = 4-ClC_6H_4-. XVI.5. m. 156-159°.[1897]

Ar = 4-$O_2NC_6H_4$-. XVI.5. m. 152-154°.[1897]

$C_6H_5PO(OCH_2CHBrCH_2Br)(NHSO_2Ar)$.

Ar = Ph. XVI.9. m. 164-167°.[1897]

Ar = 4-MeC_6H_4-. XVI.9. m. 165-167°.[1897]

Ar = 4-ClC_6H_4-. XVI.9. m. 143-145°.[1897] (by bromination of the -$OCH_2CH=CH_2$ esteramide).

$C_6H_5PO(OCH_2CBr=CH_2)(NHSO_2Ar)$.

Ar = Ph. XVI.9. m. 107-108°.[1897]

Ar = 4-$CH_3C_6H_4$-. XVI.9. m. 99-100°.[1897]

Ar = 4-ClC_6H_4-. XVI.9. m. 131-133°.[1897] (by dehydrobromination of the dibromoesteramide).

$C_6H_5PO(OCH_2C{\equiv}CH)(NHSO_2Ar)$.

Ar = Ph. XVI.9. m. 131-133°.[1897]

Ar = 4-$CH_3C_6H_4$-. XVI.9. m. 152-154°.[1897]

Ar = 4-ClC_6H_4-. XVI.9. m. 157-159°.[1897] (by dehydrobromination of the monobromoesteramide).

$C_6H_5PO(OPr)(NH_2)$. XVI.2. m. 135°.[1931]

$C_6H_5PO(OPr\text{-}i)(NHCONHAr)$.

Ar = Ph. XVI.2. m. 132-134°.[1893]

Ar = 4-$CH_3C_6H_4$-. XVI.2. m. 80-82°.[1893]

Ar = 4-BrC_6H_4-. XVI.2. m. 112-114°.[1893]

$C_6H_5PO(OPr)(NHSO_2Ar)$.

Ar = Ph. XVI.5. m. 110-112°.[1896]

Ar = 2-$CH_3C_6H_4$-. XVI.5. m. 129-130°.[1896]

Ar = $4-CH_3C_6H_4-$. XVI.5. m. 138-140°.[1896]

Ar = $4-ClC_6H_4-$. XVI.5. m. 130-134°.[1896]

Ar = $1-C_{10}H_7-$. XVI.5. m. 137-139°.[1896]

$C_6H_5PO(OBu)(NH_2)$. XVI.2. m. 104°.[1931]
$C_6H_5PO(OBu)(NHCONHPh)$. XVI.2. m. 121-123°.[1893]
$C_6H_5PO(OC_5H_{11})(NH_2)$. XVI.2. m. 82°.[1931]
$C_6H_5PO(OC_6H_4Cl-4)(NHCOR)$.

R = $-CCl_2H$. XVI.5. m. 157-158°.[1902]

R = $-CCl_3$. XVI.5. m. 155-156°.[1902]

R = $4-O_2NC_6H_4-$. XVI.5. m. 173-174°.[464]

$C_6H_5PO(OPh)(NH_2)$. XVI.1. m. 127°.[1488]
$C_6H_5PO(OPh)(NEt_2)$. XVI.3. $b_{0.03}$ 172°, n_D^{20} 1.5833, d_4^{20} 1.2155.[1637]
$C_6H_5PO(OPh)(NHPh)$. XVI.2. m. 83°, b_{25} 235°.[1260]
$C_6H_5PO(OPh)(NHAr)$.

Ar = Ph. XVI.2. m. 145-146°.[2099]

Ar = $2-MeC_6H_4-$. XVI.2. m. 132-134°.[2099]

Ar = $3-MeC_6H_4-$. XVI.2. m. 99-101°.[2099]

Ar = $4-MeC_6H_4-$. XVI.2. m. 145-147°.[2099]

Ar = $2-ClC_6H_4-$. XVI.2. m. 123-125°.[2099]

Ar = $3-ClC_6H_4-$. XVI.2. m. 94-95°.[2099]

Ar = $4-ClC_6H_4-$. XVI.2. m. 138-140°.[2099]

Ar = $-C_6H_3Cl_2-2,4$. XVI.2. m. 106-108°.[2099]

Ar = $-C_6H_2Cl_3-2,4,6$. XVI.2. m. 210-212°.[2099]

Ar = $4-BrC_6H_4-$. XVI.2. m. 151-153°.[2099]

Ar = $-C_6H_3Br_2-2,4$. XVI.2. m. 104-106°.[2099]

Ar = $-C_6H_3Br_3-2,4,6$. XVI.2. m. 202-204°.[2099]

Ar = $3-O_2NC_6H_4-$. XVI.2. m. 147-148°.[2099]

Ar = $4-O_2NC_6H_4-$. XVI.2. m. 160-170°.[2099]

Ar = $-C_6H_2(NO_2)-4-Cl_2-2,6$. XVI.2. m. 199-200°.[2099]

Ar = $4-MeOC_6H_4-$. XVI.2. m. 146-148°.[2099]

Ar = $4-EtOC_6H_4-$. XVI.2. m. 143-145°.[2099]

$C_6H_5PO(OPh)(NHCOR)$.

R = $-CCl_2H$. XVI.5. m. 143-144°.[1902]

R = $-CCl_3$. XVI.5. m. 132-133°.[1902]

$C_6H_5PO(OPh)NHSO_2Ar$.

Ar = Ph. XVI.5. m. 144-146°.[1896]

Ar = $2-MeC_6H_4-$. XVI.5. m. 130-131°.[1896]

Ar = 4-MeC_6H_4-. XVI.5. m. 155-157°.[1896]

Ar = 4-ClC_6H_4-. XVI.5. m. 124-125°.[1896]

Ar = 4-$O_2NC_6H_4$-. XVI.5. m. 120-121°.[1896]

Ar = 1-$C_{10}H_7$-. XVI.5. m. 165-166°.[1896]

$C_6H_5PO(OCH_2Ph)(NH_2)$. XVI.2. m. 121-123°.[517]

$C_6H_5PO(OCH_2Ph)(NHC_6H_{11}$-cycl.). XVI.7. m. 119-120°.[68]

$C_6H_5PO(OCH_2Ph)$(NHCONHBu). XVI.9. m. 125-126°.[517] (by reaction of $C_6H_5PO(OCH_2Ph)NH_2$ with C_4H_9NCO).

$C_6H_5PO(NH_2)_2$. XVII.1. m. 188°, solubility in H_2O, EtOH, Et_2O, $CHCl_3$, CCl_4.[1931] m. 189°.[349, 1260]

$C_6H_5PO(NH_2)$(NHAr).

Ar = Ph. XVII.1. m. 170-171°.[2101]

Ar = 2-$CH_3C_6H_4$-. XVII.1. m. 161-163°.[2101]

Ar = 4-BrC_6H_4-. XVII.1. m. 190-191°.[2101]

Ar = -$C_6H_3Cl_2$-2,4. XVII.1. m. 160-161°.[2101]

Ar = -$C_6H_2Cl_3$-2,4,6. XVII.1. m. 174-175°.[2101]

$C_6H_5PO(NHR)_2$.

R = Et. XVII.1. m. 72-74°.[733]

R = Bu. XVII.1. m. 53-55°.[733]

R = t-Bu. XVII.1. m. 188-189°.[733]

R = C_5H_{11}-. XVII.1. m. 48-49°.[733]

R = C_7H_{15}-. XVII.1. m. 44-46°.[733]

R = Ph. XVII.1. m. 211-212°.[733] XVII.2. m. 211°.[1282] XVII.1. m. 211°.[1260, 1269] m. 211.5-213°.[681] X-ray powder data. XVII.7. m. 211.5°.[1339a] m. 213-214°.[1242]

R = 2-MeC_6H_4-. XVII.1. m. 174-177°.[733]

R = 3-MeC_6H_4-. XVII.1. m. 168-169°.[733]

R = 4-MeC_6H_4-. XVII.1. m. 219-220°.[733] XVII.2. m. 220°.[1282]

R = 2-ClC_6H_4-. XVII.1. m. 192-193°.[733]

R = 3-ClC_6H_4-. XVII.1. m. 168-169°.[733]

R = 4-ClC_6H_4-. XVII.1. m. 212-218°.[733]

R = -$C_6H_3Cl_2$-2,4. XVII.1. m. 167-168°.[2101]

R = 4-$EtO_2CC_6H_4$-. XVII.1. m. 185-191°.[733]

R = 4-$O_2NC_6H_4$-. XVII.1. m. 220-221°.[2101]

R = 1-$C_{10}H_7$-. XVII.1. m. 249-251°.[733]

R = 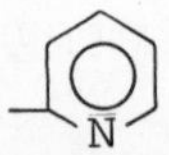 . XVII.1. m. 210-220°.[733]

R = cycl.-C_6H_{11}. XVII.1. m. 167-168°,[733] m. 166-167°, X-ray powder data.[681]

$C_6H_5PO(NHEt)(NHPh)$. XVII.1. m. 132-134°.[2101]

$C_6H_5PO(NHPh)(NHAr)$.

Ar = 2-MeC_6H_4-. XVII.1. m. 199-200°.[2101]

Ar = 4-MeC_6H_4-. XVII.1. m. 193-194°.[2101] XVII.2. m. 200°.[1282]

Ar = 4-BrC_6H_4-. XVII.2. m. 178-180°.[2101]

Ar = -$C_6H_2Br_3$-2,4,6. XVII.1. m. 218-219°.[2101]

Ar = 2-ClC_6H_4-. XVII.1. m. 207-209°.[2101]

Ar = 3-ClC_6H_4-. XVII.1. m. 160-161°.[2101]

Ar = 4-ClC_6H_4-. XVII.1. m. 175-177°.[2101]

Ar = -$C_6H_3Cl_2$-2,4. XVII.1. m. 190-191°.[2101]

Ar = -$C_6H_2Cl_3$-2,4,6. XVII.1. m. 213-215°.[2101]

Ar = 4-$O_2NC_6H_4$-. XVII.1. m. 182-184°.[2101]

Ar = -$C_6H_3(NO_2)_2$-2,4. XVII.1. m. 210-212°.[2101]

Ar = -$C_6H_2NO_2$-4-Cl_2-2,6. XVII.1. m. 199-200°.[2101]

$C_6H_5PO(NHC_6H_4Br\text{-}4)(NHC_6H_4Cl\text{-}4)$. XVII.1. m. 207-209°.[2101]

$C_6H_5PO(NEt_2)(NHAr)$.

Ar = Ph. XVII.5. m. 123°.[790] XVII.1. m. 160-162°.[2101]

Ar = 2-$CH_3C_6H_4$-. XVII.1. m. 120-123°.[2101]

Ar = 3-ClC_6H_4-. XVII.1. m. 177-178°.[2101] $b_{0.2}$ 100°.[124]

Ar = $C_6H_2Cl_3$-2,4,6. XVII.1. m. 194-196°.[2101]

$C_6H_5PO(NEt_2)$(NCH=N-CH=CH). XVII.1. m. 55-57°.[2007]

$C_6H_5PO[N(Me)Ph]$(NCH=N-CH=CH). XVII.1. m. 132-139°.[681, 138]

$C_6H_5PO(NPh_2)$(NCH=N-CH=CH). XVII.1. m. 178-186°.[1386]

C_6H_5PO(NCH=N-CH=CH)$_2$. XVII.1. m. 98-100°.[680, 1386, 1447] m. 99-104°, ^{31}P -6 ppm, X-ray powder data.[681]

$C_6H_5PO(N{=}CPh_2)_2$. XVII.8. m. 143-146°.[809]

$C_6H_5PO[N(Me)Ph]_2$. XVII.10. m. 92-93.5°.[735] m. 100-101°, X-ray powder data.[681]

$C_6H_5PO(NHPh)[N(Me)Ph]$. XVII.10. m. 148-149°.[735]

$C_6H_5PO(NC_5H_{10})_2$. XVII.1. m. 68°.[1291]

$C_6H_5PO(NHPh)(NC_5H_{10})$. XVII.2. m. 212°.[1282]

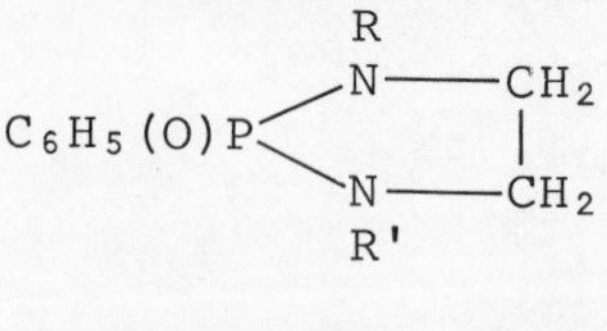

R = R' = Me. XVII.1. b_{20} 171°,[2096] $b_{0.7}$ 130°, ^{1}H.[2043]

R = Me, R' = i-Pr. XVII.1. $b_{0.03}$ 135-140°.[1792]

R = Me, R' = i-Bu. XVII.1. $b_{0.03}$ 140-142°.[1792]

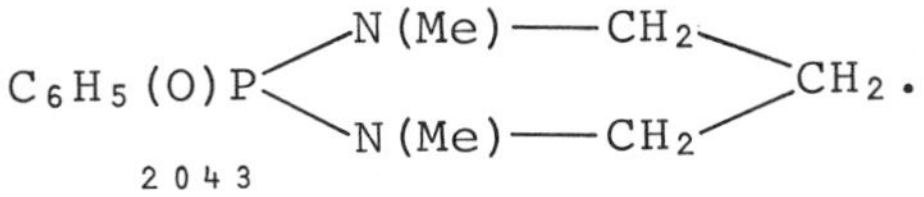

. XVII.1. $b_{0.3}$ 140-143°, ^{1}H.[2043]

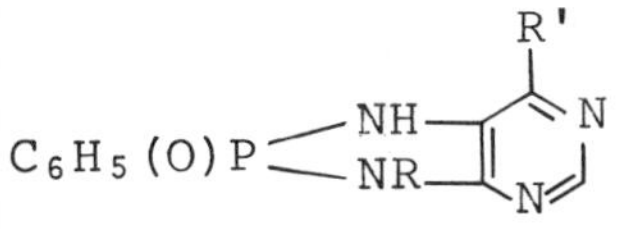

R = H, R' = Cl. XVII.1. m. >200° (dec.).[1104]

R = H, R' = $-NMe_2$. XVII.1. m. >200° (dec.).[1104]

R = Me, R' = $-NMe_2$. XVII.1. m. >200° (dec.).[1104]

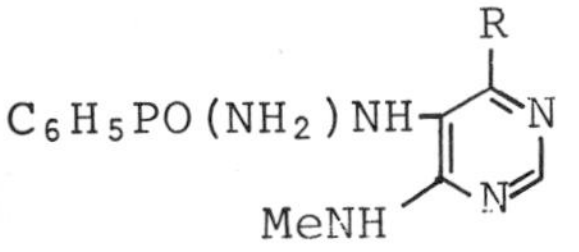

R = -OMe. XVII.1. m. >190° (dihydrate).[1105]

R = -NHMe. XVII.1. m. >170° (hexahydrate).[1105]

R = $-NH_2$. XVII.1. m. >175° (trihydrate).[1105]

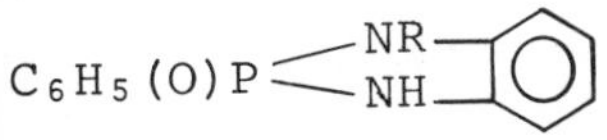

R = H. XVII.1. subl. >220° (10^{-3}mm), m. 257-270° (dec.),[731] m. 265-280° (dec.),[1947] m. 277-278°,[454] X-ray powder data.[681]

R = Ph. XVII.1. m. 214-216°.[454]

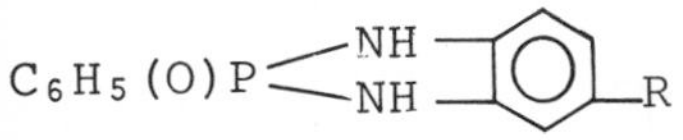

R = Cl. XVII.1. m. 233-235°.[454]

R = Me. XVII.1. m. 237-239°.[454]

R = $-CF_3$. XVII.1. m. 154-157°.[452]

R = -OMe. XVII.1. 2HCl adduct, m. 227.5-229°.[452]

R = -OEt. XVII.1. m. 2HCl adduct, m. 189-190.5°.[452]

R = $-CO_2Me$. XVII.1. m. 241-242°.[454]

R = $-NO_2$. XVII.1. m. 261-263°.[454]

R = $-NMe_2$. XVII.1. m. 196-197°.[452]

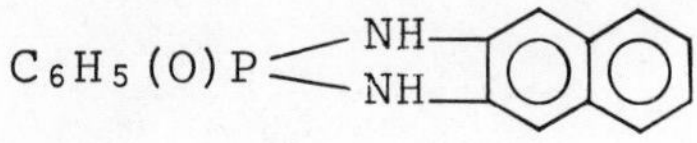

XVII.1. m. 278°.[565]

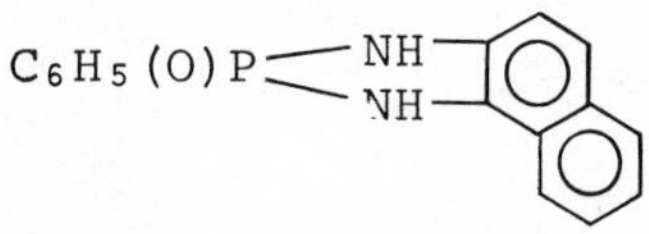

XVII.1. m. 319-322°.[454]

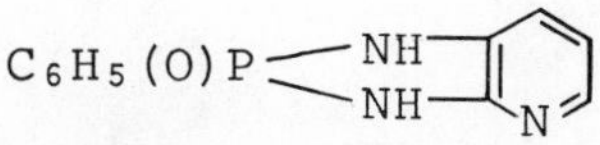

XVII.1. m. 299-301°.[454]

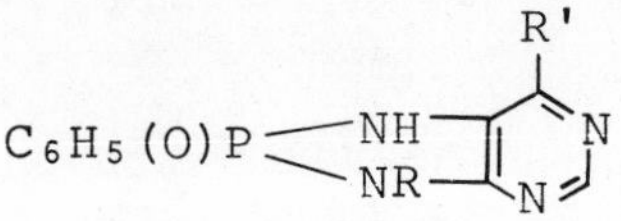

R = H, R' = $-NMe_2$. XVII.1. m. >250°.[1105]

R = H, R' = Cl. XVII.1. m. 200° (hemihydrate).[1105]

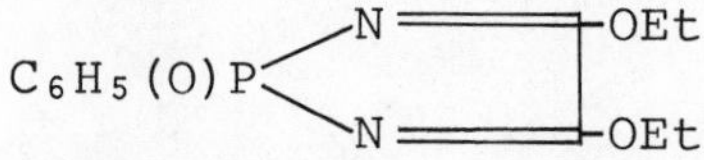

XVII.8. m. 95-97°.[466]

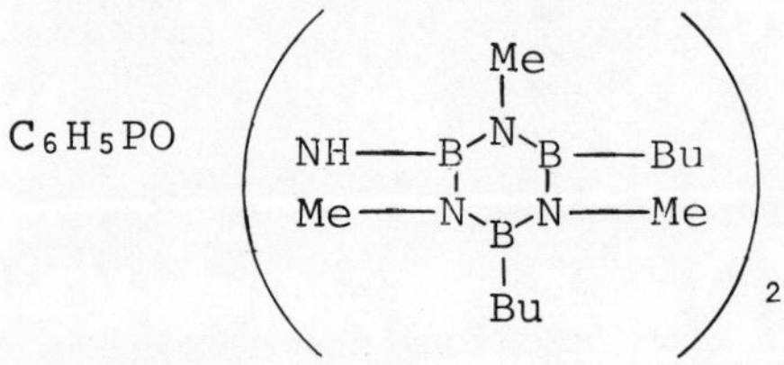

XVII.1. $b_{0.005}$ 330°, IR.[73]

$C_6H_5PO(NH_2)(NHCOAr)$.

Ar = $3-O_2NC_6H_4-$. XVII.1. m. 170-171°.[464]

Ar = $4-O_2NC_6H_4-$. XVII.1. m. 167-169°.[464]

$C_6H_5PO(NHPh)(NHCOPh)$. XVII.1. m. 163-165°.[465]

$C_6H_5PO(NHPh)(NHCOAr)$.

Ar = 4-BrC_6H_4-. XVII.1. m. 177-178°.[464]

Ar = 4-ClC_6H_4-. XVII.1. m. 182-183°.[464]

Ar = 3-$O_2NC_6H_4$-. XVII.1. m. 191-192°.[464]

Ar = 4-$O_2NC_6H_4$-. XVII.1. m. 208-209°.[464]

$C_6H_5PO(NHPh)(NHCOOR)$.

R = Me. XVII.1. m. 181-182°.[1919]

R = Et. XVII.1. m. 157-158°.[1919]

$C_6H_5PO(NHCONHR)_2$.

R = H. XVII.6. m. 230-236°, IR.[679]

R = Me. XVII.6. m. 199°.[2011]

R = Et. XVII.6. m. 194°.[2011]

R = i-Pr. XVII.6. m. 193°.[2011]

R = Bu. XVII.6. m. 221-223°,[2011] 190-193°, IR.[679]

R = t-Bu. XVII.6. m. 151-156°, IR.[679]

R = cycl.-C_6H_{11}-. XVII.6. m. 187-188°.[2013]

R = Ph. XVII.6. m. 193-195°, IR,[679] m. 175-178°.[762]

$PHPO(NHCO_2R)_2$.

R = Me. XVII.6. m. 141-149°, IR,[679] m. 161-163°.[762]

R = Et. XVII.6. m. 132-139°, IR.[679]

R = Pr. XVII.6. m. 120-128°, IR.[679]

$C_6H_5PO(NHPh)[NHCOC(CN)=CR_2]$.

R = Me. XVII.5. m. 204-205°.[1890]

R = Et. XVII.5. m. 196-197°.[1890]

R = Pr. XVII.5. m. 208-209°.[1890]

R = Bu. XVII.5. m. 216-217°.[1890]

R = -C_5H_{11}. XVII.5. m. 209-210°.[1890]

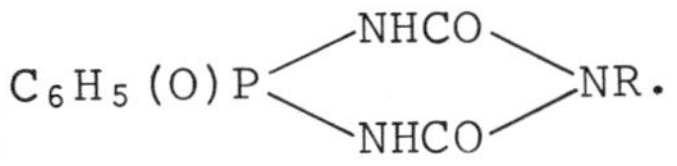

R = H. XVII.6. m. 250°.[2011]

R = Me. XVII.6. m. 236.5°.[2011]

R = Et. XVII.6. m. 225-226°.[2011]

R = i-Pr. XVII.6. m. 202-203°.[2011]

R = Bu. XVII.6. m. 224-225°,[2011] m. 221-223°.[2010]

R = Ph. XVII.6. m. 202-203°.[2011]

$C_6H_5PO(NHCONEt_2)_2$. XVII.6. m. 140-145°, IR.[679]
$C_6H_5PO(NHPh)(NHSO_2Ar)$.

Ar = Ph. XVII.5. m. 167°.[1894]

Ar = $2-CH_3C_6H_4-$. XVII.5. m. 164°.[1894]

Ar = $4-CH_3C_6H_4-$. XVII.5. m. 160°.[1894]

Ar = $2-O_2NC_6H_4-$. XVII.5. m. 165°.[1894]

Ar = $3-O_2NC_6H_4-$. XVII.5. m. 169°.[1894]

Ar = $4-O_2NC_6H_4-$. XVII.5. m. 166°.[1894]

Ar = $1-C_{10}H_7-$. XVII.5. m. 160°.[1894]

Ar = $2-C_{10}H_7-$. XVII.5. m. 164°.[1894]

$C_6H_5PO(N=PPh_3)_2$. XVII.9. m. 192-193°, IR.[202]
$C_6H_5PO(NRR')(N=PPh_3)$.

R = R' = Me. XVII.9. m. 126.5-129°, IR.[202]

R = R' = Et. XVII.9. m. 119-120.5°, IR.[202]

R = H, R' = Ph. XVII.9. m. 208-210°, IR.[202]

$C_6H_5PO(NH_2)(NHNH_2)$. XVIII.3. m. 143°.[1931]
$C_6H_5PO(NH_2)(NHN=CHC_6H_4OMe-4)$. XVIII.3. m. 163°.[1931]
$C_6H_5PO(NEt_2)(NHNHPh)$. XVIII.3. m. 138°.[459]
$C_6H_5PO(NHNH_2)_2$. XVIII.1. m. 131°.[1281, 1930]
$C_6H_5PO(NHNHPh)_2$. XVIII.1. m. 175°.[1260]

$$C_6H_5(O)P \underset{NHN=C}{\overset{NHN=C}{\diagdown}} \underset{Ph}{\overset{Ph}{\diagup}} O$$

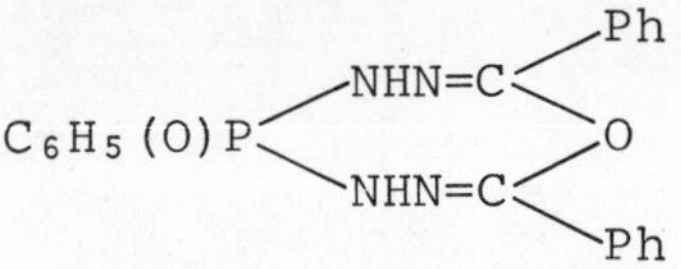

XVIII.1. m. 245-246°, IR.[1240]

$C_6H_5PO(NMe_2)N_3$. XVIII.4. $b_{0.001}$ 86-88°, n^{20} 1.5473, IR.[202]
$C_6H_5PO(N_3)_2$. XVIII.4. $b_{0.1}$ 72-74°, IR.[202D]
$C_6H_5PO(N_3)(N=PPh_3)$. XVIII.4. m. 143-145°.[202]

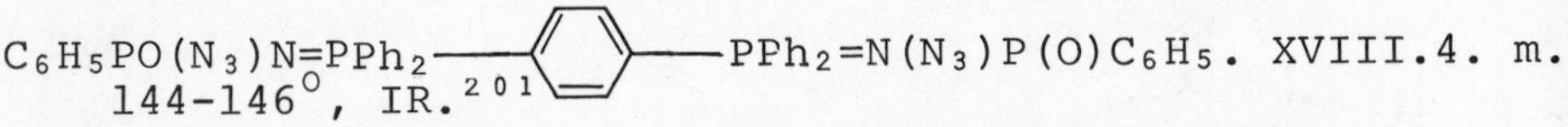

$C_6H_5PO(N_3)N=PPh_2$——⟨ ⟩——$PPh_2=N(N_3)P(O)C_6H_5$. XVIII.4. m. 144-146°, IR.[201]

$C_6H_5PO(N_3)N=PPh_2$——⟨ ⟩—⟨ ⟩——$PPh_2= N(N_3)P(O)C_6H_5$.

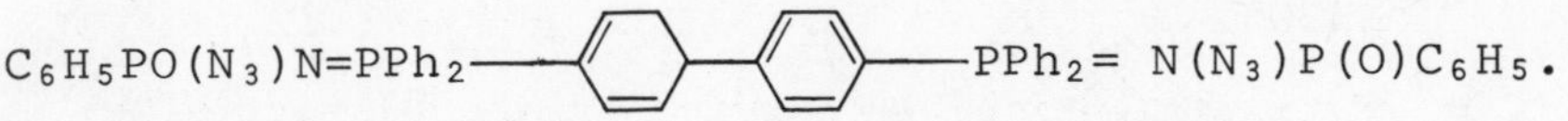

XVIII.4. m. 215-217°, IR.[201]
$4-H_2NC_6H_4PO(NHR)_2$.

R = Ph. XVII.10. m. 210-213°.[483]

R = $2-ClC_6H_4-$. XVII.10. m. 145-150° (hydrochloride).[483]

R = $3-ClC_6H_4-$. XVII.10. m. 148-151°.[483]

R = $4-ClC_6H_4-$. XVII.10. m. 202-205°.[483]

R = $3-BrC_6H_4-$. XVII.10. m. 154-157°.[483]

R = 4-MeC_6H_4-. XVII.10. m. 238.5-241.5°.[483]

R = 4-$EtO_2SC_6H_4$-. XVII.10. m. 247-248°.[483]

R = 3-$F_3CC_6H_3$-. XVII.10. m. 157-158°.[483]

R = 3-$EtOC_6H_4$-. XVII.10. m. 146-146.5°.[483]

R = 4-$EtO_2CC_6H_4$-. XVII.10. m. 124-127°.[483]

R = 4-$HO_2CC_6H_4$-. XVII.10. m. 182-185°.[483]

R = 2-$C_{10}H_7$-. XVII.10. m. 210-213°.[483]

R = 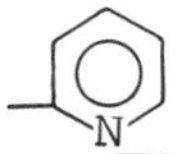 . XVII.10. m. 209-210°.[483]

4-$H_2NC_6H_4PO[N(CH_2)_3CH_2]_2$. XVII.9. m. 191-193°.[572]
4-$H_2NC_6H_4PO[N(CH_2)_4CH_2]_2$. XVII.9. m. 198-201°.[572]
$C_6H_{11}PO(NEt_2)Cl$. XV.1. m. 52°.[2007]
$C_6H_{11}PO(NH_2)F$. XV.1. m. 112°, IR, ^1H, ^{19}F, mass spect.[1761]
$C_6H_{11}PO[O(CH_2)_3OH](NEt_2)$. XVI.7. m. 42-45°.[1393]
$C_6H_{11}PO[O(CH_2)_3OH](NCH_2CH_2)$. XVI.7. $b_{0.001}$ 120-135°, n_D^{20} 1.4782.[1393]
$C_6H_{11}PO(NMe_2)_2$. XVII.1. $b_{0.45}$ 118°, IR, mass spect.[881]
$C_6H_{11}PO(NEt_2)(NCH=CH-N=CH)$. XVII.1. m. 65-67°, ^1H, IR.[2007]

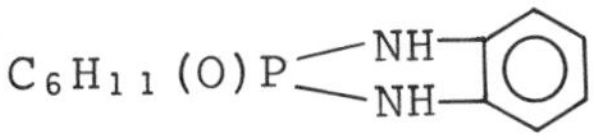 XVII.1. m. 233-235°.[454]

$C_4H_9OCH=CHPO(OBu)(NEt_2)$. XVI.1. $b_{1.0}$ 154°, n_D^{20} 1.4597, d_4^{20} 0.9691.[1501]
$(CH_2)_5C(OH)PO(OPr$-i$)(NMe_2)$. XVI.4. m. 140-141°.[28]
$C_4H_9OCH=CHPO(NR_2)_2$.

R = Me. XVII.1. b_2 136°, n_D^{20} 1.4750, d_4^{20} 0.9917.[1501]

R = Et. XVII.1. b_1 151°, n_D^{20} 1.4752, d_4^{20} 0.9880.[1501]

$CH_2=CHCH[C(CH_3)_2OH]PO(NMe_2)_2$. XVII.9. m. 52-56°, ^1H, IR.[438]
$(CH_3)_2C(OH)CH_2CH=CHPO(NMe_2)_2$. XVII.9. Oil, ^1H, IR, yields $CH_2=CHCH=CMe_2$ on thermolysis.[438]
$CH_2=CHCH[C(CH_3)_2(OH)](O)PN(Me)CH_2CH_2NMe$. XVII.9. Oil, ^1H, IR.[438]
$C_2H_5O_2CCH(CH_3)CH_2PO(OEt)(NEt_2)$. XVI.7. b_{10} 159-160°, n_D^{20} 1.4400, d_4^{20} 1.0089.[1635]
$C_2H_5O_2CCH_2CH_2CH_2PO(OEt)(NEt_2)$. XVI.4. $b_{1.5}$ 135-138°.[425]
$C_4H_9O_2CCH_2PO(OCH_2CH_2Cl)(NEt_2)$. XVI.3. b_2 153-153.5°, n_D^{20} 1.4660, d_4^{20} 1.1244, IR.[19]
$C_4H_9O_2CCH_2PO(OCHMeCH_2Cl)(NEt_2)$. XVI.3. b_3 154-156°, n_D^{20} 1.4655, d_4^{20} 1.1076, IR.[20]
i-$C_4H_9O_2CCH_2PO(OCH_2CH_2Cl)(NEt_2)$. XVI.3. b_2 149-150°, n_D^{20} 1.4652, d_4^{20} 1.1223, IR.[19]
i-$C_4H_9O_2CCH_2PO(OCHMeCH_2Cl)(NEt_2)$. XVI.3. b_3 150-151.5°, n_D^{20} 1.4644, d_4^{20} 1.1032, IR.[20]
$C_2H_5OCH_2CH_2OCH=CHPO(NBu_2)_2$. XVII.1. b_1 180°.[85]
$C_2H_5O_2CCH_2CH_2CH_2PO(NEt_2)_2$. XVII.3. $b_{0.04}$ 136-140°, n_D^{20}

1.4648.[516]

$(CH_2)_5NCH_2PO(OEt)[N(CH_2)_5]$. XVI.8. $b_{0.01}$ 125°.[284] (by reaction of $(EtO)_2POCOPh$ with $(CH_2)_5NCH_2N(CH_2)_5$).

$(C_2H_5)_2NCOCH_2PO(OEt)(NEt_2)$. XVI.3. $b_{0.5-1.7}$ 158-160° (dec.) n_D^{20} 1.4630, d_4^{20} 1.0446.[1238]

$(C_2H_5)_2NCOCH_2PO(NEt_2)_2$. XVII.3. $b_{0.4}$ 139-142°, n_D^{20} 1.4775, d_4^{20} 1.0464.[1238]

$C_2H_5OCH(CH_2OC_2H_5)PO(NMe_2)_2$. XVII.9. $b_{0.2}$ 105-106°, ^1H.[1081]

$\underline{C_7}$

$4\text{-}ClC_6H_4NHCOPO(NEt_2)_2$. XVII.9. m. 127-130°[1391] (by reaction of $(Et_2N)_2POH$ and $4\text{-}ClC_6H_4NCO$).

$C_6H_5CHClPO(NHPh)_2$. XVII.1. m. 192-194°.[924]

$3\text{-}ClC_6H_4\text{-}NHCH_2PO(NMe_2)(NHC_6H_4Cl\text{-}3)$. XVII.1. m. 93-95° (hydrochloride) (by reaction of $CH_2ClPO(NMe_2)Cl$ and $H_2NC_6H_4Cl\text{-}3$).[1242]

$C_3H_7OCH{=}C(CCl{=}CH_2)PO(NEt_2)_2$. XVII.1. $b_{0.006}$ 104-105°, n_D^{20} 1.5110, d_4^{20} 1.0652.[1920]

$C_3H_7SCH{=}C(CCl{=}CH_2)PO(NEt_2)_2$. XVII.1. $b_{0.005}$ 112°, n_D^{20} 1.5516, d_4^{20} 1.0916.[1918a]

$C_3H_7SCH{=}C(CCl{=}CH_2)PO(NCH_2CH_2)_2$. XVII.1. $b_{0.15}$ 148°, n_D^{20} 1.5996, d_4^{20} 1.2550.[2047]

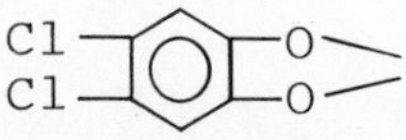
CHPO(OEt)(NEt_2). XVI.1. m. 98-102°, mass spect., ^1H.[715]

$C_6H_5COPO(OMe)(NMe_2)$. XVI.3. $b_{0.1}$ 79-80°, n_D^{20} 1.4555, d_4^{20} 1.0744.[61]

$C_6H_5COPO(OMe)(NEt_2)$. XVI.3. $b_{1.5}$ 147°, n_D^{20} 1.530, d_4^{20} 1.1322.[61]

$C_6H_5COPO(OPr)(NMeSO_2Bu)$. XVI.3. m. 106-107°.[23]

$C_6H_5COPO(OBu)(NMeSO_2Ph)$. XVI.3. m. 95-96°.[23]

$C_6H_5COPO(NEt_2)_2$. XVII.3. b_1 152-153°, b_2 159-160°, n_D^{20} 1.5390, d_4^{20} 1.0898.[61]

$C_6H_5NHCOPO(NEt_2)_2$. XVII.9. m. 112-114°.[1391] (by reaction of $(Et_2N)_2POH$ and PhNCO).

$4\text{-}CH_3C_6H_4PO(NHAr)OH$.

Ar = Ph. XIV.1. m. 150°.[1265]

Ar = $4\text{-}CH_3C_6H_4\text{-}$. XIV.1. m. 208°.[1265]

$4\text{-}CH_3C_6H_4PO(NHCONHC_4H_9)OH$. XIV.1. Isolated as Na salt, m. >310°.[517]

$4\text{-}CH_3C_6H_4PO[N(Ph)COCH_3]OH$. XIV.1. m. 162°.[1282]

$4\text{-}CH_3C_6H_4PO(OMe)(NHPh)$. XVI.2. m. 65°.[1282]

$4\text{-}CH_3C_6H_4PO(OEt)(NHC_6H_4CH_3\text{-}4)$. XVI.2. m. 53°.[1282]

$4\text{-}CH_3C_6H_4PO(OPh)(NH_2)$. XVI.1. m. 115-116°.[1265]

$4\text{-}CH_3C_6H_4PO(OPh)(NHPh)$. XVI.1. m. 59°, b_{48} 238°.[1265]

$4\text{-}CH_3C_6H_4PO(OCH_2Ph)(NH_2)$. XVI.1. m. 120-124°.[517]

$4\text{-}CH_3C_6H_4PO(OCH_2Ph)(NHCONHBu)$. XVI.9. m. 132-135°.[517]

$4\text{-}CH_3C_6H_4PO(NH_2)_2$. XVII.1. m. 176°.[1260, 1265, 1269]

$4\text{-}CH_3C_6H_4PO(NHPh)_2$. XVII.1. m. 209°.[1260, 1265, 1269] 207.5-

208.5°.[1339a] XVII.2. m. 209°.[1282]
$2\text{-}CH_3C_6H_4PO(NHPh)_2$. XVII.1. m. 230.5°.[1339a]
$4\text{-}CH_3C_6H_4PO(NHC_6H_4CH_3\text{-}4)_2$. XVII.2. m. 237°.[1265, 1282]
$4\text{-}CH_3C_6H_4PO(NC_5H_{10})_2$. XVII.1. m. 60°.[1291]
$4\text{-}CH_3C_6H_4PO(NHCONHR)_2$.

R = Bu. XVII.6. m. 182-183°.[2013]

R = cycl.-C_6H_{11}-. XVII.6. m. 191-192°.[2013]

R = Ph. XVII.6. m. 186-187°.[2013]

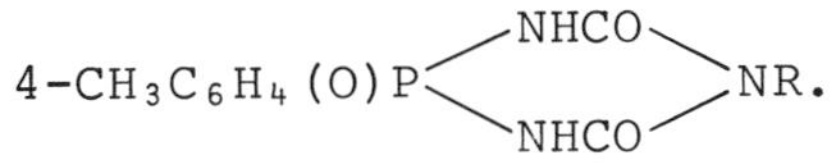

R = H. XVII.6. m. 244-245°.[2011]

R = Me. XVII.6. m. 226-227°.[2011]

$C_6H_5CH_2PO(NHCONHPh)_2$. XVII.6. m. 188-193°.[762]
$C_6H_5CH_2PO(NHCOOMe)_2$. XVII.6. m. 179-181°.[762]

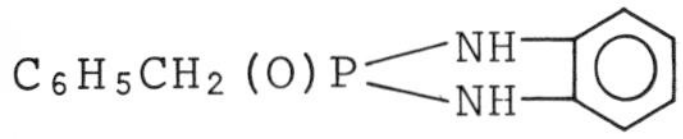

XVII.1. m. 248-254°.[1947]

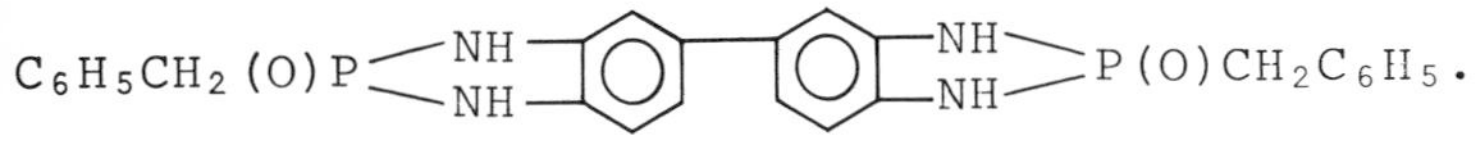

XVII.1. m. 368-378°.[1947]
$4\text{-}CH_3C_6H_4PO(OPh)(NHNHPh)$. XVIII.2. m. 173-174°.[1265]
$4\text{-}CH_3C_6H_4PO(NHNHPh)_2$. XVIII.2. m. 171°.[1265]
$C_6H_5CH(OH)PO(OEt)(NRSO_2R')$.

R = Pr, R' = Me. XVI.4. b_5 110-120°, n_D^{20} 1.4470, d_4^{20} 1.1320.[724]

R = R' = Ph. XVI.4. m. 108-109°.[724]

$C_6H_5CH(OH)PO(OPr\text{-}i)(NMe_2)$. XVI.4. m. 202-203°.[28]
$C_6H_5CH(OH)PO(OBu)(NRSO_2R')$.

R = Me, R' = Ph. XVI.4. b_5 110-120°, n_D^{20} 1.5211, d_4^{20} 1.2050.[724]

R = R' = Ph. XVI.4. m. 107-108°.[724]

$4\text{-}CH_3OC_6H_4PO(NHPh)_2$. XVII.1. m. 210°.[1339a]
$4\text{-}CH_3OC_6H_4PO(NHCONHR)_2$.

R = Ph. XVII.6. m. 183-184°.[2013]

R = cycl.-C_6H_{11}-. XVII.6. m. 187-188°.[2013]

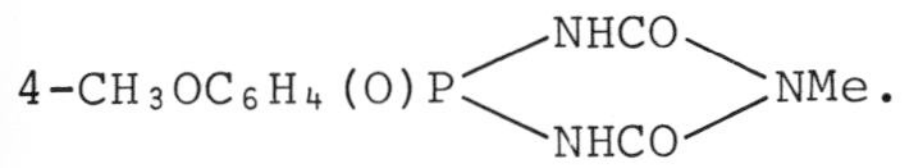

XVII.6. m. 238-239°.[2011]

$C_6H_5C(CH_3)(OH)PO(OMe)(NMe_2)$. XVI.4. m. 174-175°.[28]
$CH_2{=}CH{-}CH[C(CH_3)(C_2H_5)(OH)](O)PNMeCH_2CH_2NMe$. XVII.10. Oil, ^1H, IR.[438]
$C_4H_9O_2CCH_2CH_2PO(OEt)(NEt_2)$. XVI.4.[425]

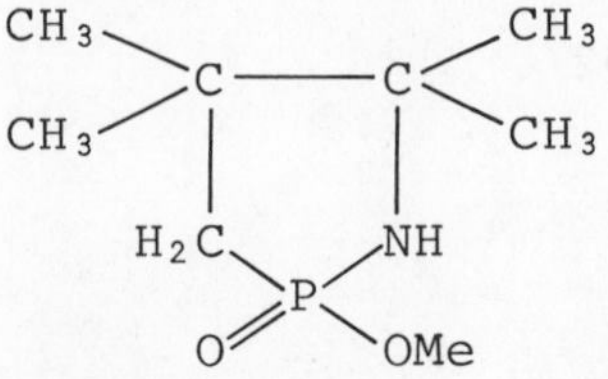

XVI.8. m. 65-66°, IR, mass spect.[750]

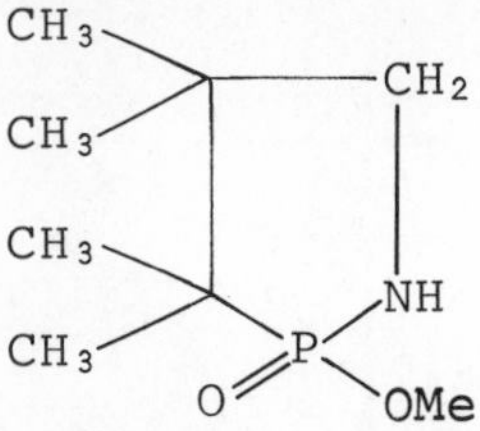

XVI.8. m. 104-106°, IR, mass spect.[750]

(Both obtained by photolysis of 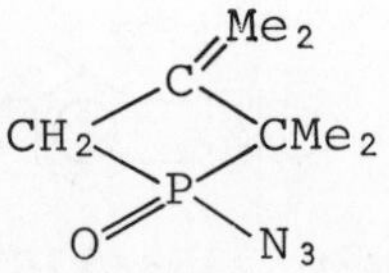 in methanol.)

$(C_2H_5)_2NCOCH_2CH_2PO(OEt)(NEt_2)$. XVI.4. $b_{0.004}$ 130-131°, n_D^{20} 1.4678, d_4^{20} 1.0351, IR.[1657]
$C_7H_{15}PO(NMe_2)_2$. XVII.7. $b_{0.4}$ 113°, n_D^{22} 1.4601, d^{25} 0.950.[1406]
$(C_2H_5)_2NCH_2CH_2CH_2PO(NEt_2)_2$. XVII.1. $b_{0.04}$ 140-141°, n_D^{20} 1.4678, d_4^{20} 0.9472, IR.[57]

C_8

$C_4H_9OCH{=}C(CCl{=}CH_2)PO(NEt_2)_2$. XVII.1. $b_{0.008}$ 108-109°, n_D^{20} 1.5100, d_4^{20} 1.0532.[1920]
$C_4H_9SCH{=}C(CCl{=}CH_2)PO(NEt_2)_2$. XVII.1. $b_{0.006}$ 118°, n_D^{20} 1.5482, d_4^{20} 1.0690.[1918a]
$C_4H_9SCH{=}C(CCl{=}CH_2)PO(NCH_2CH_2)_2$. XVII.1. $b_{0.15}$ 155°, n_D^{20} 1.5945, d_4^{20} 1.2389.[2047]
$C_6H_5NHCOCCl_2PO(NHPh)_2$. XVII.1. m. 153-155°[1887] (by reaction of $ClCOCCl_2POCl_2$ with $PhNH_2$).
$C_6H_5CH{=}CHPO(NEt_2)_2$. XVII.1. m. 103.5°.[93]
$C_6H_5CH{=}CHPO(NCH_2CH_2)_2$. XVII.1. m. 47-48°.[67]
$C_6H_5OCH{=}CHPO[N(CH_2)_4CH_2]_2$. XVII.1. b_1 202°.[82]
$C_6H_5O_2CCH_2PO(OEt)(NEt_2)$. XVI.3. $b_{0.35-0.5}$ 162-166°, n_D^{20} 1.4935.[1238]
$C_6H_5O_2CCH_2PO(NEt_2)_2$. XVII.3. $b_{0.25-0.4}$ 160-162°, n_D^{20} 1.4972, d_4^{20} 1.0552.[1238]

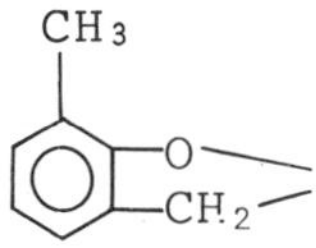 $PO(NEt_2)$. XVI.8. $b_{0.04}$ 130°, n_D^{20} 1.5270, d_4^{20} 1.1328.[833]

(By reaction of $EtOP(NEt_2)_2$ with 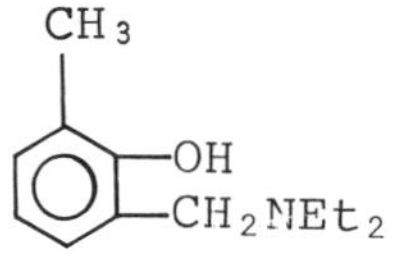)

$CH_3C(C_6H_5)(OH)PO(OMe)(NMe_2)$. XVI.4. m. 174-175°.[28]
$C_6H_5C(OH)CH_2PO(NMe_2)_2$. XVII.9. m. 87-88°, ^1H.[439]
$CH_2{=}C(CH_3)CH(CH_3)C(CH_3)_2PO(OMe)(NH_2)$. XVI.8. m. 107-115°, IR, ^1H, mass spect.[750]

(By photolysis of 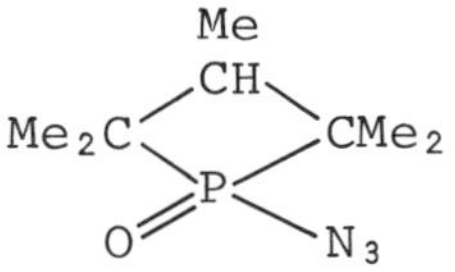 in MeOH.)

The cyclic compound 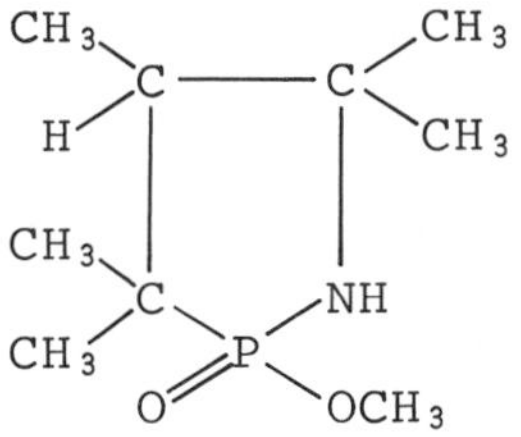 , m. 137-138°, IR, ^1H, mass spect.

is also obtained.
$C_8H_{17}PO(OEt)(NHCONC_6H_4Cl\text{-}4)$. XVI.6. m. 92-94°.[726]

$\underline{C_9}$
$C_7F_{15}CH_2CH_2PO(\overline{NCH_2CH_2})_2$. XVII.1. m. 49-50°, ^1H, IR.[1335]
$CH_3OCH{=}C(C_6H_5)PO(NHR)_2$.

R = Me. XVII.1. m. 101-102°.[1882]

R = Et. XVII.1. m. 119-120°.[1882]

R = Pr. XVII.1. m. 114-116°.[1882]

R = i-Pr. XVII.1. m. 88-89°.[1882]

R = Ph. XVII.1. m. 203-204°.[1882]

R = 4-$CH_3C_6H_4$-. XVII.1. m. 173-174°.[1882]

R = 4-$CH_3OC_6H_4$-. XVII.1. m. 173-174°.[1882]

$C_6H_5CH(OH)CH(CH_3)PO(NMe_2)_2$. XVII.9. m. 80.5-85°, diastereomers, ^1H.[439]
$C_6H_5CH[N(CH_3)_2]PO(NMe_2)_2$. XVII.4. m. 115-116°.[1390]

(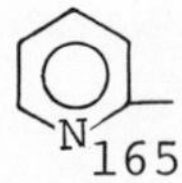)$\overline{N(CH_2)_3CH_2P}O[N($ 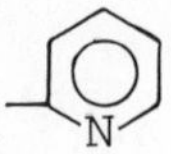 $)(CONHPh)]$. XVII.9. m. 160-165°.[768]

$\underline{C_{10}}$
$(4-ClC_6H_4)\overline{N(CH_2)_3CH_2P}O[N(C_6H_4Cl-4)(COC_6H_5)]$. XVII.9. m. 186-190°.[768] $2-C_{10}H_7PO(OMe)(NH_2)$. XVI.7. m. 103-105°.[801]

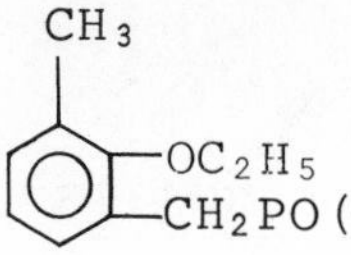
$CH_2PO(OEt)(NEt_2)$. XVI.8. $b_{0.35}$ 150-155°, n_D^{20} 1.5240, d_4^{20} 1.1258.[833]

(By reaction of $(EtO)_2PNEt_2$ with 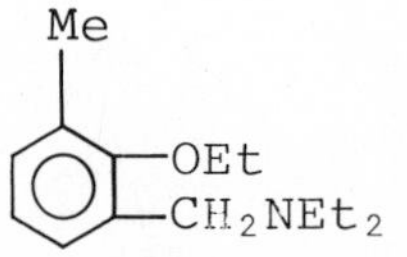 at 150-160°

$C_6H_5C(CH_3)(OH)CH(CH_3)PO(NMe_2)_2$. XVII.9. m. 99-111°, diastereomers, ^1H.[439]
cycl.-$C_6H_{11}O_2CCH_2CH_2CH_2PO(NEt_2)_2$. XVII.3. $b_{0.005}$ 137°, n_D^{2} 1.4764.[516]
$C_5H_{11}CH(C_2H_5)OCH=CHPO(NMe_2)_2$. XVII.1. b_2 171-172°.[94]

$\underline{C_{11}}$
$CH_2=CHCH[C(CH_3)(C_6H_5)(OH)](O)\overline{PNMeCH_2CH_2N}Me$. XVII.10. m. 157-157.5°, ^1H, IR.[438]
$(C_6H_5)(C_2H_5)NCOCH_2CH_2PO(NEt_2)_2$. XVII.4.[1665]
$C_6H_5CH[N(C_2H_5)_2]PO(NEt_2)_2$. XVII.4. m. 110-113°.[1390]

$\underline{C_{12}}$
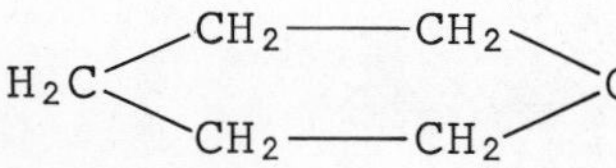
$C(OC_6H_5)PO(OPh)(NHPh)$. XVI.4. m. 157-158°.[25]

$\underline{C_{13}}$
$4-ClC_6H_4NHCH(C_6H_4Cl-4)PO(NHC_6H_4Cl-4)_2$. XVII.4. m. 129-133°.[1248]
$4-C_6H_5NHCOC_6H_4PO(NHPh)_2$. XVII.1. m. 242°.[1265]
$C_6H_5OCH(C_6H_5)PO(OPh)(NHPh)$. XVI.4. m. 149-150°.[25]
$C_6H_5NHCH(C_6H_5)PO(NHPh)_2$. XVII.4. m. 165-168°.[1248]

$\underline{C_{14}}$
$C_6H_5CH_2CH(C_6H_5)PO(NMe_2)_2$. XVII.7. m. 127-128°.[1409]
$(C_6H_5)_2C(OH)CH_2PO(NMe_2)_2$. XVII.9. m. 155-157°.[439]

$\underline{C_{15}}$

$C_6H_5COCH_2CH(C_6H_5)PO(NMe_2)_2$. XVII.7. m. 157°.[1407] XVII.4. m. 159-160°.[1248]

$(C_6H_5CH_2)_2CHPO(NH_2)OH$. XIV.1. m. 244°.[1281]

$(C_6H_5CH_2)_2CHPO(NHPh)_2$. XVII.1. m. 196°.[1281]

$(C_6H_5CH_2)_2CHPO(NHNHPh)_2$. XVIII.1. m. 164°.[1281]

$\underline{C_{16}}$

$C_6H_5COCH_2CH(COC_6H_5)PO(NRR')_2$.

R = R' = Me. XVII.4. m. 159-160°.[1248]

R = H, R' = Ph. XVII.4. m. 162-164°.[1248]

R = H, R' = 4-ClC_6H_4-. XVII.4. m. 158-160°.[1248]

$(C_6H_5)_2C(OH)C(CH_3)_2PO(NMe_2)_2$. XVII.10. m. 130-131°.[439]

$\underline{C_{20}}$

$(C_6H_5)_2POCH_2NHCH(C_6H_5)PO(NHAr)_2$.

Ar = Ph. XVII.4. m. 200-201°.[1248]

Ar = 4-ClC_6H_4-. XVII.4. m. 193-194°.[1248]

C.8. Thiophosphonic and Selenophosphonic Acid Amide Derivatvies

$\underline{C_1}$

$CClH_2PS(NHCONHC_6H_4CO_2Et\text{-}4)F$. XX.3. m. 167-169°.[855]

$CClH_2PS[NHCOCH{=}C(Ph)(NC_5H_{10})]F$. XX.3. m. 95-97°.[855]

$CClH_2PS(NHR)(OC_6H_2Cl_3\text{-}2,4,5)$.

R = i-Pr. XXI.2. $b_{0.16}$ 155°, m. 41-42°.[665]

R = i-Bu. XXI.2. $b_{0.15}$ 163-164°, m. 31-34°.[665]

$CClH_2PS(OC_6H_{11})(NMe_2)$. XXI.2. $b_{0.06}$ 105°.[1722]

$CClH_2PO(SC_6H_4NO_2\text{-}4)[N(CH_2CH_2Cl)_2]$. XXII.2. m. 95-96°.[1724]

$CClH_2PS(\overline{NCH_2CH_2})_2$. XXIII.1. $b_{0.15}$ 84°, n_D^{20} 1.5565, d_4^{20} 1.3048.[1478]

$CClH_2PS(NHPh)_2$. XXIII.1. m. 118-118.5°.[893]

$CClH_2PS(NHCONHC_6H_4CO_2Et\text{-}4)_2$. XXIII.3. m. 132-134°.[855]

$CClH_2(S)P<(NHCO)(NHCO)>C{=}C(Ph)(NC_5H_{10})$. XXIII.3. m. 185°.[855]

$CClH_2PS(SAr)[N(CH_2CH_2Cl)_2]$.

Ar = 4-ClC_6H_4-. XXV.1. m. 95-96°.[1724]

Ar = 2-MeC_6H_4-. XXV.1. m. 77-78°.[1724]

Ar = 4-MeC_6H_4-. XXV.1. m. 79-80°.[1724]

$CF_3PS(NMe_2)F$. XX.2. IR, ^1H, ^{19}F, vapor pressure, mass spect.[485]

$CF_3PS(NMe_2)_2$. XXIII.1. IR, Trouton const. 22.7, Nernst const. A = 3278, B = 0.00580, C = 7.7610.[655]

$CH_3PS(NH_2)F$. XX.1. $b_{0.02}$ 44-46°, n_D^{20} 1.5110, d_4^{20} 1.3299, IR, ^1H, ^{19}F.[1757]
$CH_3PS(NMe_2)Br$. XX.1. b_{10} 120-125°, m. 40 , IR.[1151]
$CH_3PS(NHBu\text{-}i)Cl$. XX.1. $b_{0.15}$ 96-96.5°.[665]
$CH_3PS(NR_2)Cl$.

R = Me. XX.4. b_{15} 103-105°, n_D^{20} 1.2488, d_4^{20} 1.5381.[646]

R = Et. XX.4. b_2 68-69°, n_D^{20} 1.5242, d_4^{20} 1.1450.[646]

XX.1. b_8 107-109°, n_D^{20} 1.5252, d_4^{20} 1.1612.[1372]

$CH_3PS(NHMe)F$. XX.1. b_1 58°.[1843]
$CH_3PS(NR_2)F$.

R = Me. XX.1. b_2 50°.[1843]

R = Et. XX.1. b_2 71°.[1843]

$CH_3PS(NC_5H_{10})F$. XX.1. b_1 78°.[1843]

$CH_3PS(N\langle^{CH_2CH_2}_{CH_2CH_2}\rangle O)F$. XX.1. b_1 88°.[1843]

$CH_3PS(\overline{NCH_2CH_2})F$. XX.1. b_3 46°.[1843]
$CH_3PS[\overline{N(CH_2)_3CH_2}]F$. XX.1. b_1 74°.[1843]
$CH_3PS(NHSiMe_3)F$. XX.1. $b_{0.01}$ 33°, IR, ^1H, ^{19}F.[1759]

$CH_3P(S)\langle^{NHCO}_{NHCO}\rangle C{=}C(Ph)(NC_5H_{10})$. XXIII.3. m. 177-179°.[855]

$CH_3PS(NHCHRCO_2Et)Cl$.

R = H. XX.1. $b_{0.24}$ 117-118°.[1236]

R = Me. XX.1. $b_{0.17}$ 109.5-110°.[1236]

$CH_3PS(NHCH_2CH_2CO_2Et)Cl$. XX.1. $b_{0.32}$ 132-132.5°.[1236]
$CH_3PS(OMe)(NHCSNHC_6H_4Cl\text{-}4)$. XXI.4. m. 105°.[1856]
$CH_3PS(OEt)NH_2$. XXI.1. $b_{0.4}$ 90-91°, n_D^{25} 1.5170, d_4^{25} 1.1404.[1716]

$CH_3PS(OEt)(NHR)$.

R = Me. XXI.2. $b_{0.22}$ 65.5°.[655]

R = i-Pr. XXI.2. $b_{0.2}$ 69-70°.[655]

R = sec.-Bu. XXI.2. $b_{0.18}$ 78-78.5°.[655]

$CH_3PS(OEt)(\overline{NCH_2CH_2})$. XXI.1. b_4 66-68°.[861]

$CH_3PS(OEt)(N\langle^{CH_2CH_2}_{CH_2CH_2}\rangle O)$. XXI.2. $b_{0.05}$ 91-96°, m. 38-40°.[1151]

$CH_3PS(OEt)(NHCSNHC_6H_4Cl\text{-}4)$. XXI.4. m. 119°.[1856]
$CH_3PS(OPr\text{-}i)(NMe_2)$. XXI.2. b_2 67-73°.[1151]
$CH_3PS(OPr)(\overline{NCH_2CH_2})$. XXI.1. $b_{0.05}$ 57-58°, n_D^{20} 1.4990.[351]
$CH_3PS(OBu)(\overline{NCH_2CH_2})$. XXI.1. $b_{1.5}$ 86-88°.[861]
$CH_3PS(OCH_2CO_2Et)(NHAr)$.

Ar = $C_6H_3Cl_2$-2,4. XXI.2. $b_{0.32}$ 178-180°.[1236]

Ar = C_6Cl_5. XXI.2. m. 125-126°.[1236]

$CH_3PS(OCH_2CH_2SCH_2CH_3)(NMe_2)$. XXI.2. $b_{0.01}$ 79°.[1853a]
$CH_3PS(OC_6H_{11}$-cycl.$)(NMe_2)$. XXI.2. $b_{0.6}$ 92-95°.[1722]
$CH_3PS(OPh)(NEt_2)$. XXI.2. $b_{0.05}$ 111-112°.[1372]
$CH_3PS(OC_6H_4Br$-4$)[N=C(R)NH_2]$.

   R = H. XXI.1. m. 100-102°.[1175]

   R = Me. XXI.1. m. 90-92°.[1175]

   R = Et. XXI.1. m. 72°.[1175]

$CH_3PS(OC_6H_3Cl_2$-2,4$)(NHR)$.

   R = i-Pr. XXI.5. $b_{0.2}$ 144.5-145°.[665]

   R = $-CH_2CO_2Et$. XXI.1. $b_{0.32}$ 178-180°.[1236]

   R = $-CHMeCO_2Et$. XXI.1. $b_{0.5}$ 172-172.5°.[1236]

$CH_3PS(OC_6H_2Cl_3$-2,4,5$)(NHR)$.

   R = i-Pr. XXI.2. m. 93.2-94.5°.[665]

   R = Bu. XXI.2. m. 57.5-58.5°.[665]

   R = sec.-Bu. XXI.2. m. 53-55°.[665]

   R = i-Bu. XXI.2. m. 49.5-52°.[665]

   R = $-CH_2CO_2Et$. XXI.1. m. 44-44.5°.[1236]

   R = $-(CH_2)_2CO_2Et$. XXI.1. m. 58-59°.[1236]

   R = $-(CH_2)_3CO_2Et$. XXI.1. m. 45-46°.[1236]

$CH_3PS(OC_6Cl_5)(NHR)$.

   R = i-Pr. XXI.2. m. 154.5-155.5°.[665]

   R = Bu. XXI.2. m. 114.5-115.5°.[665]

   R = sec.-Bu. XXI.2. m. 123-124°.[665]

   R = i-Bu. XXI.2. m. 132-133.5°.[665]

   R = $-CHMeCO_2Et$. XXI.1. m. 116-117.5°.[1236]

   R = $-(CH_2)_2CO_2Et$. XXI.1. m. 99.5-100°.[1236]

   R = $-(CH_2)_3CO_2Et$. XXI.1. m. 99-99.5°.[1236]

$CH_3PO(SMe)[N(Bu$-i$)_2]$. XXII.4. b_7 134-135°, n_D^{20} 1.4872, d_4^{20} 1.0299.[1470]
$CH_3PO(SEt)(NHPh)$. XXII.1. m. 155°, IR.[337]
$CH_3PO(SEt)[N(Bu)_2]$. XXII.4. b_7 126-127°, n_D^{20} 1.4870, d_4^{20} 1.0055.[1470]
$CH_3PO(SPh)(NEt_2)$. XXII.1. $b_{0.03}$ 129-130°, n_D^{20} 1.5625, d_4^{20} 1.1272, IR.[1372]
$CH_3PO(SCH_2CH_2NEt_2)(NMe_2)$. XXII.2. $b_{0.01}$ 86°.[1853a]
$CH_3PO(SC_6H_4NO_2$-4$)[N(CH_2CH_2Cl)_2]$. XXII.2. m. 62-63°.[1724]

$CH_3(O)P\underset{\displaystyle SCH_2}{\overset{\displaystyle NHCH_2}{\diagup\diagdown}}CH_2$. XXII.4. m. 86-88°.[351]

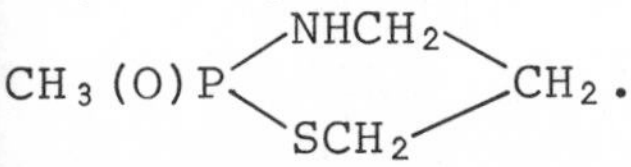

$CH_3PS(NHR)_2$.

R = i-Pr. XXIII.1. m. 75°.[1233]

R = Bu. XXIII.1. m. 47.5-48°.[1233]

R = Ph. XXIII.1. m. 177-178°.[720, 893] XXIII.2. m. 175-176°.[668]

$CH_3PS(NHMe)(NHAr)$.

Ar = Ph. XXIII.1. m. 96.5-97°.[1232]

Ar = 3-ClC_6H_4-. XXIII.1. m. 99.5-101°.[1232]

Ar = 4-ClC_6H_4-. XXIII.1. m. 122-123°.[1232]

$CH_3PS(NHEt)(NHAr)$.

Ar = Ph. XXIII.1. m. 96-98°.[1232]

Ar = 3-ClC_6H_4-. XXIII.1. m. 74-76°.[1232]

Ar = 4-ClC_6H_4-. XXIII.1. m. 83-85°.[1232]

Ar = -$C_6H_3Cl_2$-2,3. XXIII.1. m. 74-76°.[1232]

$CH_3PS(NHPr)(NHBu)$. XXIII.1. m. 47-48°.[661]

$CH_3PS(NHPr\text{-}i)(NHAr)$.

Ar = Ph. XXIII.1. m. 94-96°.[1232]

Ar = 4-ClC_6H_4-. XXIII.1. m. 116-117°.[1232]

$CH_3PS(NHBu)(NHAr)$.

Ar = Ph. XXIII.1. m. 70-72°.[1232] XXIII.2. m. 72-73°.[668]

Ar = 3-ClC_6H_4-. XXIII.1. m. 78-79°.[1232]

Ar = 4-ClC_6H_4-. XXIII.1. m. 101.5-103.5°.[1232]

Ar = -$C_6H_3Cl_2$-3,4. XXIII.1. m. 87-88.5°.[1232]

$CH_3PS(NMe_2)(NHAr)$.

Ar = Ph. XXIII.1. m. 128-128.5°.[1232]

Ar = 3-ClC_6H_4-. XXIII.1. m. 83.5-84.5°.[1232]

Ar = 4-ClC_6H_4-. XXIII.1. m. 118.5-119.5°.[1232]

Ar = 4-MeC_6H_4-. XXIII.1. m. 104.5-106.5°.[1232]

$CH_3PS(NEt_2)(NHR)$.

R = i-Pr. XXIII.1. $b_{0.2}$ 108-109°.[661]

R = Bu. XXIII.1. $b_{0.15}$ 122°, n_D^{25} 1.5045, d_4^{25} 1.0040.[66]

$CH_3PS(NMe_2)_2$. XXIII.4. b_{725} 245-247°, IR.[1152]

$CH_3PS(NEt_2)_2$. XXIII.2. m. 111.5-112.5°.[668] XXIII.1. b_{15} 149-151°.[1372]

$CH_3PS(\overline{NCH_2CH_2})_2$. XXIII.1. $b_{0.1}$ 72-73°, n_D^{20} 1.5475, d_4^{20} 1.1467.[1478]

$CH_3PS(NEt_2)[N(Pr)_2]$. XXIII.1. $b_{0.2}$ 118-119°, n_D^{25} 1.4975, d_4^{25} 0.9977.[661]

$CH_3PS(NMe_2)(\overline{NCH=NCH=CH})$. XXIII.1. n_D^{25} 1.5809.[316]

$CH_3(S)P\begin{array}{c}N=C-OR\\|\\N=C-OR\end{array}$

   R = Me. XXIII.4. m. 110-112°.[466]

   R = Et. XXIII.4. m. 85-86°.[466]

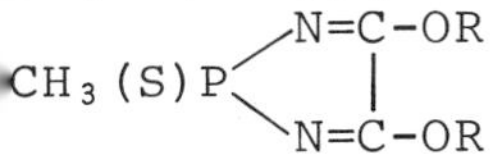

$CH_3(S)P\begin{array}{c}NHCO\\\diagdown\\NHCO\end{array}C=C(Ph)(NC_5H_{10})$. XXIII.3. m. 177-179°.[855]

$CH_3PS(NHNMe_2)F$. XXIV.1. m. 28°, IR, ^1H, ^{19}F, ^{31}P -96.2 ppm, mass spect.[1762]

$CH_3PS(OR)(NHNH_2)$.

   R = Et. XXIV.1. $b_{0.06}$ 72-76°, n_D^{20} 1.5344.[1500]

   R = i-Bu. XXIV.1. b_1 130°.[507]

   R = Ph. XXIV.1. m. 45-47°.[661]

   R = $2-ClC_6H_4-$. XXIV.1. m. 74-75°, HCl adduct, m. 158-160°.[1500]

$CH_3PS(OAr)(NHNHPh)$.

   Ar = Ph. XXIV.1. m. 73-74°.[661]

   Ar = $4-ClC_6H_4-$. XXIV.1. m. 76-77°.[661]

$CH_3PS(OAr)(NHNMe_2)$.

   Ar = Ph. XXIV.1. m. 62-63°.[661]

   Ar = $4-ClC_6H_4-$. XXIV.1. m. 54-56°.[661]

$CH_3PS(OAr)(NHN=CHPh)$.

   Ar = Ph. XXIV.1. m. 133-134°.[661]

   Ar = $4-ClC_6H_4-$. XXIV.1. m. 116-116.5^O.[661]

   Ar = $3-MeC_6H_4-$. XXIV.1. m. 104-105°.[661]

$[CH_3PS(OR)NH-]_2$.

   R = Et. XXIV.1. m. 159°.[1500]

   R = i-Bu. XXIV.1. m. 118°.[1500]

$CH_3PS(NR_2)(NHNHPh)$.

   R = Me. XXIV.1. m. 85-86°.[661]

   R = Et. XXIV.1. m. 105-106°.[661]

$CH_3PS(NEt_2)(NHNMe_2)$. XXIV.1. m. 32-34°.[661]

$CH_3PS(SMe)(NHPh)$. XXV.1. m. 116.5-117°.[720]

$CH_3PS(SPr)(NMe_2)$. XXV.1. $b_{0.15}$ 73-74°.[1722]

$CH_3PS(SPr-i)(NHAr)$.

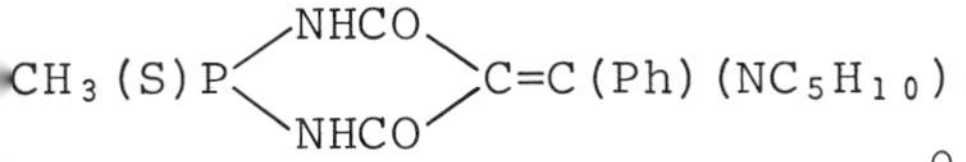

Ar = $-C_6H_2Cl_3$-2,4,6. XXV.1. m. 111-113.5°.[666]

Ar = $-C_6Cl_5$. XXV.1. m. 135-136.5°.[666]

$CH_3PS(SBu)(NHAr)$.

Ar = Ph. XXV.1. m. 68-69.5°.[666]

Ar = 4-ClC_6H_4-. XXV.1. m. 54.5-55.5°.[666]

CH_3PS(SBu-sec.)(NHAr).

Ar = 4-ClC_6H_4-. XXV.1. $b_{0.34}$ 159-160°.[666]

Ar = 4-$O_2NC_6H_4$-. XXV.1. m. 82-84°.[666]

$CH_3PS(SPh)(NMe_2)$. XXV.1. $b_{0.01}$ 102°.[1855]
$CH_3PS(SC_6H_{11}$-cycl.)(NMe_2). XXV.1. $b_{0.06}$ 92-95°.[1722]
$CH_3PS(SAr)[N(CH_2CH_2Cl)_2]$.

Ar = 4-ClC_6H_4-. XXV.2. m. 77-78°.[1724]

Ar = 4-MeC_6H_4-. XXV.2. m. 54-55°.[1724]

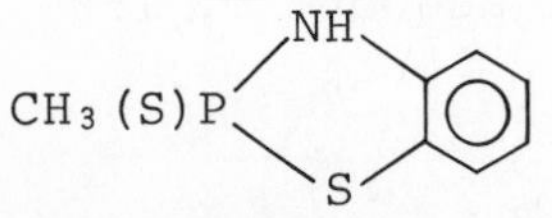 XXV.3. m. 88°, IR, ^1H.[2075]

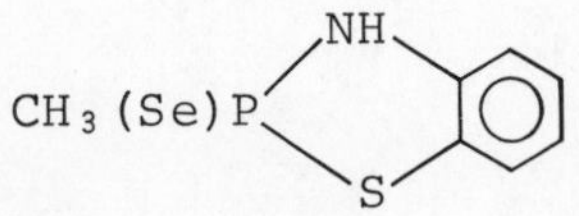 XXXV. IR, ^1H.[2075]

$\underline{C_2}$
$ClH_2CCH_2PS(OC_6H_3Cl_2$-2,4)(NHR).

R = i-Pr. XXI.2. $b_{0.28}$ 172.5-173°.[665]

R = sec.-Bu. XXI.2. $b_{0.17}$ 154.5°.[665]

CH_2=$CHPS(NR_2)F$.

R = Me. XIX.1. b_1 64°.[1843]

R = Et. XIX.1. b_1 74°.[1843]

CH_2=$CHPS[N(CH_2)_3CH_2]F$. XIX.1. b_1 84°.[1843]
CH_2=$CHPS(NC_5H_{10})F$. XIX.1. b_1 88°.[1843]
CH_2=$CHPS(N(CH_2CH_2)_2O)F$. XIX.1. b_1 98°.[1843]
CH_2=$CHPS(NCH_2CH_2)_2$. XXIII.1. $b_{0.4}$ 72-74°, n_D^{20} 1.5558, d_4^{20} 1.1476.[1478]
CH_2=$CHPS(NHPh)_2$. XXIII.1. m. 112-113°.[907]
$C_2H_5PS(NH_2)F$. XIX.1. $b_{0.01}$ 46°, n_D^{20} 1.5048, d_4^{20} 1.2498, IR, ^1H, ^{19}F.[1757]
$C_2H_5PS(NEt_2)Cl$. XIX.2. b_1 94-95°, n_D^{20} 1.5205, d_0^{20} 1.1135.[154]
$C_2H_5PS(NHMe)F$. XIX.1. b_1 60°.[1843]

$C_2H_5PS(NR_2)F$.

R = Me. XIX.1. b_2 53°.[1843]

R = Et. XIX.1. b_2 75°.[1843]

$C_2H_5PS(\overline{NCH_2CH_2})F$. XIX.1. b_3 50°.[1843]

$C_2H_5PS[\overline{N(CH_2)_3CH_2}]F$. XIX.1. b_1 78°.[1843]

$C_2H_5PS(NC_5H_{10})F$. XIX.1. b_1 85°.[1843]

$C_2H_5PS(N(CH_2CH_2)_2O)F$. XIX.1. b_1 92°.[1843]

$C_2H_5PS(NHSiMe_3)F$. XIX.1. $b_{0.01}$ 42°, IR, ^1H, ^{19}F.[1759]

$C_2H_5PS(N{=}PCl_3)F$. XIX.1. IR, ^1H, ^{19}F.[1759]

$C_2H_5PS(OR)(NEt_2)$.

R = Me. XXI.3. b_{10} 112-113°, n_D^{20} 1.4882, d_4^{20} 1.0287.[154]

R = Et. XXI.3. b_9 115-116°, n_D^{20} 1.4828, d_4^{20} 1.0042.[154]

R = Pr. XXI.3. b_1 84-85°, n_D^{20} 1.4801, d_4^{20} 0.9892.[154]

R = i-Pr. XXI.3. b_9 118.5-119.5°, n_D^{20} 1.4781, d_4^{20} 0.9858.[154]

R = Bu. XXI.3. b_1 88-89°, n_D^{20} 1.4785, d_4^{20} 0.9789.[154]

R = i-Bu. XXI.3. b_{13} 132-133°, n_D^{20} 1.4750, d_4^{20} 0.9739.[154]

R = $-C_6H_{13}$. XXI.3. $b_{1.5}$ 132-133°, n_D^{20} 1.4790, d_4^{20} 0.9642.[154]

R = $-C_7H_{15}$. XXI.3. b_2 137-137.5°, n_D^{20} 1.4789, d_4^{20} 0.9578.[154]

R = $-C_8H_{17}$. XXI.3. b_1 144-144.5°, n_D^{20} 1.4793, d_4^{20} 0.9516.[154]

R = $-C_9H_{19}$. XXI.3. b_1 151-152°, n_D^{20} 1.4796, d_4^{20} 0.9495.[154]

R = $-C_{10}H_{21}$. XXI.3. b_1 170-172°, n_D^{20} 1.4800, d_4^{20} 0.9446.[154]

$C_2H_5PS(OMe)(NHCSNHC_6H_4Cl\text{-}4)$. XXI.4. m. 105°.[1856]

$C_2H_5PS(OEt)(NHCSNHC_6H_4Cl\text{-}4)$. XXI.4. m. 100°.[1856]

$C_2H_5PS(OC_6H_4Cl\text{-}4)[N{=}C(R)NH_2]$.

R = H. XXI.1. m. 116-119°.[1175]

R = Me. XXI.1. m. 77°.[1175]

R = Et. XXI.1. m. 50-52°.[1175]

$C_2H_5PS(OC_6H_3Cl_2\text{-}2,4)(NHPr\text{-}i)$. XXI.2. $b_{0.25}$ 153-154°.[665]

$C_2H_5PS(OC_6H_2Cl_3\text{-}2,4,5)(NHPr\text{-}i)$. XXI.5. m. 90-92°.[665]

$C_2H_5PO(SMe)(NH_2)$. XXII.4. n_D^{30} 1.5507.[602a]

$C_2H_5PO(SEt)(NH_2)$. XXII.4. m. 44-46.5^O.[602a]

$C_2H_5PO(SEt)(NEt_2)$. XXII.1. b_2 106-108°, n_D^{20} 1.4892, d_0^{20} 1.0297.[1703]

$C_2H_5PO(SPr)(NH_2)$. XXII.4. m. 44-49°.[602a]

$C_2H_5PO(SPr-i)(NEt_2)$. XXII.3. b_{10} 132-133°, n_D^{20} 1.4895, d_4^{20} 1.0038.[50]

$C_2H_5PO(SCH_2CH_2NEt_2)(NMe_2)$. XXII.2. $b_{0.7}$ 151°.[1165]

$C_2H_5PO(SCH_2CH_2NEt_2)(NEt_2)$. XXII.2. $b_{0.5}$ 160°.[1165]

$C_2H_5PO(SCH_2CH_2NEt_2)(N\overset{CH_2CH_2}{\underset{CH_2CH_2}{<}}O)$. XXII.2. $b_{0.5}$ 135°.[1165]

$C_2H_5PS(NHPh)_2$. XXIII.1. m. 112-114°.[893]

$C_2H_5PS(NMe_2)_2$. XXIII.5. b_{10} 112-116°, n_D^{25} 1.5145, b_{12} 120-125°, ^{31}P -83.8 ppm.[1150]

$C_2H_5PS(NEt_2)_2$. XXIII.4. b_1 100-102°, d_4^{20} 0.9902.[1389]

$C_2H_5PS(NCH_2CRR')_2$.

R = R' = H. XXIII.1. $b_{0.5}$ 89°, n_D^{20} 1.5402, d_4^{20} 1.1267.[670]

R = H, R' = Et. XXIII.4. $b_{0.06}$ 87-88°, n_D^{20} 1.5044, d_4^{20} 1.0203, IR.[1416a]

R = R' = Me. XXIII.4. b_1 92-93°, n_D^{20} 1.5106, d_4^{20} 1.028 IR.[1416a]

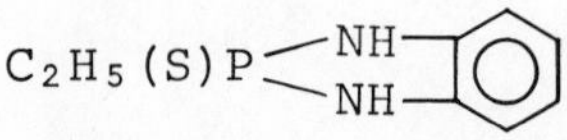

$C_2H_5(S)P\overset{NH}{\underset{NH}{<}}$⬡ XXIII.1. m. 135.5-137°.[452]

$C_2H_5(S)P\overset{N(Me)}{\underset{NH}{<}}$⬡ XXIII.1. m. 92-93° (bis-HCl adduct) [452]

$C_2H_5(S)P\overset{N=C-OEt}{\underset{N=C-OEt}{<}}$ XXIII.4. $b_{0.5}$ 140-141°.[466]

$C_2H_5PS(N=CPh_2)_2$. XXIII.4. m. 164-165°.[809a]

$C_2H_5PS(NHNMe_2)F$. XXIV.1. IR, ^1H, ^{19}F, ^{31}P -104.7 ppm, mass spect.[1762]

$C_2H_5PS(SH)(N=PPh_3)$. XXV, isolated as PPh_4^+ salt, m. 120°, IR.[1758]

$C_2H_5PS(SR)(NEt_2)$.

R = Et. XXV.3. $b_{0.025}$ 73-74°, n_D^{20} 1.5432, d^{20} 1.0550.[5]

R = i-Pr. XXV.3. $b_{0.03}$ 72-74°, n_D^{20} 1.5300, d^{20} 1.0266.[50]

R = Bu. XXV.3. $b_{0.03}$ 102-104°, n_D^{20} 1.5288, d^{20} 1.0203.[50]

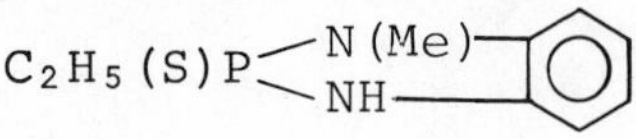

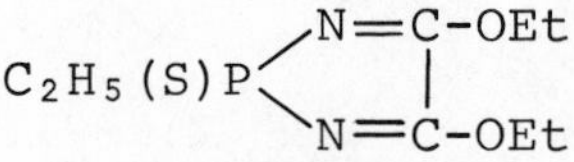

$C_2H_5PS(SAr)(NMe_2)$.

   Ar = Ph. XXV.1. $b_{0.01}$ 104°.[1855]

   Ar = 4-ClC_6H_4-. XXV.1. m. 74-75°.[1855]

$C_2H_5PS(SC_6H_4Cl-4)(NHPr-i)$. XXV.1. m. 70-73°.[666]

$C_2H_5(S)P{\displaystyle <}^{NH——CH_2}_{S———CH_2}$ XXV.1. n_D^{20} 1.6165, d_4^{20} 1.268.[313]

$\underline{C_3}$

$\overline{CH_2}$=$CHCH_2PS(NR_2)_2$.

   R = Me. XXIII.4. b_{14} 134-135°, n_D^{20} 1.5244, d_4^{20} 1.0299.
      [1701]

   R = Et. XXIII.4. $b_{0.09}$ 109-111°, n_D^{20} 1.5139, d_4^{20} 0.9971.
      [1701]

   R = Pr. XXIII.4. $b_{0.033}$ 108-111°, n_D^{20} 1.5041, d_4^{20}
      0.9614.[1701]

   R = Bu. XXIII.4. $b_{0.033}$ 151-152°, n_D^{20} 1.4955, d_4^{20}
      0.9369.[1701]

CH_2=$CHCH_2PS[\overline{N(CH_2)_4CH_2}]_2$. XXIII.4. $b_{0.038}$ 146-148°, n_D^{20}
   1.5500, d_4^{20} 1.0797.[1701]

CH_2=$CHCH_2PS(NHPh)_2$. XXIII.1. m. 97-98°.[1702]

CH_2=$CHCH_2(S)P{\displaystyle <}^{N=C-OEt}_{N=C-OEt}$ XXIII.4. $b_{0.5}$ 152-154°.[466]

CH_2=$CHCH_2PSe(NR_2)_2$.

   R = Me. XXXV. $b_{0.035}$ 87-88 ^{O}, n_D^{20} 1.5541, d_4^{20} 1.2664.
      [1701]

   R = Et. XXXV. $b_{0.053}$ 118-119°, n_D^{20} 1.5337, d_4^{20} 1.1571.
      [1701]

   R = Pr. XXXV. $b_{0.038}$ 148-150°, n_D^{20} 1.5201, d_4^{20} 1.0919.
      [1701]

$C_3H_7PS(OEt)(NHCSNHC_6H_4Cl-4)$. XXI.4. m. 95°.[1856]
$C_3H_7PS(OC_6H_3Cl_2-2,4)(NHPr-i)$. XXI.5. $b_{0.28}$ 134.5-136°.[665]
$C_3H_7PS(NHPh)_2$. XXIII.1. m. 133°.[893]
i-$C_3H_7PS(OEt)(NHCSNHC_6H_4Cl-4)$. XXI.4. m. 96°.[1856]

$\underline{C_4}$

CH_2=$CHCH$=$CHPS(NEt_2)_2$. XXIII.1. $b_{0.5}$ 142°, n_D^{20} 1.5461, d_4^{20}
   1.0375.[1629]

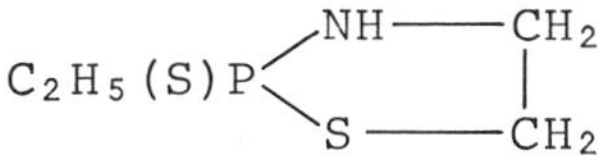

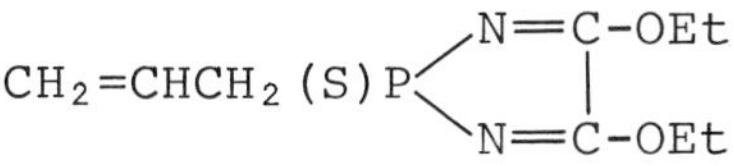

$(CH_3)_2C=CHPS(OEt)(NHC\underline{SNHC}_6H_4Cl-4)$. XXI.4. m. 90°.[1856]
$CH_3O_2CCH_2CH_2PO(SPr-i)(\overline{NCH_2CH_2})$. XXII.1. $b_{0.01}$ 100-103°, n_D^2
    1.4600, d_4^{20} 1.1480.[1475]

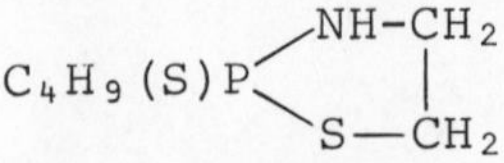

$C_4H_9(S)P\begin{matrix} \diagup NH-CH_2 \\ | \\ \diagdown S-CH_2 \end{matrix}$ XXV.1. n_D^{20} 1.5915, d_4^{20} 1.1182.[313]

sec.$-C_4H_9PS(OEt)(NHCSNHC_6H_4Cl-4)$. XXI.4. m. 86°.[1856]
sec.$-C_4H_9PS(SEt)(NRR')$.

    R = H, R' = Me. XXV.2. m. 48-48.5^O.[703]

    R = R' = Me. XXV.2. $b_{0.03}$ 152°.[703]

    R = R' = Et. XXV.2. $b_{0.015}$ 98°, n_D^{20} 1.5368, d_4^{20} 1.0336
        [703]

    R = R' = Bu. XXV.2. $b_{0.015}$ 119°, n_D^{20} 1.5231, d_4^{20}
        0.9985.[703]

sec.$-C_4H_9PS(SPr)(NR_2)$.

    R = Me. XXV.2. $b_{0.01}$ 98°.[703]

    R = Et. XXV.2. $b_{0.02}$ 130°.[703]

sec.$-C_4H_9PS(SBu)(NR_2)$.

    R = Me. XXV.2. $b_{0.01}$ 116°.[703]

    R = Et. XXV.2. $b_{0.02}$ 145°.[703]

sec.$-C_4H_9PS(SCH_2Ph)(NMe_2)$. XXV.2. $b_{0.01}$ 169-170°.[703]
sec.$-C_4H_9PS(SCH_2CO_2Et)(NR_2)$.

    R = Me. XXV.2. $b_{0.01}$ 107-109°, n_D^{20} 1.5315, d_4^{20} 1.1201.
        [703]

    R = Et. XXV.2. $b_{0.011}$ 128°, n_D^{20} 1.5263, d_4^{20} 1.1172.[703]

    R = $-CH_2CH=CH_2$. XXV.2. $b_{0.04}$ 142-143°, n_D^{20} 1.5357, d_4^{20}
        1.0884.[703]

    R = Bu. XXV.2. $b_{0.04}$ 145-147°, n_D^{20} 1.5158, d_4^{20} 1.0474.
        [703]

    R = i-Bu. XXV.2. $b_{0.02}$ 148-150°, n_D^{20} 1.5191, d_4^{20}
        1.0755.[703]

sec.$-C_4H_9PS(SCH_2CO_2Et)[\overline{N(CH_2)_4C}H_2]$. XXV.2. $b_{0.01}$ 145-147°,
    n_D^{20} 1.5415, d_4^{20} 1.1311.[703]
sec.$-\overset{O}{C}_4H_9PS(SCH_2CH_2SEt)(NR_2)$.

    R = Me. XXV.2. $b_{0.02}$ 127-128°.[706]

    R = Et. XXV.2. $b_{0.02}$ 145-148°.[706]

$\underline{C_6}$
$C_6H_5PS(NH_2)F$. XIX.1. $b_{0.01}$ 110°, IR, ^1H, ^{19}F.[1759]
$C_6H_5PS(NR_2)Cl$.

R = Me. XIX.1. $b_{0.1}$ 105°, n_D^{25} 1.6007, d_4^{25} 1.228, IR, ^{1}H.[2007]

R = Et. XIX.1. $b_{0.7}$ 137°, m. 43°, IR, ^{1}H.[2007]

R = Bu. XIX.1. $b_{0.1}$ 165°, n_D^{25} 1.5538, d_4^{25} 1.096, IR, ^{1}H.[2007]

$C_6H_5PS(OMe)(N=\underline{PPh_3})$. XXI.4. m. $121-123^{\circ}$.[202]

$C_6H_5PS(OPr-i)(\overline{NCH_2CH_2})$. XXI.1. $b_{0.05}$ $117-118^{\circ}$, n_D^{20} 1.5625.[351]

$C_6H_5PS(OPh)(NHPh)$. XXI.1. m. $93-96^{\circ}$.[1736]

$C_6H_5PS(OC_6H_4Cl-4)(NHPr-i)$. XXI.2. m. $71-72.5^{\circ}$.[666]

$C_6H_5PS(OC_6H_3Cl_2-2,4)(NHPr-i)$. XXI.5. $b_{0.2}$ $163-167^{\circ}$.[665]

$C_6H_5PO(SR)(NEt_2)$.

R = Et. XXII.3. $b_{0.08}$ $105-108^{\circ}$, n_D^{20} 1.5533, d_4^{20} 1.1045.[1747]

R = Pr. XXII.3. $b_{0.08}$ $111-113^{\circ}$, n_D^{20} 1.5455, d_4^{20} 1.0828.[1747]

R = i-Pr. XXII.3. $b_{0.1}$ $111-113^{\circ}$, n_D^{20} 1.5431, d_4^{20} 1.0775.[1747]

R = Bu. XXII.3. $b_{0.08}$ $126-128^{\circ}$, n_D^{20} 1.5390, d_4^{20} 1.0654.[1747]

R = Ph. XXII.3. $b_{0.001}$ $160-162^{\circ}$, n_D^{20} 1.5985, d_4^{20} 1.1550.[1747]

$C_6H_5PO(SBu-i)[N(Pr)_2]$. XXII.3. $b_{0.08}$ $128-129^{\circ}$, n_D^{20} 1.5308. d_4^{20} 1.0395.[1747]

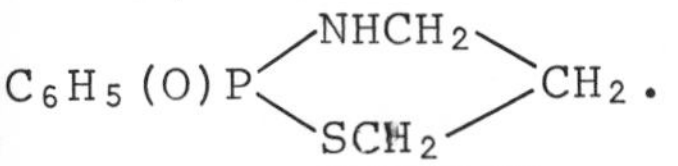

$C_6H_5(O)P\begin{smallmatrix}\diagup NHCH_2 \diagdown \\ \diagdown SCH_2 \diagup\end{smallmatrix}CH_2$. XXII.4. m. $109-110^{\circ}$.[351]

$C_6H_5(O)P\begin{smallmatrix}\diagup N(Et)CH_2 \diagdown \\ \diagdown SCH_2 \diagup\end{smallmatrix}CH_2$. XXII.4. $b_{0.02}$ $167-170^{\circ}$, n_D^{20} 1.5880.[351]

$C_6H_5PS(NH_2)_2$. XXIII.1. m. 51°.[1736, 1931]

$C_6H_5PS(NHR)_2$.

R = Me. XXIII.1. m. $69.5-70.5^{\circ}$.[2029]

R = Et. XXIII.1. m. $86-87^{\circ}$.[541] $80-81^{\circ}$.[1736]

R = $-CH_2CH=CH_2$. XXIII.1. m. $51-54^{\circ}$.[1873]

$C_6H_5PS(NHPh)_2$. XXIII.1. m. $80-81^{\circ}$,[2029] m. $175-176^{\circ}$.[893]

$C_6H_5PS(NHAr)_2$.

Ar = $2-ClC_6H_4-$. XXIII.1. m. $113-114^{\circ}$.[1736]

Ar = $4-ClC_6H_4-$. XXIII.1. m. $181-182^{\circ}$.[1736]

Ar = $-C_6H_3Cl_2-3,4$. XXIII.1. m. 145°.[1736]

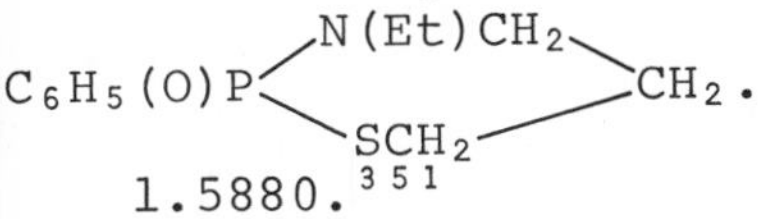

Ar = $-C_6H_3Me_2$-3,4. XXIII.1. m. 132-133°.[1736]

Ar = 4-$MeOC_6H_4$-. XXIII.1. m. 112-114°.[1736]

Ar = 4-$O_2NC_6H_4$-. XXIII.1. m. 198-200°.[1736]

C_6H_5PS 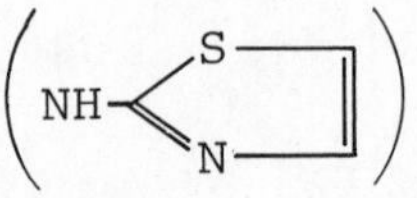 2. XXIII.1. m. 210-211°(dec.).[1873]

$C_6H_5PS(NHCH_2C_6H_3Cl_2$-2,4)$_2$. XXIII.1. m. 101-102.5°.[1873]

$C_6H_5PS(NCH_2CR_2)_2$.

R = H. XXIII.1. m. 105-107.5°.[1873]

R = Me. XXIII.4. m. 57-58°, IR.[1416a]

$C_6H_5PS(NCH_2CH_2OCH_2CH_2)_2$. XXIII.1. m. 111-112°.[1736]

$C_6H_5PS[N(CH_2)_4CH_2]_2$. XXIII.4. m. 92°.[1291]

C_6H_5PS 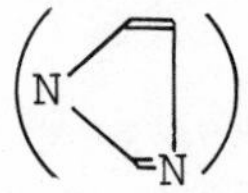 NEt_2. XXIII.4. m. 43°.[2007, 2008]

$C_6H_5PS(NHCONHAr)_2$.

Ar = Ph. XXIII.3. m. 176-178°(dec.).[1915]

Ar = 4-$EtOC_6H_4$-. XXIII.2. m. 150-151°.[668]

$C_6H_5(S)P$ 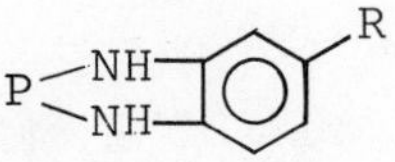

R = H. XXIII.1. m. 152-153.5°.[452]

R = Me. XXIII.1. m. 128-130 .[452]

R = Cl. XXIII.1. m. 114.5-116°.[452]

R = -OMe. XXIII.1. m. 186-188° (dihydrochloride).[452]

R = -OEt. XXIII.1. m. 113-114° (dihydrochloride).[452]

R = $-CO_2Me$. XXIII.1. m. 219-222°.[452]

R = $-NO_2$. XXIII.1. m. 230-231.5°.[452]

R = $-CF_3$. XXIII.1. m. 140-145°.[452]

$C_6H_5(S)P$ 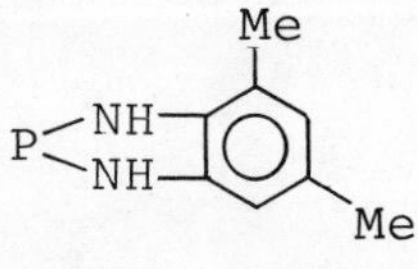 XXIII.1. m. 208-210 .[452]

$C_6H_5(S)P$ 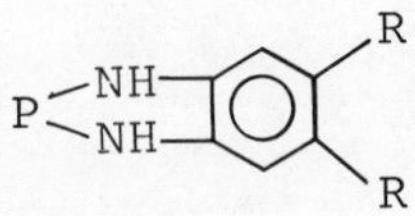

R = Me. XXIII.1. m. 198-199°.[452]

R = Cl. XXIII.1. m. 184-186°.[452]

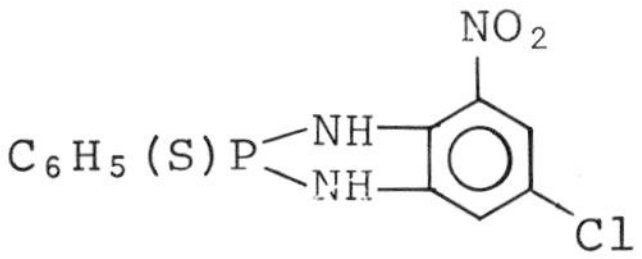

XXIII.1. m. 234-236°.[452]

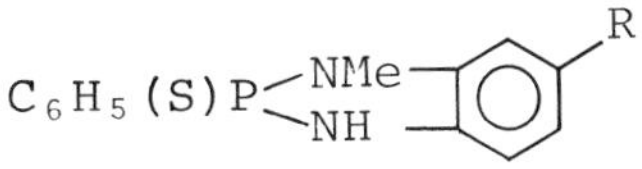

R = H. XXIII.1. m. 114-115°.[452]

R = Me. XXIII.1. m. 113-114.5°.[452]

R = -OMe. XXIII.1. m. 161.5-162.5°.[452]

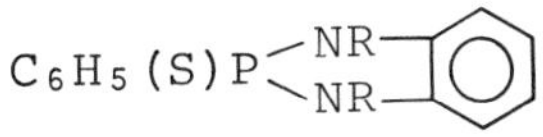

R = R'=Me. XXIII.1. m. 101-102°.[452]

R = Bu, R'=H. XXIII.1. m. 110.5-111.5° (dihydrochloride).[452]

R = cyclo-C_6H_{11}-, R' =H. XXIII.1. m. 158.5-159.5° (monohydrochloride).[452]

R = Ph, R' = H. XXIII.1. m. 188-189°.[452]

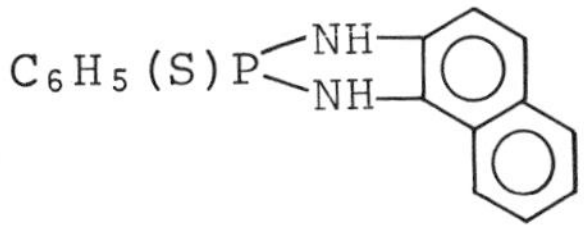

XXIII.1. m. 180-181° (dihydrochloride).[452]

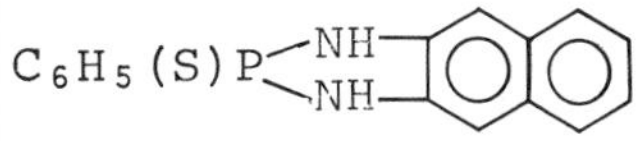

XXIII.1. m. 223-227°.[452]

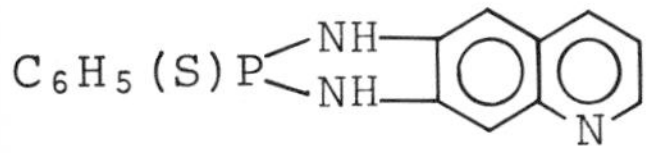

XXIII.1. m. 234-236 .[452]

$C_6H_5PS(N{=}CPh_2)_2$. XXIII.4. m. 151-152°.[809, 809a]

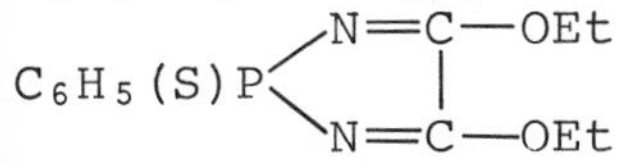

XXIII.4. m. 141-142°.[495]

$C_6H_5PS(N=PPh_3)_2$. XXIII.4. m. $195-197^O$.[202]
$C_6H_5PS(NHNH_2)_2$. XXIV.1. m. 115°.[1930]
$C_6H_5PS(NHNHPh)_2$. XXIV.1. m. $199.5-200.5^O$.[729, 1873]

$$C_6H_5(S)P \begin{array}{c} NHNH \\ \diagdown \\ NHNH \end{array} P(S)C_6H_5.\ XXIV.1.\ m.\ 168-169^O.[2006]$$

$C_6H_5PS(SR)(NEt_2)$.

R = Et. XXV.3. $b_{0.1}$ $116-118^O$, n_D^{20} 1.5942, d_4^{20} 1.1167. [1747]

R = Pr. XXV.3. $b_{0.08}$ $123-125^O$, n_D^{20} 1.5840, d_4^{20} 1.0971. [1747]

R = i-Pr. XXV.3. $b_{0.08}$ $115-117^O$, n_D^{20} 1.5824, d_4^{20} 1.0931.[1747]

R = Bu. XXV.3. $b_{0.08}$ $129-135^O$, n_D^{20} 1.5758, d_4^{20} 1.0812. [1747]

$C_6H_5PS(SPr)(NEt_2)$. XXV.3. $b_{0.08}$ $130-131^O$, n_D^{20} 1.5640, d_4^{20} 1.0524.[1747]
$C_6H_5PS(SPh)(NHPr-i)$. XXV.1. m. $56.5-59^O$.[666]

$$C_6H_5(S)P \begin{array}{c} NH—CH_2 \\ | \\ S—CH_2 \end{array}\ XXV.1.\ n_D^{20}\ 1.6807,\ d_4^{20}\ 1.321.[313]$$

$C_6H_5PSe(SR)(NEt_2)$.

R = Et. XXXV. $b_{0.06}$ $125-127^O$, n_D^{20} 1.6174, d_4^{20} 1.2868. [1747]

R = Pr. XXXV. $b_{0.04}$ $127-128^O$, n_D^{20} 1.6053, d_4^{20} 1.2527. [1747]

R = i-Pr. XXXV. $b_{0.05}$ $123-124^O$, n_D^{20} 1.6042, d_4^{20} 1.2498 [1747]

R = Bu. XXXV. $b_{0.05}$ $137-139^O$, n_D^{20} 1.5954, d_4^{20} 1.2263. [1747]

cycl.-$C_6H_{11}PS(NR_2)Cl$.

R = Me. XX.1. m. 73°.[2007]

R = Et. XX.1. $b_{0.6}$ 118°, n_D^{25} 1.5346, d_4^{25} 1.1146.[2007]

cycl.-$C_6H_{11}PS(NHPh)_2$. XXIII.1. m. $122-123^O$.[893]
cycl.-$C_6H_{11}PS(NR_2)(NCH=CHN=CH)$.

R = Me. XXIII.1. m. $53-56^O$.[2007]

R = Et. XXIII.1. m. $82-83^O$.[2007]

cycl.-$C_6H_{11}PS(SEt)(NRR')$.

R = H, R' = Me. XXV.2. m. $48-48.5^O$.[703]

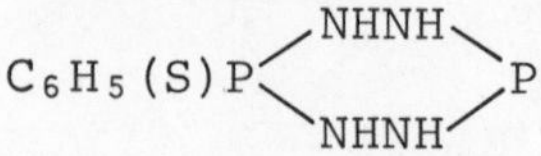

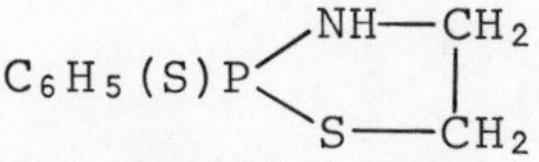

R = R' = Me. XXV.2. $b_{0.03}$ 152°, n_D^{20} 1.5630, d_4^{20} 1.0958.[703]

R = R' = Et. XXV.2. $b_{0.03}$ 130°, n_D^{20} 1.5535, d_4^{20} 1.0609.[703]

R = H, R' = Ph. XXV.2. m. 94-95°.[703]

R = R' = Bu. XXV.2. $b_{0.02}$ 135°. n_D^{20} 1.5363, d_4^{20} 1.0264.[703]

cycl.-$C_6H_{11}PS(SEt)\overline{N(CH_2)_4CH_2}$. XXV.2. m. 54-55.5°.[703]

cycl.-$C_6H_{11}PS(SPr)(NR_2)$.

R = Me. XXV.2. $b_{0.01}$ 98°, n_D^{20} 1.5550, d_4^{20} 1.0754.[703]

R = Et. XXV.2. $b_{0.02}$ 130°, n_D^{20} 1.5463, d_4^{20} 1.0522.[703]

cycl.-$C_6H_{11}PS(SBu)(NR_2)$.

R = Me. XXV.2. $b_{0.01}$ 116°, n_D^{20} 1.5490, d_4^{20} 1.0654.[703]

R = Et. XXV.2. $b_{0.02}$ 145°, n_D^{20} 1.5420, d_4^{20} 1.0412.[703]

R = Bu. XXV.2. $b_{0.01}$ 146-148°.[703]

cycl.-$C_6H_{11}PS(SCH_2Ph)(NMe_2)$. XXV.2. $b_{0.01}$ 169-170°, n_D^{20} 1.5989, d_4^{20} 1.1341.[703]

cycl.-$C_6H_{11}PS(SCH_2CO_2Et)(NR_2)$.

R = Me. XXV.2. $b_{0.01}$ 137-139°, n_D^{20} 1.5493, d_4^{20} 1.1542.[703]

R = Et. XXV.2. $b_{0.01}$ 148-149°, n_D^{20} 1.5430, d_4^{20} 1.1256.[703]

R = $-CH_2CH{=}CH_2$. XXV.2. $b_{0.01}$ 156-158°, n_D^{20} 1.5510, d_4^{20} 1.1190.[703]

R = Bu. XXV.2. $b_{0.01}$ 163-165°, n_D^{20} 1.5292, d_4^{20} 1.0706.[703]

R = i-Bu. XXV.2. $b_{0.01}$ 159-161°, n_D^{20} 1.5321, d_4^{20} 1.0812.[703]

cycl.-$C_6H_{11}PS(SCH_2CO_2Et)(NHPh)$. XXV.2. m. 82.5-83.5°.[703]

cycl.-$C_6H_{11}PS(SCH_2CO_2Et)[\overline{N(CH_2)_4CH_2}]$. XXV.2. $b_{0.01}$ 177-179°, n_D^{20} 1.5579, d_4^{20} 1.1565.[703]

cycl.-$C_6H_{11}PS(SCH_2CH_2SEt)(NR_2)$.

R = Me. XXV.2. $b_{0.03}$ 156-158°.[706]

R = Et. XXV.2. $b_{0.02}$ 158-160°.[706]

cycl.-$C_6H_{11}PS(SCH_2CH_2SEt)(NHPh)$. XXV.2. m. 76-77°.[706]

cycl.-$C_6H_{11}PS(SCH_2CH_2SEt)$(NH– 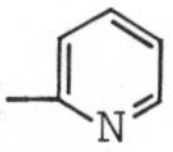). XXV.2. m. 110.5°.[706]

$\underline{C_7}$

4-$CH_3C_6H_4PS[N(CH_2)_4CH_2]_2$. XXIII.4. m. 88°.[1291]

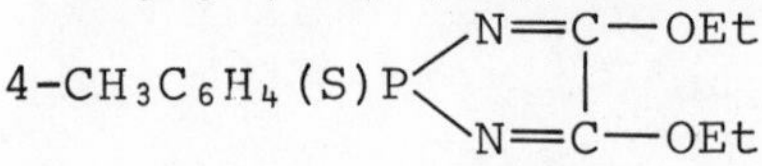

4-$CH_3C_6H_4(S)P\begin{array}{c} N{=}C{-}OEt \\ | \\ N{=}C{-}OEt \end{array}$ XXIII.4. m. $117-119^{\circ}$.[495]

C.9. Phosphonic and Thiophosphonic Acid Anhydrides and Mixed Anhydrides

$\underline{C_1}$

$\overline{(CC1H_2PO_2)}_n$. XXIX.1. ^{31}P -2.2 ppm.[1326]

$CIH_2(O)P(OEt)O(OEt)O(OEt)P(O)CH_2J$. XXVII.1. m. $75-76.5^{\circ}$.[647]

$CH_3(O)P(OH)O(OH)P(O)CH_3$. XXVI.1. m. $139-141^{\circ}$.[514] $141-142^{\circ}$.[720]

$CH_3(O)P(OR)O(OR)P(O)CH_3$.

> R = Me. XXVII.1. b_4 $137.5-138^{\circ}$.[1490]
>
> R = Et. XXVII.1. $b_{2.5}$ 122°, n_D^{20} 1.4360, d_4^{20} 1.1950.[159] $b_{0.25}$ $122-124^{\circ}$, n_D^{25} 1.4400.[1716]
>
> R = Pr. XXVII.1. b_2 $143-144^{\circ}$,[1490] $b_{1.5}$ 135°, n_D^{20} 1.4370, d_4^{20} 1.1431.[1596]
>
> R = Bu. XXVII.1. b_1 $134-135^{\circ}$, n_D^{20} 1.4304, d_4^{20} 1.0842.[268]

$CH_3(O)P(NMe_2)O(NMe_2)P(O)CH_3$. XXVIII.2. b_3 $135-140^{\circ}$, ^1H, ^{31}[875]

$(CH_3PO_2)_n$. XXIX.1. m. $122-142^{\circ}$.[383] XXIX.1. ^{31}P -13.8 ppm[1326]

$CH_3(O)P(OEt)O(OEt)P(S)CH_3$. XXX.3. $b_{0.02}$ $36-37^{\circ}$.[1153] XXX.2. $b_{0.05}$ 80°, n_D^{24} 1.4640.[337] XXX.1. $b_{0.0003}$ 73°, n_D^{20} 1.4679.[405]

$CH_3(O)P(SR)O(SR)P(O)CH_3$.

> R = Et. XXX.2. $b_{0.04}$ $120-124^{\circ}$, n_D^{20} 1.5200, d_4^{20} 1.2563.[640]
>
> R = Pr. XXX.2. $b_{0.008}$ $120-125^{\circ}$, n_D^{20} 1.5120, d_4^{20} 1.1950.[640]
>
> R = Bu. XXX.2. $b_{0.005}$ $125-130^{\circ}$, n_D^{20} 1.5060, d_4^{20} 1.1519.[640]

$CH_3(O)P(OBu)O(NEt_2)P(S)CH_3$. XXXI. $b_{0.05}$ $95-96^{\circ}$, n_D^{20} 1.4793 d_4^{20} 1.1030.[1323]

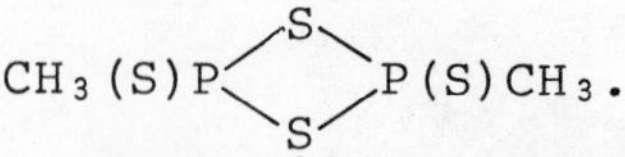

$CH_3(S)P\begin{array}{c} S \\ \diamondsuit \\ S \end{array}P(S)CH_3$. XXXII.1. m. $206-211^{\circ}$, IR, X-ray struc-

ture.[396, 1379] XXXII.3. m. 190°.[217] ^1H.[395]

$[CH_3P(S)O]_3$. XXXII.3. m. 108°, IR, ^1H, mass spect., X-ray powder data.[1758]

$CH_3PO(OR)O_2CC(Me)=CH_2$.

R = Et. XXXIII.2. n_D^{20} 1.4440, d_4^{20} 1.1562.[1596]

R = Pr. XXXIII.2. n_D^{20} 1.4450, d_4^{20} 1.1094.[1596]

$CH_3PO(OR)OC(=NOH)C_6H_4NO_2$-4.

R = Me. XXXIII.1. m. 115°, IR, ^1H, hydrolysis const.[338]

R = Et. XXXIII.1. m. 130°, IR, ^1H, hydrolysis const.[338]

R = Pr. XXXIII.1. m. 88-90°, IR, ^1H, hydrolysis const.[338]

R = i-Pr. XXXIII.1. m. 112-114°, IR, ^1H, hydrolysis const.[338]

R = Me_3CCH_2-. XXXIII.1. m. 110°, IR, ^1H, hydrolysis const.[338]

R = Me_3CCHMe-. XXXIII.1. m. 110-113°, IR, ^1H, hydrolysis const.[338]

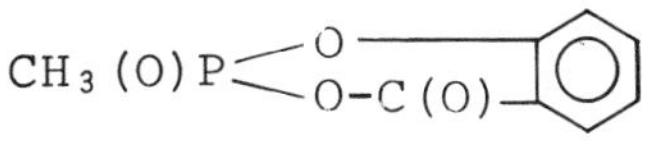

XXXIII.4. m. 99-100°.[1375, 1376]

$CH_3(O)P(OMe)OP(O)(OMe)_2$. XXXIV.1. $b_{0.001}$ 102-105°.[214, 1304]

$CH_3(O)P(NEt_2)OP(OR)_2$.

R = Pr. XXXIV.2. $b_{0.04}$ 75-76°, n_D^{20} 1.4500, d_4^{20} 1.0607.[1323]

R = Bu. XXXIV.2. $b_{0.3}$ 130-133°, n_D^{20} 1.4520, d_4^{20} 1.0321.[512, 1323]

$CH_3(O)P(NEt_2)OP(NEt_2)_2$. XXXIV.2. $b_{0.06}$ 77-80°, n_D^{20} 1.4718, d_4^{20} 1.0220.[1323]

$CH_3(O)P(NEt_2)OP(Ph)(NEt_2)$. XXXIV.2. $b_{0.05}$ 93°, n_D^{20} 1.5130, d_4^{20} 1.0820.[1323]

C_2

$C_2H_5(O)P(OR)O(OR)P(O)C_2H_5$.

R = Et. XXVII.1. b_9 154-154.5°.[1647] $b_{2.25}$ 139.5-141°, n_D^{20} 1.4280, d_4^{20} 1.1333.[1703]

R = Pr. XXVII.1. b_4 147-149°.[1647]

$C_2H_5(O)P(NMe_2)O(NMe_2)P(O)C_2H_5$. XXVIII.2. $b_{0.3}$ 117-119°.[875]

$C_2H_5(O)P(NEt_2)O(NEt_2)P(O)C_2H_5$. XXVIII.2. b_2 163.5-164.5°, n_D^{20} 1.4607, d_4^{20} 1.0619.[1703]

$C_2H_5(O)P(OEt)O(OEt)P(S)C_2H_5$. XXX.2. $[\alpha]_D$ + 10.5^O, $b_{0.07}$
76-77.5^O, n_D^{21} 1.4715.[1293, 1297] XXX.2. $[\alpha]_D^{25}$ + 41.95^O,
$b_{0.05}$ 76-78°, n_D^{23} 1.4680.[1308]
$C_2H_5(O)P(SEt)O(SEt)P(O)C_2H_5$. XXX.2. $b_{0.005}$ 115-120°, n_D^{20}
1.5127, d_4^{20} 1.1890.[640]

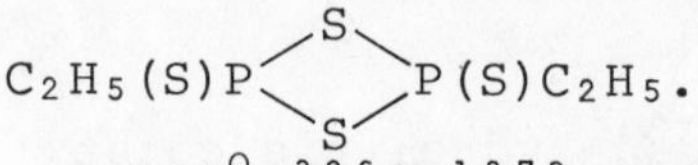

$C_2H_5(S)P\overset{\displaystyle S}{\underset{\displaystyle S}{<>}}P(S)C_2H_5$. XXXII.3. m. 142°.[217] XXXII.1. m. 14

148°.[396, 1379] XXXII.1. ^{31}P -35.4 ppm.[543]
$[C_2H_5P(S)O-]_3$. XXXII.3. m. 48°, IR, ^1H, mass spect.[1758]
$C_2H_5PO(OEt)[OC(=NOH)C_6H_4NO_2-4]$. XXXIII.1. m. 98-100°, IR,
^1H, hydrolysis const.[338]

C_3

$\overset{\displaystyle \overline{C(O)}\text{---}O}{\underset{\displaystyle CH_2\text{---}CH_2}{|}}{>}PO(OR)$.

    R = Me. XXXIII.4. $b_{0.01}$ 113-114°, n_D^{20} 1.4640, d_4^{20}
       1.392, IR.[2032]

    R = Et. XXXIII.4. $b_{0.009}$ 111-112°, n_D^{20} 1.4580, d_4^{20}
       1.305, IR.[2032]

    R = Pr. XXXIII.4. $b_{0.005}$ 108-109°, n_D^{20} 1.4575, d_4^{20}
       1.247, IR.[2032]

    R = -C(Me)$_2$CCl$_3$. XXXIII.4. $b_{0.004}$ 130-132°, n_D^{20} 1.4952
       d_4^{20} 1.4816.[504]

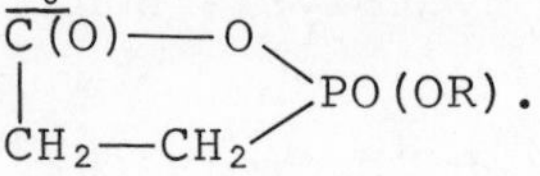

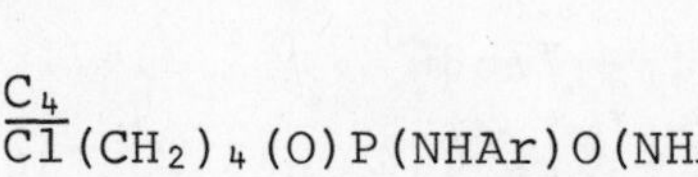

$C_3H_7(S)P\overset{\displaystyle S}{\underset{\displaystyle S}{<>}}P(S)C_3H_7$. XXXII.1. m. 97-99°.[396, 1379]

i-$C_3H_7(O)P(NMe_2)O(NMe_2)P(O)C_3H_7$-i. XXVIII.2. b_2 135-140°,
^{31}P.[875]

i-$C_3H_7(S)P\overset{\displaystyle S}{\underset{\displaystyle S}{<>}}P(S)C_3H_7$-i. XXXII.1. m. 180-181°.[396, 1379]

C_4

$Cl(CH_2)_4(O)P(NHAr)O(NHAr)P(O)(CH_2)_4Cl$.
    Ar = Ph. XXVIII.2. m. 134-138°.[767]
    Ar = 4-CH$_3$C$_6$H$_4$-. XXVIII.2. m. 132-138°.[767]
    Ar = 4-CH$_3$OC$_6$H$_4$-. XXVIII.2. m. 135-136°.[767]

$\overset{\displaystyle C(O)\text{---}O}{\underset{\displaystyle CH_3CH\text{---}CH_2}{|}}{>}PO(OR)$.

    R = Me. XXXIII.4. $b_{0.1}$ 100-102°, n_D^{23} 1.4520, d_4^{22}
       1.3004.[1492]

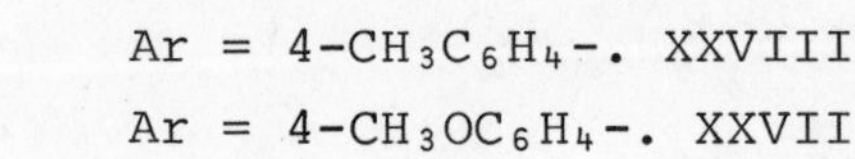

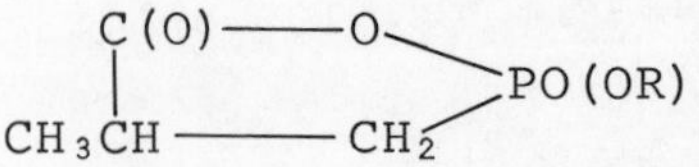

R = Et. XXXIII.4. $b_{2.5}$ 129-130°, n_D^{19} 1.4523, d_4^{20}
1.2304.[1492]

$(CH_3)_2C=CH(O)P(OEt)O(OEt)P(S)CH=C(CH_3)_2$, XXX.1. $b_{0.01}$ 116°.[114]

$(C_2H_5OCH=CHPO_2)_n$. XXIX.1. n = 4.2-4.5, m. 75-76°, dec.
140°.[1481]

$C_4H_9(O)P(OEt)O(OEt)P(O)C_4H_9$. XXVII.1. b_2 145-146°.[1647]

$C_4H_9(S)P$⟨S/S⟩$P(S)C_4H_9$. XXXII.1. m. 105-110°.[369, 1379]

XXXII.1. m. 49-52°, $b_{0.13}$ 170-173°.[708]
t-$C_4H_9PO(OEt)[OC(=NOH)C_6H_4NO_2-4]$. XXXII.1. IR, ^{1}H, hydrolysis const.[338]

$\underline{C_5}$
$(i-C_5H_{11}PO_2)n$. XXIX.1. m. 122°.[729]

$C_5H_{11}(S)P$⟨S/S⟩$P(S)C_5H_{11}$. XXXII.1. $b_{0.07}$ 178-183°, n_D^{20}
1.6273, d_4^{20} 1.1966.[708]

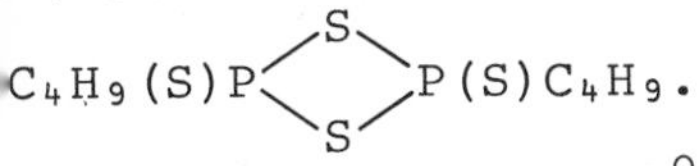

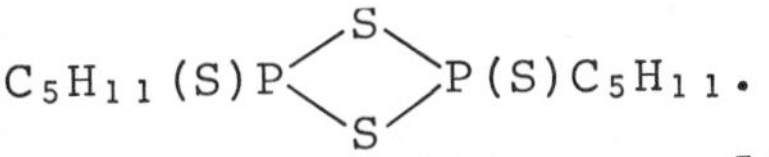

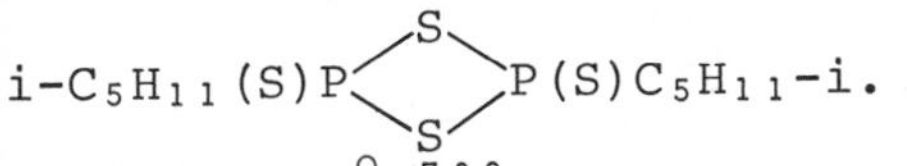

i-$C_5H_{11}(S)P$⟨S/S⟩$P(S)C_5H_{11}$-i. XXXII.1. m. 59-66°, $b_{0.16}$
184-188°.[708]

$(C_2H_5O)_2CH(O)P(OEt)O(OEt)P(O)CH(OC_2H_5)_2$. XXVII.1. $b_{0.01}$
140-141°, n_D^{22} 1.4410, ^{1}H, mass spect.[715]

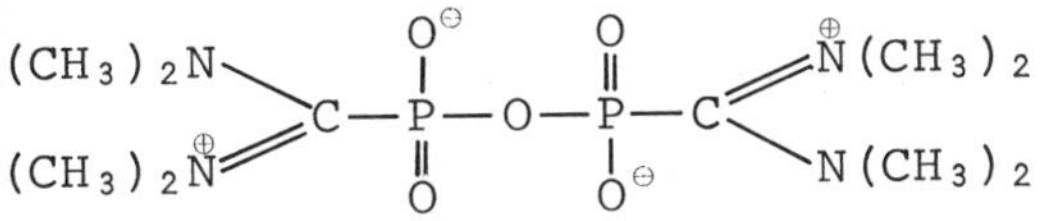

XXVI.3. Cryst. and mol. structure.[1784a]

$\underline{C_6}$

4-$FC_6H_4(O)P$⟨O/O⟩$P(O)C_6H_4F$-4. XXIX.1. m. 109-111°.[384]

4-$O_2NC_6H_4(O)P(NR_2)O(NR_2)P(O)C_6H_4NO_2$-4.

    R = Et. XXVIII.2. m. 116.5-118.5^{O}.[572]

    R = Pr. XXVIII.2. m. 77-80°.[572]

    R = i-Pr. XXVIII.2. m. >200°.[572]

    R = i-Bu. XXVIII.2. m. 125-128.5^{O}.[572]

$NR_2 = $ XXVIII.2. m. 170.5-173.5^{O}.[572]

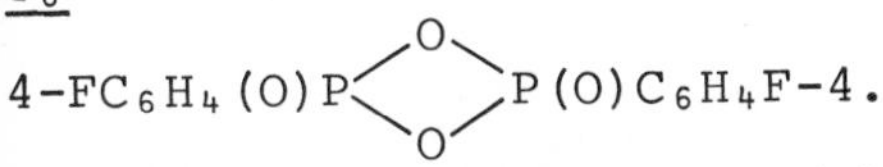

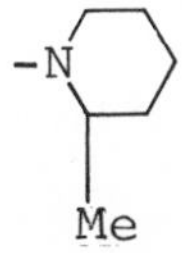

$C_6H_5(O)P(OPh)O(OPh)P(O)C_6H_5$. XXVII.1. m. 96°.[453]
$(C_6H_5PO_2)_2$. XXIX.1. m. $103-105^{O}$.[373]
$(C_6H_5PO_2)_3$. XXIX.1. m. $209-212^{O}$.[373]
$(C_6H_5PO_2)n$. XXIX.1. m. $102-103^{O}$.[509] ^{31}P 0.0 ppm[1326]

$C_6H_5(S)P\underset{S}{\overset{S}{\diagup\diagdown}}P(S)C_6H_5$. XXXII.1. m. $233-243^{O}$.[396, 1379]

    XXXII.3. m. 238°.[217] XXXII.1. ^{31}P-71.3 ppm.[543] XXXII.3.
    m. $216-226^{O}$.[1321] XXXII.3. m. 239°.[1426] XXXII.3. m.
    148°.[216] XXXII.3. m. $231-236^{O}$, X-ray structure, d_4^{20}
    1.5642.[1148, 1154]

Ph(S)P(NEt$_2$)O(NEt$_2$)P(S)Ph. XXXI. d-form, $[\alpha]_D^{25}$ + 109°, m.
    $83-85^{O}$.[1876]

$C_6H_5PO(OEt)[OC(=NOH)C_6H_4NO_2-4]$. XXXIII.1. IR, ^1H, hydrol-
    ysis const.[338]

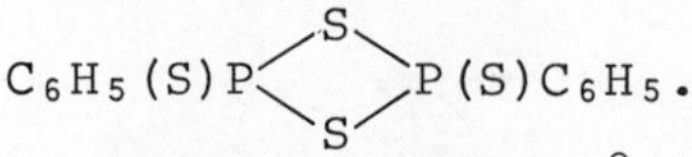

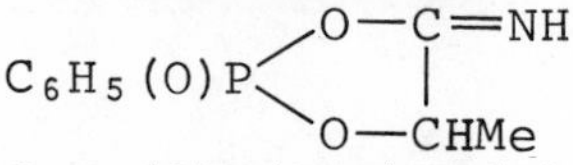

$C_6H_5(O)P\underset{O-CHMe}{\overset{O-C=NH}{\diagup\diagdown}}$ XXXIII.4. IR.[380]

$C_6H_5(O)P(OEt)OP(S)(OEt)_2$. XXXIV.3. $b_{0.05}$ $120-122^{O}$, n_D
    1.5044.[1301]

$C_6H_5(O)P(NEt_2)OP(OPr)_2$. XXXIV.2. $b_{0.0015}$ 92°, n_D^{20} 1.4904,
    d_4^{20} 1.0839.[512, 1323]

$C_6H_5(O)P(OH)O(OH)P(O)C_6H_5$. XXVI.1. m. $79.5-80^{O}$.[97] XXVI.2.
    m. $115-118^{O}$.[598] XXVI.2. Isolated as bis(morpholinium)
    salt, m. $203-207^{O}$.[597, 599] XXVI.1. Isolated as bis-
    (anilinium) salt, m. $211-213^{O}$.[556, 558] XXVI.1. Isolated
    as bis(S-benzyl-isothiuronium) salt, m. $227.5-228.5^{O}$.
    [189]

$\langle\!\rangle-(S)P\underset{S}{\overset{S}{\diagup\diagdown}}P(S)-\langle\!\rangle$.   XXXII.1. ^{31}P -43.4 ppm.[543]

cycl.-$(C_6H_{11}PO_2)_n$. XXIX.1. n = 8-9, m. $61-63^{O}$.[509]

cycl.-$C_6H_{11}(S)P\underset{S}{\overset{S}{\diagup\diagdown}}P(S)C_6H_{11}$-cycl. XXXII.1. m. $189-192^{O}$.
    [396, 1379]

$C_6H_{13}(O)P(OH)O(OH)P(O)C_6H_{13}$. XXVI.1. m. $64-66^{O}$.[326]

$C_6H_{13}(S)P\underset{S}{\overset{S}{\diagup\diagdown}}P(S)C_6H_{13}$. XXXII.1. $b_{0.02}$ $195-201^{O}$, n_D^{20}
    1.6120, d_4^{20} 1.1663.[708]

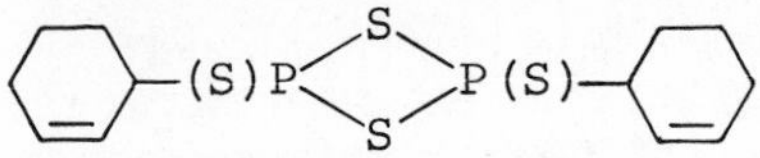

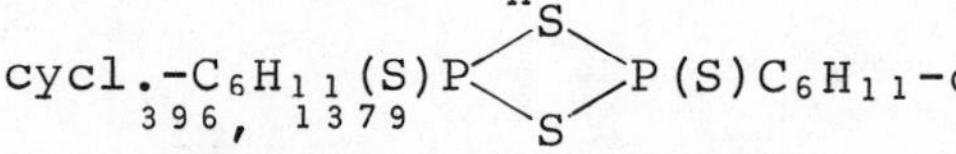

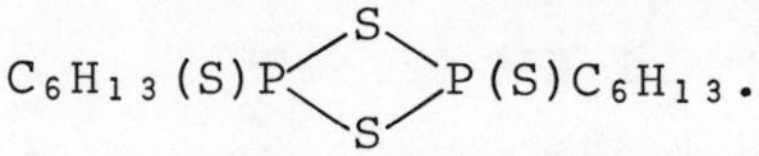

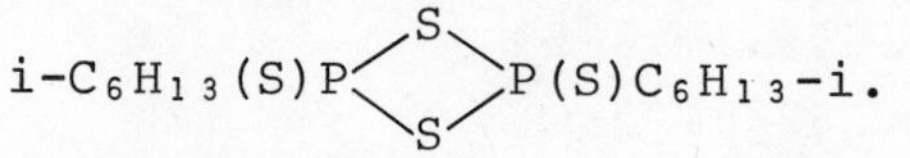

i-$C_6H_{13}(S)P\underset{S}{\overset{S}{\diagup\diagdown}}P(S)C_6H_{13}$-i. XXXII.1. $b_{0.21}$ $199-204^{O}$, n_D^{20}
    1.6132, d_4^{20} 1.1735.[708]

$\underline{C_7}$
4-$CH_3C_6H_4(O)P(OH)O(OH)P(O)C_6H_4CH_3-4$. XXVI.2. Isolated as
    bis(morpholinium) salt, m. 213-217 .[598]

4-$CH_3OC_6H_4(S)P\underset{S}{\overset{S}{\diagup\diagdown}}P(S)C_6H_4OCH_3-4$. XXXII.1. ^{31}P -14.5 ppm.[5

    XXXII.2. m. $228-229.5^{O}$.[1086]

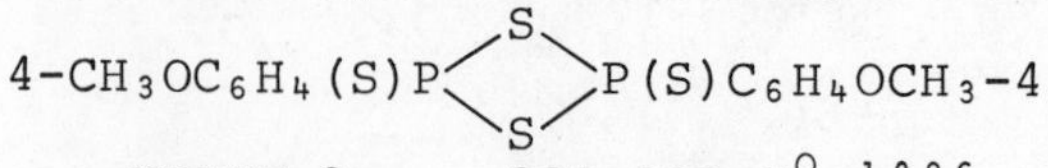

<u>C_8</u>

$4\text{-}C_2H_5OC_6H_4(S)P\langle S_2 \rangle P(S)C_6H_4OC_2H_5\text{-}4$. XXXII.1. ^{31}P -16.8 ppm.[543]

XXXII.2. m. 219-224.5°.[1086]

<u>C_{12}</u>

$(C_{12}H_{25}PO_2)_n$. XXIX.1. $\bar{n}$ = 3.7, m. 94-95°.[383]

<u>C_{15}</u>

$[(C_6H_5CH_2)_2CHPO_2]_n$. XXIX.1. m. 151°.[1281]

C.10. Iminophosphonic Acid Derivatives and Phosphonic Acid Imides.

<u>C_1</u>

$CClH_2P(=NCOR)Cl_2$.

R = $-CCl_2H$. XXXVI.1. $b_{0.08}$ 60-70° (dec.), n_D^{20} 1.5535, d_4^{20} 1.6954, IR.[1909]

R = $-CCl_3$. XXXVI.1. m. 50-52°, IR.[1909]

R = $-CF_3$. XXXVI.1. b_8 95-96°, n_D^{20} 1.4591, d_4^{20} 1.6691 IR.[1909]

$CCl_2HP(=NCOR)Cl_2$.

R = $-CCl_2H$. XXXVI.1. $b_{0.1}$ 80-90° (dec.), n_D^{20} 1.5451, d_4^{20} 1.7262, IR.[1909]

R = $-CCl_3$. XXXVI.1. m. 57-60°, IR.[1909]

R = $-CF_3$. XXXVI.1. $b_{0.07}$ 68-69°, n_D^{20} 1.4754, d_4^{20} 1.7372, IR.[1909]

$CCl_3P(=NCOR)Cl_2$.

R = $-CCl_2H$. XXXVI.1. m. 64-67°, IR.[1904]

R = $-CCl_3$. XXXVI.1. m. 67-70°, IR.[1904]

R = $-CF_3$. XXXVI.1. m. 40-44°, IR.[1904]

R = $4\text{-}ClC_6H_4-$. XXXVI.1. m. 82-84°, IR.[1904]

R = $4\text{-}O_2NC_6H_4-$. XXXVI.1. m. 100-105°, IR.[1904]

$CH_3P(=NCOCF_3)F_2$. XXXVI.1. $b_{1.5}$ 37-39°, n_D^{20} 1.3508, d_4^{20} 1.5064.[1138]

$CH_3P(=NCOR)Cl_2$.

R = $-CCl_3$. XXXVI.1. m. 81-84°, $b_{0.1}$ 120-125°, IR.[1906]

R = $-CF_3$. XXXVI.1. $b_{1.0}$ 58-60°, IR.[1906]

R = $4\text{-}NO_2C_6H_4-$. XXXVI.1. m. 100-105°, IR.[1906]

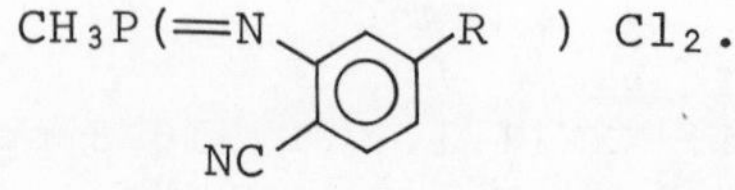

R = H. XXXVI.1. m. 71-72°, ^{1}H, ^{31}P +3.0 ppm.[1821]

R = Me. XXXVI.1. m. 50-52°.[1821]

$CH_3P(=NCO_2Me)(OR)_2$.

R = Me. XXXVI.2. $b_{0.02}$ 57-58°, n_D^{20} 1.4550, d_4^{20} 1.2111, IR.[1914]

R = Et. XXXVI.2. $b_{0.02}$ 59-60°, n_D^{20} 1.4469, d_4^{20} 1.1228, IR.[1914]

R = i-Pr. XXXVI.2. $b_{0.02}$ 59-60°, n_D^{20} 1.4450, d_4^{20} 1.0655, IR.[1914]

$CH_3P(=NCO_2Et)(OR)_2$.

R = Me. XXXVI.2. $b_{0.02}$ 69-70°, n_D^{20} 1.4520, d_4^{20} 1.1668, IR.[1914]

R = Et. XXXVI.2. $b_{0.02}$ 67-68°, n_D^{20} 1.4470, d_4^{20} 1.0944, IR.[1914]

R = i-Pr. XXXVI.2. $b_{0.02}$ 64-65°. n_D^{20} 1.4469, d_4^{20} 1.0517, IR.[1914]

$CH_3P(=NCO_2Pr\text{-}i)(OR)_2$.

R = Me. XXXVI.2. $b_{0.02}$ 64-65°, n_D^{20} 1.4483, d_4^{20} 1.1220, IR.[1914]

R = Et. XXXVI.2. $b_{0.02}$ 70-71°, n_D^{20} 1.4428, d_4^{20} 1.0600, IR.[1914]

R = i-Pr. XXXVI.2. $b_{0.02}$ 63-64°, n_D^{20} 1.4420, d_4^{20} 1.0233, IR.[1914]

$CH_3P(=NCOCF_3)(OEt)_2$. XXXVI.2. b_1 83-84°, n_D^{20} 1.4022, d_4^{20} 1.2310.[1138]

$CH_3P(=NPh)(OEt)_2$. XXXVI.2. $b_{0.001}$ 47-51°, n_D^{20} 1.5237, d_4^{20} 1.0640.[627]

$CH_3P(=NCOCCl_3)(NHPh)_2$. XXXVI.3. m. 184-186°.[1906]

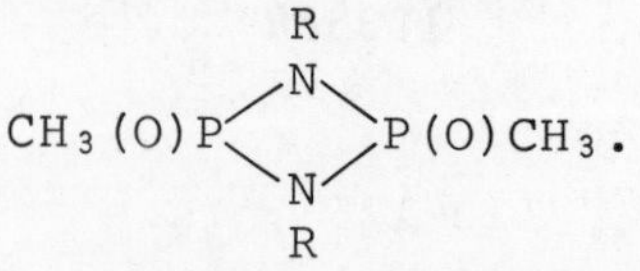

R = Bu. XXXVII. m. 123-124°.[2098]

R = Ph. XXXVII. m. 112°.[2098]

R = $-CH_2Ph$. XXXVII. m. 96-98°.[2098]

R = $4-SCNC_6H_4-$. XXXVII. m. 140-141°.[1241]

R = $3-SCN-4-ClC_6H_3-$. XXXVII. m. 147-148°.[1241]

R = $3-SCN-4-MeC_6H_3-$. XXXVII. m. 151-152°.[1241]

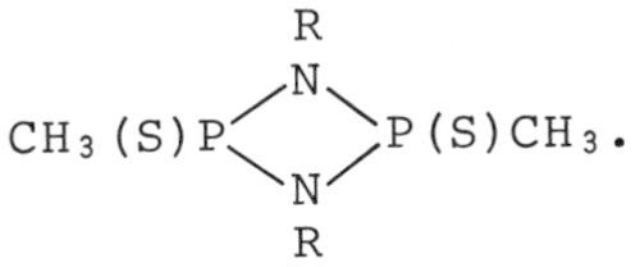

R = Me. XXXVII. m. 197-199°, IR, Raman, UV.[1233]

R = Et. XXXVII. m. 116-117.5°, IR, Raman, UV.[1233]

R = Ph. XXXVII. m. 244-245° (dec.), IR, Raman, UV.[1233]

R = $4-MeC_6H_4-$. XXXVII. m. 215-217.5°, IR, Raman, UV.[1233]

R = $3-ClC_6H_4-$. XXXVII. m. 201-203°, IR, Raman, UV.[1233]

R = $4-ClC_6H_4-$. XXXVII. m. 216-217°, IR, Raman, UV.[1233]

R = $4-MeOC_6H_4-$. XXXVII. m. 178-180°, IR, Raman, UV.[1233]

R = $4-EtOC_6H_4-$. XXXVII. m. 138.5-140°, IR, Rama, UV,[1233]

$\underline{C_2}$

$C_2H_5P(=NCOCF_3)Cl_2$. XXXVI.1. b_1 73-75°, n_D^{20} 1.4395, d_4^{20} 1.5072.[1138]

$C_2H_5P(=NCOCF_3)F_2$. XXXVI.1. $b_{1.5}$ 41-45°, n_D^{20} 1.3576, d_4^{20} 1.4509.[1138]

$C_2H_5P(=NCOCF_3)(OR)Cl$.

R = Et. XXXVI.2. b_1 78-84°, n_D^{20} 1.4175, d_4^{20} 1.3244.[1138]

R = i-Bu. XXXVI.2. b_4 103-108°, n_D^{20} 1.4210, d_4^{20} 1.2308.[1138]

$C_2H_5P(=NCOCF_3)(OEt)_2$. XXXVI.2. $b_{2.5}$ 94-95°, n_D^{20} 1.4020, d_4^{20} 1.1820.[1138]

$C_3H_5P(=NPh)(OEt)_2$. XXXVI.2. $b_{0.001}$ 47-50°, n_D^{20} 1.5200, d_4^{20} 1.0425.[627, 890]

$C_2H_5P(=NCOCF_3)(OEt)(NEt_2)$. XXXVI.3. $b_{0.5-1}$ 110-112°, n_D^{20} 1.4263, d_4^{20} 1.1405.[1138]

$\underline{C_3}$

$NCCH_2CH_2P(=NPh)(OEt)_2$. XXXVI.2. b_8 191-193°, n_D^{20} 1.5210, d_4^{20} 1.1172.[1635]

$NCCH_2CH_2P(=NAr)(OEt)_2$.

Ar = $4-MeC_6H_4-$. XXXVI.2. $b_{0.08}$ 170-171°, n_D^{20} 1.5130, d_4^{20} 1.0758.[1567]

Ar = $4-FC_6H_4-$. XXXVI.2. $b_{0.06}$ 149-150°, n_D^{20} 1.5100, d_4^{20} 1.1190.[1567]

$NCCH_2CH_2P(=NPh)(OPr)_2$. XXXVI.2. $b_{0.08}$ 149-150°, n_D^{20} 1.5160 d_4^{20} 1.0517.[1567]

$i-C_3H_7P(=NCOCF_3)F_2$. XXXVI.1. b_1 38-42°, n_D^{20} 1.3642, d_4^{20} 1.3625.[1138]

$i-C_3H_7P(=NCOCF_3)(OEt)_2$. XXXVI.2. $b_{0.2}$ 78-80°, n_D^{20} 1.4050, d_4^{20} 1.1512.[1138]

C_4

$$ClCH_2(CH_2)_3(O)P\underset{\underset{R}{\overset{|}{N}}}{\overset{\overset{R}{\overset{|}{N}}}{\diamond}}P(O)(CH_2)_3CH_2Cl.$$

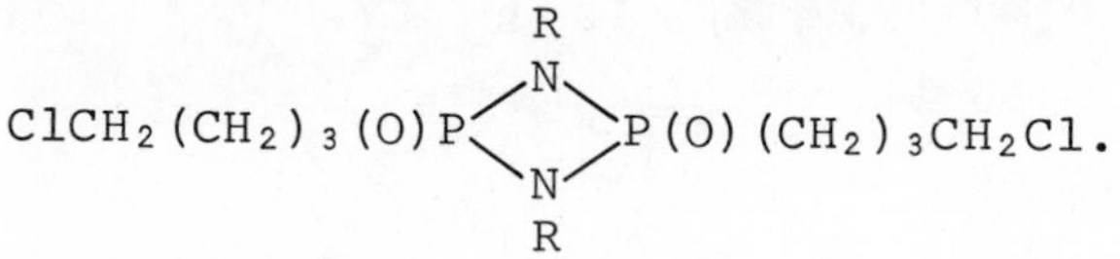

R = $3-ClC_6H_4-$. XXXVII. m. 160-163°.[768]

R = $4-ClC_6H_4-$. XXXVII. m. 183-184°.[768]

R = $4-MeSC_6H_4-$. XXXVII. m. 174-176°.[768]

R = $-CH_2Ph$. XXXVII. m. 126-130°.[768]

$NCCH(CH_3)CH_2P(=NAr)(OEt)_2$.

Ar = Ph. XXXVI.2. $b_{0.07}$ 141-142°, n_D^{20} 1.5213, d_4^{20} 1.0758.[1567]

Ar = $4-MeC_6H_4-$. XXXVI.2. $b_{0.06}$ 135-136°, n_D^{20} 1.5180, d_4^{20} 1.0878.[1567]

$CH_3O_2CCH_2CH_2P(=NPh)(OEt)_2$. XXXVI.2. b_9 187-189°, n_D^{20} 1.5070, d_4^{20} 1.1256.[1635]

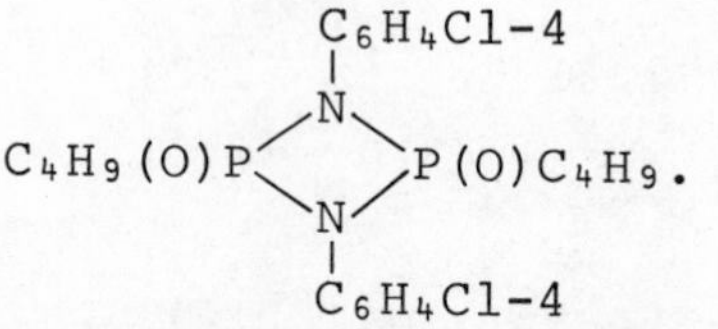

$$C_4H_9(O)P\underset{\underset{C_6H_4Cl-4}{\overset{|}{N}}}{\overset{\overset{C_6H_4Cl-4}{\overset{|}{N}}}{\diamond}}P(O)C_4H_9.$$ XXXVII. m. 162-165°.[768]

sec.$-C_4H_9P(=NCOCF_3)(OEt)_2$. XXXVI.2. b_2 100-104°, n_D^{20} 1.4078, d_4^{20} 1.1345.[1138]

C_5

$CH_3O_2CCH(CH_3)CH_2P(=NPh)(OEt)_2$. XXXVI.2. b_{10} 194-195°, n_D^{20} 1.5150, d_4^{20} 1.1110.[1635]

C_6

$4-ClC_6H_4P(=NCOAr)Cl_2$.

Ar = Ph. XXXVI.1. m. 131-133°.[471]

Ar = $4-O_2NC_6H_4-$. XXXVI.1. m. 167-169°.[471]

$C_6H_5P(=NCOAr)Cl_2$.

Ar = $3\text{-}O_2NC_6H_4$-. XXXVI.1. m. 112-113°.[471]

Ar = $4\text{-}O_2NC_6H_4$-. XXXVI.1. m. 132-133°.[471]

$C_6H_5P(=NSO_2Ar)Cl_2$.

Ar = $2\text{-}O_2NC_6H_4$-. XXXVI.1. m. 47-49°.[1889]

Ar = $3\text{-}O_2NC_6H_4$-. XXXVI.1. m. 73-75°.[1889]

Ar = $4\text{-}O_2NC_6H_4$-. XXXVI.1. m. 171-174°.[1889]

Ar = $2\text{-}C_{10}H_7$-. XXXVI.1. m. 99-102°.[1889]

$C_6H_5P(=NSO_2Ar)(OMe)_2$.

Ar = Ph. XXXVI.2. m. 48°.[1896]

Ar = $2\text{-}MeC_6H_4$-. XXXVI.2. m. 59°.[1896]

Ar = $4\text{-}MeC_6H_4$-. XXXVI.2. m. 65°.[1896]

Ar = $2\text{-}O_2NC_6H_4$-. XXXVI.2. m. 49°.[1896]

Ar = $3\text{-}O_2NC_6H_4$-. XXXVI.2. m. 64°.[1896]

Ar = $1\text{-}C_{10}H_7$-. XXXVI.2. m. 79°.[1896]

Ar = $2\text{-}C_{10}H_7$-. XXXVI.2. m. 95°.[1896]

$C_6H_5P(=NSO_2Ar)(OEt)_2$.

Ar = $2\text{-}MeC_6H_4$-. XXXVI.2. m. 100-102°.[1896]

Ar = $1\text{-}C_{10}H_7$-. XXXVI.2. m. 85°.[1896]

$C_6H_5P(=NSO_2Ph)(OCH_2CH{=}CH_2)_2$. XXXVI.2. n_D^{20} 1.5610, d_4^{20} 1.2030.[1897]

$C_6H_5P(=NSO_2Ar)(OPh)_2$.

Ar = Ph. XXXVI.2. m. 64-65°.[1896]

Ar = $2\text{-}MeC_6H_4$-. XXXVI.2. m. 56-58°.[1896]

Ar = $4\text{-}MeC_6H_4$-. XXXVI.2. m. 88-89°.[1896]

Ar = $4\text{-}ClC_6H_4$-. XXXVI.2. m. 88-89°.[1896]

Ar = $2\text{-}O_2NC_6H_4$-. XXXVI.2. m. 112-113°.[1896]

Ar = $1\text{-}C_{10}H_7$-. XXXVI.2. m. 118-120°.[1896]

Ar = $2\text{-}C_{10}H_7$-. XXXVI.2. m. 98-99°.[1896]

$C_6H_5P[=NC(CF_3)=C(CO_2Et)_2](NHPh)_2$. XXXVI.3. m. 162-163°.[279]

$C_6H_5P[=NC(CF_3)=C(CO_2Et)_2](NHC_6H_4Cl\text{-}4)_2$. XXVI.3. m. 169-173°.[279]

$C_6H_5P[=NC(CCl_3)=C(CO_2Et)_2](NHPh)_2$. XXXVI.3. m. 210-211°.[279]

$C_6H_5P[=NC(CHCl_2)=C(CO_2Me)_2]NHC_6H_3Me\text{-}2\text{-}Br\text{-}4$. XXXVI.3. m. 186-187°.[279]

$C_6H_5P(=NSO_2Ar)(NH_2)_2$.

Ar = Ph. XXXVI.3. m. 151-152°.[1895]

Ar = $2\text{-}MeC_6H_4$-. XXXVI.3. m. 140-142°.[1895]

Ar = 4-MeC$_6$H$_4$-. XXXVI.3. m. 130-131^0.[1895]

Ar = 4-ClC$_6$H$_4$-. XXXVI.3. m. 162-164^0.[1895]

Ar = 1-C$_{10}$H$_7$-. XXXVI.3. m. 149-151^0.[1895]

Ar = 2-C$_{10}$H$_7$-. XXXVI.3. m. 148-152^0.[1895]

C$_6$H$_5$P(=NSO$_2$Ar)(NHPh)$_2$.

Ar = Ph. XXXVI.3. m. 191^0.[1894]

Ar = 2-MeC$_6$H$_4$-. XXXVI.3. m. 196^0.[1894]

Ar = 4-MeC$_6$H$_4$-. XXXVI.3. m. 182^0.[1894]

Ar = 2-O$_2$NC$_6$H$_4$-. XXXVI.3. m. 212^0.[1894]

Ar = 3-O$_2$NC$_6$H$_4$-. XXXVI.3. m. 214^0.[1894]

Ar = 4-O$_2$NC$_6$H$_4$-. XXXVI.3. m. 176^0.[1894]

Ar = 1-C$_{10}$H$_7$-. XXXVI.3. m. 207^0.[1894]

Ar = 2-C$_{10}$H$_7$-. XXXVI.3. m. 215^0.[1894]

C$_6$H$_5$(O)P⟨N(R)—N(R)⟩P(O)C$_6$H$_5$.

R = Ph. XXXVII. m. 290^0.[1282]

R = 2-MeC$_6$H$_4$-. XXXVII. m. 291^0.[1282]

R = 4-MeC$_6$H$_4$-. XXXVII. m. 380^0.[1282]

R = 4-SCNC$_6$H$_4$-. XXXVII. m. 123-124^0.[1241]

R = 3-SCN-4-ClC$_6$H$_3$-. XXXVII. m. 154-155^0.[1241]

R = 3-SCN-4-MeC$_6$H$_3$-. XXXVII. m. 160-161^0.[1241]

C$_6$H$_5$(S)P⟨N(R)—N(R)⟩P(S)C$_6$H$_5$.

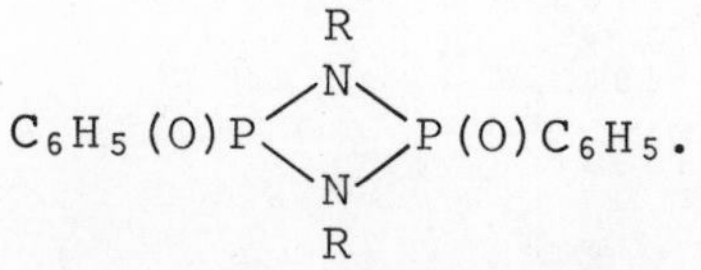

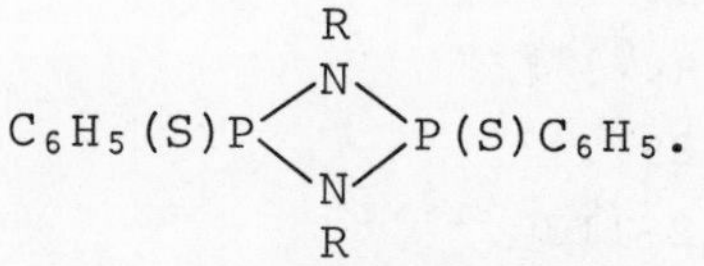

R = Me. XXXVII. m. 202-204^0, IR, Raman, UV,[1233] m.
242^0,[1793] 216-217^0, UV.[2029]

R = Et. XXXVII. m. 134^0 (cis-isomer), 142^0 (trans-
isomer), IR, ^{1}H, Raman.[541], [812]

R = i-Pr. XXXVII. m. 193-195^0, IR, Raman, UV.[1233]

R = Ph. XXXVII. m. 220-221^0, IR, Raman, UV.[1233]

R = -CH$_2$Ph. XXXVII. m. 213.5-214.5^0, UV.[2029]

R = 4-EtOC$_6$H$_4$-. XXXVII. m. >235^0, IR, Raman, UV.[1233]

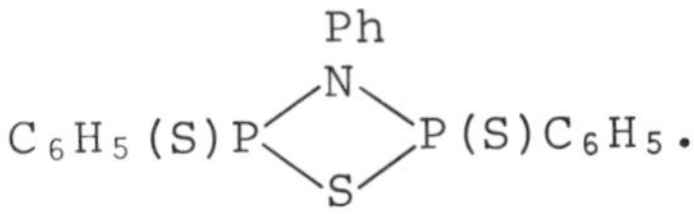

$C_6H_5(S)P \overset{\overset{\displaystyle Ph}{N}}{\underset{S}{\diamondsuit}} P(S)C_6H_5$. XXXVII. m. 183°, ^{31}P -61.0 ppm.[542]

$\underline{C_7}$

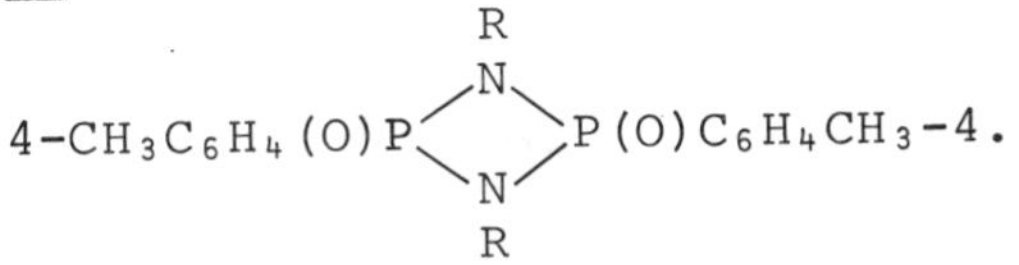

$4-CH_3C_6H_4(O)P \overset{\overset{\displaystyle R}{N}}{\underset{\underset{\displaystyle R}{N}}{\diamondsuit}} P(O)C_6H_4CH_3-4$.

R = Ph. XXXVII. m. 273°.[1282]

R = $4-MeC_6H_4-$. XXXVII. m. 312°.[1282]

$\underline{C_8}$

$O\overline{COCH(C_6H_4Br-4)P(=N)}(OEt)_2$. XXXVI.2. m. $102-105^O$.[1984]
$O\overline{COCH(C_6H_4Cl-4)P(=N)}(OEt)_2$. XXXVI.2. m. $106-108^O$.[1984]
$O\overline{COCH(C_6H_5)P(=N)}(OEt)_2$. XXXVI.2. m. $110-112^O$.[1984]

$\underline{C_{13}}$

$4-BrC_6H_4NHCH(C_6H_5)P(=NPh)(OEt)_2$. XXXVI.2. m. 192°, IR.[1568]
$4-FC_6H_4NHCH(C_6H_5)P(=NPh)(OEt)_2$. XXXVI.2. m. $192-193^O$, IR.[1568]
$4-IC_6H_4NHCH(C_6H_5)P(=NPh)(OEt)_2$. XXXVI.2. m. 187°, IR.[1568]
$C_6H_5NHCH(C_6H_4NO_2-4)P(=NNHPh)(OR)_2$.

R = Et. XXXVI.4. m. $156-157^O$.[17]

R = Bu. XXXVI.4. m. $150-152^O$.[17]

$4-O_2NC_6H_4NHCH(C_6H_5)\overline{P(=NNHPh)(OEt)_2}$. XXXVI.4. m. $143-145^O$.[17]
$C_6H_5NHCH(C_6H_4NO_2-4)P(=NNHPh)[OCH(R)CH_2O]$.

R = H. XXXVI.4. m. $138-140^O$.[17]

R = Me. XXXVI.4. m. $153-155^O$.[17]

$C_6H_5NHCH(C_6H_5)P(=NPh)(OEt)_2$. XXXVI.2. m. $195-196^O$, IR.[1568]
$C_6H_5NHCH(C_6H_5)P(=NNHPh)(OR)_2$.

R = Et. XXXVI.4. m. $148-150^O$.[17]

R = Bu. XXXVI.4. m. $148-149^O$.[17]

$C_6H_5NHCH(C_6H_5)\overline{P(=NNHPh)[OCH_2CH_2O]}$. XXXVI.4. m. $149-151^O$.[17]
$C_6H_5NHCH(C_6H_5)P(=NNHPh)[OCHMe(CH_2)_3O]$. XXXVI.4. m. $146-148^O$.[17]

$\underline{C_{14}}$

$4-ClC_6H_4NHCH(C_6H_4CH_3-4)P(=NPh)(OEt)_2$. XXXVI.2. m. $192-193^O$, IR.[1568]
$C_6H_5NHCH(C_6H_4CH_3-4)P(=NPh)(OEt)_2$. XXXVI.2. m. $188-189^O$, IR.[1568]
$2-CH_3C_6H_4NHCH(C_6H_5)P(=NNHPh)(OEt)_2$. XXXVI.4. m. $134-136^O$.[17]

$3\text{-}CH_3C_6H_4NHCH(C_6H_5)P(=NNHPh)(OR)_2$.

R = Et. XXXVI.4. m. 152-153°.[17]

R = Bu. XXXVI.4. m. 151-152°.[17]

$4\text{-}CH_3C_6H_4NHCH(C_6H_5)P(=NNHPh)(OEt)_2$. XXXVI.4. m. 139-141°.[17]

$3\text{-}CH_3C_6H_4NHCH(C_6H_5)P(=NNHPh)[OCH_2CH_2O]$. XXXVI.4. m. 151-152°.[17]

C.11. Polyphosphonic Acid Amide Derivatives

C_1

$(Me_2N)_2P(O)CH_2(O)P(NMe_2)_2$ = L, Cu^{2+} complexes of composition:

$CuL_2(ClO_4)_2$, ESR, $g_{\parallel}$ 2.5, $g_{\perp}$ 2.08 (297°K).

$CuL_3(ClO_4)_2$, ESR, $g_{\parallel}$ 2.44, $g_{\perp}$ 2.12 (88°K). X-ray powder spectra.[876]

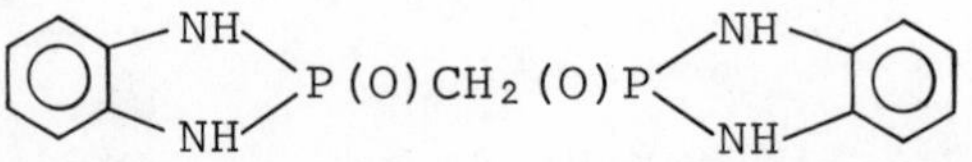

XVII.1. m. 338-341°.[1947]

C_2

$(Et_2N)_2P(O)C{\equiv}C(O)P(NEt_2)_2$. XVII.8. m. 18-20°.[1054]
$(Me_2N)_2P(S)C{\equiv}C(S)P(NMe_2)_2$. XXIII.4. m. 93°.[1054]
$(Et_2N)_2P(S)C{\equiv}C(S)P(NEt_2)_2$. XXIII.4. m. 56°.[1054]
$(Me_2N)_2P(Se)C{\equiv}C(Se)P(NMe_2)_2$. XXXV. m. 102°.[1054]
$(Et_2N)_2P(Se)C{\equiv}C(Se)P(NEt_2)_2$. XXXV. m. 60°.[1054]
$(CH_2CH_2N)_2P(O)CH_2CH_2(O)P(NCH_2CH_2)_2$. XVII.1. m. 135-136°.[1476]

$(Et_2N)_2P(S)CH_2CH_2(S)P(NCH_2CH_2)_2$. XXIII.5. m. 31-33°.[1477]
$(EtO)(Et_2N)P(O)C(OH)(CH_3)P(O)(OEt)_2$. XVI.4. b_2 142-144°, n_D^{20} 1.4335, d_4^{20} 1.1573.[1591]

C_3

$(CH_2CH_2N)_2P(O)CH(CH_3)CH_2P(O)(NCH_2CH_2)_2$. XVII.1. m. 98.5-99°.[1477]

C_4

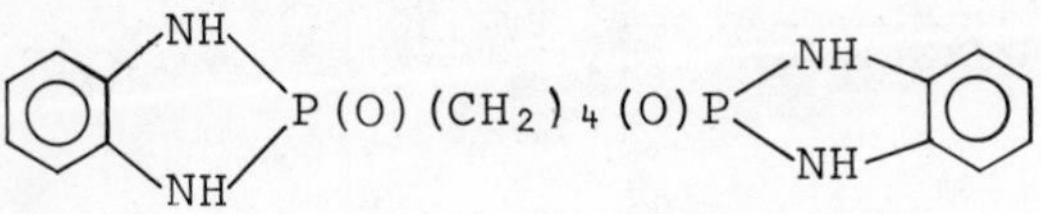

XVII.1. m. 328-334°.[1947]

C_6

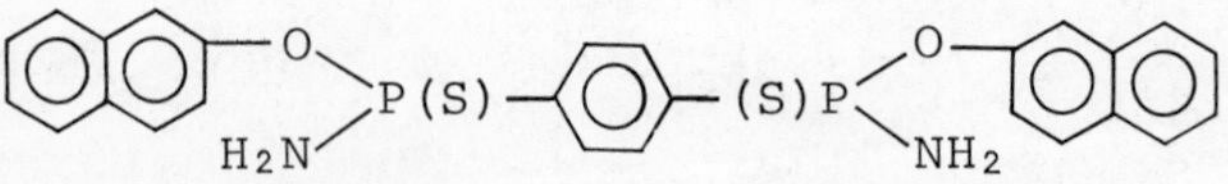

XVII.1. m. 184-185°.[272]

$\underline{C_7}$

$(EtO)(Me_2N)P(O)C(C_6H_5)(OH)P(O)(OEt)_2$. XVI.4. b_5 166-167°, n_D^{20} 1.4810, d_4^{20} 1.1130.[1591]

$\underline{C_8}$

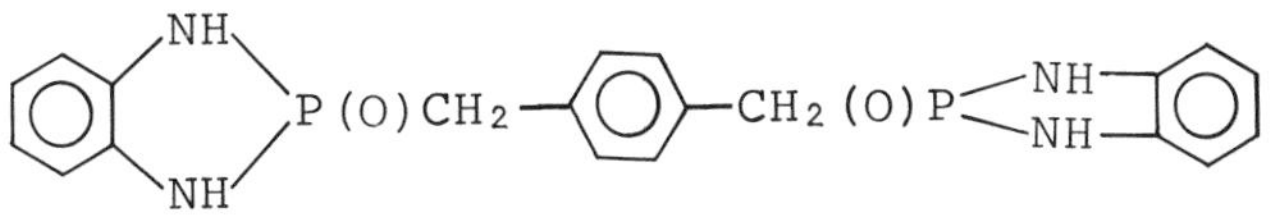

XVII.1. m. 378-382°.[1947]

$\underline{C_{11}}$

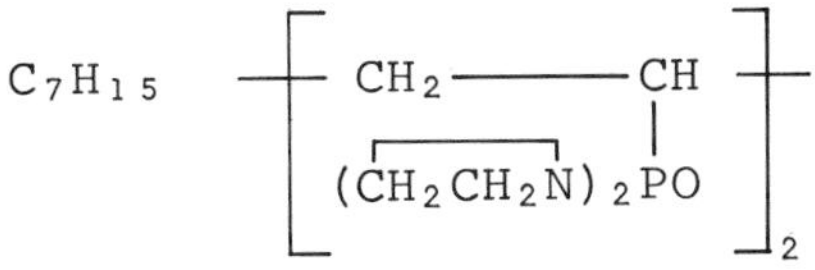

H, XVII.1. m. 103-107°, IR, ^1H.[1335]

(Received November 7, 1974)

REFERENCES

1. Aaron, H. S., T. M. Shrine, and J. I. Miller, J. Am. Chem. Soc., 80, 107 (1958).
2. Aaron, H. S., R. T. Uyeda, H. F. Frack, and J. I. Miller, J. Am. Chem. Soc., 84, 617 (1962).
3. Abduvakhabov, A. A., I. I. Agabekova, N. N. Godovikov, M. I. Kabachnik, and V. I. Rozengart, Izv. Akad. Nauk SSSR, Ser. Khim., 1968, 2480; C. A., 70, 54322v (1969).
4. Abduvakhabov, A. A., I. I. Agabekova, N. N. Godovikov, M. I. Kabachnik, and V. I. Rozengart, Izv. Akad. Nauk SSSR, Ser. Khim., 1970, 1588; C. A., 74, 61091e (1970).
5. Abduvakhabov, A. A., V. A. Avdeeva, N. N. Godovikov, M. I. Kabachnik, A. B. Katsnelson, S. S. Mickajlov, V. I. Rozengart, R. V. Sitkevich, and Ya. S. Smusin, Izv. Akad. Nauk SSSR, Ser. Khim., 1968, 2033; C. A., 69, 103316m (1968).
6. Abduvakhabov, A. A., N. N. Godovikov, and M. I. Kabachnik, Izv. Akad. Nauk SSSR, Ser. Khim., 1968, 583; C. A., 69, 59344t (1968).
7. Abduvakhabov, N. A., N. N. Godovikov, M. I. Kabachnik, V. I. Rozengart, and R. V. Sitkevich, Izv. Akad. Nauk SSSR, Ser. Khim., 1969, 1117; C. A., 71, 57021c (1969).
8. Abel, E. W., D. A. Armitage, and R. P. Bush, J. Chem. Soc., 1965, 7098.
9. Abramov, V. S., Dokl. Akad. Nauk SSSR, 95, 991 (1954); C. A., 49, 6084d (1955).

10. Abramov, V. S., Khim. Primenenie Fosforoorgan. Soedin. Akad. Nauk SSSR, Tr. 1-Konf., 1955, 71; C. A., 52, 240b (1958).
11. Abramov, V. S., Zh. Obshch. Khim., 27, 169 (1957).
12. Abramov, V. S., Zh. Obshch. Khim., 22, 647 (1952).
13. Abramov, V. S., and M. M. Azanovskaja, Zh. Obshch. Khim., 12, 270 (1942).
14. Abramov, V. S., V. I. Barabanov, and L. I. Long, Zh. Obshch. Khim., 37, 714 (1967).
15. Abramov, V. S., L. S. B. Belokon, and F. I. Machmutova Zh. Obshch. Khim., 28, 665 (1958).
16. Abramov, V. S., Yu. A. Bochkova, and A. D. Polyakova, Zh. Obshch. Khim., 23, 1013 (1953).
17. Abramov, V. S., R. Sh. Chenborisov, and A. P. Kirisova, Zh. Obshch. Khim., 39, 350 (1969).
18. Abramov, V. S., R. V. Dimitrieva, and A. S. Kapustina, Zh. Obshch. Khim., 23, 257 (1953).
19. Abramov, V. S., and Z. S. Drushina, Zh. Obshch. Khim., 37, 718 (1967).
20. Abramov, V. S., Z. S. Drushina, and T. V. Zykova, Zh. Obshch. Khim., 37, 1332 (1967).
21. Abramov, V. S., and N. A. Il'ina, Zh. Obshch. Khim., 24, 124 (1954).
22. Abramov, V. S., and M. G. Gubaidullin, Zh. Obshch. Khim., 39, 2411 (1969).
23. Abramov, V. S., and M. G. Gubaidullin, Zh. Obshch. Khim., 38, 862 (1968).
24. Abramov, V. S., and N. A. Il'ina, Zh. Obshch. Khim., 38, 677 (1968).
25. Abramov, V. S., and N. A. Il'ina, Dokl. Akad. Nauk SSSR, 125, 1027 (1959); C. A., 53, 21747 (1959).
26. Abramov, V. S., and N. A. Il'ina, Dokl. Akad. Nauk SSSR, 132, 823 (1960); C. A., 54, 22329 (1960).
27. Abramov, V. S., and N. S. Il'ina, Zh. Obshch. Khim., 41, 100 (1971).
28. Abramov, V. S., N. A. Il'ina, and I. N. Yuldaseva, Zh. Obshch. Khim., 38, 1652 (1968).
28a. Abramov, V. S., and A. S. Kapustina, Dokl. Akad. Nauk SSSR, 111, 1243 (1956); C. A., 51, 9473 (1957).
29. Abramov, V. S., and A. S. Kapustina, Zh. Obshch. Khim. 27, 173 (1957).
30. Abramov, V. S., and A. S. Kapustina, Zh. Obshch. Khim. 29, 3319 (1959).
31. Abramov, V. S., and G. A. Karp, Zh. Obshch. Khim., 24, 1823 (1954).
32. Abramov, V. S., and V. K. Khajrullin, Zh. Obshch. Khim., 29, 1599 (1959).
33. Abramov, V. S., and V. K. Khajrullin, Zh. Obshch. Khim., 29, 1222 (1959).
34. Abramov, V. S., and V. K. Khajrullin, Zh. Obshch. Khim., 26, 811 (1956).
35. Abramov, V. S., and N. S. Kuznetsov, Sbornik Statei

Obshch. Khim. Akad. Nauk SSSR, 1, 398 (1953); C. A., 49, 839a (1955).
36. Abramov, V. S., and M. N. Morosova, Zh. Obshch. Khim., 22, 257 (1952).
37. Abramov, V. S., and A. L. Salman, Zh. Obshch. Khim., 32, 827 (1962).
38. Abramov, V. S., and R. N. Savintseva, Zh. Obshch. Khim., 39, 1967 (1969).
39. Abramov, V. S., R. N. Savintseva, and F. G. Fatykhova, Zh. Obshch. Khim., 39, 2658 (1969).
40. Abramov, V. S., and L. P. Semenova, Sbornik Statei Obshch. Khim. Akad. Nauk SSSR, 1, 393 (1953); C. A., 49, 838g (1955).
41. Abramov, V. S., E. V. Sergeeva, and I. V. Chelpanova, Zh. Obshch. Khim., 14, 1030 (1940); C. A., 41, 700c (1947).
42. Ackerman, B., R. M. Chladek, and D. Swern, J. Am. Chem. Soc., 79, 6524 (1957).
43. Ackerman, B., R. M. Chladek, and D. Swern, J. Am. Chem. Soc., 81, 4335 (1959).
44. Ackerman, B., R. M. Chladek, and D. Swern, J. Am. Chem. Soc., 79, 6524 (1957).
45. Ackerman, B., T. A. Jordan, and D. Swern, J. Am. Chem. Soc., 78, 6025 (1956).
46. Ackerman, B., and D. Swern, U. S. 2,929,831; C. A., 54, 15246b (1960).
47. Agawa, T., T. Kubo, and Y. Oshiro, Synthesis, 1971, I, 27.
48. Akamsin, B. F., and N. I. Rizpolozhenskij, Izv. Akad. Nauk SSSR, Ser. Khim., 1967, 1983; C. A., 68, 114692 (1968).
49. Akamsin, V. D., and N. I. Rizpolozhenskij, Izv. Akad. Nauk SSSR, Ser. Khim., 1967, 825; C. A., 67, 100207z (1967).
50. Akamsin, V. D., and N. I. Rizpolozhenskij, Izv. Akad. Nauk SSSR, Ser. Khim., 1967, 1987; C. A., 68, 114693 (1968).
51. Akamsin, V. D., and N. I. Rizpolozhenskij, Izv. Akad. Nauk SSSR, Ser. Khim., 1966, 493; C. A., 65, 8953f (1966).
52. Akerfeldt, S., and L. Fagerlind, J. Med. Chem., 10, 115 (1967).
53. Albertson, N. F., U. S. 2,980,692 (1961); C. A., 56, 457e (1962).
54. Aleksandrova, I. A., and G. M. Vinokurova, Izv. Nauk SSSR, Ser. Khim., 1969, 1163; C. A., 71, 50084 (1969).
55. Alekseev, V. V., and M. S. Malinovskij, Zh. Obshch. Khim., 31, 3437 (1961).
56. Alekseev, V. V., and M. S. Malinovskij, Zh. Obshch. Khim., 30, 2967 (1960).
57. Aleksandrova, I. A., and G. N. Vinokurova, Izv. Akad. Nauk SSSR, Ser. Khim. 1969, 1163; C. A., 71,

50084s (1969).
58. Alexander, C. H., U. S. 2,945,052 (1955); C. A., 55, 2569 (1961).
59. Alexander, B. H., and W. F. Barthel, J. Org. Chem., 23, 101 (1958).
60. Alexander, B. H., L. S. Hafner, M. V. Garrison, and J. E. Brown, J. Org. Chem., 28, 3499 (1963).
61. Alimov, P. I., and L. A. Antokhina, Izv. Akad. Nauk SSSR, Ser. Khim., 1966, 1486; C. A., 66, 54956 (1967)
62. Allen, J. F., and O. H. Johnson, J. Am. Chem. Soc., 77, 2871 (1955).
63. Allum, K. G., and E. S. Forbes, Brit. 1,189,304 (1970 C. A., 73, 14990s (1970).
64. Almasi, L., N. Popovici, and I. Zsako, Chem. Ber., 106, 1384 (1973).
65. Almasi, L., and L. Paskucs, Rev. Roum. Chim., 10, 301 (1965); C. A., 63, 11604f (1965).
66. Almasi, I., and L. Paskucs, Chem. Ber., 97, 623 (1964).
67. Amonoo-Neizer, E. H., S. K. Ray, R. A. Shaw, and B. C. Smith, J. Chem. Soc., 1965, 6250.
68. Anand, N., and A. R. Todd, J. Chem. Soc., 1951, 1867.
69. Anderson, J. A., U. S. 2,316,089 (1941); C. A., 37, 5858 (1943).
70. Andrianov, K. A., L. M. Khanashvili, A. A. Kazakova, and A. A. Ivanow, Zh. Obshch. Khim., 31, 228 (1961).
71. Andrianov, K. A., and A. S. Shapatin, Izv. Akad. Nauk SSSR, Ser. Khim., 1962, 1753; C. A., 58, 7965a (1963).
72. Andrianov, K. A., A. S. Shapatin, and V. V. Ponomarov Izv. Akad. Nauk SSSR, Ser. Khim., 1965, 187; C. A., 62, 11846c (1965).
73. Andrianov, K. A., T. V. Vasil'eva, and T. K. Demykin, Zh. Obshch. Khim., 40, 1565 (1970).
74. Andrianov, K. A., T. V. Vasil'eva, and L. M. Khanashvili, Izv. Akad. Nauk SSSR, Ser. Khim., 1961, 1030; C. A., 55, 27019a (1961).
75. Andrianov, K. A., T. V. Vasil'eva, and L. V. Kozlova, Izv. Akad. Nauk SSSR, Ser. Khim., 1965, 381; C. A., 62, 14718e (1965).
76. Andrianov, K. A., T. V. Vasil'eva, and A. A. Minaeva, Izv. Akad. Nauk SSSR, Ser. Khim., 1963, 2227; C. A., 60, 9307b (1964).
77. Andrianov, K. A., A. A. Zhdanov, L. M. Khanashvili. and A. S. Sapatin, Zh. Obshch. Khim., 31, 224 (1961).
78. Angel, S. H., E. W. Fuller, and H. G. Berger, U. S. 2,381,377 (1945); C. A., 40, 1998 (1946).
79. Anisimov, K. N., and N. E. Kolobova, Izv. Akad. Nauk SSSR, Ser. Khim., 1956, 923 u. 927; C. A., 51, 4933f (1957).
80. Anisimov, K. N., N. E. Kolobova, and A. N. Nesmejanov Izv. Akad. Nauk SSSR, Ser. Khim., 1955, 999; C. A., 50, 11267d (1956).

81. Anisimov, K. N., N. E. Kolobova, and A. N. Nesmejanov, Izv. Akad. Nauk SSSR, Ser. Khim., 1955, 240; C. A., 50, 3217d (1956).
82. Anisimov, K. N., N. E. Kolobova, and A. N. Nesmejanov, Izv. Akad. Nauk SSSR, Ser. Khim., 1955, 665; C. A., 50, 7076 (1956).
83. Anisimov, K. N., N. E. Kolobova, and A. N. Nesmejanov, Izv. Akad. Nauk SSSR, Ser. Khim., 1955, 669; C. A., 50, 7077b (1956).
84. Anisimov, K. N., N. E. Kolobova, and A. N. Nesmejanov, Izv. Akad. Nauk SSSR, Ser. Khim., 1955, 823; C. A., 50, 9319e (1956).
85. Anisimov, K. N., N. E. Kolobova, and A. N. Nesmejanov, Izv. Akad. Nauk SSSR, Ser. Khim., 1955, 834; C. A., 50, 9320 (1956).
86. Anisimov, K. N., N. E. Kolobova, and A. N. Nesmejanov, Izv. Akad. Nauk SSSR, Ser. Khim., 1956, 23; C. A., 50, 13784i (1956).
87. Anisimov, K. N., and B. V. Kopyleva, Izv. Akad. Nauk SSSR, Ser. Khim., 1961, 277; C. A., 55, 18562h (1961).
88. Anisimov, K. N., and G. M. Kunitskaja, Izv. Akad. Nauk SSSR, Ser. Khim., 1961, 274; C. A., 55, 18562f (1961).
89. Anisimov, K. N., G. M. Kunitskaja, and I. A. Slovokhotova, Izv. Akad. Nauk SSSR, Ser. Khim., 1961, 64; C. A., 55, 18562h (1961).
90. Anisimov, K. N., and A. N. Nesmejanov, Izv. Akad. Nauk SSSR, Ser. Khim., 1955, 1006; C. A., 50, 11267h (1956).
91. Anisimov, K. N., and A. N. Nesmejanov, Izv. Akad. Nauk SSSR, Ser. Khim., 1955, 1003; C. A., 50, 11267g (1956).
92. Anisimov, K. N., and A. N. Nesmejanov, Izv. Akad. Nauk SSSR, Ser. Khim., 1956, 16; C. A., 50, 13784i (1956).
93. Anisimov, K. N., A. N. Nesmejanov, and N. E. Kolobova, Izv. Akad. Nauk SSSR, Ser. Khim., 1956, 19; C. A., 50, 13784 (1956).
94. Anisimov, K. N., and B. V. Raisbaum, Izv. Akad. Nauk SSSR, Ser. Khim., 1958, 1208; C. A., 53, 4114 (1959).
95. Anschütz, L., E. Klein, and G. Cermak, Chem. Ber., 77, 726 (1944).
96. Anschütz, L., and H. Walbrecht, J. Prakt. Chem. (2), 133, 65 (1932).
97. Anschütz, L., and H. Wirth, Chem. Ber., 89, 688 (1956).
98. Anschütz, L., and H. Wirth, Naturwissenschaften, 43, 16 (1956).
99. Antokhina, L. A., and P. I. Alimov, Izv. Akad. Nauk SSSR, Ser. Khim., 1966, 2135; C. A., 66, 95132 (1967).

100. Antokhina, L. A., and P. I. Alimov, Izv. Akad. Nauk SSSR, Ser. Khim., 1967, 1161; C. A., 68, 95899r (1968).
101. Arbuzov, A. E., Zh. Fiz. Khim., 42, 395 (1910); Chem. Zentr., 1910, II, 453.
102. Arbuzov, A. E., Zh. Fiz. Khim., 38, 687 (1906); C. A., 5, 1397 (1911).
102a. Arbuzov, A. E., Izv. Akad. Nauk SSSR, Ser. Khim., 1946, 285.
103. Arbuzov, A. E., and V. S. Abramov, Izv. Akad. Nauk SSSR, Ser. Khim., 1959, 35; C. A., 53, 15956i (1959).
104. Arbuzov, A. E., and B. A. Arbuzov, Zh. Fiz. Khim., 61, 217 (1929); C. A., 23, 3921 (1929).
105. Arbuzov, A. E., and B. A. Arbuzov, J. Prakt. Chem., 130, 103 (1931).
106. Arbuzov, A. E., and B. A. Arbuzov, Chem. Ber., 62, 1871 (1929).
106a. Arbuzov, A. E., and B. A. Arbuzov, Zh. Fiz. Khim., 61, 1923 (1929).
107. Arbuzov, A. E., B. A. Arbuzov, and B. P. Lugovkin, Izv. Akad. Nauk SSSR, Ser. Khim., 1947, 535; C. A., 42, 1886 (1948).
108. Arbuzov, A. E., and I. Arbuzova, Zh. Fiz. Khim., Sect. Khim., 62, 1533 (1930); C. A., 25, 2414 (1931).
109. Arbuzov, A. E., and M. M. Azanovskaja, Dokl. Akad. Nauk SSSR, 58, 1961 (1947); C. A., 46, 8606 (1952).
110. Arbuzov, A. E., and Ch. -L. Chang, Izv. Akad. Nauk SSSR, Ser. Khim., 1963, 1934; C. A., 60, 5539h (1964).
111. Arbuzov, A. E., and Ch. -L. Chang, Izv. Akad. Nauk SSSR, Ser. Khim., 1963, 1945; C. A., 60, 5540 (1964).
112. Arbuzov, A. E., and A. A. Dunin, Chem. Ber., 60, 291 (1927).
112a. Arbuzov, A. E., and A. A. Dunin, Zh. Fiz. Khim., 46, 295 (1914); C. A., 8, 2551 (1914).
112b. Arbuzov, A. E., and B. E. Ivanov, Zh. Fiz. Khim., 45, 690 (1913); C. A., 7, 3599 (1913).
113. Arbuzov, A. E., and G. K. Kamai, Zh. Fiz. Khim., 61, 619 (1929); C. A., 23, 4443 (1929).
114. Arbuzov, A. E., and G. K. Kamai, Zh. Fiz. Khim., 61, 2037 (1929); Chem. Zentr., 1930, I, 3550.
115. Arbuzov, A. E., and G. Kamai, Zh. Obshch. Khim., 17, 2149 (1947); C. A., 42, 4523 (1948).
116. Arbuzov, A. E., T. Konstantinova, and T. Anzyfrova, Izv. Akad. Nauk SSSR, Ser. Khim., 1946, 179; C. A., 42, 6315 (1948).
117. Arbuzov, A. E., and L. V. Nesterov, Izv. Akad. Nauk SSSR, Ser. Khim., 1954, 427; C. A., 49, 9541c (1955).
118. Arbuzov, A. E., and K. V. Nikonorov, Zh. Obshch.

Khim., 18, 1137 (1948); C. A., 43, 1346i (1949).
119. Arbuzov, A. E., and K. V. Nikonorov, Zh. Obshch. Khim., 17, 2129, 2139 (1947); C. A., 42, 4546 (1948).
119a. Arbuzov, A. E., and A. I. Razumov, Zh. Obshch. Khim., 4, 834 (1934); C. A., 29, 2145[9] (1935).
120. Arbuzov, A. E., and N. A. Razumova, Izv. Akad. Nauk SSSR, Ser. Khim., 1958, 1061; C. A., 53, 3046h (1959).
120a. Arbuzov, A. E., and F. G. Valitova, Bull. Akad. Nauk SSSR, Ser. Khim., 1940, 529; C. A., 35, 3990[1] (1941).
121. Arbuzov, A. E., and V. M. Zoroastrova, Izv. Akad. Nauk SSSR, Ser. Khim., 1951, 536; C. A., 47, 97c (1953).
122. Arbuzov, A. E., V. M. Zoroastrova, and N. I. Rizpolozhenskij, Izv. Akad. Nauk SSSR, Ser. Khim., 1948, 208; C. A., 42, 4932 (1948).
123. Arbuzov, A. E., and P. I. Alimov, Izv. Akad. Nauk SSSR, Ser. Khim., 1951, 409; C. A., 49, 159i (1955).
124. Arbuzov, A. E., and N. P. Kushkova, Zh. Obshch. Khim., 6, 283 (1936); Chem. Zentr., 1936, II, 3660; C. A., 30, 4813 (1936).
125. Arbuzov, B. A., P. I. Alimov, and O. N. Fedorova, Izv. Kazan. Fil. Akad. Nauk SSSR, 1955, 25; C. A., 52, 243 (1958).
126. Arbuzov, B. A., N. P. Bogonostseva, V. S. Vinogradova, and A. A. Musina, Izv. Akad. Nauk SSSR, Ser. Khim., 1971, 2757.
127. Arbuzov, B. A., and N. P. Bogonostseva, Zh. Obshch. Khim., 29, 2617 (1959).
128. Arbuzov, B. A., and N. P. Bogonostseva, Sbornik Statei Obshch. Khim., Akad. Nauk SSSR, 2, 1144 (1953); C. A., 49, 4556 (1955).
128a. Arbuzov, B. A., and N. P. Bogonostseva, Zh. Obshch. Khim., 26, 2419 (1956).
129. Arbuzov, B. A., and N. P. Bogonostseva, Zh. Obshch. Khim., 27, 2360 (1957).
130. Arbuzov, B. A., E. N. Dianova, and V. S. Vinogradova, Izv. Akad. Nauk SSSR, Ser. Khim., 1970, 2543; C. A., 75, 6014c (1971).
131. Arbuzov, B. A., E. N. Dianova, and V. S. Vinogradova, Izv. Akad. Nauk SSSR, Ser. Khim., 1969, 1109; C. A., 71, 50070j (1969).
132. Arbuzov, B. A., and E. N. Dianova, Izv. Akad. Nauk SSSR, Ser. Khim., 1961, 1288; C. A., 56, 3511e (1962).
133. Arbuzov, B. A., and E. N. Dianova, Izv. Akad. Nauk SSSR, Ser. Khim., 1960, 1399; C. A., 55, 410 (1961).
134. Arbuzov, B. A., E. N. Dianova, V. S. Vinogradova, and A. A. Musina, Izv. Akad. Nauk SSSR, Ser. Khim., 1969, 1530; C. A., 71, 113052c (1969).

135. Arbuzov, B. A., E. N. Dianova, and V. S. Vinogradova Izv. Akad. Nauk SSSR, Ser. Khim., 1970, 2833; C. A., 74, 112141f (1971).
136. Arbuzov, B. A., E. N. Dianova, V. S. Vinogradova, and Yu. Yu. Samitov, Dokl. Akad. Nauk SSSR, 173, 1321 (1967); C. A., 68, 12911q (1968).
137. Arbuzov, B. A., E. N. Dianova, V. S. Vinogradova, and A. V. Shamsutdinova, Dokl. Akad. Nauk SSSR, 160, 99 (1965); C. A., 62, 11848f (1965).
138. Arbuzov, B. A., and N. P. Grechkin, Zh. Obshch. Khim., 17, 2166 (1947); C. A., 42, 4522 (1948).
139. Arbuzov, B. A., and B. P. Lugovkin, Zh. Obshch. Khim., 20, 1249 (1950); C. A., 45, 1567a (1951).
140. Arbuzov, B. A., and B. P. Lugovkin, Zh. Obshch. Khim., 22, 1193 (1952).
141. Arbuzov, B. A., and B. P. Lugovkin, Zh. Obshch. Khim., 21, 1869 (1951).
142. Arbuzov, B. A., and B. P. Lugovkin, Zh. Obshch. Khim., 21, 99 (1951).
143. Arbuzov, B. A., and B. P. Lugovkin, Zh. Obshch. Khim., 22, 1193 (1952).
144. Arbuzov, B. A., and B. P. Lugovkin, Zh. Obshch. Khim., 22, 1199 (1952).
145. Arbuzov, B. A., and B. P. Lugovkin, Zh. Obshch. Khim., 21, 99 (1951).
146. Arbuzov, B. A., and O. N. Nuretdinova, Izv. Akad. Nauk SSSR, Ser. Khim., 1969, 1314; C. A., 71, 81482z (1969).
147. Arbuzov, B. A., N. A. Polezhaeva, and V. S. Vinogradova, Dokl. Akad. Nauk SSSR, 201, 91 (1971).
148. Arbuzov, B. A., L. A. Polezhaeva, and V. S. Vinogradova, Izv. Akad. Nauk SSSR, Ser. Khim., 1967, 1146; C. A., 68, 13092k (1968).
149. Arbuzov, B. A., N. A. Polezhaeva, V. S. Vinogradova, and I. I. Saidshev, Izv. Akad. Nauk SSSR, Ser. Khim., 1971, 2762. C. A. 77, 512g (1972).
150. Arbuzov, B. A., N. A. Polezhaeva, V. S. Vinogradova, and Yu. Yu. Samitov, Dokl. Akad. Nauk SSSR, 173, 93 (1967); C. A., 67, 54221z (1967).
151. Arbuzov, B. A., and A. N. Pudovik, Dokl. Akad. Nauk SSSR, 59, 1433 (1948); C. A., 47, 4281g (1953).
152. Arbuzov, B. A., and A. N. Pudovik, Zh. Obshch. Khim. 17, 2158 (1947); C. A., 42, 4522a (1948).
153. Arbuzov, B. A., and N. I. Rizpoloshenskij, Izv. Akad. Nauk SSSR, Ser. Khim., 1952, 854; C. A., 47, 9903c (1953).
154. Arbuzov, B. A., N. I. Rizpoloshenskij, and M. A. Zvereva, Izv. Akad. Nauk SSSR, Ser. Khim., 1955, 1021; C. A., 50, 11233 (1956).
155. Arbuzov, B. A., N. I. Rizpoloshenskij, and M. A. Zvereva, Izv. Akad. Nauk SSSR, Ser. Khim., 1957, 179; C. A., 51, 11237f (1957).

156. Arbuzov, B. A., T. D. Sorokina, N. P. Bogonosceva, and V. A. Vinogradova, Dokl. Akad. Nauk SSSR., 171, 605 (1966); C. A., 67, 32501p (1967).
157. Arbuzov, B. A., T. D. Sorokina, N. P. Bogonosceva, and V. S. Vinogradova, Dokl. Akad. Nauk SSSR, 171, 1097 (1966); C. A., 66, 65584a (1967).
158. Arbuzov, B. A., I. D. Sorokina, V. S. Vinogradova, and A. A. Musina, Izv. Akad. Nauk SSSR, Ser. Khim., 1969, 2727; C. A., 72, 78740x (1970).
159. Arbuzov, B. A., and V. S. Vinogradova, Izv. Akad. Nauk SSSR, Ser. Khim., 1957, 284; C. A., 51, 14587a (1957).
160. Arbuzov, B. A., and V. S. Vinogradova, Izv. Akad. Nauk SSSR, Ser. Khim., 1952, 882; C. A., 47, 10464h (1953).
161. Arbuzov, B. A., V. S. Vinogradova, and N. A. Polezhaeva, Dokl. Akad. Nauk SSSR, 137, 855 (1961); C. A., 56, 3507c (1962).
162. Arbuzov, B. A., V. S. Vinogradova, and N. A. Polezhaeva, Izv. Akad. Nauk SSSR, Ser. Khim., 1961, 71; C. A., 57, 13790h (1962).
163. Arbuzov, B. A., V. S. Vinogradova, and N. A. Polezhaeva, Izv. Akad. Nauk SSSR, Ser. Khim., 1960, 832; C. A., 54, 24454f (1960).
164. Arbuzov, B. A., V. S. Vinogradova, and N. A. Polezhaeva, Izv. Akad. Nauk SSSR, Ser. Khim., 1961, 2013; C. A., 56, 11456i (1962).
165. Arbuzov, B. A., V. S. Vinogradova, N. A. Polezhaeva, and A. K. Shamsutdinova, Izv. Akad. Nauk SSSR, Ser. Khim., 1963, 675; C. A., 59, 11551d (1963).
166. Arbuzov, B. A., V. S. Vinogradova, and O. A. Zolova, Izv. Akad. Nauk SSSR, Ser. Khim., 1968, 2290; C. A., 70, 29004k (1969).
167. Arbuzov, B. A., V. S. Vinogradova, N. A. Polezhaeva, and A. K. Shamsutdinova, Izv. Akad. Nauk SSSR, Ser. Khim., 1963, 1380; C. A., 59, 15306f (1963).
168. Arbuzov, B. A., V. S. Vinogradova, and M. A. Zwereva, Izv. Akad. Nauk SSSR, Ser. Khim., 1960, 1772; C. A., 55, 16398e (1961).
169. Arbuzov, B. A., V. S. Vinogradova, and M. A. Zwereva, Izv. Akad. Nauk SSSR, Ser. Khim., 1960, 1981; C. A., 55, 16398g (1961).
170. Arbuzov, B. A., A. O. Vizel', and K. M. Ivanovskaja, Khim. Geterotsikl. Soedin., 1967, 1130; C. A., 69, 67482b (1968).
171. Arbuzov, B. A., A. O. Vizel', K. M. Ivanovskaja, R. R. Shagidullin, V. V. Pomazanov, and E. I. Gol'dfarb, Khim. Geterotsikl. Soedin., 1971, 1329; C. A., 76, 34360 (1972).
172. Arbuzov, B. A., and D. C. Yarmukhametova, Izv. Akad. Nauk SSSR, Ser. Khim., 1962, 1405; C. A., 58, 2468f (1963).

173. Arbuzov, B. A., and D. C. Yarmukhametova, Izv. Akad. Nauk SSSR, Ser. Khim., 1960, 1767; C. A., 55, 15507c (1961).
174. Arbuzov, B. A., and M. V. Zolotova, Izv. Akad. Nauk SSSR, Ser. Khim., 1964, 1793; C. A., 62, 3921h (1965).
175. Arbuzov, B. A., O. D. Zolova, V. S. Vinogradova, and Yu. Yu. Samitov, Dokl. Akad. Nauk SSSR, 173, 335 (1967); C. A., 67,43886u (1967).
176. Arbuzov, B. A., V. M. Zoroastrova, and N. D. Ibragimova, Izv. Akad. Nauk SSSR, Ser. Khim., 1967, 704; C. A., 67, 108477h (1967).
177. Arbuzov, B. A., V. M. Zoroastrova, G. A. Tudrij, and A. V. Fuzhenkova, Dokl. Akad. Nauk SSSR, 200, 847 (1971); C. A., 76, 25367g (1972).
178. Arbuzov, B. A., and V. M. Zoroastrova, Izv. Akad. Nauk SSSR, Ser. Khim., 1960, 1030; C. A., 54, 24627 (1960).
179. Arbuzov, B. A., and V. M. Zoroastrova, Izv. Akad. Nauk SSSR, Ser. Khim., 1954, 806; C. A., 49, 13222 (1955).
180. Arbuzov, B. A., and V. M. Zoroastrova, Izv. Akad. Nauk SSSR, Ser. Khim., 1966, 254; C. A., 64, 15915e (1966).
181. Arbuzov, B. A., V. M. Zoroastrova, and L. A. Antokhina, Izv Akad. Nauk SSSR, Ser. Khim., 1961, 1016; C. A., 55, 27353b (1961).
182. Arbuzov, B. A., V. M. Zoroastrova, and N. D. Ibragimova, Izv. Akad. Nauk SSSR, Ser. Khim., 1964, 656; C. A., 61, 2959b (1964).
183. Arbuzov, B. A., V. M. Zoroastrova, and G. A. Tudrij, Izv. Akad. Nauk SSSR, Ser. Khim., 1970, 90; C. A., 72, 111564t (1970).
184. Arceneaux, R. L., J. G. Frick, E. K. Leonard, and D. Reid, J. Org. Chem., 24, 1419 (1959).
185. Arcoria, A., Ann. Chim., 56, 257 (1966).
186. Arnold, G. B. and C. S. Hamilton, J. Am. Chem. Soc., 63, 2637 (1941).
187. Ashby, E. C., and G. M. Kosolapoff, J. Am. Chem. Soc., 75, 4903 (1953).
188. Atavin, A. S., E. F. Zorina, and A.N. Mirskova, Zh. Org. Khim., 41, 229 (1971); C. A. 75, 63026g (1971).
189. Atherton, F. R., Brit. 782,267 (1954); C. A., 52, 2935 (1958).
190. Atkinson, R. E., J. I. G. Cadogan, and J. Dyson, J. Chem. Soc. (C), 1967, 2542.
190a. Auger, V., Compt. Rend., 139, 639 (1904).
191. Augustinsson, K. B., Svensk Kem. Tidskr., 64, 87 (1952).
192. Baer, E., and N. Z. Stanacey, J. Am. Chem. Soc., 87, 679 (1965).
193. Baer, E., and E. Z. Stanacey, J. Biol. Chem., 240, 3754 (1965).

194. Bai, L. I., A. Ya. Yakubovich, and L. I. Muler, Zh. Obshch. Khim., 34, 3609 (1964).
195. Baina, N. F., U.S.S.R. 221,694 (1969); C. A., 69, 106872a (1968).
196. Baker, J. W., and G. A. Saul, U. S. 2,956,919 (1955); C. A., 55, 5856 (1961).
197. Baker, J. W., and G. A. Saul, U. S. 2,924,553 (1960); C. A., 54, 13067b (1960).
198. Baker, J. W., and G. A. Saul, U. S. 2,900,296 (1959); C. A., 54, 415b (1960).
200. Bakumenko, L. A., N. V. Lebedeva, L. V. Razvodovskaja, A. F. Grapov, and N. N. Mel'nikov, Khim. Sel. Khoz., 4, 691 (1966); C. A., 66, 27927 (1967).
201. Baldwin, R. A., and M. T. Cheng, J. Org. Chem., 32, 2636 (1967).
202. Baldwin, R. A., J. Org. Chem., 30, 3866 (1965).
203. Baldwin, J. E., and J. C. Swallow, J. Org. Chem., 35, 3583 (1970).
204. Balsiger, R. W., D. G. Jones, and J. A. Montgomery, J. Org. Chem., 24, 434 (1959).
205. Banks, C. V., and R. J. Davis, Anal. Chim. Acta, 12, 418 (1955).
206. Bannard, R. A. B., J. R. Gilpin, G. R. Vavasour, and A. F. McKay, Can. J. Chem., 31, 976 (1953).
207. Bannister, B., and F. Kagan, J. Am. Chem. Soc., 82, 3363 (1960).
208. Barnes, G. H., and M. P. David, J. Org. Chem., 25, 1191 (1960).
209. Barthel, W. F., P. A. Giang, and S. A. Hall, J. Am. Chem. Soc., 76, 4186 (1954).
210. Barycki, J., P. Mastalerz, and M. Soroka, Tetrahedron Lett., 1970, 3147.
211. Bastian, J. M., A. Ebenöther, E. Jucker, E. Rissi, and A. P. Stoll, Helv. Chim. Acta, 49 (Sonderheft), 214 (1966).
212. Bataafsche (Shell), Brit. 673,451 (1950); C. A., 47, 5426 (1953).
213. Bauer, H., J. Am. Chem. Soc., 63, 2137 (1941).
214. Baudler, M. and W. Giese, Z. Anorg. Allg. Chem., 290, 258 (1957).
215. Baudler, M., Z. Anorg. Allg. Chem., 288, 171 (1956).
216. Baudler, M., K. Kipker, and H. -W. Valpertz, Naturwissenschaften, 54, 43 (1967).
217. Baudler, M., and H. W. Valpertz, Z. Naturforsch., 22b, 222 (1967).
218. Beach, L. K., and R. Drogin, U. S. 2,863,900 (1953); C. A., 53, 9059 (1959).
219. Beach, L. K., U. S. 2,908,708/709 (1954); C. A., 54, 3202 (1960).
220. Bebbington, A., and R. V. Ley, J. Chem. Soc. (C), 1966, 1410.
221. Beck, T. M., and G. I. Klein, Victor Chem. Works, DAS 1.111.629 (1961). C. A., 56, 14328g (1962).

222. Beck, T. M., and E. N. Walsh, U. S. 3,076,010 (1963); C. A., 59, 1682g (1963).
222a. McBee, E. T., O. R. Pierce, and H. M. Metz, U. S. 2,899,454 (1955); C. A., 54, 9765 (1960).
223. Behal, A., Bull. Soc. Chim. Fr., (2) 50, 632 (1888).
224. Behrenz, W., I. Hammann, H. Schlör, and G. Untersten höfer, DOS 2,017,505 (1971).
225. Behrman, E. J., M. J. Biallas, H. J. Brass, J. O. Edwards, and M. Isaks, J. Org. Chem., 35, 3069 (1970).
226. Behrman, E. J., M. J. Biallas, H. J. Brass, J. O. Edwards, and M. Isaks, J. Org. Chem., 35, 3063 (1970).
227. Beishekeev, Zh., K. Dzhundubaev, A. Aladzheva, P. I. Kozhakmetova, and T. Tokhtobekova, Zh. Obshch. Khim., 41, 2207 (1971).
228. Bell, Jr., V. L., and G. M. Kosolapoff, J. Am. Chem. Soc., 75, 4901 (1953).
229. Benezra, C., S. Nesi, and G. Ourisson, Bull. Soc. Chim. Fr., 1967, 1140.
230. Bengelsdorf, L. S., J. Am. Chem. Soc., 77, 6611 (1955).
231. Bengelsdorf, I. S., and L. B. Barron, J. Am. Chem. Soc., 77, 2869 (1955).
232. Bennett, R., A. Burger, and W. W. Umbreit, J. Med. Pharm. Chem. 1, 213 (1959).
233. Bennett, R. D., A. Burger, and W. A. Volk, J. Org. Chem., 23, 940 (1958).
234. Bennett, F. W., H. J. Emeléus, and R. N. Hazeldine, J. Chem. Soc., 1954, 3598.
235. Benschop, P. H., G. R. van den Berg, and H. L. Boter, Rec. Trav. Chim., 87, 387 (1968).
236. Benschop, H. P., G. R. van den Berg, and G. W. Kraay Rec. Trav. Chim., 89, 1025 (1970).
237. Benschop, H. P., D. H. J. M. Platenburg, S. H. Meppelder, and H. L. Boter, Chem. Commun., 1970, 33.
238. Bergmann, E., and A. Bondi, Chem. Ber., 63, 1158 (1930).
239. Bergmann, E., and A. Bondi, Chem. Ber., 64, 1455 (1931).
240. Bergmann, E., and A. Bondi, Chem. Ber., 66, 278 (1933).
241. Bergmann, E., and A. Bondi, Chem. Ber., 66, 286 (1933).
241a. Bergmann, E. D., and A. Solomonovici, Tetrahedron, 27, 2675 (1971).
242. Beriger, E., and R. Sallmann, DAS 1,098,939 (1961); C. A., 54, 15819a (1960).
243. Berlin, K. D., D. H. Burpo, and R. V. Pagilagan, Chem. Commun., 1967, 1060.
244. Berlin, K. D., D. M. Hellwege, and M. Nagabhushanan, J. Org. Chem., 30, 1265 (1965).

245. Berlin, K. D., C. Hildebrand, A. South, D. M. Hellwege, M. Petersen, E. A. Pier, and J. G. Verkade, Tetrahedron, 20, 323 (1964).
246. Berlin, K. D., and M. Nagabhushanan, Chem. Ind. (London) 1964, 974.
247. Berlin, K. D., and M. Nagabhushanan, J. Org. Chem., 29, 2056 (1964).
248. Berlin, K. D., and N. K. Ray, J. Chem. Eng. Data, 15, 579 (1970).
249. Berlin, K. D., and H. A. Taylor, J. Am. Chem. Soc., 86, 3862 (1964).
250. Bernton, A., Chem. Ber., 58, 661 (1925).
251. Bersworth, F. A., U. S. 2,841,611 (1954); C. A., 52, 17107i (1958).
252. Binder, H., and R. Heinle, DAS 1,083,818 (1960); C. A., 55, 17543 (1961).
253. Binder, H., and R. Heinle, DAS 1,084,716 (1960); C. A., 55, 21052 (1961).
254. BIOS Final Rep., 1808, 20 (1947).
255. BIOS Final Rep., 1808, 21 (1947).
255a. BIOS Final Rep., 1808, 714 (1947).
256. Birum, G. H., U. S. 3,505,431 (1970); C. A., 72, 121697c (1970).
257. Birum, G. H., and Dever, J. L., U. S. 3,029,271 (1962); C. A., 57, 8584h (1962).
258. Birum, G. H., and G. A. Richardson, U. S. 3.113,139 (1963); C. A., 60, 5551d (1964).
259. Bissey, J. E., H. Goldwhite, and D. G. Roswell, J. Org. Chem., 32, 1542 (1967).
260. Blanch, J. H., and J. Andersen, J. Chem. Soc. (B), (1968), 169.
261. Blaser, B., H. v. Freyhold, and K. -H. Worms, Ger. 1,154,028 (1963).
262. Blaser, B., H.-G. Germscheid, and K.-H. Worms, Ger. 1,148,235 (1963); C. A., 59, 11566d (1963).
262a. Blaser, B., and W. Kaiser, Ger. 1,010,965 (1957); C. A., 54, 4388e (1960).
263. Blaser, B., B. Werdelmann, and K.-H. Worms, Ger. 1,072,346 (1959); C. A., 55, 9915b (1961).
264. Blaser, B. and K.-H. Worms, Ger. 1,082,235 (1960); C. A., 55, 12722h (1961).
265. Blaser, B., K.-H. Worms, H.-G. Germscheid, and K. Wollmann, Z. Anorg. Allg. Chem., 381, 247 (1971).
266. Blaser, B., K.-H. Worms, and J. Schiefer, Ger. 1,107,207 (1961); C. A., 56, 12717a (1962).
267. Blicke, F. F., and S. Raines, J. Org. Chem., 29, 2036 (1964).
269. Bliss, A. D., and R. F. W. Rätz, U. S. 3,355,523 (1967); C. A., 68, 39656x (1968).
270. Bliznjuk, N. K., P. S. Chochlov, R. V. Strelcov, Z. N. Kvashá, and A. F. Kolomiets, Zh. Obshch. Khim., 37, 1119, (1967).

271. Bliznjuk, N. K., Z. N. Kvasha, and S. L. Varshavkij, U.S.S.R. 249,382 (1968); C. A., 72, 43862m (1970).
272. Bliznjuk, N. K., G. S. Levskaya, E. N. Matyukhina, and S. L. Varshavskij, U.S.S.R. 245,106 (1968); C. A., 72, 21781 (1970).
273. Bliznjuk, N. K., L. M. Solnceva, and Z. N. Kvasha, Zh. Obshch. Khim., 36, 1710 (1966).
274. Blum, H., and K. -H. Worms, DOS 2,165,833 (1973).
275. Blum, H., and K. -H. Worms, DOS 2,310,451 (1973).
276. Bochwic, B., and J. Michalski, Nature, 167, 1035 (1951).
277. Bochwic, B., and J. Michalski, Rocz. Chem., 25, 338 (1951); C. A., 48, 12013 (1954).
278. Bodalski, R., M. Pietrusiewicz, P. Majewski, and J. Michalski, Collect. Czech. Chem. Commun., 36, 4079 (1971).
279. Bodnarchuk, N. D., B. B. Gavrilenko, and G. I. Derkach, Zh. Obshch. Khim., 40, 1469 (1970).
280. Bodnarchuk, N. D., V. V. Malovik, and G. I. Derkach, Zh. Obshch. Khim., 40, 1210 (1970).
281. Bodnarchuk, N. V., V. V. Malovik, and G. I. Derkach, Zh. Obshch. Khim., 39, 1707 (1969).
282. Bodnarchuk, N. D., V. V. Malovik, G. I. Derkach, and A. V. Kirsanow, Zh. Obshch. Khim., 41, 1464 (1971).
283. Böhme, H., L. Koch, and E. Köhler, Chem. Ber., 95, 1849 (1962).
284. Böhme, H., and K. H. Meyer-Dulheur, Ann., 688, 78 (1965).
284a. Bogatskij, A. V., T. D. Butova, and A. A. Kolesnik, Zh. Obshch. Khim., 41, 1875 (1971).
285. Bogatskij, A. V., A. A. Kolesnik, Vu. Vu. Samitov, and T. D. Butova, Zh. Obshch. Khim., 37, 1105 (1967).
286. Bogonosceva, N. P., and T. E. Filippova, Zh. Obshch. Khim., 33, 1363 (1963).
287. Boissele, A. P., and N. A. Meinhardt, J. Org. Chem., 27, 1828 (1962).
288. Borecki, Cz., J. Michalski, and S. Musierowicz, J. Chem. Soc., 1958, 4081.
289. Borowitz, I. J., S. Firstenberg, E. W. R. Casper, and R. K. Crouch, J. Org. Chem., 36, 3282 (1971).
290. Bost, R. W., and L. D. Quin, J. Org. Chem., 18, 358 (1953).
291. Bost, R. W., L. D. Quin, and A. Roe, J. Org. Chem., 18, 362 (1953).
292. Boter, H., and G. R. van den Berg, Rec. Trav. Chim., 85, 919 (1966).
293. Boter, H. L., A. J. J. Ooms, G. R. van den Berg, and C. van Dijk, Rec. Trav. Chim., 85, 147 (1966).
294. Boter, H. L., and D. H. J. M. Platenburg, Rec. Trav. Chim., 86, 399 (1967).

295. Bott, R. W., B. F. Dowden, and C. Eaborn, J. Organometal. Chem., 4, 291 (1965).
296. Botts, M. F., and E. K. Regel, U. S. 3,485,918 (1969).
297. Bourneuf, M., Bull. Soc. Chim. Fr., (4), 33, 1808 (1923).
298. Boyd, D. R., and G. Chignell, J. Chem. Soc., 123, 813 (1923).
299. Boyd, D. R., and F. J. Smith, J. Chem. Soc., 1926, 2323.
300. Boyd, D. R., and F. J. Smith, J. Chem. Soc., 125, 1477 (1924).
301. Brace, N. O., U. S. 3,047,619 (1962).
302. Brace, N. O., J. Org. Chem., 26, 3197 (1966).
303. Brass, H. J., J. O. Edwards, and M. J. Biallas, J. Am. Chem. Soc., 92, 4675 (1970).
304. Bratkowski, T., P. Mastalerz, M. Michalewska, and B. Nitka, Rocz. Chem., 41, 471 (1967).
305. Brestkin, A. P., I. L. Brik, L. I. Ginetsinskaja, N. N. Godovikov, M. I. Kabachnik, and N. E. Teplov, Izv. Akad. Nauk SSSR, Ser. Khim., 1968, 2122; C. A., 69, 103318p (1968).
306. Brestkin, A. P., I. L. Brik, N. N. Godovikov, M. I. Kabachnik, and N. E. Teplov, Izv. Akad. Nauk SSSR, Ser. Khim., 1967, 1932; C. A., 68, 36219w (1968).
307. Brestkin, A. P., N. N. Godovikov, E. I. Godyna, M. I. Kabachnik, and E. V. Rozengart, Izv. Akad. Nauk SSSR, Ser. Khim., 1968, 2294; C. A., 70, 28997z (1969).
308. Brestkin, A. P., R. I. Volkova, N. N. Godovikov, and M. I. Kabachnik, Izv. Akad. Nauk SSSR, Ser. Khim., 1968, 2028; C. A., 69, 103315k (1968).
309. Bride, M. H., W. A. W. Cummings, and W. Pickles, J. Appl. Chem., 11, 352 (1961); C. A., 57, 1025 (1962).
310. Brocke, M. E., St. Ch. Dorman, and J. J. Menn, DOS 1,923,259 (1969); C. A., 72, 32028t (1970).
311. Brown, D. H., K. D. Crosbie, J. I. Darragh, D. S. Ross, and D. W. Sharp, J. Chem. Soc. (A), 1970, 914.
311a. Brun, G., and Chr. Blanchard, Compt. Rend. (C), 272, 2154 (1971).
312. Bruson, H. A., and T. P. O'Day, U. S. 3,037,044 (1962); C. A., 57, 12540b (1962).
313. Buchner, B., and G. G. Curtis, U. S. 3,285,999 (1966); C. A., 66, 28778 (1967).
314. Buckler, S. A., J. Am. Chem. Soc., 84, 3092 (1962).
315. Buckler, S. A., and H. Epstein, Tetrahedron, 11, 1211 (1962).
316. Budde, P. B., and H. Tolkmith, U. S. 3,466,369 (1969); C. A., 71, 124,658 (1969).

317. Büchel, K. -H. and F. Korte, DAS 1,210,837 (1966); C. A., 64, 19682a (1966).
318. Büchel, K. H., H. Röchling, and F. Korte, Ann., 685, 10 (1965).
319. Bürger, H., K. Burczyk, and A. Blaschette, Monatsh. Chem., 101, 102 (1970).
320. Bugerenko, E. F., E. A. Chernyshev, and A. D. Petrov, Izv. Akad. Nauk SSSR, Ser. Khim., 1965, 286; C. A., 62, 14721h (1965).
321. Bugerenko, E. F., E. A. Chernyshev, and A. D. Petrov, Dokl. Akad. Nauk SSSR, 143, 840 (1962). C. A., 57, 3474c (1962).
322. Bugerenko, E. F., E. A. Chernyshev, and E. M. Popov, Izv. Akad. Nauk SSSR, Ser. Khim., 1966, 1391; C. A., 66, 76078g (1967).
323. Buono, G., and G. Pfeiffer, Tetrahedron Lett., 1972, 149.
324. Burch, G. M., H. Goldwithe, and R. N. Haszeldine, J. Chem. Soc., 1964, 572.
325. Burger, A., U. S. 2,917,533 (1959); C. A., 54, 15315h (1960).
326. Burger, A., and J. J. Anderson, J. Am. Chem. Soc., 79, 3575 (1957).
327. Burger, A., J. B. Clements, N. D. Dawson, and R. B. Henderson, J. Org. Chem., 20, 1383 (1955).
328. Burger, A., and N. D. Dawson, J. Org. Chem., 16, 1250 (1951).
329. Burger, A., and W. S. Shelvers, J. Med. Pharm. Chem., 4, 225 (1961).
330. Burn, A. J., and J. I. G. Cadogan, Chem. Ind. (London), 1963, 736.
331. Burn, A. J., J. I. G. Cadogan, and P. J. Bunyan, J. Chem. Soc., 1964, 4369.
332. Burpitt, D. R., and V. W. Goodlett, J. Org. Chem., 30, 4307 (1965).
333. Cade, J. A., J. Chem. Soc., 1959, 2272.
334. Cade, J. A., J. Chem. Soc., 1959, 2266.
335. Cade, J. A., J. Org. Chem., 23, 1372 (1958).
336. Cadogan, J. I. G., J. Chem. Soc., 1957, 4154.
337. Cadogan, J. I. G., J. Chem. Soc., 1961, 3067.
338. Cadogan, J. I. G., J. A. Challis, and D. T. Eastlick, J. Chem. Soc. (B), 1971, 1988.
339. Cadogan, J. I. G., and R. K. Mackie, J. Chem. Soc. (C), 1969, 2819.
340. Cadogan, J. I. G., D. J. Sears, D. M. Smith, and J. M. Todd, J. Chem. Soc. (C), 1969, 2813.
341. McCall, M. A., and R. L. McConnell, U. S. 2,900,405 (1959); C. A., 54, 416b (1960).
342. Callot, H. J., and C. Benezra, Can. J. Chem., 48, 3382 (1970).
343. Canavan, A. E., B. F. Dowden, and C. Eaborn, J. Chem. Soc., 1962, 331.

344. Canavan, A. E., and C. Eaborn, J. Chem. Soc., 1959, 3751.
345. Carayon-Gentil, A., T. N. Thank, G. Gonzy, and P. Chabrier, Bull. Soc. Chim. Fr., 1967, 1616.
346. Cason, J., and W. N. Baxter, J. Org. Chem., 23, 1302 (1958).
347. Cassidor, J. E., and B. W. Arthur, U. S. 2,927,881 (1960); C. A., 54, 13532c (1960).
348. Cassida, J. E., and B. W. Arthur, U. S. 2,927,880 (1960); C. A., 54, 13532b (1960).
349. Cates, L. A., J. Med. Chem., 13, 301 (1970).
350. Cates, A. L., and T. E. Jones, J. Pharm. Sci., 53, 969 (1964).
351. Chabrier, P., and P. Savignac, Compt. Rend. (C), 267, 1166 (1968).
352. Chabrier, P., T. N. Thanh, and J. P. Chabrier, Compt. Rend. (C), 263, 1556 (1966).
353. Chabrier, P., N. T. Thuong, and F. Convert, Compt. Rend., 259, 2244 (1964).
354. Chabrier, P., T. N. Thuong, and D. Lemaitre, Compt. Rend. (C), 268, 1802 (1969).
355. Chadwick, D. H., Ch. H. Campbell, and S. H. Metzger, U. S. 3,064,030; C. A., 58, 9139h (1963).
356. Chalmers, M. E., and G. M. Kosolapoff, J. Am. Chem. Soc., 75, 5278 (1953).
357. Chambers, J. R., A. F. Isbell, J. Org. Chem., 29, 832 (1964).
359. Chantrell, P. G., C. A. Parce, C. R. Toyer, and R. Twaits, J. Appl. Chem., 15, 460 (1965).
360. Chao, T. H., H. Z. Lecher, and R. A. Grennwood, U. S. 2,847,458 (1955); C. A., 53, 3178b (1959).
361. Charbonnel, Y., and J. Barrans, Compt. Rend. (C), 272, 1675 (1971).
362. Charrier, C., and M. P. Simonin, Compt. Rend. (C), 264, 995 (1967).
363. Chattha, M. S., and A. S. Aguiar, J. Org. Chem., 36, 1219, 2720 (1971).
364. Chavane, V., Ann. Chim. (12), 4, 372 (1949).
365. Chavane, V., Bull. Soc. Chim. Fr. (5), 15, 774 (1948).
366. Chavane, V., Compt. Rend. 224, 406 (1947).
367. Chavane, V., Ann. Chim. (12) 4, 352 (1949).
368. Chavane, V., and P. Rumpf, Compt. Rend., 225, 1322 (1947).
369. Chelintsev, G. V., and V. K. Kuskov, Zh. Obshch. Khim., 16, 1481 (1946); C. A., 41, 5441 (1947).
370. Chen, C. H., and K. D. Berlin, J. Org. Chem., 36, 2791 (1971).
371. Chenborisov, R. Sh., and V. V. Markin, Zh. Obshch. Khim., 40, 43 (1970).
372. Cherbuliez, E., B. Baehler, F. Hunkeler, and J. Rabinowitz, Helv. Chim. Acta, 44, 1815 (1961).

373. Cherbuliez, E., B. Baehler, F. Hunkeler, and J. Rabinowitz, Helv. Chim. Acta. 44, 1812 (1961).
374. Cherbuliez, E., B. Baehler, H. Jindra, G. Weber, G. Wyss, and J. Rabinowitz, Helv. Chim. Acta, 48, 1069 (1965).
375. Cherbuliez, E., B. Baehler, and J. Rabinowitz, Helv. Chim. Acta, 44, 1820 (1961).
376. Cherbuliez, E., B. Baehler, H. Probst, and J. Rabinowitz, Helv. Chim. Acta, 45, 2656, (1962).
377. Cherbuliez, E., M. Gowhari, and J. Rabinowitz, Helv. Chim. Acta, 47, 2098 (1964).
378. Cherbuliez, E., F. Hunkeler, and J. Rabinowitz, Helv. Chim. Acta, 44, 1817 (1961).
379. Cherbuliez, E., F. Hunkeler, and J. Rabinowitz, Helv. Chim. Acta, 44, 1802 (1961).
380. Cherbuliez, E., F. Hunkeler, and J. Rabinowitz, Helv. Chim. Acta, 45, 2660 (1962).
381. Cherbuliez, E., F. Hunkeler, A. Roueche, and J. Rabinowitz, Helv. Chim. Acta, 44, 1810 (1961).
382. Cherbuliez, E., S. Jaccard, R. Prince, and J. Rabinowitz, Helv. Chim. Acta, 48, 632 (1965).
383. Cherbuliez, E., G. Weber, and J. Rabinowitz, Helv. Chim. Acta, 46, 2461 (1963).
384. Cherbuliez, E., G. Weber, and J. Rabinowitz, Helv. Chim. Acta, 45, 2665 (1962).
385. Cherbuliez, E., G. Weber, A. Yazgi, and J. Rabinowitz, Helv. Chim. Acta, 45, 2652 (1962).
386. Chernyshev, E. A., E. F. Bugerenko, E. D. Lubuzh, and A. D. Petrov, Izv. Akad. Nauk SSSR, Ser. Khim., 1962, 1001; C. A., 57, 16647h (1962).
387. Chernyshev, E. A., E. F. Bugerenko, and A. D. Petrov, Dokl. Akad. Nauk SSSR, 148, 875 (1963); C. A., 59, 3948d (1963).
388. Childs, A. F., and L. T. D. Williams, Brit. 810,930 (1954); C. A., 53, 15978 (1959).
389. Chopard, P. A., Helv. Chim. Acta, 50, 1021 (1967).
390. Chopard, P. A., V. M. Clark, R. F. Hudson, and A. J. Kirby, Tetrahedron, 21, 1961 (1965).
391. Christol, H., M. Lévy, and C. Marty, Compt. Rend. (C), 265, 1511 (1967).
392. Chung-gi, Sh., Y. Yasuchika, and Y. Juji, Bull. Chem. Soc. Jap., 44, 3488 (1971).
393. Chupp, J. P., U. S. 3,458,606 (1969); C. A., 71, 102009t (1969).
394. Chupp, J. P., and P. E. Newallis, U. S. 3,145,140 (1964); C. A., 61, 12036f (1964).
395. Chupp, J. P., and P. E. Newallis, J. Org. Chem., 27, 3832 (1962).
396. Chupp, J. O., and P. E. Newallis, U. S. 3,155,708 (1964); C. A., 62, 5298 (1965).
396a. Ciba, A. G., Fr. 1,197,647 (1956).
397. Ciba-Geigy, A. G., Belg. 772,966 (1970).

398. Ciba-Geigy, A. G., Belg. 773,066 (1970).
399. Ciba-Geigy, A. G., Belg. 762,560 (1970).
400. Cilley, W. A., D. A. Nicholson, and D. Campbell, J. Am. Chem. Soc., 92, 1685 (1970).
401. Clarence, H. R., U. S. 3,251,907 (1966); C. A., 65, 3908d (1966).
402. Claus, A., Chem. Ber., 14, 2365 (1881).
403. Clayton, J. O., and W. L. Jensen, J. Am. Chem. Soc., 70, 3880 (1948).
403a. Clovis, J. S., and F. R. Sullivan, Tetrahedron Lett., 1971, 2263.
404. Coates, H., and P. C. Crofts, Brit. 713,669 (1951); C. A., 49, 12529d (1955).
405. Coe, D. H., B. J. Perry, and R. K. Brown, J. Chem. Soc., 1957, 3604.
406. Coe, D. G., B. J. Perry, and F. S. Sherlock, J. Org. Chem., 24, 1018 (1959).
407. Cölln, R., and G. Schrader, DAS 1,132,132 (1959); C. A., 58, 1491c (1963).
408. Cölln, R., G. Schrader, DAS 1,099,535 (1959); C. A. 56, 2473a (1962).
409. Conant, J. B., J. Am. Chem. Soc., 39, 2679 (1917).
410. Conant, J. B., J. Am. Chem. Soc., 43, 1705 (1921).
411. Conant, J. B., A. H. Bump, and H. S. Holt, J. Am. Chem. Soc., 43, 1677 (1921).
412. Conant, J. B., and A. A. Cook, J. Am. Chem. Soc., 42, 830 (1920).
413. Conant, J. B., and B. B. Coyne, J. Am. Chem. Soc., 44, 2530 (1922).
414. Conant, J. B., and A. D. McDonald, J. Am. Chem. Soc., 42, 2337 (1920).
415. Conant, J. B., A. D. McDonald, and A. McB. Kinney, J. Am. Chem. Soc., 43, 1928 (1921).
416. Conant, J. B., and E. L. Jackson, J. Am. Chem. Soc., 46, 1003 (1924).
417. Conant, J. B., and V. H. Wallingford, J. Am. Chem. Soc., 46, 192 (1924).
418. Conant, J. B., V. H. Wallingford, and S. S. Gandheker, J. Am. Chem. Soc., 45, 762 (1923).
419. McConnell, R. L., M. A. McCall, and H. W. Coover, Jr., J. Org. Chem., 22, 462 (1957).
420. McConnell, R. L., and H. W. Coover, Jr., U. S. 2,875,232 (1956); C. A., 53, 11233 (1959).
421. McConnell, R. L., and H. W. Coover, Jr., J. Am. Chem. Soc., 78, 4453 (1956).
422. McConnell, R. L., and H. W. Coover, Jr., J. Am. Chem. Soc., 78, 4450 (1956).
423. McConnell, H. W., and H. W. Coover, Jr., U. S. 2,875,231 (1956); C. A., 53, 12176 (1959).
424. McConnell, R. L., and H. W. Coover, Jr., J. Org. Chem., 24, 630 (1959).
425. McConnell, R. L., and H. W. Coover, Jr., U. S.

3,121,105 (1964); C. A., 60, 13273 (1964).
426. McConnell, R. L., and H. W. Coover, Jr., J. Org. Chem., 23, 830 (1958).
427. McConnell, R. L., and H. W. Coover, Jr., J. Am. Chem. Soc., 79, 1961 (1957).
429. McConnell, R. L., and N. H. Shearer, Jr. U. S. 2,882,278 (1956); C. A., 53, 17149 (1959).
430. Cook, A. F., and J. G. Moffart, J. Am. Chem. Soc., 90, 740 (1968).
430a. Cook, A. R., and D. I. Randall, Nature, 218, 974 (1968).
431. Coover, Jr., H. W., U. S. 2,827,475 (1953); C. A., 52, 11472h (1958).
432. Coover, Jr., H. W., M. A. McCall, and J. B. Dickey, J. Am. Chem. Soc., 79, 1963 (1957).
433. Coover, Jr., H. W., and J. B. Dickey, U. S. 2,652,416 (1952); C. A., 48, 10053 (1954).
434. Coover, Jr., H. W., and J. B. Dickey, U. S. 2,787,629 (1952); C. A., 51, 10831 (1957).
435. Coover, Jr., H. W., and J. B. Dickey, U. S. 2,632,768 (1950); C. A., 48, 2084 (1954).
436. Coover, H. W., and R. L. McConnell, U. S. 2,899,455 (1959); C. A., 54, 1571i (1960).
437. Coover, H. W., and N. H. Shearer, U. S. 2,790,823 (1954); C. A., 51, 14320 (1957).
438. Corey, E. J., and D. E. Cane, J. Org. Chem., 34, 3053 (1969).
439. Corey, E. J., and G. I. Kwiatkowski, J. Am. Chem. Soc., 88, 5652 (1966).
440. Corey, E. J., and G. I. Kwiatkowski, J. Am. Chem. Soc., 90, 6816 (1968).
441. Corey, E. J., and G. Märkl, Tetrahedron Lett., 1967, 3201.
442. Corey, E. J., and J. I. Shulman, J. Org. Chem., 35, 777 (1970).
443. Cram, D. J., R. D. Trepka, and P. St. Janiak, J. Am. Chem. Soc., 86, 2731 (1964).
444. Craniadés, P., and P. Rumpf, Bull. Soc. Chim. Fr., 1955, 1194.
445. Craniadés, P., and P. Rumpf, Bull. Soc. Chim. Fr., 1954, 719.
446. Crofts, P. C., and G. M. Kosolapoff, J. Am. Chem. Soc., 75, 3379 (1953).
447. Crofts, P. C., and I. S. Fox, J. Chem. Soc., 1958, 2995.
448. Cubbon, R. C. P., and C. Hewlett, J. Chem. Soc. (C), 1970, 501.
449. Cuvigny, I., and H. Normant, Organometal. Chem. Synth., 1, 237 (1971).
449a. Daly, J. J., L. Maier, and F. Sanz, Helv. Chim. Acta, 55, 1991 (1972).
450. Daniewski, W. M., M. Gordon, and C. E. Griffin, J.

Org. Chem., 31, 2083 (1966).
451. Daniewski, W. M., and C. E. Griffin, J. Org. Chem., 31, 3236 (1966).
452. Dannley, R. L., and A. Grava, Can. J. Chem., 43, 3377 (1965).
453. Dannley, R. L., and K. R. Kabre, J. Am. Chem. Soc., 87, 4805 (1965).
454. Dannley, R. L., and P. L. Wagner, J. Org. Chem., 26, 3995 (1961).
455. Davidson, R. S., R. A. Sheldon, and S. Trippett, J. Chem. Soc. (C), 1967, 1547.
456. Davies, W. C., and C. J. O. R. Morris, J. Chem. Soc., 1932, 2880.
457. Dawson, N. D., and A. Burger, J. Org. Chem., 18, 207 (1953).
458. Dawson, T. P., and W. E. Catlin, U. S. 2,951,863 (1949); C. A., 55, 4363 (1961).
459. Debo, A., DAS 1,042,581 (1956); C. A., 54, 19600b (1960).
460. Denham, J. M., and R. K. Ingham, J. Org. Chem., 23, 1298 (1958).
461. Denney, D. B., and J. Glacin, Tetrahedron Lett., 1964, 1747.
462. Derkach, G. I., Angew. Chem., 81, 407 (1969).
463. Derkach, G. I., E. S. Gubnitskaja, and L. I. Samarai, Zh. Obshch. Khim., 36, 1949 (1966).
464. Derkach, G. I., E. S. Gubnitskaja, and V. A. Shokol, Zh. Obshch. Khim., 35, 1014 (1965).
465. Derkach, G. I., E. S. Gubnitskaja, V. A. Shokol, and A. V. Kirsanov, Zh. Obshch. Khim., 32, 1874 (1962).
466. Derkach, G. I., and Yu. V. Piven, Zh. Obshch. Khim., 36, 1087 (1966).
467. Derkach, G. I., and E. I. Slyusarenko, Zh. Obshch. Khim., 36, 1639 (1966).
468. Derkach, G. I., and E. I. Slyusarenko, Zh. Obshch. Khim., 37, 2069 (1967).
469. Derkach, G. I., and E. I. Slyusarenko, Zh. Obshch. Khim., 38, 1784 (1968).
470. Derkach, G. I., E. I. Slyusarenko, B. Ya. Libman, and N. I. Liptuga, Zh. Obshch. Khim., 35, 1881 (1965).
471. Derkach, G. I., V. A. Shokol, and E. S. Gubnitskaja, Zh. Obshch. Khim., 33, 553 (1963).
472. Dershowitz, S., and S. Proskauer, J. Org. Chem., 26, 3595 (1961).
473. Destrade, Chr., and C. Garrigon-Lagrange, J. Chim. Phys. 67, 2013 (1970).
474. Dickhäuser, H., G. Scheurer, and H. Adolphi, DAS 1,217,134 (1966); C. A., 65, 5374c (1966).
475. Dietsche, W., Ann., 712, 21 (1968).
476. Dimroth, K., and A. Nürrenbach, Angew. Chem., 70,

26 (1958).
477. Dimroth, K., and A. Nürrenbach, Chem. Ber., 93, 1649 (1960).
478. Divinskaja, L. P., V. E. Limanov, E. K. Skvortsova, G. M. Putyatina, A. V. Starkov, N. I. Grinshtein, and E. E. Nifant'ev, Zh. Obshch Khim., 36, 1244 (1966).
479. Doak, G. O., and L. D. Freedman, J. Am. Chem. Soc., 73, 5658 (1951).
480. Doak, G. O., and L. D. Freedman, J. Am. Chem. Soc., 74, 753 (1952).
481. Doak, G. O., and L. D. Freedman, J. Am. Chem. Soc., 75, 683 (1953).
482. Doak, G. O., and L. D. Freedman, J. Am. Chem. Soc., 75, 6307 (1953).
483. Doak, G. O., and L. D. Freedman, J. Am. Chem. Soc., 76, 1621 (1954).
484. Doak, G. O., and D. Freedman, J. Am. Chem. Soc., 81, 3021 (1959).
485. Dobbie, R. C., L. F. Doty, and R. G. Cavell, J. Am. Chem. Soc., 90, 2015 (1968).
486. Dobinson, B., and B. P. Stark, Brit. 1,031,369 (1966) C. A., 65, 3805e (1966).
487. Dolgov, O. N., M. G. Voronkov, and N. F. Orlov, Zh. Obshch. Khim., 40, 2259 (1970).
488. Donets University, U.S.S.R. 281, 468 (1969); C. A., 74, 100222v (1971).
489. Donohue, J., N. Mandel, W. B. Farnham. R. K. Murray, Jr., K. Mislow, and H. P. Benschop, J. Am. Chem. Soc., 93, 15 (1971).
490. Doroshenko, V. V., V. A. Stukalo, V. A. Shokol, and B. I. Kozushko, Zh. Obshch. Khim., 41, 2155 (1971).
491. Drábec, J., Z. Veselá, and M. Rupcik, Chem. Prum., 15, 675 (1965); C. A., 64, 8020d (1966).
492. Drach, B. S.,A. D. Simica, and A. V. Kirsanov, Zh. Obshch. Khim., 39, 2192 (1969).
493. Drake, L. R., and C. S. Marvel, J. Org. Chem., 2, 387 (1938).
494. Dregval, G. F., and L. V. Cherkashina, U.S.S.R. 179,315 (1966); C. A., 65, 2297b (1966).
495. Dregval', G. F., and G. I. Derkach, Zh. Obshch. Khim. 33, 2952 (1963).
496. Drozd, G. I., M. A. Pokal'skij, S. Z. Ivin, E. P. Sosova, and O. G. Strukov, Zh. Obshch. Khim., 39, 936 (1969).
497. Drozd, G. I., M. A. Sokal'skij, O. G. Strukov, and S. Z. Ivin, Zh. Obshch. Khim., 40, 2396 (1970).
498. Du Pont, Can. 848,876 (1968).
499. Eberhard, A., and F. H. Westheimer, J. Am. Chem. Soc., 87, 253 (1965).
499a. Ecker, A., and U. Schmidt, Mh. Chem., 103, 736 (1972).

500. Eckert, R., K. Hunger, and P. Tavs, Chem. Ber., 100, 639 (1967).
500a. Edgerton, L. J. and G. D. Blanpied, Nature, 219, 1064 (1968).
501. Eimers, E., DOS 1,920,293 (1970); C. A., 74, 23001t (1971).
502. Eimers, E., and H. Rudolph, DOS 1,815,946 (1970); C. A., 73, 66730b (1970).
503. Eliseenkov, V. N., and V. K. Khajrullin, Izv. Akad. Nauk SSSR, Ser. Khim., 1965, 2128; C. A., 64, 11074h (1966).
504. Eliseenkov, V. N., A. N. Pudovik, and V. K. Khajrullin, Izv. Akad. Nauk SSSR, Ser. Khim., 1966, 1582; C. A., 66, 95128x (1967).
505. Engelke, E. F., U. S. 2,377,870 (1942); C. A., 39, 4099 (1945).
506. Engelmann, M., and J. Pikl, U. S. 2,304,156 (1940); C. A., 37, 3262 (1943).
507. Englin, M. A., A. S. Filatov, Z. I. Fraer, V. K. Promonenkov, and S. Z. Ivin, Zh. Obshch. Khim., 38, 869 (1968).
508. Eto, M., K. Kishimoto, K. Matsumura, N. Obshita, and Y. Oshima, Agr. Biol. Chem. (Japan), 30, 181 (1966).
509. Evdakov, V. P., and E. I. Alipova, Zh. Obshch. Khim., 35, 1584 (1965).
509a. Evdakov, V. P., and E. I. Alipova, Zh. Obshch. Khim., 37, 2701 (1967).
510. Evdakov, V. P., and L. I. Mizrach, Zh. Obshch. Khim., 34, 1848 (1964).
512. Evdakov, V. I., L. I. Mizrach, and L. Yu. Sandalova, Zh. Obshch. Khim., 34, 3124 (1964).
513. Evdakov, V. P., L. I. Mizrach, and L. Yu. Sandalova, Dokl. Akad. Nauk SSSR, 162, 573 (1965).
514. Evdakov, V. P., L. I. Mizrach, and G. P. Sizova, Zh. Obshch. Khim., 34, 3952 (1964).
515. Falbe, J., R. Paatz, and F. Korte, Chem. Ber., 97, 2544 (1964).
516. Falbe, J., R. Paatz, and F. Korte, Chem. Ber., 98, 2312 (1965).
517. Fanshawe, W. F., V. J. Bauer, and S. R. Safir, J. Med. Chem., 10, 116 (1967).
518. Farbenfabriken Bayer, Belg. 754,312-Q (1969).
519. Farbenfabriken Bayer, Belg. 757,187-Q (1969).
520. Farbenfabriken Bayer, Ger. 1,931,052 (1969); C. A., 74, 64306q (1971).
521. Farbwerke Hoechst, Belg. 596,622 (1961).
522. Fay, P., and H. O. Lankelma, J. Am. Chem. Soc., 74, 4933 (1952).
523. Fearing, R. B., E. N. Walsh, J. J. Menn, and A. H. Freiberg, J. Agr. Food Chem., 17, 1261 (1969).
524. Fedorova, G. K., R. N. Ruba, and A. V. Kirsanov, Zh. Obshch. Khim., 39, 1471 (1969).

525. Fedorova, G. K., Ya. P. Shaturskij, L. S. Moskalevskaja, and A. V. Kirsanov, Zh. Obshch. Khim., 40, 1167 (1970).
526. Fencher, L. W., U. S. 3,591,600 (1971); C. A., 75, 77030r (1971).
527. Fest, Chr., and Schrader, G., DAS 1,445,717 (1969).
528. Fiat, D., M. Halmann, L. Kugel, and J. Reuben, J. Chem. Soc., 1962, 3837.
529. Fields, E. K., J. Am. Chem. Soc., 74, 1528 (1952).
530. Fields, E. K., U. S. 2,579,810 (1951); C. A., 46, 6140h (1952).
531. Fields, E. K., U. S. 2,635,112 (1949); C. A., 48, 7049e (1954).
532. Fields, E. K., U. S. 3,102,901 (1963); C. A., 60, 560g (1964).
533. Fields, E. K., U. S. 3,102,900 (1963); C. A., 60, 561a (1964).
534. Fields, E. K., U. S. 3,178,469 (1965); C. A., 63, 634a (1965).
535. Fields, E. K., and R. J. Rolin, Chem. Ind. (London), 1960, 999.
536. Fild, M.,Z. Anorg. Allg. Chem., 358, 257 (1965).
537. Fink, W., Angew. Chem., 73, 532 (1961).
537a. Finkelstein, J., J. Am. Chem. Soc., 68, 2397 (1946).
538. Fitch, St. J., and K. Moedritzer, J. Am. Chem. Soc., 84, 1876 (1962).
539. Fitch, St. J., and Sh. K. Liu, DAS 1,211,200 (1966); C. A., 64, 15925c (1966).
540. Fiszer, B, and J. Michalski, Rocz. Chem., 34, 1461 (1960); C. A., 55, 15331g (1961).
540a. Fleisch, H., and G. G. Russell, J. Dent. Res., 51, 324 (1972).
541. Flint, C. D., E. H. M. Ibrahim, R. A. Shaw, B. C. Smith, and C. P. Thakur, J. Chem. Soc. (A), 1971, 3513.
542. Fluck, E., and H. Binder, Angew. Chem., 79, 243 (1967).
543. Fluck, E., and H. Binder, Z. Anorg. Allg. Chem., 377, 298 (1970).
544. Fluck, E., and R. M. Reinisch, Chem. Ber., 95, 1388 (1962).
545. Fluck, E., and R. M. Reinisch, Chem. Ber., 96, 3085 (1963).
546. Fluck, F., and R. M. Reinisch, Z. Anorg. Allg. Chem., 328, 172 (1964).
547. Food Mach. Corp., Belg. 774,368 (1970).
548. Fokin, A. V., Yu. N. Studner, and A. A. Skaldner, Zh. Obshch. Khim., 33, 3366 (1963).
549. Fokin, A. V., V. I. Zimin, Yu. N. Studner, and A. I. Rapkin, Zh. Org. Khim., 7, 249 (1971); C. A., 74, 124736s (1971).
550. Fontal, B., and H. Goldwhite, J. Org. Chem., 31, 3804 (1966).

551. Ford-Moore, A. H., and B. J. Perry, Org. Synth. 31, 33, Wiley, (1951).
552. Ford-Moore, A. H., and J. H. Williams, J. Chem. Soc., 1947, 1465.
553. Fossek, W., Monatsh. Chem., 5, 121 (1884).
554. Fossek, W., Monatsh. Chem., 5, 627 (1884).
555. Fossek, W., Monatsh. Chem., 7, 20 (1886).
556. Fox, R. B., and B. J. Bailey, Publ. Board Rep., U. S. Off. Techn. Serv. 151, 809 (1959); C. A., 55, 1491a (1961).
557. Fox, R. B., and W. J. Bailey, J. Org. Chem., 25, 1447 (1960).
558. Fox, R. B., and W. J. Bailey, J. Org. Chem., 26, 2542 (1961).
559. Fox, R. B., and D. L. Venesky, J. Am. Chem. Soc., 78, 1661 (1956).
560. Frank, A. W., J. Org. Chem., 29, 3706 (1964).
561. Frank, A. W., J. Org. Chem., 30, 3663 (1965).
562. Frank, A. W., J. Org. Chem., 31, 1521 (1966).
563. Frank, A. W., J. Org. Chem., 31, 1917 (1966).
564. Frank, A. W., J. Org. Chem., 31, 1920 (1966).
565. Frank, A. W., and Ch. F. Baranauckas, U. S. 3,501,555 (1970); C. A., 73, 4031r (1970).
566. Frank, A. W., and Ch. F. Baranauckas, U. S. 3,505,433 (1970); C. A., 72, 121705d (1970).
567. Frazza, E. J., and L. Rappoport, U. S. 2,920,097 (1957); C. A., 54, 22361h (1960).
568. Freedman, L. D., J. Am. Chem. Soc., 77, 6223 (1955).
569. Freedman, L. D., and G. O. Doak, J. Am. Chem. Soc., 75, 4905 (1953).
570. Freedman, L. D., and G. O. Doak, J. Am. Chem. Soc., 77, 173 (1955).
571. Freedman, L. D., and G. O. Doak, J. Am. Chem. Soc., 77, 6221 (1955).
572. Freedman, L. D., and G. O. Doak, J. Am. Chem. Soc., 77, 6635 (1955).
573. Freedman, L. D., and G. O. Doak, Chem. Rev., 57, 479 (1957).
574. Freedman, L. D., and G. O. Doak, J. Org. Chem., 23, 769 (1958).
575. Freedman, L. D., and G. O. Doak, J. Org. Chem., 26, 2082 (1961).
576. Freedman, L. D., and H. H. Jaffé, J. Am. Chem. Soc., 77, 920 (1955).
577. Freedman, L. D., and G. O. Doak, J. Org. Chem., 29, 2450 (1964).
578. Freedman, L. D., and G. O. Doak, J. Org. Chem., 30, 1263 (1965).
579. Freedman, L. D., and G. O. Doak, J. Med. Chem., 8, 891 (1965).
580. Freedman, L. D., G. O. Doak, and E. L. Petit, J. Am. Chem. Soc., 77, 4262 (1955).
581. Freedman, L. D., G. O. Doak, and E. L. Petit, J.

Org. Chem., 25, 140 (1960).
582. Freedman, L. D., H. Tauber, G. O. Doak, and J. Magnuson, J. Am. Chem. Soc., 75, 1379 (1953).
583. Freyermuth, H. B., R. D. Jackson, and D. I. Randall, U. S. 2,845,420 (1956); C. A., 52, 21,129 (1958).
584. Freyschlag, H., H. Grassner, A. Nürrenbach, H. Pommer, W. Reif, and W. Sarnecki, Angew. Chem., 77, 277 (1965).
585. Fridland, S. V., V. S. Tsivunin, D. V. Fridland, and G. Kh. Kamai, Zh. Obshch. Khim., 40, 1993 (1970).
586. Fridland, S. V., T. V. Zykova, S. K. Chirkunova, V. A. Kataeva, and G. Kh. Kamai, Zh. Obshch. Khim., 41, 1041 (1971).
587. Friedman, L., J. Am. Chem. Soc., 86, 1885 (1964).
588. Friedmann, M., and J. A. Romersberger, J. Org. Chem., 33, 154 (1968).
589. Frost, A. E., U. S. 3,036,108 (1962); C. A., 57, 11236i; (1962).
590. Fujii, A., J. I. Dickstein, and S. I. Miller, Tetrahedron Lett., 1970, 3435.
590a. Fujii, A., and S. I. Miller, J. Am. Chem. Soc., 93, 3694 (1971).
591. Fujiwara, K., and H. Takahashi, Bull. Chem. Soc. Jap., 35, 1743 (1962).
592. Fukuto, T. R., R. L. Metcalf, and M. Y. Winton, J. Econ. Entomol., 54, 955 (1961).
593. Fuzhenkova, A. V., A. F. Zinkovski, and B. A. Arbuzov Dokl. Akad. Nauk SSSR, 201, 632 (1971).
594. Fyong, N., Zh. M. Ivanova, G. I. Derkach, and F. S. Babichev, Zh. Obshch. Khim., 41, 319 (1971).
595. Gaertner, R., U. S. 2,957,905 (1960); C. A., 55, 22136h (1961).
596. GAF Corp., Ger. 2,054,138 (1969); C. A., 75, 36321t (1971).
597. Gallagher, M. J., and J. D. Jenkins, Chem. Commun., 1965, 587.
598. Gallagher, M. J., and J. D. Jenkins, J. Chem. Soc. (C), 1971, 593.
599. Gallagher, M. J., and J. D. Jenkins, J. Chem. Soc. (C), 1966, 2176.
599a. Gardner, J. A., and G. B. Cockburn, J. Chem. Soc., 71, 1157 (1897); 73, 704 (1898).
600. Garner, A. Y., U. S. 2,953,591 (1958); C. A., 55, 5346 (1961).
602. Garrigon-Lagrange, Ch., and Chr. Destrade, J. Chim. Phys., 67, 1646 (1970).
603. Gavrilin, G. F., and B. A. Voosi, U.S.S.R. 162,140 (1964); C. A., 61, 9528e (1964).
604. Gazizov, T. Kh., A. P. Pashinkin, and A. N. Pudovik, Zh. Obshch. Khim., 41, 2418 (1971).
605. Gazizov, T. Kh., A. P. Pashinkin, and A. N. Pudovik, Zh. Obshch. Khim., 40, 31 (1970).

606. Gazizov, T. Kh., and A. N. Pudovik, Zh. Obshch. Khim., 40, 1202 (1970).
607. Gazizov, M. B., and A. I. Razumov, Zh. Obshch. Khim., 39, 2600 (1969).
608. Gazizov, M. B., D. B. Sultanova, L. P. Ostanina, and A. M. Rusalkina, Zh. Obshch. Khim., 41, 2575 (1971).
609. Gazizov, M. B., D. B. Sultanova, L. P. Ostanina, T. V. Zykova, R. A. Salakhutdinova, and A. I. Razumov, Zh. Obshch. Khim., 41, 2167 (1971).
610. Gazizov, T. Kh., M. A. Vasyanina, A. P. Pashinkin, N. P. Anoshina, Z. I. Gol'dfarb, and A. N. Pudovik, Zh. Obshch. Khim., 41, 1957 (1971).
611. Geffers, H., U. -W. Hendricks, M. Quadvlieg, and R. Schliebs, DOS 1,767,391 (1971).
612. Geffers, H., W. Radt, R. Schliebs, and H. Schulz, DOS 2,061,838 (1972).
613. Geffers, H., W. Radt, R. Schliebs, and H. Schulz, DOS 2,015,068 (1971).
614. Gefter, E. L., Zh. Obshch. Khim., 31, 949 (1961).
615. Gefter, E. L., Zh. Obshch. Khim., 31, 3316 (1961).
616. Gefter, E. L., Plast. Massy, 1959, 42.
617. Gefter, E. L., Plast Massy, 11, 38 (1961).
618. Gefter, E. L., Zh. Obshch. Khim., 33, 3548 (1963).
619. Gefter, E. L., Zh. Obshch. Khim., 28, 1908 (1958).
620. Gefter, E. L.,N. I. Bondar, and M. I. Ermlina, U.S.S.R. 306,133 (1970); C. A., 75, 98676w (1971).
621. Gefter, E. L., V. N. Borbunov, and D. -M. Filippenko, Zh. Obshch. Khim., 39, 1739 (1969).
622. Gefter, E. L., and M. I. Kabachnik, Izv. Akad. Nauk SSSR, Ser. Khim., 1957, 194.
623. Gefter, E. L., and P. A. Moshkin, Plast. Massy, 1960, 54; C. A., 54, 24345e (1960).
624. Gefter, E. L., and I. A. Rogacheva, Zh. Obshch. Khim., 32, 3962 (1962).
625. Gefter, E. L., and I. A. Rogacheva, Zh. Obshch. Khim., 32, 964 (1962).
626. Gefter, E. L., and I. A. Rogacheva, Zh. Obshch. Khim., 36, 79 (1966).
626a. Geike, F., Naturwiss. Rundschau, 24, 335 (1971).
627. Genkina, G. K., V. A. Gilyarov, E. I. Matrosov, and M. I. Kabachnik, Zh. Obshch. Khim., 40, 1496 (1970).
628. George, E. F., and A. J. Davidson, DOS 2,106,168 (1971).
629. Germscheid, H. -G., and J. Schiefer, Ger. 1,148,551 (1963); C. A., 59, 11566c (1963).
630. Geterocych. Cpds. Inst. Akad. Sci. SSSR, U.S.S.R. 287,938 (1969); C. A., 74, 125845p (1971).
631. Gilbert, E. E., J. A. Otto, and J. J. Donleavry, U. S. 2,929,759 (1960); C. A., 54, 18362g (1960).
632. Giljazov, M. M., T. A. Zjablikova, E. Ch. Muchametz- janova, and I. M. Shermegorn, Izv. Akad. Nauk SSSR,

Ser. Khim., 1970, 1177.
633. Gilman, H., and R. K. Abbott, Jr., J. Am. Chem. Soc. 71, 659 (1949).
634. Gilman, H., and J. Robinson, Rec. Trav. Chim., 48, 328 (1929).
635. Ginsberg, A. E., K. H. Buechel, and A. G. K. Korte, DAS 1,214,681 (1966); C. A., 65, 2297d (1966).
636. Ginsburg, V. A., and A. Ya. Yakubovich, Zh. Obshch. Khim., 30, 3979 (1960).
637. Ginsburg, V. A., and A. Ya. Yakubovich, Zh. Obshch. Khim., 28, 728 (1958).
638. Ginsburg, V. A., and A. Ya. Yakubovich, Zh. Obshch. Khim., 30, 3987 (1960).
639. Gladshtein, B. M., B. L. Zakharov, M. M. Sosina, and A. A. Spitsyn, Zh. Obshch. Khim., 40, 1245 (1970).
640. Gladshtein, B. M., B. L. Zakharov, and I. A. Achkasov, Zh. Obshch. Khim., 39, 1001 (1969).
641. Glamkowski, E. J., G. Gal, R. Purick, A. J. Dawidson and M. Sletzinger, J. Org. Chem., 35, 3510 (1970).
642. Gluesenkamp, E. W., and T. M. Patrick, Jr., U. S. 2,612,513 (1950); C. A., 47, 9343h (1953).
643. Goddard, L. E., and J. D. Odenweller, U. S. 3,006,94 (1961); C. A., 56, 2474e (1962).
644. Godovikov, N. N., E. I. Godyna, I. M. Kabachnik, M. Ja. Mickel'son, E. V. Rozengart, and V. A. Jakovlev, Dokl. Akad. Nauk SSSR, 151, 1104 (1963); C. A., 60, 1995b (1964).
645. Godovikov, N. N., N. E. Teplov, and M. I. Kabachnik, Izv. Akad. Nauk SSSR, Ser. Khim., 1966, 164; C. A., 64, 12716f (1966).
646. Godovikov, N. N., and M. I. Kabachnik, Zh. Obshch. Khim., 31, 1638 (1961).
646a. Gösele, H., Diss., University of Munich (1954).
647. Gold, A. M., J. Org. Chem., 26, 3991 (1961).
648. Goldwhite, H., and D. G. Rowsell, J. Am. Chem. Soc., 88, 3572 (1966).
649. Goldwhite, H., D. G. Rowsell, and C. Valdez, J. Organometal. Chem. (Amsterdam), 12, 133 (1968).
649a. Golobov, Yu. G., and V. V. Semidetko, Zh. Obshch. Khim., 36, 950 (1966).
650. Golobov, Yu. G., and L. Z. Soborovskij, Zh. Obshch. Khim., 33, 2955 (1963).
651. Golobov, Yu. G., I. A. Zaishlova, and A. S. Buntyako Zh. Obshch. Khim., 35, 1240 (1965).
652. Good, M. L., and T. H. Sidall, Inorg. Nucl. Chem. Lett., 2, 337 (1966).
653. Gordon, M., and C. E. Griffin, J. Org. Chem., 31, 333 (1966).
654. Gordon, M., V. A. Notaro, and C. E. Griffin, J. Am. Chem. Soc., 86,1898 (1964).
655. Gosling, K., and A. B. Burg, J. Am. Chem. Soc., 90, 2011 (1968).

655a. Grabenstetter, R. J., O. T. Quimby, and T. J. Flautt, J. Phys. Chem., 71, 4194 (1967).
656. Gradsten, M. A., U. S. 3,223,514 (1965); C. A., 64, 8239f (1966).
657. Graf, R., Chem. Ber. 85, 9 (1952).
658. Graf, R., Ger. 910,649 (1943); Chem. Zentr., 1954, 6589.
659. Grandberg, I. I., and A. N. Kost, Zh. Obshch. Khim., 31, 129 (1961).
660. Granoth, I., A. Kalir, Z. Pelah, and E. D. Bergmann, Isr. J. Chem., 8, 613 (1970).
661. Grapov, A. F., O. B. Mikhailova, L. V. Razvodovskaja, and N. N. Mel'nikov, Zh. Obshch. Khim., 41, 1441 (1971).
662. Grapov, A. F., N. N. Mel'nikov, S. L. Portnova, and L. V. Razvodovskaja, Zh. Obshch. Khim., 36, 2222 (1966).
663. Grapov, A. F., N. N. Mel'nikov, and N. K. Kozlovskaja, U.S.S.R. 274,111 (1969); C. A., 74, 13266 (1971).
664. Grapov, A. F., N. V. Lebedeva, and N. N. Mel'nikov, Zh. Obshch. Khim., 38, 1751 (1968).
665. Grapov, A. F., N. V. Lebedeva, and N. N. Mel'nikov, Zh. Obshch. Khim., 38, 2260 (1968).
666. Grapov, A. F., N. V. Lebedeva, and N. N. Mel'nikov, Zh. Obshch. Khim., 38, 2658 (1968).
667. Grapov, A. F., L. V. Razvodovskaja, and N. N. Mel'nikov, Zh. Obshch. Khim., 37, 828 (1967).
668. Grapov, A. F., L. V. Razvodovskaja, and N. N. Mel'nikov, Zh. Obshch. Khim., 39, 165 (1969).
669. Grayson, M., and C. F. Farley, J. Org. Chem., 32, 236 (1967).
670. Grechkin, N. P., and G. S. Bobchenko, Dokl. Akad. Nauk SSSR, 129, 569 (1959); C. A., 54, 7671 (1960).
671. Grechkin, N. P., and L. K. Nikonorova, Izv. Akad. Nauk SSSR, Ser. Khim., 1967, 939; C. A., 67, 108,159 (1967).
672. Grechkin, N. P., and I. A. Nuretdinov, Izv. Akad. Nauk SSSR, Ser. Khim., 1965, 1105; C. A., 63, 8290 (1964).
673. Grechkin, N. P., and I. A. Nuretdinov, Izv. Akad. Nauk SSSR, Ser. Khim., 1963, 302; C. A., 59, 5194 (1963).
674. Grechkin, N. P., and R. R. Shagidullin, Izv. Akad, Nauk SSSR, Ser. Khim., 1960, 2135; C. A., 55, 16409 (1961).
675. Green, M., J. Chem. Soc., 1963, 1324.
676. Green, M., R. Haszeldine, B. R. Iles, and D. G. Roswell, J. Chem. Soc. 1965, 6879.
677. Green, M., and R. F. Hudson, J. Chem. Soc., 1958, 3129.
678. Green, M., and R. F. Hudson, J. Chem. Soc., 1963, 3883.

678a. Green, M., and B. Saville, J. Chem. Soc., 1956, 3887.
679. Greenhalgh, R., J. E. Newberry, R. Woodcock, and R. F. Hudson, Chem. Commun. 1969, 22.
680. Greenley, R. Z., and M. L. Nielsen, U. S. 3,227,727 (1966); C. A., 64, 11250 (1966).
681. Greenley, R. Z., M. L. Nielsen, and L. Parts, J. Org Chem., 29, 1009 (1964).
682. Greenwood, R. A., M. Scalera, and H. Z. Lecher, Ger. 954,876 (1954); C. A., 53, 10128c (1959).
683. Greenwood, R. A., M. Scalera, and H. Z. Lecher, U. S. 2,814,645 (1953); C. A., 52, 5464 (1958).
684. Grell, W., and H. Machleidt, Ann., 699, 53 (1966).
685. Griffin, C. E., Chem. Ind. (London), 1958, 415.
686. Griffin, C. E., J. Org. Chem., 25, 665 (1960).
687. Griffin, C. E., and J. T. Brown, J. Org. Chem., 26, 853 (1961).
688. Griffin, B. S., and A. Burger, J. Am. Chem. Soc., 78, 2336 (1956).
689. Griffin, C. E., and N. T. Castelluci, J. Org. Chem., 26, 629 (1961).
690. Griffin, C. E., R. B. Davison, and M. Gordon, Tetrahedron, 22, 561 (1966).
691. Griffin, C. E., and S. K. Kundu, J. Org. Chem., 34, 1532 (1969).
692. Griffin, C. E., and T. D. Mitchell, J. Org. Chem., 30, 1935 (1965).
693. Griffin, C. E., and T. D. Mitchell, J. Org. Chem., 30, 2829 (1965).
694. Griffin, C. E., R. P. Peller, and J. A. Peters, J. Org. Chem., 30, 91 (1965).
695. Griffin, C. E., E. H. Uhing, and A. D. F. Toy, J. Am. Chem. Soc., 87, 4757 (1965).
696. Griffin, C. E., and H. J. Wells, J. Org. Chem., 24, 2049 (1959).
696a. Griffin, M. J., Ph. D. Dissertation, University of Chicago (1960).
697. Grinev, G. V., G. I. Khervenjuk, and A. V. Dombrovskij, Zh. Obshch. Khim., 39, 1253 (1969).
698. Grinstein, E. I., A. B. Bruker, and L. Z. Soborovskij, Zh. Obshch. Khim., 36, 1138 (1966).
699. Grishina, O. N., and L. M. Bezzubova, Zh. Obshch. Khim., 37, 470 (1967).
700. Grishina, O. N., and L. M. Bezzubova, Izv. Akad. Nauk SSSR, Ser. Khim., 1965, 1619; C. A., 64, 3587 (1966).
701. Grishina, O. N., and L. M. Bezzubova, Izv. Akad. Nauk SSSR, Ser. Khim., 1965, 2140; C. A., 64, 11240 (1966).
702. Grishina, O. N., and L. M. Bezzubova, Izv. Akad. Nauk SSSR, Ser. Khim., 1966, 1617; C. A., 66, 76095 (1967).

703. Grishina, O. N., and I. A. Elfimova, Zh. Obshch. Khim., 36, 1655 (1966).
704. Grishina, O. N., and I. A. Elfimova, Zh. Obshch. Khim., 40, 579 (1970).
705. Grishina, O. N., and L. M. Kosova, Zh. Obshch. Khim., 37, 2276 (1967).
706. Grishina, O. N., L. M. Kosova, and I. A. Elfimova, Zh. Obshch. Khim., 38, 853 (1968).
707. Grishina, O. N., L.M. Kosova, I. P. Lipatova, and R. R. Shagidullin, Zh. Obshch. Khim., 40, 66 (1970).
708. Grishina, O. N., and M. I. Potekhina, Neftekhimiya, 8, 111 (1968); C. A., 69, 2980 (1968).
709. Grisley, Jr., D. W., J. Org. Chem., 26, 2544 (1961).
710. Grisley, Jr., D. W., U. S. 3,033,891 (1962); C. A., 58, 3458h (1963).
711. Grisley, D. W., G. H. Birum, and S. A. Heininger, U. S. 3,005, 008 (1961); C. A., 56, 4796h (1962).
712. Groß, H., B. Costisella, and L. Haase, J. Prakt. Chem., 311, 577 (1969).
713. Groß, H., and B. Costisella, J. Prakt. Chem., 311, 571 (1969).
714. Groß, H., and B. Costisella, J. Prakt. Chem., 311, 925 (1969).
715. Groß, H., and B. Costisella, J. Prakt. Chem., 313, 265 (1971).
715a. Groß, H., and B. Costisella, Ann., 750, 44 (1971).
715b. Groß, H., and B. Costisella, J. Prakt. Chem., 314, 87 (1972).
715c. Groß, H., B. Costisella, and L. Brennecke, J. Prakt. Chem., 314, 969 (1972).
715d. Groß, H., B. Costisella, and L. Brennecke, Phosphorus 4, 241 (1974).
716. Groß, H., B. Costisella, and W. Bürger, J. Prakt. Chem., 311, 563 (1969).
717. Groß, H., G. Engelhardt, J. Freiberg, W. Burger, and B. Costisella, Ann., 707, 35 (1967).
718. Groß, H., J. Freiberg, and B. Costisella, Chem. Ber., 101, 1250 (1968).
719. Groß, H., and H. Seibt, J. Prakt. Chem., 312, 475 (1970).
720. Gryszkiewicz-Trochimowski, E., Bull. Soc. Chim. Fr., 1967, 2232.
721. Gryszkiewicz-Trochimowski, E., Bull. Soc. Chim. Fr., 1967, 4289.
722. Gryszkiewicz-Trochimowski, E., and A. Chmelevski, Bull. Soc. Chim. Fr., 1966, 2043.
723. Gryszkiewicz-Trochimowski, E., J. Quinchon, and O. Gryszkiewicz-Trochimowski, Bull. Soc. Chim. Fr., 1960, 1794.
724. Gubaidullin, M. G., and V. V. Pilyagina, Zh. Obshch. Khim., 40, 1966 (1970).
725. Gubaidullin, M. G., and V. V. Pilyagina, Zh. Obshch.

Khim., 40, 1962 (1970).
726. Gubnitzkaja, E. S., and G. I. Derkach, Zh. Obshch. Khim., 38, 1530 (1968).
727. Gubnitzkaja, E. S., I. M. Loseva, A. A. Kropacheva, and G. I. Derkach, Pharm. Chem. J., 6, 308 (1970); C. A., 73, 75432h (1970).
729. Guichard, F., Chem. Ber., 32, 1572 (1899).
730. Guseinov, I. I., Zh. Obshch. Khim., 39, 1737 (1969).
731. Gutmann, V., D. E. Hagen, and K. Utvary, Monatsh. Chem., 93, 627 (1962).
732. Gutmann, V., D. E. Hagen, and K. Utvary, Monatsh. Chem., 93, 747 (1962).
733. Gutmann, V., D. E. Hagen, and K. Utvary, Monatsh. Chem., 91, 836 (1960).
734. Gutmann, V., A. Meller, and R. Schlegel, Monatsh. Chem., 94, 733 (1963).
735. Gutmann, V., G. Mörtl, and K. Utvary, Monatsh. Chem. 92, 1258 (1961).
736. Gutmann, V., G. Mörtl, and K. Utvary, Monatsh. Chem. 94, 897 (1963).
737. Haage, K., D. Jobsky, and H. Reinheckel, DOS 2,007,7 (1970); C. A., 73, 131130x (1970).
738. Haas, H., Ger. 938,186 (1953); C. A., 50, 12096i (1956).
739. Hackley, B. E., and O. O. Owens, J. Org. Chem., 24, 1120 (1959).
740. Hafner, L. S., J. D. Fellman, and G. C. Briney, J. Med. Chem., 13, 1025 (1970).
741. Hafner, L. S., M. V. Garrison, J. E. Brown, and B. H. Alexander, J. Org. Chem., 30, 677 (1965).
742. Hagemeyer, H. J., U. S. 2,596,679; C. A., 46, 7279 (1952).
743. Hagemeyer, H. J., and A. Bell, U. S. 2,566,194 (1948 C. A., 46, 2564 (1952).
744. Hamilton, L. A., U. S. 2,365,466 (1942); C. A., 39, 4619 (1945).
745. Hamilton, L. A., U. S. 2,382,309 (1944); C. A., 39, 4619 (1945).
746. Hamilton, L. A., and R. H. Williams, U. S. 2,957,931 (1960); C. A., 55, 10317i (1961).
747. Hanford, W. E., and R. M. Joyce, U. S. 2,478,390 (1947); C. A., 44, 1126c (1950).
748. Hardy, E. E., U. S. 2,923,729 (1950); C. A., 54, 12995 (1960).
749. Hardy, E. M., U. S. 3,017,321 (1962); C. A., 56, 11622d (1962).
750. Harger, M. J. P., Chem. Commun., 1971, 442.
751. Harvey, R. G., Tetrahedron, 22, 2561 (1966).
752. Harvey, R. G., H. J. Jacobson, and E. V. Jensen, J. Am. Chem. Soc., 85, 1618 (1963).
753. Harvey, R. G., and E. V. Jensen, J. Org. Chem., 28, 470 (1963).

754. Harvey, T. B., G. M. Omietanski, K. C. -T. Sze, and J. J. Zuckermann, Inorg. Chim. Acta (Padova), 4, 235 (1970).
754a. Harvey, R. G., and E. R. de Sombre, Topics in Phosphorus Chemistry, 1S, 57-111, Interscience (1964).
755. Harwood, H. J., U. S. 3,089,892 (1963); C. A., 59, 11562b (1963).
756. Hassan, A., S. M. A. D. Zayed, and F. M. Abdel-Hamit, Biochem. Pharmacol. 14, 1577 (1965).
757. Hasserodt, V., F. Korte, DAS 1,159,443 (1963); C. A., 60, 1071f (1964).
758. Hassner, A., and J. E. Galle, J. Am. Chem. Soc., 92, 3733 (1970).
759. Hatt, H. H., J. Chem. Soc., 1929, 2412.
760. Hatt, H. H., J. Chem. Soc., 1933, 776.
761. Haven, A. C., J. Am. Chem. Soc., 78, 842 (1956).
762. Haven, A. C., U. S. 2,835,652 (1954); C. A., 53, 221 (1959).
763. Hays, H. R., J. Org. Chem., 36, 98 (1971).
764. Hays, H. R., and T. J. Logan, J. Org. Chem., 31, 3391 (1966).
765. Heins, A., Blum, H., and K. -H. Worms, DOS 2,217,692 (1973).
766. Helferich, B., and E. Aufderhaar, Ann., 658, 100 (1962).
767. Helferich, B., and U. Curtius, Ann., 655, 59 (1962).
768. Helferich, B., and L. Schröder, Ann., 670, 48 (1963).
769. Hellwege, D. M., M. Petersen, E. A. Pier, and J. G. Verkade, Tetrahedron, 20, 323 (1964).
769a. Hendlin, D., E. O. Stapley, M. Jackson, H. Wallick, A. K. Miller, F. J. Wolf, T. W. Miller, L. Chaiet, F. M. Kahan, E. L. Foltz, and H. B. Woodruff, Science, 166, IV, 123 (1969).
770. Henning, H., and D. Gloyna, Z. Chem., 6, 28 (1966).
771. Henning, H. G., G. Hilgetag, and G. Busse, J. Prakt. Chem. (4), 33, 188 (1966).
772. Henning, H. G., and Petzold, G., Z. Chem., 5, 419 (1965).
773. Henning, H. G., and G. Petzold, Z. Chem., 7, 183 (1967).
774. Herring, D. L., Chem. Ind. (London), 1960, 717.
775. Herring, D. L., and C. M. Douglas, Inorg. Chem., 4, 1012 (1965).
776. Hersman, M. F., and L. F. Audrieth, J. Org. Chem., 23, 1889 (1958).
777. Herweh, J., J. Org. Chem., 31, 2422 (1966).
778. Hewertson, W., R. A. Shaw, and B. C. Smith, J. Chem. Soc., 1963, 1670.
779. Hieronymus, E., and R. Wirth, DAS 1,197,882 (1965); C. A., 63, 13318c (1965).
780. Hilgetag, G., K. Zieloff, and H. Paul, Angew. Chem., 77, 261 (1965).

781. Hindersinn, R. R., and R. S. Ludington, J. Org. Chem., 30, 4020 (1965).
782. Hindersinn, R. R., and I. I. Miltiades, Can. 853,735 (1970).
783. Hodan, J. J., and H. Tieckelmann, J. Org. Chem., 26, 4429 (1961).
784. Hoff, M. C., and P. Hill, J. Org. Chem., 24, 356 (1959).
784a. Hoffmann, F. W., B. Kagan, and J. H. Canfield, J. Am. Chem. Soc., 81, 148 (1959).
785. Hoffmann, F. W., and T. R. Moore, J. Am. Chem. Soc., 80, 1150 (1958).
786. Hoffmann, F. W., and A. M. Reeves, J. Org. Chem., 26, 3040 (1961).
787. Hoffmann, F. W., T. C. Simmons, and L. J. Glunz, J. Am. Chem. Soc., 79, 3570 (1957).
788. Hoffmann, F. W., D. H. Wadsworth, and H. D. Weiss, J. Am. Chem. Soc., 80, 3945 (1958).
789. Hoffmann, H., and H. Förster, Monatsh. Chem., 99, 380 (1968).
790. Hoffmann, H., R. Grünewald, and L. Horner, Chem. Ber., 93, 861 (1960).
791. Hofman, A. W., Chem. Ber., 6, 303 (1873).
792. Hofmann, A. W., Chem. Ber., 5, 104 (1872).
793. Hogg, D. R., and J. H. Smith, Mechanisms of Reactions of Sulfur Compounds, 3, 63 Intra Sci. Res. Found., St. Monica, Cal. 1968.
794. Holiday, E. R., J. S. L. Philpot, and L. A. Stocken, Biochem. J. 47, 637 (1950).
795. Holy, A., Collect. Czech. Chem. Commun., 32, 3713 (1967); C. A., 68, 49955h (1968).
796. Holy, A., Tetrahedron Lett., 1967, 881.
797. Holy, A., Ng. D. Hong, Collect. Czech. Chem. Commun., 36, 316 (1971); C. A., 74, 76623d (1971).
798. Hoke, D. I., U. S. 3,108,119 (1963); C. A., 60, 3014f (1964).
798a. Hook, E. O., and G. A. Loughran, U. S. 2,954,379 (1957); C. A., 55, 4030 (1961).
799. Hooper, J. F., and R. A. Massy-Westrop, Aust. J. Chem., 17, 946 (1964).
799a. Horner, L., H. Hoffmann, and H. G. Wippel, Chem. Ber., 91, 61 (1958).
799b. Horiguchi, M., and M. Kandatsu, Nature, 184, 901 (1959).
800. Hotten, B. W., and F. O. Johnson, U. S. 2,862,882 (1956); C. A., 53, 6602 (1959).
801. Houalla, D., R. Miquel, and R. Wolf, Bull. Soc. Chim. Fr., 1963, 1152.
802. Hudson, R. F., and P. A. Chopard, Helv. Chim. Acta, 45, 1137 (1962).
803. Hudson, R. F., and G. Salvadori, Helv. Chim. Acta, 49, (Sonderheft), 96 (1966).

804. Hudson, R. F., and R. J. G. Searle, Chem. Commun., 23, 1249 (1967).
805. Hudson, R. F., R. J. G. Searle, and F. H. Dewitt, J. Chem. Soc. (B), 1966, 789.
806. Hudson, R. F., and R. Woodcock, Chem. Commun., 1971, 1050.
807. Huller, T. L., Tetrahedron Lett., 1967, 4921.
808. Humiec, F. S., and I. I. Bezman, J. Am. Chem. Soc., 83, 2210 (1961).
809. Hunger, K., Brennstoffchem., 49, 201 (1968); C. A., 69, 106813 (1968).
809a. Hunger, K., DOS 1,811,352 (1970).
810. Hunt, C. D., and E. H. M. Ibrahim, Tetrahedron Lett., 1969, 3061.
811. ICI Ltd., Brit. 797,603 (1958); C. A., 53, 2157h (1959).
812. Ibrahim, E. H. M., and R. A. Shaw, Chem. Commun., 1967, 244.
813. Il'ina, N. A., I. N. Yuldasheva, and E. P. Trutneva, Zh. Obshch. Khim., 41, 2173 (1971).
814. Iliopulos, M. J., Tetrahedron Lett., 1968, 609.
815. Iliopulos, M. J., Z. Naturforsch. B, 18, 767 (1963).
816. Iliopulos, M. I., and H. Bredereck, DOS 1,618,539 (1971).
817. Iliopulos, M. J., and H. Wieder, Angew. Chem., 77, 618 (1965).
818. Il'yashevich, I. I., V. N. Podkhainova, N. V. Sebryakova, L. G. Egarova, and G. N. Tyurenkova, Zh. Obshch. Khim., 41, 758 (1971).
819. Imaev, M. G., A. M. Shakirova, and R. A. Galeeva, Zh. Obshch. Khim., 36, 1230 (1966).
820. Imaev, M. G., A. M. Shakirova, E. P. Sirmanova, and E. K. Kas'janova, Zh. Obshch. Khim., 34, 3950 (1964).
821. Imaev, M. G., A. M. Shakirova, and M. Kh. Yuferova, Zh Obshch. Khim., 36, 1142 (1966).
822. Inukai, K., T. Ueda, and H. Muramatsu, J. Org. Chem., 29, 2224 (1964).
823. Ionin, B. I., V. B. Lebedev, and A. A. Petrov, Dokl. Akad. Nauk SSSR, 152, 1354 (1963); C. A., 60, 1560d (1964).
823a. Ionin, B. I., and A. A. Petrov, Zh. Obshch. Khim., 32, 2387 (1962).
824. Ionin, B. I., and A. A. Petrov, Zh. Obshch. Khim., 34, 1174 (1964).
825. Ionin, B. I., and A. A. Petrov, Zh. Obshch. Khim., 33, 432 (1963).
826. Ionin, B. E., and A. A. Petrov, Zh. Obshch. Khim., 33, 2863 (1963).
827. Ionin, B. I., and A. A. Petrov, Zh. Obshch. Khim., 35, 1917 (1965).
828. Ionin, B. I., and A. A. Petrov, Zh. Obshch. Khim., 35, 2255 (1965).

829. Irani, R. R., Belg. 665,004 (1965); C. A., 64, 12248 (1966).
830. Irkutsk Org. Chem. Inst. Sib. Sect. Acad. Sci. SSSR, U.S.S.R. 283,220 (1969); C. A., 75, 63974g (1971).
831. Ismail, R. M., Z. Anorg. Allg. Chem., 372, 252 (1970).
832. Ismail, R. M., Ann., 732, 107 (1970).
833. Ivanov, B. E., A. B. Ageeva, and R. R. Shagidullin, Izv. Akad. Nauk SSSR, Ser. Khim., 1967, 1994; C. A., 68, 69079 (1968).
834. Ivanov, B. E., Yu. A. Gorin, and S. S. Krokhina, Izv. Akad. Nauk SSSR, Ser. Khim., 1970, 2627.
835. Ivanov, B. E., and T. I. Karpova, Izv. Akad. Nauk SSSR, Ser. Khim., 1964, 1230; C. A., 60, 15913d (1964).
836. Ivanov, B. E., and S. S. Krokhina, Izv. Akad. Nauk SSSR, Ser. Khim., 1967, 424; C. A., 67, 21975v (1967).
837. Ivanov, B. E., and S. S. Krokhina, Izv. Akad. Nauk SSSR, Ser. Khim., 1971, 2773.
838. Ivanov, B. E., and S. S. Krokhina, Izv. Akad. Nauk SSSR, Ser. Khim., 1971, 2493; C. A. 76, 127085 (1971
839. Ivanov, B. E., and S. S. Krokhina, Izv. Akad. Nauk SSSR, Ser. Khim., 1970, 2629 (1970); C. A., 75, 6034j (1971).
840. Ivanov, B. E., and I. A. Kudrjavtseva, Izv. Akad. Nauk SSSR, Ser. Khim., 1971, 125; C. A., 75, 36253x (1971).
841. Ivanov, B. E., and L. A. Kudrjavtseva, Izv. Akad. Nauk SSSR, Ser. Khim., 1970, 1180; C. A., 73, 66670g (1970).
842. Ivanov, B. E., L. A. Kudrjavtseva, and T. G. Bykova, Izv. Akad. Nauk SSSR, Ser. Khim., 1970, 2063; C. A., 74, 31811x (1971).
843. Ivanov, B. E., L. A. Kudrjavtseva, I. G. Bykova, and T. A. Zjablikova, Izv. Akad. Nauk SSSR, Ser. Khim., 1969, 1851; C. A., 72, 31922t (1970).
844. Ivanov, B. E., L. A. Kudrjavtseva, and T. A. Zjablikova, Izv. Akad. Nauk SSSR, Ser. Khim., 1970, 96; C. A., 72, 111574w (1970).
845. Ivanova, Zh. M., N. I. Liptuga, Stukalova, E. A., and Derkach, G. I., Zh. Obshch. Khim., 36, 162 (1966
846. Ivanov, B. E., and L. A. Valitova, Izv. Akad. Nauk SSSR, Ser. Khim., 1963, 1049; C. A., 59, 7555c (1963).
847. Ivanov, B. E., and L. A. Valitova, Izv. Akad. Nauk SSSR, Ser. Khim., 1967, 1090; C. A., 68, 39724t (1968).
848. Ivanov, B. E., and L. V. Valitova, Izv. Akad. Nauk SSSR, Ser. Khim., 1967, 1087; C. A., 68, 39723s (1968).
849. Ivanov, B. E., and T. G. Vasilova, U.S.S.R. 169,116 (1965).

850. Ivanova, Zh. M., and A. V. Kirsanov, Zh. Obshch. Khim., 31, 3991 (1961).
851. Ivanova, Zh. M., and A. V. Kirsanov, Zh. Obshch. Khim., 35, 1974 (1965).
852. Ivanova, Zh. M., N. I. Liptuga, E. A. Stukalo, and G. I. Derkach, Zh. Obshch. Khim., 36, 162 (1966).
853. Ivanova, Zh. M., S. K. Mikhailik, V. A. Shokol, and G. I. Derkach, Zh. Obshch. Khim., 39, 1504 (1969).
854. Ivanova, Zh. M., S. K. Mikhailik, and G. I. Derkach, Zh. Obshch. Khim., 39, 1037 (1969).
855. Ivanova, Zh. M., S. K. Mikhailik, and G. I. Derkach, Zh. Obshch. Khim., 40, 1473 (1970).
856. N. L. Ivanova, E. E. Nifant'ev, G. K. Genkina, S. E. Chesnokova, V. A. Giljarov, and M. I. Kabachnik, Izv. Akad. Nauk SSSR, Ser. Khim., 1969, 1393; C. A., 71, 124586 (1969).
857. Ivanova, Zh. M., E. A. Stukalo, and G. I. Derkach, Zh. Obshch. Khim., 37, 1144 (1967).
858. Ivashchenko, Ya. N., L. S. Sologub, S. D. Moshchitskij, and A. V. Kirsanov, Zh. Obshch. Khim., 39, 1695 (1969)
859. Ivasyuk, N. V., and I. M. Shermergorn, Zh. Obshch. Khim., 41, 2199 (1971).
860. Ivanova, Zh. M., S. K. Mikhailik, and G. I. Derkach, Zh. Obshch. Khim., 38, 1334 (1968).
861. Ivin, S. Z., Yn. A. Kondrat'ev, O. D. Shelakova, I. A. Zaishlova, and I. I. Gubenko, Zh. Obshch. Khim., 35, 1218 (1965).
862. Ivin, S. Z., and V. K. Promonenkov, U.S.S.R. 166,696 (1965); C. A., 62, 10461h (1965).
863. Ivin, S. Z., and V. K. Promonenkov, U.S.S.R. 166,690 (1965); C. A., 62, 10461b (1965).
864. Ivin, S. Z., and V. K. Promonenkov, U.S.S.R. 166,693 (1965); C. A., 62, 10461a (1965).
865. Jackson, H. L., U. S. 2,722,538 (1951); C. A., 50, 4218 (1956).
866. Jacobson, H. J., M. J. Griffin, S. Preis, and E. V. Jensen, J. Am. Chem. Soc., 79, 2608 (1957).
867. Jaffé, H. H., L. D. Freedman, and G. O. Doak, J. Am. Chem. Soc., 76, 1548 (1954).
868. Jagodić, V., Chem. Ber., 93, 2308 (1960).
869. Jagupol'skij, L. M., and R. V. Belinskaja, Zh. Obshch. Khim., 30, 2014 (1960).
870. Jagupol'skij, L. M., and Zh. M. Ivanova, Zh. Obshch. Khim., 29, 3766 (1959).
871. Jagupol'skij, L. M., and Zh. M. Ivanova, Zh. Obshch. Khim., 30, 4026 (1960).
872. Jagupol'skij, L. M., and Zh. M. Ivanova, Zh. Obshch. Khim., 30, 1284 (1960).
873. Jean, H., Bull. Soc. Chim. Fr., 1957, 783.
874. Jelinek, A. G., U. S. 2,503,390 (1948); C. A., 44, 6435d (1950).
875. Joesten, M. D., and Y. T. Chen, Inorg. Chem., 11, 429 (1972).

876. Joesten, M. D., R. C. Koch, T. W. Martin, J. H. Venable, J. Am. Chem. Soc., 93, 1138 (1971).
877. Johnston, F., U. S. 2,616,918 (1949); C. A., 47, 8772 (1953).
879. Jones, G. H., H. P. Albrecht, N. P. Damodaran, and J. G. Moffat, J. Am. Chem. Soc., 92, 5510 (1970).
880. Jones, G. H., and J. G. Moffat, J. Am. Chem. Soc., 90, 5337 (1968).
881. Jones, J. B., and P. W. Marr, Can. J. Chem., 49, 1300 (1971).
882. De Jong, L. P. A., and H. P. Benschop, Rec. Trav. Chim., 89, 1038 (1970).
883. Jugelt, W., and D. Schmidt, Tetrahedron, 25, 5569 (1969).
884. Jurzenko, T. I., and B. I. Kaspruk, Dokl. Akad. Nauk SSSR, 168, 113 (1966).
885. Ju-Yu, Ch., and Ch. Kwei-Fen, Acta Chim. Sinica, 24, 172 (1957).
886. Ju-Yu, Ch., and Ch. Kwei-Fen, Acta Chim. Sinica, 24, 203 (1958).
887. Kabachnik, M. I., Izv. Akad. Nauk SSSR, Ser. Khim., 1947, 631; C. A., 42, 5845 (1948).
888. Kabachnik, M. I., Izv. Akad. Nauk SSSR, Ser. Khim., 1947, 233; C. A., 42, 413 (1948).
889. Kabachnik, M. I., U.S.S.R. 162,143 (1964); C. A., 61 16094h (1964).
890. Kabachnik, M. I., and V. A. Giljarov, Dokl. Akad. Nauk SSSR, 96, 991 (1954); C. A., 49, 8842 (1955).
891. Kabachnik, M. I., V. A. Giljarov, and W. M. Popov, Izv. Akad. Nauk SSSR, Ser. Khim., 1961, 1022; C. A., 55, 27014 (1961).
892. Kabachnik, M. I., V. A. Giljarov, and M. M. Jusupov, Dokl. Akad. Nauk SSSR, 164, 812 (1965); C. A., 64, 3465 (1966).
893. Kabachnik, M. I., and N. N. Godovikov, Dokl. Akad. Nauk SSSR, 110, 217 (1956); C. A., 51, 4982 (1957).
894. Kabachnik, M. I., and N. N. Godovikov, Zh. Obshch. Khim., 33, 1941 (1963).
895. Kabachnik, M. I., N. N. Godovikov, and E. I. Godyna, Zh. Obshch. Khim., 33, 1335 (1963).
896. Kabachnik, M. I., N. N. Godovikov, D. M. Paikin, M. P. Shabanova, N. M. Gamper, and L. F. Efimova, Zh. Obshch. Khim., 28, 1568 (1958).
897. Kabachnik, M. I., E. I. Golubeva, D. M. Paikin, M. P. Shabanova, N. M. Gamper, and L. F. Efimova, Zh. Obshch. Khim., 29, 1671 (1959).
898. Kabachnik, M. I., Ch. Jung-Yu, and E. N. Tsvetkov, Zh. Obshch. Khim., 32, 3351 (1962).
899. Kabachnik, M. I., and K. A. Kovyrzina, Zh. Obshch. Khim., 24, 307 (1954).
899a. Kabachnik, M. I., R. P. Lastovskij, T. Ya. Medved, V. V. Medyntsev, I. D. Kolpakova, and N. M. Dyatlova

Dokl. Akad. Nauk SSSR, 177, 582 (1967); C. A., 69, 5682h (1968).
900. Kabachnik, M. I., and T. A. Mastryukova, Izv. Akad. Nauk SSSR, Ser. Khim., 1953, 163; C. A., 48, 3243e (1954).
901. Kabachnik, M. I., T. A. Mastryukova, and N. I. Kurochkin, Izv. Akad. Nauk SSSR, Ser. Khim., 1956, 193; C. A., 50, 13727f (1956).
902. Kabachnik, M. I., T. A. Mastryukova, N. I. Kurochkin, N. P. Rodionova, and E. M. Popov, Zh. Obshch. Khim., 26, 2228 (1956).
903. Kabachnik, M. I., T. A. Mastryukova, and T. A. Melent'eva, Zh. Obshch. Khim., 33, 382 (1963).
904. Kabachnik, M. I., and T. Ya. Medved', Izv. Akad. Nauk SSSR, Ser. Khim., 1959, 2142; C. A., 54, 10834 (1960).
905. Kabachnik, M. I., and T. Ya. Medved', Izv. Akad. Nauk SSSR, Ser. Khim., 1953, 868; C. A., 49, 840c (1955).
906. Kabachnik, M. I., and T. Ya. Medved', Izv. Akad. Nauk SSSR, Ser. Khim., 1952, 540; C. A., 47, 4848 (1953).
906a. Kabachnik, M. I., and T. Ya. Medved', Dokl. Akad. Nauk SSSR, 99, 765 (1954); C. A., 49, 14664 (1955).
907. Kabachnik, M. I., and T. Ya. Medved', Izv. Akad. Nauk SSSR, Ser. Khim., 1961, 604; C. A., 55, 23319 (1961).
908. Kabachnik, M. I., I Ya. Medved', Izv. Akad. Nauk SSSR, Ser. Khim., 1966, 1365; C. A., 66, 54802 (1967).
909. Kabachnik, M. I., and T. Ya. Medved', Sbornik Statei Obshch. Khim., Akad. Nauk SSSR, 2, 12 (1952); C. A., 48, 564 (1954).
910. Kabachnik, M. I., and T. Ya. Medved', Izv. Akad. Nauk SSSR, Ser. Khim., 1951, 95; C. A., 46, 421 (1952).
911. M. I. Kabachnik, and T. Ya. Medved', Izv. Akad. Nauk SSSR, Ser. Khim., 1950, 635; C. A., 45, 8444 (1951).
912. Kabachnik, M. I., and T. Ya. Medved', Izv. Akad. Nauk SSSR, Ser. Khim., 1953, 1126; C. A., 49, 2306f (1955).
913. Kabachnik, M. I., and T. Ya. Medved', Dokl. Akad. Nauk SSSR, 83, 689 (1952); C. A., 47, 2724h (1953).
914. Kabachnik, M. I., T. Ya. Medved', G. K. Kozlova, V. S. Balabukha, E. A. Mironova, and L. I. Tikhonova, Izv. Akad. Nauk SSSR, Ser. Khim., 1960, 651; C. A., 54, 22329b (1960).
915. Kabachnik, M. I., T. Ya. Medved', G. K. Kozlova, V. S. Balabukha, M. M. Senyawin, and L. I. Tinkhonov, Izv. Akad. Nauk SSSR, Ser. Khim., 1958, 1070; C. A., 53, 3039 (1959).
916. Kabachnik, M. I., T. Ya. Medved', and T. A. Mastryukova, Dokl. Akad. Nauk SSSR, 92, 959 (1953); C. A.,

49, 839 (1955).
917. Kabachnik, M. I., and P. A. Rossijakaja, Izv. Akad. Nauk SSSR, Ser. Khim., 1946, 515; C. A., 42, 7242 (1948).
917a. Kabachnik, M. I., and P. A. Rossijskaja, Izv. Akad. Nauk SSSR, Ser. Khim., 1947, 389.
917b. Kabachnik, M. I., and P. A. Rossijskaja, Izv. Akad. SSSR, Ser. Khim., 1947, 631; C. A., 42, 5846e (1948)
918. Kabachnik, M. I., P. A. Rossijskaja, and E. S. Shepeleva, Izv. Akad. Nauk SSSR, Ser. Khim., 1947, 163; C. A., 42, 4132 (1948).
919. Kabachnik, M. I., and P. A. Rossijskaja, Izv. Akad. Nauk SSSR, Ser. Khim., 1948, 95; C. A., 42, 5846 (1948).
920. Kabachnik, M. I., and P. A. Rossijskaja, Izv. Akad. Nauk SSSR, Ser. Khim., 1946, 403; C. A., 42, 7242 (1948).
921. Kabachnik, M. I., and P. A. Rossijskaja, Izv. Akad. Nauk SSSR, Ser. Khim., 1945, 597; C. A., 41, 88 (1947).
922. Kabachnik, M. I., and P. A. Rossijskaja, Izv. Akad. Nauk SSSR, Ser. Khim., 1945, 364; C. A., 40, 4688 (1946).
923. Kabachnik, M. I., P. A. Rossijskaja, and N. N. Novikova, Izv. Akad. Nauk SSSR, Ser. Khim., 1947, 97; C. A., 42, 4132 (1948).
924. Kabachnik, M. I., and E. S. Shepeleva, Izv. Akad. Nauk SSSR, Ser. Khim., 1950, 39; C. A., 44, 7257 (1950).
925. Kabachnik, M. I., and E. S. Shepeleva, Izv. Akad. Nauk SSSR. Ser. Khim., 1951, 185; C. A., 45, 10191 (1951).
926. Kabachnik, M. I., and E. S. Shepeleva, Dokl. Akad. Nauk SSSR, 75, 219 (1950); C. A., 45, 6569i (1951).
928. Kabachnik, M. I., and E. N. Tsvetkov, Dokl. Akad. Nauk SSSR, 117, 817 (1957); C. A., 52, 8070c (1958).
929. Kabachnik, M. I., and E. N. Tsvetkov, Zh. Obshch. Khim., 30, 3227 (1960).
930. Kabachnik, M. I., E. N. Tsvetkov, and Ch. Jung-Yu, Dokl. Akad. Nauk SSSR, 131, 1334 (1960); C. A., 54, 20845a (1960).
931. Kabachnik, M. I., E. N. Tsvetkov, and Ch. Jung-Yu. Zh. Obshch. Khim., 32, 3340 (1962).
932. Kagan, F., R. D. Birkenmeyer, and R. D. Strube, J. Am. Chem. Soc., 81, 3026 (1959).
933. Kamai, G., Dokl. Akad. Nauk SSSR, 55, 219 (1947); C. A., 41, 5863 (1947).
934. Kamai, G., and L. P. Egorova, Zh. Obshch. Khim., 16, 1521 (1946); C. A., 41, 5439 (1947).
935. Kamai, G., and V. A. Kukhtin, Dokl. Akad. Nauk. SSSR 102, 283 (1955); C. A., 50, 1993 (1956).

936. Kamai, G., and V. A. Kukhtin, Zh. Obshch. Khim., 27, 949 (1957); C. A., 52, 3665 (1958).
936a. Kamai, G., and V. A. Kukhtin, Dokl. Akad. Nauk SSSR, 112, 868 (1957); C. A., 51, 13742 (1957).
936b. Kamai, G., and V. A. Kukhtin, Zh. Obshch. Khim., 27, 2372, 2376 (1957).
937. Kamai, G., V. S. Tsivunin, and S. Kh. Nurtdinov, Zh. Obshch. Khim., 35, 1817 (1965).
938. Karayannis, N. M., C. M. Mikulski, M. J. Stroko, L. L. Pytlewski, and M. M. Labes, Inorg. Chim. Acta, 4, 557 (1970).
939. Karayannis, N. M., C. M. Mikulski, M. J. Stroko, L. L. Pytlewski, and M. M. Labes, Z. Anorg. Allg. Chem., 384, 267 (1971).
940. Kashman, Y., and M. Sprecher, Tetrahedron, 27, 1331 (1971).
941. Kataev, E. G., F. R. Tantasheva, and E. G. Yarkova, Zh. Obshch. Khim., 35, 759 (1965).
942. Kaufmann, S., U. S. 2,843,617 (1956); C. A., 53, 2090f (1959).
943. McKay, A. F., R. A. B. Bannard, R. O. Braun, and R. L. Benness, J. Am. Chem. Soc., 76, 3546 (1954).
944. McKay, A. F., R. O. Braun, and G. R. Vavasour, J. Am. Chem. Soc., 74, 5540 (1952).
945. Kazan Chem. Technol. Inst., U.S.S.R. 289,094 (1969); C. A., 75, 6091a (1971).
946. Kazuo, K., Y. Hiroshi, O. Tsuyoski, and I. Saburo, Bull. Chem. Soc. Jap., 42, 3245 (1969).
947. Keat, R., W. Sim, and D. S. Payne, J. Chem. Soc. (A), 1970, 2715.
948. Keay, L., Can. J. Chem., 43, 2637 (1965).
949. Keay, L., J. Org. Chem., 28, 1426 (1963).
950. Keay, L., J. Org. Chem., 28, 329 (1963).
951. Keeber, W. H., and H. W. Pott, J. Org. Chem., 21, 509 (1956).
952. Kelbe, W., Chem. Ber., 9, 1051 (1876), 11, 1499 (1878).
953. Kelso, C. D., and L. W. Mixon, U. S. 2,316,090 (1941); C. A., 37, 5858 (1943).
954. Kemp, R. H., W. A. Thomas, M. Gordon, and C. E. Griffin, J. Chem. Soc. (B), 1969,527.
955. Kennard, K. C., and C. S. Hamilton, J. Am. Chem. Soc., 77, 1156 (1955).
956. Kennedy, J., Chem. Ind. (London), 1956, 1348.
957. Kennedy, J., E. S. Lane, and B. K. Robinson, J. Appl. Chem. (London), 8, 459 (1958).
958. Kennedy, J., E. S. Lane, and J. L. Willans, J. Chem. Soc., 1956, 4670.
959. Kenyon, G. L., and F. H. Westheimer, J. Am. Chem. Soc., 88, 3557 (1966).
959a. Khayrullin, V. K., R. M. Kondrat'eva, and A. N.

Pudovik, Izvest. Akad. Nauk SSSR, Ser. Khim., 1967, 2097; C. A., 68, 43714 (1968).
960. Khajrullin, V. K., and T. I. Sobchuk, Izv. Akad. Nauk SSSR, Ser. Khim., 1965, 2010; C. A., 64, 8062f (1966).
961. Khajrullin, V. K., A. I. Lebedeneva, and V. S. Abramov, Zh. Obshch. Khim., 29, 2551 (1959).
962. Khamitov, R. N., and N. P. Grechkin, Dokl. Akad. Nauk SSSR, 172, 1099 (1967); C. A., 66, 115768m (1967).
963. Kharasch, M. S., R. A. Mosher, and I. S. Bengelsdorf J. Org. Chem., 25, 1000 (1960).
964. Kharrasova, F. M., G. Kamai, and R. R. Shagidullin, Zh. Obshch. Khim., 35, 1993 (1965).
965. Kinnear, A. M., and E. A. Perren, J. Chem. Soc., 1952, 3437.
966. Kipping, W., Chem. Tech., 16, 208 (1964).
967. Kirillova, L. M., and V. A. Kukhtin, Zh. Obshch. Khim., 35, 1146 (1965).
968. Kirillova, K. M., V. A. Kukhtin, and T. M. Sudakova, Dokl. Akad. Nauk SSSR, 149, 316 (1963); C. A., 59, 7556c (1963).
969. Kirilov, M., M. Arnandov, G. Petrov, and M. Shishkova Chem. Ber., 103, 3190 (1970).
970. Kirilov, M., and G. Petrov, Monatsh. Chem., 99, 166 (1968).
971. Kirilov, M., and G. Petrov, Chem. Ber., 100, 3139 (1967).
972. Kirsanov, A. V., and L. P. Zuravleva, Zh. Obshch. Khim., 30, 3038 (1960).
973. Klamann, D., H. Wache, and P. Weyerstahl, Monatsh. Chem., 98, 911 (1967).
974. Klauke, E., E. Kühle, I. Hammann, and W. Lorenz, Ger 1,210,833 (1966); C. A., 64, 15925b (1966).
974a. Klebe, J. F., H. Finkbeiner, and D. M. White, J. Am. Chem. Soc. 88, 3390 (1966).
975. Kleine-Weischede, K., DAS 1,231,801 (1966).
976. Knollmueller, K. O., U. S. 3,534,125 (1970); C. A., 74, 53989p (1971).
977. Knott, E. B., J. Chem. Soc., 1965, 3793.
978. Knunjants, I. L., and E. G. Bykhovskaya, Izv. Akad. Nauk SSSR, Ser. Khim., 1966, 1571; C. A., 66, 94647 (1967).
979. Knunjants, I. L., E. Ja. Pervova, and V. V. Tjuleneva, Izv. Akad. Nauk SSSR, Ser. Khim., 1963, 1576; C. A., 59, 15306b (1963).
980. Knunjants, I. L., E. Ja. Pervova, and V. V. Tjuleneva, Dokl. Akad. Nauk SSSR, 129, 576 (1959); C. A., 54, 7536e (1960).
981. Knunjants, I. L., and R. N. Sterlin, Dokl. Akad. Nauk SSSR, 56, 49 (1947); C. A., 42, 519 (1948).
982. Knunjants, I. L., R. W. Sterlin, V. V. Tjuleneva,

and L. N. Pinkina, Izv. Akad. Nauk SSSR, Ser. Khim. 1963, 1123; C. A., 59, 8784e (1963).

983. Knunjants, I. L., V. V. Tjulaneva, E. Ja. Pervova, and R. N. Sterlin, Izv. Akad. Nauk SSSR, Ser. Khim., 1964, 1797; C. A., 62, 2791d (1965).

984. Koe, B. K., I. A. Seto, A. R. English, and T. J. McBride, J. Med. Chem., 6, 653 (1963).

985. Köhler, H., and A. Michaelis, Chem. Ber., 10, 807 (1877).

986. Köhler, H., and A. Michaelis, Chem., Ber., 9, 1053 (1876).

987. Kohn, G. K., U. S. 3,069, 312 (1962); C. A., 59, 666a (1963).

988. Koketsu, J., S. Kojima, and Y. Ishii, Bull. Chem. Soc. Jap., 43, 3232 (1970).

989. Kolobova, N. E., and K. N. Anisimov, Izv. Akad. Nauk SSSR, Ser. Khim., 1962, 726; C. A., 57, 15144a (1962).

990. Kolobova, N. E., and K. N. Anisimov, Izv. Akad. Nauk SSSR, Ser. Khim., 1962, 1117; C. A., 58, 547e (1963).

991. Kormachev, V. V., V. S. Tsivunin, and N. A. Koren', Zh. Obshch. Khim., 40, 1711 (1970).

992. Kormachev, V. V., V. S. Tsivunin, and N. A. Koren', Zh. Obshch. Khim., 40, 1989 (1970).

993. Kormachev, V. V., V. S. Tsivunin, and N. A. Koren, U.S.S.R. 283,215 (1969); C. A., 75, 63969s (1971).

994. Korshak, V. V., I. A. Gribova, and V. K. Shitikov, Izv. Akad. Nauk SSSR, Ser. Khim., 1958, 210; C. A., 52, 12804e (1958).

995. Korshak, V. V., I. A. Gribova, and M. A. Andreeva, Izv. Akad. Nauk SSSR, Ser. Khim., 1957, 631; C. A., 51, 14621g (1957).

996. Korte, F., and F. -F. Wiese, Chem. Ber., 97, 1963 (1964).

997. Kosolapoff, G. M., U. S. 2,389,576 (1943); C. A., 40, 1536 (1946).

998. Kosolapoff, G. M., J. Am. Chem. Soc., 66, 109 (1944).

999. Kosolapoff, G. M., J. Am. Chem. Soc., 66, 1511 (1944).

1000. Kosolapoff, G. M., J. Am. Chem. Soc., 67, 1180 (1945).

1001. Kosolapoff, G. M., J. Am. Chem. Soc., 67, 2259 (1945).

1002. Kosolapoff, G. M., U. S. 2,397,422 (1945); C. A., 40, 3468 (1946).

1003. Kosolapoff, G. M., J. Am. Chem. Soc., 68, 1103 (1946).

1003a. Kosolapoff, G. M., J. Am. Chem. Soc., 70, 1283 (1948).

1004. Kosolapoff, G. M., J. Am. Chem. Soc., 69, 1002

(1947).
1005. Kosolapoff, G. M., J. Am. Chem. Soc., 69, 2112 (1947).
1006. Kosolapoff, G. M., J. Am. Chem. Soc., 69, 2248 (1947).
1007. Kosolapoff, G. M., J. Am. Chem. Soc., 70, 1971 (1948).
1008. Kosolapoff, G. M., J. Am. Chem. Soc., 70, 3465 (1948).
1009. Kosolapoff, G. M., J. Am. Chem. Soc., 71, 4021 (1949).
1010. Kosolapoff, G. M., U. S. 2,594,455 (1949); C. A., 47, 1182 (1953).
1011. Kosolapoff, G. M., J. Am. Chem. Soc., 72, 5508 (1950).
1012. Kosolapoff, G. M., Organophosphorus Compounds, Wiley (1950).
1013. Kosolapoff, G. M., J. Am. Chem. Soc., 73, 4040 (1951).
1014. Kosolapoff, G. M., "The Synthesis of Phosphonic and Phosphinic Acids," in Organic Reactions, VI, Wi (1951).
1015. Kosolapoff, G. M., J. Am. Chem. Soc., 75, 1500 (1953).
1016. Kosolapoff, G. M., J. Am. Chem. Soc., 76, 3222 (1954).
1017. Kosolapoff, G. M., J. Chem. Soc., 1955, 3092.
1018. Kosolapoff, G. M., and L. B. Payne, J. Org. Chem., 21, 413 (1956).
1018a. Kosolapoff, G. M., and J. J. Duncan, J. Am. Chem. Soc., 77, 2419 (1955).
1019. Kosolapoff, G. M., and W. F. Huber, J. Am. Chem. Soc., 68, 2540 (1946).
1020. Kosolapoff, G. M., and W. F. Huber, J. Am. Chem. Soc., 69, 2020 (1947).
1021. Kosolapoff, G. M., and G. G. Priest, J. Am. Chem. Soc., 75, 4847 (1953).
1022. Kosolapoff, G. M., and C. H. Roy, J. Org. Chem., 26, 1895 (1961).
1023. Kost, D., and M. Sprecher, Tetrahedron Lett., 1970, 2535.
1024. Kotzschmar, A., and W. Wolf, DAS 1,032,240 (1956); C. A., 54, 19490g (1960).
1025. Kovalev, L. S., N. A. Razumov, and A. A. Petrov, Zh. Obshch. Khim., 38, 2277 (1968).
1026. Kovalev, L. S., N. A. Razumova, and A. R. Petrov, Zh. Obshch. Khim., 38, 2277 (1968).
1027. Kozlov, N. S., V. D. Pak, and E. S. Elin, Zh. Obshch. Khim., 39, 2407 (1969).
1028. Krasilnikova, E. A., A. M. Potapov, and A. I. Razumov, Zh. Obshch. Khim., 38, 609 (1968).
1029. Krasil'nikova, E. A., A. M. Potapov, and A. I. Razumov, Zh. Obshch. Khim., 37, 1409 (1967).

1030. Krasil'nikova, E. A., A. N. Potapov, and A. I. Razumov, Zh. Obshch. Khim., 37, 2365 (1967).
1031. Krasil'nikova, E. A., A. M. Potapov, and A. I. Razumov, Zh. Obshch. Khim., 38, 1098 (1968).
1032. Krawiecka, B., and J. Michalski, Bull. Acad. Pol. Sci., Ser. Sci. Chim., 19, 377 (1971); C. A., 75, 140944j (1971).
1033. Kreutzberger, A., and D. Abel, Arch. Pharm., 303, 715 (1970).
1034. Kreutzkamp, N., Chem. Ber., 88, 195 (1955).
1035. Kreutzkamp, N., Naturwissenschaften, 43, 81 (1956).
1036. Kreutzkamp, N., and G. Cordes, Ann., 623, 103 (1959).
1037. Kreutzkamp, N., and G. Cordes, Arch. Pharm., 295, 276 (1962).
1038. Kreutzkamp, N., and E. M. Gemser, Arch. Pharm., 295, 188 (1962).
1039. Kreutzkamp, N., and G. Kayser, Chem. Ber., 89, 1614 (1956).
1040. Kreutzkamp, N., and W. Mengel, Ann., 657, 19 (1962).
1041. Kreutzkamp, N., and W. Mengel, Arch. Pharm., 295, 773 (1962).
1042. Kreutzkamp, N., and W. Mengel, Arch. Pharm., 300, 389 (1967).
1043. Kreutzkamp, N., and W. Mengel, Chem. Ber., 100, 709 (1967).
1044. Kreutzkamp, N., and J. Pluhatsch, Arch. Pharm., 292, 159 (1959).
1045. Kreutzkamp, N., J. Pluhatsch, H. Schindler, and H. Kayser, Arch. Pharm., 295, 81 (1962).
1046. Kreutzkamp, N., and H. Schindler, Chem. Ber., 92, 1695 (1959).
1047. Krilov, M., and G. Petrov, Monatsh. Chem., 99, 166 (1968).
1048. Krilov, M., and J. Petrova, Chem. Ber., 103, 1047 (1970).
1049. Krivun, S. V., S. N. Baranov, and O. F. Voziyanova, Dokl. Akad. Nauk SSSR, 196, 600 (1971); C. A., 75, 35599c (1971).
1050. Krüger, F., and Bauer, L., Fr. 2,089,675 (1971).
1051. Krüger, F., and Bauer, L., DAS 2,013,371 (1971).
1052. Krupnov, V. V., and A. N. Pudovik, U.S.S.R. 249,384 (1968); C. A., 72, 43863n (1970).
1053. Kuchen, W., and H. Buchwald, Chem. Ber., 91, 2296 (1958).
1054. Kuchen, W., and K. Koch, Z. Naturforsch., B, 25, 1189 (1970).
1055. Kudriavtsev, B. V., and D. Kh. Yarmukhametova, Izvest. Akad. Nauk SSSR, Ser. Khim., 1970, 1583; C. A., 74, 13229j (1971).
1055a. Kühle, E., Angew. Chem., 81, 18 (1969).
1055b. Kukhar, V. P., V. I. Pasternak, and A. V. Kirsamov,

Zh. Obshch. Khim., 42, 1169 (1972).
1056. Kukhtin, V. A., V. S. Abramov, and K. M. Orekhova, Dokl. Akad. Nauk SSSR, 128, 1198 (1959); C. A., 54, 7536b (1960).
1057. Kukhtin, V. A., G. Kamai, and L. A. Sinchenko, Dokl. Akad. Nauk SSSR, 118, 505 (1958); C. A., 52, 10956 (1958).
1058. Kukhtin, V. A., and G. Kamai, Zh. Obshch. Khim., 28, 1196 (1958).
1059. Kukhtin, V. A., and K. M. Kirillova, Zh. Obshch. Khim., 31, 2226 (1961).
1060. Kukhtin, V. A., and K. M. Orechova, Zh. Obshch. Khim., 30, 1208 (1960).
1061. Kukhtin, V. A., and K. M. Orechova, Zh. Obshch. Khim., 28, 2790 (1958); C. A., 53, 7972 (1959).
1062. Kukhtin, V. A., Ju. Ju. Samitov, and K. M. Kirillova, Izv. Akad. Nauk SSSR, Ser. Khim., 1967, 356; C. A., 67, 21361s (1967).
1063. Kumial Chem. Ind. Co. Ltd., Jap. 7,109,480 (1964).
1064. Kunbazarov, A., A. M. Sorochan, and M. M. Senyavin, Zh. Neorg. Khim., 16, 651 (1971); C. A., 74, 116323q (1971).
1065. Kuskov, V. K., and G. F. Bebich, Dokl. Akad. Nauk SSSR, 136, 354 (1961); C. A., 55, 16456b (1961).
1066. Kuznetsov, E. V., and M. I. Bakhitov, Zh. Obshch. Khim., 31, 3015 (1961).
1067. Kuznetsov, E. V., and M. I. Bakhitov, Dokl. Akad. Nauk SSSR, 141, 1105 (1961); C. A., 57, 4577a (1962).
1068. Kuznetsov, E. V., and M. I. Bakhitov, Zh. Obshch. Khim., 32, 278 (1962).
1069. Labarre, M. -Ch., D. Voigt, and F. Gallais, Bull. Soc. Chim. Fr., 1967, 3328.
1070. Lacoste, R. G., J. W. Wisner, and H. D. Kluge, U. S. 2,935,505 (1958); C. A., 54, 21746 (1960).
1071. Ladd, E. C., and M. P. Harvey, U. S. 2,609,376 (1949); C. A., 47, 7540 (1953).
1072. Ladd, E. C., and M. P. Harvey, U. S. 2,651,656 (1950); C. A., 48, 10052 (1954).
1073. Ladd, E. C., and M. P. Harvey, U. S. 2,971,019 (1961); C. A., 55, 16427i (1961).
1074. Ladd, E. C., and M. P. Harvey, U. S. 3,048,613 (1962); C. A., 57, 15154h (1962).
1075. Landau, M. A., V. V. Sheluchenko, and S. S. Dubov, Zh. Strukt. Khim., 11, 513 (1970); C. A., 73, 65601 (1970).
1076. Landauer, S. R., and H. N. Rydon, J. Chem. Soc., 1953, 2224.
1077. Laskorin, B. N., E. P. Buchikhin, V. V. Shatalov, and S. I. Ponomareva, Radiokhimiya, 13, 809 (1971); C. A., 76, 117985h (1972).
1078. Laughlin, R. G., U. S. 3,138,629 (1964); C. A., 61, 8529h (1964).

1079. Laughlin, R. G., J. Org. Chem., 27, 1005 (1962).
1080. Laughlin, R. G., J. Org. Chem., 27, 3644 (1962).
1081. Lavielle, G., and D. Reisdorf, Compt. Rend. (C), 272, 100 (1971).
1082. Lavielle, G., G. Sturtz, and H. Normant, Bull. Soc. Chim. Fr., 1967, 4186.
1083. Lea-Ronal Inc., DOS 2,023,304 (1969); C. A., 74, 82554y (1971).
1084. Lechner, H. Z., T. H. Chao, and K. C. Whitehouse, U. S. 2,717,906 (1953); C. A., 50, 6508 (1956).
1085. Lecher, H. Z., and R. A. Greenwood, U. S. 2,870,204 (1955) American Cyanamid Co.; C. A., 53, 11306 (1959).
1086. Lecher, H. Z., R. A. Greenwood, K. C. Whitehouse, and T. H. Chao, J. Am. Chem. Soc., 78, 5018 (1956).
1088. Lecher, H. Z., T. H. Chao, K. C. Whitehouse, and R. A. Greenwood, J. Am. Chem. Soc., 76, 1045 (1954).
1089. Lee, J. D., G. W. Goodace, S. C. Peace, M. Fild, and R. Schmutzler, Naturwissenschaften, 57, 195 (1970).
1090. Lemmon, N. E., E. N. Roberts, and A. R. Sabol, U. S. 2,843,579 (1956); C. A., 53, 1701 (1959).
1091. Lember, A. L., and H. Tieckelmann, Tetrahedron Lett., 1964, 3053.
1092. Lenard-Borecka, B., J. Michalski, and S. Musierowicz, Rocz. Chem., 32, 1301 (1958); C. A., 53, 17884d (1959).
1093. Lenzi, M., G. Sturtz, and G. Lavielle, Compt. Rend. (C), 264, 1425 (1967).
1094. Lerch, I., and Kottler, A., DAS 1,002,355 (1957); C. A., 53, 21814e (1959).
1094a. Letts, E. A., R. F. Blake, Trans. Roy. Soc. Edinburgh, 35, 527 (1889).
1095. Levin, Ya. A., I. P. Gozman, and E. E. Sidorova, Izv. Akad. Nauk SSSR, Ser. Khim., 1970, 173; C. A., 72, 111565u (1970).
1096. Levy, D., and M. Sprecher, Tetrahedron Lett., 1971, 1909.
1097. Lewis, A. H., and R. D. Stayner, U. S. 2,655,533 (1951); C. A., 48, 10055 (1954).
1098. Liebig, H., Chem.-Ztg., 95, 301 (1971).
1099. Lieske, C. N., J. W. Havanec, G. M. Steinberg, J. N. Pikulin, W. J. Lennox, A. B. Ash, and P. Blumbergs, J. Arg. Food Chem., 17, 255 (1969).
1100. Lindner, J., and M. Strecker, Monatsh. Chem., 53/54, 274 (1929).
1100a. Lindner, J., W. Wirth, and B. Zaunbauer, Monatsh. Chem., 701, (1937).
1101. Linville, R. G., DAS 1,090, 210 (1960); C. A., 53, 1147d (1959).
1102. Liorber, B. G., Z. M. Khammatova, and A. I. Razumov, Zh. Obshch. Khim., 39, 1551 (1969).

1103. Lira, E. P., and C. W. Huffmann, J. Org. Chem., 31, 2188 (1966).
1104. Lister, S. H., and G. M. Timmis, Chem. Ind. (London), 20, 819 (1963).
1105. Lister, J. H., and G. M. Timmis, J. Chem. Soc. (C), 1966, 1242.
1106. Litthauer, S., Chem. Ber., 22, 2144 (1889).
1107. Loane, C. M., and J. W. Gaynor, U. S. 2,316,078-82 (1941); C. A., 37, 5858 (1943).
1108. Lomakina, V. I., Ja. A. Mandel'baum, and N. N. Mel'nikov, Zh. Obshch. Khim., 36, 447 (1966).
1109. Lomakina, V. I., V. V. Voronkova, Ja. A. Mandel'-baum, and N. N. Mel'nikov, Zh. Obshch. Khim., 35, 1752 (1965).
1110. Lorenz, W., and Chr. Fest, DAS 1,187,622 (1965); C. A. 63, 3000b (1965).
1111. Lorenz, W., and I. Hammann, DOS 2,003,141 (1971); C. A., 75, 88500v (1971).
1112. Lorenz, W., A. Henglein, and G. Schrader, J. Am. Chem. Soc., 77, 2554 (1955).
1113. Lorenz, W., and G. Schrader, DAS 1,105,656 (1959); C. A., 55, 277566 (1961).
1113a. Lorenz, W., and G. Schrader, DAS 1,067,017 (1958); C. A., 56, 1482 (1962).
1113b. Lorenz, W., and G. Schrader, DAS 1,124,034 (1960); C. A., 57, 3485b (1962).
1113c. Lorenz, W., and G. Schrader, DAS 1,114,478 (1960); C. A., 56, 10045e (1962).
1114. Lorenz, W., and G. Schrader, DAS 1,141,489 (1962); C. A., 59, 2860 (1963).
1115. Lubuzh, E. D., and A. D. Petrov, Izv. Akad. Nauk SSSR, Ser. Khim., 1962, 1001.
1116. Ludvik, G. F., J. P. Chupp, and P. E. Newallis, U. S. 3,165,441 (1965); C. A., 62, 11854c (1965).
1117. Lugovkin, B. P., Zh. Obshch. Khim., 27, 1524 (1957).
1118. Lugovkin, B. P., Zh. Obshch. Khim., 30, 2427 (1960).
1119. Lugovkin, B. P., Zh. Obshch. Khim., 31, 3406 (1961).
1120. Lugovkin, B. P., Zh. Obshch. Khim., 31, 3408 (1961).
1121. Lugovkin, B. P., Khim. Geterotsikl. Soedin., 1968, 117; C. A., 70, 4213g (1969).
1122. Lugovkin, B. P., Khim. Geterotsikl. Soedin., 1969, 694; C. A., 72, 67051x (1970).
1123. Lugovkin, B. P., Khim., Geterotsikl. Soedin, 1969, 900; C. A., 72, 111422v (1970).
1124. Lugovkin, B. P., Zh. Obshch. Khim., 40, 1050 (1970).
1125. Lugovkin, B. P., Zh. Obshch. Khim., 40, 2391 (1970).

1125a. Lugovkin, B. P., Zh. Obshch. Khim., 41, 815 (1971).
1126. Lugovkin, B. P., Zh. Obshch. Khim., 41, 2079 (1971).
1127. Lugovkin, B. P., and B. A. Arbuzov, Dokl. Akad. Nauk SSSR, 59, 1301 (1948); C. A., 42, 7265 (1948).
1128. Lugovkin, B. P., and B. A. Arbuzov, Izv. Akad. Nauk SSSR, Ser. Khim., 1950, 56; C. A., 44, 7256e (1950).
1129. Lukin, A. M., N. A. Bolotina, T. V. Tsernysheva, and G. B. Zavarachina, Dokl. Akad. Nauk SSSR, 173, 361 (1967); C. A., 67, 55143n (1967).
1130. Lukin, A. M., and I. D. Kalinina, Dokl. Akad. Nauk SSSR, 137, 873 (1961); C. A., 56, 1474f (1962).
1131. Lukin, A. M., and I. D. Kalinina, Zh. Obshch. Khim., 30, 1597 (1960).
1131a. Lukin, A. M., and G. S. Petrova, Zh. Obshch. Khim., 27, 2171 (1957).
1132. Lutsenko, I. F., and M. Kirilov, Dokl. Akad. Nauk SSSR, 132, 842 (1960); C. A., 54, 20842a (1960).
1133. Lutsenko, I. F., M. Kirilov, and G. A. Ovcinnikova, Zh. Obshch. Khim., 31, 2028 (1961).
1134. Lutsenko, I. F., M. Kirilov, and G. B. Postnikova, Zh. Obshch. Khim., 31, 2034 (1961).
1135. Lutsenko, I. F., M. Kirilov, and G. B. Postnikova, Zh. Obshch. Khim., 32, 263 (1962).
1136. Lutsenko, I. F., and Z. S. Krajc, Dokl. Akad. Nauk SSSR, 1960, 860.
1137. Lynch, E. R., J. Chem. Soc., 1962, 3729.
1138. Lysenko, V. V., S. Z. Ivin, K. V. Karavanov, and V. V. Fedotova, Zh. Obshch. Khim., 37, 1096 (1967).
1139. Machleidt, H., and G. Strehlke, Ann., 681, 21 (1965).
1140. Machleidt, H., and G. U. Strehlke, Angew. Chem., 76, 494 (1964).
1141. Machleidt, H., and R. Wessendorf, Ann., 674, 1 (1964).
1142. Makhsumov, A. G., A. M. Sladkov, and V. V. Korsak, Izv. Akad. Nauk SSSR, Ser. Khim., 1964, 733; C. A., 61, 3142a (1964).
1143. Magerlein, B. J., and F. Kagan, J. Am. Chem. Soc., 82, 593 (1960).
1144. Maguire, M. H., and G. Shaw, J. Chem. Soc., 1955, 1756.
1145. Maguire, M. H., R. K. Ralph, and G. Shaw, J. Chem. Soc., 1958, 2299.
1146. Maguire, M. H., G. Shaw, and C. G. Greenham, Chem. Ind. (London); 1953, 668.
1147. Mahler, W., and A. B. Bury, J. Am. Chem. Soc., 80, 6161 (1958).
1148. Maier, L., Helv. Chim. Acta, 46, 1812 (1963).

1149. Maier, L., Helv. Chim. Acta, 46, 2667 (1963).
1150. Maier, L., Helv. Chim. Acta, 47, 27 (1964).
1151. Maier, L., Helv. Chim. Acta, 47, 1448 (1964).
1152. Maier, L., U. S. 3,137,692 (1964); C. A., 61, 7045 (1964).
1153. Maier, L., U. S. 3,188,339 (1965); C. A., 63, 8405 (1965).
1154. Maier, L., U. S. 3,336,378 (1967); C. A., 68, 22056 (1968).
1154a. Maier, L., Helv. Chim. Acta, 52, 858 (1969).
1154b. Maier, L., Helv. Chim. Acta, 52, 827 (1969).
1155. Maier, L., Helv. Chim. Acta, 53, 1948 (1970).
1156. Maier, L., Helv. Chim. Acta, 53, 2069 (1970).
1157. Maier, L., DOS 1,945,146 (1970); C. A., 72, 100881t (1970).
1158. Maier, L., Phosphorus, 1, 67 (1971).
1158a. Maier, L., Phosphorus, 2, 229 (1973).
1158b. Maier, L., Phosphorus, 3, 19 (1973).
1159. Majoral, J.-P., R. Pujol, J. Naveck, and F. Mathis, Tetrahedron Lett., 1971, 3755.
1160. Makljaev, F. I., N. K. Bliznjuk, and G. I. Fremin, Zh. Obshch. Khim., 30, 4053 (1960).
1161. Makljaev, F. L., M. I. Douzin, and I. V. Palagina, Zh. Obshch. Khim., 31, 2012 (1961).
1162. Makljaev, F. L., N. V. Kirillov, A. V. Fokin, and L. S. Rudnitzkaja, Zh. Obshch. Khim., 40, 1014 (1970).
1163. Malatesta, L., Gazz. Chim. Ital., 77, 509 (1947).
1164. Malatesta, L., Gazz. Chim. Ital., 80, 527 (1950).
1165. Malatesta, P., and G. Migliaccio, Ann. Chim. (Rome), 50, 1158 (1960).
1166. Malatesta, L., and R. Pizzoti, Chim. Ind. (Milan), 27, 6 (1945).
1167. Malatesta, B. L., and R. Pizzoti, Gazz. Chim. Ital., 76, 167 (1946).
1168. Malatesta, L., A. Sacco and L. Ormezzano, Gazz. Chim. Ital., 80, 658 (1950).
1169. Malatesta, P., Farm. Ed. Sci., 18, 714 (1963); C. A., 60, 5540g (1964).
1170. Malatesta, P., and A. Ciamarella, Ann. Chimica, 51, 230 (1961).
1171. Malatesta, P., A. Ciamarella, and G. Migliaccio, Farm. Ed. Sci., 17, 65 (1962); C. A., 57, 16648a (1962).
1172. Malevannaya, R. A., E. N. Tsvetkov, and M. I. Kabachnik, Zh. Obshch. Khim., 41, 1426 (1971).
1173. Malinoskij, M. S., and V. V. Alekseev, Zh. Obshch. Khim., 30, 2965 (1960).
1174. Malinowskij, M. S., Z. F. Solomko, and E. I. Ectushenko, Zh. Obshch. Khim., 30, 2591 (1960).
1175. Malz, H., and G. Hermann, DAS 1,223,381 (1966); C. A., 65, 15272 (1966).

1176. Maneke, G., and H. Heller, DAS 1,175,242 (1964); C. A., 61, 13346f (1964).
1177. Maneke, G., and H. Heller, Chem. Ber., 95, 2700 (1962).
1177a. Marie, C., Compt. Rend., 133, 219 (1901).
1178. Marie, C., Compt. Rend., 135, 106 (1902).
1178a. Marie, C., Compt. Rend., 134, 286 (1902).
1179. Marie, C., Compt. Rend., 134, 847 (1902).
1180. Marie, C., Compt. Rend., 134, 994 (1902).
1181. Marie, C., Compt. Rend., 135, 1118 (1902).
1182. Marie, C., Compt. Rend., 136, 48 (1903).
1183. Marie, C., Compt. Rend., 137, 124 (1903).
1184. Marie, C., Compt. Rend., 136, 508 (1903).
1185. Marie, C., Compt. Rend., 136, 234 (1903).
1186. Marie, C., Ann. Chim. Phys. (8), 3, 335 (1904).
1187. Mark, V., Tetrahedron Lett., 1961, 295.
1188. Mark, V., Tetrahedron Lett., 1962, 281.
1189. Mark, V., U. S. 3,197,497 (1965); C. A., 63, 13318 (1965).
1189a. Mark, V., C. H. Dungan, M. M. Crutchfield, and J. R. Van Wazer, in Topics in Phosphorus Chemistry, 5, Wiley (1967).
1190. Mark, V., J. R. Van Wazer, J. Org. Chem. 29, 1006 (1964).
1190a. Markowska, A., and J. Michalski, Rocz. Chem., 34, 1675 (1960); C. A., 56, 7345i (1962).
1191. Marmor, R. S., and D. Seyferth, J. Org. Chem., 36, 128 (1971).
1192. Marsden, C. G., U. S., 2,963,503 (1960); C. A., 55, 8292d (1961).
1193. Marsi, K. L., C. A. Van der Werf, and W. E. McEven, J. Am. Chem. Soc., 78, 3063 (1956).
1194. Marsh, J. E., and J. A. Gardner, J. Chem. Soc., 65, 35 (1894).
1195. Martin, D. J., DOS 2,118,223 (1971).
1196. Martin, D. J., M. Gordon, and C. E. Griffin, Tetrahedron, 23, 1831 (1967).
1197. Martin, D., and H. -J. Niclas, Chem. Ber., 100, 187 (1967).
1198. Martin, H., C. Richter, R. Egli, and P. Sieber, Swiss. 339,925 (1959).
1199. Martynov, I. V., Y. U. L. Kruglyak, and S. I. Malekin, U.S.S.R. 245,097 (1968); C. A., 72, 67095q (1970).
1200. Martynov, I. V., S. I. Malekin, N. F. Privesentseva, U.S.S.R. 228,022 (1969); C. A., 71, 50215k (1969).
1201. Martynov, V. F., and V. E. Timofeev, Zh. Obshch. Khim., 32, 3449 (1962).
1202. Martynov, V. F., and V. L. Timofeev, Zh. Obshch. Khim., 34, 3890 (1964).
1203. Maruszewska-Wieczorkowska, E., and J. Michalski,

J. Org. Chem., 23, 1886 (1958).
1204. Maruszewska-Wieczorkowska, E., and J. Michalski, Rocz. Chem., 33, 405 (1959); C. A., 53, 21936a (1959).
1205. Maruszewska-Wieczorkowska, E., and J. Michalski, Rocz. Chem., 37, 1315 (1963); C. A., 60, 5546a (1964).
1206. Maruszewska-Wiezcorkowska, E., and J. Michalski, Rocz. Chem., 38, 625 (1964); C. A., 61, 10702h (1964).
1207. Mashlyakovskij, L. N, and B. I.Ionin, Zh. Obshch. Khim., 35, 1577 (1965).
1208. Mashlyakovskij, L. N., T. A. Zagudaeva, B. I. Ionin, and I. S. Okhrimenko, Zh. Obshch. Khim., 41, 330 (1971).
1209. Maslennikov, V. P., and V. P. Sergeeva, Zh. Obshch. Khim., 40, 1906 (1970).
1210. May, R. L., U. S. 2,356,073 (1943); C. A., 39, 614 (1945).
1213. McMaster, E. L., and W. K. Glesner, U. S. 2,980,721 (1961); C. A., 56, 506b (1962).
1214. Mastrjukova, T. A., G. M. Baranov, V. V. Perek, and M. I. Kabachnik, Dokl. Akad. Nauk SSSR, 171, 1341 (1966); C. A., 66, 54959a (1967).
1215. Maurer, E. W., A. J. Shirton, W. C. Ault, and K. J. Weil, J. Am. Oil Chem. Soc., 41, 205 (1964).
1216. Mauret, P., J. -P. Fayet, and M. -Cl. Labarre, Compt. Rend. (C), 265, 65 (1967).
1217. Maynard, J. A., and J. M. Swan, Aust. J. Chem., 16, 609 (1963).
1218. Maynard, J. A., and J. M. Swan, Aust. J. Chem., 16, 596 (1963).
1219. Medved, T. Ya., and M. I. Kabachnik, Dokl. Akad. Nauk SSSR, 84, 717 (1952); C. A., 47, 3226c (1953).
1220. Medved, T. Ya., and M. I. Kabachnik, Izv. Akad. Nauk SSSR, Ser. Khim., 1954, 314; C. A., 48, 10541b (1954).
1220a. Medved, T. Ya., and M. I. Kabachnik, Izv. Akad. Nauk SSSR, Ser. Khim., 1956, 684; C. A., 51, 1817 (1957).
1221. Medved, T. Ya., and M. I. Kabachnik, Izv. Akad. Nauk SSSR, Ser. Khim., 1957, 1357; C. A., 52, 7316b (1958).
1222. Medved, T. Ya., and M. I. Kabachnik, Izv. Akad. Nauk SSSR, Ser. Khim., 1958, 1212; C. A., 53, 4111a (1959).
1223. Medved, I. Ya., and M. I. Kabachnik, Izv. Akad. Nauk SSSR, Ser. Khim., 1961, 270; C. A., 55, 20922 (1961).
1224. Medved, T. Ya., M. V. Rudomino, E. A. Mironova, V. S. Balabukha, and M. I. Kabachnik, Izv. Akad. Nauk SSSR, Ser. Khim., 1967, 351; C. A., 67, 21973t (1967).

1225. Meek, J. S., and J. S. Towler, J. Org. Chem., 33, 985 (1968).
1226. Meisters, A., and J. M. Swan, Aust. J. Chem., 16, 725 (1963).
1227. Meisters, A., and J. M. Swan, Aust. J. Chem., 18, 163 (1965).
1228. Meisters, A., and J. M. Swan, Aust. J. Chem., 18, 155 (1965).
1229. Melchiker, P., Chem. Ber., 31, 2915 (1898).
1230. Mel'nikov, N. N., and A. F. Grapov, Zh. Obshch. Khim., 36, 1841 (1966).
1231. Mel'nikov, N. N., A. F. Grapov, L. V. Razvodovskaja, and S. L. Portnova, Zh. Obshch. Khim., 35, 1771 (1965).
1232. Mel'nikov, N. V., A. F. Grapov, and L. V. Razvodovskaja, Zh. Obshch. Khim., 36, 269 (1966).
1233. Mel'nikov, N. N., A. F. Grapov, A. V. Razvodovskaja, and T. M. Ivanova, Zh. Obshch. Khim., 37, 239 (1967).
1234. Mel'nikov, N. N., A. F. Grapov, and N. V. Lebedeva, Zh. Obshch. Khim., 36, 457 (1966).
1235. Mel'nikov, N. N., A. F. Grapov, and N. V. Lebedeva, Zh. Obshch. Khim., 36, 450 (1966).
1236. Mel'nikov, N. N., N. V. Lebedeva, N. K. Daragan, and A. F. Grapov, Zh. Obshch. Khim., 37, 2760 (1967).
1237. Mel'nikov, N. N., Ja. A. Mandel'baum, and Z. M. Bakanova, Zh. Obshch. Khim., 31, 3953 (1961).
1238. Mel'nikov, N. N., Ya. A. Mandel'baum, V. I. Lomakina, and V. S. Livsiv, Zh. Obshch. Khim., 31, 3949 (1961).
1239. Mel'nikov, N. N., Ya. A. Mandelbaum, and V. I. Lomakina, Zh. Obshch. Khim., 29, 3289 (1959).
1240. Mel'nikov, N. N., O. B. Mikhailova, and A. F. Grapov, Zh. Obshch. Khim., 38, 2099 (1968).
1241. Mel'nikov, N. N., K. D. Shvecova-Silovskaya, L. V. Nifant'ev, Khim. Geterotsikl. Soedin., 3, 465 (1966); C. A., 65, 15255 (1966).
1242. Mel'nikov, N. N., L. V. Razvodovskaya, and A. F. Grapov, Zh. Obshch. Khim., 38, 2648 (1968).
1243. Melton, T. M., U. S. 3,479,418 (1969); C. A., 72, 55653g (1970).
1244. Melton, Th. M., and D. A. Brown, U. S. 3,476,836 (1969); C. A., 72, 32017p (1970).
1245. Melton, T. M. and H. A. Kaufmann, U. S. 3,428,655.
1246. Melton, Th. M., U. S. 3,297,797 (1967); C. A., 66, 46481 (1967).
1247. Menn, J. J., and K. Szabo, J. Econ. Entomol., 58, 734 (1965).
1248. Messinger, P., Arch. Pharm., 304, 842 (1971).
1249. Metcalf, R. F., Fr. 2,002,164 (1969).
1250. Metcalf, R. L., and T. R. Fukuto, J. Econ. Entomol., 53, 127 (1960).

1251. Metzger, S. H., A. F. Isbell, and O. H. Basedow, J. Org. Chem., 29, 627 (1964).
1252. Metzger, S. H., and A. F. Isbell, J. Org. Chem., 29, 623 (1964).
1254. Meyer, F., and C. Hackmann, Ger. 963,876 (1953), Chem. Zentr., 1958, 2815; C. A., 53, 11312b (1959).
1255. Meyers, T. C., K. Nakamura, and A. B. Danielzaden, J. Org. Chem., 30, 1517 (1965).
1256. Michaelis, A., Chem. Ber., 6, 816 (1873).
1256a. Michaelis, A., Chem. Ber., 12, 1009 (1879).
1257. Michaelis, A., Ann., 181, 265 (1876).
1258. Michaelis, A., Ann., 181, 294 (1876).
1259. Michaelis, A., Ann., 181, 335-339 (1876).
1259a. Michaelis, A., Chem. Ber., 13, 2174 (1880).
1260. Michaelis, A., Ann., 293, 193 (1896).
1260a. Michaelis, A., Chem. Ber., 17, 1273 (1884); 18, 898 (1885); 19, 1009 (1886).
1261. Michaelis, A., Ann., 293, 207 (1896).
1262. Michaelis, A., Ann., 293, 229 (1896).
1263. Michaelis, A., Ann., 293, 242 (1896).
1264. Michaelis, A., Ann., 293, 254 (1896).
1265. Michaelis, A., Ann., 293, 261 (1896).
1266. Michaelis, A., Ann., 293, 263 (1896).
1267. Michaelis, A., Ann., 293, 295 (1896).
1268. Michaelis, A., Ann., 293, 319 (1896).
1269. Michaelis, A., Ann., 294, 1 (1897).
1270. Michaelis, A., Ann., 294, 8 (1896).
1271. Michaelis, A., Ann., 294, 40 (1896).
1272. Michaelis, A., Ann., 315, 43 (1901).
1273. Michaelis, A., Ann., 315, 51 (1901).
1274. Michaelis, A., Ann., 315, 54 (1901).
1275. Michaelis, A., Ann., 326, 129 (1903).
1276. Michaelis, A., Ann., 407, 316 (1915).
1277. Michaelis, A., and J. Ananoff, Chem. Ber., 7, 1688 (1874).
1278. Michaelis, A., and T. Becker, Chem. Ber., 30, 1003 (1897).
1279. Michaelis, A., and E. Benzinger, Chem. Ber., 8, 1310 (1875).
1280. Michaelis, A., and E. Benzinger, Ann., 188, 275 (1877).
1281. Michaelis, A., and A. Flemming, Chem. Ber., 34, 1291 (1901).
1282. Michaelis, A., B. V. Gaza, and W. Rehse, Ann., 407, 316 (1915).
1283. Michaelis, A., and R. Kaehne, Chem. Ber., 31, 1048 (1898).
1284. Michaelis, A., and F. Kammerer, Chem. Ber., 8, 1306 (1875).
1285. Michaelis, A., and H. Köhler, Chem. Ber., 9, 519 (1876).
1285a. Michaelis, A., and C. Mathias, Chem. Ber., 7, 1070 (1874).

1286. Michaelis, A., and C. Paneck, Ann., 212, 203 (1882).
1287. Michaelis, A., and C. Paneck, Chem. Ber., 14, 405 (1881).
1288. Michaelis, A., and R. Pasternak, Chem. Ber., 32, 2411 (1899).
1289. Michaelis, A., and F. Rothe, Chem. Ber., 25, 1747 (1892).
1290. Michaelis, A., and A. Schenk, Ann., 260, 1 (1890).
1291. Michaelis, A., and G. Schlüter, Chem. Ber., 31, 1037 (1898).
1292. Michailov, B. M., and N. F. Kucherova, Dokl. Akad. Nauk SSSR, 78, 709 (1951); C. A., 46, 2004 (1952).
1293. Michalski, J., Rocz. Chem., 29, 960 (1955); C. A., 50, 10641 (1956).
1294. Michalski, J., and Cz. Krawiecki, Rocz. Chem., 31, 715 (1957); C. A., 52, 5284g (1958).
1295. Michalski, J., and W. Mikolajczyk, Chem. Commun., 1963, 35.
1296. Michalski, J., and M. Mikolajczyk, Tetrahedron Lett., 1966, 3055.
1297. Michalski, J., M. Mikolajczyk, and A. Skowronska, Chem. Ind. (London), 24, 1053 (1962).
1298. Michalski, J., and M. Mikolajczyk, Chem. Ind. (London), 16, 661 (1964).
1299. Michalski, J., and B. Pliszka-Krawiecka, J. Chem. Soc., (C), 1966, 2249.
1300. Michalski, J., and A. Ratayczak, Rocz. Chem., 36, 911 (1962); C. A., 59, 3946a (1963).
1301. Michalski, J., and A. Skowronska, J. Chem. Soc. (C), 1970, 703.
1302. Michalski, J., and A. Skowronska, Rocz. Chem., 31, 301 (1957); C. A., 51, 16278 (1957).
1303. Michalski, J., and J. Wieczorkowski, Rocz. Chem., 33, 105 (1959); C. A., 53, 15956e (1957).
1304. Michalski, J., and A. Zwierzak, Proc. Chem. Soc., 1964, 80.
1305. Middleton, W. J., and W. H. Sharkey, J. Org. Chem., 30, 1384 (1965).
1306. Mikhailova, N., Uchenye Zapiski Kazan. Gosudarst, Univ. V. I. Ul'yanova-Lenina, 101, Nr. 3 Sbornik Studencheskikh Rabot Nr. 2, 58 (1941); C. A., 40, 55 (1946).
1307. Mikolayczyk, M., Chem. Ind. (London), 1966, 2059.
1308. Mikolajczyk, M., Chem. Ber., 99, 2083 (1966).
1309. Mikolajczyk, M., Angew. Chem., 78, 393 (1966).
1310. Mikolajczyk, M., Tetrahedron, 23, 1543 (1967).
1311. Mikulski, C. M., N. M. Karayannis, J. V. Minkiewicz, L. L. Pytlewski, and M. M. Labes, Inorg. Chim. Acta (Padova), 3, 523 (1969).
1312. Mikulski, C. M., N. M. Karayannis, M. Y. Strocko,

L. L. Pytlewski, and M. M. Labes, Inorg. Chem., 9, 2053 (1970).
1313. Miller, L. A., U. S. 3,093, 672 (1963); C. A., 59, 13823d (1963).
1314. Miller, J. A., Tetrahedron Lett., 1970, 3427.
1315. Milobendzki, T., and K. Szulgin, Chem. Polski, 15, 66 (1917); Chem. Zentr., 1918, I, 914.
1316. Minich, D., N. I. Rizpolozhenskij, and V. D. Akamsin, Izv. Akad. Nauk SSSR, Ser. Khim., 1968, 1792; C. A., 70, 37096g (1969).
1317. Minich, D., N. I. Rizpoloshenskij, V. D. Akamsin, and O. A. Raevskij, Izv. Akad. Nauk SSSR, Ser. Khim., 1969, 876; C. A., 71, 49447f (1969).
1318. Miranova, R. I., Yu. G. Golobov, Yu. M. Zinov'ev, and L. S. Soborovskij, Zh. Obshch. Khim., 39, 384 (1969).
1319. Misra, A. K., and J. B. Lal, Ind. J. Appl. Chem., 28, 187 (1965).
1320. Miyata, H., Bull. Chem. Soc. Jap., 36, 127 (1963).
1321. Mizrach, L. I., and V. P. Evdakov, Zh. Obshch. Khim., 36, 469 (1966).
1322. Mizrach, L. I., Mamonov, V. I. Svergun, and L. N. Kozlova, Zh. Obshch. Khim. 41, 1406 (1971).
1323. Mizrach, L. I., V. P. Evdakov, and L. Yu. Sandalova, Zh. Obshch. Khim., 35, 1871 (1965).
1324. Mizrach, K. I., V. G. Yakovlev, E. M. Yukno, V. I. Mamonov, and V. I. Sergun, Zh. Obshch. Khim., 41, 2654 (1971).
1325. Mobil Oil Corp., Brit. 1,097,634 (1968); C. A., 69, 10353f (1968).
1326. Moedritzer, K., J. Am. Chem. Soc., 83, 4381 (1961).
1327. Moedritzer, K., U. S. 3,242,236 (1966); C. A., 64, 15925h (1966).
1328. Moedritzer, K., and R. R. Irani, J. Org. Chem., 31, 1603 (1966).
1329. Moedritzer, K., and R. R. Irani, J. Inorg. Nucl. Chem., 22, 297 (1961).
1329a. Moedritzer, K., L. Maier, and L. C. D. Groenweghe, J. Chem. Eng. Data, 7, 307 (1962).
1330. Monsanto Chem. Comp., Belg. 774,349 (1971).
1331. Monsanto Chem. Comp., Brit. 989,443 (1965); C. A., 63, 3000d (1965).
1332. Monsanto Chem. Comp., Brit. 1,209,486 (1966); C. A., 74, 23000s (1971).
1333. Monsanto Chem. Comp., Brit. 1,247,405 (1971); C. A., 75, 129936n (1971).
1334. Moreau, J. P., and L. H. Chance, J. Chem. Eng. Data, 15, 581 (1970).
1335. Moreau, J. P., and L. H. Chance, J. Chem. Eng. Data, 14, 403 (1969).
1336. Morris, R. C., and J. L. van Winkle, U. S. 2,815,361 (1953); C. A., 52, 5465 (1958).

1337. Morrison, D. C., J. Org. Chem., 22, 444 (1957).
1338. Morrison, D. C., J. Org. Chem., 21, 705 (1956).
1339. Morrison, D. C., J. Am. Chem. Soc., 73, 896 (1951).
1339a. Morrison, D. C., J. Am. Chem. Soc., 73, 5896 (1951).
1340. Morrison, A. L., and F. R. Atherton, Brit. 682,706 (1951); C. A., 47, 11223f (1953).
1341. Moschel, W., H. Jonas, and W. Noll, Ger. 832,499 (1950); C. A., 52, 6850c (1958).
1342. Mosher, W. A., and R. R. Irino, J. Am. Chem. Soc., 91, 756 (1969).
1343. Moshkina, T. M., and A. N. Pudovik, Zh. Obshch. Khim., 35, 2042 (1965).
1344. Moshkina, T. M., and A. N. Pudovik, Zh. Obshch. Khim., 32, 1671 (1962).
1345. Moskva, V. V., V. M. Ismailov, and A. I. Razumov, Zh. Obshch. Khim., 41, 90 (1971).
1346. Moskva, V. V., V. M. Ismailov, and A. I. Razumov, Zh. Obshch. Khim., 41, 1495 (1971).
1347. Moskva, V. V., V. M. Ismailov, T. V. Zykova, and A. I. Razumov, Zh. Obshch. Khim., 41, 1676 (1971).
1348. Moskva, V. V., V. M. Ismailov, T. V. Zykova, and A. I. Razumov, Zh. Obshch. Khim., 41, 96 (1971).
1349. Moskva, V. V., A. I. Maikova, and A. I. Razumov, Zh. Obshch. Khim., 39, 2451 (1969).
1350. Moskva, V. V., G. F. Nazdurova, I. V. Zykova, A. I. Razumov, and L. A. Chemodonova, Zh. Obshch. Khim., 41, 1680 (1971).
1351. Moss, R. D., U. S. 2,959,590 (1960); C. A., 55, 7448b (1961).
1352. Moss, R. D., U. S. 3,012,054 (1961); C. A., 57, 4698d (1962).
1353. Mountcastle, Jr., W. R., W. H. Martin, P. D. Ballard, and D. H. Miles, J. Inorg. Nucl. Chem., 33, 3537 (1971).
1354. Müller, E., and H. G. Padeken, Chem. Ber., 100, 521 (1967).
1355. Mukaiyama, T., and Y. Yokota, Bull. Chem. Soc. Jap., 38, 858 (1965).
1356. Mukmenev, E. T., and G. Kamai, Dokl. Akad. Nauk SSSR, 153, 605 (1963); C. A., 60, 6737c (1964).
1357. Muratova, A. A., I. Ya. Kuramshim, E. G. Yarkova, and A. N. Pudovik, Zh. Obshch. Khim., 41, 1967 (1971).
1358. Murayama, K., S. Morimura, Y. Nakamura, and G. Sumagawa, J. Pharm. Soc. Jap., 85, 757 (1965).
1359. Murray, A. W., and K. Vaughn, Chem. Commun., 24, 1282 (1967).
1360. Mustafa, A., M. M. Sidky, and F. M. Soliman, Tetrahedron, 23, 99 (1967).
1361. Mustafa, A., M. M. Sidky, and F. M. Soliman, Ann., 698, 109 (1966).
1362. Mustafa, A., M. M. Sidky, and F. M. Soliman, Tetra-

hedron, 22, 393 (1966).
1363. Myannik, A. O., E. S. Shepeleva, and P. I. Sanin, Neftechimija, 4, 899 (1964); C. A., 62, 11675g (1965).
1364. Myers, T. C., R. G. Harvey, and E. V. Jensen, J. Am. Chem. Soc., 77, 3101 (1955).
1365. Myers, T. C., K. Nakamura, and J. W. Flesher, J. Am. Chem. Soc., 86, 3292 (1963).
1366. Myers, T. C., S. Preis, and E. V. Jensen, J. Am. Chem. Soc., 76, 4172 (1954).
1367. Myers, T. C., and L. N. Simons, J. Org. Chem., 30, 443 (1965).
1368. Nakayama, Sh., M. Yoshifuyi, R. Okazaki, and N. Inamoto, Chem. Commun., 1971, 1186.
1369. Neidlein, R., W. Haussmann, and E. Henkelbach, Chem. Ber., 99, 1252 (1966).
1370. Neimysheva, A. A., I. V. Savchuk, and I. L. Knunjants, Zh. Obshch. Khim., 36, 500 (1966).
1371. Nesterov, L. V., and A. Ya. Kessel, Zh. Obshch. Khim., 37, 728 (1967).
1372. Nesterov, L. V., A. Ya. Kessel, Yu. Yu. Samitov, and A. A. Musina, Zh. Obshch. Khim., 40, 1237 (1970)
1373. Nesterov, L. V., N. E. Krepysheva, and R. I. Mutalapova, Zh. Obshch. Khim., 35, 2050 (1965).
1374. Nesterov, L. V., N. E. Krepysheva, R. A. Sabirova, and G. N. Romanova, Zh. Obshch. Khim., 41, 2449 (1971).
1375. Nesterov, L. V., and R. A. Sabirova, Zh. Obshch. Khim., 31, 2358 (1961).
1376. Nesterov, L. V., and R. A. Sabirova, Zh. Obshch. Khim., 35, 2006 (1965).
1377. Nesterov, L. V., R. A. Sabirova, and N. E. Krepyshev Zh. Obshch. Khim., 39, 1943 (1969).
1378. Newallis, P. E., and J. P. Chupp, U. S. 3,058,878 (1962); C. A., 58, 6864e (1963).
1379. Newallis, P. E., J. P. Chupp, and L. C. D. Groenweghe, J. Org. Chem., 27, 3829 (1961).
1380. Newallis, P. E., J. P. Chupp, and H. L. Nufer, J. Chem. Eng. Data, 15, 455 (1970).
1381. Nguyen, H. P., N. T. Thuong, and P. Chabrier, Compt. Rend. (C), 271, 1465 (1970).
1382. Nicholson, D. A., Phosphorus, 2, 143 (1972).
1383. Nicholson, D. A., W. A. Cilley, and O. T. Quimby, J. Org. Chem., 35, 3149 (1970).
1384. Nicholson, D. A., and H. Vaughn, J. Org. Chem., 36, 3843 (1971).
1384a. Nicholson, D. A., and H. Vaughn, J. Org. Chem., 36, 1875 (1971).
1385. Nielsen, M. L., U. S. 3,296,301 (1967); C. A., 66, 85856 (1967).
1386. Nielsen, M. L., and R. Z. Greenley, U. S. 3,359,276 (1967); C. A., 68, 39621 (1968).

1387. Nifant'ev, E. E., I. P. Gudkova, and N. K. Kochetkov, Zh. Obshch. Khim., 40, 460 (1970).
1388. Nifant'ev, E. E., and N. L. Ivanova, Zh. Obshch. Khim., 40, 1492 (1970).
1389. Nifant'ev, E. E., N. L. Ivanova, I. P. Gudkova, and I. V. Shilov, Zh. Obshch. Khim., 40, 1420 (1970).
1390. Nifant'ev, E. E., and I. V. Shilov, Zh. Obshch. Khim., 41, 2372 (1971).
1391. Nifant'ev, E. E., and I. V. Shilov, Zh. Obshch. Khim., 41, 2104 (1971).
1392. Nifant'ev, E. E., and A. I. Zavalishina, Zh. Obshch. Khim., 37, 1854 (1967).
1393. Nifant'ev, E. E., A. I. Zavalishina, and I. V. Komlev, Zh. Obshch. Khim., 37, 2497 (1967).
1394. Nihon Kagaku Kogyo Co. Ltd., Jap. 7,008,220 (1970); C. A., 73, 14991t (1970).
1395. Nijk, D. R., Rec. Trav. Chim., 41, 461 (1922).
1396. Nikitina, V. I., and A. N. Pudovik, Zh. Obshch. Khim., 29, 1219 (1959).
1397. Nikonorov, K. V., E. A. Gurilev, Izv. Akad. Nauk SSSR, Ser. Khim., 1965, 2136; C. A., 64, 11245b (1966).
1398. Nikonorov, K. V., and V. A. Nikonenko, Izv. Akad. Nauk SSSR, Ser. Khim., 1962, 1882; C. A., 58, 7968h (1963).
1399. Nikonorov, K. V., and V. A. Nikonenko, Izv. Akad. Nauk SSSR, Ser. Khim., 1963, 1662; C. A., 59, 15308d (1963).
1400. Nikonorov, K. V., and V. A. Nikonenko, Izv. Akad. Nauk SSSR, Ser. Khim., 1966, 1373; C. A., 66, 54958z (1967).
1401. Nishiwaki, T., Tetrahedron, 21, 3043 (1965).
1402. Nishiwaki, T., Tetrahedron, 22, 711 (1966).
1403. Nishiwaki, T., Tetrahedron, 23, 2181 (1967).
1404. Nishiwaki, T., and T. Saito, J. Chem. Soc. (C), 1971, 3021.
1405. Noble, A. M., and J. M. Winfield, J. Chem. Soc. (A), 1970, 501.
1406. Normant, H., Fr. 1,509,091 (1968); C. A., 70, 47592 (1969).
1407. Normant, H., Bull. Soc. Chim. Fr., 1966, 3601.
1408. Normant, H., J. -F. Brault, Compt. Rend., 264, 707 (1967).
1409. Normant, H., T. Cuvigny, J. Normant, and B. Angelo, Bull. Soc. Chim. Fr., 1965, 3441.
1410. Normant, H., and G. Sturtz, Compt. Rend., 253, 2366 (1961).
1411. Novikova, Z. S., and M. A. Kramovskaja, Zh. Obshch. Khim., 39, 1060 (1969).
1412. Novikova, Z. S., and I. F. Lutsenko, Zh. Obshch. Khim., 40, 2129 (1970).
1413. Novikova, Z. S., S. N. Mashoshina, and I. F. Lutsenko,

Zh. Obshch. Khim., 41, 2110 (1971).
1414. Novikova, Z. S., S. N. Mashoshina, T. A. Saposhnikov
and I. F. Lutsenko, Zh. Obshch. Khim., 41, 2622 (1971).
1415. Novikova, Z. S., M. V. Proskurnina, L. I. Petrovska
I. V. Bogdanova, N. P. Galitskova, and I. F. Lutsenko, Zh. Obshch. Khim., 37, 2080 (1967).
1416. Nuretdinova, O. N., and B. A. Arbuzov, Izv. Akad. Nauk SSSR, Ser. Khim., 1971, 353; C. A., 75, 36256a (1971).
1416a. Nuretdinov, S. Kh., and N. P. Grechkin, Izv. Akad. Nauk SSSR., Ser. Khim., 1967, 436; C. A., 67, 21750t (1967).
1417. Nuretdinov, S. Kh., N. M. Ismagilova, T. V. Zykova, R. A. Salakhutdinov, V. S. Tsivunin, and G. Kh. Kamai, Zh. Obshch. Khim., 41, 2486 (1971).
1418. Nylen, P., Chem. Ber., 57, 1023 (1924).
1419. Nylen, P., Chem. Ber., 59, 1119 (1926).
1420. Nylen, P., Studien über Org. Phosphorverbindungen, Diss. Uppsala 1930.
1421. Obrycki, R., and C. E. Griffin, Tetrahedron Lett., 1966, 5049.
1422. Obrycki, R., and C. E. Griffin, J. Org. Chem., 33, 632 (1968).
1423. Oertel, G., and H. Holtschmidt, DAS 1,129,149 (1962)
C. A., 57, 11238h (1962).
1424. Ogata, J., and H. Tomioka, J. Org. Chem., 35, 596 (1970).
1425. Okamoto, Y., and H. Sakurai, Bull. Chem. Soc. Jap., 43, 2613 (1970).
1425a. Oldbury-Electrochemical Co., U. S. 2,500,022 (1946); C. A., 44, 4923 (1950).
1426. Olin Mathieson Chem. Co., Brit. 1,032,504 (1966); C. A., 65, 8961 (1966).
1428. Omelanczuk, J., and M. Mikolajczyk, Tetrahedron, 27, 5587 (1971).
1429. Opitz, G., A. Griesinger, and H. W. Schubert, Ann. Chem., 665, 91 (1963).
1430. Orlov, N. F., and M. A. Belokrinitzkij, Zh. Obshch. Khim., 40, 504 (1970).
1431. Orlov, N. F., B. I. Ionin, and V. P. Shvedov, Zh. Obshch. Khim., 35, 2046 (1965).
1432. Orlov, N. F., and V. P. Mileshkevich, U.S.S.R. 172,789 (1966); C. A., 64, 2128e (1966).
1433. Orlov, N. F., and V. P. Mileshkevich, Dokl. Akad. Nauk SSSR, 164, 344 (1965); C. A., 63, 18141f (1965).
1433a. Orlov, N. F., and V. P. Mileshkevich, Zh. Obshch. Khim., 36, 892 (1966).
1434. Orlov, N. F., and V. P. Mileshkevich, Zh. Obshch. Khim., 36, 518 (1966).
1435. Orlov, N. F., and V. P. Mileshkevich. Zh. Obshch.

Khim., 36, 699 (1966).
1436. Orlov, N. F., V. P. Mileshkevich, and E. S. Andronov, Zh. Obshch. Khim., 35, 2193 (1965).
1437. Orlov, N. F., V. P. Mileshkevich, and E. L. Gefter, Zh. Obshch. Khim., 35, 1312 (1965).
1438. Orlov, N. F., V. P. Mileshkevich, and E. L. Gefter, Zh. Obshch. Khim., 35, 590 (1965).
1439. Orlov, N. F., V. P. Mileshkevich, and V. M. Vajnburg, Zh. Obshch. Khim., 36, 1075 (1966).
1440. Orlov, N. F., L. N. Slesar, and M. S. Sorokin, Zh. Obshch. Khim., 40, 2585 (1970).
1441. Orlov, N. F., O. F. Victorov, and N. S. Elkina, Dokl. Akad. Nauk SSSR, 166 378 (1968).
1442. Orlov, H. F., and M. S. Voronkov, Zh. Obshch. Khim., 32, 608 (1962).
1442a. Orthner, L., M. Reuter, and E. Wolf, Ger. 1,045,373 (1958); C. A., 54, 23129b (1961).
1443. Oswald, A. A., and G. N. Schmit, DOS 1,925,763 (1969); C. A., 72, 79221r (1970).
1444. Overberger, C. G., and E. Sarlo, J. Org. Chem., 26, 4711 (1961).
1445. Owens, C., N. M. Karayannis, L. L. Pytlewski, and M. M. Labes, J. Phys. Chem., 75, 637 (1971).
1446. Page, H. J., J. Chem. Soc., 101, 423 (1912).
1447. Parts, L., U. S. 3,227,728 (1966); C. A., 64, 8239 (1966).
1448. Parts, L., M. L. Nielsen, and J. T. Miller, Inorg. Chem., 3, 1261 (1964).
1449. Pastushkov, V. N., Yu. A. Kondrat'ev, S. Z. Ivin, E. S. Vdovina, and A. S. Vasil'ev, Zh. Obshch. Khim., 38, 1407 (1968).
1450. Pastushkov, V. N., E. S. Vdovina, Yu. A. Kondrat'ev, S. Z. Ivin, and V. V. Tarasov, Zh. Obshch. Khim., 38, 1408 (1968).
1451. Patandva, I. M., R. G. Kuzovleva, and V. A. Belaev, U.S.S.R. 252,335 (1968); C. A., 72, 67099u (1970).
1452. Pattison, V. A., J. G. Colson, and R. L. K. Carr, J. Org. Chem., 33, 1084 (1968).
1453. Paul, L., and K. Zieloff, Chem. Ber., 99, 1431 (1966).
1454. Pelchowicz, Z., J. Chem. Soc., 1961, 238.
1455. Pelchowicz, Z., S. Brukson, and E. D. Bergmann, J. Chem. Soc., 1961, 4348.
1456. Pelchowicz, Z., and H. Leader, J. Chem. Soc., 1963, 3320.
1457. Pelchowicz, Z., H. Leader, S. Cohen, and D. Balderman, J. Chem. Soc., 1962, 3824.
1458. Peppard, D. F., J. R. Ferraro, and G. W. Mason, J. Inorg. Nucl. Chem., 1959, 60.
1459. Perkow, W., K. Ullerich, and F. Meyer, Naturwissenschaften, 39, 353 (1952).
1460. Pernert, J. C., Ger. 1,003,750 (1955); C. A., 53,

18972d (1959).
1461. Peterlein, K., DAS 1,194,414 (1965); C. A., 63, 7044c (1965).
1462. Petersen, H., Ann., 726, 89 (1969).
1463. Petersen, H., J. Plückhahn, and F. Fuchs, DOS 2,024,280 (1971).
1464. Petrov, K. A., R. A. Baksova, and L. V. Khorkhoyanu, Zh. Obshch. Khim., 35, 732 (1965).
1465. Petrov, K. A., R. A. Baksova, L. V. Khorkhoyanu, B. Ya. Libman, and L. P. Sinogejkina, Zh. Prikl. Khim., 39, 2798 (1966); C. A., 66, 76088t (1967).
1466. Petrov, K. A., R. A. Baksova, L. V. Korkhoyanu, L. P. Sinogeikina, and T. V. Skudina, Zh. Obshch. Khim., 35, 723 (1965).
1467. Petrov, K. A., A. A. Basyuk, V. P. Evdakov, and L. I. Mizrach, Zh. Obshch. Khim., 34, 2226 (1964).
1468. Petrov, K. A., N. K. Bliznjuk, M. A. Korshunov, F. L. Makljaev, and A. N. Voronkov, Zh. Obshch. Khim., 29, 3407 (1959).
1469. Petrov, K. A., N. K. Bliznjuk, and I. Yu. Mansurov, Zh. Obshch. Khim., 31, 176 (1961).
1470. Petrov, K. A., N. K. Bliznjuk, and V. A. Savostenok, Zh. Obshch. Khim., 31, 1361 (1961).
1471. Petrov, K. A., N. K. Bliznjuk, Yu. N. Studnev, and A. F. Kolomiets, Zh. Obshch. Khim., 31, 179 (1961).
1472. Petrov, K. A., V. P. Evdakov, and L. I. Mizrach, Zh. Obshch. Khim., 33, 1246 (1963).
1473. Petrov, K. A., V. P. Evdakov, L. I. Mizrach, and V. P. Romodin, Zh. Obshch. Khim., 32, 3062 (1962).
1474. Petrov, K. A., L. G. Gachenko, and A. A. Neymysheva, Zh. Obshch. Khim., 29, 1827 (1959).
1475. Petrov, K. A., A. I. Gavrilova, and M. M. Butilov, Zh. Obshch. Khim., 35, 1856 (1965).
1476. Petrov, K. A., A. I. Gavrilova, and A. M. Kopylov, Zh. Obshch. Khim., 30, 2863 (1960).
1477. Petrov, K. A., A. I. Gavrilova, and V. P. Korotkova, Zh. Obshch. Khim., 32, 1978 (1962).
1478. Petrov, K. A., A. I. Gavrilova, V. K. Shatunov, and V. P. Korotkova, Zh. Obshch. Khim., 31, 3081 (1961).
1479. Petrov, K. A., B. I. Ionin, and V. M. Ignatyev, Tetrahedron Lett., 1968, 15.
1480. Petrov, K. A., L. V. Khorkhoyanu, Ih. Beisheev, and K. Dzhundubaev, Zh. Obshch. Khim., 41, 110 (1971).
1481. Petrov, K. A., L. V. Khorkhoyanu, K. Dzhundubaev, and A. Sulaimanov, Zh. Obshch. Khim., 40, 1968 (1970).
1482. Petrov, K. A., L. V. Khorkhoyanu, and A. S. Sulaimanov, U.S.S.R. 297,645 (1970); C. A., 75, 88761f (1971).
1483. Petrov, K. A., V. A. Kravchenko, V. P. Evdakov, and L. I. Mizrach, Zh. Obshch. Khim., 34, 2586 (1964).

1484. Petrov, K. A., F. L. Makljaev, and N. K. Bliznjuk, Zh. Obshch. Khim., 29, 588 (1959).
1484a. Petrov, K. A., F. L. Makljaev, and N. K. Bliznjuk, Zh. Obshch. Khim., 29, 591 (1959).
1485. Petrov, K. A., F. L. Makljaev, and N. K. Bliznjuk, Zh. Obshch. Khim., 30, 1602 (1960).
1486. Petrov, K. A., F. L. Makljaev, and N. K. Bliznjuk, Zh. Obshch. Khim., 30, 1608 (1960).
1487. Petrov, K. A., F. L. Makljaev, and N. K. Bliznjuk, Zh. Obshch. Khim., 30, 1960 (1960).
1487a. Petrov, K. A., F. L. Makljaev, and M. A. Korshunov, Zh. Obshch. Khim., 29, 301 (1959).
1488. Petrov, K. A., F. L. Makljaev, A. A. Neimysheva, and N. K. Bliznjuk, Zh. Obshch. Khim., 30, 4060 (1960).
1489. Petrov, K. A., and A. A. Neimysheva, Zh. Obshch. Khim., 29, 1819 (1959).
1490. Petrov, K. A., and A. A. Neimysheva, Zh. Obshch. Khim., 29, 1822 (1959).
1491. Petrov, K. A., and A. A. Neimysheva, Zh. Obshch. Khim., 29, 3026 (1959).
1492. Petrov, K. A., A. A. Neimysheva, and E. V. Smirnov, Zh. Obshch. Khim., 29, 1491 (1959).
1493. Petrov, K. A., E. E. Nifant'ev, and R. G. Gol'cova, Zh. Obshch. Khim., 33, 1485 (1963).
1494. Petrov, K. A., E. E. Nifant'ev, and R. G. Gol'cova, Zh. Obshch. Khim., 31, 2370 (1961).
1495. Petrov, K. A., E. E. Nifant'ev, and R. G. Gol'cova, Zh. Obshch. Khim., 31, 3174 (1961).
1496. Petrov, K. A., E. E. Nifant'ev, R. G. Gol'cova, M. A. Belvencev, and S. M. Korneev, Zh. Obshch. Khim., 32, 1277 (1962).
1497. Petrov, K. A., E. E. Nifant'ev, R. G. Gol'cova, and G. V. Gubin, Zh. Obshch. Khim., 31, 2732 (1961).
1498. Petrov, K. A., E. E. Nifant'ev, and R. F. Nikitina, Zh. Obshch. Khim., 31, 1705 (1961).
1499. Petrov, K. A., N. A. Pokatilo, and L. V. Solomina, Zh. Obshch. Khim., 40, 2192 (1970).
1500. Petrov, K. A., V. A. Parshina, and G. Shefer, Zh. Obshch. Khim., 40, 1234 (1970).
1501. Petrov, K. A., M. A. Raksha, and V. L. Vinogradov, Zh. Obshch. Khim., 37, 2047 (1967).
1502. Petrov, K. A., M. A. Raksha, V. P. Korotkova, and E. Shmidt, Zh. Obshch. Khim., 41, 324 (1971).
1503. Pfeiffer, G., A. Guillemonat, and J. C. Traynard, Bull. Soc. Chim. Fr. 1966, 1652.
1503a. Philippot, E., J. C. Jumas, G. Brun, and M. Maurin, Cryst. Struct. Commun. 1, 103 (1972).
1504. Phytopathology Inst., U.S.S.R. 186,464 (1967); C. A., 66, 85601g (1967).
1505. Phytopathology Inst., U.S.S.R. 186,466 (1967); C. A., 66, 94699x (1967).

1506. Phytopathology Inst., DOS 1,956,187 (1971); C. A., 75, 77028w (1971).
1507. Phytopathology Inst., U.S.S.R. 189,846 (1967); C. A.,68, 13176r (1968).
1508. Phytopathology Inst., U.S.S.R. 189,851 (1967); C. A., 68, 13177s (1968).
1509. Phytopathology Inst., U.S.S.R. 192,204 (1967); C. A., 68, 59717w (1968).
1510. Phytopathology Inst., U.S.S.R. 193,504 (1967); C. A., 69, 27524k (1968).
1511. Phytopathology Inst., U.S.S.R. 218,864 (1969).
1512. Phytopathology Inst., U.S.S.R. 239,946 (1966); C. A., 72, 43853v (1970).
1513. Phytopathology Inst., U.S.S.R. 243,616 (1968); C. A., 71, 81523p (1969).
1514. Phytopathology Inst., U.S.S.R. 248,673 (1968); C. A., 72, 100889b (1970).
1515. Phytopathology Inst., U.S.S.R. 251,576 (1968); C. A., 72, 55655j (1970).
1516. Phytopathology Inst., U.S.S.R. 256,767 (1968); C. A., 72, 132964p (1970).
1517. Phytopathology Inst., U.S.S.R. 259,881 (1968); C. A., 73, 14995x (1970).
1518. Phytopathology Inst., U.S.S.R. 268,420 (1969); C. A., 73, 88021y (1970).
1519. Phytopathology Inst., U.S.S.R. 287,016 (1969); C. A., 75, 49334z (1971).
1520. Phytopathology Inst., U.S.S.R. 289,096 (1969); C. A., 75, 6092b (1971).
1521. Phytopathology Inst., U.S.S.R. 292,985 (1969); C. A., 75, 20620d (1971).
1522. Pilgram, K., and H. Ohse, Angew, Chem., 78, 820 (1966).
1523. Pikl. K., U. S. 2,328,358 (1941); C. A., 38, 754 (1944).
1523a. Pistschimuka, P., J. Prakt. Chem.(2), 84, 746 (1911)
1524. Pitré, D., and E. B. Grabitz, J. Prakt. Chem. (4), 32, 317 (1966).
1525. Plazek, E., and Z. E. Golubski, Rocz. Chem., 44, 681 (1970); C. A., 74, 3044f (1971).
1526. Plieninger, H., and A. Müller, Synthesis, 1970, 586.
1526a. Plöger, W., unpublished.
1527. Plöger, W., N. Schindler, K. Wollmann, and K. H. Worms, Z. Anorg. Allg. Chem., 389, 119 (1972).
1528. Plumb, J. B., and C. E. Griffin, J. Org. Chem., 27, 4711 (1962).
1529. Plumb, J. B., R. Obrycki, and C. E. Griffin, J. Org. Chem., 31, 2455.
1530. Ponomarev, V. V., S. A. Golubcov, K. A. Andrianov, and E. A. Khuprova, Izv. Akad. Nauk SSSR, Ser. Khim., 1969, 1551.

1531. Popoff, I. C., U. S. 3,294,628 (1966); C. A., 66, 65632 (1967).
1532. Popoff, I. C., U. S. 3,321,516 (1967); C. A., 67, 73678 (1967).
1533. Popov, G. L., N. F. Baina, and L. Z. Soborovskij, U.S.S.R. 253062 (1968); C. A., 72, 132966r (1970).
1534. Poroshin, K. T., and V. K. Burichenko, Dokl. Akad. Nauk SSSR, 156, 386 (1964); C. A., 61, 3192d (1964).
1535. Prat, J., L. Bourgeois, and P. Ragon, Mem. Serv. Chim., 34, 393 (1948); C. A., 44, 5800 (1950).
1536. Preis, S., T. C. Myers, and E. V. Jensen, J. Am. Chem. Soc., 77, 6225 (1955).
1537. Preston, J., and H. G. Clark, U. S. 2,928,859 (1958); C. A., 55, 3522j (1961).
1538. Prokof'eva, A. F., N. N. Melnikov, I. L. Vladimirova, and L. I. Einisman, Zh. Obshch. Khim., 41, 1702 (1971).
1540. Promonenko, V. K., Yu. E. Utkin, and N. A. Utkina, U.S.S.R. 247,301 (1968); C. A., 72, 3567n (1970).
1541. Protopopov, I. S., and M. J. Kraft, Zh. Obshch. Khim., 34, 1446 (1964).
1542. Pudovik, A. N., Dokl. Akad. Nauk SSSR, 73, 499 (1950); C. A., 45, 2856d (1951).
1543. Pudovik, A. N., Dokl. Akad. Nauk SSSR, 80, 65 (1951); C. A., 50, 4143 (1956).
1544. Pudovik, A. N., Dokl. Akad. Nauk SSSR, 85, 349 (1952); C. A., 47, 5351 (1953).
1545. Pudovik, A. N., Zh. Obshch. Khim., 22, 462 (1952).
1546. Pudovik, A. N., Zh. Obshch. Khim., 22, 473 (1952).
1547. Pudovik, A. A., Dokl. Akad. Nauk, SSSR, 83, 865 (1952); C. A., 47, 4300g (1953).
1548. Pudovik, A. N., Izv. Akad. Nauk SSSR, Ser. Khim., 1952, 926; C. A., 47, 10467e (1953).
1549. Pudovik, A. N., Zh. Obshch. Khim., 22, 1143 (1952).
1550. Pudovik, A. N., Zh. Obshch. Khim., 22, 1371 (1952).
1551. Pudovik, A. N., Zh. Obshch. Khim., 22, 2047 (1952).
1552. Pudovik, A. N., Zh. Obshch. Khim., 25, 2173 (1955).
1553. Pudovik, A. N., and M. M. Aladzheva, Zh. Obshch. Khim., 31, 2052 (1961).
1554. Pudovik, A. N., and I. M. Aladzheva, Zh. Obshch. Khim., 32, 2005 (1962).
1555. Pudovik, A. N., and Aladzheva, I. M., Zh. Obshch. Khim., 33, 707 (1963).
1556. Pudovik, A. N., and I. M. Aladzheva, Dokl. Akad. Nauk SSSR, 151, 1110 (1963); C. A., 59, 13798d (1963).
1557. Pudovik, A. N., and I. M. Aladzheva, Zh. Obshch. Khim., 33, 1816 (1963).
1558. Pudovik, A. N., and I. M. Aladzheva, Zh. Obshch. Khim., 33, 3096 (1963).
1559. Pudovik, A. N., I. M. Aladzheva, and L. N. Jakovenko, Zh. Obshch. Khim., 35, 1210 (1965).

1560. Pudovik, A. N., I. M. Aladzheva, and N. A. Petruseva Zh. Obshch. Khim., 34, 2907 (1964).
1561. Pudovik, A. N., I. M. Aladzheva, I. A. Sokolova, and G. A. Kozlova, Zh. Obshch. Khim., 33, 102 (1963)
1562. Pudovik, A. N., and B. A. Arbuzov, Dokl. Akad. Nauk SSSR, 73, 327 (1950); C. A., 45, 2853b (1951).
1563. Pudovik, A. N., and B. A. Arbuzov, Izv. Akad. Nauk SSSR, Ser. Khim., 1949, 522; C. A., 44, 1893 (1950).
1564. Pudovik, A. N., and B. A. Arbuzov, Zh. Obshch. Khim., 21, 382 (1951).
1565. Pudovik, A. N., and B. A. Arbuzov, Zh. Obshch. Khim., 21, 1837 (1951).
1566. Pudovik, A. N., and V. P. Averjanova, Zh. Obshch. Khim., 26, 1426 (1956).
1567. Pudovik, A. N., and E. S. Batyeva, Zh. Obshch. Khim., 39, 334 (1969).
1568. Pudovik, A. N., E. S. Batyeva, and O. A. Raevskij, Zh. Obshch. Khim., 39, 1235 (1969).
1569. Pudovik, A. N., E. S. Batyeva, R. R. Shagidullin, O. A. Raevskij, and M. A. Pudovik, Zh. Obshch. Khim., 40, 1195 (1970).
1570. Pudovik, A. N., and E. G. Chebotareva, Zh. Obshch. Khim., 28, 2492 (1958).
1571. Pudovik, A. N., and G. M. Denisova, Zh. Obshch. Khim., 23, 263 (1953).
1572. Pudovik, A. N., and O. S. Durova, Zh. Obshch. Khim., 36, 1460 (1966).
1573. Pudovik, A. N., and G. I. Estav'ev, Zh. Obshch. Khim., 34, 890 (1964).
1574. Pudovik, A. N., and E. M. Fajzullin, Zh. Obshch. Khim., 32, 231 (1962).
1575. Pudovik, A. N., E. M. Fajzullin, and V. P. Zhukov, Zh. Obshch. Khim., 36, 310 (1966).
1576. Pudovik, A. N., and M. M. Frolova, Zh. Obshch. Khim., 22, 2052 (1952).
1577. Pudovik, A. N., and R. D. Gareev, Zh. Obshch. Khim., 33, 3441 (1963).
1578. Pudovik, A. N., and R. D. Gareev, Zh. Obshch. Khim., 34, 3942 (1964).
1579. Pudovik, A. N., and R. D. Gareev, Zh. Obshch. Khim., 40, 1025 (1970).
1580. Pudovik, A. N., R. D. Gareev, and A. V. Aganov, Zh. Obshch. Khim., 41, 1017 (1971).
1581. Pudovik, A. N., R. D. Gareev, A. V. Aganov, and L. A. Stabrovskaja, Zh. Obshch. Khim., 41, 1173 (1971).
1581a. Pudovik, A. N., R. D. Gareev, A. V. Aganov, and L. A. Stabrovskaja, Zh. Obshch. Khim., 41, 1232 (1971).
1582. Pudovik, A. N., R. D. Gareev, and L. I. Kusnetsova, Zh. Obshch. Khim., 39, 1536 (1969).
1583. Pudovik, A. N., R. D. Gareev, and O. E. Raevskaja, Zh. Obshch. Khim., 40, 1189 (1970).

1584. Pudovik, A. N., T. Ch. Gazizov, and A. P. Pashinkin, Zh. Obshch. Khim., 36, 563 (1966).
1585. Pudovik, A. N., T. C. Gazizov, and A. P. Pashinkin, Zh. Obshch. Khim., 36, 951 (1966).
1586. Pudovik, A. N., T. C. Gazizov, J. J. Samitov, and T. V. Zykova, Dokl. Akad. Nauk SSSR, 166, 615 (1966); C. A., 64, 12716b (1966).
1587. Pudovik, A. N., T. C. Gazizov, Yu. Yu. Samitov, and T. V. Zykova, Zh. Obshch. Khim., 37, 706 (1967).
1588. Pudovik, A. N., and G. A. Golitsyna, Zh. Obshch. Khim., 34, 876 (1964).
1589. Pudovik, A. N., I. P. Gozman, and V. I. Nikitina, Zh. Obshch. Khim., 33, 3201 (1963).
1590. Pudovik, A. N., I. V. Gur'janova, and S. P. Perevezencova, Zh. Obshch. Khim., 37, 1090 (1967).
1591. Pudovik, A. N., I. V. Gur'janova, G. V. Romanov, and L. G. Rakhimova, Zh. Obshch. Khim., 41, 1485 (1971).
1592. Pudovik, A. N., I. V. Gur'janova, and M. G. Zimin, Zh. Obshch. Khim., 37, 2580 (1967).
1592a. Pudovik, A. N., and M. G. Imaev, Izv. Akad. Nauk SSSR, Ser. Khim., 1952, 916; C. A., 47, 10463 (1953).
1593. Pudovik, A. N., and E. A. Ishmaeva, Zh. Obshch. Khim., 35, 358 (1965).
1594. Pudovik, A. N., E. A. Ishmaeva, R. S. Akhmerova, and I. M. Aladzheva, Zh. Obshch. Khim., 36, 161 (1966).
1595. Pudovik, A. N., and B. E. Ivanov, Izv. Akad. Nauk SSSR, Ser. Khim., 1952, 947; C. A., 47, 10464 (1953).
1596. Pudovik, A. N., E. I. Kashevarova, and V. M. Gorchakova, Zh. Obshch. Khim., 34, 2213 (1964).
1597. Pudovik, A. N., E. I. Kashevarova, and N. G. Khusainova, Dokl. Akad. Nauk SSSR, 145, 818 (1962); C. A., 58, 547b (1963).
1598. Pudovik, A. N., E. I. Kashevarova, and Yu. P. Rudnev, Dokl. Akad. Nauk SSSR, 140, 841 (1961); C. A., 56, 3506e (1962).
1599. Pudovik, A. N., N. G. Khusainova, Zh. Obshch. Khim., 34, 3938 (1964).
1599a. Pudovik, A. N., N. G. Khusainova, T. V. Timoshina, and O. E. Raevskaya, Zh. Obshch. Khim., 41, 1476 (1971).
1600. Pudovik, A. N., and N. G. Khusainova, Zh. Obshch. Khim., 36, 1345 (1966).
1601. Pudovik, A. N., and N. G. Khusainova, Zh. Obshch. Khim., 39, 1646 (1969).
1602. Pudovik, A. N., N. G. Khusainova, and A. B. Ageeva, Zh. Obshch. Khim., 34, 3938 (1964).
1603. Pudovik, A. N., and N. G. Khusainova, Zh. Obshch. Khim., 36, 1236 (1966).
1604. Pudovik, A. N., and N. G. Khusainova, Zh. Obshch. Khim., 39, 2426 (1969).
1605. Pudovik, A. N., and N. G. Khusainova, Zh. Obshch.

Khim., 40, 1419 (1970).
1606. Pudovik, A. N., N. G. Khusainova, and I. M. Aladzhev Zh. Obshch. Khim., 34, 2470 (1964).
1607. Pudovik, A. N., N. G. Khusainova, and I. M. Aladzhev Zh. Obshch. Khim., 33, 1045 (1963).
1608. Pudovik, A. N., N. G. Khusainova, and T. I. Frolova, Zh. Obshch. Khim., 41, 2420 (1971).
1609. Pudovik, A. N., N. G. Khusainova, and R. G. Galeeva, Zh. Obshch. Khim., 36, 69 (1966).
1610. Pudovik, A. N., N. G. Khusainova, and T. V. Timoshin Zh. Obshch. Khim., 40, 1040 (1970).
1611. Pudovik, A. N., and Yu. P. Kitaev, Zh. Obshch. Khim. 22, 467 (1952).
1612. Pudovik, A. N., and I. V. Konovalova, Zh. Obshch. Khim., 27, 1617 (1957).
1613. Pudovik, A. N., and I. V. Konovalova, Zh. Obshch. Khim., 28, 1208 (1958).
1614. Pudovik, A. N., and I. V. Konovalova, Zh. Obshch. Khim., 29, 3342 (1959).
1615. Pudovik, A. N., and I. V. Konovalova, Zh. Obshch. Khim., 30, 2348 (1960).
1616. Pudovik, A. N., and I. V. Konovalova, Zh. Obshch. Khim., 31, 1693 (1961).
1617. Pudovik, A. N., and I. V. Konovalova, Zh. Obshch. Khim., 32, 467 (1962).
1618. Pudovik, A. N., and I. V. Konovalova, Zh. Obshch. Khim., 33, 98 (1963).
1619. Pudovik, A. N., and I. V. Konovalova, Dokl. Akad. Nauk SSSR, 149, 1091 (1963); C. A., 59, 6434g (1963).
1620. Pudovik, A. N., and I. V. Konovalova, Zh. Obshch. Khim., 33, 3100 (1963).
1621. Pudovik, A. N., and I. V. Konovalova, Zh. Obshch. Khim., 33, 3442 (1963).
1622. Pudovik, A. N., and I. V. Konovalova, Dokl. Akad. Nauk SSSR, 143, 875 (1962); C. A., 57, 3480a (1962).
1623. Pudovik, A. N., I. V. Konovalova, and L. V. Banderova, Zh. Obshch. Khim., 35, 1206 (1965).
1624. Pudovik, A. N., I. V. Konovalova, and L. A. Burnaeva Zh. Obshch. Khim., 41, 2413 (1971).
1625. Pudovik, A. N., I. V. Konovalova, and L. V. Dedova, Zh. Obshch. Khim., 34, 2902 (1964).
1626. Pudovik, A. N., I. V. Konovalova, and L. V. Dedova, Zh. Obshch. Khim., 34, 2905 (1964).
1627. Pudovik, A. N., I. V. Konovalova, and O. S. Durova, Zh. Obshch. Khim., 31, 2656 (1961).
1628. Pudovik, A. N., I. V. Konovalova, and A. A. Guryleva Zh. Obshch. Khim., 33, 2924 (1963).
1629. Pudovik, A. N., I. V. Konovalova, and E. A. Ishmaeva Zh. Obshch. Khim., 33, 2509 (1963).
1630. Pudovik, A. N., I. V. Konovalova, and E. A. Ishmaeva Zh Obshch. Khim., 32, 237 (1962).

1631. Pudovik, A. N., I. V. Konovalova, and R. E. Krivonosova, Zh. Obshch. Khim., 26, 3110 (1956).
1632. Pudovik, A. N., and M. V. Korchemkina, Izv. Akad. Nauk SSSR, Ser. Khim., 1952, 940; C. A., 47, 10468f (1953).
1633. Pudovik, A. N., and K. A. Kovyrzina, Zh. Obshch. Khim., 24, 307 (1954).
1634. Pudovik, A. N., and G. P. Krupnov, Zh. Obshch. Khim., 31, 4053 (1961).
1635. Pudovik, A. N., and G. P. Krupnov, Zh. Obshch. Khim., 34, 1157 (1964).
1636. Pudovik, A. N., and V. K. Krupnov, Zh. Obshch. Khim., 38, 1406 (1968).
1637. Pudovik, A. N., and V. K. Krupnov, Zh. Obshch. Khim., 39, 1890 (1969).
1638. Pudovik, A. N., and V. V. Krupnov, Zh. Obshch. Khim., 39, 2415 (1969).
1639. Pudovik, A. N., and A. V. Kusnetsova, Zh. Obshch. Khim., 25, 1369 (1955).
1640. Pudovik, A. N., and R. G. Kuzovleva, Zh. Obshch. Khim., 33, 2755 (1963).
1641. Pudovik, A. N., and R. G. Kuzovleva, Zh. Obshch. Khim., 34, 1031 (1964).
1642. Pudovik, A. N., and R. G. Kuzovleva, Zh. Obshch. Khim., 35, 354 (1965).
1643. Pudovik, A. N., and T. N. Moshkina, Zh. Obshch. Khim., 27, 1611 (1957).
1644. Pudovik, A. N., T. M. Moshkina, and I. V. Konovaleva, Zh. Obshch. Khim., 29, 3338 (1959).
1646. Pudovik, A. N., A. A. Muratova, T. I. Konova, T. Feoktistova, and L. N. Levkova, Zh. Obshch. Khim., 30, 2624 (L960).
1647. Pudovik, A. N., A. A. Muratova, I. I. Konnova, T. Feoktistova, and L. N. Levkova, Zh. Obshch. Khim., 30, 2606 (1960).
1648. Pudovik, A. N., A. A. Muratova, T. I. Konnova, T. Feoktistova, and L. N. Levkova, Zh. Obshch. Khim., 30, 2624 (1960).
1649. Pudovik, A. N., A. A. Muratova, and V. A. Savel'eva, Zh. Obshch. Khim., 34, 2582 (1964).
1650. Pudovik, A. N., V. I. Nikitina, and V. V. Evdokimova, Zh. Obshch. Khim., 40, 294 (1970).
1651. Pudovik, A. N., V. I. Nikitina, and A. M. Kurguzova, Zh. Obshch. Khim., 40, 291 (1970).
1652. Pudovik, A. N., and N. I. Plakatina, Sbornik Statei Obshch. Khim., Akad. Nauk SSSR, 2, 831 (1953); C. A., 49, 6821c (1955).
1653. Pudovik, A. N., and M. A. Pudovik, Zh. Obshch. Khim., 33, 3353 (1963).
1654. Pudovik, A. N., and M. A. Pudovik, Zh. Obshch. Khim., 36, 1467 (1966).
1655. Pudovik, A. N., and M. A. Pudovik, Zh. Obshch. Khim., 39, 1645 (1969).

1656. Pudovik, A. N., M. A. Pudovik, N. P. Anoshima, and L. S. Andreeva, Zh. Obshch. Khim., 39, 1719 (1969).
1657. Pudovik, A. N., M. A. Pudovik, and R. R. Shagidullin Zh. Obshch. Khim., 39, 1973 (1969).
1658. Pudovik, A. N., M. A. Pudovik, and S. A. Terent'eva, Zh. Obshch. Khim., 40, 33 (1970).
1659. Pudovik, A. N., R. D. Sabirova, and T. A. Terner, Zh. Obshch. Khim., 24, 1026 (1954).
1660. Pudovik, A. N., and M. K. Sergeeva, Zh. Obshch. Khim., 25, 1759 (1955).
1661. Pudovik, A. N., and L. I. Sidnichina, Zh. Obshch. Khim., 24, 1193 (1954).
1662. Pudovik, A. N., and F. N. Sitdikova, Dokl. Akad. Nauk SSSR, 125, 826 (1959); C. A., 53, 19850c (1959).
1663. Pudovik, A. N., and S. N. Sitdikova, Zh. Obshch. Khim., 34, 1682 (1964).
1664. Pudovik, A. N., S. A. Terent'eva, and M. A. Pudovik, Izv. Akad. Nauk SSSR, Ser. Khim., 1969, 2584; C. A., 72, 67055b (1970).
1664a. Pudovik, A. N., S. A. Terent'eva, and M. A. Pudovik, Zh. Obshch. Khim., 40, 1707 (1970).
1665. Pudovik, A. N., S. A. Terent'eva, and M. A. Pudovik, Izv. Akad. Nauk SSSR, Ser. Khim., 1971, 645; C. A., 75, 36245 (1971).
1666. Pudovik, A. N., and R. I. Tarasova, Zh. Obshch. Khim., 34, 293 (1964).
1667. Pudovik, A. N., and R. I. Tarasova, Zh. Obshch. Khim., 34, 1151 (1964).
1668. Pudovik, A. N., and R. I. Tarasova, Zh. Obshch. Khim., 34, 3946 (1964).
1669. Pudovik, A. N., R. I. Tarasova, and R. A. Bulgakova, Zh. Obshch. Khim., 33, 2560 (1963).
1670. Pudovik, A. N., and D. C. Yarmukhametova, Izv. Akad. Nauk SSSR, Ser. Khim., 1954, 636; C. A., 49, 8789a (1955).
1671. Pudovik, A. N., and D. C. Yarmukhametova, Izv. Akad. Nauk SSSR, Ser. Khim., 1952, 721; C. A., 47, 10467a (1953).
1672. Pudovik, A. N., G. E. Yastrebova, and V. I. Nikitina Zh. Obshch. Khim., 36, 1232 (1966).
1673. Pudovik, A. N., G. E. Yastrebova, and Pudovik, O. A., Zh. Obshch. Khim., 40, 499 (1970).
1674. Pudovik, A. N., and G. A. Zametaeva, Izv. Akad. Nauk SSSR, Ser. Khim., 1952, 932; C. A., 47, 10467i (1953).
1675. Pudovik, A. N., M. G. Zimin, and A. M. Kurguzova, Zh. Obshch. Khim., 41, 1964 (1971).
1676. Pushkus, A. C., and J. E. Herweh, J. Org. Chem., 29, 2567 (1964).
1677. Quesnel, G., M. de Botton, A. Chambolle, and R. Dubon, Compt. Rend., 251, 1074 (1960).

1677a. Quimby, O. T., and J. B. Prentice, U. S. 3,400,149 (1968).
1677b. Quimby, O. T., J. B. Prentice, and D. A. Nicholson, J. Org. Chem., 32, 4111 (1967).
1677c. Quimby, O. T., J. D. Curry, D. A. Nicholson, and J. B. Prentice, J. Organometal. Chem., 13, 199 (1968).
1677d. Quimby, O. T., U. S. 3,387,024 (1968).
1677e. Quin, L. D., Topics in Phosphorus Chemistry, 4, 23, Interscience (1967).
1678. Rabilloud, M. G., Compt. Rend., 259, 3009 (1964).
1679. Rabinowitz, R., U. S. 3,175,998 (1965); C. A., 63, 757a (1965).
1680. Rabinowitz, R., J. Am. Chem. Soc., 82, 4564 (1960).
1681. Rabinowitz, R., J. Org. Chem., 26, 5152 (1961).
1682. Rabinowitz, R., J. Org. Chem., 28, 2975 (1963).
1683. Rätz, R., J. Am. Chem. Soc., 77, 4170 (1955).
1684. Rätz, R., and A. D. Bliss, J. Heterocycl. Chem., 3, 14 (1966).
1685. Rafikov, S. R., and M. E. Ergebekov, Zh. Obshch. Khim., 34, 2230 (1964).
1686. Rajan, K. S., L. Murase, and A. E. Martell, J. Am. Chem. Soc., 91, 4408 (1969).
1687. Rakov, A. P., Zh. Obshch. Khim., 40, 2129 (1970).
1688. Ralston, P. H., DOS 1,517,399 (1969).
1689. Ramirez, F., O. P. Madan, and C. P. Smith, J. Org. Chem., 30, 2284 (1965).
1690. Ramirez, F., O. P. Madan, and S. R. Heller, J. Am. Chem. Soc., 87, 731 (1965).
1691. Ramirez, F., A. V. Patwardhan, and S. R. Heller, J. Am. Chem. Soc., 86, 514 (1964).
1692. Ramirez, F., H. Yamanaka, and O. H. Basedov, J. Am. Chem. Soc., 83, 173 (1961).
1693. Ramsey, W. M., and C. Kezerian, U. S. 2,917,528 (1959); C. A., 54, 24399i (1960).
1694. Ramsey, W. M., and C. Kezerian, U. S. 2,964,549 (1960); C. A., 55, 13319f (1961).
1695. Randall, D. I., and R. W. Wynn, DOS 2,050,247 (1971); C. A., 75, 20614e (1971).
1695a. Ranneva, Yu. I., I. S. Temnova, E. S. Petrov, A. I. Shatenshtein, E. N. Tsvetkov, and M. I. Kabachnik, Izv. Akad. Nauk SSSR, Ser. Khim., 1967, 2129; C. A., 68, 43714 (1968).
1696. Razumov, A. I., and N. P. Kashurina, Trans. Kirov's Inst. Chem. Technol. Kazan, 8, 45 (1940); C. A., 35, 2474 (1941).
1697. Razumov, A. I., and B. G. Liorber, Dokl. Akad. Nauk SSSR, 135, 1150 (1960); C. A., 55, 13298b (1961).
1698. Razumov, A. I., and B. G. Liorber, Zh. Obshch. Khim., 32, 4063 (1962).
1699. Razumov, A. I., B. G. Liorber, M. B. Gazizov, and Z. M. Khammatova, Zh. Obshch. Khim., 34, 1851 (1964).

1700. Razumov, A. I., B. G. Liorber, M. P. Sokolov, and T. V. Zykova, Zh. Obshch. Khim., 41, 2106 (1971).
1701. Razumov, A. I., B. G. Liorber, and M. P. Sokolov, Zh. Obshch. Khim., 40, 1252 (1970).
1702. Razumov, A. I., B. G. Liorber, M. B. Gazizov, and Z. M. Khammatova, Zh. Obshch. Khim., 34, 1851 (1964).
1703. Razumov, A. I., E. A. Markovich, and O. A. Mukhacheva, Khim. Primenenie Fosforoorgan Soedin. Akad. Nauk SSSR, Kazan. Filial. Tr. 1-Konf. 1955, 194; C. A., 52, 237 (1958).
1704. Razumov, A. I., E. A. Markovich, and O. A. Mukhacheva, Zh. Obshch. Khim., 28, 194 (1958); C. A., 52, 11733 (1958).
1704a. Razumov, A. I., E. A. Markovich, and A. D. Reshetnikova, Zh. Obshch. Khim., 27, 2394 (1957).
1705. Razumov, A. I., and V. V. Moskva, Zh. Obshch. Khim., 34, 2589 (1964).
1706. Razumov, A. I., and V. V. Moskva, Zh. Obshch. Khim., 34, 3125 (1964).
1707. Razumov, A. I., and V. V. Moskva, Zh. Obshch. Khim., 35, 1149 (1965).
1708. Razumov, A. I., and V. V. Moskva, Zh. Obshch. Khim., 35, 1595 (1965).
1709. Razumov, A. I., and V. V. Moskva, U.S.S.R. 175,961 (1966); C. A., 64, 9596h (1966).
1710. Razumov, A. I., O. A. Mukhacheva, and S. Do-Khen, Izv. Akad. Nauk SSSR, Ser. Khim., 1952, 894; C. A., 10466c (1953).
1711. Razumov, A. I., O. A. Mukhacheva, and O. A. Markovich, Zh. Obshch. Khim., 27, 2389 (1957).
1711a. Razumov, A. I., O. A. Mukhacheva, and E. A. Markovich, Zh. Obshch. Khim., 28, 194 (1958).
1712. Razumova, N. A., K. I. Novickij, and M. V. Kivi, Zh. Obshch. Khim., 37, 1136 (1967).
1713. Razumov, A. I., and N. G. Zabusova, Zh. Obshch. Khim., 30, 1307 (1960).
1714. Razumova, N. A., A. A. Petrov, A. K. Vosnesenskaja, and V. I. Nevickij, Zh. Obshch. Khim., 36, 244 (1966).
1715. Razvodovskaja, L. V., A. F. Grapov, and N. N. Mel'nikov, Zh. Obshch. Khim., 39, 1260 (1969).
1716. Razvodovskaja, L. V., N. N. Mel'nikov, N. V. Lebedeva, and A. F. Grapov, Zh. Obshch. Khim., 41, 1446 (1971).
1717. Reddy, G. S., and R. Schmutzler, Z. Naturforsch., B, 25, 1199 (1970).
1718. Redmore, D., J. Org. Chem., 35, 4114 (1970).
1719. Rednik, V. S., and Yu. S. Shvetsov, Izv. Akad. Nauk SSSR, Ser. Khim., 1971, 2006; C. A., 76, 14648 (1972).
1720. Reetz, Th., U. S. 2,909,558 (1959); C. A., 54, 3204a (1960).

1721. Reetz, Th., U. S. 2,935,518 (1958); C. A., 54, 19598c (1960).
1722. Reetz, Th., U. S. 3,010,986 (1961); C. A., 56, 4476 (1962).
1723. Reetz, T., D. H. Chardwick, E. E. Hardy, and S. Kaufman, J. Am. Chem. Soc., 77, 3813 (1955).
1724. Reetz, Th., and W. D. Dixon, U. S. 3,253,062 (1966); C. A., 65, 5488 (1966).
1725. Regel, E. K., U. S. 3,193,372 (1965); C. A., 63, 8976f (1965).
1726. Regel, E. A., U. S. 3,342,583 (1967); C. A., 69, 59371z (1968).
1727. Regel, E. A., U. S. 3,463,839 (1969); C. A., 71, 124649d (1969).
1728. Regel, E. K., and M. F. Botts, Can. 841,055 (1970).
1729. Regel, E. K., and M. F. Botts, U. S. 3,416,912 (1968); C. A., 71, 22185g (1969).
1730. Regel, E. K., and M. F. Botts, U. S. 3,467,736 (1969); C. A., 72, 32037s (1970).
1731. Regel, E. K., and M. F. Botts, U. S. 3,481,731 (1969); C. A., 72, 32022m (1970).
1731a. Regel, E. K., and M. F. Botts, U. S. 3,529,041 (1970); C. A., 73, 131133a (1970).
1732. Regitz, M., and W. Anschütz, Ann., 730, 194 (1969).
1733. Reichel, L., and H. J. Jahns, Ann., 751, 69 (1971).
1734. Reiff, L. P., and H. S. Aaron, J. Am. Chem. Soc., 92, 5275 (1970).
1735. Reiff, L. P., L. J. Szafraniec, and H. S. Aaron, Chem. Commun., 1971, 366.
1736. Reist, E. J., I. G. Junga, and B. R. Baker, J. Org. Chem., 25, 666 (1960).
1737. Reynolds, H. T., R. L. Metcalf, and T. R. Fukuto, J. Econ, Entomol., 58, 293 (1966).
1738. Reznik, V. S. and Yu. S. Shvetsov, Izv. Akad. Nauk SSSR, Ser. Khim., 1971, 2006; C. A., 76, 14648y (1972).
1739. Reznik, V. S., and Yu. S. Shetsov, Izv. Akad. Nauk SSSR, Ser. Khim., 1970, 2254; C. A., 75, 118372z (1971).
1740. Richard, J. J., K. E. Burke, J. W. Laughlin, and C. V. Banks, J. Am. Chem. Soc., 83, 1722 (1961).
1741. Richardson, G. A., and G. H. Birum, U. S. 3,190,892 (1965); C. A., 63, 11613d (1965).
1742. Richardson, G. A., and G. H. Birum, U. S. 3,235,567 (1966); C. A., 64, 14219e (1966).
1743. Richter, S. B., U. S. 2,914,439 (1959); C. A., 54, 6587c (1960).
1744. Richter, S. B., U. S. 3,459,836 (1969); C. A., 71, 81519s (1969).
1745. Rimmer, R. V., Belg. 633,648 (1963); C. A., 61, 575h (1964).
1746. Rizpolozhenskij, N. I., M. A. Zvereva, Izv. Akad. Nauk SSSR, Ser. Khim., 1959, 358; C. A., 53, 17888

(1959).
1747. Rizpolozhenskij, N. I., V. D. Alkamsin, and T. M. Dorova, Izv. Akad. Nauk SSSR, Ser. Khim., 1970, 622; C. A., 73, 14929 (1970).
1748. Rizpolozhenskij, N. I., L. V. Bojko, and M. A. Zvereva, Dokl. Akad. Nauk SSSR, 155, 1137 (1964); C. A., 61, 1817e (1964).
1749. Rizpolozhenskij, N. I., and F. S. Mukhametov, Izv. Akad. Nauk SSSR, Ser. Khim., 1967, 210; C. A., 66, 115648x (1967).
1750. Rizpoloxhenskij, N. I., and F. S. Mukhametov, Izv. Akad. Nauk SSSR, Ser. Khim., 1970, 1087; C. A., 73, 66679s (1970).
1751. Rizpolozhenskij, N. I., and A. A. Muslinkin, Izv. Akad. Nauk SSSR, Ser. Khim., 1961, 1600; C. A., 56, 4791d (1962).
1752. Rochlitz, F., DAS 1,300,296 (1969).
1753. Rockett, J., U. S. 2,922,791 (1960); C. A., 54, 8857d (1960).
1755. Roedig, A., F. Hagedorn, and G. Märkl, Chem. Ber., 97, 3322 (1964).
1756. Roesky, H. W., Chem. Ber., 101, 3679 (1968).
1757. Roesky, H. W., Z. Naturforsch., B, 24, 5 (1969).
1758. Roesky, H. W., and D. Bormann, Chem. Ber., 101, 630 (1968).
1759. Roesky, H. W., and W. Grosse Böwing, Z. Naturforsch. B, 24, 1250 (1969).
1760. Roesky, H. W., and M. Dietl., Z. Anorg. Allg. Chem., 376, 230 (1970).
1761. Roesky, H. W., and W. Kloker, Z. Anorg. Allg. Chem., 375, 140 (1970).
1762. Roesky, H. W., O. Petersen, Z. Naturforsch., B, 26, 1232 (1971).
1763. Rolih, R. J., and E. K. Fields, U. S. 3,054,821 (1962); C. A., 58, 4599a (1963).
1764. De Roos, A. M., and H. J. Toet, Rec. Trav. Chim., 78, 59 (1959).
1765. Rosenheim, A., and J. Pinsker, Chem. Ber., 43, 2003 (1910).
1766. Rosenheim, A., and M. Pritze, Chem. Ber., 41, 2708 (1908).
1767. Rosenheim, A., W. Stadler, and F. Jocobsohn, Chem. Ber., 39, 2837 (1906).
1768. Rosenthal, A. F., J. Chem. Soc., 1965, 7345.
1769. Rosenthal, A. F., and R. P. Geyer, J. Am. Chem. Soc., 80, 5240 (1958).
1770. Rosenthal, A. F., G. M. Kosolapoff, and R. P. Geyer, Rec. Trav. Chim., 83, 1273 (1964).
1771. Rosenthal, A. F., and M. Pousada, Rec. Trav. Chim., 84, 833 (1965).
1772. Rosin, J., and J. Haus, U. S. 2,899,456 (1959); C. A., 54, 2168b (1960).

1773. Rossijskaja, P. A., and M. I. Kabachnik, Izv. Akad. Nauk SSSR, Ser. Khim., 1947, 389; C. A., 42, 1558 (1948).

1774. Rozen, S., I. Shahack, and F. D. Bergmann, Synthesis, 1971, 646.

1775. Rueggeberg, W. H. C., J. Chernak, and I. M. Rose, J. Am. Chem. Soc., 72, 5336 (1950).

1776. Rubtsova, J. K., and V. J. Kirilovich, Plast. Massy, 1965, 59; C. A., 63, 4329d (1965).

1776a. Rumpf, P., and V. Chavane, Compt. Rend., 224, 919 (1947).

1776b. Russo, J. L., H. C. Dostal, and A. C. Leopold, Bioscience, 18, 109 (1968).

1777. Sabol, A. R., U. S. 2,806,022 (1955); C. A., 52, 2048 (1958).

1778. Sachs, H., Chem. Ber., 25, 1514 (1892).

1779. Sänger, A., Ann., 232, 1 (1886).

1780. Sallmann, R., DAS 1,082,262 (1960); C. A., 53, 222c (1959).

1781. Samarai, L. I., O. I. Kolodyazhnyi, and G. I. Derkach, Zh. Obshch. Khim., 39, 1712 (1969).

1782. Samuel, G., and R. Weiss, Tetrahedron, 26, 3951 (1970).

1783. Sandalova, L. Yu., L. I. Mizrach, and V. P. Evdakov, Zh. Obshch. Khim., 35, 1314 (1965).

1784. Sander, M., Angew. Chem., 73, 67 (1961).

1784a. Sanz, F., and J. J. Daly, J. Chem. Soc. Dalton, 1972, 2267.

1785. Sasin, R., R. M. Naumann, and D. Swern, J. Am. Chem. Soc., 80, 6336 (1958).

1786. Sasin, R., R. M. Naumann, and D. Swern, J. Am. Chem. Soc., 81, 4335 (1959).

1787. Sasin, R., W. F. Olszewski, J. R. Russell, and D. Swern, J. Am. Chem. Soc., 81, 6275 (1959).

1787a. Sasse, K., Methoden der organischen Chemie Band XII/1, Georg Thieme Verlag Stuttgart (1963).

1788. Saunders, B. C., and P. Simpson, J. Chem. Soc., 1963, 3351.

1789. Saunders, B. C., and P. Simpson, J. Chem. Soc., 1963, 3464.

1790. Saunders, B. C., G. J. Stacey, F. Wild, and I. G. E. Wilding, J. Chem. Soc., 1948, 699.

1791. Savel'eva, N. I., A. S. Kostyuk, Vu. I. Bankov, and I. F. Lutsenko, Zh. Obshch. Khim., 41, 485 (1971).

1792. Savignac, Ph., J. Chenault, Compt. Rend. (C), 270, 2164 (1970).

1793. Shaw, R. A., and E. H. M. Ibrahim, Angew. Chem., 79, 575 (1967).

1794. Shaw, R. A., and C. Stratton, Chem. Ind. (London), 1959, 52.

1795. Schegk, E., H. Schlör, and G. Schrader, DAS 1,058,992

(1957); C. A., 54, 12998b (1960).
1796. Schegk, E., and G. Schrader, DAS 1,044,812 (1956); C. A., 55, 1532e (1961).
1797. Schegk, E., and G. Schrader, DAS 1,078,124 (1957); C. A., 58, 1492b (1963).
1798. Schegk, E., and G. Schrader, DAS 1,089,376 (1958); C. A., 57, 13805a (1962).
1799. Schenk, A., and A. Michaelis, Chem. Ber., 21, 1497 (1888).
1800. Schicke, H. G., DAS 1,183,494 (1964); C. A., 62, 7684g (1965).
1801. Schicke, H. G., DAS 1,183,506 (1964); C. A., 62, 7798c (1965).
1802. Schicke, H. G., DAS 1,198,359 (1965); C. A., 63, 18155f (1965).
1803. Schicke, H. G., and W. Lorenz, DAS 1,144,264 (1963); C. A., 60, 4184e (1964).
1804. Schicke, H. G., W. Lorenz, and G. Schrader, U. S. 3,228,999 (1966); C. A., 57, 666c (1962).
1805. Schicke, H. G., and G. Schrader, DAS 1,129,484 (1962); C. A., 57, 13806b (1962).
1806. Schicke, H. G., and G. Schrader, DAS 1,206,900 (1965); C. A., 64, 9641c (1966).
1807. Schicke, H. G., and G. Schrader, U. S. 3,076,012 (1963); C. A., 57, 15156b (1962).
1808. Schimmelschmidt, K., and W. Denk, Ger. 1,023,033 (1956); C. A., 54, 5466e (1960).
1809. Schindelbauer, H., Chem. Ber., 100, 3432 (1967).
1809a. Schindler, N., and W. Plöger, Chem. Ber., 104, 2021 (1971).
1810. Schliebs, R., DAS 1,131,670 (1960); C. A., 57, 16660h (1962).
1811. Schlör, H., E. Schegk, and G. Schrader, Belg. 586,314 (1960).
1812. Schlör, H., E. Schegk, and G. Schrader, DAS 1,057,112 (1957); C. A., 54, 12998b (1960).
1813. Schlör, H., E. Schegk, and G. Schrader, DAS 1,099,53 (1959); C. A., 56, 8748e (1962).
1814. Schlör, H., E. Schegk, and G. Schrader, DAS 1,143,20 (1963); C. A., 59, 2861a (1963).
1815. Schlör, H., and G. Schrader, DAS 1,183,082 (1964); C. A., 62, 6514f (1965).
1816. Schlör, H., G. Schrader, and H. Scheinpflug, DOS 1,902,928 (1970); C. A., 73, 119693a (1970).
1817. Schmidt, P., Swiss 321,397 (1953); C. A., 52, 2067 (1958).
1818. Schmidt, P., Swiss 316,731 (1953); C. A., 52, 2066h (1958).
1819. Schmidt, P., U. S. 2,719,167 (1954); C. A., 50, 7843 (1956).
1820. Schmidt, K-J., and I. Hammann, DAS 1,221,634 (1966); C. A., 65, 13760h (1966).

1821. Schmidpeter, A., and N. Schindler, Chem. Ber., 102, 2201 (1969).
1822. Schmidpeter, A., and N. Schindler, DOS 1,916,844 (1969); C. A., 74, 13268 (1971).
1823. Schmidpeter, A., and N. Schindler, Angew. Chem., 80, 1030 (1968).
1824. Schmutzler, R., and G. S. Reddy, Inorg. Chem., 4, 191 (1965).
1825. Schrader, G., Belg. 597,651 (1960).
1826. Schrader, G., Ger. 938,187 (1952); C. A., 53, 2158c (1959).
1827. Schrader, G., DAS 1,032,247 (1954); C. A., 54, 22363a (1960).
1828. Schrader, G., DAS 1,051,851 (1957); C. A., 55, 17503b (1961).
1829. Schrader, G., DAS 1,051,852 (1957); C. A., 55, 17502e (1961).
1830. Schrader, G., DAS 1,053,503 (1957); C. A., 54, 10964f (1960).
1831. Schrader, G., DAS 1,063,166 (1958); C. A., 55, 17502c (1961).
1832. Schrader, G., DAS 1,064,510 (1960); C. A., 55, 14381f (1961).
1833. Schrader, G., DAS 1,064,512 (1958); C. A., 55, 17505a (1961).
1834. Schrader, G., DAS 1,071,701 (1958); C. A., 55, 12301h (1961).
1834a. Schrader, G., Ger. 1,071,702 (1958); C. A., 55, 12357h (1961).
1835. Schrader, G., DAS 1,072,245 (1957); C. A., 54, 7559a (1960).
1836. Schrader, G., DAS 1,074,042 (1957); C. A., 54, 7559a (1960).
1837. Schrader, G., DAS 1,076,129 (1957).
1837a. Schrader, G., DAS 1,080,108 (1958); C. A., 55, 16424b (1961).
1838. Schrader, G., DAS 1,104,506 (1959); C. A., 56, 1482h (1962).
1839. Schrader, G., DAS 1,081,012 (1958); C. A., 55, 19862i (1961).
1840. Schrader, G., DAS 1,081,458 (1960); C. A., 55, 15419a (1961).
1841. Schrader, G., DAS 1,087,600 (1959); C. A., 55, 19789c (1961).
1842. Schrader, G., DAS 1,094,745 (1960); C. A., 55, 22132 (1961).
1843. Schrader, G., DAS 1,099,532 (1959); C. A., 56, 3515c ( ).
1844. Schrader, G., DAS 1,099,534 (1959); C. A., 56, 1482b (1962).
1845. Schrader, G., Ger. 1,104,520 (1961); C. A., 56, 14328e (1962).

1846. Schrader, G., DAS 1,105,413 (1957).
1846a. Schrader, G., DAS 1,108,216 (1959); C. A., 56, 4795i (1962).
1847. Schrader, G., DAS 1,109,680 (1957).
1848. Schrader, G., DAS 1,116,223 (1957); C. A., 56, 8748g (1962).
1849. Schrader, G., DAS 1,116,224 (1961); C. A., 56, 14328c (1962).
1849a. Schrader, G., DAS 1,124,946 (1959); C. A., 58, 5724g (1963).
1850. Schrader, G., DAS 1,129,954 (1962); C. A., 58, 5724h (1963).
1851. Schrader, G., Aust. 210,438 (1960).
1852. Schrader, G., and H. -G. Schicke, DAS 1,122,060 (1962); C. A., 57, 12325h (1962).
1853. Schrader, G., DAS 1,134,373 (1962); C. A., 58, 3316c (1963).
1853a. Schrader, G., DAS 1,136,333 (1959); C. A., 58, 6864g (1963).
1854. Schrader, G., DAS 1,137,012 (1957); C. A., 57, 15155a (1962).
1855. Schrader, G., DAS 1,139,119 (1962); C. A., 58, 12601 (1963).
1856. Schrader, G., DAS 1,139,494 (1961); C. A., 55, 12601 (1961).
1857. Schrader, G., DAS 1,139,495 (1962); C. A., 58, 12601e (1963).
1858. Schrader, G., DAS 1,142,606 (1963); C. A., 59, 1683a (1963).
1859. Schrader, G., DAS 1,145,615 (1963); C. A., 59, 11565d (1963).
1860. Schrader, G., DAS 1,146,882 (1963); C. A., 59, 10121 (1963).
1861. Schrader, G., DAS 1,156,409 (1963); C. A., 60, 4184a (1964).
1862. Schrader, G., DAS 1,169,925 (1964); C. A., 61, 6959e (1964).
1863. Schrader, G., DAS 1,240,850 (1967); C. A., 67, 53729j (1967).
1864. Schrader, G., and R. Cölln, DAS 1,101,417 (1959); C. A., 57, 4698f (1962).
1865. Schrader, G., and I Hammann, DOS 1,906,954 (1970); C. A., 73, 99046m (1970).
1866. Schrader, G., I. Hamman, and W. Behrenz, DAS 1,262,266 (1968); C. A., 68, 114735c (1968).
1866a. Schrader, G., and W. Lorenz, DAS 1,103,324 (1959); C. A., 56, 5888c (1962).
1867. Schroeder, H., J. Am. Chem. Soc., 81, 5658 (1959).
1868. Schulze, W., H. Willitzer, and H. Fritzsche, Chem. Ber., 100, 2640 (1967).
1869. Schumann, H., P. Schwabe, and N. Schmidt, Inorg. Nucl. Chem. Lett., 2, 309 (1966).

1870. Schumann, H., P. Schwabe, and N. Schmidt, Inorg. Nucl. Chem. Lett., 2, 313 (1966).

1870a. Schwarzenbach, G., H. Ackermann, and P. Ruskstuhl, Helv. Chim. Acta, 32, 1175 (1949).

1871. Schwarzenbach, G., P. Ruckstuhl, and J. Zurc, Helv. Chim. Acta, 34, 455 (1951).

1872. Schwarzenbach, G., and J. Zurc, Monatsh. Chem., 81, 202 (1950).

1873. Scola, D. A., and D. W. Grisley, Jr., J. Chem. Eng. Data, 13, 571 (1968); C. A., 69, 106808 (1968).

1874. Scotti, F., and E. J. Frazza, J. Org. Chem., 29, 1800 (1964).

1875. Sedykh-Zadse, S. I., a.o., U.S.S.R. 172,320 (1966); C. A., 63, 16385b (1965).

1876. Seiber, J. N., and H. Tolkmith, Tetrahedron Lett., 1967, 3333.

1877. Seux, R., and A. Foucaud, Compt. Rend., (C) 273, 842 (1971).

1878. Seyferth, D., J. Y. P. Mui, and G. Singh, J. Organometal. Chem., 9, 185 (1966).

1878a. Seyferth, D., R. S. Marmor, and P. Hilbert, J. Org. Chem., 36, 1379 (1971).

1879. Seyferth, D., and R. S. Marmor, Tetrahedron Lett., 1970, 2493.

1880. Shahak, I., and E. D. Bergmann, Isr. J. Chem., 4, 225 (1966); C. A., 66, 55552 (1967).

1881. Sharov, V. N., A. L. Klebanskij, and V. A. Bartashev, Zh. Obshch. Khim., 40, 2011 (1970).

1882. Shaturskij, Ya. P., Yu. S. Grushin, G. K. Fedorova, and A. V. Kirsanov, Zh. Obshch. Khim., 39, 1467 (1969).

1883. Shealy, J. F., R. F. Struck, J. D. Clayton, and J. A. Montgomery, Inorg. Chem., 26, 4433 (1961).

1884. Sheinkman, A. K., G. V. Samoilenko, and S. N. Baranov, Dokl. Akad. Nauk SSSR, 196, 1377 (1971); C. A., 75, 20514x (1971).

1885. Shenhav, H., Z. Rappoport, and S. J. Pateu, J. Chem. Soc. (B), 1970, 469.

1886. Shepeleva, E. S., and P. I. Sanin, Dokl. Akad. Nauk SSSR, 109, 555 (1956); C. A., 51, 4934i (1957).

1887. Shevchenko, V. I., and N. D. Bodnarchuk, Zh. Obshch. Khim., 36, 1645 (1966).

1888. Shevchenko, V. I., P. D. Bodnarchuk, and A. V. Kirsanov, Zh. Obshch. Khim., 32, 2994 (1962).

1889. Shevchenko, V. I., and Z. V. Merkulova, Zh. Obshch. Khim., 29, 1005 (1959).

1890. Shevchenko, V. I., Mohamed el Dik, and A. M. Pinchuk, Zh. Obshch. Khim., 39, 1514 (1969).

1891. Shevchenko, V. I., A. S. Shtepanek, and A. V. Kirsanov, Zh. Obshch. Khim., 31, 3062 (1961).

1892. Shevchenko, V. I., A. S. Shtepanek, and A. V.

Kirsanov, Zh. Obshch. Khim., 32, 150 (1962).
1893. Shevchenko, V. I., A. S. Shtepanek, and A. V. Kirsanov, Zh. Obshch. Khim., 33, 3693 (1963).
1894. Shevchenko, V. I., and V. T. Stratienko, Zh. Obshch. Khim., 29, 3458 (1959).
1895. Shevchenko, V. I., and V. T. Stratienko, Zh. Obshch. Khim., 29, 3757 (1959).
1896. Shevchenko, V. I., and V. T. Stratienko, Zh. Obshch. Khim., 30, 1561 (1960).
1897. Shevchenko, V. I., V. P. Tkach, and A. V. Kirsanov, Zh. Obshch. Khim., 35, 992 (1965).
1898. Shevchenko, V. I., V. P. Tkach, and A. V. Kirsanov, Zh. Obshch. Khim., 35, 1224 (1965).
1899. Shikiev, I. A., M. I. Aliev, and S. Z. Israfilova, Zh. Obshch. Khim., 32, 2686 (1962).
1900. Shner, S. M., L. P. Bocharova, and I. K. Rubtsova, Zh. Obshch. Khim., 37, 418 (1967).
1901. Shner, S. M., I. K. Rubtsova, L. P. Bocharova, and V. S. Kolganova, Plast. Massy, 8, 25 (1971); C. A. 75, 144197e (1971).
1902. Shokol, V. A., G. I. Derkach, and A. V. Kirsanov, Zh. Obshch. Khim., 32, 166 (1962).
1903. Shokol, V. A., V. V. Doroshenko, and G. I. Derkach, Zh. Obshch Khim., 40, 1458 (1958).
1904. Shokol, V. A., V. F. Gamaleya, and G. I. Derkach, Zh. Obshch. Khim., 37, 1147 (1967).
1905. Shokol, V. A., V. F. Gamaleya, and G. I. Derkach, Zh. Obshch. Khim., 37, 2528 (1967).
1906. Shokol, V. A., V. F. Gamaleya, and G. I. Derkach, Zh. Obshch. Khim., 38, 1867 (1968).
1907. Shokol, V. A., V. F. Gamaleya, and G. I. Derkach, Zh. Obshch Khim., 39, 856 (1969).
1908. Shokol, V. A., V. F. Gamaleya, and G. I. Derkach, Zh. Obshch. Khim., 39, 1703 (1969).
1909. Shokol, V. A., V. F. Gamaleya, and V. P. Kukhar, Zh. Obshch. Khim., 40, 554 (1970).
1910. Shokol, V. A., G. A. Golik, and G. I. Derkach, Zh. Obshch. Khim., 41, 545 (1971).
1911. Shokol, V. A., G. A. Golik, and G. I. Derkach, Zh. Obshch. Khim., 39, 2197 (1969).
1912. Shokol, V. A., G. A. Golik, B. Ya. Libman, and G. I. Derkach, Zh. Obshch. Khim., 36, 1636 (1966).
1913. Shokol, V. A., G. A. Golik, V. T. Tsyba, Yu. P. Egorov, and G. I. Derkach, Zh. Obshch. Khim., 40, 1680 (1970).
1914. Shokol, V. A., L. I. Molyavko, and G. I. Derkach, Zh. Obshch. Khim., 40, 998 (1970).
1915. Shokol, V. A., L. I. Molyavko, A. G. Matyusha, N. K. Mikhailyuchenko, and G. I. Derkach, Zh. Obshch. Khim., 41, 2380 (1971).
1916. Short, J. N., U. S. 2,818,406 (1954); C. A., 52, 6407 (1958).

1917. Shostakovsky, M. F., A. S. Atavin, B. A. Trofimov, A. V. Gusarov, V. M. Nikitin, and V. I. Skorobogatova, Zh. Obshch. Khim., 40, 70 (1970).
1918. Shostakovskij, M. F., I. I. Gusejnov, and G. S. Vasil'ev, Zh. Obshch. Khim., 32, 375 (1962).
1918a. Shostakovskij, M. F., I. I. Gusejnov, and G. S. Vasil'ev, Zh. Obshch. Khim., 32, 380 (1962).
1919. Shtepanek, A. S., V. I. Shevchenko, and A. V. Kirsanov, Zh. Obshch. Khim., 35, 1023 (1965).
1920. Shostakovskij, M. F., L. I. Shmonina, and I. I. Gusejnov, Zh. Obshch. Khim., 31, 734 (1961).
1921. Sidky, M. M., F. M. Soliman, and R. Shabana, Tetrahedron, 27, 3431 (1971).
1922. Sieper, H., Tetrahedron Lett., 1967, 1987.
1923. Silcox, C. M., and J. J. Zuckerman, J. Am. Chem. Soc., 88, 168 (1966).
1924. Simon, A., and W. Schulze, Chem. Ber., 94, 3251 (1961).
1925. Simonnin, M. P., J. J. Basselier, and C. Charrier, Bull. Soc. Chim. Fr., 1967, 3544.
1926. Simov, D., M. Kirilov, L. Kamenov, and G. Petrov, Zh. Obshch. Khim., 40, 2131 (1970).
1927. Sisler, H. H., J. Weiss, Inorg. Chem., 4, 1514 (1965).
1928. Smith, B. E., and A. Burger, J. Am. Chem. Soc., 75, 5891 (1953).
1929. Smith, P. L., and Ch. W. McGary, U. S. 3,236,863 (1966); C. A., 64, 14363g (1966).
1930. Smith, W. C., R. Gher, and L. F. Audrieth, J. Org. Chem., 21, 113 (1956).
1931. Smith, W. C., and L. F. Audrieth, J. Org. Chem., 22, 265 (1957).
1932. Snyder, C. D., W. E. Bondinell, and H. Rapoport, J. Org. Chem., 36, 3951 (1971).
1933. Soborovskij, L. Z., and N. F. Baina, Zh. Obshch. Khim., 29, 1144 (1959).
1934. Soborovskij, L. Z., Yu. G. Golobova, and V. V. Fedotova, Zh. Obshch. Khim., 30, 2586 (1960).
1935. Soborovskij, L. Z., Yu. M. Zinov'ev, and L. I. Muler, Zh. Obshch. Khim., 29, 3947 (1959).
1936. Sokolovskij, M. A., and P. M. Zavlin, Zh. Obshch. Khim., 30, 3562 (1960).
1937. Sommer, K., Z. Anorg. Allg. Chem., 376, 37 (1970).
1938. Songstadt, J., Acta Chem. Scand., 21, 1681 (1967).
1940. Speziale, A. J., U. S. 3,066,140 (1962); C. A., 58, 9139d (1963).
1941. Speziale, A. J., and R. C. Freeman, J. Org. Chem., 23, 1883 (1958).
1942. Spivack, J. D., U. S. 3,270,091 (1966); C. A., 67, 64529v (1967).
1943. Spivack, J. D., U. S. 3,281,505 (1966); C. A., 66, 2638f (1967).

1944. Stauffer, Chem. Comp., Brit. 1,194,349 (1967); C. A., 73, 56234d (1970).
1945. Stauffer Chem. Co., U. S. 3,597,439 (1971); C. A., 75, 88757j (1971).
1946. Stayner, R. D., U. S. 2,652,426 (1948); C. A., 48, 10052 (1954).
1946a. Steinberg, G. M., and J. Bolger, J. Org. Chem., 24, 1120 (1959).
1947. Steininger, E., and H. Deibig, Monatsh. Chem., 97, 1326 (1966).
1948. Stern, Ch. J., U. S. 3,179,676 (1965); C. A., 63, 2999h (1965).
1949. Stetter, H., and W. -D. Last, Chem. Ber., 102, 3364 (1969).
1950. Stewart, W. E., and T. H. Siddall, J. Inorg. Nucl. Chem., 32, 3599 (1970).
1951. Stewart, A. P., and S. Trippett , Chem. Commun., 1970, 1279.
1952. Stiles, A. R., and F. Rust, U. S. 2,724,718 (1955); C. A., 50, 10124d (1956).
1953. Stiles, A. R., F. F. Rust, and W. E. Vaughn, J. Am. Chem., Soc., 74, 3282 (1952).
1954. Stiles, A. R., W. E. Vaughn, and F. F. Rust, J. Am. Chem. Soc., 80, 714 (1958).
1955. Stölzer, C., A. Simon, Chem. Ber., 93, 1323 (1960).
1956. Strube, R. E., R. D. Birkenmeyer, and F. Kagan, U. S. 3,052,709 (1962); C. A., 58, 10240e (1963).
1957. Struck, R. F., J. Med. Chem., 9, 231 (1966).
1958. Sturtz, G., Bull. Soc. Chim. Fr., 1964, 2333.
1959. Sturtz, G., Bull. Soc. Chim. Fr., 1964, 2340.
1960. Sturtz, G., Compt. Rend., 260, 1984 (1965).
1961. Sturtz, G., Bull. Soc. Chim. Fr., 1967, 1345.
1962. Sturtz, G., and C. Charrier, Compt. Rend., 261, 1019 (1965).
1963. Sturtz, G., C. Charrier, and H. Normant, Bull. Soc. Chim. Fr., 1966, 1707.
1964. Le Suer, W. M., U. S. 3,158,640 (1964); C. A., 62, 9174a (1965).
1965. Sundberg, R. J., J. Org. Chem., 30, 3604 (1965).
1966. Swern, D., and H. B. Knight, U. S. 2,989,562 (1961); C. A., 55, 25764h (1961).
1967. Szabo, K., U. S. 3,272,892 (1966); C. A., 66, 11047y (1967).
1968. Szabo, K., U. S. 3,170,944 (1965); C. A., 62, 13180 (1965).
1969. Szabo, K., and J. G. Brady, U. S. 2,988,474 (1961); C. A., 55, 23918i (1961).
1970. Szabo, K., and J. G. Brady, U. S. 3,117,908; C. A., 60, 6869e (1964).
1971. Szabo, K., and J. G. Brady, U. S. 3,172,902 (1965); C. A., 63, 1818g (1965).
1972. Szabo, K., and J. G. Brady, U. S. 3,236, 919 (1966); C. A., 64, 14219a (1966).

1973. Szabo, K., and J. J. Menn, J. Agr., Food Chem., 17, 863 (1969).
1974. Szafraniec, L. J., L. P. Reiff, and H. S. Aaron, J. Am. Chem. Soc., 92, 6391 (1970).
1975. Szmuszkovicz, J., U. S. 2,849,454 (1957); C. A., 53, 1251 (1959).
1976. Szmuszkovicz, J., U. S. 2,877,234 (1957); C. A., 53, 13174 (1959).
1977. Szmuszkovicz, J., J. Am. Chem. Soc., 80, 3782 (1958).
1978. Tahori, S., and S. Sternberg, J. Econ. Entomol., 59, 196 (1966).
1979. Takamizava, A., K. Hiras, and Y. Hamashima, Tetrahedron Lett., 1967, 5081.
1980. Takamizawa, A., and Y. Sato, Chem. Pharm. Bull. (Japan), 12, 398 (1964); C. A., 61, 3104f (1964).
1981. Taljanker, E. G., S. L. Lihina, and E. L. Gefter, Zh. Obshch. Khim., 36, 1473 (1966).
1982. Tammelin, L. E., Acta Chem. Scand., 11, 859 (1957).
1983. Tammelin, L. E., and L. Fagerlind, Acta Chem. Scand. 14, 1353 (1960).
1984. Tarasova, R. I., N. M. Kislitsyna, and A. N. Pudovik, Zh. Obshch. Khim., 41, 1972 (1971).
1985. Tavs, P., Brit. 1,200,273 (1970).
1986. Tavs, P., Chem. Ber., 100, 1571 (1967).
1987. Tavs, P., Chem. Ber., 103, 2428 (1970).
1988. Tavs, P., and F. Korte, Tetrahedron, 23, 4677 (1967).
1989. Tavs, P., and H. Weitkamp, Tetrahedron, 26, 5529 (1970).
1990. Tesi, G., and P. J. Slota, Jr., Proc. Chem. Soc., 1960, 404.
1991. Tesoro, G. C., St. B. Sello, D. R. Moore, and R. F. Wurster, U. S. 3,551,422 (1970); C. A., 74, 65509v (1971).
1991a. Thiele, J., Chem. -Ztg., 36, 657 (1912).
1992. Thompson, J. E., J. Org. Chem., 30, 4276 (1965).
1993. Thuong, N. T., and J. P. Chabrier, Bull. Soc. Chim. Fr., 1970, 780.
1994. Thuong, N. T., F. Convert, G. Martin, and P. Chabrier, Bull. Soc. Chim. Fr., 1965, 1925.
1995. Tiemann, Ch. H., U. S. 3,189,635 (1965); C. A., 63, 5678d (1965).
1996. Tikhonova, L. I., Zh. Fiz.-Khim., 44, 3118 (1970); C. A., 74, 91818q (1971).
1997. Tikhonova, L. I., Radiokhimiya, 12, 519 (1970); C. A., 73, 134383t (1970).
1998. Tilemans, M., J. H. T. Ledrut, and G. Combes, Prod. Pharmacol., 18, 569 (1963).
1999. Timmler, H., R. Wegler, G. Unterstenhöfer, and I. Hammann, DAS 1,212,967 (1966); C. A., 64, 15925e (1966).
2000. Timofeeva, T. M., V. M. Ignat'ev, B. I. Ionin,

and A. A. Petrov, Zh. Obshch. Khim., 39, 2446 (1969).
2001. Tömösközi, I., and G. Janzso, Chem. Ind. (London), 1962, 2085.
2002. Tolkmith, H., U. S. 2,654,781 (1951); C. A., 48, 10050 (1954).
2002a. Tolkmith, H., U. S. 2,668,838 (1953); C. A., 49, 5518 (1955).
2003. Tolkmith, H., U. S. 2,668,846 (1953); C. A., 49, 5520 (1955).
2004. Tolkmith, H., U. S. 2,680,760 (1953); C. A., 49, 10359 (1955).
2005. Tolkmith, H., U. S. 2,686,197 (1954); C. A., 49, 11000 (1955).
2006. Tolkmith, H., and E. C. Britton, J. Org. Chem., 24, 705 (1959).
2007. Tolkmith, H., P. B. Budde, D. R. Mussell, and R. A. Nyquist, J. Med. Chem., 10, 1074 (1967).
2008. Tolkmith, H., and D. R. Mussell, World Rev. Pest Control, 6, 74 (1967); C. A., 68, 86362 (1968).
2009. Toma, S., and E. Koluzayova, Chem. Zvesti., 23, 540 (1969); C. A., 73, 14960g (1970).
2010. Tomaschewski, G., and B. Breitfeld, J. Prakt. Chem., 311, 256 (1969).
2011. Tomaschewski, G., C. Berseck, and G. Hilgetag, Chem. Ber., 101, 2037 (1968).
2012. Tomaschewski, G., A. Otto, and D. Zanke, Arch. Pharm., 301, 520 (1968).
2013. Tomaschewski, G., R. Pannier, L. Ausner, A. Otto, and G. Hilgetag, Arch. Pharm., 301, 525 (1968).
2014. Tomchina, L. F., P. M. Zavlin, and V. V. Razumovskij Zh. Obshch. Khim., 39, 1256 (1969).
2015. Tomchina, L. F., P. M. Zavlin, and V. V. Razumovskij Zh. Obshch. Khim., 38, 564 (1968).
2016. Tong, B. L., and A. S. Dalwa, Chem. Ind. (London), 1967, 582.
2017. Torralba, A. F., and T. C. Myers, J. Org. Chem., 22, 972 (1957).
2018. Torya Kogyo Inc., Jap. 7,017,164 (1967); C. A., 73, 66728g (1970).
2019. Toy, A. D. F., J. Am. Chem. Soc., 70, 186 (1948).
2020. Toy, A. D. F., U. S. 2,382,622 (1944); C. A., 40, 604 (1946).
2021. Toy, A. D. F., U. S. 2,400,577 (1944); C. A., 40, 4745 (1946).
2021a. Toy, A. D. F., U. S. 2,425,766 (1947); C. A., 42, 596h (1948).
2022. Toy, A. D. F., and R. S. Cooper, J. Am. Chem. Soc., 76, 2191 (1954).
2023. Toy, A. D. F., and K. H. Rattenburg, U. S. 2,863,903 (1954); C. A., 53, 9151 (1959).
2024. Toy, A. D. F., and K. H. Rattenburg, U. S. 2,922,810 (1953); C. A., 54, 9848 (1960).

2026. Toy, A. D. F., and K. H. Rattenburg, U. S. 2,960,522 (1960); C. A., 55, 19861h (1961).
2027. Toy, A. D. F., E. Walsh, and J. R. Froli, U. S. 3,094,405 (1963); C. A., 59, 12844e (1963).
2028. Treckner, D. J., and J. P. Henry, J. Am. Chem., Soc., 85, 3204 (1963).
2029. Trippett, S., J. Chem. Soc., 1962, 4731.
2030. Trofimov, B. A., A. S. Atavin, G. M. Garilova, and G. A. Kalabin, Izv. Akad. Nauk SSSR, Ser. Khim., 1969, 1200; C. A., 71, 50117e (1969).
2031. Trowbridge, D. B., G. L. Kenyon, J. Am. Chem. Soc., 92, 2181 (1970).
2032. Tsivunin, V. S., and N. I. D'yakonova, Zh. Obshch. Khim., 40, 1995 (1970).
2033. Tsivunin, V. S., S. V. Fridland, T. V. Zykova, and G. Ch. Kamai, Zh. Obshch. Khim., 36, 1424 (1966).
2034. Tsivunin, V. S., G. Kh. Kamai, and S. V. Fridland, Zh. Obshch. Khim., 36, 436 (1966).
2035. Tsivunin, V. S., G. Kh. Kamai, and S. V. Fridland, Zh. Obshch. Khim., 33, 2146 (1963).
2037. Tsivunin, V. S., G. Kh. Kamai, and V. V. Kormachev, Zh. Obshch. Khim., 36, 1663 (1966).
2038. Tsvetkov, E. N., R. A. Malevannaya, and M. I. Kabachnik, Zh. Obshch. Khim., 39, 1520 (1969).
2039. Tuchtenhagen, G., and K. Rühlmann, Ann., 711, 174 (1968).
2040. Tyka, R., Tetrahedron Lett., 1965, 3071.
2041. Ueda, T., K. Inukai, and H. Maramatsu, Bull. Chem. Soc. Jap., 42, 1684 (1969).
2041a. Uhlig, E., and W. Achilles, J. Prakt. Chem., 311, 529 (1969).
2042. Ulrich, H., Fr. 1,457,452 (1966); C. A., 67, 100,244 (1967).
2043. Ulrich, H., B. Tucker, and A. A. R. Sayigh, J. Org. Chem., 32, 1360 (1967).
2044. Upjohn Co., Brit. 866,409 (1961); C. A., 55, 2468h (1961).
2045. Utvary, K., Inorg. Nucl. Chem. Lett., 1, 77 (1965).
2046. Utvary, K., E. Freundlinger, and V. Gutmann, Monatsh. Chem., 97, 679 (1966).
2047. Vasil'ev, G. S., I. I. Gusejnov, and M. F. Shostakovskij, Zh. Obshch. Khim., 34, 1216 (1964).
2048. Vasil'ev, G. S., E. N. Prilezhaeva, V. F. Bistrov, and M. F. Shostakovskij, Zh. Obshch. Khim., 35, 1350 (1965).
2049. Verheyden, J. P. H., and J. G. Moffatt, J. Org. Chem., 35, 2319 (1970).
2050. Verheyden, J. P. H., and J. G. Moffat, J. Org. Chem., 35, 2868 (1970).
2051. Verizhnikov, L. V., O. V. Voskresenskaja, V. Kh. Kadyrova, P. A. Kirpichnikov, and E. T. Mukmenov, Zh. Obshch. Khim., 41, 2162 (1971).
2052. Verkade, J., T. Huttmann, M. Fung, and R. King,

Inorg. Chem., 4, 83 (1965).
2053. Viout, A., and P. Rumpf, Compt. Rend., 239, 1291 (1954).
2054. Vizgert, R. V., and M. P. Voloshin, Zh. Obshch. Khim., 41, 1991 (1971).
2055. Vizgert, R. V., and M. P. Voloshin, Zh. Obshch. Khim., 41, 2576 (1971).
2056. Voronkov, M. G., and Yu. I. Skorik, Zh. Obshch. Khim., 35, 106 (1965).
2057. Voziyanova, O. F., S. N. Baranov, and S. V. Krivun, Zh. Obshch. Khim., 40, 1905 (1970).
2058. Wadsworth, W. S., U. S. 3,047,606 (1962); C. A., 58, 4599b (1963).
2059. Wadsworth, Jr., W. S., and W. D. Emmons, J. Am. Chem. Soc., 84, 610 (1962).
2060. Walling, C., F. R. Stacey, E. J. Saunders, and E. S. Huyser, J. Am. Chem. Soc., 80, 4546 (1958).
2061. Walling, C., F. R. Stacey, E. J. Saunders, and E. S. Huyser, J. Am. Chem. Soc., 80, 4543 (1958).
2061a. Walsh, E. N., J. Am. Chem. Soc., 81, 3023 (1959).
2062. Walsh, E. N., T. M. Beck, and A. D. F. Toy, J. Am. Chem. Soc., 78, 4455 (1956).
2063. Walsh, E. N., T. M. Beck, and W. H. Woodstock, J. Am. Chem. Soc., 77, 929 (1955).
2063a. Warner, H. L., and A. C. Leopold, Bioscience, 17, 722 (1967).
2064. Warren, S., and M. R. Stuart, J. Chem. Soc. (B), 1971, 618.
2065. Weber, C. W., U. S. 2,882,242 (1953); C. A., 53, 17904 (1959).
2066. Weil, E. D., E. Dorfman, and J. Linder, U. S. 3,515,537 (1970); C. A., 73, 35509g (1970).
2067. Weil, E. D., and K. J. Smith, U. S. 3,202,692 (1965); C. A., 63, 14907d (1965).
2068. Welch, C. M., E. J. Gonzales, and J. D. Guthrie, J. Org. Chem., 26, 3270 (1961).
2069. Weller, J., Chem. Ber., 20, 1718 (1887).
2070. Weller, J., Chem. Ber., 21, 1492 (1888).
2071. v. d. Westeringh, C., and H. Veldstra, Rec. Trav. Chim., 77, 1096 (1958).
2072. Whetstone, R. R., W. J. Raab, and W. E. Hall, U. S. 2,867,646 (1956); C. A., 53, 12237 (1959).
2073. Whitehouse, K. C., and H. Z. Lecher, U. S. 2,799,70 (1957); C. A., 51, 18003f (1957).
2074. Whitehouse, K. C., and H. Z. Lecher, U. S. 2,894,02 (1959); C. A., 54, 1424e (1960).
2075. Wieber, M., and J. Otto, Chem. Ber., 100, 974 (1967).
2076. Wieber, M., J. Otto, and M. Schmidt, Angew. Chem., 76, 648 (1964).
2077. Wiers, B. H., Inorg. Chem., 10, 2581 (1971).
2078. Willstätter, R., and E. Sonnenfeld, Chem. Ber.,

47, 2801 (1914).
2079. Van Winkle, J. L., and R. C. Morris, U. S. 2,681,920 (1951); C. A., 49, 6989g (1955).
2080. Van Winkle, J. L., and R. C. Morris, U. S. 2,874,184 (1957); C. A., 53, 13056c (1959).
2081. Woodstock, W. H., U. S. 2,471,472 (1945); C. A., 43, 7499 (1949).
2082. Worms, K. H., and H. Blum, umpublished.
2083. Worms, K. H., and K. Wollmann, Z. Anorg. Allg. Chem., 381, 260 (1971).
2084. Wozniak, M., J. Nicole, and G. Tridot, Compt. Rend. (C), 272, 635 (1971).
2085. Yakshin, V. V., and L. I. Sokal'skaja, Zh. Obshch. Khim., 41, 484 (1971).
2086. Yakubovich, A. Ya., and V. A. Ginsburg, Zh. Obshch. Khim., 22, 1534 (1952).
2087. Yakubovich, A. Ya., and V. A. Ginsburg, Dokl. Akad. Nauk SSSR, 82, 273 (1952); C. A., 47, 2685 (1953).
2088. Yakubovich, A. Ya., and V. A. Ginsburg, Zh. Obshch. Khim., 24, 1465 (1954).
2089. Yakubovich, A. Ya., and V. A. Ginsburg, Zh. Obshch. Khim., 24, 2250 (1954).
2090. Yakubovich, A. Ya., and V. A. Ginsburg, Zh. Obshch. Khim., 22, 1534 (1952).
2091. Yakubovich, A. Ya., V. A. Ginsburg, and S. P. Makarov, Dokl. Akad. Nauk SSSR, 71, 303 (1950); C. A., 44, 8320 (1950).
2092. Yamashita, M., H. Yashida, T. Ogata, and S. Inokawa, J. Synth. Org. Chem. Jap., 27, 984 (1969).
2093. Yarmukhametova, D. Kh., and B. V. Kudryavtsev, U.S.S.R. 282,317; C. A., 75, 20616g (1971).
2094. Yarmukhametova, D. Kh., and Z. G. Speranskaja, Izv. Akad. Nauk SSSR, Ser. Khim., 1969, 1393; C. A., 71, 70689w (1969).
2095. Yarmukhametova, D. Kh., and I. V. Cheplanova, Izv. Akad. Nauk SSSR, Ser. Khim., 1966, 1260; C. A., 65, 16995g (1966).
2096. Yoder, C. H., J. J. Zuckerman, J. Am. Chem. Soc., 88, 2170 (1966).
2097. Zavlin, P. M., A. S. Fedoseeva, N. K. Ryasinskaja, and V. M. Shek, Zh. Obshch. Khim., 39, 1887 (1969).
2098. Zavlin, P. M., V. A. Zamora, and A. S. Fedoseeva, Zh. Obshch. Khim., 41, 481 (1971).
2098a. Zeffert, B. M., P. B. Coulter, and H. Tannenbaum, J. Am. Chem. Soc., 82, 3843 (1960).
2099. Zhmurova, I. N., Zh. Obshch. Khim., 33, 549 (1963).
2100. Zhmurova, I. N., and A. V. Kirsanov, Zh. Obshch. Khim., 33, 182 (1963).
2101. Zhmurova, I. N., and I. Yu. Voitsekkovskaja, Zh. Obshch. Khim., 33, 1349 (1963).
2102. Zhmurova, I. N., and I. Yu. Voitsekkovskaja, Zh. Obshch. Khim., 35, 2197 (1965).

2103. Zieloff, K., H. Paul, and G. Hilgetag, Chem. Ber., 99, 357 (1966).
2104. Zimmer, H., and P. J. Bercz, Ann., 686, 107 (1965).
2105. Zimmer, H., P. J. Bercz, and E. Heuer, Tetrahedron Lett., 1968, 171.
2106. Zieloff, K., H. Paul, and G. Hilgetag, Z. Chemie, 4, 148 (1964).
2107. Zinov'ev, Yu. M., V. N. Kulakova, R. I. Mironova, and L. Z. Soborovskij, Zh. Obshch. Khim., 39, 606 (1969).
2108. Zinov'ev, J. M., V. N. Kulakova, and L. Z. Soborovskij, Zh. Obshch. Khim., 28, 1551 (1958).
2109. Zinov'ev, Yu. M., and L. Z. Soborovskij, Zh. Obshch Khim., 29, 3954 (1959).
2110. Zinov'ev, J. M., T. G. Spiridonova, and L. Z. Soborovskij, Zh. Obshch. Khim., 29, 3594 (1959).
2111. Zolch, L., and W. Kipping, Brit. 903,339 (1962); C. A., 58, 1492a (1963).
2112. Zolotova, M. V., and T. V. Konstantinova, Zh. Obshch. Khim., 40, 2131 (1970).
2113. Zyablikova, T. A., I. M. Magdeev, and I. M. Shermergorn, Izv. Akad. Nauk SSSR, Ser. Khim., 1969, 694; C. A., 71, 50081p (1969).

Chapter 19. Organic Derivatives of Thio (Seleno, Telluro) Phosphoric Acid.

DAVID E. AILMAN AND RICHARD J. MAGEE

American Cyanamid Company, Princeton, New Jersey

A. INTRODUCTION

Organic derivatives of thio- and selenophosphoric acid have been developed rapidly since the appearance of the first edition of "Organophosphorus Compounds" in 1950. Phosphorotelluronates remain relatively little explored.

An important segment of the development of phosphorothioates and -selenoates was concerned with agricultural products, particularly the phosphorothioate and -dithioate tertiary esters as pesticides. Commercially important products in this field receive frequent review in the literature. Thus only incidental mention of these products will be found in this chapter. For more detailed accounts, the reader should consult "Pesticide Manual," H. Martin, British Crop Protection Council, 2nd. Ed. 1971; "Commercial and Experimental Organic Insecticides," E. E. Kenaga and W. E. Allison, Bulletin of the Entomological Society of America, 16 68 (1970); "Die Entwicklung neuer insektizider Phosphorsäure-Ester," G. Schrader, Verlag Chemie GMBH, Weinheim Bergstr. (1963)[1168a]; "The Chemistry of Organophosphorus Pesticides," C. Fest and K.-J. Schmidt, Springer Verlag, Berlin-Heidelberg-New York (1973); and "Pesticides '72," Chemical Week, McGraw-Hill, June 21, pp. 34-66 and July 26, pp. 18-46, 1972

Current reviews of phosphorothioate chemistry are to be found in "Organophosphorus Chemistry," S. Trippett, senior reporter, The Chemical Society, London, Chapter 6, N. K. Hamer, vol. 3, 1972.

Halogen analogs of the primary and secondary acids, such as $ROP(S)Cl_2$, $(RO)_2P(S)Cl$, $(RO)_2P(Se)Cl$, are useful intermediates for the preparation of secondary and tertiary esters. The methods of preparation of the halogen analogs are designated with capital letters A through P. The products are to be found in Sections F.1 and F.2 of the List of Compounds.

Primary and secondary esters, more generally referred to as phosphorodithio-, -thio-, or -seleno- acids, such as $(RO)_2(S)SH$, $ROP(S)(OH)_2$, $(RO)_2POSH$, $(RO)_2POSeH$, are widely used as intermediates. The free acids derived from pure alcohols, phenols, thiols and selenols can often be distilled or crystallized. Their salts with alkali metals or organic bases may be isolated and purified, or they may be prepared in situ in reaction mixtures. Their heavy metal salts were a subject of considerable interest at the turn of the century when the subject of tautomeric forms, such as $P(O)SM \rightleftharpoons P(S)OM$, was being studied. These heavy metal salts have been omitted from this review.

Alkali metal and organic base salts are included in the tables of primary and secondary esters in Sections F.3.2, F.4.1, F.4.2, F.5.7, F.5.8, and F.6 of the List of Compounds. The corresponding methods of preparation are given Roman numerals I through VII.

Secondary esters and their salts derived from mixed grades of higher alcohols or "cresylic acids" find wide commercial application as oil additives, lubrication improvers, rubber additives, mineral-flotation agents, and the like. Since these are usually used in a crude form without isolation of pure compounds, these have had to be excluded from this review.

Acknowledgments: The authors wish to thank the American Cyanamid Company for the use of library and duplicating facilities. We also thank Mrs. Mildred Scott and Miss Nancy Hoffman for assistance in preparing the manuscript.

B. PHYSICAL PROPERTIES AND PHYSICAL METHODS

B.1. Dipole Moments

Arbuzov and Shavsha (1950)[60] measured dipole moments of trialkyl phosphorothionates. They calculated the value of the dipole moment of the P(S) link as 3.55 D. Brown, Verkade, and Piper (1961)[133] studied the effect on dipole moment of adding S to a constrained phosphite ester:

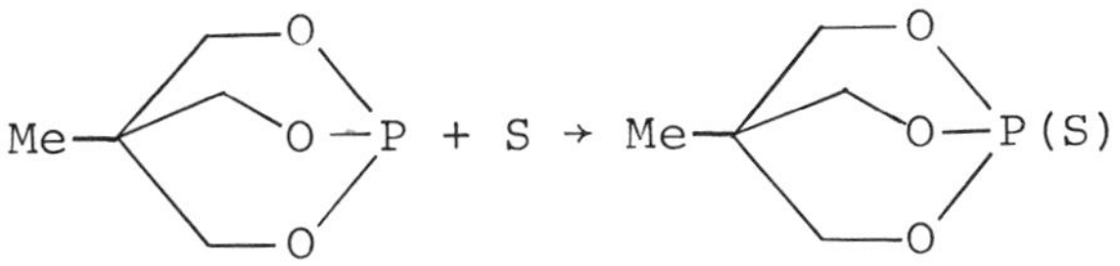

μ 4.15 D μ 6.77 D Δμ 2.62 D

In comparison, the effect of addition of S to $(PhO)_3P$, μ 2.03 D, to give $(PhO)_3P(S)$, μ 2.58 D, was only Δμ 0.55 D. Arbuzov and Arshinova (1970)[55] calculated dipole moments and Kerr constants for several dioxaphosphorinanes, including 2-chloro-2-thiono-1,3,2-dioxaphosphorinane and its 4-methyl homolog:

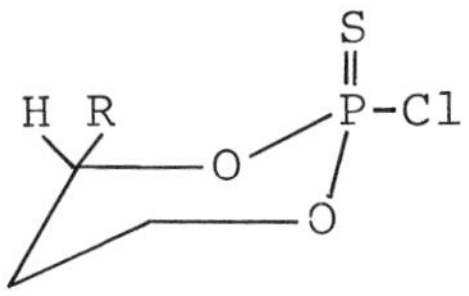

R	μ		mK x 10^{12}	
	exptl.	calc.	exptl.	calc.
H	5.30 D	5.46 D	1356	positive only for conformation given
Me(axial)	5.55 D	5.46 D	1195	positive only for conformation given

B.2. Infrared Spectra

In the infrared, the P(O) group is a strong absorber between 1300 - 1250 cm^{-1} [Gore (1950)].[374] Chen and Dority (1967)[162] made the assignments for P(O): 1256 cm^{-1} in $(EtO)_2P(O)SCH_2SC_6H_4Cl$-4 and $(EtO)_2P(O)SCH_2S(O)_2C_6H_4Cl$-4, and 1259 in $(EtO)_2P(O)SCH_2S(O)C_6H_4Cl$-4.

Bellamy and Beecher (1952)[80] made the following assignments for P-O-(Ar): 1232 cm^{-1} in $(EtO)_2P(S)OC_6H_4NO_2$-4, and 1205 cm^{-1} in $EtOP(S)(OC_6H_4NO_2\text{-}4)_2$.

The P(S) bands in the infrared are generally weak and poorly characterized in the 690 - 640 cm^{-1} region [Gore (1950)].[374] However, the following assignments have been made:

$\nu_{P(S)}$ cm^{-1}	Compound(s)	Ref.
763	$(EtO)_2P(S)OC_6H_4NO_2$-4	Bellamy, Beecher (1952)[80]
766	$EtOP(S)(OC_6H_4NO_2\text{-}4)_2$	Bellamy, Beecher (1952)[80]

685	$(RS)_3P(S)$	Menefee et al. (1957)[764]
700	$EtOP(S)Cl_2$	Popov et al. (1959)[968]
660	$(RO)_2P(S)Cl$	Popov et al. (1959)[968]
650	$(EtO)_2P(S)SH$	Popov et al. (1959)[968]
660	$(EtO)_2P(S)SEt$	Popov et al. (1959)[968]
690-675	Alkyl-NH_3^+ salts	Vasil'ev, Khaskin (1964)[1348]
652	$(EtO)_2P(S)SCH_2SC_6H_4Cl$-4	Chen, Dority (1967)[162]
651	$(EtO)_2P(S)SCH_2S(O)_2C_6H_4Cl$-4	Chen, Dority (1967)[162]
653	$(EtO)_2P(S)SCH_2S(O)C_6H_4Cl$-4	Chen, Dority (1967)[162]
709	$Cl_3CCSSP(S)F_2$	Roesky, Dietl (1970)[1045]
699	$Cl_3CCOSP(S)F_2$	Roesky, Dietl (1970)[1045]
728,701	$(EtO)P(S)Cl_2$	Nyquist et al. (1971)[902]
727,699	$MeCD_2OP(S)Cl_2$	Nyquist et al. (1971)[902]
688,675	$CD_3CH_2OP(S)Cl_2$	Nyquist et al. (1971)[902]

Herail and Viossat (1964)[429] found that $(MeO)_3P(Se)$ showed two distinct bands of equal intensity at 558 and 504 cm^{-1}, which were attributed to $\nu_{P(Se)}$. Turkevich and Viblii (1966)[1334] found that the P(Se) absorption occurred at 595 - 590 cm^{-1} in $(RO)_3P(Se)$, and at 585-580 in $(RO)_2P(Se)SR$.

Infrared spectra have been reproduced in the following references: McIvor et al. (1956),[673] (1958)[674] on ROP(S)-Cl_2, $(RO)_2P(S)Cl$, $(RO)_2P(S)SH$, $(RO)_2POSH$, $(RO)_2P(S)SR$, $(RO)_2P(O)SR$, $(RO)_3P(S)$ for R = Me, Et, i-Pr, Pr, and Bu; Menefee et al. (1957)[764] on $(RS)_3P(S)$; Alford et al. (1959)[13] on $(RS)_2P(S)SH$ (the SH region: broad due to H-bonding in the pure liquid; sharp 2545 cm^{-1} in CCl_4 solution); Popov et al. (1959)[968] on $EtOP(S)Cl_2$, $EtSP(O)Cl_2$, $(RO)_2P$-(S)Cl, $(EtO)_2POSNa$, $(EtO)_2P(S)SH$, and $(RO)_2P(S)S(CH_2)_n$-SR; Vasil'ev and Khaskin on alkylammonium salts of phosphorodithioates and their disulfides; Chen and Dority (1967)[162] on "Trithion" and its possible metabolic products: and Gore et al. (1971),[375] IR and UV spectra of pesticides including nine phosphorodithio- or -thioates.

B.2.1. Raman Spectra

Zemlyanskii and Klimovskaya (1960)[1443] reported that

the P(S) bond produced a Raman line in the region of 598 - 662 cm^{-1} in several acids of the types $(RO)_2P(S)SH$ and $(RO)_2P(S)OH$. Nyquist et al. (1970)[902] reported two isomers, with the Raman lines in cm^{-1} for each: $EtOP(S)Cl_2$, 730, 706; $MeCD_2OP(S)Cl_2$, 720, 697; $CD_3CH_2OP(S)Cl_2$, 685, 677.

B.3. Molar Refraction

The Eisenlohr molar refraction, Mn_D^{20} or MR_E, was extended to phosphorus compounds by Sayre (1958).[1078] The precision and convenience of the use of Mn_D^{20} were compared with the Lorenz-Lorentz MR both on the basis of atom and bond refractions. Some 38 phosphorodithio- and -thioates were given. Data from that paper have been included in the tables under the designation "mol. ref." The Eisenlohr molar refraction was used by Stoelzer and Simon (1963)[1271] on $EtOP(S)F_2$ and $(EtOP(S)F)_2O$ and compared with the Lorenz-Lorentz molecular refraction.

B.4. Parachor

Surface tensions of trialkyl phosphorothionates, $(RO)_3P(S)$, for R = Et, Pr, Bu, C_6H_{13}, and C_8H_{17} were determined by Arbuzov and Vinogradova (1947, 1952).[61, 62] The parachors as reported by these authors will be found in the List of Compounds.

B.5. Nuclear Magnetic Resonance, ^{1}H-NMR, ^{19}F, and ^{31}P

The past two decades have seen the widespread publication of results and interpretation of proton magnetic resonance spectra of organophosphorus compounds. No attempt has been made to include these results in the lists of compounds, but references to the ^{1}H-NMR spectra of the compounds are given there. High-resolution ^{1}H-NMR spectra of pesticides have been published by Keith, Garrison and Alford (1968-1970),[524-6] and by Babad, Herbert and Goldberg (1968).[74] Stothers and Spencer (1961)[1275] used ^{1}H-NMR to establish the structures of cis- and trans- "thiono-Phosdrin," $(MeO)_2P(S)OCMe{=}CHCOOMe$. Verkade and King (1962)[1355] studied the ^{1}H- and ^{31}P-NMR spectra of polycyclic P compounds with particular attention to the effect of the P-O-C-H angle on the spectra:

Compound	^{31}P (in DMSO)	$J_{P-O-C-H}$
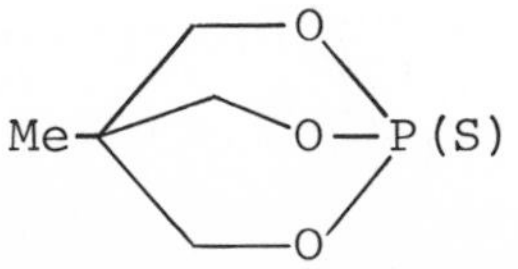	-57.4 ppm	6 cps

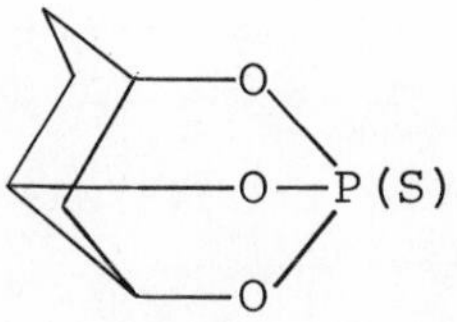

-64.0 ppm 19cps

Martin and Besnard (1963)[722] studied the effect of dilution in nonpolar solvents on variations of the effect of coupling J_{H--P} on $(MeS)_3P(O)$ in hexane, and $MeSP(O)Cl_2$ and other organic P compounds in $CHCl_3$, CCl_4 and hexane. The influence of dilution on J could be attributed to variation in the bond angles. Mavel and Martin (1963)[734] using sensitive equipment determined the ^{13}CH coupling constants (J) in various methyl compounds: $(MeO)_3P(S)$, $(MeO)_3P(Se)$, $(MeS)_3P(O)$, $(MeS)_3P(S)$, and $MeSP(O)Cl_2$. Takahashi et al (1966)[1299] observed long-range coupling of P-H through four bonds in $EtOP(S)Cl_2$, $(EtO)_2P(S)Cl$, and $(EtO)_3P(S)$. Miyazima et al. (1967)[816] constructed a frequency-swept heteronuclear decoupling system to observe the proton resonance under irradiation with a phosphorus reasonance around 24.29 Mcps. The coupling constants were tabulated for: $EtOP(S)Cl_2$, $(EtO)_2P(S)Cl$, $(EtO)_3P(S)$, $(EtO)_3P(Se)$, and $(EtO)_2P(S)SH$. Magnetic nonequivalence of the methyls in the ethoxy groups in (1), which contains an asymmetric

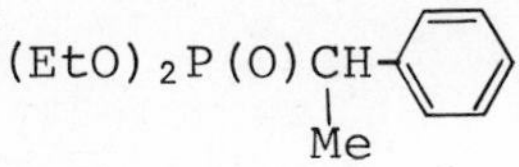

(1)

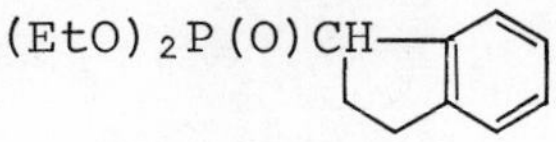

(2)

center, was shown by Moen and Mueller (1966).[818] The separation of the triplets was greatest in nonpolar solvents. However, (2) did not show magnetic nonequivalence. Kaplan and Schulz (1967)[515] demonstrated long-range P-H spin-spin coupling through four and five bonds to the *o*- and *m*-protons of an aromatic system in $2,4,5\text{-}Cl_3C_6H_2OP(S)Cl_2$. Edmundson and Mitchell (1968) reported ^1H-NMR data for

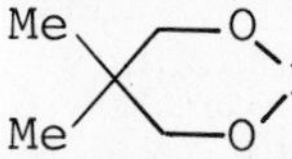

P(S)OMe in $CDCl_3$ and in benzene. Data were also assigned for the Me_{eq} and Me_{ax} (gem-dimethyl) groups in both solvents. Williamson and Griffin (1968)[1395] determined 3- and 4-bond ^{31}P - ^1H coupling constants and geminal proton nonequivalence in Et esters: $(EtO)_3P(S)$, $(EtO)_2$-$P(S)Cl$, $[(EtO)_2P(S)]_2O$, $EtOP(S)Cl_2$, $(EtO)_3P(Se)$, and $(EtO)_2$ $P(S)SH$. Kainosho (1970)[500] reported P-H spin-spin coupling constants in open-chain compounds: $(MeO)_3P(X)$ and $(EtO)_3$-$P(X)$ for X = S or Se. Tseng and Chan (1971)[1325] gave ^1H-NMR data on the esters: $(MeO)_2P(S)SEt$, $MeOP(O)(SMe)SEt$, and $EtOP(O)(SMe)_2$.

The results of measurements for ^{19}F and ^{31}P have been included in the lists of compounds wherever possible. Unless otherwise stated, it may be assumed that ^{31}P measurements are relative to phosphoric acid. ^{19}F and ^{31}P-NMR measurements on P-F compounds, including $(MeO)_2P(S)F$, $(PhO)_2P(S)F$, $MeOP(S)F_2$, and $PhOP(S)F_2$, by Reddy and Schmutzler (1970)[1013] revealed no meaningful substitution rules, either for chemical shifts or for P-F coupling constants.

Of general interest on ^{31}P-NMR are papers by Muller et al. (1956);[850] Jones and Katritzky (1962);[488] Mark, Van Wazer et al. (1967)[715]; Moedritzer, Maier and Groenweghe (1962);[817] and Van Wazer et al. (1956).[1345] Ross and Biros have measured ^{31}P-NMR on pesticides with P_4O_6 as external standard (1970).[1047]

The following publications on ^{31}P-NMR are addressed to limited areas or to more specific problems: Finegold (1958)[326] on $(EtO)_2POSH$, $(EtO)_2P(S)SH$, $(EtO)_2P(S)Cl$, $PrOP(S)Cl_2$, $BuOP(S)Cl_2$, $(RO)_2P(O)SR'$, $(RS)_3P(O)$, $(RO)_3P(S)$, $(RO)_2P(S)SR'$, and $(RS)_3P(S)$; Stothers and Robinson (1964)[1274] on $(RO)_2P(O)SC_6Cl_5$, $(RO)_2P(S)OC_6Cl_5$, and $(RO)_2P(S)SC_6Cl_5$ where R = Me and Et; Fluck et al. (1966)[339] on

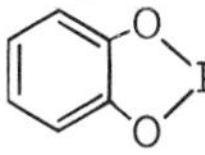
P(S)X where X = Br and Cl; Harris et al. (1967)[403]

on $((EtO)_2P(S))_2O$ with P_4O_6 external standard; Dwek, Richards, et al. (1966, 1970)[240, 241] have interpreted qualitatively enhancements at 12,500 gauss of P-S and other P compounds in solutions containing tri-t-butylphenoxy radical, TTBB, on $(EtO)_3P(S)$, $(4\text{-}MeC_6H_4O)_3P(S)$, $(4\text{-}MeC_6H_4S)_3P(O)$, $(4\text{-}MeC_6H_4S)_3P(S)$, $(PhCH_2S)_3P(O)$, and $(PhS)_3P(S)$; Mikolajczyk and Schiebel (1969)[797, 800] on geometric isomerism (see Section B.6); Dale and Hobbs (1971)[195] on spin-lattice relaxation times of ^{31}P in $(MeO)_3P(S)$ and other P compounds; and Potenza et al. (1969)[970] on ^{31}P enhancements at 74 gauss in solution with the free radical bis(diphenylenephenylallyl) on $(EtO)_3P(S)$, $(EtO)_2P(S)Cl$, and their P(O) analogs.

B.6. Stereochemistry

Donninger and Hutson (1968)[221] resolved tetramethylammonium methyl-1-naphthylphosphorothionate (3), with (-) ephedrine to obtain (-) ephedrine (+) methyl-1-naphthyl-

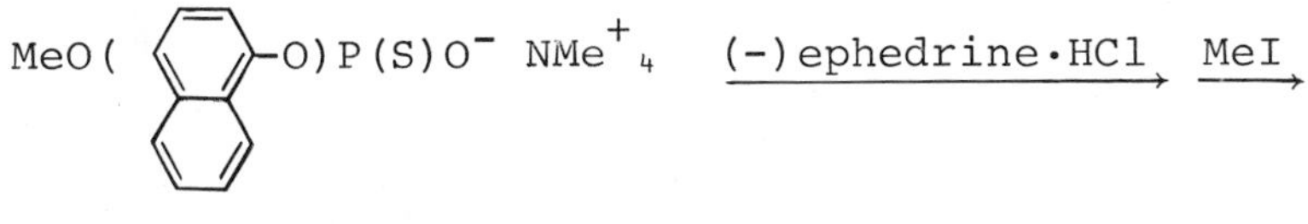

(3)

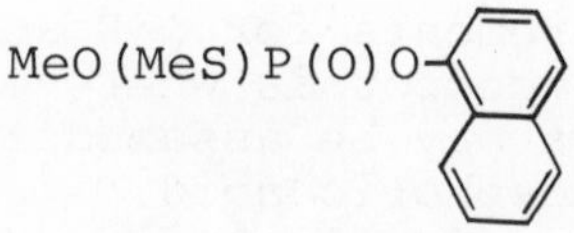

(4)

phosphorothiolate, $[\alpha]^{24}_D$(4) 33° (c 1.44, $CHCl_3$) and (-) ephedrine (-) methyl-1-naphthylphosphorothiolate, $[\alpha]^{24}_D$ -45.9° (c 9.20, $CHCl_3$). The ephedrine salts were reacted with MeI in MeOH to give (+) and (-) -O-methyl-S-methyl-1-naphthylphosphorothiolate, (4) $[\alpha]^{24}_D$ 53.4° (c 6.34, $CHCl_3$), and $[\alpha]^{24}_D$ -51° (c 1.75, $CHCl_3$), respectively.

Teichmann and Lam (1969)[1303] treated $(MeO)_2P(S)SMe$ in MeOH with strychnine and isolated the (+) and (-) methostrychninium salts. These were converted to (+) and (-) MeO(MeS)P(S)OH by ion exchange. Neutralization of the acids with potassium carbonate and reaction with EtBr gave (+) and (-) MeO(MeS)P(S)OEt, respectively.

Mikolajczyk and Schiebel (1969)[800] have prepared the geometrical isomers of imidazolium propylenethionophosphate, but have been unable to assign cis- and trans-

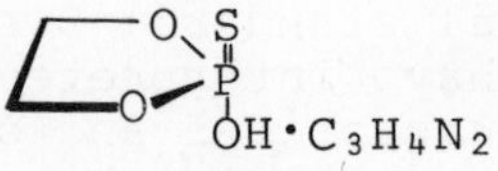

cis

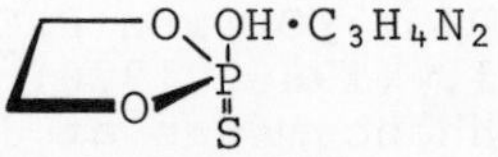

trans

forms to the products, m. 122-125° and 78.5-83°, respectively. Mikolajczyk (1969)[797] accomplished the stereospecific synthesis of cis-2-methoxy-4-methyl-1,3,2-dioxaphosphor^V-inane-2-thione. This ester with Me_3N gave the Me_4N^+ salt of 2-hydroxy-4-methyl-1,3,2-dioxaphosphor^V-inane-2-thione and permitted the assignment of cis- and trans-forms to 6-ring products from Ref. 800:

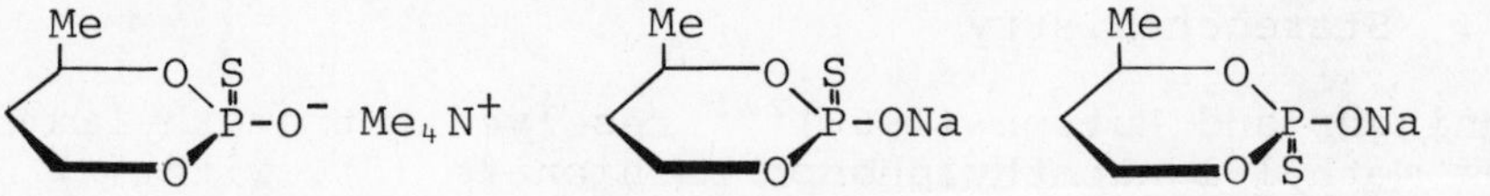

cis ^{31}P -112.8 ppm | cis ^{31}P -112.9 ppm | trans ^{31}P -120 ppm

Bodkin and Simpson (1969)[112] and Simpson and Bodkin (1971)[1248] have shown that the more stable stereoisomers of 2-alkoxy-4-methyl-2-thiono-1,3,2-dioxaphosphorinanes (5), have the trans- configuration and assume a chair con-

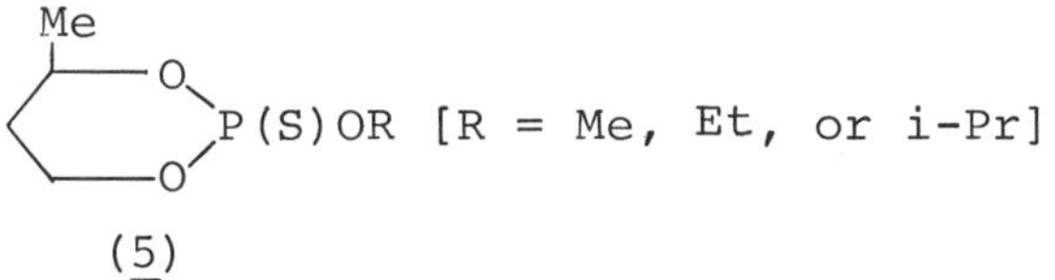

(5)

formation with equatorial methyl and axial alkoxy groups. The less-stable isomers have the cis- configuration, and at room temperature adopt a rapidly flipping chair conformation. Configurational stability is solvent dependent.

Usher, Richardson, and Eckstein (1970)[1344] have reported "an unambiguous determination of the geometry of the second step" of ribonuclease action following the isolation by Eckstein, J. Am. Chem. Soc. 92 4718 (1970), of the two diastereoisomers of uridine-2',3'-cyclic phosphorothioate:

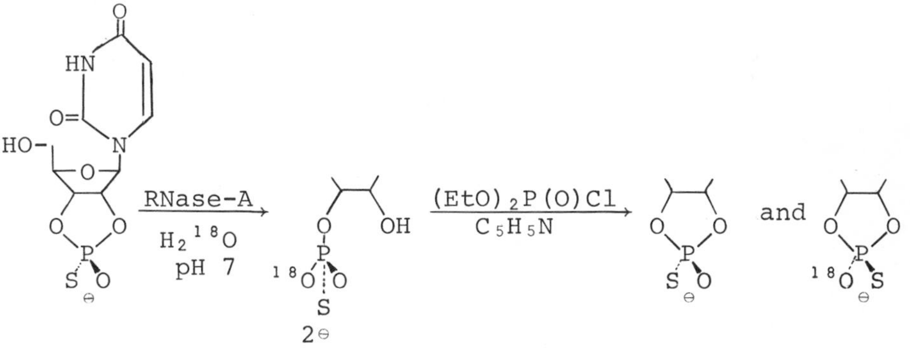

B.7. Mass Spectra

Damico (1966)[196] has published mass spectra for 23 organophosphorus pesticides and related compounds: phosphorodithioates, -thiolates, -thionates, and phosphates. Molecular ions were observed in all but two cases. The spectra are discussed in detail.

Cooks and Gerrard (1968)[184] published mass spectra for $(PhO)_2P(S)Cl$ and $(PhO)_2P(S)OMe$ and some related amides and mass spectral data for $(MeO)_2P(S)Cl$ and $(PhO)_2P(S)OEt$. The fragmentation processes upon electron impact include the following bond-forming reactions: (a) molecular ion rearrangement of the thiono-thiol type, $P(S)OR \rightarrow P(O)SR$; (b) bond formation between substituents (especially aryl groups) on phosphorus; (c) loss of SH from the molecular ion where the H can be derived from an alkyl or an aryl group; (d) H rearrangement to give the phenol molecular ion (or an isomer) in compounds having a phenoxy substituent.

Pritchard (1970)[979] has shown that $(MeO)_2P(S)Cl$ yields CH_2O in the initial step in the decomposition of the positive ion. For the ethyl ester, $(EtO)_2P(S)Cl$, MeCHO is principally produced along with a $C_2H_3\cdot$ radical. With the analogous $(EtO)_2P(O)Cl$ ester, the aldehyde is not observed but the $C_2H_3\cdot$ radical is produced along with 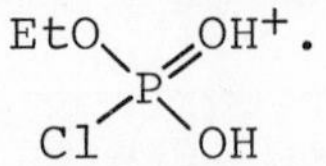.

Rearrangements occurring in the ion fragmentations of O-(3,6-dichloro-2-pyridyl)-O,O-diethylphosphorothioate (6) were studied by Tou (1971)[1317] and compared with similar

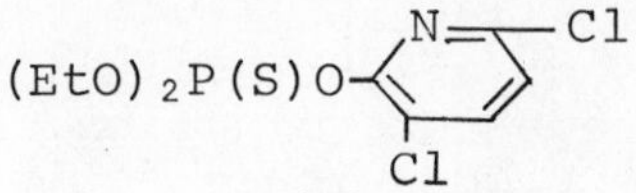

(6)

studies of O-2,3,6-trichlorophenyldimethylthiocarbamate, $Me_2NC(S)OC_6H_2Cl_3$-2,3,6. In both compounds, thermal reaction and ion fragmentation give rise to thion-thiol isomerization. The theoretical aspects of the reactions are discussed and equations are derived for the rate ratio of production of rearranged ions compared with those generated by simple bond scission.

B.8. Electron Spin Resonance

The ESR spectrum of (7) has been studied by Pannell (1962)

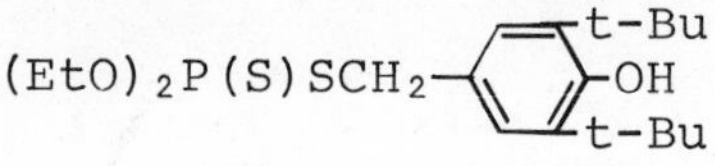

(7)

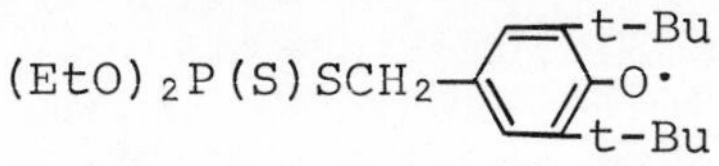

(8)

[924] and shown to consist of a triplet of triplets with coupling parameters for the p-substituent proton, 6.37 gauss, and the m-ring protons, 1.80 gauss. The radical species was (8).

B.9. Nuclear Quadrupole Resonance

Semin and Babushkina (1968)[1218]: N.Q.R. of ^{35}Cl data on numerous organophosphorus compounds containing P-Cl are summarized. The average for $(\nu_{P(S)} - \nu_{P(O)})$ is 1.23 MHz and for $(\nu_{P(S)} - \nu_{P(III)})$ is 1.96 MHz.

Whitehead and Hart (1971)[1389]: N.Q.R. data at 77 °K for $ROP(S)Cl_2$ (R = Me, Et) and for $(RO)_2P(S)Cl$ (R = Me, Et, i-Pr) shows that the ^{35}Cl N.Q.R. frequencies of thiophosphoryl chlorides all exceed those of their P(O) analogs. The frequencies are used to predict possible molecular structures, and are related to σ- and π-bond effects within the molecules, by use of BEEM (bond electronegativity equalization method).

B.10. Magnetic Susceptibility

Data on diamagnetic susceptibilities (specific (χ) and molar (χ_M) susceptibilities) for $(RS)_3P(O)$ for R = Et, Pr, and Bu [Voigt, Labarre; and Fournes (1966)][1365] and for $(EtO)_2P(S)SH$ and $(EtO)_2P$-(S)SEt[Zgadzai and Maklakov (1964)][1455] will be found in the List of Compounds.

B.11. Bond Magnetic Rotation

Bond magnetic rotations have been measured by Labarre, Voigt, and Gallais (1967)[580] for $ROP(S)Cl_2$, $(RO)_2P(S)Cl$, $(RO)_3P(S)$, $(RS)_3P(O)$, and $(RS)_3P(S)$, where R is Et, Pr, and Bu. Data on magnetic rotation $(\rho)_M$ and on molecular magnetic rotation $(\Omega)_M$ will be found in the List of Compounds.

B.12. Molecular Polarizability

Aroney, Le Febre, and Saxby (1963)[69] calculated major and minor axes for the group $O_3P(S)$ to be 0.869 and 0.492 x 10^{-23} cc based on polarizability ellipsoids of revolution. Polarizations, dipole moments, and molar Kerr constants are given for (9), where (X) is (:), (O), (S), and (As).

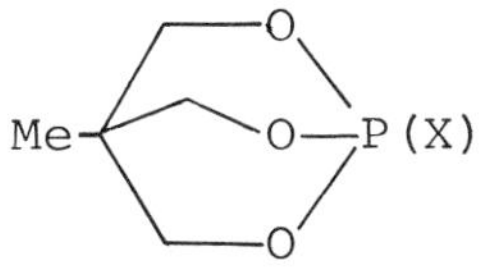

(9)

For the compound where (X) is (S), μ(D) is 6.77.

B.13. Viscosity

In a study of viscosity and structure of tertiary esters [Arbuzov and Vinogradova (1952)][63] the observed solution viscosities of $(C_6H_{13}O)_3P(S)$ and $(C_8H_{17}O)_3P(S)$ in CCl_4 and in benzene indicated the compounds in solution existed in the extended form rather than the parallel-chain form. Data on specific viscosity η_{sp} and Z_η in solution are given in the List of Compounds.

B.14. Crystallography

Hanic and Cakajdova (1958)[401] reported observations on crystallography of bis(nitrophenyl) alkylphosphorothioates: $MeOP(S)(OR)_2$, $MeSP(O)(OR)_2$, $EtOP(S)(OR)_2$, and $EtSP(O)(OR)_2$, where R is $-C_6H_4NO_2$. The thiolo compounds differed very little from their O-analogs, while P(S) and P(O) compounds showed considerable differences in structure type and dimensions.

The crystal and molecular structure of 1-chloro-1-thiono-2,5,1-dithiaphospha(V) cyclopentane, 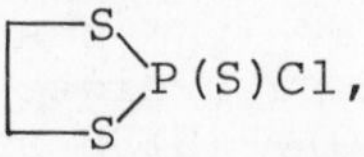, which crystallizes in the orthorhombic system, was determined by Lee and Goodacre (1971).[597] Complete data are given for bond lengths and angles, and distortions due to P(S) and 5-membered ring are discussed.

B.15. Cryoscopy

The absolute mole percent purity of the pesticides, malathion, dimethoate, Zinophos®, and phorate, was determined cryoscopically by Best and Hersch (1964).[93] Time-temperature melting curves were analyzed to determine melting points of the pure compounds, T_o, and cryoscopic constants, K, were established by adding known impurities and measuring the new melting points. T_o for the compounds above were 3.70 ± 0.09^o, 50.25 to 50.56^o, -1.67 ± 0.1^o, and -43.7^o, respectively. Mole percent purities of high purity samples of the products can be determined by comparison with temperature vs. mole % purity curves given in the publication.

C. METHODS OF PREPARATION OF PHOSPHOROHALIDOTHIOATES AND -SELENOATES

METHOD A. REACTION OF HYDROXY OR MERCAPTO COMPOUNDS WITH PHOSPHORYL, PHOSPHOROTHIOYL, OR PHOSPHOROSELENOYL HALIDES

Thiophosphoryl chloride reacts with primary alcohols to give chlorothionophosphates:

$$P(S)Cl_3 + ROH \rightarrow ROP(S)Cl_2 + HCl$$

$$P(S)Cl_3 + 2ROH \rightarrow (RO)_2P(S)Cl + 2HCl$$

Its reactivity is lower than that of phosphoryl chloride allowing the reaction to be restricted to the formation of primary products, $ROP(S)Cl_2$, if desired. Manske and Kulka (1951)[713] ran the reaction with ethanol in refluxing benzene, t-BuCl, etc. Bakanova et al. (1956)[77] ran the reaction in benzene at 50° with $Al(OEt)_3$ as catalyst. Gay and Hamer (1970)[360] treated equimolar quantities of ethylene chlorohydrin and $P(S)Cl_3$ at reflux temperature without solvent. The products were purified by distillation.

Charlton and Cavell (1968)[159a] treated thiophosphoryl iodofluoride with methanol or methyl mercaptan on warming to room temperature to give $MeOP(S)F_2$ and $MeSP(S)F_2$ respectively with SPF_2H as the by-product.

In place of $P(S)Cl_3$, Dye (1956)[242] added alcohol in benzene to a mixture of sulfur and PCl_3 in refluxing benzene to obtain $(EtO)_2P(S)Cl$ and $(EtO)_3P(S)$.

With temperature control (MeOH 0°, EtOH 25°), Olah and Oswald (1957)[908] replaced one Cl from $P(S)Cl_2F$ to give ROP(S)ClF; in the presence of pyridine (Method B below) both Cl's were readily replaced.

Methanol in chlorobenzene at room temperature for 18 hr replaced one Cl from $ArOP(S)Cl_2$ to give MeO(ArO)P(S)Cl, Dow Chemical (1959).[224]

Bliznyuk et al. (1967)[108] used an excess of acrylonitrile as an acid acceptor in preparing $ROP(S)Cl_2$, where R was isoamyl and higher:

$$P(S)Cl_3 + ROH + 2CH_2{=}CHCN \rightarrow ROP(S)Cl_2 + ClCH_2CH_2CN$$

METHOD B. REACTION OF OH AND SH COMPOUNDS WITH HALO-PHOSPHORUS COMPOUNDS IN THE PRESENCE OF BASES

In this review, no distinction is made between the use of active-H compounds plus base (inorganic or organic), as compared with the use of preformed salts (RONa, RSK, $RONH_4$, etc. Frequently both methods are used in a given publication where advantages may be discussed for particular cases.

A few generalized equations are:

$$P(S)X_3 \text{ (excess)} + ROH + B \rightarrow ROP(S)X_2 + B{\cdot}HX$$
$$ROP(S)X_2 + ROH + B \rightarrow (RO)_2P(S)X + B{\cdot}HX$$
$$P(S)X_3 + RONa \rightarrow (RO)P(S)X_2 + NaX$$

Representative examples from the literature include:

a. $P(S)Cl_3$ + ROH + tert. amine → $(RO)P(S)Cl_2$

CD_3CH_2OH pyridine [Nyquist (1970)][902]
$MeOCH_2CH_2OH$ picoline [Kishikawa (1969)][539]
$4\text{-}MeCOC_6H_4OH$ Et_3N [Yagnyukova (1971)][1402]
cholesterol pyridine [Cremlyn, Olson (1969)][191]

$P(S)Cl_3$ + 2ROH + tert. amine → $(RO)_2P(S)Cl$

$MeCHOHCH_2OH$ pyridine [B. A. Arbuzov (1954)][56]

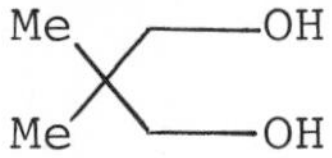

pyridine [Edmundson (1965)][247] [Wagner (1973)][1369]

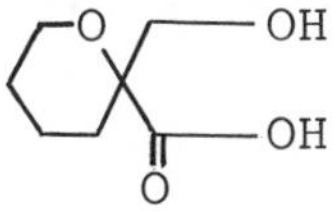

pyridine [Lanham (1958)][585]

b. $P(S)Cl_3 + EtOCH_2CH_2ONa \rightarrow EtOCH_2CH_2OP(S)Cl_2$ [Strivastava (1966)][1260]

$^{32}P(S)Cl_3 + 2MeOH + 2NaOH \rightarrow (MeO)_2{}^{32}P(S)Cl$ [Mandel'baum (1954)][699]

$${}^{32}P(S)Cl_3 + 2EtONa \rightarrow (EtO)_2{}^{32}P(S)Cl$$
$$[Fukuto, Metcalf (1954)][345]$$

c. excess $P(O)Cl_3 + RSH + pyridine \rightarrow RSP(O)Cl_2$
   [R = Me to Bu][Kobayashi (1970)][543]

d. $P(O)Cl_3 +$ $\rightarrow$

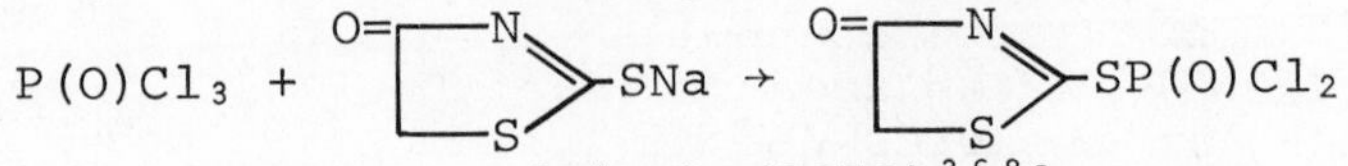

$$[Ginak (1970)[368]]$$

e. $P(S)Cl_2F + PhSH + pyridine \rightarrow PhSP(S)ClF$
   [Roesky (1968)][1040, 1041]

f. $P(O)Cl_2F + 2EtSH + 2Me_2NPh \rightarrow (EtS)_2P(O)F$
   [Chapman, Saunders (1948)][159]

g. $EtOP(S)F_2 + ROH + Et_3N \rightarrow EtO(RO)P(S)F$
   [Farbenfabriken-Bayer (1957)][278]

$$FCH_2CH_2OP(S)Cl_2 + EtOH + Et_3N \rightarrow$$
$$EtO(FCH_2CH_2O)P(S)Cl$$
$$[Kabachnik (1959)][490]$$

$-OP(S)Cl_2 + ArOH + Et_3N \rightarrow$ $-O(ArO)P(S)C$

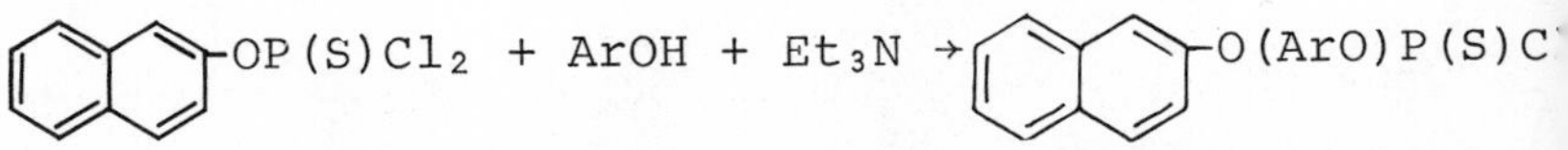

$$[Buchner, Jacoves (1967)][135]$$

$$EtOP(S)Cl_2 + 3\text{-}ClC_6H_4OH + Et_3N \xrightarrow{PhCl}$$
$$EtO(3\text{-}ClC_6H_4O)P(S)Cl$$
$$[Richter (1967)][1028]$$

h. $ROP(S)Cl_2 + MeONa \rightarrow MeO(RO)P(S)Cl$
   [Schrader (1964)][1174]

$ROP(S)Cl_2 + R'OH + NaOH \rightarrow RO(R'O)P(S)Cl$
   [Schrader (1968)][1187]

$$EtOP(S)Cl_2 + 2\text{-}ClC_6H_4OH + NaOH \rightarrow$$
$$EtO(2\text{-}ClC_6H_4O)P(S)Cl$$
$$[Schrader (1969)][1188]$$

$$EtOP(S)Cl_2 + 2\text{-}HOC_6H_4COO\text{-}i\text{-}Pr + NaOH \rightarrow$$
$$EtO(2\text{-}i\text{-}PrO_2CC_6H_4O)P(S)Cl$$
$$[Farbenfabriken-Bayer (1970)][314]$$
$$[Richter (1969)][1030]$$

METHOD C. HALOGEN EXCHANGE REACTIONS OF PHOSPHORO-CHLORIDOTHIOATES

Exchange of chlorine for fluorine occurs readily in a variety of compounds on gentle warming with inorganic fluorides. Dihalides may give mixtures of chlorofluorides and difluorides [Booth (1948)]:[118]

$$EtOP(S)Cl_2 + SbF_3 \rightarrow EtOP(S)ClF + EtOP(S)F_2$$

Seel and co-workers (1962)[1211] used KSO_2F at 50°:

$$EtOP(S)Cl_2 + KSO_2F \rightarrow EtOP(S)F_2$$

Monochlorides are replaced cleanly by the use of a variety of fluorides:

$$EtO(EtS)P(O)Cl + SbF_3 \rightarrow EtO(EtS)P(O)F$$ [Petrov (1956)][947]

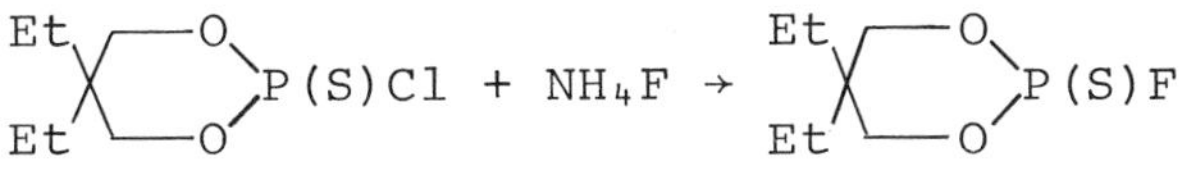
[Lanham (1957)][584]

$$(RO)_2P(S)Cl + NaF \xrightarrow[\text{UV 5 hr}]{\text{boiling PhMe}} (RO)_2P(S)F$$ [Olah (1955)][909]

METHOD D. THIONO - THIOLO ISOMERIZATION

Heating $MeOP(S)Cl_2$ for 5 hr at 100° [Hilgetag (1960)][434] isomerized the material to $MeSP(O)Cl_2$. Heating $(MeO)_2$-$P(S)Cl$ at 130° gave only MeCl and resinous residue.

Aksenov et al. (1971)[11] showed that phenylphosphorodichloridothioate, $PhOP(S)Cl_2$, was very resistant to thermal isomerization. In the vapor phase at 550° in helium carrier gas, the products contained about 38% unchanged starting material, about 1% isomer, $PhSP(O)Cl_2$, a number of volatile products and six unidentified high-boiling substances.

METHOD E. ADDITION OF SULFUR OR SELENIUM TO HALOPHOSPHITES

Elemental sulfur and selenium add readily to halophosphites to form thiono or seleno halophosphates. The reaction is frequently run in a sealed tube or in solution-suspension at temperatures above 100°. Sulfur was added to $RSPCl_2$ at 140-160° in preparing $PhSP(S)Cl_2$ [Michaelis and Linke (1907)][773] and $EtSP(S)Cl_2$ [Pistschimuka (1912)].[960] Malatesta (1953)[684] prepared $ClCH_2CH_2OP(S)Cl_2$ in a sealed tube at 120° for 24 hr. B. A. Arbuzov (1970)[58] added sulfur to $MeO(ClCH_2CH_2CHMeO)PCl$ at 140-160° to prepare MeO-$(ClCH_2CH_2CHMeO)P(S)Cl$.

Kosolapoff (1951)[559] caused PCl_3, ethylene oxide, and sulfur to react in solvents such as benzene, toluene, or xylene at 70-80°:

$$PCl_3 + C_2H_4O + S \rightarrow (ClCH_2CH_2O)_2P(S)Cl$$

The use of $P(S)Cl_3$ as the source of the sulfur was demonstrated by Gottlieb (1932):[376]

$$(PhO)_2PCl + P(S)Cl_3 \xrightarrow{\text{reflux}} (PhO)_2P(S)Cl + PCl_3$$

Tolkmith (1962)[1316] ran the following reaction at 125-

155° with the temperature rising as PCl_3 distilled:

$$2,4,5\text{-}Cl_3C_6H_2OPCl_2 + P(S)Cl_3 \rightarrow 2,4,5\text{-}Cl_3C_6H_2OP(S)Cl_2 + PCl_3$$

A similar procedure, attributed to Gottlieb,[376] was employed by Kalashnikov and Zemlyanskii (1969)[508] at bath temperatures of 150-180° and 200-220°:

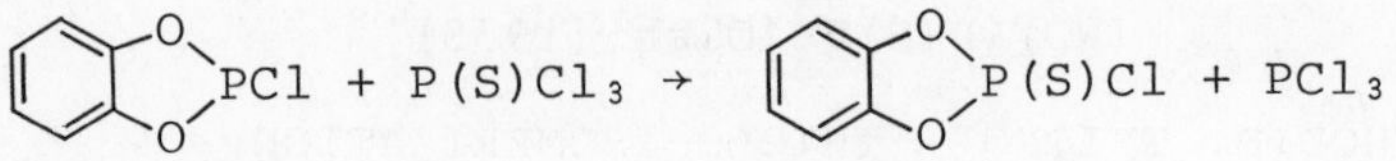

Schrader (1964)[1173] used sulfur monochloride in benzene to prepare phosphorodichloridothioates from phosphoryldichlorides in rapid reactions at 15-20°:

$$3\ s\text{-}BuOPCl_2 + S_2Cl_2 \rightarrow 2\ s\text{-}BuOP(S)Cl_2 + P(O)Cl_3 + s\text{-}BuCl$$

Zemlyanskii (1966)[1433] added selenium to $(EtO)_2PCl$ in toluene at 105-110° to obtain $(EtO)_2P(Se)Cl$.

METHOD F. REACTIONS OF P_2S_5 WITH DERIVATIVES OF PHOSPHORODICHLORIDIC ACIDS

Heating $EtOP(O)Cl_2$ or $EtSP(O)Cl_2$ with P_2S_5 at 150° under nitrogen for 5 hr produced $EtSP(S)Cl_2$ [Godovikov and Kabachnik (1961)].[371] $EtOP(S)Cl_2$ is probably formed first and isomerizes before undergoing the second step to give the final product. In the case of $ArOP(O)Cl_2$ the reaction produces only $ArOP(S)Cl_2$. Resistance to isomerization (see Method D) with aryl derivatives prevents the second step reaction.

METHOD G. DISPLACEMENT REACTION OF TERTIARY PHOSPHOROTRITHIOATES

In the reaction of $(EtS)_3P(O)$ with acetyl chloride [Divinskii (1948)][213] at 160° for 3 hr, the initial attack is probably at the S linkage giving rise to the intermediate shown below and eliminating ethyl thioacetate to give the product:

$$(EtS)_3P(O) + MeCOCl \rightarrow ((EtS)_2P(O)S^{\oplus}(COMe)EtCl^{\ominus}) \rightarrow (EtS)_2P(O)Cl + MeCOSEt$$

An excess of acetyl chloride can carry the product to further stages of degradation to give $RSP(O)Cl_2$ and finally $P(O)Cl_3$.

METHOD H. HALOGENATION OF PHOSPHORODITHIOIC AND -THIOIC ACIDS AND THEIR DERIVATIVES

The action of chlorine or chlorinating agents at moderate temperatures (usually 0-50°) on O,O-dialkyl phosphorodithioic acids provided a method for preparing dialkyl phosphorochloridothionates without the use of $P(S)Cl_3$. The processes were developed during the 1950s to provide important intermediates for the rapidly developing pesticides industry. Fletcher et al. (1950)[334] chlorinated crude phosphorodithioic acids in benzene solution at 25-30°:

$$(RO)_2P(S)SH + 2Cl_2 \rightarrow (RO)_2P(S)Cl + SCl_2 + HCl$$

The products were purified by washing and distilling. Hu (1956)[464] used chlorine in $CHCl_3$ at -5°. Ishida (1968)[477] used chlorine in $PhPCl_2$.

Hechenbleikner (1949)[421] used chlorine on sulfides of the type $[(RO)_2P(S)]_2S_n$. Lippman (1966)[606] studied the chlorination of the neutral fractions from the preparation of dimethyl and diethyl phosphorodithioic acid. It was found that all of the constituents except $(RO)_3P(S)$ were converted to $(RO)_2P(S)Cl$.

Sulfuryl chloride was used by Ashbolt and Coates (1951)[70] on the dithioic acids, their alkali metal salts or their sulfide derivatives, $[(RO)_2P(S)]_2S_n$.

Delepine (1912)[197a] observed a poor yield of $(MeO)_2P(S)Cl$ from $(MeO)_3P(S)$ plus PCl_5. PCl_5 with PCl_3 as diluent was used by Toy and McDonald (1955)[1319] with the dithioic acids or their alkali metal salts. Mühlmann and Schrader (1960)[845] used PCl_5 on acids or salts of the type $RO(RS)P(O)OH$ at 30° to prepare $RO(RS)P(O)Cl$. With PCl_5 on $(EtO)_2P(S)ONH_4$ in CCl_4, Pesin and Khaletskii (1961)[935] obtained $(EtO)_2P(S)Cl$.

METHOD J. HALOGENATION OF SECONDARY PHOSPHOROTHIOITES AND -SELENOITES

Martynov (1969)[723] chlorinated 2-chloro-1,3,2-oxathiaphospholane in ether at 15° with ring-opening to give S-2-chloroethyldichlorothiophosphate:

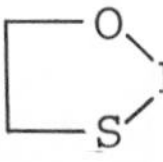 $PCl + Cl_2 \rightarrow ClCH_2CH_2SP(O)Cl_2$

Delventhal and Kuchen (1971)[198] carried out the bromination of the bisthiophosphite of pentaerythritol in acetic acid at $10°$:

$$H(S)P\left\langle\begin{array}{c}O\\O\end{array}\right\rangle\!\!\times\!\!\left\langle\begin{array}{c}O\\O\end{array}\right\rangle P(S)H \ + \ 2Br_2 \rightarrow$$

$$Br(S)P\left\langle\begin{array}{c}O\\O\end{array}\right\rangle\!\!\times\!\!\left\langle\begin{array}{c}O\\O\end{array}\right\rangle P(S)Br \ + \ 2HBr$$

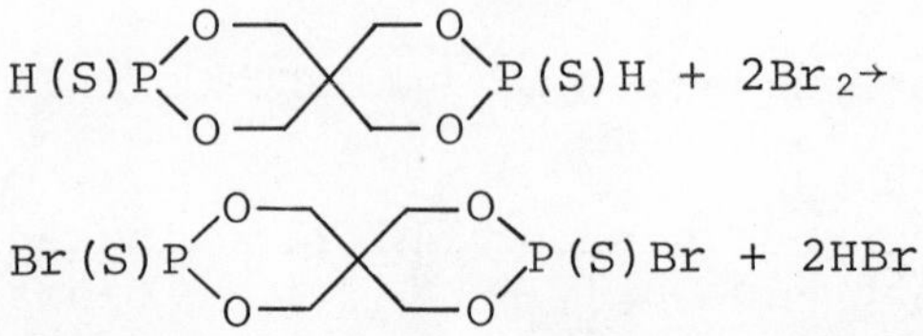

A stable phosphoroiodidothionate was obtained by Edmundson (1965)[246] on treatment of 5,5-dimethyl-2-thiono-1,3,2-dioxaphosphorinane with iodine in alcohol with KHCO$_3$ at $40°$:

$$\begin{array}{c}Me\\Me\end{array}\!\!\times\!\!\left\langle\begin{array}{c}O\\O\end{array}\right\rangle P(S)H \ + \ I_2 \ \rightarrow \ \begin{array}{c}Me\\Me\end{array}\!\!\times\!\!\left\langle\begin{array}{c}O\\O\end{array}\right\rangle P(S)I \ + \ HI$$

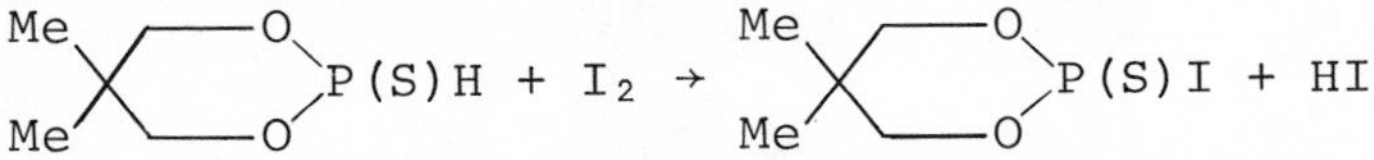

The other possible product, RSSR, was not observed.

Thiophosphites treated with CCl$_4$ in benzene in the presence of an amine gave phosphorochloridothionates [Lorenz and Schrader (1959)]:[635]

$$(MeO)_2P(S)H \ + \ CCl_4 \ \xrightarrow[\text{benzene} <60°]{\text{Et}_3\text{N (cat.)}} \ (MeO)_2P(S)Cl \ + \ CHCl_3$$

Sodium diethylthiophosphite added to excess CCl$_4$ at $-10°$ and kept 24 hr at room temperature gave diethyl phosphorochloridothionate and a new trichloromethylphosphonothionate [Almasi and Paskucz (1964)]:[31]

$$(EtO)_2PSNa \ + \ CCl_4 \ \rightarrow \ (EtO)_2P(S)Cl \ + \ (EtO)_2P(S)CCl_3$$

Krawiecki et al. (1969)[562] treated dialkyl phosphoroselenoites with sulfuryl chloride in benzene at $5°$ to prepare dialkyl phosphorochloridoselenoates:

$$(EtO)_2P(Se)H \ + \ SO_2Cl_2 \ \rightarrow \ (EtO)_2P(Se)Cl \ + \ HCl \ + \ SO_2$$

Earlier work [Michalski and Krawiecki (1957)][776] with the same reactants in the absence of solvent gave a violent reaction from which, however, a good yield of product was obtained.

METHOD K. ACTION OF HALOGEN ACIDS ON PHOSPHORODITHIOATES AND -THIOATES

Dialkyl phosphorodithioic acids in the presence of H$_2$S acceptors such as acetonitrile were converted by gaseous HCl to the chlorides:

$$(RO)_2P(S)SH + HCl + MeCN \rightarrow (RO)_2P(S)Cl + MeCSNH_2$$

Meinhardt (1960)[737] ran the reaction in ether at reflux temperature. Schicke and Schrader (1961)[1104] gassed the solution in acetonitrile with no mention of temperature or inert solvent.

Mühlmann and Schrader (1960)[845] used gaseous HCl in benzene at ice temperature to convert sodium O,S-dimethyl phosphorothioate to the chloride:

$$MeO(MeS)P(O)ONa + 2HCl \rightarrow MeO(MeS)P(O)Cl + NaCl + H_2O$$

Tetraethyl dithionopyrophosphate was cleaved by slowly adding it to anhydrous HF at ice bath temperature and warming to 50-55° [White and Hood (1953)]:[1388]

$$[(EtO)_2P(S)]_2O + H_2F_2 \rightarrow 2(EtO)_2P(S)F + H_2O$$

METHOD L. REACTIONS OF HALOPHOSPHITES OR ELEMENTAL PHOSPHORUS WITH ALKYL SULFENYL CHLORIDES

Addition of dialkyl chlorophosphites to alkyl sulfenyl chlorides in toluene at -30 to -50° gave alkyl chlorides and O,S-dialkyl phosphorochloridothioates [Lippman (1965)]:[605]

$$(RO)_2PCl + RSCl \rightarrow RO(RS)P(O)Cl + RCl$$

$$[R = Me \text{ or } Et]$$

Hoffmann (1968-1970)[451, 454] added MeSCl at 10-30° to a CCl_4 solution of 2-chloro-1,3,2-dioxaphospholane to obtain the open-chain thiophosphoryl chloride:

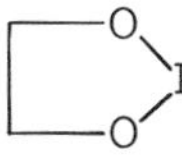PCl + MeSCl → $MeS(ClCH_2CH_2O)P(O)Cl$

Similarly, using bromine and Me_2S at -10 to -15° gave $MeS(BrCH_2CH_2O)P(O)Cl$.[452] Reaction with 4-chlorobenzenesulfenyl chloride in CH_2Cl_2 at 30° gave $ClCH_2CH_2S(4\text{-}ClC_6H_4S)\text{-}P(O)Cl$.[453]

Schrader (1963)[1172] prepared $C_6Cl_5SP(O)Cl_2$ in benzene at 0°:

$$MeOPCl_2 + C_6Cl_5SCl \rightarrow C_6Cl_5SP(O)Cl_2 + MeCl$$

Shitov and Gladshtein (1968)[1236] ran analogous reactions without solvent at -15 to -10°:

$$ROPF_2 + PhSCl \rightarrow PhSP(O)F_2 + RCl$$

Petrov et al. (1961)[945] treated red phosphorus with EtSCl in liquid SO_2 at -50 to -30° to form a mixture of Et_2S_2, $EtSP(O)Cl_2$, and $(EtS)_2P(O)Cl$. The following reaction was also observed at -50°:

$$PCl_3 + RSCl \rightarrow [RSPCl_4] \rightarrow SO_2 \rightarrow RSP(O)Cl_2 + SOCl_2$$

Aichenegg and Gillen (1969)[4] prepared 1,2,2-trichloroethylthiophosphoryl dichloride in CCl_4 at $30-32^O$:

$$EtOPCl_2 + Cl_2CHCHClSCl \rightarrow Cl_2CHCHClSP(O)Cl_2 + EtCl$$

METHOD M. REACTIONS OF PHOSPHITES, HALOPHOSPHITES, OR HALOPHOSPHOLANES WITH α-HALONITROSOALKANES

Martynov, Kruglyak, et al. (1969-70)[565, 723, 724] treated thiophosphites, halothiophosphites, 1,3,2-dioxa- and 1,3,2-oxathiaphospholanes in ether at -10 to 10° with α-halonitrosoalkanes to obtain chlorothiophosphates. The P-S linkages remained intact:

$$EtO(EtS)_2P + CCl_2FNO \rightarrow ClFC=NOP(O)(SEt)_2 + EtCl$$

$$EtO(EtS)PCl + CCl_2FNO \rightarrow ClFC=NO(EtS)P(O)Cl + EtCl$$

$$\ce{\underset{S}{\overset{O}{[}}PCl} + CCl_2FNO \rightarrow ClFC=NO(ClCH_2CH_2S)P(O)Cl$$

$$\ce{\underset{S}{\overset{O}{[}}PCl} + MeEtCClNO \rightarrow MeEtC=NO(ClCH_2CH_2S)P(O)Cl$$

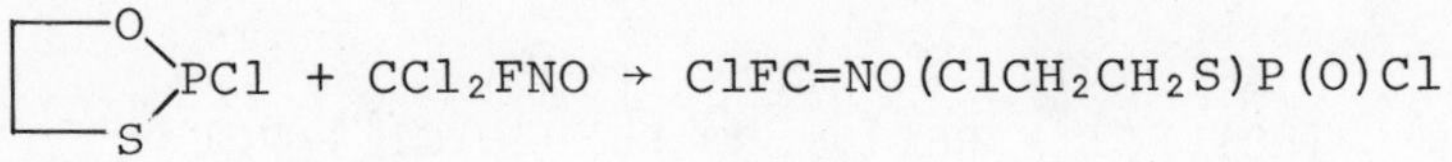

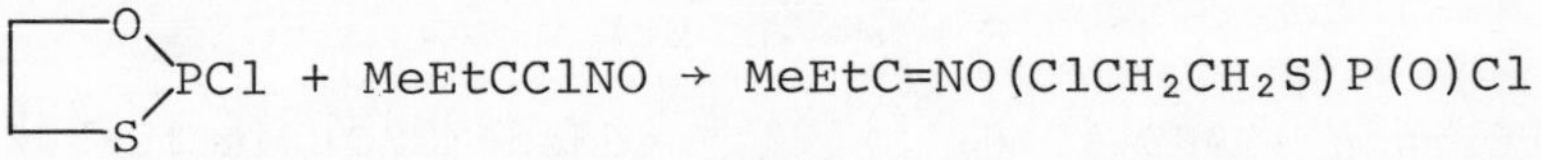

METHOD N. MISCELLANEOUS REACTIONS OF PCl_3, $P(O)Cl_3$, and $P(S)X_3$

Regel and Potts (1966, 1970)[1014, 1015] studied the reaction of PCl_3 with dialkyl disulfides in the presence of $AlCl_3$ at 0°. Treatment of the complex with ice water gave the thiophosphoryl chlorides:

$$RSSR + PCl_3 + AlCl_3 \rightarrow (RS)_2P(O)Cl$$

Petrov et al. (1961)[942] treated PCl_3 with sulfenamides in liquid SO_2 at -50° to obtain complexes that interacted with SO_2 to form alkyl dichlorophosphorothioates. These were difficult to separate from the mixture of other products:

$$PCl_3 + Et_2NSBu \xrightarrow[-50^O]{SO_2} \left[\underset{NEt_2}{BuSPCl_3} \right] \xrightarrow{SO_2} BuSP(O)Cl_2 + Et_2NSOCl$$

Abel, Armitage, and Bush (1964)[1] heated $P(O)Cl_3$ with ethylthiotrimethylsilane to evolve trimethylchlorosilane and leave a residue of ethyl phosphorodichloridothioate, which was difficult to purify by distillation because of the presence of $(EtS)_2P(O)Cl$ and $(EtS)_3P(O)$.

$$P(O)Cl_3 + EtSSiMe_3 \rightarrow EtSP(O)Cl_2 + Me_3SiCl$$

With a reactant ratio of 1:3, the main product was $(EtS)_3$-$P(O)$.

Gololobov (1967)[373] prepared vinyl esters of thiophosphorus acids by heating $P(S)Cl_3$ with carbonyl compounds in the presence of Et_3N. In the case of MeCHO, the mixing was done at 0° and the mixture held at room temperature for 2 days:

$$P(S)Cl_3 + MeCHO + Et_3N \rightarrow CH_2{=}CHOP(S)Cl_2$$

Kosolapoff (1951)[559] added ethylene oxide to PCl_3 in cold xylene. Sulfur was then added and the mixture stirred and heated to 75°:

$$PCl_3 + 2\overset{\text{O (ring)}}{CH_2{-}CH_2} \xrightarrow[\text{cold}]{\text{xylene}} \xrightarrow[75^{\circ}]{S} (ClCH_2CH_2O)_2P(S)Cl$$

Kolka (1959)[546a] treated 1,2-oxides or -sulfides with $P(S)X_3$ in the presence of a catalytic amount of Et_3N to obtain phosphorothioic halides:

$$P(S)Br_3 + 2Me\overset{\text{S (ring)}}{CH{-}CH_2} \xrightarrow{Et_3N} (MeCHBrCH_2S)_2P(S)Br$$

METHOD O. MISCELLANEOUS PREPARATIONS OF HALOPHOSPHOROTHIOATES

Birum (1958)[95] studied the reaction of phosphorus sesquisulfide with various aliphatic, aromatic, or heterocyclic sulfenyl chlorides and obtained mixtures of thiophosphates and trithiophosphates:

$$P_4S_3 + Cl_3CSCl \xrightarrow{80^{\circ}}$$
$$Cl_3CSPCl_2 + Cl_3CSP(S)Cl_2 + (Cl_3CS)_2P(S)Cl$$

Roesky (1969)[1042] treated metallic dithiophosphoryl difluorides in tetramethylene sulfone at 70-80° for 3 hr with organic halides to obtain alkyl, benzyl, or cyclohexenyl phosphorodifluoridodithioates:

$$CsS_2P(S)F_2 + RBr \rightarrow RSP(S)F_2 + CsBr$$

Mel'nikov and co-workers (1957)[752] have made use of magnesium ethylate in alcohol at 45^0 to convert alkyl phosphorodichloridothioates to O-ethyl O-alkyl phosphoro-chloridothionates:

$$2ROP(S)Cl_2 + (EtO)_2Mg \rightarrow 2EtO(RO)P(S)Cl + MgCl_2$$

Addition of a chlorophosphoryl sulfenyl chloride to vinyl propionate in ether at 40^0 formed an acyloxyalkyl phosphorochloridothioate [Vegter (1965)]:[1351]

$$EtO(ClS)P(O)Cl + CH_2=CHOCOEt \rightarrow EtO(EtOCOCHClCH_2S)P(O)Cl$$

Aichenegg and Gillen (1969)[4] converted haloalkylphos-phorodichloridothioates to the corresponding haloalkyl sulfinyl phosphoryl dichlorides using ozone in $CHCl_3$ at $5-15^0$:

$$Cl_2CHCHClSP(O)Cl_2 + O_3 \rightarrow Cl_2CHCHCl\overset{\displaystyle O}{\overset{\displaystyle \|}{S}}P(O)Cl_2$$

METHOD P. PREPARATION OF THIO- AND OXOPHOSPHORANE SUL-FENYL CHLORIDES

O,O-Dialkyl thiophosphoranesulfenyl chlorides may be produced by chlorination of phosphorodithioic or -thioic acids under mild conditions. They are stable enough to be distilled, but on heating they deposit sulfur and form the more stable dialkyl phosphorochloridothioates or dialkyl phosphorochloridates, respectively.

Hu (1956)[464] treated diethylphosphorodithioic acid with chlorine in $CHCl_3$ at -5^0 overnight to obtain diethyl thiophosphorane sulfenyl chloride:

$$(EtO)_2P(S)SH + Cl_2 \rightarrow (EtO)_2P(S)SCl + HCl$$

The same product was also obtainable from $[(EtO)_2P(S)S]_2S$ in CCl_4 in a violent reaction with chlorine.

Another method of preparing the above product involves treating piperidinium or morpholinium salts of the dithi-oic acid with dry HCl in CCl_4 [Almasi and co-workers (1964 1965)].[19,32] The amine hydrochloride is filtered off and the product distilled:

$$(EtO)_2P(S)SH \cdot HN\bigcirc \quad + HCl \rightarrow (EtO)_2P(S)SCl + \bigcirc NH \cdot HCl$$

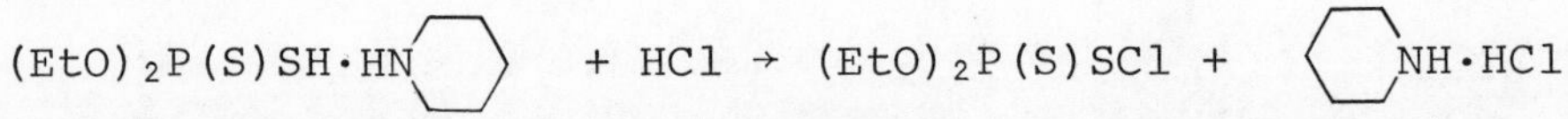

In a study of the preparation of O,O-diethyl oxophos-phorane sulfenyl chloride, Lenard-Borecka and Michalski (1957)[601] found that 76-78% yields were obtained using sulfuryl chloride in benzene at -5^0 on O,O-diethyl phos-phorothioic acid, its K salt or its disulfide:

$$(EtO)_2P(O)SH + SO_2Cl_2 \rightarrow (EtO)_2P(O)SCl + HCl + SO_2$$

The action of chlorine in CCl_4-benzene at -5° gave a 59% yield.

Continuing the above studies, Michalski (1959)[774] simplified the procedure by treating dialkyl phosphites with sulfur monochloride in benzene at -5 to 0°, and without isolating the product, introducing sulfuryl chloride at the same temperature (60% yield):

$$(EtO)_2P(O)H \xrightarrow{S_2Cl_2} \xrightarrow{SO_2Cl_2} (EtO)_2P(O)SCl$$

Michalski and co-workers (1963)[785] have also obtained the oxophosphorane sulfenyl halides (Br or Cl) in excellent yields by treating trialkyl phosphorothionates with halogens or sulfuryl chloride in benzene at -5 to 0°:

$$(RO)_3P(S) + SO_2Cl_2 \rightarrow (RO)_2P(O)SCl + RCl + SO_2$$

D. METHODS OF PREPARATION OF PRIMARY AND SECONDARY ESTERS AND THEIR SALTS

I. DISPLACEMENT OF HALOGEN FROM HALOTHIONOPHOSPHATES

Halothionophosphates are more resistant to hydrolysis than the corresponding oxo analogs. Thus they may be washed with dilute aqueous alkali without significant loss by hydrolysis.

Alcoholic alkalies are used to produce the phosphorothioate salts [Mastin (1945)]:[725]

$$ROP(S)Cl_2 + 4KOH \xrightarrow{\text{alcohol}} ROP(S)(OK)_2 + 2KCl + 2H_2O$$

$$(RO)_2P(S)Cl + 2KOH \rightarrow (RO)_2P(S)OK + KCl + H_2O$$

Reaction of KSH with $(RO)_2P(O)Cl$ gave the same $(RO)_2P(S)OK$ salt as was obtained from the thiono chloride and KOH. Analogous reactions with the thiono chlorides and KSH gave $ROP(S)(SK)_2$ and $(RO)_2P(S)SK$.

Delventhal and Kuchen (1971)[198] carried out the reaction:

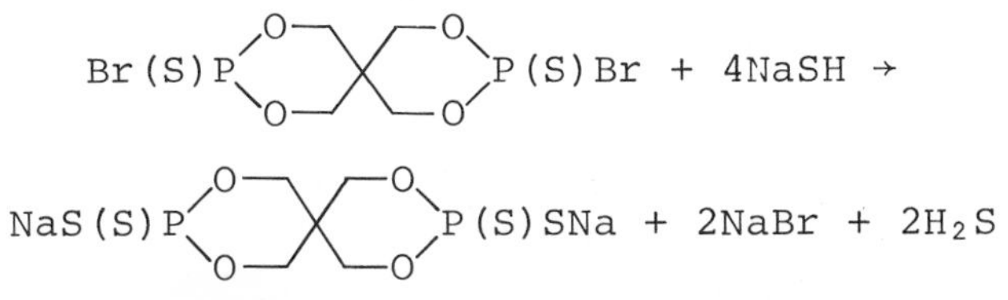

Aryl chloridothionophosphates are more resistant to hydrolysis than alkyl types. Miller (1960)[801] demonstrated a convenient procedure for accomplishing the hydrolysis of diphenylchloridothionophosphate using an aqueous-dioxane solution of potassium acetate. Evidently an intermediate mixed anhydride was formed which immediately reacted with another mole of potassium acetate to produce acetic anhydride and the phosphorothioate salt:

$$(PhO)_2P(S)Cl + AcOK \xrightarrow{-KCl} [(PhO)_2P(S)OOAc] \xrightarrow{AcOK} Ac_2O + (PhO)_2P(S)OK$$

Second-order kinetics were observed with $k = 3.53 \pm 0.09 \times 10^{-3}$ moles/liter/sec at 58^0 C. The analogous reaction with t-BuCOOK was also carried out with $k = 2.73 \pm 0.12 \times 10^{-3}$ moles/liter/sec at 58^0.

Hydrolysis of cyclic phosphorochloridothionates has been carried out using amines in place of alkalies. Cierpka (1968)[181] used triethylamine and acetic acid in the following reaction, in which the R's were H, halogen, or were paired in benzo rings:

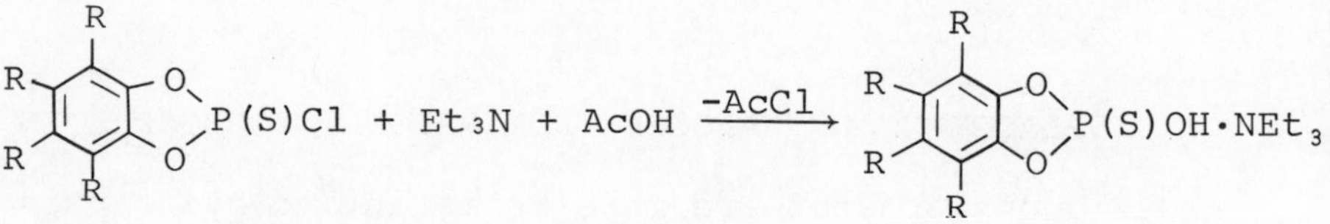

The reactions of cyclic phosphorochloridothioates with imidazole formed unstable imidazolides, which were treated with water to give stable imidazolium salts [Mikolajczyk and Schiebel (1969)].[800] The reactions were used in the preparation of geometrical isomers of 5- and 6-membered ring phosphorothioates:

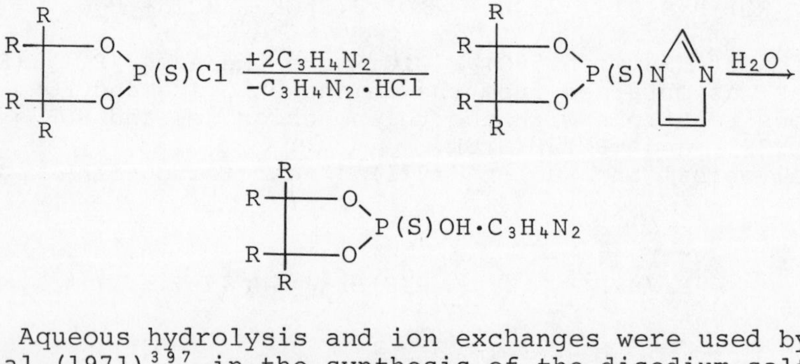

Aqueous hydrolysis and ion exchanges were used by Haga et al.(1971)[397] in the synthesis of the disodium salt of inosine 5'-phosphorothionate:

$$ROP(S)Cl_2 + H_2O \xrightarrow{NH_3} (NH_4)_2 \text{ salt} \rightarrow (NaO)_2P(S)OR$$

The analogous thiolo compound, $(NaO)_2P(O)SR$, was prepared from Na_3PO_3S and iodoinosine (see Sections D.VI and F.4.1).

Fife and Milstien (1969)[325] prepared S-(2-carboxyphenyl) phosphorothioate using the reactions:

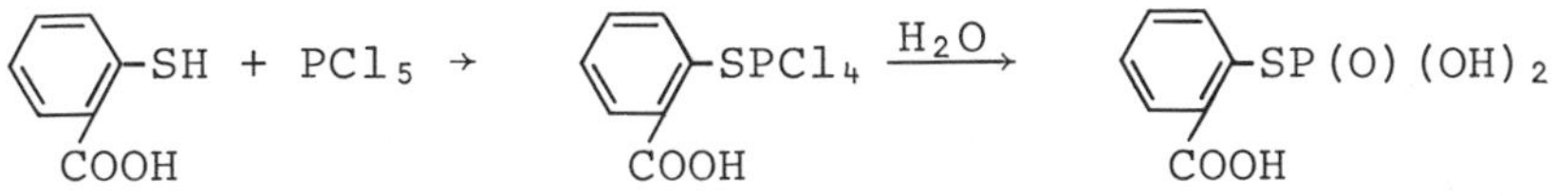

Hydrolysis of the final product was studied and compared with the 2-i-Pr and 4-Me analogs to demonstrate carboxyl-group participation in phosphorothioate hydrolysis.

Blair and Britton (1960)[104] used successive reactions of alcohols and water on phosphorodichloridothioates to obtain O-aryl O-alkyl O-hydrogen phosphorothioates:

$$ArOP(S)Cl_2 \xrightarrow{ROH} RO(ArO)P(S)Cl \xrightarrow{H_2O} RO(ArO)P(S)OH$$

The final products were characterized as amine salts.

Zemlyanskii et al.(1965)[1432, 1433] have used reactions of potassium hydrosulfide with dialkyl or diaryl phosphorochloridoselenoates to obtain phosphoroselenothioates:

$$(RO)_2P(Se)Cl + 2KSH \rightarrow (RO)_2P(Se)SK + KCl + H_2S$$

The reactions were carried out in dry toluene at reflux.

II. PARTIAL HYDROLYSIS OF TERTIARY ESTERS

Hydrolysis of tertiary esters is difficult to control in acidic media. Prolonged exposure yields phosphoric acid with total loss of sulfur. Thus Cowen (1956)[190] applied the acidic hydrolysis (also referred to in this case as a devinylation) to diethyl dithiophosphatosuccinic anhydride to prepare mercaptosuccinic acid. The phosphorothiomoiety was rejected as phosphoric acid and H_2S:

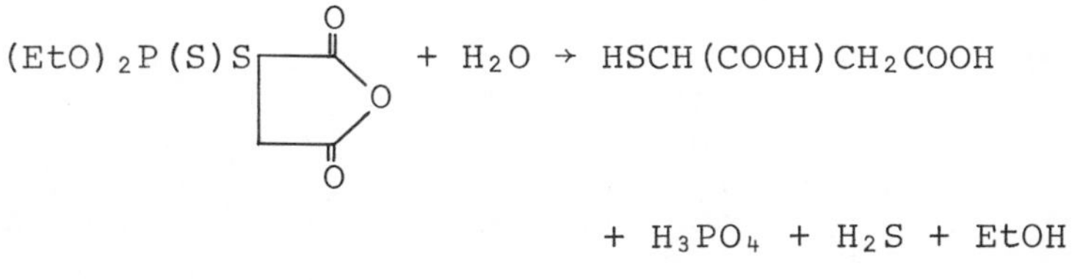

A novel approach to the synthesis of primary S-alkyl phosphorothioates involves preparation of di-t-butyl S-alkylphosphorothioates and removal of the t-Bu groups with

HCl in chloroform [Zwierzak and Gramze (1971)].[1457] The products were isolated as the dicyclohexylammonium salts:

$$(t\text{-}BuO)_2P(S)OH\cdot NEt_3 + RI \rightarrow (t\text{-}BuO)_2P(O)SR \xrightarrow[CHCl_3]{HCl}$$

$$\xrightarrow{(C_6H_{11})_2NH} 2t\text{-}BuCl + RSP(O)(OH)_2\cdot HN(C_6H_{11})_2$$

For preservation of the sulfur content and control of the number of groups removed from tertiary esters, the reactions are carried out in alcoholic alkalies or alkali hydrosulfides [Emmett and Jones (1911)].[253] [Pistschimuka (1912)].[961]

A solution of sodium hydroxide in ethanol was saturated with hydrogen sulfide and used to treat phosphorodithioates to prepare dithioate salts [Sumitomo Chemical (1969)]:[1285]

$$(EtO)_2P(S)SCH_2Ph + NaOH \xrightarrow{H_2S} EtO(PhCH_2S)P(S)ONa + EtOH$$

Vitenberg, Pesin, and Preser (1967)[1360] prepared potassium O,O-bis-p-nitrophenylphosphorothioate by the partial hydrolysis of O,O,O-tris-p-nitrophenylphosphorothioate in boiling 4:1 aqueous alcohol:

$$(4\text{-}NO_2C_6H_4O)_3P(S) + 2KOH \rightarrow$$

$$(4\text{-}NO_2C_6H_4O)_2P(O)SK + 4\text{-}NO_2C_6H_4OK + H_2O$$

Chernaya and Zemlyanskii (1966)[164] hydrolyzed O,O-diethylpropylphosphoroselenothioate, which contained a preponderance of the thiolic isomer, by refluxing the tertiary ester with 10% alcoholic potassium hydroxide for 30-35 hr:

$$(EtO)_2P(Se)SPr + KOH \rightarrow (EtO)_2P(Se)OK + PrSH$$

III. DEALKYLATION AND REALKYLATION OF TERT.-ESTERS OF PHOSPHORODITHIOIC OR -THIOIC ACIDS

Tertiary esters containing one or more methoxy groups attached to phosphorus (V) are readily demethylated by nucleophilic agents such as iodides or phosphorodithioates Itskova et al. (1968)[480] treated dimethyl salts with trimethyl esters with complete scrambling of methyl groups:

$$(MeO)_2P(S)SK + (MeO)_2P(S)SMe \rightarrow$$

$$(MeO)_2P(S)SMe + MeO(MeS)POSK$$

$$MeO(MeS)POSK + MeOP(O)(SMe)_2 \rightarrow$$

MeO(MeS)P(O)SMe + (MeS)$_2$P(O)OK

The reaction of sodium dimethylphosphorodithioate with O,O-dimethyl-S-4,6-diamino-s-triazin-2-yl phosphorodithioate was used to prepare the demethylated intermediate,(10), which was re-alkylated to give S-alkyl derivatives,(11) [Floyd (1966)].[336]

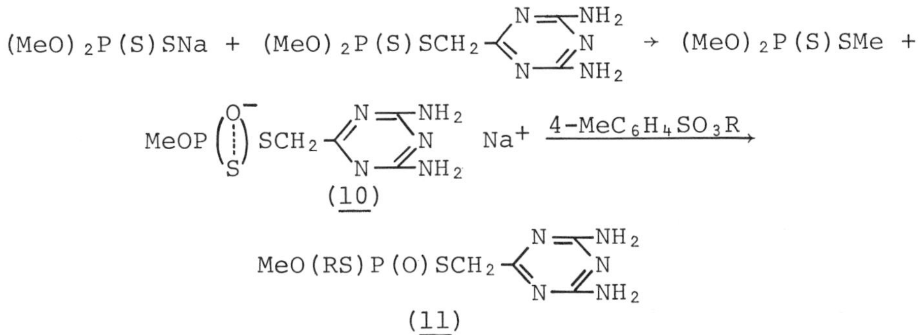

(MeO)$_2$P(S)SNa + (MeO)$_2$P(S)SCH$_2$–⟨triazine N=NH$_2$, N, N–NH$_2$⟩ → (MeO)$_2$P(S)SMe +

MeOP(O$^-$)(S)SCH$_2$–⟨N=NH$_2$, N, N–NH$_2$⟩ Na$^+$ $\xrightarrow{4\text{-MeC}_6\text{H}_4\text{SO}_3\text{R}}$

(10)

MeO(RS)P(O)SCH$_2$–⟨N=NH$_2$, N, N–NH$_2$⟩

(11)

When the nucleophile contained only groups larger than methoxy, reaction was easily controlled. The reaction is exemplified by trimethylphosphorodithioate plus potassium diisopropylphosphorodithioate [DuBreuil et al.(1964)].[192, 235] Negligible scrambling of alkyl groups occurs with the bulky isopropyl group in the salt:

(MeO)$_2$P(S)SMe + (i-PrO)$_2$P(S)SK →

MeO(MeS)POSK + (i-PrO)$_2$P(S)SMe

Remethylation using dimethyl sulfate gives S-methylation:

MeO(MeS)POSK + Me$_2$SO$_4$ → MeOP(O)(SMe)$_2$

Analogous reactions involving demethylating esters such as malathion or dimethioate with the diisopropyl phosphorodithioate salt and remethylating with Me$_2$SO$_4$ were used in preparing isomalathion, isodimethoate, and their homologs[192, 235], (see Sections E.XVII and F.3.2). The intermediate demethylated salts were noncrystalline.

Tertiary phosphorodithioates and -thioates react with tertiary amines to form quaternary ammonium salts of phosphorodithioic and -thioic acids. Hilgetag and Teichmann (1959),[437] Thuong (1966),[1307, 1308] and Mel'nikov et al. (1968)[754] used trimethyl amine, dimethyl aniline or N-substituted morpholine on various trimethyl esters:

(MeO)$_3$P(S) + Me$_2$NR → MeO(MeS)P(O)O$^-$ NRMe$_3$$^+$

(MeO)$_2$P(O)SMe + Me$_2$NR → MeO(MeS)P(O)O$^-$ NRMe$_3$$^+$

(MeS)$_3$P(O) + Me$_3$N → (MeS)$_2$P(O)S$^-$ NMe$_4$$^+$

$$(MeO)_3P(S) + PhCOCH_2CH_2N\overset{\frown}{\underset{\smile}{}}O \rightarrow$$

$$(MeO)_2P(O)S^- \;\; PhCOCH_2CH_2\overset{Me}{\overset{|}{N}}{}^+\overset{\frown}{\underset{\smile}{}}O$$

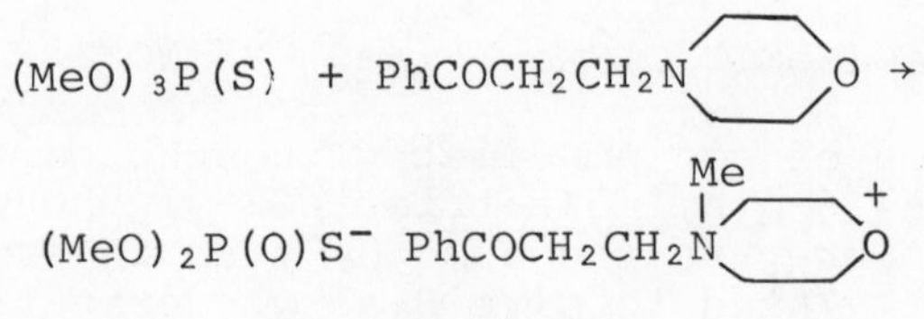

[R = Me or Ph]

The reaction has been applied to ring opening of dioxa-phospholanes and -phosphorinanes [Nguyen et al. (1970)]:[870]

$$RO(X)P\overset{O-}{\underset{O-}{}}\rangle + Me_3N \rightarrow ROP\overset{O}{\underset{X}{(\overset{|}{})}}{}^- \;\; OCH_2CH_2NMe_3{}^+$$

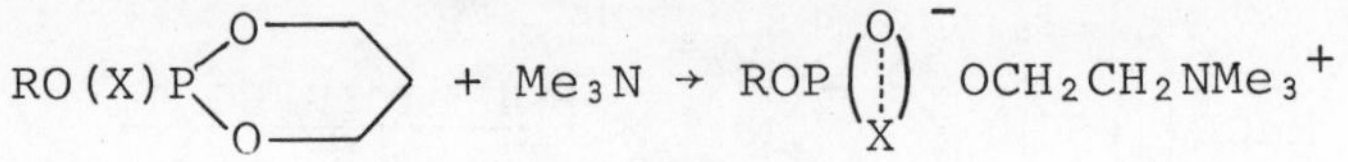

[X = S or Se]

Matyuk et al. (1967)[733] treated O,O,S-trimethylphosphorothioate with 2-(dialkylamino)ethyl alkyl carbamates to form oily salts:

$$(MeO)_2P(O)SMe + MeNHCOOCH_2CH_2NR_2 \rightarrow$$

$$MeO(MeS)P(O)O^- \;\; MeNHCOOCH_2CH_2NMeR_2{}^+$$

Thiono-thiolo isomerization has been observed in reactions with pyridine, quinoline, or hydroxyethylpiperidine in petroleum ether at room temperature for 30 days [Mel'-nikov and co-workers (1968)][744] or on heating with pyridine-N-oxide [Khaskin et al. (1971)]:[534]

$$(MeO)_2P(S)OEt + C_5H_5N \rightarrow EtO(MeS)P(O)O^- \;\; C_5H_5NMe^+$$

$$(MeO)_2P(S)SMe + C_9H_8N \rightarrow (MeS)_2P(O)O^- \;\; C_9H_8NMe^+$$

$$(MeO)_2P(S)SMe + HOCH_2CH_2N\overset{\frown}{\underset{\smile}{}}\rangle \longrightarrow$$

$$(MeS)_2P(O)O^- \;\; HOCH_2CH_2\overset{Me}{\overset{|}{N}}{}^+\overset{\frown}{\underset{\smile}{}}\rangle$$

$$(MeO)_2P(S)OR + \overset{\frown}{\underset{\smile}{}}N=O \rightarrow MeS(RO)P(O)O^- \;\; \overset{\frown}{\underset{\smile}{}}NOMe^+$$

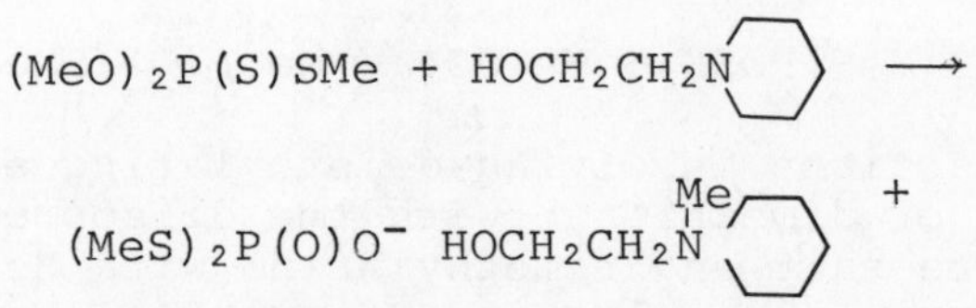

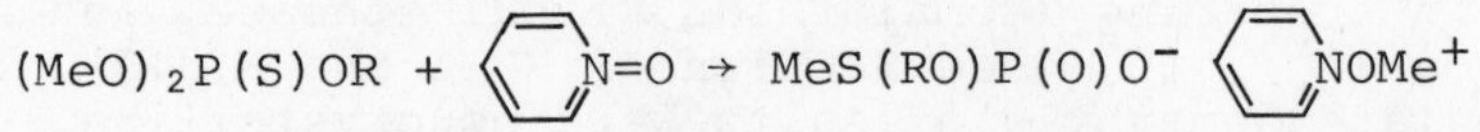

In this connection, Khaskin, Mel'nikov, and Sablina (1968)[533] observed that when excess of the thionophosphate was used, dialkylation of 4,4'-dipyridyl was accompanied by thiono-thiolo isomerization, but when molar quantities of the reactants were used to give monoalkylation, the isomerization did not occur:

$$2(MeO)_3P(S) + N\overset{\frown}{\underset{\smile}{}}-\overset{\frown}{\underset{\smile}{}}N \rightarrow$$

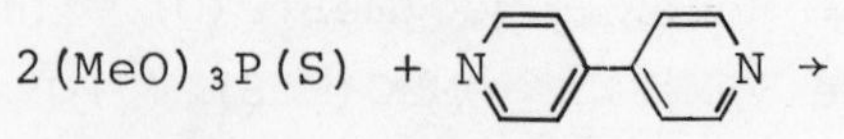

$[MeO(MeS)P(O)O^-]_2$ $[MeN\langle\bigcirc\rangle\langle\bigcirc\rangle NMe]^{++}$

$(MeO_3P(S)$ $+$ $N\langle\bigcirc\rangle\langle\bigcirc\rangle N$ $\rightarrow$

$(MeO)_2P(O)S^-$ $N\langle\bigcirc\rangle\langle\bigcirc\rangle NMe^+$

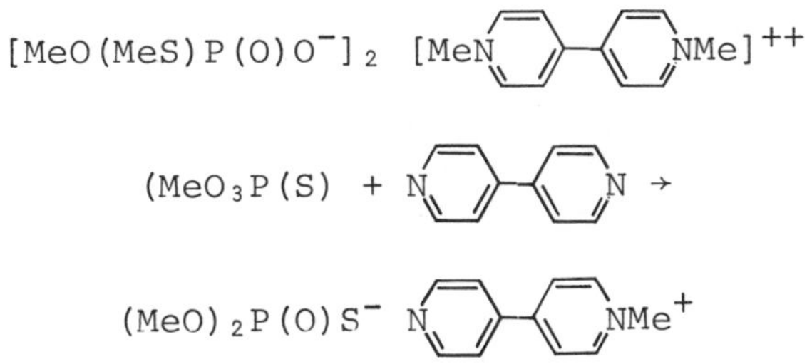

Tertiary phosphorodithioates and -thioates react with dimethyl sulfide to form sulfonium salts with thiono-thiolo isomerization [Hilgetag and co-workers (1959)]:[435][436]

$$(MeO)_3P(S) + Me_2S \xrightarrow[12\ hr]{90^O} MeO(MeS)P(O)O^- SMe_3^+$$

$$(MeO)_2P(O)SMe + Me_2S \rightarrow MeO(MeS)P(O)O^- SMe_3^+$$

$$(MeO)_2P(S)OPh + Me_2S \rightarrow PhO(MeS)P(O)O^- SMe_3^+$$

Sulfonium salts may also be produced from silver phosphorothioates and sulfonium iodides.[435]

$$(ArO)_2P(O)SAg + Me_3S^+ I^- \rightarrow (ArO)_2P(O)S^- SMe_3^+$$

O,O,O-Trimethylphosphorothionate reacts with thiourea to give a sulfonium salt [Hilgetag and Teichmann (1939)]:[437]

$$(MeO)_3P(S) + CS(NH_2)_2 \rightarrow MeO(MeS)P(O)O^- MeSC(NH_2)_2^+$$

Amino esters reacted with tertiary phosphorodithioates and -thioates [Shvetsova-Shilovskaya et al. (1964)].[1240] Thiono-thiolo isomerization was being investigated, but the thiono forms were reported:

$$(MeO)_3P(S) + MeNHCH_2COOEt \rightarrow (MeO)_2P(S)O^- Me_2NHCH_2COOEt^+$$

Reaction was much faster and occurred under milder conditions with methyl esters than with higher homologs. The products were oils.

Tertiary esters with lithium chloride and amines eliminate alkyl chloride and form amine salts [Lecocq and Todd (1954)]:[596]

$$(PhO)_2P(S)OPr + LiCl + C_6H_{11}NH_2 \xrightarrow[-LiCl]{+\ HCl}$$

$$PrCl + (PhO)_2P(O)SH \cdot H_2NC_6H_{11}$$

Amines can be used to cleave phosphorothioic anhydrides [Edmundson (1965)]:[247]

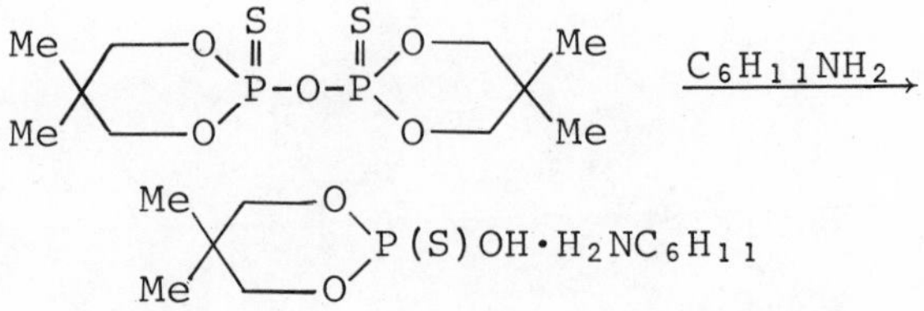

Ammonolysis of tetraethyl trisulfanebis(thiophosphoric acid) produced ammonium diethyl phosphorodithioate [Klement et al. (1964)]:[540]

$$[(EtO)_2P(S)]_2S_3 + NH_3\ (liq.) \xrightarrow[4\ wk]{R.T.} (EtO)_2P(S)SNH_4$$

IV. ADDITION OF SULFUR OR SELENIUM TO SECONDARY PHOSPHITES

Alkali metal salts of secondary phosphites readily add sulfur or selenium in anhydrous media [Foss (1947)]:[343]

$$(RO)_2PONa + S \xrightarrow{ether} (RO)_2POSNa$$

The use of metallic sodium or potassium to form the metal salts can be avoided by employing metallic salts of phenols. In the presence of sulfur the phosphorothioates are formed [Pesin and Khaletskii (1969)]:[934]

$$(EtO)_2P(O)H + ArONa \xrightarrow{-ArOH} (EtO)_2PONa \xrightarrow{S} (EtO)_2POSNa$$

The reaction with sulfur can also be carried out using sodium, potassium, or ammonium carbonates or acetates, where the formation of the intermediate phosphite salts was not observed.

Geometrical isomers of dioxaphosphorinanes have been prepared by Mikolajczyk (1969)[798] by adding sulfur or selenium to the phosphites in the presence of dicyclohexylamine or imidazole:

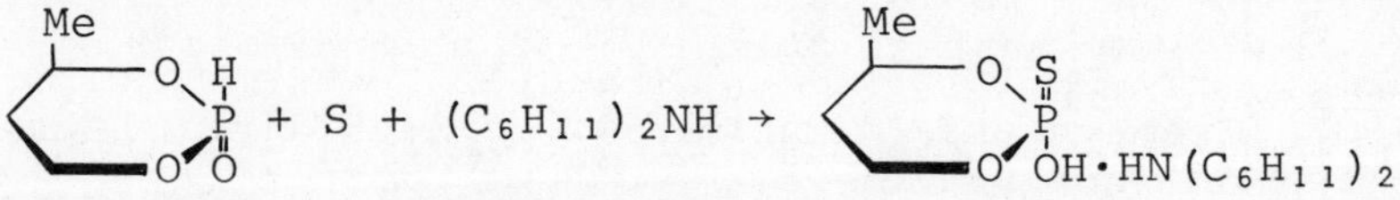

V. REACTIONS OF PHOSPHORUS SULFIDES AND SELENIDES WITH HYDROXY COMPOUNDS OR THIOLS

The early work (to 1945)[148, 725, 960, 1068, 1370] on the reactions of hydroxy and mercapto compounds with P_2S_5 was reviewed in the first edition (1950) "Organophosphorus Compounds," Chapter 9, Reaction XXII.

The commercially important reaction with alcoholic and phenolic materials is:

$$4ROH + P_2S_5 \rightarrow 2(RO)_2P(S)SH + H_2S$$

The reaction may be carried out by mixing the ingredients, with or without solvents; by adding P_2S_5 to hydroxy compounds, with or without solvent; or by adding the hydroxy compound to P_2S_5 in a stirred suspension in inert solvent, in batch or continuous processes. Most of the acids of Section F.3.2 were obtained as products of this reaction. The salts were prepared by neutralization with appropriate alkalies or amines, their bicarbonates or carbonates, with or without solvents.

Pudovik and Sergeeva (1955)[1003] prepared aromatic amine salts of O,O-diethylphosphorodithioic acid by mixing the components. Anils also reacted to give amine salts, the anil apparently being hydrolyzed by moisture:

$$(EtO)_2P(S)SH + 4\text{-}ClC_6H_4N=CHPh \xrightarrow[-PhCHO]{+ H_2O}$$

$$(EtO)_2P(S)SH \cdot H_2NC_6H_4Cl\text{-}4$$

Zemlyanskii and Glushkov (1967)[1438] found that yields in the reaction of P_2S_5 with di and trichlorophenols or tribromophenol were improved by the use of triethylamine in the phenol at reaction temperatures below 100°. Reactant mole ratios of P_2S_5:phenol:Et_3N were 4:1:2 or 2:1:2. The products were isolated as triethylamine salts:

$$4Cl_2C_6H_3OH + P_2S_5 + 2Et_3N \xrightarrow{-H_2S} 2(Cl_2C_6H_3O)_2P(S)SH \cdot NEt_3$$

The use of inert solvent (ether) and temperature control (35°) is recommended in preparing unsaturated phosphorodithioic acids [Mel'nik and Zemlyanskii (1970)].[739] The products were characterized as potassium and heavy metal salts:

$$4CH_2=CHCH_2OH + P_2S_5 \xrightarrow{ether} \xrightarrow{K_2CO_3} 2(CH_2=CHCH_2O)_2P(S)SK$$

When a solution of cholesterol in carbon disulfide was treated with P_2S_5, and the suspension boiled for 3 hr and cooled overnight at 0°, O,O-dicholesteryl phosphorodithioic acid, $(RO)_2P(S)SH$, was obtained as needles melting at $192\text{-}193^\circ$ [Cremlyn and Olson (1969)].[191] Earlier claims of formation of $(RS)_2P(O)OH$[1370] were not substantiated.

Glycols (with at least one secondary OH) added to suspensions of P_2S_5 in toluene at $95\text{-}100^\circ$ gave dioxaphospholane dithioic acids [Lanham (1956)].[583, 585, 587] The reaction can also be conducted in the absence of solvent:

$$2MeCHOHCH_2OH + P_2S_5 \rightarrow 2 \begin{array}{c} Me \\ \end{array} \overset{O}{\underset{O}{\boxed{}}} P(S)SH + H_2S$$

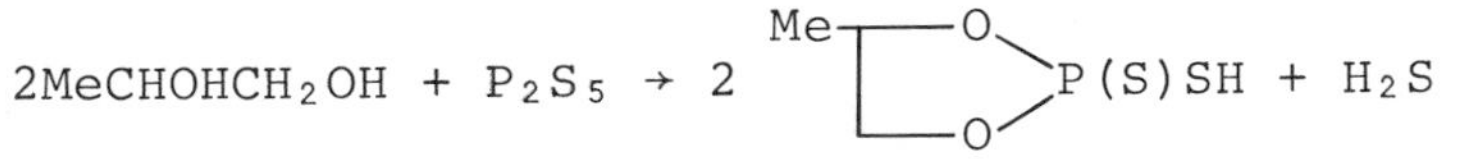

The 1,2 or 1,3 di-primary glycols, ethylene glycol and

trimethylene glycol, with P_2S_5 gave only polymers [Pudovik (1967)].[987] Yields of dioxaphospholanes more than doubled in going from the mono-secondary glycol, $MeCHOHCH_2OH$, to the di-secondary glycol, MeCHOHCHOHMe. The latter also reacted with $(EtO)_2P(S)SH$ to give 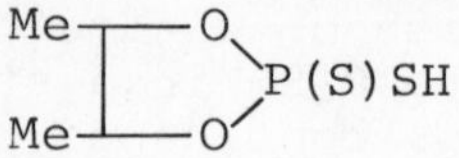 in unstated yield.

For the reactions with 1,3-butanediol and 2,5-hexanediol, the glycol and P_2S_5 were stirred and heated in toluene at 50-60° to obtain substituted phosphorinanes and phosphepanes [Zemlyanskii (1967)]:[1441]

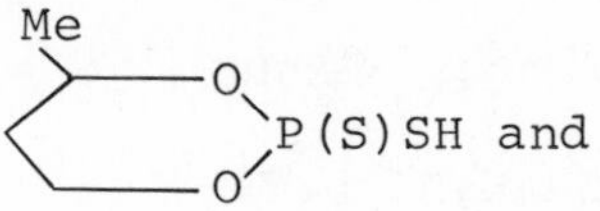

and

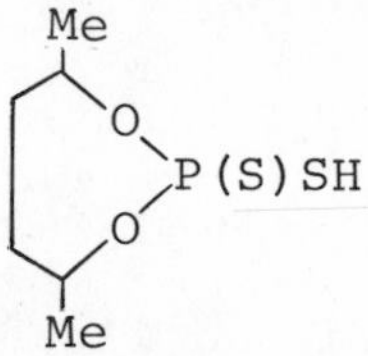

The reactions of aminoalcohols with P_2S_5 produce inner salts of dithioic and trithioic acids [Gupalo and Zemlyanskii (1967)]:[387, 398]

$$R_2N(CH_2)_nOH + P_2S_5 \rightarrow \xrightarrow{KOH}$$

$$[R_2N(CH_2)_nO]_2P(S)SK \text{ and } R_2N(CH_2)_nOP(S)(SK)_2$$

The products were separated by fractional crystallization of their potassium salts. Weakly basic aminoalkanols formed dithioates while increasing basicity increased the yield of (less soluble) trithioates.

Tertiary alcohols and benzyl alcohol with P_2S_5 at 80-100° gave olefins and sulfides. However, at controlled temperatures of 45-60° the unstable acids, $(t\text{-}BuO)_2P(S)SH$ and $(PhCH_2O)_2P(S)SH$, were isolated in the form of their lead salts [Hu and Chen (1958)].[463]

Adding an alcohol plus Et_3N to P_2S_5 in the mole ratio of 3:3:1 in boiling ether or benzene gave good yields of primary trithioates, isolated as potassium salts, ROP(S)-$(SK)_2$ [Kotovich et al. (1967)].[561] These were alkylated with alkyl bromides or iodides to give the expected tertiary esters, $ROP(S)(SR)_2$. They were also treated with alcohols or phenols and gaseous HCl to give secondary acids:

$$ROP(S)(SK)_2 + R'OH + 2HCl \rightarrow RO(R'O)P(S)SH + 2KCl + H_2S$$

Rosnati (1946)[1046] studied the reactions of thio-p-cresol and thiophenol with P_2S_5. The products were H_2S, aryl trithiometaphosphates, $ArSPS_2$ which is a dimer,[679b], and triaryltetrathiophosphates, $(ArS)_3P(S)$. The former reacted with phenols to liberate thiophenols and form diarylphosphorodithioic acids:

$$2ArOH + ArSPS_2 \rightarrow (ArO)_2P(S)SH + ArSH$$

The reactions offered no preparative advantages over the method of Cambi (1944),[148] using ArOH and P_2S_5.

Heating P_2S_5 with excess MeSH at 200-240° (pressure reaction) gave $RSPS_2$ [Scott (1956)].[1209] This in turn was caused to react with PrSH to give MeS(PrS)P(S)SH (not isolated) but which was treated with n-octene to give MeS-(PrS)$(C_8H_{17}S)$P(S).

The reactions of alcohols with P_4S_7 are of limited synthetic value. For alcohols above methyl, at least three liquid products may be expected: $(RO)_2P(S)H$, $(RO)_2P(S)SH$, and $(RO)_2P(S)SR$. With MeOH, McIvor, McCarthy, and Grant (1956)[675] prepared $(MeO)_2P(S)SMe$ with $(MeO)_2P(S)H$ as the major co-product.

In the presence of Et_3N in ether with the mole ratio P_4S_7:ROH:Et_3N of 1:5:5, Murav'ev, Zemlyanskii, and Panov (1968)[852] found that the primary products were $(RO)_2P(S)H$ and $ROP(S)(SH{\cdot}NEt_3)_2$. Alkylation with the corresponding RBr then gave good yields of the phosphorotrithioates, $(RS)_2P(S)OR$:

R	%$(RO)_2P(S)H$	%$(RS)_2P(S)OR$
Me	50% $(MeO)_2P(O)H$	73
Et	92% $(RO)_2P(S)H$	74
i-Pr	74	48
Bu	80	72

Alcohols have been shown by Kudchadker et al. (1968)[566] to react with P_2Se_5 to give unstable phosphorodiselenoic acids, $(RO)_2P(Se)SeH$. With exclusion of air and moisture, these were converted to potassium salts and chromium complexes. When the salts of the acids were not isolated from the reaction mixtures, derivatives having bridged selenium atoms, such as $(i\text{-}PrO)_4P_2Se_4$, $(C_6H_{11}O)_4P_2Se_4$, and $(EtO)_4P_2Se_5$ were obtained.

VI. REACTIONS OF Na_3PO_3S

Reactions of trisodium phosphorothioate with varied aminoalkylene bromides have been used in the studies on antiradiation agents [Johnston, Piper, and co-workers (1966-1971)].[486, 487, 957, 958] The reaction may be exemplified by:

$$Na_3PO_3S + H_2N(CH_2)_nNHCH_2CH_2Br \cdot 2\ HBr \rightarrow$$
$$3NaBr + H_2N(CH_2)_nCH_2CH_2SP(O)(OH)_2$$

Final products may be isolated in the form of mono- or disodium salts.

Airapetyan et al. (1967)[8] cleaved β-lactones with trisodium phosphorothioate to prepare S-substituted β-mercapto acids:

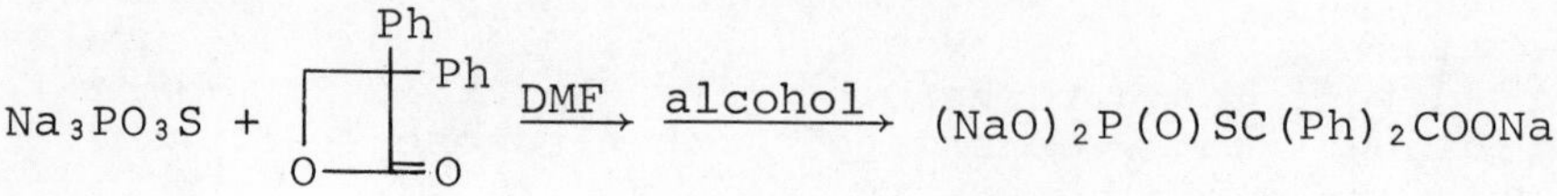

Haga et al. (1971)[397] reacted trisodium phosphorothioate with 5'-deoxy-5'-iodoinosine to give the disodium salt of 5'-deoxy-5'-thioinosine-5'-phosphorothioate (Section F.4.1):

$$Na_3PO_3S + RI \rightarrow NaI + (NaO)_2P(O)SR \quad (R = \text{inosine})$$

VII. MISCELLANEOUS PREPARATIONS OF SECONDARY ESTERS

(a) A reaction of possible synthetic interest was observed by Michalski and co-workers (1959)[795] involving sodium dialkyl phosphites and dialkyl or diaryl disulfides. The products were isolated as cyclohexylamine salts:

$$(EtO)_2PONa + RSSR \xrightarrow[\text{benzene}]{\text{boiling}} \xrightarrow{C_6H_{11}NH_2}$$
$$EtO(RS)P(O)OH \cdot C_6H_{11}NH_2$$

(b) Some reactions that have as a primary purpose the preparation of $[(RO)_2P(S)S]_2S$, (12), also give amine salts of dialkyl phosphorodithioic acids as co-products [Almasi and Hantz (1966)]:[22, 23]

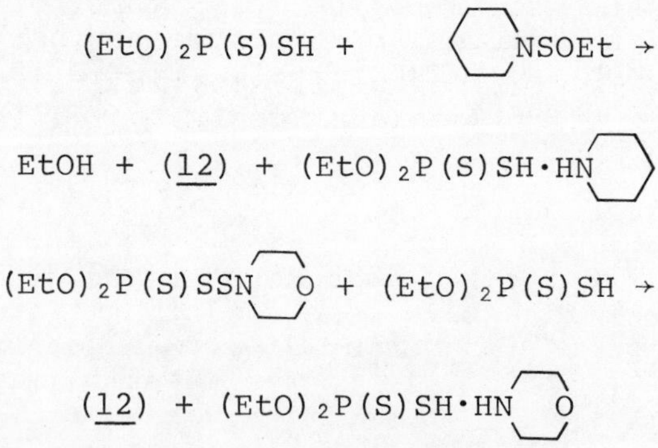

(c) Reactions of dipotassium O-alkyl phosphorotrithioates with o-, m-, and p-dihydric phenols in the presence of HCl in ether gave two types of products: 1:1 and 2:1 substitutions on the phenols [Kolesnikova, Zemlyanskii, and Kotovich (1971)]:[546]

$$\mathrm{EtOP(S)(SK)_2 + C_6H_4(OH)_2 \rightarrow EtO(HOC_6H_4O)P(S)SK + KSH}$$

$$\mathrm{2EtOP(S)(SK)_2 + C_6H_4(OH)_2 \rightarrow}$$
$$\mathrm{EtO(KS)P(S)OC_6H_4OP(S)(SK)OEt + 2KSH}$$

With the indicated reactant ratios, both types of products were obtained. Treatment with potassium carbonate and fractional crystallization from alcohol-acetone separated the products. The 1:1 products were hygroscopic and soluble in alcohol.

(d) Nucleoside phosphorothioates and a dinucleoside phosphorothionate have been prepared by Cook, Holman, and Nussbaum (1969)[183] and by Eckstein (1966)[243, 244] from pyridinium S-ethyl phosphorothioate and a variety of protected nucleosides using dicyclohexylcarbodiimide (DCC) as condensing agent (Sections F.4.1 and F.4.2).

(e) Truchlik and Masek (1956)[1322] have demonstrated the transesterification of dimethyl- or diethyl phosphorodithioic acid with alcohols of higher boiling point:

$$\mathrm{(MeO)_2P(S)SH + 2ROH \rightarrow 2MeOH + (RO)_2P(S)SH}\ [R = i\text{-Pr, Pr}]$$

$$\mathrm{(EtO)_2P(S)SH + 2ROH \rightarrow 2EtOH + (RO)_2P(S)SH}\ [R = \text{Bu, Lauryl}]$$

(f) It has been shown that diethyl phosphorodithioic acid in solution in alcohol at 50° for several hours reacts to form diethyl phosphorothioic acid and O,O,S-triethylphosphorodithioate in approximately 1:1 mole ratio [Ailman (1963)].[5] Some $\mathrm{(EtO)_3P(S)}$ was also observed:

$$\mathrm{2(EtO)_2P(S)SH + EtOH \rightarrow (EtO)_2POSH + (EtO)_2P(S)SEt + H_2S}$$

(g) One equivalent of p-nitrophenyl dihydrogen phosphate and excess 2-mercaptopyridine were treated with three equivalents each of triphenylphosphine and 2,2'-dipyridyl disulfide in acetonitrile at room temperature for 3 hr to give a 95% yield of p-nitrophenyl S-(2-pyridyl) phosphorothioate [Mukaiyama and Hashimoto (1971)]:[848]

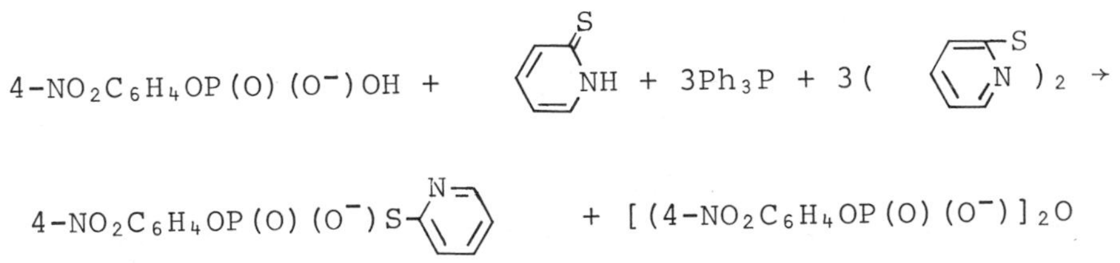

$+ Ph_3P(O) +$ 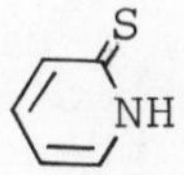

(h) The distillation of crude tri-t-butylphosphorothionate was rendered difficult due to evolution of isobutylene and formation of di-t-butylphosphorothioic acid in an exothermic reaction above about 85° [Mark and Van Wazer (1964)]:[716]

$$(t\text{-}BuO)_3P(S) \xrightarrow{85^o} Me_2C{=}CH_2 + (t\text{-}BuO)_2POSH$$

E. METHODS OF PREPARATION OF TERTIARY ESTERS

VIII. REACTION OF ACTIVE-H COMPOUNDS WITH $P(S)Cl_3$, $ROP(S)Cl_2$, OR $(RO)_2P(S)Cl$ IN THE PRESENCE OF ACID ACCEPTORS

This commercially important reaction is widely used (especially with $(RO)_2P(S)Cl$) to prepare a large portion of the products of Section F.4.4. The active-H compounds may be alcohols, phenols, thiols, or selenols. The acid acceptors may be preformed alkali metal or ammonium salts, or reaction may be conducted in the presence of a base. The reactions may be run in solution or in solution-suspension in heterogeneous systems usually at temperatures from about -20° to 100°C. The products are usually tertiary thiono esters or their dithio or seleno analogs. In some instances, particularly at higher temperatures, thiono-thiolo isomerization (Reaction XXXI) may occur to produce thiolo compounds of Section F.4.3.

Rates of displacement of chloride in O,O-diphenyl phosphorochloridothioate by anions in t-BuOH at 58° were measured by Miller, (1962).[803] Rates were related to the nucleophilicity of attacking anions:

$$(PhO)_2P(S)Cl + HA \rightarrow (PhO)_2P(S)A + HCl$$

Summary: A and k_2 at 58° in t-BuOH at 58°: liter/mole/sec x 10^2: t-BuO, 121; AmS, 35.7; PhO, 26.7; 2-pyridyloxy, 14.5; $4\text{-}MeC_6H_4O$, 9.20; $4\text{-}NCC_6H_4O$, 7.75; PhS, 4.48; 2-pyrazinyloxy, 3.11; 2-pyridylthio, 0.517; Me_3CCO, 0.283; MeCOO, 0.343. The rates for the following oximes were measured at 20°: $Me_2C{=}NO$, 33.3; $Ph_2C{=}NO$, 81.3. Rates increased with basicity of attacking anion without regard to S or O anions. The high reactivities of the oximes were attributed to bifunctional catalysis.

The following examples demonstrate the variety of reactions and products included in Method VIII.

Fletcher (1950):[331]

$(EtO)_2P(S)Cl + 4-Cl-2-NO_2C_6H_3OH \xrightarrow[Na_2CO_3]{MeCOEt}$

$(EtO)_2P(S)OC_6H_4-4-Cl-2-NO_2 + HCl$

Schrader (1952):[1125]

$(EtO)_2P(S)Cl + NaOC_6H_4NO_2-4 \xrightarrow[Na_2CO_3\ 80^0]{PhCl,\ Cu,\ KBr}$

$(EtO)_2P(S)OC_6H_4NO_2-4 + NaCl$

Bakanova, Mandel'baum, Mel'nikov (1956):[76] The following reaction was described as a disproportionation and transesterification:

$(MeO)_2P(S)Cl + HOC_6H_4NO_2-4 \xrightarrow[125-130^0]{C_5H_5N\cdot HCl\ (cat.)}$

$(4-NO_2C_6H_4O)_2P(O)SMe + HCl$

Schrader, Mühlmann (1953):[1201]

$(RO)_2P(S)Cl + HOC_6H_4CH_2CH=CH_2-4 \xrightarrow[K_2CO_3]{MeCOEt,\ Cu}$

$(RO)_2P(S)OC_6H_4CH_2CH=CH_2-4 + HCl$

Ketelaar, Gersmann (1950):[527]

$(4-NO_2C_6H_4O)_2P(S)Cl + EtONa \xrightarrow{acetone}$

$(4-NO_2C_6H_4O)_2P(S)OEt + NaCl$

Yamasaki (1954):[1406]

$(EtO)_2P(S)Cl + ROCSSM \rightarrow (EtO)_2P(S)SC(S)OR + MCl$

Schrader, Mühlmann (1955):[1200]

$3(MeO)_2P(S)Cl + HOCH_2CHOHCH_2OH \xrightarrow[40^0]{C_5H_5N}$

$[(MeO)_2P(S)OCH_2]_2CHOP(S)(OMe)_2 + 3HCl$

Schrader (1956):[1133]

$(EtO)_2P(S)Cl + Et_2NCH_2CH_2SNa \xrightarrow{benzene}$

$(EtO)_2P(S)SCH_2CH_2NEt_2 + NaCl$

Sallman (1957):[1066]

Me—O
    ⟩P(S)Cl + NaSCH_2CH_2SC_6H_11 $\xrightarrow{PhH}$
—O

Me—O
    ⟩P(S)SCH_2CH_2SC_6H_11 + NaCl
—O

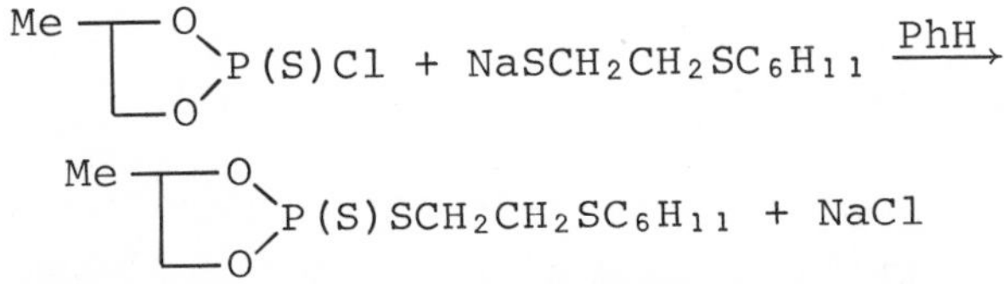

Nishizawa et al. (1960):[884, 887]

$(MeO)_2P(S)Cl + Cl_2CHCOONa \rightarrow (MeO)_2P(S)OCOCHCl_2 + NaCl$

Tichy et al. (1957):[1309]

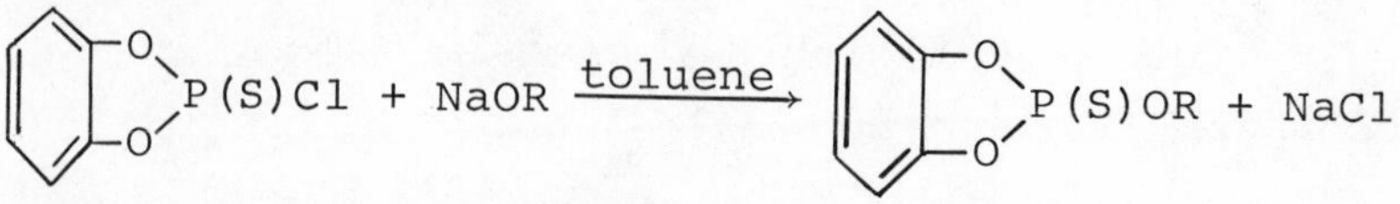

Mel'nikov et al. (1958):[757]

$$(\text{RO})_2\text{P(S)Cl} + \text{O}= \xrightarrow[\text{K}_2\text{CO}_3]{\text{benzene}} (\text{RO})_2\text{P(S)O}- + \text{HCl}$$

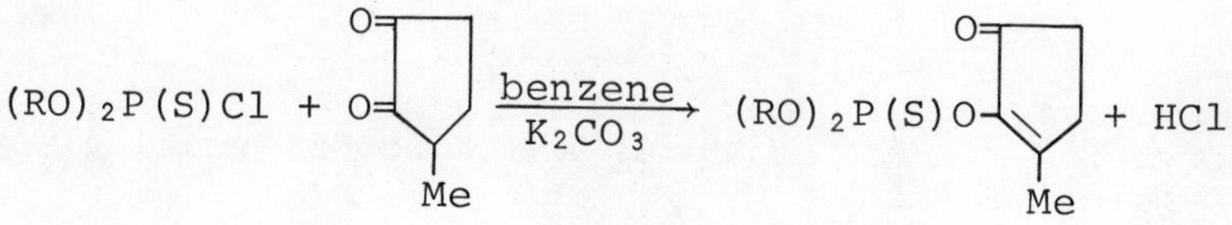

Nishizawa (1961):[882]

$$(EtO)_2P(S)Cl + NaOC(Ph){=}CHCOMe \rightarrow$$
$$(EtO)_2P(S)OC(Ph){=}CHCOMe + NaCl$$

Schrader (1963):[1170]

$$(EtO)P(S)Cl + HOCH_2P(S)(OEt)_2 \xrightarrow[40-50^\circ]{C_5H_5N}$$
$$(EtO)_2P(S)OCH_2P(OEt)_2 + HCl$$

Farbenfabriken Bayer (1959):[287]

$$(EtO)_2PCl + S + HOCH_2CH_2ONO_2 \xrightarrow[\text{toluene, } 25^\circ]{C_5H_5N}$$
$$(EtO)_2P(S)OCH_2CH_2ONO_2 + HCl$$

Nuretdinov (1970):[891] EtSNa or EtSH + Et_3N:

$$(PhO)_2P(Se)Cl + EtSNa \rightarrow (PhO)_2P(Se)SEt + NaCl$$

Schrader (1962):[1167] reaction mixture treated with H_2O_2:

$$(EtO)EtS)P(O)Cl + HOC_6H_4Cl\text{-}4 \xrightarrow[\text{toluene, } 40^\circ]{C_5H_5N}$$
$$EtO(EtS)P(O)OC_6H_4Cl\text{-}4 + HCl$$

Drake, Erbel (1952):[229a]

$$2,4,6\text{-}Cl_3C_6H_2OP(S)Cl_2 + 2ROH \xrightarrow{C_5H_5N}$$
$$(RO)_2P(S)OC_6H_2Cl_3\text{-}2,4,6 + 2HCl$$

Tolkmith (1954):[1313]

$$2,4,5\text{-}Cl_3C_6H_2OP(S)Cl + EtOH + HC{\equiv}CCH_2OH \xrightarrow{Et_3N}$$
$$EtO(HC{\equiv}CCH_2O)P(S)OC_6H_2Cl_3\text{-}2,4,5 + 2HCl$$

Britton, Blair (1960):[126]

$$ArOP(S)Cl_2 + 2NaOH \text{ in } MeOH \xrightarrow[-4^\circ]{MeOH} (MeO)_2P(S)OAr + 2NaCl$$

Farbenfabriken Bayer (1963):[298]

$$3\text{-Me-4-(MeS)}C_6H_3OP(S)Cl_2 + NaOEt + NaSEt \rightarrow$$
$$EtO(EtS)P(S)OC_6H_3\text{-3-Me-4-SMe} + 2NaCl$$

Edmundson, Lambie (1966):[250]

$$4\text{-}NO_2C_6H_4OP(S)Cl_2 + HOCH_2CH_2CH_2OH \xrightarrow[\text{toluene}\ 100^{\circ}]{Et_3N}$$

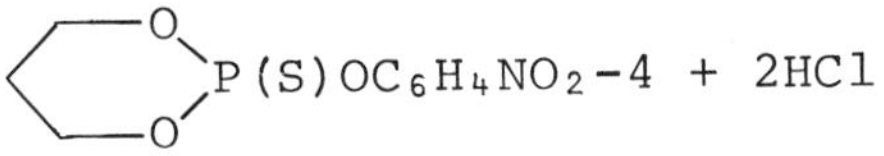

Buchner, Jacoves (1967):[135]

$$2\text{-}C_{10}H_7OP(S)Cl_2 + 2MeOH \xrightarrow[\text{at 200 mm}]{\text{reflux}}$$
$$(MeO)_2P(S)OC_{10}H_7\text{-2} + 2HCl$$

Sumitomo (1967):[1284]

$RSP(S)Cl_2 +$ $\xrightarrow[\text{aq. NaOH}]{\text{toluene}}$ $+ 2HCl$

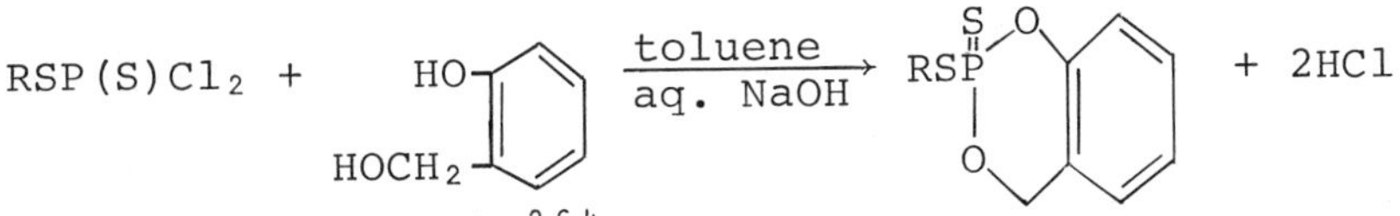

Eto et al. (1968):[264]

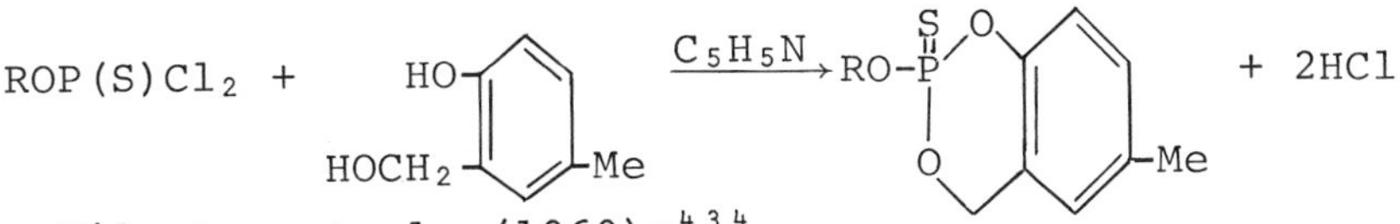

Hilgetag et al. (1960):[434]

$$MeSP(O)Cl_2 + 2Na \text{ in } MeOH \xrightarrow{0^{\circ}} (MeO)_2P(O)SMe + 2NaCl$$

Markley, Larson (1955):[719]

$P(S)Cl_3 + MeCH\text{-}CH_2 \xrightarrow[60\text{-}80^0]{PCl_3 \text{ (cat.)}} (MeCHClCH_2O)_3P(S)$

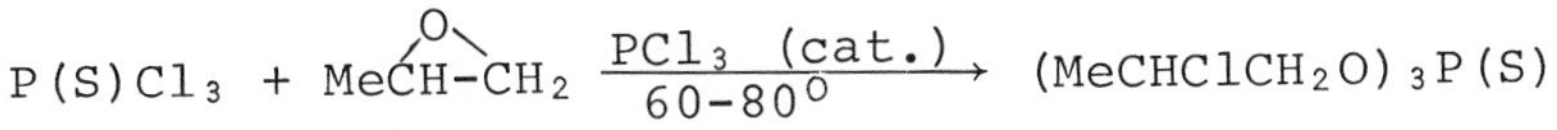

Blackman, Dewar (1957):[99]

$$P(S)Cl_3 + C_{12}H_{25}SNa \xrightarrow[\text{(trace)}]{(C_{12}H_{25}S)_3P} (C_{12}H_{25}S)_3P(S) + 3NaCl$$

Reznik et al. (1967):[1017]

$P(S)Cl_3 + 2NaO$ $+ MeONa \xrightarrow{\text{benzene}}$

$MeOP(S)(O$ $)_2 + 3NaCl$

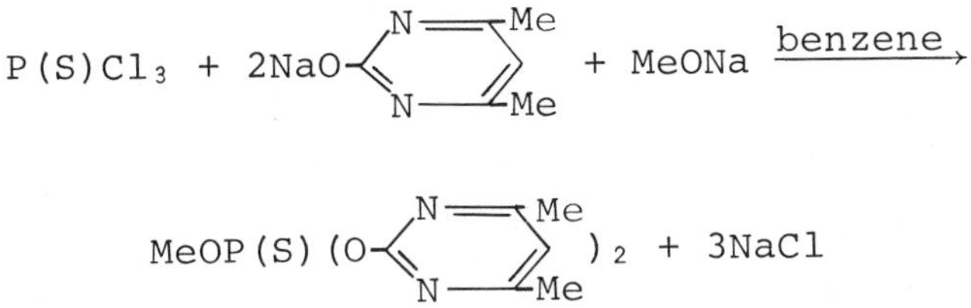

Dye (1946):[242] EtOH added to PCl_3 plus sulfur in benzene forms $(EtO)_2P(S)Cl$ and $(EtO)_3P(S)$.

IX. REACTIONS OF THIOLS WITH $(RO)_2P(O)Cl$, $ROP(O)Cl_2$, OR $P(O)Cl_3$

The oxophosphoryl halides are generally much more reactive than their thiono analogs of Method VIII. Solvents may be used to dilute the reactants and control the temperature. Gradual addition of one of the components also helps to control the reactions. With $P(O)Cl_3$ or $ROP(O)Cl_2$, incomplete displacement of the halogen produces mixtures that are frequently difficult to separate by distillation.

Examples of Method IX include:

Ghosh (1956):[366]

$$(EtO)_2P(O)Cl + Et_2NCH_2CH_2SNa \xrightarrow[\text{reflux}]{\text{benzene}} (EtO)_2P(O)SCH_2CH_2NEt_2 + NaCl$$

Sallmann (1957):[1064, 1065]

$(EtO)_2P(O)Cl$ + NaS– 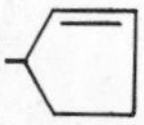 $\xrightarrow[20-30^{\circ}]{\text{benzene}}$ $(EtO)_2P(O)S$– 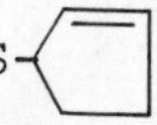 + NaCl

$$(EtO)_2P(O)Cl + NaSCH_2CH_2C_6H_{11} \xrightarrow[10-20^{\circ}]{\text{toluene}} (EtO)_2P(O)SCH_2CH_2C_6H_{11} + NaCl$$

Malinowskii, Solomko (1960):[685]

$$(EtO)_2P(O)Cl + NaSC(S)OR \xrightarrow{\text{ether}} (EtO)_2P(O)SC(S)OR + NaCl$$

Fiszer et al. (1953):[327]

$$(EtO)_2P(O)Cl + (EtO)_2POSNa \xrightarrow{\text{ether}} (EtO)_2P(O)OP(S)(OEt)_2 + NaCl$$

Sirrenberg, Coelln (1968):[1250]

$$Cl_2C{=}CHOP(O)Cl_2 + ROH + R'SH \xrightarrow[-10\text{ to }-5^{\circ}]{\text{toluene}} RO(R'S)P(O)OCH{=}CCl_2 + 2HCl$$

Farbenfabriken Bayer (1967);[310] Inoue et al. (1969):[472]

$$EtOP(O)Cl_2 + 2NaSAr \xrightarrow[5-10^{\circ}]{\text{benzene}} EtOP(O)(SAr)_2 + 2NaCl$$

Aichenegg, Gillen (1969):[3, 4]

$$EtOPCl_2 + Cl_2C{=}CHSCl \xrightarrow[30-32^{\circ}]{CCl_4} Cl_2C{=}CHSP(O)Cl_2 + EtCl$$

$$Cl_2C{=}CHSP(O)Cl_2 + 2\ 4\text{-}ClC_6H_4SH \xrightarrow[CCl_4,\ 25-30^{\circ}]{Et_3N}$$

$$(4\text{-}ClC_6H_4S)_2P(O)SCH{=}CCl_2 + 2HCl$$

Schrader, Scheinpflug (1969):[1202]

$$\text{cyclo-}C_6H_{11}OP(O)Cl_2 + 2PhSH \xrightarrow[K_2CO_3]{MeCN}$$

$$(PhS)_2P(O)OC_6H_{11}\text{-cyclo} + 2HCl$$

Chapman, Saunders (1948):[159]

$$P(O)Cl_2F + 3EtSNa \xrightarrow{\text{cool}} (EtS)_3P(O) + 2NaCl + NaF$$

Blackman, Dewar (1957):[99]

$$P(O)Cl_3 + 3C_{12}H_{25}SNa \xrightarrow{\text{ether}} (C_{12}H_{25}S)_3P(O) + 3NaCl$$

Abel et al. (1964):[1]

$$P(O)Cl_3 + 3Me_3SiSEt \xrightarrow{\text{heat}} (EtS)_3P(O) + 3Me_3SiCl \text{ (gas)}$$

X. REACTIONS OF OLEFIN OXIDES AND SULFIDES

De Witt (1958)[204] added propylene oxide to PCl_3 at temperatures below 35°, followed by gradual addition of sulfur flowers at 60°.

$$PCl_3 + MeCH\overset{O}{-}CH_2 \longrightarrow \xrightarrow{S} (MeCHClCH_2O)_3P(S)$$

XI. REACTIONS OF DIALKYL PHOSPHORODITHIOIC ACIDS WITH ETHYLENE OXIDE OR ETHYLENE SULFIDE

O,O-Dialkyl-S-hydroxyethylphosphorodithioates are formed when ethylene oxide is passed into the dithioic acids at about 30°. Analogous products are obtained when ethylene sulfide or ethylene imine are used in place of the oxide.

Examples include:

Kabachnik et al. (1955):[493]

$$(RO)_2P(S)SH + CH_2\overset{O}{-}CH_2 \xrightarrow{30^\circ} (RO)_2P(S)SCH_2CH_2OH$$

Payne et al. (1962)[927] prepared a "dimer of β-(O,O-dimethylphosphorodithioate) of lactaldehyde:"

$$(MeO)_2P(S)SH + CH_2\overset{O}{-}CHCHO \xrightarrow[10\text{-}15^\circ]{CHCl_3} [(MeO)_2P(S)SCHOHCH_2CHO]_2$$

XII. REACTIONS OF P_2S_5 UNDER FORCING CONDITIONS

P_2S_5 can be made to react at high temperatures (above

100°) in the presence of catalysts and in pressurized vessels with various intermediates to produce phosphorotetrathioates and other products. The following examples demonstrate the variety of reactions.

Brannock (1951):[123]

$$P_2S_5 + HCOOEt \xrightarrow[90^\circ]{15\ min} \xrightarrow[150^\circ]{30\ min} (EtO)_2P(S)SEt + HCSOEt$$

$$P_2S_5 + HC(SEt)_3 \xrightarrow[100^\circ]{1.5\ hr} (EtS)_3P(S) + HCSSEt$$

Th. Goldschmidt (1966)[1054] used H_3PO_4, H_2SO_4, I_2, or $SbCl_3$ as catalysts for the reactions of ethers with P_2S_5:

$$P_2S_5 + R_2O \xrightarrow[140^\circ]{5\ hr} (RS)_3P(S) + P_2O_5$$

Worrel (1969):[1400]

$$P_2S_5 + \alpha\text{-Pinene} + H_2S \xrightarrow[112-140^\circ]{220\ min} (C_{10}H_{17}S)_3P(S)$$

Blagoveschenskii et al. (1969)[100] treated tetraalkoxysilanes with P_2S_5 in the ratio of 3:4. They reported that the use of excess P_2S_5 may cause an explosion.

$$4P_2S_5 + 3(RO)_4Si \xrightarrow[105-110^\circ]{reflux} \xrightarrow[160-170^\circ]{reflux} (RS)_3P(S)$$

Trialkylborates behaved similarly:[101]

$$P_2S_5 + (RO)_3B \xrightarrow[170-175^\circ]{} \xrightarrow[250-255^\circ]{} (RS)_3P(S)$$

Reactions of P_2S_5 with thenoate esters (13) in xylene and in pyridine have been studied [Brevilet *et* al. (1971)].[124] Thieno-oxathiaphosphorins (14) and thienodithioles (15) are the major products. Thienodithioles only are obtained in pyridine.

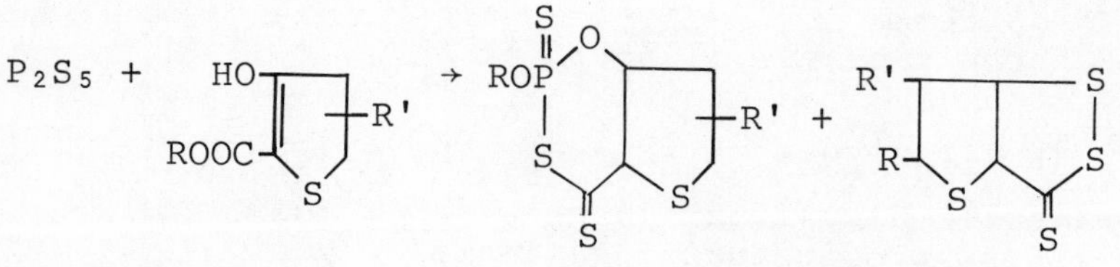

(13) (14) (15)

Maier and Van Wazer (1962)[680] have studied solution and reaction of P_2S_5 with $(RS)_3P(S)$. Metastable equilibria were shown by ^{31}P-NMR to occur immediately upon dissolution of the P_2S_5 and upon further heating for a period as long as 50 hr at temperatures in the range of 155-175°.

Scott, Menefee, Alford (1956-1957):[1209, 1210]

$$P_2S_5 + MeSH \xrightarrow[125-150^\circ,\ 100-110\ psi]{toluene} MeSPS_2$$

$$MeSPS_2 + PrSH + 1\text{-octene} \xrightarrow{reflux} MeS(PrS)P(S)SC_8H_{17}\text{-}n$$

XIII. REACTIONS OF PHOSPHITES WITH ALKANETHIOSULFONATES

Trialkyl phosphites reacted at low temperatures with organic thiosulfonates to produce thiolic esters and organic sulfenates, according to Michalski and co-workers (1958)[794], (1960)[780]:

$$(BuO)_3P + BuSO_2SEt \xrightarrow[20-25^\circ]{} (BuO)_2P(O)SEt + BuSOOBu$$

Similarly, dialkyl thiophosphites reacted with sodium ethylthioethylthiosulfonate in water plus Et_3N to form phosphorodithioates [Lorenz, Schrader (1963)]:[646]

$$(MeO)_2P(S)H + EtSCH_2CH_2SSO_2Na + Et_3N \xrightarrow[10-20^\circ]{water}$$

$$(MeO)_2P(S)SCH_2CH_2SEt + NaSO_2H$$

XIV. OXIDATION OF PHOSPHOROTHIOITES

Triphenylphosphorotrithioate was oxidized by H_2O_2 or by $KMnO_4$ and acetic acid to obtain S,S,S-triphenylphosphorothioate [Michaelis, Linke (1907)]:[773]

$$(PhS)_3P + (O) \rightarrow (PhS)_3P(O)$$

XV. ADDITION OF SULFUR OR SELENIUM TO PHOSPHITES

This method offers convenient procedures for preparing many of the products of Sections F.4.4, F.5.8, and F.6 when the phosphites are available. The additions occur readily on warming in inert solvents such as CS_2 or hydrocarbons. Other sources of sulfur, such as $P(S)Cl_3$, may be used instead of sulfur. Examples include:

Pistschimuka (1911):[960, 961]

$$(MeO)_3P + S \text{ or } Se \rightarrow (MeO)_3P(S) \text{ or } (MeO)_3P(Se)$$

Arbuzov, Razumova (1956):[51]

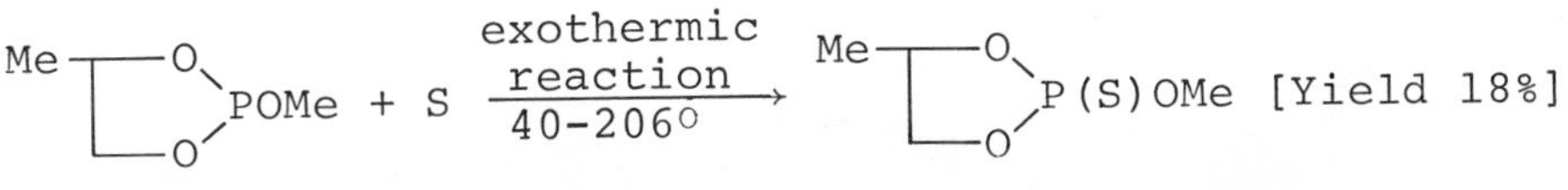

The selenium analog was obtained in 4.7% yield. It deposited Se on standing.

Markley, Larson (1958):[720]

$$(MeCHClCH_2O)_3P + S \rightarrow (MeCHClCH_2O)_3P(S)$$

3 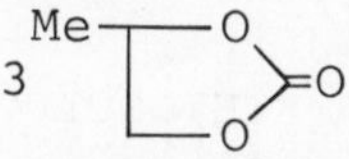 $+ PCl_3 + S \rightarrow (MeCHClCH_2O)_3P(S) + 3CO_2$

Schrader (1960):[1156]

$$(EtO)_2P(O)H + S + ClCH_2SC_6H_{11} \rightarrow (EtO)_2P(S)OCH_2SC_6H_{11} + HCl$$

Schrader (1957):[1127]

$$(EtO)_2P(O)H + S + MeOSOCl \xrightarrow{110^\circ} (EtO)_2P(S)OSO_2Me + HCl$$

$$(EtO)_2P(O)H + S + Me_2NCOCl \rightarrow (EtO)_2P(S)OCONMe_2 + HCl$$

Gottlieb (1932):[376]

$$(PhO)_3P + P(S)Cl_3 \rightarrow (PhO)_3P(S) + PCl_3$$

Kuznetsov, Valetdinov (1956):[579a]

$$(NCCMe_2O)_3P + P(S)Cl_3 \xrightarrow{120\text{-}130^\circ} (NCCMe_2O)_3P(S) + PCl_3$$

XVI. REACTIONS OF TERTIARY PHOSPHOROTHIONATES WITH ORGANIC HALIDES

The general reaction

$$(RO)_3P(S) + R'X \rightarrow (RO)_2(R'S)P(O) + RX$$

was explored by Pistschimuka (1912).[961] For the cases where R = R', the reaction becomes an isomerization (see Method XXXI).

Walling, Rabinowitz (1959):[1372]

$$(EtO)_3P(S) + BuI \xrightarrow[43\ hr]{reflux} (EtO)_2P(O)SBu + EtI$$

XVII. REACTIONS OF PHOSPHORODITHIOATE OR -THIOATE SALTS WITH ORGANIC HALIDES

This commercially important reaction has been used to prepare most of the products of Sections F.3.3, -dithioates, F.4.3, -thiolates, F.5.7, -trithioates, F.5.8, -tetrathioates, and F.6, -selenoates.

Generalized equations take the forms:

$$(RO)_2P(S)SH + R'X + B \rightarrow (RO)_2P(S)SR' + B \cdot HX$$

$$(RO)_2P(O)SM + R'X \rightarrow (RO)_2P(O)SR' + MX$$

In these equations, B is a base (organic or inorganic), M is alkali metal or ammonium (occasionally a heavy metal), and X is a halogen (frequently Br). Reactions are usually carried out in polar solvents or heterogeneous systems at or slightly above room temperature.

The following examples illustrate the variety of reactants, reaction conditions and products included in Method XVII:

Norman, LeSuer, Mastin (1952):[890]

$$(EtO)_2P(S)SNa + \text{2-Br-octane} \xrightarrow[\text{reflux}]{\text{EtOH}} (EtO)_2P(S)SCHMe(CH_2)_5Me + NaBr$$

Diveley (1958)[210] observed simultaneous condensation and dehydrohalogenation in the following reaction:

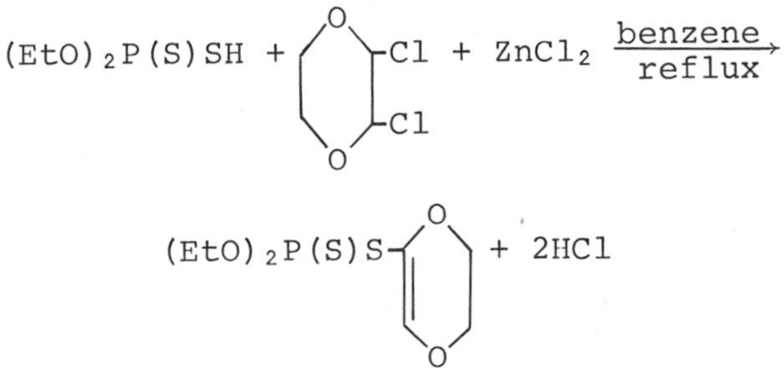

Diveley (1959),[212] Haubein (1959):[419]

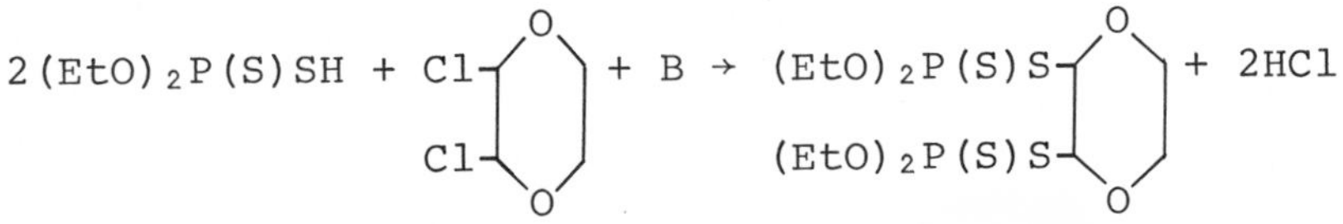

Malinovskii, Solomko, Evtushenko (1960):[686]

$$(EtO)_2P(S)SH + ClCH_2CH_2OSO_2Cl \xrightarrow{Na_2CO_3} (EtO)_2P(S)SCH_2CH_2OSO_2Cl + HCl$$

Addor, Levy, Magee (1964):[2]

$$(EtO)_2P(S)SK + ClCH_2CHO \xrightarrow[\text{reflux}]{CHCl_3,\ H_2O} (EtO)_2P(S)SCH_2CHO + KCl$$

The aldehyde was subsequently condensed with $HOCH_2CH_2OH$ or $HSCH_2CH_2SH$ in the presence of p-toluene sulfonic acid to give:

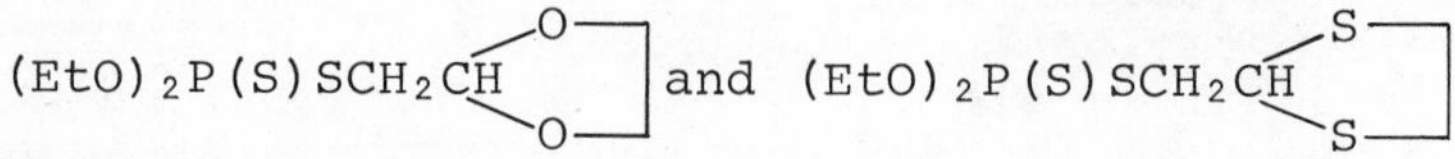

Ruefenacht (1968):[1053]

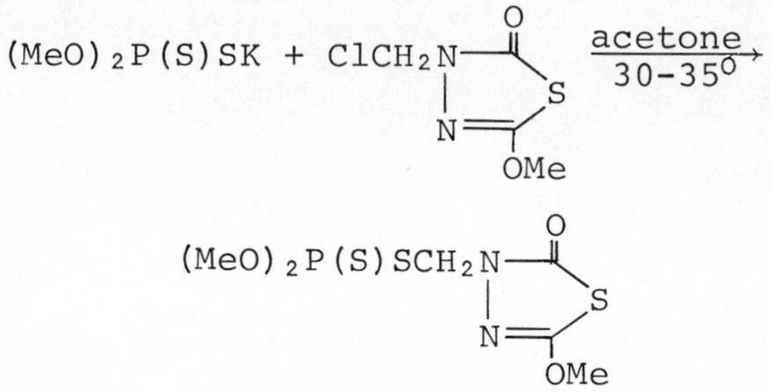

Richter (1967):[1028]

$$EtO(ArO)P(S)Cl + KSH \rightarrow EtO(ArO)P(S)SK\ (\underline{16}) + HCl$$

$$(\underline{16}) + ClCH_2CON(OMe)\text{-}i\text{-}Pr \xrightarrow[\text{reflux}]{\text{benzene}} EtO(ArO)P(S)SCH_2CON(OMe)\text{-}i\text{-}Pr + KCl$$

Nakagawa, Sakamoto (1967):[857]

$$2(EtO)_2P(S)SNa + CH_2Cl_2 + KBr \xrightarrow[\text{autoclave}]{\text{water } 85^\circ \text{ 12 hr}} [(EtO)_2P(S)S]_2CH_2 + 2NaCl$$

Thuong, Demozay (1966):[1308]

$$(MeS)_2P(O)S^- \ NMe_4^+ + ClCH_2CONHMe \xrightarrow[80^\circ \text{ pressure}]{CHCl_3} (MeS)_2P(O)SCH_2CONHMe + Me_4NCl$$

Gupalo, Zemlyanskii (1967):[386]

$$Me_2NCH_2CH_2OP(S)(SK)_2 + (CH_2Br)_2 \xrightarrow{50^\circ}$$

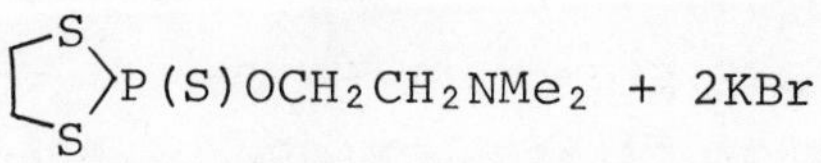

Raetz (1967):[1005]

$$(RO)_2P(S)SNH_4 + ClR' \xrightarrow[\text{R.T.}]{\text{acetone}}$$

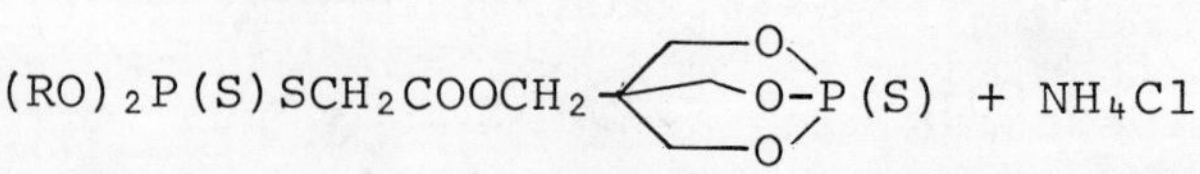

Shvetsova-Shilovskaya et al. (1961):[1242]

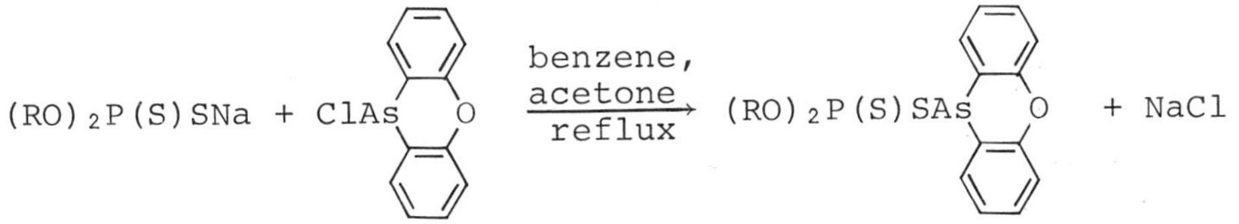

Schrader (1961):[1157]

$$EtO(EtS)P(O)SNa + ClCH_2CN \xrightarrow[40-45^{\circ}]{EtOH} EtO(EtS)P(O)SCH_2CN + NaCl$$

Farbenfabriken Bayer (1958):[279] in the absence of base:

$$(MeO)_2P(S)SH + ClCH_2SBu \xrightarrow[heat]{(CH_2Cl)_2} (MeO)_2P(S)SCH_2SBu + HCl\ (gas)$$

Feher, Blümcke (1957):[319]

$$(MeO)_2P(S)SK + Me_3SiCl\ (HCl\text{-}free) \xrightarrow[acetone]{ether} (MeO)_2P(S)SSiMe_3 + KCl$$

Truchlik, Tichy (1957):[1324]

$$(EtO)_2P(S)SK + Et_2NCSCl \rightarrow (EtO)_2P(S)SC(S)NEt_2 + KCl$$

Nuretdinov, Arbuzov (1971):[897]

$$(EtO)_2P(O)SH{\cdot}NEt_3 + ClCH_2\overset{S}{CH-CH_2} \xrightarrow{benzene} (EtO)_2P(O)SCH_2\overset{S}{CH-CH_2} + Et_3N{\cdot}HCl$$

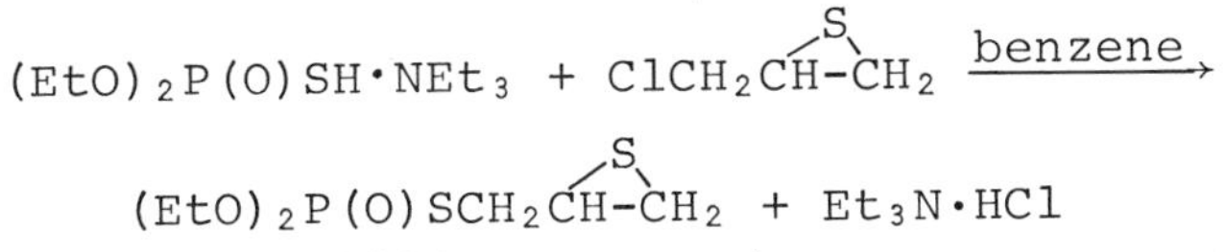

Schrader (1952):[1123]

$$(EtO)_2PONa + S + EtSCH_2Cl \xrightarrow{toluene} (EtO)_2P(O)SCH_2SEt + NaCl$$

Schrader (1961):[1175] in the absence of base:

$$(EtO)_2P(O)H + S + EtSCH_2Cl \xrightarrow[125-130^{\circ}]{} (EtO)_2P(S)OCH_2SEt + HCl\ (gas)$$

Kabachnik et al. (1956):[494]

$$(EtO)_2POSNa + RCOCl \xrightarrow[100^{\circ}]{} (EtO)_2P(S)OCOR + NaCl$$

Hilgetag, Lehmann (1957)[433] demethylated $(MeO)_2P(S)$-$OC_6H_4NO_2$-4 with strychnine to separate the optical isomers; these were treated with $HClO_4$ to liberate the unstable acids that were isolated as Pb salts; they were caused to react with MeI to prepare d- and l-$MeO(MeS)P(O)OC_6H_4NO_2$-4.

Chabrier, Thuong (1964):[154]

$$(MeO)_3P(S) + Me_3N \rightarrow (MeO)_2P\left(\begin{smallmatrix}O^-\\ \vdots\\ S\end{smallmatrix}\right) NMe_4^+ \xrightarrow{RX}$$

$$(MeO)_2P(O)SR + Me_4NCl$$

Floyd (1966):[336] demethylation with $(MeO)_2P(S)SNa$ to form (17):

$$MeOP\left(\begin{smallmatrix} O^- \\ \| \\ S \end{smallmatrix}\right)SCH_2-\left\langle \begin{smallmatrix} N = -NH_2 \\ N \\ N - NH_2 \end{smallmatrix} \right. + RI \xrightarrow[cold]{MeONa}$$

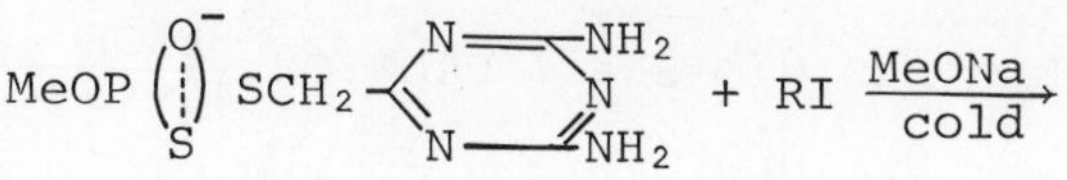

(17)

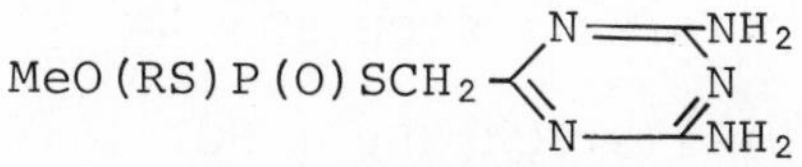

Curry et al. (1967):[192] demethylation with $(i\text{-}PrO)_2P$-(S)SK to form (18):

$$MeO(MeS)P(O)SK + ClCH_2CONHMe \rightarrow MeO(MeS)P(O)SCH_2CONHMe$$

(18)

In preparing "methyl acetophos" and its ethyl homolog, Zaks et al. (1966)[1426] noted the large difference in yields: 35% for Me vs. 85% for Et:

$$(RO)_2P(S)ONH_4 + ClCH_2COOEt \xrightarrow{-NH_4Cl}$$

$$(RO)_2P(O)SCH_2COOEt \quad [R= Me \text{ or } Et]$$

They then studied the scrambling of methyl groups (involving $(MeO)_2P(S)SMe$, and $(MeS)_2P(S)OMe$), which accounted for the reduced yield of methylacetophos.

Uchiyama, Kosaka (1968):[1340] demethylation with KOCSS-Et:

$$MeOP\left(\begin{smallmatrix}O^-\\ \vdots\\ S\end{smallmatrix}\right) SCH_2CH_2NHCOMe + MeI \rightarrow MeO(MeS)P(O)SCH_2CH_2NHCOMe + I^-$$

Donninger, Hutson (1968):[221] Demethylation with -ephedrine and methylation with MeI gave optical isomers:

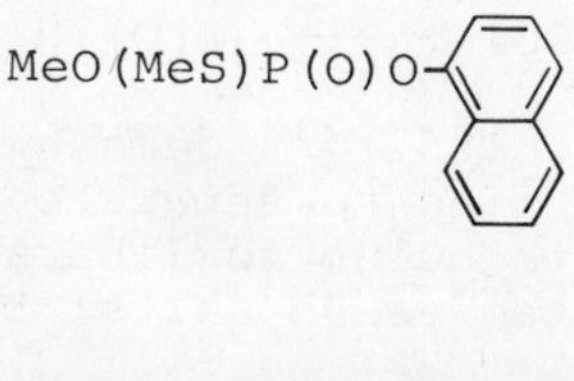

XVIII. REACTION OF PHOSPHORODITHIOIC ACIDS WITH ALDEHYDES AND KETONES

The formation of methylene bis(dithiophosphates) has been reported by Bachinskii and Zemlyanskii (1968):[75]

$$2(RO)_2P(S)SH + (CH_2O)_x \xrightarrow[50^\circ]{40\ hrs} [(RO)_2P(S)S]_2CH_2 + H_2O$$

Benzaldehyde gave benzylidene bis(dithiophosphates) [Zemlyanskii, Bachinskii (1968)]:[1428]

$$2(RO)_2P(S)SH + PhCHO \xrightarrow[25-40^\circ]{HCl\ (gas)} [(RO)_2P(S)S]_2CHPh + H_2O$$

With diethyl mesoxalate, Pudovik and Cherkasov (1967)[985] obtained a quantitative yield of an undistillable, monobasic acid, which they characterized as diethyl [(diethoxyphosphinothioyl)thio]tartronate. The acidic properties of the hydroxy group were attributed to the presence of electronegative groups on the C carrying the OH:

$$(EtO)_2P(S)SH + (O)C(COOEt)_2 \xrightarrow[35-40^\circ]{benzene}$$

$$(EtO)_2P(S)SC(OH)(COOEt)_2$$

XIX. REACTIONS OF PHOSPHITES OR PHOSPHOROTHIOITES WITH SULFENYL HALIDES

Secondary phosphites and phosphorothioites either as alkali metal salts or in the presence of organic bases react readily with sulfenyl halides. Tertiary phosphites also react with elimination of alkyl halide. The products are phosphorothiolates or -dithioates.

The following examples illustrate the reactions:

Birum (1960):[96]

$$(EtO)_3P + Cl_2CHSCl \xrightarrow[2-5^\circ]{CH_2Cl_2} (EtO)_2P(O)SCHCl_2 + EtCl$$

Kharasch (1960):[531]

$$(MeO)_3P + C_6Cl_5SCl \xrightarrow{(CH_2Cl)_2} (MeO)_2P(O)SC_6Cl_5 + MeCl$$

Almasi, Hantz (1960),[15] Schrader (1962):[1161]

$$(MeO)_2P(S)H + ArSCl \xrightarrow[10-20^\circ]{CCl_4} (MeO)_2P(S)SAr + HCl$$

Fearing, McClellan (1968):[318]

$$(EtO)_2P(S)H + Cl_3CSCl + Et_3N \xrightarrow[15^\circ]{benzene} (EtO)_2P(S)SCCl_3 + HCl$$

In a study of reactions of dialkylaminosulfenyl chlorides with di- and trialkyl phosphites to prepare dialkoxy oxophosphoranesulfenamides, Michalski and Pliszka-Krawiecka (1966)[784] obtained the better yields using the tertiary phosphites:

$$(EtO)_3P + Et_2NSCl \xrightarrow[0-10^o]{benzene} (EtO)_2P(O)SNEt_2 \ (74\%) + EtCl$$

$$(EtO)_2P(O)H + Et_2NSCl + Et_3N \xrightarrow[ice\ temp.]{benzene}$$

$$(EtO)_2P(O)SNEt_2 \ (35\%) + HCl$$

Reaction of N-chlorodiethylamine with sodium diethylphosphorothioate gave a 36% yield:

$$(EtO)_2POSNa + Et_2NCl \xrightarrow[R.T.]{petr.\ ether\ (30-60^o)}$$

$$(EtO)_2P(O)SNEt_2 + NaCl$$

Almasi, Paskucz (1965):[32]

$$(EtO)_2P(S)H + ClSN\langle\ \rangle O \xrightarrow[C_5H_5N]{CCl_4} (EtO)_2P(S)SN\langle\ \rangle O + HCl$$

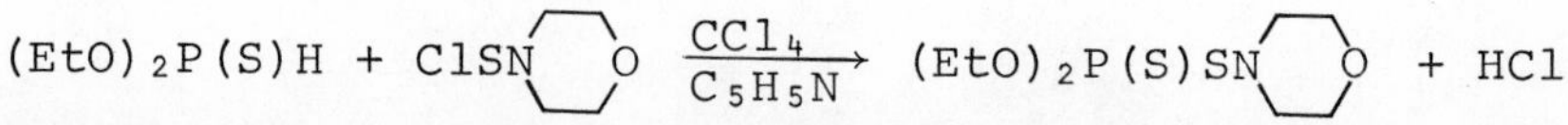

Reactions of phosphites with sulfonyl chlorides give products similar to those from sulfenyl chlorides. Portions of the phosphites are converted to phosphates:

Gilbert and McGough (1954):[367]

$$3(EtO)_3P + BuSO_2Cl \xrightarrow[30^o]{ether}$$

$$(EtO)_2P(O)SBu + 2(EtO)_3P(O) + EtCl$$

$$3(EtO)_3P + ArSO_2Cl \xrightarrow[15^o]{CH_2Cl_2}$$

$$(EtO)_2P(O)SAr + 2(EtO)_3P(O) + EtCl$$

Hoffman et al. (1956):[450]

$$3(EtO)_3P + 4-BrC_6H_4SO_2Cl \rightarrow$$

$$(EtO)_2P(O)SC_6H_4Br-4 + 2(EtO)_3P(O) + EtCl$$

Lorenz (1958):[612]

$$(EtO)_2P(S)H + NO_2C_6H_4SO_2Cl + B \rightarrow$$

$$(EtO)_2P(S)SC_6H_4NO_2-(2,3,\ or\ 4)$$

Unstable products (disulfides) were obtained from dialkyl phosphites and sulfur monochloride [Ettel and Zbirovsky (1956)]:[265]

$$(RO)_2P(O)H + S_2Cl_2 \xrightarrow[-10\ to\ -5^o]{ether} \xrightarrow{aeration}$$

$$(RO)_2P(O)SSP(O)(OR)_2 + 2HCl$$

Almasi and Hantz (1966)[32] obtained a stable product in the reaction:

$$(EtO)_2P(S)H + 4-MeC_6H_4SSCl \xrightarrow[0^0; \text{ aeration}]{\text{petr. ether}}$$

$$(EtO)_2P(S)SSC_6H_4Me-4 + HCl$$

Lorenz (1963)[620] treated a suspension of bis(2-N-methylcarbamoylphenyl)disulfide in $CHCl_3$ with SO_2Cl_2 at 20-25^0, and then added dimethyl phosphite at 20-30^0. The process probably formed the sulfenyl chloride as an intermediate:

$$(SC_6H_4CONHMe-2)_2 + SO_2Cl_2 \xrightarrow{CHCl_3} ClSC_6H_4CONHMe-2 \ (\underline{19}) + SO_2$$

$$(MeO)_2P(O)H + (\underline{19}) \xrightarrow{CHCl_3} (MeO)_2P(O)SC_6H_4CONHMe-2 + HCl$$

Yoshido et al. (1967)[1417] used a similar process with benzyl disulfide:

$$(ArCH_2S)_2 + SO_2Cl_2 \xrightarrow{CCl_4}_{-5^0} 2ArCH_2SCl + SO_2$$

$$(RO)_2P(O)H + ArCH_2SCl \xrightarrow{CCl_4}_{50^0} (RO)_2P(O)SCH_2Ar + HCl$$

XX. REACTIONS OF PHOSPHATES, PHOSPHORODITHIOATES, AND -THIOATES WITH SULFENYL CHLORIDES

The reactions are similar to Method XIX.

Almasi, Paskucz (1961):[28]

$$(EtO)_2P(S)SH + ClSC_6H_4NO_2-4 \xrightarrow{CHCl_3}_{\text{aeration}}$$

$$(EtO)_2P(S)SSC_6H_4NO_2-4 + HCl$$

Schrader, Lorenz (1961):[1197]

$$(EtO)_2P(S)SH + RSCl \xrightarrow{CCl_4}_{10^0} (EtO)_2P(S)SSR + HCl$$

$$(EtO)_2P(O)SH + RSCl \rightarrow (EtO)_2P(O)SSR + HCl$$

Almasi, Hantz (1966):[23]

$$(EtO)_2P(S)SH \cdot C_5H_5N + ClSN \underset{\underset{\smile}{}}{\frown} O \xrightarrow{\text{petr. ether}}_{10^0}$$

$$(EtO)_2P(S)SSN \underset{\underset{\smile}{}}{\frown} O + HCl \cdot NC_5H_5$$

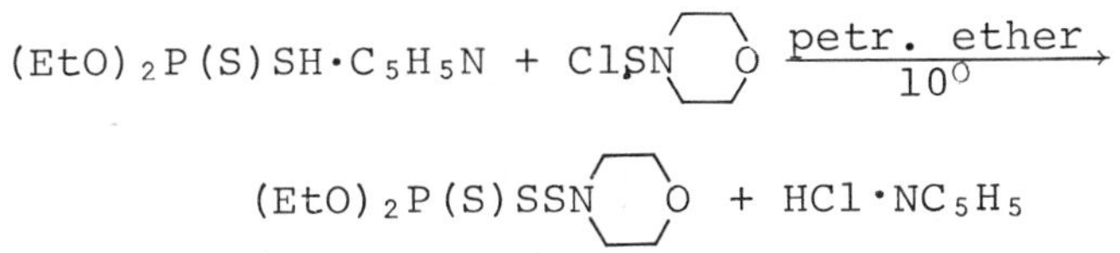

XXI. REACTION OF PHOSPHORODITHIOIC ACIDS WITH ALDEHYDES AND ACTIVE-H COMPOUNDS

The generalized equation is:

$$(RO)_2P(S)SH + CH_2O + R'SH \rightarrow (RO)_2P(S)SCH_2SR' + H_2O$$

Preformed $HOCH_2SR'$ may be employed in place of CH_2O and R'SH. Other lower aldehydes may be used in place of CH_2O. Other active-H compounds include R'OH, $R'NH_2$, H_2NCOR', etc. The reactions are usually conducted at room temperature in aqueous or aqueous-organic solvent systems. In reactions involving CH_2O, methanol-free formalin or paraformaldehyde must be used to avoid contamination with $(RO)_2P(S)SCH_2OMe$.

Hook, Moss (1952):[458a]

$$(EtO)_2P(S)SH + CH_2O\ (aq.) + EtSH \rightarrow (EtO)_2P(S)SCH_2SEt + H_2O$$

Shvetsova-Shilovskaya et al. (1956):[1246]

$$(RO)_2P(S)SH + CH_2O + HNR'COOR'' \rightarrow$$

$$(RO)_2P(S)SCH_2NR'COOR'' + H_2O$$

Ruefenacht (1968):[1053]

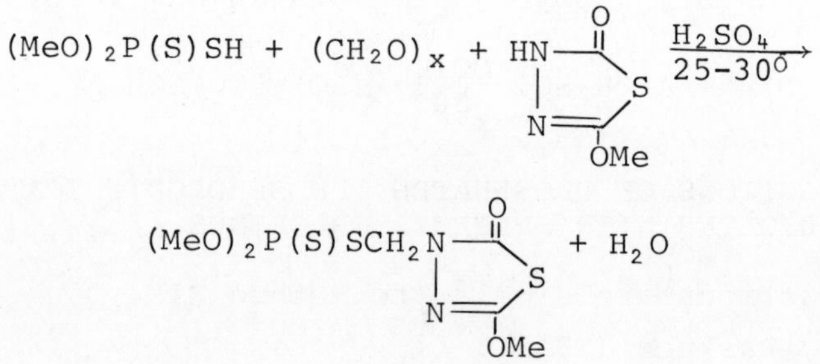

XXII. ADDITION OF PHOSPHORODITHIOIC AND -THIOIC ACIDS TO OLEFINIC COMPOUNDS

Phosphorodithioic acids add readily to many types of olefinic compounds. The reactions are sometimes carried out in the presence of tertiary amines, but in most cases no catalyst is necessary.

With α-olefins, addition is to the 2-position as shown in the studies by Norman, Le Suer, and Mastin (1952).[890] The reactions with liquid olefins proceeded readily at temperatures below about 100°. Gaseous olefins, ethylene and propylene, were added at 100-115° under pressure. Structures were confirmed by alternate syntheses.

$$(RO)_2P(S)SH + CH_2{=}CHR' \rightarrow (RO)_2P(S)SCHMeR'$$

In the production of malathion by addition of O,O-dimethylphosphorodithioic acid to diethyl maleate and fumarate [Cassaday (1951)],[150] the use of hydroquinone to suppress polymerization of the unsaturated ester was recommended. Triethylamine was also mentioned as a catalyst for the addition. Subsequent studies have revealed that while the addition to maleate is quite rapid, the isomerization of maleate to fumarate is also rapid in the presence of phosphorodithioic acids. A general equation for the reaction is:

$$(MeO)_2P(S)SH + EtOOCCH{=}CHCOOEt \xrightarrow{65^\circ}$$

$$(MeO)_2P(S)SCH(COOEt)CH_2COOEt$$

Pudovik and Cherkasov (1968)[986] reported rate constants at 45° in 3:1 heptane:xylene for addition of O,O-diethyl phosphorodithioic acid to 1,3-conjugated systems containing a hetero atom: k_2 = liter/mole/sec x 10^4: $CH_2{=}CHCN$, 8.36; $CH_2{=}CMeCN$, 7.23; $CH_2{=}CHCOOH$, 1.89; $CH_2{=}CMeCOOH$, 1.43; $CH_2{=}CHCOOMe$, 0.86; $CH_2{=}CMeCOOMe$, 0.43; $CH_2{=}C(COOEt)_2$, very fast; $MeCH{=}C(COOEt)_2$, 9.37; $EtCH{=}C(COOEt)_2$, 4.57; $PrCH{=}C(COOEt)_2$, 3.60; i-$PrCH{=}C(COOEt)_2$, 1.50. For $ArCH{=}C(COOEt)_2$, the rates were much slower and were measured at 75°; for Ar = Ph, 4.01. Substitutions in the Ar group had little effect on rates. Diethyl maleate (isomeric with the very reactive $CH_2{=}C(COOEt)_2$) was inert in the solvent system, as were maleic and fumaric acids and diethyl fumarate.

Oswald, Mueller, and Daniker (1971)[921] have studied additions to conjugated dienes: piperylene, isoprene, chloroprene, cyclopentadiene, butadiene, and 2,5-dimethyl-2,4-hexadiene. The mono- addition of O,O-dimethylphosphorodithioic acid to piperylene gave 75% yield of adduct with the composition 90% 4,1-, $(RO)_2P(S)SCHMeCH{=}CHMe$, isomer and 5% 1,4- isomer, $(RO)_2P(S)SCH_2CH{=}CHEt$. With isoprene the product ratio was about 63% 4,1- and 27% 1,4- isomer.

For the mono-addition of O,O-dimethylphosphorothioic acid to butadiene, a radical initiator, methyl hydroquinone, had a mild catalytic effect. A cationic catalyst, $HClO_4$, with a pK_a greater than $(MeO)_2POSH$, proved to be a strong catalyst for both the 2,1- addition and the subsequent isomerization to the 1,4- isomer. The cationic catalyst was also stated to inhibit polymerization of the diene and to suppress addition of a second mole of the phosphorothioate to the allylic phosphate ester:

$$(MeO)_2POSH + H_2C{=}CHCH{=}CH_2 \xrightarrow{HClO_4} (MeO)_2P(O)SCH_2CH_2CH{=}CH_2 \overset{\sim}{\rightarrow}$$

$$(MeO)_2P(O)SCH_2CH{=}CHMe$$

Other examples include:

Miller, Birum (1962):[810] catalyst: $PhCH_2NMe_3^+\ OMe^-$:

$$(EtO)_2P(S)SH + HC{\equiv}CCOOMe \xrightarrow[\text{reflux}]{\text{MeOH, benzene}}$$

$$(EtO)_2P(S)SCH{=}CHCOOMe$$

Nishiwaki (1967):[880]

$$(RO)_2P(S)SH + H_2C{=}CHOCOMe \xrightarrow[40^\circ]{\text{toluene}} (RO)_2P(S)SCHMeOCOMe$$

Haubein (1967):[420]

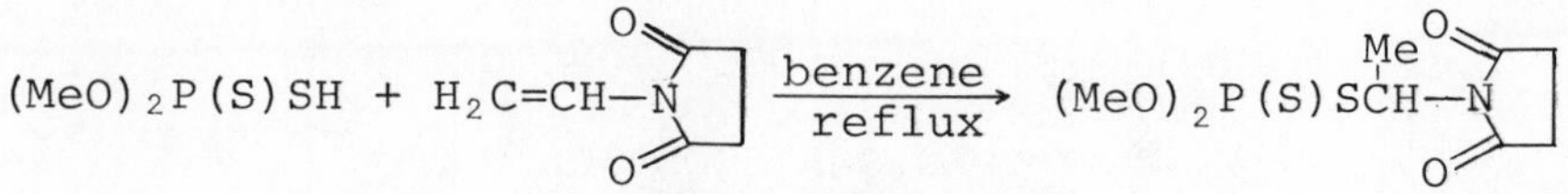

Pudovik, Sergeeva (1955):[1003]

$$(EtO)_2P(S)SH + 4\text{-}ClC_6H_4N{=}CHAr \xrightarrow[\text{moisture}]{\text{exclude}}$$

$$(EtO)_2P(S)SCH(Ar)NHC_6H_4Cl\text{-}4$$

Knunyants, Bykhovskaya (1966):[541]

$$(i\text{-}PrO)_2P(S)OH + F_2C{=}C(CF_3)_2 + Et_3N \xrightarrow[-10\text{ to }-5^\circ]{\text{ether}}$$

$$(i\text{-}PrO)_2P(S)OCF_2CH(CF_3)_2$$

Michalski and Tulimowski (1966)[791] found that $(EtO)_2P(Se)SH$ added to isolated and conjugated double bonds in the presence or absence of catalyst. C-S rather than C-Se bonds were formed. The direction of addition was consistent with a free radical mechanism for the uncatalyzed reaction.

$$(EtO)_2P(Se)SH + CH_2{=}CHCN \xrightarrow{\text{no catalyst}} (EtO)_2P(Se)SCHMeCN$$

$$\xrightarrow{\text{MeONa}} (EtO)_2P(Se)CH_2CH_2CN$$

XXIII. OXIDATION OF PHOSPHORODITHIOATES TO DISULFIDES AND POLYSULFIDES

Mild oxidizing agents convert phosphorodithioic acids or their salts to disulfides and polysulfides:

Rudel, Boyle (1950):[1050]

$$2(s\text{-}BuO)_2P(S)SNa + 2NaNO_2 \rightarrow$$

$$(s\text{-}BuO)_2P(S)SS(S)P(O\text{-}s\text{-}Bu)_2 + 2NO + Na_2O$$

Hu, Li, Chen (1956):[464]

$$2(EtO)_2P(S)SH + Br_2 \xrightarrow{CCl_4} (EtO)_2P(S)SS(S)P(OEt)_2 + 2HBr$$

$$(EtO)_2P(S)SH + Br_2 \xrightarrow[\text{R.T., 1 wk}]{CHCl_3} (EtO)_2P(S)SSS(S)P(OEt)_2$$

Hu, Chen (1956):[462]

$$[(RO)_2P(S)S]_2Hg + NaOH + Br_2 \xrightarrow[\text{R.T.}]{-HgBr_2} (RO)_2P(S)SS(S)P(OR)_2$$

Edmundson (1965):[247]

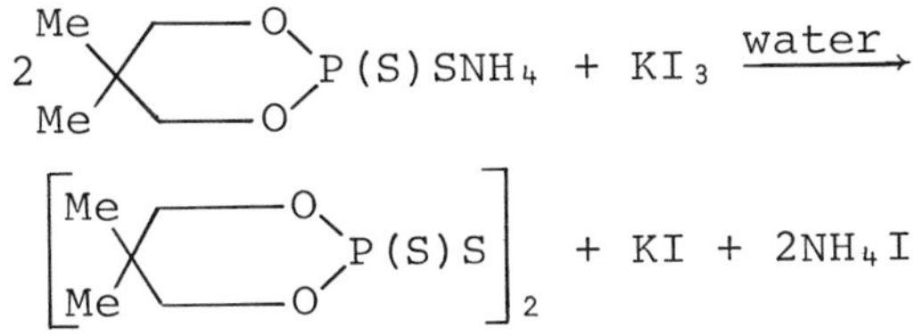

Use of NH_4 salt plus H_2O_2 or Na salt plus $NaNO_2$ gave lower yields.[247]

Cremlyn, Olson (1969):[191]

$$2(RO)_2P(S)SNa + Br_2 \xrightarrow[\text{warm}]{\text{EtOH}} (RO)_2P(S)SS(S)P(OR)_2 + 2NaBr$$

[R = cholesteryl]

Zemlyanskii, Chernaya (1966):[1430]

$$2(PhO)_2P(S)SeK + I_2 \xrightarrow[\text{acetone, R.T.}]{\text{EtOH}}$$

$$(PhO)_2PSSeSeSP(OPh)_2 + 2KI$$

IR showed both P=S and P=Se. The corresponding Et analog was too unstable to obtain an IR spectrum.

XXIV. REACTION OF PHOSPHITES WITH THIOCYANATES

Thiocyanates react at moderate temperatures with secondary phosphites or phosphorothioites and with tertiary phosphites to form tertiary esters with elimination of HCN, NaCN, or RCN.

Schrader (1952):[1123]

$$(EtO)_2PONa + EtSCH_2CH_2SCN \xrightarrow[40^{\circ}]{\text{toluene}}$$

$$(EtO)_2P(O)SCH_2CH_2SEt + NaCN$$

Vitenberg, Pesin (1969):[1358]

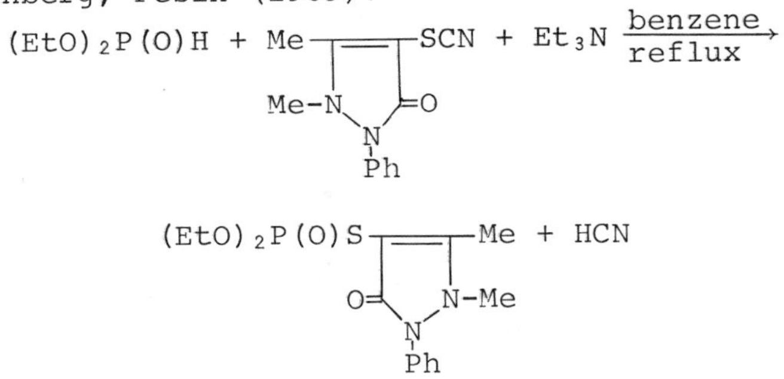

Calderbank et al. (1959):[147]

$$(EtO)_2PSNa + EtPhNCH_2CH_2SCN \xrightarrow[35-40^\circ]{benzene} (EtO)_2P(S)SCH_2CH_2NEtPh + NaCN$$

Lorenz, Schrader (1960):[641]

$$(EtO)_2P(S)H + EtSCN + NaOMe \xrightarrow[20^\circ]{benzene} (EtO)_2P(S)SEt + NaCN + MeOH$$

Raetz, Bliss (1967):[1006]

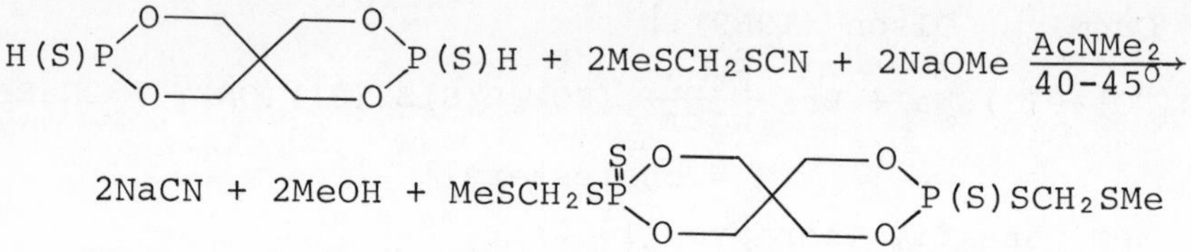

Schrader, Lorenz (1955):[1191]

$$(EtO)_3P + EtSCH_2CH_2SCN \xrightarrow[110-140^\circ]{xylene} (EtO)_2P(O)SCH_2CH_2SEt + EtCN$$

Uchiyama et al. (1967):[1339]

$$(MeO)_3P + ClCH_2SCN \xrightarrow[0-90^\circ]{toluene} (MeO)_2P(O)SCH_2Cl + MeCN$$

Pilgram and Korte (1964)[954] observed that with nitroaryl thiocyanates and tertiary phosphites reaction took a different course. RCN was not eliminated. Instead, the liberated R group consumed an equivalent of thiocyanate to produce the nitroaryl thioether as the by-product:

$$2(EtO)_3P + 2(2\text{-}MeO\text{-}5\text{-}NO_2\text{-}C_6H_3SCN) \xrightarrow[90^\circ]{}$$

$$(EtO)_2P(O)SC_6H_3\text{-}2\text{-}MeO\text{-}5\text{-}NO_2 + EtSC_6H_3\text{-}2\text{-}MeO\text{-}5\text{-}NO_2$$

$$+ EtCN + [(EtO)_2\overset{O}{\overset{\|}{P}}CN \xrightarrow{(EtO)_3P} EtNC + (EtO)_2\overset{O}{\overset{\|}{P}}\text{-}\overset{O}{\overset{\|}{P}}(OEt)_2]$$

Schrader, Lorenz (1951):[1190]

$$(EtO)_2PONa + RSeCN \xrightarrow[30-60^\circ]{toluene} (EtO)_2P(O)SeR + NaCN$$

XXV. ADDITION OF PHOSPHORANESULFENYL HALIDES TO OLEFINIC MATERIALS

Phosphoranesulfenyl halides add easily to olefinic or enolic unsaturation with formation of halogen-substituted tertiary esters.

Michalski, Lenard (1956):[777]

$$(EtO)_2P(O)SCl + CH_2{=}CH_2 \xrightarrow{\text{R.T.}} (EtO)_2P(O)SCH_2CH_2Cl$$

(20)

$$(\underline{20}) + MeCOOCMe{=}CH_2 \text{ or } Me_2CO \rightarrow (EtO)_2P(O)SCH_2COMe$$

$$(\underline{20}) + CH_2{=}CHOEt \xrightarrow[\text{below } 5^\circ]{\text{benzene}} (EtO)_2P(O)SCH_2CHClOEt$$

Petrov, Neimysheva (1959):[944]

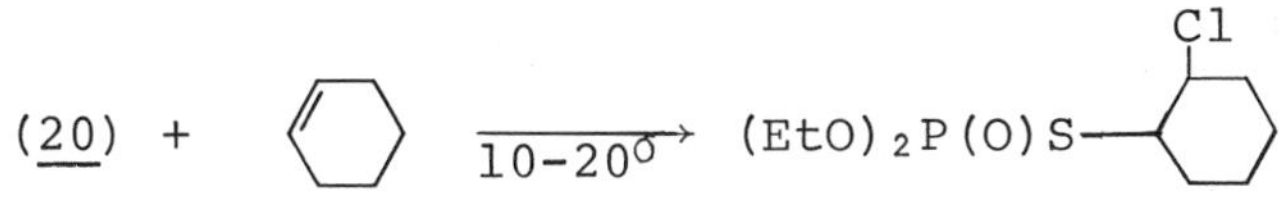

Almasi, Paskucz (1965):[32]

$$(EtO)_2P(S)SCl + CH_2CO \xrightarrow[-15^\circ]{\text{ether}} (EtO)_2P(S)SCH_2COCl$$

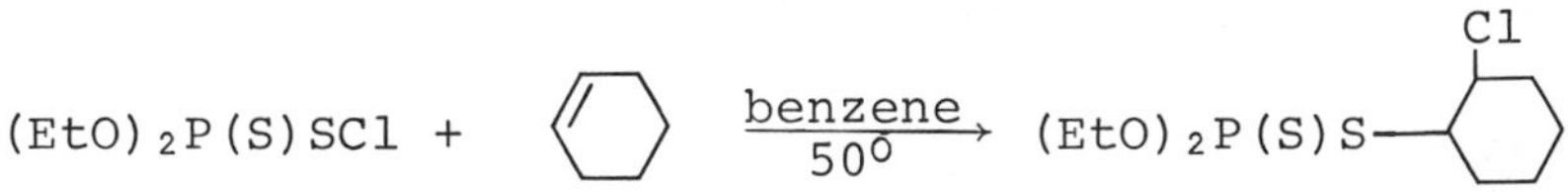

Mueller, Rubin, and Butler (1966)[842] found both Markovnikov and anti-Markovnikov additions to a variety of olefins and dienes. Thus with propylene the ratio of $(MeO)_2P(O)SCHMeCH_2Cl$ to $(MeO)_2P(O)SCH_2CHClMe$ was 51:49; for 1-pentene the analogous products were in the ratio of 65:35. With butadiene, the products were $-SCH_2CHClCH{=}CH_2$ and $-SCH(CH_2Cl)CH{=}CH_2$ with the ratio 73:27. With allene, only one product was observed: $(MeO)_2P(O)SC(CH_2Cl){=}CH_2$ (anti-Markovnikov addition).

XXVI. REACTIONS OF PHOSPHORANESULFENYL HALIDES WITH N COMPOUNDS, ALKYLATING AGENTS, AND THIOPHOSPHITES

The phosphoranesulfenyl halides react easily with amines and with salts of arylsulfonamides:

Michalski, Markowska, Strzelecka (1959):[779]

$$(EtO)_2P(O)SCl + 2Et_2NH \xrightarrow[\text{below } 5^\circ]{\text{ether}}$$

$$(EtO)_2P(O)SNEt_2 + Et_2NH \cdot HCl$$

Reaction with diethylaniline gave a product with ring substitution:

$$(EtO)_2P(O)SCl + 2PhNEt_2 \rightarrow$$

$$(EtO)_2P(O)SC_6H_4NEt_2\text{-}4 + PhNEt_2 \cdot HCl$$

Almasi, Paskucz, Fey (1970):[34]

$$(RO)_2P(O)SCl + NaNHSO_2Ar \xrightarrow[\text{R.T.}]{\text{ether}} (EtO)_2P(O)SNHSO_2Ar + NaCl$$

Reactions with diazoalkyls retain the halogen in the S-ester group:

Petrov, Neimysheva (1959):[944]

$$(EtO)_2P(O)SCl + CH_2N_2 \xrightarrow[\text{ice-salt bath}]{\text{ether}} (EtO)_2P(O)SCH_2Cl + N_2$$

Almasi, Paskucz (1965):[32]

$$(EtO)_2P(S)SCl + MeCHN_2 \xrightarrow[\text{R.T.}]{\text{ether}} (EtO)_2P(S)SCHClMe + N_2$$

Reactions with Grignard reagents replace the halogen of the sulfenyl halide:

Semidetko, Gololobov (1967):[1217]

$$(EtO)_2P(O)SCl + CH_2{=}CHMgBr \xrightarrow{\text{THF}} (EtO)_2P(O)SCH{=}CH_2 + MgBrCl$$

Almasi, Hantz (1968)[25]

$$(EtO)_2P(S)SCl + RMgBr \xrightarrow[5^0]{\text{ether}}$$

$$(EtO)_2P(S)SR + MgBrCl \ (+ R{-}R \text{ for } R = CH_2Ph)$$

$$(EtO)_2P(S)SCl + PhC{\equiv}CMgBr \xrightarrow[5^0]{\text{benzene,THF}}$$

$$(EtO)_2P(S)SC{\equiv}CPh + MgBrCl$$

Reactions with ketones eliminated HCl to form esters [Michalski, Lenard (1956)]:[777]

$$(EtO)_2P(O)SCl + Me_2CO \xrightarrow[\text{R.T.}]{} (EtO)_2P(O)SCH_2COMe$$

$$(EtO)_2P(O)SCl + CH_2(COMe)_2 \rightarrow (EtO)_2P(O)SCH(COMe)_2$$

Reaction with a phosphite to form a sulfide [Petrov, Neimysheva (1959)]:[944]

$$(EtO)_2P(O)SCl + (EtO)_2P(O)H \xrightarrow[-10 \text{ to } 0^0]{} [(EtO)_2P(O)]_2S + HCl$$

Reaction with a thiophosphite eliminated HCl to form sulfides:

Almasi, Hantz (1964);[19] Almasi, Paskucz (1965):[32]

$$(EtO)_2P(S)SCl + (EtO)_2P(S)H \xrightarrow[-5 \text{ to } 0^0]{CCl_4}$$

$$(EtO)_2P(S)S(S)P(OEt)_2 + HCl$$

XXVII. ALTERNATIVE ESTERIFICATIONS

The conventional method of esterifying dithioate and thioate salts has been presented as Method XVII. Other means for esterification of phosphorodithioic or -thioic acids or their salts are given in the following examples.

Reactions of phosphorodithioic acids with methyl or benzyl phosphorous, phosphonous, phosphinous esters result in methylation of the dithioate,

Pudovik, Krupnov (1968):[1001]

$$(MeO)_2P(S)SH + (MeO)_3P \rightarrow [(MeO)_3PH^+ (MeO)_2P(S)S^-] \rightarrow$$

$$(MeO)_2P(S)SMe + (MeO)_2P(O)H$$

Reaction was run at 0-5° with dropwise addition of the acid to phosphite. Analogous reactions were observed with $(MeO)_2PPh$ and $MeO(Et)PPh$.

Mandel'baum et al. (1967):[709]

$$(RO)_2POSNH_4 + (MeO)_2P(O)H \xrightarrow[\text{reflux}]{\text{benzene}}$$

$$(RO)_2P(O)SMe + (MeO)(NH_4O)P(O)H$$

Hechenbleikner et al. (1958):[422]

$$(EtO)_2P(S)SH + \left[\begin{array}{c}O\\O\end{array}\right]POCH_2Ph \xrightarrow[120°]{\text{toluene}}$$

$$(EtO)_2P(S)SCH_2Ph + \left[\begin{array}{c}O\\O\end{array}\right]P(O)H$$

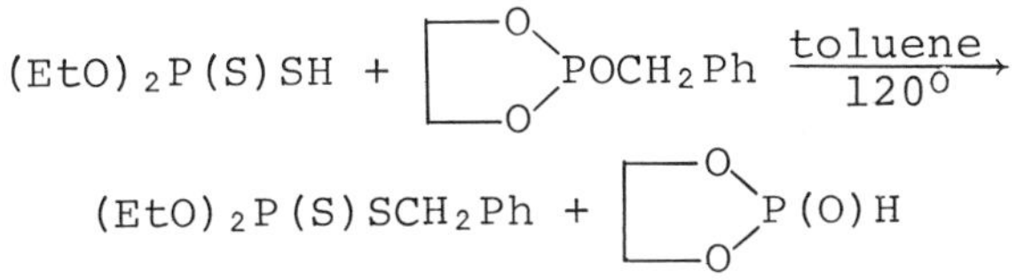

Schrader, Schliebs (1967):[1204]

$$(EtO)_2P(S)SH + (MeO)_2P(O)C\equiv CMe \xrightarrow[50°]{Et_3N\ cat.}$$

$$(EtO)_2P(S)SMe + MeO(HO)P(O)C\equiv CMe$$

While alkylation of diethylphosphorodithioate salts and higher homologs proceeds cleanly, the scrambling of methyl groups when dimethylphosphorodithioate salts are refluxed in acetone with tertiary trimethyl esters has been demonstrated by Itskova et al. (see also Method III):

$$(MeO)_2P(S)SK + (MeO)_2P(S)SMe \rightarrow (MeO)_2P(S)SMe + MeO(MeS)POSK$$

$$MeO(MeS)POSK + (MeO)_2P(S)SMe \rightarrow (MeS)_2P(O)OMe + MeO(MeS)POSK$$

$$MeO(MeS)POSK + (MeS)_2P(O)OMe \rightarrow (MeS)_2P(O)OMe + (MeS)_2P(O)OK$$

Kabachnik, Ioffe, and Mastryukova (1955),[491] in an investigation of tautomerism of O,O-diethyl phosphorothioic acid, demonstrated the use of diazomethane to prepare the methyl esters. Reaction of the acid in ether, by the conventional procedure, gave two products (91.4% total yield) in a 4:1 ratio of thiolo:thiono ester. On the other hand, reaction of the sodium salt in a benzene-water system gave only the thiolo ester (38% yield):

$$(EtO)_2POSH + CH_2N_2 \xrightarrow{\text{ether}}$$

$$(EtO)_2P(O)SMe + (EtO)_2P(S)OMe;\ 4:1$$

Shafik, Bradway, Biros, and Enos (1970)[1223a] used gas chromatography and mass spectral data to identify the products of alkylation of potassium O,O-diethylphosphorothioate with diazoalkanes. With CH_2N_2 the thiolo:thiono ratio was 85:15; with $MeCHN_2$, 62.5:37.5. With $RCHN_2$, O-alkylation (to produce thiono compound) increased as R was increased.

The use of ethyl orthoformate was demonstrated by McConnell and Coover (1961):[669]

$$(EtO)_2P(S)SH + HC(OEt)_3 \xrightarrow{55\text{-}100^\circ}$$

$$(EtO)_2P(S)SEt + EtOH + HCOOEt$$

Gololobov et al. (1965)[372] used divinyl mercury to form vinyl esters:

$$(RO)_2P(S)SH + (CH_2{=}CH)_2Hg \xrightarrow{70\text{-}80^\circ}$$

$$(RO)_2P(S)SCH{=}CH_2 + CH_2{=}CH_2 + Hg$$

Vogel and Meinhardt (1959)[1364] used diethylphosphorochloridothioate as an alkylating agent in the following reaction:

$$(BuO)_2P(S)SH + Et_3N + (EtO)_2P(S)Cl \xrightarrow[85^\circ]{\text{dioxane}} (BuO)_2P(S)SEt$$

Diazotized aromatic amines can be used with phosphorodithioate or -thioate salts to prepare S-aryl esters. Reactions were conducted by conventional methods at 0-2° with Cu powder catalyst followed by warming to 40-50° until N_2 evolution ceased:

Mel'nikov et al. (1957-8),[742, 750] Bianchetti (1957):[94]

$$(RO)_2P(S)SK + ArN_2X \rightarrow (RO)_2P(S)SAr + N_2 + KX$$

$$(RO)_2P(O)SNa + ArN_2X \rightarrow (RO)_2P(O)SAr + N_2 + NaX$$

Aromatic sulfonate esters have been used to esterify alkali salts of phosphorodithioic and -thioic acids:

Arbisman et al. (1967):[48]

$$(RO)_2POSNa + 4\text{-}MeC_6H_4SO_3CH_2C{\equiv}CH \xrightarrow[\text{reflux}]{\text{acetone}}$$

$$(RO)_2P(O)SCH_2C{\equiv}CH + 4\text{-}MeC_6H_4SO_3Na$$

Prib, Glushkova, Sendega (1968):[972]

$$(RO)_2P(S)SK + 4\text{-}MeC_6H_4SO_3CH_2CH{=}CHPh \xrightarrow[\text{R.T.}]{\text{acetone}}$$

$$(RO)_2P(S)SCH_2CH{=}CHPh + 4MeC_6H_4SO_3K$$

Glushkova et al. (1968):[370]

$$(ArO)_2P(S)SK + PhSO_3C_6H_3(NO_2)_2\text{-}2,6 \xrightarrow[\text{reflux}]{\text{benzene}}$$

$$(ArO)_2P(S)SC_6H_3(NO_2)_2\text{-}2,6 + PhSO_2K$$

Note: $PhSO_3C_6H_3(NO_2)_2$-2,4 gave no reaction.

Alkyl- and arylthiosulfonic S-alkyl or S-aryl esters have also been used [Lorenz, Schrader (1961)]:[642]

$$(EtO)_2P(S)SH + PhSO_2SR + Et_3N \xrightarrow[0\text{-}10^\circ]{\text{MeOH}}$$

$$(EtO)_2P(S)SR + PhSO_2SH$$

Michalski and Tulimowski (1966)[790] investigated the ambident reactivity of the organophosphorus selenothio anion. The results of alkylation and acylation were interpreted in terms of "hard" and "soft" acids and bases. Reactions of $(EtO)_2PSSeNa$ with organic halides gave the following product ratios:

Halide	Thiono	Selenono
EtI	$(EtO)_2P(S)SeEt$	--
$ClCH_2OMe$	60% $(EtO)_2P(S)SeCH_2OMe$	40% $(EtO)_2P(Se)SCH_2OMe$
$Et_3O^+ BF_4^-$	20% $(EtO)_2P(S)SeEt$	80% $(EtO)_2P(Se)SEt$
PhCOCl	--	90% $(EtO)_2P(Se)SCOPh$
$(EtO)_2P(O)Cl$	$(EtO)_2P(S)OP(Se)(OEt)_2$	

XXVIII. CLEAVAGE OF DISULFIDES AND DISELENIDES BY PHOSPHITES

Reaction of trialkyl phosphites with dialkyl disulfides in the absence of radical initiators at temperatures above 100° for long periods (10-40 hr) produced thiolophosphate esters and dialkyl sulfides. The reaction was formulated as an Arbuzov-type process by Jacobson, Harvey, and Jensen (1955);[481] Harvey et al. (1963):[417]

$$(EtO_3P + EtSSEt \rightarrow EtS^- \; EtSP(OEt)_3^+ \rightarrow$$

$$(EtO)_2P(O)SEt + Et_2S$$

Irradiation with a General Electric R S sunlamp at 60° or the use of di-t-butyl peroxide at 120-125° for 3 hr gave thionophosphate esters and dialkyl sulfides. A radical mechanism was given by Walling and Rabinowitz (1957-1959):[1371, 1372]

$$(EtO)_3P + RSSR \rightarrow (EtO)_3P(S) + RSR$$

While not ruling out the earlier reaction mechanism, the

suggestion was made that thermal isomerization may have occurred during the prolonged heating periods in the earlier work (see Method XXXI):

$$(RO)_3P(S) \overset{\sim}{\rightarrow} (RO)_2P(O)SR$$

Pilgram and Korte (1964):[954]

$$(MeO)_3P + (2\text{-}Me\text{-}3\text{-}NO_2\text{-}C_6H_3S)_2 \xrightarrow[\text{R.T.-}60^{\circ}]{\text{exotherm.}}$$
$$MeSC_6H_3\text{-}2\text{-}Me\text{-}3\text{-}NO_2 + (MeO)_2P(O)SC_6H_3\text{-}2\text{-}Me\text{-}3\text{-}NO_2$$

Bis(carbalkoxyalkyl) disulfides were readily cleaved by trialkyl phosphites without catalyst at 100-115°. Reactivities were apparently related to the pK_a values of the parent mercaptans:

Ailman (1965):[6]

$$(MeO)_3P + (SCH_2COOEt)_2 \rightarrow (MeO)_2P(O)SCH_2COOEt + MeSCH_2COOEt$$

$$(MeO)_3P + [SCH(COOEt)CH_2COOEt]_2 \rightarrow MeSCH(COOEt)CH_2COOEt$$
$$+ (MeO)_2P(O)SCH(COOEt)CH_2COOEt \text{ (Malaoxon)}$$

Michalski and Wieczorkowski (1957)[793] cleaved thiocyanogen and diacyldisulfides with trialkyl phosphites at room temperature to obtain thionophosphates:

$$(RO)_3P + (SCN)_2 \xrightarrow[\text{R.T.}]{} (RO)_3P(S) + S(CN)_2$$

$$(EtO)_3P + (SCOMe)_2 \rightarrow (EtO)_3P(S) + S(COMe)_2$$

Alkyl (dialkoxyphosphinyl) disulfides were cleaved by phosphites to give mixtures of thiono and thiolo phosphates:[793]

$$2(EtO)_3P + 2BuSSP(O)(OEt)_2 \rightarrow$$
$$(EtO)_2P(O)SEt + (EtO)_3P(S) + 2(EtO)_2P(O)SBu$$

Sulfenylmonothioates or -dithioates reacted with trialkyl phosphites in solution [Schrader (1961)]:[1159]

$$(EtO)_3P + (EtO)_2P(S)SSC_6H_4Cl\text{-}4 \xrightarrow[40^{\circ}]{\text{benzene}}$$
$$(EtO)_2P(O)SC_6H_4Cl\text{-}4 + (EtO)_2P(S)SEt$$

Methyl trichloromethyl disulfide reacted with trialkyl phosphites to form phosphorothiolate esters and $R_7CO_9P_3S$ [Birum (1961)]:[97]

$$(EtO)_3P + MeSSCCl_3 \xrightarrow[155^{\circ}]{5\text{ hr}} (EtO)_2P(O)SMe + Et_7CO_9P_3S$$

The use of dialkyl phosphites or -thiophosphites or their alkali metal salts proceeded redily at 20° in solvents [Michalski et al. (1959)]:[795]

$$(EtO)_2PONa + RSSR \xrightarrow[20^\circ]{benzene} (EtO)_2P(O)SR + RSNa$$

$$(EtO)_2PSNa + BuSSBu \rightarrow (EtO)_2P(S)SBu + BuSNa$$

Petrov, Neimysheva (1959):[944a]

$$(EtO)_2PONa + (MeCHClCH_2S)_2 \xrightarrow[125\text{-}130^\circ]{toluene}$$

$$(EtO)_2P(O)SCH_2CHClMe + MeCHClCH_2SNa$$

Petrov et al. (1961)[941] demonstrated the use of a catalytic amount of sodium with secondary phosphites or thiophosphites in nitrogen atmosphere under reduced pressure for removal of mercaptan as formed:

$$(RO)_2P(O)H + R'SSR' \xrightarrow[110\text{-}140^\circ]{Na,\ N_2} (RO)_2P(O)SR' + R'SH$$

$$(RO)_2P(S)H + R'SSR' \rightarrow (RO)_2P(S)SR' + R'SH$$

Farbenfabriken Bayer (1963);[297] Lorenz, Schrader (1963):[647]

$$(EtO)_2P(S)H + PhCH_2SeSeCH_2Ph + NaOMe \rightarrow$$

$$(EtO)_2P(S)SeCH_2Ph + HSeCH_2Ph$$

Mel'nikov et al. (1960):[759]

$$(MeO)_2P(O)H + (RO)_2P(S)SS(S)P(OR)_2 + Et_2N \rightarrow$$

$$2(RO)_2P(S)SMe + Et_3N\cdot HPO_3$$

[R = Me, Et, i-Pr, Pr, Bu]

XXIX. REACTIONS OF PHOSPHOROTHIOITES WITH ALDEHYDES AND KETONES

Secondary phosphorothioites, $(RO)_2P(S)H$, react with chloral and other activated aldehydes and ketones to form intermediate phosphonates which rearrange to phosphorothiolates or -thionates.

Thiolo analogs of the insecticide dichlorvos are obtainable from dialkyl phosphorothioites and chloral [Pelchowicz (1961)].[929] From the work of Lorenz and Schrader (1963),[648] it appeared that, in benzene a condensation product was formed which then lost HCl:

$$(RO)_2P(S)H + Cl_3CCHO \xrightarrow[50\text{-}60^\circ]{benzene}$$

(intermediate) $\xrightarrow[-HCl]{}$ $(RO)_2P(O)SCH=CCl_2$

In the reactions of dialkyl phosphorothioites with butyl glyoxalate, evidence for an intermediate hydroxy-(carbobutoxy)methylphosphonic ester (21) was presented. The intermediate, in the presence of sodium alcoholate, rearranged to the phosphorothionate (22) [Pudovik and Gur'yanova (1967)]:[994]

$$(RO)_2P(S)H + BuOCOCHO \xrightarrow[50-100^\circ]{} (RO)_2P(S)CHOHCOOBu \qquad (21)$$

$$\xrightarrow[\sim]{EtONa} (RO)_2P(S)OCH_2COOBu \qquad (22)$$

The reaction of ethyl benzoyl formate was similar [Pudovik et al. (1968)].[995] In this case the intermediate phosphonate (23) was isolated and heated with or without sodium ethylate to give the phosphorothionate (24):

$$(RO)_2P(S)H + PhCOCOOEt \xrightarrow[50-110^\circ]{} (RO)_2P(S)CPhOHCOOEt \qquad (23)$$

$$\xrightarrow[110^\circ]{\pm\ EtONa} (RO)_2P(S)OCHPhCOOEt \qquad (24)$$

Nishimura, Okuda, and Yoshido (1971)[879] treated di-alkyl phosphorothioites with 2-pyridone in the presence of tertiary amines to prepare phosphorothionates:

$$(RO)_2P(S)H + O=\overset{HN-}{\bigcirc} + Et_3N \xrightarrow[reflux]{CCl_4}$$

$$(RO)_2P(S)O-\overset{N=}{\bigcirc} + CHCl_3 + Et_3N \cdot HCl$$

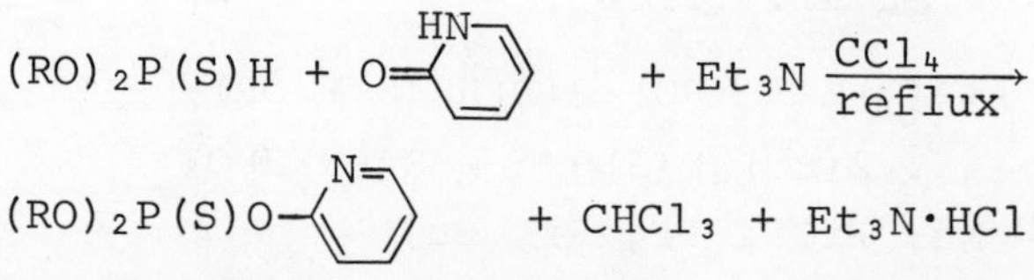

Schrader (1959)[1154] treated tertiary phosphorothioites with chloral to eliminate RCl and obtain dichlorovinyl-phosphorothionates:

$$(EtO)_2PSEt + Cl_3CCHO \xrightarrow[40^\circ]{benzene} (EtO)_2P(S)OCH=CCl_2 + EtCl$$

The analogous reaction, in which a glycolphosphoro-thioite was used, resulted in ring opening and introduction of the dichlorovinyloxy esterifying group to give a phosphorothiolate:

$$\overset{\overset{O}{\shortmid}}{\underset{\underset{O}{\shortmid}}{\boxed{}}}PSMe + Cl_3CCHO \xrightarrow[40^\circ]{benzene} ClCH_2CH_2O(Cl_2C=CHO)P(O)SMe$$

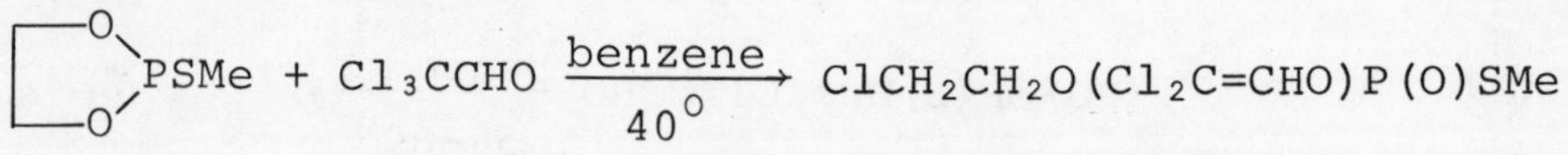

XXX. REACTIONS OF PHOSPHORODITHIOATES WITH ISOCYANATES

Addition of dialkyl phosphorodithioates to isocyanates at temperatures below 40° afforded isolable thiocarbamates [Ottmann and Hooks (1966)].[923] These, on heating, gave isothiocyanates:

$$(EtO)_2P(S)SH + RNCO \xrightarrow[20-25^\circ]{}$$

$$(EtO)_2P(S)SCONHR \xrightarrow[125-130^\circ]{} RNCS + (EtO)_2P(S)OH$$

[R = Me, et, i-Pr, Pr, Bu]

Addition of O,O-dimethyl phosphorodithioic acid to p-thiocyanatophenyl isocyanate in hexane in an atmosphere of argon gave O,O-dimethyl-S-carbo-p-thiocyanatoanilidophosphorothioate [Nifant'eva et al. (1968)]:[873]

$$(MeO)_2P(S)SH + OCNC_6H_4SCN\text{-}4 \rightarrow (MeO)_2P(S)SCONHC_6H_4SCN\text{-}4$$

XXXI. ISOMERIZATION OF TERTIARY PHOSPHOROTHIONATES TO -THIOLATES

Thiono-thiolo isomerization may be carried out thermally, catalytically, or by reactions with organic halides under the conditions of Method XVI.

Thermal isomerization was studied by Fukuto and Metcalf (1954)[345] on ^{32}P Systox and found to be complete in 4 hr at 120-130°:

$$(EtO)_2{}^{32}P(S)OCH_2CH_2SEt \rightarrow (EtO)_2{}^{32}P(O)SCH_2CH_2SEt$$

Isomerizations of parathion, methyl parathion, and malathion were carried out at 150° by Metcalf and March (1953).[766]

Schrader and Gönnert (1956)[1185] accomplished an isomerization rapidly at 200-210° under reduced pressure:

$$(4\text{-}EtOOCC_6H_4O)_2P(S)OMe \xrightarrow[1-2\text{ mm, }10\text{ min}]{200-210^\circ} (4\text{-}EtOOCC_6H_4O)_2P(O)SMe$$

With crotyl or methallyl esters, Pudovik and Aladzheva (1960)[981] observed that isomerization was accompanied by inversion:

$$(RO)_2P(S)OCH_2CH{=}CHMe \xrightarrow[3-6\text{ hr}]{120-140^\circ} (RO)_2P(O)SCHMeCH{=}CH_2$$

$$(RO)_2P(S)OCHMeCH{=}CH_2 \longrightarrow (RO)_2P(O)SCH_2CH{=}CHMe$$

Isomerization can also be conducted with catalysts such as HgI_2, $FeCl_3$, Ph_3P, or Et_4NI:
Pistschimuka (1912):[961]

$$(MeO)_3P(Se) + HgI_2 \rightarrow (MeO)_2P(O)SeMe \cdot HgI_2$$

When $FeCl_3$ is used, an adduct may be obtained, according to Hilgetag, Teichmann, Krueger (1965);[439] (1970):[440]

$$3(MeO)_3P(S) + 2FeCl_3 \xrightarrow[\text{below } 25]{\text{petr. ether}} [(MeO)_2P(O)SMe]_3 \cdot (FeCl_3$$

$$(MeO)_2P(S)OPh + FeCl_3 \xrightarrow{\text{benzene}} MeO(MeS)P(O)OPh$$

The isomerization may also be catalyzed by NaOMe or KOH in MeOH [Hilgetag, Teichmann (1959)]:[438]

$$(MeO)_3P(S) + MeONa \xrightarrow[\text{reflux}]{\text{MeOH}} (MeO)_2P(O)SMe$$

Ford-Moore and Wood (1960)[340] used $NaOCH_2CH_2SEt$ as catalyst:

$$(EtO)_2P(S)OCH_2CH_2SEt \xrightarrow[\text{5 hr}]{\text{to } 130^{\circ}} (EtO)_2P(O)SCH_2CH_2SEt$$

Nguyen, Thuong, Chabrier (1971):[871,872]

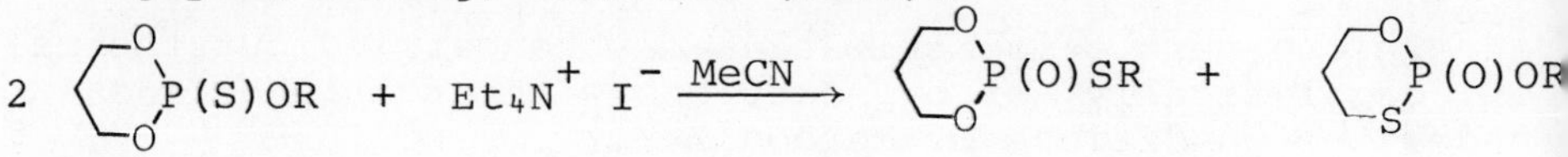

The isomerization may be performed with alkyl halides (see Method XVI) [McIvor, McCarthy and Grant (1956)]:[675]

$$(EtO)_3P(S) + EtI \xrightarrow{\text{reflux}} (EtO)_2P(O)SEt$$

Frequently isomerization of O,O-dimethylphosphorodithioic or -thioic acid derivatives is accomplished through a process of demethylation, using potassium O,O-diisopropylphosphorodithioate, and remethylation with methyl sulfate or methyl iodide. The remethylation was included in Method XVII. The overall process (using malathion ⟶ isomalathion as an example) is [Du Breuil, Curry et al. (1964-1967):[192,235]

$$(MeO)_2P(S)SCH(COOEt)CH_2COOEt + (i\text{-}PrO)_2P(S)SK \rightarrow$$

$$K^+ \; MeOP\binom{O^-}{S}SCH(COOEtCH_2COOEt \; (\underline{25}) + (i\text{-}PrO)_2P(S)SMe$$

$$(\underline{25}) + (MeO)_2SO_2 \rightarrow MeO(MeS)P(O)SCH(COOEt)CH_2COOEt$$

The demethylation may also be accomplished with a tertiary amine with formation of a quaternary ammonium salt which is then treated with RX or $(RO)_2SO_2$ [Chabrier and Thuong (1964)]:[154]

$$(MeO)_3P(S) + Me_3N \xrightarrow{\text{acetone}} (MeO)_2P(O)S^- \; NMe_4^+ \xrightarrow{RX}$$

$$(MeO)_2P(O)SR + Me_4NX$$

The reaction is also applicable to $(MeO)_3P(Se)$.[154]

The preparation of optically active thiophosphoric acid esters [Hilgetag and Lehmann (1957)][433] by the following series of reactions involves thiono-thiolo isomerization: [R = $C_6H_4NO_2$-4]

$$(MeO)_2P(S)OR \xrightarrow{AgNO_3} MeO(RO)P(S)OAg \xrightarrow{\text{strychnine}\cdot MeI}$$

$$(MeO)_2P(S)OR\cdot\text{strychnine} \xrightarrow{HClO_4} \xrightarrow{Pb^{2+}} [(MeO)(RO)P(S)(O)]_2Pb$$

(2 diastereomeric forms)

$$\xrightarrow{MeI} \text{d- and l-}MeO(MeS)P(O)OR$$

Demethylation and rearrangement occur on treatment with sodium iodide [Schrader and Scheinpflug (1968)].[1202a] The resulting salt can be alkylated or acylated by procedures of Method XVII:

$$(MeO)_2P(S)OC_6H_2Cl_3\text{-}2,4,5 \xrightarrow[40-50^\circ]{\text{NaI, acetone}} MeS(O^-)P(O)OC_6H_2Cl_3$$

$$\text{-}2,4,5\text{-}Na^+ \xrightarrow[60^\circ]{ClCOO\text{-}i\text{-}Pr \text{ and } C_5H_5N} MeS(2,4,5\text{-}Cl_3C_6H_2O)P(O)$$

$$OCOO\text{-}i\text{-}Pr$$

(The final product was not characterized.)

Onium salts are obtainable from thionophosphates and nucleophiles such as Me_2S, $(H_2N)_2CS$, and $PhNMe_2$. These decompose to form thiolophosphates [Hilgetag et al. (1959)]:[435]

$$(PhO)_2P(S)OMe + Me_2S \xrightarrow[R.T.]{} (PhO)_2P(S)O^- \; Me_3S^+ \xrightarrow{90^\circ}$$

$$Me_2S + (PhO)_2P(O)SMe$$

The isomerization of parathion has been accomplished using ethyl sulfenyl chloride by Jaras (1956):[484]

$$(EtO)_2P(S)OC_6H_4NO_2\text{-}4 \xrightarrow{EtSCl} EtO(EtS)P(O)OC_6H_4NO_2\text{-}4$$

XXXII. PREPARATION OF SALTS AND COMPLEXES OF TERTIARY ESTERS

A mixture of the known methylstrychninium salts was obtained by refluxing methyl parathion with strychnine in methanol or acetonitrile without the use of $AgNO_3$ (Method XXXI) [Hilgetag and Lehmann (1959)]:[432]

$$(MeO)_2P(S)OC_6H_4NO_2\text{-}4 \ + \ strychnine \ \rightarrow \ MeOP\overset{O^-}{\underset{S}{\big(\!\!\big|\!\!\big)}}OC_6H_4NO_2$$

$$\text{-4}\cdot methylstrychninium^+$$

Thionophosphates formed adducts with $FeCl_3$, $AlCl_3$, $SnCl_4$, $SnBr_4$, or $TiCl_4$ in petroleum ether. Hydrolysis of the adducts with $CaCl_2$ in water produced thiolophosphates (see also Method XXXI) [Hilgetag, Teichmann, and Krueger (1970)]:[440]

$$(MeO)_3P(S) \ + \ MX_n \ \longrightarrow \ complexes \ \xrightarrow{\ CaCl_2\ }{water} \ (MeO)_2P(O)SMe$$

Reactions of tertiary thiono or dithio esters or anhydrosulfides with Mannich bases containing tertiary nitrogen at room temperature or in boiling benzene resulted in alkylation of the base and the formation of salts of the residual phosphates [Mel'nikov et al. (1968)]:[754]

$$(MeO)_3P(S) \ + \ PhCOCH_2CH_2N\!\!\bigcirc\!\!O \ \xrightarrow[reflux]{benzene} \ (MeO)_2P\overset{O^-}{\underset{S}{\big(\!\!\big|\!\!\big)}}$$

$$PhCOCH_2CH_2\overset{+}{\underset{Me}{N}}\!\!\bigcirc\!\!O$$

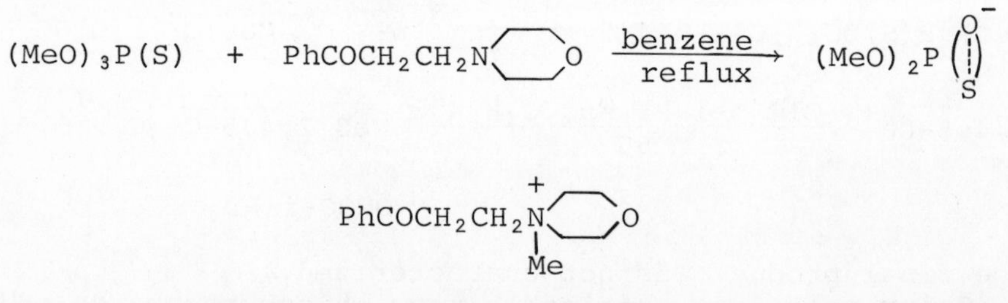

Stec and Michalski (1970)[1265] obtained a distillable complex, $(EtO)_2P(O)SEt\cdot HOOCCCl_3$, from the components. IR and NMR spectra showed that the complex involved H bonding between the basic O of the phosphorothioate and the OH group of trichloroacetic acid.

XXXIII. Miscellaneous Syntheses

In the category of miscellaneous synthese are listed methods for preparing anhydrides (Sections F.5.1 and F.5.4) anhydrosulfides (Sections F.5.2 and F.5.5), and polysulfide (Sections F.5.3 and F.5.6). Some recent preparative method that have not yet found broad application may also be found here.

Note: The methods listed here always involve substitution on phosphorus or on one of the O, S, or Se atoms attached to phosphorus. Preparations involving substitution in a remote part of the molecule by the usual methods of organic chemistry, such as $(RO)_2P(S)OCH_2CH_2OH \longrightarrow (RO)_2P(S)OCH_2CH_2OR$, will be found in the tables of compounds under the product in question.

Toy (1951):[1318]

$$2(EtO)_2P(S)Cl + H_2O + C_5H_5N \xrightarrow{38-40^\circ} (EtO)_2P(S)OP(S)(OEt)_2 + 2HCl$$

Farbenfabriken Bayer (1959):[292]

$$2(RO)_2P(S)H + Me_2NS(O)Cl \xrightarrow{benzene} (RO)_2P(S)\overset{O}{\overset{\|}{S}}P(S)(OR)_2 + Me_2NH\cdot HCl$$

Stoelzer, Simon (1963):[1270]

$$EtOP(S)ClF + (EtO)_2P(S)F \xrightarrow[3.5\ hr]{185-195^\circ} [(EtO)P(S)F]_2O + EtCl$$

Edmundson (1965):[247]

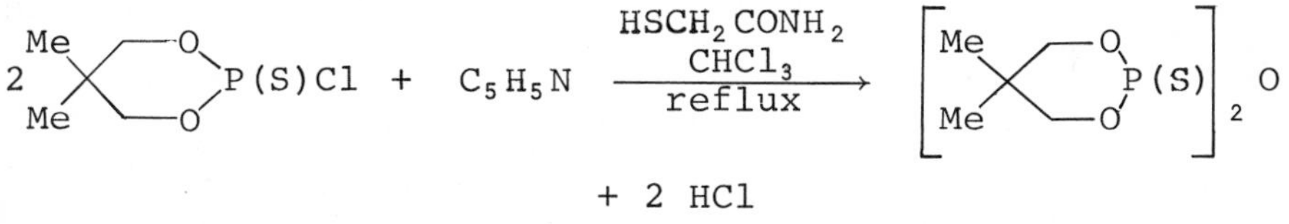

Almasi, Hantz (1966):[22]

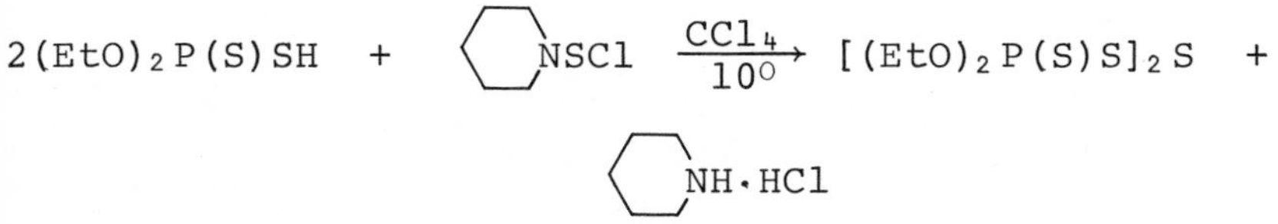

Almasi, Hantz (1966):[23]

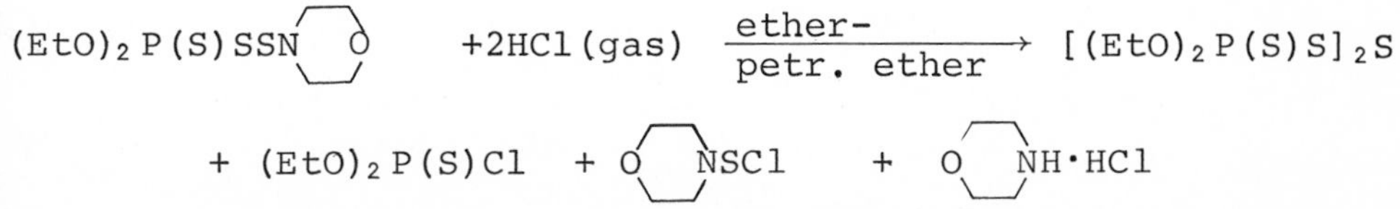

Mel'nikov et al. (1960):[760]

$$[(RO)_2P(S)S]_2 \quad + \quad (ArO)_3P \xrightarrow[\text{reflux}]{\text{benzene}} [(RO)_2P(S)]_2S \quad + \quad (ArO)_3P(S)$$

Edmundson (1965):[247]

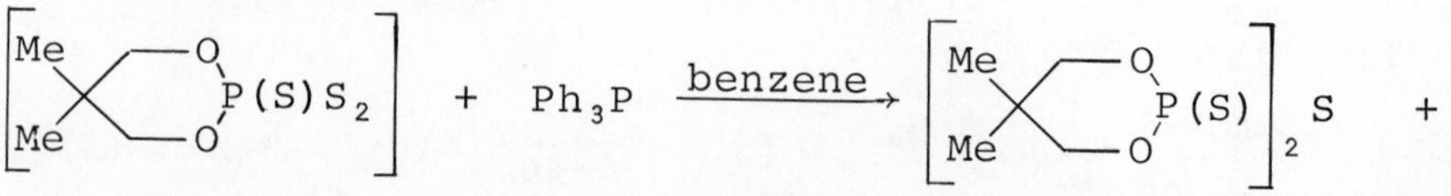

$$Ph_3P(S)$$

Michalski, Wieczorkowski (1957):[593]

$$[(EtO)_2P(O)S]_2 \quad + \quad (PhO)_3P \xrightarrow[\text{R.T.}]{} [(EtO)_3P(O)]_2S \quad + \quad (PhO)_3P(S)$$

Stoffey (1968):[1272]

$$(RO)_2P(S)SH \quad + \quad MeCONHCH_2CH_2SO_2SCH_2CH_2NHCOMe \xrightarrow[\text{R.T. 5 days}]{\text{acetone}}$$

$$(RO)_2P(S)SSCH_2CH_2NHCOMe \quad + \quad MeCONHCH_2CH_2SO_2H$$

Lorenz (1961):[614]

$$(EtO)_2P(S)SH \quad + \quad NaOSO_2SEt \xrightarrow[\text{R.T.}]{\text{water}} (EtO)_2P(S)SSEt \quad + \quad NaHS$$

Lorenz (1962):[615]

$$(EtO)_2P(S)SH \quad + \quad MeOSO_2SCH_2COOEt \xrightarrow[\text{R.T.}]{\text{water}}$$

$$(EtO)_2P(S)SSCH_2COOEt \quad + \quad SO_2 \quad + \quad MeOH$$

Scherer, Wokulat (1968):[1089]

$$[(EtO)_2P(S)S]_2 \quad + \quad NaN(SiMe_3)_2 \xrightarrow[80^{\circ}\ 12\ \text{hr}]{\text{benzene}}$$

$$(EtO)_2P(S)SN(SiMe_3)_2 \quad + \quad (EtO)_2P(S)SNa$$

Malz et al. (1959):[690]

$$(RO)_2P(S)SNa \quad + \quad ClN(C_6H_{11})_2 \xrightarrow[\text{water}]{\text{acetone-}}$$

$$(RO)_2P(S)SN(C_6H_{11})_2 \quad + \quad NaCl$$

Pudovik, Krupnov (1968):[1002]

$$(EtO)_2P(S)SH \ + \ (EtO)_2POCOMe \ \xrightarrow[70°]{6 \ hr} \ (EtO)_2P(S)SCOMe$$

$$+ \ (EtO)_2P(O)H$$

Truchlik, Masek, Drabek (1957):[1323]

$$2(RO)_2P(S)SH \ + \ HCl \ + \ ClN \underset{O}{\overset{O}{\diagup}} \ \longrightarrow \ [(RO)_2P(S)S]_2 \ +$$

$$2HCl \ + \ HN \underset{O}{\overset{O}{\diagup}}$$

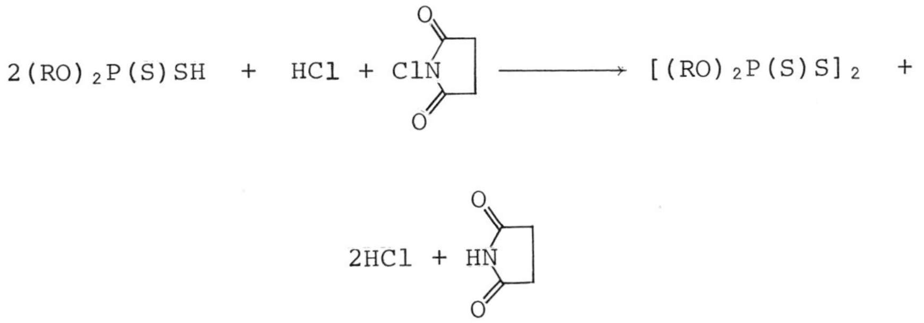

Miller (1960-1964)[802,804,805,807] has demonstrated the cleavage of dithiophosphoryl disulfides in ether or ether-hexane with metal alkyls or Grignard reagents:

$$[(RO)_2P(S)S]_2 \ + \ R'M \longrightarrow (RO)_2P(S)SR' \ + \ (RO)_2P(S)S^- \ M^+$$

where R'M is: $MeC\equiv CLi$;[804] $PhCH=CHMgBr$ or $CH_2=CHMgBr$;[805] BuLi or PhMgBr.[807]

Gay and Hamer (1970)[350] prepared sodium O,S-ethylene phosphorothioate by treatment of 2-chloroethyl phosphorodichloridothioate with a suspension of barium carbonate in dioxane followed by treatment with aqueous sodium sulfate:

$$ClCH_2CH_2OP(S)Cl_2 \ \xrightarrow{BaCO_3} \ \xrightarrow{Na_2SO_4} \ \begin{array}{c} O \\ \diagup \\ S \end{array} P(O)ONa$$

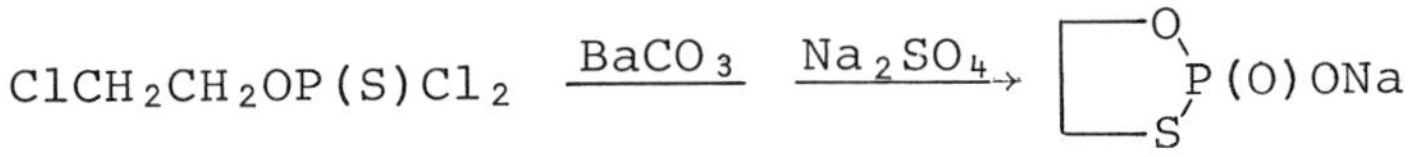

Its base solvolysis was studied in excess sodium hydroxide solution in aqueous ethanol 3:1 (v/v) at 25°. Second-order kinetics was observed with $k_2 = 0.70$ liter/mole/min. The solvolysis reaction was:

$$\begin{array}{c} O \\ \diagup \\ S \end{array} P(O)O^- \ \xrightarrow{NaOH} \ HOCH_2CH_2SP(O)(O^-)_2$$

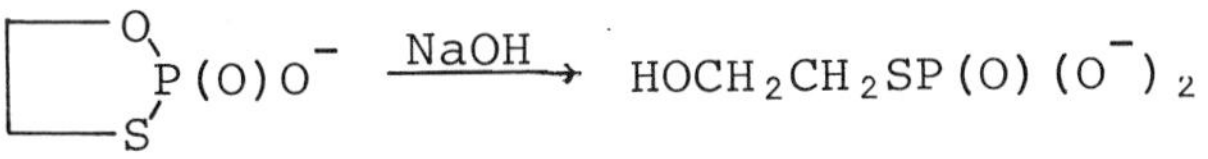

The rates and mechanism were compared with open-chain phosphorylated derivatives of 2-mercaptoethanol.

Mel'nikov et al. (1958)[758] (9.5% yield):

$(EtO)_2P(S)Cl$ + [furfuryl-CH_2OH] + C_5H_5N $\xrightarrow[\text{R.T.}-100^\circ]{\text{benzene}}$

$(EtO)_2P(S)OCH_2$—[furfuryl ring]

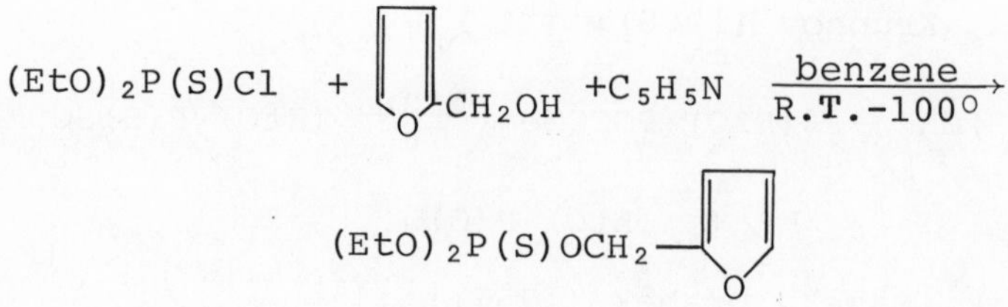

Lorenz and Schrader (1962):[643]

$(MeO)_2P(O)OH$ + $NaSCH_2CH_2SEt$ + $NaOMe$ $\xrightarrow[0-5^\circ]{\text{MeOH-benzene}}$

$(MeO)_2P(O)SCH_2CH_2SEt$ + $NaOH$

Bunyan and Cadogan (1962):[138]

$2(EtO)_3P$ + $BuSH$ + $2BrCCl_3$ $\xrightarrow{\text{heat}}$ $(EtO)_2P(O)SBu$ +

$(EtO)_2P(O)CCl_3$ + $2EtBr$ + $CHCl_3$

Murdock and Hopkins (1968):[853]

$2(MeO)_3P$ + $2BrCCl_3$ + $2PhSH$ $\xrightarrow[\text{reaction}]{\text{exothermic}}$

$(MeO)_2P(O)SPh$ + $2CHCl_3$ + $2CH_3Br$

The above reactions reportedly involve a radical-chain transfer mechanism.[853] Best yields were obtained with a ratio of $(MeO)_3P:BrCCl_3:PhSH=1:2:4$.

Shlenkova and Evdakov (1967):[1237]

$(RO)_2POCOMe$ + $2R'SH$ $\xrightarrow[40-50^\circ]{R''_3N}$ $(RO)_2P(S)SR'$ + $R'H$ +

$MeCOOH \cdot R''_3N$

H. Hoffmann, Scheinpflug (1968):[453]

[Me—O / Me—O ring]PCl + ClC_6H_4SCl-4 $\xrightarrow[25^\circ]{CH_2Cl_2}$ $\xrightarrow[\text{heat}]{\text{MeOH}}$

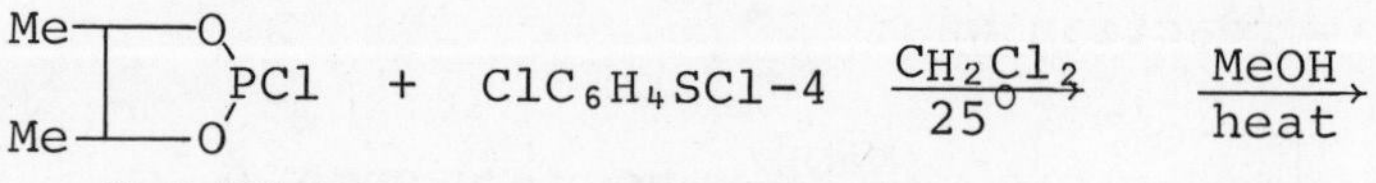

$MeO(ClCHMeCHMeO)P(O)SC_6H_4Cl-4$ + HCl

Nishizawa (1961).[882]

$(EtO)_2P(S)H$ + $PhCOCHClCOMe$ $\longrightarrow$ $(EtO)_2P(S)OC(Ph)=CHCOMe$

+ HCl

Petrov et al. (1961):[942]

$(EtO)_2P(O)H$ + Et_2NSBu $\xrightarrow[\text{1-2 hr}]{\text{100-120}^\circ}$ $(EtO)_2P(O)SBu$ + Et_2NH

$(BuO)_2P(S)H$ + Et_2NSBu $\longrightarrow$ $(BuO)_2P(S)SBu$ + Et_2NH

Shell (1969):[1232]

$\overset{O}{\underset{O}{\diagup}}P(S)H$ + $Cl_2CHCOC_6H_3Cl_2$-2,4 + NH_3 $\xrightarrow[\text{10}^\circ]{\text{EtOH}}$

$\overset{O}{\underset{O}{\diagup}}P(S)OC(=CHCl)C_6H_3Cl_2$-2,4 + NH_4Cl

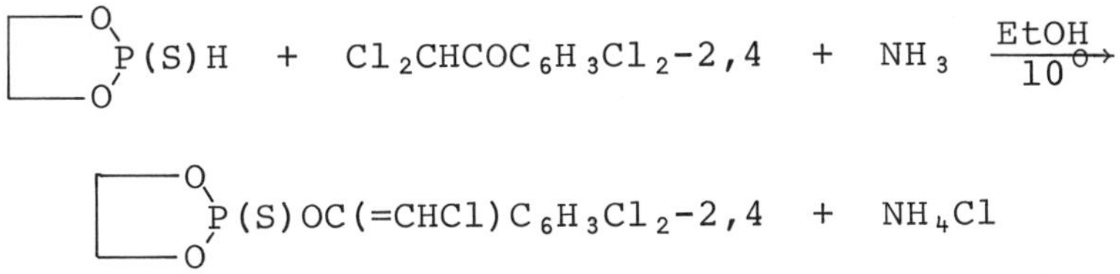

Demarcq and Sleziona (1966):[199]

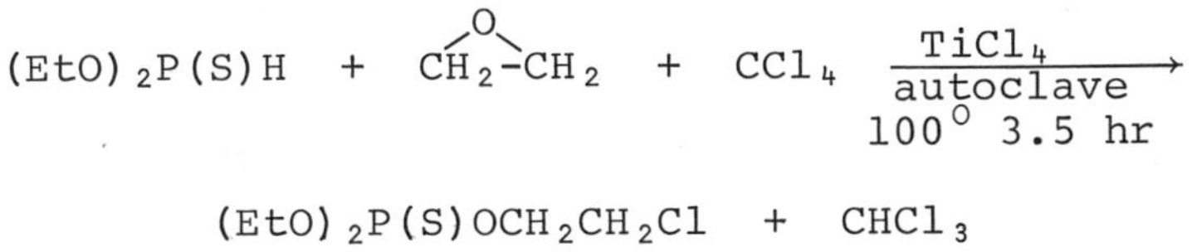

$(EtO)_2P(S)H$ + $CH_2\text{-}CH_2$ (epoxide) + CCl_4 $\xrightarrow[\text{100}^\circ \text{ 3.5 hr}]{\genfrac{}{}{0pt}{}{TiCl_4}{\text{autoclave}}}$

$(EtO)_2P(S)OCH_2CH_2Cl$ + $CHCl_3$

Diveley (1958):[210]

$(EtO)_2P(S)S\text{--}$ $\xrightarrow[\text{0.2 mm}]{\text{162}^\circ}$ $(EtO)_2P(S)S\text{--}$ +

$(EtO)_2P(S)SH$

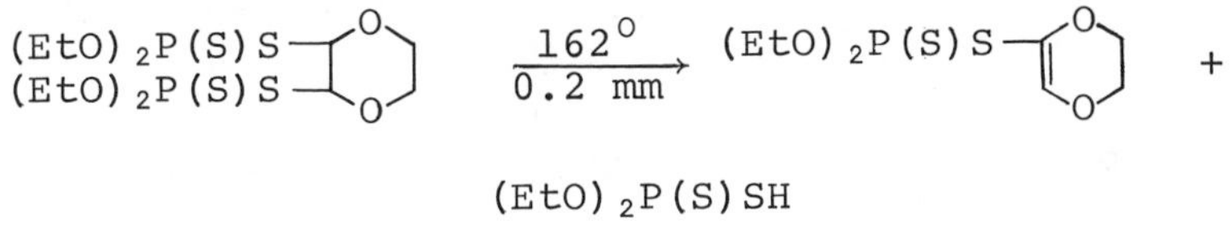

Also the oxathian analog: 150° (0.5 mm.): $(EtO)_2P(S)S\text{--}$

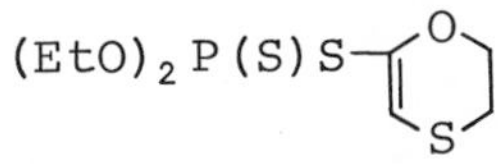

Johnson (1955):[485] HNO_3. Berkelhammer, Dauterman, O'Brien (1963):[87] N_2O_4:

$(MeO)_2P(S)SCH(COOEt)CH_2COOEt$ $\xrightarrow[\text{25-31}^\circ]{(O)}$
Malathion

$(MeO)_2P(O)SCH(COOEt)CH_2COOEt$ (+S)
Malaoxon

Also N_2O_4 for[87]:

$(EtO)_2P(S)OC_6H_4NO_2$-4 $\longrightarrow$ $(EtO)_2P(O)OC_6H_4NO_2$-4
Parathion   Paraoxon

Schrader (1961):[1160]

$$(EtS)_2P(S)OEt + EtSK + EtS(CH_2)_nCl \xrightarrow[70-80^o]{MeOH}$$

$$(EtS)_2P(O)S(CH_2)_nSEt + KCl + Et_2S$$

for n = 1 or 2.

Price, Walsh, Hallett (1966):[977]

$$(EtO)_2P(O)SCl + EtSCH_2CH_2OH + Et_3N \xrightarrow[10^o]{ether}$$

$$(EtO)_2P(O)SOCH_2CH_2SEt + Et_3N{\cdot}HCl$$

Blagoveshchenskii et al. (1969):[101]

$$(BuS)_3P(S) + PhOH \xrightarrow{180^o} (BuS)_2P(S)OPh + BuSH$$

Michalski, Tulimowski (1966):[790]

$$(EtO)_2PSSeNa + (EtO)_2P(O)Cl \longrightarrow$$

$$(EtO)_2P(S)OP(Se)(OEt)_2 + NaCl$$

Nussbaum and Tiberi (1965)[899] described a method for phosphorylating acid- and base-sensitive materials of biological interest. An alkylthio group on the phosphorylating agent was later removed by iodine at pH 7: R = a steroidal alcohol, 11-desoxycorticosterone:

$$ROH + EtSP(O)(O^-)_2 \xrightarrow[pyridine]{DCC} RO(EtS)P(O)O^-$$

$$\xrightarrow[H_2O]{I_2} ROP(O)(O^-)_2 + I^- + EtSSEt$$

(DCC = dicyclohexylcarbodiimide.)

Miscellaneous Reactions

In this section are listed a few reactions that gave a number of products. At least one of the products is to be found in the List of Compounds. However, due to the complexities of the reactions, these reactions are generally not to be considered as useful methods of preparation. No attemps have been made to balance the equations.

Almasi, Paskucz (1964):[31]

$$(EtO)_2PSNa + CCl_4 + EtOH \xrightarrow{EtOH} (EtO)_3P(S) +$$

$$[(EtO)_2PS]_2 + (EtO)_2P(S)Cl + CHCl_3$$

Zwierzak (1969):[1456]

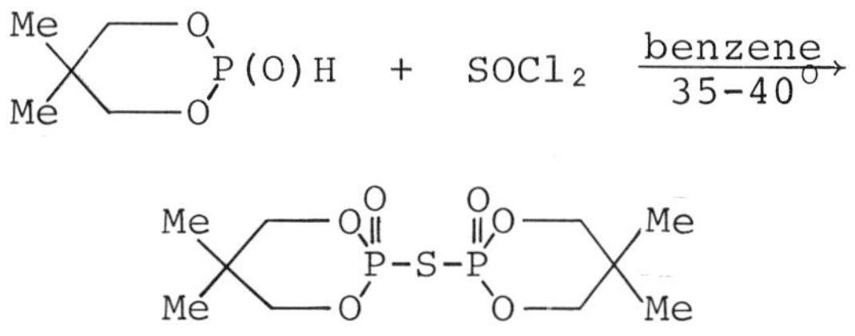

$+ :P(O)Cl,\ :P(O)O(S)P:,\ :P(O)O(O)P:,\ :P(O)OH,\ :P(S)OH$

plus three unidentified products.

Michalski, Skowronska (1970):[789]

$$(EtO)_2P(O)SCl + (EtO)_3P \xrightarrow[0-5^{\circ}]{benzene}$$

$$\underset{15\%}{(EtO)_2P(S)OP(O)OEt)_2} + \underset{40\%}{(EtO)_2P(O)Cl} + \underset{70\%}{(EtO)_3P(S)}$$

$$(EtO)_2P(O)SCl + (PhO)_3P \longrightarrow$$

$$\underset{72\%}{(EtO)_2P(O)Cl} + \underset{74\%}{(PhO)_3P(S)}$$

Almasi, Hantz (1965):[21]

$$(EtO)_2P(S)SNa + 4\text{-}ClC_6H_4SOCl \xrightarrow[5-10^{\circ}]{benzene}$$

$$[(EtO)_2P(S)S]_2 + 4\text{-}ClC_6H_4SO_2H + (EtO)_2P(S)SSC_6H_4Cl\text{-}4$$

$$+ 4\text{-}ClC_6H_4SO_2SC_6H_4Cl\text{-}4$$

Zemlyanskii, Prib, Drach (1961):[1451]

$$(RO)_2P(S)SK + ArSO_2Cl \xrightarrow[2-3^{\circ}]{acetone} [(RO)_2P(S)S]_2 +$$

$$ArSO_2K + ArSO_2SK + (ArSO_2)_2 + KCl$$

List of Compounds

The List of Compounds in sections places the dihalides in Section F.1 (order of halides is alphabetical -- Br, Cl, F, I), and monohalides in Section F.2.

Section F.3, phosphorodithioates, has three subdivisions: Section F.3.1, primary dithioates (nonexistent); Section F.3.2, secondary dithioates (the phosphorodithioic acids and their salts used as synthetic intermediates); Section F.3.3, the tertiary esters of wide theoretical and commercial interest.

Section F.4, phosphorothioates, is similarly divided: Sections F.4.1, primary esters; F.4.2, secondary esters; F.4.3, tertiary phosphorothiolates, $(RO)_2P(O)SR'$; and F.4.· tertiary phosphorothionates, $(RO)_2P(S)OR'$.

Section F.5 contains miscellaneous phosphorodithio and -thio derivatives.

Section F.6 contains the phosphoroselenoates, arranged approximately as given above for thioates.

Arrangement: In the smaller Sections: F.1, F.2, F.3.2 F.4.1, F.4.2, F.5, and F.6 and their subsections, the compounds are listed in order of increasing C-content and chain length of the R groups in $ROP(S)X_2$, $(RO)_2P(S)X$, etc.

In the large Sections: F.3.3, F.4.3, and F.4.4, the R ("substrate") group in $(RO)_2P(A)SR'$ (A is O or S) determin the order of listing. Thus, $(RO)_2P(A)SMe$ appears first, followed by non-C substitutions on the CH group in alphabetical order, such as Br, Cl, F, I, NH_2, 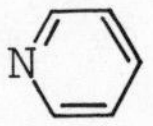 , NHR, OH, OR, SH, SR, SeH, SeR, SiR_3, etc.

When the C-content of the R' group is increased to Et, the same order of substitutions on CH_2CH_2 is observed. Further increases in C-content of the R' group appear in increasing C-chain length: i-Pr, Pr, t-Bu, i-Bu, s-Bu, Bu . . . C_n (aliphatic).

The aromatic and heterocyclic R' groups appear appropriately as 3-rings, 4-rings, 5-rings, 6-rings, . . ., n-rings.

The benzyl types follow the aromatic and heterocyclic O- and S-substituents in the order CH_2Ph, CH_2-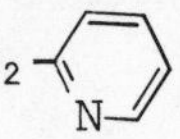 , etc. CH_2CH_2Ph, . . ., $CHMeCH_2Ph$, etc.

Groups and subgroups are indicated in the Contents by further subheadings.

"Homologs and Analogs": This entry in the List of Com· pounds indicates an author's preparation of a number of structurally related compounds that space does not permit including in the lists. "Homologs" is used here usually to refer to increasing C-content in the R groups of $(RO)_2P(A)$-SR'. "Analogs" refers either to increasing C-content in th R' group or to substitution, such as halogens, NO_2, etc., in the R' group.

Thiono and Oxo Groups: For convenience in typing, the forms P(S) and P(O) are used throughout to represent thion and oxo forms, respectively. With the phosphorothioic acid: $(RO)_2POSH$, and their salts, the parentheses are usually omitted. The thiono, P(S), form of the acids has been shown by Kabachnik and co-workers (1955-1960)[491,496] to predominate in solutions in 80% EtOH-water while the oxo, P(O), form predominates in 7% EtOH-water solution.

Occasionally the form MeOP(O/S)SR (O above and S below, joined by a dashed bond) is introduced to represent a phosphorothioate anion.

Molar refraction: In the List of Compound the term molar refraction or "mol. ref." refers to Mn_D^{20}, the Eisenlohr molar refraction [Sayre (1958)].[1078]

Contents

F.1. Phosphorodihalidodithioates, -thioates, and -selenoat

F.1.1. Phosphorobromidofluoridothioates, ROP(S)BrF

EtOP(S)BrF. A. Liq. b_{12} 35°, n_D^{20} 1.4800, d_{20}^{20} 1.6221.[1269]

F.1.2. Phosphorochloridofluoridodithioates, RSP(S)ClF

MeSP(S)ClF. B. Liq. b_{12-14} 42-45°, IR, ^{19}F-NMR 1.26 ppm, Mass spec.[1041]

EtSP(S)ClF. B. Liq. b_{12-14} 57°, IR, ^{19}F -3.02 ppm.[1041]

PhSP(S)ClF. B. Liq. $b_{0.3}$ 73-74°, IR, ^{19}F 2.2 ppm, ^{31}P -84.8 ppm.[1040]

F.1.3. Phosphorochloridofluoridothioates, ROP(S)ClF

MeOP(S)ClF. A.[908] Liq. b_{50} 37°, $n_D^{20.5}$ 1.4429,[908] mol. ref. 214.31.[1078]

EtOP(S)ClF. A.[908,1269] C.[118,1044] Liq. b_{15} 26°,[1044] b_{20} 26.2°,[118] b_{50} 50-51°,[908] b_{60} 52°,[1269] n_D^{20} 1.4493,[1044] n_D^{27} 1.4390,[908] mol. ref. 233.92,[1078] IR, ^{19}F 14.45 ppm.[1044]

i-PrOP(S)ClF. A.[908] Liq. b_{80} 64-65°, n_D^{20} 1.4363,[908] mol. ref. 253.64.[1078]

F.1.4. Phosphorodibromidothioates, $ROP(S)Br_2$

EtOP(S)Br_2. A. Liq. b_{20} 105°.[961]

PhOP(S)Br_2. E. Liq. b_{11} 156-157°.[1277]

F.1.5. Phosphorodichloridodithioates, $RSP(S)Cl_2$

$Cl_3CSP(S)Cl_2$. O. Liq. n_D^{25} 1.6142.[95]

EtSP(S)Cl_2. E.[960,961] F.[371] Liq. b_3 61-62°,[371] b_{10} 92°,[961] n_D^{20} 1.6904,[371] d_0^0 1.4453,[961] d^{20} 1.4246.[371]

PhSP(S)Cl_2. E. Liq. b_{16} 168-170°.[773]

F.1.6. Phosphorodichloridothioates

F.1.6.1. $ROP(S)Cl_2$

$MeOP(S)Cl_2$. A.[675,959,961,1296] Liq. b_{11} 45°,[675] b_5 48°,[1296] b_{40} 70°,[961] n_D^{25} 1.5124,[675] d_0^0 1.4949,[961] IR,[427,673,1347,1348] ^{31}P -59.2,[715] ^{35}Cl NQR at 77°K ν 27.626, 27.834.[1389]

$CH_2DOP(S)Cl_2$. IR.[1349]

$CD_3OP(S)Cl_2$. IR.[1349]

$EtOP(S)Cl_2$. A.[77,118,675,713,1296] Liq. b_{11} 52°,[713] b_{12} 56.6-56.9°,[1211] b_{14} 57-59°,[675] b_{20} 67°,[1296] b_{20} 68°,[77,118,517] n_D^{20} 1.5030,[517] n_D^{25} 1.5026,[675] 1.5030,[77] d_0^0 1.3966,[961] 1.3968,[77] d^{20} 1.3970,[517] IR,[673,674] Raman[902,968] ^1H-NMR,[816,1299,1395] ^{31}P -55.9 ± 0.2,[715] ^{35}Cl NQR at 77°K ν 27.748,[1389] bond magn. rotation.[580]

$EtO^{32}P(^{35}S)Cl_2$. A. Liq. b_{10} 68°.[708]

$MeCD_2OP(S)Cl_2$. IR, Raman.[902]

$CD_3CH_2OP(S)Cl_2$. B. Liq. b_{25} 68-69°, IR, Raman.[902]

$ClCH_2CH_2OP(S)Cl_2$. A.[360] B.[356] E.[684] Liq. b_1 78-80°,[684] b_{14} 104-108°,[356] b_{15} 98-99°,[360] n_D^{20} 1.5362, d^{20} 1.4671.[356]

$FCH_2CH_2OP(S)Cl_2$. A. Liq. b_8 67-68°, n_D^{20} 1.5041, d^{20} 1.5108.[490]

$MeOCH_2CH_2OP(S)Cl_2$. A. Liq. b_8 81-82°.[539]

$EtOCH_2CH_2OP(S)Cl_2$. B.[108,356,1260] Liq. b_1 85-86°,[108] b_{10} 93-96°,[1260] b_{23} 108-114°,[356] n_D^{20} 1.4868,[108] 1.4910,[356] d^{20} 1.2974,[108] 1.317.[356]

$BuOCH_2CH_2OP(S)Cl_2$. A. Liq. b_1 95-96°, n_D^{20} 1.4787, d^{20} 1.2224.[108]

$Cl_2P(S)OCH_2CH_2OP(S)Cl_2$. B. Liq. b_4 78° n_D 1.4938.[1408]

CH_2=$CHOP(S)Cl_2$. N. Liq. b_{30} 53-55°, n_D^{20} 1.5180, d^{20} 1.4263.[373]

i-$PrOP(S)Cl_2$. A.[1296] B.[522] Liq. b_{10} 59°,[1296] b_{10} 59-60°.[522]

$PrOP(S)Cl_2$. A.[959,961,1296] Liq. b_{20} 80°,[961] b_{20} 84°,[959,1296] d_0^0 1.3344,[959,961] ^{31}P -56.3,[326,715] IR,[674] bond magn. rotation.[580]

CH_2=$CHCH_2OP(S)Cl_2$. A.[961b] E.[961a] Liq. b_{25} 74°.[961a,961b]

i-$BuOP(S)Cl_2$. A.[959,961] Liq. b_{20} 88°,[961] b_{20} 91°,[959] d_0^0 1.2724.[959]

s-$BuOP(S)Cl_2$. B.[522] E.[1173] Liq. b_{10} 64°,[522] b_{12} 75°.[1173]

$BuOP(S)Cl_2$. A.[725] B.[517] Liq. b_{10} 67°,[517] b_{10} 81-82°,[725] n_D^{2} 1.4951,[517] ^{31}P -56.4 ppm,[326,488,715] bond magn. rotatio[580]

i-$AmOP(S)Cl_2$. A.[108,197a] Liq. b_1 67-70°,[108] b_{15} 108-109°,[197a] n_D^{20} 1.4811,[108] d_4^0 1.2370,[197a] d_4^{17} 1.2188,[197a] d^{20} 1.1895.[108]

$AmOP(S)Cl_2$. A. Liq. b_1 74-75° n_D^{20} 1.4845, d^{20} 1.1920.[108]

MeCH=C(Et)$OP(S)Cl_2$. N. Liq. b_5 69-71°, n_D^{20} 1.5110, d^{20} 1.2466.[373]

t-BuCH(Me)$OP(S)Cl_2$. E. Liq. b_2 74°.[1173]

$PhOP(S)Cl_2$. A.[153] B.[72,73,529,700] E.[47,1405] F.[371] Liq. b_8 107-107.5°,[371] b_{11} 119-120°,[47] b_{13} 129°,[700] b_{15} 133-135°,[73] n_D^{20} 1.5730,[700] 1.5738,[371] 1.5785,[11] d^{20} 1.4050,[700] 1.4064,[1405] 1.4095,[371] d_4^{20} 1.4084,[11] ^{31}P -26 ppm,[11] -53.7 ppm.[715]

4-$BrC_6H_4OP(S)Cl_2$. B. Liq. b_4 120-122°, n_D^{20} 1.6080.[980]

4-$ClC_6H_4OP(S)Cl_2$. B.[980] E.[1277] Liq. b_5 98-100°,[980] b_{11} 143-145°,[1277] n_D^{20} 1.5850.[980]

4-$IC_6H_4OP(S)Cl_2$. B. Solid. m. 30-31°.[980]

2,4,5-$Cl_3C_6H_2OP(S)Cl_2$. B.[229] E.[1316] Liq. b_1 110°,[229] b_{10} 174-176°,[1316] n_D^{20} 1.6084, d^{20} 1.6653,[229] ^1H-NMR.[515]

2,4,6-$Cl_3C_6H_2OP(S)Cl_2$. B.[1314] Solid. b_3 140-144°,[224] b_6 166°, m. 47-48°.[1314]

$C_6Cl_5OP(S)Cl_2$. B. Solid. m. 99-101° (from petroleum ether).[1412]

4-$MeOC_6H_4OP(S)Cl_2$. N. Liq. b_3 127-134°.[1316]

4-$NO_2C_6H_4OP(S)Cl_2$. B.[700,980] Solid. $b_{0.15}$ 130-135°,[700] m. 50-51.5°,[980] 54°.[700]

2-$MeC_6H_4OP(S)Cl_2$. B.[980] E.[1277] Liq. b_6 95-96°,[980] b_{15} 130-131°,[1277] n_D^{20} 1.5672.[980]

3-$MeC_6H_4OP(S)Cl_2$. E. Liq. b_{12} 138°.[1277]

4-$MeC_6H_4OP(S)Cl_2$. B.[73,980] E.[1277] F.[371] Liq. b_3 91-92°,[371] b_6 104-105°,[980] b_{11} 135-136°,[1277] b_{12} 138°,[73] n_D^{20} 1.5615,[371] 1.5660,[980] d^{20} 1.3575.[371]

3,4-$Me(MeS)C_6H_4OP(S)Cl_2$. Liq. $b_{0.01}$ 98°.[298]

3-$EtC_6H_4OP(S)Cl_2$. N. Liq. b_9 141-148°.[1316]

2-$MeCOC_6H_4OP(S)Cl_2$. B. Liq. $b_{0.1}$ 125-127°, n_D^{20} 1.5810.[1402]

3-$MeCOC_6H_4OP(S)Cl_2$. B. Liq. $b_{0.1}$ 143-145°, n_D^{20} 1.5830.[1402]

4-$MeCOC_6H_4OP(S)Cl_2$. B. Liq. $b_{0.25}$ 135-137°, n_D^{20} 1.5860.[1402]

4-t-Bu-2-$ClC_6H_3OP(S)Cl_2$. B. Solid. b_{10} 175°, m. 37-38°(from petroleum ether).[1314]

2-s-$BuC_6H_4OP(S)Cl_2$. N. Liq. b_7 129-136°.[1316]

2-s-Bu-4,6-$(NO_2)_2C_6H_2OP(S)Cl_2$. B.[227,550] Solid. m. 67-69.5° (from heptane),[227] 68-70°.[550]

2-Br-4-$PhC_6H_3OP(S)Cl_2$. N. Liq. b_1 198°.[1316]

4-$Me_3CCH_2C(Me)_2C_6H_4OP(S)Cl_2$. E. Liq. $b_{0.1}$ 138°, n_D^{19} 1.5392.[1454]

Cholesteryl-$OP(S)Cl_2$. B. Solid. m. 145° (dec.)(from petroleum ether). UV. ν_{max} 1020 (P-O-C), 740, 710 (P=S), 537, 428 (P-Cl) cm^{-1}.[191]

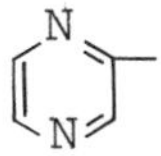—$OP(S)Cl_2$. B.[37,808a] Solid. b_5 100-105°,[37] b_5 101-105°,[808a] m. 37-38° (from petroleum ether).[37]

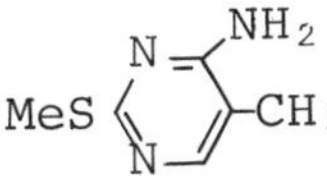$CH_2OP(S)Cl_2$. B. Solid. m. 231-234°.[442]

F.1.6.2. $RSP(O)Cl_2$

$MeSP(O)Cl_2$. A.[543] D.[434] Liq. b_{10} 74.5-75.5°,[434] b_{15} 95-108°,[543] n_D^{20} 1.5294,[434] IR[426,429,901] ^{1}H-NMR.[722,734]

$EtSP(O)Cl_2$. A.[543] L.[945] N.[1] Liq. $b_{0.1}$ 32°,[1] b_{15} 87-88°,[945] b_{20} 95-105°,[543] n_D^{20} 1.5279,[945] n_D^{23} 1.5145,[1] d^{20} 1.4040,[945] IR.[968]

$ClCH_2CH_2SP(O)Cl_2$. J. Liq. $b_{0.02}$ 91-92°, n_D^{20} 1.5405, d^{20} 1.552.[723]

$Cl_2CHCHClSP(O)Cl_2$. O. Liq. $b_{0.02}$ 75°, n_D^{25} 1.5565.[4] Prepared from $EtOPCl_2$ + $Cl_2CHCHClSCl$ in CCl_4 at 30-32°.[4]

$Cl_3CCH_2SP(O)Cl_2$. O. Liq. $b_{0.05}$ 75-76°, n_D^{25} 1.5500.[4]

$Cl_2C{=}CHSP(O)Cl_2$. O. Liq. $b_{0.12}$ 60-65°, n_D^{25} 1.5711.[4]

i-$PrSP(O)Cl_2$. A. Liq. b_5 70-80°.[543]

$PrSP(O)Cl_2$. A. Liq. b_{10} 105-110°.[543]

i-$BuSP(O)Cl_2$. L. Liq. b_6 96-97.5°, n_D^{20} 1.5002, d_4^{20} 1.2192.[945]

$BuSP(O)Cl_2$. A.[543] L.[945] N.[942] Liq. b_{15} 107-110°,[942] b_{15} 113-114°,[942] b_{17} 115-120°,[543] n_D^{20} 1.5037,[945] 1.5110[942] d_4^{20} 1.2395,[945] d^{20} 1.2900.[942]

i-$AmSP(O)Cl_2$. L. Liq. b_8 114-115°, n_D^{20} 1.5044, d_4^{20} 1.2020.[945]

$PhSP(O)Cl_2$. D.[11] L.[1236] Solid.[1236] b_1 95-97°,[1236] b_5 120-121°,[11] m. 38°,[1236] n_D^{20} 1.5989,[11] d_4^{20} 1.4444,[11] ^{31}P -27 ppm.[11] Prepared in 1% yield by thermal isomerizati[on] of $PhOP(S)Cl_2$ at 550°.[11]

4-$ClC_6H_4SP(O)Cl_2$. O.[1172,1236] Solid. b_1 117-118°,[1236] b_3 145°,[1172] m. 65°,[1236] 66°.[1172]

$C_6Cl_5SP(O)Cl_2$. O. Solid. m. 98°.[1172]

F.1.6.3. $R\overset{O}{\overset{\|}{S}}P(O)Cl_2$

$Cl_2CHCHCl\overset{O}{\overset{\|}{S}}P(O)Cl_2$. O. Liq. n_D^{25} 1.5285. Prepared from Cl_2CH $CHClSP(O)Cl_2$ and ozone in $CHCl_3$ at 5°.[4]

$Cl_3CCH_2\overset{O}{\overset{\|}{S}}P(O)Cl_2$. O. Liq. n_D^{25} 1.5335.[4]

$ClCH{=}CCl\overset{O}{\overset{\|}{S}}P(O)Cl_2$. O. Liq. n_D^{25} 1.5340. Prepared from Cl_2CH $CHCl\overset{O}{\overset{\|}{S}}P(O)Cl_2$ and Et_3N in $CHCl_3$ at 15°.[4]

$Cl_2C{=}CH\overset{O}{\overset{\|}{S}}P(O)Cl_2$. O. Liq. b_{12} 60-65°, n_D^{22} 1.5711.[4]

F.1.7. Phosphorodifluoridodithioates, $RSP(S)F_2$

$MeSP(S)F_2$. A.[159a] Liq. Vap. pressure -20.0° 2.9 mm, 0° 11.[?] mm, 24.8° 45.5 mm, ^1H-NMR, ^{19}F 26.7 ppm., mass spect.[159a] Prepared from $P(S)F_2I$ plus MeSH.[159a]

$EtSP(S)F_2$. O. Liq. b_{760} 124°, prepared from $CsSP(S)F_2$ +

EtBr in tetramethylene cyclic sulfone at 70-80°.[1042]

i-PrSP(S)F_2. O. Liq. b_{760} 137°.[1042]

$PhCH_2SP(S)F_2$. O. Liq. b_8 106-109°.[1042]

cyclo-$C_6H_{10}[SP(S)F_2]_2$-1.2. O. Liq. $b_{0.1}$ 98°.[1042]

$Cl_3CCOSP(S)F_2$. IR. Prepared from $HSP(S)F_2$ and Cl_3CCOCl.[1045]

$Cl_3CCSSP(S)F_2$. IR. Prepared from $HSP(S)F_2$ and Cl_3CCSCl.[1045]

F.1.8. Phosphorodifluoridothioates

F.1.8.1. ROP(S)F_2

MeOP(S)F_2. A.[159a] C.[1013] Liq. b_{760} 47°,[1013] vap. pressure -36.8° 10.3 mm, -8.1° 58.8 mm, 24.6° 259.5 mm.,[159a] IR, Raman,[236] ^{1}H-NMR,[159a] ^{19}F 48.0 ppm,[159a] 48.4 ppm,[1013] ^{31}P -52.5 ppm,[1013] mass spect.[159a]

$CD_3OP(S)F_2$. IR, Raman.[236]

EtOP(S)F_2. B[906] C.[118,1211] Liq. b_{741} 77.5-77.7°,[1211] b_{760} 78.4°,[118] 78-78.5°,[906] f.p. -124° ± 0.5°,[118] n_D^{20} 1.3942,[906] d_0^0 1.3019,[118] mol. ref. 201.21.[1271]

i-PrOP(S)F_2. B. Liq. b_{760} 88-89°, n_D^{20} 1.3813.[906]

t-BuOP(S)F_2. B. Liq. b_{20} 47-78°, n_D^{20} 1.3889.[906]

cyclo-$C_6H_{11}OP(S)F_2$. B. Liq. b_5 42-43°, n_D^{20} 1.4408.[906]

PhOP(S)F_2. B.[906] C.[1013] Liq. b_{11} 50°,[1013] b_{14} 52-52.5°,[906] n_D^{24} 1.4860,[1013] n_D^{20} 1.4869,[906] ^{1}H-NMR, ^{19}F 45.6 ppm,[1013] ^{31}P -43.3 ppm.[1013]

2-$ClC_6H_4OP(S)F_2$. B. Liq. b_4 52°, n_D^{20} 1.5033.[906]

2-$NO_2C_6H_4OP(S)F_2$. B. Liq. b_8 118-120°, n_D^{20} 1.5389.[906]

2-$MeC_6H_4OP(S)F_2$. B. Liq. b_{10} 59-60°, n_D^{20} 1.4849.[906]

1-Naphthyl-OP(S)F_2. B. Liq. b_2 82-83°, n_D^{20} 1.5704.[906]

F.1.8.2. RSP(O)F_2

MeSP(O)F_2. B. Liq. b_{760} 105-106°, n_D^{20} 1.3979,[1043] IR,[1208] μ_{PO} 5.76 D.[1208]

EtSP(O)F_2. B. Liq. b_{760} 111-112°, n_D^{20} 1.4076.[1043]

PhSP(O)F_2. B.[1040] L.[1236] Solid. $b_{0.6}$ 62°,[1040] b_{20} 88-89°,[1236] m. 23.5-24°,[1040] m. 28-29°,[1236] IR,[1040] ^{31}P -11.9 ± 0.5 ppm.[1040]

4-$ClC_6H_4SP(O)F_2$. L. Solid. b_4 88-90°, m. 39-40°.[1236]

F.2. Phosphorohalidodithioates, -thioates, -trithioates, selenoates, and phosphoranesulfenyl halides

F.2.1. Phosphorobromidothioates and -selenoates, $(RO)_2P(S)Br$ and $(RO)_2P(Se)Br$

$(PhO)_2P(S)Br$. E. Needles. b_{11} 200°, m. 72.5° (from EtOH).[1277]

$(4\text{-}Me_2NC_6H_4O)_2P(S)Br$. B. White crystals m. 158-164° (from CCl_4 and toluene-petroleum ether). Forms a dimethiodide dihydrate, pale yellow solid, m. 137-141° (dec.).[328]

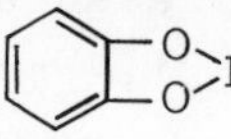P(S)Br. ^{31}P -55.1 ± 1.0 ppm.[339,715]

Br(S)P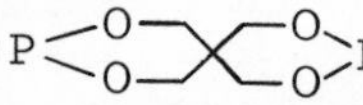P(S)Br. J. Colorless crystals m. 200-202°, from the corresponding thiophosphite with bromine in acetic acid.[198]

$(PhO)_2P(Se)Br$. E. Crystals. m. 64-65° (from ligroin).[1277]

F.2.2. Phosphorobromidotrithioates, $(RS)_2P(S)Br$

$(MeCHBrCH_2S)_2P(S)Br$. N. No physical properties given, from $P(S)Br_3$ and 1,2-propylene sulfide with tertiary amine catalyst.[546a]

F.2.3. Phosphorochloridodithioates

F.2.3.1. $(RS)_2P(O)Cl$

$(MeS)_2P(O)Cl$. N. Liq. $b_{0.04}$ 74°, n_D^{23} 1.5780.[1015]

$(EtS)_2P(O)Cl$. B.[725] F.[213] L.[945] N.[1015] Liq. $b_{0.03}$ 80°,[1015] b_6 129-131.5°,[945] b_{11} 125°,[213] b_{22} 145-150°,[725] n_D^{20} 1.5193,[945] n_D^{22} 1.5485,[1015] d^{20} 1.4030,[945] Raman.[968]

$(PrS)_2P(O)Cl$. N. Liq. $b_{0.05}$ 103°, n_D^{24} 1.5324.[1015]

$(BuS)_2P(O)Cl$. L.[945] N.[1014,1015] Liq. $b_{0.05}$ 105°,[1015] $b_{0.2}$ 112°,[1014] b_3 160-161°,[945] n_D^{25} 1.5210,[1015] n_D^{20} 1.5241,[1014] 1.5790,[945] d^{20} 1.1424.[945]

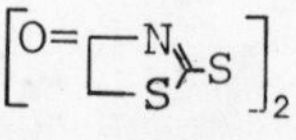P(O)Cl. B. Amorphous solid. m. 209° (dec.)(from $CHCl_3$).[368]

F.2.3.2. RS(R'S)P(O)Cl

MeS($BrCH_2CH_2S$)P(O)Cl. L. Liq. n_D^{25} 1.6041.[452]
PhS($BrCH_2CH_2S$)P(O)Cl. L. Liq. n_D^{25} 1.5905.[452]
4-ClC_6H_4S($ClCH_2CH_2S$)P(O)Cl. L. Liq. b_1 180-185°, n_D^{20} 1.6254.[451]

F.2.4. Phosphorochloridoselenoates, $(RO)_2P(Se)Cl$

$(EtO)_2P(Se)Cl$. E.[1433] From $(EtO)_2PCl$ + Se in toluene at 105-110°, J.[562,776] from $(EtO)_2P(Se)H$ + SO_2Cl_2 in benzene at 5°,[562] liq. $b_{0.8}$ 52-53°,[562] b_1 62°,[776] b_{33} 86°,[1433] n_D^{25} 1.4974,[776] 1.4994,[562] n_D^{20} 1.5402,[1433] d^{25} 1.4411,[562,776] d^{20} 1.5939.[1433]
$(PhO)_2P(Se)Cl$. E. Needles. m. 59-59.5° (from MeOH), b_{11} 200°.[1277]
$(4\text{-}ClC_6H_4O)_2P(Se)Cl$. E. Crystals. m. 59-61° (from ligroin), b_{11} 245°.[1277]
$(2\text{-}MeC_6H_4O)_2P(Se)Cl$. E. Liq. b_{11} 224-227°.[1277]
$(4\text{-}MeC_6H_4O)_2P(Se)Cl$. E. Needles. m. 48-49°, b_{11} 235°.[1277]

F.2.5. Phosphorochloridothioates

F.2.5.1. $(RO)_2P(S)Cl$

$(MeO)_2P(S)Cl$. A.[753] B.[399] H.[36,70,197a,334,447,599,606] J.[635] Liq. b_1 40°,[36] b_9 55°,[635] b_{16} 66°,[197a] b_{23} 73-74°,[606] $n_D^{25.5}$ 1.4774,[606] n_D^{25} 1.4795,[334] 1.4799,[399] 1.4834,[753] d_4^0 1.3414, d_4^{17} 1.3217,[197a] d_4^{20} 1.3351,[753] IR.[427,673,900,968,1346,1348] Raman.[238] ^{31}P -72.9 ppm,[715,817] ^{35}Cl NQR at 77°K, ν 25.943,[1389] mass spect.[184,979]
$(MeO)_2{}^{32}P(S)Cl$. B. Liq. b_{16} 66°, n_D^{20} 1.4820, d_0^0 1.3350.[699]
$(EtO)_2P(S)Cl$. A.[77,242] B.[333,725] H.[36,70,334,421,464,477,683,935,1319] J.[31] K.[1104] Liq. b_1 49-50°,[421] b_2 60°,[31] b_8 75°,[683] $b_{11.5}$ 80.5-81.5°,[70] b_{20} 94-96°,[334] n_D^{19} 1.4693,[935] n_D^{20} 1.4705,[1078] n_D^{25} 1.4685,[334] Mol. ref. 277.35 [1078] d_0^0 1.2015,[77] d_4^{20} 1.2015,[517] IR.[374,427,673,674,900,968,1346,1348] Raman,[968] ^1H-NMR,[816,1299,1395] ^{31}P -67.7,[817] -68.1,[326,488,850] -68.1 ± 0.3 ppm,[715] ^{35}Cl NQR at 77°K

ν 25.409,[1389] EPR,[970] mass spect.[979] Bond magn. rotation.[580]

$(EtO)_2{}^{32}P(S)Cl$. B. Liq. b_{15} 74°.[345]

$(EtO)_2{}^{32}P(^{35}S)Cl$. B. Liq. b_{25} 96-99°, n_D^{20} 1.4678, d^{20} 1.201[708]

$(EtO)_2P(^{35}S)Cl$. B. Liq. b_{15} 84-87°.[443]

$(ClCH_2CH_2O)_2P(S)Cl$. B.[356] E.[559] b_{17} 130°, n_D^{20} 1.5641, d^{20} 1.5135.[356] Also prepared from PCl_3, ethylene oxide, and S in xylene at 70-80°.[559]

$(EtOCH_2CH_2O)_2P(S)Cl$. B. Liq. b_5 160-162°.[1260]

$(i\text{-}PrO)_2P(S)Cl$. H.[36,70,334] K.[737,1104] Liq. $b_{0.55}$ 55°,[36] b_3 61°,[737] b_{14} 91°,[70] n_D^{25} 1.4601,[334] 1.4612,[737] IR.[673,968] ^{31}P -65.1 ppm,[715] ^{35}Cl NQR at 77°K ν 25.732.[1389]

$((Me(ClCH_2)CHO)_2P(S)Cl$. E. b_3 131-132°, n_D^{20} 1.5012, d_4^{20} 1.3424.[992]

$(PrO)_2P(S)Cl$. H.[334,1319] K.[737] Liq. b_1 70-75°,[334] b_2 72°,[737] n^{25} 1.4671,[1319] 1.4673,[334] 1.4701,[737] IR.[673,968] Bond magn. rotation.[580]

$(i\text{-}BuO)_2P(S)Cl$. H. Liq. $b_{0.5}$ 76-82°, n_D^{25} 1.4624.[334]

$(s\text{-}BuO)_2P(S)Cl$. K. Liq. $b_{0.1}$ 54°, n_D^{25} 1.4692.[737]

$(BuO)_2P(S)Cl$. B.[725] H.[36,334,1319] K.[737] Liq. $b_{0.05}$ 73°,[517] $b_{0.5}$ 76°,[737] b_2 95-98°,[725] n_D^{20} 1.4687,[517] n_D^{25} 1.4670,[334] 1.4674,[1319] d_4^{20} 1.4844,[517] IR.[673,968]

$(AmO)_2P(S)Cl$. E. Liq. $b_{0.9}$ 116-118°, n_D^{20} 1.4700, d^{20} 1.0487[16]

$(PhO)_2P(S)Cl$. A.[73] B.[72,73] E.[47,256,376,1277,1405] F.[371] H.[36] Needles. b_1 180-183°,[376] b_{11} 194°,[73] m. 66-67°,[72,73,371] 67-68°,[596] 68° (from alcohol),[256,1405] 68-71°,[36] ^{31}P -59.0[715] Mass spect.[184]

$(4\text{-}ClC_6H_4O)_2P(S)Cl$. E. Crystals. b_{11} 243-245°, m. 43-44°.[1277]

$(3\text{-}Me_2NC_6H_4O)_2P(S)Cl$. B. Liq. n_D^{25} 1.6135.[328,329]

$(4\text{-}Me_2NC_6H_4O)_2P(S)Cl$. B. Solid. m. 165.5-166.5° (from benzene - iPrOH), forms a dimethiodide hydrate, m. 165-167°.[328,329]

$(2\text{-}MeC_6H_4O)_2P(S)Cl$. E. Liq. b_{11} 212°.[1277]

$(3)MeC_6H_4O)_2P(S)Cl$. E. Needles. b_{11} 218°, m. 33-34°.[127]

$(4\text{-}MeC_6H_4O)_2P(S)Cl$. B.[72,73] E.[1277] F.[371] Crystals. m. 51-52°,[371] 53°,[72,73] 54-55°.[1277]

$(2\text{-}s\text{-}Bu\text{-}4,6(NO_2)_2C_2H_6O_2)_2P(S)Cl$. B. Solid. m. 105-107°.[550]

$(2\text{-}C_{10}H_7O)_2P(S)Cl$. B. Solid. m. 112° (from alcohol).[833]

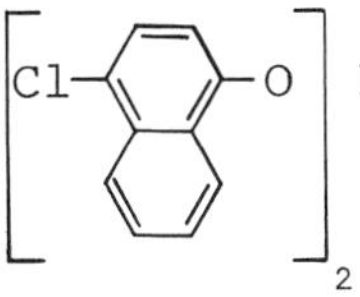
P(S)Cl. B. Crystals. m. 113°.[834]

F.2.5.2. RO(R'O)P(S)Cl

MeO(EtO)P(S)Cl. A.[751,753] O.[752] Liq. b_{17} 79-80°,[752] 80-81°,[753] n_D^{20} 1.4740,[752,753] d_4^{20} 1.2506.[752,753]

MeO(i-PrO)P(S)Cl. A.[751,1174] B.[1174] Liq. b_3 52°,[1174] b_{22} 87-88°,[751] n_D^{20} 1.4680,[751] d^{20} 1.1879.[751]

MeO(PrO)P(S)Cl. A.[753] Liq. b_{25} 107-110°, n_D^{20} 1.4650, d^{20} 1.1852.[753]

MeO(s-BuO)P(S)Cl. A. B. Liq. b_1 61°.[1174]

$MeO(ClCH_2CH_2CHMeO)P(S)Cl$. E. Liq. $b_{0.25}$ 80-82°, n_D^{20} 1.5018, d_4^{20} 1.2953.[58]

MeO(BuO)P(S)Cl. A.[751,753] Liq. $b_{0.3}$ 45-47°,[753] b_9 89-90°,[751] n_D^{20} 1.4750,[751] 1.4765,[753] d^{20} 1.1828,[751] d^{20} 1.1830.[753]

MeO(i-AmO)P(S)Cl. A.[751,753] Liq. b_{16} 107-110°,[751] b_{32} 138-140°,[753] n_D^{20} 1.4672,[753] 1.4676,[751] d^{20} 1.1259,[751] 1.1261.[753]

$MeO(4\text{-}ClC_6H_4O)O(S)Cl$. A. Liq. n_D^{25} 1.5581.[1315]

$MeO(3\text{-}MeCOC_6H_4O)P(S)Cl$. A. Liq. $b_{0.00046}$ 97-100°, n_D^{20} 1.5610.[1402]

$MeO(4\text{-}MeCOC_6H_4O)P(S)Cl$. A. Liq. $b_{0.00038}$ 94-96°, n_D^{20} 1.5639, d_4^{20} 1.3345.[1402]

$MeO(2\text{-}Cl\text{-}4\text{-}NO_2C_6H_2O)P(S)Cl$. A. Liq. n_D^{25} 1.5888.[1315]

$MeO(2,4,5\text{-}Cl_3C_6H_2O)P(S)Cl$. A. Solid. m. 38°.[1315]

$MeO(2,4,6\text{-}Cl_3C_6H_2O)P(S)Cl$. A. Solid. m. 64-66°.[224,1315]

$EtO(FCH_2CH_2O)P(S)Cl$. B. Liq. $b_{1.5}$ 62-63°, n_D^{20} 1.4712, d^{20} 1.3309.[490]

EtO(i-PrO)P(S)Cl. O.[752] Liq. b_1 56-58°,[1085] b_{10} 84-85°, n_D^{2} 1.4620, d_4^{20} 1.1561.[752]

EtO(BuO)P(S)Cl. O. Liq. b_{22} 126-127°, n_D^{20} 1.4660, d_4^{20} 1.1432.[752]

EtO(i-AmO)P(S)Cl. O. Liq. b_{20} 131-132°, n_D^{20} 1.4620, d_4^{20} 1.0885.[752]

EtO(PhO)P(S)Cl. B.[701,749] Liq. $b_{0.5}$ 94-96°,[749] $b_{0.2}$ 95-100°,[701] n_D^{20} 1.5390,[749] 1.542,[701] d^{20} 1.2499,[749] d_4^{20} 1.2666.[701]

$EtO(2-ClC_6H_4O)P(S)Cl$. B.[1028,1188] Liq. $b_{0.01}$ 90°,[1188] $b_{0.2}$ 140-150°,[1028] n_D^{20} 1.5545.[1028]

$EtO(3-ClC_6H_4O)P(S)Cl$. B. Liq. b_{3-4} 140-150°, n_D^{25} 1.5480.[10

$EtO(4-ClC_6H_4O)P(S)Cl$. B.[108,749,1028] Liq. $b_{0.2}$ 113-116°,[74 b_1 130-132°,[108] n_D^{20} 1.5520,[749] 1.5553,[108] n_D^{28} 1.5492,[1028] d^{20} 1.3625,[108] 1.3664.[749]

$EtO(4-NO_2C_6H_4O)P(S)Cl$. B.[108,701,749] Liq. $b_{0.2}$ 160°,[749] $b_{0.4}$ 165-170°,[701] n_D^{20} 1.5655,[108] 1.5740,[749] 1.5810,[710] d_4^{20} 1.4151,[701] d^{20} 1.4247,[108] d_4^{20} 1.4343.[749]

$EtO(2-MeC_6H_4O)P(S)Cl$. B. Liq. n_D^{23} 1.4422.[1030]

$EtO(3-MeC_6H_4O)P(S)Cl$. B. Liq. n_D^{28} 1.5394.[1030]

$EtO(4-MeC_6H_4O)P(S)Cl$. B. Liq. n_D^{23} 1.5400.[1030]

$EtO(2-EtOCOC_6H_4O)P(S)Cl$. B. Liq. $b_{0.01}$ 110°.[314]

$EtO(2-i-PrOCOC_6H_4O)P(S)Cl$. B. Liq. $b_{0.01}$ 120°.[314,1187]

$EtO(3-MeCOC_6H_4O)P(S)Cl$. A. Liq. $b_{0.00012}$ 102-104°, n_D^{20} 1.5500, d_4^{20} 1.3139.[1402]

$EtO(4-MeCOC_6H_4O)P(S)Cl$. A. Liq. $b_{0.00012}$ 93-95°, n_D^{20} 1.5531 d_4^{20} 1.3038.[1402]

$EtO(2,4-Cl_2C_6H_3O)P(S)Cl$. A.[1315] B.[108,749,1028] Liq. $b_{0.4}$ 133-137°,[749] n_D^{20} 1.5540,[108] 1.5600,[749] n_D^{25} 1.5888,[1315] n_D^{29} 1.5750,[1028] d^{20} 1.4280,[108] 1.4386.[749]

$EtO(2,4,5-Cl_3C_6H_2O)P(S)Cl$. A.[1315] B.[108,749,1028] Liq. $b_{0.7}$ 163-166°,[749] n_D^{20} 1.5705,[108] 1.5981,[748] n_D^{25} 1.5689,[1315] n_D^{28} 1.5866,[1028] d^{20} 1.5039,[108] 1.5548.[749]

PrO(3-MeCOC$_6$H$_4$O)P(S)Cl. A. Liq. $b_{0.00011}$ 115-117°, n_D^{20} 1.5495.[1402]

PrO(4-MeCOC$_6$H$_4$O)P(S)Cl. A. Liq. $b_{0.000052}$ 84-87°, n_D^{20} 1.5472, d_4^{20} 1.2579.[1402]

4-MeC$_6$H$_4$O(2-C$_{10}$H$_7$O)P(S)Cl. B. Solid m. 76-78°.[135]

2,4-Cl$_2$C$_6$H$_4$O(2-C$_{10}$H$_7$O)P(S)Cl. B. Liq. n_D^{20} 1.168.[135]

F.2.5.3. 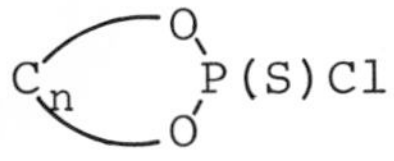

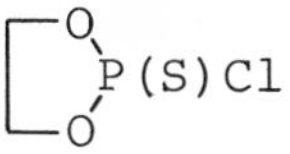 A. B.[1407] Solid. $b_{3.5}$ 88°, m. 30.5°.[1407]

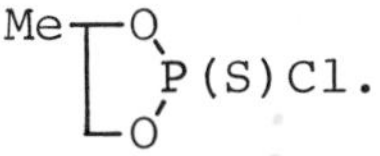 A.[1407] B.[56,1407] E.[249] Liq. $b_{0.1-0.2}$ 61-62°,[249] b_3 82-84°,[56] b_3 96-98°,[1407] b_{13} 111-115°,[1066] n_D^{16} 1.5112,[1407] n_D^{20} 1.5090,[56] 1.5102,[249] d^{20} 1.3902.[66]

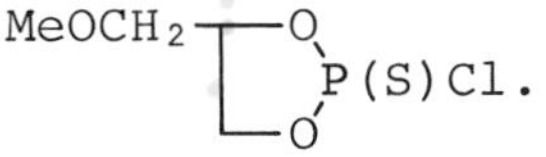 A. B. Liq. b_4 114°, n_D^{18} 1.5085.[1407]

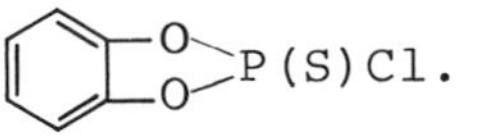 E.[43,508] Solid. b_1 106°,[43] m. 49-50° (from benzene),[43,508] ^{31}P -76.5 ± 1.0 ppm.[339,715]

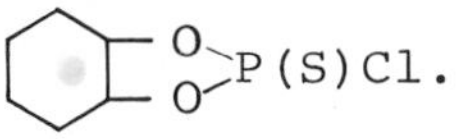 A. Solid. m. 54-56°.[1439]

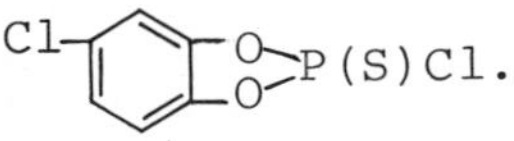 Liq. b_{14} 134°, n_D^{23} 1.5933.[181]

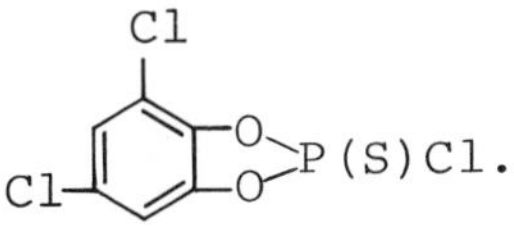 Solid. m. 69°.[181]

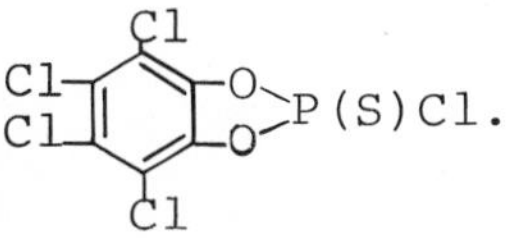 Solid. $b_{0.3}$ 140-142°, m. 184-186°.[181]

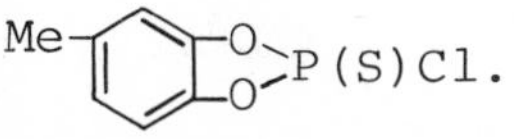 E. Liq. b_{11} 142-143°.[46]

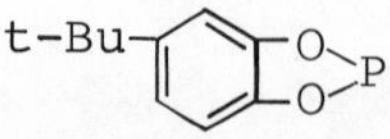 P(S)Cl. E. Liq. $b_{0.2}$ 100-101°, n_D^{25} 1.5523.[181]

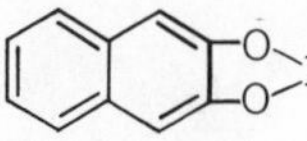 P(S)Cl. Solid. m. 114-116°.[181]

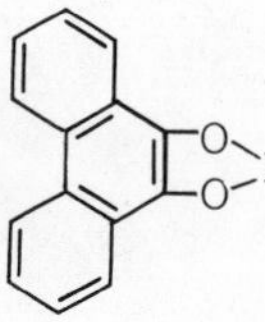 P(S)Cl. Solid. m. 170-172° (sublimed at 150° at 0.2 mm).[161]

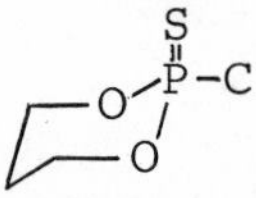 -Cl. A. Solid. m. 39-40°.[1439] μ 5.30 D. Kerr const, mK·10^{12} 1356, rigid chair conformation with axial P-Cl

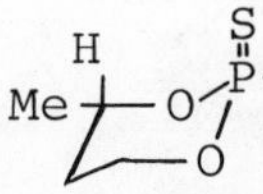 -Cl. A.[1439] B.[661] Liq. b_2 110-112°, n_D^{20} 1.5255, d^{2} 1.3623.[1439] IR.[681] μ 5.55 D, Kerr const., mK·10^{12} 1195, calc. show rigid chair conformation with axial P-Cl and 4-Me.[55]

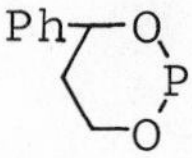 P(S)Cl. B. Liq. n^{30} 1.5852.[591]

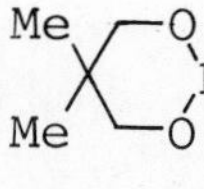 P(S)Cl. B.[247,248,681,1369] Solid. m. 89-90° (from hexane),[1369] m. 90-91.5° (from ligroin),[247,248] m. 91-92°, IR.[681] ^{31}P -58 ppm (in $CHCl_3$).[679a]

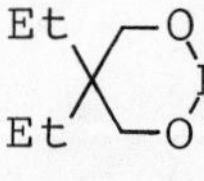 P(S)Cl. B. Solid. m. 56-62°.[583a]

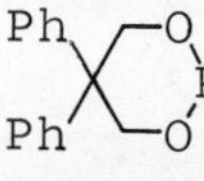 P(S)Cl. B. Solid. m. 135°. IR.[681]

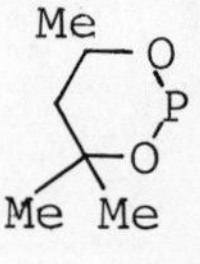 P(S)Cl. B.[250,682] Solid. n_D^{20} 1.5129,[250] m. 51-52°.[682]

IR, ^{1}H-NMR.[682]

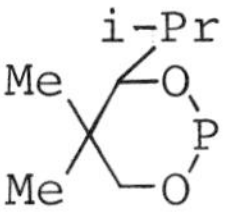 P(S)Cl. B. Solid. m. 75-76°, IR.[682]

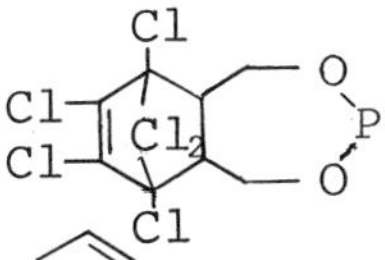 P(S)Cl. B. Solid. m. 66-68°.[380]

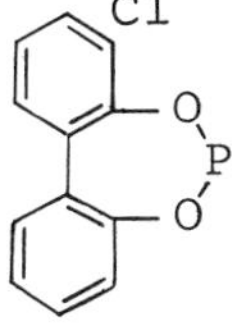 P(S)Cl. E. Solid. m. 107° (from benzene-petroleum ether).[44]

 P(S)Cl. A. Solid. m. 86-89°.[588]

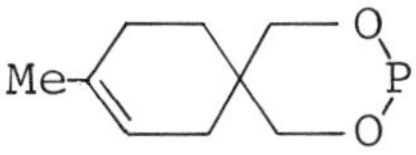 P(S)Cl. A. Solid. m. 51-57°.[588]

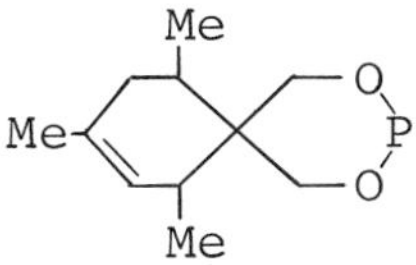 P(S)Cl. A. Solid. m. 77-79°.[585]

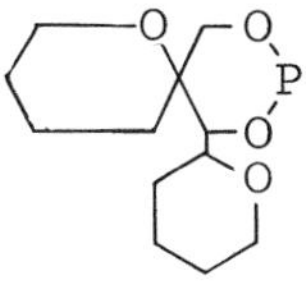 P(S)Cl. B. Solid m. 146-156°.[585]

F.2.5.4. RO(RS)P(O)Cl

MeO(MeS)P(O)Cl. A. B.[846] H. K.[845] L.[605] Liq. b_1 52°,[845] b_2 58°,[845,846] b_3 63-66°,[605] n_D^{24} 1.5145,[845] n_D^{25} 1.4928,[605] ^{1}H-HMR, ^{31}P -37.1 ppm.[605]

EtO(EtS)P(O)Cl. H. K.[845] L.[605,945] Liq. b_1 67-71°,[1010] $b_{1.5}$ 78°,[605] b_5 95-96°,[945] n_D^{20} 1.4844,[945] n_D^{25} 1.4803,[605] d_4^{20} 1.2234,[945] ^{1}H-NMR, ^{31}P -34.1 ppm.[605]

EtO(PhS)P(O)Cl. B. Liq. $b_{0.5}$ 117-119°, n_D^{20} 1.5494, d^{20} 1.2812.[108]

$BrCH_2CH_2O$(MeS)P(O)Cl. L. Liq. b_2 130°.[452,454]

$BrCH_2CH_2O$(EtS)P(O)Cl. L. Liq. n_D^{25} 1.5172.[452]

$BrCH_2CH_2O(PhS)P(O)Cl$. L. Liq. n_D^{25} 1.5905.[452]

$BrCH_2CH_2O(PhCH_2S)P(O)Cl$. L. Liq. n_D^{25} 1.5655.[452]

$ClCH_2CH_2O(MeS)P(O)Cl$. L. Liq. b_1 106°.[451,453]

$ClCH_2CH_2O(OCHCMe_2S)P(O)Cl$. L. Liq. $b_{0.4}$ 150°, n_D^{20} 1.5158.[45

$BrCH_2CHMeO(ClCH_2CH_2S)P(O)Cl$. L. Liq. n_D^{25} 1.5260.[452]

$BrCH_2(ClCH_2)CHO(MeS)P(O)Cl$. L. Liq. n_D^{25} 1.5317.[452]

$(ClCH_2)_2CHO(4\text{-}ClC_6H_4S)P(O)Cl$. L. Liq. b_1 194°, n_D^{20} 1.5765.[451]

$MeCHBrCHMeO(MeS)P(O)Cl$. L. Liq. n_D^{25} 1.5002.[452]

$MeCHClCH_2CH_2O(MeS)P(O)Cl$. L. Liq. b_1 124°.[451]

$MeCHClCHMeO(4\text{-}ClC_6H_4S)P(O)Cl$. L. Liq. b_1 162°.[451]

$PhO(EtS)P(O)Cl$. L. Liq. b_5 144-145°, n_D^{20} 1.5508, d_4^{20} 1.2920[945]

$ClFC{=}NO(EtS)P(O)Cl$. M. Liq. b_1 99-101°, n_D^{20} 1.4979, d_4^{20} 1.4790.[724]

$ClFC{=}NO(ClCH_2CH_2S)P(O)Cl$. M. Liq. $b_{0.01}$ 101-105°, n_D^{20} 1.519 d_4^{20} 1.6760.[723]

$ClFC{=}NO(ClCOCH_2S)P(O)Cl$. M. Liq. $b_{0.004}$ 93-95°, n_D^{20} 1.5220, d_4^{20} 1.6830.[565]

$ClFC{=}NO(BuS)P(O)Cl$. M. Liq. b_2 107-108°, n_D^{20} 1.4925, d_4^{20} 1.3692.[724]

$ClF_2CCF{=}NO(EtS)P(O)Cl$. M. Liq. $b_{0.01}$ 57-59°, n_D^{20} 1.4582, d_4^{20} 1.5131.[724]

$Me_2C{=}NO(MeS)P(O)Cl$. M. Liq. n_D^{20} 1.5079, d_4^{20} 1.2950.[724]

$Me_2C{=}NO(EtS)P(O)Cl$. M. Liq. n_D^{20} 1.5052, d_4^{20} 1.2640.[724]

F.2.6. Phosphorochloridotrithioates, $(RS)_2P(S)Cl$

$(Cl_3CS)_2P(S)Cl$. O. Liq. n_D^{25} 1.6462.[95]

$(EtS)_2P(S)Cl$. B. Liq. b_2 110-113°.[725]

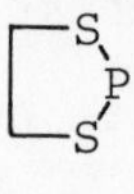
P(S)Cl. Crystal and molecular structure.[597]

F.2.7. Phosphorofluoridodithioates

F.2.7.1. RO(RS)P(S)F

$MeO(EtS)P(S)F$. A. IR. ^{19}F 27.8 ppm, mass spect.[1041]

$MeO(PhS)P(S)F$. A. Liq. $b_{0.3}$ 95-97°, IR. ^{31}P -90.7 ppm.[1040]

EtO(MeS)P(S)F. A. Liq. b_{12} 70-73°, IR. ^{19}F 28.6 ppm. Mass spect.[1041]

EtO(EtS)P(S)F. A. Liq. $b_{0.15}$ 28-30°, IR. ^{19}F 24.9 ppm. Mass spect.[1041]

EtO(PhS)P(S)F. B. Liq. $b_{0.3}$ 104-107°, IR. ^{31}P -89.1 ppm. Mass spect.[1041]

F.2.7.2. $(RS)_2P(O)F$

$(MeS)_2P(O)F$. B. Liq. $b_{0.03}$ 30-31°, n_D^{20} 1.5223.[1043] With Ph_4PCl gives $MeS(S^-)P(O)F.Ph_4P^+$ m. 208°, IR.[1041] With Ph_4AsCl gives $MeS(S^-)(P(O)F.Ph_4As^+$ m. 164°, IR.[1041]

$(EtS)_2P(O)F$. A.[667] B.[159] Liq. b_{15} 104-107°.[159,667] With Ph_4PCl gives $EtS(S^-)P(O)F.Ph_4P^+$ m. 111°, IR.[1041]. With Ph_4AsCl gives $EtS(S^-)P(O)F.Ph_4As^+$ m. 126°, IR.[1041]

$(PhS)_2P(O)F$. B. Solid. m. 88°.[1043] With Ph_4PCl gives $PhS(S^-)P(O)F.Ph_4P^+$ m. 179.5 , IR.[1040] With Ph_4AsCl gives $PhS(S^-)P(O)F.Ph_4As^+$ m. 171°, IR.[1040]

F.2.8. Phosphorofluoridothioates

F.2.8.1. $(RO)_2P(S)F$

$(MeO)_2P(S)F$. B.[908] C.[909] Liq. b_{10} 68-70°,[909] b_{20} 75-77°,[908] n_D^{20} 1.4528,[908,1078] mol. ref. 209.36,[1078] ^{19}F 48.5 ppm, ^{31}P -65.5 ppm.[1013]

$(EtO)_2P(S)F$. B.[278,908] C.[909] O.[1388] Liq. b_{10} 49°,[278] 79-81°,[908] 80-82°,[909] $b_{11.5-12}$ 55.5-55.8°.[1388]

$(i\text{-}PrO)_2P(S)F$. B. Liq. b_{10} 88-89°.[908]

$(PhO)_2P(S)F$. B. Liq. b_1 112-113°.[907] ^{19}F 41.7, ^{31}P -49.5.[1013]

$(4\text{-}Me_2NC_6H_4O)_2P(S)F$. B. Solid. m. 110.5-112° (from EtOH). Forms dimethiodide dihydrate, m. 159-162°.[328]

F.2.8.2. RO(R'O)P(S)F

MeO(EtO)P(S)F. B. Liq. b_{10} 32°, n_D^{20} 1.4250.[908] Mol. ref. 225.35.[1078]

MeO(i-PrO)P(S)F. B. Liq. b_{10} 34-35°, n_D^{20} 1.4225.[908] Mol. ref. 244.90.[1078]

$EtO(EtOCH_2CH_2O)P(S)F$. B. Liq. b_{10} 95°.[278]

$EtO(EtSCH_2CH_2O)P(S)F$. B. Liq. $b_{0.4}$ 69°.[278]

$EtO(Cl_3CCH_2O)P(S)F$. B. Liq. b_3 75°.[278]

$EtO(i\text{-}PrO)P(S)F$. B.[278,908] Liq. b_{10} 37-38°,[908] b_{12} 68°,[278] n_D^{20} 1.4237,[908,1078] mol. ref. 265.08.[1078]

$EtO(MeEtCHO)P(S)F$. B. Liq. b_{13} 81-82°.[278]

$EtO(C_6H_{13}O)P(S)F$. B. Liq. b_3 90-92°.[278]

$EtO(t\text{-}BuMeCHO)P(S)F$. B. Liq. b_4 65°.[278]

$PhO(4\text{-}ClC_6H_4O)P(S)F$. B. Liq. $b_{1.5}$ 130-131°.[907]

$PhO(4\text{-}MeC_6H_4O)P(S)F$. B. Liq. $b_{1.5}$ 123-124°.[907]

$PhO(2\text{-}C_{10}H_7O)P(S)F$. B. Liq. $b_{1.5}$ 155-157°.[907]

F.2.8.3. 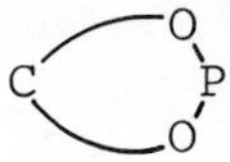 P(S)F

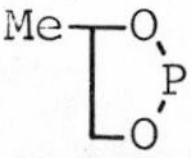 P(S)F. E. Liq. b_1 84°, n_D^{20} 1.4550, d^{20} 1.3398.[1012]

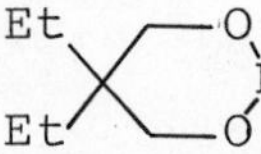 P(S)F. C. Solid. m. 33-34° (from ether).[584]

F.2.8.4. RO(RS)P(O)F

$EtO(EtS)P(O)F$. L. Liq. b_6 70.5-71°, d_0^0 1.2120.[947]

$EtO(ClCH_2CH_2S)P(O)F$. C. Liq. b_{18} 131-133°, d_{20}^{20} 1.3581.[947]

F.2.8.5. Phosphoroidothioates

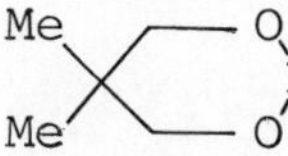 P(S)I. J. Solid. m. 142-143° (from EtOH), IR.[246]

F.2.9. Phosphorofluoridotrithioates, $(RS)_2P(S)F$

$(MeS)_2P(S)F$. A. Liq. $b_{0.01}$ 44-46°, IR. ^{19}F 29.5 ppm. Mass spect.[1041]

$(EtS)_2P(S)F$. A. Liq. $b_{0.07}$ 62-65°, IR. ^{19}F 20.5 ppm. Mass spect.[1041]

$(PhS)_2P(S)F$. A. Liq. $b_{0.2}$ 168-172°, IR. ^{31}P -112 ppm.[1040]

F.2.10. Phosphoranesulfenyl Halides

F.2.10.1. $(RO)_2P(S)SCl$

$(EtO)_2P(S)SCl$. P.[32,464] Liq. $b_{0.06}$ 52-53°, $b_{0.4}$ 65-66°,[32] b_1 95-100°,[464] n_D^{20} 1.5270, d^{20} 1.2687.[32]

$(i\text{-}PrO)_2P(S)SCl$. P. Liq. $b_{0.4}$ 78°, n_D^{20} 1.5145, d^{20} 1.1890.[3

$(PrO)_2P(S)SCl$. P. Liq. $b_{0.6}$ 93°, n_D^{20} 1.5175, d^{20} 1.2015.[33]

$(i\text{-}BuO)_2P(S)SCl$. P. Liq. $b_{0.5}$ 95°, n_D^{20} 1.5020, d^{20} 1.1385.[33]

$(BuO)_2P(S)SCl$. P. Liq. $b_{0.5}$ 102°, n_D^{20} 1.5000, d^{20} 1.1375.[33]

F.2.10.2. $(RO)_2P(O)SCl$

$(MeO)_2P(O)SCl$. P. Liq. $b_{0.6}$ 41-42°, n_D^{20} 1.4820.[785]

$(EtO)_2P(O)SCl$. P.[601,774,777,785,788,944,1456] Liq. $b_{0.12}$ 45-46°,[785] $b_{0.2}$ 53-55°,[1456] $b_{0.4}$ 61°,[601] 61-62°,[774] $b_{3.5}$ 90-91°,[944] n_D^{20} 1.4668,[785] 1.4672,[774] 1.4692,[1456] d^{17} 1.2834,[944] d^{20} 1.2706,[601] d^{25} 1.2703.[777]

$(i\text{-}PrO)_2P(O)SCl$. P. Liq. $b_{0.8}$ 71°, n_D^{20} 1.4587, d^{20} 1.1774.[601]

$(PrO)_2P(O)SCl$. P.[601,785] Liq. $b_{0.5}$ 80°,[785] 84°,[601] n_D^{20} 1.4672,[785] 1.4679,[601] d^{20} 1.1978.[601]

$(BuO)_2P(O)SCl$. P.[601,785,788] Liq. $b_{0.07}$ 73-74°,[788] $b_{0.8}$ 71°,[601,785] n_D^{20} 1.4665,[788] 1.4677,[601,785] d_4^{20} 1.1440.[601]

$EtO(PhO)P(O)SCl$. P.[785,788] Liq. $b_{0.3}$ 97-98°, n_D^{20} 1.5335.[785,788]

F.3. Phosphorodithioic Acid Esters

F.3.1. Primary Phosphorodithioates, $ROP(S)(SH)OH$, $RSP(S)(OH)_2$: Nonexistent

F.3.2. Secondary Phosphorodithioates

F.3.2.1. $(RO)_2P(S)S^-$

$(MeO)_2P(S)SH$. III.[675] V.[342,496,714] Liq. $b_{0.5}$ 42-44°,[714] b_4 56-57°,[496] b_5 70.5°,[675] n_D^{20} 1.5340,[496] 1.5343.[1078] n_D^{25} 1.5346,[675] mol. ref. 242.70,[1078] d^{20} 1.2869,[496] IR.[673,674,900] Raman.[1443] pK_a 1.55 (in 7% EtOH), 2.64 (in 80% EtOH),[496] potassium salt; m. 171-172°(dec.).[1451] 2,2'-Bipyridinium salt (1:1) m. 43-45°,[533] $(MeO)_2P(S)S^{\ominus}$ $PhCOCH_2CH_2$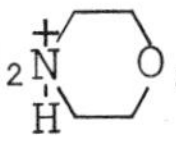, m. 107-109° (from dry benzene),[754] $(MeO)_2P(S)SH \cdot H_2NCH_2CH_2SP(S)(OMe)_2$, m. 74-75°.[128]

$(EtO)_2P(S)SH$. I.[725] III.[675] IV.[683] V.[496,961] Liq. $b_{0.75}$ 65°,[675] b_4 77-78°,[496] b_{4-5} 85-90°,[683] n_D^{20} 1.5070,[1078] 1.5073,[496] n_D^{25} 1.5096,[675] mol. ref. 280.66,[1078] d^{20} 1.1651,[496]

IR.[374,673,674,900,968] Raman,[968,1443] ^{1}H-NMR.[816,1395] ^{31}P -85.7 ppm.[326,488,715] Magn. suscept. $-\chi\cdot 10^6$ 112.7,[1455] pK_a 1.62 (in 7% EtOH), 2.56 (in 80% EtOH).[496] Potassium salt, m. 152-153° (from EtOH-ether),[725] m. 195° [117,1451] IR.[374] Ammonium salt, m. 180-182° (from EtOH-ether),[540] ^{31}P -110.5 ppm.[715] Methyl ammonium salt, m. 76.5-77°, IR.[1348] Piperidinium salt, m. 57°.[22] Morpholinium salt, m. 89-90°.[23] Aniline salt, m. 86°; 4-chloroaniline salt, m. 119-120°, p-toluidine salt, m. 87-88°; 2-naphthyl amine salt, m. 116-117°.[1003] 2,2' Bipyridinium salt (1:1), m. 73.5-75.5°.[533] $(EtO)_2P(S)S^\ominus$

$PhCOCH_2CH_2$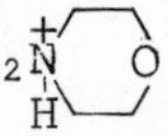, m. 128-129° (from benzene).[754] $(EtO)_2$

$P(S)SH\cdot H_2NCH_2CH_2SP(S)(OEt)_2$, m. 100-101°, $(EtO)_2P(S)SH\cdot MeNHCOCH_2CH_2SP(S)(OEt)_2$, m. 74-75°. $(EtO)_2P(S)SH\cdot EtNHCOCH_2CH_2SP(S)(OEt)_2$, m. 75°.[128]

$(ClCH_2CH_2O)_2P(S)SH$. V. Liq. $b_{0.008}$ 75-80°, n_D^{20} 1.5520, d_4^{20} 1.4227.[1450] Potassium salt, m. 117-118° (from acetone-benzene).[1450]

$(MeOCH_2CH_2O)_2P(S)SH$. Potassium salt, V. m. 174-175° (from acetone-ether).[1450]

$(EtOCH_2CH_2O)_2P(S)SH$. Potassium salt, V. m. 133-134° (from acetone-ether).[1450]

$(PhOCH_2CH_2O)_2P(S)SH$. Potassium salt. V. m. 192-193° (from acetone-ether).[1450]

$(Me_2NCH_2CH_2O)_2P(S)SH$. Potassium salt. V. m. 168-169°.[389]

$(EtPhNCH_2CH_2O)_2P(S)SH$. Potassium salt. V. m. 85-87°.[389]

(2-pyridyl-$CH_2CH_2O)_2P(S)SH$. Potassium salt. V. m. 40-41°.[387]

(2-piperidinyl-$CH_2CH_2O)_2P(S)SH$. Potassium salt. V. m. 166-167.5°.[387]

(2-morpholinyl-$CH_2CH_2O)_2P(S)SH$. Potassium salt. V. m. 175-176.5°.[387]

$(i\text{-}PrO)_2P(S)SH$. V.[496,1409] Also from $(MeO)_2P(S)SH$ + i-PrOH.[1322] Liq. b_2 84-87°,[1409] b_3 70.5-71.5°,[1322] 71-72°,[18,496] n_D^{20} 1.4915,[18] 1.4918,[496,1078] 1.4920,[1322] n_D^{25} 1.4906

[1409] mol. ref. 319.68,[1078] d^{20} 1.0900,[18] 1.0911,[498] 1.0913,[1322] d^{25} 1.0779,[1409] IR.[674] ^{31}P -82,[715] pK_a 1.82 (in 7% EtOH), 2.65 (in 80% EtOH).[496] Potassium salt, m. 193°,[117,1451] ^{31}P -107.4 ppm.[715] As a primary standard for sulfide generation.[107]

$((ClCH_2)_2CHO)_2P(S)SH$. Potassium salt. V. m. 137-138° (from acetone-benzene).[1450]

$(PrO)_2P(S)SH$. V.[496,1409] Also from $(MeO)_2P(S)SH$ + PrOH.[1322] Liq. b_2 82°,[18] b_3 85-86°,[496] n_D^{20} 1.4980,[18] 1.4987,[496,1078] n_D^{25} 1.4976,[1409] mol. ref. 321.16,[1078] d^{20} 1.1020,[18] 1.1040,[496] 1.1209,[1322] 1.0899,[1409] IR.[674] Raman.[1443] pK_a 1.75 (in 7% EtOH), 2.57 (in 80% EtOH).[496] Potassium salt, m. 163-164°,[117] 164-165°,[548] 165°,[1451] Ethanolamine salt, liq., n^{20} 1.5322, d_4^{20} 1.1565; diethanolamine salt, n^{20} 1.5325, d_4^{20} 1.1706; triethanolamine salt, n_D^{20} 1.5335, d_4^{20} 1.1828.[548]

$(CH_2{=}CHCH_2O)_2P(S)SH$. V. Liq. n_D^{20} 1.5330, d_4^{20} 1.1656.[910] Potassium salt, m. 126° (from acetone-ether).[739,1445]

$(HC{\equiv}CCH_2O)_2P(S)SH$. V. Liq. (Free acid liberated from potassium salt with HCl.) $b_{0.0001}$ 75°, n_D^{20} 1.5620, d^{20} 1.2703. Potassium salt, m. 138°.[739,1447]

$(t\text{-}BuO)_2P(S)SH$. V. (unstable liquid) n_D^{25} 1.4761. Isolated as lead salt, m. 109° (dec.).[463]

$(i\text{-}BuO)_2P(S)SH$. V.[496] Liq. $b_{0.1}$ 100°,[18] $b_{1.5}$ 77.5-78°,[496] n_D^{20} 1.4880,[18] 1.4921,[496] 1.4889,[1078] mol. ref. 360.83,[1078] d^{20} 1.0550,[18] 1.0620,[496] potassium salt, m. 189-190°.[117]

$(Me_2C(NO_2)CH_2O)_2P(S)SH$. V. Solid. m. 103.8-104°.[516] Pyridine salt, m. 103-104°.[516]

$(BuO)_2P(S)SH$. I.[725] V.[496,1409] Also from $(EtO)_2P(S)SH$ + BuOH,[1322] Liq. $b_{0.2}$ 103°,[18] $b_{1.2}$ 117.5-118°,[1409] b_2 99-99.5°,[496] n_D^{20} 1.4861,[1322] 1.4950,[18] 1.4971,[496] n_D^{25} 1.4916,[1409] d^{20} 1.0620,[496] 1.0685,[18] 1.1348,[1322] d^{25} 1.0605,[1409] IR.[674] ^{31}P -86.[715] pK_a 1.83 (in 7% EtOH), 2.64 (in 80% EtOH).[496] Potassium salt, m. 142.5-143°,

[548] 147-147.5°.[117] $(BuO)_2P(S)S^{\ominus}$ $PhCOCH_2CH_2$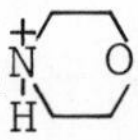 m. 80-81.5°.[754] $(BuO)_2P(S)S^{\ominus}$ $PhCOCH_2CH_2\overset{+}{N}HMe_2$, oil, n^{22} 1.5463. Ethanol amine salt, liq. n_D^{20} 1.5200, d_4^{20} 1.1063; triethanolamine salt, n_D^{20} 1.5230, d_4^{20} 1.1379.[548]

$(MeCH{=}CHCH_2O)_2P(S)SH$. V. Liq. (Free acid liberated from potassium salt with HCl) n_D^{20} 1.5300, d_4^{20} 1.1262.[1446] Potassium salt, m. 135°.[739,1447]

$(CH_2{=}CMeCH_2O)_2P(S)SH$. V. Liq. (Free acid liberated from potassium salt with HCl) n_D^{20} 1.5260, d^{20} 1.1240.[1446] Potassium salt, m. 115°.[739,1447]

$(i\text{-}AmO)_2P(S)SH$. Liq. $b_{0.3}$ 103°, n_D^{20} 1.4900, d^{20} 1.0365.[18]

$(AmO)_2P(S)SH$. Liq. $b_{0.3}$ 112°, n_D^{20} 1.4920, d^{20} 1.0391.[18] Potassium salt, m. 144-145°.[117,548] Ethanolamine salt, liq., n_D^{20} 1.5185, d_4^{20} 1.0870; diethanolamine salt, n_D^{20} 1.5190, d_4^{20} 1.1040; triethanolamine salt, n^{20} 1.5208, d_4^{20} 1.1214.

$(Me_2CHCH_2CHMeO)_2P(S)SH$. ^{31}P -82.4 ppm.[715]

$(RO)_2P(S)SH$ with R= C_6H_{13} to $C_{10}H_{21}$. Salts:

R	Potassium, m.	Ethanolamine	n_D^{20}	d_4^{20},[548]
C_6H_{13}	149.5-150.5°,[548]	Mono	1.5115	1.0590
	154-155.5°,[117]	Di	1.5125	1.0753
		Tri	1.5140	1.0936
C_7H_{15}	151-152°,[548]	Mono	1.5085	1.0417
	154-154.5°,[117]	Di	1.5100	1.0599
		Tri	1.5100	1.0754
C_8H_{17}	152-153°,[548]	Mono	1.5040	1.0244
	153.5-154.5°,[117]	Di	1.5050	1.0425
		Tri	1.5062	1.0563
C_9H_{19}	150-153°,[117]	Mono	1.502	1.0111
	156-157°,[548]	Tri	1.503	1.0393
$C_{10}H_{21}$	157-158°,[548]	Mono	1.4985	0.9970
	158-158.5°,[117]	Tri	1.5020	1.0319

$(C_{12}H_{25}O)_2P(S)SH$. V.[814a] Also from $(EtO)_2P(S)SH + C_{12}H_{25}OH$.

[1322] Liq. n_D^{20} 1.4782, d^{20} 0.9467.[1322] Pyridine salt, m. 43-45°.[814a]

$(PhO)_2P(S)SH$. III.[148] V.[148,334,496,549,1435] Solid. m. 60-61°,[496] 61°,[148] 61-62°,[549] 62.5-63.5° (from benzene-hexane).[334] Potassium salt, m. 187-188°, (from acetone-benzene),[911,1435] 191.5-192°.[117] Ethanolamine salt, m. 101-102°; diethanolamine salt, m. 86-87°; triethanolamine salt, m. 85.5-86°; triethylamine salt, m. 99-99.5°.[549] o-Toluidine salt, m. 121-123° (from CCl_4); m-toluidine salt, m. 106-108° (from CCl_4); p-toluidine salt, m. 110-112° (from water); 1-naphthyl amine salt, m. 116-118° (from benzene-ligroin); 2-naphthyl amine salt, m. 151-153° (from $CHCl_3$).[1435] Pyridine compound, m. 93-95°.[814a]

$(2\text{-}ClC_6H_4O)_2P(S)SH$. V. Potassium salt, m. >180°(dec.)(from acetone-benzene).[1435]

$(4\text{-}ClC_6H_4O)_2P(S)SH$. V. Very unstable as the free acid; decomp. to give H_2S, $4\text{-}ClC_6H_4OH$, and H_3PO_4.[1437] Potasium salt, m. 210-212° (from acetone-benzene).[1435] Sodium salt, decomp. at 250°; forms a trihydrate, m. 97-99° (sealed capillary).[1435] Aromatic amine salts: aniline, m. 163-165° (from water); o-toluidine, m. 145-147° (from water); m-toluidine, m. 132-133° (from CCl_4); p-toluidine, m. 140-142° (from water); 1-naphthyl amine, m. 144-149° (from CCl_4); 2-naphthyl amine, m. 154-156° (from water).[1435]

$(2,4\text{-}Cl_2C_6H_3O)_2P(S)SH$. V. Potassium salt m. 185-186°.[1438] Triethylamine salt, m. 108-109°.[1438]

$(2,4,5\text{-}Cl_3C_6H_2O)_2P(S)SH$. IR at 29°, 0°, -25°, -50°, -75°, -100°.[900]

$(2,4,6\text{-}Br_3C_6H_2O)_2P(S)SH$. V. Triethylamine salt, m. 200-202°.[1438]

$(2,4,6\text{-}Cl_3C_6H_2O)_2P(S)SH$. V. Triethylamine salt, m. 175-177°.[1438]

$(3\text{-}MeC_6H_4O)_2P(S)SH$. V. Solid. m. 45-47°. Potassium salt, m.

205-207°.[1437]

(4-MeC$_6$H$_4$O)$_2$P(S)SH. V.[1046,1222,1437] Solid. m. 52-54°.[1437] Potassium salt, m. 217-218°.[1437]

(4-t-BuC$_6$H$_4$O)$_2$P(S)SH. V. Solid. m. 114-115°.[1222] Potassium salt, m. 320° (dec.).[117]

(4-t-AmC$_6$H$_4$O)$_2$P(S)SH. Potassium salt, m. 300° (dec.)[117]

(4-t-C$_8$H$_{17}$C$_6$H$_4$O)$_2$P(S)SH. V. Liq. n_D^{20} 1.5419, d^{20} 1.1132.[122] Potassium salt, m. 181-182°.[117]

(4-Me$_2$PhCC$_6$H$_4$O)$_2$P(S)SH. Potassium salt, m. 204-205°.[117]

(2,4(t-Am)$_2$C$_6$H$_3$O)$_2$P(S)SH. V. Potassium salt, m. 192-193° (from EtOH-ether).[467,469]

(2-C$_{10}$H$_7$O)$_2$P(S)SH. V. Solid. m. 112-113°.[549]

(PhCH$_2$O)$_2$P(S)SH. V. Liq. n_D^{25} 1.597].[463] Potassium salt, m. 152-154°.[117]

(PhCH=CHCH$_2$O)$_2$P(S)SH. V. Potassium salt, m. 140°.[739]

CH$_2$O)$_2$P(S)SH. V. Potassium salt, m. 188-189°.[1450]

(Cholesteryl-O)$_2$P(S)SH. V. Solid. m. 192-193°, UV ν_{max} 980 cm^{-1} (P-O-C), 740, 680 cm^{-1} (P=S). Sodium salt (from the acid plus Na in xylene), m. 267°.[191]

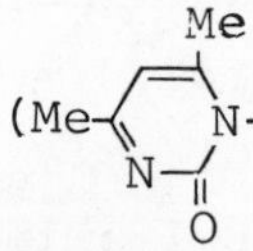
N-CH$_2$CH$_2$O)$_2$P(S)SH. V. Solid. m. 159-161°.[1017]

F.3.2.1. RO(R'O)P(S)S$^-$

MeO(EtO)P(S)SH. VII. From MeOP(S)(SK)$_2$ + EtOH + HCl gas. Liq. b$_2$ 52-54°, n_D^{20} 1.5260, d_4^{20} 1.2451. Potassium salt, m. 172-173°.[561] Salt with Me$_2$NCH$_2$COOEt, liq. n_D^{20} 1.5428, d_4^{20} 1.2315.[1240]

MeOP(S)(SH)OCH$_2$CH$_2$OP(S)(SH)OEt. VII. Dipotassium salt, m. 162-163°.[545]

MeO(PrO)P(S)SH. VII. Liq. b$_1$ 62-64°, n_D^{20} 1.5090, d_4^{20} 1.1541 Potassium salt, m. 153-154°.[561]

MeO(CH$_2$=CHCH$_2$O)P(S)SH. VII. Potassium salt, m. 149-151°.[561]

MeO(BuO)P(S)SH. VII. Potassium salt, m. 159-161°.[561]

MeO(HO$(CH_2)_4$O)P(S)SH. VII. Potassium salt, m. $138-139^{\circ}$.[545]

MeOP(S)(SH)O$(CH_2)_4$OP(S)(SH)OMe. VII. Dipotassium salt, m. $161-162^{\circ}$.[545]

MeO(PhO)P(S)SH. VII. Potassium salt. m. $176-178^{\circ}$.[560]

EtO(HOCH_2CH_2O)P(S)SH. VII. Potassium salt, m. 212° (dec.).[545]

EtOP(S)(SH)OCH_2CH_2OP(S)(SH)OEt. VII. Dipotassium salt, m. $164-165^{\circ}$.[545]

i-PrOP(S)(SH)OCH_2CH_2OP(S)(SH)O-i-Pr. VII. Dipotassium salt, m. $156-157^{\circ}$.[545]

EtO(PhO)P(S)SH. VII. Potassium salt, m. $166-168^{\circ}$.[560]

EtO(2-HOC_6H_4O)P(S)SH. VII. Potassium salt, m. >198° (dec.).[546]

EtO(3-HOC_6H_4O)P(S)SH. VII. Potassium salt, m. >180° (dec.).[546]

EtO(4-HOC_6H_4O)P(S)SH. VII. Potassium salt, m. >210° (dec.).[546]

EtO(PrO)P(S)SH. VII. Liq. b_2 $72-74^{\circ}$, n_D^{20} 1.5053, d_4^{20} 1.1581. Potassium salt, m. $168-168.5^{\circ}$.[561]

EtOP(S)(SH)OCH_2CH_2CHMeOP(S)(SH)OEt. VII. Dipotassium salt, m. $176-178^{\circ}$.[545]

EtO(BuO)P(S)SH. VII. Liq. b_1 $75-77^{\circ}$, n_D^{20} 1.4960, d_4^{20} 1.1056. Potassium salt, m. $146-147^{\circ}$.[561]

EtO(C_8H_{17}O)P(S)SH. VII. Potassium salt, m. $137-139^{\circ}$.[561]

EtO($C_{10}H_{21}$O)P(S)SH. VII. Potassium salt, m. $167-168^{\circ}$.[561]

1,2-(EtOP(S)(SH)O$)_2C_6H_4$. VII. Dipotassium salt, m. $182-183^{\circ}$.[546]

1,3-(EtOP(S)(SH)O$)_2C_6H_4$. VII. Dipotassium salt, m. $184-185^{\circ}$.[546]

1.4-(EtOP(S)(SH)O$)_2C_6H_4$. VII. Dipotassium salt, m. $179-180^{\circ}$.[546]

EtO(4-MeC_6H_4O)P(S)SH. VII. Potassium salt, m. $165-167^{\circ}$.[560]

PrO(BuO)P(S)SH. VII. Potassium salt, m. $142-142.5^{\circ}$.[561]

F.3.2.3. 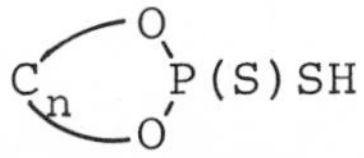

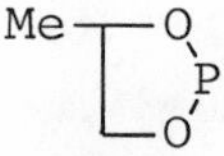 P(S)SH. V.[583,987] Liq. $b_{0.2}$ 100°,[563] $b_{0.4}$ 96-96.5°,[987] n_D^{20} 1.5610,[987] n_D^{30} 1.5592,[583] d^{20} 1.3812.[987]

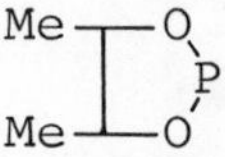 P(S)SH. V.[78,987] Liq. $b_{0.2}$ 97-98°,[987] b_3 120-122°,[78] n_D^{20} 1.5482, d^{20} 1.2941.[987] Sodium salt, m. 87-87.5°.[987]

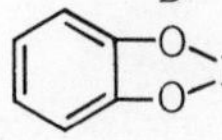 P(S)SH. V. Triethylamine salt, m. 137-138°, IR.[507]

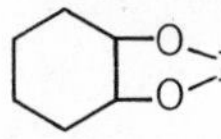 P(S)SH. V. Potassium salt, m. >200° (dec.).[1442]

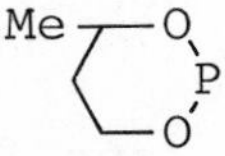 P(S)SH. V.[987,1441] Liq. $b_{0.4}$ 130-131°, n_D^{20} 1.5669, d^{20} 1.3270.[987] Ammonium salt, m. 187-188°. Potassium salt, m. >185° (dec.).[1441] Sodium salt, m. 120-121°.[987]

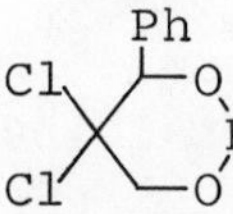 P(S)SH. V. Solid. m. 134-137°.[591]

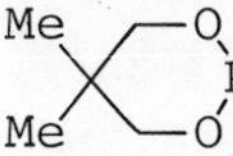 P(S)SH. V.[78,247] Solid. b_1 126-140°,[78] m. 81-82° (from ether or petroleum ether).[247] Cyclohexylamine salt, m. 216-217° (from $CHCl_3$-petroleum ether).[247]

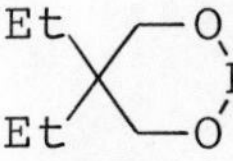 P(S)SH. V. Solid. m. 99-101°.[583]

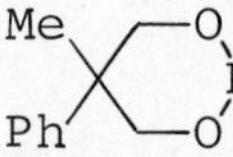 P(S)SH. V. Solid. m. 136-138°.[591]

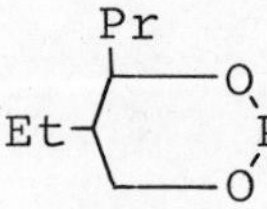 P(S)SH. V.[583,659] Liq. $b_{0.2}$ 124°, n_D^{30} 1.5341.[583] Ammonium salt, m. 209-210°. Potassium salt, m. >265°.[659]

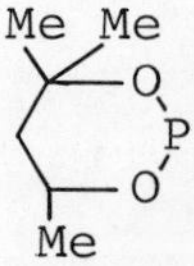 P(S)SH. V.[583,659] Solid. m. 42-44°.[583] Ammonium salt,

m. 206-207°, Potassium salt, m. 250-260° (dec.).[659]

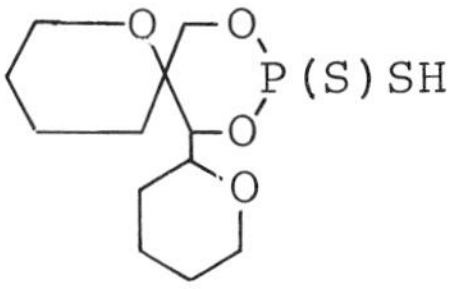

V. Solid. m. 125-130°.[585]

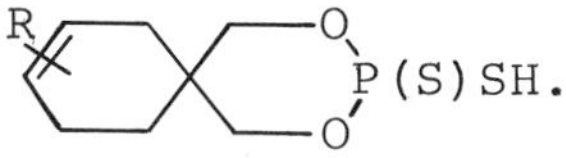

V.[587] Solids:

R	m.
H	98-100°
1'-Cl	144-147°
5'-Me	223 (dec.)
1',3',5'-Me_3	124-130°

I. Solid. m. 179-181°, from sodium salt by cation exchange. Dipotassium salt, m. 260° (dec.). Disodium salt, browns at 245-250° without melting.[198]

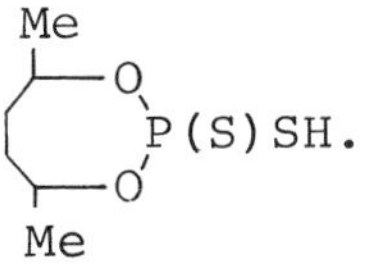

V. Ammonium salt, m. 150-152°. Potassium salt, m. 270° (dec.).[1441]

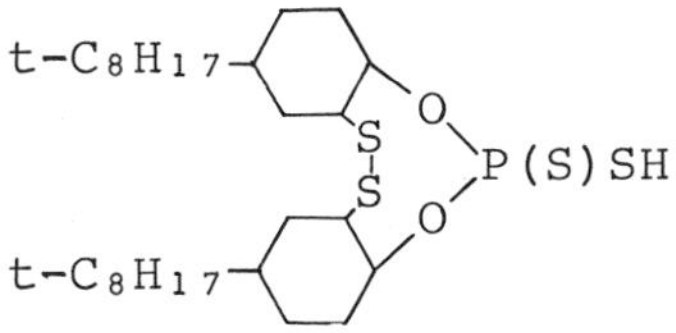

V. Potassium salt, m. 118-119° (from EtOH).[469]

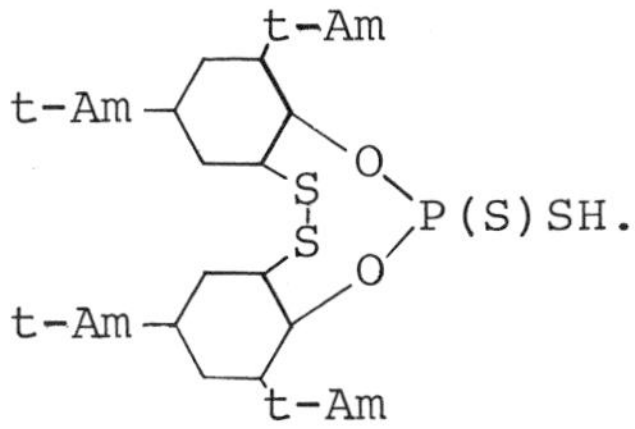

V.[467,469] Potassium salt, m. 223-225° (from BuOH). Sodium salt, m. 218-220° (from acetone).[467,469]

F.3.2.4. RO(RS)POSH

MeO(MeS)POSH. III.[192,235,480] Potassium salt, m. 152.5-153.5°,[192,235] 110-112°(from ether).[480] 4,4'-Bipyridyl salt (1:1). VII. m. 106-107.5° (dec.).[533]

MeO(EtS)POSH. VII. Diethylmethylamine salt, m. 61-62°.[745] Also several analogs.

MeO($MeNHCOCH_2S$)POSH. III. Dicyclohexylamine salt, m. 111-114°.[192,235]

MeO(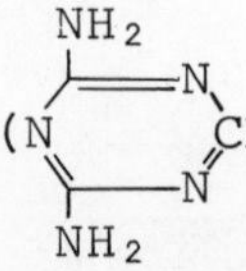CH_2S)POSH. II. Inner salt, m. 195-204° (dec.).[336]

EtO(EtS)POSH. II. Sodium salt.[961]

EtO($EtSCH_2CH_2S$)POSH. II. Potassium salt, m. 145-146°.[1285]

EtO(MeCH=CHS)POSH. II. Ethyltrimethylammonium salt, m. 70-71°.[922] Also several homologs and analogs.

EtO(BuS)POSH. II. Potassium salt, m. 168-170°.[1285]

EtO($PhCH_2S$)POSH. II. Sodium salt, m. 180-181°.[1285]

EtO(PhCHMeS)POSH. II. Potassium salt, m. 143-145°.[1285]

EtO($PhCH_2CH_2S$)POSH. II. Potassium salt, m. 156-158°.[1285]

PrO($PhCH_2S$)POSH. II. Potassium salt, m. 169-171°.[1285]

BuO($PhCH_2S$)POSH. II. Potassium salt, m. 175-176°.[1285]

F.3.2.5. $(RS)_2P(O)O^{\ominus}$

$(MeS)_2P(O)OH$. III.[192,235,480] Potassium salt, m. 214-215°,[192,235] >250°.[480] N-(2-hydroxyethyl)-N-methylpiperidinium salt, liq. n_D^{20} 1.5485, d^{20} 1.2608.[744] N-Methylquinolinium salt, m. 130-131.5°.[744] N,N'-Dimethyl-4,4'-bipyridylium salt (2:1), m. >138° (dec.).[533]

MeS(EtS)P(O)OH. VII. Triethylmethylammonium salt, m. 42-47°.[745] N-Methyloctadecylamine salt, m. 37-40°.[746] Trimethyloctadecylammonium salt, m. 78-82°.[746] N,N'-Dimethyl-4,4'-bipyridylium salt (2:1), m. >118° (dec.).[533]

MeS($MeNHCHCH_2S$)P(O)OH. VII. N-Methoxypyridinium salt, liq. n_D^{20} 1.5650, d_4^{20} 1.3500.[534] N-(2-hydroxyethyl)-N-Methylpiperidinium salt, liq., n_D^{20} 1.5508, d_4^{20} 1.3021.[744] Trimethyloctadecylammonium salt, liq., n_D^{20} 1.5047, d_4^{20} 1.0288.[746]

MeS(PrS)P(O)OH. VII. N-Methoxypyridinium salt, liq., n_D^{20} 1.5480, d_4^{20} 1.3057.[534]

F.3.3. Tertiary Phosphorodithioates

F.3.3.1. $(RO)_2P(S)SR$

$(MeO)_2P(S)SMe$. VIII.[606,890] XVII.[890] XXII.[197a,890] XXVII.[606,1001] XXVIII.[759] XXXIII.[675,890] Liq. b_1 50-51^0,[1001] b_2 55^0,[606] b_8 86^0,[890] b_{10} 93-95^0.[1001] n_D^{20} 1.5290,[1001] 1.5292,[1078] n_D^{25} 1.5273,[606] mol. ref. 263.34,[1078] d_4^0 1.2587,[197a] d_4^{20} 1.2390,[1001] IR.[673,674] ^{31}P -99.5 ppm.[715] Methostrychninium salts.[1303]

$(EtO)_2P(S)SEt$. VIII.[606] IX.[960,961] XII.[123] XVII.[960,961] XXII.[890,960,961] XXVII.[25,606,641,642,669] XXVIII.[941] XXXIII.[890] Liq. b_1 74-77^0,[890] b_2 83^0,[641] b_{10} 115-116^0,[25,123] b_{20} 128^0,[960,961] n_D^{20} 1.5013,[1078] 1.5014,[937] n_D^{25} 1.5008,[25] mol. ref. 321.71,[1078] d_0^0 1.1341,[960,961] d_4^{20} 1.1109,[941] d^{20} 1.1174.[25] IR.[673,674,1348] IR and Raman.[968] ^{31}P -94.2,[715,850] -94.9,[715] -95.[488,715] Magn. susceptibility $-\chi \times 10^6 = 135.7$.[1455]

$(i\text{-}PrO)_2P(S)S\text{-}i\text{-}Pr$. Liq. n_D^{20} 1.4843, mol. ref. 380.53,[1078] ^{31}P -91.0.[715]

$(PrO)_2P(S)SPr$. Liq. n_D^{20} 1.4955, mol. ref. 383.40.[1078]

$(i\text{-}BuO)_2P(S)S\text{-}i\text{-}Bu$. Liq. n_D^{20} 1.4859, mol. ref. 443.47.[1078]

$(BuO)_2P(S)SBu$. XXVII.[942] XXXIII.[1237] Liq. $b_{0.15}$ 73^0,[1237] b_5 160-163^0,[942] n_D^{20} 1.4831,[942] 1.4865,[1237] d_4^{20} 1.0124,[942] 1.0152.[1237]

$(PhO)_2P(S)SPh$. VIII. Solid. m. 71.5-73^0 (from MeOH).[803]

$(PhCH_2O)_2P(S)SCH_2PH$. XVII. Solid. n_D^{30} 1.6127, m. 28^0 (from ether).[463]

F.3.3.2. $(RO)_2P(S)SR'$

$(RO)_2P(S)SMe$

$(EtO)_2P(S)SMe$. XXVII.[1204] XXVIII.[759] Liq. $b_{0.01}$ 48^0,[1204] $b_{0.08}$ 63.5-64^0,[759] n_D^{20} 1.5100, d_4^{20} 1.1951.[759] From $(EtO)_2P(S)SH$ plus $(MeO)_2P(O)C{\equiv}CMe$ with Et_3N catalyst.[1204]

$Me_3\overset{+}{N}CH_2CH_2O$
$\qquad\qquad\qquad P(S)SMe.I^{\ominus}$. From L plus MeI. Solid. m. 216-
$Me_2NCH_2CH_2O$

218°.[388]

$(Me_3N^{+}CH_2CH_2O)_2P(S)SMe.2I^{\ominus}$. From L plus 2MeI. Solid. m. 166.5-167°.[388]

$(EtN^{+}(Me)_2CH_2CH_2O)_2P(S)SMe.2I^{\ominus}$. From L plus 2MeI. Solid. m. 143-144°.[388]

(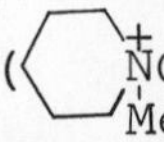$NCH_2CH_2O)_2P(S)SMe.2I^{\ominus}$. From L plus 2MeI. Solid. m. 152-153°.[388]

($NCH_2CH_2O)_2P(S)SMe.2I^{\ominus}$. From L plus 2MeI. Solid. m. 149-151°.[388]

(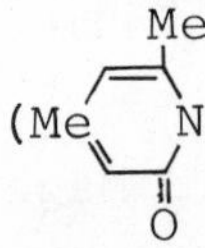$NCH_2CH_2O)_2P(S)SMe$. XVII. Solid. m. 215-217°.[1017]

$(RO)_2P(S)SMe$.

R	Method	State	n_D^{20}	d_4^{20}
i-Pr	XXVIII.[759]	Liquid. $b_{0.07}$ 60-65°	1.4950	1.073
From $[(RO)_2P(S)S]_2$ plus $(MeO)_2P(O)H$ plus Et_3N.[759]				
Pr	XXVIII.[759]	Liquid. $b_{0.1}$ 68-70°	1.5008	1.080
$CH_2=CHCH_2$	XVII.[1445]	Liquid. $b_{0.0001}$ 26°	1.5298	1.142
$HC\equiv CCH_2$	XVII.[1447]	Liquid. $b_{0.0001}$ 58°	1.5490	1.220
i-Bu	XXVIII.[759]	Liquid. $b_{0.07}$ 75-76°	1.4930	1.048
$CH_2=CMeCH_2$	XVII.[1447]	Liquid. $b_{0.0001}$ 42°	1.5224	1.100
Bu	XXVIII.[759]	Liquid. $b_{0.08}$ 89-90°	1.4960	1.054
$MeCH=CHCH_2$	XVII.[1447]	Liquid. $b_{0.0001}$ 55°	1.5240	1.104
$3\text{-}Me\text{-}4\text{-}NO_2C_6H_4$	XVII.[926]	Solid. m. 84-85°.		
Cholesteryl	XVII.[191]	Solid m. 188°. ^1H-NMR.[191]		

$(MeO)_2P(S)SCH_2Cl$. XVII.[949,1087] Liq. b_{10} 99-103°,[949] 100°,[1087] n_D^{20} 1.5268.[949]

$(EtO)_2P(S)SCH_2Cl$. XVII.[949,1087] Liq. b_1 93-95°,[1087] $b_{2.5}$ 113-115°, n_D^{20} 1.5200.[949]

$(EtO)_2P(S)SCCl_3$. XVII. Liq. n_D^{25} 1.5304.[318]

$(MeO)_2P(S)SCH_2NCO$. XVII. Liq. $b_{0.1}$ 82-83°, n_D^{20} 1.5398, d_4^{20} 1.3428, IR.[1238]

$(EtO)_2P(S)SCH_2NCO$. XVII. Liq. $b_{0.06}$ 77-78°, n_D^{20} 1.5170, d_4^{20} 1.2276.[1238]

$(PrO)_2P(S)SCH_2NCO$. XVII. Liq. $b_{0.1}$ 103-105°, n_D^{20} 1.5090, d_4^{20} 1.1683. IR.[1238]

$(RO)_2P(S)SCH_2NH$-

$(RO)_2P(S)SCH_2NHCOOMe$. From the isocyanates plus MeOH.[1238]

R	State	n_D^{20}	d_4^{20}
Me	Liq.	1.5254	1.3031
Et.	Liq.	1.5157	1.2413
Pr	Liq.	1.5093	1.1850

$(RO)_2P(S)SCH_2NHCOOEt$.

R	Methods	State	n_D^{20}	d_4^{20}
Me	XVII,[741] XXI[1246]	Liq. $b_{0.2}$107-110°	1.5091	1.3498[741,1246]
	(*)[1238]		1.5282	1.2978[1238]
(*) From the isocyanate plus EtOH.				
Et	XVII,[741] XXI[1246]	Liq. $b_{0.65}$64-68°	1.4990	1.1904[741,1246]
	(*)[1238]		1.5118	1.2051[1238]
Pr	XVII[741]	Liq. $b_{0.075}$82°	1.4912	1.0866[741]
i-Bu	XVII	Solid. $b_{0.175}$122-124°(solidified).[741]		
Bu	XVII	Liq. $b_{0.2}$100°	1.5000	1.2523[741]
$(RO)_2P(S)SCH_2NMeCOOEt$.				
Et	XXI	Liq. $b_{0.1}$74-83°	1.4896	1.1592[1246]
i-Pr	XXI	Liq. $b_{0.4}$90-93°	1.4744	1.1008[1246]
i-Bu	XXI	Liq. $b_{0.3-0.4}$95-114°	1.4906	1.0845[1246]

$(ArO)_2P(S)SCH_2N^+Me_3.X^{\ominus}$.	Method	State	$X^{\ominus}$=Br	$X^{\ominus}$=I
	XXXII (from L. plus MeX)			
Ph	XXXII	Solid	m.110-115.5°	m.105-110°[535,536]
4-BrC_6H_4	XXXII	Solid	m.76-79°	m.171-173°[537]
4-ClC_6H_4	XXXII	Solid	m.139-141°	m.155-157.5°[537]

$(RO)_2P(S)SCH_2$-(N-ring)

$(RO)_2P(S)SCH_2N$ 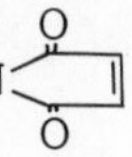

	Method	State	
Me	XVII	Solid	m.46-48° (from petroleu[m] ether)[625]
Et	XVII	Liq.	n_D^{21} 1.45[5] [625]

$(RO)_2P(S)SCH_2N$ 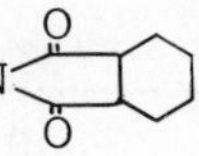

Me XVII.[268,608,676] XXI.[1261] Solid. m. 63-64°(from petroleum ether-benzene),[676] m. 70°,[608] 72-73°.[1261] "Imida liq. n_D^{28} 1.5959.[268] ^{1}H-NMR.[74,526] ^{31}P 18.6 (P_4O_6 external).[1047] Mass spect.[196]

Et XVII.[268,608,676] XXI.[1261] Solid. m. 59-60°,[1261] m. 63 - 64° (from petroleum ether-benzene),[676] 63-65° (from i-PrOH),[268] 64°.[608]

i-Pr XVII.[268,608] Solid. m. 113°,[608] 114-115.5° (from MeOH).[268]

$(RO)_2P(S)SCH_2N$ 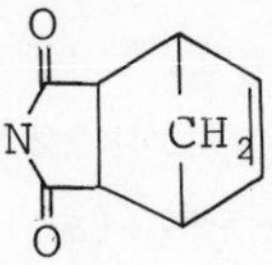 .

	Method	State	n_D^{20}
Et	XVII	Liq.	1.5233[1287]
i-Pr	XVII	Liq.	1.5165
Pr	XVII	Liq.	1.5095
i-Bu	XVII	Liq.	1.5060

$(RO)_2P(S)SCH_2N$ 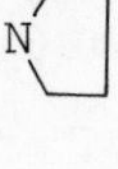 .

	Method	State	
Me	XVII	Liq.	1.5980[967]
Et	XVII	Liq.	1.5718
Bu	XVII	Liq.	1.5519

$(EtO)_2P(S)SCH_2N$ 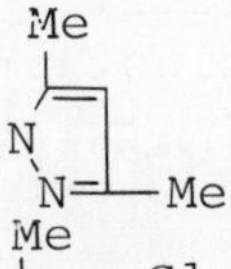 XVII. Liq. $b_{0.1}$ 120°. Also 12 analog[s][363]

$(RO)_2P(S)SCH_2N$ 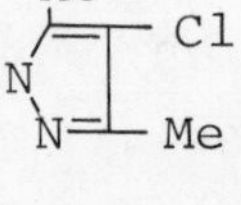

Me   XVII. Liq. $b_{0.2}$ 115°.[363,1052]

Et   XVII. Liq. $b_{0.2}$ 115°.[363,1052]

$(EtO)_2P(S)SCH_2N$ XVII. Liq. $b_{0.05}$ 135°.[1052]

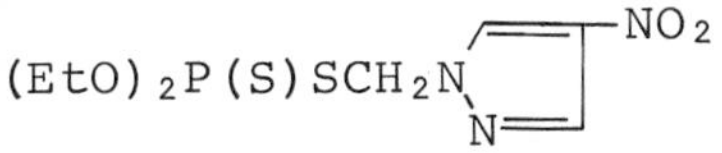

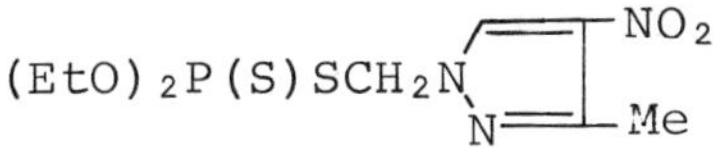

$(EtO)_2P(S)SCH_2N$ XVII. Liq. $b_{0.05}$ 120°.[363]

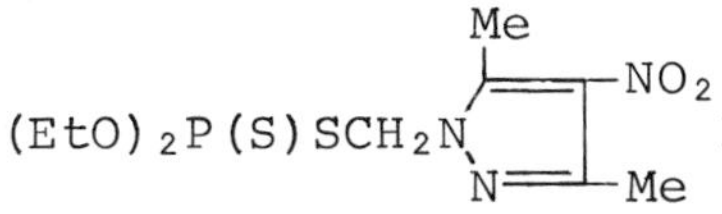

$(EtO)_2P(S)SCH_2N$ XVII. Solid. m. 50-51°.[1052]

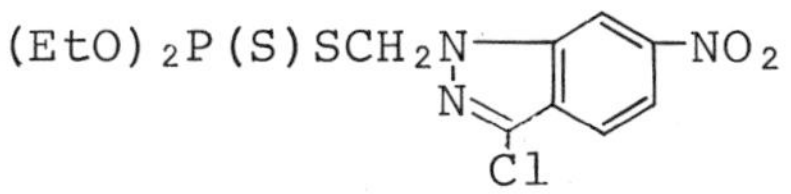

$(EtO)_2P(S)SCH_2N$ Solid. m. 90-95°.[814]

$(RO)_2P(S)SCH_2N$

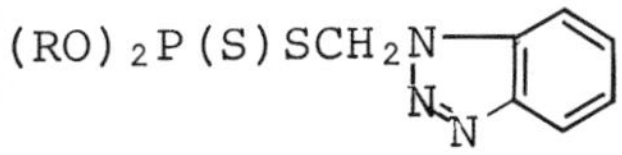

Me   XXI. Solid. m. 70-72°.[1261]

Et   XXI. Solid. m. 45-47°.[1261]

$(RO)_2P(S)SCH_2N$

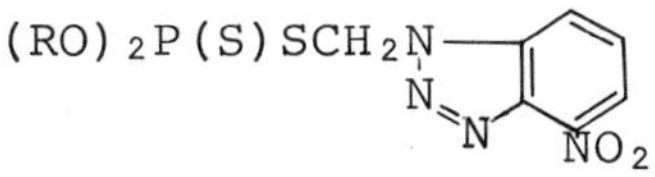

Me   XVII. Solid. m. 69°.[609]

Et   XVII. Solid. m. 103-104°.[609]

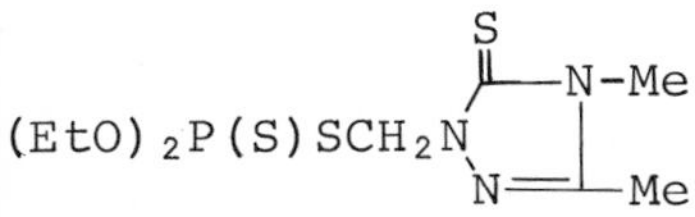

$(EtO)_2P(S)SCH_2N$ XVII. Solid. m. 77°.[1311]

$(RO)_2P(S)SCH_2N$

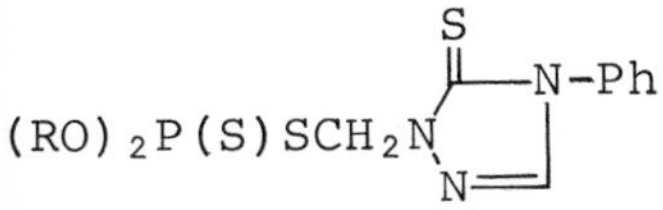

Me   XVII. Solid. m. 61°.[1311]

Et   XVII. Solid. m. 84°.[1311]

$(MeO)_2P(S)SCH_2N$ XVII. Solid. m. 91°.[1311]

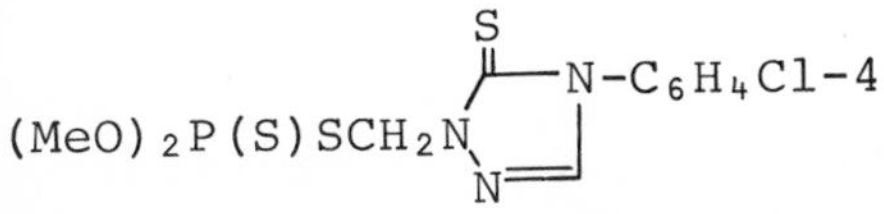

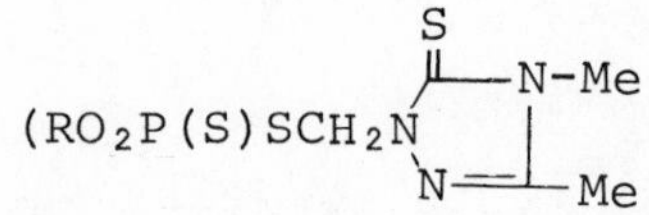

Me — XVII. Solid. m. 113°.[1311]

Et — XVII. Solid. m. 77°.[1311]

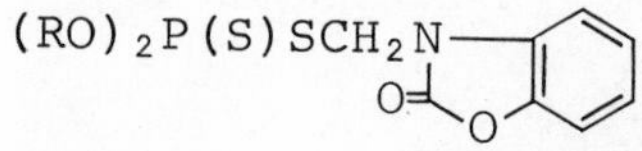

Me — XVII.[122,161,1022] Solid. m. 68-72° (from benzene ligroin),[122,161] m. 70° (from cyclohexane).[1022]

Et — XVII.[122,161,770,1020] Solid. m. 120-128° (from benzene-ligroin)[122,161] (also given as m. 30-31°[770,1020]).

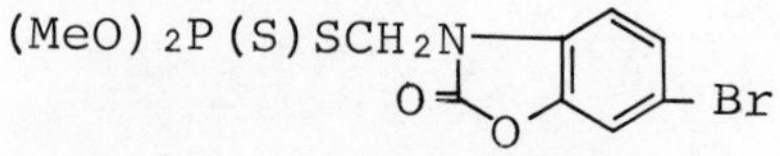

XVII. Solid. m. 87-90°.[122,161]

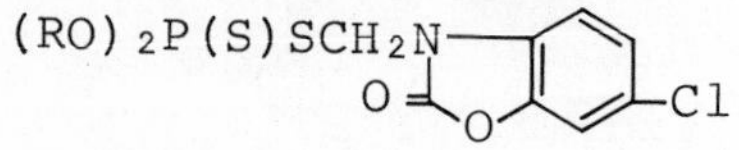

Me — XVII.[122,161] XXI.[1261] Solid. m. 55-57° (from benzene-petroleum ether),[122,161] m. 95-96°.[1261]

Et — XXI.[1261] Solid. m. 39-41°,[1261] 47.5-48°.[182a] "Phosalone, Zolon

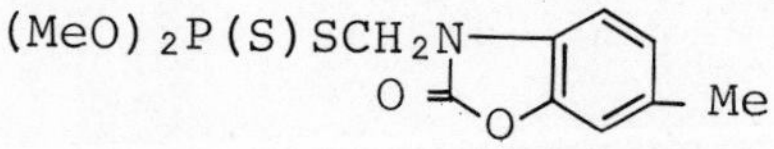

XVII. Solid. m. 95-96° (from benzene).[1020]

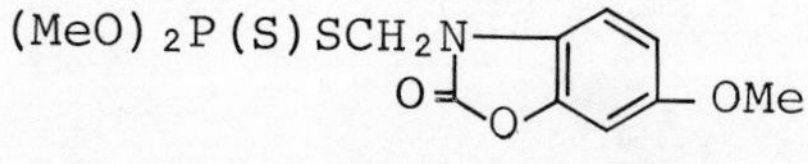

XVII. Solid. m. 62° (from cyclohexane).[1020]

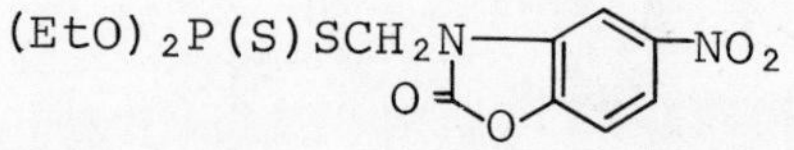

XVII. Solid. m. 54-56° (from MeOH).[626]

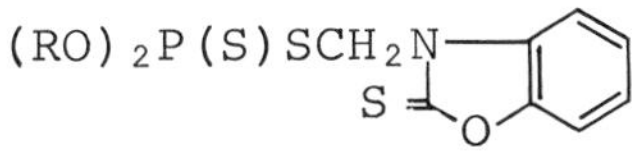

Me XVII. Solid. m. 104° (from benzene),[626] 105° (from MeOH).[1058]

Et XVII. Solid. m. 73° (from MeOH),[626] 70-71° (from EtOH).[1058]

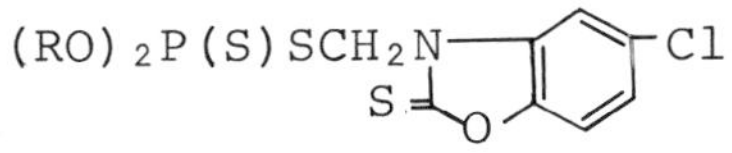

Me XVII. Solid. m. 95-98°.[1058]

Et XVII. Solid. m. 68-71°.[1058]

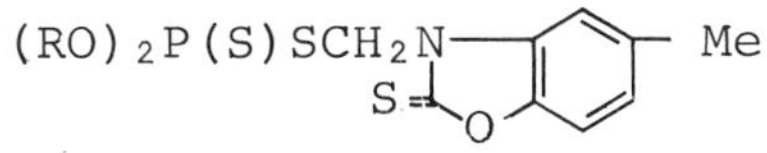

Me XVII. Solid. m. 79-81°.[1058]

Et XVII. Solid. m. 67-68°.[1058]

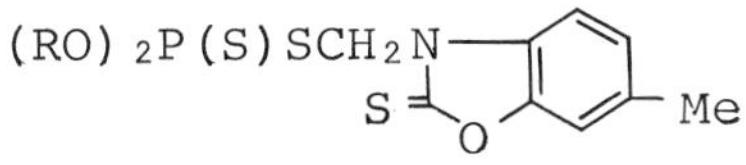

Me XVII. Solid. m. 104-107°.[1058]

Et XVII. Solid. m. 85-86°.[1058]

Me XVII. Liquid. $b_{0.005}$ 110°. Also 22 analogs.[364]

Et XVII. Liquid. $b_{0.001}$ 105°.[364]

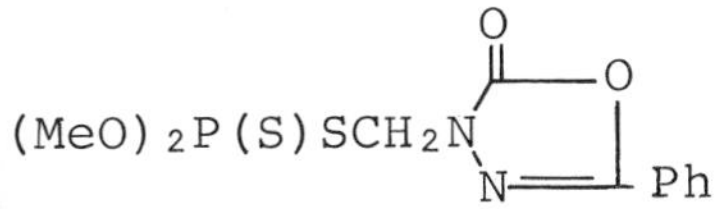

XXI. Solid m. 78-80°.[1261]

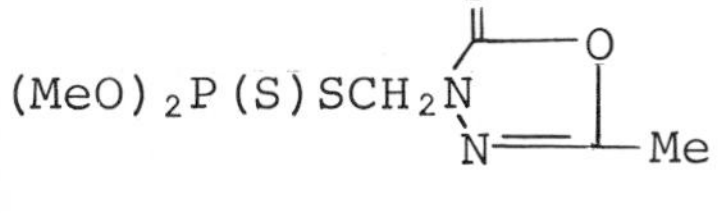

XVII. Liq. $b_{0.001}$ 145°.[364]

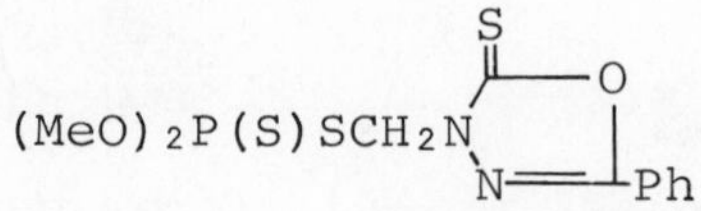

$(MeO)_2P(S)SCH_2N$ [structure]   XVII. Solid. m. 100-101°.[364]

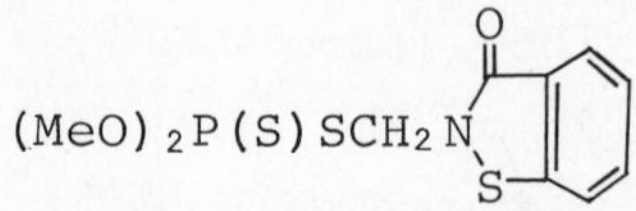

$(MeO)_2P(S)SCH_2N$ [structure]   XVII. Solid. m. 70° (from i-PrOH).

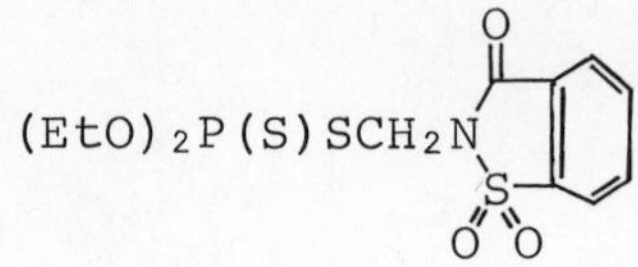

$(EtO)_2P(S)SCH_2N$ [structure]   XVII. Solid. m. 77°.[608]

$(RO)_2P(S)SCH_2N$ [structure, OMe]

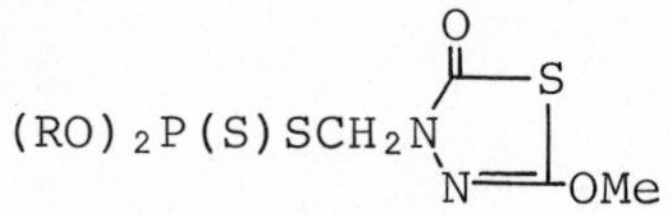

Me   XVII.[1051,1053] XXI.[1053,1261]
                    Solid. $b_{0.0001}$ 130°,[1051] m.
                    39-40°.[1053,1261] [1H]-NMR.[525]
                    Also 40 analogs.[1051] "Supra-
                    cide."

Et   XVII.[1051,1053] XXI.[1261] Solid.
                    $b_{0.2}$ 130°,[1051] m. 43-44°.[1261]

$(RO)_2P(S)SCH_2N$ [structure, OEt.]

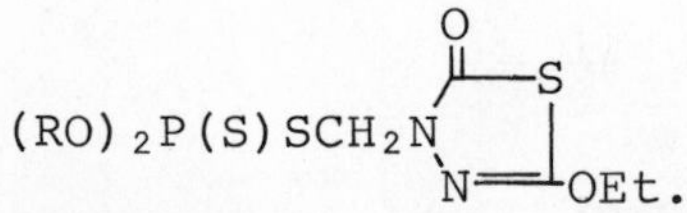

Me   XXI. Solid. m. 50-51°.[1261]
ClCH$_2$CH$_2$ XXI. Solid. m. 43-44°.[1261]
EtOCH$_2$CH$_2$ XXI. Solid. m. 32-33°.[1261]

$(RO)_2P(S)SCH_2N$ [structure, SMe]

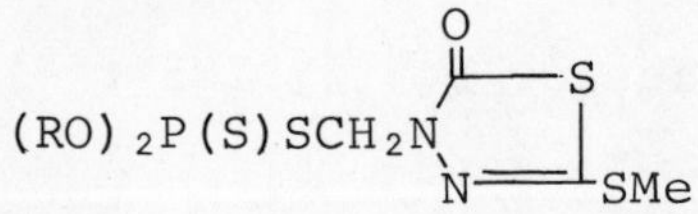

Me   XVII.[1051] XXI.[1261] Solid. $b_{0.1}$
                    130°,[1051] m. 28-29°.[1261]

Et   XVII. Liq. $b_{0.1}$ 140°.[1051]

$(EtO)_2P(S)SCH_2N$ [structure, Me]   XVII. Solid. m. 52-53°.[364]

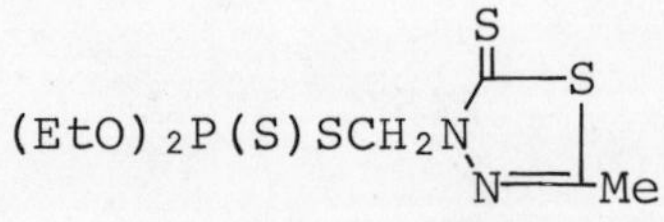

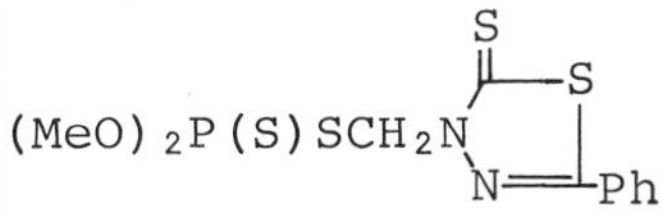

XVII. Solid. m. 75-76°.[364]

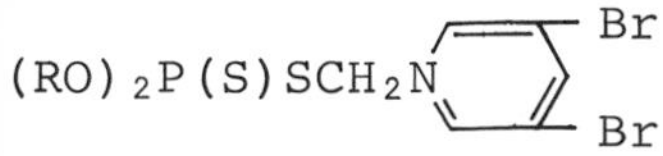

Me XVII. Solid. m. 109-111°.[1320]

Et XVII. Solid. m. 64-66°.[1320]

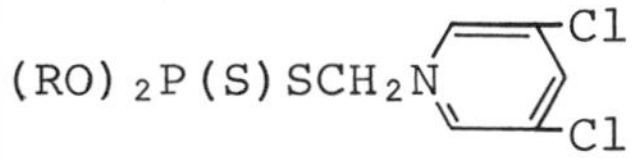

Me XVII. Solid. m. 72-75°.[1320]

Et XVII. Solid. m. 51-53°.[1320]

Me XVII. Solid. m. 86-88°.[1320]

Et XVII. Solid. m. 62-64°.[1320]

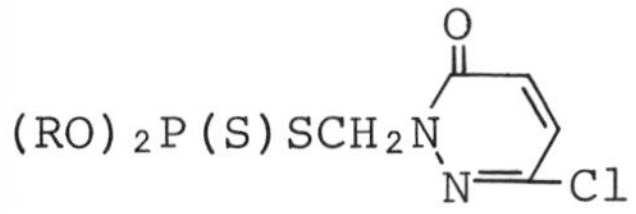

Me XVII. Liq. n_D^{25} 1.5870.[233]

Et XVII. Solid. b_1 115-120°, n_D^{25} 1.5363, m. 52.5-53.5°.[233]

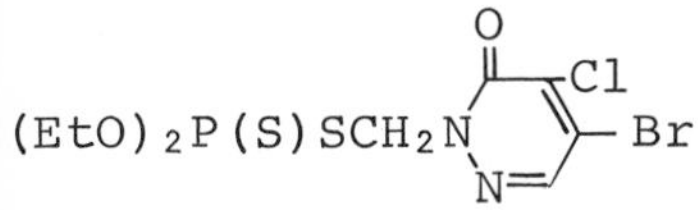

XVII. Liq. n_D^{25} 1.5980.[233]

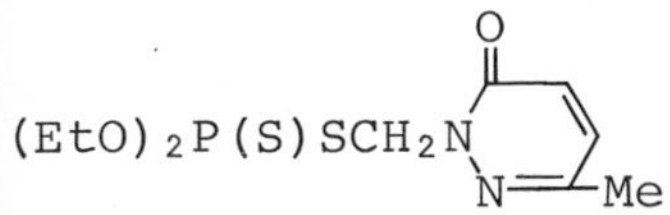

XVII. Liq. n_D^{25} 1.5535.[233]

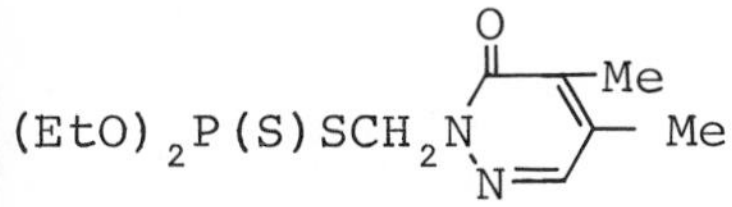

XVII. Solid. m. 105.5-106.5°.[233]

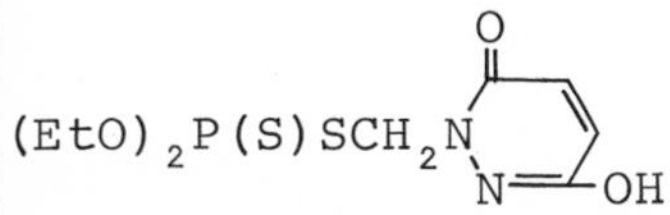

XVII. Solid. m. 104-104.5°.[233] With Ac_2O gives acetoxy analog. Liq. n_D^{25} 1.5525.[233]

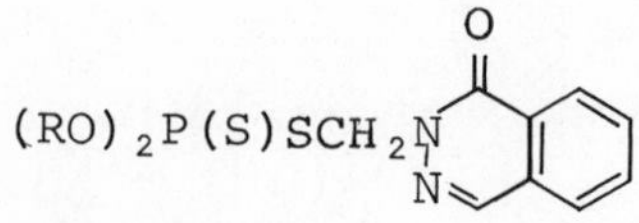

Me — VIII. Solid. m. 87°.[617]

Et — VIII. Solid. m. 58° (from dilu i-PrOH).[617]

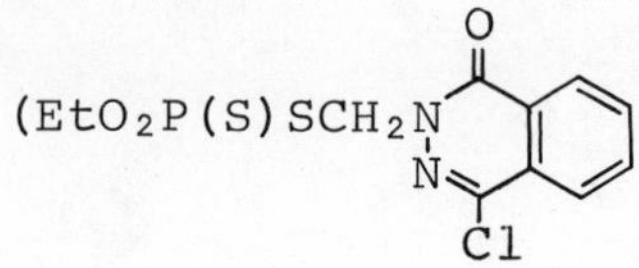

XVII. Solid. m. 78.5-79°.[233]

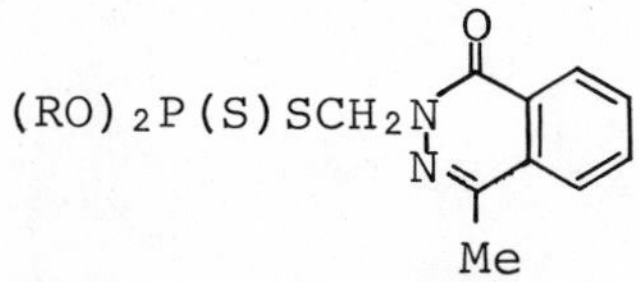

Me — XVII. Solid. m. 113-115.5° (fr (MeOH).[1062]

Et — XVII. Solid. m. 79-82°.[1062]

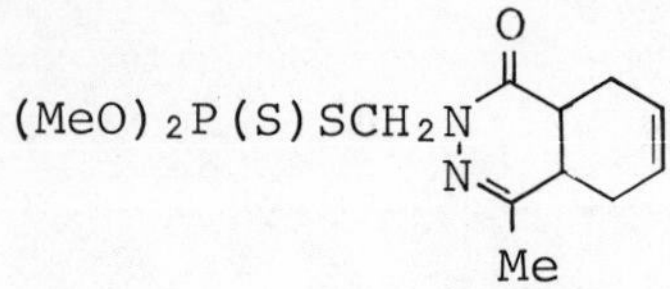

XVII. Solid. m. 105-110° (from MeOH).[1069]

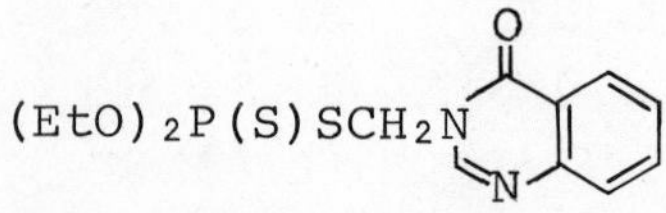

XVII. Solid. m. 71°. (from ben-zene-ligroin).[613]

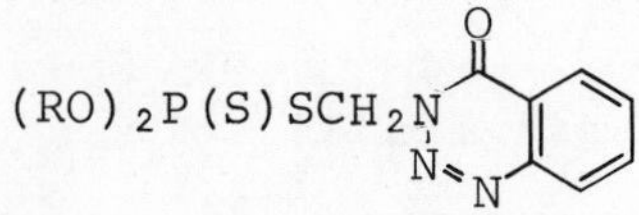

Me — XVII.[610] XXI.[311] Solid. m. 63-66°,[311] 72° (from MeOH).[610] IR and UV.[375] ^{1}H-NMR.[74,526] ^{31}P 18.2 (P_4O_6 external).[1047] Mass spect.[196] "Guthion, Azinphos-methyl."

Et — XVII.[610] XXI.[311] Solid. m. 49-50° (from MeOH).[610] ^{1}H-NMR.[74] "Azinphosethyl"

i-Pr — XVII. Solid. m. 56° (from MeOH).[610]

Pr — XVII. Solid. m. 53° (from MeOH).[610]

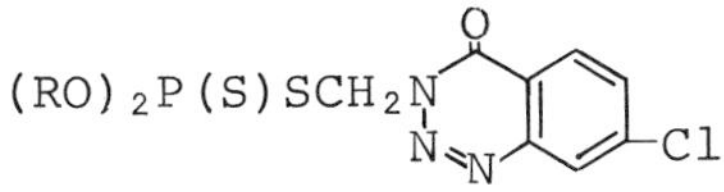

Me — XVII. Solid. m. 79-80°.[1035]

Et — XVII. Solid. m. 64-65° (from MeOH).[1035] Also about 30 analogs similarly prepared.

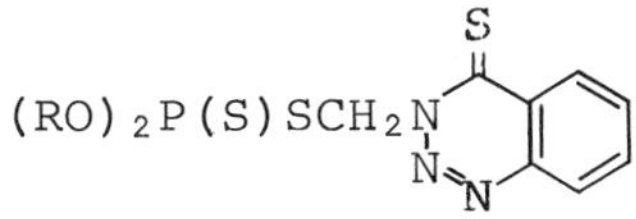

Me — From C=O analog plus P_2S_5. Solid. m. 102-103° (from benzene-ligroin).[1097]

Et — From C=O analog plus P_2S_5. Solid. m. 84° (from i-PrOH)[1097]

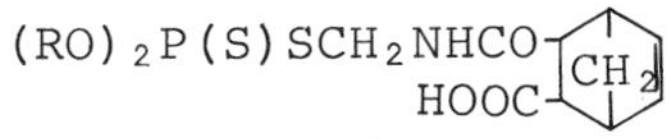 From 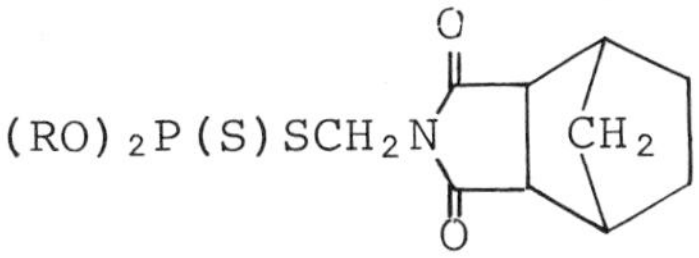 plus 10% NaOH.[1287]

Et	Liq.	n_D^{20}	1.5435
Pr	Liq.		1.5273
Bu	Liq.		1.5268

$(EtO)_2P(S)SCH_2NHCO$ CH_2 $SP(S)(OEt)_2$ HOOC

XXII. Liquid. n_D^{20} 1.5513.[1287] Reacts with CH_2N_2 to give the COOMe ester. Liquid. n_D^{20} 1.5264.[1287]

$((EtO)_2P(S)SCH_2NH)_2CO$. XXI. Solid. m. 68-78°.[458]

$(RO)_2P(S)SCH_2NHCOOR'$

$(RO)_2P(S)SCH_2NHCOOEt$

	Methods		State		n_D^{20}	d^{20}
Me	XVII,[741]	XXI[1246]	Liq.	$b_{0.2}$107-110°,	1.5091	1.3
Et	XVII	XXI	Liq.	$b_{0.065}$64-68°	1.4990	1.1
Pr	XVII	XXI	Liq.	$b_{0.075}$82°	1.4912	1.08
i-Bu	XVII	XXI	Solid m. 22°			

Also 8 analogs.[1246]

$(RO)_2P(S)SCH_2NHCOO$-i-Pr. From the isocyanates plus i-PrOH.

Me	Liq.	1.5015	1.20
Pr	Liq.	1.5008	1.13

$[(EtO)_2P(S)SCH_2NH]_2CS$. XXI. Solid. m. 66-68°.[458]

$(EtO)_2P(S)SCR_2N{=}NCR_2SP(S)(OEt)_2$. XVII. Solids. R, Me, m. 33-35° (from MeOH); Et, m. 41°.[689]

$(EtO)_2P(S)SC({=}NH)NH_2$. XVII. Solid. m. 83-85°.[741]

(i-BuO)$_2$P(S)SC(=NH)NH$_2$. XVII. Solid. m. 146-148°.[741]

$(MeO)_2P(S)SC({=}NC_6H_3Et_2$-2,6)$NHC_6H_3Et_2$-2,6. XXII. Solid. m. 72-74° (from MeOH).[691]

$(MeO)_2P(S)SC({=}NC_6H_3$-i-Pr$_2$-2,6)NHC_6H_3-i-Pr$_2$-2,6. XXII. Solid m. 116-118° (from MeOH).[691]

$(EtO)_2P(S)SC({=}NC_6H_2Et_2Cl$-2,6,4)$NHC_6H_2Et_2Cl$-2,6,4. XXII. Solid. m. 104°.[691]

$(RO)_2P(S)SCH_2OR'$

(i-PrO)$_2$P(S)SCH$_2$OMe. XXI. Liq. $b_{0.26-0.28}$ 83-87°.[1298]

$(EtO)_2P(S)SCH_2OBu$. XXI. Liq. $b_{0.09-0.11}$ 113-119°.[1298]

$(EtO)_2P(S)SCH_2OEt$. XVII.[279,741,1192] Liq. $b_{0.01}$69°,[279,1192] $b_{0.18}$73-74°, n_D^{20} 1.5029, d^{20} 1.1484.[741]

$(EtO)_2P(S)SCH_2O$-i-Pr. XVII.[279,741,1192] Liq. $b_{0.01}$65°,[279,1192] $b_{0.15}$ 71-73°, n_D^{20} 1.5021, d^{20} 1.1427.[741]

$(EtO)_2P(S)SCH_2OCH_2CH_2Br$. XVII. Liq. $b_{0.01}$ 88°.[1164]

$(EtO)_2P(S)SCH_2OCH_2CH_2OC_6H_3Cl_2$-2,4. XXI. Liq. n_D^{30} 1.5552.[273] 20 analogs.

$(EtO)_2P(S)SCH_2OCH_2CH_2SEt$. From $(EtO)_2P(S)SCH_2OCH_2CH_2Br$ plus EtSH and EtONa in benzene. Liq. $b_{0.01}$ 96°.[1164]

$(MeO)_2P(S)SCH_2OCOMe$. XVII. Liq. $b_{0.02}$ 100°.[1351]

$(EtO)_2P(S)SCH_2OCOCH_2SMe$. From $(EtO)_2P(S)SCH_2OH$ plus $MeSCH_2COCl$ and pyridine. Liq. n_D^{30} 1.5592.[392]

$(MeO)_2P(S)SCH_2OCOCH_2S$-i-Pr. From $(MeO)_2P(S)SCH_2OCOCH_2Cl$ plus i-PrSNa. Liq. n_D^{30} 1.5282.[392] Also SBu, 1.5426; S-i-Am, 1.5150; SAm, 1.4540; SC_7H_{15}, 1.4940; SPh, 1.5892.[392]

$(EtO)_2P(S)SCH_2OCHMeCOOEt$. XXI. Liq. $b_{0.3}$ 105-110°, n_D^{32} 1.5020.[888]

$(EtO)_2P(S)SCH_2OC_6H_2Cl_3$-2,4,5. XVII. Liq. n_D^{25} 1.5795.[1293]

$(EtO)_2P(S)SCH_2OC_6H_4NO_2$-4. XVII. Liq. n_D^{25} 1.5781.[1376] Also several analogs.

$(RO)_2P(S)SCH_2O$- 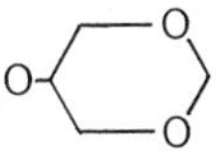

Me	XXI. Liq. n_D^{30} 1.5402.[391]
Et	XXI. Liq. n_D^{30} 1.5128.[391]

$(RO)_2P(S)SCH_2OCH$- 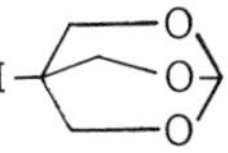 P(S).

Me	XVII. Solid. m. 94.5° (from MeOH).[1009]
Et	XVII. Solid. m. 69° (from CCl_4).[1009]
i-Pr	XVII. Solid. m. 116.5-117° (from i-PrOH).[1009]

$(RO)_2P(S)SCH_2OCH_2CH_2NHCOCH_2OC_6H_3Cl_2$-2,4.

Me	XXI. Liq. n_D^{30} 1.5753.[274]
Et	XXI. Liq. 1.5700.
i-Pr	XXI. Liq. 1.5540.

Also 28 similar compounds.[274]

$(MeO)_2P(S)SCH_2O$- 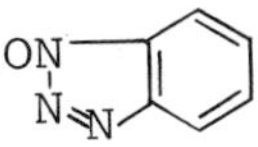 XVII. Solid. m. 93-84° (from MeOH).[302] Also 5-Cl, m. 79°; 6-CF_3, m. 85°.[302]

<u>$(RO)_2P(S)SCH_2SR'$</u>

$(i\text{-}PrO)_2P(S)SCH_2SMe$. XXI. Liq. $b_{0.3}$ 112-116°.[466a]

$[(RO)_2P(S)S]_2CH_2$.

R	Methods	State	n_D^t	d^{20}
Me	XVII[294,1393]	Liq. $b_{0.01}116^{\circ}$[294]	1.5247^{25},[1393]	1.2162[75]
Et	XVII,[75,294,857,1243,1391,1393] XVIII.[75] Liq. $b_{0.01}125^{\circ}$ $b_{0.3}$ 164-165°,[1243] n_D^{20} 1.5478,[1243] n_D^{25} 1.5325,[857] d^{20} 1.2277,[1243] IR and UV,[375] ^{1}H-NMR,[74,524,526] ^{31}P 20.0 (P_4O_6 external),[1047] mass spect.[196] "Ethion"			
i-Pr	XVII[75,1243,1393]	Liq.	1.5206^{20},[75]	1.1360[75]
	XVIII[75]		1.5163^{25},[1243]	1.1539[124
Pr	XVII.[75,294,1243,1393]	Liq. $b_{0.01}148^{\circ}$,[294]	1.5276^{20},[1243]	
	XVIII.[75]	$b_{0.1}$175-176°[1243]	1.5220^{25},[1393]	
i-Bu	XVII.[1243,1393]	Liq.	1.5103^{20},[1243]	1.0919[124
			1.5088^{25},[1393]	
Bu	XVII.[1243,1393]	Liq.	1.5153^{20},[1243]	1.1084[124
			1.5110^{25},[1393]	

[(cyclo-$C_6H_{11}O)_2P(S)S]_2CH_2$. XVII.[1391,1393] Liq. n_D^{25} 1.5290.

[(4-$NO_2C_6H_4O)_2P(S)S]_2CH_2$. XVII. Liq. n_D^{20} 1.5753.[1391]

[(MeO)$_2$P(S)$SCH_2]_2$S. XVII. Liq. n_D^{20} 1.6084.[169]

[(i-PrO)$_2$P(S)$SCH_2]_2$S. XVII. Liq. n_D^{20} 1.5580.[169]

(EtO)$_2$P(S)SCH_2SC(S)SEt. From (EtO)$_2$P(S)SCH_2Cl plus EtSCSSNa Liq. n_D^{25} 1.5964.[1375] Also several analogs.[1375]

((EtO)$_2$P(S)S)$_2$C=NPh. XVII. Solid. m. 72° (from light gasoline). Analogs.[693]

(RO)$_2$P(S)SCH_2SEt.

Me XVII.[279,1192] XXI.[633] Liq. $b_{0.01}78^{\circ}$,[279,1192] $b_{1.5}$114-116°.[633]

Et XVII.[279,1192] XXI.[458a,633] Liq. $b_{0.01}$ 75-78°,[279,1192] b_2 125-127°.[633] ^{1}H-NMR.[74,524,526] Cryoscopic m. p. T_o -43.7 (100% pure).[93] "Phorate."

i-Pr XVII.[40] XXI.[466a,633] From (i-PrO)$_2$P(S)SCH_2OMe plus

EtSH and HCl.[1297] Liq. $b_{0.3}$ 109-112°,[1297] 112-116°,[466a] b_2 126-128°,[633] n_D^{25} 1.5168.[40]

Pr XXI.[633] Liq. $b_{1.5}$ 132-133°.[633]

$(EtO)_2P(S)SCH_2SCH_2CH_2Cl$. XVII. Liq. $b_{0.01}$ 105°.[1198]

$(MeO)_2P(S)SCH_2SCH_2CH_2OCOCl$. (*) Liq. n_D^{30} 1.5500.[393]

$(EtO)_2P(S)SCH_2SCH_2CH_2OCOCl$. (*) Liq. n_D^{30} 1.5342.[393]

(*) From $(RO)_2P(S)SCH_2SCH_2CH_2OH$ plus $COCl_2$.[393]

$(EtO)_2P(S)SCH_2SCH_2CH_2OCONH_2$. From -OCOCl plus NH_3. Liq. n_D^{30} 1.5460.[735]

$(EtO)_2P(S)SCH_2SCH_2CH_2OCONMe_2$. From -COCl plus Me_2NH. Liq. n_D^{30} 1.5508.[735]

$(EtO)_2P(S)SCH_2SCH_2CH_2OCONHEt$. From $(EtO)_2P(S)SCH_2SCH_2CH_2OH$ plus EtNCO. Liq. n_D^{30} 1.5320.[735] Also, RNCO give: -CONH-i-Pr, 1.5263; -CONHBu, 1.5182; -CONHPh, 1.5678.[735]

$(EtO)_2P(S)SCH_2SCH_2CH_2OCOSBu$. (*) Liq. n_D^{25} 1.5295.[1373,1374]

$(EtO)_2P(S)SCH_2SCH_2CH_2OCOSPh$. (*) Liq. n_D^{25} 1.5770.[1373,1374]

(*) From $(EtO)_2P(S)SCH_2SCH_2CH_2OH$ plus RSCOCl and Et_3N.[1373,1374]

$(EtO)_2P(S)SCH_2SCH_2CH_2OP(S)(OEt)_2$. From $(EtO)_2P(S)SCH_2CH_2CH_2OH$ plus $(EtO)_2P(S)Cl$ and Et_3N. Liq. n_D^{25} 1.5205.[1373]

$(EtO)_2P(S)SCH_2SCH_2CONH_2$. (*) Liq. n_D^{25} 1.5426.[1292]

$(EtO)_2P(S)SCH_2SCH_2CONHMe$, (*) Liq. b_{20} 140°, n_D^{25} 1.5424.[1292]

(*) From $(EtO)_2P(S)SCH_2Cl$ plus $NaSCH_2CONHR$.[1292]

$(MeO)_2P(S)SCH_2SCH_2COOMe$. XVII. Liq. b_2 116-120°, n_D^{20} 1.5292.[309] Also -COOEt, $b_{0.6}$ 105-106°, n_D^{20} 1.5250; -COOPr, 1.5102; -COO-i-Bu, 1.5180$^{21.5°}$; -COO-cyclo-C_6H_{11}, 1.4982.[309]

$(EtO)_2P(S)SCH_2SCOOMe$. XVII. Liq. n_D^{20} 1.5257.[309]

$(EtO)_2P(S)SCH_2SCOOEt$. XVII.[309] XXI.[888] $b_{0.4}$ 140-142°,[888] $b_{0.5}$ 121-125°,[309] n_D^{20} 1.5052,[309] n_D^{31} 1.5159.[888]

$(MeO)_2P(S)SCH_2SCH_2CONHMe$. (*) Liq. n_D^{23} 1.5212.[858]

$(EtO)_2P(S)SCH_2SCH_2CONHMe$. (*) Liq. n_D^{32} 1.5203.[858]

(*) From $(RO)_2P(S)SCH_2SCH_2COOMe$ plus $MeNH_2$.[858]

$(MeO)_2P(S)SCH_2SBu$. XVII.[279,1192] Liq. $b_{0.01}$ 93°.[279,1192]

$(EtO)_2P(S)SCH_2SBu$. XVII.[279,1192] Liq. $b_{0.01}$ 95°.[279,1192]

$(i\text{-}PrO)_2P(S)SCH_2SBu$. XVII. Liq. $b_{0.01}$ 95°.[279,1192]

$(MeO)_2P(S)SCH_2SC_6H_{13}$. XVII.[279,1192] Liq. $b_{0.01}$ 109°.[279,11

$(i\text{-}PrO)_2P(S)SCH_2SC_6H_{13}$. XVII. Liq. $b_{0.01}$ 115°.[279,1192]

$(EtO)_2P(S)SCH_2S$–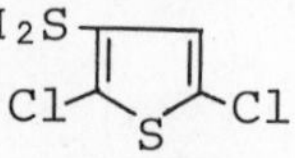 From $(EtO)_2P(S)SCH_2Cl$, RSH and KOH. Liq. n_D^{30} 1.5720.[257] Also several analogs.[257]

$(EtO)_2P(S)SCH_2S$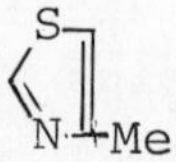 From $(EtO)_2P(S)SCH_2Cl$ and NaSR. Liq. $b_{0.01}$ 122°.[624]

$(RO)_2P(S)SCH_2SPh$. XVII.[279,353,1192,1245] Liq.

R	b./mm	n_D^{20}	d^{20}
Me	100°/0.001	1.6100.[353]	
Et	128-130°/0.01.[279,1192]		
	128°/0.03	1.5909	1.2044[1245]
i-Pr	133°/0.18	1.5720	1.1691
Pr	139-142°/0.08	1.5726	1.1670
i-Bu	151-152°/0.18	1.5673	1.1214
Bu	175°/0.15	1.5583	1.1227

$[(RO)_2P(S)S]_2CPh_2$. XVIII. Liq.[1429]

R	n_D^{20}	d_4^{20}
Et	1.5945	1.2374[1429]
i-Pr	1.5862	1.2014
Pr	1.5884	1.2102

$(MeO)_2P(S)SCH_2SC_6H_4Cl\text{-}4$. ^{1}H-NMR.[74,526] ^{31}P 15.9 (P_4O_6 external).[1047] Mass spect.[196] "Methyl Trithion."

$(EtO)_2P(S)SCH_2SC_6H_4Cl\text{-}4$. XVII. Liq. n_D^{26} 1.6198.[269] IR.[162,375] UV.[375] ^{1}H-NMR.[74,524,526] Mass spect.[196] "Trithion."

$(i\text{-}PrO)_2P(S)SCH_2SC_6H_4Cl\text{-}4$. XVII. Liq. n_D^{25} 1.6200.[269]

$(EtO)_2P(S)SCH_2SC_6H_3Cl_2\text{-}2,5$. XVII.[353] Liq. $b_{0.001}$ 120°, n_D^{21} 1.6007, d_4^{21} 1.3507.[1168a] ^{1}H-NMR.[526] "Phenkapton."

$(EtO)_2P(S)SCH_2S$–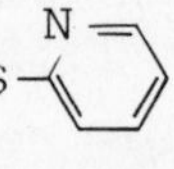 (*). Liq. $b_{0.01}$ 124°.[303]

$(EtO)_2P(S)SCH_2S$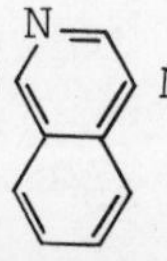 Me. (*) Solid. m. 105-106°.[303]

$(EtO)_2P(S)SCH_2S$-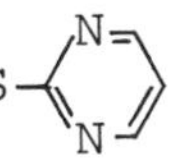 (*). Liq. $b_{0.01}$ 136-137°, n_D^{20} 1.5872,[629] n_D^{25} 1.5392.[1306]

$(EtO)_2P(S)SCH_2S$-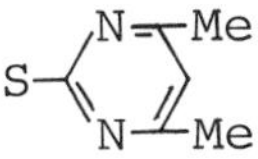 (*). Solid. m. 72-74°,[1306] 77° (from ligroin).[629]

(*) From $(EtO)_2P(S)SCH_2Cl$ plus NaSR.[303,629,1306]

$(EtO)_2P(S)SCH_2SCH_2Ph$. XVII. Liq. $b_{0.01}$ 138-140°.[639]

$(MeO)_2P(S)SCH_2SCH_2C_6H_4Cl$-2. XVII. Liq. $b_{0.01}$ 130°.[639]

$(EtO)_2P(S)SCH_2SCH_2C_6H_4Cl$-2. XVII. Liq. $b_{0.01}$ 127°.[639]

<u>$(RO)_2P(S)SCH_2S(O)_nR'$</u>

$(EtO)_2P(S)SCH_2S(O)Et$. Liq. $b_{0.1}$ 150°, n_D^{30} 1.5365.[765]

$(EtO)_2P(S)SCH_2S(O)_2Et$. Liq. $b_{0.1}$ 150°, n_D^{30} 1.5251.[765]

$(i\text{-}PrO)_2P(S)SCH_2S(O)Et$. (*) Liq. n_D^{25} 1.5232.[40]

$(i\text{-}PrO)_2P(S)SCH_2S(O)_2Et$. (*) Liq. n_D^{25} 1.5135.[40]

(*) From $(i\text{-}PrO)_2P(S)SCH_2SEt$ and 1 or 2 moles of perbenzoic acid.[40]

$(EtO)_2P(S)SCH_2S(O)C_6H_4Cl$-4. (*) Liq. n_D^{30} 1.5916.[277]

$(EtO)_2P(S)SCH_2S(O)_2C_6H_4Cl$-4. (*) Liq. n_D^{30} 1.5733.[277]

(*) From $(EtO)_2P(S)SCH_2SC_6H_4Cl$-4 plus $KMnO_4$.[277]

<u>$(RO)_2P(S)SCH_2SeR'$</u>

$(EtO)_2P(S)SCH_2SePh$. XVII. Liq. $b_{0.01}$ 124°.[1152]

$(EtO)_2P(S)SCH_2SeC_6H_{11}$-cyclo. XVII. Liq. $b_{0.01}$ 123°.[1152]

<u>$(RO_2P(S)SCH_2SiR'_3$</u>

$(EtO)_2P(S)SCH_2SiMe_2OEt$. XVII. Liq. b_1 126°, n_D^{20} 1.4863, d^{20} 1.0715.[41]

$(BuO)_2P(S)SCH_2SiMe_2OEt$. XVII. Liq. b_{2-3} 166-168°, n_D^{20} 1.4821, d^{20} 1.0267.[41]

$[(EtO)_2P(S)SCH_2SiMe_2)]_2O$. XVII. Liq. $b_{0.0001}$ 100°, n_D^{20} 1.4915, d^{20} 1.1134.[41]

$(EtO)_2P(S)SCH_2SiMe(OEt)_2$. XVII. Liq. b_6 159-160°, n_D^{20} 1.4815, d^{20} 1.0919.[41]

$(BuO)_2P(S)SCH_2SiMe(OEt)_2$. XVII. Liq. b_2 170-171°, n_D^{20} 1.4770, d^{20} 1.0514.[41]

$(RO)_2P(S)SCH_2SiMe_2OSiEt_3$.		State	n_D^{20}	d_4^{20}
Et	XVII. (*).	Liq. b_{3-4} 168-170°	1.4823	1.028[42]
Bu	XVII. (*).	Liq. $b_{0.01}$ 100-103°	1.4780	0.9971[4

(*) Structures confirmed by alternative syntheses: $(RO)_2P(S)SCH_2SiMe_2OEt$ plus Et_3SiOH.[42]

$(RO)_2P(S)SEt$

$(RO)_2P(S)SEt$	Methods	State	n_D^{20}	d_4^{20}
Me	XXVIII[759, 1159]	Liq. $b_{0.05}$35°[1159]	1.5080[759]	1.1795[75
	XXXIII[643]	b_2 79°[643]	^{1}H-NMR[1325]	
i-Pr	XXVIII[759]	Liq. $b_{1.7}$59-60°	1.4900	1.0720[75
Pr	XXVIII[759]	Liq. $b_{0.07}$72-74°,[759]	1.4945[759]	1.0638[75
	XXXIII[1237]	b_1 78°[1237]	1.4960[1237]	1.0679[12
$CH_2=CHCH_2$	XVII	Liq. $b_{0.0001}$36°	1.5238	1.1164[14
$HC\equiv CCH_2$	XVII	Liq. $b_{0.0001}$61°	1.5414	1.1850[14
$CH_2=CMeCH$	XVII	Liq. $b_{0.0001}$48°	1.5175	1.0793[14
Bu	XXVII[1364]	Liq. $b_{0.08}$86-88°[759]	1.4895[1237]	1.0368[12
	XXVIII[759,941]			
	XXXIII[1237]	$b_4$141-142°[941]	1.4923[759]	1.0400[75
$MeCH=CHCH_2$	XVII	Liq. $b_{0.0001}$68°	1.5240	1.0819[14

$(MeO)_2P(S)SCH_2CH_2Br$. XVII. Liq. $b_{0.01}$ 55°.[1144]

$(EtO)_2P(S)SCH_2CH_2Br$. XVII. Liq. $b_{0.01}$ 60°.[1144]

$(EtO)_2P(S)SCH_2CH_2Cl$. XVII.[1144] XXVIII.[1159] From $(EtO)_2P(S)SCH_2CH_2OH$ plus PCl_5.[493] Liq. $b_{0.01}$ 86°,[1159] $b_{0.1}$ 110°,[1144] $b_{1.5}$ 103-104°,[493] n_D^{20} 1.5230, d^{20} 1.2270.[493]

$(EtO)_2P(S)SCHClMe$. XXVI. Liq. $b_{0.2}$ 82°, n_D^{20} 1.5130, d^{20} 1.2048.[32]

$(EtO)_2P(S)SCH(SEt)CH_2Cl$. XVII. Liq. $b_{0.01}$ 104°.[1177]

$(EtO)_2P(S)SCH_2CHBrCl$. XVII. Liq. b_2 120-122°.[313,1113]

$(RO)_2P(S)SCHOHCCl_3$. XXI. Liq. R, n_D^{20}, d_4^{20}: Et, 1.5583, 1.4827; i-Pr, 1.5310, 1.3294; Pr, 1.5202, 1.3268; i-Bu, 1.4840, 1.2139; Bu, 1.5110, 1.2675.[572]

$(EtO)_2P(S)SCH(OMe)CCl_3$. XVII. Liq. $b_{0.01}$ 80°.[217]

$(MeO)_2P(S)SCH(OEt)CCl_3$. XVII. Liq. $b_{0.01}$ 79°.[217]

$(EtO)_2P(S)SCH(OEt)CCl_3$. XVII. Liq. $b_{0.01}$ 83-86°.[217]

$(MeO)_2P(S)SCH(SEt)CCl_3$. XVII. Liq. $b_{0.01}$ 111°.[1153]

$(EtO)_2P(S)SCH(SEt)CCl_3$. XVII. Liq. $b_{0.01}$ 108°.[1153]

<u>$(RO)_2P(S)SCH_2CH_2NR'_2$</u>

$(i\text{-}PrO)_2P(S)SCH_2CH_2NH_2$. XVII. Liq. $b_{0.045}$ 90°, n_D^{20} 1.5126, d^{20} 1.102.[134]

$(EtO)_2P(S)SCH_2CH_2NMe_2$. XVII. Liq. $b_{0.1}$ 96-102°.[362]

$(EtO)_2P(S)SCH_2CH_2NEt_2$. IX.[1133] XVII.[362,741] $b_{0.2}$ 108-109°,[362] b_1 136°.[1133]

$(EtO)_2P(S)SCH_2CH_2NEtPh$. XXIV.[147] Liq. $b_{0.0003-0.0004}$ 135-139°,[147] $b_{0.01}$ 135°.[1165]

$(EtO)_2P(S)SCH_2CH_2NEtC_6H_4Cl\text{-}2$. Liq. $b_{0.01}$ 144°.[1165]

$(RO)_2P(S)SCH_2CH_2NHCHOHCCl_3$. From $(RO)_2P(S)SH$ plus $(CH_2)_2NH$ and Cl_3CCHO. Liq. R. $b_{0.5}$, n_D^{20}: Me, 50-60°, 1.5653; Et, 50-60°, 1.5520; i-Pr, --, 1.5103; Pr, --, 1.5218.[514]

$(MeO)_2P(S)SCH_2CH_2NHCOOEt$. XVII. Liq. n_D^{30} 1.4695. Also several analogs.[1273]

$(EtO)_2P(S)SCH_2CH_2NHCOOMe$. XVII. Liq. $b_{0.01}$ 150°, n_D^{25} 1.5160.[177]

$(EtO)_2P(S)SCH_2CH_2NHCOOEt$. XVII.[177,1273] Liq. $b_{0.04}$ 150°, n_D^{26} 1.5092.[177]

$(EtO)_2P(S)SCH_2CH_2NHCOOR$. XVII. Liq. R, b_{mm}, n_D^t: CH_2Ph, $b_{0.01}$ 175°, 1.5460$^{26°}$; i-Bu, $b_{0.04}$ 175°, 1.4993$^{22°}$; CH_2CH_2OPr, $b_{0.01}$ 175°; CH_2COOBu, $b_{0.01}$ 175°.[177]

$(MeO)_2P(S)SCH_2CH_2N{=}C$. VIII. Liq. $b_{0.02}$ 75° (dec.).[1342]

$(RO)_2P(S)SCH_2C{\equiv}N$. XVII. Liq. R, n_D^{25}: Me, 1.5306; Et, 1.5153; Am, 1.4984; Ph, 1.5180.[152]

$(MeO)_2P(S)SCH(SMe)C{\equiv}N$. XVII. Liq. $b_{0.01}$ 94°.[1186]

$(EtO)_2P(S)SCH(SMe)C{\equiv}N$. XVII. Liq. $b_{0.01}$ 94°.[1186]

$(MeO)_2P(S)SCH(SC_6H_4Me\text{-}4)C{\equiv}N$. XVII. Liq. n_D^{25} 1.5820.[1108]

$(EtO)_2P(S)SCH(SC_6H_4Cl\text{-}4)C{\equiv}N$. XVII. Solid. m. 51°.[1108]

<u>$(RO)_2P(S)SCH_2CH_2$-(N-ring)</u>

$(EtO)_2P(S)SCH_2CH_2N$ 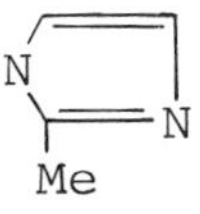 XVII. Picrate. Solid. m. 110-111°.[1077]

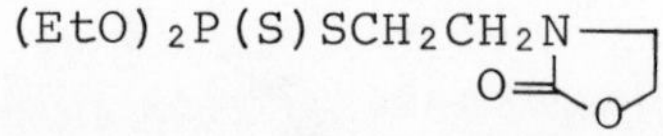

$(EtO)_2P(S)SCH_2CH_2N$⟶ (ring with O)   XVII. Liq. $b_{0.005}$ 188-192°.[187]

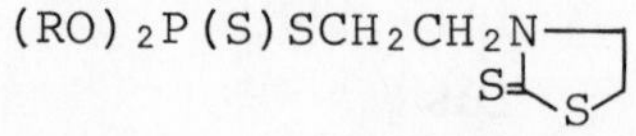

$(RO)_2P(S)SCH_2CH_2N$⟶ (ring with S)   XVII. Solids. R, m: Me, 53-54°; Et
63-64° (from acetic acid).[187]

$(EtO)_2P(S)SCH_2CH_2N$ (ring) XVII. Liq. $b_{0.3}$ 133-135°.[362]

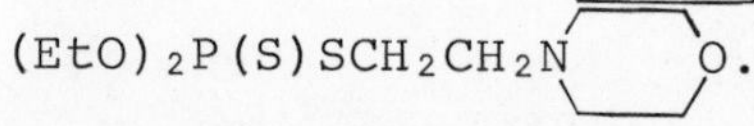

$(EtO)_2P(S)SCH_2CH_2N$ (ring with O). XVII.[362,602,662] Liq. $b_{0.08}$ 148-
152°,[602,662] b_2 140-142°.[362]

$\underline{(RO)_2P(S)SCH_2CH_2NHSO_2R'}$

$(EtO)_2P(S)SCH_2CH_2NHSO_2Me$. XVII. Liq. n_D^{30} 1.5248.[1263]

$(EtO)_2P(S)SCH_2CH_2NHSO_2Ar$. XVII. Liq. Ar, n_D^{30}; Ph, 1.5468;
4-MeC_6H_4, 1.5496; 2,5-$Me_2C_6H_3$, 1.5563; 4-ClC_6H_4, 1.5661;
3,4-$Cl_2C_6H_4$, 1.5729; 2,5-$Cl_2C_6H_3$, 1.5685; 3-$NO_2C_6H_4$,
1.5708.[203,271]

$(i-PrO)_2P(S)SCH_2CH_2NHSO_2Ar$. XVII. Liq. Ar, n_D^{30}: Ph, 1.5428,
[203,271] ^{1}H-NMR,[74] ^{31}P 21.3 (P_4O_6 external),[1047] "Be-
tasan;" 4-MeC_6H_4, 1.5411; 2,5-$Me_2C_6H_3$, 1.5423; 3,4-Cl_2
C_6H_3, 1.5590; 2,5-$Cl_2C_6H_3$, 1.5595; 3-$NO_2C_6H_4$, 1.5501.
[203,271]

$(PrO)_2P(S)SCH_2CH_2NHSO_2Ar$. XVII. Liq. Ar, n_D^{30}: Ph, 1.5445;
2,5-$Cl_2C_6H_3$, 1.5583.[203,271]

$(BuO)_2P(S)SCH_2CH_2NHSO_2Ph$. XVII. Liq. n_D^{30} 1.5377.[203,271]

$\underline{(RO)_2P(S)SCH_2CH_2OR'}$

$(RO)_2P(S)SCH_2CH_2OH$. XI. Liq.

	b_2	n_D^{20}	d^{20}
Me	--	1.5380	1.2911[493]
Et	119-120°	1.5250	1.2042
i-Pr	118.5-120°	1.5083	1.1323
Pr	124-126°	1.5140	1.1440
i-Bu	135-138°	1.5045	1.0965

$(RO)_2P(S)SCH_2CH_2OMe$. XXI. Liq.

	b_{mm}	n_D^{25}	d^{25}
Pr	b_8 114-117°	1.5139	1.1297[1409]
Bu	$b_{0.00006}$104°	1.5090	1.0876
$ClCH_2CH_2$	$b_{0.00008}$117°	1.5416	1.3013

$(EtO)_2P(S)SCH_2CH_2OEt$. XVII. Liq. $b_{0.08}$ 75-75.5°, n_D^{20} 1.4978

d^{20} 1.1044.[741]

$(RO)_2P(S)SCH_2CH_2OCH_2CH_2OCOMe$. From $(RO)_2P(S)SCH_2CH_2OH$ plus $(MeCO)_2O$ in Pyridine.[493] Liq.

	b_{mm}	n_D^{20}	d^{20}
Et	$b_{1.5}$135.5-136.5°	1.5010	1.1845[493]
i-Bu	$b_{0.5}$140-141°	1.4890	1.0948

$(MeO_2P(S)SCH_2CH_2OCOMe$. XXII. Liq. b_2 100-103°, n_D^{20} 1.5252, d^{20} 1.1558.[756]

$(EtO)_2P(S)SCH_2CH_2OCOMe$. XXII. Liq. $b_{1.5}$ 115-117°, n_D^{20} 1.4948, d^{20} 1.1517.[756]

Also several homologs.[756]

$(MeO)_2P(S)SCH_2CH_2OPh$. XVII. Liq. $b_{0.28}$ 140-142°, n_D^{20} 1.5698, d^{20} 1.2408.[761]

$(EtO)_2P(S)SCH_2CH_2OSO_2Et$. XVII. Liq. $b_{0.01}$ 108°.[289]

$(EtO)_2P(S)SCH_2CH_2OSO_2Cl$. XVII. Liq. b_7 84-85°, n_D^{20} 1.4995, d^{20} 1.3652.[686]

$(RO)_2P(S)SCH_2CH_2SR'$

$(RO)_2P(S)SCH_2CH_2SH$. XI. Liq.

	b./mm	n_D^{20}	d_4^{20}
Me	105-106°/1	1.5738	1.2915[727]
Et	110-112°/1.5	1.5485	1.2041
i-Pr	110-112°/1.5	1.5365	1.1506
Pr	113-115°/1	1.5370	1.1522
i-Bu	129-131°/1.5	1.5265	1.1088

$(RO)_2P(S)SCH_2CH_2SMe$. XI. Liq.

	b./mm	n_D^{20}	d_4^{20}
Et	111-113°/1	1.5388	1.1708[727]
i-Pr	115°/1.5	1.5179	1.1087
Pr	128-129°/1.5	1.5270	1.1229

$(EtO)_2P(S)SCH(OR)CH_2SCCl_3$. XXII. Liq.

	b./mm	n_D^{20}
Et	40-42°/0.02	1.5370[501]
Bu	77-78°/0.04	1.5288
i-Am	53-54°/0.02	1.5250

$(EtO)_2P(S)SCH(OAr)CH_2SCCl_3$. Liq.

		n_D^{20}	d^{20}
Ph	XVII[502,547] XXII[501,547]	1.5762	1.3567[502,547]
2-MeC_6H_4	XVII[502]	1.5794	1.3139[502]

3-MeC_6H_4	XVII	1.5774	1.3242
4-MeC_6H_4	XVII	1.5713	1.3263
4-ClC_6H_4	XVII	1.5798	1.4052
2,4,5-$Cl_3C_6H_2$	XVII	1.5868	1.4803

$(MeO)_2P(S)SCH_2CH_2SEt$. VIII.[666] XIII.[646] XVII.[279,666] XXVIII.[297,647] XXXIII.[595] (*)[295,1140] Liq. $b_{0.01}$ 77°,[297,647] $b_{0.01}$ 78°,[295,1140] b_2 95°,[279] n_D^{20} 1.5501,[297,647] 1.5516.[595]

$(EtO)_2P(S)SCH_2CH_2SEt$. VIII.[666] XVII.[279,633,666] XXVII.[642] XXVIII.[595] XXXIII.[595] (*)[295,1140] Liq. $b_{0.01}$ 86°,[642] 87°,[297,647] $b_{0.1}$ 105-106°,[595] n_D^{20} 1.5328,[666] 1.5330,[297,595,647] IR and Raman.[968] ^{1}H-NMR.[74,524,526] ^{31}P 18.2 (P_4O_6 external).[1047] Mass spect.[196] "Disulfoton," "Di-Syston," (*) From $(EtO)_2P(S)SCH_2CH_2Br$ plus EtSH and EtONa.[295,1140]

$(MeO)_2P(S)SCH_2CH_2SCH=CH_2$. XXII. Liq. $b_{0.01}$ 90°.[1158]

$(EtO)_2P(S)SCH_2CH_2SCH=CH_2$. XXII. Liq. $b_{0.02}$ 90°.[1158]

$(EtO)_2P(S)SCH_2CH_2SCH_2CH=CH_2$. XVII. Liq. b_1 150°, n_D^{20} 1.5395[666]

$(EtO)_2P(S)SCH_2CH_2SBu$. From $(EtO)_2P(S)SCH_2CH_2Br$ plus EtSH and EtONa. Liq. $b_{0.01}$ 107°.[1140]

$(MeO)_2P(S)SCH_2CH_2SCHMeCONHMe$. XVII. Solid. m. 97-99°.[1019]

$(EtO)_2P(S)SCH_2CH_2SCH_2SEt$. XVII. Liq. $b_{0.01}$195°.[1146]

$(EtO)_2P(S)SCH_2CH_2S$- 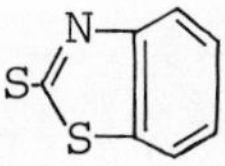 From $(EtO)_2P(S)SCH_2CH_2Br$ plus RSH and EtONa. Liq. $b_{0.01}$ 170°.[1140]

$(EtO)_2P(S)SCH_2CH_2SPh$. From $(EtO)_2P(S)SCH_2CH_2Br$ plus PhSH and EtONa. Liq. $b_{0.02}$ 128°.[295,1140]

$(EtO)_2P(S)SCH(OPh)CH_2SPh$. XXII. Liq. n_D^{20} 1.5925.[501]

$(EtO)_2P(S)SCH_2CH_2SC_6H_4SMe$-4. XVII. Liq. n_D^{20} 1.6115, d^{20} 1.235.[1083]

$(i\text{-}PrO)_2P(S)SCH_2CH_2SC_6H_4SMe$-4. XVII. Liq. n_D^{20} 1.5880, d^{20} 1.170.[1083]

$(MeO)_2P(S)SCH_2CH_2SCH_2Ph$. XVII. Liq. $b_{0.4}$ 180°.[1143]

$(EtO)_2P(S)SCH_2CH_2SCH_2Ph$. XVII. Liq. $b_{0.05}$ 171°.[1143]

$(EtO)_2P(S)SCH_2CH_2SC(=NH)NH_2$. Solid. m. 85°.[741]

$(EtO)_2P(S)SCH_2CH_2SCOMe$. (*) Liq. $b_{2.5}$ 135-137°, n_D^{20} 1.5351, d_4^{20} 1.1994.[727]

$(i-BuO)_2P(S)SCH_2CH_2SCOMe$. (*) Liq. $b_{2.5}$ 143-144°, n_D^{20} 1.5170, d_4^{20} 1.1074.[727]

(*) From $(RO)_2P(S)SCH_2CH_2SH$ plus $(MeCO)_2O$ and pyridine.[727]

$[(RO_2P(S)S]_2(CH_2)_2$. XVII. Liq.	b./mm	n_D^{20}	d^{20}
Et	187°/0.25	1.5427	1.2188[1243]
i-Pr	--	1.5268	1.1790
Pr	185-186°/0.2	1.5265	1.1638
i-Bu	--	1.5143	1.1006
Bu	--	1.5156	1.1045

$(EtO)_2P(S)SC(SEt)=CHSEt$. XXII. Liq. n_D^{20} 1.5643.[209]

$(RO)_2P(S)SCH_2CH_2S(O)_nR'$

$(MeO)_2P(S)SCH_2CH_2S(O)Et$. XXII. Liq. n_D^{20} 1.5570, d^{20} 1.2891.[1223]

$(EtO)_2P(S)SCH_2CH_2S(O)Et$. XVII.[289,627] XXII.[1223] Liq. $b_{0.01}$ 108°,[289,627] n_D^{20} 1.5390.[1223] Also several analogs.[1223]

$(EtO)_2P(S)SCH_2CH_2S(O)Bu$. XXII. Liq. n_D^{20} 1.5305, d^{20} 1.1655.[1223]

$(EtO)_2P(S)SCH_2CH_2S(O)_2Et$. XVII.[289] XXII.[1239] Liq. $b_{0.01}$ 109°,[289] $b_{0.04}$ 133-137°.[1239]

$(MeO)_2P(S)SCH_2CH_2S(O)_2CH_2CH_2SP(S)(OMe)_2$. XXII. Liq. n_D^{20} 1.5600.[1024]

$(EtO)_2P(S)SCH_2CH_2SO_2CH_2CONH_2$. XVII. Solid. m. 72°.[1021]

$(EtO)_2P(S)SCH_2CH_2SO_2CH_2CONHMe$. XVII. Solid. m. 55°.[1021]

$(EtO)_2P(S)SCH_2CH_2SO_2C_6H_4Cl-4$. XVII. Solid. m. 20-22°.[1072]

$(EtO)_2P(S)SCH_2CH_2SO_2C_6H_4Me-4$. XVII. Solid. m. 45-50°.[1072]

$(EtO)_2P(S)SCH_2CH_2SO_2NMe_2$. XVII. Liq. b_1 60°.[1114]

$(i-PrO)_2P(S)SCH_2CH_2SO_2NMe_2$. XVII. Solid. m. 37-39°.[1114]

$(MeO)_2P(S)SCH_2CH_2SO_2OPh$. XXII. Liq. n_D^{23} 1.5432.[1112]

$(EtO)_2P(S)SCH_2CH_2SO_2OPh$. XXII. Liq. n_D^{24} 1.5464.[1112]

$(RO)_2P(S)SCHMe-(N-ring)$

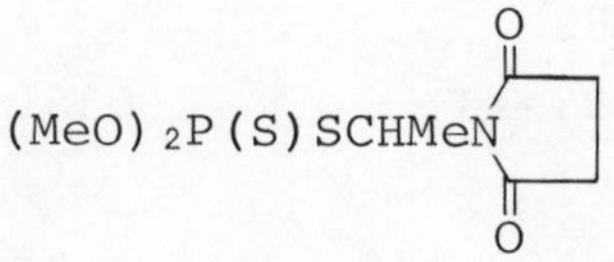

XXII. Solid. m. 95-97° (from benzene-petroleum ether).[420]

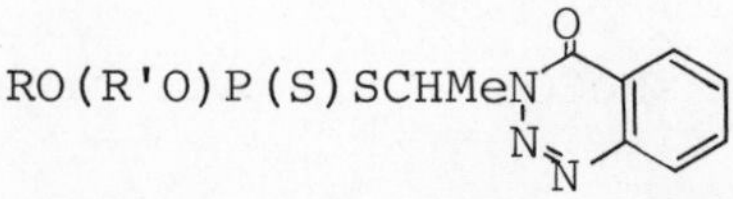

XVII. Solids. R, R', m.: Me, Me, 94°; Et, Et, 56°; Me, Pr, 70-71°.[305]

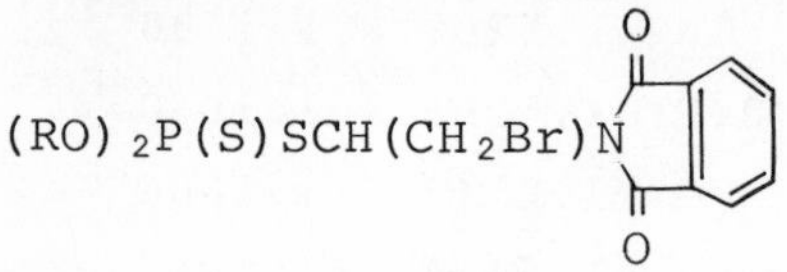

XVII. Solids. R, m.: Me, 102-103° (from toluene-petroleum ether); Et, 73-74° (from toluene-petroleum ether).[430]

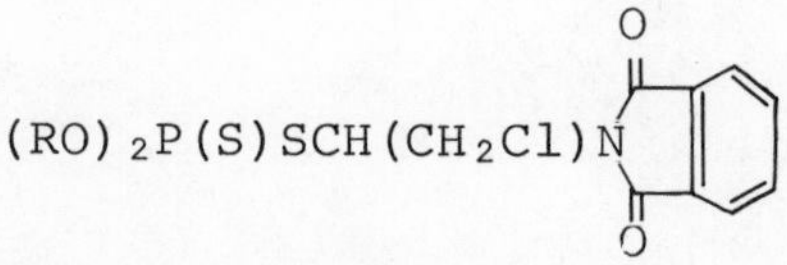

XVII. Solids. R, m.: Me, 76-77° (from toluene-hexane); Et, 62-64° (from toluene-hexane).[430,482] "Torak."

$(RO)_2P(S)SCHMeOR'$

$(EtO)_2P(S)SCHMeOEt$. XXII. Liq. $b_{0.025}$ 50-52°, n_D^{20} 1.4960, d_4^{20} 1.1284.[978]

$(RO)_2P(S)SCHMeOPh$. XXII.[503,506,978] Liq.

	b./mm	n_D^{20}	d_4^{20}
Me	78-79°/0.1	1.5589	1.2151[503]
Et	86-87°/0.1	1.5410	1.1550[506]
	96-97°/0.027	1.5450	1.1566[978]
i-Pr	125-129°/0.3	1.5302	1.1101[503]

Also several analogs and homologs.[503,504,506]

$(RO)_2P(S)SCH(CH_2Cl)OPh$. XVII.[505] XXII.[501] Liq.

	b./mm	n_D^{20}	d_4^{20}
Me	117°/0.015	1.5695	1.2984[505]
Et	127-128°/0.017	1.5543[501]	

Also several analogs and homologs.[501,505]

(EtO)$_2$P(S)SCH(CHCl$_2$)OPh. XXII. Liq. $b_{0.03}$ 130-133°, n_D^{20} 1.5580.[501]

(RO)$_2$P(S)SCHMeOCOMe. XXII. Products of cationic addition (Markownikoff).[880]

		b./mm	n_D^{20}	
Et	Liq.	119-120°/2	1.4938,	^{1}H-NMR, mass spect.
Bu		145-147°/2	1.4850	
i-Am		158-159°/3	1.4811	

(RO)$_2$P(S)CHMeSR'

(RO)$_2$P(S)SCHMeSEt. XVII.[728] XXI.[57] XXII.[728] Liq.

	b./mm	n_D^{20}	d_4^{20}
Me	82-83°/4	1.5320	1.1788[57]
Et	109-110°/2.5	1.5290	1.1392[728]
	118-119°/3	1.5240	1.1347[57]
Pr	122-128°/4	1.5130	1.0929[57]
i-Bu	124-126°/3	1.5012	1.0422[728]
	--	1.5060	1.0522[57]
Bu	--	1.5085	1.060[57]

(RO)$_2$P(S)SCHMeSCH$_2$CH$_2$OBu. XXII. Liq.

	b./mm	n_D^{20}	d^{20}
Et	123-125°/3	1.5125	1.0940[728]
i-Bu	124-126°/3	1.5012	1.0422[728]

(RO)$_2$P(S)SCHMeSBu. XXII. Liq.

Et	109-110°/2	1.5198	1.0965[728]
i-Bu	121-122°/2	1.5052	1.0384[728]

(EtO)$_2$P(S)SCHMeSPh. XXII. Liq. $b_{0.045}$ 110°, n_D^{20} 1.5751, d_4^{20} 1.1851.[978]

((RO)$_2$P(S)S)$_2$CHMe. XXI. Liq.

		n_D^{20}	d_4^{20}
Et	Solid. m. 62-62.5°.[75]		
i-Pr		1.5254	1.1358[75]
Pr		1.5259	1.1395[75]

(EtO)$_2$P(S)SCHMeS(O)Et. XXII. Liq. n_D^{20} 1.5342, d_4^{20} 1.1916.[978]

(EtO)$_2$P(S)SCHMeS(O)Ph. XXII. Liq. n_D^{20} 1.5760, d_4^{20} 1.2996.[978]

(RO)$_2$P(S)SCH$_2$COR'

(MeO)$_2$P(S)SCH$_2$CHO. XVII. Liq. $b_{0.02}$ 45-55°.[2]

$(RO)_2P(S)SCH_2COPh$. XVII. Liq.

	n_D^{20}	d_4^{20}
Et	1.5689	1.2329[1336]
i-Pr	1.5486	1.1733
Pr	1.5592	1.1772
Bu	1.5430	1.1426

$(RO)_2P(S)SCH_2COC_6H_4Br$-4. XVII. Liq.

	n_D^{20}	d_4^{20}
Et	1.5919	1.4429[1336]
i-Pr	1.5735	1.2675
Pr	1.5768	1.3454
Bu	1.5664	1.2131

$(RO)_2P(S)SCH_2COC_6H_4Cl$-4. XVII. Liq.

	n_D^{20}	d_4^{20}
Et	1.5720	1.3039[1336]
i-Pr	1.5590	1.2186
Pr	1.5670	1.2266
Bu	1.5519	1.1911

$[(EtO)_2P(S)S]_2CHCONH_2$. XVII. Liq. n_D^{25} 1.5017.[1392]

$(MeO)_2P(S)SCH_2COCl$. XXXIII. From $(MeO)_2P(S)SCH_2COOH$ plus $(COCl)_2$. Liq. n_D^{23} 1.537.[1090]

$(EtO)_2P(S)SCH_2COCl$. XXV. From $(EtO)_2P(S)SCl$ plus CH_2=CO. Liq. $b_{0.5}$ 108°, n_D^{25} 1.5185, d^{25} 1.2828.[32]

$(EtO)_2P(S)SCH_2COF$. XVII. Liq. $b_{0.1}$ 84-90°.[149]

<u>$(RO)_2P(S)SCH_2CONHR'$</u>

$(MeO)_2P(S)SCH_2CONH_2$. XVII.[447] From $(MeO)_2P(S)SCH_2COOCOMe$ plus NH_3.[149] Solid. m. 58-59° (from CCl_4),[149] 62-63°.[447]

$(EtO)_2P(S)SCH_2CONH_2$. XVII.[151,447] From $(EtO)_2P(S)SCH_2COCl$ plus NH_3.[32,149] From $(EtO)_2P(S)SCH_2COOCOR$ plus NH_3.[581] Solid. m. 58-59° (from CCl_4),[581] 59.5-60.5 (from water; $i-Pr_2O$).[149]

$(MeO)_2P(S)SCH_2CONHMe$. XVII.[88,819,932,1422,1424] XXVII.[646] XXXIII.[643] From $(MeO)_2P(S)SCH_2COOR$ plus $MeNH_2$: R= : H,[655] Me,[930] $(CH_2)_2$,[114] COMe,[149] COPh.[1214] From $(MeO)_2P(S)SCH_2COOH$ plus MeNCO.[904,1281,1419] From $(MeO)_2P(S)SCH_2COOH$ plus 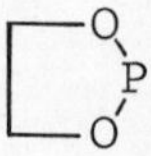 PCl, Et_3N and $MeNH_2$.[1423] Solid. m.

49-50.5°,[1281,1419] 50-51° (from toluene-hexane),[88] 51-52° (from ether),[1214] Cryoscopic m.p. T_o = 50.25-50.56° (100% pure).[93] IR and UV.[375] ^{1}H-NMR.[74,526] ^{31}P 15.3 (P_4O_6 external).[1047] Mass spect.[196] "Dimethoate."

$(MeO)_2{}^{32}P(S)SCH_2CONHMe$. XVII. Solid. m. 50-51°.[382]

$(EtO)_2P(S)SCH_2CONHMe$. XVII. Liq. n_D^{25} 1.5292.[447]

$(FCH_2CH_2O)_2P(S)SCH_2CONHMe$. XVII. Liq. n_D^{20} 1.5229, d^{20} 1.386.[828,931]

$(MeO)_2P(S)SCH(SMe)CONHMe$. XVII. Solid. m. 84° (from benzene-ether).[631]

$(EtO)_2P(S)SCHSMe)CONHMe$. XVII. Solid. m. 85° (from ether).[631]

$(MeO)_2P(S)SCH(SEt)CONHMe$. XVII. Solid. m. 112°.[631]

$(EtO)_2P(S)SCHSEt)CONHMe$. XVII. Solid. m. 95°(from benzene-petroleum ether).[631]

$(EtO)_2P(S)SCH_2CONHCH_2NHCOOEt$. XVII. Liq. n_D^{30} 1.5218.[1264]

$(EtO)_2P(S)SCH_2CONHCH_2NHCOOPh$. XVII. Liq. n_D^{30} 1.5485.[1264]

$(EtO)_2P(S)SCH_2CONHCH_2NEt_2$. From $(EtO)_2P(S)SCH_2CONHCH_2Cl$ plus Et_2NH. Liq. n_D^{25} 1.5282.[1291]

$(MeO)_2P(S)SCH_2CONHCH_2NMeCOOCH_2CH_2SC_6H_4Cl$-4. XVII. Liq. n_D^{20} 1.5629.[1264]

$(MeO)_2P(S)SCH_2CONHCH_2OMe$. XVII.[176,824] Also from $(MeO)_2P(S)SCH_2CONHCH_2OH$ plus MeOH and HCl.[824] Solid. m. 40-42°,[824] 43-44°.[176]

$(EtO)_2P(S)SCH_2CONHCH_2OMe$. XVII. Liq. n_D^{20} 1.5269.[952]

$(EtO)_2P(S)SCH_2CONHCH_2OR$. XVII. Liq. R, n_D^{20}; Et, 1.5392; i-Pr, 1.5208; $CH_2CH{=}CH_2$, 1.5258; Bu, 1.5200; cyclo-C_6H_{11}, solid. m. 50-51.5°.[952]

$(MeO)_2P(S)SCH_2CONHCH_2SMe$. (*). Solid. m. 39-40° (from ether-petroleum ether).[821,1048]

$(EtO)_2P(S)SCH_2CONHCH_2SMe$. (*). Solid. m. 49-49.5°.[821,1048]

(*) From $(RO)_2P(S)SCH_2CONHCH_2OH$ plus MeSH and HCl.[821,1048]

$(FCH_2CH_2O)_2P(S)SCH_2CONHCH_2SMe$. XVII. Solid. m. 43.5-45°.[828,831,931]

$(MeO)_2P(S)SCH_2CONHCH_2SR$. (*) Solids. R, m. : Et, 41-42° (from ether-petroleum ether); i-Pr, 59.5-60.5°; t-Bu, 66° (from alcohol).[1048]

$(EtO)_2P(S)SCH_2CONHCH_2SR$. (*) Solids. R, m.: Et, 28.5-29.5°; i-Pr, 42-43°; t-Bu, 32°; Ph, 60-61°.[1048]

(*) From $(RO)_2P(S)SCH_2CONHCH_2OH$ plus mercaptan and HCl.[1048]

$(EtO)_2P(S)SCH_2CONHCH_2SC_6H_4Cl$-4. From $(EtO)_2P(S)SCH_2CONHCH_2C$ plus HSC_6H_4Cl-4. Liq. n_D^{24} 1.5920.[1291]

$[(MeO)_2P(S)S]_2CHCONHEt$. XVII. Solid. m. 59-61°.[1392]

$[(EtO)_2P(S)S)]_2CHCONHEt$. XVII. Liq. n_D^{25} 1.5020.[1392]

$(MeO)_2P(S)SCH_2CONHEt$. XVII.[88,1422,1424] From $(MeO)_2P(S)SCH_2COOH$: with $EtNH_2$,[655] with EtNCO.[904] Solid. m. 65-66°,[655,904] 67-68°.[88,1422,1424]

$(MeO)_2P(S)SCH_2CONHCH_2CH_2Cl$. XVII.[1338] From $(MeO)_2P(S)SCH_2COCl$ plus $(CH_2)_2NH$.[1337] Solid. m. 59-60° (from acetone-ligroin),[1338] 63.5-64°.[1337]

$(EtO)_2P(S)SCH_2CONHCH_2CH_2Cl$. XVII.[1338] From $(EtO)_2P(S)SCH_2COCl$ plus $(CH_2)_2NH$.[1337] Liq. n_D^{30} 1.5337,[1337] n_D^{31} 1.5332.[1338]

$(i\text{-}PrO)_2P(S)SCH_2CONHCH_2CH_2Cl$. XVII. Liq. n_D^{31} 1.5171.[1338]

$(MeO)_2P(S)SCH_2CONHCH(NMe)CCl_3$. XVII. Solid. m. 99-101° (from benzene).[215]

$(EtO)_2P(S)SCH_2CONHCH(NMe)CCl_3$. XVII. Solid. m. 83-85° (from benzene-ether).[215]

$(EtO)_2P(S)SCH_2CONHCH(NHC_6H_4Cl\text{-}3)CCl_3$. XVII.[215] From $(EtO)_2P(S)CH_2CONH_2$ plus CCl_3CHO and $3\text{-}ClC_6H_4NCO$.[220] Solid. m. 64-66°,[220] 69-72° (from benzene).[215]

$(EtO)_2P(S)SCH_2CONHCHOHCCl_3$. From $(EtO)_2P(S)SCH_2CONH_2$ plus CCl_3CHO. Solid. m. 51-53°.[220]

$((EtO)_2P(S)SCH_2CONHCH_2)_2$. XVII. Solid. m. 103-105° (from MeOH).[350]

$(MeO)_2P(S)SCH_2CONHCH_2CH_2OMe$. XVII. Liq.[176]

$(MeO)_2P(S)SCH_2CONHR$. XVII.[88,176,932,1422,1424] From $(MeO)_2P(S)SCH_2COOH$ plus RNCO.[904] Solids. R, m.: CH_2CH_2Cl,

66-67°;[904] i-Pr, 72-73°,[932] 72-74°,[904] 76-77° (from MeOH-ether);[88,1422,1424] Pr, 63.5-64° (from MeOH):[88,1422,1424] $CH_2CH_2CH_2OMe$, 43-44°;[176] $CH_2CH{=}CH_2$, 57.5-58°;[1422,1424] t-Bu, 58-60.5°;[1422,1424] i-Bu, 68.5-69°;[1422,1424] Bu, 31-31.5° (from ether);[88] 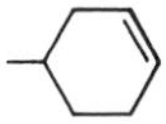, 55-57°;[904] cyclo-C_6H_{11}, 98-99°.[904]

$(EtO)_2P(S)SCH_2CONHAr$. XVII.[151,447] Solids. Ar, m.: Ph, 85-87° (from heptane);[151,447] $C_6H_4NO_2$-4, 109-110°;[151,447] $C_6H_4SO_2NH_2$-4, 145-146°.[447] Also several homologs and analogs.[151,447]

$(EtO)_2P(S)SCH_2CONHC_6H_4Cl$-4. From $(EtO)_2P(S)SCH_2COOCOR$ plus 4-$ClC_6H_4NH_2$. Solid. m. 81-83° (from EtOH-water).[581]

$[(EtO)_2P(S)SCH_2CO]_2NH$. XVII.[201,1116] Solid. m. 41-43° (from ether-petroleum ether or alcohol-water),[201] 54-56° (from benzene-petroleum ether).[1116]

$[(i\text{-}PrO)_2P(S)SCH_2CO]_2NH$. XVII. Solid. m. 46-48°.[1116]

$(EtO)_2P(S)SCH_2CONHCOAr$. XVII. Solids. Ar, m. : Ph, 79° (from Ac_2O); 4-ClC_6H_4, 127-128°.[283]

$(MeO)_2P(S)SCH_2CONHCONH_2$. XVII. Solid. m. 117-118°(from CCl_4).[444,447]

$(EtO)_2P(S)SCH_2CONHCONH_2$. XVII. Solid. m. 93-95° (from CCl_4).[444] Also several analogs.[444]

$(RO)_2P(S)SCH_2CONHCONH$-i-Pr. XVII. Solids. R, m. (from CCl_4): Me, 72-74°; Et, 101-103°; i-Pr, 82-85°.[201]

$(MeO)_2P(S)SCH_2CONHCONHPh$. XVII. Solid. m. 92-94° (from MeOH).[202]

$(MeO)_2P(S)SCH_2CONHCONHC_6H_4Br$-4. XVII. Solid. m. 129-130° (from CCl_4).[201]

$(MeO)_2P(S)SCH_2CONHCONHC_6H_4Cl$-4. XVII. Solid. m. 108-110° (from MeOH-benzene).[202] Also 2 homologs or analogs.[201,212]

$(MeO)_2P(S)SCH_2CONHCONHAr$. XVII. Solids. Ar, m.: C_6H_4OMe-4, 118-120°(from CCl_4); C_6H_4Me-2, 125-127° (from CCl_4); C_6H_4Me-4, 125-127° (from CCl_4); 1-$C_{10}H_7$, 167-168° (from acetone); 2-$C_{10}H_7$, 136-137.5° (from acetone).[201] Also $(EtO)_2$ and $(i\text{-}PrO)_2$ homologs.[201]

$(MeO)_2P(S)SCH_2CONHCH_2C{\equiv}N$. XVII. Solid. m. 38-40°.[656]

$(EtO)_2P(S)SCH_2CONHCH_2C{\equiv}N$. XVII. Solid. m. 71-72°.[656]

$(MeO)_2P(S)SCH_2CONHCOOMe$. XVII. Solid. m. 50-55° (from ether).[201]

$(EtO)_2P(S)SCH_2CONHCOOMe$. XVII. Solid. m. 66-69° (from alcohol).[201]

$(i\text{-}PrO)_2P(S)SCH_2CONHCOOMe$. XVII. Solid. m. 80-83° (washed with CCl_4).[201]

$(RO)_2P(S)SCH_2CONHCOOEt$. XVII. Solids. R, m.: Me, 45°;[1139] Et, 70-73° (from MeOH),[201] 73°;[1139] i-Pr, 40°,[1139] 46-48° (from alcohol-water).[201]

$(i\text{-}PrO)_2P(S)SCH_2CONHCOO\text{-}i\text{-}Pr$. XVII. Solid. m. 60-65° (from petroleum ether).[201]

$(MeO)_2P(S)SCH_2CONHCOOAr$. XVII. Solids. Ar, m.: Ph, 65-68° (from ether);[201] 1-$C_{10}H_7$, 107-109° (from MeOH).[202]

$(i\text{-}PrO)_2P(S)SCH_2CONHCOOPh$. XVII. Solid. m. 41-45° (washed with CCl_4).[201]

$(EtO)_2P(S)SCH_2CONHCOOC_6H_4Cl\text{-}4$. XVII. Solid. m. 53-57° (washed with CCl_4).[201]

$(i\text{-}PrO)_2P(S)SCH_2CONHCOOC_6H_4Cl\text{-}4$. XVII. Solid. m. 68-72° (from ether).[201]

<u>$(RO)_2P(S)SCH_2CONH\text{-}i\text{-}Pr$</u>

$(MeO)_2P(S)SCH_2CONH\text{-}i\text{-}Pr$. XVII.[634,819] XXXIII.[643] From $(MeO)_2P(S)SCH_2COOMe$ plus $i\text{-}PrNH_2$.[1213] Solid. m. 72-73°,[819] 77-78° (from ether).[634,643] Liq. n_D^{20} 1.4861.[1213]

$(EtO)_2P(S)SCH_2CONH\text{-}i\text{-}Pr$. XVII.[819,932] From $(EtO)_2P(S)SCH_2COOMe$ plus $i\text{-}PrNH_2$.[1213] Liq. $b_{0.1-0.2}$ 152-153°,[819,932] n_D^{20} 1.5079,[1213] 1.5182,[819] d^{20} 1.1561,[819] Solid. m. 23-24°.[932]

$(MeO)_2P(S)SCH_2CONHCH_2CH_2CH_2OMe$. XVII. Solid. m. 43-44°.[82,176]

$(EtO)_2P(S)SCH_2CONHCH_2CH_2CH_2OMe$. XVII.[176] From $(EtO)_2P(S)SCH_2COOMe$ plus $H_2NCH_2CH_2CH_2OMe$.[82] Solid. m. 40°.[82,176]

$(MeO)_2P(S)SCH_2CONHCH_2CH{=}CH_2$. XVII. Solid. m. 55-56.6°.[1094]

$(RO)_2P(S)SCH_2CON(Me)CH_2C{\equiv}CH$. XVII. Liq. R, n_D^{25}: Me, 1.548; Et, 1.532; i-Pr, 1.519.[207] Also 18 analogs.[207]

$(MeO)_2P(S)SCH_2CONHCMe_2C{\equiv}N$. XVII. Solid. m. 57-59°.[656]

$(EtO)_2P(S)SCH_2CONHCMe_2C{\equiv}N$. XVII.[407,656,827] Solid. m. 82-83°,[407] 83-83.5° (from water).[656,827]

$(MeO)_2{}^{32}P(S)SCH_2CONHBu$. Solid. m. 50-51°.[382]

$(RO)_2P(S)SCH_2CONH$— 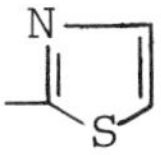

XVII. Solids. R, m.: Me, 113-115°; Et, 121-124°.[270] 23 analogs similarly prepared.[270]

$(EtO)_2P(S)SCH_2CONH$— 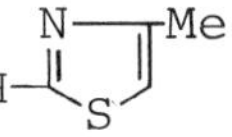

VIII. Solid. m. 114°.[1333]

$(EtO)_2P(S)SCH_2CONH$— 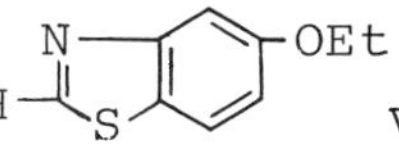

VIII. Solid. m. 113°.[1333]

<u>$(RO)_2P(S)SCH_2CONHAr$</u>

$(MeO)_2P(S)SCH_2CONHC_6H_4Br$-4. XVII. Solid. m. 65-67° (from CCl_4-hexane).[1401]

$(RO)_2P(S)SCH_2CONHC_6H_4Cl$-4. XVII.[554,1401] Solids. R, m.: Me, 63-64° (from hexane);[1401] Et, 79-81° (from hexane).[554]

$(MeO)_2P(S)SCH_2CONHC_6H_4OMe$-4. XVII. Solid. m. 68-69° (from MeOH).[1401]

$(MeO)_2P(S)SCH_2CONHC_6H_4NO_2$-2. XVII. Solid. m. 51.5-52.5° (from CCl_4).[1401]

$(EtO)_2P(S)SCH_2CONHC_6H_4NO_2$-3. XVII. Solid. m. 102°.[705]

$(RO)_2P(S)SCH_2CONHC_6H_4NO_2$-4. XVII.[554,705,1401] Solids. R, m.: Me, 102-103° (from CCl_4-hexane);[1401] Et, 100-102° (from dilute EtOH),[554] 112°.[705]

$(RO)_2P(S)SCH_2CONHC_6H_4CF_3$-2. XVII. Solids. R, m.: Me, 68-69° (from hexane); Et. 54-55° (from hexane).[554] Also several analogs.[554]

$(EtO)_2P(S)SCH_2CONHC_6H_3Me_2$-2,6. XVII. Solid. m. 58-60°.[197]

$(MeO)_2P(S)SCH_2CONHC_6H_3Et_2$-2,6. XVII. Solid. m. 61-63°.[197]

$(EtO)_2P(S)SCH_2CONHC_6H_{11}$-cyclo. From $(EtO)_2P(S)SCH_2COOH$

plus cyclo-$C_6H_{11}NCO$. Solid. m. 81°.[904]

<u>$(RO)_2P(S)SCH_2CONR'_2$</u>

$(EtO)_2P(S)SCH_2CONMe_2$. XVII.[705] From $(EtO)_2P(S)SCH_2COOMe$ plus Me_2NH.[1213] Liq. n_D^{20} 1.5093,[1213] 1.5210,[705] d^{20} 1.1985.[705]

$(EtO)_2P(S)SCH_2CONEt_2$. XVII.[705] Liq. n_D^{20} 1.5136, d^{20} 1.1478.[705]

$(MeO)_2P(S)SCH_2CONRCHO$. XVII. Liq. R, n_D^{20}: Me, 1.5522; Et, 1.5380; i-Pr, 1.5286.[665]

$(MeO)_2P(S)SCH_2CONMeCOOMe$. XVII. Liq. n_D^{20} 1.5292.[1206]

$(EtO)_2P(S)SCH_2CONMeCOOMe$. XVII.[951,1206] Liq. n_D^{20} 1.5218,[1206] 1.5307.[951]

$(MeO)_2P(S)SCH_2CONMeCOOEt$. XVII. Liq. n_D^{20} 1.5154.[1206]

$(EtO)_2P(S)SCH_2CONMeCOOEt$. XVII.[948,951,1206] Liq. $b_{0.02}$ 144°[948] n_D^{20} 1.5049,[1206] 1.5091,[951] 1.5138,[948] d_{20}^{20} 1.223. "Mecarbam."

$(EtO)_2P(S)SCH_2CONMeCOOBu$. XVII. Liq. n_D^{20} 1.5060.[951]

$(EtO)_2P(S)SCH_2CONMeCOOC_{10}H_7$-1. Solid. m. 88-89°.[951]

$(RO)_2P(S)SCH_2CONEtCOOMe$. XVII. Liq. R, n_D^{20}: Me, 1.5194; Et, 1.5118.[1206]

$(RO)_2P(S)SCH_2CONEtCOOEt$. XVII. Liq. R, n_D^{20}: Me, 1.5118; Et, 1.5048,[1206] 1.5071.[951]

$(EtO)_2P(S)SCH_2CONEtCOOR$. XVII. Liq. R, n_D^{20}: i-Pr, 1.4980; $CH_2CH{=}CH_2$, 1.5122; Bu, 1.4949.[951]

$(EtO)_2P(S)SCH_2CONMeSO_2Me$. XVII. Liq. n_D^{20} 1.5315, d_4^{20} 1.3261[480]

$(RO)_2P(S)SCH_2CONEtSO_2Me$. XVII. Liq. R, n_D^{20}, d_4^{20}: Me, 1.5340 1.3581; Et, 1.5170, 1.2710.[480]

$(EtO)_2P(S)SCH_2CONBuSO_2Et$. XVII. Liq. n_D^{20} 1.5050, d_4^{20} 1.1926[480]

$(RO)_2P(S)SCH_2CON(CH_2CH{=}CH_2)_2$. XVII. Liq. R, n_D^{21}: Me, 1.5313 Et, 1.5253.[1094]

$(EtO)_2P(S)SCH_2CON(CH_2CH{=}CH_2)CH_2NMeCOOEt$. XVII. Liq. n_D^{30} 1.4997.[1264]

$(RO)_2P(S)SCH_2CONMe$-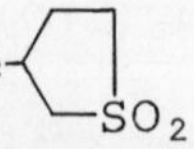 XVII. Solids. R, m.: Me, 104-105°;

Et, 83-84.5°.[1228]

$(RO)_2P(S)SCH_2CON(t\text{-}Bu)$- XVII. Solids. R, m.: Me, 138-139°; Et, 95°.[1228]

$(RO)_2P(S)SCH_2CON(cyclo\text{-}C_6H_{11})$- XVII. Solids. R, m.: Me, 94-95°; Et, 74°.[1228]

$(RO)_2P(S)SCH_2CONMeC_6H_4Cl$-4. XVII. Liq. R, n_D^{20}: Me, 1.5189; Et, 1.5669.[1207]

$(RO)_2P(S)SCH_2CONMeC_6H_3Cl_2$-2,5. XVII. Solids. R, m.: Me, 79°; Et, 72-73°.[1207]

$(RO)_2P(S)SCH_2CONMeC_6H_3Cl_2$-3,4. XVII. Solids. R, m.: Me, 64°; Et, 71-72°.[1207]

$(RO)_2P(S)SCH_2CONMeC_6H_2Cl_3$-2,4,5. XVII. Liq. R, n_D^{20}: Me, 1.5953; Et, 1.5760.[1207]

<u>$(RO)_2P(S)SCH_2CONHNR'_2$</u>

$(MeO)_2P(S)SCH_2CONHNH_2 \cdot HCl$. From $(MeO)_2P(S)SCH_2COOC_6H_4NO_2$-4 plus N_2H_4; product precipitated as HCl salt. Solid. m. 130-140°.[1049]

$(RO)_2P(S)SCH_2CONHNH_2$. From $(RO)_2P(S)SCH_2COOEt$ plus N_2H_4. Liq. R, n_D^{25}: Et, 1.5370; i-Pr, 1.5220.[317]

$(EtO)_2P(S)SCH_2CONHNMe_2$. XVII. Solid. m. 66°.[189]

$(i\text{-}PrO)_2P(S)SCH_2CONHNMe_2$. From $(i\text{-}PrO)_2P(S)SCH_2COOEt$ plus NH_2NMe_2. Liq. n_D^{25} 1.4860.[317]

<u>$(RO)_2P(S)SCH_2CONHSO_2R'$</u>

$(PrO)_2P(S)SCH_2CONHSO_2Et$. XVII. Solid. m. 55-57° (from ether)[480]

$(RO)_2P(S)SCH_2CONHSO_2Ph$. XVII. Solids and liquids.[698]

	m	n_D^{40}	d^{40}
Me	96-98°	--	--
Et	48-50°	--	--
Pr	--	1.5477	1.2654
Bu	--	1.5369	1.2343

Also $-SO_2C_6H_4Cl$ and $-SO_2C_6H_4Me$.[698]

$(PrO)_2P(S)SCH_2CONHSO_2C_6H_3Cl_2$-3,4. XVII. Solid. m. 53-54° (from petroleum ether).[480]

$(RO)_2P(S)SCH_2CON(OR')R''$

$(MeO)_2P(S)SCH_2CON(OH)Me$. XVII. Solid. m. 67°.[175]

$(MeO)_2P(S)SCH_2CON(OH)C_6H_3Cl_2$-3,4. XVII. Solid. m. 87-88°.[564]

$(MeO)_2P(S)SCH_2CON(OMe)H$. XVII. Solid. m. 47-$48^\circ$.[175]

$(MeO)_2P(S)SCH_2CON(OMe)Ar$. XVII. Liq. Ar, n_D^{27}: Ph, 1.5712[280]; C_6H_4Cl-4, 1.5744; $C_6H_3Cl_2$-3,4, 1.5932; C_6H_4Me-3, 1.5744.[564]

$(RO)_2P(S)SCH_2CO$-(N-ring)

$(EtO)_2P(S)SCH_2CON$ 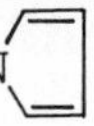 From $(EtO)_2P(S)SCH_2COOCOMe$ plus pyrrole. Liq. n_D^{20} 1.5149.[1213]

$(RO)_2P(S)SCH_2CON$ 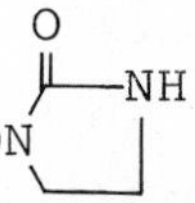 XVII. Solids. R, m.: Me, 115-122°; Et, 64-66° (from toluene-petroleum ether).[483]

$(RO)_2P(S)SCH_2CON$  XVII. Liq. R, n_D^{20}: Me, 1.5632; Pr, 1.5386.[1312] Also several analogs.[1312]

$(RO)_2P(S)SCH_2CON$ 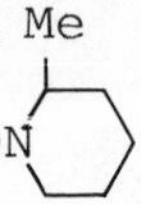 XVII. Liq. R, n_D^{20}: Me, 1.5446; Et, 1.5332; Pr, 1.5258; $CH_2{=}CHCH_2$, 1.5430; Bu, 1.5200; $MeOCH_2CH_2$, 1.5262; $EtOCH_2CH_2$, 1.5204.[1312]

$(RO)_2P(S)SCH_2CON$ 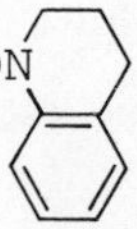 XVII. Liq. R, n_D^{20}, d_4^{20}: Me, 1.5980, 1.2965; Et, 1.5835, 1.2424.[480]

$(MeO)_2P(S)SCH_2CON$ 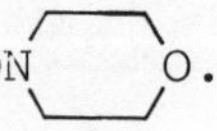. XVII.[903] From $(MeO)_2P(S)SCH_2COCl$ plus morpholine.[149] From $(MeO)_2P(S)SCH_2COOCOR$ plus morpholine.[149,581] Solid. m. 62° (from EtOH),[903] 65° (from CCl_4).[149,581]

$(MeO)_2P(S)SCH_2CON$ 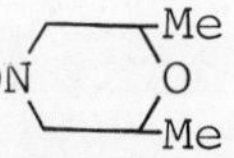 XVII. Liq. n_D^{24} 1.5424.[85a]

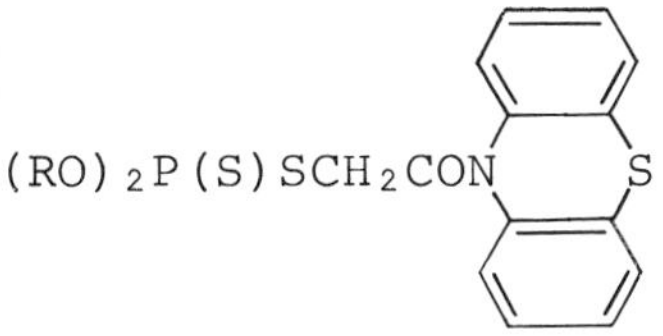

$(RO)_2P(S)SCH_2CON$⟨S⟩ XVII.[66,1414] Solids. R, m.: Me, 103–104°;[66] Et, 101–103°,[1414] 113–114°; [66] Pr. 104–106°;[1414] i-Bu, 128–130°.[1414]

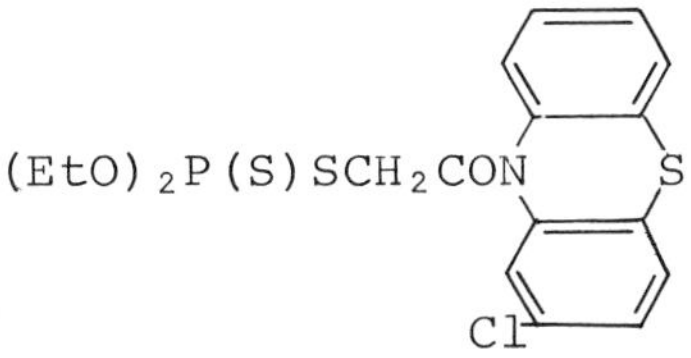

$(EtO)_2P(S)SCH_2CON$⟨S⟩ XVII. Solid. m. 108–109°.[1413]

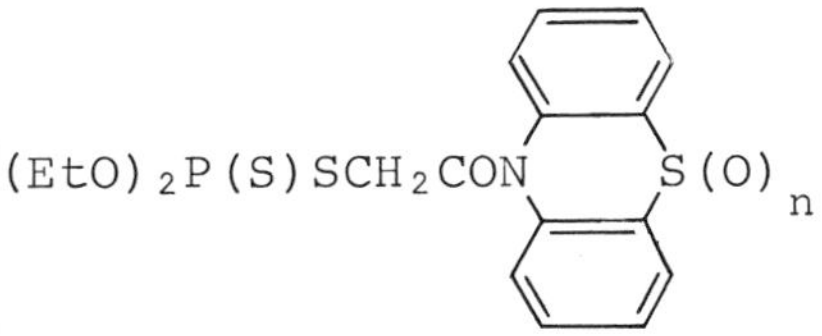

$(EtO)_2P(S)SCH_2CON$⟨S(O)$_n$⟩ XVII. Solids. n, m.: 0, 94–95° (from EtOH); 2, 161–163°. [413]

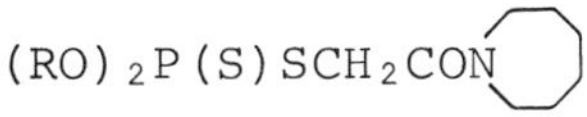

$(RO)_2P(S)SCH_2CON$⟨⟩ XVII. Liq. R, n_D^{20}: Et, 1.5381; Pr, 1.5297; $CH_2=CHCH_2$, 1.5481; Bu, 1.5239; $MeOCH_2CH_2$, 1.5327; $EtOCH_2CH_2$, 1.5240.[1312]

$\underline{(RO)_2P(S)SCH_2COOR'}$

$(MeO)_2{}^{32}P(S)SCH_2COOH$. XVII. Solid. m. 43.5°.[382]

$(MeO)_2P(S)SCH_2COOMe$. XVII.[933] From $(MeO)_2P(S)SCH_2COONa$ plus MeI.[113] From $(MeO)_2P(S)SCH_2COOH$ plus Me_2SO_4 and $NaHCO_3$.[1420] From $(MeO)_2P(S)SCH_2COOH$ plus MeOH and HCl (gas). [1305] Liq. $b_{0.01}$ 85°,[113,1305] $b_{0.1}$ 110°,[933] n_D^{20} 1.5207, d_{20}^{20} 1.3094.[933]

$(EtO)_2P(S)SCH_2COOMe$. XVII. Liq. $b_{0.15}$ 122°, n_D^{20} 1.4994, d_{20}^{20} 1.1880.[933,1259]

$(FCH_2CH_2O)_2P(S)SCH_2COOMe$. XVII. Liq. $b_{0.8}$ 154–156°.[352,830]

$(EtO)_2P(S)SCH_2COOCH_2SMe$. From $(EtO)_2P(S)SCH_2COOH$ plus $ClCH_2SMe$ and Et_3N. Liq. n_D^{30} 1.5214.[530]

$(MeO)_2P(S)SCH_2COOEt$. XVII.[933] From $(MeO)_2P(S)SCH_2COOH$ plus EtOH and HCl (gas); or EtI and EtONa.[1305] Liq. $b_{0.01}$ 87–90°,[1305] b_2 115–116°,[661] n_D^{20} 1.5153,[661] 1.5200,[933] d_4^{20} 1.2469,[661] d_{20}^{20} 1.3064.[933]

$(EtO)_2P(S)SCH_2COOEt$. XVII.[280,933] XXXIII.[643] Liq. $b_{0.08}$ 104.5-105.5°,[661] $b_{0.1}$ 110-112°,[933] n_D^{20} 1.5003,[933] 1.5011,[661] d_4^{20} 1.1823,[661] d_{20}^{20} 1.1857.[933] Acethion.

$(FCH_2CH_2O)_2P(S)SCH_2COOEt$. XVII. Liq. $b_{0.2}$ 149-151°.[352,830]

$(MeOCH_2CH_2O)_2P(S)SCH_2COOEt$. XVII. Liq. $b_{0.07}$ 166-169°.[851]

$(RO)_2P(S)SCH_2COOEt$. Liquids. (Structure-biological activity[661]

R	b./mm	n_D^{20}	d_4^{20}
i-Pr	135-136°/2	1.4908	1.1280
Pr	148-150°/2	1.4969	1.1367
i-Bu	146-148°/2	1.4895	1.0926
Bu	167-169°/2	1.4931	1.1009

$(EtO)_2P(S)SCH_2COO$-i-Bu. XVII. Liq. $b_{0.2}$ 120°, n_D^{20} 1.4910, d_{20}^{20} 1.1341.[933]

$(EtO)_2P(S)SCH_2COOCMe_2C{\equiv}CH$. XVII. Liq. $b_{0.4}$ 129-134°.[654]

$(EtO)_2P(S)SCH_2COOCMeEtC{\equiv}CH$. XVII. Liq. $b_{0.1}$ 132-134°.[654]

$(EtO)_2P(S)SCH_2COOC_6H_4Cl$-4. XVII. Liq. $b_{0.15}$ 170°, n_D^{20} 1.5540, d^{20} 1.2820.[705]

$(EtO)_2P(S)SCH_2COOC_6H_4NO_2$-4. XVII. Liq. n_D^{20} 1.5640, d^{20} 1.3135.[705]

$(EtO)_2P(S)SCH_2COO$– 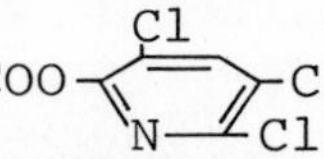

. XVII. Solid. m. 39-42°.[105]

$(EtO)_2P(S)SCH_2COSEt$. XVII.[705] Liq. $b_{0.1}$ 140-143°,[705] $b_{0.15}$ 114.5-115.5°,[661] n_D^{20} 1.5370,[705] 1.5393,[661] d^{20} 1.1961,[705] d_4^{20} 1.2081.[661]

$(EtO)_2P(S)SCH_2COSPh$. XVII. Liq. n_D^{20} 1.5845, d^{20} 1.2321.[705]

$(EtO)_2P(S)SCH_2COSC_6H_4Cl$-4. XVII. Liq. n_D^{20} 1.5890, d^{20} 1.3060.[705]

$(MeO)_2P(S)SCH(SEt)COOEt$. XVII. Liq. $b_{0.015}$ 120-125°, n_D^{23} 1.5215.[877,878]

$(EtO)_2P(S)SCH(SEt)COOEt$. XVII. Liq. $b_{0.05}$ 125-130°, n_D^{23} 1.5171,[877,878]

$(MeO)_2P(S)SCH(SC_6H_4Me$-4$)COOEt$. XVII. Liq. n_D^{20} 1.5735.[878]

$(EtO)_2P(S)SCH(SC_6H_4Me$-4$)COOEt$. XVII. Liq. $b_{0.01}$ 120°, n_D^{20} 1.5608. Also 20 analogs.[878]

$[(EtO)_2P(S)SCH(COOR)]_2S$. XVII. Liq. R, n_D^t: Me, $1.5320^{25°}$: Et. $1.5256^{24°}$.[1099]

$(RO_2P(S)SCH=CH_2$

$(MeO)_2P(S)SCH=CH_2$. Liq. XXII.[925] XXVII.[802] $b_{0.7}$ 62-63°,[925] b_2 74-77°,[802] n_D^{25} 1.533,[802] n_D^{27} 1.527.[925]

$(EtO)_2P(S)SCH=CH_2$. XXII.[802] XXVII.[372,802] XXXIII.[805] Liq. $b_{0.1}$ 62-64°,[925] $b_{0.7}$ 108-112°,[802] b_3 87-89°,[372] n_D^{20} 1.5010,[372] n_D^{25} 1.512,[802,805,925] d^{20} 1.1221.[372]

$(i\text{-}PrO)_2P(S)SCH=CH_2$. XXII. Liq. $b_{0.1}$ 66-69°.[925]

$(i\text{-}BuO)_2P(S)SCH=CH_2$. XXVII. Liq. $b_{0.001}$ 88-92°, n_D^{20} 1.4900, d^{20} 1.0466.[372]

$(EtO)_2P(S)SCBr=CMeBr$. From $(EtO)_2P(S)SC{\equiv}CMe$ plus Br_2 in $CHCl_3$. Liq. n_D^{25} 1.5618.[806]

$(MeO)_2P(S)SCH=CClOAr$. From $(MeO)_2P(S)SH$ plus $ClCH=CClOAr$ in $CHCl=CCl_2$ at 80-100°. Liq. Ar, b/mm: Ph, 147°/0.2; $4\text{-}ClC_6H_4$, 163-167°/0.1-0.2.[1404]

$(MeO)_2P(S)SCH=CClSEt$. XXII. Liq. $b_{0.1}$ 100.7°, n_D^{27} 1.5746.[1403]

$(EtO)_2P(S)SCH=CClSEt$. XXII. Liq. $b_{0.1}$ 113°, n_D^{27} 1.5520.[1403]

$(EtO)_2P(S)SCH=CClSC_6H_4Me\text{-}4$. XXII. Liq. $b_{0.1-0.2}$ 150-155°, n_D^{20} 1.5950.[1404]

$(EtO)_2P(S)SR$. XXII. From $(EtO)_2P(S)SH$ plus fluorinated acetylenes in the presence of Et_3N. Liq.

R	b./mm	n_D^{20}	d_4^{20}
$CF=CH_2$	88-91°/3	1.4961	1.1650[361]
$CBr=CHCF_3$	122°/2	$1.5198^{25°}$	$1.5962^{25°}$
$CCl=CHCF_3$	84-86°/1	1.4855	1.3175
$CH=CHCF_3$	78°/2	1.4657	1.2159
$C(CF_3=CHCF_3)$	87°/4	1.4438	1.3524

$(i\text{-}PrO)_2P(S)SCF=CH_2$. XXII. Liq. b_3 93-95°, n_D^{20} 1.4500, d_4^{20} 1.0870.[361]

$(EtO)_2P(S)SC{\equiv}CMe$. XXXIII. Liq. $b_{0.8}$ 95-100°, n_D^{25} 1.5236.[804]

$(MeO)_2P(S)SC{\equiv}CPh$. XXXIII. Liq. n_D^{25} 1.6104.[804]

$(EtO)_2P(S)SC{\equiv}CPh$. XXXIII. Liq. n_D^{25} 1.5903.[804]

$(EtO_2P(S)SCH=CHEt$. XXXIII. Liq. n_D^{25} 1.520.[805]

$(RO)_2P(S)S$-i-Pr

$(EtO)_2P(S)S$-i-Pr. XXII. Liq. b_1 73-77°.[890]

$(EtO)_2P(S)SCHMeCH_2NHSO_2Me$. XVII. Liq. n_D^{25} 1.4957.[1263]

$(MeO)_2P(S)SCHMeCH_2OCO$-

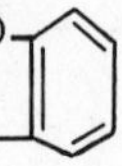

$CH_2=CHCH_2OCO$- XXII. Liq. $b_{0.1}$ 120°, n_D^{24} 1.5480.[1026]

$(MeO)_2P(S)SCHMeCH_2SCH_2CHMeSP(S)(OMe)_2$. XXII. Liq. n_D^{20} 1.5648.[1025]

$[(MeO)_2P(S)SCHMeCH_2]_2S$. XXII. Liq. n_D^{20} 1.5648.[1353]

$[(MeO)_2P(S)SCHMeCH_2]_2SiMePh$. XXII. From $(MeO)_2P(S)SH$ plus $MePhSi(CH_2CH=CH_2)_2$. Liq. n_D^{20} 1.5550, d^{20} 1.1799.[940]

$(RO)_2P(S)SCHMeCONMeCHO$. XVII. Liq. R, n_D^{20}: Me, 1.5314; Et, 1.5257.[665]

$(EtO)_2P(S)SCHMeCONH_2$. XVII. Solid. m. 70-71°.[820]

$(EtO_2P(S)SCHMeCONHEt$. XVII. Solid. m. 73-74°.[820]

$(EtO)_2P(S)SCHMeCONH$-i-Pr. XVII. Solid. m. 41-43°.[820]

$(EtO)_2P(S)SCHMeCONEt_2$. XVII. Liq. n_D^{20} 1.5040, d_4^{20} 1.1173.[1425]

$(EtO)_2P(S)SCHMeC{\equiv}N$. XVII. Liq. $b_{0.01}$ 77-78°.[1150]

$(EtO)_2P(S)SCHMeCOOEt$. XVII. Liq.[280,1425] b_2 134°,[280] $b_{0.3}$ 116-118°, n_D^{20} 1.4975, d_4^{20} 1.1490.[1425]

$(EtO)_2P(S)SCH(CH_2OMe)CH_2SEt$. XVII. Liq. n_D^{26} 1.5175.[1326]

$(MeO)_2P(S)SCH(CH_2OEt)CH_2SEt$. XVII. Liq. n_D^{23} 1.5252.[1326]

$(EtO)_2P(S)SCH(CH_2SEt)_2$. XVII. Liq. n_D^{18} 1.5457.[666]

$(EtO)_2P(S)SCH(CH_2OMe)CONHMe$. XVII. Solid. m. 77-78°.[962]

$(EtO)_2P(S)SCH(CH_2OEt)CONHR$. XVII. Solids. R, m.: H, 61-62°; Me, 77-78°; Et, 72-73°; i-Pr, 78-79°.[962]

$(MeO)_2P(S)SCH(CONHMe)_2$. XVII.[206,644] Solid. m. 135°,[644] 137-138°.[206]

$(EtO)_2P(S)SCH(CONHMe)_2$. XVII.[206,644] Solid. m. 159° (from MeCN),[644] 162-163° (from MeOH).[206]

$(EtO)_2P(S)SCH(COOEt)_2$. XVII. Liq. $b_{0.5}$ 146-152°, n_D^{25} 1.4920.[163]

(i-PrO)$_2P(S)SCH(COOEt)_2$. XVII. Liq. $b_{0.5}$ 152-154°, n_D^{25} 1.4818.[163]

$(EtO)_2P(S)SC(OH)(COOEt)_2$. XVIII. Liq. n_D^{20} 1.4900, d_4^{20}

1.2270.[985]

<u>(RO)$_2$P(S)SPr</u>

(CH$_2$=CHCH$_2$O)$_2$P(S)SPr. XVII. Liq. $b_{0.0001}$ 40°, n_D^{20} 1.5195, d_4^{20} 1.0948.[1445]

(HC≡CCH$_2$O)$_2$P(S)SPr. XVII. Liq. $b_{0.0001}$ 65°, n_D^{20} 1.5360, d_4^{20} 1.1574.[1447]

(MeCH=CHCH$_2$O)$_2$P(S)SPr. XVII. Liq. $b_{0.0001}$ 76°, n_D^{20} 1.5198, d_4^{20} 1.0645.[1447]

(MeO)$_2$P(S)SCHBrCHBrMe. From (MeO)$_2$P(S)SCH=CHMe plus Br_2 in $CHCl_3$. Liq. n_D^{25} 1.5895.[806]

(BuO)$_2$P(S)SCH$_2$CH$_2$CH$_2$OH. XXII. Liq. $b_{0.03}$ 90-95°, n_D^{20} 1.4835, d^{20} 1.0491.[756]

(RO)$_2$P(S)SCH$_2$CH$_2$CH$_2$OP(S)(SPr)$_2$. XXII. Liq. R, $b_{0.002}$, n_D^{20}, d_4^{20}: Et, 132°, 1.5598, 1.1904; Pr, 140°, 1.5534, 1.1565. IR.[740]

(EtO)$_2$P(S)SCH$_2$CH$_2$CH$_2$SEt. XXII. Liq. $b_{0.1}$ 126-127°.[381]

(EtO)$_2$P(S)SCH$_2$CH(SEt)CH$_2$SEt. XVII. Liq. $b_{0.01}$ 113°.[1203]

[(EtO)$_2$P(S)S]$_2$(CH$_2$)$_3$. XVII.[1243] XXII.[1248] Liq. $b_{0.0001}$ 110°,[1448] $b_{0.35}$ 192°,[1243] n_D^{20} 1.5402,[1243,1448] d^{20} 1.1922,[1243] d_4^{20} 1.1931.[1448]

[(i-PrO)$_2$P(S)S]$_2$(CH$_2$)$_3$. XXII.[1448] Liq. $b_{0.0001}$ 118°, n_D^{20} 1.5200, d_4^{20} 1.1242, IR.[1448]

[(PrO)$_2$P(S)S]$_2$(CH$_2$)$_3$. XXII. Liq. $b_{0.0001}$ 129°, n_D^{20} 1.5260, d_4^{20} 1.1383.[1448]

(RO)$_2$P(S)SCH$_2$CH$_2$CH$_2$SP(S)(OR')$_2$. XXII. Liq.[1448]

R	R'	$b_{0.0001}$	n_D^{20}	d_4^{20}	
Me	Pr	106°	1.5464	1.2142	
Et	i-Pr	116°	1.5300	1.1629	
Et	Pr	118°	1.5336	1.1699	IR
Et	Bu	132°	1.5275	1.1330	
i-Pr	Pr	139°	1.5226	1.1309	
i-Pr	Bu	134°	1.5180	1.1115	
Pr	Bu	136°	1.5216	1.1172	IR

<u>(RO)$_2$P(S)SCH$_2$CH$_2$C(O)R'</u>

(RO)$_2$P(S)SCH$_2$CH$_2$CHO. XXII. Liq.[756]

R	b./mm	n_D^{20}	d^{20}
Et	80-81°/0.05	1.5070	1.0984
Pr	74°/0.15	1.5040	1.1348
Bu	75-77°/0.025	1.4955	1.0756

$(MeO)_2P(S)SCHOHCH_2CHO$ - dimer. XI. From $(MeO)_2P(S)SH$ plus CH_2—CHCHO (epoxide, O bridging). Solid. m. 98-101°.[927]

$(EtO)_2P(S)SCH_2CH_2CO$- =X: 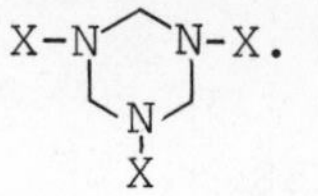 XXII. From $(EtO)_2P(S)S$ plus intermediate wher X= CH_2=CHCO-. Solid. m 58°.[1397]

$(EtO)_2P(S)SCH_2CH_2CONEt_2$. XVII. Liq. n_D^{20}. 1.5150, d_4^{20} 1.1364 [1425]

$(EtO)_2P(S)SCH_2CH_2CONHPh$. XVII. Liq. n_D^{20} 1.5790, d_4^{20} 1.2261 [1425]

$(EtO)_2P(S)SCH_2CH(SEt)CONHMe$. XVII. Solid. m. 64-65°.[632]

$(EtO)_2P(S)SCH_2CH_2CN$. XVII.[988] XXII.[756,988] Liq. $b_{0.5}$ 111-113°,[988] $b_{3.5}$ 137-142°,[756] n_D^{20} 1.5174,[988] 1.5195,[756] d_4^{20} 1.1704,[756] 1.1831.[988] Rate of formation: $(EtO)_2P(S)$SH plus CH_2=CHCN in 3:1 heptane-xylene at 45°: k_2 = 8.36 liters/mole/sec.[986] For the homolog: $(EtO)_2P(S)SC$ CHMeCN, k_2 = 7.23 liters/mole/sec.[986]

$(RO)_2P(S)SCH_2CH_2CN$. XXII.[756] Liq.

	b./mm	n_D^{20}	d^{20}
i-Pr	98-99°/0.05	1.5020	1.0182[756]
Pr	116-120°/0.05	1.5068	1.0505
Bu	121-123°/0.03	1.5010	1.0986

$(PhO)_2P(S)SCH_2CH_2CN$. XXII. Liq. n_D^{20} 1.6030, d^{20} 1.2448.[741]

$(EtO)_2P(S)SCH_2CHRCOOH$. XXII. Rates of formation: $(EtO)_2P(S)$ plus CH_2=CHCOOH in 3:1 heptane-xylene at 45°: R, k_2 liter/mole/sec.: H, 1.89; Me, 1.43.[986]

$(MeO)_2P(S)SCH_2CH_2COOMe$. XXII. Liq. b_3 145°, n_D^{20} 1.5160, d^2 1.2625.[756]

$(EtO)_2P(S)SCH_2CH_2COOMe$. XVII.[890] XXII.[756,890,986] Liq. $b_{0.}$ 65-66°,[890] $b_{1.5}$ 167°,[756] n_D^{20} 1.5050,[756] d^{20} 1.1911.[756]

Rates of formation: $(EtO)_2P(S)SH$ plus $CH_2=CHCOOMe$ in 3:1 heptane-xylene at 45°: k_2 = 0.86 liter/mole/sec. For the homolog, $(EtO)_2P(S)SCH_2CHMeCOOMe$, k_2 = 0.43 liter/mole/sec.[986]

$(PrO)_2P(S)SCH_2CH_2COOMe$. XXII. Liq. $b_{0.04}$ 85-86°.[890]

$(i\text{-}BuO)_2P(S)SCH_2CH_2COOMe$. XXII. Liq. $b_{0.2}$ 124-129°, n_D^{20} 1.4915, d^{20} 1.1172.[756]

$(BuO)_2P(S)SCH_2CH_2COOMe$. XXII. Liq. $b_{0.03}$ 96-97°.[890]

$(PhO)_2P(S)SCH_2CH_2COOMe$. XXII. Liq. n_D^{20} 1.5862, d^{20} 1.2577.[741]

$(EtO)_2P(S)SCH_2CH_2COOCH_2SMe$. From $(EtO)_2P(S)SCH_2CH_2COOH$ plus Et_3N and $ClCH_2SMe$. Liq. n_D^{30} 1.5212.[530]

$(EtO)_2P(S)SCH_2CH_2COOEt$. XVII. Liq. $b_{0.2}$ 114°, n_D^{20} 1.4980, d_4^{20} 1.1653.[1425]

$(EtO)_2P(S)SCH_2CH_2COOC_6H_4Cl\text{-}4$. XVII. Liq. n_D^{20} 1.5500, d_4^{20} 1.2651.[1425]

$(EtO)_2P(S)SCH_2CH_2COSEt$. XVII. Liq. $b_{0.3}$ 140°, n_D^{20} 1.5298, d_4^{20} 1.1861.[1425]

<u>$(RO)_2P(S)SCH_2CHORMe$</u>

$(MeO)_2P(S)SCH_2CHOHMe$. XI. Liq. n_D^{20} 1.5320, d^{20} 1.2541.[993]

$(EtO)_2P(S)SCH_2CHOHMe$. XI. Liq. n_D^{20} 1.5130, d^{20} 1.1682.[993]

$(MeO)_2P(S)SCH_2CHOHCH_2Cl$. XI. Liq. n_D^{20} 1.5480, d^{20} 1.3774.[993]

$(EtO)_2P(S)SCH_2CHOHCH_2Cl$. XI. Liq. $b_{0.02}$ 120-123°, n_D^{20} 1.5038, d^{20} 1.2393.[993]

$(PrO)_2P(S)SCH_2CHOHCH_2Cl$. XI. Liq. n_D^{20} 1.5130, d^{20} 1.2112.[993]

$(EtO)_2P(S)SCH(OAr)CH_2CCl_3$. XVII. Liq. Ar, n_D^{20}, d_4^{20}: Ph, 1.5560, 1.3277; $4\text{-}ClC_6H_4$, 1.5630, 1.3708; $3\text{-}MeC_6H_4$, 1.5543, 1.2976.[505]

$(PhO)_2P(S)SCH(OPh)CH_2CCl_3$. XVII. Solid. m. 80-81°.[505]

$(MeO)_2P(S)SCH_2CHOHCH_2OH$. XI. Liq. n_D^{20} 1.5460, d^{20} 1.3378.[993]

$(EtO)_2P(S)SCH_2CHOHCH_2OH$. XI. Liq. n_D^{20} 1.5270, d^{20} 1.2435.[993]

$(EtO)_2P(S)SCH_2CH(SEt)Me$. From $(EtO)_2P(S)SCH=CHMe$ plus EtSH under UV light at 17° for 24 hr. Liq. $b_{0.5}$ 100-102°.[919]

<u>$(RO)_2P(S)SCH_2COMe$</u>

$(MeO)_2P(S)SCH_2COMe$. XVII.[2,1336] Liq. $b_{0.25}$ 113-115°,[1336] n_D^{20} 1.5320,[1336] n_D^{25} 1.5293,[2] d_4^{20} 1.2731.[1336]

$(EtO)_2P(S)SCH_2COMe$. XVII.[2,1336] Liq. $b_{0.25}$ 120-121°,[1336] $b_{0.3-0.5}$ 125-128°,[2] n_D^{20} 1.5138,[1336] n_D^{25} 1.5126,[2] d_4^{20} 1.1876.[1336]

$(i\text{-}PrO)_2P(S)SCH_2COMe$. XVII.[193,1336] Liq. $b_{0.08}$ 115°,[1336] $b_{0.3-0.5}$ 105-106°,[193] n_D^{20} 1.4950,[1336] d_4^{20} 1.1215.[1336]

$(RO)_2P(S)SCH_2COMe$. XVII. Liq. R, b/mm, n_D^{20}, d_4^{20}: Pr, 125-128°/0.15, 1.5060, 1.1236; i-Bu, 123°/0.075, 1.4940, 1.0913; Bu, 128°/0.075, 1.5013, 1.0963.[1336]

$(EtO)_2P(S)SCH_2C(SEt)_2Me$. From $(EtO)_2P(S)SCH_2COMe$ plus EtSH and HCl gas, Liq. n_D^{25} 1.5420.[1377] Also several analogs.

$[(EtO)_2P(S)SCH]_2CHNMe_2.(COOH)_2$. XVII. Solid. m. 91-92° (dec.). IR. ^{1}H-NMR.[556]

$\underline{(RO)_2P(S)SCH_2CH{=}CH_2}$

$(RO)_2P(S)SCH_2CH{=}CH_2$. XVII. Liq.

	b./mm	n_D^{20}	d^{20} [14
Et	95-96°/3	1.5159	1.1056
i-Pr	97-98°/2	1.5015	1.0499
Pr	118-119°/3	1.5074	1.0700
Bu	140-141°/3	1.4991	1.0333

$(HC{\equiv}CCH_2O)_2P(S)SCH_2CH{=}CH_2$. XVII. Liq. $b_{0.0001}$ 76°, n_D^{20} 1.5510, d_4^{20} 1.1842.[1447]

$(MeCH{=}CHCH_2O)_2P(S)SCH_2CH{=}CH_2$. XVII. Liq. $b_{0.0001}$ 78°, n_D^{20} 1.5308, d_4^{20} 1.0800.[1447]

$(CH_2{=}CMeCH_2O)_2P(S)SCH_2CH{=}CH_2$. XVII. Liq. $b_{0.0001}$ 62°, n_D^{20} 1.5257, d_4^{20} 1.0787.[1447]

$(PhO)_2P(S)SCH_2CH{=}CH_2$. XVII. Liq. b_1 172-173°, n_D^{20} 1.6070, d_4^{20} 1.217, IR.[383]

$(RO)_2P(S)SCH{=}CHMe$. XXII. Liq. R, b/mm: Me, 60-64°/0.3; Et, 74-78°/0.4; i-Pr, 68-69°/0.05.[919]

$(HC{\equiv}CCH_2O)_2P(S)SCH_2C{\equiv}CH$. XVII. Liq. $b_{0.0001}$ 80°, n_D^{20} 1.5600 d_4^{20} 1.2212.[1447]

$(MeCH{=}CHCH_2O)_2P(S)SCH_2C{\equiv}CH$. XVII. Liq. $b_{0.0001}$ 84°, n_D^{20} 1.5384, d_4^{20} 1.1077.[1447]

$(CH_2{=}CMeCH_2O)_2P(S)SCH_2C{\equiv}CH$. XVII. Liq. $b_{0.0001}$ 72°, n_D^{20} 1.5321, d_4^{20} 1.1067.[1447]

$(EtO)_2P(S)SCH{=}CHCHO$. XXII. Liq. $b_{0.025}$ 107°, n_D^{20} 1.5515, d^2

1.2027.[989]

$(PrO)_2P(S)SCH{=}CHCHO$. XXII. Liq. $b_{0.025}$ 122-123°, n_D^{20} 1.5410, d^{20} 1.1444.[989]

$(PrO)_2P(S)SCH{=}CHCH{=}NNHC_6H_3(NO_2)_2$-2,4. From $(PrO)_2P(S)SCH{=}CHCHO$ plus 2,4-dinitrophenylhydrazine. Solid. m. 121°.[989]

$(EtO)_2P(S)SCH{=}CHCOOMe$. XXII. From $HC{\equiv}CCOOMe$ plus $(EtO)_2P(S)SH$ with $PhCH_2NMe_3^+$ OMe^- as catalyst. Liq. $b_{0.05}$ 133-135°.[810]

$(EtO)_2P(S)SCH_2C{\equiv}CH$. XVII.[1155,1436] Liq. $b_{0.1}$ 66°,[1155] b_3 104-105°,[1436] n_D^{20} 1.5182, d_4^{20} 1.1262.[1436]

$(RO)_2P(S)SCH_2CH{=}CH$. XVII.[973,1436] Liq.

R	b./mm	n_D^{20}	d_4^{20}
i-Pr	80-83°/0.001	1.5075	1.0892[973]
	113-114°/3	1.5088	1.0881[1436]
Pr	70°/0.01	1.5158	1.0995[973]
	119-120°/3	1.5147	1.0978[1436]
Bu	90-92°/0.01	1.5085	1.0602[973]
	138-139°/3	1.5060	1.0624[1436]
Am	100-103°/0.01	1.5040	1.036 [973]
$4\text{-}ClC_6H_4$	143-148°/0.001	1.6108	1.5104[973]

$(MeO)_2P(S)SC{\equiv}CMe$. XXXIII. Liq. $b_{2.3}$ 115-117°, n_D^{25} 1.539.[802]

$(EtO)_2P(S)SC{\equiv}CMe$. XXXIII. Liq. $b_{0.8}$ 95-99°, n_D^{25} 1.524.[802]

$(EtO)_2P(S)SC{\equiv}CPh$. XXXIII. Liq. $b_{0.01}$ 85-89°, n_D^{25} 1.592.[802]

<u>$(RO)_2P(S)SBu$</u>

$(BuO)_2P(S)S$-t-Bu. XXXIII. Liq. $b_{0.001}$ 58-60°, n_D^{20} 1.4825, d_4^{20} 1.0081.[1237]

$(EtO)_2P(S)S$-i-Bu. XXVIII.[941] XXXIII.[802] Liq. $b_{0.01}$ 65-75°,[802] b_5 117-119°,[941] n_D^{20} 1.4834,[941] n_D^{25} 1.497,[802] d_4^{20} 1.0646.[941]

$(CH_2{=}CHCH_2O)_2P(S)S$-i-Bu. XVII.[910,1445] Liq. $b_{0.0001}$ 44°,[1445] n_D^{20} 1.5254.[910] 1.5155,[1445] d_4^{20} 1.1037,[910] 1.0773.[1445]

$(BuO)_2P(S)S$-i-Bu. XXVIII. Liq. b_5 151-152°, n_D^{20} 1.4808, d_4^{20} 1.0044.[941]

$(MeO)_2P(S)SCH_2CHMeCH_2Cl$. XXII. Liq. $b_{0.07}$ 115-116°.[411]

$(EtO)_2P(S)SCH_2CHMeCH_2Cl$. XXII. Liq. $b_{0.15}$ 120-123°.[411]

$(MeO)_2P(S)SCH_2CHMeCCl_3$. XXII. Liq. $b_{0.06}$ 126-128°.[414]

$(EtO)_2P(S)SCH_2CHCMeCCl_3$. XXII. Liq. $b_{0.07}$ 141-144°.[414]

$(EtO)_2P(S)SCH_2CH_2C(SEt)MeCONH_2$. XVII. Solid. m. 68-70°.[1103]

$(RO)_2P(S)SCH_2CHMeCOOMe$. XXII. Liq.

	b./mm	n_D^{20}	d_4^{20} [756]
Me	133-135°/2	1.5100	1.2330
Et	154.5°/5	1.4995	1.1577
i-Pr	84-85°/0.02	1.4935	1.1203
i-Bu	115-116°/0.04	1.4915	1.1138
Bu	110°/0.025	1.4918	1.1132

$(EtO)_2P(S)SCH_2C(SMe)MeCOOMe$. XVII. Liq. $b_{0.4}$ 142°.[1103]

$(EtO)_2P(S)SCH_2CHMeCOOCH_2SMe$. From $(EtO)_2P(S)SCH_2CHMeCOOH$ plus $ClCH_2SMe$ in the presence of Et_3N. Liq. n_D^{30} 1.5160.[530]

$(EtO)_2P(S)SCH_2CHMeCOOCH_2S(O)Me$. From $(EtO)_2P(S)SCH_2CHMeCOOC$ SMe plus H_2O_2 in acetic acid. Liq. n_D^{30} 1.5260.[530]

$(EtO)_2P(S)SCH_2CH(COOEt)_2$. XXII. Liq. $b_{0.6}$ 158-160°, n_D^{20} 1.4710, d^{20} 1.1448.

$(EtO)_2P(S)S$-s-Bu. XXXIII. Liq. $b_{0.01}$ 65-75°, n_D^{25} 1.499.[802]

$(MeO)_2P(S)SCHMeCH_2COOMe$. XXII. Liq. $b_{0.04}$ 102-104°.[337]

$(EtO)_2P(S)SCHMeCH_2COOMe$. XXII. Liq. $b_{0.02}$ 101-102°.[337] Also several analogs.[337]

$(RO)_2P(S)SCHMeCOMe$. XVII. Liq.

	b./mm	n_D^{20}	d_4^{20} [1336]
Me	89-92°/0.07	1.5220	1.2140
Et	95-97°/0.06	1.5069	1.1503
i-Pr	102-105°/0.16	1.4941	1.0910
Pr	100-104°/0.07	1.5008	1.1919
i-Bu	99.5-101°/0.12	1.4950	1.0678
Bu	132.5-134°/0.21	1.4970	1.0748

$(EtO)_2P(S)SCHMeCH(SMe)COOMe$. XVII. Liq. $b_{0.01}$ 108°.[630]

$(i\text{-}PrO)_2P(S)SCHMeCH(SMe)COOMe$. XVII. Liq. $b_{0.01}$ 110°.[630]

$(EtO)_2P(S)SCHMeCH(SEt)COOMe$. XVII. Liq. $b_{0.03}$ 115-116°.[630]

$(i\text{-}PrO)_2P(S)SCHMeCH(SEt)COOMe$. XVII. Liq. $b_{0.01}$ 121°.[630]

<u>$(RO)_2P(S)SCH(CONHR')CH_2COOR''$</u>

$(EtO)_2P(S)SCH(CONHEt)CH_2COOH$. XXII. Liq. n_D^{20} 1.5252.[348]

$(EtO)_2P(S)SCH(CONH\text{-}i\text{-}Pr)CH_2COOH$. XXII. Liq. n_D^{20} 1.5243.[348]

$(EtO)_2P(S)SOH(CONHEt)CH_2COOMe$. From the acid plus CH_2N_2. Liq. n_D^{20} 1.5204.[348]

$(EtO)_2P(S)SCH(CONH\text{-}i\text{-}Pr)CH_2COOMe$. From the acid plus CH_2N_2. Liq. n_D^{20} 1.5124.[348]

$(EtO)_2P(S)SCH(CONHEt)CH_2COOEt$. From the acid plus $SOCl_2$ and EtOH. Liq. n_D^{25} 1.5281.[339]

$(EtO)_2P(S)SCH(CONHC_6H_4Cl\text{-}4)CH_2COOEt$. XXII. Solid. m. 89-91° (from alcohol).[339]

$(EtO)_2P(S)SCH(COOCH_2SMe)CH_2CONMe_2$. From the amido-acid plus $ClCH_2SMe$ and K_2CO_3. Liq. n_D^{30} 1.5250.[530] Ten analogs.[530]

$(EtO)_2P(S)SCH(COOCH_2S(O)Me)CH_2CONHMe_2$. From the sulfide plus H_2O_2. Liq. n_D^{30} 1.5248.[530] Six analogs.[530]

$(EtO)_2P(S)SCH(COOEt)CH_2CONMeCOOEt$. XXII. Solid. m. 68-70° (from benzene-hexane).[557]

$(EtO)_2P(S)SCH(COOEt)CH_2CONEtCOOEt$. XXII. Solid. m. 71-72° (from benzene-hexane).[557]

$(MeO)_2P(S)SCH(CON(OMe)Pr)CH_2CON(OMe)Pr$. XXII. Solid. m. 65.5-69°.[77a]

$(MeO)_2P(S)SCH(CONMe(O\text{-}i\text{-}Pr))CH_2CONMe(O\text{-}i\text{-}Pr)$. XXII. Solid. m. 91.5-94°.[77a]

<u>$(RO)_2P(S)SCH(COOR')CH_2COOR'$</u>

$(MeO)_2P(S)SCH(COOH)CH_2COOH$. XXII. Solid. m. 115-118° (from $CHCl_3$).[714]

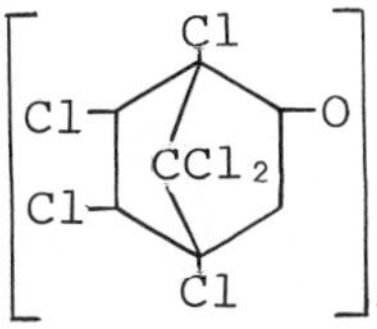

$P(S)SCH(COOH)CH_2COOH$. XXII. Solid. m. 143°.[324]

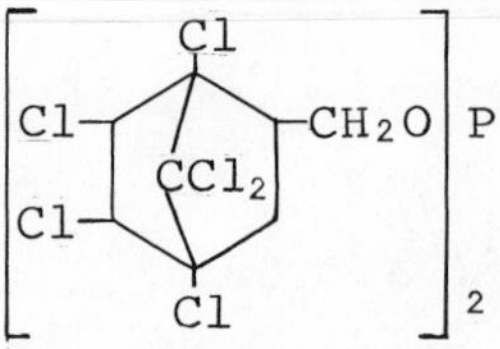
P(S)SCH(COOH)CH_2COOH. XXII. Liq. n_D^{20} 1.5539.[324]

(MeO)$_2$P(S)SCH(COOMe)CH_2COOMe. XXII.[150,756] Liq. $b_{0.2}$ 134.5°[756] n_D^{20} 1.5070,[756] n_D^{25} 1.5080,[150] d^{20} 1.2864.[746] ^{31}P -95.[715]

(EtO)$_2$P(S)SCH(COOMe)CH_2COOMe. XXII.[150,756] Liq. $b_{0.025}$ 116-120°,[756] n_D^{20} 1.4985,[756] n_D^{25} 1.4970,[150] d^{20} 1.2237.[756] ^{31}P -88.[715]

(RO)$_2$P(S)SCH($COOCH_2$SMe)CH_2COOCH_2SMe. XXII. Liq. R, n_D^{30}: Me, 1.5580; Et, 1.5344.[530]

(RO)$_2$P(S)SCH(COOEt)CH_2COOCH_2SMe. XXII. Liq. R, n_D^{30}: Me, 1.5169; Et, 1.5120.[530] Three analogs.[530]

(EtO)$_2$P(S)SCH(COOR)CH_2COOCH_2S(O)Me. From (EtO)$_2$P(S)SCH(COOR)CH_2COOCH_2SMe and H_2O_2. Liq. R, n_D^{30}: Et, 1.5195; Ph, 1.5456; 4-ClC_6H_4, 1.5517.[530]

(MeO)$_2$P(S)SCH(COOEt)CH_2COOEt. XV. (From (MeO)$_2$PCl plus HSCH(COOEt)CH_2COOEt and S).[1056] XXII.[150,714,756] Liq. $b_{0.7}$ 155°,[1056] $b_{3.5}$ 160-170°,[756] n_D^{20} 1.4960,[756] 1.4991,[766] n_D^{24} 1.5149,[714] n_D^{25} 1.4970,[150] d^{20} 1.2076,[756] IR and UV.[375] ^1H-NMR.[74,524,526] ^{31}P -94.[715] ^{31}P 17.5 (P_4O_6 external).[1047] Mass spect.[196] Cryoscopic m.p. 3.70° ± 0.09° (100% pure).[93] Thermal isomerization.[766] "Malathion."

(EtO)$_2$P(S)SCH(COOEt)CH_2COOEt. XXII.[150,756,986] Liq. $b_{0.5}$ 139-141°,[986] b_3 157-162°,[756] n_D^{20} 1.4910,[756] 1.4920,[986] n_D^{25} 1.4895,[150] d_4^{20} 1.1736,[986] d^{20} 1.1742,[756] ^{31}P -89.[715]

(RO)$_2$P(S)SCH(COOEt)CH_2COOEt. XXII.[150,756] Liq.

	b./mm	n_D^{20}	d^{20}
i-Pr	161°/4	1.5440	1.2076[756]
	n_D^{25}	1.4811[150]	
Pr	145°/0.1	1.4880	1.1706[756]

i-Bu	117-130°/0.04	1.4855	1.0642
Bu	125-128°/0.025	1.4861	1.1708
$C_{10}H_{21}$	n_D^{25}	1.4769[150]	
Ph		1.5532[150]	

$(EtO)_2P(S)SCH(COOCH_2CH_2Cl)CH_2COOCH_2CH_2Cl$. XXII. Liq. n_D^{25} 1.5110.[150]

$(MeO)_2P(S)SCH(COO\text{-}i\text{-}Pr)CH_2COO\text{-}i\text{-}Pr$. XXII. Liq. n_D^{25} 1.4862.[150]

$(EtO)_2P(S)SCH(COO\text{-}i\text{-}Pr)CH_2COO\text{-}i\text{-}Pr$. XXII. Liq. n_D^{25} 1.4805.[150]

$(PrO)_2P(S)SCH(COO\text{-}i\text{-}Pr)CH_2COO\text{-}i\text{-}Pr$. XXII. Liq. $b_{0.075}$ 125-128°, n_D^{20} 1.4785, d^{20} 1.1146.[756]

$(MeO)_2P(S)SCH(COOPr)CH_2COOPr$. XXII. Liq. n_D^{25} 1.4890.[150]

$(EtO)_2P(S)SCH(COOPr)CH_2COOPr$. XXII. Liq. n_D^{25} 1.4852.[150]

$(EtO)_2P(S)SCH(COOCH_2CH{=}CH_2)CH_2COOCH_2CH{=}CH_2$. XXII. Liq. n_D^{25} 1.5000.[150]

$(RO)_2P(S)SCH(COO\text{-}i\text{-}Bu)CH_2COO\text{-}i\text{-}Bu$. XXII.[150,756] Liq.

	b./mm	n_D^{20}	d^{20}
Me	127.5-128°/0.03	1.4835	1.1483[756]
Et	124-128°/0.02	1.4775	1.1098
	n_D^{25}	1.4793[150]	
Pr	143-145°/0.05-0.06	1.4822	1.0957[756]

$(EtO)_2P(S)SCH(COOBu)CH_2COOBu$. XXII.[150,1055] Liq. n_D^{25} 1.4814,[150] n_D^{35} 1.4814.[1055] Also higher homologs.[150]

$(MeO)_2P(S)SCH(COOMe)CHClCOOMe$. XVII. Liq. $b_{0.02}$ 135-140°.[410]

$(EtO)_2P(S)SCH(COOMe)CHClCOOMe$. XVII. Liq. $b_{0.03}$ 145-151°.[410]

$(MeO)_2P(S)SCH(COOEt)CHClCOOEt$. XXII. Liq. $b_{0.18}$ 136°, n_D^{20} 1.5110, d^{20} 1.2902.[741]

$(EtO)_2P(S)SCH(COOEt)CHClCOOEt$. XXII. Liq. $b_{0.15}$ 125-128°, n_D^{20} 1.5010, d^{20} 1.2151.[741]

$(MeO)_2P(S)SCHCOOEt$. XXII. Liq. n_D^{25} 1.5321.[1024a]
$(MeO)_2P(S)SCHCOOEt$ (the two CH groups joined by a bond)

$(EtO)_2P(S)SBu$. XVII.[807,1070] XXVIII.[795,941] XXXIII.[802] Liq. $b_{0.01}$ 60-70°,[802,807] b_6 124-126°,[941] b_{12} 145°,[795] n_D^{20}

1.4961,[795] 1.4970,[1070] n_D^{25} 1.495,[807] d_4^{20} 1.0664,[941] 1.0766.[1070]

$(PrO)_2P(S)SBu$. XXXIII. Liq. b_1 87-88°, n_D^{20} 1.4920, d_4^{20} 1.0377.[1237]

$(CH_2=CHCH_2O)_2P(S)SBu$. XVII. Liq. $b_{0.0001}$ 46°, n_D^{20} 1.5156, d_4^{20} 1.0760.[1445]

$(HC{\equiv}CCH_2O)_2P(S)SBu$. XVII. Liq. $b_{0.0001}$ 71°, n_D^{20} 1.5312, d_4^{20} 1.1382.[1447]

$((EtO)_2P(S)S)_2(CH_2)_4$. XVII. Liq. $b_{0.22}$ 204°, n_D^{20} 1.5335, d_4^{20} 1.1783.[1243]

$((RO)_2P(S)S)_2CHCH_2CH_2Me$. XXI. Liq. R, n_D^{20}, d_4^{20}: Et, Solid, m. 60-61°; i-Pr, 1.5148, 1.1079; Pr, 1.5161, 1.1099.[75]

$(RO)_2P(S)SCH_2CH_2COMe$. XVII.[1336] XXII.[445] Liq.

	b./mm	n_D^{20}	d_4^{20}
Et	114-118°/0.12	1.4990	1.1300[1336]
	n_D^{25}	1.5074[445]	
i-Pr	125-127°/0.08	1.4980	1.1102[1336]
Pr	130-131.5°/0.2	1.5036	1.1162
I-Bu	135-138°/0.22	1.4968	1.1151
Bu	--	1.5010	1.0870

$(RO)_2P(S)SCH_2COCH_2Me$. XVII. Liq.

	b./mm	n_D^{20}	d_4^{20} [1336]
Et	93-96°/0.05	1.5091	1.1987
i-Pr	95-97°/0.03	1.4970	1.0973
Pr	111-113°/0.15	1.5035	1.1078

$(EtO)_2P(S)SCH_2CH_2CH_2CONEt_2$. XVII. Liq. n_D^{20} 1.5125, d_4^{20} 1.1136.[1425]

$(EtO)_2P(S)SCH_2CH_2CH_2COOEt$. XVII. Liq. $b_{0.5}$ 140° n_D^{20} 1.5000 d_4^{20} 1.1446.[1425]

$(EtO)_2P(S)SCH_2CH_2CH_2COOC_6H_4Cl$-4. XVII. Liq. n_D^{20} 1.5490, d_4^{20} 1.2486.[1425]

$(EtO)_2P(S)SCH_2CH_2CH_2COSEt$. XVII. Liq. $b_{0.4}$ 145°, n_D^{20} 1.530 d_4^{20} 1.1640.[1425]

$(EtO)_2P(S)SCH_2CH_2CH_2COSC_6H_4Cl$-4. XVII. Liq. n_D^{20} 1.5870, d_4^{2} 1.2781.[1425]

(RO)$_2$P(S)S-butenyls

$(CH_2=CHCH_2O)_2P(S)SCH_2CMe=CH_2$. XVII. Liq. $b_{0.0001}$ 54°, n_D^{20} 1.5302, d_4^{20} 1.0975.[1447]

$(HC\equiv CCH_2O)_2P(S)SCH_2CMe=CH_2$. XVII. Liq. $b_{0.0001}$ 79°, n_D^{20} 1.5460, d_4^{20} 1.1609.[1447]

$(CH_2=CMeCH_2O)_2P(S)SCH_2CMe=CH_2$. XVII. Liq. $b_{0.0001}$ 73°, n_D^{20} 1.5230, d_4^{20} 1.0651.[1447]

$(MeCH=CHCH_2O)_2P(S)SCH_2CMe=CH_2$. XVII. Liq. $b_{0.0001}$ 82°, n_D^{20} 1.5282, d_4^{20} 1.0696.[1447]

$(PhO)_2P(S)SCH_2CMe=CH_2$. XVII. Liq. b_1 182-183°, n_D^{20} 1.5960, d_4^{20} 1.192.[383]

$(RO)_2P(S)SCH_2CH=CHMe$. XXII. Liq.

	b./mm	% 1,4 adduct	% 4,1*adduct.[921]
Me	77-79°/0.2	94	6
Et	--	93	7
i-Bu	95-97°/0.1	100	-

* $(RO)_2P(S)SCHMeCH=CH_2$. Product ratio estimated by NMR. For R = Me, a series of experiments with free radical initiator (methylhydroquinone) or cationic catalyst ($HClO_4$), the latter catalyzed addition and/or rearrangement to the 1,4 adduct.[920,921]

$(PhO)_2P(S)SCH_2CH=CHMe$. XVII. Liq. b_1 178-179°, n_D^{20} 1.6010, d_4^{20} 1.193, IR.[383]

$(RO)_2P(S)SCH_2CCl=CHMe$. XXII. Liq.

	b./mm	% 1,4 adduct	% 4,1*adduct.[921]
Me	80-82°/0.05	87	13
Et	89-92°/0.05	83	17
i-Pr	89-92°/0.01	82	18

*$(RO)_2P(S)SCHMeCCl=CH_2$. Product ratios estimated by NMR.

$(RO)_2P(S)SCH_2CH=CMeCl$. XVII. Liq.

	b./mm	n_D^{20}	d^{20} [726]
Me	115-116°/4	1.5457	1.2631
Et	113-114°/3	1.5293	1.1917
i-Pr	132-133°/5	1.5138	1.1284

$(MeO)_2P(S)SCCl=CMeCH_2Cl$. XXII. Liq. $b_{0.04}$ 118-123°.[415]

$(EtO)_2P(S)SCCl{=}CMeCH_2Cl$. XXII. Liq. $b_{0.03}$ 108-116°.[415]

$(EtO)_2P(S)SCH_2CH_2CF{=}CF_2$. XVII. Liq. n_D^{30} 1.4730.[129a]

$(MeOCH_2CH_2O)_2P(S)SCH_2CH_2CF{=}CF_2$. XVII. Liq. n_D^{30} 1.4782.[129a]

$(EtO)_2P(S)SCH_2CH{=}CHCN$. XVII. Liq. $b_{0.01}$ 108-109°.[290]

$(EtO)_2P(S)SCH_2CMe{=}CHCN$. XVII. Liq. $b_{0.01}$ 112°.[290]

$(EtO)_2P(S)SCH(CH_2OH)CH{=}CH_2$. XI. Liq. $b_{0.04}$ 123-124°, n_D^{20} 1.4918, d^{20} 1.0974.[993]

$(PrO)_2P(S)SCH(CH_2OH)CH{=}CH_2$. XI. Liq. $b_{0.02}$ 112-116°, n_D^{20} 1.4855, d^{20} 1.0723.[993]

$(MeO)_2P(S)SCMe{=}CHCOOMe$. XXII. Liq. $b_{0.02}$ 128-130°.[83,174]

$(EtO)_2P(S)SCMe{=}CHCOOMe$. XXII. Liq. $b_{0.08}$ 108-115°.[83,174]

<u>$(RO)_2P(S)SAm$</u>

$(EtO)_2P(S)SCMe(COOEt)CH_2COOEt$. XVII. Liq. $b_{0.07}$ 127°, n_D^{20} 1.5113, d^{20} 1.1911.[1363]

$(MeO)_2P(S)SCH(COOEt)CH(CN)Me$. XXII. Liq. $b_{0.5}$ 145-149°.[1279]

$(EtO)_2P(S)SCHMeCH(COOEt)_2$. XXII. Liq. $b_{0.3}$ 149-150°, n_D^{20} 1.4910, d^{20} 1.1587.[986] Rate of formation.[986]

$(EtO)_2P(S)SCMePrN{=}NCMePrSP(S)(OEt)_2$. XVII. Solid. m. 61°.[689]

$(MeO)_2P(S)SCEt_2N{=}NCEt_2SP(S)(OEt)_2$. XVII. Solid. m. 86-88°.[689]

$(RO)_2P(S)SCH(COOMe)CHMeCOOMe$. XXII. Liq.

	b./mm	n_D^{20}	d^{20}.[741,1362]
Me	137°/0.22	1.5240	1.2737
Et	117-119°/0.07	1.5042	1.2004
i-Bu	141°/0.1	1.4880	1.1301
Bu	120-122°/0.1	1.4950	1.0910

Also several analogs.[741,1362]

$(EtO)_2P(S)SCH(COOEt)CHMeCOOEt$. XXII. Liq. $b_{0.12}$ 120-122°, n_D^{20} 1.4950, d^{20} 1.2038.[1363] Also several analogs.[1363]

$(RO)_2P(S)SCHMeCH_2COMe$. XXII. Liq. R, n_D^{25}: Et, 1.5087; i-Pr, 1.4998; $Me(CH_2)_{13}$, 1.4805; Ph, 1.5878.[445]

$(MeO)_2P(S)SCH_2CH(COOEt)COMe$. XXII. Liq. n 1.4960.[1294] Also several homologs and analogs.

$(CH_2{=}CHCH_2O)_2P(S)S\text{-}i\text{-}Am$. XVII.[910,1445] Liq. $b_{0.0001}$ 47°,

[1445] n_D^{20} 1.5113,[1445] 1.5238,[910] d_4^{20} 1.0582,[1445] 1.0878.[910]

$(CH_2=CHCH_2O)_2P(S)SAm$. XVII. Liq. $b_{0.0001}$ 69°, n_D^{20} 1.5118, d_4^{20} 1.0595.[1447]

$(HC\equiv CCH_2O)_2P(S)SAm$. XVII. Liq. $b_{0.0001}$ 80°, n_D^{20} 1.5262, d_4^{20} 1.1167.[1447]

$(i\text{-}AmO)_2P(S)SAm$. XVII. Liq. b_1 147-148°, n_D^{20} 1.4845, d_4^{20} 0.9872.[1070]

$(PhO)_2P(S)SAm$. VIII. Liq. $b_{0.0008}$ 110-115°, n_D^{25} 1.5698.[803]

$(RO)_2P(S)S(CH_2)_4CCl_3$. XVII. Liq.

	b./mm	n_D^{20}	d_4^{20}.[1070]
Et	164-165°/1	1.5270	1.2895
i-Pr	151-152°/1	1.5135	1.2282
i-Bu	168-169°/1.5	1.5110	1.1932
i-Am	179-180°/1	1.5092	1.1742
$CH_2CHEtBu$	--	1.5010	1.1032
$C_{16}H_{33}$	Solid. m. 61.5-62.5°		

$(RO)_2P(S)SCHMeCH=CH=CHMe$. XXII. Liq.

	b./mm	% 1,4 adduct	% 4,1* adduct[921]
Me	80-84°/0.1	92	(8)
Et	81-83°/0.01	73	(27)

* $(RO)_2P(S)SCH_2CH=CHCH_2Me$.

$(RO)_2P(S)SCH_2CHMeCH=CH_2$ (*). XXII. Liq.

	b./mm	n_D^{20}	d_4^{20}.[1444]
Et	76-80°/0.04	1.5184	1.4610
Pr	--	1.5072	1.436
i-Bu	--	1.5015	1.429

(*) 1,2 or 3,4 addition to isoprene not established. IR data.

<u>$(RO)P(S)SC_6$ - Higher Alkyls</u>

$(EtO)_2P(S)SCHEtCH(COOEt)_2$. XXII. Liq. $b_{0.3}$ 152-153°, n_D^{20} 1.4902, d^{20} 1.1555.[986] Rate of formation.[986]

$(MeO)_2P(S)SCMe_2CH_2COMe$. XXII. Liq. $b_{0.05}$ 93-96°, n_D^{20} 1.5265, d^{20} 1.1979.[756]

$(EtO)_2P(S)SC(CN)(COOMe)CHMe_2$. XXII. Liq. $b_{0.06}$ 73-76°.[860]

$(EtO)_2P(S)SC(CN)(COOMe)CH_2CH_2Me$. XXII. Liq. $b_{0.06}$ 84-88°.[860]

$(EtO)_2P(S)SCH(CONH_2)CH_2CH_2CH(CONH_2)SP(S)(OEt)_2$. XVII. Solid m. 145° (from EtOH).[1105]

$(EtO)_2P(S)SCH_2COC(CN)Me_2$. XVII. Liq. $b_{0.8}$ 175-185°.[407]

$(MeO)_2P(S)SC_6H_{13}$. XXII. Liq. $b_{2.2}$ 110°, n_D^{20} 1.5128, d^{20} 1.0967.[573]

$(CH_2{=}CHCH_2O)_2P(S)SC_6H_{13}$. XVII. Liq. $b_{0.0001}$ 80°, n_D^{20} 1.5100 d_4^{20} 1.0465.[1447]

$(MeO)_2P(S)SCH{=}CHCMe(OH)CH{=}CH_2$. XXII. Liq. n_D^{20} 1.5685.[1289] Also several homologs.[1289]

$(EtO)_2P(S)SC(C{\equiv}CH)(Me)CH{=}CH_2$. XXII. Liq. n_D^{20} 1.5460.[1289]

$(MeO)_2P(S)SCMe(CH{=}CH_2)C{\equiv}CH$. XXII. Liq. n_D^{20} 1.5684.[1289] Also several homologs.[1289]

$(MeO)_2P(S)SC(CH{=}CH_2){=}CHCH{=}CH_2$. XXII. Liq. n_D^{20} 1.5670.[289] Also several homologs.[1289]

$(RO)_2P(S)SC({=}CHCH{=}CH_2)CH_2CH_2SP(S)(OR)_2$. XXII. Liq. R, n_D^{20}: Me, 1.5724; Et, 1.5494; cyclo-C_6H_{11}, 1.5560.[1289]

$(MeO)_2P(S)SC_7H_{15}$. XXII. Liq. $b_{1.4}$ 114-115°, n_D^{20} 1.5112, d^{20} 1.0905.[573]

$(EtO)_2P(S)SCH_2CH_2CH_2COC(CN)Me_2$. XVII. Liq. $b_{0.2}$ 140°.[407]

$(EtO)_2P(S)SCHMe(CH_2)_5Me$. XVII. XXII. Liq. $b_{0.04}$ 68-70°, n_D^{20} 1.4918, d_{20}^{20} 1.0187.[890]

$(PrO)_2P(S)SCHMe(CH_2)_5Me$. XVII. XXII. Liq. $b_{0.03}$ 80°, n_D^{20} 1.4893, d_{20}^{20} 0.9970.[890]

$(BuO)_2P(S)SCHMe(CH_2)_5Me$. XVII. XXII. Liq. $b_{0.03}$ 98-99°, n_D^{20} 1.4870, d_{20}^{20} 0.9830.[890]

$(MeO)_2P(S)SC_8H_{17}$. XXII. Liq. $b_{1.1}$ 118-119°, n_D^{20} 1.5090, d^{20} 1.0840.[573]

$(EtO)_2P(S)SC_8H_{17}$. VIII, XVII. Liq. $b_{0.02}$ 75-76°, n_D^{20} 1.4925 d_{20}^{20} 1.0184.[890]

$(MeO)_2P(S)SCH(i\text{-}Pr)CH{=}CMe_2$. XXII. Liq. $b_{0.05}$ 80-83°. 90% 1,2 adduct by NMR.[921]

$(MeO)_2P(S)SC_9H_{19}$. XXII. Liq. $b_{1.3}$ 128°, n_D^{20} 1.4978, d^{20} 1.0478.[573]

$(BuO)_2P(S)SC_9H_{19}$. XVII. Liq. b_1 173-174°, n_D^{20} 1.4681, d_4^{20} 0.9767.[1070]

$(EtO)_2P(S)S(CH_2)_8CCl_3$. XVII. Liq. b_1 184-185°, n_D^{20} 1.5140, d_4^{20} 1.2010.[1070]

$(MeO)_2P(S)SC_{10}H_{21}$. XXII. Liq. $b_{1.2}$ 132°, n_D^{20} 1.4944, d^{20} 1.0329.[573]

<u>$(RO)_2P(S)S$-4- and 5-ring</u>

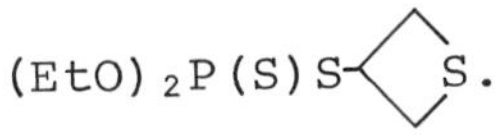

$(EtO)_2P(S)S$⟨◇⟩S. XVII. Liq. $b_{0.07}$ 100.5-102°.[898]

$(RO)_2P(S)S$—⟨⬠⟩ XXII. Liq. R, $b_{0.3}$ *: Me, 94-96°; Et, 110-111°; i-Pr, 115-117°. * Some decomp. during distillation.[921]

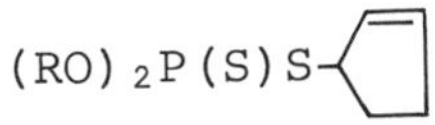

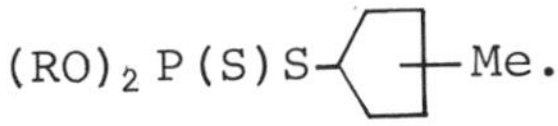

$(RO)_2P(S)S$—⟨⬠$-$Me⟩. XXII. Liq.

	b./mm	n_D^{20}	d^{20}.[571]
Et	110-112°/0.1	1.5141	1.1049
i-Pr	118-120°/0.5	1.5055	1.0661
Pr	75-76°/0.1	1.4940	1.0803
i-Bu	116-120°/0.1	1.5024	1.0474
Bu	118-120°/0.2	1.4957	1.0302

$(MeO)_2P(S)S$—[⟩=O, NEt, =O]. XXII. Solid. m. 59-60°.[446] Liq. n_D^{20} 1.5210.[348]

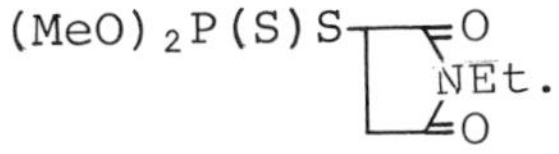

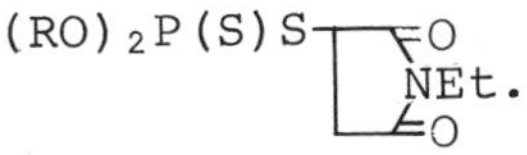

$(RO)_2P(S)S$—[⟩=O, NEt, =O]. XXII. R = Et. Liq. n_D^{20} 1.4925.[348] R = i-Pr. Solid. m. 40°. R = Pr. Solid. m. 47°.

$(EtO)_2P(S)S$—[⟩=O, NCH_2CH=CH_2, =O]. XXII. Liq. n_D^{25} 1.5312.[446]

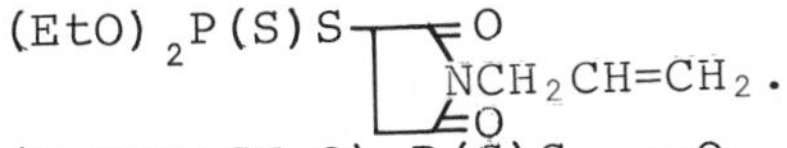

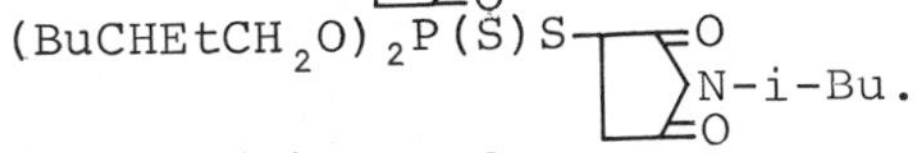

$(BuCHEtCH_2O)_2P(S)S$—[⟩=O, N-i-Bu, =O]. XXII. Liq. n_D^{25} 1.4958.[446]

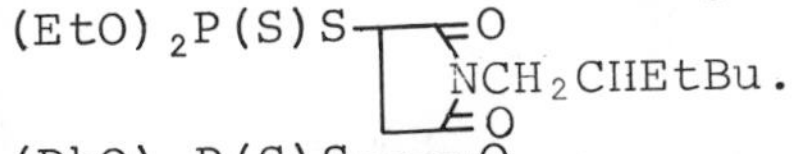

$(EtO)_2P(S)S$—[⟩=O, NCH_2CHEtBu, =O]. XXII. Liq. n_D^{25} 1.5083.[446]

$(PhO)_2P(S)S$—[⟩=O, NPh, =O]. XXII. Liq. n_D^{25} 1.5055.[446]

$(EtO)_2P(S)S$—[⟩=O, NC_6H_4R-4, =O]. XXII. Solids. R, m.: Cl, 90.5-

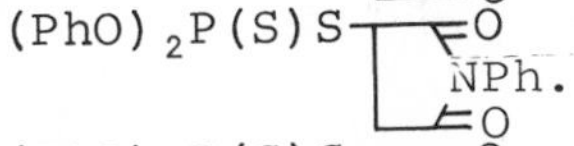

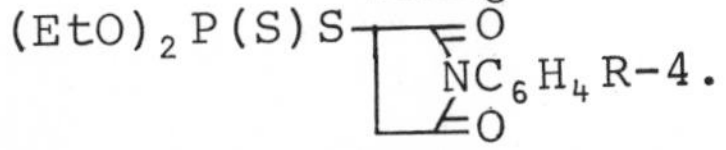

92°; NO_2, 118.5-119.5°; SO_2NH_2, 199-200° (dec.).[446]

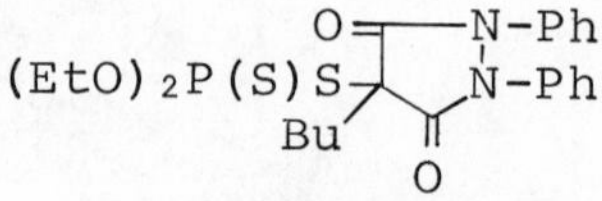

XVII. Solid. m. 129-130° (from alcohol).[938]

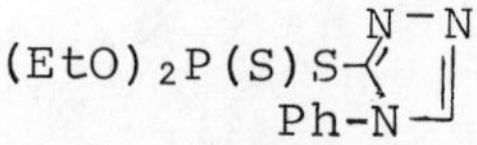

VIII. Liq. n_D^{23} 1.471.[1310]

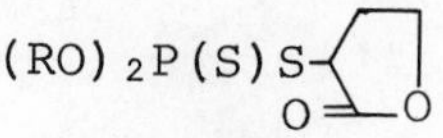

XVII. Liq. R, n_D^t: Me, $1.5432^{20°}$; Et, $1.5302^{14°}$; Pr, $1.5159^{19°}$.[470]

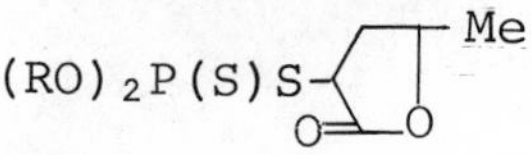

XVII. Liq. R, n_D^t: Me, $1.5311^{25°}$; Et, $1.5198^{23°}$; Pr, $1.5090^{21°}$.[471]

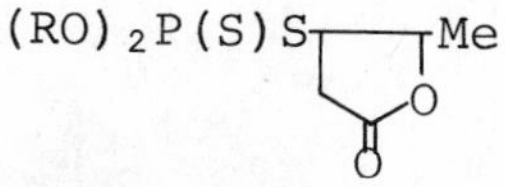

XXII. Liq. R, n_D^t: Me, $1.5326^{21°}$; Et, $1.5199^{21°}$; Bu, $1.5028^{24°}$.[338]

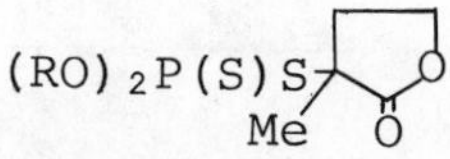

XVII. Liq. R, n_D^t: Me, $1.5221^{24°}$; Et, $1.5167^{22°}$.[471]

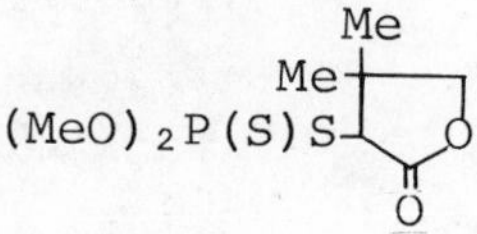

XVII. Liq. n_D^{25} 1.5362.[471]

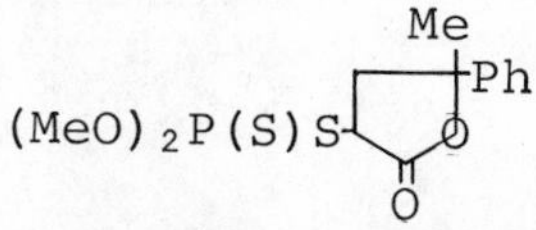

XVII. Liq. n_D^{20} 1.5347.[471]

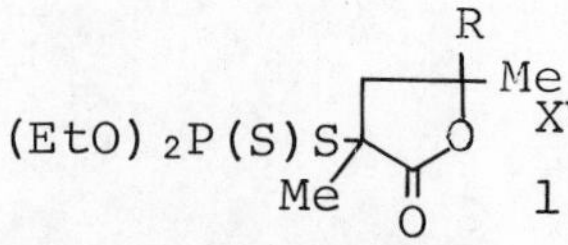

XVIII. Liq. R, n_D^t: H, $1.5203^{21°}$; Me, $1.5196^{20°}$.

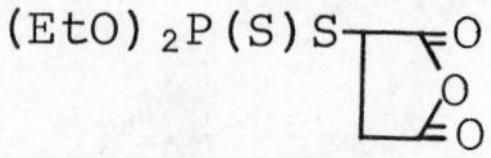

XXII. Brown liq. Refluxing with aqueous HCl gave 83% of $HSCH(COOH)CH_2COOH$.[190]

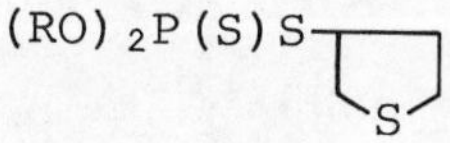

XXII. Liq. R, $b_{0.02}$: Me, 50-55°; Et, 80-82°.[378]

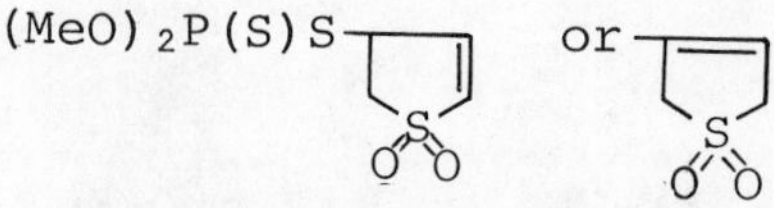

XVII. Solid. m. 66-67° (from MeOH).[738]

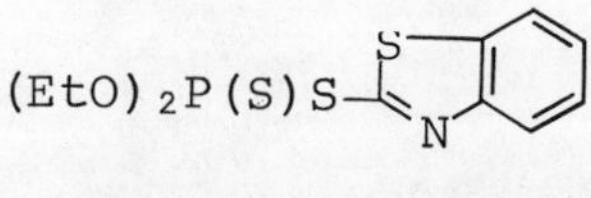

VIII. Solid. m. 46-48°.[1278]

<u>$(RO)_2P(S)S$-6-ring</u>

$(MeO)_2P(S)SPh$. XV.[813] XIX.[1161,1182] XXVII. (from $(MeO)_2P(S)$ SH plus PhN_2X.[742] Liq. $b_{0.04}$ 93-94°,[813] $b_{0.1}$ 95-97°,[742] n_D^{20} 1.5927,[742] 1.5911,[813] d_4^{20} 1.2460,[742] d^{20} 1.2529,[813] IR.[813]

$(EtO)_2P(S)SPh$. XIX.[15] XXVII.[25,94,742] XXXIII.[802,807] Liq. $b_{0.01}$ 88-95°,[802] $b_{0.4}$ 109°,[25] $b_{1.4}$ 140-145°,[807] n_D^{20} 1.5629,[742] 1.5660,[25] n_D^{25} 1.558,[802,807] d^{20} 1.1770,[25] d_4^{20} 1.1823.[742]

$(i\text{-}PrO)_2P(S)SPh$. XV.[813] XXVII.[743] Liq. $b_{0.04}$ 102.5-104°,[743] 106-107°,[813] n_D^{20} 1.5440,[813] 1.5487,[743] d^{20} 1.1174,[743] 1.1188.[813] IR.[813]

$(PrO)_2P(S)SPh$. XV.[813] XXVII.[743] Liq. $b_{0.04}$ 115-116°,[813] $b_{0.1}$ 125.5-127.5,[743] n_D^{20} 1.5487,[743] 1.5532,[813] d^{20} 1.1262,[743] 1.1330.[813] IR.[813]

$(MeO)_2P(S)SC_6H_4Br$-4. XIX. Solid. $b_{0.01}$ 110°, m. 62°.[1161]

$(EtO)_2P(S)SC_6H_4Br$-4. XIX. Liq. $b_{0.5}$ 144°, n_D^{20} 1.5910, d^{20} 1.4280.[15]

$(MeO)_2P(S)SC_6H_4Cl$-2. XIX. Solid. m. 81°.[1161]

$(EtO)_2P(S)SC_6H_4Cl$-2. XXV.[32] Liq. $b_{0.3}$ 128°, n_D^{20} 1.5340, d^{20} 1.2050.[32]

$(EtO)_2P(S)SC_6H_4Cl$-3. XXVII.[94,742] Liq. $b_{0.08}$ 104.5-106.5°,[742] $b_{0.4}$ 142-145°,[94] n_D^{20} 1.5730, d_4^{20} 1.2538.[742]

$(MeO)_2P(S)SC_6H_4Cl$-4. IX.[1253] XIX.[1161] XXVII.[743] Liq. Solid. $b_{0.01}$ 97°,[1161] m. 42-43°,[1161] $b_{0.15}$ 115.5-116°,[743] n_D^{20} 1.5948, n_D^{25} 1.6013,[1253] d^{20} 1.3235,[743] d^{25} 1.3918.[1253]

$(EtO)_2P(S)SC_6H_4Cl$-4. IX.[1253] XIX.[15,20] XXVII.[742] Liq. $b_{0.2}$ 122-123°,[742] 128°,[20] n_D^{20} 1.5742,[20] 1.5750,[742], n_D^{25} 1.5697,[1253] d^{20} 1.2578,[20] d_4^{20} 1.2627,[742] d^{25} 1.2455.[1253]

$(i\text{-}PrO)_2P(S)SC_6H_4Cl$-4. XXVII. Liq. $b_{0.12}$ 121.5-123.5°, n_D^{20} 1.5612, d^{20} 1.1904.[743]

$(MeO)_2P(S)SC_6H_4F$-4. XIX.[1161] Liq. $b_{0.01}$ 72°, n_D^{21} 1.5726.[1161]

$(EtO)_2P(S)SC_6H_4F$-4. XIX. Liq. $b_{0.1}$ 111°, n_D^{20} 1.5462, d^{20} 1.2310.[15]

$(EtO)_2P(S)SC_6H_3Br_2$-2,5. XIX. Liq. $b_{0.3}$ 146°, n_D^{20} 1.6310.[24]

$(EtO)_2P(S)SC_6H_3Cl_2$-2,5. IX.[1253] XIX.[17] $b_{0.25}$ 133-134°,[17] n_D^{20} 1.5822,[17] n_D^{25} 1.5663,[1253] d^{20} 1.3343,[17] d^{25} 1.3234.[1253]

$(EtO)_2P(S)SC_6H_3Cl_2$-3,4. XIX. Liq. $b_{0.25}$ 138°, n_D^{20} 1.5859, d^{20} 1.3379.[17]

$(EtO)_2P(S)SC_6H_2Cl_3$-2,4,5. IX. Liq. n_D^{25} 1.5511, d^{25} 1.3422.[1253]

$(MeO)_2P(S)SC_6Cl_5$. XIX.[1411] Solid. m. 99-101°.[1411] ^{31}P -90.6.[715,1274]

$(EtO)_2P(S)SC_6Cl_5$. XIX.[1411] Solid. m. 84-86°.[1411] ^{31}P -85.6.[715,1274]

(i-PrO)$_2$P(S)SC_6Cl_5. XIX. Solid. m. 94-96°.[1411]

$(PrO)_2P(S)SC_6Cl_5$. XIX. Liq. n_D^{20} 1.5948, d_4^{20} 1.4758.[1411]

$(MeO)_2P(S)SC_6H_4NH_2$-4. XIX. Solid. m. 79°.[618]

$(EtO)_2P(S)SC_6H_4NMe_2$-4. XIX. Liq. $b_{0.01}$ 152°.[618]

$(EtO)_2P(S)SC_6H_4NO_2$-2. IX.[291] XIX.[28,612] XX.[28] Solid. $b_{0.05}$ 105-108°,[291] m. 31.5-32°,[28] 32°.[612]

$(EtO)_2P(S)SC_6H_4NO_2$-3. XIX. Liq. $b_{0.01}$ 117°.[612]

$(MeO)_2P(S)SC_6H_4NO_2$-4. XIX. Solid. m. 102°.[618]

$(EtO)_2P(S)SC_6H_4NO_2$-4. VIII.[520] IX.[291] XIX.[28,612,618] XX.[28] XXVII.[742] Solid. $b_{0.1}$ 137-139°,[742] m. 49-50°.[291,612,618]

$(RO)_2P(S)S$- 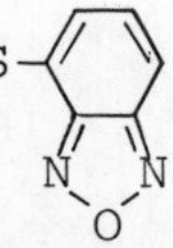 XVII. Solids. R, m.: Me, 106-107°, Et, 99-100°.[1229]

$(RO)_2P(S)S$ 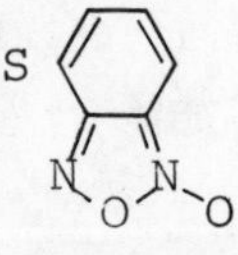 XVII. Solids. R, m.: Me, 105.5-106°; Et, 70-71°.[1229]

$(EtO)_2P(S)S$- 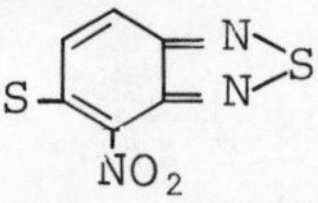 XVII. Solid. m. 66-67° (from alcohol).[938]

$(RO)_2P(S)SC_6H_3(NO_2)_2$-2,4. XVII. Liq. and solids. R, b./mm or m.: Et, 139-140°/0.001; i-Pr, 29-31°; Pr, 46-48°; Bu, 48-50°; Am, 150°/0.04.[1452]

$(RO)_2P(S)SC_6H_3(NO_2)_2$-2,6. XVII. Solids. R, m.: Et, 47-49°;

i-Pr, 64-66°; Pr, 36-38°; Bu, 52-53.5°; Am, Liq. $b_{0.001}$ 135-140°.[1452]

$(ArO)_2P(S)SC_6H_3(NO_2)_2$-2,6. XXVII. Solids. Ar, m.: Ph, 84-85°; 4-ClC_6H_4, 92-94°; 3-MeC_6H_4, 107-109°; 4-MeC_6H_4, 91-92.5°.[370]

$(EtO)_2P(S)SC_6H_4(OMe)$-2. IX Liq. b_3 140-170°.[1126]

$(MeO)_2P(S)SC_6H_4(OMe)$-4. XIX. Liq. $b_{0.01}$ 105°.[618,1161,1182]

$(EtO)_2P(S)SC_6H_4(OMe)$-4. VIII.[1081] XIX.[15] XXVII.[94,743] Liq. $b_{0.15}$ 129.5-130°,[743] $b_{0.4}$ 154-158°, n_D^{20} 1.5680,[743] 1.5690,[15] d^{20} 1.2003,[743] 1.2010.[15]

$(EtO)_2P(S)SC_6H_4(OEt)$-2. IX. Liq. b_1 155°.[1126]

$(EtO)_2P(S)SC_6H_4(OEt)$-4. IX. Liq. b_1 175°.[1126]

$(EtO)_2P(S)SC_6H_4(SMe)$-4. XIX. Liq. $b_{0.02}$ 140°.[296]

$(RO)_2P(S)S$-cyclo-C_6H_{11}. XXII. Liq.

	b./mm	n_D^{20}	d^{20}.[571]
Et	133-134°/13	1.5231	1.1121
i-Pr	106-107°/0.1	1.5095	1.0662
Pr	105-106°/1	1.5155	1.0779
i-Bu	135-138°/0.4	1.5066	1.0409
Bu	134-135°/0.1	1.5090	1.0506

$(MeO)_2P(S)SC_6H_4Me$-2. XIX.[1161] XXVII.[742] Solid. m. 83°,[1161] liq. $b_{0.1}$ 101-102.5°, n_D^{20} 1.5660, d_4^{20} 1.1644.[742]

$(EtO)_2P(S)SC_6H_4Me$-2. XXVII.[94,742] Liq. $b_{0.2}$ 118-118.5°,[742] $b_{0.2}$ 126-128°,[94] n_D^{20} 1.5642, d_4^{20} 1.1696.[742]

$(MeO)_2P(S)SC_6H_4Me$-3. XXVII. Liq. $b_{0.1}$ 102.5-102.8°, n_D^{20} 1.5662, d_4^{20} 1.1729.[742]

$(EtO)_2P(S)SC_6H_4Me$-3. XXVII.[94,742] Liq. $b_{0.09}$ 105-107°,[742] $b_{0.3}$ 138°,[94] n_D^{20} 1.5624, d_4^{20} 1.1781.[742]

$(MeO)_2P(S)SC_6H_4Me$-4. XIX.[1161,1182] XXVII.[742] Liq. $b_{0.01}$ 103°,[1161,1182] $b_{0.15}$ 104.5-105°, n_D^{20} 1.5829, d_4^{20} 1.2136.[742]

$(EtO)_2P(S)SC_6H_4Me$-4. XIX.[15] XXVII.[94,742] Liq. $b_{0.12}$ 110-112°,[742] $b_{0.3}$ 130-131°,[94] b_2 150-151°,[15] n_D^{20} 1.5624,[15] 1.5639,[742] d^{20} 1.1723,[15] d_4^{20} 1.1779.[742]

$(PhO)_2P(S)SC_6H_4Me$-4. XXVII. Liq. $b_{0.8}$ 215-218°.[94]

$(MeO)_2P(S)SC_6H_4CN$-2. XIX. Solid. m. 74°.[618]

$(EtO)_2P(S)SC_6H_4CN$-2. XIX. Solid. m. 62°.[618] Also 36 homologs and analogs.[618]

$(MeO)_2P(S)SC_6H_4COOMe$-2 XIX. Solid. m. 67°.[618]

$(EtO)_2P(S)SC_6H_4COOMe$-2. XXVII. Liq. $b_{0.2}$ 144-145°, n_D^{20} 1.5648, d^{20} 1.2330.[743]

$(EtO)_2P(S)SC_6H_3BrMe$-4,2. XIX.[17,24] Liq. $b_{0.1}$ 153°,[17] $b_{0.4}$ 163°,[24] n_D^{20} 1.5882,[17] d_4^{20} 1.3871.[17,24]

$(EtO)_2P(S)SC_6H_3BrMe$-4,3. XIX.[17,24] Liq. $b_{0.15}$ 141.5°,[17] $b_{0.8}$ 160°,[24] n_D^{20} 1.5840,[17] d_4^{20} 1.3822.[17,24]

$(EtO)_2P(S)SC_6H_3BrMe$-5,2. XIX.[17,24] Liq. $b_{0.6}$ 147°,[17] $b_{0.7}$ 148°,[24] n_D^{20} 1.5826,[17] d_4^{20} 1.3809.[17,24]

$(EtO)_2P(S)SC_6H_3ClMe$-4,2. XIX.[24] Liq. $b_{0.4}$ 142°, d_4^{20} 1.2346.[24]

$(EtO)_2P(S)SC_6H_3ClMe$-4,3. XIX. Liq. $b_{0.35}$ 140°, n_D^{20} 1.5719, d^{20} 1.2346.[17]

$(EtO)_2P(S)SC_6H_3ClMe$-5,2. XIX.[17,24] Liq. $b_{0.25}$ 138°,[17] $b_{0.5}$ 147-149°,[24] n_D^{20} 1.5711,[17] d_4^{20} 1.2334.[17,24]

$(MeO)_2P(S)SC_6H_3ClMe$-2,3. XIX. Liq. $b_{0.01}$ 97°.[1161]

$(MeO)_2P(S)SC_6H_3ClMe$-3,4. XIX. Liq. $b_{0.01}$ 106°.[1161]

$(EtO)_2P(S)SC_6H_3$-2-Me-4-NO_2. XIX. Liq. $b_{0.01}$ 130°.[616]

$(MeO)_2P(S)SC_6H_3$-3-Me-4-NO_2. XIX. Solid. m. 35°.[616]

$(EtO)_2P(S)SC_6H_3$-3-Me-4-NO_2. XIX. Liq. $b_{0.01}$ 130°.[616]

$(EtO)_2P(S)SC_6H_3Me_2$-2,4. XIX. Liq. $b_{0.5}$ 155-156°, n_D^{20} 1.5620, d_4^{20} 1.1411.[17,24]

$(EtO)_2P(S)SC_6H_3Me_2$-2,5. XIX. Solid. m. 45°.[17,24]

$(EtO)_2P(S)SC_6H_3Me_2$-3,4. XIX.[17,24] Liq. $b_{0.1}$ 132°,[17] b_1 155°,[24] n_D^{20} 1.5662,[17] d_4^{20} 1.1473.[17,24]

$(MeO)_2P(S)SC_6H_4CONHMe$-2. XXVII. Solid. m. 112-113°.[620]

$(EtO)_2P(S)SC_6H_4CONHMe$-2. XXVII. Solid. m. 76-78°.[620] Also 4 analogs.[620]

$(EtO)_2P(S)S$- 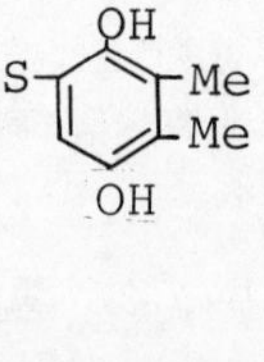 XVII. Solid. m. 90-91°.[357] Also several analogs.[357]

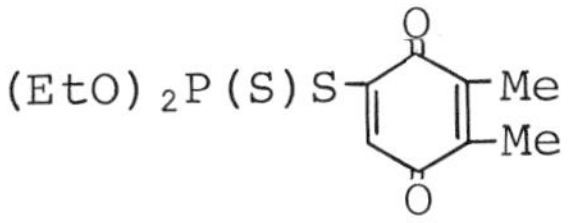

XVII. Solid. m. 61-62°.[357]

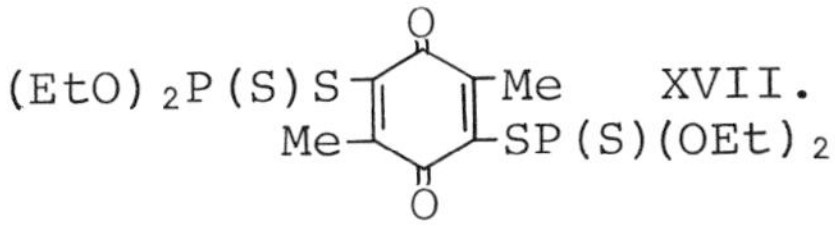

XVII. Solid. m. 51-52° (from alcohol).[358]

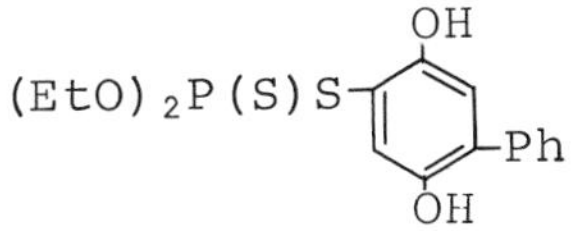

From $(EtO)_2P(S)SH$ plus 2-phenyl-p-quinone. Liq. n_D^{20} 1.6112.[671]

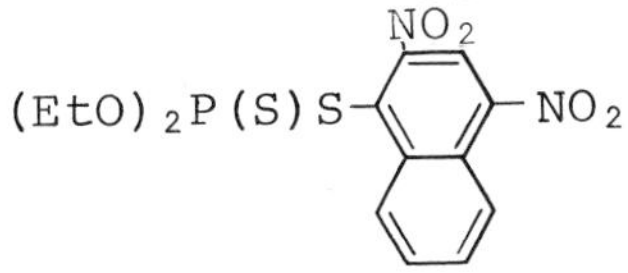

XVII. Solid. m. 133-134° (from alcohol).[938]

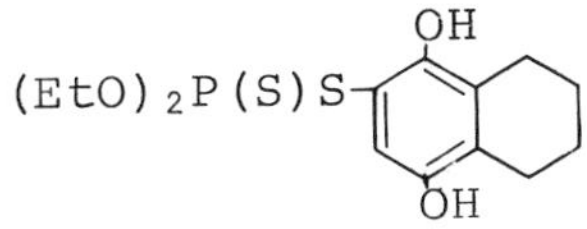

XVII. Solid. m. 102-108°.[357]

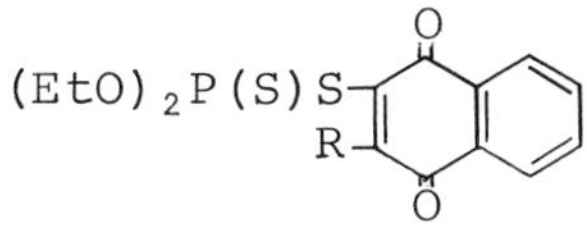

XVII. Solids. R, m. H, 102-102.5° (from alcohol); Me, 103.5-104.5° (from alcohol).[357]

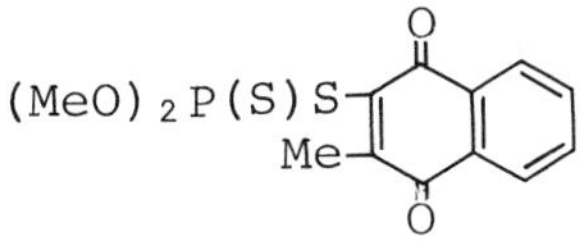

XVII. Solid. m. 94.5-95° (from MeOH).[358]

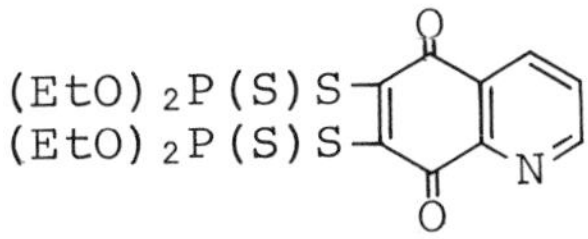

XVII. Solid. m. 103-104° (from EtOAc-petroleum ether).[357,358]

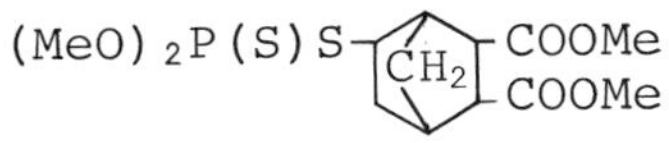

XXII. Solid. m. 88-89°.[160]

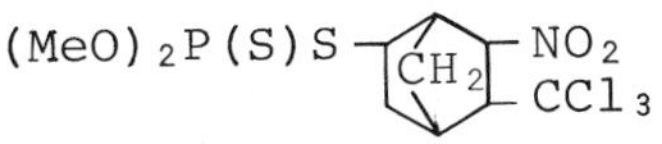

XXII. Solid. m. 70-73°.[160]

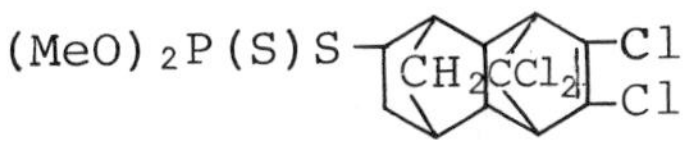

XXII. Solid. m. 108-109°.[160]

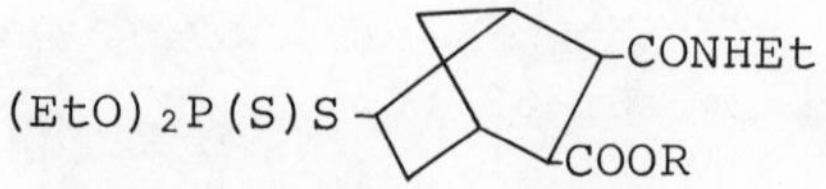

XXII. Liq. R, n_D^{20}: H, 1.5357; Me, 1.5077. Endo-cis configurations.[1288]

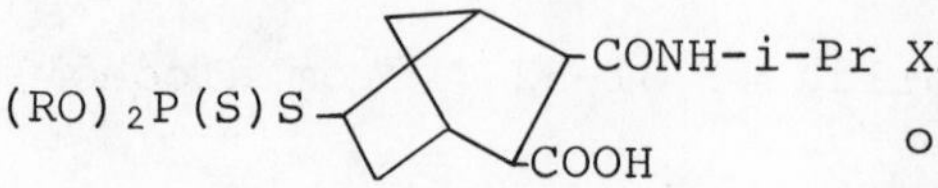

XXII. Liq. or solid. R, n_D^{20}, or m.: Et, 1.5250; i-Pr, 32°.[1288] Endo-cis configurations.[1288]

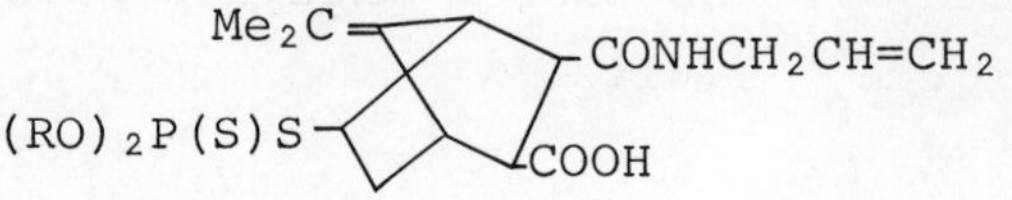

XXII. Solids. R, configuration, m.: Et, endo-cis, 83° i-Pr, exo-cis, 63°.[1288]

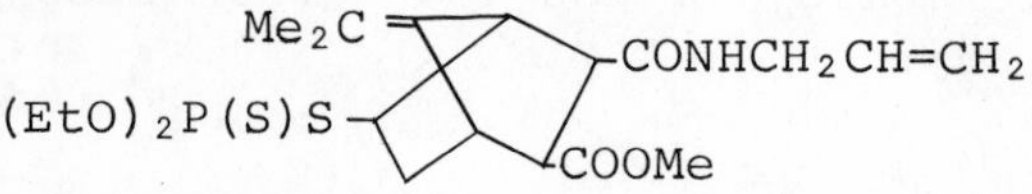

From R-COOH plus CH_2N_2. Solid. m. 70-72°.[1288]

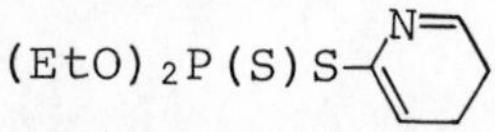

XVII. Liq. $b_{0.01}$ 97-98°.[621] Also 20 analogs.[621]

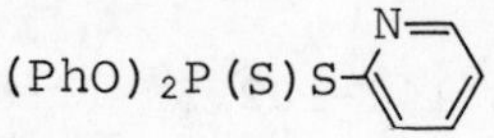

VIII. Oil. λ_{max} 268 mu, 5520.[803]

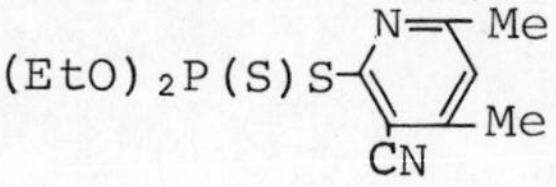

XVII. Solid. m. 56-58°.[258]

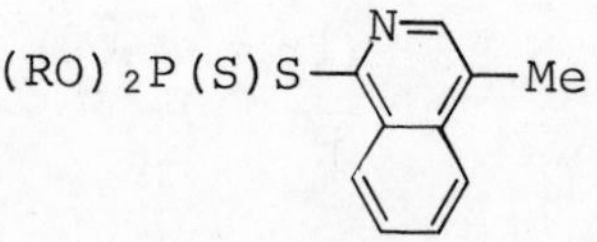

XVII. Solids. R, m. Me, 100°; Et, 63°.[621]

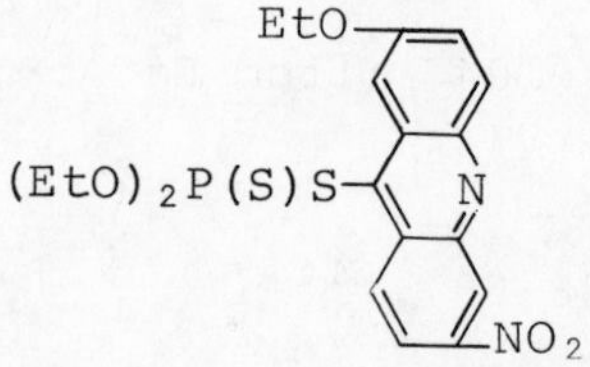

XVII. Solid. m. 164-165° (from acetone)[938]

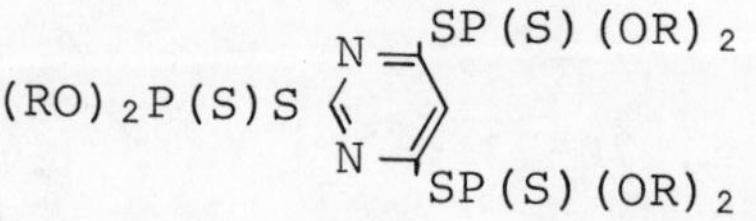

XVII. Solids. R, m.: Et, 46-47° (from EtOH); i-Pr, 65.5° (from EtOH).[30]

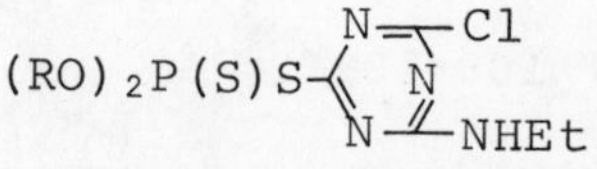

From 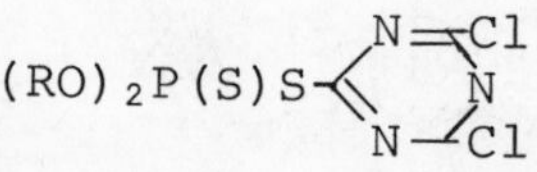 plus $EtNH_2$.

Solids. R, m. (from hexane): Me, 82-84°; Et, 88-90°; i-Pr, 115-116°.[915]

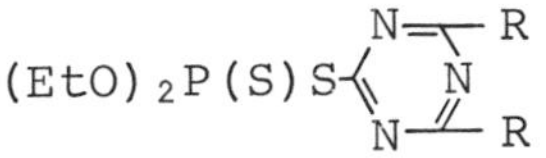

XVII. Liq. and solid. R = H, n_D^{30} 1.5620; R = Me, m. 43-44°.[1247]

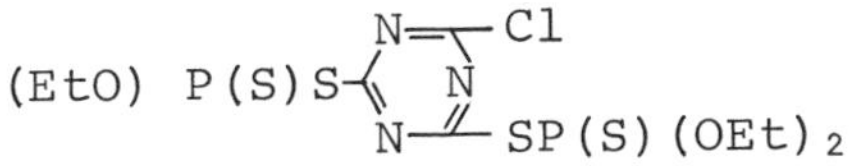

XVII. Solid. m. 41-42° (from light petroleum).[402]

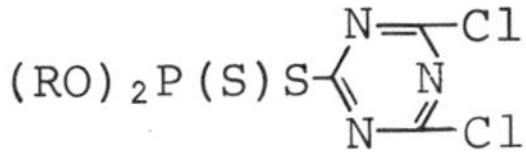

XVII. Solid. R, m. (from hexane): Me, 97-98°; Et, 71°; i-Pr, 39-40°.[915]

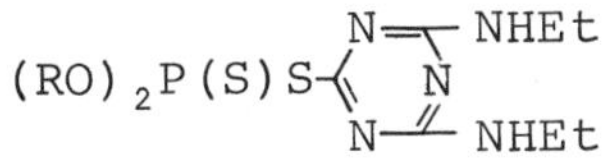

From 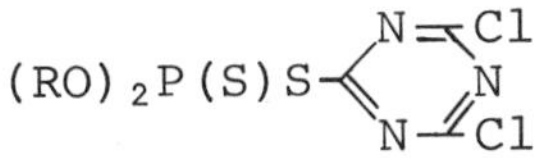 plus $EtNH_2$. Solids. R, m.: Et, 95-97° (from hexane); i-Pr, 84-85° (from hexane).[915]

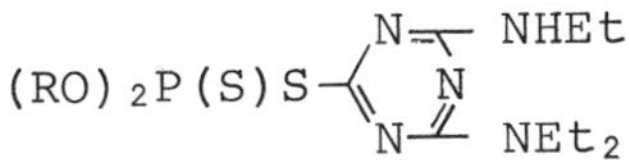

From 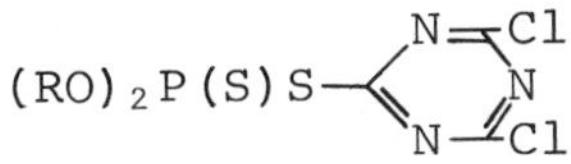 plus $EtNH_2$ followed by Et_2NH. Solids. R, m. (from hexane): Me, 83-84°; Et, 47-49°; i-Pr, 62-64°.[915]

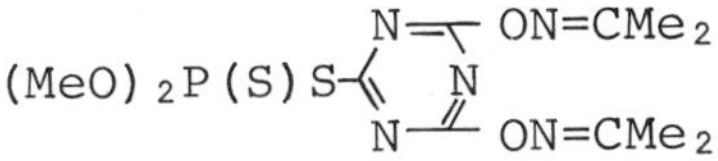

XVII. Solid. m. 108-109°.[170]

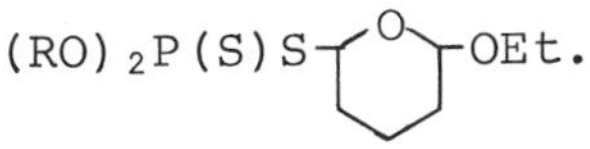

XXII. Liq. R, n_D^{20}: Me, 1.5332; Et, 1.5115.[670]

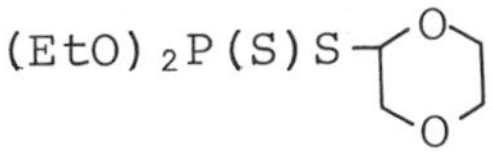

XXII. Liq. n_D^{20} 1.5242.[212]

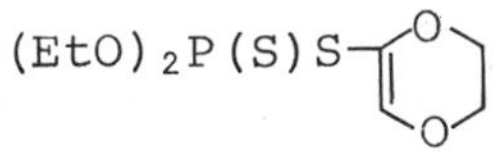

XVII. XXXIII. $b_{0.25}$ 118-122 , $b_{0.3-0.5}$ 136-142°.[210]

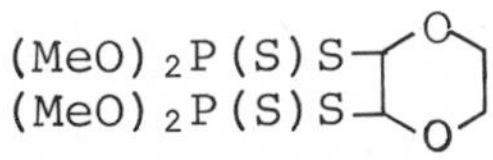

XVII. XXXIII. Cis, liq. Trans, m. 80-81°.[212]

(EtO)$_2$P(S)S–
(EtO)$_2$P(S)S– 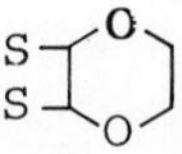 XVII. XXII. Liq. n_D^{20} 1.5409. IR, absorption bands 9.12 μ cis and 8.86 μ trans were used to calculate ratio of isomers, about 2:3.[211,212] ^{1}H-NMR.[74,526] Mass spect.[196] "Dioxathion."

(EtO)$_2$P(S)S– 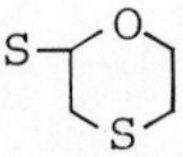 XXII. Liq. n_D^{20} 1.5232.[209a]

(RO)$_2$P(S)S–  –SP(S)(OR)$_2$ XVII. Solids. R, m.: Me, 126-128°; Et, 54-57°.[1385]

(MeO)$_2$P(S)S–  From the sulfide plus H_2O_2. Solid. m. 119-132° (from benzene-hexane).[461]

(EtO)$_2$P(S)S– 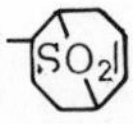 From the sulfide plus H_2O_2. Solid. m. 70-71°.[461]

<u>(RO)$_2$P(S)SCH$_2$-3-ring</u>

(EtO)$_2$P(S)SCH$_2\overset{O}{\widehat{CH—CH_2}}$. From (EtO)$_2$P(S)SCH$_2$CHOHCH$_2$Cl plus KOH. Liq. $b_{0.02}$ 66=66.5°, n_D^{20} 1.4605, d^{20} 1.1113.[993]

(i-PrO)$_2$P(S)SCH$_2\overset{O}{\widehat{CH—CH_2}}$. XVII. Liq. n_D^{20} 1.5180.[194]

(C$_{13}$H$_{27}$O)$_2$P(S)SCH$_2\overset{O}{\widehat{CH—CH_2}}$. XVII. Liq. n_D^{20} 1.4885.[194]

(i-PrO)$_2$P(S)SCH$_2\overset{O}{\widehat{CH—CH}}CH_2$SP(S)(OEt)$_2$. XVII. Liq. n_D^{20} 1.5278[194]

(MeO)$_2$P(S)SCH$_2\overset{S}{\widehat{CH—CH_2}}$. XVII. Liq. $b_{0,01}$ 64°.[1196]

(EtO)$_2$P(S)SCH$_2\overset{S}{\widehat{CH—CH_2}}$. XVII.[898,1196] Liq. $b_{0.01}$ 73°,[1196] $b_{0.18}$ 112°,[898] n_D^{20} 1.5555, d_4^{20} 1.2227.[898]

<u>(RO)$_2$P(S)SCH$_2$-5-ring</u>

(EtO)$_2$P(S)SCH$_2$– 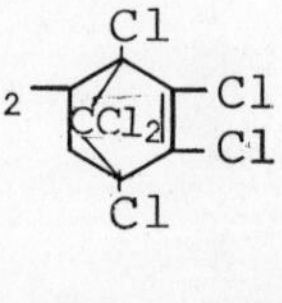 From (EtO)$_2$P(S)SCH$_2$CH=CH$_2$ and hexachlorocyclopentadiene at 105-110°

for 10-12 hr. Liq. n_D^{20} 1.4859, d^{20} 1.5660.[568a]

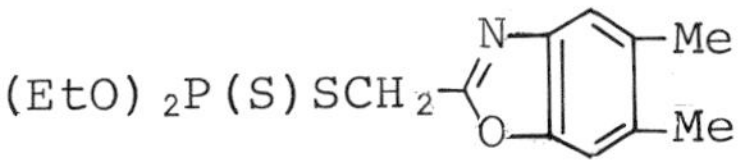

XVII. Solid. m. 37-38°.[316]

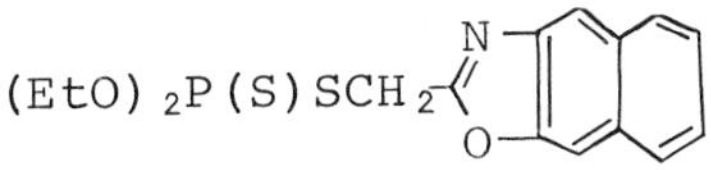

XVII. Solid. m. 51-52°.[316]

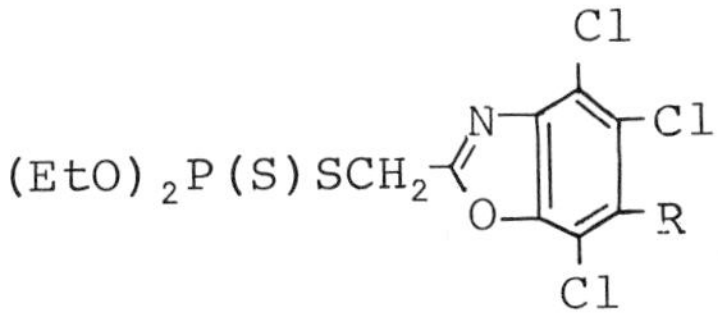

XVII. Solids. R, m.: H, 98-99°; Cl, 133-134°.[316]

From $(EtO)_2P(S)SCH_2CNS$ plus oxalyl chloride. Solid. m. 100-102° (from benzeneligroin).[1092]

XVII. Liq. $b_{0.01}$ 125°.[323]

XVII. Solid. m. 107-108°.[1234]

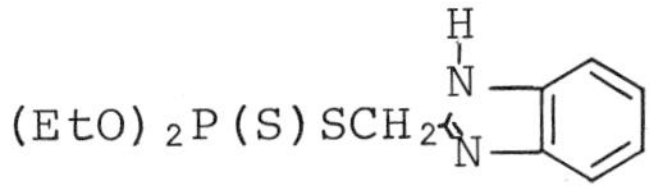

XVII. Solid. m. 112° (from ether, toluene, EtOAc).[950] Also 5(6) Cl or NO_2 analogs.[950]

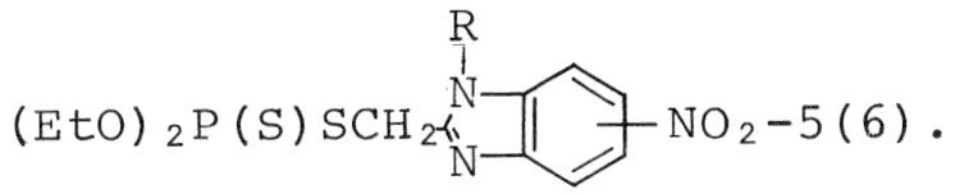

Solids. R, m.: CH_2OMe, oil n_D^{22} 1.6082; COSMe, 112-116° (from MeOH); COOEt, (mixed isomers) 103-104° (from MeOH), 94-95° (from light petroleum); COO-i-Pr, 88-90° (from MeOH); COOBu, 102-106° (from MeOH).[950]

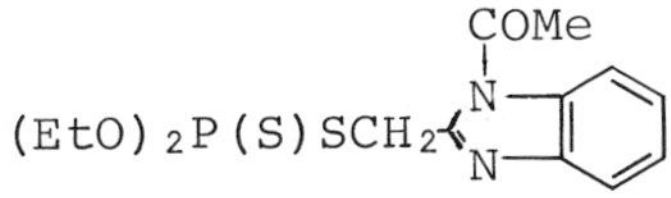

XVII. Solid. m. 112.5°.[950]

$(MeO)_2P(S)SCH_2$– 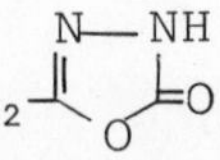 XVII. Solid. m. 65°.[1073]

$(MeO)_2P(S)SCH_2$– 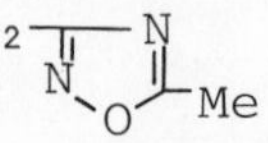 XVII. Liq. n_D^{19} 1.5390.[1399]

$(MeO)_2P(S)SCH_2$– 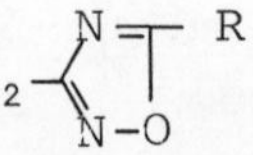 XVII. Liq. R, n_D^t: Me, 1.5390[19°]; Pr, 1.5154[20°]; COOEt, 1.5206[21°]; $CONH_2$, solid, m. 78-85°.[1399]

$(MeO)_2P(S)SCH_2$– 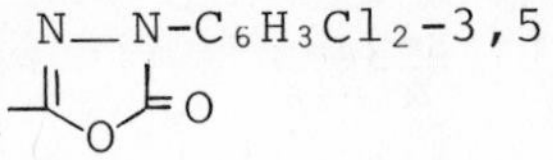 XVII. Solid. m. 104°.[1073]

$(RO)_2P(S)SCH_2$– 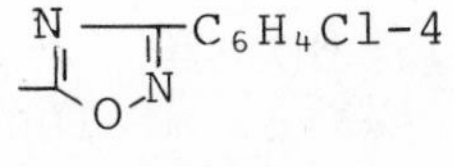 XVII. R, solidification point Me, 36°; Et, 26-27°.[772]

$(MeO)_2P(S)SCH_2$– 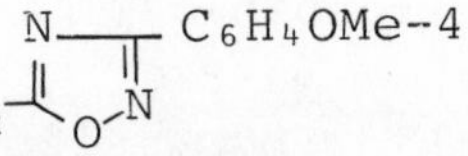 XVII. Solid. m. 52°.[772]

$(EtO)_2P(S)SCH_2$– 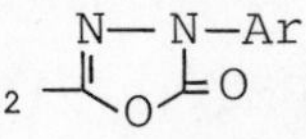 XVII. Solids. Ar, m.: $4\text{-}ClC_6H_4$, 72°; $2,4\text{-}Cl_2C_6H_3$, 70°.[1073]

$(EtO)_2P(S)SCH_2$– 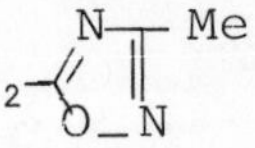 XVII. Liq. $b_{0.4}$ 140-143°.[772]

$(MeO)_2P(S)SCH_2$– 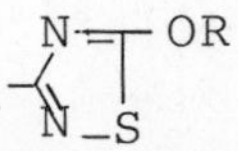 XVII. Liq. R, n_D^{20}: Me, 1.5610; Et, 1.5409.[1008]

$(i\text{-}PrO)_2P(S)SCH_2$– 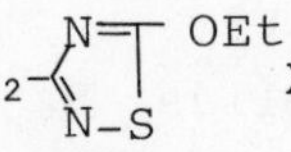 XVII. Liq. b_{14} 80°.[1008]

$(EtO)_2P(S)SCH_2$– 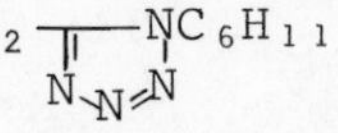 XVII. Solid. m. 73° (from ether).[1234]

$(EtO)_2P(S)SCH_2$– 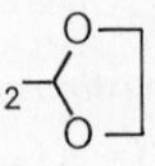 From $(EtO)_2P(S)SCH_2CHO$ plus $HOCH_2CH_2OH$ and p-toluene sulfonic acid. Liq. $b_{0.01}$ 85-86°.[2]

$(EtO)_2P(S)SCH_2$– 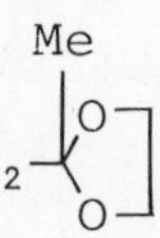 From $(EtO)_2P(S)SCH_2COMe$ plus $HOCH_2CH_2OH$ and p-toluene sulfonic acid. Liq. $b_{0.002}$ 60-70°.[2]

$(EtO)_2P(S)SCH_2$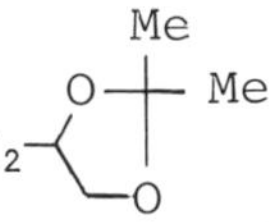 IX. Liq. $b_{0.4}$ 134°.[171]

$(EtO)_2P(S)SCH_2$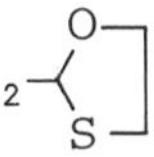 XVII. Liq. $b_{0.001}$ 100°.[2]

$(RO)_2P(S)SCH_2$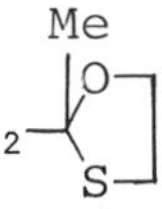 XVII. Liq. R, b./mm: Me, 70-75°/0.001; Et, 70°/0.002.[2]

$(EtO)_2P(S)SCH_2$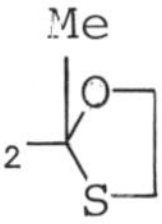 From $(EtO)_2P(S)SCH_2COMe$ plus $HSCH_2CH_2OH$ and $Et_2O \cdot BF_3$. Liq. nK 1.5430.[1378]

$(EtO)_2P(S)SCH_2$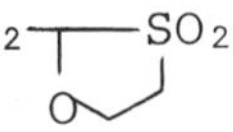 From $(EtO)_2P(S)SCH_2$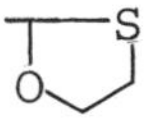 plus $KMnO_4$ and AcOH. Liq. n_D^{25} 1.5366.[603]

$(EtO)_2P(S)SCH_2$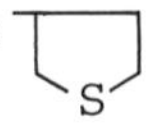 XVII. Liq. n_D^{30} 1.5262.[257]

$(RO)_2P(S)SCH_2$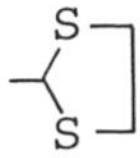 XVII. Liq. R, b./mm: Me, 81-103°/0.003; Et, 90-105°/0.001.[2]

$(RO)_2P(S)SCH_2$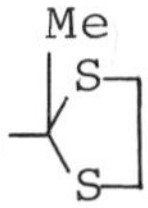 XVII. Liq. R: Me, n_D^{25} 1.5957; Et, $b_{0.001}$ 70°.[2]

$(EtO)_2P(S)SCH_2$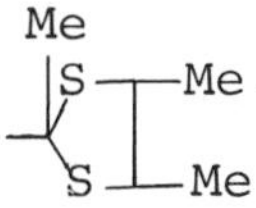 From $(EtO)_2P(S)SCH_2COMe$ plus HSCHMe CHMeSH. Liq. $b_{0.002}$ 60-90°; n_D^{25} 1.5533.[2]

$(EtO)_2P(S)SCHMe$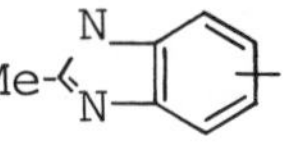 NO_2-5(6). XVII. Solid. m. 86-88°.[950]

<u>$(RO)_2P(S)SCH_2$-6-ring</u>

$(MeO)_2P(S)SCH_2Ph$. XVII.[761] XXXIII.[643] Liquid. $b_{0.01}$ 89°,[643] $b_{0.11}$ 117-117.5°, n_D^{20} 1.5818, d^{20} 1.2217.[761]

$(EtO)_2P(S)SCH_2Ph$. XVII.[1194] XXVII.[25,422] XXXIII.[643,802] Liq. $b_{0.01}$ 90°,[643] $b_{0.1}$ 117-118°, b_2 150°,[422] n_D^{20} 1.5597, n_D^{25} 1.555,[802] d^{20} 1.1563.[25]

$(RO)_2P(S)SCH_2Ph$. XVII. Liq.

R	$b_{0.0001}$	n_D^{20}	d_4^{20}.[1447]
CH_2=$CHCH_2$	95°	1.5706	1.1516
HC≡CCH_2	110°	1.5870	1.2139
CH_2=$CMeCH_2$	105°	1.5575	1.1187
MeCH=$CHCH_2$	108°	1.5562	1.1166

$(EtO)_2P(S)SCH(NHC_6H_4Cl$-4)Ph. XXI. From $(EtO)_2P(S)SH$ plus 4-ClC_6H_4N=CHPh with exclusion of moisture. Solid. m. 96-97°.[1003]

$(EtO)_2P(S)SCH(OH)Ph$. XXI. Solid. m. 104°.[1003]

$(EtO)_2P(S)SCH(SMe)Ph$. XVII. Liq. $b_{0.25}$ 135°.[578]

$(MeO)_2P(S)SCH(SEt)Ph$. XVII. Liq. n_D^{23} 1.5933.[578]

$(EtO)_2P(S)SCH(SEt)Ph$. XVII. Liq. $b_{0.18}$ 140°, n_D^{25} 1.5703.[578]

$((RO)_2P(S)S)_2CHPh$. XXI. Solids. R, m.: Me 61.2° (from ether pentane); Et, 37-38°.[1276] Also 14 analogs with R = Me, substitutions on Ph.[1276]

$((RO)_2P(S)S)_2CHPh$. XVIII. Solids or liq. Et, m. 38.5-39.5°; i-Pr, n_D^{20} 1.5486, d^{20} 1.1457; Pr, 1.5545, 1.1602; Ph, m. 80-81°.[1428]

$(EtO)_2P(S)SCH_2C_6H_4Cl$-2. XVII. Liq. b_2 169-171°.[1194]

$(RO)_2P(S)SCH_2C_6H_4Cl$-4. XVII.[1193,1194,1245] Liq. R, b/mm, n_D^{20}, d^{20}: Me, 98-100°/0.01;[1193,1194] Et, 105-107°/0.01,[1193,1194] 143°/0.06,[1245] 1.5932, 1.2763;[1245] i-Pr, 120°/0.04,[1193,1194] 1.5775, 1.2259;[1245] Pr, 107°/0.01,[1193,1194] 180-182°/0.25,[1245] 1.5808, 1.2269;[1245] Bu, 1.5685, 1.1721.[1245]

$(EtO)_2P(S)SCH_2C_6H_3Cl_2$-3,4. XVII. Liq. b_2 180°.[1193,1194]

$(EtO)_2P(S)SCH_2C_6H_3Cl_2$-2,6. XVII. Liq. b_2 182°.[1193,1194]

$(RO)_2P(S)SCH_2C_6H_3(OMe)Cl$-2,4. XVII. Liq. R, b./mm, n^{20}, d^{20} Et, 167-168°/0.4, 1.5640, 1.2516; i-Pr, 1.5565, 1.2205; Pr, 161-162°/0.2, 1.5542, 1.2023; i-Bu, 1.5424, 1.1542; 1.5470, 1.1658.[755] Also analogs with $C_6H_3(OEt)Cl$.[755]

$(RO)_2P(S)SCH_2C_6H_3(OMe)Cl$-4,3. XVII. Liq. R, b./mm, n_D^{20}, d^{20} Et, 145-147°/0.1, 1.5651, 1.2518; i-Pr, 1.5570, 1.2086; pr, 167-170°/0.2, 1.5534, 1.2101; i-Bu, 1.5435, 1.1734; Bu, 1.5510, 1.1695.[755]

$(EtO)_2P(S)SCH_2C_6H_4NO_2$-4. XXXIII. Liq. $b_{0.01}$ 152-153°.[643]

$(EtO)_2P(S)SCH_2C_6H_4SMe$-4. XVII. Liq. $b_{0.01}$ 125°.[637]

$(EtO)_2P(S)SCH(NHC_6H_4Cl-4)C_6H_4Me$-4. XXI. From $(EtO)_2P(S)SH$
plus 4-$ClC_6H_4N=CHC_6H_4Me$-4 with exclusion of moisture.
Solid. m. 96°.[1003]

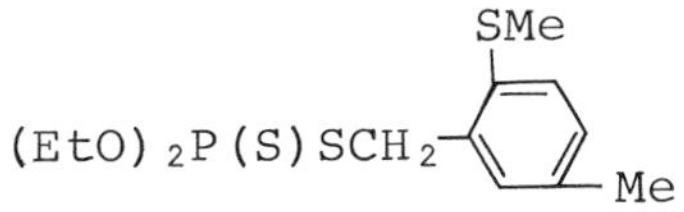

$(EtO)_2P(S)SCH_2-$ XVII. Liq. $b_{0.01}$ 127-130°.[285]

$(RO)_2P(S)SCH_2C_6H_4CN$-2. XVII. Solids. R, m. (from ether-petroleum ether): Me, 36-36.3°; Et, 38-39°.[89]

$(MeO)_2P(S)SCH_2C_6H_4CN$-3. XVII. Liq. n_D^{25} 1.5863.[89]

$(RO)_2P(S)SCH_2C_6H_4CN$-4. XVII. Solids. R, m.: Me, 63-64°
(from ether-petroleum ether); Et, 33-33.5° (from petroleum ether).[89]

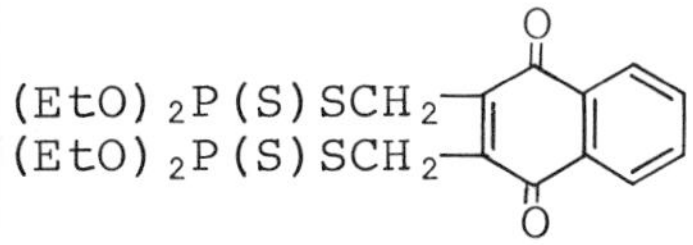

$(EtO)_2P(S)SCH_2-$
$(EtO)_2P(S)SCH_2-$ XVII. Solid. m. 87-88°.[359]

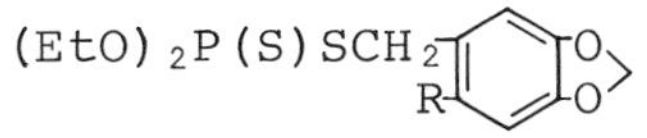

$(EtO)_2P(S)SCH_2-$ Liq. and solids. R, b./mm, or m.: H,
170-185°/0.08; Cl, 180-190°/0.06;
NO_2, m. 60-63°.[416]

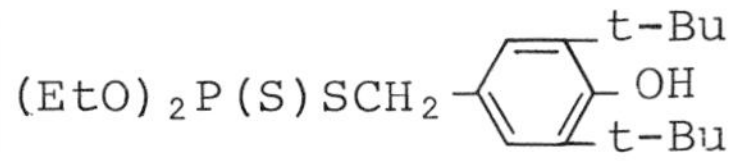

$(EtO)_2P(S)SCH_2-$ ESR.[924]

$(EtO)_2P(S)SCH_2-$ XVII. Liq. $b_{0.25}$ 159°, n_D^{20} 1.5325, d_4^{20}
1.1009.[1357]

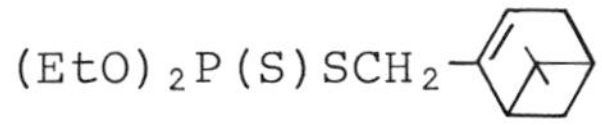

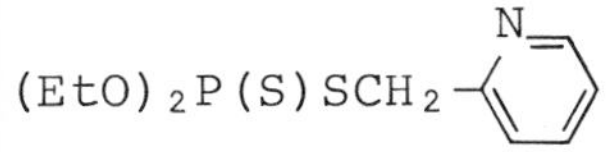

$(EtO)_2P(S)SCH_2-$ XVII.[258,650] Liq. $b_{0.01}$ 105-106°,[650]
n_D^{30} 1.5164,[258] L.HCl, solid, m. 54°
(from EtOAc).[650]

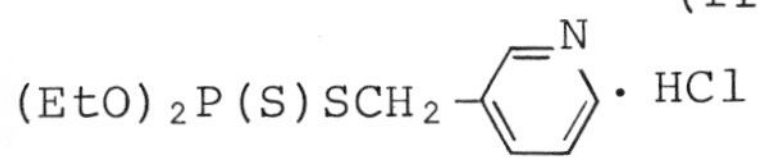

$(EtO)_2P(S)SCH_2-$ · HCl   XVII. Solid. m. 100° (dec.)(from
EtOAc).[650]

$(RO)_2P(S)SCH_2-$ ·HCl   XVII. Solids. R, m.: Me, 100° (dec.);
Et, 127° (from EtCOMe).[650]

$(EtO)_2P(S)SCH_2-$ XVII. Solid. m. 122-123°.[1234]

$(MeO)_2P(S)SCH_2$– 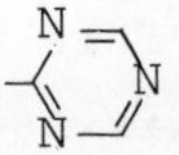 XVII. Liq. n_D^{25} 1.5715.[91]

$(MeO)_2P(S)SCH_2$– 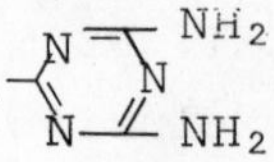 XVII. Solid. m. 164-166°.[145] "Menazon." Forms a sulfonate. Solid. m. 161-162°.[145] Heating Menazon with 1:1 HCl produces unexpected products (hydrolysis of NH_2 groups).[144] Menazon-^{32}P. Solid. m. 145-147°.[1011]

$(EtO)_2P(S)SCH_2$– 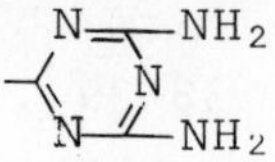 XVII. Solid. m. 168-169°.[145]

$(MeO)_2P(S)SCH_2$– 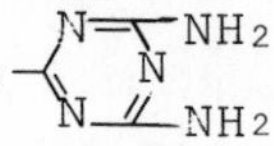 XVII. Solid. m. 128-129°.[678]

$(MeO)_2P(S)SCH_2$– 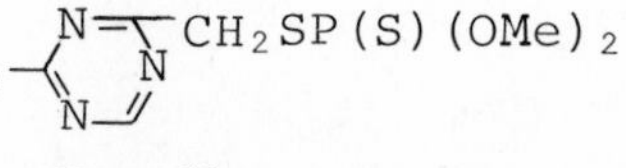 XVII. Liq. n_D^{25} 1.5838.[9

$(RO)_2P(S)SCH_2$ – 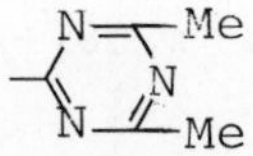 XVII. Solids. R, m.: Me, 28.5-30°; Et, 18-19° (from petroleum ether).[9

$(MeO)_2P(S)SCH_2$ – 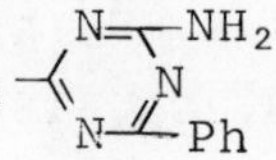 XVII. Solid. m. 97-98° (from benzen hexane and aqueous EtOH).[678]

$(EtO)_2P(S)SCH_2$– 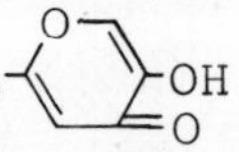 XVII. Solid. m. 94° (from CCl_4).[767, 1018]

$(RO)_2P(S)SCH_2$– 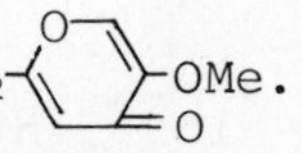 XVII. Solids. R, m.: Me, 90°; Et, 62-64° (from CCl_4); i-Pr, 78-80°; Pr, oil.[767,1018]

$(RO)_2P(S)SCH_2$– 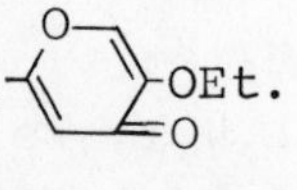 XVII. Solids. R, m.: Me, 76-78°; E 46-48°.[767,1018]

$(RO)_2P(S)SCHMe$-6-ring

$(RO)_2P(S)SCHMePh$. XVII. XXII. Liq. R, b./mm: Et, 85°/0.02; Pr, 99.5-101°/0.03; Bu, 119-121°/0.16.[890]

$(RO)_2P(S)SCH(CH_2OH)Ph$. XI. Liq. R, n_D^{20}, d^{20}: Me, 1.5561, 1.2383; Et, 1.5612, 1.1976.[993]

$(EtO)_2P(S)SCH(CONHR)Ph$. XVII. Solids. R, m.: Et, 104-105°; i-Pr, 109°.[820]

$(RO)_2P(S)SCH(CONH\text{-}i\text{-}Pr)Ph$. XVII. Solids. R, m.: Me, 90°; i-Pr, 115-116°.[820] Also several analogs.[820]

$(EtO)_2P(S)SCH(CN)Ph$. XVII. Liq. $b_{0.22}$ 135°.[1150]

$(EtO)_2P(S)SCH(COOH)Ph$. XVII. Solid. m. 83-85.5°.[832]

$(MeO)_2P(S)SCH(COOMe)Ph$. XVII. Liq. n_D^{20} 1.5631, d_{20}^{20} 1.2656.[822]

$(MeO)_2P(S)SCH(COOEt)Ph$. XVII.[280,822,1421] Liq. $b_{0.01}$ 122-125°,[280] n_D^{20} 1.5449,[822] n_D^{80} 1.5490,[1421] d_{20}^{20} 1.2635.[822] "Phenthoate." Phenthoate-^{32}P. Liq. n_D^{20} 1.5538.[230]

$(EtO)_2P(S)SCH(COOEt)Ph$. XVII.[822,832] Liq. $b_{0.05}$ 149-150°,[832] n_D^{20} 1.5371, d_{20}^{20} 1.1790.[822]

$(i\text{-}PrO)_2P(S)SCH(COOEt)Ph$. XVII. Liq. n_D^{20} 1.5269, d_{20}^{20} 1.1355.[822]

$(FCH_2CH_2O)_2P(S)SCH(COOEt)Ph$. XVII.[352,830] Solid. m. 41.4° (from ligroin),[830] m. 41-43° (from petroleum ether).[352]

$(EtO)_2P(S)SCH(COO\text{-}t\text{-}Bu)Ph$. XVII. Solid. m. 45.5-46° (from MeOH).[351]

$(MeO)_2P(S)SCH(COOCMe_2CN)Ph$. XVII. Liq. $b_{0.6}$ 124°.[823]

$(EtO)_2P(S)SCH(COOCMe_2C{\equiv}CH)Ph$. XVII. Solid. m. 43-44°.[654]

$(EtO)_2P(S)SCH(COOEt)C_6H_4Cl\text{-}4$. XVII. Liq. $b_{0.05}$ 158-160°, n_D^{20} 1.5470, d_{20}^{20} 1.2381.[822]

$(i\text{-}PrO)_2P(S)SCH(COOEt)C_6H_4Cl\text{-}4$. XVII. Liq. n_D^{20} 1.5332, d_{20}^{20} 1.1914.[822]

$(EtO)_2P(S)SCH(COOEt)C_6H_4NO_2\text{-}4$. XVII. Liq. n_D^{20} 1.5351, d_{20}^{20} 1.3244.[822]

$(MeO)_2P(S)SCHMe$— 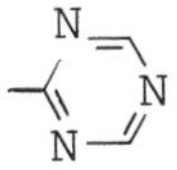 . XVII. Liq. n_D^{25} 1.5557.[91]

$[(RO)_2P(S)S]_2CMePh$. XVIII. Liq. R, n_D^{20}, d_4^{20}: Et, 1.6262, 1.3214; i-Pr, 1.5704, 1.1964, IR; Pr, 1.5960, 1.2357.[1429]

$[(RO)_2P(S)S]_2CMeC_6H_4Br\text{-}4$. XVIII. Liq. R, n_D^{20}, d_4^{20}: Et, 1.6222, 1.4369; i-Pr, 1.6022, 1.3689; Pr, 1.6038, 1.3699.[1429]

$(RO)_2P(S)SCH(C_6H_4Cl\text{-}4)_2$. XVII. Liq. R, n_D^{30}: Me, 1.6213; Et, 1.5884.[272]

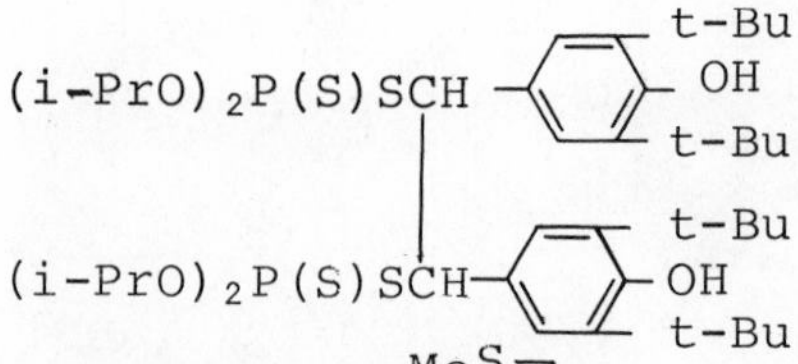

XXII. Solid. m. 180-181° (from MeOH).[369]

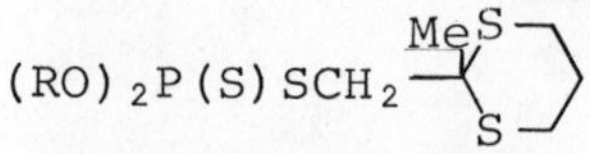

XVII. R: Me, solid, m. 42-43°; Et, liq., n_D^{25} 1.5720.[2]

<u>$(RO)_2P(S)SCH_2CH_2$-6-ring</u>

$(MeO)_2P(S)SCH_2CH_2Ph$. XXII. Liq. $b_{0.1}$ 139-140°,[761], $b_{0.45}$ 128-132°,[756] n_D^{20} 1.5644,[761] 1.5705,[756] d^{20} 1.1712,[761] 1.2108.[756]

$(EtO)_2P(S)SCH_2CH_2Ph$. VIII.[890] XVII.[890,1151] XXII.[756] Liq. $b_{0.02}$ 85-86° (mol. still),[890] $b_{0.01}$ 143°,[1151] $b_{0.15}$ 135-137°,[756] n_D^{20} 1.5498, d^{20} 1.1144.[756]

$(RO)_2P(S)SCH_2C(=NNHC_6H_3(NO_2)_2$-2,4)Ph. XVII. Solids. R, m.: Et, 131-132°; i-Pr, 109-110°; Pr, 94-95°; Bu, 90-91°.[1336]

$(RO)_2P(S)SCH_2C(=NNHC_6H_3(NO_2)_2$-2,4)$C_6H_4Br$-4. XVII. Solids. R, m.: Et, 108°; i-Pr, 125-126°; Pr, 102-104°; Bu, 117.5-118°.[1336]

$(RO)_2P(S)SCH_2CH_2C_6H_4Cl$-4. XVII. Liq. R, $b_{0.01}$: Me, 135°; Et 144°.[1151]

$(RO)_2P(S)SCH_2C(=NNHC_6H_3(NO_2)_2$-2,4)$C_6H_4Cl$-4. XVII. Solids. R, m.: Et, 108-109°; i-Pr, 125-125.5°; Pr, 108-110°; Bu, 108°.[1336]

$(EtO)_2P(S)SCH_2CH_2C_6H_4Me$-4. XVII. Liq. $b_{0.01}$ 141°.[1151]

$(MeO)_2P(S)SCH{=}CHPh$. XXXIII. Liq. n_D^{25} 1.6190.[805]

$(EtO)_2P(S)SCH{=}CHPh$. XXVII.[802] XXXIII.[805] Liq. $b_{0.01}$ 110-122°,[802] n_D^{25} 1.582-3.[802,805]

$(RO)_2P(S)SC{\equiv}CPh$. XXVI. Liq. R, b./mm, n_D^{20}, d^{20}: Et, 130°/0.15, 1.5935, 1.1634; Pr, 134°/0.1, 1.5795, 1.1272; Bu, 147°/0.15, 1.5696, 1.1039.[25]

XXII. From $(MeO)_2P(S)SH$ plus isosafrole). Liq. $b_{0.05}$ 128°, n_D^{20} 1.5743.[110]

$(EtO)_2P(S)SC(cyclo\text{-}C_6H_{11})(CN)_2$. XXII. Liq. $b_{0.2-0.3}$ 85-90°.[860]

$(EtO)_2P(S)SCH_2CH_2$— 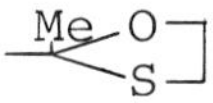 From $(EtO)_2P(S)SCH_2CH_2COMe$ plus $HOCH_2CH_2SH$ and p-$MeC_6H_4SO_3H$. Liq. n_D^{25} 1.5375.[2]

$(EtO)_2P(S)SCH_2CH_2$—  From $(EtO)_2P(S)SCH_2CH_2COMe$ plus $HSCH_2CH_2SH$ and p-$MeC_6H_4SO_3H$. Liq. n_D^{25} 1.5848.[2]

$(EtO)_2P(S)SCH(C_6H_4Cl\text{-}2)CH_2COOR$. XVII. Liq. R, n_D^t: Me, 1.5530 25°; Et, 1.5518 16°; i-Pr, 1.5422 17°; s-Bu, 1.5423 17°.[337]

$(RO)_2P(S)SCH_2CH_2CH_2C_6H_4OH\text{-}2$. XXII. Liq. Me and Et, $b_{0.2}$>100°. From additions to 2-allylphenol in the presence of AIBN catalyst and Hg vapor light.[1384]

$(MeO)_2P(S)SCH_2CH_2CH_2$— 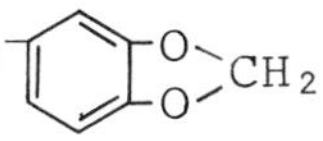 XXII. From $(MeO)_2P(S)SH$ plus safrole. Liq. $b_{0.01}$ 124°, n_D^{20} 1.5734.[110]

$(EtO)_2P(S)SCH(CH(COOEt)_2)Ar$. XXII. From $(EtO)_2P(S)SH$ plus $ArCH{=}C(COOEt)_2$. Rates of formation, $k_2 \times 10^4$ liter/mole/sec at 75° in xylene. Solids. Ar, m., k_2: Ph, 47-48°, 4.01; 4-BrC_6H_4, 62.5-63 , 3.57; 4-ClC_6H_4, 58-58.5°, 4.64; 4-FC_6H_4, 48-48.5°, 3.20.[986]

$(RO)_2P(S)SCH_2CH{=}CHPh$. XXVII. Liq. $b_{0.0001}$, n_D^{20}, d^{20}: Et, 105-110°, 1.5650, 1.1400; i-Pr, 128-130°, 1.5659, 1.1122; Pr, 125-128°, 1.5655, 1.1206; Bu, 140-145°, 1.5760, 1.1226; Am, 150-155°, 1.5480, 1.0639; Ph, 170-172°, 1.6300, 1.2162; cyclo-C_6H_{11}, 135-140°, 1.5674, 1.0305; C_8H_{17}, 185-187°, 1.5140, 0.9982; $C_{10}H_7$, solid, m. 57-59°.[972]

$(PhO)_2P(S)SR$. XVII. Liq. R, b_1, n_D^{20}, d_4^{20}: $CH_2CH{=}CHC_6H_4NO_2$-4, 175-175.5°, 1.6020, 1.240, IR; $CH_2CMe{=}CHC_6H_4NO_2$-4, 183-183.5°, 1.5877, 1.208; $CH_2CH{=}CMeC_6H_4NO_2$-4, 182°, 1.5972, 1.221.[383]

$(PhO)_2P(S)SR$. XVII. Liq. R, b_1, n_D^{20}, d_4^{20}: $CH_2CH{=}CHC_6H_4Me$-4,

177-178°, 1.6060, 1.175; $CH_2CH=CMeC_6H_4Me-4$, 176°, 1.6040, 1.173.[383]

F.3.3.3. $C\underset{n}{\overset{O}{\langle\rangle}}\overset{O}{O}P(S)SR$

$\overset{O}{\underset{O}{\square}}P(S)SR$

_____

$\overset{O}{\underset{O}{\square}}P(S)SEt$. XV.[249] Liq. $b_{0.4}$ 114°,[249] $b_{0.7}$ 92-94°,[478] n_D^{20} 1.5340,[478] n_D^{21} 1.5540,[249] d_4^{20} 1.3656.[478] μ 4.1 D.[478]

$\overset{O}{\underset{O}{\square}}P(S)SR$. VIII. Liq. R, n_D^{20}, d_4^{20}: Pr, 1.5478, 1.2462; Bu 1.5439, 1.2204; i-Am, 1.4856, 1.0921.[508]

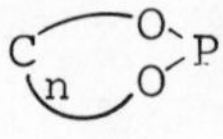

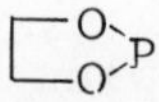

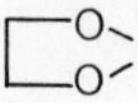

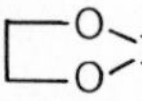

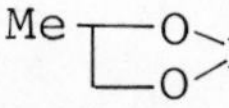

$Me\overset{O}{\underset{O}{\square}}P(S)SEt$. XV. Liq. $b_{0.35-0.45}$ 102-108°, n_D^{22} 1.4407.[24]

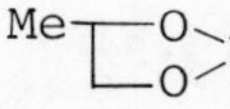

$Me\overset{O}{\underset{O}{\square}}P(S)SCH_2CH_2CN$. XXII. Liq. $b_{0.4}$ 162-163°, n_D^{20} 1.5605, d_4^{20} 1.3291.[987]

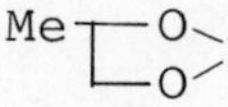

$Me\overset{O}{\underset{O}{\square}}P(S)SR$. VIII. Liq. R, n_D^{20}, d_4^{20}: Bu, 1.5300, 1.1681; i-Am, 1.4712, 1.0437.[508]

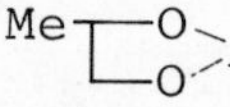

$Me\overset{O}{\underset{O}{\square}}P(S)SR$. XVII. Liq. R, b./mm, n_D^{20}, d^{20}: CH_2Ph, 157-158°/0.4, 1.5974, 1.2765; CH_2CH_2Ph, 145-147° 0.25, 1.5796, 1.2311.[987]

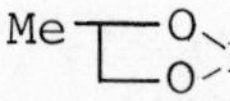

$Me\overset{O}{\underset{O}{\square}}P(S)SCH_2CH_2C_6H_{11}$-cyclo. VIII. Yellowish oil.[1066]

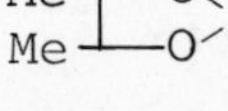

$\underset{Me}{Me}\overset{O}{\underset{O}{\square}}P(S)SEt$. XV.[249] XVII.[987] Liq. $b_{0.25}$ 106°,[249] $b_{0.4}$ 110-111°,[987] n_D^{20} 1.5330,[987] n_D^{23} 1.5327,[249] d^{20} 1.2010.[987]

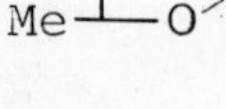

$\underset{Me}{Me}\overset{O}{\underset{O}{\square}}P(S)SR$. XVII. Liq. R, $b_{0.4}$, n_D^{20}, d^{20}: CH_2Ph, 165-166°, 1.5839, 1.2299; CH_2CH_2Ph, 167-168°, 1.5730, 1.2028.[987]

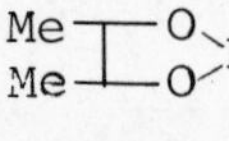

$\underset{Me}{Me}\overset{O}{\underset{O}{\square}}P(S)SCH_2CH_2CN$. XXII. Liq. $b_{0.4}$ 188-189°, n_D^{20} 1.5410, d^{20} 1.2650.[987]

$\underset{Me}{Me}\overset{O}{\underset{O}{\square}}P(S)SCH_2CONHMe$. VIII.[248] XVII.[249] Solid. m. 106-108°

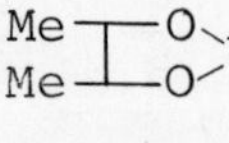

[248] (from i-PrOH).[249]

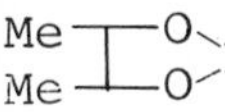P(S)SCH$_2$CON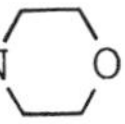O. VIII.[248] XVII.[249] Solid. m. 105-107° (from i-PrOH).[248,249]

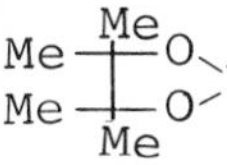P(S)SEt. XV. Liq. $b_{0.15}$ 99-106°, n_D^{24} 1.5222.[249]

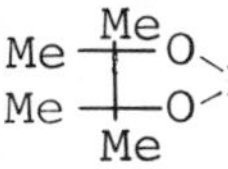P(S)SCH$_2$CONHMe. VIII. Solid. m. 105-106°.[248]

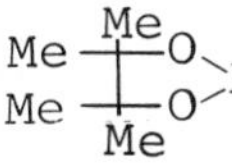P(S)SCH(COOEt)CH$_2$COOEt. XXII. Liq. n_D^{30} 1.5018.[589]

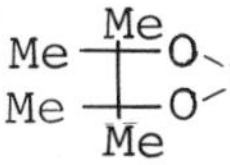P(S)SCHPhCH$_2$COMe. XXII. Solid. m. 105-106°.[590]

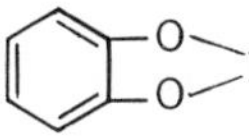P(S)SR. VIII.[508] XVII.[507,1440] Liq. and solids. R, m., or n_D^{20}, d_4^{20}: Me, 49°;[507,1440] Et, 28-29°;[507,1440] Pr, 1.5751, 1.2753;[508] i-Am, 1.5520, 1.1768.[507,1440]

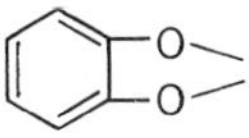P(S)SCOC$_6$H$_4$NO$_2$-4. XVII. Solid. m. 163° (dec.).[507]

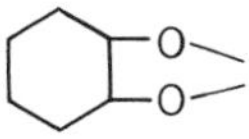P(S)SMe. XVII. Solid. m. 42-45°.[1442]

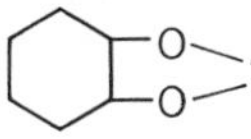P(S)SR. VIII.[508] XVII.[1442] Liq. R, n_D^{20}, d_4^{20}: CH$_2$-COOEt, 1.5515, 1.2901;[1442] Pr, 1.5480, 1.2229.[508]

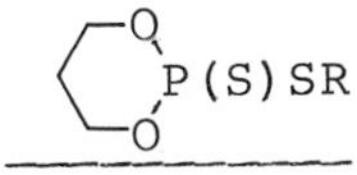

P(S)SMe. XVII. Liq. n_D^{20} 1.5578, d_4^{20} 1.2708.[1442]

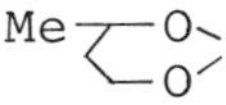P(S)SEt. XVII.[987,1442] Liq. $b_{0.6}$ 111-112°,[987] n_D^{20} 1.5470,[987] 1.5480,[1442] d_4^{20} 1.2314,[1442] d^{20} 1.2329.[987]

P(S)SCH$_2$COOMe. XVII. Liq. n_D^{20} 1.5265, d_4^{20} 1.2846.[1440]

P(S)S-i-Pr. XVII. Liq. n_D^{20} 1.5365, d_4^{20} 1.1838.[1440,1442]

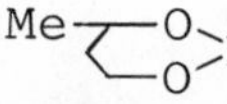P(S)SR. XVII. Liq. R, n_D^{20}, d_4^{20}: Pr, 1.5400, 1.1977; $CH_2CH{=}CH_2$, 1.5565, 1.2238; $CH_2C{\equiv}CH$, 1.5680, 1.2602.[1441]

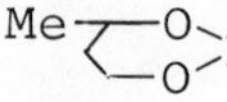P(S)SCH_2CH_2CN. XXII. Liq. $b_{0.4}$ 192-193°, n_D^{20} 1.5561, d^{20} 1.2950.[987]

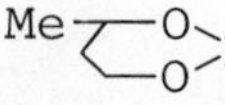P(S)SR. VIII.[508] Liq. R, n_D^{20}, d_4^{20}: Bu, 1.5312, 1.161 i-Am, 1.4411, 0.9753; Am, 1.5118, 1.1055.[508]

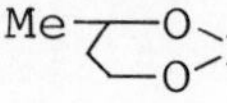P(S)SCH_2CH_2Ph. XVII. Liq. $b_{0.1}$ 160-161°, n_D^{20} 1.5723, d^{20} 1.2011.[987]

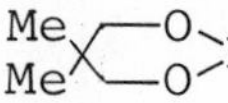P(S)SR. XVII. Solids. R, m.: Me, 85.5-86°; Et, 69.5-70.5°; i-Pr, 108.5°.[245]

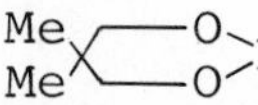P(S)SCH_2CONH_2. VIII.[248] XVII.[250] Solid. m. 96-98° (from i-PrOH).[250]

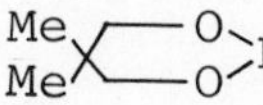P(S)$SCH_2CONHMe$. VIII.[248] XVII.[250] Solid. m. 136-137°.[248]

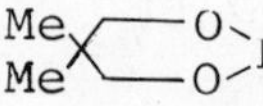P(S)SCH_2CONO. VIII.[248] XVII.[250] Solid. m. 123-124° (from i-PrOH).[250]

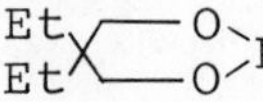P(S)SCH_2CH_2OCOMe. XXII. Liq. n_D^{20} 1.5220.[586] Also nine homologs and ring-substituted analogs.[586]

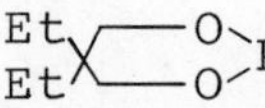P(S)SCH_2CH_2COOEt. XXII. Solid. m. 36-39°.[589] Also several analogs.[589]

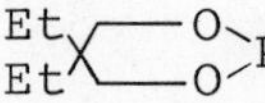P(S)$SCH_2CHMeCOOMe$. XXII. Liq. $b_{0.2}$ 135°.[79]

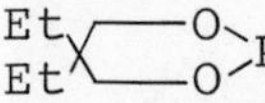P(S)$SCMe_2CH_2COMe$. XXII. Solid. m. 65-70°.[590]

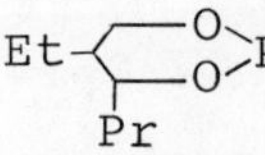P(S)SCH_2CH_2CN. XXII. Liq. n_D^{30} 1.5315.[79]

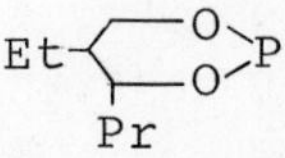P(S)SCH(COOEt)CH_2COOEt. XXII. Liq. n_D^{30} 1.5009.[79,589]

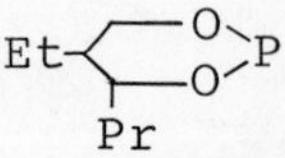P(S)$SCHPhCH_2COMe$. XXII. Liq. n_D^{30} 1.5600.[590,1343]

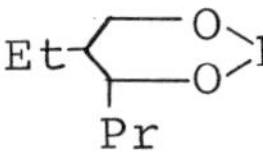P(S)SCHPhCH_2COPh. XXII. Solid. m. 133-135°.[1343]

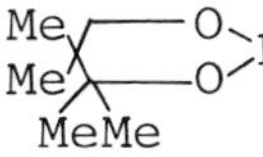P(S)SCHPhCH_2COMe. XXII. Solid. m. 105-106°.[1343] Also four analogs.[1343]

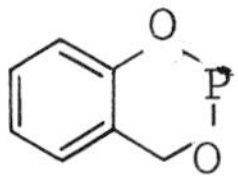P(S)SR. VIII.[542,1284] Solids. R, m.: Me, 69-70°; Ph, 79-80°.[542,1284] Liq. R, b./mm, n_D^{25}: Et, 145-147°/0.2, 1.6221; i-Pr, 140-143°/0.1, 1.5920; Pr, 145-150°/0.25, 1.5998; CH_2CH=CH_2, 140-147°/0.3, 1.6305; Bu, 160-167°/0.25.[542,1284]

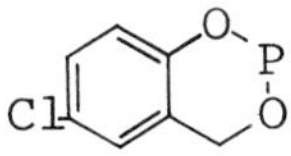P(S)SMe. VIII.[264,1284] Liq. $b_{0.2}$ 160-170°,[264] 160-165°.[1284]

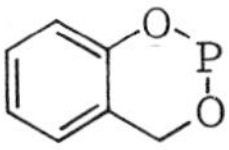P(S)SCH_2CH_2OR. VIII. From $ROCH_2CH_2SP(S)Cl_2$ plus saliginin and 20% NaOH (Schotten-Bauman conditions) 10-50°, 2-3 hr. Liq. R, b./mm: Me, 149-151°/0.02; Et, 160-165°/0.03.[544]

MeS(S)PP(S)SMe. XVII. Solid. m. 212.5° (from EtOAc).[198]

$MeS(CH_2)_nS$(S)PP(S)$S(CH_2)_n$SMe. VIII, XXIV. Solids. n, m.: 1, 169-170° (from acetone); 2, 187-188° (from acetone); 3, 169-170° (from acetone).[1006] Also analogs with n = 2 or 3.[1006]

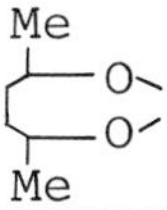P(S)SR. XVII. Liq.

R	n_D^{20}	d_4^{20}
Me	1.5418	1.1987[1442]
Et	1.5352	1.1674
i-Pr	1.5272	1.1327
$C_{10}H_{21}$	1.5050	1.0321
CH_2CH_2OH	1.5345	1.2098[1440]

$CHMeCH_2OH$	1.5259	1.1945
CH_2COOMe	1.5330	1.2452
Pr	1.5255	1.1306[1441]
$CH_2CH{=}CH_2$	1.5389	1.1558

F.3.3.4. RO(R'O)P(S)SR"

MeO(R'O)P(S)SR"

MeO(EtO)P(S)$SCH_2OC_6H_3Cl_2$. XVII. Liq. n_D^{25} 1.5762.[1293] Also five analogs (position of Cl not given).[1293]

MeO(EtO)P(S)SCH_2N 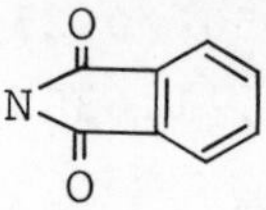 XVII.[275,866] Solid. m. 73°.[866] Liq. (crude). n_D^{30} 1.5860.[275]

(MeO(EtO)P(S)$SCH_2CONHMe$. VIII. Solid. m. 61-62°.[598]

MeO(EtO)P(S)$SCH_2CONHCH_2SR$. From MeO(EtO)P(S)$SCH_2CONHCH_2OH$ plus RSH in EtOAc with HCl (gas) in benzene, 45° for 6 hr. Liq. and solid: R= Me, n_D^{20} 1.5610, d^{20} 1.283; Et, n_D^{20} 1.5525, d^{20} 1.254; i-Pr, solid, m. 42-44°.[829]

MeO(EtO)P(S)$SCH_2CONHCH_2OR$. From MeO(EtO)P(S)$SCH_2CONHCH_2OH$ plus ROH and HCl. Liq. R, n_D^{20}, d_4^{20}: Me, 1.5300, 1.265; Et, 1.5245, 1.230.[658]

MeO(EtO)P(S)$SCH_2CONHCH_2OR$. XVII. Liq. R, n_D^{20}: Me, 1.5300; Et, 1.5245; i-Pr, 1.5253; Pr, 1.5273; $CH_2CH{=}CH_2$, 1.5387; $CH_2C{\equiv}CH$, 1.5360; CH_2CH_2OMe, 1.5231; CH_2CH_2SMe, 1.5529.[824]

MeO(EtO)P(S)$SCH_2CONEtCH_2CN$. XVII. Liq. n_D^{20} 1.5262, d^{20} 1.200.[656]

MeO(EtO)P(S)$SCH_2CONMeCH_2CH_2CN$. XVII. Liq. n_D^{20} 1.5353, d^{20} 1.248.[656]

MeO(EtO)P(S)$SCH_2CONHCMe_2CN$. XVII. Liq. n_D^{20} 1.5233, d^{20} 1.223.[656]

MeO(EtO)P(S)SBu. VIII, XVII. Liq. b_1 114-116°, n_D^{20} 1.5050, d_4^{20} 1.0965.[561]

MeO(EtO)P(S)SCH(COOR)Ph. XVII. Liq. R, n_D^{20}, d^{20}: Me, 1.5536 1.227; Et, 1.5470, 1.210; i-Pr, 1.6370, 1.175.[657]

MeO(EtO)P(S)$SCH(COOCH_2CN)Ph$. XVII. Liq. n_D^{20} 1.5520, d^{20} 1.259.[823]

MeO(EtO)P(S)SCH(COOCHMeCN)Ph. XVII. Liq. n_D^{20} 1.5503, d^{20} 1.237.[823]

MeO(EtO)P(S)SCH(COOCMe$_2$CN)Ph. XVII. Solid. m. 52-53.5° (from MeOH).[823]

MeO(EtO)P(S)SCH(COOCMeEtCN)Ph. XVII. Liq. n_D^{20} 1.5379, d^{20} 1.189.[823]

MeO(EtO)P(S)SCH(COOEt)C$_6$H$_4$Cl-4. XVII. Liq. n_D^{20} 1.5413, d^{20} 1.248.[657]

MeO(EtO)P(S)SCH(COOEt)C$_6$H$_4$NO$_2$-4. XVII. Liq. n_D^{20} 1.5629, d^{20} 1.288.[657]

MeO(EtO)P(S)S 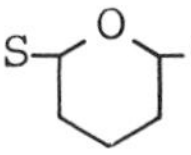 XXII. Liq. n_D^{20} 1.5186.[670]

MeO(EtO)P(S)S 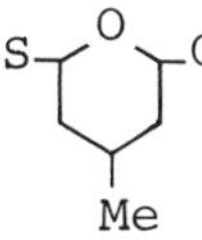 XXII. Liq. n_D^{20} 1.5000.[670]

MeO(EtSCH$_2$CH$_2$O)P(S)SMe. VIII. Liq. $b_{0.01}$ 82°.[1163]

MeO(i-PrO)P(S)SCH$_2$SEt. XVII. Liq. $b_{0.1}$ 79°.[304]

MeO(i-PrO)P(S)SCH$_2$ 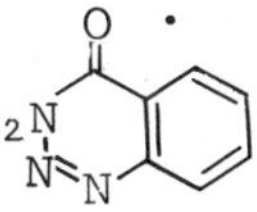 XVII.[623] XXI.[311] Solid. m. 64°,[311] 68°.[623]

MeO(i-PrO)P(S)SCH$_2$CH$_2$SEt. XVII. Liq. $b_{0.01}$ 88°.[304]

MeO(i-PrO)P(S)S 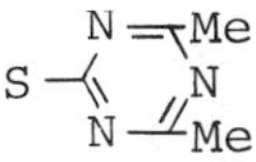 XVII. Liq. n_D^{30} 1.5418.[1247]

(MeO(PrO)P(S)SCH$_2$ 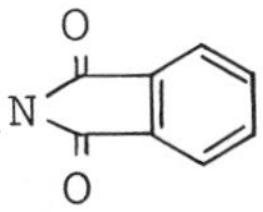 XVII. Liq. n_D^{30} 1.5809.[275]

MeO(PrO)P(S)SCH$_2$OCOCH$_2$SMe. From MeO(PrO)P(S)SCH$_2$OCOCH$_2$Cl plus MeSNa. Liq. n_D^{30} 1.5496.[392]

MeO(CH$_2$=CHCH$_2$O)P(S)SEt. XVII. Liq. n_D^{20} 1.5238, d_4^{20} 1.1524.[560]

MeO(s-BuO)P(S)SCH$_2$SEt. XVII. Liq. $b_{0.01}$ 94°.[304]

MeO(s-BuO)P(S)SCH$_2$CH$_2$SEt. XVII. Liq. $b_{0.01}$ 92°.[304]

MeO(Me$_3$CCHMeO)P(S)SCH$_2$ 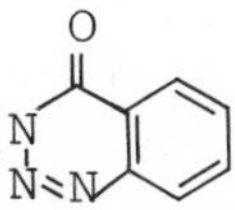 XVII.[623]

$MeO(cyclo\text{-}C_6H_{11}O)P(S)SCH_2N$ 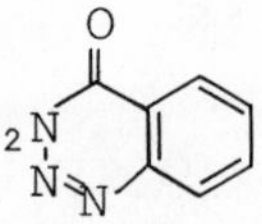 XVII. Solid. m. 78°.[623]

$MeO(PhO)P(S)SCH_2CON(OMe)$-i-Pr. XVII. Liq. n_D^{25} 1.5720.[1029]

<u>EtO(R'O)P(S)SR"</u>

$EtO(ClCH_2CH_2O)P(S)SEt$. XV. Liq. b_1 102°.[1162]

$EtO(FCH_2CH_2O)P(S)SCH(COOEt)CH_2COOEt$. VIII. Liq. $b_{0.000001}$ 110-115°, n_D^{20} 1.4840, d^{20} 1.2297.[490]

$EtO(HOCH_2CH_2O)P(S)Et$. XVII. Liq. n_D^{20} 1.4704, d_4^{20} 1.0986.[545]

$EtO(EtSCH_2CH_2O)P(S)SMe$. VIII. Liq. $b_{0.01}$ 85°.[1163]

$EtO(EtSCH_2CH_2O)P(S)SEt$. VIII. Liq. $b_{0.01}$ 87°.[1163]

$EtO(EtSCH_2CH_2O)P(S)S$-i-Pr. VIII. Liq. $b_{0.01}$ 92°.[1163]

$EtO(EtSCH_2CH_2O)P(S)S$-i-Bu. VIII. Liq. $b_{0.01}$ 96°.[1163]

$EtO(O{=}CHCH_2O)P(S)SEt$. From $EtO(HOCH_2CH_2O)P(S)SEt$ plus H_2O_2. Liq. n_D^{20} 1.5109, d_4^{20} 1.1968.[545]

$EtO(i\text{-}PrO)P(S)SCH_2N$ 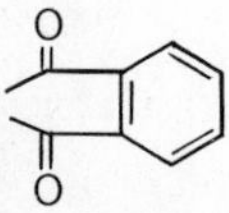 XVII.[275,866] Solid. m. 75-85°,[275] 78-82°.[866]

$EtO(i\text{-}PrO)P(S)SCH_2SEt$. XVII. Liq. $b_{0.01}$ 87°.[304]

$EtO(i\text{-}PrO)P(S)SCH_2CH_2SEt$. XVII. Liq. $b_{0.01}$ 92°.[304]

$EtO(ClCH_2MeCHO)P(S)SEt$. XV. Liq. b_1 104°.[1162]

$EtO((ClCH_2)_2CHO)P(S)SEt$. XV. Liq. $b_{0.01}$ 82°.[1162]

$EtO(PrO)P(S)SCH_2N$ 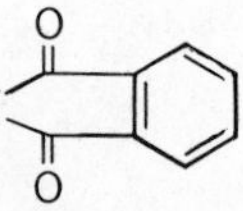 XVII. Liq. n_D^{30} 1.5746.[275]

$EtO(PrO)P(S)S$ 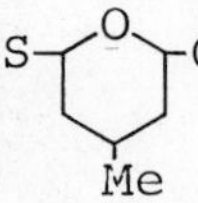 OMe. XXII. Liq. n_D^{20} 1.5030.[670]

$EtO(s\text{-}BuO)P(S)SCH_2SEt$. XVII. Liq. $b_{0.01}$ 94°.[304]

$EtO(s\text{-}BuO)P(S)SCH_2CH_2SEt$. XVII. Liq. $b_{0.01}$ 103°.[304]

$EtO(BuO)P(S)SCH_2CN$. XVII. Liq. n_D^{20} 1.5094, d_4^{20} 1.1372.[560]

$EtO(BuO)P(S)SPr$. XVII. Liq. n_D^{20} 1.5022, d_4^{20} 1.0642.[560]

$EtO(BuO)P(S)SCH_2Ph$. XVII. Liq. n_D^{20} 1.5428, d_4^{20} 1.0991[560]

$EtO(PhO)P(S)SEt$. XVII. Liq. n_D^{20} 1.5550, d^{20} 1.1754.[560]

$EtO(PhO)P(S)SCH_2CON(OMe)Me$. XVII. Liq. n_D^{25} 1.5804.[1029]

EtO(PhO)P(S)SCH$_2$CON(OMe)-i-Pr. XVII. Liq. n_D^{25} 1.5615.[1029]

EtO(PhO)P(S)SCH$_2$CON(OMe)Ph. XVII. Liq. n_D^{27} 1.5929.[564]

EtO(2-ClC$_6$H$_4$O)P(S)SCH$_2$CON(OMe)-i-Pr. XVII. Liq. n_D^{29} 1.5375.[1028]

EtO(3-ClC$_6$H$_4$O)P(S)SCH$_2$CON(OMe)-i-Pr. XVII. Liq. n_D^{29} 1.5340.[1028]

EtO(4-ClC$_6$H$_4$O)P(S)SCH$_2$CON(OMe)-i-Pr. XVII. Liq. n_D^{22} 1.5391.[1028]

EtO(2,4-Cl$_2$C$_6$H$_3$O)P(S)SCH$_2$CON(OMe)-i-Pr. XVII. Liq. n_D^{25} 1.5498.[1028]

EtO(2,4,5-Cl$_3$C$_6$H$_2$O)P(S)SCH$_2$CON(OMe)-i-Pr. XVII.[1028]

EtO(4-ClC$_6$H$_4$O)P(S)SEt. VIII. Liq. $b_{0.01}$ 95°.[1163]

EtO(2,4-Cl$_2$C$_6$H$_3$O)P(S)SEt. VIII. Liq. $b_{0.01}$ 99°.[1163]

EtO(HOC$_6$H$_4$O)P(S)SEt. XVII. Liq. Position of HO, n_D^{20}, d_4^{20}: 2, 1.4644, 1.0348; 3, 1.4950, 1.0918; 4, 1.4545, 1.0072.[546]

EtO(4-HOC$_6$H$_4$O)P(S)S-i-Pr. XVII. Liq. n_D^{20} 1.4391, d_4^{20} 1.0143.[546]

EtO(4-NO$_2$C$_6$H$_4$O)P(S)SEt. VIII. Liq. $b_{0.01}$ 140°.[298]

EtO(4-MeSC$_6$H$_4$O)P(S)SEt. VIII. Liq. $b_{0.01}$ 105°.[1163]

EtO(4-MeSC$_6$H$_4$O)P(S)S-i-Pr. VIII. Liq. $b_{0.01}$ 128°.[1163]

EtO(4-MeSC$_6$H$_4$O)P(S)S-i-Bu. VIII. Liq. $b_{0.01}$ 136°.[1163]

EtO(4-MeC$_6$H$_4$O)P(S)SEt. XVII. Liq. n_D^{20} 1.5548, d^{20} 1.1567.[560]

EtO(4-MeC$_6$H$_4$O)P(S)SCH$_2$CN. XVII. Liq. n_D^{20} 1.5645, d_4^{20} 1.2262.[560]

EtO(MeC$_6$H$_4$O)P(S)SCH$_2$CON(OMe)-i-Pr. VIII. Liq. Position of Me, n_D^t: 2, 1.5295$^{27°}$; 3, 1.5326$^{25°}$; 4, 1.5314$^{20°}$.[1030]

EtO(3-Me-4-MeSC$_6$H$_3$O)P(S)SEt. VIII. Liq. $b_{0.01}$ 142°.[298]

EtO(4-NCC$_6$H$_4$O)P(S)SEt. VIII. Liq. $b_{0.01}$ 98°.[1163]

i-PrO(PhO)P(S)SCH$_2$CON(OMe)-i-Pr. XVII. Liq. n_D^{25} 1.5523.[1029]

PrO(BuO)P(S)SCH$_2$CN. XVII. Liq. n_D^{20} 1.4990, d_4^{20} 1.0981.[560]

PrO(PhO)P(S)SCH$_2$CON(OMe)-i-Pr. XVII. Liq. n_D^{25} 1.5587.[1029]

MeOP(S)(SEt)O(CH$_2$)$_4$OP(S)(SEt)OMe. XVII. Liq. n_D^{20} 1.5232, d_4^{20} 1.1405.[545]

EtOP(S)(SEt)OCH$_2$CH$_2$CHMeOP(S)(SEt)OEt. XVII. Liq. n_D^{20} 1.4616,

d_4^{20} 1.0422.[545]

ROP(S)(S-i-Pr)OCH_2CH_2OP(S)(S-i-Pr)OR. XVII. Liq. R, n_D^{20}, d_4^{20}: Et, 1.5033, 1.1233; i-Pr, 1.5127, 1.1176.[545]

EtOP(S)(SEt)OC_6H_4OP(S)(SEt)OEt. XVII. Liq. Position on C_6H_4, n_D^{20}, d_4^{20}: 1,2, 1.4761, 1.0588; 1,3, 1.5012, 1.1072; 1,4, 1.5087, 1.1151.[546]

EtOP(S)S(S-i-Pr)OC_6H_4OP(S)(S-i-Pr)OEt-1,2. XVII. n_D^{20} 1.450 d_4^{20} 0.9954.[546]

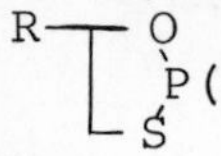

XVII. Liq.

R	b./mm	n_D^{20}	d^{20}
H	106-107°/1	1.5280	1.2801[993]
Me	92-94°/0.02	1.5359	1.2342
$ClCH_2$	105-106°/0.02	1.5400	1.3434

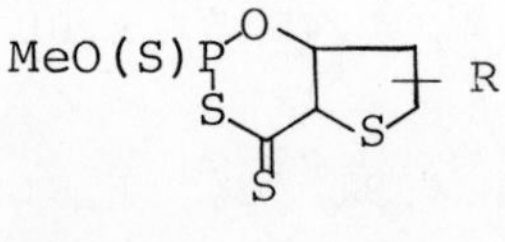

XII. Solids. R, m.: H, 145°; 6-Me, 125°; 7-Me, 101°; 6,7-Me_2, 116°; 7-Et-6-Me, 12 6-Ph, 130°.[124]

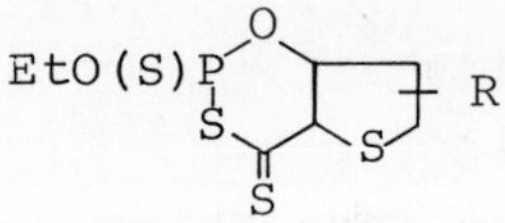

XII. Solids. R, m.: 6-Me, 91°, 7-Me, 69° 6,7-Me_2, 112°; 7-Et-6-Me, 84°.[124]

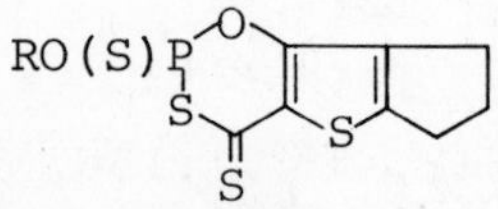

XII. Solids. R, m.: Me, 121°; Et, 111 [124]

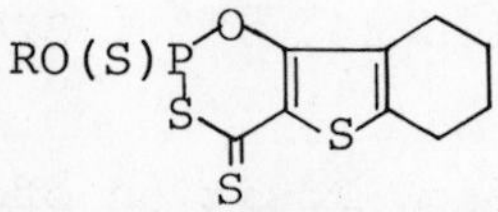

XII. Solids. R, m.: Me, 109°; Et, 72° [124]

F.3.3.5. ROP(O)$(SR')_2$

MeOP(O)$(SMe)_2$. XXVII. Liq. $b_{0.05}$ 60-62°, n_D^{20} 1.5340, d_4^{20} 1.2506.[480]

MeOP(O)$(SEt)_2$. XIV. Liq. $b_{0.055}$ 62.5-63°, n_D^{20} 1.5225, d^{20} 1.1831.[812]

MeOP(O)$(SPr)_2$. XIV. Liq. b_8 138-150°.[1394]

MeOP(O)$(SBu)_2$. XIV. Liq. $b_{0.065}$ 82.5-83.5°, n_D^{20} 1.5035, d^{20} 1.0823.[812]

$)_2$. IX. Amorphous solid. m. 185° (dec.)(fr

alcohol).[368]

EtOP(O)(S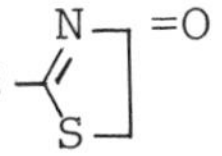)$_2$. IX. Amorphous solid. m. 173° (dec.) (from alcohol).[368]

EtOP(O)(SEt)$_2$. IX. XVII. XXI.[960,961] Liq. b_{20} 148°, d_0^0 1.1623.[960,961] Mol. ref. 321.97.[1078]

EtOP(O)(SPr)$_2$. XIV.[1394] Pale yellow liq. "Mocap."

EtOP(O)(SPh)$_2$. VIII.[472] IX.[310] Liq. $b_{0.01}$ 152-154°.[310,472]

EtOP(O)(SAr)$_2$. IX. Solids. Ar, m.: 4-ClC$_6$H$_4$, 77°; 4-NO$_2$C$_6$H$_4$, 106°.[310]

EtOP(O)(SAr)$_2$. IX. Liq. Ar, n_D^{20}: 4-MeOC$_6$H$_4$, 1.6152; 2-MeC$_6$H$_4$, 1.6078; 4-C$_6$H$_4$, 1.6067.[310]

Cl$_2$C=CHOP(O)(SR)$_2$. IX. Liq. R, $b_{0.01}$, n_D^{20}: Et, 112°, 1.5472; Bu, 146°.[1250]

cyclo-C$_6$H$_{11}$OP(O)(SAr)$_2$. IX. Solids. Ar, m.: Ph, 65° (from ligroin); 4-ClC$_6$H$_4$, 84° (from petroleum ether); 4-MeC$_6$H$_4$, 69°.[1202]

4-Me-cyclo-C$_6$H$_{10}$OP(O)(SPh)$_2$. IX. Solid. m. 65°.[1202]

F.3.3.6 ROP(O)(SR)(SR')

MeOP(O)(SMe)SCH$_2$CH$_2$NHCOMe. XVII. Liq. n_D^{20} 1.5442.[1340]

MeOP(O)(SMe)SCH$_2$CH$_2$SEt. XVII. Liq. $b_{0.01}$ 70°.[1166]

MeOP(O)(SMe)SCH$_2$CONHMe. XVII. Liq. n_D^{25} 1.5511.[192,235]

MeOP(O)(SMe)SCH$_2$CONEtSO$_2$Me. XVII. Liq. n_D^{20} 1.5385, d_4^{20} 1.3654.[480]

MeOP(O)(SMe)SCH$_2$CO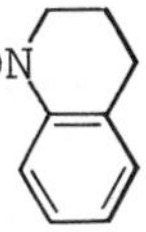 XVII. Liq. n_D^{20} 1.5860, d_4^{20} 1.2763.[480]

MeOP(O)(SMe)SCH$_2$CH$_2$CN. XVII. Liq. $b_{0.01}$ 82°.[1157]

MeOP(O)(SMe)SCHMeCOOEt. XVII. Liq. $b_{0.01}$ 66°.[1157]

MeOP(O)(SMe)SCH(COOEt)CH$_2$COOEt. XVII. Liq. n_D^{25} 1.4982.[192,235]

MeOP(O)(SMe)SCH$_2$CH=CHMe. XVII. Liq. $b_{0.3}$ 91-93°.[922] Also seven homologs and analogs.[922]

MeOP(O)(SMe)SCH$_2$Ph. XVII. Liq. n_D^{21} 1.5861.[1327] Also 52 analogs.[1327]

MeOP(O)(SMe)SCH_2– 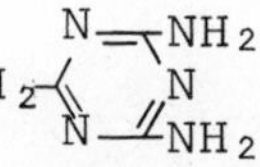 XVII. Solid. m. 180° (dec.).[336]

EtOP(O)(SMe)SCH_2CH_2SEt. VIII. Liq. $b_{0.05}$ 102°.[282]

EtOP(O)(SEt)SCH_2SEt. VIII. Liq. $b_{0.02}$ 85-87°.[282]

EtOP(O)(SEt)$SCH_2CH_2NEt_2$. XVII. Liq. $b_{0.01}$ 88°.[1157]

EtOP(O)(SEt)SCH_2CN. XVII. Liq. $b_{0.01}$ 82°.[1157]

EtOP(O)(SEt)SCH_2CH_2SEt. XVII.[1166,1327] Liq. $b_{0.01}$ 87°,[1166] n_D^{21} 1.5560,[1327]

EtOP(O)(SEt)SCH_2COOEt. VIII. Liq. $b_{0.01}$ 126°.[282]

EtOP(O)(SEt)SCH_2CH_2CN. XVII. Liq. $b_{0.01}$ 97°.[1157]

EtOP(O)(SEt)SCH=CHMe. XVII. Liq. $b_{0.7}$ 82-85°.[922]

EtOP(O)(SEt)SPh. VIII. Liq. $b_{0.01}$ 90°.[1167]

EtOP(O)(SEt)SC_6H_4Cl-4. VIII. Liq. $b_{0.01}$ 105°.[1167]

EtOP(O)(SEt)$SCH_2CHPhMe$. XVII. Liq. n_D^{27} 1.5511.[1332]

EtOP(O)(SCH_2CH_2Br)SC_6H_4Cl-4. VIII. Liq. n_D^{23} 1.5560.[454] Also five analogs.[454]

EtOP(O)(SCH_2CH_2Cl)SPh. XVII. Liq. n_D^{23} 1.5710.[1327]

EtOP(O)(SCH_2CH_2Cl)$SCH_2CH_2CH_2Ph$. XVII. Liq. n_D^{27} 1.5599.[1332] Also several homologs and analogs.[1332]

EtOP(O)($SCH_2C{\equiv}CH$)$SCH_2CH_2CH_2Ph$. XVII. Liq. n_D^{27} 1.5627.[1332]

EtOP(O)(SBu)SCH_2Ph. XVII. Liq. n_D^{18} 1.5510.[1286,1327]

EtOP(O)(SBu)$SCH_2CHPhMe$. XVII. Liq. n_D^{27} 1.5283.[1332]

EtOP(O)(SBu)$SCH_2CH_2CH_2Ph$. XVII. Liq. n_D^{27} 1.5379.[1332]

$BrCH_2CH_2OP$(O)(SMe)SPh. VIII. Liq. n_D^{23} 1.6023, 1.5798.[454]

$BrCH_2(ClCH_2)CHOP$(O)(SMe)SC_6H_4Cl-4. VIII. Liq. n_D^{23} 1.5993.[45

$ClCH_2MeCHOP$(O)(SMe)$SC_6H_4NO_2$-4. VIII.[453]

BuOP(O)(SEt)$SCH_2CH_2CH_2Ph$. XVII. Liq. n_D^{27} 1.5379.[1332]

$Me_2CHCH_2MeCHOP$(O)(SMe)SPh. IX. Liq. n_D^{21} 1.5528.[455] Also three analogs.[455]

$Me_2CHCH_2MeCHOP$(O)(SMe)SC_6H_4Me-4. IX. Liq. $b_{0.1}$ 160°.[455]

F.4. Phosphorothioates

F.4.1. Phosphorothioic Acid: Primary Esters

F.4.1.1. ROP(S)$(OH)_2$

MeOP(S)$(OH)_2$. II. Disodium salt. MeOP(S)$(ONa)_2 \cdot 6H_2O$. m.

49°.[253]

EtOP(S)(OH)$_2$. I. Disodium salt. Crystals from MeOH-acetone.[360] Silver salt.[961]

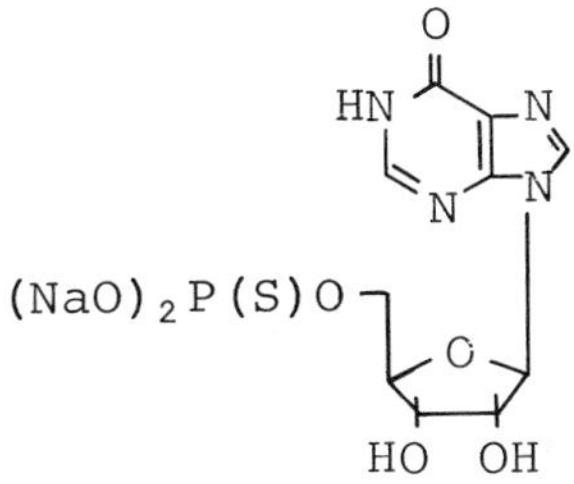

I. m. 205° (dec.) ^{31}P -43.1 ppm.[397]

F.4.1.2. RSP(O)(OH)$_2$

RSP(O)(OH)O^- $H_2N(C_6H_{11})_2^+$ salts. II. Solids. R, m.: Me, 159-160°; Et, 157-158°; i-Pr, 156°; Pr, 161-162°; $CH_2CH{=}CH_2$, 164°; i-Bu, 154-155°; s-Bu, 146-147°; Bu, 157°; $PhCH_2$, 151-152°.[1457]

EtSP(O)(OLi)$_2$. XVII. Solid (from water plus EtOH), From Li_3PO_3S and EtBr.[9]

$H_2NCH_2CH_2SP(O)(OH)_2$. XXV. Solid. Inner salt, m. 156-157°.[129] Also about 40 similar compounds.[129]

t-BuNHCH$_2$CH$_2$SP(O)(OH)$_2$. XXV. Solid. Inner salt, m. 190-192°.[129]

(NaO)$_2$P(O)SCH$_2$CH$_2$NHCSSNa. Infusible solid. From NaO(HO)P(O)CH$_2$CH$_2$NH$_2$ plus CS$_2$ and NaOH.[8]

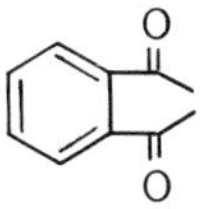

$N(CH_2)_mNH(CH_2)_nSP(O)(OH)_2 \cdot xH_2O$. VI. Solids.[486]

m	n	x	m.
2	2	1	131-132° (dec.)
3	3	1.3	158-160°
4	3	1.3	151-152°
5	2	1	148-150° (dec.)
5	3	1.3	179-180°

Also several homologs and analogs.[486,487]

$H_2N(CH_2)_mNH(CH_2)_nSP(O)(OH)_2 \cdot xH_2O$. VI. Solids.[957]

m	n	x	m.
2	2	1	139-141°

3	2	1	$160-161^{\circ}$
-	3	-	$293-295^{\circ}$ (dec.)
2	3	-	$168-170^{\circ}$
3	3	2	$140-143^{\circ}$
4	3	2	$171-172^{\circ}$
-	4	-	200° (dec.)

$CH_2SP(O)(OH)_2 \cdot H_2O$. VI. Solid. m. 235° (dec).[958]
NH_2

$H_2N-CH_2SP(O)(OH)_2$

$\quad\quad\quad\quad\quad\quad$ VI. Solid. m. 260° (dec).[958]

$H_2N-CH_2SP(O)(OH)_2$

$(NaO)_2P(O)S$ $\cdot 1.5\ H_2O$. VI. Solid. m. 208° (dec).
$\quad\quad\quad\quad\quad\quad\quad\quad$ ^{31}P -16.1 ppm.[397]

$SP(O)(OH)_2 \cdot H_2O$. I. Solid. m. $101-103^{\circ}$ (from acetone,
$COOH$ $\quad\quad\quad\quad\quad\quad\quad$ benzene).[325]

F.4.2. Phosphorothioic Acid: Secondary Esters

F.4.2.1. $(RO)_2POSH$

Thiono-thiol equilibrium, acidity and structure. K_{TS} = tautomeric equilibrium constant in solvent system S.[491,49]

R	7% EtOH in water			80% EtOH		
	pK_{H_2O}	K_{H_2O}	K_{TS}	pK_{EtOH}	K_{EtOH} x 10^3	K_{TS}
Me	1.18	6.61	0.38	2.50	3.16	3.32
Et	1.49	3.24	0.42	2.84	1.45	3.76
i-Pr	1.59	2.57	0.35	2.90	1.26	3.01
Pr	1.55	2.82	0.31	2.82	1.51	2.67
i-Bu	1.65	2.24	0.32	2.94	1.15	2.8

From values of K_{TS}: The thiol form predominates in aqueous solution; the thiono form predominates in EtOH solution.

$(MeO)_2POSH$. IV.[491,496,934] Liq. Characterized as ammonium or amine salts: NH_4 salt, m. $70-78^{\circ}$.[934] Et_3NH salt, m.

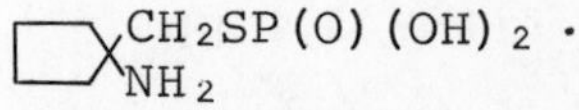

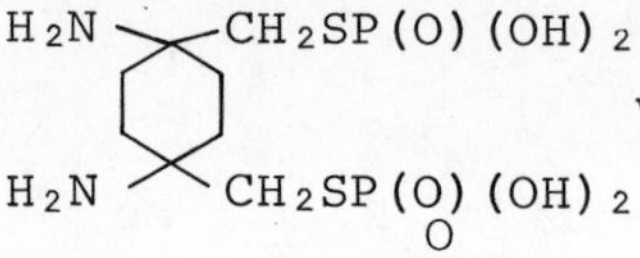

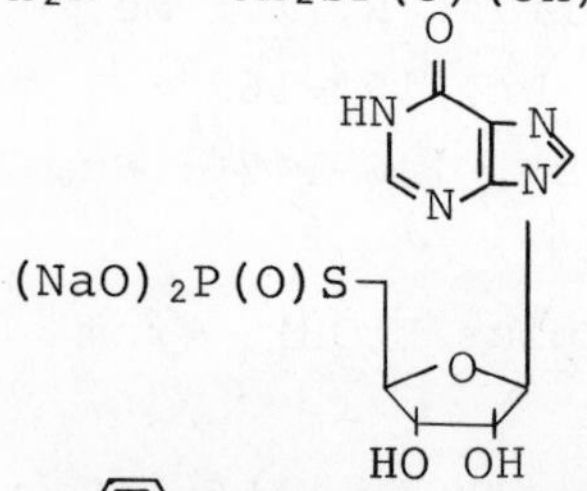

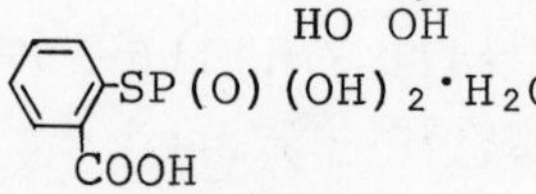

156^{0} (dec).[491,496] Na and K salts.[342] Amine salts: From $(MeO)_3P(S)$ plus amine $\rightarrow$ $(MeO)_2POS^{-}$ $MeNR_3^{+}$:
Solid salts:

$MeNR_3^{+}$		m.
NMe_4^{+}		$\sim 98^{0}$[1307]
$(C_6H_{11})_2NH_2^{+}$		$184\text{-}185^{0}$[120]
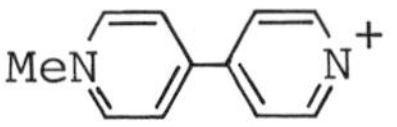	"Thiono"	$>210^{0}$ (dec)[533]
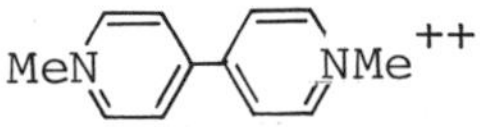	"Thiolo"	$59\text{-}61.5^{0}$[533]
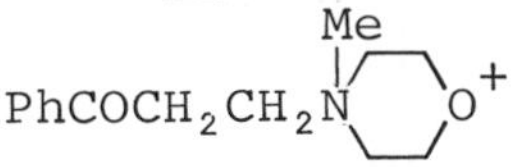		$113\text{-}115^{0}$ (from benzene).[754]

Liquid salts:

$MeNR_3^{+}$	n_D^{20}	d_4^{20}
$MeNH_2CH(CH_2OH)Et^{+}$	1.5291	1.215[1240]
$Me_2NHCH_2COOEt^{+}$	1.4961	1.2145.[1240]
$Et_2NMeCH_2CH_2OCONHC_6H_4Cl\text{-}3^{+}$	1.5683	1.2681.[1241]

$(EtO)_2POSH$. IV.[491,496] From $[(EtO)_2P(O)S]_2$ plus EtMgBr $\rightarrow$ $(EtO)_2POSH$ + $(EtO)_2P(O)SEt$.[721] Liq. $b_{0.05}$ $76\text{-}78^{0}$,[721] $b_{2.5}$ $106\text{-}107^{0}$.[491,496] n_D^{20} 1.4702,[721] 1.4719, d_4^{20} 1.1806.[491,496] IR.[528,673,674] ^{31}P -24.0.[326] μ 2.9 D.[466] NH_4 salt: IV. m. $144\text{-}145.5^{0}$,[934] ^{31}P -55.0.[715] Potassium salt: I.[725] VI.[5] Solid. m. $197\text{-}197.5^{0}$,[141,725] $197\text{-}201^{0}$ (from acetone).[5] IR.[374] ^{31}P -56.[488] Sodium salt: IV.[327,934] Solid. m. $196\text{-}198^{0}$,[492,934] 203^{0} (from $CHCl_3$-ether).[327] IR.[374] ^{31}P -57 ±1.[715,1345] $(C_6H_{11})_2NH_2$ salt: Solid. m. $155\text{-}156^{0}$.[120] $(EtO)_2POS^{-}$ $4\text{-}ClC_6H_4CH_2S{=}C(NH_2)_2^{+}$ (from $(EtO)_2POSNa$ plus $4\text{-}ClC_6H_4CH_2S{=}C(NH_2)_2^{+}$ Cl^{-}). Solid. m. $81\text{-}82^{0}$ (from ether).[418]
Amine salts: From $(EtO)_3P(S)$ plus amine $\rightarrow$ $(EtO)_2POS^{-}$ $EtNR_3^{+}$: Liquids:

$EtNR_3^{+}$	n_D^{20}	d_4^{20}
$EtNH_2CH(CH_2OH)Et^{+}$	1.5261	1.198.[1240]
$Me_2NHCH_2COOEt^{+}$	1.4922	1.1812.[1240]
$Bu_2NEtCH_2CH_2OCONHPh^{+}$	1.4945	1.0659.[1241]

$(FCH_2CH_2O)_2POSH$. IV. Liq. $b_{0.03}$ 97-100°, n_D^{20} 1.4595, d^{20} 1.4238.[490]

$(EtOCH_2CH_2O)_2POSHN_4$. IV. Solid. m. 106-110°.[934]

$(EtSCH_2CH_2POSNa$. IV. Solid. m. 109-110° (needles from ether-petroleum ether).[736]

$(i\text{-}PrO)_2POSH$. IV.[342,491,496] Liq. $b_{1.5}$ 89-90°, n_D^{20} 1.4592, d^{20} 1.0906.[491,496] Raman.[1443] Ammonium salt: IV. Solid. m. 156-157°.[934] Sodium and potassium salts.[342] $(C_6H_{11})_2NH_2^+$ salt: Solid. m. 199-200°.[120]

$(PrO)_2POSH$. I.[961] IV.[342,491,496] Liq. $b_{0.09}$ 108.5-109.5°.[491] Sodium and potassium salts.[342] Raman.[1443]

$(CH_2{=}CHCH_2O)_2POSNH_4$. IV. Solid. m. 133-140°.[934]

$(t\text{-}BuO)_2POSH$. Formed on attempted distillation of $(t\text{-}BuO)_3P(S)$; an exothermic reaction at 80° gave the product. Solid. m. 90-93° (dec.). IR.[716] ^{31}P -25.0 ppm.[715,716] Et_3NH^+ salt: IV. Solid. m. 80-81° (from ligroin).[1457]

$(i\text{-}BuO)_2POSH$. II.[961] IV.[342,491,496] Liq. $b_{0.05}$ 82.5-84°, n_D^{20} 1.4570, d^{20} 1.0386.[491,496] Raman.[1443] Ammonium salt. IV. Solid. m. 172-176°.[934] Potassium salt.[961] Silver salt, m. 160° (from MeOH-ether).[961]

$(BuO)_2POSH$. IV.[342,491,496] Liq. $b_{0.008}$ 75-76°,[491] $b_{0.08}$ 88-89°,[496] n_D^{20} 1.4654,[496] 1.4670,[491] d^{20} 1.0670,[491] 1.0672,[496] μ 2.8 D.[466] Ammonium salt: IV. Solid. m. 147-150°.[934]

$(i\text{-}AmO)_2POSH$. IV.[342,934] Ammonium salt. Solid. m. 167°.[934] Potassium salt. Needles from $CHCl_3$-ether.[342]

$(C_6H_{13}O)_2POSH$ μ 2.7 D.[466]

$(C_8H_{17}O)_2POSH$. μ 2.6 D.[466]

$(PhO)_2POSH$. IV. Solid. m. 86-87°.[491,496] Potassium salt: Solid. m. 165-167° (from acetone-benzene).[801,1360] Silver salt, m. 222-224° (dec.).[801] cyclo-$C_6H_{11}NH_3^+$ salt: From $(PhO)_2P(S)OPr$ + LiCl + $C_6H_{11}NH_2$. Solid. m. 127-128°.[596] $(PhO)_2POS^-$ Me_3S^+: From $(PhO)_2P(S)OMe$ plus Me_2S. Solid. m. 57-59° (from $CHCl_3$).[435]

$(4\text{-}NO_2C_6H_4O)_2POSK$. Solid. m. 232-233° (from alcohol).[1360]

$(Me_3SiO)_2POSH$. IV. Liq. b_3 93-95°, n_D^{20} 1.4200, d_4^{20} 0.9778.[1367]

F.4.2.2. RO(R'O)POSH

$(MeO(EtO)P(S)O^-)_2$ 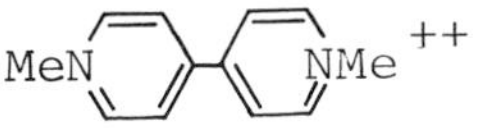 VII. From $2(MeO)_2P(S)$ OEt plus 4,4'-dipyridyl. Solid. m. 78-80°.[533]

$MeO(2,4,5\text{-}Cl_3C_6H_2O)POSH$. Amine salts. I. Solids.[104]

Amine	m.
NH_3	195°
NH_2NH_2	185-186°
$MeNH_2$	154-156°
$EtNH_2$	128-132°
$i\text{-}PrNH_2$	150-151°
$PhNH_2$	175-177°
$cyclo\text{-}C_6H_{11}NH_2$	199-200°
$(cyclo\text{-}C_6H_{11})_2NH$	176-179°

$MeO(2,4,5\text{-}Cl_3C_6H_2O)POS^-$. Ammonium salts. VII. From $(MeO)_2P(S)OC_6H_2Cl_3\text{-}2,4,5$ plus tert. amine. Solids.

Ammonium ion	m.
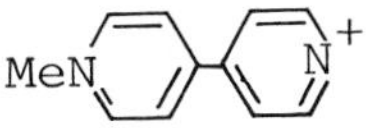	84-85° [533]
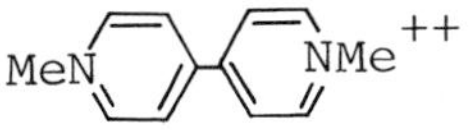	166° (dec.)[533]
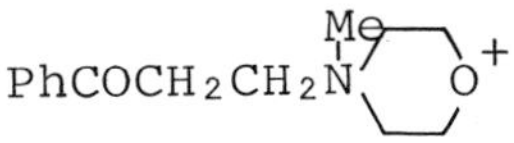	122-124° (from benaene)[754]

$MeO(4\text{-}NO_2C_6H_4O)POS^-$ Methyl strychninium$^+$. XXXII. Solids.[432]

d- salt, m. 217° (dec).$[\alpha]_D^{21}$ 13.6° (c. 0.25, MeCN).

l- salt, m.218° (dec).$[\alpha]_D^{21}$ -13.7° (c. 0.25, MeCN).

$MeO(4\text{-}t\text{-}Am\text{-}2\text{-}ClC_6H_3O)POSH$. $cyclo\text{-}C_6H_{11}NH_2$ salt. I. Solid. m. 97-98°.[104]

$MeO(1\text{-}C_{10}H_7O)POS^-$ -Methyl ephedrine$^+$ salts. XXXII. Solids.[22]

d- salt, m. 169-171°. $[\alpha]_D^{24}$ 33° (c. 1.44, $CHCl_3$)

l- salt. --- $[\alpha]_D^{24}$ -45.9° (c. 9.20, $CHCl_3$)

$MeO(4\text{-}Br\text{-}1\text{-}C_{10}H_6O)POS^-$ -Methyl ephedrine$^+$ salts. XXXII.

Solid. m. 149-157° $[\alpha]_D^{24}$ 27° (c. 1.028, $CHCl_3$).[221]

EtO(i-PrO)POSH. IV. Liq. $b_{0.05}$ 84°, n_D^{22} 1.4618.[786] Quinine salts: (i) m. 108-110°, $[\alpha]_D^{20}$ -122.8° (c. 0.378, MeOH); (ii) m. 80-82°, $[\alpha]_D^{20}$ -125° (c. 0.400, MeOH).[787]

EtO(PrO)POSH. IV. Liq. $b_{0.015}$ 84°, n_D^{22} 1.4649.[786]

EtO(i-BuO)POSH. IV. Liq. $b_{0.01}$ 90°, n_D^{22} 1.4651.[786] Silver salt, m. 149°.[961]

EtO(BuO)POSH. IV. Liq. $b_{0.08}$ 97-98°, n_D^{22} 1.4632.[786] Cinchonidine salts: (i) m. 140-142°, $[\alpha]_D^{22}$ -84.61° (c. 0.524, MeOH); (ii) m. 108-109°, $[\alpha]_D^{22}$ -80.5° (c. 0.631, MeOH).[787]

EtO(2,4,5-$Cl_3C_6H_2O$)POSH. cyclo-$C_6H_{11}NH_2$ salt. I. Solid. m. 125-126°.[104]

EtO(2,4,5-$Cl_3C_6H_2O$)POS^-. Ammonium salts. VII.

Ammonium ion	m.	n_D^{20}
	93.5-94.5[744]	--
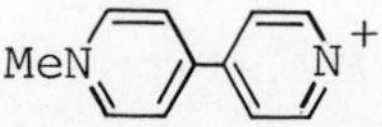	--	1.6141[533]
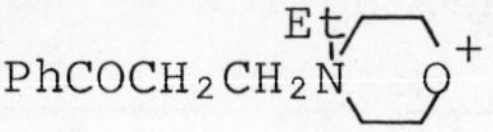	--	1.5765[754]

EtO(4-$NO_2C_6H_4O$)POSH. Amine salts. Solids. cyclo-$C_6H_{11}NH_2$ salt, m. 106-108°; (cyclo-C_6H_{11})$_2$NH salt, m. 160-161°.[192]

i-PrO($Me_3N^+CH_2CH_2CH_2O$)POS^-. Inner salt. From Me_3N plus

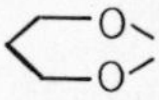 P(S)O-i-Pr. Solid. m. 218°.[870]

PhO($Me_3\overset{+}{N}CH_2CH_2CH_2O$)$POS^-$. Inner salt. From Me_3N plus 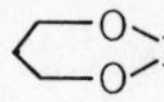 P(S)OPh. Solid. m. 169-170°.[870]

F.4.2.3. 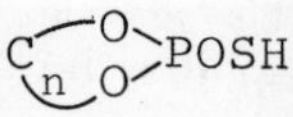

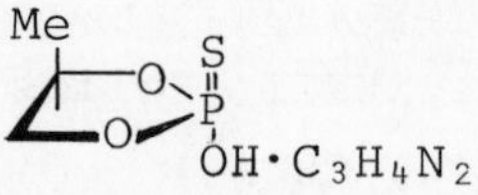 and 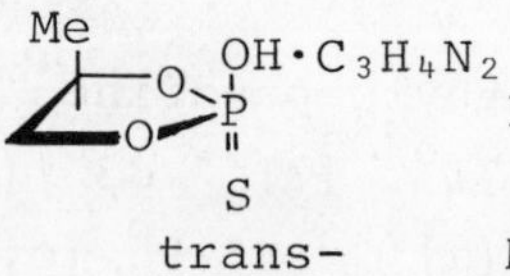 Imidazole salts. From

cis- trans-

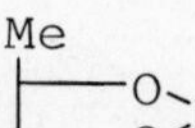 P(S)Cl plus

2 $C_3H_4N_2$. Two solids. m. 122-125° and 78.5-83°. ^{1}H-NMR. *cis*- and *trans*- assignments not made.[800]

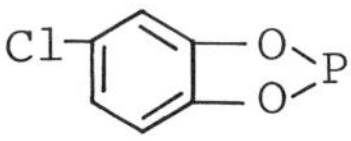POSH. Et_3N salt. From 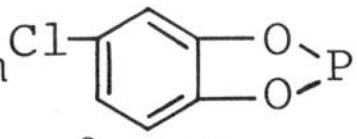P(S)Cl plus Et_3N and AcOH. Solid. m. 73.5-74.5°.[181]

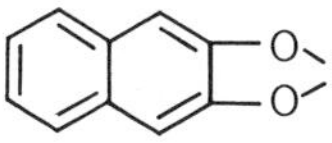POSH. Et_3N salt. From :P(S)Cl plus Et_3N and AcOH. Solid. m. 127-128.5°.[181]

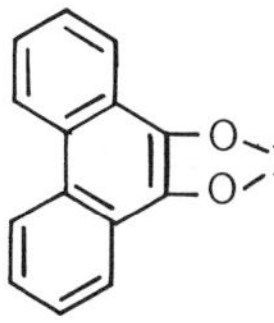POSH. Et_3N salt. From :P(S)Cl plus Et_3N and AcOH. Solid. m. 207-208°.[181]

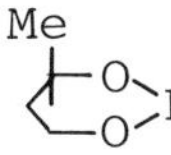POSNa. *cis*- and *trans*-. ^{1}H-NMR.[800]

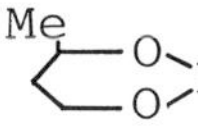POSH. Amine salts. Solids: $C_3H_4N_2$ salt, m. 99-100°. Mixture of cis-, trans-, and meso- forms; (cyclo-C_6H_{11})$_2$NH salt, m. 208-211°.[798]

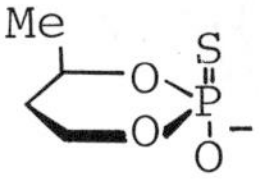 NMe_4+. From cis- $C_5H_{11}O_3PS$ plus Me_3N. Solid. m. 205-208°. ^{1}H-NMR.[797]

cis-

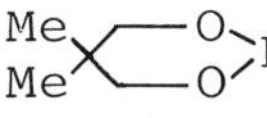POSH. Amine salts. Solids: Et_3N salt, m. 157-158° (from EtOAc-$CHCl_3$); cyclo-$C_6H_{11}NH_2$ salt, m. 240° (from i-PrOH).[247]

F.4.2.4. RO(RS)P(O)OH

MeO(MeS)P(O)O$^-$ Me_4N^+. VII.[156,437] Solid. m. 195-201° (from acetone-i-PrOH).[437] m. 110°.[156] Isomerization and kinetics.[156]

MeO(MeS)P(O)O$^-$ Et_3MeN^+. VII. Solid. m. 33-36°.[745]

MeO(MeS)P(O)O$^-$ Me_3PhN^+. VII. Solid. m. 123-124° (from acetone-i-PrOH).[437]

$MeO(MeS)P(O)O^{-}$ $RNHCOOCH_2CH_2\overset{+}{N}MeR'_2$. VII. Liq.[733]

R'	R = Me n^{20}	R = Me d_4^{20}	R = Et n_D^{20}	R = Et d_4^{20}
Et	1.5051	1.2241	1.5000	1.1872
i-Pr	1.4950	1.2106	1.4940	1.2016
Bu	1.4859	1.0885	1.4910	1.1230
i-Am	1.4810	1.0636	1.4878	1.1277
Am	--	--	1.4815	1.0927
C_6H_{13}	1.4832	1.0635	1.4800	1.0355

$MeO(MeS)P(O)O^{-}$ $MeSC(NH_2)_2^{+}$. VII. Solid. m. 145-146.8° (from ether-dioxane).[437]

$MeO(MeS)P(O)O^{-}$ Me_3S^{+}. VII. Solid. m. 116-117.5°.[436]

$RO(R'N^{+}CH_2CH_2S)P(O)O^{-}$. Inner salts. From 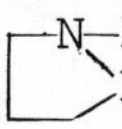 P(O)OR plus H_2O. Solids.[157]

R	R' = H m.	R' = Me m.
Me	164°	152-153°
Et	177	166-167
Pr	185	138-140
Ph	163	--

$EtO(MeS)P(O)O^{-}$ Me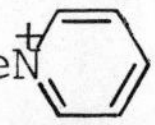 VII. Liq. n_D^{20} 1.5355, d_4^{20} 1.2445. Also several analogs.[744]

$EtO(MeS)P(O)O^{-}$ MeO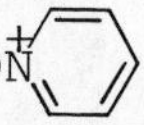 VII. Liq. n_D^{20} 1.5310, d_4^{20} 1.2837.[534]

$EtO(RS)P(O)O^{-}$ $4\text{-}ClC_6H_4CH_2S{=}C(NH_2)_2^{+}$. From $(EtO)_2P(O)SR$ plus NaSEt plus $4\text{-}ClC_6H_4CH_2S{=}C(NH_2)_2^{+}$ Cl^{-} (for identification of organophosphorus acids and esters). R, m.: Et, 151-152° (from water); Ph, 174-176° (from water).[418]

EtO(BuSP(O)OH. cyclo-$C_6H_{11}NH_2$ salt. VII.[596,795] Solid. m. 125-126°.[795]

EtO(PhS)P(O)OH. cyclo-$C_6H_{11}NH_2$ salt. VII. Solid. m. 130-131°.[795]

EtO 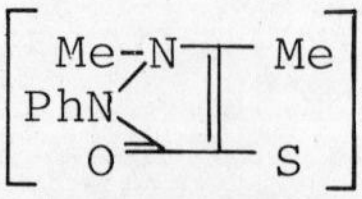 P(O)ONa. From $(EtO)_2P(O)S$-antipyrinyl plus

NaI. Solid. m. 248-249° (from i-PrOH).[1358]

PrO(MeS)P(O)O⁻ 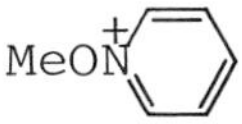 VII. Liq. n_D^{20} 1.5120, d_4^{20} 1.2731.[534]

PhO(MeS)P(O)O⁻ Me_3S^+. VII. Solid. m. 99° (from acetone).[435]

2,4,5-$Cl_3C_6H_2O$(RS)P(O)O⁻ tert-ammonium salts. VII. Solids.

Ammonium ion	R = Me m.	R = Et m.
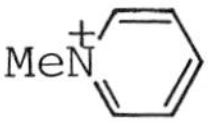	95-96°.[744]	--
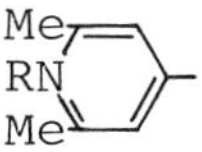 Me⁺	126.5-127.5°	70-71.5°[744]
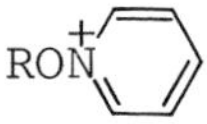	d_4^{20} 1.5319	74-76°[534]

4-$NO_2C_6H_4O$(MeS)P(O)O⁻ 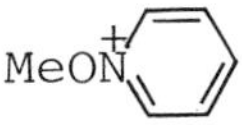 VII. Liq. n_D^{20} 1.5850, d_4^{20} 1.4250.[534]

4-$NO_2C_6H_4O$(MeS)P(O)O⁻ 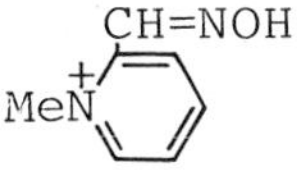 VII. Solid. m. 143-144° (from MeO -ether).[435]

4-$NO_2C_6H_4O$(MeS)P(O)O⁻ Me_3S^+. VII. Solid. m. 114-116°.[435]

4-$NO_2C_6H_4O$(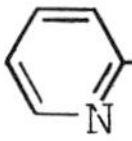 -S)P(O)OH. cyclo-$C_6H_{11}NH_2$ salt. From 4-NO_2 $C_6H_4OP(O)(OH)O^-$ + 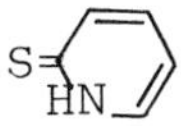 + Ph_3P + ( -S)$_2$; isolated as cyclo-$C_6H_{11}NH_2$ salt. Solid. m. 135° λ_{max} 275 mu (ε 17760) at pH 7.[848]

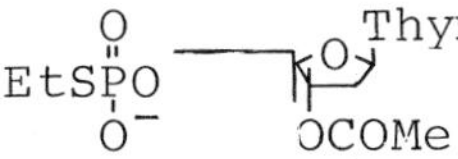 $Na^- \cdot H_2O$. White solid. $\lambda_{max}^{H_2O}$ 265 mu (ε 9300). IR.[183]

RO(RS)P(O)OH. R = cholesteryl. V., Solid, m. 188°.[1370]

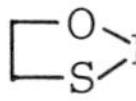 P(O)ONa. I. Solid. From $ClCH_2CH_2OP(S)Cl_2$ plus $BaCO_3$ in dioxane, plus Na_2SO_4. Crystals (from MeOH-ether).[360]

F.4.3. Phosphorothioic Acid: Tertiary thiolo Esters

F.4.3.1. $(RO)_2P(O)SR$

$(MeO)_2P(O)SMe$. VIII.[434] XVII.[154,253,438] XIX.[216] XXI.[961]

XXVII.[709] XXXI.[154,439,1302] XXXII.[440] Liq. $b_{0.01}$ 37°,[1302] $b_{0.15}$ 42°,[438] b_2 68-69°,[216] b_{17} 103°,[439,440] n_D^{20} 1.4654,[434] d_4^{20} 1.2484.[81] IR.[426,673,674] Alkaline hydrolysis, kinetics.[81]

$(EtO)_2P(O)SEt$. VIII.[959,961] XIII.[780] XVII.[142,729,935,1456] XIX.[947] XXI.[961] XXIV. (From $(EtO_3P$ + EtSCN → EtCN + $(EtO)_2P(O)SEt$).[792] XXVII. (from $(EtO)_2POSNa$ plus $Et_3O^+ BF_4^-$).[731] XXVII.[1372] XXVIII.[481,941] XXXI. (From $(EtO)_3P(S)$ plus EtBr,[492] or EtI,)[675] XXXIII. (From $(EtO)_2P(O)SCl$ or $[(EtO)_2P(O)S]_2$ plus EtMgBr.) From $(EtO)_2P(S)SK$ + 2,4-$(NO_2)_2C_6H_3Cl$ → $(EtO)_2P(O)SEt_2$ + some $(EtO)_2P(S)SEt$ + $(EtO)_2POSH$ + $(2,4-(NO_2)_2C_6H_3)_2S$.[937] Liq. $b_{3.5}$ 65-66°,[729] b_5 83°,[675] b_8 105-106°,[731] b_{12} 115°,[780] b_{760} 237° (sl. dec.).[961] n_D^{20} 1.4577,[1078] n_D^{25} 1.4553,[142,675] Mol. ref. 288.96,[1078] d^{20} 1.1063,[492] d_4^{20} 1.1082,[941] $d^{31.3}$ 1.0957, d^{60} 1.0681,[125] IR.[374,528,673,674] Raman.[968] ^{31}P -26.4.[850]

$(EtO)_2P(O)SEt \cdot HOOCCCl_3$. XXXII. Liq. complex. $b_{0.01}$ 82° IR and ^1H-NMR showed H bonding between the basic O of the P(O) group and OH of trichloroacetic acid.

$(PrO)_2P(O)SPr$. XVII, XXI. Liq. b_{20} 156°, d_0^0 1.0532.[961]

$(i\text{-}BuO)_2P(O)S\text{-}i\text{-}Bu$. XVII, XXI. Liq. b_{20} 170°, d_0^0 1.0101.[961]

$(PhO)_2P(O)SPh$. VIII. Liq. b_{25} 275-282°. (See Kosolapoff, p. 262.)

$(PhCH_2O)_2P(O)SCH_2Ph$. XXVIII. Liq. (decomp. on distillation)[1417]

F.4.3.2. $(RO)_2P(O)SR'$

$(EtO)_2P(O)SMe$. XVII.[1123] XIX.[836,1123] XXIV.[1233] XXVII.[491] [491,709] XXVIII.[97] (From $(EtO)_3P$ plus $MeSSCCl_3$ or MeS SPh).[417] Liq. $b_{0.3}$ 65°,[97] b_4 89°,[417] b_8 106°,[140] b_{18} 122°,[1233] n_D^{20} 1.4607,[81,491] n_D^{25} 1.4559,[1233] d^{20} 1.1418,[491] d^{25} 1.1168.[836] Alkaline hydrolysis, kinetics.[81]

$(i\text{-}PrO)_2P(O)SMe$. XXVIII. Liq. $b_{0.1}$ 55°.[97] $b_{0.06}$ 63-64°, n_D^{20} 1.4555, d_4^{20} 1.0767.[81]

$(PrO)_2P(O)SMe$. XXVII.[709] Liq. $b_{0.15}$ 84-85°,[709] b_{20} 143°,[81]

n_D^{20} 1.4565,[709] 1.4603,[81] d^{20} 1.0744,[709] d_4^{20} 1.0864.[81] Alkaline hydrolysis, kinetics.[81]

$(i\text{-}BuO)_2P(O)SMe$. Liq. $b_{0.035}$ 88-89°, n_D^{20} 1.4558, d_4^{20} 1.0392. Alkaline hydrolysis, kinetics.[81]

$(BuO)_2P(O)SMe$. XXVII.[709] Liq. $b_{0.25}$ 90°,[709] b_{11} 148-149°,[81] n_D^{20} 1.4590,[709] 1.4599,[81] d_4^{20} 1.0481,[81] d^{20} 1.0495.[709] Alkaline hydrolysis, kinetics.[81]

$(PhO)_2P(O)SMe$. XXXI. Liq. $b_{0.01}$ 145-147°, n_D^{20} 1.5765.[435]

$(4\text{-}NO_2C_6H_4O)P(O)SMe$. VIII.[434] XVII.[1360] XXXI.[435] XXXIII. (From $(MeO)_2P(S)Cl$ plus $4\text{-}NO_2C_6H_4OH$ in the presence of C_5H_5N catalyst: disprop. and transesterification).[76] Solid. m. 108-109° (from alcohol).[434,1360] Crystallography.[401]

$(4\text{-}EtOCOC_6H_4O)_2P(O)SMe$. XXXI. (From $(RO)_2P(S)OMe$ at 200-210°, 1-2 mm pressure, 10 min). Solid. m. 89-90° (from petroleum ether).[1185]

$(MeO)_2P(O)SCH_2Cl$. XXIV. Liq. $b_{0.25}$ 69.5-70°, n_D^{20} 1.4825.[1339]

$(EtO)_2P(O)SCH_2Cl$. XVII.[949] XXIV.[1339] XXVI. (From $(EtO)_2P(O)SCl$ plus CH_2N_2).[944] Liq. $b_{0.35-0.4}$ 85.5-86.5°,[1339] $b_{1.3}$ 95-98°,[949] $b_{3.5}$ 113-115°,[944] n_D^{20} 1.4752,[949] d^{18} 1.2610.[944]

$(RO)_2P(O)SCHCl_2$. XIX. Liq. R, n_D^{25}: Me, 1.4893; Et, $b_{0.1}$ 82-83°, 1.4816; $ClCH_2CH_2$, 1.5166.[96]

$(RO)_2P(O)SCCl_3$. XX. Liq. R, b./mm, n_D^{20}: Me, 110°/0.5, 1.5110; Et, 122°/0.7, 1.4992; i-Pr, 126°/0.4, 1.4876; Pr, 133°/0.5, 1.4923; i-Bu, 144°/0.5, 1.4818; Bu, 151°/0.5, 1.4848.[266]

$(MeO)_2P(O)SCCl_2F$. XIX. Liq. $b_{0.1}$ 63-65°.[694]

$(EtO)_2P(O)SCCl_2F$. XIX. Liq. $b_{0.1}$ 66-69°.[694]

$(EtO)_2P(O)SCH_2NHCOOR$. From $(EtO)_2P(O)SCH_2NCO$ plus ROH. Liq. R, b./mm, n_D^{20}, d_4^{20}: Me, 115-118°/0.05, 1.4750, 1.2006; Et, 118-120°/0.08, 1.4698, 1.1438; i-Pr, 112-115°/0.05, 1.4675, 1.1378.[1238]

$(EtO)_2P(O)SCH_2NCO$. XVII. Liq. $b_{0.06}$ 94-96°, n_D^{20} 1.4720,

d_4^{20} 1.2226, IR.[1238]

$(EtO)_2P(O)SC(=NC_6H_3Et_2\text{-}2,6)NHC_6H_3Et_2\text{-}2,6$. XXII. Solid. m. 75° (from ligroin).[691]

$(EtO)_2P(O)SCH_2OC_6H_2Cl_3\text{-}2,4,5$. XVII. Liq. n_D^{25} 1.5520.[1293]

$(EtO)_2P(O)SCH_2OCH_2$ 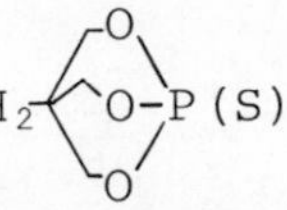 . XVII. Solid. m. 95° (from EtOH).[1009]

$(EtO)_2P(O)SCH_2O$ 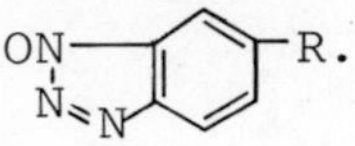 R. XVII. Solids. R, m.: H, 74-76° OMe, 106-108°.[302]

$(EtO)_2P(O)SCH_2SC(S)SEt$. From $(EtO)_2P(O)SCH_2Cl$ plus EtSCSSN Liq. n_D^{25} 1.5622. Also homologs.[1375]

$(EtO)_2P(O)SCH_2SEt$. XVII. (From $(EtO)_2PONa$ plus S and $EtSCH_2Cl$.)[1123] XXI.[1145] Liq. $b_{0.01}$ 75-80°,[1145] $b_{1.5}$ 129-130°.[1123]

$(EtO)_2P(O)SCH_2SCH_2CH_2Cl$. XVII. Liq. $b_{0.01}$ 108°.[1198]

$(EtO)_2P(O)SCH_2SR$. XXI. Liq. R, $b_{0.01}$: i-Pr, 84°; Pr, 91-92°; Bu, $b_{0.04}$ 100°; CH_2CH_2SEt, 115°.

$(EtO)_2P(O)SCH_2SC_6H_4Me\text{-}4$. XVII. Liq. $b_{0.5}$ 178°.[1123]

$(EtO)_2P(O)SCH_2SC_6H_4Et\text{-}4$. XVII. Liq. b_1 136°.[1123]

$(EtO)_2P(O)SCH_2SC_6H_4COMe\text{-}4$. XVII. Liq. $b_{0.01}$ 160°.[1080]

$(EtO)_2P(O)SCH_2S$ 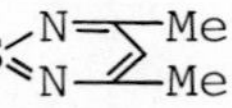 From $(EtO)_2P(O)SCH_2Cl$ plus

NaS 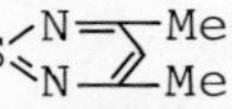 . Liq. n_D^{24} 1.5128,[629] n_D^{25} 1.5273.[1306]

$(EtO)_2P(O)SCH_2SCH_2Ph$. XVII. Liq. $b_{0.01}$ 138°.[639]

$(EtO)_2P(O)SCH_2SCH_2C_6H_4Cl\text{-}2$. XVII. Liq. $b_{0.01}$ 135°.[639]

$[(EtO)_2P(O)S]_2CH_2$. XVII. Liq. n_D^{25} 1.4678.[1391,1393]

$(EtO)_2P(O)SCH_2S(O)Et$. Liq. $b_{0.2}$ 150°, n_D^{30} 1.4981.[765]

$(EtO)_2P(O)SCH_2SO_2Et$. Liq. From $(EtO)_2P(O)SCH_2SEt$ plus $KMnO_4$.[281,843] $b_{0.01}$ 106°,[281,843] $b_{0.2}$ 150°.[765]

$(MeO)_2P(O)SCH_2S(O)C_6H_4Cl\text{-}3$. From $(MeO)_2P(O)SCH_2SC_6H_4Cl\text{-}3$ plus $KMnO_4$. Liq. n_D^{30} 1.5756.[277]

$(EtO)_2P(O)SCH_2SePh$. XVII. Liq. $b_{0.01}$ 118°.[1152]

<u>$(RO)_2P(O)SEt$</u>

$(MeO)_2P(O)SEt$. XIX. Liq. $b_{0.05}$ 32-35°.[216]

$(i\text{-}PrO)_2P(O)SEt$. XXVIII. Liq. b_{12} 113-116°, n_D^{20} 1.4503,

d_4^{20} 1.0452.[941]

$(PrO)_2P(O)SEt$. XIX.[836] XXVIII.[941] Liq. b_8 127-128°,[941] b_{25} 141-142°,[836] n_D^{20} 1.4564, d_4^{20} 1.0516,[941] d^{25} 1.0588.[836]

$(i\text{-}BuO)_2P(O)SEt$. XXVIII. Liq. b_7 131-134°, n_D^{20} 1.4510, d_4^{20} 1.0152.[941]

$(BuO)_2P(O)SEt$. XIII.[780,794] XIX.[836] XXIV.[792] XXVIII.[941] Liq. $b_{0.3}$ 87-89°,[780] b_4 132°,[836] b_7 143-145°,[941] n_D^{20} 1.4548,[792] n_D^{25} 1.4524,[794] d_4^{20} 1.0273.[941]

$(PhO)_2P(O)SEt$. XVII.[1360] From $(PhO)_2PCl$ plus EtSCl in liq. SO_2 at -50°.[945] Liq. b_1 156-159°,[1360] b_9 222-224°,[945] n_D^{20} 1.5685,[1360] d_4^{20} 1.2290.[945]

$(4\text{-}NO_2C_6H_4O)_2P(O)SEt$. XVII. Solid. m. 111° (from EtOH). IR.[1360] Crystallography.[401]

$(4\text{-}MeOCOC_6H_4O)_2P(O)SEt$. XXXI. Solid. m. 104° (from petroleum ether).[1185]

$(EtO)_2P(O)SCH_2CH_2Br$. XVII.[1144] XXV.[788] Liq. $b_{0.02}$ 69-70°,[1135] $b_{0.02}$ 75°,[1144] $b_{0.2}$ 86°,[788] n_D^{23} 1.4900.[788]

$(MeO)_2P(O)SCH_2CH_2Cl$. XVII.[154,1307] XIX.[947] XXV.[785,842] XXXI.[154] Liq. $b_{0.03}$ 69-72°,[785] $b_{0.3}$ 82.5,[842] b_4 119°,[947] n_D^{20} 1.4952,[947] n_D^{24} 1.4892,[785] d^{20} 1.2941,[947] ^1H-NMR.[842]

$(EtO)_2P(O)SCH_2CH_2Cl$. VIII. (from $ClCH_2CH_2OP(O)Cl_2$ plus 2 EtOH and Et_3N).[723] XV. (From $(EtO)_2P(O)Cl$ plus
O
CH_2-CH_2 and S.)[729] XVII.[729,935] XIX.[947] XXV.[602,775,777,785,836,944] Liq. $b_{0.06}$ 68°,[602] $b_{0.1}$ 88°,[723] b_4 120°,[947] n_D^{20} 1.4785,[602,775] 1.4825,[723] d^{20} 1.2178,[777] 1.2260,[723] d^{25} 1.2036.[836]

$(i\text{-}PrO)_2P(O)SCH_2CH_2Cl$. XXV. Liq. $b_{0.09}$ 100°, n_D^{20} 1.4680, d^{20} 1.1462.[602,775]

$(PrO)_2P(O)SCH_2CH_2Cl$. XXV. Liq. $b_{0.005}$ 72°, n_D^{20} 1.4751, d^{20} 1.1657.[602,775]

$(BuO)_2P(O)SCH_2CH_2Cl$. XXV. Liq. $b_{0.01}$ 95.2°, n_D^{20} 1.4731, d^{20} 1.1208.

$(PhO)_2P(O)SCH_2CH_2Cl$. XXV. Liq. $b_{0.01}$ 145-150°, n_D^{23} 1.5735.[785]

$(EtO)_2P(O)SCH(SEt)CH_2Cl$. XVII. Liq. $b_{0.01}$ 108°.[1177]

$(MeO)_2P(O)SCHClCHCl_2$. VIII. Liq. n_D^{25} 1.5095.[4]

$(MeO)_2P(O)SCH_2CCl_3$. VIII.[4] XIX.[424] Liq. $b_{0.012}$ 92-93°,[424] $b_{0.04}$ 100-102°,[4] n_D^{25} 1.5070,[4] 1.5101,[424] d_{20}^{20} 1.523.[424]

$(EtO)_2P(O)SCH_2CCl_3$. XIX. Liq. $b_{0.019}$ 106-108°, n_D^{25} 1.4490, d_{20}^{20} 1.406.[424] Also several analogs.[424]

$(PhO)_2P(O)SCH_2CCl_3$. XIX. Liq. $b_{0.018}$ 182-183°, n_D^{25} 1.5749, d_{20}^{20} 1.421.[424]

$(EtO)_2P(O)SCH(OMe)CCl_3$. XVII. Liq. $b_{0.01}$ 72°,[217]

$(EtO)_2P(O)SCH(OEt)CCl_3$. XVII. Liq. $b_{0.01}$ 86-88°.[217]

$(i\text{-}PrO)_2P(O)SCH(OEt)CCl_3$. XVII. Liq. $b_{0.01}$ 87°.[217]

$(EtO)_2P(O)SCH(SEt)CCl_3$. XVII. Liq. $b_{0.01}$ 92°.[1153]

$(EtO)_2P(O)SCH(SPr)CCl_3$. XVII. Liq. $b_{0.05}$ 115°.[1153]

$(EtO)_2P(O)SCH_2CH_2NMe_2$. VIII.[330,365] Liq. $b_{0.225}$ 70°,[365] $b_{0.8}$ 105-112°,[330] n_D^{21} 1.4680.[365]

$(EtO)_2P(O)SCH_2CH_2NEt_2$. VIII.[366] XXIV.[366,1134] Liq. $b_{0.04}$ 88°,[143] $b_{0.2}$ 97°,[365] b_2 134°,[1134] n_D^{21} 1.4732,[365] n_D^{25} 1.4666,[143] "Amiton" high-temperature decomp.[141a,143]

$(EtO)_2P(O)SCH_2CH_2N$ VIII. Liq. $b_{0.01}$ 108°, n_D^{21} 1.4875.[365]

$(RO)_2P(O)SCH_2CH_2N$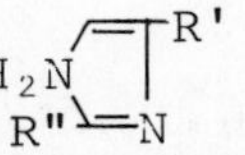 XVII.[1077] XXXIII. (from $(RO)_2P(S)S$-CH_2CH_2N-ring plus NHO_3). Liq. Solid picrates:

R	R'	R"	Picrate, m.
Me	H	Me	101-102°[1075]
Et	H	Me	99-102°[1075,1077]
Et	Me	Et	81-82°[1077]

$(EtO)_2P(O)SCH_2CH_2NHCOOEt$. XVII. Liq. n_D^{30} 1.4547.[1273] Also several analogs.[1273]

<u>$(RO)_2P(O)SCH_2CH_2Ar$</u>

$(EtO)_2P(O)SCH_2CH_2NMeC_6H_4R$. XVII. Liq.[1304]

R	b./mm	n_D^{20}	d^{20}	MeOSO$_3$H salt n^{20}
H	132-133°/0.001	1.5424	1.1500	1.5215
3-Me	140-142°/0.001	1.5400	1.1322	1.5205
4-Me	135-137°/0.001	1.5385	1.1235	1.5026

3-Cl	138-140°/0.0001	1.5517	1.2167	1.5237
4-Cl	140-142°/0.0001	1.5507	1.2177	m. 70-71°

$(EtO)_2P(O)SCH_2CH_2NMeSO_2Ph$. XVII. Liq. n_D^{30} 1.5332.[1263]

$(EtO)_2P(O)SCH_2CH_2NEtC_6H_4Cl$-2. XVII. Liq. $b_{0.01}$ 140°.[1165]

$(EtO)_2P(O)SCH_2CH_2NEtC_6H_4Me$-4. XXIV. Liq. $b_{0.0004}$ 135-140°.[147] Also several analogs.[147]

$(MeO)_2P(O)SCH_2CH_2NMePh$. XVII. Liq. $b_{0.01}$ 102°.[1165]

$(MeO)_2P(O)SCH_2CH_2NEtPh$. XVII. Liq. $b_{0.01}$ 128°.[1165]

$(MeO)_2P(O)SCH_2CH_2NEtC_6H_4Cl$-2. XVII. Liq. $b_{0.01}$ 146°.[1165]

$(EtO)_2P(O)SCH_2CH_2N$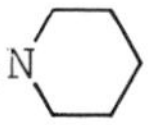 VIII.[365,663] IX.[366] Liq. $b_{0.01}$ 108-110°,[366] $b_{0.08}$ 136-140°,[663] n_D^{21} 1.4941.[365]

$(EtO)_2P(O)SCH_2CH_2N$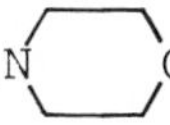O. XVII. Liq. $b_{0.08}$ 136-140°.[662]

$(EtO)_2P(O)SCH_2CH{=}NNHCONH_2$. By standard procedures for semicarbazones. Solid. m. 160°.[781]

$(EtO)_2P(O)SCH_2CN$. XVII.[65,152] Liq. b_3 132-135°, n_D^{20} 1.4700, d^{20} 1.1970.[65]

$(EtO)_2P(O)SCH(SC_6H_4Me\text{-}4)CN$. XVII. Liq. n_D^{25} 1.5489.[1108]

$(i\text{-}PrO)_2P(O)SCH_2CH_2OMe$. XVII. Liq. $b_{0.5}$ 94-95°.[360]

$(RO)_2P(O)SCH_2CH_2OCH{=}CH_2$. XXVIII. Liq.[1321]

R	b_1	n_D^{20}	d_4^{20}	
Et	104-105°	1.4783	1.1310	IR
Pr	111-113°	1.4773	1.0944	
Bu	126°	1.4700	1.0545	IR

$(MeO)_2P(O)SCH_2CH_2OCOMe$. VIII. Liq. $b_{0.09}$ 83°.[1351]

$(EtO)_2P(O)SCH_2CH_2OCOMe$. VIII. Liq. $b_{0.01}$ 92-94°.[1351]

<u>$(RO_2P(O)SCH_2CH_2SR'$</u>

$(EtO)_2P(O)SCH_2CH_2SMe$. XVII. XIX. Liq. b_2 134-138°.[1123]

$(EtO)_2P(O)SCH_2CH_2SCH_2SEt$. XVII. Liq. $b_{0.05}$ 150°.[1146]

$(MeO)_2P(O)SCH_2CH_2SEt$. XIX.[216] XXVII.[642] XXVIII.[297,647] XXXIII.[595,643] Liq. $b_{0.01}$ 75-76°,[643] $b_{0.1}$ 115°,[595] n_D^{20} 1.5052.[595]

$(EtO)_2P(O)SCH_2CH_2SEt$. XVII.[492,1115,1123] XXIV.[1191] XXVII.[646] XXVIII.[297,647] XXXI.[340,449] XXXIII.[595] From $(EtO)_2P(O)SCH_2CH_2Br$ plus EtSNa.[286,1135] Liq. $b_{0.01}$ 85°,[297,

[647] b_2 132-134°,[1191] n_D^{20} 1.4982,[492] 1.5027,[647] n_D^{25} 1.4922,[449] Raman.[968] (1/3 component of "Systox," "Demeton" - see Section F.4.4.) $(EtO)_2P(S)OCH_2CH_2SEt)$. L. Me_2SO_4. Liq. n_D^{29} 1.5023.[347]

EtO(i-PrO)P(O)SCH_2CH_2SEt. XVII. Liq. $b_{0.5}$ 118°.[1115]

EtO(BuO)P(O)SCH_2CH_2SEt. XVII. Liq. $b_{0.01}$ 98°.[1115]

(BuO)P(O)SCH_2CH_2SEt. XXVIII. Liq. $b_{4.5}$ 193-195°, n_D^{20} 1.495 d_4^{20} 1.0643.[941]

$(EtO)_2P(O)SCH(OCOMe)CH_2SEt$. XVII. Liq. $b_{0.12}$ 125-127°.[1093]

$(EtO)_2P(O)SCH_2CH_2SR$. From $(EtO)_2P(O)SCH_2CH_2Br$ plus RSNa. Liq.

R	b./mm
CH_2CH_2OH	140°/0.01
C(S)OEt	140°/0.1
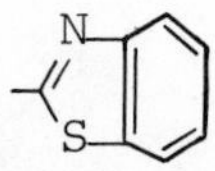	190°/0.05

$(EtO)_2P(O)SCH_2CH_2SCOMe$. XVII. Liq. $b_{0.01}$ 101°.[1141]

$(EtO)_2P(O)SCH_2CH_2SCH{=}CH_2$. XXII. Liq. $b_{0.01}$ 86°.[1158]

$(EtO)_2P(O)SCH_2CH_2SP(O)(OEt)_2$. XXIV (from 2 $(RO)_2PONa$ plus $(CH_2SCN)_2$).[1123] Liq. $b_{0.5}$ 190°.[1123]

$(EtO)_2P(O)SCH_2CH_2S$-i-Pr. XXXI. Liq. $b_{0.04}$ 96-97°, n_D^{25} 1.4881, d^{25} 1.0164.[449]

$(MeO)_2P(O)SCH_2CH_2SCHMeCONHMe$. XVII.[771,1019] Solid. m. 32-33°,[1019] Optical isomers: l-$[\alpha]_D^{21}$ -79° (c. 10.0, water d-$[\alpha]_D^{23}$ 73° (c. 5.0, water).[771]

$(EtO)_2P(O)SCH_2CH_2SAr$. From $(EtO)_2P(O)SCH_2CH_2Br$ plus ArSNa. Liq. Ar, b./mm: Ph, 137°/0.01; 4-MeC_6H_4, 148°/0.1; 4-ClC_6H_4, 163°/0.2; 3,4-$Cl_2C_6H_3$, 171°/0.2.[1135]

$(EtO)_2P(O)SCH_2CH_2SC_6H_{11}$-cyclo. IX. Liq. $b_{0.1}$ 150-155°.[1064]

$(MeO)_2P(O)SCH_2CH_2S(O)Et$. By oxidation of $(MeO)_2P(O)SCH_2CH_2$ SEt with H_2O_2,[582] or Br_2-water,[628] Liq. $b_{0.01}$ 105-106° [628] ^{1}H-NMR.[74,524,526] "Meta-Systox R*."[582]

$(EtO)_2P(O)SCH_2CH_2S(O)Et$. XVII.[289,627] By oxidation of $(EtO)_2P(O)SCH_2CH_2SEt$ with H_2O_2,[582] or Br_2-water.[628] liq. $b_{0.01}$ 108°,[627,628] 115°,[289] $b_{0.05}$ 150°.[765]

$(MeO)_2P(O)SCH_2CH_2SO_2Et$. From $(MeO)_2P(O)SCH_2CH_2SEt$ plus $KMnO_4$. Solid. m. 52° (from benzene), $b_{0.01}$ 115°.[281,843]

$(EtO)_2P(O)SCH_2CH_2SO_2Et$. XVII.[289] From $(EtO)_2P(O)SCH_2CH_2SEt$ plus $KMnO_4$.[281,843] $b_{0.01}$ 115°,[289] $b_{0.1}$ 170°.[765]

$(EtO)_2P(O)SCH_2CH_2SO_2Ph$. From $(EtO)_2P(O)SCH_2CH_2SPh$ plus $KMnO_4$. Liq. $b_{0.01}$ 148°.[843]

$(EtO)_2P(O)SCH_2CHO$. XXV. Liq. $b_{0.04}$ 82°, n_D^{25} 1.4708.[781]

$(EtO)_2P(O)SCH_2CH(OEt)_2$. IX.[172,782] XXV.[781,782] Liq. $b_{0.06}$ 80-80.5°,[781,782] $b_{0.2}$ 118°,[172] n_D^{25} 1.4531.[781]

<u>$(RO)_2P(O)SCH_2CONHR'$</u>

$(MeO)_2P(O)SCH_2CONHMe$. XVII.[88,825] XXVIII.[7] Liq. $b_{0.005}$ 105°,[7] $b_{0.1-0.2}$ 150-152°,[825] n_D^{20} 1.4987,[825] n_D^{25} 1.4984.[88] ^1H-NMR.[526] Mass spect.[196] ("Dimethoate" - O-analog.)

$(MeO)_2{}^{32}P(O)SCH_2CONHMe$. XVII.[382]

$(EtO)_2P(O)SCH_2CONHCH_2SMe$. From $(EtO)_2P(O)SCH_2CONHCH_2SH$ plus MeSH and HCl. Liq. n_D^{20} 1.5079.[821]

$(EtO)_2P(O)SCH_2CONHCOOEt$. XVII. Solid. m. 44°.[1139]

$[(i\text{-}PrO)_2P(O)SCH_2CONH]_2CH_2$. XVII.[1350,1398] Solid. m. 90° (from benzene-petroleum ether).[1398]

$[(EtO)_2P(O)SCH_2CONH]_2(CH_2)_2$. XVII.[1350,1398] Solid. m. 61° (from ether).[1398]

$(MeO)_2P(O)SCH_2CONHCH_2CH_2OMe$. XVII. Liq. $b_{0.12}$ 133-135°.[82,176]

$(MeO)_2P(O)SCH_2CONHCH_2CH=CH_2$. XVII. Liq. n_D^{19} 1.5040.[1094]

$(MeO)_2P(O)SCH_2CONHCONHMe$. XVII. Solid. m. 87-89° (from benzene-petroleum ether).[395]

$(RO)_2P(O)SCH_2CONHSO_2Ph$. XVII. Solids. R, m.: Me, 114-115°; Et, 78-80°; Pr, 56-57°; Bu, 46-47°.[698] Also several analogs.[698]

$(RO)_2P(O)SCH_2CONH$ XVII. Solids. R, m.: Me, 75°; Et, 101°; i-Pr, 77-78°.[1333]

$(RO)_2P(O)SCH_2CONH$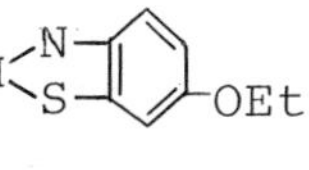 VIII. Solids. R, m.: Me 115°; Et, 120°; i-Pr, 146°.[1333]

$(EtO)_2P(O)SCH_2CONHCOPh$. XVII. Solid. m. 56-58°.[283]

$(EtO)_2P(O)SCH_2CONHCOC_6H_4Cl$-4. XVII. Solid. m. 72-75°.[283]

$(EtO)_2P(O)SCH_2CONHC_6H_4NO_2$-4. VIII, XVII. Solid. m. 110-111°.[568]

$(MeO)_2P(O)SCH_2CONMeOMe$. XVII. Liq. $b_{0.12}$ 131-135°.[84,175]

$(EtO)_2P(O)SCH_2CONMeOMe$. XVII. Liq. $b_{0.5}$ 155-159°.[84,175]

$(EtO)_2P(O)SCH_2CONMeCHO$. XVII. Liq. n_D^{20} 1.4928.[665]

$(EtO)_2P(O)SCH_2CONMeCOOMe$. XVII. Liq. n_D^{20} 1.4828.[1206]

$(EtO)_2P(O)SCH_2CONMeCOOEt$. XVII. Liq. n_D^{20} 1.4775.[1206]

$(MeO)_2P(O)SCH_2CONMeCH_2C{\equiv}CH$. XVII. Liq. n_D^{25} 1.505.[207]

$(EtO)_2P(O)SCH_2CON$ 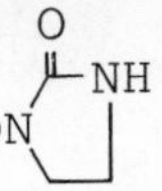 XVII. Solid. m. 64-66°.[483]

$(EtO)_2P(O)SCH_2CON$ 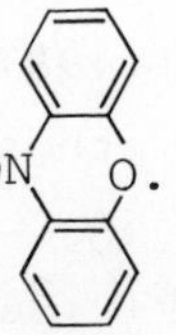 XVII. Solid. m. 81-82°.[1414]

$(RO)_2P(O)SCH_2CON$ 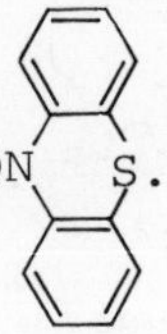 XVII. Solids. R, m.: Et, 98-99°; Pr, 75-76°; i-Pr, 99-99.5°; i-Bu, 89-90°; Bu, 82-82.5°.[1414]

<u>$(RO)_2P(O)SCH_2COOR'$</u>

$(EtO)_2P(O)SCH_2COOMe$. XVII.[65,288] Liq. $b_{0.01}$ 86°,[288] b_3 129°[65] n_D^{20} 1.4645, d^{20} 1.2192.[65]

$(MeO)_2P(O)SCH_2COOEt$. XVII.[154,288,696,697,1307] XXVIII.[6] XXXI.[154] Liq. $b_{0.05}$ 86°,[154,1307] $b_{0.35}$ 116-120°,[696] n_D^2 1.4740,[696] d^{20} 1.2500.[696] IR.[426,673,697] "Methylacetophos."

$(EtO)_2P(O)SCH_2COOEt$. XVII.[65,696,933,1115] Liq. $b_{0.01}$ 78-81°,[1115] $b_{0.03}$ 95°,[696] b_3 138°,[65] n_D^{20} 1.4624,[696] d^{20} 1.1840,[696] "Acetophos."

$(i\text{-}PrO)_2P(O)SCH_2COOEt$. XVII. Liq. b_3 134°, n_D^{20} 1.4568, d^{20} 1.1203.[65]

$(BuO)_2P(O)SCH_2COOEt$. XVII. Liq. $b_{2.5}$ 150°, n_D^{20} 1.4642, d^{20} 1.0984.[65]

$(EtO)_2P(O)SCH_2COO$-i-Pr. XVII. Liq. $b_?$ 136-138°.[851]

$(EtO)_2P(O)SCH_2COOC_6H_{11}$. XVII. Liq. $b_?$ 163-164°.[851]

$(EtO)_2P(O)SCH_2COOBu$. XVII. Liq. $b_{0.5}$ 138-140°, n_D^{20} 1.4616, d_{20}^{20} 1.1144.[933]

$(EtO)_2P(O)SCH_2COO$-i-Am. XVII. Liq. $b_{0.5}$ 143-145°, n_D^{20} 1.4600, d_{20}^{20} 1.1132.[933]

$(EtO)_2P(O)SCH_2COOCMe_2C{\equiv}CH$. XVII. Liq. $b_{0.1-0.2}$ 124-128°.[654]

$(EtO)_2P(O)SCH_2COOCMeEtC{\equiv}CH$. XVII. Liq. $b_{0.1-0.2}$ 140°.[654]

$(MeO)_2P(O)SCH(SEt)COOEt$. XVII. Liq. $b_{0.02}$ 125-130°.[877] n_D^{23} 1.4940.[878]

$(EtO)_2P(O)SCH(SEt)COOEt$. XVII. Liq. $b_{0.01}$ 130-135°, n_D^{20} 1.4880.[878]

$(MeO)_2P(O)SCH(SPh)COOEt$. XVII. Liq. $b_{0.02}$ 100°, n_D^{25} 1.5493.[878]

$(EtO)_2P(O)SCH(SPh)COOEt$. XVII. Liq. $b_{0.01-0.02}$ 125-130°, n_D^{25} 1.5330.[877,878]

$(EtO)_2P(O)SCH{=}CH_2$. XXVI. Liq. $b_{0.02}$ 50°, n_D^{20} 1.4713, d_4^{20} 1.1431.[1217]

$(MeO)_2P(O)SCH{=}CCl_2$. VIII.[4] XIX.[424] XXIX.[648,929] Liq. $b_{0.007}$ 71-73°,[424] b_2 113-115°,[929] n_D^{25} 1.5146, d_4^{22} 1.4442.[929]

$(EtO)_2P(O)SCH{=}CCl_2$. XIX.[424] XXIX.[648,929] Liq. $b_{0.007}$ 84-86°,[424] b_5 130°, n_D^{25} 1.5020, d_4^{22} 1.3119.[929]

$(i\text{-}PrO)_2P(O)SCH{=}CCl_2$. XXIX. Liq. $b_{0.01}$ 67°.[648]

$(PrO)_2P(O)SCH{=}CCl_2$. XXIX.[648,929] Liq. $b_{0.01}$ 78°,[648] b_2 135°,[929] n_D^{25} 1.4870, d_4^{23} 1.2350.[929]

$(PhO)_2P(O)SCH{=}CCl_2$. XIX. Liq. $b_{0.015}$ 171-172°, n_D^{25} 1.5821, d_{20}^{20} 1.362.[424]

$(RO)_2P(O)SCCl{=}CHCl$. XIX. Liq.[424]

R	b./mm	n_D^t	d_{20}^{20}
Me	87°/0.018	1.5174^{20}	--
Et	81-82°/0.008	1.4980^{25}	1.3208
Bu	143-147°/0.045	1.4855^{27}	1.1828
Ph	199-200°/0.08	1.5829^{27}	1.3699

$(EtO)_2P(O)SCH{=}CHNCO$. XXV. Liq. $b_{0.06}$ 127-130°, n_D^{20} 1.5020, d_4^{20} 1.2558.[223]

$(EtO)_2P(O)SCCl{=}CHNCO$. From $(EtO)_2P(O)SCH{=}CHNCO$ plus Cl_2 in CCl_4. Liq. $b_{0.01}$ 120-122°, n_D^{20} 1.5055, d_4^{20} 1.3093.[233]

$(i\text{-}PrO)_2P(O)SCF{=}CH_2$. XXII. Liq. b_3 93-95°, n_D^{20} 1.4500, d_4^{20} 1.0870.[361]

$(EtO)_2P(O)S(HOCH_2CH_2S)C{=}CMeN(CHO)CH_2$- 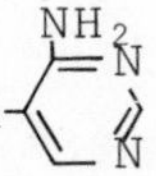 Solid. m. 133-134°.[876]

<u>$(RO)_2P(O)S\text{-}i\text{-}Pr$</u>

$(MeO)_2P(O)S\text{-}i\text{-}Pr$. XIX. Liq. b_2 85-86°.[216]

$(MeO)_2P(O)SCH(Me)CH_2Cl$:$(MeO)_2P(O)SCH_2CHClMe$ in the ratio 51:49. XXV. $b_{0.3}$ 83.5°. Ratio determined by ^1H-NMR.[842]

$(MeO)_2P(O)SCHMeCOOEt$. XVII.[280] XIX.[216] Liq. $b_{0.01}$ 76-78°.[280]

$(EtO)_2P(O)SCHMeCOOEt$. XVII.[280,1115] XXIV.[1132] Liq. $b_{0.01}$ 72-75°,[1115] b_1 137°.[1132]

$(MeO)_2P(O)SCHMeCOOPr$. XIX. Liq. $b_{0.01}$ 88°.[216]

$(MeO)_2P(O)SCHMeCOOC_6H_{13}$. XVII.[280] XIX.[216] Liq. $b_{0.01}$ 98-100°,[280] 100-102°.[216]

$(MeO)_2P(O)SCHMeCH_2SEt$. XVII. Liq. $b_{0.01}$ 83°.[293,1178]

$(EtO)_2P(O)SCHMeCH_2SEt$. XVII. Liq. $b_{0.01}$ 92°.[293,1178]

$(EtO)_2P(O)SCH(CH_2NMe_2)CH_2SEt$. XVII. Liq. $b_{0.01}$ 111-112°.[1184]

$(EtO)_2P(O)SCH(CH_2OMe)CH_2NEt_2$. XXIV. Liq. $b_{0.003}$ 104-105°.[366]

$(EtO)_2P(O)SCH(CH_2OPr)CH_2NEt_2$. XXIV. Liq. $b_{0.004}$ 117°.[366]

$(MeO)_2P(O)SCH(CH_2OMe)CH_2SMe$. XVII. Liq. n_D^{26} 1.5058.[1326]

$(MeO)_2P(O)SCH(CH_2OMe)CH_2SEt$. XVII. Liq. n_D^{25} 1.4991.[1326]

$(MeO)_2P(O)SCH(CH_2OEt)CH_2SEt$. XVII. Liq. n_D^{21} 1.4950.[1326]

$(EtO)_2P(O)SCH(CH_2OEt)CH_2SEt$. XVII. Liq. n_D^{21} 1.4865.[1326]

$(EtO)_2P(O)SCH(CH_2SEt)_2$. VIII. Liq. $b_{0.004}$ 104-110°.[1007]

$(MeO)_2P(O)SCH(CONHMe)_2$. XVII. Solid. m. 113°.[644]

$(EtO)_2P(O)SCH(CONHMe)_2$. XVII.[206,644] Solid. m. 102-103°,[206] 111°.[644] Also several analogs.[644]

$(MeO)_2P(O)SCH(COOEt)_2$. XVII. Liq. $b_{0.05}$ 132° n_D^{20} 1.464.[154,1307] IR.[426,673]

MeO(EtO)P(O)SCH(COOEt)$_2$. XVII. Liq. b$_{0.05}$ 136°.[1307]

(EtO)$_2$P(O)SCHMeCN. XVII. Liq. b$_{0.01}$ 77-78°.[1150]

(MeO)$_2$P(O)SC(Me)=CH(Cl):(MeO)$_2$P(O)SC(Me)=CCl(H) in the ration 71:29. XXV. b$_{0.22}$ 74-77°. Ratio determined by ^{1}H-NMR.[842]

(RO)$_2$P(O)SPr

(MeO)$_2$P(O)SPr. XIX.[216] XXVIII.[97] Liq. b$_{0.3}$ 75°,[97] b$_2$ 83-84°,[216] ^{31}P -31 ± 1.[1345]

(EtO)$_2$P(O)SPr. XVII.[1123] XIX.[1123] XXVIII.[417] XXXIII.[721] Liq. b$_{1.2}$ 90°,[417] b$_1$ 103-105°,[1123] n_D^{20} 1.4581,[721] n_D^{25} 1.4560,[417] ^{31}P -26.5.[850]

(MeO)$_2$P(O)SCH$_2$CHClMe. XIX.[947] XXV.[600] Liq. b$_{0.01}$ 68°,[600] b$_3$ 121°,[947] n_D^{20} 1.4840,[600] 1.4880,[947] d^{20} 1.1212.[947]

(EtO)$_2$P(O)SCH$_2$CHClMe. XXV.[600] XXVIII.[944a] Liq. b$_{0.02}$ 60°,[600] b$_1$ 115-116°,[944a] n_D^{19} 1.4730,[944a] n_D^{20} 1.4818,[600] d^{19} 1.1794.[944a]

(BuO)$_2$P(O)SCH$_2$CHClMe. XXV. Liq. b$_{0.01}$ 90°, n_D^{20} 1.4712.[600]

(RO)$_2$P(O)SCH$_2$CH$_2$CF$_3$. XXVIII. Liq.[12]

R	b./mm	n_D^{20}	d_4^{20}
Me	101-102°/8	1.4199	1.3535
Et	115°/9	1.4210	1.2515
i-Pr	115-115.5°/9	1.4189	1.1732
Pr	127-128°/7.5	1.4254	1.1893
i-Bu	133-134°/7.5	1.4261	1.1401
s-Bu	147-147.5°/16	1.4261	1.1438
Bu	149.5-150°/9	1.4290	1.1487
Me$_3$CCHMe	134-135°/3.5	1.4362	1.0956

(EtO)$_2$P(O)SCH$_2$CH(SMe)Me. From (EtO)$_2$P(O)SCH=CHMe plus MeSH under UV irradiation. Liq. b$_{0.15}$ 100-102°.[260,841]

(EtO)$_2$P(O)SCH$_2$CH(SEt)CH$_2$SEt. XVII. Liq. b$_{0.01}$ 114°.[1203]

(EtO)$_2$P(O)SCH$_2$CH$_2$CN. XVII. Liq. b$_{0.01}$108°.[1115]

(EtO)$_2$P(O)SCH$_2$CH$_2$COOMe. XVII. Liq. b$_3$ 137°, n_D^{20} 1.4640, d^{20} 1.1867.[65]

(MeO)$_2$P(O)SCH$_2$CH$_2$COOEt. XXVIII. Liq. b$_{0.12}$ 108-113°, n_D^{25} 1.4770.[6]

$(MeO)_2P(O)SCH_2COMe$. XVII.[154,1307,1335] Liq. $b_{0.05}$ 109-110°,[154,1307] $b_{0.09}$ 88-90°.[1335] IR.[426,673]

$(EtO)_2P(O)SCH_2COMe$. XVII.[1335] XXV.[781] XXXIII. (From $(EtO)_2$-P(O)SCl plus acetone.)[777] $b_{0.03}$ 94-95°,[777] $b_{0.7}$ 82°,[781] $b_{0.13}$ 110-111°,[1335] n_D^{20} 1.4712,[777] 1.4745,[1335] n_D^{25} 1.4685,[781] d^{20} 1.1863,[777] d_4^{20} 1.1888.[1335] p-Nitrophenyl hydrazone, solid, m. 92-93°.[781]

$(RO)_2P(O)SCH_2COMe$. XVII. Liq.[1335]

R	b./mm	n_D^{20}	d_4^{20}
i-Pr	96-97°/0.12	1.4620	1.1173
Pr	104-105°/0.14	1.4671	1.1313
i-Bu	122°/0.22	1.4645	1.0818
Bu	108-109°/0.21	1.4742	1.0889

$(MeO)_2P(O)SCH_2CH{=}CH_2$. XXXI. Liq. $b_{2.5}$ 90-91°, n_D^{20} 1.4798, d_4^{20} 1.1848.[981]

$(EtO)_2P(O)SCH_2CH{=}CH_2$. XXII.[259,840] XXXI.[981] XXXIII.[721] Liq. $b_{0.01}$ 53-54°,[721] $b_{0.25}$ 70-70.5°,[529,840] b_5 107-108°,[981] n_D^{20} 1.4710,[981] 1.4742,[721] d_4^{20} 1.1067.[981]

$(BuO)_2P(O)SCH_2{=}CH_2$. XXXI. Liq. b_2 138-140°, n_D^{20} 1.4673, d_4^{20} 1.0349.[981]

$(EtO)_2P(O)SCH{=}CHMe$. XXII. Liq. $b_{0.2}$ 69.5-70°, n_D^{20} 1.4756.[260,841]

$(i\text{-}PrO)_2P(O)SCH{=}CHMe$. XXII. Liq. $b_{0.4}$ 79-80°, n_D^{20} 1.4651.[260,841]

$(MeO)_2P(O)SCH_2CCl{=}CH_2$. XXV. Liq. $b_{0.24}$ 84.5-86°. ^{1}H-NMR.[842]

$(EtO)_2P(O)SCH_2C{\equiv}CH$. XVII.[1155] XXVII.[48] Liq. $b_{0.01}$ 55°,[1155] b_1 89-90°, n_D^{20} 1.4815, d^{20} 1.1522. IR.[48]

$(RO)_2P(O)SCH_2C{\equiv}CH$. XXVII. Liq. IR.[48]

R	b_1	n_D^{20}	d^{20}
i-Pr	95°	1.4701	1.0799
Pr	102-105°	1.4781	1.0835
i-Bu	118-120°	1.4705	1.0321
Bu	128°	1.4760	1.0607

<u>$(RO)_2P(O)S$-t-Bu</u>

$(EtO)_2P(O)S$-t-Bu. IX.[344] XXII.[839] Liq. $b_{0.25}$ 64.5°,[839] $b_{1.5}$ 98-100°,[344] n_D^{20} 1.4583,[839] IR.[344] ^{1}H-NMR.[344,839]

$(EtO)_2P(O)SCMe_2CN$. XV. (<u>Note</u>: The expected thiono product apparently isomerized to the thiolo form.) Liq. $b_{0.017}$ 71-74°, n_D^{20} 1.4450, d_0^{20} 1.1023. IR shows P(O) at 1276 $cm.^{-1}$.[68]

$(EtO)_2P(O)S$-i-Bu. VIII.[961] XVI.[961] XVII.[1123] XIX.[1123] XXVIII.[941] Liq. $b_{0.5}$ 105°,[1123] b_{11} 129-131°, n_D^{20} 1.4588, d_4^{20} 1.0591.[941]

$(EtO)_2P(O)SCH_2CMe_2NMe_2$. VIII. XVII. Liq. $b_{0.12}$ 85-89°, $b_{0.25}$ 102-106°. Picrate. Solid. m. 112-113° (from EtOH). The expected thiono compound of Method VIII must have isomerized to give the same product as Method XVII.[146]

$(MeO)_2P(O)SCH_2CMe(SMe)CONH_2$. XVII. Solid. m. 84-85°.[1103]

$(EtO)_2P(O)SCH_2CMe(SMe)COOMe$. XVII. Liq. $b_{0.16}$ 127-129°.[1103]

$(MeO)_2P(O)SCH_2CClMe_2:(MeO)_2P(O)SC(CH_2Cl)Me_2:(MeO)_2P(O)SCH_2CMe{=}CH_2$ in the ratio of 57:19:24. XXV. (From $(MeO)_2P(O)SCl$ plus $Me_2C{=}CH_2$.) Liq. $b_{0.003}$ 69.5-72°. Ratio determined by ^{1}H-NMR.[842]

$(EtO)_2P(O)SCH_2CH{=}CMe_2$. XXII. Liq. $b_{0.15}$ 89-90°, n_D^{20} 1.4795.[839]

$(EtO)_2P(O)SCHMeCH(SMe)COOMe$. XVII. Liq. $b_{0.05}$ 119-120°.[630]

$(EtO)_2P(O)SCHMeCH(SEt)COOMe$. XVII. Liq. $b_{0.01}$ 110-111°.[630]

$(EtO)_2P(O)SCHMeCOMe$. XIX. Liq. $b_{0.2}$ 100-102°.[71]

$(EtO)_2P(O)SCHMeCOEt$. XIX. Liq. $b_{0.2}$ 110-112°.[71]

$(MeO)_2P(O)SCH(COOMe)CH_2COOMe$. XXVIII. Liq. $b_{0.12}$ 126-132°, n_D^{25} 1.4690.[6]

$(MeO)_2P(O)SCH(COOEt)CH_2COOEt$. XIX.[6] XXVIII.[6] XXXIII. (From $(MeO)_2P(S)SCH(COOEt)CH_2COOEt$ plus HNO_3.)[485] Liq. $b_{0.1}$ 132°,[6] n_D^{25} 1.4666. IR.[6] ^{1}H-NMR.[524,526] Mass spect.[196] "Malaoxon."

$(EtO)_2P(O)SCH(COOEt)CH_2COOEt$. XVII.[1131] XXIV.[1132] XXVIII.[6]

Liq. $b_{0.01}$ 112°,[1132] $b_{0.02}$ 114-115°,[1131] $b_{0.15}$ 133-137°.[6]

$(RO)_2P(O)SCHMeCH=CH_2$. XXXI. Liq.[981]

R	b./mm	n_D^{20}	d_4^{20}
Me	95-96°/3	1.4770	1.1402
Et	110-112°/2	1.4730	1.0807

$(RO)_2P(O)SCMe=CHCOOMe$. XXII. Liq. R, b./mm: Me, 62-64°/0.01, Et, 123-129°/0.06.[174]

$(EtO)_2P(O)SCHEt=CH(OPh)_2$. XXV. (From $(EtO)_2P(O)SCl$ plus EtCH=CHOPh plus PhOH.) Liq. $b_{0.01}$ 110-111°, n_D^{21} 1.5046.[782]

$(EtO)_2P(O)SBu$. XVI.[1372] XXIV.[792] XXVIII,[795,941] XXXIII.[138,367,450,942] Liq. $b_{0.04}$ 74°,[138] $b_{1.5}$ 102.5-103°,[1372] b_{15} 137°,[792] n_D^{20} 1.4587,[795] n_D^{25} 1.4550.[450] ^{31}P -26.6.[450]

$(RO)_2P(O)SBu$. XXXIII. Liq.[942]

R	b./mm	n_D^{20}	d_4^{20}
i-Pr	143-145°/15	1.4533	1.0096
Pr	151-153°/8	1.4567	1.0091
Bu	160°/8	1.4582	1.0013

$(EtO)_2P(O)SCH_2CH_2CH_2CONEt_2$. XVII. Liq. n_D^{20} 1.4770, d_4^{20} 1.1287.[1425]

$(EtO)_2P(O)SCH_2CH_2CH_2COOEt$. XVII. Liq. $b_{0.55}$ 135°, n_D^{20} 1.4640, d_4^{20} 1.1403.[1425]

$(EtO)_2P(O)SCH_2CH_2CH_2COOC_6H_4Cl$-4. XVII. Liq. $b_{0.25}$ 170-173°, n_D^{20} 1.5210, d_4^{20} 1.2397.[1425]

$(MeO)_2P(O)SCH_2CH=CHMe$. VIII. Liq. b_3 110°, n_D^{20} 1.4835, d_4^{20} 1.1538.[981]

$(EtO)_2P(O)SCH_2CH=CHMe$. VIII.[981] XVII.[981] XXII.[840] Liq. $b_{0.18}$ 79.5-80°,[840] b_5 124-125°,[981] n_D^{20} 1.4759,[840] 1.4750-1.4782,[981] d_4^{20} 1.0886-1.0890.[981]

$(i\text{-}PrO)_2P(O)SCH_2CH=CHMe$. XXII. Liq. $b_{0.07}$ 68-69°, n_D^{20} 1.4670.[840]

$(MeO)_2P(O)SCH_2CHClCH=CH_2$: $(MeO)_2P(O)SCH(CH_2Cl)CH=CH_2$ in the ratio 73:27. XXV. (From $(MeO)_2P(O)SCl$ plus butadiene.)

Liq. $b_{0.25}$ 104-106°. Ratio determined by ^{1}H-NMR.[842]

$(EtO)_2P(O)SCH_2CHClCH=CH_2$. XXV. (From $(EtO)_2P(O)SCl$ plus butadiene.) Liq. $b_{0.5}$ 98-100°, n_D^{20} 1.4830, d_4^{20} 1.1730. (Note: 1,2 addition; 1,4 and 3,4 addition discussed; discarded on the basis of IR and of hydrolysis products.)[996] See above.

$(EtO)_2P(O)SCH_2CH=CMeCl$. XVII. Liq. b_3 114-114.5°, n_D^{20} 1.4909, d^{20} 1.1812.[726]

$(i\text{-}PrO)_2P(O)SCH_2CH=CMeCl$. XVII. Liq. b_2 104-104.5°, n_D^{20} 1.4820, d^{20} 1.1263.[726]

$(EtO)_2P(O)SCH_2CH=CHCN$. XVII. Liq. $b_{0.01}$ 94°.[290]

$(PrO)_2P(O)SCH_2CH=CHCN$. XVII. Liq. $b_{0.01}$ 116-118°.[290]

$(EtO)_2P(O)SAm$. XXII.[840] XXXIII. (From $(EtO)_2P(O)SCl$ plus AmMgBr.)[721] Liq. $b_{0.18}$ 88-89°,[840] b_{26} 170°,[721] n_D^{20} 1.4583,[840] 1.4590,[721] ^{31}P -26.6.[850]

$(i\text{-}PrO)_2P(O)SAm$. XXII. Liq. $b_{0.15}$ 85-86°, n_D^{20} 1.4525.[840]

$(MeO)_2P(O)SCH(CH_2Cl)Pr$:$(MeO)_2P(O)SCH_2CHClPr$ in the ratio of 65:35. XXV. Liq. $b_{0.004}$ 94°. Ratio determined by ^{1}H-NMR.[842]

$(EtO)_2P(O)SCH(COMe)_2$. XXXIII. (from $(EtO)_2P(O)SCl$ plus $CH_2(COMe)_2$). Liq. $b_{0.04}$ 81-82°, n_D^{20} 1.5000, d^{20} 1.2086.[7]

$(RO)_2P(O)S(CH_2)_4CCl_3$. By "known methods." Liq.[855]

R	b_2	n_D^{20}	d^{20}
Et	165-167°	1.4340	1.2892
Pr	186-187°	1.4890	1.2226
i-Bu	183-184°	1.4850	1.1875

$(MeO)_2P(O)SCH_2CHClCMe=CH_2$:$(MeO)_2P(O)SCH(CH_2Cl)CMe=CH_2$ in the ratio of 73:27. XXV. (From $(MeO)_2P(O)SCl$ plus isoprene.) Liq. $b_{0.005}$ 117-120°. Ratio determined by ^{1}H-NMR.[842]

$(MeO)_2P(O)SCHMeCH=CHMe$. XXII. Liq. $b_{0.05}$ 70-71°.[921]

$(MeO)_2P(O)SCH_2CH=CMe_2$. XXII. Liq. $b_{0.2}$ 65-70°. At least 90% 4,1 adduct by ^{1}H-NMR.[921]

$(EtO)_2P(O)SCH_2CMe=CHCN$. XVII. Liq. $b_{0.01}$ 108°.[290]

$(MeO)_2P(O)SC_6H_{13}$. XVII.[1123] XIX.[216,1123] Liq. $b_{0.01}$ 71-

73°,[216] b_1 125-130°.[1123]

$(EtO)_2P(O)SCMe_2CMe_2Cl$. XXV. (From $(EtO)_2P(O)SCl$ plus $Me_2C{=}CMe_2$.) Liq. $b_{0.02}$ 67°, n_D^{20} 1.4840, d^{20} 1.1106.[602,775,777]

$(PrO)_2P(O)SCMe_2CMe_2Cl$. XXV. Liq. $b_{0.01}$ 86°, n_D^{25} 1.4858, d^{25} 1.0908.[602,775]

$[(RO)_2P(O)SCH(CONH_2)CH_2]_2$. XVII. Solids. R, m.: Et, 152-153° (from acetone-EtOH); i-Pr, 144° (from MeOH).[1105]

$(EtO)_2P(O)SCH(i\text{-}Pr)_2$. IX. Liq. $b_{0.05}$ 86-87°, ^{1}H-NMR.[344]

$(EtO)_2P(O)S(CH_2)_8CCl_3$. By "known methods." Liq. b_2 217-218°, n_D^{20} 1.4885, d^{20} 1.1900.[855]

$(EtO)_2P(O)SC_9H_{19}$. By "known methods." Liq. b_2 177-178°, n_D^{20} 1.4600, d^{20} 0.9985.[855]

$(EtO)_2P(O)SC_{12}H_{25}$. XVII, XIX.[1123] Liq. $b_{0.5}$ 185-190°.[1123]

<u>$(RO)_2P(O)S$-5-ring</u>

$(EtO)_2P(O)SC_5H_7$-cyclo, IX. Liq. b_2 105-108°. (Position of unsaturation not given).[1065]

$(EtO)_2P(O)SC_5H_9$-cyclo. XXII. Liq. $b_{0.04}$ 79°, n_D^{20} 1.4808.[839]

$(EtO)_2P(O)S$– 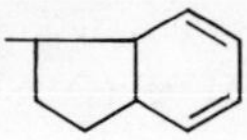 XXII. Liq. n_D^{20} 1.5385.[839] ^{1}H-NMR.[839,818]

$(EtO)_2P(O)S$– 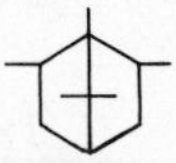 –Cl. XXV. Liq. $b_{0.1}$ 108-110°, n_D^{20} 1.4968, d^{20} 1.1254. n_D^{20} -8.6° (liq.).[111]

$(EtO)_2P(O)S$– 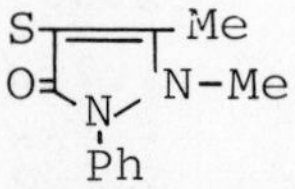 XXIV. Solid. m. 100-101° (from benzene-petroleum ether).[936,1358]

$(EtO)_2P(O)S$– 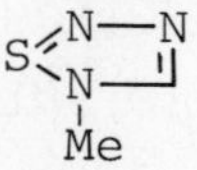 IX. Solid. m. 76°.[1310]

$(EtO)_2P(O)S$– 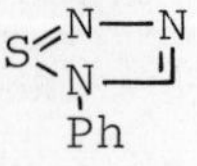 IX. Solid. m. 65° (from MeOH).[1310]

$(EtO)_2P(O)S$– 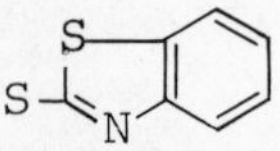 VIII. Liq. $b_{0.035}$ 85-89°.[1278]

$(Et)_2P(O)S$– 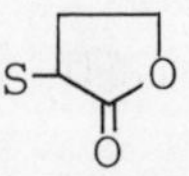 XVII. Liq. n_D^{20} 1.4680.[470]

$(EtO)_2P(O)S$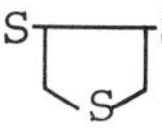SEt. XXV. Liq. $b_{0.05}$ 118-120°.[1383]

$(EtO)_2P(O)S$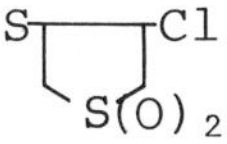Cl. XXV.[379]

$(EtO)_2P(O)S$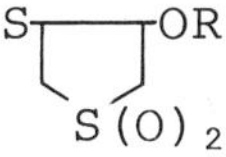OR. From $(EtO)_2P(O)S$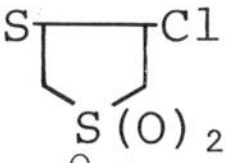Cl plus RONa. Liq. R, b./mm: Et, 120°/0.0015; Ph, 175°/0.001.[379]

<u>$(RO)_2P(O)S$-6-ring</u>

$(MeO)_2P(O)SPh$. XV.[813] XIX.[955] XXXIII.[853] Liq. $b_{0.046}$ 93-94°,[813] $b_{0.04}$ 118-119,[955] n_D^{20} 1.5443,[955] 1.5450,[813] d_4^{20} 1.2586,[813] IR.[853]

$(EtO)_2P(O)SPh$. VIII.[596,1189] XIX.[836,955] XXIV.[1233] XXVII.[94,750] XXVIII.[417,795] XXXIII. (From $(EtO)_2P(O)SCl$ or $[(EtO)_2P(O)S]_2$ plus PhMgBr.)[721] (From $(EtO)_3P$ plus PhS_2O_2Ph.)[780] (from $(EtO)_3P$ plus 2 $BrCCl_3$ plus 4 PhSH.)[853]. Liq. $b_{0.01}$ 63°,[721] $b_{0.04}$ 87-88°,[780] $b_{0.75}$ 120°,[1233] $b_{1.5}$ 148°, n_D^{16} 1.5240,[596] n_D^{20} 1.5227,[1223] n_D^{26} 1.5213,[417] d^{20} 1.1633,[750] d^{25} 1.0941.[836] IR.[853] ^{31}P -22 ± 1.[1345]

$(i\text{-}PrO)_2P(O)SPh$. XV.[813] Liq. $b_{0.04}$ 93-94°, n_D^{20} 1.5098, d_4^{20} 1.1144.[813]

$(BuO)_2P(O)SPh$. XV. Liq. $b_{0.04}$ 127-128°, n_D^{20} 1.5106, d_4^{20} 1.0925.[813]

$(EtO)_2P(O)SC_6H_4Br$-4. XIX. (From $(EtO)_3P$ plus 4-$BrC_6H_4SO_2Cl$.)[450] XXVII.[450,853] Liq. $b_{0.04}$ 115-117°, n_D^{25} 1.5445.[450] IR.[853]

$(MeO)_2P(O)SC_6H_4Cl$-2. XIX.[956,1169] Liq. $b_{0.005}$ 120-122°,[955] $b_{0.01}$ 92°,[1169] n_D^{20} 1.5573.[955]

$(EtO)_2P(O)SC_6H_4Cl$-2. XIX.[1169] XXVII.[853] . Liq. $b_{0.01}$ 103°.[1169] IR.[853]

$(MeO)_2P(O)SC_6H_4Cl$-3. XIX.[955,1169] Liq. $b_{0.01}$ 91°,[1169] $b_{0.018}$ 118°, n_D^{20} 1.5503.[955]

$(EtO)_2P(O)SC_6H_4Cl$-3. XIX.[1169] XXVII.[94] Liq. $b_{0.01}$ 95°,[1169] $b_{0.8}$ 137-139°.[94]

$(MeO)_2P(O)SC_6H_4Cl$-4. XIX. Liq. $b_{0.005}$ 124°, n_D^{20} 1.5550.[955]

$(EtO)_2P(O)SC_6H_4Cl$-4. XIX.[1169] XXVII.[750,853] XXVIII.[97,1159] Liq. $b_{0.01}$ 98°,[1169] 103°,[1159] $b_{0.6}$ 160-166,[750] n_D^{20} 1.5298, d^{20} 1.2568.[750] P(O):P(S) ratio 63:37.[750] IR.[853] ^{31}P -21 ± 1.[1345]

$(ClCH_2CH_2O)_2P(O)SC_6H_4Cl$-4. XXVIII. Liq. $b_{0.01}$ 136°.[1159]

$(EtO)_2P(O)SC_6H_4F$-4. XXVII. IR.[853]

$(EtO)_2P(O)SC_6H_4NO_2$-4. XXVII. IR.[853]

$(EtO)_2P(O)SC_6H_4OMe$-4. XXVII.[94,750] XXVIII.[417] Liq. $b_{0.2}$ 152-155°,[94] n_D^{25} 1.5296.[417]

$(MeO)_2P(O)SC_6H_3ClNO_2$-4,2. XIX. Liq. n_D^{20} 1.5384.[955]

$(MeO)_2P(O)SC_6H_3ClOH$-5,2. XXV. Liq. d^{20} 1.389.[1390]

$(EtO)_2P(O)SC_6H_3ClOH$-5,2. XXV. Liq. n_D^{20} 1.531.[1390]

$(MeO)_2P(O)SC_6H_3Cl_2$. XIX. Solids. Position of Cl's, m.: 2,4, 49-50° ($b_{0.01}$ 134°);2,5, 67-68°; 2,6, 66-67°; 3,4, 11-12° (n_D^{20} 1.5663). Also several analogs.[955]

$(MeO)_2P(O)SC_6HCl_3OH$-3,5,6,2. XXV. Liq. n_D^{20} 1.541.[1390]

$(EtO)_2P(O)SC_6HCl_3OH$-3,5,6,2. XXV. Liq. n_D^{20} 1.539.[1390]

$(MeO)_2P(O)SC_6Cl_5$. XIX.[98,531,1411] Solid. m. 144-146°,[1411] 145-147°.[98] ^{31}P -21.4.[1274]

$(EtO)_2P(O)SC_6Cl_5$. XIX.[98,1411] Solid. m. 87-89°,[1411] 91-92°.[98] ^{31}P -17.9.[1274]

$(RO)_2P(O)SC_6Cl_5$. XIX. Solids. R, m.: i-Pr, 78-80°; Pr, 49-53°; Bu, 49-52°.[1411]

$(MeO)_2P(O)SC_6H_4NH_2$-4. XXIV. Solid. m. 64°.[618]

$(MeO)_2P(O)SC_6H_4NHCOMe$-4. XXIV. Solid. m. 104°.[618]

$(MeO)_2P(O)SC_6H_4NMe_2$-4. XXIV. Solid. m. 45-46°.[618]

$(EtO)_2P(O)SC_6H_4NMe_2$-4. XXIV. Liq. $b_{0.01}$ 123°.[618]

$(EtO)_2P(O)SC_6H_4NEt_2$-4. XXVI. Liq. $b_{0.05}$ 140°, n_D^{25} 1.5532. Picrate. Solid. m. 92°.[779]

$(MeO)_2P(O)SC_6H_4NO_2$-2. XIX. Solid. m. 44-45°.[955]

$(EtO)_2P(O)SC_6H_4NO_2$-2. XIX. Liq. n_D^{21} 1.5395.[955]

$(MeO)_2P(O)SC_6H_4NO_2$-3. XIX. Solid. m. 57°.[955]

$(EtO)_2P(O)SC_6H_4NO_2$-3. XIX. Solid. m. 32-33°.[955]

$(MeO)_2P(O)SC_6H_4NO_2$-4. VIII.[1189] XIX.[519,955] Solid. m. 55-56°,[519,955] 59°.[1189]

$(EtO)_2P(O)SC_6H_4NO_2$-4. VIII.[119,1189] XIX.[519,955] Solid. m. 40-41°,[119,519] 42-43°,[1189] $b_{0.15}$ 158°.[119]

$(EtO)_2P(O)S$ 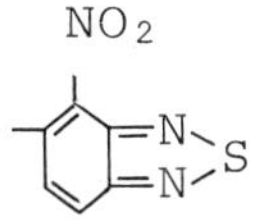 XIX.[1358] XXIV.[936] Amorph. yellow powder. m. 159-161° (dec.) (From benzene-petroleum ether.)[936,1358]

$(MeO)_2P(O)SC_6H_3(NO_2)_2$-2,4. XIX. Solid. m. 62-63° (from MeOH).[953]

$(MeO)_2P(O)SC_6H_3NO_2OH$-5,2. XXV. Liq. n_D^{20} 1.533.[1390]

$(EtO)_2P(O)SC_6H_3NO_2OH$-5,2. XXV. Liq. d^{20} 1.335.[1390]

$(MeO)_2P(O)SC_6H_3NO_2OMe$-4,2. XIX. Solid. m. 82°.[955]

$(MeO)_2P(O)SC_6H_3NO_2OMe$-5,2. XIX. Solid. m. 92°. λ_{max} 200, 233, 297; log ε 4.19, 3.98, 3.97.[955]

$(EtO)_2P(O)SC_6H_3NO_2OMe$-5,2. XIX. Solid. m. 74-75°.[954]

$(MeO)_2P(O)SC_6H_4OH$-2. XXV. Liq. n_D^{20} 1.523.[1390]

$(EtO)_2P(O)SC_6H_4OMe$-2. VIII. Liq. b_3 155°.[1189]

$(EtO)_2P(O)SC_6H_4OMe$-4. VIII. Liq. n_D^{20} 1.5332, d_4^{20} 1.169.[1081]

$(EtO)_2P(O)SC_6H_4OEt$-2. VIII. Liq. b_2 160°.[1189]

$(EtO)_2P(O)SC_6H_4OEt$-4. VIII. Liq. b_3 185°.[1189]

$(EtO)_2P(O)SC_6H_4SMe$-4. XIX. Liq. $b_{0.03}$ 160°.[296]

$(EtO)_2P(O)SC_6H_{11}$-cyclo. IX. Liq. $b_{0.2}$ 99-102°.[1065]

$(EtO)_2P(O)S$-cyclo-$C_6H_{10}Cl$-2. XXV.[602,775,783,944] Liq. $b_{0.05}$ 103°,[783] b_1 145-146°,[944] b_3 160-161°,[944] n_D^{19} 1.4918,[944] n_D^{20} 1.4963,[775] d^{18} 1.2162,[944] d^{20} 1.1948.[775]

$(EtO)_2P(O)S$ 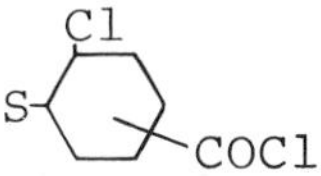 XXV. Liq. $b_{0.02}$ 126-129°, n_D^{20} 1.5115, IR.[109]

$(MeO)_2P(O)SC_6H_4Me$-2. XIX. Liq. $b_{0.01}$ 80°.[1169]

$(EtO)_2P(O)SC_6H_4Me$-2. XIX.[1169] XXVII.[94,750] Liq. $b_{0.01}$ 94°,[1169] $b_{0.3}$ 92-96°,[750] $b_{0.4}$ 87-88°,[94] n_D^{20} 1.4930, d^{20} 1.1319.[750] P(O):P(S) ratio 63:37.[750]

$(EtO)_2P(O)SC_6H_4Me$-3. XXVII.[94,750] Liq. $b_{0.12}$ 75-80°,[750] $b_{0.3}$ 92°,[94] n_D^{20} 1.5094, d^{20} 1.1253. P(O):P(S) ratio 56:44.[750]

$(MeO)_2P(O)SC_6H_4Me$-4. VIII.[1189] XIX.[955] Liq. $b_{0.01}$ 115-116°,

[955] $b_{1.5}$ 182°,[1189] n_D^{20} 1.5429.[955]

$(EtO)_2P(O)SC_6H_4Me$-4. VIII.[1189] XIX.[955] XXVII.[94,853] XXVIII.[721] Liq. $b_{0.01}$ 92°,[721] b_1 144°,[1189] n_D^{20} 1.5189,[721] 1.5382.[955] IR.[853]

$(MeO)_2P(O)SC_6H_4MeCl$-4,3. XIX.[1169] Liq. $b_{0.01}$ 103°.[1169]

$(EtO)_2P(O)SC_6H_4MeCl$-4,3. XIX. Liq. $b_{0.01}$ 125°.[1169]

$(MeO)_2P(O)SC_6H_4CF_3$-2. XIX.[955,956] Solid. m. 31.5°,[955] $b_{0.03}$ 112-113°,[955] $b_{0.5}$ 116-117°, n_D^{25} 1.5925.[955]

$(EtO)_2P(O)SC_6H_4CF_3$-2. XIX. Liq. $b_{0.01}$ 116°, n_D^{25} 1.4789.[955]

$(MeO)_2P(O)SC_6H_3MeNO_2$-2,3. XIX.[954, 955] XXIV.[954] Solid. m. 56-57°,[954] 59-60°.[955]

$(MeO)_2P(O)SC_6H_3MeNO_2$-2,4. XIX. Solid. m. <37°.[616]

$(EtO)_2P(O)SC_6H_3MeNO_2$-2,4. XIX. Liq. $b_{0.01}$ 128-130°.[616]

$(MeO)_2P(O)SC_6H_3MeNO_2$-2,5. XIX. Solid. m. 69°.[955]

$(MeO)_2P(O)SC_6H_3MeNO_2$-2,6. XIX. Solid. m. 54-55°.[955]

$(EtO)_2P(O)SC_6H_3MeNO_2$-3,4. XIX. Liq. $b_{0.01}$ 140°.[616]

$(MeO)_2P(O)SC_6H_4CONHMe$-2. XIX. Solid. m. 75°.[620] Also three analogs.[620]

$(EtO)_2P(O)SC_6H_4CN$-2. XXIV. Solid. m. 44°.[618]

$(EtO)_2P(O)SC_6H_4COOEt$-2. VIII. Liq. b_1 185°.[1189]

$(PhCH_2O)_2P(O)SC_6H_4$-i-Pr-2. XIX. Solid. m. 39° (from ether).[325]

$(EtO)_2P(O)S$- 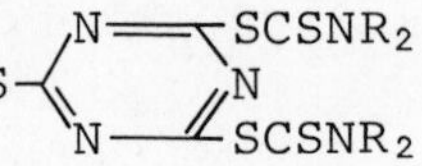 XVII. Liq.[276]

R	n_D^{30}
Et	1.5802
Pr	1.5834
Bu	1.5672
C_7H_{15}	1.5418
$CH_2CH=CH_2$	1.5267

<u>$(RO)_2P(O)SCH_2$-3-Ring</u>

$(MeO)_2P(O)SCH_2CH-CH_2$ (S bridging CH and CH_2). XVII. Liq. $b_{0.02}$ 71°.[1196]

$(EtO)_2P(O)SCH_2\overset{S}{CH-CH_2}$. XVII.[896, 897, 1196] (From $(EtO)_2$-$POSH \cdot NEt_3$ plus $\overset{S}{CH_2-}CHCH_2Cl$, or from $(EtO)_2P(S)SH \cdot NEt_3$ plus $\overset{O}{CH_2-}CHCH_2Cl$.[896] Liq. $b_{0.02}$ 75°,[1196] $b_{0.25}$ 117-119°,[896] $b_{0.3}$ 116-118°.[897]

$(RO)_2P(O)SCH_2$-5-ring

$(EtO_2)P(O)SCH_2N$ 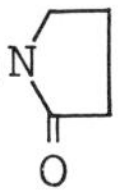 XVII. Liq. n_D^{25} 1.5400.[967]

$(EtO)_2P(O)SCH_2N$ 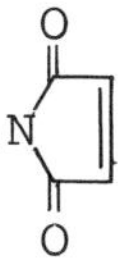 XVII. Liq. n_D^{21} 1.5120.[625]

$(RO)_2P(O)SCH_2N$ 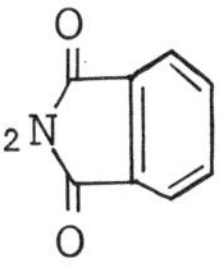 XVII.[608, 866] Solids. R, m.: Me, 70°,[866] 74°,[608]; Et, 71°.[608]

$(EtO)_2P(O)SCH_2N$  $-N-C_6H_4Cl-4$ XVII. Solid. m. 86-88°.[1249] 43 analogs.[1249]

$(RO_2P(O)SCH_2N$ 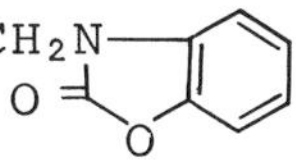 XVII.[122, 161, 1020] Solids. R, m.: Me, 52-55°,[122, 161] 66° (from cyclohexane);[1020] Et. 28-30°.[1020]

$(MeO)_2P(O)SCH_2N$ 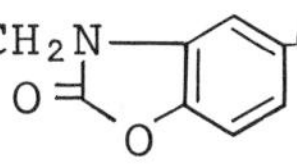 Cl. XVII. Solid. m. 95° (from cyclohexane).[1020]

$(RO)_2P(O)SCH_2N$ 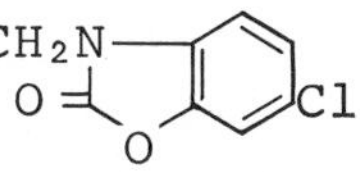 XVII. Solids. R, m.: Me, 110° (from benzene); Et, 66° (from benzene).[1020] Also several analogs.

$(EtO)_2P(O)SCH_2N$ 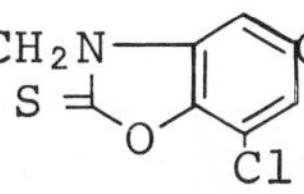 Cl. XVII. Solid. m. 104.5-105°.[626]

$(MeO)_2P(O)SCH_2$ 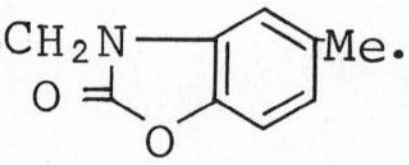 XVII. Solid. m. 90-92° (from benzene-ligroin). Also several analogs.[626]

$(EtO)_2P(O)SCH_2$ 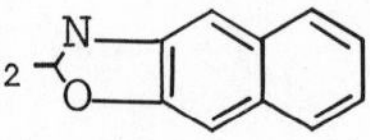 XVII. Solid Softens 25-26°.[316]

$(EtO)_2P(O)SCH_2$ 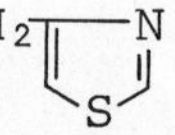 XVII. Liq. $b_{0.07}$ 130°.[1234]

$(EtO)_2P(O)SCH_2$ 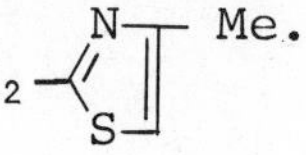 XVII. Liq. $b_{0.01}$ 112°.[323]

$(EtO)_2P(O)SCH_2$ 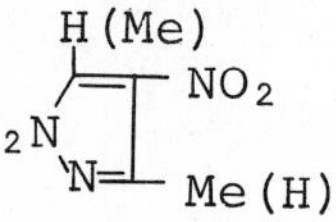 XVII. Liq. $b_{0.1}$ 130°.[363, 1052]

$(EtO)_2P(O)SCH_2$ 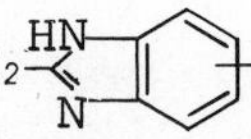 $-NO_2$-5(6). XVII. Solid. m. 127-129° (from benzene, i-PrOH).[950]

$(EtO)_2P(O)SCH_2$ 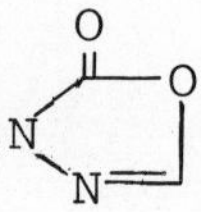 XVII. Liq. $b_{0.001}$ 105°. Also about 13 analogs.[364]

$(EtO)_2P(O)SCH_2$ 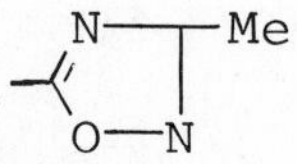 XVII. Liq. $b_{0.5}$ 145-150°.[772]

$(MeO)_2P(O)SCH_2$ 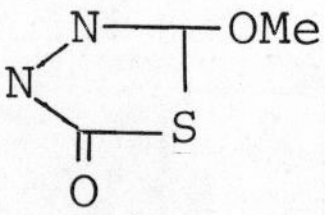 XXI. Solid m. 50-51°.[1261]

$(MeO)_2P(O)SCH_2$ 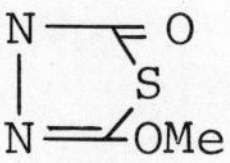 XVII. Solid. m. 49-50°.[1053]

$(EtO)_2P(O)SCH_2$ 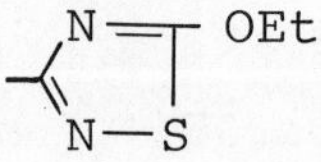 XVII. Liq. b_1 187°, n_D^{20} 1.5050.[1008]

$(EtO)_2P(O)SCH_2$ 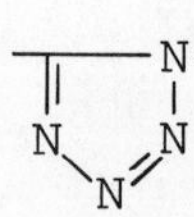 $-C_6H_5NO_2$-4. XVII. Solid. m. 54-55°.[1234]

$(EtO)_2P(O)SCH_2$ 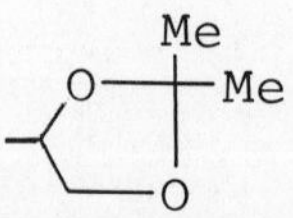 IX. Liq. $b_{0.25}$ 129-135°.[173]

<u>$(RO)_2P(O)SCH_2$-6-ring</u>

$(MeO)_2P(O)SCH_2Ph$. XVII.[288] XIX.[216, 1417] Liq. $b_{0.01}$ 73°,[216] 98°,[288] $b_{0.5}$ 113-115°.[1417]

$(EtO)_2P(O)SCH_2Ph$. XVII.[142, 288, 492, 1123, 1194] XIX.[1123, 1417] XXIV.[792, 1233] Liq. $b_{0.01}$ 99°,[288] $b_{0.6}$ 128-130°,[1233] b_1 133-134°,[792] b_3 160°,[1194] n^{20} 1.5252,[792] 1.5258,[492] n_D^{25} 1.5234,[140, 142] 1.5245,[1233] d^{20} 1.1540,[792] 1.1569.[492]

$(i\text{-}BuO)_2P(O)SCH_2Ph$. XIX. Liq. n_D^{18} 1.5167.[905]

$(s\text{-}BuO)_2P(O)SCH_2Ph$. XIX. Liq. n_D^{18} 1.5167.[905]

$(BuO)_2P(O)SCH_2Ph$. XIX. Liq. $b_{0.01}$ 142-145°, n_D^{26} 1.5056.[905]

$(PhO)_2P(O)SCH_2Ph$. XXVIII. Solid. m. 65-67°.[1417]

$(EtO)_2P(O)SCHClPh$. XXV. Solid. m. 98-99°.[602]

$(EtO)_2P(O)SCH_2C_6H_4Cl$-2. XVII.[1193, 1194] b_2 171°,[1193] 180°.[1194]

$(MeO)_2P(O)SCH_2C_6H_4Cl$-4. XVII. Liq. $b_{0.01}$ 110°.[288]

$(EtO)_2P(O)SCH_2C_6H_4Cl$-4. XVII.[1115, 1193, 1194] Liq. $b_{0.01}$ 120°,[1115] b_3 177°.[1193, 1194]

$(ClCH_2CH_2O)_2P(O)SCH_2C_6H_4Cl$-4. XXVIII. Liq. $b_{0.01-0.02}$ 191-194 .[1417]

$(EtO)_2P(O)SCH_2C_6H_3Cl_2$-2,6. XVII. Liq. b_3 183°.[1194]

$(MeO)_2P(O)SCH_2C_6H_3Cl_2$-3,4. XVII. Liq. $b_{0.01}$ 125°.[288]

$(EtO)_2P(O)SCH_2C_6H_3Cl_2$-3,4. XVII.[288, 1194] Liq. $b_{0.01}$ 123°,[288] b_2 190°.[1194]

$(MeO)_2P(O)SCH_2C_6H_2Cl_3$-2,4,5. XVII. Liq. $b_{0.01}$ 135°.[288]

$(EtO)_2P(O)SCH_2C_6H_2Cl_3$-2,4,5. XVII. Liq. $b_{0.01}$ 145°.[288]

$(MeO)_2P(O)SCH_2C_6H_4F$-4. XVII. Liq. $b_{0.01}$ 109°.[636]

$(MeO)_2P(O)SCH_2C_6H_4NO_2$-4. XXVIII. Liq. $b_{0.01-0.02}$ 45-48°.[1417]

$(EtO)_2P(O)SCH_2C_6H_4NO_2$-4. XXVIII. Liq. $b_{0.01-0.02}$ 155-160°.[1417]

$(RO)_2P(O)SCH_2C_6H_4OMe$-4. XVII. Liq.[755]

R	b./mm	n_D^{20}	d^{20}
Me	140-146°/0.2	1.5410	1.2475
Et	150-155°/0.3	1.5270	1.1863

i-Pr	140-145°/0.14	1.5188	1.1358
Pr	165-170°/0.18	1.5210	1.1389
i-Bu	155-160°/0.1	1.5120	1.1042
Bu	155-160°/0.1	1.5148	1.1115

Also analogs with $CH_2C_6H_4OEt$-4.

$(EtO)_2P(O)SCH_2C_6H_4SMe$-4. XVII. Liq. $b_{0.01}$ 124°.[637]

$(EtO)_2P(O)SCH_2$– 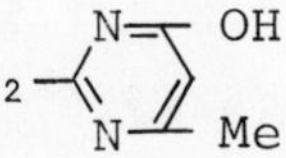 XVII. Solid. m. 90° (from EtOH).[1234]

$(RO)_2P(O)SCH_2$– 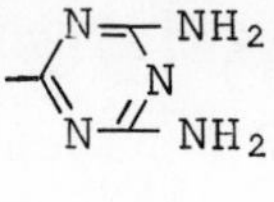 XVII. Solids. R, m.: Me 198°; Et, 169°, hydrochloride, 144-146° (dec).[145]

$(EtO)_2P(O)SCH_2$ 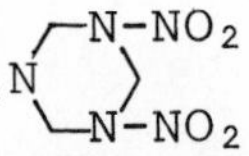 XVII. Solid. m. 32.5-34.2°.[1381]

$(EtO)_2P(O)SCH_2$– 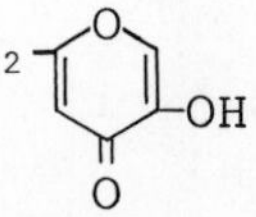 XVII. Solid. m. 70° (from CCl_4).[767, 1018]

$(RO)_2P(O)SCH_2$– 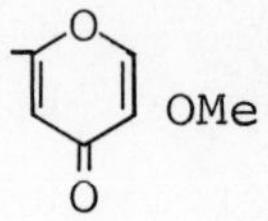 XVII.[767, 768, 1018] Solids. R, m.: Me, 74-76° (from benzene); Et, 70-72°; i-Pr, 66-68°; Bu 60°.

$(EtO)_2P(O)SCH_2$– 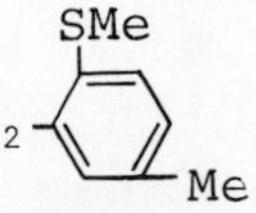 XVII. Liq. $b_{0.01}$ 144°.[285]

$(EtO)_2P(O)SCH_2$– 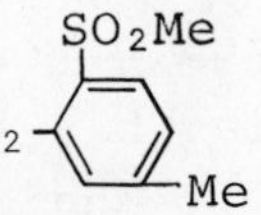 XVII. Solid. m. 63°.[285]

$(MeO)_2P(O)SCH_2C_6H_4CN$-4. XVII. Solid. m. 54-55°.[89]

$(EtO)_2P(O)SCH_2$–  XVII. Liq. $b_{2.5}$ 165-166.5°, n_D^{20} 1.5042, d_4^{20} 1.0922.[1357]

$(RO)_2P(O)SCH_2C_{10}H_7$-1. XVII. Liq. R, n_D^t: Me, $1.5898^{21.5°}$; Et, $1.5784^{23°}$; i-Pr, $1.5649^{21.5°}$.[182]

$(RO)_2P(O)SCH_2C_{10}H_7$-2. XVII. Liq. R, $n_D^{23.5°}$: Me, 1.5942; Et, (solid) m. 47-48°; i-Pr, 1.5650.[182]

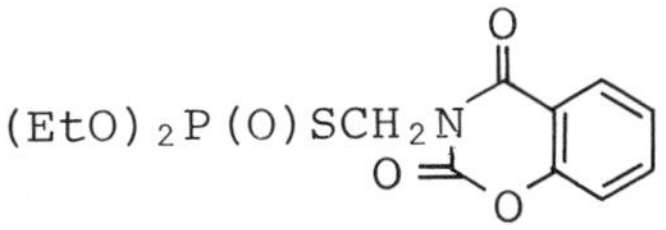

XVII. Solid. m. 79-82° (from $i\text{-}Pr_2O$-EtOH).[864]

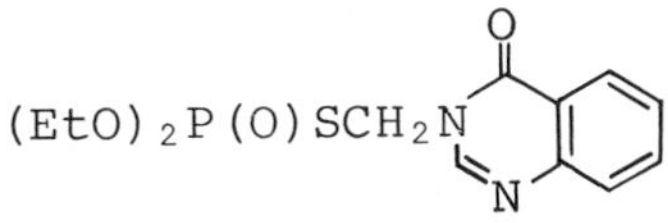

XVII. Solid. m. 74° (from benzene-ligroin).[613]

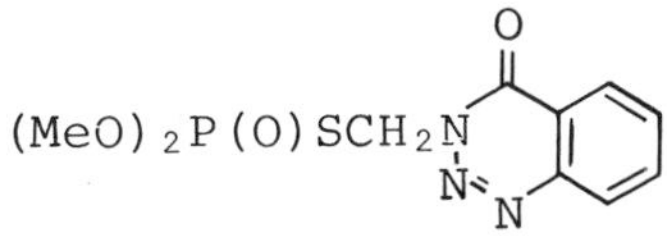

XVII. Solid. m. 81-83° (from i-PrOH).[611] ^1H-NMR.[526] Mass spect.[196] "Guthion-O-analog."

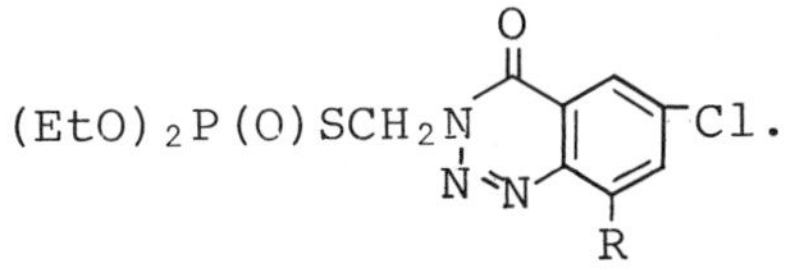

XVII. R: H, liq., n_D^{25} 1.5825, R: Cl, solid, m. 88-90°.[1035]

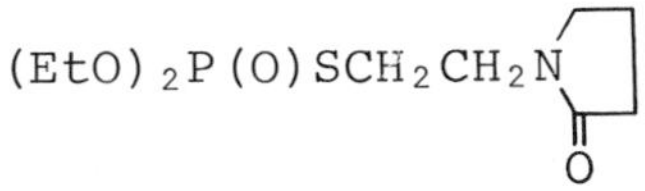

XVII. Liq. n_D^{20} 1.5204.[967]

$(MeO)_2P(O)SCH_2CH_2Ph$. XVII. Liq. $b_{0.5}$ 115-118°.[499]

$(MeO)_2P(O)SCH_2CH_2CH_2Ph$. XVII. Liq. $b_{0.005}$ 115-118°.[499]

$(EtO)_2P(O)S(CH_2)_nPh$. XVII. Liq.[499, 1418] n, b./mm: 2, 150-155°/0.4; 3, 138°/0.01; 4, 152-156°/0.15; 5, 153-157°/0.03; 10, 162-165°/0.08.[499, 1418] Also several analogs.

$(MeO)_2P(O)SCH_2CHClPh$. XXV. Liq. $b_{0.02}$ 95°, n_D^{20} 1.5459.[600]

$(EtO)_2P(O)SCH_2CHClPh$. XXV. Liq. $b_{0.01}$ 105°, n_D^{20} 1.5292.[600]

XVII. Solid. m. 167-169°. L.HCl, solid, m. 155-156°.[145a]

$(EtO)_2P(O)SCH_2CH_2SC_6H_4SMe\text{-}4$. XVII. Liq. n_D^{20} 1.5719, d^{20} 1.205.[1083]

$(RO)_2P(O)SCH_2COC_6H_4Cl\text{-}4$. XVII. Liq. 2,4-dinitrophenylhydrazones. Solids.[1355]

R	n_D^{20}	d_4^{20}	2,4-DNPH, m.
Me	1.5584	1.2479	110-112°
Et	1.5502	1.2613	78-80
i-Pr	1.5360	1.2204	102-104
Pr	1.5425	1.2861	73-75
Bu	1.5412	1.2334	55-57
Et, i-Pr	1.5355	1.3122	83-85

$(EtO)_2P(O)SCHMePh$. XVII. Liq. $b_{0.05}$ 123-132°.[499, 1418] ^{1}H-NMR.[818]

$(EtO)_2P(O)SCH(CN)Ph$. XVII. Liq. $b_{0.01}$ 124-126°.[1150]

$(MeO)_2P(O)SCH(COOEt)Ph$. XXIV. Liq. $b_{0.01}$ 123°.[1132] ^{32}P, XVII. Liq., n_D^{20} 1.5220.[230]

$(EtO)_2P(O)SCH(COOEt)Ph$. XVII.[280] XXIV.[1132] Liq. $b_{0.01}$ 127°,[280] $b_{0.1}$ 140°.[1132]

$(EtO)_2P(O)SCHPhCOOCMe_2C{\equiv}CH$. XVII. Liq. $b_{0.8}$ 175°.[654]

$(EtO)_2P(O)SCHPhCOOCMeEtC{\equiv}CH$. XVII. Liq. $b_{0.8}$ 180°.[654]

F.4.3.3. 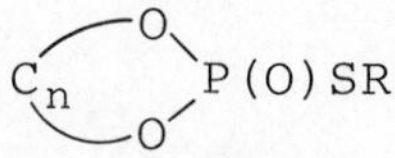

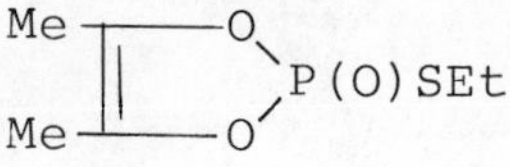 IX. Liq. $b_{0.035}$ 76-68°, n_D^{20} 1.4870, d_4^{20} 1.2210.[377]

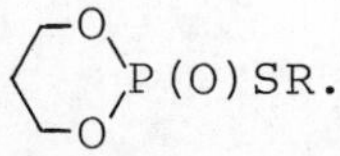 XXXI. Liq. R, $b_{0.05}$, n_D^{20}: Me, 103°, 1.5088; Et, 105°, --; Pr, 115°, 1.4988.[872]

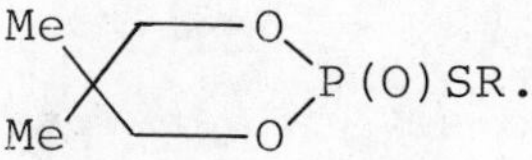 XVII. Solids. R, m.: Me, 81-81.5°; Et, 42-44°.[245]

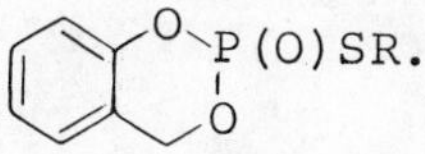 IX.[543, 917] Liq. R, b./mm: Me, 144-145°/0.1; Et, 140-145°/0.04; i-Pr, 155-158°/0.1; Pr, 145-147°/0.07; Bu, 157-160°/0.02; Ph, 115°/0.04,[917] solid, m. 88-89°;[543] $CH_2{=}CHCH_2$, 160-163°/0.15;[917] $PhCH_2$, 130-132°/0.04.[917]

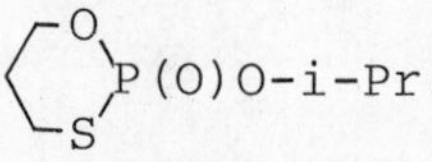 XXXI.[870, 872] Liq. $b_{0.05}$ 109°,[872] n_D^{20} 1.4860. ^{1}H-NMR.[870]

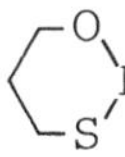P(O)OPh. XXXI. Solid. m. 82°. ^{1}H-NMR.[870]

F.4.3.4. RO(R'O)P(O)SR"

MeO(PhO)P(O)SMe. XXXI. XXXII. Liq. $b_{0.01}$ 104-106°,[440] $b_{0.1}$ 104-106°.[439]

MeO(4-$NO_2C_6H_4O$)P(O)SMe. VIII.[605] XVII.[192, 433] XXXI-XXXII.[439, 440] Liq. n_D^{20} 1.5612-1.5626,[440] n_D^{25} 1.5616,[192] ^{1}H-NMR.[605] ^{31}P -25.2 ppm.[605] Optical isomers - see table below.

MeO(4-$NO_2C_6H_4O$)P(O)SR. XVII. Liq. Optical isomers.[433]

R	n_D^{22} (l-) (-)	$[\alpha]_D^{22}$	c.(MeOH)	n_D^{22} (d-) (+)	$[\alpha]_D^{22}$	c.(MeOH)
Me	1.5639	-30.2 ± 9°	0.5	1.5628	35.1 ± 1.5°	0.4
Et	1.5576	-10.4 ±0.8°	0.6	1.5553	13.0 ± 0.8	0.6
Pr	1.5501	-9.1 ±1.4	0.4	1.5504	13.6 ± 1.5	0.3

MeO(1-$C_{10}H_7O$)P(O)SMe. XVII. Solids. Optical isomers.[221] (+), m. 57-59°, $[\alpha]_D^{24}$ 53.4° (c 6.34, $CHCl_3$); (-), m. 55-58°, $[\alpha]_D^{24}$ -51° (c 1.75, $CHCl_3$).[221]

MeO(4-Br-1-$C_{10}H_6O$)P(O)SMe. XVII. Liq. $[\alpha]_D^{24}$ 43.4° (c 2.44, $CHCl_3$).[221]

MeO(EtO)P(O)SCH_2Cl. XVII. Liq. n_D^{20} 1.4785.[949]

MeO(RO)P(O)SCH_2SEt. XVII. Liq. R, $b_{0.01}$ or n_D^{22}: i-Pr, 1.468; Bu, 81°; $Me_3CCHMeO$, 96°; cyclo-$C_6H_{11}O$, 106°.[1171, 1176]

MeO(EtOP(O)SCH_2CH_2SEt. XVII. Liq. $b_{0.01}$ 94°.[1115]

MeO(RO)P(O)SCH_2CH_2SEt. XVII. Liq. R, $b_{0.01}$: i-Pr, 76°,[1171] $b_{0.5}$ 118°,[1115]; $Me_3CCHMeO$, 98°; cyclo-$C_6H_{11}O$, 112°.[1171, 1176]

MeO(RO)P(O)SCH_2CN. Liq. R, n_D^{21} or $b_{0.01}$: i-Pr, 1.4774; s-Bu, 93°; $Me_3CCHMeO$, 92°; cyclo-$C_6H_{11}O$, 104°.[1171, 1176]

MeO(RO)P(O)SCH_2CH_2CN. XVII. Liq. R, n_D^{21} or $b_{0.01}$: i-Pr, 1.4787; s-Bu, 98°; $Me_3CCHMeO$, 96°; cyclo-$C_6H_{11}O$, 108°.[1171, 1176]

MeO(i-PrO)P(O)SC_6H_4Cl-4. XIX. Liq. $b_{0.01}$ 104°.[1169]

MeO(ClCHMeCHMeO)P(O)SC_6H_4Cl-4. XXXIII. Liq. $b_{0.5}$ 148-150°. Also about 20 analogs.[312, 453]

MeO(cyclo-$C_6H_{11}O$)P(O)SC_6H_4Cl-4. XIX. Colorless oil.[1180]

MeO(EtO)P(O)$SC_6H_4NO_2$-4. VIII. m. 37° (from petroleum ether).[520]

MeOP(O) 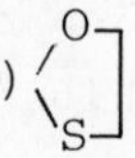 XIV. Liq. $b_{0.01}$ 73-75°.[400]

MeO(i-PrO)P(O)$SCH_2C_{10}H_7$-1. XVII. Liq. $n_D^{21.5}$ 1.5831.[182]

EtO(i-PrO)P(O)SMe. XVII. Liq. $b_{0.05}$ 53-55°, n_D^{25} 1.4426. Optical isomers: (+)$[\alpha]_D^{25}$ 2.1° (neat); (-)$[\alpha]_D^{25}$ -2.0° (neat).[787]

EtO(BuO)P(O)SMe. XVII. Liq. $b_{0.03}$ 61°, n_D^{25} 1.4553. Optical isomers: (+)$[\alpha]_D^{25}$ 0.8° (neat); (-) -1.1° (neat).[787]

EtO(FCH_2CH_2O)P(O)SEt. XXXI. Liq. $b_{0.5}$ 81-82°, n_D^{20} 1.4596, d_D^{20} 1.2012.[490]

EtO(Me_3SiO)P(O)SEt. XXXI. Liq. b_{14} 118.5-119.5°, n_D^{20} 1.4495, d_4^{20} 1.0505.[137]

EtO(RO)P(O)SEt. VIII. Liq.[1167] R, $b_{0.01}$: $EtSCH_2CH_2$, 84°; 4-ClC_6H_4, 96°; 2,4-$Cl_2C_6H_3$, 108°; 2,4,5-$Cl_3C_6H_2$, 112°; 4-$MeSC_6H_4$, 104°; 3,4-Me(MeS)C_6H_3, 110°.[1167]

EtO(4-$NO_2C_6H_4O$)P(O)SEt. VIII.[605, 846] XVII.[192] XXXI.[484] Liq. $b_{0.01}$ 102°,[846] $b_{0.01-0.025}$ 140-148°, n_D^{20} 1.5495,[484] n_D^{25} 1.5492,[192] d^{20} 1.2945.[484] ^{31}P -24.0 ppm.[605]

EtO(EtS)P(O)O- 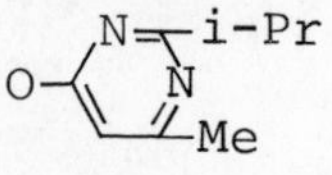 VIII. Liq. "S-ethyl diazinon." Oil with poor storage stability; characterized by IR and ^1H-NMR.[1010]

EtO(PhO)P(O)SR. XVII. Liq. R, n_D^t: $ClCH_2CH_2$, 1.5324$^{18°}$; CH_2=$CHCH_2$, 1.5310$^{20°}$; $PhCH_2CH_2$, 1.5567$^{22°}$; $PhSCH_2CH_2$, 1.5790$^{23°}$;

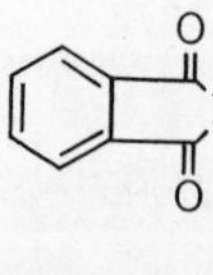 NCH_2, 1.5850$^{21°}$.[1331]

$EtO(PhO)P(O)SCH_2CH_2Cl$. XXV. Liq. $b_{0.3}$ 124-125°, n_D^{23} 1.5340.[788]

$EtO(ClCH_2CH_2O)P(O)SBu$. IX. Liq. $b_{in\ vacuo}$ 110°.[1361]

$EtO(PhO)P(O)SBu$. XVII. Liq. n_D^{20} 1.5125. About 80 analogs.[1331]

$EtO(EtSCH_2CH_2O)P(O)SC_6H_4Cl$-4. XXVIII. Liq. $b_{0.01}$ 114°.[1159]

$EtO(PhO)P(O)SC_6H_4Cl$. ^{31}P -39 ± 1.[1345]

$EtO(cyclo\text{-}C_6H_{11}O)P(O)SC_6H_4Cl$-4. XIX. Liq. n_D^{22} 1.5322.[1180]

$EtO(RC_6H_4O)P(O)SCH_2CH_2Ph$. XVII. Liq. R, n_D^t: 3-Me, 1.5522 25°; 4-Me, 1.5536 25°; 4-t-Bu, 1.5412 29°; 3,4-di-Me, 1.5530 28°; 4-Cl, 1.5610 26°.[1331] About 80 analogs.[1331]

$EtO(3\text{-}Me\text{-}4\text{-}MeSC_6H_3O)P(O)SCH_2CH_2Ph$. VIII. Liq. $b_{0.01}$ 120°.[1188] About 15 homologs and analogs.[1188]

i-BuO(MeEtC=NO)P(O)SCH_2COCl. From 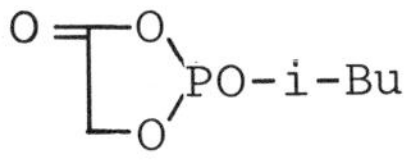 plus CCl_2FNO. Liq. n_D^{20} 1.4850, d_4^{20} 1.1830.[565]

$ClCH_2CH_2O(CCl_2{=}CHO)P(O)SR$. XXIX. Liq. R, b./mm: Me, 138°/3; Et, 134°/2; Ph, 174°/2.[1136]

$PhO(CCl_2{=}CHO)P(O)SEt$. IX. Liq. $b_{0.01}$ 130-135°, n_D^{20} 1.5448.[1250]

$BuO(2,4,5\text{-}Cl_3C_6H_2O)P(O)SMe$. VIII. Liq. $b_{0.2}$ 165-167°.[315]

F.4.4. Phosphorothioic Acid: Tertiary thiono Esters

F.4.4.1. $(RO)_3P(S)$

$(MeO)_3P(S)$. VIII.[753] XV.[154, 959, 961] Liq. b_3 75°,[253] b_{12} 80°,[253] b_{20} 82°,[959, 961] $n_D^{10.5}$ 1.4583,[253] n_D^{20} 1.4545,[1078] 1.4599,[753] mol. ref. 227.12,[1078] d_0^0 1.2192,[959, 961] d_0^{20} 1.2192,[753] IR.[238, 674, 1348] IR at 77°K,[425, 427, 429] Raman.[238, 1443] ^{1}H-NMR.[500, 734] ^{31}P -73.0[488] -72.9 ± 0.1.[715] L.NEt_3, n_D^{20} 1.5010.[762]

$(MeO)_3{}^{32}P(S)$. VIII. Liq. b_{13} 73°, n_D^{20} 1.4590, d_0^0 1.2190.[699]

$(EtO)_3P(S)$. VIII.[242, 959, 961] XV.[141, 142, 1456] XXVII.[1372] XXVIII.[417, 481, 793] From $(EtO)_2PSNa$ plus CCl_4 plus EtOH.[31] From $(EtO)_3P$ plus sulfenyl halides.[789] Liq. $b_{1.3}$ 62°,[417] b_8 88-89°,[855, 959, 961] b_{12}

94-95°,[60, 62, 789] n_D^{20} 1.4488,[60, 62, 1078] n_D^{25} 1.4464,[142] n_D^{30} 1.4454.[417] Mol. ref. 287.20.[1078] d_0^{20} 1.0756,[60, 61, 62] d^{20} 1.0746,[855] $d^{31.3}$ 1.0783, d^{60} 1.0491.[125] IR.[673, 674] ^{1}H-NMR.[500, 816, 970, 1299, 1395] Raman.[968, 1443] ^{31}P -68.0 ± 0.1,[715, 817] -68.1.[850] μ 2.91 D,[60] 2.9 D.[466] Electric moment in benzene at 25°, 2.82.[261] γ^{20} 29.65 dynes/cm.[61] Parachor, 431.2.[62] Bond magn. rotation.[580] L.NEt$_3$, liq. n_D^{20} 1.4909.[762]

$(ClCH_2CH_2O)_3P(S)$. VIII.[356] X, XV.[204] Liq. b_9 142-150°,[356] n_D^{20} 1.5080,[204] 1.5650,[356] d_4^{20} 1.4778.[356]

$(Me_2NCH_2CH_2O)_3P(S)$. VIII. Liq. (dec.∿125°), n_D^{25} 1.4708.[330]

$(i\text{-}PrO)_3P(S)$. ^{31}P -65.1.[715]

$[Me(ClCH_2)CHO]_3P(S)$. XV. Liq. $b_{0.04}$ 117-119°, n_D^{20} 1.4885, d_4^{20} 1.2832.[992]

$[RSCH_2(CH_2Cl)CHO]_3P(S)$. XV. Liq.[570] R, n_D^{20}, d_4^{20}: n-C_6H_{13}, 1.5089, 1.0971; n-C_7H_{15}, 1.5078, 1.0951; n-C_8H_{17}, 1.5074, 1.0950; n-C_9H_{19}, 1.5056, 1.0733; n-$C_{10}H_{21}$, 1.5053, 1.0728.[570]

$(PrO)_3P(S)$. IX.[961] XV.[60, 62] XXIX. (Transesterification $(PrO)_2P(S)OCHPhCOOEt$ plus PrONa in PrOH.)[995] Liq. b_{10} 123.5-124.5°,[60, 62] b_{20} 130-131°,[995] n_D^{20} 1.4502,[60, 62, 1078] mol. ref. 348.50,[1078] d_0^0 1.0409,[961] d_0^{20} 1.0177,[60, 62] d_4^{20} 1.0412.[995] ^{31}P -68.2.[715] γ^{20} 28.47, Parachor 545.4.[60, 62] μ 2.99 D.[60, 62] Bond magn. rotation.[580]

$(MeCHClCH_2O)_3P(S)$. X. (From $P(S)Cl_3$ plus $MeCH\overset{O}{-}CH_2$ with PCl_3 catalyst.[204, 719] XV.[204, 720] Liq. b_1 140-160°,[204, 720] $n_D^{21.5}$ 1.4875,[719] 1.4910,[204] 1.4916.[720]

$(CH_2\overset{O}{-}CHCH_2O)_3P(S)$. XV. Liq. $b_{0.005}$ 180-182°, n_D^{20} 1.4937, d^{20} 1.3156.[1039]

$(t\text{-}BuO)_3P(S)$. XV.[716] Liq. n_D^{25} 1.4482. ^{1}H-NMR, IR.[716] ^{31}P - 41.2.[715, 716] Pure material was distilled, $b_{0.4}$ 80°, but the bulk of the sample decomp. at a pot temp. above 85° to give $(t\text{-}BuO)_2POSH$.

$(NCCMe_2O)_3P(S)$. XV. (from $(NCCMe_2O)_3P$ plus $P(S)Cl_3$). Solid.

m. 81° (from xylene).[579a]

$(i\text{-}BuO)_3P(S)$. IX. Liq. b_{20} 155°, d_0^0 0.9907.[961]

$(BuO)_3P(S)$. XV.[60, 62] XXVIII.[793] Liq. $b_{0.06}$ 104°,[793] b_{11} $158\text{-}159^{\circ}$,[60, 62] n_D^{20} 1.4515,[60, 62, 1078] Mol. ref. 409.89,[1078] d_0^{20} 0.9871,[60, 62] ^{31}P -68.5,[715] -69.2,[715, 850] μ 3.04 D.[60] 3.02 D,[261] Elect. moment in benzene at 25°: 2.84.[261] Dielectric const. 6.82 ± 0.03.[261] γ28.36 dynes/ cm, Parachor 660.2.[61, 62] Bond magn. rotation.[580]

$(Me_3CCH_2O)_3P(S)$. ^{31}P -68.5.[715, 717]

$(i\text{-}AmO)_3P(S)$. VIII. Liq. d_0^{12} 0.849 (Chevrier, 1869).

$(C_6H_{13}O)_3P(S)$. XV.[60, 62, 63] Liq. $b_{2.5}$ $188\text{-}188.5^{\circ}$, n_D^{20} 1.4568, d_0^{20} 0.9501. γ^{20} 28.68, Parachor, 892.6[62] Mol. ref. 533.40.[1078] μ 3.18 D.[60] Viscosity.[63]

$(C_6H_{13}MeCHO)_3P(S)$. ^{31}P -69.5.[715]

$(i\text{-}C_8H_{17}O)_3P(S)$. ^{31}P -69.6.[715]

$(C_8H_{17}O)_3P(S)$. XV.[60, 62, 63] Liq. $b_{0.5}$ $224\text{-}226^{\circ}$, n_D^{20} 1.4592, d_0^{20} 0.9293 γ^{20} 29.11 dynes/cm. Parachor 1126.4.[62] Mol. ref. 657.68.[1078] μ 3.08 D.[60] Viscosity.[63]

$(PhO)_3P(S)$. VIII.[73, 371, 1226] IX.[72, 376] XV.[1111, 1405] XXVIII.[793] Solid b_1 $148\text{-}150^{\circ}$,[793] m. 48°,[376, 793] $52.5\text{-}53^{\circ}$,[72, 604, 961, 1405] (from petroleum ether),[1226] ^{31}P -53.4,[715, 604] μ 2.58 D.[133, 529, 604]

$(2\text{-}BrC_6H_4O)_3P(S)$. VIII. Solid. m. $89\text{-}90^{\circ}$.[747]

$(4\text{-}BrC_6H_4O)_3P(S)$. VIII. Solid. m. $89\text{-}99^{\circ}$.[747]

$(2\text{-}ClC_6H_4O)_3P(S)$. VIII.[747] XV.[513] Liq. $b_{0.025}$ $160\text{-}170^{\circ}$,[747] b_3 $250\text{-}251^{\circ}$,[513] n_D^{20} 1.5992,[747] d_0^{20} 1.458,[513] d^{20} 1.3899.[747]

$(3\text{-}ClC_6H_4O)_3P(S)$. VIII. Liq. $b_{0.05}$ $170\text{-}180^{\circ}$, n_D^{20} 1.6002, d^{20} 1.4318.[747]

$(4\text{-}ClC_6H_4O)_3P(S)$. VIII.[747, 1226] XV.[513] XXXIII. (From $(4\text{-}ClC_6H_4O)_3P$ plus $[(RO)_2P(S)]_2S$.)[760] Solid. $b_{0.15}$ $190\text{-}200^{\circ}$,[760] b_4 $262\text{-}263^{\circ}$,[513] m. $84\text{-}86^{\circ}$,[747, 1226] $108\text{-}108.5^{\circ}$,[760] 113°.[513]

$(2,4\text{-}Cl_2C_6H_3O)_3P(S)$. XV. Solid. m. 106°.[509]

$(2,4\text{-}ClNO_2C_6H_3O)_3P(S)$. VIII. Solid. m. 186-187°.[747]

$(4,2\text{-}ClNO_2C_6H_3O)_3P(S)$. VIII. Solid. m. 136-137°.[747]

$(2,6,4\text{-}Cl_2NO_2C_6H_2O)_3P(S)$. VIII. Solid. m. 212-213°.[747]

$(2\text{-}IC_6H_4O)_3P(S)$. VIII. Solid. m. 147-148°.[747]

$(4\text{-}IC_6H_4O)_3P(S)$. VIII. Solid. m. 123-124°.[747]

$(4\text{-}NH_2C_6H_4O)_3P(S)$. From $(4\text{-}NO_3C_6H_4O)_3P(S)$ plus H_2 and Raney Ni in MeOH. Solid. m. 156°.[456, 457]

$(3\text{-}Me_2NC_6H_4O)_3P(S)$. VIII. Solid. m. 55.5-56.5° (from EtOH-acetone). Trimethiodide·$2H_2O$, m. 178.5-181° (dec).[328]

$(4\text{-}Me_2NC_6H_4O)_3P(S)$. VIII. Solid. m. 127.5-129°. Trimethiodide hexahydrate. m. 162-164° (dec.).[328]

$(3\text{-}Et_2NC_6H_4O)_3P(S)$. VIII. Liq. $b_{0.002}$ about 325°, n_D^{25} 1.5948.[328]

$(4\text{-}OCNC_6H_4O)_3P(S)$. From $(4\text{-}NH_2C_6H_4O)_3P(S)$ plus $COCl_2$. Solid. m. 84-86°.[456]

$[4\text{-}(2,6\text{-}Cl_2C_6H_3CH{=}NO_2CNH)C_6H_4O]_3P(S)$. From $(4\text{-}OCNC_6H_4O)_3$-P(S) plus $2,6\text{-}Cl_2C_6H_3CH{=}NOH$. Solid. m. 90-100°.[208]

[s-BuNH— 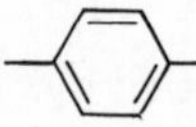 —N(s-Bu)CONH— 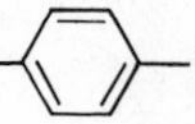 —O]$_3$P(S). From $(4\text{-}OCNC_6\text{-}H_3O)_3P(S)$ plus (s-BuNH—  —NH-s-Bu). Solid. m. 137-139°.[299]

$(2\text{-}NO_2C_6H_4O)_3P(S)$. XV. Solid. m. 118°.[513]

$(4\text{-}NO_2C_6H_4O)_3P(S)$. IX.[456, 527] Solid. m. 174°,[527] 181-183°.[456] IR.[528] μ 3.33 D.[529]

$(3\text{-}PhOC_6H_3O)_3P(S)$. VIII. Liq. $b_{0.05}$ 304-308°.[811]

$(2\text{-}MeC_6H_4O)_3P(S)$. VIII.[376, 1277] XV.[1405] Solid. b_1 260-265°,[376] m. 45-46°,[376, 1277, 1405] ^{31}P -52.2.[715, 850]

$(3\text{-}MeC_6H_4O)_3P(S)$. XV.[127, 1405] Solid. b_{12} 270-272°,[127] m. 40-41° (from EtOH).[127, 1405]

$(4\text{-}MeC_6H_4O)_3P(S)$. VIII.[371, 1277] IX.[73] Solid. m. 87-88°,[73, 371] 93-94° (from EtOH).[1277] ^{31}P -53.5,[715, 1345] -54.[715, 850]

$[2\text{-}s\text{-}Bu\text{-}4,6\text{-}(NO_2)_2C_6H_2O]_3P(S)$. VIII. Solid m. 211-215°.[551]

$[3,5,4\text{-}(t\text{-}Bu)_2HOC_6H_2O]_3P(S)$. XV. Solid. m. 194-197° (from i-PrOH).[262, 763]

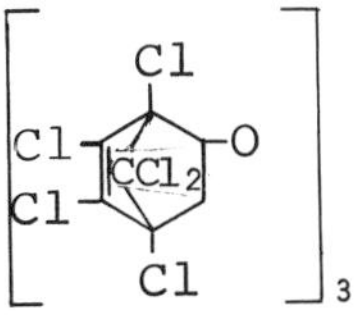

P(S). VIII. Liq. n_D^{20} 1.5378.[324]

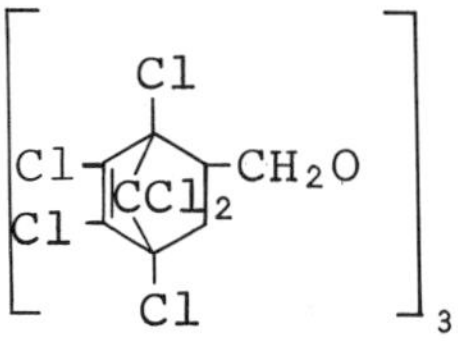

P(S). VIII. Liq. n_D^{20} 1.5240.[324]

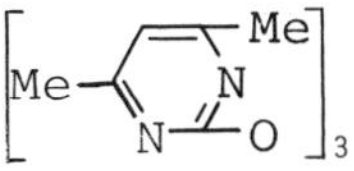

P(S). IX. Solid. m. 129-131°.[1017]

$(Me_3SiO)_3P(S)$. XV. Liq. b_1 79-80°, n_D^{20} 1.4350, d_4^{20} 0.9678.[913, 914]

$(MeEt_2SiO)_3P(S)$. XV. Liq. b_2 148°, n_D^{20} 1.4562, d_4^{20} 0.9703.[913]

$(Et_3SiO)_3P(S)$. XV. Liq. b_1 159-160°, n_D^{20} 1.4610, d_4^{20} 0.9656.[913, 914]

F.4.4.2. 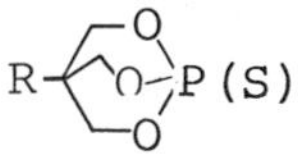

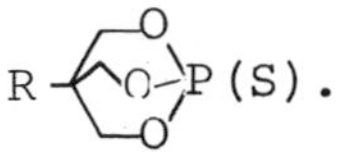 Solids. (Q = 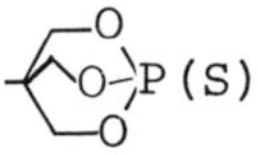 in tables below).

R	Method
Me	XV.[1356, 1386] m. 224-226°.[1356, 1386] ^{31}P -57.3,[715] -51.4.[1355] μ 6.77 D.[133] Molecular polarizability.[69]
$ClCH_2$	From $HOCH_2$-Q plus DMF and Cl_2. (Also with PCl_5 in low yield.[1009] m. 216° (sublimed).
$HOCH_2$	VIII.[1004, 1009] m. 160.5-161° (from xylene).
$ClCH_2OCH_2$	From $HOCH_2Q$ plus CH_2O and HCl.[1009] m. 157-158° (from xylene).
$ClCH_2COOCH_2$	From $HOCH_2Q$ plus $ClCH_2COCl$.[1004, 1005, 1009] m.

	142° (from o-$Cl_2C_6H_4$).
$Cl_2CHCOOCH_2$	From $HOCH_2Q$ plus $Cl_2CHCOCl$.[1004, 1005, 1009] m. 164.5° (from o-$Cl_2C_6H_4$).
$Cl_3CCOOCH_2$	From $HOCH_2Q$ plus Cl_3CCOCl.[1004, 1005, 1009] m. 148° (from benzene).
$MeNHCOOCH_2$	From $HOCH_2Q$ plus MeNCO.[1009] m. 170-172° (from MeOH).
$EtNHCOOCH_2$	From $HOCH_2Q$ plus EtNCO.[1009] m. 124° (from EtOH)
$BuNHCOOCH_2$	From $HOCH_2Q$ plus BuNCO.[1009] m. 102°.
Et	XV.[459, 869, 1368, 1386, 1396] m. 176-179° (from EtOH).[869, 1368] m. 181.2-185°.[1396]
Ph	VIII. m. 207-209°.[158]
$C_6Br_5OCH_2$	VIII. m. 234-239°.[158]
$C_6Cl_5OCH_2$	VIII. m. 218-222°.[158]

$(RO)_2P(S)SCH_2COOCH_2$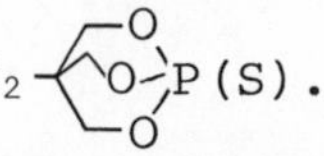. XVII. Solids. R, m.: Me, 119° (from MeOH); Et, 86° (from EtOH); i-Pr, 116° (from i-PrOH).

$ArOCH_2COOCH_2$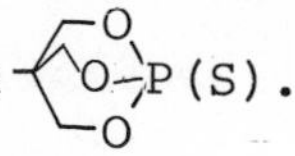. From $ClCH_2COOCH_2$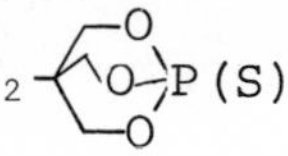 plus ArOH).[1004, 1009] Solids. R, m.: Ph, 202° (from o-Cl_2-C_6H_4 and acetone); 2,4-$ClFC_6H_3$, 159.5-160° (from o-$Cl_2C_6H_4$); 2,4-$Cl_2C_6H_3$, 187-187.5°; 2,4,5-$Cl_3C_6H_2$, 204.5-205.5°; Cl_5C_6, 228° (from MeCN); 4-FC_6H_4, 146-148° (from o-$Cl_2C_6H_4$).

$(S)P(OCH_2$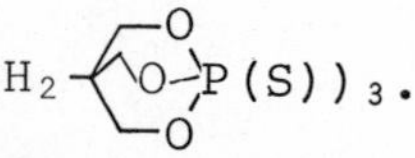$)_3$. VIII. Solid. m. 270°.[1009]

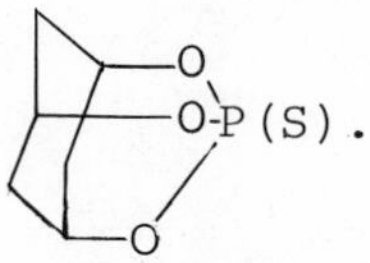. IX, XV.[1267] Solid. m. 250-251°. ^{31}P -64.0[715, 1355]

F.4.4.3. $(RO)_2P(S)OR'$

$(EtO)_2P(S)OMe$. XXVII. Liq. b_{26} 103-104.5°, n_D^{20} 1.4523, d^{20} 1.1071.[491] ^{31}P -69.6.[715]

$(\overset{O}{CH_2-CH}CH_2O)_2P(S)OMe$. Liq. $b_{0.5}$ 125-126°, n_D^{20} 1.4890, d_4^{20} 1.3002.[1036]

$(Cl_3CCMe_2O)_2P(S)OMe$. XV. Liq. $b_{0.01}$ 134.5-136°, n_D^{20} 1.5220, d^{20} 1.4844.[252]

$(PhO)_2P(S)OMe$. Mass spect.[184]

$(4\text{-}NO_2C_6H_4O)_2P(S)OMe$. VIII. Solid m. 96°. Crystallography.[401]

$(4\text{-}EtOCOC_6H_4O)_2P(S)OMe$. VIII. Solid m. 48° (from petroleum ether).[1185]

$(3,4\text{-}MeNO_2C_6H_3O)_2P(S)OMe$. VIII. Solid. m. 92.5-93.5°.[926]

$(2\text{-}s\text{-}Bu\text{-}4,6\text{-}(NO_2)_2C_6H_2O)_2P(S)OMe$. VIII. Liq. $b_{0.1}$ ∿70°.[552]

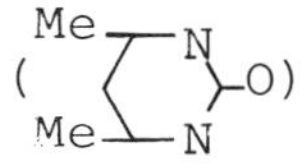
$O)_2P(S)OMe$. VIII. Solid. m. 115.5-116.5°.[1017]

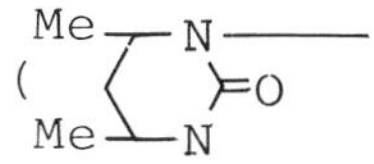
$CH_2CH_2O)_2P(S)OMe\cdot HCl\cdot H_2O$. VIII. Solid. m. 200° (dec).[1016]

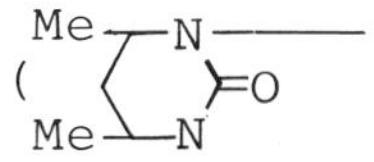
$CH_2CHMeO)_2P(S)OMe$. 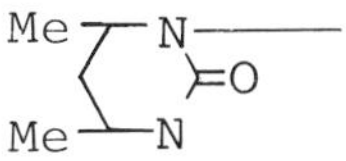 $CH_2CHMeOH.HCl$. VIII. Solid. m. 203-204° (dec.).[1016]

$(EtO)_2P(S)OCH_2Cl$. XVII. Liq. $b_{1.5}$ 118-122°.[1087]

$(PhO)_2P(S)OCH_2NMe_3^+\ X^-$. XXXII. (from $(PhO)_2P(S)OCH_2NMe_2$ plus MeX). Solids. X, m.: Br, 115-118°; I, 97-99°.[535]

$(RO)_2P(S)OEt$

$(EtO)_2P(S)OCH_2SEt$. XVII. Liq. $b_{0.01}$ 75-78°.[1175]

$(EtO)_2P(S)OCH_2SC_6H_4Cl\text{-}4$. XVII. Liq. $b_{0.1}$ 120°.[1175]

$(EtO)_2P(S)OCH_2SC_6H_4Me\text{-}4$. XVII. Liq. $b_{0.01}$ 130-132°.[1175]

$(EtO)_2P(S)OCH_2SC_6H_{11}$-cyclo. XV. Liq. $b_{0.01}$ 99-102°.[1156]

$(EtO)_2P(S)OCH_2P(S)(OEt)_2$. VIII. Liq. $b_{0.01}$ 104°.[1170]

$(EtO)_2P(S)OCH_2SiMe_2OEt$. VIII. Liq. b_{15} 89-94°, n_D^{20} 1.4450, d^{20} 1.0581.[354]

$(MeO)_2P(S)OEt$. VIII.[432, 753] XV.[510] Liq. b_{10} 72°,[432] b_{17} 79.5-80.5°,[510] b_{25} 96°.[753] n_D^{20} 1.4520,[753] 1.4530,[432, 510] d^{20} 1.1527,[753] 1.1631.[510]

$(FCH_2CH_2O)_2P(S)OEt$. XV. Liq. b_3 86-88°, n_D^{20} 1.4580, d^{20} 1.2686.[490]

$(EtOCH_2CH_2O)_2P(S)OEt$. VIII. Liq. b_3 170°.[1260]

$(CH_2{=}CHO)_2P(S)OEt$. XV. Liq. b_7 72-73°, n_D^{20} 1.4634, d^{20} 1.1017.[868]

$(RO)_2P(S)OEt$. XV.[468, 510] Liq.

R	b./mm	n_D^{20}	d^{20}	d_4^{20}
i-Pr	106-108°/17	1.4425	1.1631[510]	
Pr	105-107°/13	1.4475	1.0388[510]	
	226-227°/atm.	1.4505		1.0443[468]
i-Bu	134-135°/16	1.4450	0.9942[510]	
Bu	135-137°/13	1.4486	0.9969[510]	
	129-130°/14	1.4506		1.0041[468]
i-Am	127-129°/5	1.4530	0.9881[510]	
	135-148°/12	1.4510		0.9856[468]

$(\overset{O}{CH_2\text{-}CH}CH_2O)_2P(S)OEt$. XV. Liq. b_2 134-135°, n_D^{20} 1.4850, d_4^{20} 1.2531.[1036]

$(t\text{-}BuO)_2P(S)OEt$. ^{31}P -50.2.[715, 717]

$(CH_2{=}CHCHClCH_2O)_2P(S)OEt$. XV. Liq. $b_{0.07}$ 124-126°, n_D^{20} 1.4940, d_4^{20} 1.2912.[990]

$(Et_3SiO)_2P(S)OEt$. XV. Liq. b_1 138-140°, n_D^{20} 1.4588, d_4^{20} 0.9778.[913]

$(PhO)_2P(S)OEt$. VIII.[596] IX.[153] Liq. $b_{0.02}$ 140°,[596] b_{16} 220-230°.[153] ^{31}P -42.[488] Mass spect.[184]

$(2\text{-}BrC_6H_4O)_2P(S)OEt$. VIII. Liq. $b_{0.25}$ 165°, n_D^{20} 1.5958, d^{20} 1.6246.[747]

$(4\text{-}BrC_6H_4O)_2P(S)OEt$. VIII. Liq. $b_{0.025}$ 157-160°, n_D^{20} 1.5940, d^{20} 1.6165.[747]

$(Cl_nC_6H_{5-n}O)_2P(S)OEt$. VIII. Liq. and solids.[747]

n	Position	m.	b./mm	n_D^{20}	d^{20}
1	2	--	128-132°/0.025	1.5693	1.2987
1	3	--	135-140°/0.05	1.5645	1.2966
1	4	--	140-146°/0.05	1.5713	1.3353
2	2,6	53-54°			
3	2,4,5	68°			
3	2,4,6	112-113°			
5	2,3,4,5,6	156-157°			

$(2\text{-}IC_6H_4O)_2P(S)OEt$. VIII. Solid. m. 47-48°.[747]

$(4\text{-}IC_6H_4O)_2P(S)OEt$. VIII. Solid. m. 55-56°.[747]

$(4\text{-}NO_2C_6H_4O)_2P(S)OEt$. VIII. Solid. m. 125-126°.[527] IR.[80, 528] μ 4.97 D.[529] Crystallography.[401]

$(4\text{-}NO_2C_6H_4O)_2{}^{32}P({}^{35}S)OEt$. VIII. Solid. m. 125° (from EtOH).[708]

$(4\text{-}MeOCOC_6H_4O)_2P(S)OEt$. VIII. Solid. m. 128° (from petroleum ether).[1185]

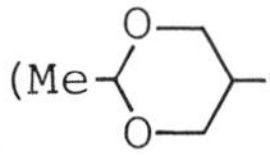 $)_2P(S)OEt$. XV. Liq. $b_{0.5}$ 155-157°, n_D^{20} 1.4820.[64]

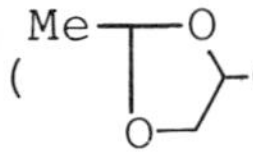 $CH_2O)_2P(S)OEt$. XV. Liq. $b_{0.5}$ 138°, n_D^{20} 1.4740, d^{20} 1.2133.[64]

$(EtO)_2P(S)OCH_2CH_2Cl$. XXXIII. (From $(EtO)_2P(S)H$ plus 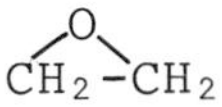 and CCl_4, $TiCl_4$, autoclave at 100°, 3.5 hr.) Liq. $b_{0.45-0.5}$ 61-69°.[199]

$(EtO)_2P(S)OCH_2CCl_3$. "Known methods." Liq. b_3 130-131°, n_D^{20} 1.4840, d^{20} 1.3414.[855]

$(EtO)_2P(S)OCH(OR)CCl_3$. XV. Liq.[875]

R	b./mm	n_D^{20}	d_4^{20}
Me	84-86°/0.035	1.4810	1.3406
Et	80-82°/0.025	1.4766	1.3069
Pr	92-94°/0.025	1.4760	1.2767
Bu	95-97°/0.025	1.4745	1.2505

$(EtO)_2P(S)OCH(SR)CCl_3$. XV. Liq.[1149] R, b./mm: Et, 100°/0.01; Ph, 140°/0.05; $4\text{-}ClC_6H_4$, 160°/0.05; $3,4\text{-}Cl_2C_6H_3$, 150°/0.01.

$(BuO)_2P(S)OCH(SBu)CCl_3$. XV. Liq. $b_{0.03}$ 145-148°, n_D^{20} 1.4993, d_4^{20} 1.1958.[874]

$(EtO)_2P(S)OCH_2CH_2F$. XV. Liq. $b_{0.5}$ 65-66°, n_D^{20} 1.4570, d^{20} 1.1876.[490]

$(MeO)_2P(S)OCH_2CH_2NMe_2$. VIII. Liq. $b_{0.4}$ 64-65°.[330]

$(EtO)_2P(S)OCH_2CH_2NMe_2$. VIII.[330, 362, 663, 1300] Liq. $b_{0.2}$ 83-86°,[663] $b_{0.3}$ 108-109°,[362] n_D^{20} 1.4590.[1300] Oxalate salt, solid, m. 132° (from acetone).[1300]

$(RO)_2P(S)OCH_2CH_2NMe_2$. VIII. Liq. R, b./mm: i-Pr, 83-93°/

0.6; Pr, 92-112°/0.3; Bu, 103-120°/0.3.[330] Also analogs and salts.

$(EtO)_2P(S)OCH_2CH_2NEt_2$. VIII.[330, 362, 663] Liq. $b_{0.3}$ 98-100°,[663] 103-106°,[330] b_{12} 153-156°.[362]

$(EtO)_2P(S)OCH_2CH_2N_3$. VIII. Liq. $b_{0.01}$ 71°.[1137]

$(EtO)_2P(S)OCH_2CH_2NO_2$. VIII. Liq. $b_{0.01}$ 87°.[287]

$(EtO)_2P(S)OCH_2CH_2CN$. VIII. Liq. $b_{0.01}$ 62-64°.[218]

$(EtO)_2P(S)OCH_2CH_2$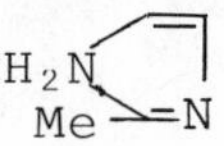 VIII. Picrate, solid. m. 88-89°.[1076]

$(EtO)_2P(S)OCH_2CH_2$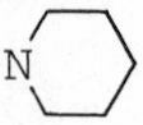 VIII. Liq. $b_{0.35}$ 151-153°.[362]

$(EtO)_2P(S)OCH_2CH_2OH$. VIII.[1200, 1408] b_2 108-109°,[1408] 110-115°,[1200] n_D^{18} 1.4695, d^{18} 1.767.[1408]

$(EtO)_2P(S)OCH_2CH_2OEt$. VIII. Liq. $b_{0.42}$ 57-58°, n_D^{20} 1.4510, d^{20} 1.030.[741]

$(4\text{-}NO_2C_6H_4O)_2P(S)OCH_2CH_2OEt$. VIII. Solid. m. 165°.[1260]

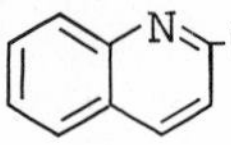$O)_2P(S)OCH_2CH_2OEt$. VIII. Solid. m. 209°.[1260]

$(EtO)_2P(S)OCH_2CH_2OCH_2CH_2OH$. VIII. Liq. $b_{0.5}$ 90°.[1200]

$(EtO)_2P(S)OCH_2CH_2OCH_2CH_2SMe$. VIII. Liq. $b_{2.5}$ 168°.[1122]

$(EtO)_2P(S)OCH_2CH_2OCOCMe{=}CH_2$. XV. Liq. $b_{1.5}$ 134-135°, n_D^{20} 1.4701, d^{20} 1.1401.[998]

$(EtO)_2P(S)OCH_2CH_2OCONRR'$. VIII. Liq. R, R', n_D^{20} : H, H, 1.4904; H, Me, 1.4804; Me, Me, 1.4740.[475]

$(EtO)_2P(S)OCH_2CH_2OCOMe$. By acetylation of $(EtO)_2P(S)OCH_2$-CH_2OH. Liq. $b_{0.5}$ 117°, n_D^{20} 1.4622, d^{20} 1.1585.[1408]

$(EtO)_2P(S)OCH_2CH_2OSO_2Cl$. XVII. Liq. b_4 112°, n_D^{20} 1.4495, d^{20} 1.0632.[686]

$(i\text{-}PrO)_2P(S)OCH_2CH_2OSO_2Cl$. XVII. Liq. b_7 120-121°, n_D^{20} 1.4770, d^{20} 1.0028.[686]

$((EtO)_2P(S)O)_2(CH_2)_2$. VIII. Liq. $b_{0.02}$ 110°.[1200]

$(MeO)_2P(S)OCH_2CH_2SMe$. VIII.[1183] XV.[1130] Liq. $b_{0.07}$ 57°,[1183] b_2 115°.[1122]

$(EtO)_2P(S)OCH_2CH_2SMe$. VIII.[1183] XV.[1130] Liq. $b_{0.04}$ 73-75°.

[1183] $b_{0.05}$ 73°.[1130]

$(EtO)_2P(S)OCH_2CH_2SCH_2OR$. VIII.[1147, 1148] Liq. R, b./mm: Me, 132°/2; Et, 138°/2; i-Pr, 115°/0.01; s-Bu, 120°/0.01.

$(EtO)_2P(S)OCH_2CH_2SCH_2OCOR$. VIII. Liq. R, n_D^{18}: Me, 1.5102; Et, 1.5017; i-Pr, 1.4938; Pr, 1.4945[19°].[1352]

$(MeO)_2P(S)OCH_2CH_2SEt$. VIII.[703, 741, 1122, 1183] XV.[449] Liq. $b_{0.06}$ 62°,[1183] $b_{0.1}$ 79°,[703, 741] b_2 134°,[1122] n_D^{20} 1.4980,[703, 741] n_D^{25} 1.4949,[449] d^{20} 1.1848,[703, 741] d^{25} 1.1771.[449]

$MeO(RO)P(S)OCH_2CH_2SEt$. VIII.[703, 741] Liq.

R	b./mm	n_D^{20}	d^{20}
Et	80-82°/0.1	1.4930	1.1526
i-Pr	88-92°/0.05	1.4870	1.1434
Bu	86°/0.25	1.4930	1.1105
i-Am	125-130°/0.13	1.4980	1.0842

$(EtO)_2P(S)OCH_2CH_2SEt$. VIII.[741] XV.[340, 449, 943, 969] Liq. $b_{0.002}$ 85-86°, $b_{0.2}$ 124-128°,[969] n_D^{20} 1.4875,[943] n_D^{25} 1.4865,[449] d^{20} 1.1190,[741] d^{25} 1.1114.[449] IR and UV.[375] ^1H-NMR.[74] ^{31}P -67.7.[715] Raman.[968] "Systox" = 2:1 P(S): P(O) isomer.[1122, 1123] Isomerization.[345]

$(EtO)_2P(S)OCH_2CH_2SCH_2CN$. VIII. Liq. $b_{0.01}$ 96°.[1107]

$(EtO)_2P(S)OCH_2CH_2SR$. VIII.[1141] Liq. R, b./mm: COMe, 95°/0.01; $COCHCl_2$, 108°/0.01; $COCCl_3$, 110°/0.01; $CONMe_2$, 127°/0.01; COPh, 160°/0.04.

$(EtO)_2P(S)OCH_2CH_2SCH_2COOEt$. VIII. Liq. b_2 170°.[1147]

$(EtO)_2P(S)OCH_2CH_2S$-i-Pr. XV. Liq. $b_{0.00002}$ 57°, n_D^{25} 1.4805, d^{25} 1.0853.[449]

$(i\text{-}PrO)_2P(S)OCH_2CH_2S$-i-Pr. XV. Liq. $b_{0.0001}$ 65°, n_D^{25} 1.4678, d^{25} 0.9612.[449]

$(EtO)_2P(S)OCH_2CH_2SCH_2CH_2SR$. VIII. Liq.[1122] R, b./mm: Pr, 140-146°/2; Bu, 158-162°/3; Ph, 177-178°/2. Also several analogs.[1122]

$(EtO)_2P(S)OCH_2CH_2SC_6H_{11}$-cyclo. VIII. Liq. $b_{0.01}$ 146-148°.[1064]

$(MeO)_2P(S)OCH_2CH_2S(O)Et$. From $(MeO)_2P(S)OCH_2CH_2SEt$ plus Br_2 and aqueous Na_2CO_3. Liq. $b_{0.01}$ 95-96°.[628]

$(EtO)_2P(S)OCH_2CH_2SO_2Et$. From $(EtO)_2P(S)OCH_2CH_2SEt$ plus $KMnO_4$. Liq. $b_{0.01}$ 104°.[281, 843]

$(MeO)_2P(S)OCH_2CONHMe$. VIII. Liq. b_{1-2} 140°, n_D^{20} 1.4910, d_4^{20} 1.3779.[697]

$(EtO)_2P(S)OCH_2CONHMe$. VIII. Liq. $b_{0.5}$ 130-133°, n_D^{20} 1.4812, d_4^{20} 1.1827.[697]

$(EtO)_2P(S)OCH_2CONHC_6H_4NO_2$-4. VIII. XVII. Solid. m. 61-62°.[568]

$(MeO)_2P(S)OCH_2COOEt$. VIII. Liq. b_2 118-121°, n_D^{20} 1.4630, d_4^{20} 1.2498, IR.[697]

$(EtO)_2P(S)OCH_2COOEt$. VIII. Liq. b_{15} 123-127°, n_D^{20} 1.4599, d_4^{20} 1.2501.[697]

$(PrO)_2P(S)OCH_2COOBu$. XXIX. Liq. b_{15} 185°, n_D^{20} 1.4490, d_4^{20} 1.0643.[994]

$(BuO)_2P(S)OCH_2COOBu$. XXIX. Liq. b_{12} 192°, n_D^{20} 1.4540, d_4^{20} 1.0532.[994]

$(EtO)_2P(S)OCH_2COOC(CN)Me_2$. XVII. Liq. $b_{0.04}$ 165-170°.[407]

$(EtO)_2P(S)OCH_2COO$-i-Pr. XVII. Liq. $b_{0.45}$ 130-131°.[851]

$(EtO)_2P(S)OCH_2COO$-i-Bu. XVII. Liq. $b_{0.6}$ 141-142°.[851]

$(RO)_2P(S)OCH=CH_2$. XV. Liq.[868]

R	b./mm	n_D^{20}	d^{20}
Et	82°/7.5	1.4562	1.0904
Pr	92°/6	1.4581	1.0505
Bu	126-127°/8	1.4575	1.0195
Ph	163-164°/2	1.5655	1.2164

$(MeO)_2P(S)OCH=CCl_2$. XXIX. Liq. b_1 70-71°.[1154]

$(EtO)_2P(S)OCH=CCl_2$. XXIX. Liq. b_1 87-88°.[1154]

$(MeO)_2P(S)O$-i-Pr. VIII. (From MeO(i-PrO)P(S)Cl plus MeOH and NaOH.) Liq. b_{12} 79-80°, n_D^{20} 1.4540, d^{20} 1.1031.[704]

$(EtOCH_2CH_2O)_2P(S)O$-i-Pr. VIII. Liq. b_5 174°.[1260]

$(\overset{\frown O \frown}{CH_2-CH}CH_2O)_2P(S)O$-i-Pr. XV. Liq. b_1 130-132°, n_D^{20} 1.4791, d^{20} 1.2211.[1039]

$(4\text{-}Me_2NC_6H_4O)_2P(S)$-i-Pr. VIII. Solid. m. 74-76° (from i-PrOH). Dimethiodide·$3H_2O$. Solid. m. 177-177.5° (dec.) [328]

$(EtO)_2P(S)OCHMeCH_2Cl$. XV. Liq. $b_{0.04}$ 73-75°, n_D^{20} 1.4613, d_4^{20} 1.1504.[992]

$(EtO)_2P(S)OCHMeCH_2NMe$. VIII. Liq. $b_{0.5}$ 77-78°.[330]

$(EtO)_2P(S)OCHMeCH_2N$  VIII. Liq. $b_{0.06}$ 108-112°.[663]

$(EtO)_2P(S)OCHMeCH_2NO_2$. VIII. Liq. $b_{0.01}$ 85°.[287]

$(EtO)_2P(S)OCHMeCN$. VIII. Liq. $b_{0.01}$ 57-58°.[218]

$(MeO)_2P(S)OCH(CH_2OMe)CH_2SEt$. VIII. Liq. $b_{0.5}$ 128-132°, n_D^{20} 1.4945.[664]

$(EtO)_2P(S)OCH(CH_2OMe)CH_2SEt$. VIII. Liq. $b_{0.3}$ 121-127°, n_D^{20} 1.4861.[664]

$(MeO)_2P(S)OCH(CH_2OEt)CH_2SEt$. VIII. Liq. $b_{0.1}$ 108-112°, n_D^{20} 1.4901.[664]

$(EtO)_2P(S)OCH(CH_2OEt)CH_2SEt$. VIII. Liq. $b_{0.3}$ 130-132°, n_D^{20} 1.4858.[664]

$(EtO)_2P(S)OCH(CH_2SMe)CH_2SEt$. VIII. Liq. $b_{0.004}$ 102-104°.[1067]

$(EtO)_2P(S)OCH(CH_2SMe)CH_2S$-i-Pr. VIII. Liq. $b_{0.005}$ 108°.[1067]

$(MeO)_2P(S)OCH(CH_2SEt)_2$. VIII. Liq. $b_{0.01}$ 83-89°.[1067]

$(EtO)_2P(S)OCH(CH_2SEt)_2$. VIII. Liq. $b_{0.001}$ 107 °.[1067]

$(MeO)_2P(S)OCH(CH_2OP(S)(OMe)_2)_2$. VIII. (from $(MeO)_2P(S)Cl$ plus glycerol.) Liq. $b_{0.05}$ 60°.[1200]

$(EtO)_2P(S)OCHMeCOOEt$. XVII.[280] XXIX.[1000] Liq. $b_{0.01}$ 69°,[280] b_1 117°, n_D^{20} 1.4500, d_4^{20} 1.1230.[1000]

(i-PrO)$_2$P(S)OCHMeCOOEt. XXIX. Liq. $b_{1.5}$ 113°, n_D^{20} 1.4610, d_4^{20} 1.0863.[1000]

$(RO)_2P(S)OCH(COOEt)_2$. XXIX. Liq.[999]

R	b_2	n_D^{20}	d_4^{20}
Et	135-136°	1.4541	1.1704
i-Pr	145-147°	1.4530	1.1315
Pr	149-151°	1.4498	1.1329

$(MeO_2P(S)OCMe=CHSMe$. VIII. Liq. $b_{0.3}$ 103°.[222]

$(MeO)_2P(S)OCMe{=}CHS\text{-}i\text{-}Pr$. VIII. Liq. $b_{0.1}$ 92-93°.[222]

$(MeO)_2P(S)OCMe{=}C(SMe)_2$. VIII. Liq. $b_{0.1}$ 92-93°.[222]

$(EtO)_2P(S)OCMe{=}C(SMe)_2$. VIII. Liq. $b_{0.2}$ 119-120°.[222]

$(EtO)_2P(S)OCHMeCH_2OP(S)(OEt)_2$. XV. Liq. b_3 193-195°, n_D^{20} 1.4764, d_4^{20} 1.1576.[983]

$(RO)_2P(S)OPr$. VIII.[753] XV.[468] Liq.

R	Method	b./mm	n_D^{20}	d_4^{20}	d^{20}
Me	XIII[753]	104-105°/22	1.4571	--	1.1507
	XV[468]	213-214°/atm.	1.4530	1.1053	--
Et	XV[468]	95-97°/13	1.4507	1.0544	--
Bu	XV[468]	259-260°/atm.	1.4460	0.9837	--

$(EtOCH_2CH_2O)_2P(S)OPr$. VIII. Liq. b_8 193°.[1260]

$(CH_2\text{-}CHCH_2O)_2P(S)OPr$ (epoxy). XV. Liq. b_1 144-146°, n_D^{20} 1.4800, d_4^{20} 1.2192.[1036]

$(Et_3SiO)_2P(S)OPr$. XV. Liq. b_1 140-142°, n_D^{20} 1.4598. d_4^{20} 0.9770.[913]

$(PhO)_2P(S)OPr$. VIII. Liq. n_D^{16} 1.5578.[596]

$(EtO)_2P(S)OCH_2CH_2CH_2OH$. VIII. Liq. $b_{0.02}$ 72°.[1200]

$(EtO_2P(S)OCH_2CHOHMe$. VIII. Liq. b_4 119-121°, n_D^{20} 1.4659, d^{20} 1.361.[1408]

$(EtO)_2P(S)OCH_2CHOHCH_2Cl$. From $(EtO)_2P(S)OCH_2CH\text{-}CH_2$ (epoxy) plus HCl. Liq. $b_{0.002}$ 108-110°, n_D^{20} 1.4845, d_4^{20} 1.2487.[1038]

$(EtO)_2P(S)OCH_2CHOHCH_2OMe$. VIII. Liq. b_1 130-132°, n_D^{23} 1.4658, d^{23} 1.1557.[1408]

$(EtO)_2P(S)OCH_2CH_2CH_2OP(S)(OEt)_2$. VIII. Liq. $b_{0.02}$ 120°.[1200]

$(RO)_2P(S)OCH_2CH{=}CH_2$. VIII. Liq.[981]

R	b./mm	n_D^{20}	d_4^{20}
Me	88°/10	1.4683	1.1451
Et	82°/4	1.4605	1.0773
Bu	106-107°/1	1.4635	1.0208

$(CH_2\text{-}CHCH_2O)_2P(S)OCH_2CH{=}CH_2$ (epoxy). XV. Liq. b_1 144-145°, n_D^{20} 1.4989, d_4^{20} 1.2664.[1036]

$(PhO)_2P(S)OCH_2CH{=}CH_2$. XV. Liq. b_3 82-83°, n_D^{20} 1.5651, d^{20} 1.2095.[511]

$(MeO)_2P(S)OCH{:}CClCOOEt$. VIII. Liq. $b_{0.05}$ 117°.[180]

$(EtO)_2P(S)OCH{:}CClCOOEt$. VIII. Liq. $b_{0.04}$ 117-118°.[180]

$(EtO)_2P(S)OCH_2C{\equiv}CH$. VIII.[886, 964] XV.[984] Liq. $b_{0.03}$ 59-64°,[964] $b_{0.2}$ 70-71°,[984] $b_{0.5}$ 81-83°,[886] n_D^{20} 1.4688, d_4^{20} 1.1112.[984]

$(PrO)_2P(S)OCH_2C{\equiv}CH$. XV. Liq. b_1 85-87°, n_D^{20} 1.4680, d_4^{20} 1.0692.[984]

$(EtO)_2P(S)O$-t-Bu. ^{31}P -59.0, -64.6.[715, 717]

$(\overset{O}{CH_2-CH}CH_2O)_2P(S)O$-i-Bu. XV. Liq. $b_{0.5}$ 128-129°, n_D^{20} 1.4792, d^{20} 1.1911.[1039]

$(EtO)_2P(S)OCH_2C(OH)(CH_2Cl)CH_2NO_2$. XVII. Liq. $b_{0.04}$ 87-88°.[558]

$(i\text{-}PrO)_2P(S)OCF_2CH(CF_3)_2$. XXII. Liq. b_2 65°, n_D^{20} 1.3820, d_2^{22} 1.3215.[541]

$(EtO)_2P(S)OCH_2C(-N{=}C)Me_2$. VIII. (From $(EtO)_2P(S)Cl$ plus $HOCH_2CMe_2NHCHO$, Et_3N, and $COCl_2$.) Liq. $b_{0.004}$ 70-75°.[1342]

$(MeO)_2P(S)OCH{:}CMeCOOEt$. VIII. Liq. $b_{0.08}$ 115°.[180]

$(EtO)_2P(S)OCH{:}CMeCOOEt$. VIII. Liq. $b_{0.05}$ 120°.[180]

$(EtO)_2P(S)OCH_2CMe_2NMe_2$. VIII. Liq. $b_{0.25}$ 102-106°. Picrate. Solid. m. 112-113° (from EtOH).[146]

$(EtO)_2P(S)OCH(CN)Et$. VIII. Liq. $b_{0.01}$ 62-66°.[218]

$(EtO)_2P(S)OCHMeCOMe$. XXIX. Liq. $b_{0.5}$ 116°, n_D^{20} 1.4610, d_4^{20} 1.1472.[1000]

$(EtO)_2P(S)OCMe{:}CHCOOMe$. XV. Liq. b_1 106-109°, n_D^{20} 1.4830, d^{20} 1.1614.[497]

$(MeO)_2P(S)OCHMeCCl{:}CCl_2$. VIII. Liq. $b_{0.05}$ 50-52°.[460, 1110]

$(EtO)_2P(S)OCHMeCCl{:}CCl_2$. VIII. Liq. $b_{0.1}$ 85-87°.[460, 1110]

$(EtO)_2P(S)OCH(CH_2OMe)CCl{:}CH_2$. VIII. Liq. $b_{0.001}$ 96°.[965]

$(MeO)_2P(S)OCH(CH_2SEt)CCl{:}CH_2$. VIII. Liq. $b_{0.5}$ 95-100°.[965]

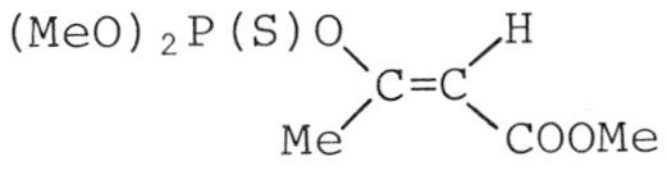

VIII. Solid. m. 49-50°, separated from cis isomer by crystallization. ^1H-NMR.[1275]

trans-"Thiono Phosdrin."

$(EtO)_2P(S)OCH(CH_2SEt)CH{:}CH_2$. VIII. Liq. $b_{0.5}$ 130-135°.[966]

$(MeO)_2P(S)OCMe{:}CPhCN$. VIII. Liq. b_1 143-146°, n_D^{26} 1.5447.[883]

$(EtO)_2P(S)OCMe{:}CPhCN$. VIII. Liq. $b_{0.6}$ 150-155°, $n_D^{26.5}$ 1.5334.[883]

$(MeO)_2P(S)OBu$. VIII.[704, 753] XV.[510] Liq. b_{10} 89-91°,[704] b_{13} 100-102°,[510] b_{20} 114°,[753] n_D^{20} 1.4560,[753] d^{20} 1.0941.[753]

$(EtO)_2P(S)OBu$. XV. Liq. b_{14} 116-117°, n_D^{20} 1.4502, d^{20} 1.0354.[510]

$(i\text{-}AmO)_2P(S)OBu$. XV. Liq. b_9 126-127°, n_D^{20} 1.4544, d^{20} 0.9695.[510]

$(EtOCH_2CH_2O)_2P(S)OBu$. VIII. Liq. b_2 174°.[1260]

$(CH_2{:}CHO)_2P(S)OBu$. XV. Liq. b_9 96°, n_D^{20} 1.4654, d^{20} 1.0617.[868]

$(PrO)_2P(S)OBu$. XV. Liq. $b_{0.5}$ 79-80°, n_D^{20} 1.4489.[267]

$(\overset{\frown O \frown}{CH_2-CH}CH_2O)_2P(S)OBu$. XV. Liq. $b_{0.1}$ 128-131°, n_D^{20} 1.4791, d_4^{20} 1.1910.[1036]

$(EtO)_2P(S)O(CH_2)_4OP(S)(OEt)_2$. XV. Liq. b_2 170-171°, n_D^{20} 1.4767, d_4^{20} 1.1742.[983]

$(MeO)_2P(S)OCH_2CH{:}CHMe$. VIII. Liq. b_3 82-83°, n_D^{20} 1.4706, d_4^{20} 1.1325.[981]

$(EtO)_2P(S)OCH_2CH{:}CHMe$. VIII. Liq. b_2 99-100°, n_D^{20} 1.4672, d_4^{20} 1.0643.[981]

$(EtO)_2P(S)OCH_2CH{:}CMeCl$. VIII. Liq. b_3 101-102°, n_D^{20} 1.4830, d^{20} 1.1689.[726]

$(EtO)_2P(S)OCH_2CHClCH{:}CH_2$. XV. Liq. b_4 111-112°, n_D^{20} 1.4760, d_4^{20} 1.1571[990]

$(MeO)_2P(S)OCH{:}CHCOMe$. VIII. Liq. $b_{0.01}$ 105°.[180]

$(EtO)_2P(S)OCH{:}CHCOMe$. VIII. Liq. $b_{0.01}$ 105-110°.[180]

$(EtO)_2P(S)OCHMeC{\equiv}CH$. VIII. Liq. $b_{0.05}$ 101-110°.[964]

$(MeO)_2P(S)O\text{-}i\text{-}Am{\cdot}NEt_3$. By mixing the components. Liq. n_D^{20} 1.4892.[762]

$(\overset{O}{CH_2-CHCH_2O})_2P(S)O$-i-Am. XV. Liq. $b_{0.005}$ 122-124°, n_D^{20} 1.4792, d^{20} 1.1702.[1039]

$(EtO)_2P(S)OAm$. XV. Liq. b_3 109-113°, n_D^{20} 1.4508.[49]

$(EtO)_2P(S)O(CH_2)_4CCl_3$. "Known methods." Liq. b_4 153-154°, n_D^{20} 1.4873, d^{20} 1.2647.[855]

$(RO)_2P(S)OCH_2CHMeCOMe$. XV. Liq.[1037]

R	b./mm	n_D^{20}	d^{20}
Et	92-83°/0.01	1.4617	1.1034
i-Pr	85-87°/0.01	1.4537	1.0528
Pr	94-96°/0.08	1.4600	1.0666
i-Bu	112-113°/0.05	1.4532	1.0256
Bu	116-118°/0.08	1.4595	1.0394

$(MeO)_2P(S)OCMe_2CCl{:}CCl_2$. VIII. Liq. $b_{0.05}$ 55-59°.[460, 1110]

$(EtO)_2P(S)OCMe_2CCl{:}CCl_2$. VIII. Liq. $b_{0.05}$ 75-80°.[460, 1110]

$(EtO)_2P(S)OCMe{:}CHCOMe$. XV. Liq. b_1 92.5-93° n_D^{20} 1.4807, d^{20} 1.1216.[497]

$(MeO)_2P(S)OCHMeCMe_3$. VIII. Liq. $b_{0.01}$ 40°.[1142]

$(EtO)_2P(S)OCHMeCMe_3$. VIII. Liq. $b_{0.01}$ 61°.[1142]

$(EtO)_2P(S)OCMe{:}CHCHMe_2$. VIII. Liq. $b_{0.05}$ 60°.[222]

$(MeO)_2P(S)OCMeEtC{\equiv}CH$. VIII. Liq. $b_{0.5}$ 94-95°, n_D^{25} 1.5008.[886]

$(EtO)_2P(S)OCMeEtC{\equiv}CH$. VIII. Liq. $b_{0.8}$ 100-101°.[886]

$(i\text{-}PrO)_2P(S)O$ 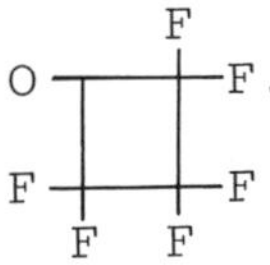 XXII. Liq. b_3 67°, n_D^{20} 1.3752, d_4^{20} 1.3206.[541]

$(EtO)_2P(S)O$  VIII.[254] XV.[59] Liq. $b_{0.1}$ 86°,[59] $b_{0.6}$ 125-135°,[254] n_D^{20} 1.5080, d^{20} 1.2022.[59]

$(RO)_2P(S)O$ 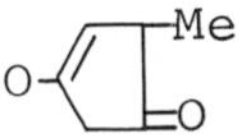 VIII. Liq. R, n_D^{21}: Me, 1.5152; Et, $b_{0.005}$ 99°, 1.5040.[92]

RO(R'O)P(S)O 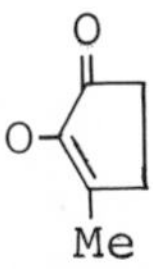 VIII. Liq.[757]

R	R'	b./mm	n_D^{20}	d^{20}
Me	Me	116-117°/0.2	1.5155	1.2562
Me	Et	125-127°/0.1	1.5060	1.2254
Et	Et	126-128°/0.2	1.5025	1.1715

$(EtO)_2P(S)O$  SEt. VIII. Liq. $b_{0.2}$ 120-121°.[222]

$(MeO)_2P(S)O$ 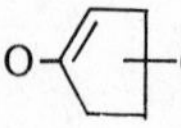 COOEt. VIII. Liq. $b_{1.2}$ 140-143°, $n_D^{23.5}$ 1.5348.[881]

$(EtO)_2P(S)ON$  VIII. Solid. m. 55°.[652]

$(RO)_2P(S)ON$ 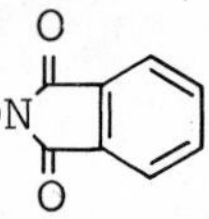 VIII. Solids. R, m.: Me, 122° (from MeOH); Et, 108° (from MeOH).[652]

$(EtO)_2P(S)ON$ 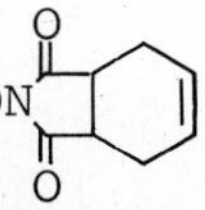 VIII. Solid. m. 48°.[652]

$(EtO)_2P(S)ON$ 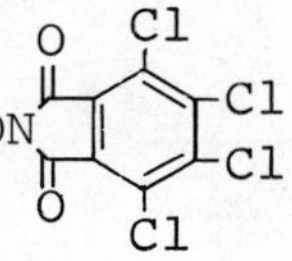 VIII. Solid. m. 171°.[652]

$(EtO)_2P(S)ON$ 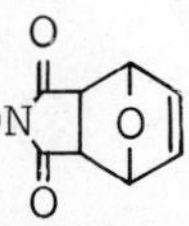 VIII. Solid. m. 80.5-82°.[1032]

$(EtO)_2P(S)ON$ 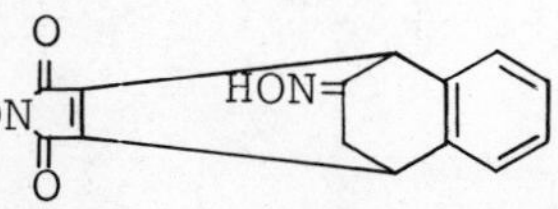 VIII. Solid. m. 91°.[652]

$(EtO)_2P(S)O$ 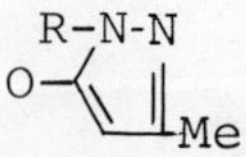 VIII. Liq.[741]

R	b./mm	n_D^{20}	d^{20}
H	102-104°/0.08	1.4980	1.0878
Ph	125-126°/0.1	1.5572	1.1717

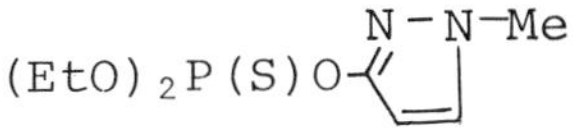

VIII. Liq. $b_{0.08}$ 102-104°.[758]

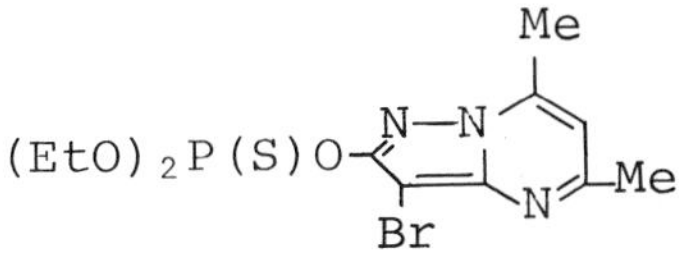

VIII. Solid. m. 95°.[1101]

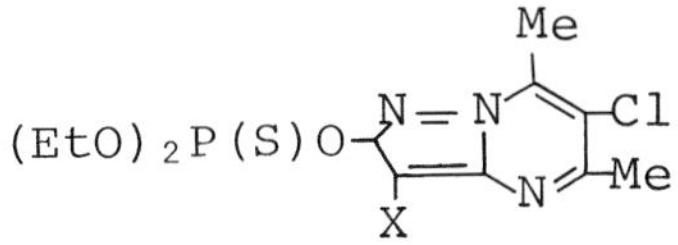

VIII. Solids.[308] X, m.: H, 83-84° (from ligroin); Cl, 87-88°. Also nine analogs.

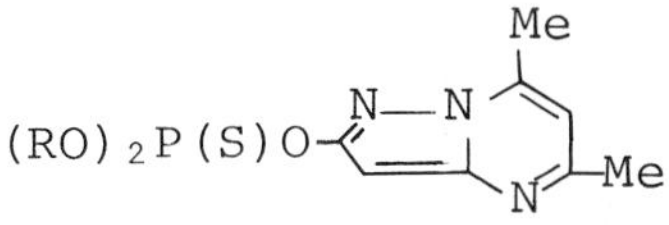

VIII. Solids. R, m.: Me, 72-74°;[1088] Et, 42-43° (from benzene-ligroin),[1101] 45° (from benzene-ether).[1088]

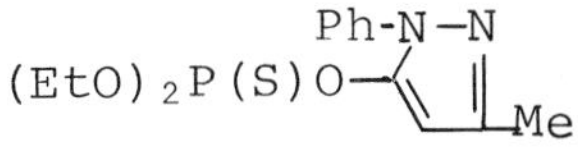

VIII. Liq. $b_{0.1}$ 125-126°.[758]

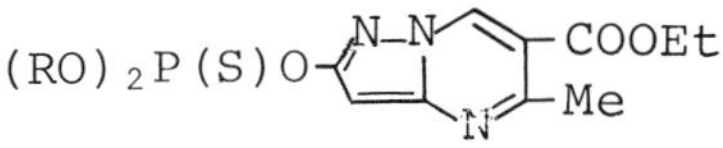

VIII. Solids.[1088] R, m.: Me, 25°; Et, 38-40°. "Afugan," m. 50-51°.

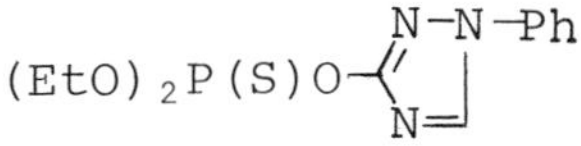

"Hostathion."

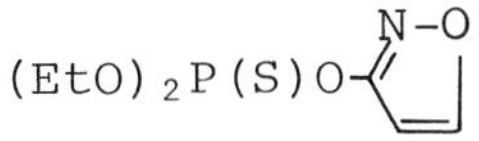

VIII. Liq. $b_{0.2}$ 90°.[815, 1071]

VIII. Liq.[815, 1071] R, b./mm: Me, 120° /0.2; OMe, 125°/0.1; COOMe, 125°/0.1; $CH(OEt)_2$, 120°/0.1.

VIII. Liq.[815, 1071] R, b./mm: Me, 120-123°/0.2; Pr, 120°/0.2; $C_{10}H_{21}$, 160°

/0.2; COOEt, 150°/0.2; Ph, 160°/0.15; C_6H_4CN-4, solid, m. 60-63°.

$(EtO)_2P(S)O$- 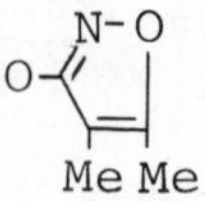 VIII. Liq. $b_{0.2}$ 110°.[815]

RO(R'O)P(S)O 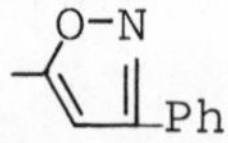 VIII.[622] R, R', properties: Me, Me, solid, m. 188°; Et, Et, liq., $b_{0.01}$ 90-92°, n_D^{24} 1.5298; me, i-Pr, liq., $b_{0.01}$ 132°.

$(EtO)_2P(S)O$ 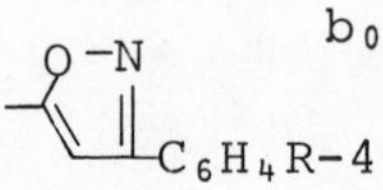 VIII.[622] R, properties: OMe, liq., n_D^{21} 1.5314; NO_2, solid, m. 65°.

$(EtO)_2P(S)O$- 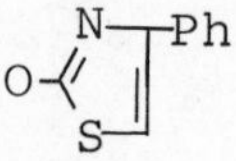 VIII. Solid. m. 33-34° (from PrOH).[301]

$(RO)_2P(S)O$- 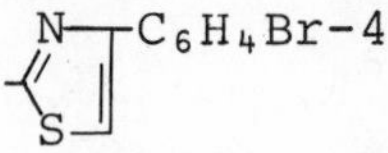 VIII. Solids. R, m.: Me, 46-47.5°; Et, 54-55°.[301]

$(EtO)_2P(S)OPh$. VIII.[700, 701, 807, 971] IX.[153] Liq. $b_{0.8}$ 120-122°,[701] b_{16} 153°,[153] n_D^{20} 1.5110,[700] 1.5155,[701] d_4^{20} 1.1763.[701]

$(CH_2{:}CHO)_2P(S)OPh$. XV. Liq. b_{10} 134-135°, n_D^{20} 1.5268, d^{20} 1.1719.[868]

$(i\text{-}PrO)_2P(S)OPh$. XV. Liq. b_{12} 151-151.5°, n_D^{20} 1.5005, d^{20} 1.0932.[511]

$(PrO)_2P(S)OPh$. XV. Liq. b_6 147-148°, n_D^{20} 1.5052, d^{20} 1.1040.[511]

$(\overline{CH_2\text{-}CHCH_2O})_2P(S)OPh$ (epoxide O bridging CH_2–CH). XV. Liq. $b_{0.005}$ 195-198°, n_D^{20} 1.5326, d_4^{20} 1.2805.[1036]

$(4\text{-}NO_2C_6H_4O)_2P(S)OPh$. VIII. (From $PhOP(S)Cl_2$ plus 2 $NaOC_6H_4NO_2$-4.) Solid. m. 85° (from EtOH). μ 4.78 D.[529] IR.[528]

$(4\text{-}MeC_6H_4O)_2P(S)OPh$. IX. Solid. m. 54° (from EtOH).[73]

$(2\text{-}PhC_6H_4O)_2P(S)OPh$. VIII. Liq. b_{10} 320-330°.[837]

$(EtO)_2P(S)OC_6H_4Br$-2. VIII. Liq. $b_{0.01}$ 100-104°, n_D^{20} 1.5401,

d^{20} 1.4036.[747]

$(EtO)_2P(S)OC_6H_4Br$-4. VIII. Liq. $b_{0.01}$ 111-115°, n_D^{20} 1.5356, d^{20} 1.3902.[747]

$(MeO)_2P(S)OC_6H_2BrCl_2$-4,2,5. VIII.[116, 1216] Solid. $b_{0.01}$ 140-142°, m. 51° (from MeOH). "Bromophos," m. 53-54°; v. p. 1.3 x 10^{-4} at 20° (Pesticide Manual).

$(MeO)_2P(S)OC_6T_2BrCl_2$-4,2,5. VIII. Solid. m. 59-60°. "Bromophos-T_2."[1268]

$(EtO)_2P(S)OC_6T_2BrCl_2$-4,2,5. VIII. Liq. $b_{0.001}$ 122-123°. "Bromophos-ethyl-T_2."[1268]

$(EtO)_2P(S)OC_6H_3BrOEt$-2,4. VIII. Liq. $b_{0.02}$ 142-150°, n_D^{20} 1.5410, d^{20} 1.3596.[355]

$(PrO)_2P(S)OC_6H_3BrOEt$-2,4. VIII. Liq. $b_{0.05}$ 136-140°, n_D^{20} 1.5290, d^{20} 1.3113.[355]

$(MeO)_2P(S)OC_6H_2Br_2Cl$-2,4,5. VIII. Solid. m. 46-47°.[85]

$(EtO)_2P(S)OC_6H_2Br_2Cl$-2,4,5. VIII. Liq. n_D^{24} 1.5672.[85]

$(BuO)_2P(S)OC_6H_2Br_2NO_2$-2,6,4. VIII. Liq. b_3 82°.[1220]

$(EtO)_2P(S)OC_6H_4Cl$-2. VIII.[121, 747, 1220] Liq. $b_{0.2}$ 93-99°,[747] $b_{0.48}$ 112-113.5°,[121] n_D^{20} 1.5208, d^{20} 1.2223.[747]

$(BuO)_2P(S)OC_6H_4Cl$-2. VIII. Liq. b_8 110°.[1220]

$(EtO)_2P(S)OC_6H_4Cl$-3. VIII.[121, 747] Liq. $b_{0.1}$ 95°,[747] $b_{0.25}$ 106.5-108°,[121] n_D^{20} 1.5218, d^{20} 1.2242.[747]

$(EtO)_2P(S)OC_6H_4Cl$-4. VIII.[121, 701, 747, 895, 1220] Liq. $b_{0.02}$ 98.5-99°,[895] $b_{0.1}$ 110-111.5°, $b_{0.25}$ 123-127°,[701] n_D^{20} 1.5219,[121] d_4^{20} 1.2277.[747]

$(BuO)_2P(S)OC_6H_4Cl$-4. VIII. Liq. b_8 138°.[1220]

$(EtO)_2P(S)OC_6H_3ClOEt$-2,4. VIII. Liq. $b_{0.15}$ 110-126°, n_D^{20} 1.5285, d^{20} 1.2391.[355]

$(PrO)_2P(S)OC_6H_3ClOEt$-2,4. VIII. Liq. $b_{0.05}$ 121-126°, n_D^{20} 1.5285, d^{20} 1.1883.[355]

$(MeO)_2P(S)OC_6H_3ClNO_2$-2,4. VIII. Solid. m. 51-52°.[332] ^{1}H-NMR.[74, 524, 526] "Dicapthon."

$(EtO)_2P(S)OC_6H_3ClNO_2$-2,4. VIII. Liq. $b_{0.06}$ 115-120°, n_D^{20} 1.5482, d^{20} 1.3612.[747]

$(EtO)_2P(S)OC_6H_3ClNO_2$-4,2. VIII.[331, 747] $b_{0.03}$ 105-111°,

n_D^{20} 1.5308, d^{20} 1.3115.[747]

$(MeO)_2P(S)OC_6H_3ClNO_2$-4,3. VIII. Liq. $b_{0.01}$ 122°.[1199]

$(RO)_2P(S)O$ VIII. Liq. R, n_D^{25}: Me, 1.5408,[1027] Et, 1.5170.

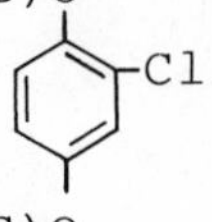

$(RO)_2P(S)O$

$(EtO)_2P(S)OC_6H_3$-2-Cl-4-SMe. VIII. Liq. $b_{0.0001}$ 92°, n_D^{25} 1.5597, d_4^{25} 1.2798.[867]

$(MeO)_2P(S)OC_6H_3$-3-Cl-4-SMe. VIII. Liq. $b_{0.05}$ 116-118°.[885]

$(EtO)_2P(S)OC_6H_3$-3-Cl-4-SMe. VIII. Liq. $b_{0.0001}$ 98°, n_D^{25} 1.5604, d_4^{25} 1.2716.[867]

$(RO)_2P(S)OC_6H_3$-3-Cl-4-SCH_2CN. VIII. Liq. R, n_D^{24}: Me, 1.5729; Et, 1.5567.[1098]

$(MeO)_2P(S)OC_6H_3$-3-Cl-4-SO_2NHMe. VIII. Solid. m. 120-121°.[574, 1059, 1283]

$(EtO)_2P(S)OC_6H_3$-3-Cl-4-SO_2NHMe. VIII. Solid. m. 52-53° (from EtOH).[1059]

$(MeO)_2P(S)OC_6H_3$-3-Cl-4-SO_2NMe_2. VIII. Solid. m. 60-62° (from MeOH).[575]

$(MeO)_2P(S)OC_6H_3Cl_2$-2,4. VIII. Liq. $b_{0.2}$ 120-125°, n_D^{29} 1.5478.[121]

$(EtO)_2P(S)OC_6H_3Cl_2$-2,4. VIII.[121, 700, 1256] XV.[895] Liq. $b_{0.02}$ 100-101°,[895] $b_{0.12-0.25}$ 126-131°,[121] $b_{0.3}$ 150-154°,[700] n_D^{20} 1.5370, d_4^{20} 1.3206.[895] ^{31}P 49.7 (P_4O_6 ext.).[1047] "VC-13."

(i-PrO)$_2P(S)OC_6H_3Cl_2$-2,4. VIII. Liq. $b_{0.2}$ 116-118°, n_D^{29} 1.5179.[121]

$(BuO)_2P(S)OC_6H_3Cl_2$-2,4. VIII. Liq. $b_{0.1}$ 140-145°, n_D^{29} 1.5161.[121]

$(EtO)_2P(S)OC_6H_3Cl_2$-2,5. VIII. Liq. $b_{0.13}$ 113.5-115°, n_D^{27} 1.5300.[121]

$(MeO)_2P(S)OC_6H_3Cl_2$-3,4. VIII. Liq. n_D^{25} 1.5547, d^{25} 1.418.[126]

$(MeO)_2P(S)OC_6H_2Cl_2I$-2,5,4. VIII. Solid. m. 72-73°.[178]

$(EtO)_2P(S)OC_6H_2Cl_2I$-2,5,4. VIII. Solid. m. 47-48°.[178]

$(MeO)_2P(S)OC_6H_2Cl_2NO_2$-2,4,5. VIII. Solid. m. 80-82°.[1199]

$(EtO)_2P(S)OC_6H_2Cl_2NO_2$-2,6,4. VIII. Liq. $b_{0.075}$ 135-140°, n_D^{20} 1.5528, d^{20} 1.4112.[747]

$(MeO)_2P(S)OC_6H_2Cl_2NO_2$-3,5,4. VIII. Solid. m. 52° (from ligroin).[219]

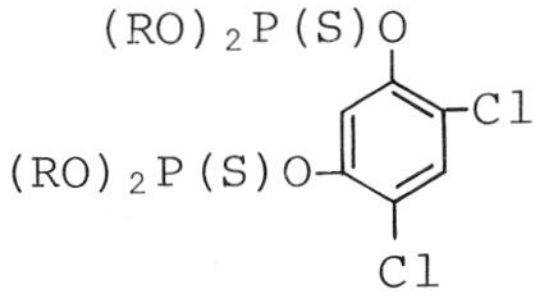
VIII. R, properties: Me, solid, m. 70.5-73°; Et, liq., n_D^{25} 1.5230.[1027]

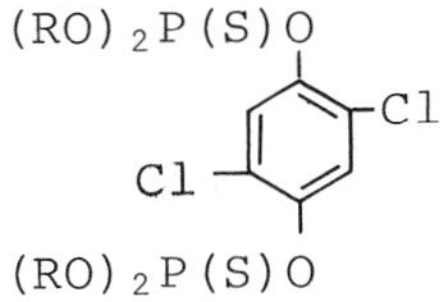
VIII. Solids. R, m.: Me, 105-107.5°; Et, 40.5-44 .[1027]

$(MeO)_2P(S)OC_6H_2Cl_2SMe$-2,5,4. VIII. Liq. $b_{0.001}$ 139-140°.[1212]

$(EtO)_2P(S)OC_6H_2Cl_2SMe$-2,5,4. VIII. Liq. $b_{0.001}$ 150-151°.[1212]

$(EtO)_2P(S)OC_6H_2Cl_2SMe$-2,6,4. VIII. Solid. m. 45-47°.[867]

$(MeO)_2P(S)OC_6H_2Cl_3$-2,4,5. VIII.[103, 126, 688, 838, 1255] XV.[943] Solid. $b_{1-1.5}$ 148-157°,[688] n_D^{35} 1.5597,[838] m. 39-42°,[103, 126] IR and UV.[375] ^1H-NMR.[74, 524, 526] ^{31}P 46.6 (P_4O_6 ext.).[1047] Mass spect.[196] "Ronnel."

$(EtO)_2P(S)OC_6H_2Cl_3$-2,4,5. VIII.[121, 838] XV.[895] Liq. $b_{0.025}$ 108-109°,[895] $b_{0.11}$ 116-117°,[121] n_D^{20} 1.5468, d_4^{20} 1.3892.[895]

MeO)PrO)P(S)$OC_6H_2Cl_3$-2,4,5. VIII. Liq. n_D^{25} 1.5525, d^{25} 1.4081.[103]

$(RO)_2P(S)OC_6H_2Cl_3$-2,4,6. VIII. (From 2,4,6-$Cl_3C_6H_2OP(S)Cl_2$ plus ROH and pyridine.) Solids. R, m.: Me, 59-60°; Et, 49-51 .[229a]

$(EtO)_2P(S)OC_6H_3Cl_3$-2,4,6. VIII. Solid. $b_{0.01}$ 122-124°, m. 55.5-56.5°.[121]

$(EtO)_2P(S)OC_6HCl_4$-2,3,4,6. VIII. Liq. $b_{0.6}$ 153-155°, n_D^{27} 1.5582.[121]

$(MeO)_2P(S)OC_6Cl_5$. VIII. Solid. m. 82-83°.[441] ^{31}P -66,[715, 1274]

$(EtO)_2P(S)OC_6Cl_5$. VIII.[441, 1410] XV.[1410] Solid. m. 96-98° (from EtOH).[1410] ^{31}P -62.4.[715, 1274]

$(PrO)_2P(S)OC_6Cl_5$. VIII. Solid. m. 39-40°.[441]

$(EtO)_2P(S)OC_6H_4F$-2. VIII. Liq. $b_{0.5}$ 120°.[489]

$(EtO)_2P(S)OC_6H_4F$-4. VIII. Liq. $b_{0.4}$ 110°.[489]

$(EtO)_2P(S)OC_6H_2FI_2$-4,2,6. VIII. Liq. b_3 125°.[489] Also several analogs.[489]

$(EtO)_2P(S)OC_6H_4I$-2. VIII. Liq. $b_{0.05}$ 125-130°, n_D^{20} 1.5691, d^{20} 1.5733.[747]

$(EtO)_2P(S)OC_6H_4I$-4. VIII. Liq. $b_{0.03}$ 116-118°, n_D^{20} 1.5680, d^{20} 1.5817.[747]

$(EtO)_2P(S)O$ 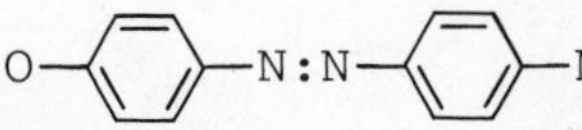

NO_2. VIII. Solid. m. 70-71°.[653]

$(EtO)_2P(S)OC_6H_4NR_2$. From $(EtO)_2P(S)C_6H_4NH_2$ plus RSO_2Cl. Solids. R, m.: $MeSO_2$, 136°; $ClCH_2SO_2$, 102°; $PhSO_2$, 85°; 4-$ClC_6H_4SO_2$, 148°.[1082]

$(EtO)_2P(S)OC_6H_4NHCSNH_2$-n. VIII. Solids (from benzene-ligroin). n, m.: 2, 119°; 3, 69°; 4, 109°.[1138]

$(MeO)_2P(S)OC_6H_4NO_2$-3. VIII. Liq. $b_{0.01}$ 102°.[1179]

$(MeO)_2P(S)OC_6H_4NO_2$-4. VIII.[35, 334, 399, 700, 766, 1125] Solid. b_2 158°.[700, 1125] m. 37-38°,[334] ^1H-NMR,[74, 524, 526] ^{31}P -65.5,[715] 47.2 (P_4O_6 ext.).[1047] Mass spect.[196] "Methyl Parathion."

$(MeO)_2{}^{32}P(S)OC_6H_4NO_2$-4. VIII. Solid. m. 35-36°.[699]

$MeO(EtO)P(S)OC_6H_4NO_2$-4. VIII. Liq. b_1 120-121°, n_D^{20} 1.5470, d^{20} 1.3182

$(EtO)_2P(S)OC_6H_4NO_2$-4. VIII.[35, 239, 333, 700, 766, 971] XV.[943] Liq. $b_{0.03}$ 115-117°,[943] $b_{0.6}$ 157-162°,[333] b_2 196°.[971] n_D^{20} 1.5381,[1078] 1.5395,[700] Mol. ref. 448.00.[1078] n_D^{25} 1.5370,[35, 333] d_4^{20} 1.2704.[701] m. 3.4°,[239] IR.[80] IR and UV.[375] ^1H-NMR.[74, 525, 526] ^{31}P -42,[715] -42 ± 1,[1345] 46.0 (P_4O_6 ext.).[1074] μ 4.98 D.[529] Mass spect.[196] v.p. 3.78 x 10^{-5} at 20°.[125] "Parathion."

$(EtO)_2{}^{32}P({}^{35}S)OC_6H_4NO_2$-4. VIII. Liq. $b_{0.04}$ 115°, n_D^{20} 1.5374, d^{20} 1.2704.[708]

$(RO)_2P(S)OC_6H_4NO_2$-4. VIII.[35, 334] Liq. and solids.

R	m.	b./mm	n_D^{25}
i-Pr	56-57°[334]	--	--
Pr	--	164°/0.5	1.5259[35]
i-Bu	--	165-175°/0.4	1.5155[334]
Bu	--	--	1.5195[334]
2-Et-hexyl	--	--	1.5052[35]
$C_{10}H_{21}$	--	--	1.4940[35]
Ph	64-65°[35, 334]	--	--

$(EtOCH_2CH_2O)_2P(S)OC_6H_4NO_2$-4. VIII. Liq. b_5 183°.[1260]

$(PhO)_2P(S)OC_6H_4NO_2$-4. VIII. Solid. m. 47° (from EtOH). μ 4.52 D.[529] IR.[528]

$(EtO)_2P(S)O$— 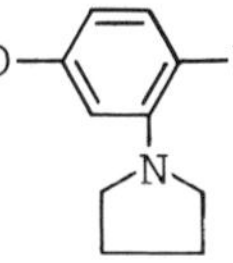 —NO_2. VIII. Solid. m. 52-54°.[710]

$(MeO)_2P(S)OC_6H_3(NO_2)_2$. VIII. Solid. m. 94-95°.[926]

$(MeO)_2P(S)OC_6H_3$-2-OMe-4-NO_2. VIII. Solid. m. 54°.[300]

$(EtO)_2P(S)OC_6H_3$-2-OMe-4-NO_2. VIII. Liq. $b_{0.01}$ 132°.[645]

$(MeO)_2P(S)OC_6H_3$-3-OMe-4-NO_2. VIII. Solid. m. 72°.[300]

$(EtO)_2P(S)OC_6H_3$-3-OMe-4-NO_2. VIII. Solid. m. 42-43°.[300, 645]

$(MeO)_2P(S)OC_6H_3$-2-OEt-4-NO_2. VIII. Solid. m. 67-69° (from hexane).[555]

$(EtO)_2P(S)OC_6H_3$-2-NO_2-4-SMe. VIII. Liq. n_D^{25} 1.5627, d_4^{25} 1.2888.[867]

$(MeO)_2P(S)OC_6H_3$-2-NO_2-5-SMe. VIII. Solid. m. 62°.[423]

$(MeO)_2P(S)OC_6H_3$-3-NO_2-4-SMe. VIII. Solid. m. 80-82°.[423]

$(MeO)_2P(S)OC_6H_3$-3-NO_2-4-SEt. VIII. Solid. m. 57-59°.[423]

$(EtO)_2P(S)OC_6H_3$-4-NO_2-3-SMe. VIII. Solid. m. 61-62° (from cyclohexane).[423]

$(EtO)_2P(S)OC_6H_3$-4-NO_2-3-SEt. VIII. Solid. m. 42-44°.[423]

$(EtO)_2P(S)OC_6H_3$-4-NO_2-3-SAr. VIII. Solids. Ar, m.: 2-MeC_6H_4, 65-66°; 4-MeC_6H_4, 69-71°; 4-ClC_6H_4, 63-65°.[423]

$(EtO)_2P(S)OC_6H_3(NO_2)_2$-2,4. VIII. Solid. m. 44-45° (from benzene-petroleum ether).[1359]

$(EtO)_2P(S)OC_6H_4OMe-2$. VIII. Liq. b_8 170-174°, n_D^{20} 1.5990, d^{20} 1.1672.[355]

$(EtO)_2P(S)OC_6H_4OMe-3$. VIII. Liq. $b_{0.02}$ 97°, n_D^{20} 1.5050, d^{20} 1.1483.[355]

$(EtO)_2P(S)OC_6H_4OMe-4$. VIII.[355, 1081] XXVII.[750] Liq. $b_{0.01}$ 130°,[1081] $b_{0.1}$ 117-122°, n_D^{20} 1.5180, d^{20} 1.1910.[355] Also several homologs and analogs.[355]

$(MeO)_2P(S)OC_6H_4OEt-3$. VIII. Liq. $b_{0.02}$ 100-107°, n_D^{20} 1.5280, d^{20} 1.1970.[355]

$((EtO)_2P(S)O)_2C_6H_4$. Liq.

Isomer	Method	b./mm	n_D^{20}	d_4^{20}
1,2	VIII[1309]	110°/0.1	1.5110	1.2610
1,2	XV[982]	151-153°/0.027	1.5161	1.2183
1,3	XV[982]	166-167°/0.025	1.5149	1.2144
1,3	VIII[707]	130°/0.15	1.5113	1.1991
1,4	VIII[707]	134°/0.2	1.5158	1.2496
1,4	XV[982]	186-188°/0.023	1.5161	1.2207

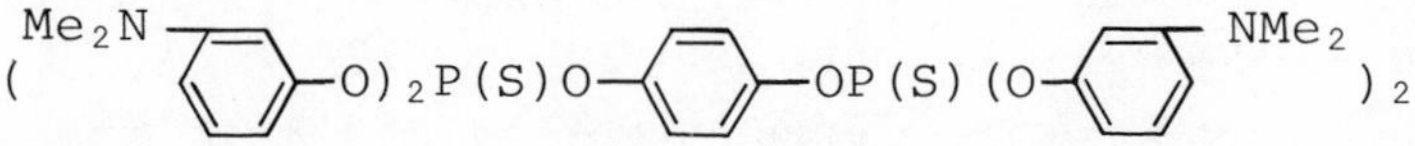

$(\text{Me}_2\text{N}-\langle\text{ring}\rangle-O)_2P(S)O-\langle\text{ring}\rangle-OP(S)(O-\langle\text{ring}\rangle-\text{NMe}_2)_2$. VIII. Solid.

m. 130-138°. Tetramethiodide·$4H_2O$, solid, m. 133-136° (dec.).[329]

$(MeO)_2P(S)OC_6H_4SMe-4$. VIII.[885, 1079] Liq. $b_{0.01}$ 108°,[1079] $b_{0.05}$ 101-103°,[885] n_D^{20} 1.5710, d_4^{20} 1.266.[1079]

$(EtO)_2P(S)OC_6H_4SMe-4$. VIII. Liq. n_D^{25} 1.5462, d_4^{25} 1.1947.[867]

$(RO)_2P(S)OC_6H_4SCH_2Cl-4$. From $(RO)_2P(S)OC_6H_4SMe$ plus SO_2Cl_2. [1095] Liq. R, n_D^t : Me, 1.5780[26°]; Et, 1.5540[27°]. Also 12 analogs.[1095]

$(MeO)_2P(S)OC_6H_4SCF_3-4$. VIII. Liq. $b_{0.01}$ 84°.[1023]

$(EtO)_2P(S)OC_6H_4SCF_3-4$. VIII. Liq. $b_{0.01}$ 94°.[1023]

$(RO)_2P(S)OC_6H_4-SCH_2CH_2OMe$. VIII. Liq. R, b./mm: Me, 168-172°/0.7; Et, 177°/1.5. Also several analogs.[473]

$(MeO)_2P(S)OC_6H_4SCH_2CN-4$. VIII. Liq. n_D^{24} 1.5596.[1098]

$(EtO)_2P(S)OC_6H_4SCH_2CN-4$. VIII. Liq. n_D^{23} 1.547.[1098]

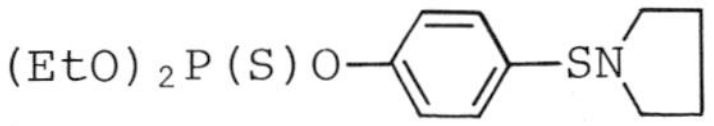. From 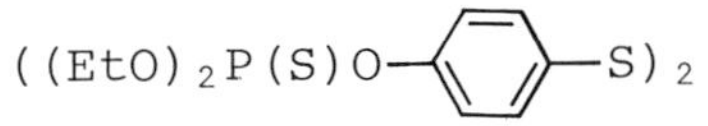 plus SO_2Cl_2 and pyrrolidine. Liq. n_D^{24} 1.5618.[1109] Also analogs from morpholine, 1.5621; and diallylamine, 1.5599.[1109]

$(EtO)_2P(S)OC_6H_4SPh$-*n*. VIII. Liquids. *n*, b./mm: 2, 124°/0.001; 3, 145°/0.01; 4, 145°/0.01.[1252]

4-$(MeO)_2P(S)OC_6H_4SC_6H_4NO_2$-4. VIII. Solid. m. 58-59° (from MeOH).[1341]

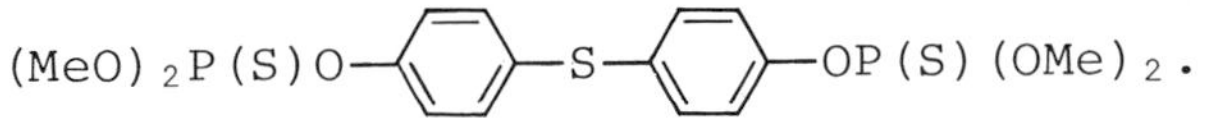. VIII. Liq. n_D^{25} 1.5883.[660] ^{1}H-NMR.[74] ^{31}P 46.6 (P_4O_6 ext.).[1047] "Abate."

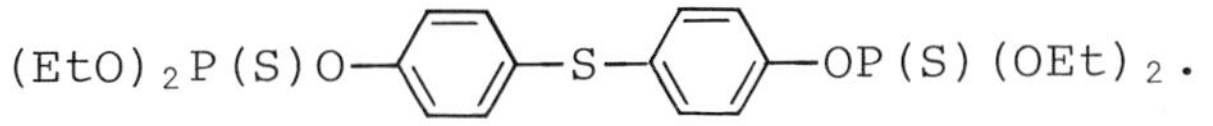. VIII.[660, 1251, 1252] Liq. b_1 100°,[1251, 1252] n_D^{25} 1.5610[660]

$(MeO)_2P(S)OC_6H_4S(O)Me$-4. From $(MeO)_2P(S)OC_6H_4SMe$-4 plus H_2O_2. Liq. $b_{0.01}$ 100° (bath temp.), n_D^{20} 1.5459, d_4^{20} 1.217.[1079]

$(EtO)_2P(S)OC_6H_4S(O)Me$-4. From $(EtO)_2P(S)OC_6H_4SMe$-4 plus H_2O_2. Liq. $b_{0.01}$ 138-141°.[1079, 1084] ^{31}P 50.0 (P_4O_6 ext.).[1047] "Dasanit."

$(EtO)_2P(S)OC_6H_4$-S(O)Ph-2. Liq. b_1 100°.[1252]

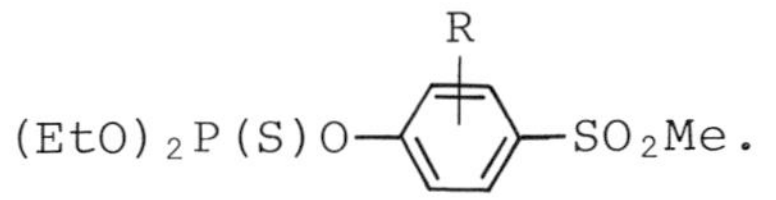. VIII. Solids and liq.[867]

R	m.	n_D^{25}	d_4^{25}
H	42-44°	--	--
2-Cl	66-67	--	--
2,6-Cl_2	95-97	--	--
2-Me	--	1.5313	1.2600
3-Me	--	1.5348	1.2628
2-NO_2	65-66	--	--

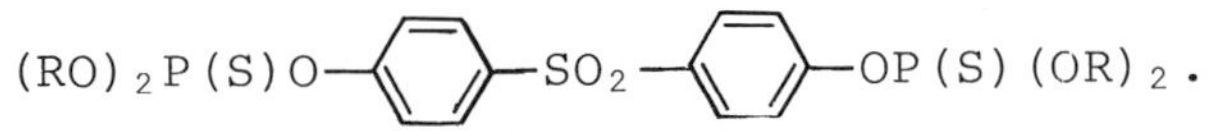. VIII. R, properties:

Me, solid, m. 66-68° (from EtOH); Et, liq., n_D^{25} 1.5524.[660]

$(MeO)_2P(S)OC_6H_4$-SO_2NH_2-4. VIII.[86, 431] Solids. m. 42.5-43° and 70-71°,[431] m. 70-71° (from toluene).[86]

$(EtO)_2P(S)OC_6H_4$-SO_2NH_2-4. VIII.[86, 431, 474] Liq. n_D^{25} 1.5346,[86, 431] $n_D^{32.3}$ 1.5330.[474]

$(MeO)_2P(S)OC_6H_4$-SO_2NHMe-4. VIII.[431, 1290] Liq. or solid. n_D^{25} 1.5352,[1290] m. 36-38°.[431]

$(MeO)_2P(S)OC_6H_4$-SO_2NHR-4. VIII. Liq. R, n_D^t: Et, $1.5383^{23°}$; i-Pr, $1.5313^{31°}$; Pr, $1.5329^{31°}$.[1290]

$(EtO)_2P(S)OC_6H_4$-SO_2NHR-4. VIII.[474, 1290] Liq. R, n_D^t: i-Pr, $1.5203^{30°}$,[1290] $1.5184^{32.2°}$;[474] $CH_2CH{=}CH_2$, $1.5292^{32°}$; Bu, $1.5166^{32°}$; Ph, $1.5620^{32.2°}$.[474]

$(MeO)_2P(S)OC_6H_4$-SO_2NMe_2-4. VIII.[86, 431, 1282, 1290] Solid. m. 52.5-53°.[86, 431] Also several homologs and analogs.[1290]

$(EtO)_2P(S)OC_6H_4$-SO_2NMe_2-4. VIII.[474, 1290] Solid. m. 68°.[1290]

$(MeO)_2P(S)OC_6H_4$-SO_2NEt_2-4. VIII.[474, 1282, 1290] Solid. m. 77-80°.

$(EtO)_2P(S)OC_6H_4$-SO_2NEt_2-4. VIII. Solid. m. 33-35°.[1290]

$(MeO)_2P(S)OC_6H_4$-SO_2NHCOR-4. VIII. Solids. R, m.: Me, 142-144°; i-Pr, 103-104.5°; Ph, 102-104°.[39]

$(EtO)_2P(S)OC_6H_4$-SO_2OR-4. VIII. Liq. R, n_D^t: Me, $1.5252^{27°}$; Et. $1.5177^{27.3°}$; CH_2CH_2OMe, $1.5123^{28°}$.[474]

$(EtO)_2P(S)OC_6H_4$-SO_2OAr-4. VIII. Liq. Ar, n_D^{32}: Ph, 1.5418; 4-ClC_6H_4, 1.5474; 4-$MeSC_6H_4$, 1.5649; 4-$NCSC_6H_4$, 1.5682; 4-$SCNC_6H_4$, 1.5878.[474]

$(RO)_2P(S)O$ 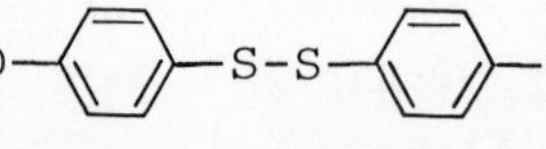 $OP(S)(OR)_2$. VIII.[677, 1102] Liq. R, n_D^t: Me, $1.5929^{24°}$,[1102] $1.5814^{25°}$;[677] Et, $1.5747^{24°}$.[1102]

$(MeO)_2P(S)O$ 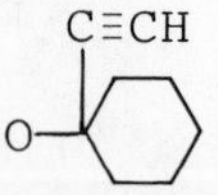 . VIII. Liq. n_D^{25} 1.533.[1221]

$(EtO)_2P(S)OC_8H_{15}$-cyclo. VIII. Liq. $b_{0.01}$ 125-126°.[963]
$(MeO)_2P(S)OC_6H_4Me$-2. VIII.[577, 926] $b_{0.06}$ 112-113°,[577] $b_{0.07}$ 87.5-89°,[926] n_D^{20} 1.5269, d^{20} 1.1941.[926]
$(MeO)_2P(S)OC_6H_4Me$-3. VIII.[577, 926] $b_{0.05}$ 109-110°,[577] $b_{0.1-0.15}$ 115-120°,[926] n^{20} 1.5262, d^{20} 1.1907.[926]
$(MeO)_2P(S)OC_6H_4Me$-4. VIII.[103, 577, 926] $b_{0.07-0.08}$ 97-100°,[926] $b_{0.08}$ 113-114°,[577] n_D^{20} 1.5255, d^{20} 1.1903.[926]
$(PhO)_2P(S)OC_6H_4Me$-4. IX. Solid. m. 69° (from EtOH).[73]
$(EtO)_2P(S)OC_6H_3MeCl$-2,4. VIII. Liq. $b_{0.26}$ 118°, n_D^{30} 1.500.[121]
$(EtO)_2P(S)OC_6H_3MeCl$-3,4. VIII. Liq. $b_{0.2}$ 113-118°, $n_D^{25.5}$ 1.5211.[121]
$(EtO)_2P(S)OC_6H_4MeCl$-2,6. VIII. Liq. $b_{0.15}$ 116-117.5°, n_D^{29} 1.5215.[121]
$(EtO)_2P(S)OC_6H_3$-2-Me-5-N:C. VIII. Liq. $b_{0.02}$ 142-145° (dec).[1342]
$(MeO)_2P(S)OC_6H_3MeNO_2$-2,4. VIII. Solid. m. 55°.[645]
$(MeO)_2P(S)OC_6H_3MeNO_2$-2,6. VIII. Solid. m. 47-48°.[926]
$(MeO)_2P(S)OC_6H_3MeNO_2$-3,4. VIII.[889, 926] Liq. $b_{0.1}$ 140-145° (dec.),[889] $b_{0.15}$ 159.5°,[926] n_D^{25} 1.5528,[889] d^{20} 1.3404.[926] Mass spect.[196] "Sumithion."
$(MeO)_2P(S)OC_6H_3MeNO_2$-4,2. VIII. Liq. $b_{0.01}$ 140-144°, n_D^{25} 1.5420, d^{20} 1.3118.[926]
$(EtO)_2P(S)OC_6H_3MeNO_2$-4,2. VIII. Liq. $b_{0.5-1.0}$ 175-185°, n_D^{25} 1.5283.[331]
$(MeO)_2P(S)OC_6H_2Me(NO_2)_2$-3,2,4. VIII. Liq. n_D^{20} 1.5530, d^{20} 1.4189.[926]
$(MeO)_2P(S,OC_6H_2Me(NO_2)_2$-3,4,6. VIII. Liq. n_D^{20} 1.5642.[926]
$(MeO)_2P(S)OC_6H_3$-2-Me-4-SMe. VIII. Liq. $b_{0.01}$ 86-88°.[885]
$(MeO)_2P(S)OC_6H_3$-3-Me-4-SMe. VIII.[885, 1079, 1084] Liq. $b_{0.01}$ 105°.[1084] ^{1}H-NMR.[74, 524, 526] ^{31}P 45.9 (P_4O_6 ext.).[1047] "Fenthion," "Baytex."
$(EtO)_2P(S)OC_6H_3$-2-Me-4-SMe. VIII. Liq. $b_{0.0001}$ 92°, n_D^{25} 1.5487, d_4^{25} 1.1780.[867]
$(EtO)_2P(S)OC_6H_3$-3-Me-4-SMe. VIII.[867, 1084] Liq. $b_{0.0002}$

88-90°, n_D^{25} 1.5499, d_4^{25} 1.1782.[867]

$(MeO)_2P(S)OC_6H_3$-3-Me-4-SCH_2CH_2OMe. VIII. Liq. n_D^{32} 1.5487.[473] Also about 10 analogs.[473]

$(MeO)_2P(S)OC_6H_3$-3-Me-4-SCH_2CN. VIII. Liq. n^{24} 1.5652.[1098] Et homolog, 1.5475.[1098]

$(MeO)_2P(S)OC_6H_3$-3-Me-4-CF_3. VIII. Liq. $b_{0.01}$ 93°.[1023]

$(MeO)_2P(S)OC_6H_2$-2-Cl-5-Me-4-SCN. VIII. Solid. m. 54-56° (from EtOH).[398]

$(MeO)_2P(S)OC_6H_3$-3-Me-4-S(O)Me. From $(MeO)_2P(S)OC_6H_3$-3-Me-4-SMe plus H_2O_2. Liq. $b_{0.01}$ 110°. Et homolog, n^{20} 1.5471, d_4^{20} 1.2319.[1084] Also several analogs.[1084]

$(RO)_2P(S)OC_6H_3$-3-Me-4-SO_2NH_2. VIII. Solids. R, m.: Me, 84-85°; Et, 80-81.5° (from benzene).[38]

$(MeO)_2P(S)OC_6H_4$-3-Me-4-SO_2NHMe. VIII. Solid. m. 85.5-86.5°.[38] Et homolog, liq., n_D^{25} 1.5365.[38]

$(MeO)_2P(S)OC_6H_4$-3-Me-4-$SO_2NHCOMe$. VIII. Solid. m. 120-121°.[39]

$(MeO)_2P(S)OC_6H_3$-3-Me-4-SO_2NMe_2. VIII. Solid. m. 72.5-73°.[255] Et homolog. Liq. $b_{0.001}$ 110°, n_D^{25} 1.5349-5360.[255]

$(MeO)_2P(S)OC_6H_4CF_3$-3. VIII. Liq. $b_{0.15}$ 87-88°, n_D^{20} 1.4751.[228]

$(EtO)_2P(S)OC_6H_4CF_3$-3. VIII. Liq. $b_{0.1}$ 104-107°.[228]

$(MeO)_2P(S)OC_6H_3$-4-Cl-3-CF_3. VIII. Liq. $b_{0.2}$ 103-104°, n_D^{20} 1.4970. Et homolog. Liq. $b_{0.2}$ 111-112°, n_D^{20} 1.4870.[228]

$(MeO)_2P(S)OC_6H_2Cl_2CF_3$-2,4,5. Liq. $b_{0.12}$ 126-128°, n_D^{20} 1.5150. Et homolog. Liq. $b_{0.1}$ 124-128°.[228]

$(MeO)_2P(S)OC_6H_4CONH_2$-2. VIII. Liq. n_D^{20} 1.5238.[1060]

$(EtO)_2P(S)OC_6H_4CONHPh$-2. VIII. Liq. n_D^{20} 1.5359.[1060]

$(RO)_2P(S)OC_6H_4CONH_2$-4. VIII. Solids. R, m.: Me, 89°; Et, 72°.[1181]

$(EtO)_2P(S)OC_6H_4CONHMe$-2. VIII. Solid. m. 53°.[1181]

$(RO)_2P(S)OC_6H_4CONHMe$-4. VIII. Solids. R, m.: Me, 86°; Et, 55°.[1181]

$(MeO)_2P(S)OC_6H_3$-2-Cl-4-CONHMe. VIII. Solid. m. 82°.[1061]

$(MeO)_2P(S)OC_6H_4CONMe_2$-4. VIII. Liq. n_D^{20} 1.5449.[1060]

$(MeO)_2P(S)OC_6H_4CSNH_2$-4. VIII. Solid. m. 107-108°.[1106]

$(EtO)_2P(S)OC_6H_4CSNH_2$-4. VIII.[732, 809]. Solid. m. 92° (from petroleum ether-benzene).[809]

$(i\text{-}PrO)_2P(S)OC_6H_4CSNH_2$-4. VIII. Solid. m. 112-114° (from benzene-hexane).[809]

$(EtO)_2P(S)OC_6H_3$-2-X-4-$CSNH_2$. VIII. X, properties: Br, solid, m. 71-73° (from benzene-hexane); Cl, liq., n_D^{25} 1.6060.[809]

$(RO)_2P(S)O$— 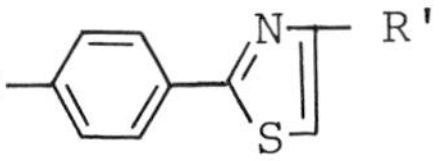 VIII. Liq. R, R', n_D^{25}: Me, H, 1.6057; Et, H, 1.5732; Me, Me, 1.6002; Et, Me, 1.5815.[808]

$(MeO)_2P(S)OC_6H_4CN$-2. VIII. Liq. $b_{0.01}$ 91°.[619]

$(MeO)_2P(S)OC_6H_4CN$-4. VIII.[579, 889] Liq. n_D^{21} 1.5457.[579] "Cyanox."

$(RO)_2P(S)OC_6H_4CN$-4. VIII. Liq. R, b./mm, n_D^t: i-Pr, 124-125°/0.15, 1.5137$^{24°}$; Pr, 129-131°/0.03, 1.5199$^{23°}$: Bu, 140°/0.04; 1.5123$^{20°}$; Am, --, 1.5083$^{25°}$.[576]

$(RO)_2P(S)OC_6H_3ClCN$. VIII.[711, 1280, 1415] Solids and liq.

R	Cl	CN	m.	b./mm
Me	2	4	57-60°[1415]	--
Et	2	4	56°[1280]	131-135°/0.2[1280, 1415]
Et	3	4	--	150-160°/0.001[711]
Me	4	2	35-38°[1415]	--
Et	4	2	--	132-136°/0.2[1280, 1415]
Et	4	3	--	145-148°/0.45[1415]

$(MeO)_2P(S)OC_6H_3$-4-CN-2-OMe. VIII.[1057, 1262] Solid. m. 79-81° (from hexane).[1057] Also six analogs (position isomers or homologs).[1057]

$(MeO)_2P(S)OC_6H_3$-2-CN-4-OMe. VIII. Liq. $b_{0.17}$ 154-162°. Also five position isomers.[553]

$(EtO)_2P(S)OC_6H_3$-2-CN-4-OMe. VIII. Liq. $b_{0.2}$ 161-169°.[1057]

$(MeO)_2P(S)OC_6H_2$-3-Cl-4-CN-6-OMe. VIII. Solid. m. 133.5-134.5° (from hexane-benzene). Also several analogs.[553]

$(MeO)_2P(S)OC_6H_3$-4-CN-2-OEt. VIII.[553, 1057] Solid. m. 87-88°

(from hexane).[553]

$(EtO)_2P(S)OC_6H_3$-4-CN-2-OEt. VIII. Solid. m. 47-48°.[1057]

$(MeO)_2P(S)OC_6H_3$-4-CN-3-OEt. VIII. Solid. m. 74-75°.[553]

$(EtO)_2P(S)OC_6H_4CH{=}X$. VIII and appropriate procedures for derivatives. Solids and liq. X, n_D^{30}: O, 1.5239; NOH, 1.5460; NOCONHMe, 1.5394. Also 196 homologs and analogs.[390]

$(RO)_2P(S)OC_6H_4$-4-CH:$NNHC_6H_4$-X-4. From $(RO)_2P(S)OC_6H_4CHO$-4 plus $ArNHNH_2$.[1235]

R	X	m.
Me	H	77-79°
Et	H	49-52
Me	NO_2	164-166
Et	NO_2	140-143
Me	$CONH_2$	163-166
Et	$CONH_2$	121-124

$(MeO)_2P(S)OC_6H_4COMe$-2. VIII. Liq. $b_{0.14}$ 120-126°, n_D^{20} 1.5372, d_4^{20} 1.2465.[385]

$(EtO)_2P(S)OC_6H_4COMe$-2. VIII. Liq. $b_{0.09}$ 110-114°, n_D^{20} 1.5271, d_4^{20} 1.1911.[385]

$(EtO)_2P(S)OC_6H_4COMe$-4. VIII.[385, 1091] Liq. $b_{0.08}$ 127-130°, n_D^{20} 1.5280, d^{20} 1.1822.[385]

$(RO)_2P(S)OC_6H_3$-2-Cl-4-COMe. VIII. Liq. R, n_D^{25}; Me, 1.555; Et, 1.534.[1091]

$(MeO)_2P(S)OC_6H_3$-4-Cl-2-COMe. VIII. Liq. $b_{0.15}$ 136-143°, n^{20} 1.5510, d_4^{20} 1.3519.[385] Also 10 isomers and analogs.[385] Et homolog, liq. $b_{0.13}$ 126-130°, n_D^{20} 1.5295, d_4^{20} 1.2531.[385]

$(MeO)_2P(S)OC_6H_3$-2-Cl-4-COEt. VIII. Liq. n_D^{25} 1.546.[1091]

$(EtO)_2P(S)OC_6H_4$-4-CH:$CHNO_2$. VIII. Liq. n_D^{25} 1.5719.[1330]

$(EtO)_2P(S)OC_6H_4CH_2CH{:}CH_2$-4. VIII. Liq. b_2 146°.[1201]

$(EtO)_2P(S)OC_6H_3$-4-Cl-2-$CH_2CH{:}CH_2$. VIII. Liq. b_3 161°.[1201]

$(EtO)_2P(S)OC_6H_3$-4-$CH_2CH{:}CH_2$-2-OMe. VIII.[1201, 1220] Liq. b_2 173 .[1201]

$(MeO)_2P(S)OC_6H_4$-4-CH:$CMeNO_2$. VIII. Liq. $n_D^{18.5}$ 1.5970.[476]

$(EtO)_2P(S)OC_6H_4-4-CH:CMeNO_2$. VIII. Liq. $n_D^{19.5}$ 1.5718.[476]

$(RO)_2P(S)OC_6H_4-4-CH:C(CN)_2$. From $(RO)_2P(S)OC_6H_4CHO-4$ plus $CH_2(CN)_2$, AcOH, and AcONH_4. R, m.: Me, 80-81°; Et, 76-77.5°.[1328, 1329]

$(EtO)_2P(S)OC_6H_3-2-Cl-5-t-Bu$. VIII. Liq. n_D^{20} 1.5157.[179a]

$(EtO)_2P(S)OC_6H_3-2-I-5-t-Bu$. VIII. Liq. n_D^{20} 1.5420.[179a]

$(PhO)_2P(S)OC_6H_4-4-t-Bu$. XV. Liq. b_7 252-267°.[837]

$(EtO)_2P(S)OC_6H_4-4-CH:CHCOMe$. VIII. Liq. n_D^{24} 1.5700.[882]

$(EtO)_2P(S)OC_6H_4-4-CMe_2CH_2CMe_3$. XV. Liq. $b_{0.4}$ 188-192°.[1454]

$(RO)_2P(S)OC_6H_4Ph-\underline{n}$. VIII.[523] XV.[837] Liq.

R	$\underline{n}$	b./mm	n_D^{25}	d^{25}
Me	2	--	1.5822	1.221[523]
Et	2	--	1.5649	1.640[523]
Ph	2	275-281°/6.[837]	--	--
4-t-BuC_6H_4	2	316-330°/7.5[837]	--	--
Me	3	--	1.5899	1.218[523]
Et	3	--	1.5760	1.119[523]
Et	4	Solid. m. 118-124°.[523]	--	--

$(MeO)_2P(S)OC_6H_3-2-Br-4-Ph$. VIII.[103, 523] Solid. m. 65-65.5°.[103]

$(EtO)_2P(S)OC_6H_3-2-Br-4-Ph$. VIII. Liq. n_D^{25} 1.5818, d^{25} 1.234.[523]

$(MeO)_2P(S)OC_6H_3-2-Cl-4-Ph$. VIII. Solid. m. 64-65°.[523]

$(EtO)_2P(S)OC_6H_3-2-Cl-4-Ph$. VIII. Liq. n_D^{25} 1.5886, d^{25} 1.334.[523]

$(RO)_2P(S)OC_6H_4-4-t-Bu-2-Ph$. VIII. Liq. R, n_D^{25}, d^{25}; Me, 1.5648, 1.41; Et, 1.5517, 1.111.[523]

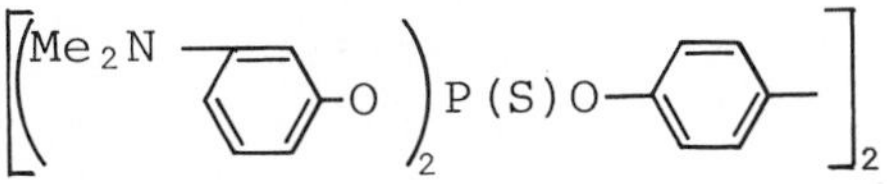

$$\left[\left(Me_2N-\!\!\bigcirc\!\!-O\right)_2 P(S)O-\!\!\bigcirc\!\!-\right]_2 \cdot nMeI.$$ VIII. Solids. n, m.:

3, 122-125° (dec.); 4·(4 H_2O), 181-185° (dec.).[329]

$(RO)_2P(S)OC_6H_4-2-C_6H_{11}-cyclo$. VIII. Liq. R, n_D^{25}: Me, 1.5353; Et, 1.5316.[521]

$(RO)_2P(S)OC_6H_3-Cl-C_6H_{11}-cyclo$. VIII. Liq.[521]

R	Cl	cyclo-C_6H_{11}	n_D^{25}
Me	2	4	Solid, m. 57-58°
Et	2	4	1.5312
Me	4	2	1.5415
Et	4	2	1.5339

$(MeO)_2P(S)OC_6H_4$-4-CH_2Ph. VIII. Liq. n_D^{25} 1.5219, d^{25} 1.109.[1254]

$(EtO)_2P(S)OC_6H_4$-4-CH_2Ph. VIII. Liq. n_D^{20} 1.5522.[1086]

$(RO)_2P(S)OC_6H_4$-4-$CH_2C_6H_4$-4-Cl. VIII. Liq. R, n_D^{25}, d^{25}: Me, 1.5758, 1.268; Et, 1.5591, 1.207.[1254]

$(RO)_2P(S)OC_6H_3$-2-NO_2-4-CH_2Ph. VIII. Liq. R, n_D^{20}: Me, 1.5798; Et, 1.5622.[1086]

$(RO)_2P(S)OC_6H_4$-4-CMe_2Ph. VIII. Liq. R, n_D^{20}: Me, 1.5668; Et. 1.5650.[1086]

$(MeO)_2P(S)OC_6H_3Me_2$. VIII. Liq.[577]

Me positions	b./mm	n_D^t
2,3	108-110°/0.1	1.5310[25°]
2,4	113-115°/0.1	1.5290[23]
2,5	121-122°/0.1	1.5289[24]
2,6	110-113°/0.07	1.5297[24]
3,4	116-118°/0.08	1.5332[23]
3,5	116-118°/0.1	1.5296[23]

$(EtO)_2P(S)OC_6H_3Me_2$-3,5. VIII. Liq. $b_{0.04}$ 117-119°, n_D^{22} 1.5098.[577]

$(MeO)_2P(S)OC_6H_2Me_2NO_2$-3,5,4. VIII. Solid. m. 24-25°.[219]

$(MeO_2P(S)OC_6H_3$-2-Me-4-CN. VIII. Solid. m. 60-60.5° (from benzene).[711]

$(MeO)_2P(S)OC_6H_3$-3-Me-4-CN. VIII. Liq. $b_{0.05}$ 124-128°.[711]

$(MeO)_2P(S)OC_6H_3$-2-Me-4-COMe. VIII. Liq. $b_{0.18}$ 153-156°, n_D^{20} 1.5465, d_4^{20} 1.2400.[385]

$(EtO)_2P(S)OC_6H_3$-2-Me-4-COMe. VIII. Liq. $b_{0.18}$ 158-160°, n_D^{20} 1.5296, d_4^{20} 1.1695. Also six isomers and analogs.[385]

$(PhO)_2P(S)OC_6H_3$-2-Me-5-iPr. VIII. Liq. $b_{7.5}$ 240-248°.[837]

$(3\text{-}ClC_6H_4O)_2P(S)OC_6H_3\text{-}2\text{-}Me\text{-}5\text{-}i\text{-}Pr$. VIII. Liq. $b_{7.5}$ 274-282°.[837]

$(MeO)_2P(S)OC_6H_2Me_3$. VIII. Liq. Isomer, n_D^t: 2,4,6, 1.5165 $^{23.5°}$; 3,4,5, 1.5350$^{22.5°}$.[577]

$(MeO)_2P(S)OC_6Me_5$. VIII. Liq. n_D^{27} 1.5557.[577]

$(RO)_2P(S)OC_6H_2Me_2COMe\text{-}3,5,2$. VIII. Liq. R, $b_{0.05}$, n_D^{20}, d_4^{20}: Me, 150-155°, 1.5330, 1.1407; Et, 145-147°, 1.5194, 1.1977. Also five analogs.[385]

$(RO)_2P(S)OC_6H_2(i\text{-}Pr)_2CN\text{-}2,6,4$. VIII. Solids. R, m.: Me, 89°; Et, 86°.[1168]

$(MeO)_2P(S)O$– 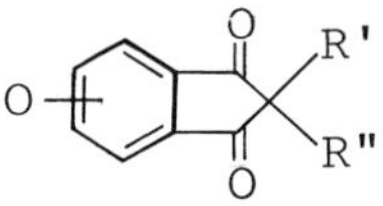 VIII. Solids.

R'	R"	m.
Cl	Me	49-51°[538]
Cl	Cl	57-59°
benzylidene		122-124°
phenylhydrazone		146-149°
Me	Ph	86-88.5°
Ph	Ph	175-177°
Ph	$PhCH_2$	52-56°

$(MeO)_2P(S)O$– 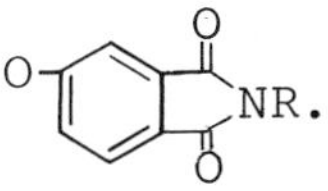 VIII.[712, 1227] Solids. R, m.: H, 118-120° (from PrOH)[712]; Me, 63-64°[712]; Et, 43.5-44° (from hexane)[1227]; i-Pr, 64.6° (from hexane)[1227]; Pr, 42-44° (from petroleum ether)[1227]; $CH_2CH{:}CH_2$, 35°[712]. Also several homologs.[1227]

$(EtO)_2P(S)O$– 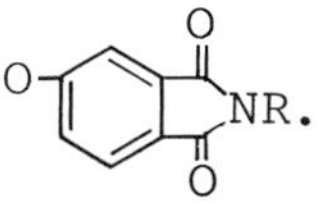 VIII. Solids.[712] R, m.: H, 113-115° (from PrOH); Me, 78-79°; Et, 72-74°; Pr, 35°; $CH_2CH{:}CH_2$, 54-55°.[712]

$(i\text{-}PrO)_2P(S)O$– 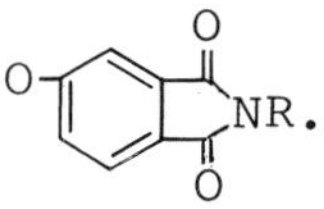 VIII. Solids. R, m.: Me, 54-56°; $CH_2CH{:}CH_2$, 59-61°.[712]

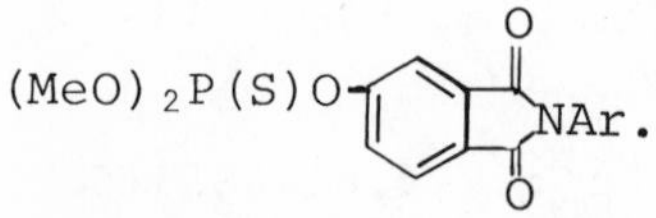

VIII. Solids. Ar, m. (from i-PrOH): Ph, 111.5-114°; 2-MeC_6H_4, 112-114°; 3-MeC_6H_4, 95-96°; 4-MeC_6H_4, 97.5-99°; 3,4-$Me_2C_6H_3$, 88-90°. Also 30 analogs.[396, 1227] Et homolog with NPh, m. 93-94.5°.[396]

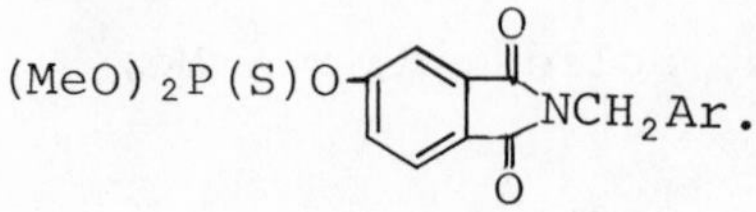

VIII. Solids. Ar, m. (from MeOH): 2,4-$Me_2C_6H_3$, 91.5-99°; 2-ClC_6H_4, 113.5-115°; 2,4-$Cl_2C_6H_3$, 102-106°; 3,4-$Cl_2C_6H_3$, 79-81°; 4-$NO_2C_6H_4$, 72-74° (from MeOH-hexane).[1227]

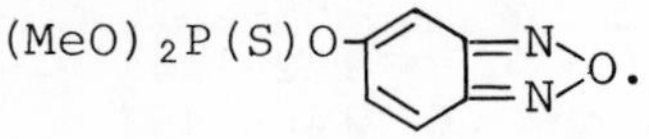

VIII. Solid. m. 47.5-49°.[1229]

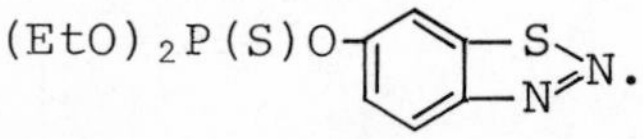

VIII. Solid. m. 39°.[1230]

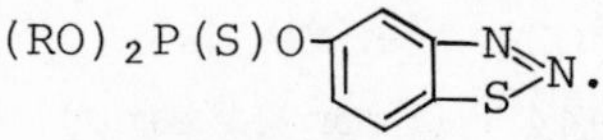

VIII. Solids. R, m.: Me, 60-61°; Et, 47-49°. Also several homologs.[1230]

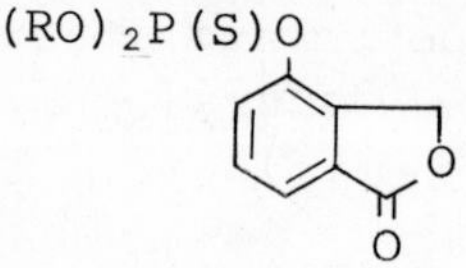

VIII. Solids. R, m.: Me, 86-88°; Et, 36-38°.[1231]

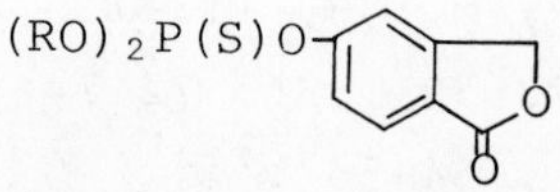

VIII. Solids. R, m.: Me, 59.5-61.5° (from hexane); Et, 38.5-40.5°.[1231]

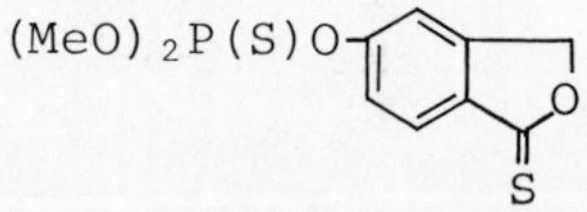

VIII. Solid. m. 39-40.5°.[1231]

$(EtO)_2P(S)O$– 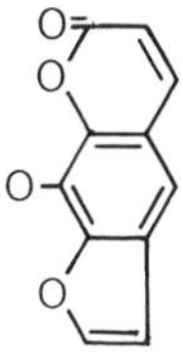 VIII. Solid. m. 131-133° (from ligroin).[854]

$(EtO)_2P(S)O$– 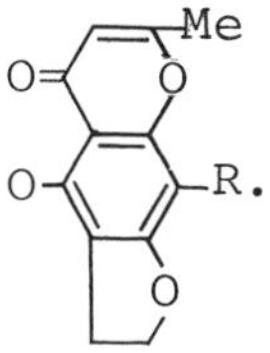 VIII. Solids. R, m.: H, 151-153° (from ligroin 110-120°); OMe, 103-105°.[854]

$(PrO)_2P(S)OC_{10}H_7$-1. XV. Liq. b_1 199-202°, n_D^{20} 1.5585, d_4^{20} 1.3161.[512]

$(MeO)_2P(S)OC_{10}H_7$-2. VIII. Liq. n_D^{20} 1.5997.[135]

$(EtO)_2P(S)OC_{10}H_7$-2. VIII. Liq. $b_{0.8-1.0}$ 151-154°, n_D^{20} 1.5728.[135]

$(MeO)_2P(S)O$-1-NO_2-2-$C_{10}H_7$. VIII. Solid. m. 31-32°.[479]

$(EtO)_2P(S)O$-6-NO_2-2-$C_{10}H_7$. VIII. Solid. m. 51-52.5° (from Skellysolve B).[346]

$(MeO)_2P(S)O$-6-tetralin. VIII. Liq. $b_{0.03}$ 137°.[115, 1215]

$(EtO)_2P(S)O$-6-tetralin. VIII. Liq. $b_{0.03}$ 132°.[115, 1215]

$(MeO)_2P(S)O$-7-Cl-6-tetralin. VIII. Liq. $b_{0.001}$ 144°.[115, 1215]

$(EtO)_2P(S)O$– 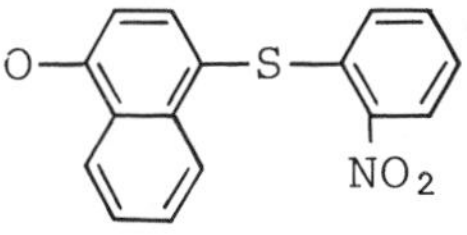 VIII. Solid. m. 102°.[1252]

$(RO)_2P(S)O$– 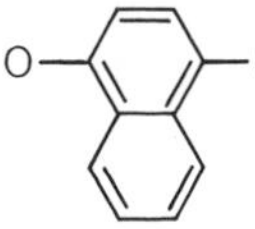 –COMe. VIII. Solids. R, m.: Me, 48-49°; Et, 40-42°.[1091]

$(EtO)_2P(S)O$– 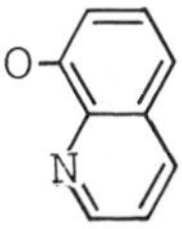 VIII. Liq. $b_{0.5}$ 160°.[405]

$(EtO)_2P(S)O$– 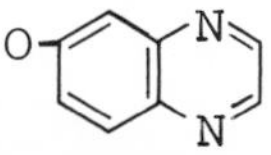 VIII. Liq. n_D^{28} 1.5669.[1118]

$(RO)_2P(S)O$- 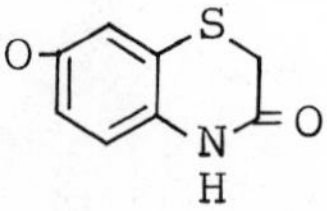 VIII. Solids. R, m.: Me, 153-154°; Et, 113-114° (from benzene-petroleum ether).[1100]

$(RO)_2P(S)O$- 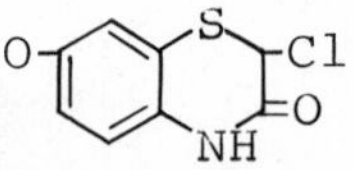 From H analogs above plus SO_2Cl_2. Solids. R, m.: Me, 157°; Et, 140°.[1096]

(RO) P(S)O- 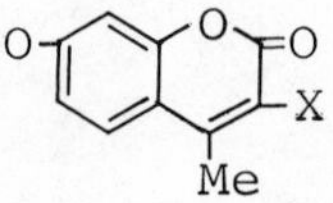 VIII. Solids.

R	X	m.
Me	H	77° (from MeOH)[1125]
Et	H	38°[1125]
$ClCH_2CH_2$	H	48°[131, 185]
Et	Br	105°[1128]
Et	Cl	95°.[1128] ^{1}H-NMR.[74, 524, 526]

^{31}P 50.3 (P_4O_6ext.).[1047] Mass spect.[196]

"Co-Ral"

$ClCH_2CH_2$	Cl	74°.[185]
$ClCH_2CH_2$	Me	74°.[131]
Me	CH_2COOMe	67-68°.[188]

$(EtO)_2P(S)O$- 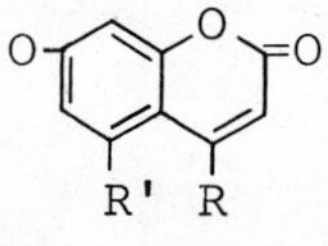 VIII. Liq.[1219]

R	R'	b./mm
H	H	170°/8
Me	Cl	165°/10
Me	Et	112°/10
Me	Pr	145°/8
Me	Bu	135°/8

$(RO)_2P(S)O$- 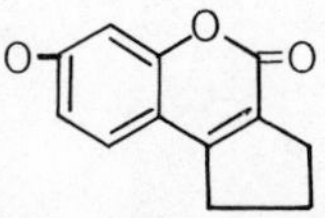 VIII. Solids. R, m.: Me, 88-89°;[349] Et, 85-86°.[349]

$(EtO)_2P(S)O$- 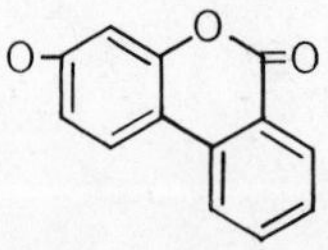 VIII. Solid. m. 103° (from benzene-ligroin).[826]

$(RO)_2P(S)O$- 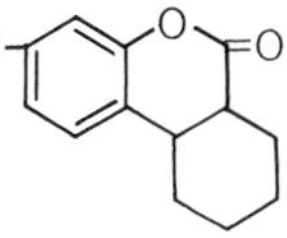 =O VIII. Solids.[349]

R	m.
Me	99-100°
Et	84-86°
i-Pr	112-114°

$(RO)_2P(S)O$- 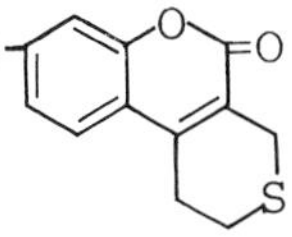 =O VIII. Solids.[826]

R	m.(from EtOH)
Me	100-101°
Et	77-78°

$(EtO)_2P(S)O$-norbornyl. XXII. Liq. $b_{0.08}$ 113-116°, n_D^{20} 1.4941.[839]

$(EtO)_2P(S)O$-cholesteryl. IX. Solid. m. 100°.[191]

$(RO)_2P(S)O$- 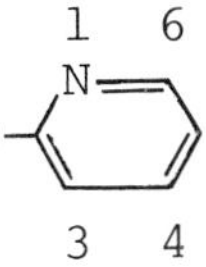 5. XXIX. Liq. R, properties: Et, b_2 40°,[879] i-Pr, $n_D^{17.5}$ 1.4720.[879] VIII. Liq. Ph, λ_{max} 260 mμ ε4330.

R	Substitutions	Methods	Properties
Et	6-Cl	VIII	Liq. n_D^{25} 1.5277.[1031]
Et	4,6-Cl_2	XXIX	Liq. $n_D^{15.5}$ 1.5356.[879]
i-Pr	4,6-Cl_2	XXIX	Liq. $n_D^{16.5}$ 1.4751.[879]
Et	3,6-Cl_2	--	Mass spect.[1317]
Et	3,5-Cl_2	VIII	Liq. n_D^{25} 1.5336.[1031] Mass spect.[1317]
Me	3,5,6-Cl_3	VIII	Liq. n_D^{25} 1.5745.[1031]
Et	3,5,6-Cl_3	VIII	Solid. m. 41-42°.[1031] ^1H-NMR.[74] ^{31}P 52.2 (P_4O_6 ext.).[1047] "Dursban."
Et	3,4,5,6-Cl_4	VIII	Solid. m. 47-49°.[1031]
Et	5-SMe	VIII	Liq. n_D^{25} 1.5022.[1034]
	5-SEt	VIII	Liq. 1.5486

	5-S-i-Pr	VIII	Liq. 1.5258
	5-Ph	VIII	Liq. 1.5380
Et	5-SO_2Me	VIII	Solid. m. 56.5-58° (from MeOH).[1034]
Et	5-SO_2-i-Pr	VIII	Liq. n_D^{25} 1.5221.[1034]
Et	6-SO_2Me	VIII	Liq. n_D^{25} 1.5332.[1034]
Me	5-CN	VIII	Solid. m. 85-86°.[1033]
Et	5-CN	VIII	Liq. n_D^{25} 1.5238.[1033]
Et	3-Cl-5-CN	VIII	Liq. n_D^{25} 1.5310.[1033]
Me	5- 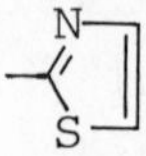	VIII	Liq. n_D^{22} 1.5902.[384]
Et	5- 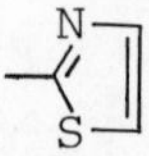	VIII	Solid. m. 57-59 (from MeOH).[384] L.HCl, m. 84-87° L.H_2SO_4, m. 92-94°
Pr	5- 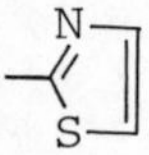	VIII	Solid. m. 20-25°.[384]
Et	3-Br-5- 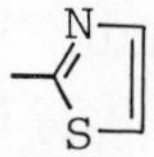	VIII	Solid. m. 86-87°.[384]
Et	3-Cl-5- 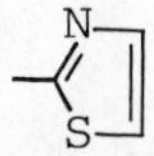	VIII	Solid. m. 58-60°.[384]
Et	3-CN-4,6-Me_2	VIII	Solid. m. 46-47°.[1117]
Et	5-Br-3-CN-4,6-Me_2	VIII	Solid. m. 67-68°.[1117]
Et	5-Cl-3-CN-4,6-Me_2	VIII	Solid. m. 53-55°.[1117]
Et	3-CN-6-Me-4-CH_2OEt	VIII	Solid. m. 38-40°.[1118]
Et	3-CN-6-Me-4-COOEt	VIII[404,1117]	Solid. m. 91°.[1117]
Et	5-Br-3-CN-6-Me-4-COOEt	VIII	Solid. m. 39°.[1117]
Et	3-CN-6-Me-4-Ph	VIII	Solid. m. 50°.[1117]
Et	1-oxo-5-COOEt	VIII	Liq. $b_{0.06}$ 145-155°.[406]

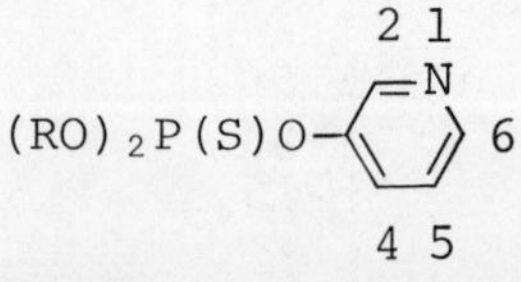

VIII. R, substitutions, properties: Et, 2-Cl, liq., n_D^{25} 1.5145.[1031] Et,

2-CN, liq., n_D^{25} 1.5160.[1033] Et, 6-Me, liq., $b_{0.08}$ 110°.[200] i-Pr, 6-Me, liq., $b_{0.05}$ 110-112°.[200]

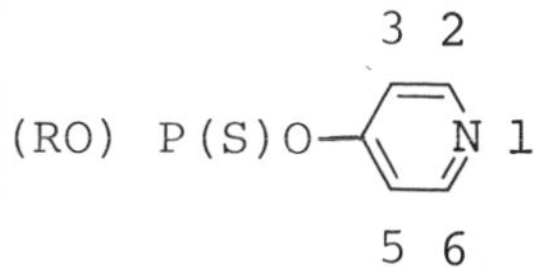

VIII. R, substitutions, properties:

Et, 3,5-Cl_2, liq., n_D^{25} 1.5210.[1031]
Et, 2,6-$(SEt)_2$, liq., n_D^{25} 1.4422.[1034]
Me, 1,2-Me_2-6-oxo, liq., n_D^{20} 1.5370,[132] n_D^{25} 1.5550.[1205]
Et, 1,2-Me_2-6-oxo, solid, m. 62-63°.[132, 1205]
i-Pr, 1,2-Me_2-6-oxo, solid, m. 82-83°.[132]
Et, 1-Pr-2-Me-6-oxo, solid, m. 30-32°.[1205]
Et, 1-$CH_2CH:CH_2$-2-Me-6-oxo, liq., or solid. n_D^{20} 1.5370,[132] n_D^{25} 1.5336,[1205] m. 54-57°.[1205]

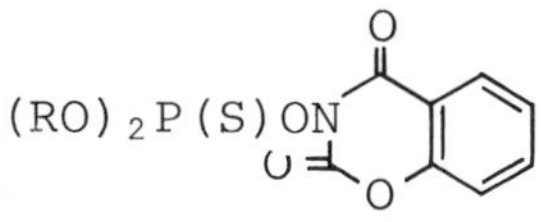

XVII.[861, 862] VIII.[865] Solids. R, m.: Me, 101-103°; Et, 107-108.5° (from EtOH).

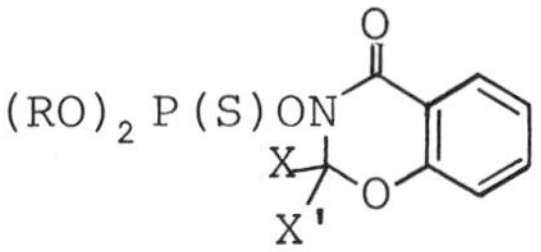

VIII. R, X, X', properties:
Me, H, Et, solid, m. 78-79°.[865]
Me, Me, Me, liq., n_D^{28} 1.5470.[847]
Et, Me, Me, liq., n_D^{21} 1.5322.[847]
Me, Me, Et, liq., n_D^{25} 1.5475.[847]
Et, Me, Et, liq., n_D^{25} 1.5333.[847]

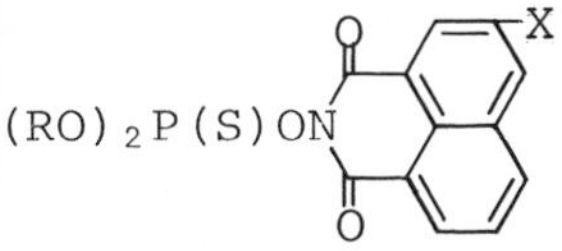

VIII. Solids.[1387] R, X, m. (from EtOH): Me, H, 170-172°; Et, H, 162-164°; i-Pr, H, 168-170°; Et, NO_2, 173-178°.[1387]

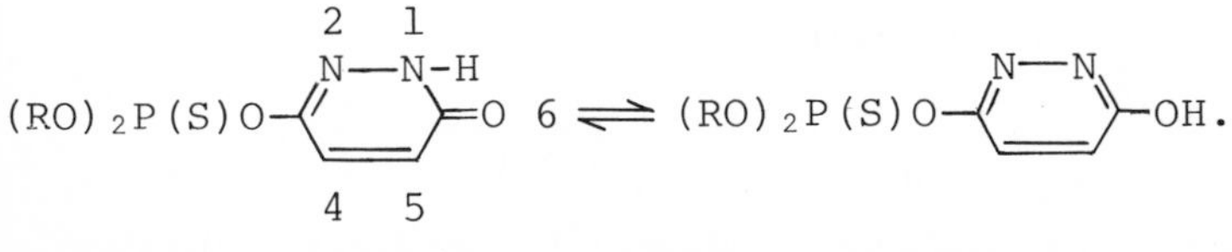

R, Substituents, methods, properties:
Me, H, VIII.[232, 234] Solid. m. 71.5-73° (from ether).

[234]

Et, H, VIII.[232, 234, 1295] Solid. m. 90.5-91.5° (from ether-heptane).[234]

i-Pr, H. VIII.[232, 234] Solid. m. 86.5-87° (from benzene-heptane).

Pr, H. VIII.[232, 234] Solid. m. 113-114° (from benzene-heptane).

Bu, H, VIII. Liq. n_D^{25} 1.4832.[232]

Et, 4,5-Br_2. VIII. Solid. m. 121-122°.[232]

Et, 4-Cl. VIII. Solid. m. 144-145°.[232, 234]

Et, 5-Cl. VIII. Solid. m. 91-92°.[232, 234]

Et, 4-Me. VIII. Solid. m. 109-109.5°.[232, 234]

Et, 5-Me. VIII. Solid. m. 88-89°.[232]

Et, 4-Ph. VIII. Solid. m. 162.5-163.5°.[232, 234]

Et, 5-Ph. VIII. Solid. m. 135.5-136.5°.[232, 234]

Et, 1-Me. VIII. Liq. n_D^{25} 1.5133.[231]

Et, 1-$C_{12}H_{25}$. VIII. Liq. n_D^{25} 1.4843.[231]

Me, 1-Ph. VIII. Solid. m. 46.5-47°.[231]

Et, 1-Ph. VIII. Liq. n_D^{25} 1.5374.[231]

Et, 4-Cl-1-Ph. VIII. Solid. m. 82.5-83°.[231]

Et, 5-Cl-1-Ph. VIII. Liq. n_D^{25} 1.5753.[231]

Et, 1-(C_6H_4Br-4). VIII. Solid. m. 67-67.5°.[231]

Et, 1-($C_6H_4NO_2$-4). VIII. Solid. m. 32-32.5°.[231]

i-Pr, 1-Me. VIII. Solid. m. 37.5-38.5°.[231]

Et, 4,5-Br_2-1-Me. VIII. Solid. m. 67-68°.[231]

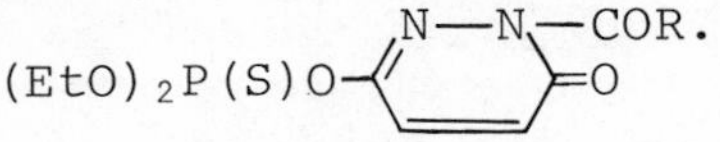

From $(EtO)_2P(S)O$ 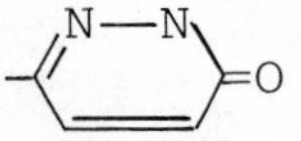

plus RCOCl. Solids. R, m.: $C_{11}H_{23}$, 40-47°; $C_{15}H_{31}$, 50-58°; $C_6H_4NO_2$-4, 108-110°.[412]

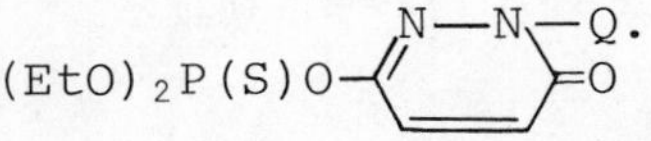

From $(EtO)_2P(S)O$ 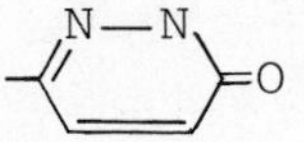

plus QCl and KOH.

Q	Compounds[1120, 1121]
CH_2CONR_2	Solids and liq. R, m. or n_D^{20}: H, 115-

	116°;
	Me, 1.5302; Et, 1.5185; i-Pr, 88-89°; Pr, 1.5122.
CH_2COOR	Me, 55-56°; Et, 36-37°.
$CH_2CH_2CONR_2$	H, 68-69°; Me, 1.5166; Et, 1.5054.
CH_2CH_2CN	51-53°.
CH_2CH_2COOR	Me, 1.5125; Et, 1.5057.

$(EtO)_2P(S)O$– 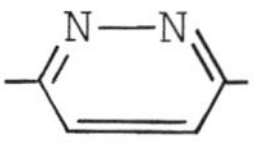 –$OP(S)(OEt)_2$. VIII. Solid. m. 78°.[322]

$(EtO)_2P(S)O$– 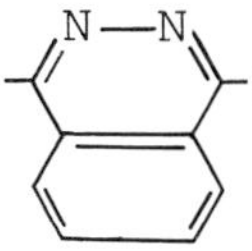 –OH. VIII. Solid. m. 99-100°.[232]

$(RO)_2P(S)O$– 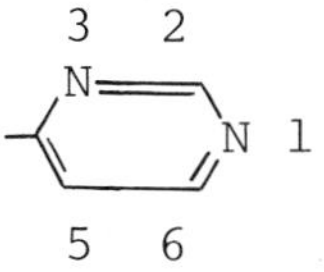 VIII. Solids and liquids. R, substituents, properties:

Me, 2-NH_2-6-Me. Solid. m. 106-108° (from benzene-petroleum ether).[1257, 1258]

Et, 2-NH_2-6-Me. Solid. m. 103° (from benzene-petroleum ether).[1257]

Et, 2-NH_2-6-Pr. Solid. m. 84° (from 60-80° light petroleum).[1258]

Et, 2-NH_2-5-$CH_2CH{:}CH_2$-6-Me. Solid. m. 84° (from light petroleum).[1258]

Et, 2-NMe_2-6-Me. Liq. $b_{0.04}$ 128-132°.[672]

Bu, 2-NMe_2-6-Me. Liq. $b_{0.3}$ 160-164°.[672]

Et, 2-NMe_2-5-Et-6-Me. Liq. $b_{0.002}$ 143-148°, n_D^{26} 1.5168, ^1H-NMR, UV.[1258]

Et, 2-NMe_2-5-Am-6-Me. Liq. n_D^{22} 1.5142, ^1H-NMR, UV.[1258]

Me, 2-NEt_2-6-Me. Liq. n^{25} 1.5291.[1225] "Actellic."

Et, 2-NEt_2-6-Me. "Primicid."

Et, 2-NHCOMe-6-Me. Solid. m. 74° (from i-PrOH).[1258]

Et, 2-NHBu-6-Me. Liq. n_D^{22} 1.5227.[1258]

Et, 2-(=NH)-6-Me. Solid. m. 107-108° (from benzene-petroleum ether).[67]

Et, 2-SH-6-Me. Liq. $b_{0.3}$ 132-133°.[758]

Et, 2,6-Me_2. Liq. $b_{0.13}$ 93-95°,[758, 1244] b_4 152-154°,[67] n_D^{20} 1.5010,[758] 1.5060,[67] d^{20} 1.1455,[758] d_4^{20} 1.1507.[67]

Et, 2-Et-6-Me. Liq. $b_{0.06-0.07}$ 87-92°,[758, 1244] $b_{0.3}$ 115-117°,[394] n_D^{20} 1.4983, d^{20} 1.1350.[758, 1244]

Me, 2-i-Pr-6-Me. Liq. $b_{0.4}$ 99-101°.[394]

Et, 2-i-Pr-6-Me. Liq. $b_{0.002}$ 92-94°,[394] $b_{0.05}$ 85-90°, n_D^{20} 1.4922, d^{20} 1.1088.[758, 1244] ^{1}H-NMR.[74, 524, 526] ^{31}P 52.0 (P_4O_6 ext.).[1047] Mass spect.[196] "Diazinon."

Et, 2-Pr-6-Me. Liq. $b_{0.05}$ 78°, n_D^{20} 1.4941, d^{20} 1.1111.[758, 1244]

Et, 2-CH_2F-6-Me. Liq. $b_{0.01}$ 95°.[321]

Et, 2-Me-6-Ph. Solid. m. 46-48° (from petroleum ether).[67]

$(RO)_2P(S)O$ 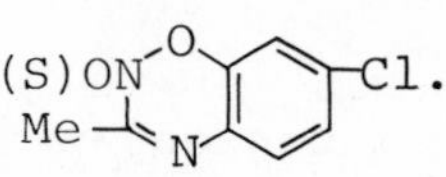 VIII. Solids. R, m.: Me, 87-89°; Et, 83-86°.[803]

$(EtO)_2P(S)O$ 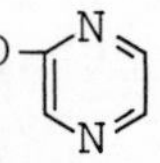 VIII. Liq. n_D^{25} 1.5131, d^{25} 1.207.[214] m. -1.67 ± 0.1° (cryoscopy).[93] ^{1}H-NMR.[526] ^{31}P 51.2 (P_4O_6 ext.).[1047] "Zinophos."

$(RO)_2P(S)O$ 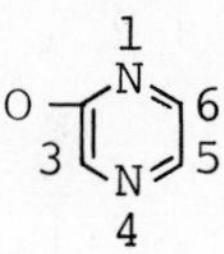 VIII.

R, substituents, properties:

Ph, H. Liq. λ_{max} 270 mu, ε 6040.[803]

Et, 6-Cl. Liq. n_D^{25} 1.5187.[214]

Et, 5-Ph. Solid. m. 87-88°.[214]

Et, 5,6-Me_2. Liq. n_D^{25} 1.5078.[214]

Et, 3,5,6-Me . Liq. n_D^{25} 1.5048.[214]

$(RO)_2P(S)O$ 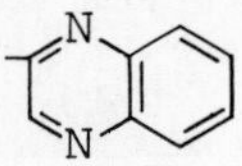 VIII.[307, 1069] Solids. R, m.: Me, 50-51°; Et, 31°.[1069] Me, i-Pr, liq., n_D^{25} 1.5603.[307]

R, substituents, properties:
Me, 6 or 7-Cl. Solid. m. 100°.[1069]
Et, 6 or 7-Cl. Solid. m. 66-67°.[1069]
Et, 3-Me. Liq. $b_{0.0001}$ 110°.[1069]
Et, 3-COOEt. Solid. m. 43-44°.[1069]
Et, 6,7-Me_2. Solid. m. 44-45°.[1069]

$(EtO)_2P(S)O$– 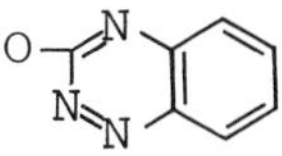 VIII. Solid. m. 56°.[1119]

$(EtO)_2P(S)O$– 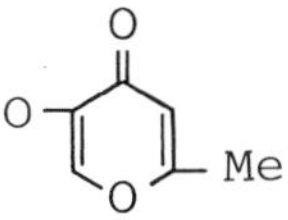 VIII. Solid. m. 51° (from petroleum ether).[769]

$(RO)_2P(S)O$– 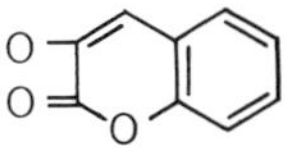 VIII. Solids. R, m.: Me, 93-94°; Et, 61-63 .[186]

$(EtO)_2P(S)O$– 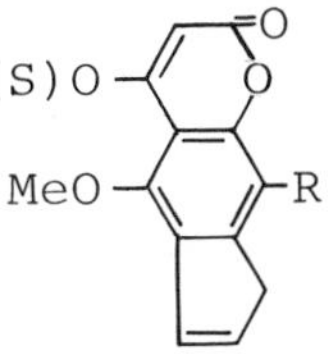 VIII. Solids. R, m. (from ligroin): H, 85-87°; OMe, 70-80°.[854]

$(EtO)_2P(S)O$– 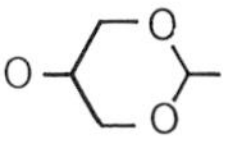 Me. XV. Liq. b_1 113-114°, n_D^{20} 1.4685, d^{20} 1.1651.[64]

$(RO)_2P(S)OCH_2CH-CH_2$ (O bridging CH–CH₂). XV.[1036, 1039] Liq.

R	$b_{0.5}$	n_D^{20}	d_4^{20}
Et	83-85°	1.4671	1.1700[1036]
i-Pr	82	1.4594	1.1038[1039]
Pr	91-93	1.4654	1.1163[1036]
i-Bu	102-103	1.4617	1.0732[1039]
Bu	107-108/0.2	1.4637	1.0807[1036]

$(EtO)_2P(S)OCH_2CH-CH_2$ (S bridging CH–CH₂). XV. Liq. $b_{0.3}$ 101-103°, n_D^{20} 1.4950, d_4^{20} 1.1891.[896]

$(RO)_2P(S)OCH_2$– 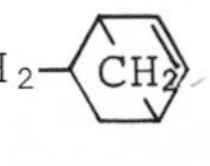 VIII. Liq. R, n_D^{20}: Me, 1.5033; Et, 1.4932.[225]

$(MeO)_2P(S)OCH_2$– 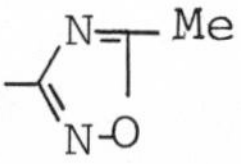 XVII. Liq. n_D^{18} 1.4872.[1399]

$(EtO)_2P(S)OCH_2$–  XXXIII. Liq. $b_{0.1}$ 89-92°, n_D^{20} 1.5001, d^{20} 1.0820.[741, 758]

$(EtO)_2P(S)OCH_2$– 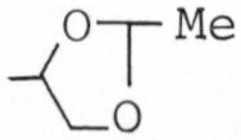 XV. Liq. b_1 113-114°, n_D^{20} 1.4636, d^{20} 1.1575.[64]

$(EtO)_2P(S)OCH_2Ph$. VIII.[139, 492] Liq. $b_{0.2}$ 130°,[139] $b_{2.5}$ 122.5-123.5°,[492] n_D^{20} 1.5152, d^{20} 1.1301.[492] Reaction with PrI in a sealed tube, 20 hr at 105°, gave debenzylation/deethylation in the ratio of 5:2.[139]

$(PhO)_2P(S)OCH_2Ph$. XV. Solid. m. 62.5-63.5°.[511]

$(EtO)_2P(S)OCH_2C_6H_3$-2-Me-5-CH:CRNO$_2$. XVII. Liq. R, b_1: H, 130°; Me, 127°; Pr, 140°/0.5.[563]

$(EtO)_2P(S)OCHPh_2$. VIII. Liq. n_D^{30} 1.5813.[272]

$(EtO)_2P(S)OCH(C_6H_4Cl\text{-}4)_2$. VIII. Liq. n_D^{30} 1.5895.[272]

$(EtO)_2P(S)OCH_2$– 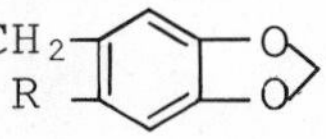 VIII. Liq. R, b./mm: H, 170-180°/0.06; Br, 150-157°/0.02; Cl, 160-180°/0.07; NO$_2$, 200°/0.04.[416]

$(EtO)_2P(S)OCH_2$– 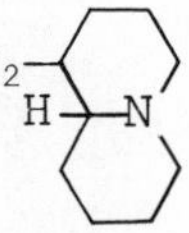 VIII. Liq. $b_{0.3}$ 152°, n_D^{20} 1.5025.[569]

$(RO)_2P(S)OCH_2$– 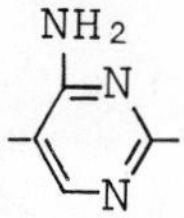 –SMe·HCl. VIII. Solids. R, m, λ_{max} in 95% EtOH, logε : Me, 149-152°, 252, 4.26; Et, 157-159°, 249, 4.22.[442]

$(EtO)_2P(S)OCCOOEt$. VIII. Liq. $b_{0.02}$ 110-115°, n_D^{20} 1.5023.[136]

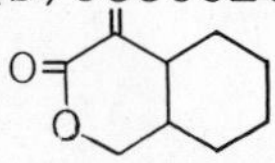

$(EtO)_2P(S)OC$= 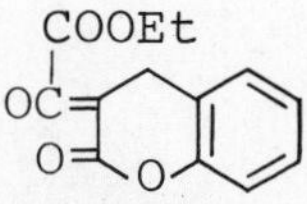 VIII. Liq. $b_{0.04}$ 130-135°, n_D^{20} 1.5082.[136]

(RO)$_2$P(S)OCH$_2$ 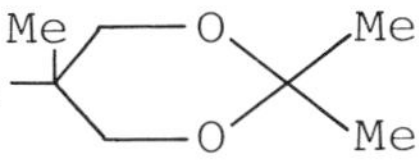 XV. Liq. R, n^{20}, d_4^{20}: Ph, 1.5523, 1.2497; $C_{10}H_{21}$, 1.4706, 1.0119.[946]

(RO)$_2$P(S)OCH(CCl$_3$) 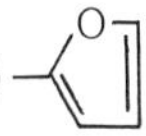 VIII. Liq. R, $b_{0.06}$: Et, 125-135°; Pr, 140-150°; Bu, 180°/0.1.[409]

(EtO)$_2$P(S)O 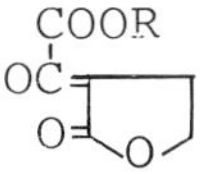 VIII. Liq. R, $b_{0.01}$: Me, 120-124°; Et, 130-135°, n_D^{20} 1.4345.[136]

(EtO)$_2$P(S)O 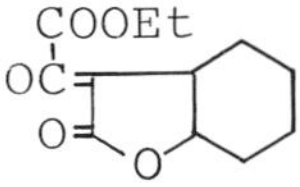 VIII. Liq. $b_{0.01}$ 130-140°, n_D^{20} 1.4985.[136]

(EtO)$_2$P(S)O 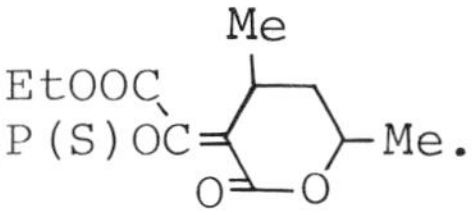 VIII. Liq. R, $b_{0.02}$, n_D^{20}: H, 140-148°, 1.4882; Me, 150-160°, 1.4894.[136]

(EtO)$_2$P(S)OCH$_2$CH$_2$ 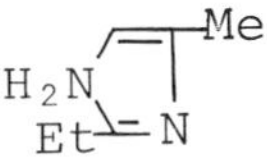 VIII. Picrate. Solid. m. 60-64°.[1076]

(EtO)$_2$P(S)OCH$_2$CH$_2$ 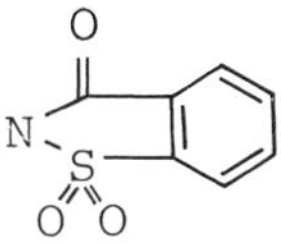 VIII. Solid. m. 27°.[1129]

(RO)$_2$P(S)OCH$_2$CO 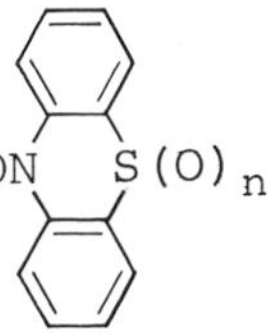 XVII. Solids. R, n, m.: Me, O, 125-128° (from EtOH-acetone); Et, 1, 130-133° (from EtOH); Et, 2, 152-154°.[413]

(EtO)$_2$P(S)OCClPhMe. XV. Liq. b_1 144-146°, n_D^{20} 1.5240, d_4^{20} 1.1975.[990]

(EtO)$_2$P(S)OCHPhCCl$_3$. VIII. Liq. $b_{0.1}$ 200°.[409]

(EtO)$_2$P(S)OCH(4-ClC$_6$H$_4$)CCl$_3$. VIII.[409] XV.[14] Liq. $b_{0.05}$ 145-154°,[409] b_1 162-165°, n_D^{20} 1.5415, d^{20} 1.3806.[14]

(i-PrO)$_2$P(S)OCH(4-ClC$_6$H$_4$)CCl$_3$. XV. Solid. m. 87°.[14]

$(i\text{-}BuO)_2P(S)OCH(4\text{-}ClC_6H_4)CCl_3$. XV. Liq. b_1 185-187°, n_D^{20} 1.5224, d^{20} 1.2753.[14]

$(EtO)_2P(S)OCH(4\text{-}MeOC_6H_4)CCl_3$. VIII. Liq. $b_{0.09}$ 135-142°.[409]

$(EtO)_2P(S)OCH(3,4\text{-}CH_2O_2C_6H_3)CCl_3$. VIII. Liq. $b_{0.05}$ 180°.[409]

$(EtO)_2P(S)OCH(Ph)COOEt$. XXIX. Liq. $b_{0.03}$ 132-133°, n_D^{20} 1.5018, d^{20} 1.1514.[995]

$(PrO)_2P(S)OCH(Ph)COOEt$. XXIX. Liq. b_2 175-176°, n_D^{20} 1.4986, d^{20} 1.1219.[995]

$(MeO)_2P(S)OC(:CHCl)C_6H_3Br_2\text{-}2,4$. VIII. Solid. m. 55-56°.[1380]

$(MeO)_2P(S)OC(:CHCl)C_6H_3Cl_2\text{-}2,5$. VIII. Solid. m. 81-82°.[1379]

$(EtO)_2P(S)OC(:CHCl)C_6H_3Cl_2\text{-}2,5$. VIII.[1379, 1380] Solid. m. 27.5-28.5°. "Akton." Also 15 homologs and analogs.

$(RO)_2P(S)OC(:CHCl)C_6H_2Cl_3\text{-}2,4,5$. VIII. Solids. R, m.: Me, 75-76°; Et, 55-56°.[1380]

$(EtO)_2P(S)OC(Ph):CHCOOEt$. VIII. Liq. $b_{0.5}$ 161°, $n_D^{23.5}$ 1.5348.[881]

$(EtO)_2P(S)OC(C_6H_4NO_2\text{-}4):CHCOOEt$. VIII. Liq. $b_{0.005}$ 180-183°.[408]

$(EtO)_2P(S)OC(Ph):CHCOMe$. XXXIII. From $(EtO)_2P(S)H$ plus $PhCOCHClCOMe$. Liq. $b_{0.2}$ 150-152°, n_D^{26} 1.5451.[882]

$(MeO)_2P(S)OSiMe_3$. XVII. Liq. b_{12} 83-84°, n_D^{20} 1.4460, d^{20} 1.071.[319]

$(EtO)_2P(S)OSiMe_3$. XV.[137] XVII.[319] Liq. b_{12} 96-97°, n_D^{20} 1.4430, d^{20} 1.027.[137, 319] Raman.[319]

$(BuO)_2P(S)OSiMe_3$. XV. Liq. b_{10} 123-124°, n_D^{20} 1.4450, d_4^{20} 0.9687.[913]

$(EtO)_2P(S)OSiEt_3$. XV.[137, 913] Liq. b_1 96-98°,[913] b_2 89-90°,[137] n_D^{20} 1.4540,[137] 1.4560,[913] d_4^{20} 0.9984,[913] 1.0129.[137]

F.4.4.4. 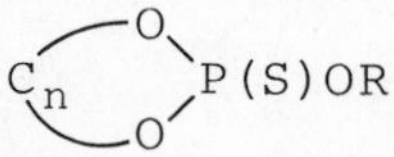

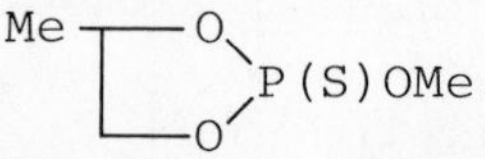 VIII.[56] XV.[51] Liq. $b_{2.5}$ 136-138°, n_D^{20} 1.4930, d_0^{20} 1.3001.[51]

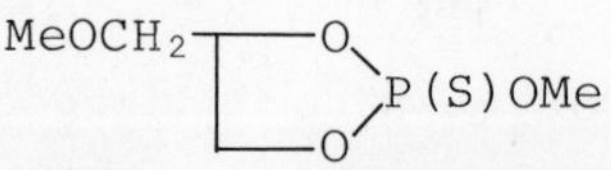 VIII. Liq. $b_{1.5}$ 111-112.5°, n_D^{20} 1.4889, d_0^{20} 1.2877.[54]

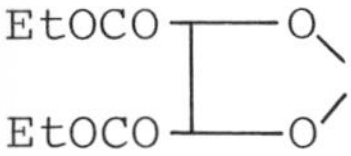 P(S)OMe. XV. Liq. $b_{4-4.5}$ 163-164°, n_D^{20} 1.4695, d_0^{20} 1.3048. $[\alpha]_D^{20}$ -56.6.[53]

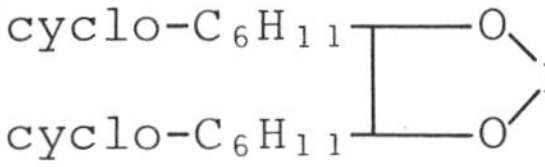 P(S)OMe. XV. Solid. m. 115.5-116.2° (from MeOH).[50]

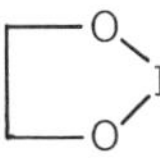 P(S)OEt. VIII.[1407] XV.[249] Liq. b_3 98-99°, n_D^{21} 1.4849, d^{21} 1.2737.[1407]

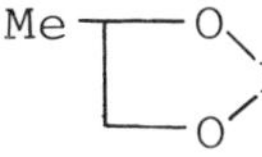 P(S)OEt. VIII.[56, 1407] XV.[51, 249] Liq. $b_{0.5}$ 102-104°,[51] b_3 96°,[1407] b_6 107-108°,[56] n_D^{20} 1.4770,[56, 249] d_0^{20} 1.2072.[51]

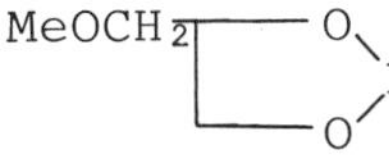 P(S)OEt. VIII.[54, 1407] Liq. b_2 121.5-122°,[54] b_3 119-120°,[1407] n_D^{21} 1.4790, d^{21} 1.2310.[1407]

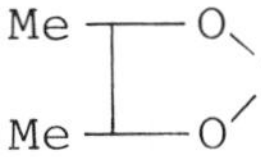 P(S)OEt. XV. Liq. $b_{0.06}$ 66°, n_D^{19} 1.4785.[249]

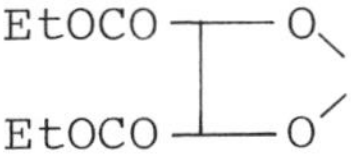 P(S)OEt. XV. Liq. $b_{3.5}$ 159-160°, n_D^{20} 1.4680, d_0^{20} 1.2603. $[\alpha]_D^{20}$ -37.61°.[53]

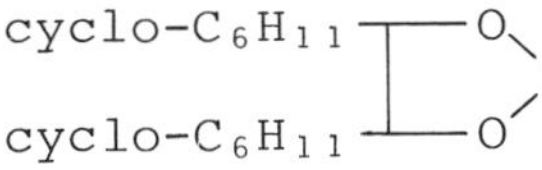 P(S)OEt. XV. Solid. m. 65-66° (from petroleum ether).[50]

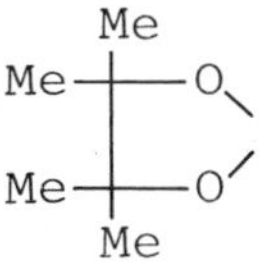 P(S)OEt. XV. Liq. n_D^{19} 1.4775.[249]

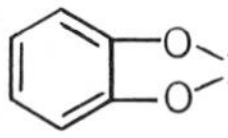 P(S)OEt. VIII. Liq. n_D^{20} 1.5622, d^{20} 1.2954.[1309]

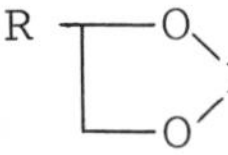 P(S)OCH$_2$CH$_2$Cl. XV. Liq. R, b./mm, n_D^{20}, d_4^{20}: Me, 116-118°/2, 1.4954, 1.3445; MeOCH$_2$, 142-143°/1.5, 1.4960, 1.3234.[991]

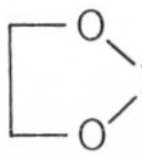 P(S)OCH$_2$CH$_2$OCH:CH$_2$. XV. Liq. $b_{0.5}$ 105-106°, n_D^{20} 1.5035, d^{20} 1.2850.[532]

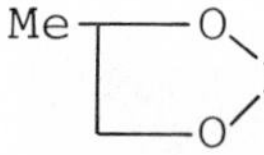
P(S)O-i-Pr. VIII. Liq. b_3 104-106°, n_D^{20} 1.4740, d^{20} 1.188.[56]

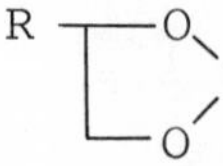
$P(S)OCHMeCH_2Cl$. XV. Liq. R, b/mm., n_D^{20}, d_4^{20}: H, 129-130°/3, 1.4876, 1.2881; Me, 126-128°/2, 1.4841, 1.2433.[991]

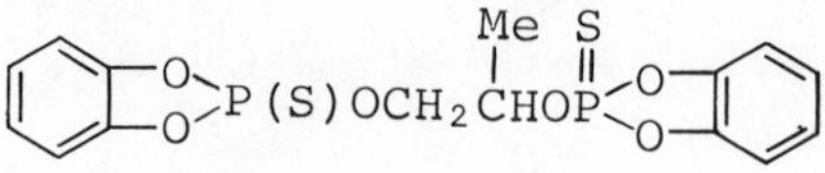
XV. Liq. $b_{5.5}$ 216-219°.[849]

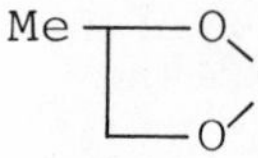
P(S)OPr. VIII.[56] XV.[51] Liq. $b_{0.2}$ 99-101°,[51] b_5 110-112°,[56] n_D^{20} 1.4662,[51] 1.4741,[56] d_0^{20} 1.1221,[51] d^{20} 1.187.[56]

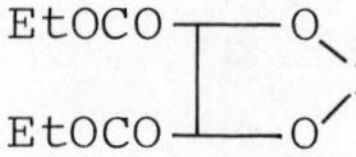
P(S)OPr. XV. Liq. $b_{1.5-2}$ 158-160°, n_D^{20} 1.4685, d_0^{20} 1.2282. $[\alpha]_D^{20}$ -38.43°.[53]

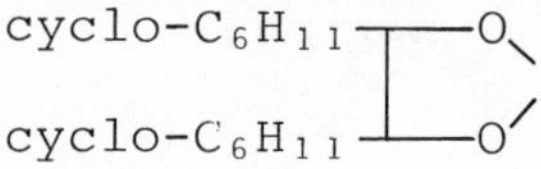
P(S)OPr. XV. Solid. m. 50.5-52° (from petroleum ether).[50]

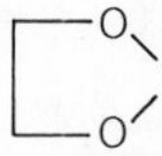
$P(S)OCH_2CH:CH_2$. XV. Liq. b_4 130-132°, n_D^{20} 1.5025, d_0^{20} 1.2619.[52]

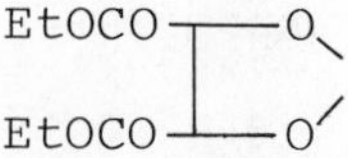
$P(S)OCH_2CH:CH_2$. XV. Liq. b_2 170-172°, n_D^{20} 1.4723, d_0^{20} 1.2670. $[\alpha]_D^{20}$ -38.7°.[53]

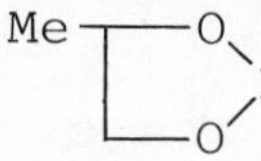
P(S)C i-Bu. VIII.[56] XV.[51] Liq. $b_{0.12}$ 86.5-88°,[51] b_3 112-115°,[56] n_D^{20} 1.4705,[51] 1.4720,[56] d_0^{20} 1.1232,[51] d^{20} 1.138.[56]

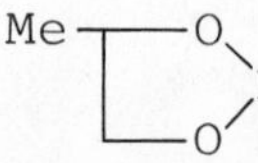
P(S)OBu. VIII. Liq. b_3 114-116°, n_D^{20} 1.4730, d^{20} 1.1435.[56]

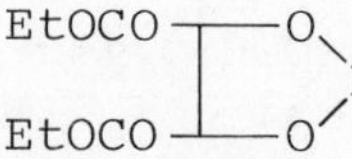
P(S)OBu. XV. Liq. $b_{3.5}$ 177-178°, n_D^{20} 1.4690, d_0^{20} 1.2060, $[\alpha]_D^{20}$ -36.75°.[53]

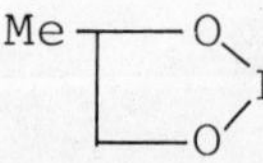
$P(S)OC_6H_{13}$. XV. Liq. $b_{0.15}$ 118.5-119°, n_D^{20} 1.4725, d_0^{20} 1.0891.[51]

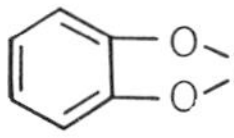P(S)OPh. VIII.[1309] XV.[45] Solid. m. 71-72°.[45]

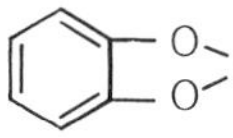$P(S)OC_6H_4OMe$-2. XV. Solid. m. 93-94°.[45]

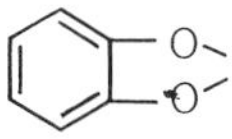P(S)OAr. VIII. Solids. Ar, m.: 2-ClC_6H_4, 125.5°; 2,4-$Cl_2C_6H_3$, 80°; 2-$NO_2C_6H_4$, 142°; 4-$NO_2C_6H_4$, 88°.[1309]

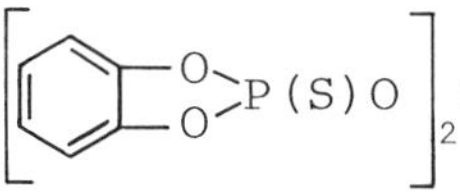$[P(S)O]_2C_6H_4$. XV. Liq. Position on C_6H_4, $b_{3.5}$: 1,2, 263-264°; 1,3, 267-268°; 1,4, 277-278°.[849]

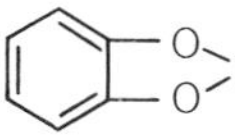$P(S)OC_6H_4Me$-2. VIII.[1309] XV.[45] Solid. m. 87-88°.[45] b_{14} 197-200°.[45] C_6H_4Me-4 isomer, solid, m. 71-72°.[45]

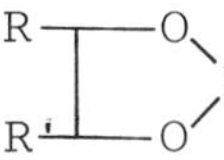$P(S)OCH_2CH-CH_2$. XV. Liq.[1266]

R	R'	$b_{0.005}$	n_D^{20}	d_4^{20}
H	H	109-110°	1.5081	1.3856
H	Me	132-135	1.4926	1.2978
H	$ClCH_2$	136-137	1.5198	1.4342
H	$MeOCH_2$	127-130	1.4960	1.3208
Me	Me	144-146	$1.4940^{25°}$	$1.2948^{25°}$

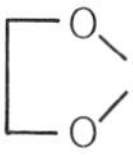$P(S)OC(CHCl)C_6H_3Cl_2$-2,4. XXXIII. (From 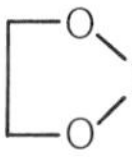P(S)H plus $Cl_2CHCOC_6H_3Cl_2$-2,4 and NH_3.) Solid. m. 110-111° (from (EtOH).[1232]

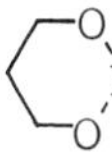P(S)OR. XV.[870, 872] Liq.

R	$b_{0.05}$	n_D^{20}	
Me	100°	1.5045	
Et	103-105	$1.4995^{25°}$	
i-Pr	100	$1.4845^{25°}$	^1H-NMR[870]
Pr	108	1.4900	
HC≡CCH_2	115	1.5170	

Ph Solid, m. 45° -- ^{1}H-NMR[870]

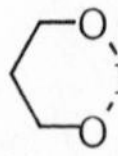P(S)$OC_6H_4NO_2$-4. VIII. Solid. m. 165.5-166.5°.[250]

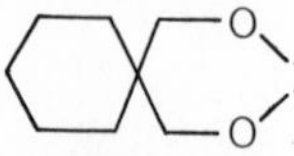P(S)OEt. VIII. Solid. m. 73-77°.[592]

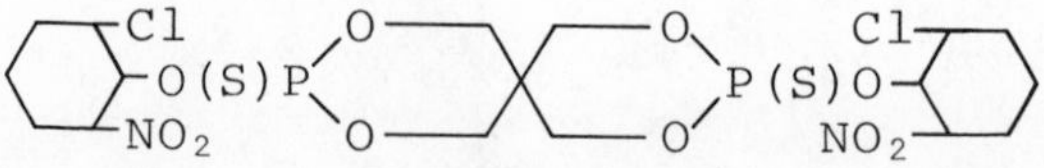 VIII. Solid. m. 254-255°. Also 17 analogs.[1007]

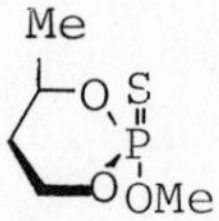 XV. (From "cisoid" 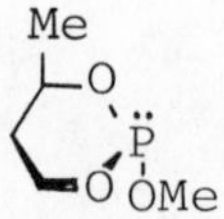 plus S). Liq. $b_{0.3}$

cis

78-80°, n_D^{20} 1.4922. ^{1}H-NMR. ^{31}P -148.6 ppm.[797] Cf. 1248 below.

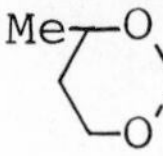P(S)OR. XV.[1248] Liq. ^{1}H-NMR, Dipole moments, configurations and conformations:

R	isomer	b./mm	n_D^t
Me	cis	68-70°/0.02	1.4913[20°]
	trans	74-76°/0.02	1.4958[19°]
i-Pr	cis	62-65°/0.03	1.4822[19°]
	trans	68°/0.01	1.4823[20°]

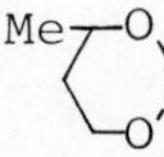P(S)OEt. ^{1}H-NMR.[112]

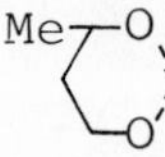P(S)$OCH_2CH:CH_2$. XV. Liq. b_{4-5} 158-160°, n_D^{20} 1.5025, d_0^{20} 1.2111.[52]

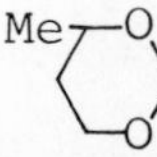P(S)OPh. VIII. Liq. Molecular distn. at 25° at 5 x 10^{-5} mm. IR.[681]

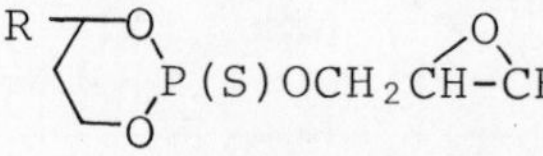P(S)OCH_2CH-CH_2. XV. Liq. R, $b_{0.005}$, n_D^{20}, d_4^{20}: H, 135-140°, 1.5081, 1.3307; Me, 143-145°, 1.5040, 1.2860.[1266]

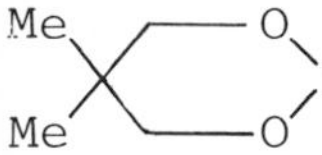
P(S)OR.

R	Method	Properties
Me	XV	Solid. m. 93.5-94.5°.[245] ^{31}P -63.0.[715]
Et	XV	Solid. m. 62-63°.[245]
Pr	XV	Liq. $b_{0.25}$ 90-94°, n_D^{18} 1.4754.[245]
Ph	VIII	Solid. m. 98-99°. IR.[681] ^{31}P -54.1 ppm (in CH_3OH).[679a]
$C_6H_4NO_2$-4	VIII	Solid. m. 157.5-159.5° (from PhCl).[248, 250]

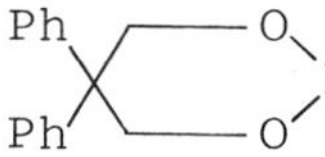
P(S)OPh. VIII. Solid. m. 165-166°. IR.[681]

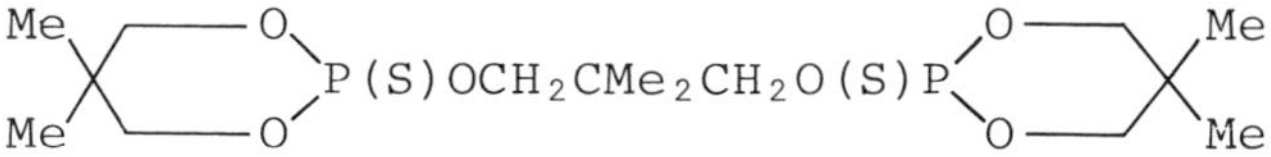

XV. Solid. m. 163-164° (from benzene).[668]

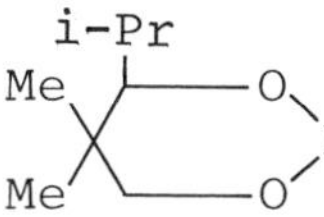
P(S)OPh. VIII. Solid. m. 75-76°. IR.[681]

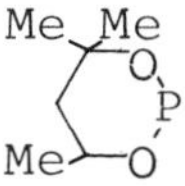
P(S)OPh. VIII. Solid. m. 112-113.5°. IR. ^{1}H-NMR.[682]

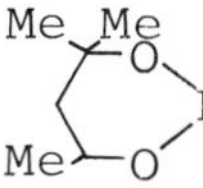
P(S)$OC_6H_4NO_2$-4. VIII. Solid. m. 107.5-109° (from PhCl-ligroin).

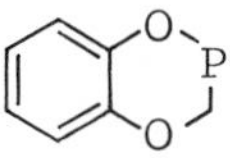
P(S)OR. VIII. (From ROP(S)Cl_2 plus 2-$HOC_6H_4CH_2OH$ and C_5H_5N):

R	Properties
Me	Solid. m. 51-53° (from MeOH. "Salithion."[916, 1284]
Et	Liq. n_D^{25} 1.5495.[263, 916]
CH_2CH_2Cl	Liq. $b_{0.08}$ 160-162°.[544]
CH_2CH_2OMe	Liq. $b_{0.02}$ 149-151°.[544]
CH_2CH_2OEt	Liq. $b_{0.03}$ 160-165°.[544]
Ph	Solid. m. 36°.[263]

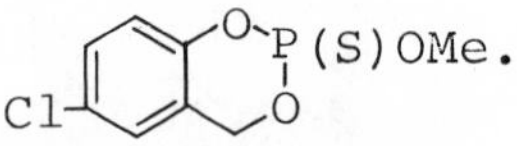 VIII. Liq. $b_{0.2}$ 170-178°.[264]

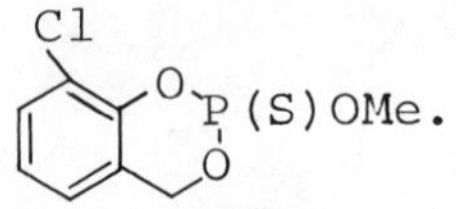 VIII. Solid. m. 72-73°.[1284]

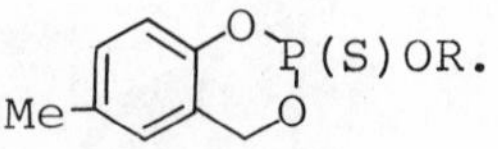 VIII. Solids. R, m.: Me, 34-35°; Et, 71-72°; Pr, liq., $b_{0.2}$ 158-160°.[264] Also isomers with 7-Me and 8-Me.[264]

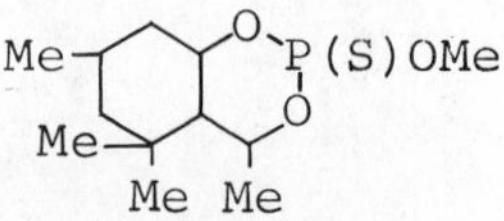 or 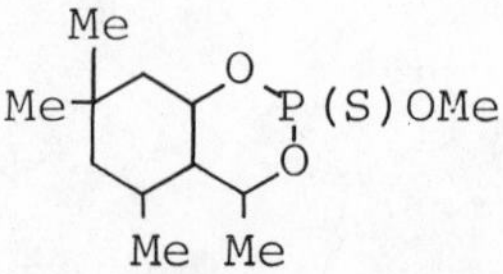 VIII. Liq. n_D^{30} 1.5010.[593]

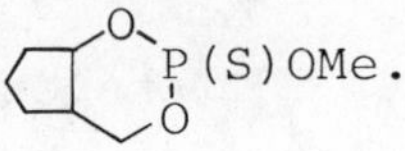 VIII. Liq. n_D^{30} 1.5063.[593]

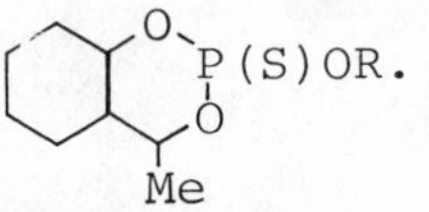 VIII. Liq. R, n_D^{30}: Me, 1.5065; $MeOCH_2CH_2$-OCH_2CH_2, 1.5047; , 1.5356; $EtSCH_2CH_2$, 1.5280.[593]

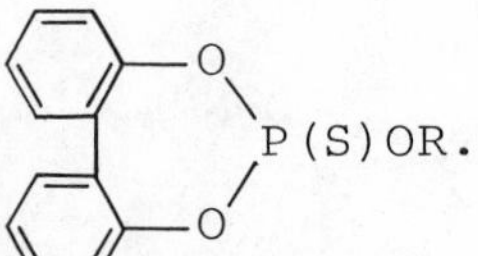 XV. Liq.[1354]

R	b./mm	n_D^{20}	d_4^{20}
C_6H_{13}	182-183°/0.007	1.5770	1.1908
$C_{10}H_{21}$	227-228°/0.009	1.5517	1.1194

F.4.4.5. RO(R'O)P(S)OR"

MeO(EtO)P(S)OR. VIII. Liq.[704]

R	$b_{0.3}$	n_D^{20}	d^{20}
Pr	58-61°	1.4540	1.1224
i-Bu	64-67	1.4470	1.0540
Bu	60-64	1.4490	1.0625
i-Am	70-72/0.25	1.4560	1.0645

$MeO(EtO)P(S)OCH_2CH_2NEt_2$. VIII. Liq. b_1 70°, n_D^{20} 1.4545, d_4^{20} 1.0907.[687]

$MeO(EtO)P(S)OC_6H_2Br_2Cl$-2,4,5. VIII. Liq. n_D^{24} 1.5843.[85]

$MeO(EtO)P(S)OC_6H_2Cl_2I$-2,5,4. VIII. Liq. n_D^{18} 1.5941.[178]

$MeO(RO)P(S)OC_6H_2Cl_3$-2,4,5. VIII. Liq. R, n_D^{25}: Et, 1.5518; i-Pr, 1.5456; Pr, 1.5525; $MeEtCHCH_2$, 1.5498.[126a]

$MeO(EtO)P(S)OC_6H_4NO_2$-4. VIII. Liq. $b_{0.12}$ 116°, n_D^{20} 1.5480, d^{20} 1.3182.[748]

$MeO(EtO)P(S)OC_6H_2$-2,4-Cl_2-5-CF_3. VIII. Liq. $b_{0.15}$ 130°.[228]

$MeO(EtO)P(S)OC_6H_4$-4-$CSNH_2$. VIII. Solid. m. 70-72° (from petroleum ether).[809]

$MeO(EtO)P(S)OC_6H_3$-3-Me-4-NO_2. VIII. Liq. n_D^{20} 1.5425, d^{20} 1.2824.[926]

$MeO(EtO)P(S)OC_6H_3$-3-Me-4-SMe. VIII. Liq. $b_{0.01}$ 128°.[1085]

$MeO(EtO)P(S)OC_6H_4SMe$-4. VIII. Liq. $b_{0.01}$ 120°, n_D^{20} 1.5625, d^{20} 1.250.[1085]

$MeO(EtO)P(S)OC_6H_4S(O)Me$-4. From sulfide plus H_2O_2. Liq. n_D^{20} 1.5638, d_4^{20} 1.288.[1085]

MeO(EtO)P(S)O 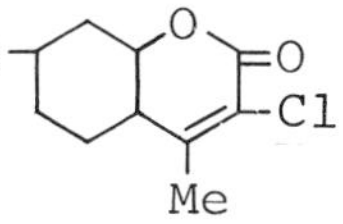 VIII. Solid. m. 108°.[1128]

$MeO(EtO)P(S)OC_6H_3$-2-Br-5-t-Bu. VIII. n_D^{24} 1.5217.[179a]

$MeO(EtOCH_2CH_2O)P(S)ON{:}C(CN)Ph$. VIII. Liq. n_D^{20} 1.5383. Also several analogs.[539]

$MeO(EtSCH_2CH_2O)P(S)OAr$. VIII. Liq. Ar, n_D^{20}, d^{20}: Ph, 1.5540, 1.2150; 4-ClC_6H_4, 1.5690, 1.3548; 4-$NO_2C_6H_4$, 1.5880, 1.3655.[702]

$MeO(i\text{-}PrO)P(S)OC_6H_4NO_2$-4. VIII. Liq. n_D^{24} 1.5400. Also 17 analogs.[1174]

$MeO(s\text{-}BuO)P(S)OC_6H_4NO_2$-4. VIII. Solid. m. 70°. Also 11 analogs.[1174]

$EtO(FCH_2CH_2O)P(S)OCH_2CH_2SEt$. VIII. Liq. $b_{0.0001}$ 72-74°, n_D^{20} 1.4831, d^{20} 1.1970.[490]

$EtO(FCH_2CH_2O)P(S)OC_6H_4NO_2$-4. VIII. Liq. $b_{0.0001}$ 134-135°,

n_D^{20} 1.5290, d^{20} 1.3469.[490]

$EtO(MeOCH_2CH_2O)P(S)OC_6H_4NO_2$-4. VIII. Liq. n_D^{28} 1.5280. Also several analogs.[539]

$EtO(EtSCH_2CH_2O)P(S)OAr$. VIII. Liq. Ar, n_D^{20}, d^{20}: Ph, $b_{0.5-0.6}$ 163-166°, 1.5400, 1.1910; 4-ClC_6H_4, 1.5510, 1.2723; 4-$NO_2C_6H_4$, 1.5680, 1.3282.[702]

$EtO(i\text{-}PrO)P(S)OC_6H_4NO_2$-4. VIII. Solid. m. 28-29°, n_D^{33} 1.5209.[226]

$EtO(i\text{-}PrO)P(S)OC_6H_4SMe$-4. VIII. Liq. $b_{0.01}$ 105°, n_D^{20} 1.504, d^{20} 1.170.[1085]

$EtO(i\text{-}PrO)P(S)OC_6H_4S(O)Me$-4. From the sulfide plus H_2O_2. Liq. n_D^{20} 1.5426, d_4^{20} 1.220.[1085]

$EtO(HC{\equiv}CCH_2O)P(S)OC_6H_2Cl_3$-2,4,5. VIII. (From 2,4,5-$Cl_3C_6H_2OP(S)Cl_2$ plus EtOH plus $HC{\equiv}CCH_2OH$ and Et_3N.) Liq. n_D^{20} 1.5585, d^{20} 1.3754.[1313]

$EtO(BuO)P(S)OCH_2CH_2NMe_2$. VIII. Liq. $b_{0.2}$ 86-96°.[330]

$EtO(CCl_3CMe_2O)P(S)OCMeEtCCl_3$. XV. Liq. $b_{0.007}$ 132-133°, n_D^{20} 1.5158, d^{20} 1.4265.[252]

$EtO(PhO)P(S)OCH_2N^+Me_3$ X^-. XXXII. Solids. X, m.: Br, 103-106°; I, 98-101.5°.[535]

EtO(PhO)P(S)O 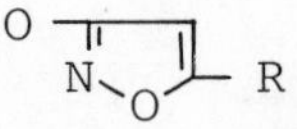 VIII. Liq. R, b./mm: H, 140°/0.1; Me, 135-140°/0.1; $CH(OEt)_2$, 163°/0.04; COOMe, 155°/0.04.[1071]

F.5. Miscellaneous Phosphorodithioate and -thioate Derivatives

F.5.1. Phosphorodithioate Anhydrides

F.5.1.1. $(RO)_2P(S)SCOR$

$(EtO)_2P(S)SCOCl$. XVII. Liq. $b_{0.05}$ 80°, n_D^{25} 1.5198.[141]

$(MeO)_2P(S)SCONHMe$. XVII. Solid. m. 50°.[903]

$(EtO)_2P(S)SCONH$-i-Pr. XVII.[903] XXX.[923] Solid. m. 70.5-71.5°.[923]

$(MeO)_2P(S)SCONHAr$. XXX.[594, 873] Solids. Ar, m.: Ph, 91.5-92° (from benzene);[594] 4-ClC_6H_4, 80° (from CCl_4-petroleum ether);[594] 3,4-$Cl_2C_6H_3$, 116° (from CCl_4);[594]

C_6H_4SCN-4, 72° (dec.);[873] C_6H_3-3-Cl-4-SCN, 109-110°.[873]

$(EtO)_2P(S)SCONHAr$. XXX.[594, 923] Solids. Ar, m.: Ph, 56°;[594, 923] 3-MeC_6H_4, 45-46°;[923] 4-ClC_6H_4, 69-70°;[923] 3,4-$Cl_2C_6H_3$, 76-77°;[923] $C_6H_4NO_2$-3, 80-81°.[923]

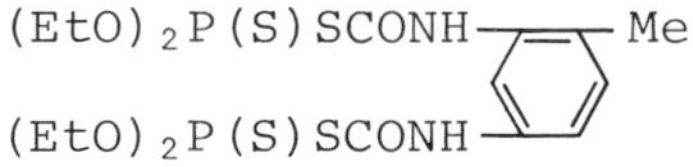

XXX. Solid. m. 116-117°.[923]

$(EtO)_2P(S)SCON(CHCl_2)CCl_3$. XVII. Solid. m. 102° (from ether).[695]

$(EtO)_2P(S)SCOOMe$. XVII. Liq. $b_{0.05}$ 80°, n_D^{25} 1.5040,[141] n_D^{20} 1.5063, Mol. ref. 367.96.[1078]

$(EtO)_2P(S)SCOMe$. XXXIII. (From $(EtO)_2P(S)SH$ plus $(EtO)_2P$-(O)Ac or $(PrO)_2P(O)Ac$.) Liq. $b_{0.5}$ 65-67°, n_D^{20} 1.5060, d^{20} 1.1845.[1002] n_D^{20} 1.5154, Mol. ref. 345.94.[1078]

$(ROCH_2CH_2O)_2P(S)SCOMe$. XVII. Liq. R, n^{20}, d_4^{20}: Me, 1.5138, 1.2335; Et, 1.5008, 1.1704; Ph, 1.5798, 1.2550;

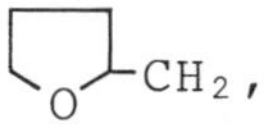

1.5378, 1.2587.[1450]

$(CH_2{:}CHCH_2O)_2P(S)SCOR$. XVII. Liq. R, n_D^{20}, d_4^{20}: Me, 1.5470, 1.2069; i-Pr, 1.5565, 1.1877; i-Bu, 1.5342, 1.1519; Ph, 1.5780, 1.2121.[910]

$(ClCH_2CH_2O)_2P(S)SCOCH_2Br$. XVII. Liq. n_D^{20} 1.5700, d_4^{20} 1.6540.[1450]

$(cyclo\text{-}C_6H_{11}O)_2P(S)SCOCCl_3$. XVII. Solid. m. 48° (from ether).[1449]

$(CH_2{:}CHCH_2O)_2P(S)SCO(CH_2)_nCOSP(S)(OCH_2CH{:}CH_2)_2$. XVII. Liq. n, n_D^{20}, d_4^{20}: 0, 1.5450, 1.2454; 1, 1.5445, 1.2205; 2, 1.5440, 1.2235; 3, 1.5432, 1.2123.[910]

$(PhO)_2P(S)SCO(CH_2)_nCOSP(S)(OPh)_2$. XVII. Liq. n, n_D^{25}, d_4^{25}: 0, 1.5542, 1.1933; 2, 1.6135, 1.3361; 3, 1.5767, 1.3016; 4, solid, m. 75-76°.[911]

$(RO)_2P(S)SCOCH{:}CH_2$. XVII. Liq. R, n_D^{20}, d_4^{20}: Et, 1.5347, 1.2147; i-Pr, 1.5139, 1.1364, IR data; i-Bu, 1.5117, 1.0966, IR data; Bu, 1.5171, 1.1115, IR data.[1444]

$(RO)_2P(S)SCOCMe{:}CH_2$. XVII. Liq. R, n_D^{20}, d_4^{20}: Et, 1.5296,

1.1877; Pr, 1.5207, 1.1321; s-Bu, 1.5100, 1.0958; Bu, 1.5130, 1.0967. IR data.[1444]

$(EtO)_2P(S)SCOCH_2CH_2Me$. From $(EtO)_2P(S)SH$ plus $(MeCH_2CH_2CO)_2O$. Liq. b_1 85-87°, n_D^{20} 1.5070.[918]

$(RO)_2P(S)SCOCH{:}CHCOSP(S)(OR)_2$. XVII. Liq. R, n_D^{20}, d_4^{20}: Et, $b_{0.0001}$ 76-77°, 1.556, 1.294; i-Pr, 1.544, 1.217; Pr, 1.546, 1.220; Bu, --, 1.163. IR data.[1444]

$(EtO)_2P(S)SCOPh$. From $(EtO)_2P(S)SH$ plus $(PhCO)_2O$. Liq. $b_?$ 100° (dec).[918]

$(ROCH_2CH_2O)_2P(S)SCOPh$. XVII. Liq. R, n_D^{20}, d_4^{20}: Me, 1.5663, 1.2483; Et, 1.5528, 1.1914; Ph, 1.6128, 1.2617.[1450]

$(PhO)_2P(S)SCOPh$. XVII. Liq. n_D^{20} 1.5861, d^{20} 1.4130.[912]

$(PhO)_2P(S)SCOC_6H_4NO_2$-4. XVII. Liq. n_D^{20} 1.6130, d^{20} 1.5210.[912]

(cyclo-$C_6H_{11}O)_2P(S)SCOAr$. XVII. Solids. Ar, m.: Ph, 57.5° (from EtOH); $C_6H_4NO_2$-4, 100° (from acetone).[1449]

$(ROCH_2CH_2O)_2P(S)SCOC_6H_4NO_2$-4. XVII. Liq. R, n_D^{20}, d_4^{20}: Me, 1.5790, 1.3329; Et, 1.5650, 1.2751; Ph, solid, m. 53-54° (from ether); 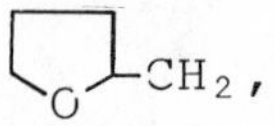 1.5858, 1.3390.[1450]

$(ROCH_2CH_2O)_2P(S)SCO(CH_2)_3COSP(S)(OCH_2CH_2OR)_2$. XVII. Liq. R, n_D^{20}, d_4^{20}: Me, 1.5352, 1.2811; Et, 1.5095, 1.1994; Ph, 1.5884, 1.2818; 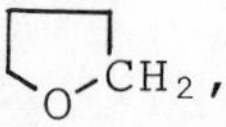 1.5190, 1.2252.[1450]

$(MeO)_2P(S)SCOCH_2C_6H_3Cl_2$-2,4. XVII. Liq. n_D^{23} 1.5915.[179]

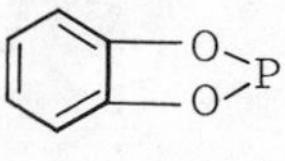 $P(S)SCOC_6H_4NO_2$-4. XVII. Solid. m. 163° (dec).[507, 1440]

$(EtO)_2P(S)SO_2S$–
 XVII. Liq. n_D^{25} 1.5667.[1382]
$(EtO)_2P(S)SO_2S$–

$(EtS)_2P(O)ON{:}CClF$. XXXIII. Liq. b_2 108°, n_D^{20} 1.5218, d_4^{20} 1.3070.[724]

F.5.2. Phosphorodithioate Anhydrosulfides

F.5.2.1. $(RO)_2P(S)SC(S)R'$

$(EtO)_2P(S)SC(S)NMe_2$. VIII. Liq. $b_{0.35}$ 63-64°, n_D^{30} 1.4646.[859]

$(EtO)_2P(S)SC(S)NEt_2$. VIII.[859] XX. (From $(EtO)_2P(S)SK$ plus $Et_2NC(S)Cl$.)[1324] Liq. $b_{0.35}$ 73-77°,[859] n_D^{20} 1.5706, d^{20} 1.1737.[1324]

$(EtO)_2P(S)SC(S)NBu_2$. VIII. Liq. $b_{0.2}$ 95-98°, n_D^{30} 1.5035.[859]

$(EtO)_2P(S)SC(S)$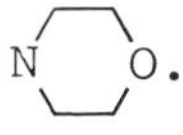. VIII. Liq. $b_{0.1}$ 81-83°, n_D^{30} 1.4854.[859]

$(RO)_2P(S)SC(S)OR'$. VIII. IX. Liq.[1406]

R	R'	b./mm	d_{18}^{18}
Me	Me	63-67°/12	1.2531
Me	Et	65-72°/10	1.1764
Pr	Et	105-109°/10	0.9653
Et	i-Pr	60-63°/3.5	1.0507
Et	Pr	80-84°/10	1.0169
Et	Bu	92-95°/12	0.9872

F.5.2.2. $(RO)_2P(S)SP(S)(OR)_2$

$(RO)_2P(S)SP(S)(OR)_2$.

R	Method	m.	b./mm	n_D^{20}	d_4^{20}	^{31}P[715]
Me	XXVIII[606]	--	--	--	--	-83.5[606]
	XXXIII[760]	34°	108.5-111°/0.07	--	--	
Et	XXVI[19]	43-44°	--	--	--	-78.4
	XXVIII[606]	42.5-43°				
	XXXIII[760]	40°	105-112°/0.07			
i-Pr	XXVI[19]	--	140°/0.5	1.5113	1.1284	-75.2
	XXXIII[760]		107-108°/0.08			
Pr	XXVI[19]		150°/0.4	1.5163	1.1361	
i-Bu	XXVI[19]		156°/0.2	1.5020	1.0815	

$(EtO)_2P(S)SP(S)(O\text{-}i\text{-}Pr)_2$. XX. Liq. $b_{0.4}$ 138°, n_D^{20} 1.5203, d^{20} 1.1663.[32]

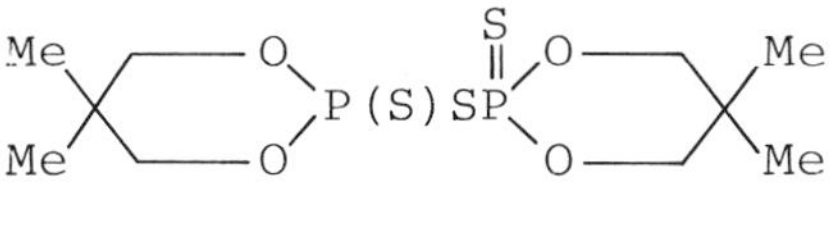

XXXIII. (From

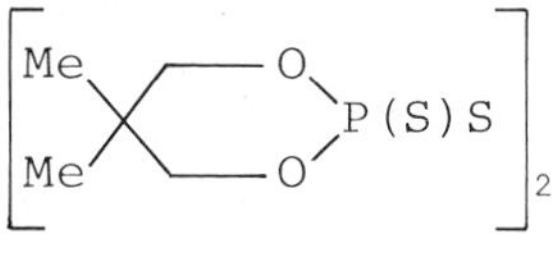

plus Ph_3P.) Solid. m. 223°.[247]

$(RO)_2P(S)S(O)P(S)(OR)_2$. XXXIII. (From sulfide plus H_2O_2; or from $2(RO)_2P(S)H$ plus $Me_2NS(O)Cl$.: Liq. R, $b_{0.01}$:

Me, 55-60°; Et, 91°; i-Pr, 112-115°; Pr, 118°.[292, 1195]

F.5.3. Phosphorodithiosulfenates

F.5.3.1. $(RO)_2P(S)SOR'$

$(MeO)_2P(S)SOCH(CH_2Br)_2$. XXXIII. (From $(MeO)_2P(S)SCl$ plus $(BrCH_2)_2CHOH$ and Et_3N.) Liq. n_D^{25} 1.5259.[977]

$(EtO)_2P(S)SOCH(CH_2Br)_2$. XXXIII. Liq. n_D^{25} 1.5230.[977]

$(MeO)_2P(S)SOC_6Cl_5$. XXXIII. Liq. n_D^{25} 1.5730.[977]

$(EtO)_2P(S)SOC_6Cl_5$. XXXIII. Liq. n_D^{25} 1.5569.[977]

F.5.3.2. $(RO)_2P(S)SNR'_2$

$(EtO)_2P(S)SNHCOOEt$. XVII. Solid. m. 71-73° (from benzene).[692]

$(EtO)_2P(S)SNHSO_2Ar$. From $(RO)_2P(S)SN(SO_2Ar)SP(S)(OR)_2$ plus HI. Solids. Ar, m.: Ph, 170°; 4-ClC_6H_4, 80-81°; 4-FC_6H_4 85°; 4-$MeOC_6H_4$, 98°; 4-MeC_6H_4, 100°.[26] IR of 4-ClC_6H_4 compound shows dimerization (broad band at 3200 cm^{-1}, neat); monomeric in CCl_4 (sharp band at 3355 cm^{-1}).[27]

$(PrO)_2P(S)SNHSO_2Ar$. Liq. Ar, n_D^{25}: 4-$MeOC_6H_4$, 1.5580; 4-MeC_6H_4, 1.5568. (Cf. preceding entry.)[26]

$(EtO)_2P(S)SN(SO_2Ar)SP(S)(OEt)_2$. XXVI. Solids or liq. Ar, m. or n_D^{25}: Ph, 75-76°; 4-BrC_6H_4, 1.5874; 4-ClC_6H_4, 1.5826; 4-FC_6H_4, 1.5670; 4-$MeOC_6H_4$, 62-63°; 4-MeC_6H_4, 87°.[26]

$(PrO)_2P(S)S(SO_2Ar)SP(S)(OPr)_2$. XXVI. Liq. Ar, n_D^{25}: 4-MeO-C_6H_4, 1.5618; 4-MeC_6H_4, 1.5537.[26]

$(MeO)_2P(S)SNMeCOMe$. XVII. Liq. $b_{0.6}$ 96-101°.[1416]

$(EtO)_2P(S)SNMeSO_2Ph$. XVII. Solid. m. 65° (from petroleum ether).[692]

$(PhO)_2P(S)SNMeSO_2Ph$. XVII. Solid. m. 92-94° (from MeOH).[692]

$(EtO)_2P(S)SNEtPh$. XX. Liq. n_D^{23} 1.5655, d^{23} 1.1479.[32]

$(EtO)_2P(S)SNPhCHO$. XVII. Solid. m. 53-54° (from EtOH).[692]

$(EtO)_2P(S)SNPhCOMe$. XVII.[692, 1416]. Solid. m. 60-62° (from petroleum ether),[692] 74-75°.[1416]

$(EtO)_2P(S)SN(C_6H_{11}\text{-cyclo})_2$. XXXIII. (From $(EtO)_2P(S)SNa$ plus $(C_6H_{11}\text{-cyclo})_2NCl$.) Solid. m. 70-71°.[690]

$(PhO)_2P(S)SN(C_6H_{11}\text{-cyclo})_2$. XXXIII. Solid. m. 128-130°.[690]

$(EtO)_2P(S)SN$ 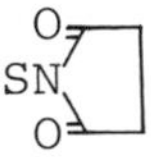 XVII. Solid. m. 94.5-96° (from EtOH).[692]

$(EtO)_2P(S)SN$ 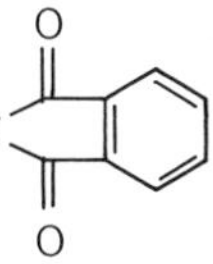 XVII. Solid. m. 123°.[692]

$(EtO)_2P(S)SN$ 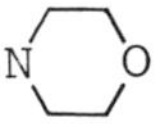 XX. Solid. m. 30° (from ligroin 60-80°).[32]

$(RO)_2P(S)SN$ 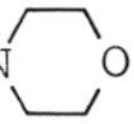 XX. Solids or liq. R, m. or n_D^{20}, d^{20}: i-Pr, 31°; Pr, 30°; i-Bu, 1.5120, 1.1073; Bu, 1.5127, 1.1077.[33]

$(EtO)_2P(S)SN(COOMe)NHCOOMe$. XXII. Solid. m. 51-53.5°.[498]

$(RO)_2P(S)SN(COOEt)NHCOOEt$. XXII. Solids. R, m.: Me, 40-45°; Et, 57.5-59° (from petroleum ether); Bu, 68-70°.[498]

$(i\text{-}PrO)_2P(S)SN{:}NC_6H_4Cl\text{-}4$. XXVII. Solid. m. 33-35°. Warming with aqueous Cu_2Cl_2 gives $(i\text{-}PrO)_2P(S)SC_6H_4Cl\text{-}4$ plus N_2).[743]

$(PhO)_2P(S)SN{:}NAr$. XVII. Solids. Ar, m. (dec.): $C_6H_4NO_2\text{-}4$, 50°; $C_6H_4COOH\text{-}2$, 70°; $C_6H_4COOH\text{-}3$, 75°; $C_6H_4COOH\text{-}4$, 81°. IR.[1453]

$(EtO)_2P(S)SNPhOSOC_6H_4Me\text{-}4$. XXXIII. Liq. n_D^{25} 1.5497.[976]

$(EtO)_2P(S)SN(SiMe_3)_2$. XXXIII. (From $((EtO)_2P(S)S)_2$ plus $NaN(SiMe_3)_2$). Liq. b_1 120-123°.[1089]

$(RO)_2P(S)SAs$ 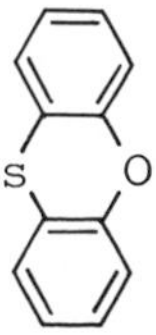 XVII. Solids. R, m.: Me, 90-93°; Et, 63-65°.[1242]

F.5.3.3. $(RO)_2P(S)SSR'$

$(EtO)_2P(S)SSMe$. XX.[29, 1197] Liq. $b_{0.1}$ 98°,[1197] $b_{2.5}$ 100.5-101.5°, n_D^{20} 1.5500, d_4^{20} 1.2142.[29]

$(i\text{-}PrO)_2P(S)SSMe$. XX. Liq. b_2 99-100°, n_D^{20} 1.5297, d_4^{20}

1.1471.[29]

$(MeO)_2P(S)SSEt$. XIX. Liq. b_1 95°, n_D^{25} 1.5542.[975]

$(EtO)_2P(S)SSEt$. XIX.[975] XX.[29, 1197] XXXIII.[614] Liq. $b_{0.01}$ 62°,[614] $b_{0.1}$ 106°,[1197] b_3 106-107°,[29] n_D^{20} 1.5431, d_4^{20} 1.1810.[29]

$(i\text{-}PrO)_2P(S)SSEt$. XX. Liq. $b_{3.5}$ 117-118°, n_D^{20} 1.5240, d_4^{20} 1.1189.[29]

$(i\text{-}BuO)_2P(S)SSEt$. XX. Liq. $b_{2.5}$ 129-130°, n_D^{20} 1.5183, d_4^{20} 1.0866.[29]

$(RO)_2P(S)SSCH_2CH_2NHCOMe$. XXXIII. (From $(RO)_2P(S)SH$ plus $MeCONHCH_2CH_2S(O)_2SCH_2CH_2NHCOMe$.) Liq. R, n_D^{25}: i-Pr, 1.5375 (solidified on standing, m. 40-42°); Pr, 1.5420; Bu, 1.5345.[1272]

$(MeO)_2P(S)SSCH_2CH_2SEt$. XXXIII. Liq. n_D^t 1.5772[240].[614]

$(EtO)_2P(S)SSCH_2CH_2SEt$. XIX.[975] XXXIII.[614] Liq. $b_{0.01}$ 96-98°, n_D^{20} 1.5635.[614]

$(EtO)_2P(S)SSCH_2COOEt$. XIX.[975] XXXIII. Liq. $b_{0.01}$ 88°.[615]

$(EtO)_2P(S)SSPr$. XX. Liq. $b_{0.5}$ 106-107°, n_D^{20} 1.5370, d_4^{20} 1.1547. IR.[29]

$(EtO)_2P(S)SS\text{-}t\text{-}Bu$. XIX. Liq. n_D^{25} 1.5489.[975]

$(EtO)_2P(S)SS\text{-}i\text{-}Bu$. XX. Liq. $b_{0.2}$ 104°, n_D^{20} 1.5310, d_4^{20} 1.1370.[29]

$(MeO)_2P(S)SSBu$. XIX. Liq. n_D^{25} 1.5425.[975]

$(MeO)_2P(S)SSCH(COOEt)CH_2COOEt$. XIX. Liq. n_D^{25} 1.5225.[975]

$(EtO)_2P(S)SSBu$. XIX.[975] XX.[29] Liq. $b_{0.2}$ 139.5-140°, n_D^{20} 1.5306, d_4^{20} 1.1246.[29]

$(EtO)_2P(S)SSPh$. XX.[29, 1197] Liq. $b_{0.001}$ 118°.[1197] b_2 153°, n_D^{20} 1.5993, d_4^{20} 1.2263.[29]

$(PhO)_2P(S)SSPh$. XIX. Solid. m. 55-56°.[29]

$(EtO)_2P(S)SSC_6H_4Br\text{-}4$. XX. Solid. m. 24-25°.[29]

$(PhO)_2P(S)SSC_6H_4Br\text{-}4$. XIX. Solid. m. 81-82°.[29]

$(MeO)_2P(S)SSC_6H_4Cl\text{-}4$. XX. Liq. $b_{0.01}$ 116°.[1197] ^{31}P -74.[488]

$(EtO)_2P(S)SSC_6H_4Cl\text{-}4$. XX,[29, 1197] XXXIII. (From R. P. of $(EtO)_2P(S)SNa$ with $4\text{-}ClC_6H_4SOCl$.)[21] Liq. $b_{0.01}$ 124°,[1197] $b_{0.6}$ 146-148°,[29] $b_{0.8}$ 154°,[21] n_D^{20} 1.6062, d_4^{20}

1.2970.[29]

$(RO)_2P(S)SSC_6H_4Cl$-4. XX.[18] Solids and liq. R, properties: i-Pr, solid, m. 39°,[18] ^{31}P -74,[715] -74 ± 2;[1345] Pr, solid, m. 29-30°; i-Bu, solid, m. 35°; Bu, liq., $b_{0.1}$ 158° (dec.); i-Am, $b_{0.12}$ 165°, n_D^{20} 1.5555, d^{20} 1.1541; Am, n_D^{20} 1.5610, d^{20} 1.1571.[18]

$(PhO)_2P(S)SSC_6H_4Cl$-4. XX. Solid. m. 63-64°.[29]

$(EtO)_2P(S)SSC_6H_4F$-4. XX. Liq. $b_{0.3}$ 126°, n_D^{20} 1.5827, d_4^{20} 1.2749. IR.[29]

$(PhO)_2P(S)SSC_6H_4F$-4. XX. Solid. m. 62-63°. IR.[29]

$(EtO)_2P(S)SSC_6H_4Me$-4. XX. Liq. $b_{0.2}$ 138.5°, m. 10-11°, n_D^{20} 1.5936, d_4^{20} 1.1975.[29] Also several analogs.[22, 29]

$(PhO)_2P(S)SSC_6H_4Me$-4. XX. Solid. m. 65°. Also several analogs.[29]

$(RO)_2P(S)SSC_6H_4NO_2$-4. XX. Solids. R, m.: Et, 53-54°; Ph, 96-98° (from ether-petroleum ether).[28]

$(RO)_2P(S)SSC_6H_4(NO_2)_2$-2,4. XX. Solids. R, m.: Et, 66-67° (from cyclohexane); Ph, 132-133° (from benzene, ether).[28]

F.5.3.4. $[(RO)_2P(S)S]_2$

$[(MeO)_2P(S)S]_2$. XXIII.[462, 606, 807] Solid m. 52°.[606] IR.[1348] ^{31}P -88.3.[606]

$[MeO(EtO)P(S)S]_2$. XXIII. Liq. n_D^{20} 1.5785, d_4^{20} 1.3200.[560]

$[(EtO)_2P(S)S]_2$. XXIII.[462, 464, 606, 807] XXXIII.[21, 1323, 1451] Liq. $b_{0.2}$ 135°,[21] b_{1-2} 170-178°,[1323] m. 23-24°,[1451] n_D^{20} 1.5580,[21] n_D^{25} 1.5600,[1451] d_4^{20} 1.2539.[21] IR.[1348] ^{31}P -83.7,[606, 715] -84.2.[715]

$[EtO(RO)P(S)S]_2$. XXIII. Liq. R, n_D^{20}, d_4^{20}: Pr, 1.5465, 1.1998; Ph, 1.6108, 1.3000; $C_{10}H_{21}$, 1.5206, 1.0670.[560]

$[(i\text{-}PrO)_2P(S)S]_2$. XXIII.[462] XXXIII.[292, 1323, 1451] Solid. m. 90-91° (from EtOH),[462] 92-93°,[292] IR.[1348] ^{31}P -81.15.[715]

$[(Me_2CHOCH_2CHMeO)_2P(S)S]_2$. ^{31}P -81.8.[715]

$[(PrO)_2P(S)S]_2$. XXXIII.[1323, 1451] Liq. n_D^{20} 1.5468, d_4^{20} 1.1652.[1451]

$[(CH_2{:}CHCH_2O)_2P(S)S]_2$. XXIII. Liq. n_D^{20} 1.5760, d^{20} 1.2385.[1446]

$[(HC{\equiv}CCH_2O)_2P(S)S]_2$. XXIII. Liq. n_D^{20} 1.6062, d^{20} 1.3429.[1446]

$[(i\text{-}BuO)_2P(S)S]_2$. XXIII.[462] XXXIII.[1323] Solid. m. 90° (from EtOH).[462] Liq. n_D^{20} 1.5242, d^{20} 1.1110.[1323] Also several homologs.[462, 1323]

$[(CH_2{:}CMeCH_2O)_2P(S)S]_2$. XXIII. Liq. n_D^{20} 1.5588, d^{20} 1.1793.[1446]

$[(MeCH{:}CHCH_2O)_2P(S)S]_2$. XXIII. Liq. n_D^{20} 1.5652, d^{20} 1.1702.[1446]

$[(PhO)_2P(S)S]_2$. XXIII. Solid. m. 74.6-75°.[807]

$[(3\text{-}MeC_6H_4O)_2P(S)S]_2$. XXIII. Solid. m. 42-43° (from hexane-petroleum ether).[1437]

$[(4\text{-}MeC_6H_4O)_2P(S)S]_2$. XXIII. Solid. m. 100.5-102° (from benzene-EtOH).[1437]

$[(2,4\text{-}(t\text{-}Am)_2C_6H_3O)_2P(S)S]_2$. XXIII. Solid. m. 157-158°.[469]

$[(PhCH_2O)_2P(S)S]_2$. XXIII. Liq. n_D^{25} 1.6109.[463]

$[(Cholesteryl\text{-}O)_2P(S)S]_2$. XXIII. Solid. m. 203-204°.[191]

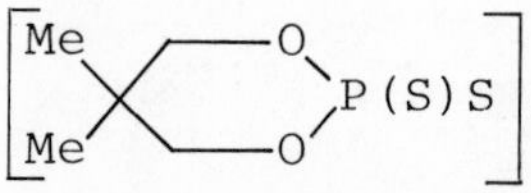

2. XXIII. Solid. m. 133.5-134° (from EtOH).[247]

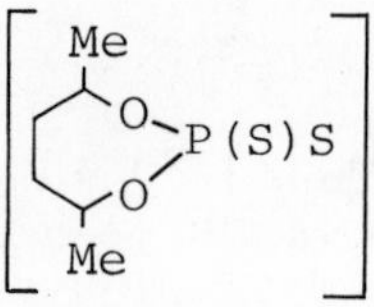

2. XXIII. Solid. m. 103-104°.[1440]

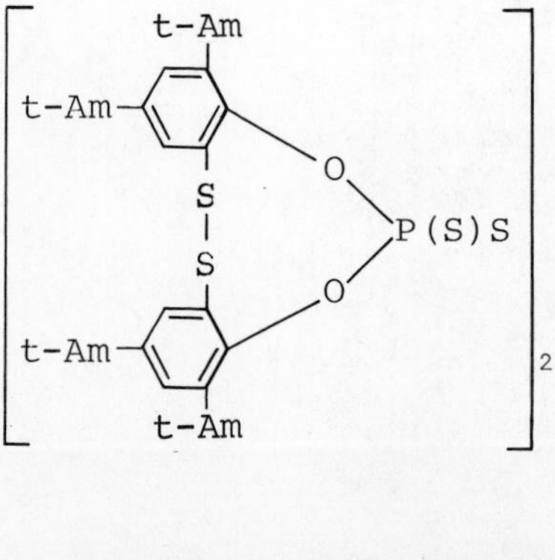

XXIII. Solid. m. 178-180°.[469]

$[MeO(O^-)P(S)S]_2 \cdot 2\ (Et_2NHMe)^+$. IR.[1348] Also several homologs.

$[RO(O^-)P(S)S]_2 \cdot 2\ PhCOCH_2CH_2$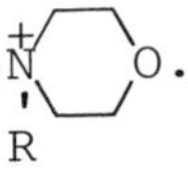O. XXXII. R, properties:

Me, liq., d_4^{50} 1.1644; Et, solid, m. 120-121.5° (from benzene); i-Pr, solid, m. 115-116° (from benzene).[754]

$[i\text{-}PrO(O^-)P(S)S]_2 \cdot 2\ ArCOCH_2CH_2\overset{+}{N}Me_2$ (N-i-Pr). XXXII. Solids. Ar, m.:

Ph, 88-90° (from hot benzene); $4\text{-}ClC_6H_4$, 107-108° (from benzene); $C_{10}H_7$, liq., d_4^{50} 1.2456.[754]

$[(EtO)_2P(S)S]_2S$. VIII. (From $[Br_2P(S)S]_2S$ plus EtOH.)[540]

XXIII.[464] XXXIII. (From $(EtO)_2P(S)SH$ plus NSCl.)

[22, 23] Solid. m. 73.5-75° (from EtOH).[22, 540] ^{31}P -83.0.[715]

$[(RO)_2P(S)S]_2S$. XXXIII. Liq. R, n_D^{20}: Pr, 1.5604; i-Bu, 1.5403; Bu, 1.5454; i-Am, 1.5388.[23]

$[(EtO)_2P(S)S]_2S_2$. ^{31}P -83.4.[715]

$EtOP(S)[SP(S)(OEt)_2]_2$. IR.[673]

F.5.4. Phosphorothiolate and -thionate Anhydrides

F.5.4.1. RO(RS)P(O)OCOR

MeO(MeS)P(O)OCOOEt. XVII. Liq. $b_{0.05}$ 103°, n_D^{20} 1.4579.[1307]

$2,4,5\text{-}Cl_3C_6H_2O(MeS)P(O)OCOO\text{-}i\text{-}Pr$. XVII.[1202a]

$(EtO)_2P(O)SCN$. VIII. Liq. b_{13} 40°.[1074]

F.5.4.2. $(RO)_2P(S)OCOR$

$(EtO)_2P(S)OCONMe_2$. XV. (From $(EtO)_2P(O)Cl$ plus C_5H_5N, Me_2NCOCl, and S.) Liq. b_2 128°.[1127]

$(EtO)_2P(S)OCOOMe$. XVII. Liq. b_4 115.5-117°, n_D^{20} 1.4603, d^{20} 1.2181.[494]

$(MeO)_2P(S)OCOOEt$. XVII. Liq. $b_{0.03}$ 70-73°, n_D^{20} 1.4519.[1307]

$(EtO)_2P(S)OCOOEt$. XVII. Liq. $b_{2.5}$ 114-116°, n_D^{20} 1.4560, d^{20} 1.1851.[494]

$(MeO)_2P(S)OCOMe$. XVII. Liq. $b_{0.05}$ 64°, n_D^{20} 1.459.[1307]

$(EtO)_2P(S)OCOMe$. XV.[49, 495] XVII.[494] XXII.[799] Liq. $b_{0.9}$

65-66°,[799] b_2 81-82°,[495] b_3 93-96°,[49] n_D^{20} 1.4545, d^{20} 1.1596.[495]

(i-PrO)$_2$P(S)OCOMe. XXII. Liq. $b_{0.8}$ 65-66°, n_D^{20} 1.4480.[799]

(BuO)$_2$P(S)OCOMe. XXII. Liq. $b_{0.2}$ 79-80°, n_D^{20} 1.4548.[799]

(4-Cl-1-$C_{10}H_6O$)$_2$P(S)OCOMe. VIII. Solid. m. 160-161°.[834]

(2-$C_{10}H_7O$)$_2$P(S)OCOMe. VIII. Solid. m. 150-152°.[833]

(MeO)$_2$P(S)OCOCH$_2$Cl. XVII. Liq. $b_{0.05}$ 75-76°.[1307]

(MeO)$_2$P(S)OCOCHCl$_2$. VIII. Liq. $b_{1.3}$ 121-122°, n_D^{28} 1.4891.[884]

(EtO)$_2$P(S)OCOEt. XV, XVII.[494] Liq. b_4 94-94.5°, n_D^{20} 1.4563, d^{20} 1.1337.[494]

(EtO)$_2$P(S)OCOCCl$_2$Me. VIII. Liq. $b_{0.4}$ 100-108°, n_D^{30} 1.4643.[887]

(4-Cl-1-$C_{10}H_6O$)$_2$P(S)OCOPr. VIII. Solid. m. 158-159°.[834]

(4-Cl-1-$C_{10}H_6O$)$_2$P(S)OCOPh. VIII. Solid. m. 157-158°.[834]

(2-$C_{10}H_7O$)$_2$P(S)OCOC$_6$H$_4$NO$_2$-3. VIII. Solid. m. 150° (dec.).[833]

($C_{10}H_7O$)$_2$P(S)OCOCH:CHCOOP(S)(O$C_{10}H_7$)$_2$. VIII. Solids. 1-Naphthyl, m. 76°; 2-naphthyl, m. 129°.[835]

(EtO)$_2$P(S)OCOC$_5$H$_{11}$. XV. Liq. b_3 109-113°, n_D^{20} 1.4508.[49]

($C_{10}H_7O$)$_2$P(S)OCO(CH$_2$)$_7$COOP(S)(O$C_{10}H_7$)$_2$. VIII. Solids. 1-Naphthyl, m. 114°; 2-naphthyl, m. 117°.[835]

($C_{10}H_7O$)$_2$P(S)OCOCH:CHPh. VIII. Solids. 1-Naphthyl, m. 80°; 2-naphthyl, m. 120°.[835]

(EtO)$_2$P(S)OCOPh. XVII. Liq. $b_{0.009}$ 60-62°, n_D^{20} 1.5182, d_4^{20} 1.1930.[730]

(EtO)$_2$P(S)OCOC$_6$H$_4$NO$_2$-4. XVII. Solid. m. 59-60°, IR, ^{1}H-NMR.[730]

(EtOP(S)F)$_2$O. XXXIII. (From EtOP(S)ClF plus (EtO)$_2$P(S)F.) Liq. $b_{0.001}$ 80-85°, n_D^{20} 1.478, d_{20}^{20} 1.388.[1270] Mol. ref. 390.17.[1271]

(EtO)$_2$P(S)OP(S)(OEt)$_2$. XV.[1127] XXXIII. (From$_2$ (EtO)$_2$POSK plus SOCl$_2$.)[856] (From (EtO)$_2$P(S)Cl plus H$_2$O and C$_5$H$_5$N.)[1318] Liq. $b_{0.2}$ 110-113°,[1318] b_3 140-141°,[1127] n_D^{25} 1.4753, d_4^{25} 1.189.[1318] IR.[528, 673] ^{1}H-NMR.[74, 403, 1395]

^{31}P 60.1 (P_4O_6 ext.).[403] "Sulfotepp."

$(i\text{-}PrO)_2P(S)OP(S)(O\text{-}i\text{-}Pr)_2$. XXXIII. Liq. n_D^{25} 1.4620, d_4^{25} 1.093.[1318]

$(PrO)_2P(S)OP(S)(OPr)_2$. XXXIII. Liq. $b_{0.01}$ 104°, n_D^{25} 1.4712, d_{20}^{20} 1.12.[1318] ^1H-NMR.[74, 524, 526] "Aspon."

$(BuO)_2P(S)OP(S)(OBu)_2$. XXXIII. Liq. n_D^{25} 1.4690, d_4^{25} 1.068.[1318]

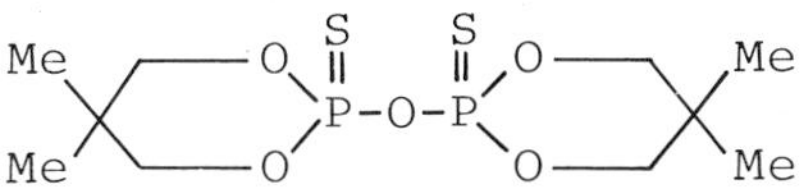

XXXIII. Solid. m. 233° (from benzene).[247]

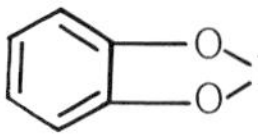

$P(S)OP(S)(OR)_2$. VIII. Liq. R, n_D^{20}, d^{20}: Et, 1.5368, 1.3073; Pr, 1.5311, 1.2514.[1309]

$(RO)_2P(S)OSO_2OP(S)(OR)_2$. XXXIII. (From $(RO)_2POSK$ plus SO_2Cl_2.) Liq. R, b./mm: Me, 120°/25; Et, 112.5-117.5°/24.[856] (Note: These b. p.'s appear to be too low.)

F.5.5. Phosphorothioate Anhydrosulfides

F.5.5.1. $(RO)_2P(O)SC(S)OR'$

$(EtO)_2P(O)SC(S)OR$. IX. Liq.[685]

R	b./mm	n_D^{10}	d^{10}
Me	51°/5	1.4915	1.1792
Et	59-60°/4	1.4715	1.1043
Pr	80°/4	1.4860	1.1072

F.5.5.2. $(RO)_2P(O)SP(O)(OR)_2$; probably $(RO)_2P(O)OP(S)(OR)_2$[789]

$(MeO)_2P(O)SP(O)(OMe)_2$. IX. Liq. $b_{0.02}$ 68-69°, b_2 128-130°, n_D^{25} 1.4519, d^{25} 1.3360.[327]

$(MeO)_2P(O)SP(O)(OEt)_2$. IX. Liq. $b_{0.5}$ 116°, n_D^{25} 1.4499, d^{25} 1.2536.[327]

$(EtO)_2P(O)SP(O)(OEt)_2$. Formed at ice temperature from $(EtO)_2P(O)SCl$ and $(Et)_3P$. Detected by TLC. Unstable at room temperature, forming $(EtO)_2P(O)OP(S)(OEt)_2$.[789]

Liq. IX.[327] XIX.[944] XXXIII.[793] $b_{0.04-0.05}$ 82-84°,[327,793] b_2 139°,[944] n_D^{18} 1.4500.[944]

$(EtO)_2P(O)SP(O)(OPr)_2$. IX. Liq. $b_{0.001}$ 103.5°, n_D^{25} 1.4429, d^{25} 1.1347.[327]

$(RO)_2P(O)SP(O)(OR)_2$. IX. Liq.[327]

R	$b_{0.01}$	n_D^{25}	d^{25}
i-Pr	82-84°	1.4370	1.0885
Pr	94-95	1.4363	1.1075
i-Bu	96-98	1.4463	1.0488
Bu	112-114	1.4517	1.0374

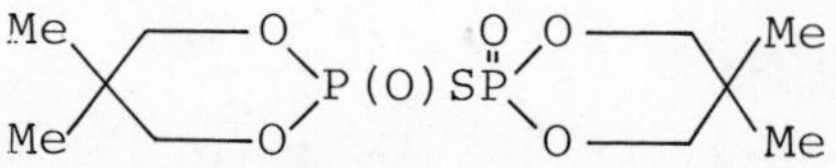 Isolated from 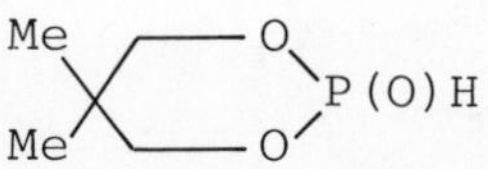 plus $SOCl_2$. Solid. m. 186-187° (from benzene-$CHCl_3$, 1:1).[1456]

F.5.6. Miscellaneous Phosphorothioate Derivatives

$(EtO)_2P(O)SCOC_6H_4NO_2$-4. XVII. Solid. m. 56-57° (from petroleum ether or EtOH).[937]

$(EtO)_2P(O)SCOOMe$. XVII. Liq. $b_{0.05}$ 86°, n_D^{23} 1.4560.[141]

F.5.6.1. $(RO)_2P(O)SNR'_2$

$(EtO)_2P(O)S(O)NMe_2$. XIX. Liq. $b_{0.05}$ 59-63°, n_D^{20} 1.4278.[679]

$(MeO)_2P(O)SNMeCOOEt$. XVII. Liq. $b_{0.1}$ 137-140°.[692]

$(EtO)_2P(O)SNEt_2$. XIX.[784] XXVI.[779] Liq. $b_{0.01}$ 54°,[784] $b_{0.12}$ 52-53°,[779] n_D^{22} 1.4605,[779] 1.4657,[784] d^{22} 1.0085.[779]

$(PrO)_2P(O)SNEt_2$. XIX. Liq. $b_{0.03}$ 70-71°, n_D^{22} 1.4640.[784]

$(BuO)_2P(O)SNEt_2$. XIX. Liq. n_D^{22} 1.4629.[784]

$(MeO)_2P(O)SC(COOMe)NHCOOMe$. XXII. Solid. m. 39-42°.[498]

$(i\text{-}PrO)_2P(O)SC(COOMe)NHCOOMe$. XXII. Solid. m. 56-59.5°.[498]

$(RO)_2P(O)SNHSO_2Ar$. XXVI. Solids.[34]

Ar	R = Et m.	R = Pr m.
Ph	97°	87°
4-MeC_6H_4	96-97	87-88
4-$MeOC_6H_4$	84-85	85-86
4-BrC_6H_4	105-106	95-96

4-ClC_6H_4	107-108	96-97
4-FC_6H_4	124-125	103-104
4-IC_6H_4	118-119	115-116

$(EtO)_2P(O)SN(OSOPh)C_6H_4Cl$-4. XXVI. Liq. n_D^{25} 1.5430.[976]
$(EtO)_2P(O)SN(OSOPh)C_6H_4Me$-4. XXVI. Liq. n_D^{25} 1.5369.[976]
$(EtO)_2P(S)ONMeCOMe$. XVII. Liq. $b_{0.25}$ 87-95°.[1416]
$(EtO)_2P(S)ON(COMe)C_6H_4NO_2$-4. XVII. Solid. m. >150° (dec.).[1416]

F.5.6.2. $(RO)_2P(O)SSR'$

$(EtO)_2P(O)SSEt$. XX.[492] XXVI.[975] Liq. $b_{3.5}$ 82-85°, n_D^{20} 1.4721, d^{20} 1.1224.[492] Also several analogs.[975]
$(EtO)_2P(O)SSCH_2SEt$. XXVI. Liq. n_D^{25} 1.5186.[975]
$(MeO)_2P(O)SSCH_2COOEt$. XXVI. Liq. n_D^{25} 1.4960.[975]
$(EtO)_2P(O)SSCH_2COOEt$. XXVI. Liq. n_D^{25} 1.4744.[975]
$(EtO)_2P(O)SSBu$. XX. Liq. b_4 132-134°, n_D^{20} 1.4885, d^{20} 1.1248.[492]
$(EtO)_2P(O)SSPh$. XX.[1197] XXVI.[975] Liq. $b_{0.001}$ 116°,[1197] n_D^{25} 1.5126.[975]
$(EtO)_2P(O)SSC_6H_4Cl$-4. XX.[1197] XXVI.[975] Liq. $b_{0.001}$ 124°,[1197] n_D^{25} 1.4863.[975]
$(MeO)_2P(O)SSC_6H_4NO_2$-2. XX. Solid. m. 72° (from CS_2).[343]
$(RO)_2P(O)SSC_6H_3$-5-Me-2-NO_2. XX. Solids. R, m.: Me, 93° (from CS_2); Et, 53° (from ether); i-Pr, 44° (from ether or petroleum ether).[343]
$(MeO)_2P(O)SSeC_6H_4NO_2$-2. XX. Solid. m. 79°.[341]

F.5.6.3. $(RO)_2P(O)SON{:}CR'_2$

$EtO(C_8H_{17}O)P(O)SON{:}CMePh$. XXVI. Liq. n_D^{25} 1.5066.[974]
$(C_8H_{17}O)_2P(O)SON{:}CMePh$. XXVI. Liq. n_D^{25} 1.4874.[974]
$(EtO)_2P(S)ON{:}CMe_2$. VIII. Liq. $b_{0.005}$ 57-60°, m. -20°, n_D^{20} 1.4770.[205]
$(MeO)_2P(O)SON{:}CMePh$. XXVI. Liq. n_D^{25} 1.5248.[974]
$(EtO)_2P(O)SON{:}CMeC_6H_4Cl$-4. XXVI. Liq. n_D^{25} 1.5322.[974]
$(MeO)_2P(S)ON{:}CMeC_6H_4I$-4. VIII. Solid. m. 63-65°.[465]
$(MeO)_2P(S)ON{:}CMeC_6H_4NO_2$-4. VIII. Solid. m. 70-71° (from petroleum ether).[465] Also several analogs.[465]

$(EtO)_2P(S)ON{:}CMeC_6H_4NO_2$-4. VIII. Solid. m. 47-48°.[465]

$(EtO)_2P(S)ON{:}CH(Ph)CN$. VIII. Liq. $b_{0.01}$ 102°, n_D^{22} 1.5395, d^{20} 1.176.[306] Also 11 analogs.[306]

$(ClCH_2CH_2O)_2P(S)ON{:}CH(C_6H_4Cl\text{-}2)CN$. VIII. Liq. n_D^{21} 1.5641.[306] Also 12 analogs.[306]

$(RO)_2P(S)ON{:}C(CN)$ 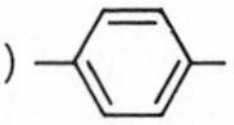 $C(CN){:}NOP(S)(OR)_2$. VIII. Solids. R, m.: Me, 117-119°; Et, 78°.[651]

$(MeO)_2P(O)S$ 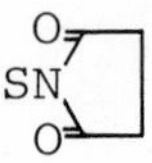 XVII. Solid. m. 90-93° (from benzene).[692]

$(RO)_2P(S)O$ 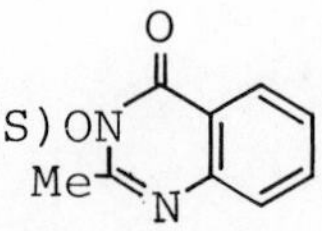 VIII. Solids. R, m. Me, 97-100°; Et, 76-79° (from petroleum ether).[863]

$(EtO)_2P(O)SON{:}CH$ 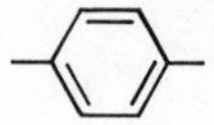 $OSP(O)(OEt)_2$. Liq. n_D^{25} 1.4925.[974]

F.5.6.4. $(RO)_2P(O)S(O)_nOR'$

$(EtO)_2P(O)SOCH_2CH_2SEt$. XXXIII. (From $(EtO)_2P(O)SCl$ plus $EtSCH_2CH_2OH$.) Liq. n_D^{25} 1.4820.[977] Also 32 analogs.[977]

$(EtO)_2P(O)SOCH(CH_2Br)_2$. XXXIII. Liq. n_D^{25} 1.5026.[977]

$(EtO)_2P(O)SO$ 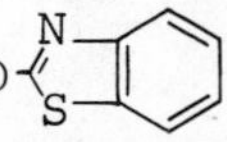 XXXIII. Liq. n_D^{25} 1.5418.[977]

$(RO)_2P(O)S(O)OCH_2CH_2OEt$ and $(RO)_2P(S)OS(O)OCH_2CH_2OEt$. XIX. Liq.[1224]

R	P(O) comp'd. b./mm	P(S) comp'd. b./mm
Me	94-97°/9	119-122°/12
Et	113-115°/9	125-130°/6
i-Pr	153-155°/12	107-111°/18
Pr	142-144°/5	117-121°/10
i-Bu	109-111°/2.5	97-101°/5
Bu	154-157°/10	146-149°/6

F.5.6.5. $(RO)_2P(O)SSP(O)(OR)_2$

$(RO)_2P(O)SSP(O)(OR)_2$. XIX. (From $(RO)_2P(O)H$ plus S_2Cl_2.) Liq.[265]

R	n_D^{20}	d_4^{20}
Me	1.4968	1.4084
Et	1.4870	1.2550
i-Pr	1.4761	1.1462
Pr	1.4808	1.1591
i-Bu	1.4725	1.0932
Bu	1.4770	1.0988

F.5.6.6. $(RO)_2P(S)OSi(OR')_3$

$(RO)_2P(S)OSi(OSiMe_3)_3$. XVII. Liq.[320]

R	$b_{0.001}$	n_D^{20}	d^{20}
Et	82-84°	1.4205	0.990
i-Pr	84-86	1.4200	0.973
Pr	91-94	1.4240	0.982

F.5.7. Phosphorotrithioates

F.5.7.1. $(RS)_2POS^-$ (acids and salts)

$(EtS)_2POSH$. II. Lead salt. Glass[961]

$ROP(S)(SK)_2$. From ROH plus P_2S_5 and Et_3N; boil; treat with KOH. R = Me, i-Pr, Pr, i-Bu, Bu. Anal. for P and S.[561]

$Me_2NCH_2CH_2OP(S)(SK)_2$. V. Solid. m. 180° (dec.)(from MeOH-acetone).[389]

$Et_2NCH_2CH_2OP(S)(SK)_2$. V. Solid. m. 195° (dec.).[389]

$RCH_2CH_2OP(S)(SK)_2$. V. Solids. R, m. (dec.): 2-pyridyl, 165°; 2-piperidinyl, 190°; 2-morpholinyl, 175°.[387]

$Et_2NCH_2CH_2CH_2OP(S)(SK)_2$. V. Solid. m. 165° (dec.).[389]

$Me_2NCH_2CH_2CH_2CH_2OP(S)(SK)_2$. V. Solid. m. 128° (dec.).[389]

$(MeS)_2POS^-\ NMe_4^+$. XXXII. (From $(MeS)_3P(O)$ plus Me_3N.) Solid. m. 132°.[1308]

F.5.7.2. $(RS)_3P(O)$

$MeSP(S)_2$. XII. (From MeSH plus P_2S_5.) Solid. m. 112°.[1209]

$(MeS)_3P(O)$. XVII. Liq. n_D^{20} 1.6000, d_4^{20} 1.2747.[387] IR,[426, 428] ^1H-NMR.[722, 734]

$(EtS_3P(O)$. IX.[1, 159, 675, 1366] XVII.[387] Liq. $b_{0.001}$ 107°,[1] b_3 124.5°,[675] b_{18} 175°,[159] n_D^{20} 1.5700,[1366] n_D^{25} 1.5664,[675] d^{20} 1.2146.[1366] IR.[673, 674] ^{31}P -61.3 ppm.[850] Magn.

susceptibility.[1365, 1366] Bond magn. rotation.[580]

$(PrS)_3P(O)$. IX. Liq. $b_{0.9}$ 131.5°, n_D^{20} 1.5497, d^{20} 1.1034.[1366] Magn. susceptibility.[1365]

$(BuS)_3P(O)$. IX.[718, 1366] Liq. $b_{0.3}$ 150°,[718] b_1 166.5°,[1366] n_D^{20} 1.5345, d^{20} 1.0552,[1366] d 1.532.[718] ^{1}H-NMR.[526] Bond magn. rotation.[580] Magn. susceptibility.[1365, 1366] "Def."

$(C_{12}H_{25}S)_3P(O)$. IX.[99, 1068] Solid. m. 26-27° (from acetone)[99] L. HgI_2. m. 162-164°.[99]

$(PhS)_3P(O)$. XIV. Solid. m. 115° (from EtOH).[773]

$(C_6F_5S)_3P(O)$. IX. (From $P(O)Cl_3$ plus $(C_6F_5S)_2Pb$.) Solid. m. 145-146° (vacuum sublimed).[928]

$(4\text{-}MeC_6H_4S)_3P(O)$. (Isolated in 9% yield from $4\text{-}MeC_6H_4SO_2Cl$ plus PH_3 in C_5H_5N). Solid. m. 138-140°.[135a] ^{31}P Enhancement at 12,500 gauss, +3.[241]

$(PhCH_2S)_3P(O)$. ^{31}P 50.8 ± 0.8 (P_4O_6 ext.).[241] ^{31}P Enhancement at 12,500 gauss, +2.[241]

F.5.7.3. $(RS)_2P(O)SR'$

$(MeS)_2P(O)SCHClCHCl_2$. VIII. Liq. $b_{0.01}$ 129-130°, n_D^{25} 1.6205.[3, 4] Also 20 analogs.[3, 4]

$(MeS)_2P(O)CH_2CCl_3$. VIII. Liq. n_D^{25} 1.6038.[3, 4]

$(MeS)_2P(O)SCH{:}CCl_2$. VIII. Liq. $b_{0.04}$ 120-123°, n_D^{26} 1.6303.[3, 4]

$(4\text{-}ClC_6H_4S)_2P(O)CH{:}CCl_2$. IX. Liq. n_D^{26} 1.6692.[3, 4]

$(PrS)_2P(O)SCH_2CH_2CMe_2$. XVII. Liq. $b_{0.02}$ 77°, n_D^{20} 1.5482, d_4^{20} 1.1050.[386]

$(MeS)_2P(O)SCH_2CH_2NEt_2$. L.MeBr. Solid. m. 112°. L.$(COOH)_2$. m. 124°.[167]

$(PrS)_2P(O)SCH_2CH_2N$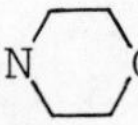O. XVII. Liq. n_D^{20} 1.5410, d_4^{20} 1.1242.[386]

$(EtS)_2P(O)SCH_2SEt$. XXXIII. Liq. $b_{0.01}$ 83°.[1160]

$(MeS)_2P(O)SCH_2CH_2SEt$. XVII. Liq. $b_{0.01}$ 92°.[1160]

$(EtS)_2P(O)SCH_2CH_2SEt$. XXXIII. (From $(EtS)_2P(S)OEt$ plus

EtSK and $EtSCH_2CH_2Cl$.) Liq. $b_{0.01}$ 108°.[1160]

$(MeS)_2P(O)SCH_2CONHMe$. XVII. Solid. m. 80°.[1308]

$(MeS)_2P(O)SCH_2CONHC_6H_3Cl_2$-2,4. XVII. Solid. m. 90°.[155]

$(MeS)_2P(O)SCH_2COOEt$. IX.[1015] XVII.[1308] Liq. $b_{0.01}$ 112-115°,[1308] $b_{0.02}$ 135°,[1015] n_D^{20} 1.5705,[1308] n_D^{23} 1.5685.[1015]

$(RS)_2P(O)SCH_2COOEt$. IX. Liq.[1015]

R	b./mm	n_D^t
Et	143°/0.15	$1.5525^{20°}$
Pr	146°/0.01	$1.5395^{25°}$
Bu	150°/0.03	$1.5285^{26°}$

$(MeS)_2P(O)SCHMeCOOEt$. XVII. Liq. $b_{0.01}$ (0.07 ?) 116°, n_D^{20} 1.5564.[155, 1308]

$(MeS)_2P(O)SCH(COOEt)_2$. XVII. Liq. $b_{0.01}$ 148-150°.[155, 1308]

$(MeS)_2P(O)SCH_2CH_2COOEt$. XVII. Liq. $b_{0.05}$ 130-132°, n_D^{20} 1.5577.[1308]

$(MeS)_2P(O)SCH_2C_6H_3Cl_2$-3,4. XVII. Liq. $b_{0.05}$ 179°.[1308]

$MeS(BrCH_2CH_2S)P(O)SPh$. VIII. Liq. n_D^{23} 1.5798, 1.6023.[454] Also 24 analogs.[454]

$EtS(BrCH_2CH_2S)P(O)SPh$. VIII. Liq. n_D^{23} 1.5880.[454]

F.5.7.4. $(RS)_2P(S)OR'$

$(MeS)_2P(S)OMe$. XVII.[561] From P_4S_7 plus MeOH and Et_3N (mole ratio 5:1:5).[852] Liq. b_1 76-77°,[852] 79-80°,[561] n_D^{20} 1.5948,[561] 1.5953,[852] d_4^{20} 1.2827,[852] 1.2840.[561]

$(RS)_2P(S)OMe$. XV. Liq.[812]

R	b./mm	n_D^{20}	d^{20}
Et	61-62°/0.045	1.5743	1.1916
i-Pr	67-68°/0.04	1.5435	1.1150
Pr	76.5-77°/0.04	1.5550	1.1391
i-Bu	84-84.5°/0.045	1.5373	1.0890
Bu	87.5-89°/0.06	1.5417	1.0939

$(H_2NCOCH_2S)_2P(S)OMe$. IX. Solid. m. 133-135°.[706]

$(i\text{-}PrNHCOCH_2S)_2P(S)OMe$. IX. Solid. m. 76-78°.[706]

$(EtS)_2P(S)OEt$. IX.[961] From EtOH plus Et_3N and P_4S_7 in the mole ratio of 1:5:5).[852] Liq. b_1 98-100°,[852] b_{20} 155°,

[961] n_D^{20} 1.5570, d_4^{20} 1.1515,[852] L.2 $HgCl_2$, m. 81°; L.2 HgI_2, m. 112°.[961]

$(RNHCOCH_2S)_2P(S)OEt$. IX. Solids. R, m.: H, 147-150°; Me, 108-110°; Et, 93-94°; i-Pr, 70-72°; Pr, 66-68°; Bu, 53-55°.[706] Also homologs with i-PrO, PrO, and BuO substituted for EtO.[706]

$(PrS)_2P(S)OCH_2CH_2NMe_2$. XVII. Liq. $b_{0.02}$ 70°, n_D^{20} 1.5390, d_4^{20} 1.0834.[386]

$(PrS)_2P(S)OCH_2CH_2NEt_2$. XVII. Liq. $b_{0.02}$ 90°, n_D^{20} 1.5350, d_4^{20} 1.0610.[386]

$(MeS)_2P(S)OCH_2CH_2NMeR_2^+ I^-$. XXXIII. (From L plus MeI.) R_2, m.: Me_2, 39-40°; piperidino, 129-130°; morpholino, 106-108°.[388]

$(PrS)_2P(S)OCH_2CH_2NMe_2Pr^+ I^-$. XXXIII. (From L plus MeI.) Solid. m. 75-76°.[388]

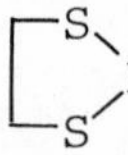$P(S)OCH_2CH_2NR_2$. XVII. Solids. R_2, m.: Me_2, 50-52°; Et_2, 66-68°; morpholino, 55-57°.[386]

$(i\text{-}PrS)_2P(S)O\text{-}i\text{-}Pr$. XVII.[561] From i-PrOH plus Et_3N and P_4S_7).[852] Liq. b_1 101-103°, n_D^{20} 1.5312-1.5314, d_4^{20} 1.0685-1.0693.[561, 852]

$(PrS)_2P(S)OPr$. XVII. Liq. b_1 119-120°, n_D^{20} 1.5380, d_4^{20} 1.0887.[561]

$(PrS)_2P(S)OCH_2CH_2CH_2NEt_2$. XVII. Liq. $b_{0.02}$ 95°, n_D^{20} 1.5284, d_4^{20} 1.0518.[386]

$(MeS)_2P(S)OCH_2CH_2CH_2NEt_2Me^+ I^-$. XXXIII. (From L. plus MeI.) Solid. m. 130-131°.[388]

$(RS)_2P(S)OCH_2CH:CH_2$. XVII. Liq.[740]

R	$b_{0.001}$	n_D^{20}	d_4^{20}	
Me	42°	1.5938	1.2348	
Et	51	1.5682	1.1625	
Pr	56	1.5516	1.1156	IR
$CH_2CH:CH_2$	58	1.5838	1.1533	
$CH{\equiv}CCH_2$	69	1.6034	1.2316	
Bu	58	1.5422	1.0842	

(i-BuS)$_2$P(S)O-i-Bu. XVII. Liq. b_1 104-106°, n_D^{20} 1.5270, d_4^{20} 1.0508.[561]

(EtS)$_2$P(S)OBu. XV. Liq. $b_{0.05}$ 100-101°, n_D^{20} 1.5472, d_4^{20} 1.1098.[812]

(BuS)$_2$P(S)OBu. XVII.[561] (From BuOH plus Et_3N and P_4S_7 in the mole ratio of 1:5:5).[852] Liq. b_1 153-154°, n_D^{20} 1.5275, d_4^{20} 1.0428.[852]

(MeS)$_2$P(S)O(CH$_2$)$_4$NMe$_3^+$ I^-. XXXIII. (From L. plus MeI.) Solid. m. 50-51°.[388]

(BuS)$_2$P(S)OPh. XXXIII. (From (BuS)$_3$P(S) plus PhOH.) Liq. b_1 180-183°, n_D^{20} 1.5814, d^{20} 1.1367.[101]

(EtS)$_2$P(S)OC$_6$H$_3$-2-Cl-5-t-Bu. VIII. Liq. n_D^{24} 1.5887.[179a]

MeO(MeS)P(S)SCH$_2$CH$_2$SEt. IX. Liq. $b_{0.01}$ 95°.[282]

EtO(EtS)P(S)SPh. VIII. Solid. m. 106°.[1163]

EtO(EtS)P(S)OC$_6$H$_4$Cl-4. VIII. Solid. m. 119°.[1163]

F.5.8. Phosphorotetrathioates

F.5.8.1. (RS)$_2$P(S)SH (acids and salts)

MeS(RS)P(S)SH. IR for R = Me, Et, i-Pr, Pr.[13]

(MeS)$_2$P(S)$^-$ NMe $^+$. XXXII. (From (MeS)$_3$P(S) plus Me$_3$N.) Solid. m. 171°.[1308]

F.5.8.2. (RS)$_3$P(S)

(MeS)$_3$P(S). XII.[100, 102, 1054, 1210] Solid. m. 29-30°.[100] $b_{0.05}$ 96-104°,[1054] $b_{0.2}$ 126-130°.[1210] IR.[426, 764] ^{1}H-NMR.[734] ^{31}P -98.[715]

(EtS)$_3$P(S). VIII.[675, 961, 1366] XII.[100, 102, 123, 961, 1054, 1210] Liq. $b_{0.15}$ 110°,[1210] $b_{0.6}$ 117°,[1366] $b_{1.5}$ 124-127°,[100, 102] n_D^{20} 1.6180,[102, 1366] 1.6201,[1078] Mol. ref. 399.23,[1078] d_4^{20} 1.1893.[102] IR.[673] ^{31}P -92.9,[488, 715, 850] -91.7.[715] Bond magn. rotation.[580]

(RNHCOCH$_2$S)$_3$P(S). VIII. Solids. R, m.: H, 170-172°; Me, 169-171°; i-Pr, 151-153°; Pr, 114-116°.[706]

(i-PrS)$_3$P(S). XII. Liq. $b_{0.3}$ 123-125°.[1210]

(PrS)$_3$P(S). XII.[100, 102, 680, 1210, 1366] Liq. $b_{0.5}$ 131-132°,[1366] $b_{1.8-2}$ 140-142°,[100, 102] n_D^{20} 1.5870-1.5885,

[100, 102, 1366] d_4^{20} 1.1160.[102] ^{31}P -93.1, -92.5.[715]

$(t\text{-}BuS)_3P(S)$. XII. Liq. $b_{0.1}$ 115-120°.[1054]

$(BuS)_3P(S)$. XII.[100, 102, 680, 1054, 1366] Liq. $b_{0.05}$ 150-160°,[1054] $b_{2.5}$ 164-166°,[100] n_D^{20} 1.5675-1.5772,[100, 102] d_4^{20} 1.0765.[102] ^{31}P -92.6.[715] Bond magn. rotation.[580] Also homologs to $C_{10}H_{21}$.[101, 102]

$(C_{12}H_{25}S)_3P(S)$. VIII. Liq. m. 15°, n_D^{20} 1.5051. L. HgI_2. Solid, m. 162-163°.[99]

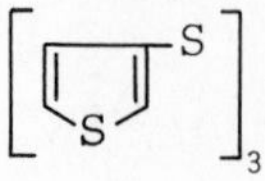 P(S). VIII. (From $P(S)Cl_3$ plus 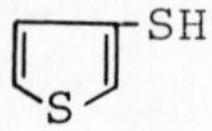 .) Solid. m. 76.7-78°.[130]

$(PhS)_3P(S)$. VIII. IX.[733] XII. (Solution and reaction of $(RS)_3P(S)$ with P_2S_5 followed by ^{31}P-NMR.)[680] Solid. m. 86-88°,[680, 776] ^{31}P 20.5 ± 2.4 (P_4O_6 ext.),[240, 241] -91.1, -93.[715, 817]

$(4\text{-}MeC_6H_4S)_3P(S)$. VIII.[73] XII.[1046] Solid. m. 119-120°,[1046] m. 121-122° (from EtOH).[73] ^{31}P Enhancement at 12,500 gauss, +3.[241]

$(PhCH_2S)_3P(S)$. XII. Liq. m. -13° (from EtOH).[1046]

(Bornyl-S)$_3$P(S). XII. (From α-pinene plus P_2S_5 and H_2S.) Solid. m. 237.5-239.5°.[1400]

F.5.8.3. $(RS)_2P(S)SR'$

MeS(RS)P(S)SR'. XII. (From $MeSP(S)_2$ plus RSH and olefin.)[1210]

R	R'	b./mm[1210]	
Me	Et	135-139°/1.0	
Me	i-Pr	125-130°/0.8	
Et	i-Pr	105-115°/0.2	IR[764]
Et	Et	109-129°/0.5	IR[764]
i-Pr	i-Pr	130-135°/0.6	IR[764]

$(MeS)_2P(S)SCH_2CH_2NEt_2Me^+ Br^-$. XXXII. (From L. plus MeBr.) Solid. m. 113°.[167]

$(MeS)_2P(S)SCH_2CH_2NEt_2$. XXXII. Oxalate salt. Solid. m. 118°.[167]

$(MeS)_2P(S)SCH_2C_6H_2Cl_3$-2,4,5. XXXII. Solid. m. 55°. L. $(COOH)_2$, m. 118°. L. MeBr, m. 113°.[1308]

$MeS(PrS)P(S)SCH_2(CH_2)_6Me)$. XII. (From $MeSP(S)_2$ plus PrSH and n-octene.) Liq. $b_{0.2}$ 120-128°.[1209]

F.6. Phosphoroselenoates and -telluroates

F.6.1. $(RO)_2PSeSeH$

$(RO)_2PSeSeK$. V. Solids.[566]

R	m. (from ligroin-EtOH)		
Et	132-136°	IR	^{1}H-NMR
Pr	92-95		
Bu	92-93	IR	^{1}H-NMR
Am	132-133 (monohydrate)		
cyclo-C_6H_{11}	202-208 (dec.)	IR	^{1}H-NMR

F.6.2. $(RO)_2PSSeH$

$(EtO)_2PSSeK$. I. Solid. m. 156° (dec).[1433]

$(EtO)_2PSSeH$. IV. cyclo-$C_6H_{11}NH_2$ salt. Solid. m. 98-99°.[562]

$(PhO)_2PSSeK$. I. Solid. m. 190-191° (from acetone-water).[1431, 1432]

F.6.3. $(RO)_2POSeH$

$(MeO)_2POSeNa$. IV. Needles (from MeOAc).[342]

$(EtO)_2POSeNa$. II.[961] IV.[342] Needles. m. 146°.[961]

$(EtO)_2POSeK$. II. Solid. m. 170° (dec).[164]

$(EtO)_2POSeH$. XXXII. cyclo-$C_6H_{11}NH_2$ salt. m. 82-83°.[791]

$(RO)_2POSeH$. IV. Na and K salts described as needles from various solvents. R = i-Pr, Pr, i-Bu, Bu, i-Am, $MeCH_2$-$CHMeCH_2$.[342]

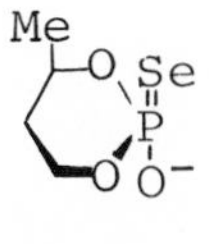

$(C_6H_{11})_2N^+H_2$ IV. (cyclo-$C_6H_{11})_2NH$. IV. Solid. m. 203-207°.[798]

PhOP($\overset{O^-}{\vdots}$ / Se)$OCH_2CH_2NMe_3^+$. XXXII. (From 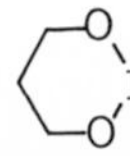 P(Se)OPh plus Me_3N.)

Solid. m. 172°.[870]

F.6.4. $(RO)_2P(Se)SeR'$

	R	n_D^{20}	d_4^{20}
$(RO)_2P(Se)SeMe$. XVII. Liq.[557a]	Me	1.6020	1.7789
	Et	1.5675	1.6164
	i-Pr	1.5440	1.4312
	Pr	1.5470	1.4458
	Bu	1.5320	1.3401
$(RO)_2P(Se)SeEt$.	Me	1.5820	1.6662
	Et	1.5580	1.5056
	i-Pr	1.5360	1.3784
	Pr	1.5385	1.3845
	Bu	1.5255	1.3118
	Ph	1.6142	1.4546
$(RO_2P(Se)SePr$.	Me	1.5650	1.6160
	Et	1.5495	1.4462
	i-Pr	1.5302	1.3392
	Pr	1.5340	1.3460
$(RO)_2P(Se)Se$-i-Bu	Me	1.5520	1.5050
	Et	1.5380	1.3950
	i-Pr	1.5240	1.3120
	Pr	1.5295	1.3210
$(RO)_2P(Se)SeC_8H_{17}$-n	Me	1.5460	1.3660
	Et	1.5230	1.2789

F.6.5. $(RO)_2P(O)SeR'$

$(MeO)_2P(O)SeMe$. XVII.[154] XXXI.[961] Liq. b_{15} 107°, n_D^{20} 1.496.[154] IR.[426, 427]

$(EtO)_2P(O)SeMe$. XXIV. Liq. $b_{2.5}$ 95°, b_4 100-103°.[1190]

$(PhO)_2P(O)SeMe$. XVI. Liq. $b_{0.003}$ 145-146°, n_D^{20} 1.5920, d_4^{20} 1.3923. IR. ^{31}P -13 ppm.[892]

$(EtO)_2P(O)SeEt$. XXIV.[1190] XVI, XXXI.[961] Liq. b_2 102-105°,[1190] b_{20} 140°, d_0^0 1.3593.[961]

$(MeO)_2P(O)SeCH_2CH_2Cl$. XVII. XXXI. Liq. $b_{0.05}$ 105°.[154]

$(EtO)_2P(O)SeCH_2CH_2NH_2\cdot(COOH)_2$. XI. Solid. m. 94-95°.[10]

$(EtO)_2P(O)SeCH_2CH_2NEt_2$. XI. Liq. n_D^{25} 1.4830, d_4^{22} 1.230.[10]

$(EtO)_2P(O)SeCH_2CH(OEt)_2$. XXV. (From $(EtO)_2P(O)SeCl$ plus CH_2:CHOEt and EtOH.) Liq. $b_{0.05}$ 81-81.5°, n_D^{20} 1.4710.[778]

$(EtO)_2P(O)SeCH_2CH_2SMe$. XIX.[1190] XXIV.[1124] Liq. b_3 152°.[1124, 1190]

$(EtO)_2P(O)SeCH_2CH_2SEt$. XIX.[1190] XXIV.[1124] Liq. b_2 153°.

$(EtO)_2P(O)SeCH_2CH_2SC_6H_4Me$-4. XXIV. Liq. b_1 189°.[1124]

$(EtO)_2P(O)SePr$. XXIV. Liq. b_2 105°.[1190]

$(PhO)_2P(O)SeCH_2CH:CH_2$. XXXI. ^1H-NMR. ^{31}P -6.5 ppm.[893]

$(PhO)_2P(O)SeCHMeCH:CH_2$. XXXI. (From $(PhO)_2P(Se)OCH_2CH$:CHMe by thermal isomerization during distillation.) ^1H-NMR. ^{31}P -10.5.[893]

$(MeO)_2P(O)SeBu$. XIX. Liq. b_2 86°, n_D^{25} 1.4856, d_4^{25} 1.3565.[518]

$(EtO)_2P(O)SeBu$. XIX.[518] XXIV.[1190] Liq. b_3 106°,[518] 122°,[1190] n_D^{25} 1.4763, d_4^{25} 1.2572.[518] Also higher homologs.[1190]

$(MeO)_2P(O)SePh$. XIX.[518, 939] Liq. $b_{0.005}$ 91-92°,[939] $b_{0.04}$ 109°,[518] n_D^{25} 1.5677,[518] n_D^{26} 1.5606,[939] d_4^{25} 1.4956.[518] IR.[518] UV λ_{max}, log ε : 248, 3.39; 271, 2.84.[518]

MeO(EtO)P(O)SePh. VIII, XIX. Liq. $b_{0.5}$ 138°, n_D^{25} 1.5509, d^{25} 1.4170.[520]

$(EtO)_2P(O)SePh$. XIX.[518, 939] Liq. $b_{0.005}$ 102°,[939] $b_{0.012}$ 105°,[518] n_D^{25} 1.5369,[518] n_D^{26} 1.5408,[939] d_4^{25} 1.3597.[518] λ_{max}, log ε: 248, 3.36; 271, 2.85.[518]

EtO(i-PrO)P(O)SePh. VIII, XIX.[520] Liq. $b_{0.2}$ 133°, n_D^{25} 1.5350, d^{25} 1.3309.[520]

EtO(BuO)P(O)SePh. VIII, XIX. Liq. $b_{0.3}$ 146°, n_D^{25} 1.5460, d^{25} 1.3200.[520]

$(i\text{-}PrO)_2P(O)SePh$. XIX.[518, 939] Liq. $b_{0.005}$ 101°,[939] $b_{0.02}$ 110°,[518] n_D^{25} 1.5258,[518] n_D^{26} 1.5215,[939] d_4^{25} 1.2824,[518] UV. λ_{max}248 Log ε 3.38.[518]

$(PrO)_2P(O)SePh$. XIX.[518, 939] Liq. $b_{0.005}$ 109°,[939] $b_{0.04}$ 130°,[518] n_D^{25} 1.5364,[518] n_D^{26} 1.5296,[939] d_4^{25} 1.3039.[518]

UV. λ_{max}, log ε : 250, 3.35; 272, 2.85.[518]

$(i\text{-}BuO)_2P(O)SePh$. XIX. Liq. $b_{0.5}$ 140°, n_D^{25} 1.5290, d_4^{25} 1.2561.[519]

$(s\text{-}BuO)_2P(O)SePh$. XIX. Liq. $b_{0.005}$ 115°, n_D^{25} 1.5172.[939]

$(BuO)_2P(O)SePh$. XIX.[518, 939] Liq. $b_{0.005}$ 123°,[939] $b_{0.02}$ 140°,[518] n_D^{25} 1.5274,[518] n_D^{26} 1.5185,[939] d_4^{25} 1.2499.[518] UV. λ_{max} 248, Log ε 3.45.[518]

$(MeO)_2P(O)SeC_6H_4NO_2$-2. XIX. Solid. m. 37-38.5°.[519]

$(MeO)_2P(O)SeC_6H_4NO_2$-4. XIX. Solid. m. 56°.[519]

$(EtO)_2P(O)SeC_6H_4NO_2$-4. XIX. Solid. m. 49°.[519]

$(EtO)_2P(O)Se$— 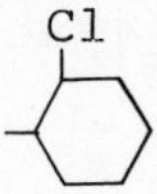 XXV.[721a, 778] Liq. $b_{0.05}$ 93-94°,[721a] $b_{0.07}$ 93-94°,[778] n_D^{18} 1.5166,[778] n_D^{22} 1.5160.[721a]

$(MeO)_2P(O)SeSeP(O)(OMe)_2$. XXIII. Liq. m. 3-4°.[342]

$(EtO)_2P(O)SeSeP(O)(OEt)_2$. XXIII. Liq. m. <-70°.[342]

F.6.6. $(RO)_3P(Se)$

$(MeO)_3P(Se)$. XV.[154, 894, 960, 961] Liq. b_{11} 73-74°,[894] b_{20} 95°,[960, 961] n_D^{20} 1.4487,[894] d_0^0 1.5387,[960, 961] d_4^{20} 1.5112.[894] IR.[425, 427] Conformation and spectrum.[528, 529] ^{1}H-NMR.[500, 734] ^{31}P -78.4 ppm.[715]

$(EtO)_3P(Se)$. XV.[894, 960, 961] Liq. b_9 85°,[894] b_{20} 117°,[960, 961] n_D^{20} 1.4730,[894] d_0^0 1.3189,[960, 961] d_4^{20} 1.3040.[894] IR.[1334] ^{1}H-NMR.[500, 717, 816, 1395] ^{31}P -71.[488, 715] -71 ± 1.[1345] μ 3.1 D.[466]

$(i\text{-}PrO)_3P(Se)$. XV. Liq. $b_{0.2}$ 63°, n_D^{20} 1.4557, d_4^{20} 1.1615.[894] ^{31}P -67.9.[715, 717]

$(PrO)_3P(Se)$. XV.[894] Liq. $b_{0.09}$ 67-68°, n_D^{20} 1.4680, d_4^{20} 1.1958.[894] IR.[1334]

$(t\text{-}BuO)_3P(Se)$. XV.[716] ^{31}P -31.1 ppm.[715, 716]

$(i\text{-}BuO)_3P(Se)$. XV. Liq. $b_{0.01}$ 77-78°, n_D^{20} 1.4620, d_4^{20} 1.1227.[894]

$(BuO)_3P(Se)$. XV. Liq. $b_{0.04}$ 83-84°, n_D^{20} 1.4663, d_4^{20} 1.1363.[894] IR.[1334] ^{31}P -73.0 ppm.[715] μ 2.6 D.[466]

$(Me_3CCH_2O)_3P(Se)$. ^{31}P -72.6 ppm.[715, 717]

$(PhO)_3P(Se)$. XV.[1277, 1301, 1405] Solid. m. 73-74°.[1277, 1405] ^{31}P -58.0.[715, 832a] X-ray photoelectron spectrum.[832a]

$(4\text{-}ClC_6H_4O)_3P(Se)$. XV. Solid. m. 88° (from EtOH).[1277]

$(2\text{-}MeC_6H_4O)_3P(Se)$. XV. Solid. m. 50-51°.[1277]

$(4\text{-}MeC_6H_4O)_3P(Se)$. XV. Solid. m. 111-112° (from EtOH).[1277]

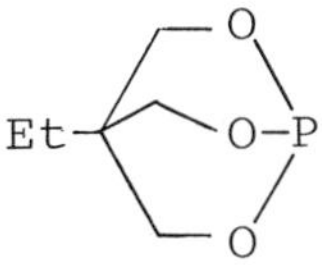
(Se). XV.[158, 459, 1386] Solid. m. 207-210° (from $MeOCH_2CH_2OH$).[158, 1386]

F.6.7. $(RO)_2P(Se)OR'$

$(PhO)_2P(Se)OMe$. VIII. Liq. n_D^{20} 1.5780, d_4^{20} 1.3677. IR. ^{31}P -65 ppm.[892]

$(t\text{-}BuO)_2P(Se)OEt$. ^{31}P -45.2.[715, 717]

$(PhO)_2P(Se)OEt$. VIII, XV. Liq. $b_{0.02}$ 133-134°, n_D^{20} 1.5765, d_4^{20} 1.3709. IR. ^{31}P -62 ppm.[892]

P(Se)OEt. XV. Liq. $b_{0.125}$ 107°, n_D^{20} 1.4923. Deposited red Se on standing.[51]

$(EtO)_2P(Se)OCOMe$. XXII. Liq. $b_{0.4}$ 60-62°, n_D^{20} 1.4772.[799]

$(PhO)_2P(Se)OCOR$. VIII. Solids. R, m.: Me, 72° (from MeOH); CH_2Cl, 61°; Ph 70-71° (from $CHCl_3$).[1427]

$(PhO)_2P(Se)OCOCOOP(Se)(OPh)_2$. VIII. Solid. m. 62°.[1427]

$(PhO)_2P(Se)OPr$. VIII. Liq. $b_{0.002}$ 141-143°, n_D^{20} 1.5703, d_4^{20} 1.3354. IR. ^{31}P -62 ppm.[892]

$(PhO)_2P(Se)OCH_2CH{:}CH_2$. VIII. Liq. $b_{0.02}$ 141-142°, n_D^{20} 1.5895, d_4^{20} 1.3748. IR. ^{31}P -83.5 ppm. Isomerizes on distillation to $(PhO)_2P(O)SeCH_2CH{:}CH_2$.[893]

$(PhO)_2P(Se)OCH_2CH{:}CHMe$. VIII. Liq. $b_{0.015}$ 167-168.5°, n_D^{20} 1.5810, d_4^{20} 1.3339. IR. ^{31}P -81.2 ppm. Isomerizes on distillation to $(PhO)_2P(O)SeCHMeCH{:}CH_2$.[893]

$(PhO)_2P(Se)OCOCH{:}CH_2$. VIII. Solid. m. 58-59°.[997]

$(PhO)_2P(Se)OCOCMe{:}CH_2$. VIII. Solid. m. 74-75° (from EtOH), b_1 191°.[997]

$(EtO)_2P(Se)OAr$. XV. Liq.[895]

Ar	b./mm	n_D^{20}	d_4^{20}
Ph	100-102°/0.06	1.5330	1.3459
$4-ClC_6H_4$	114-115°/0.09	1.5435	1.4222
$2,4-Cl_2C_6H_3$	118-119°/0.002	1.5528	1.4893
$2,4,5-Cl_3C_6H_2$	129-130°/0.005	1.5618	1.5563

F.6.8. $(RO)_2P(Se)SR'$

$(EtO)_2P(Se)SCH_2OMe$. XVII. Liq. $b_{0.01}$ 45°, n_D^{20} 1.4962.[790]

$(PhO)_2P(Se)SEt$. VIII.[891, 1431, 1432] Liq. $b_{0.002}$ 144-147° $b_{0.005}$ 152-153°, n_D^{20} 1.6125 - 1.6223, d_4^{20} 1.3874-1.3986.[891] These constants differ appreciably from the earlier work (Ref. 1431, 1432) where reaction with excess EtSNa is suggested.[891] IR.[1334]

$(EtO)_2P(Se)SPr$. VIII. Liq. $b_{0.02}$ 62-64°, n_D^{20} 1.5018, d^{20} 1.3205.[1433] IR.[1334]

$(EtO)_2P(Se)SEt$. XXVII. (From $(EtO)_2PSSeNa$ plus Et_3O^+ PF_4^- to give 80% P(Se) and 20% P(S) isomers.) Liq. b_{10} 71°, n_D^{20} 1.4868.[790] IR.[1334]

$(EtO)_2P(Se)CH_2CH_2CN$. XXII. Liq. $b_{0.05}$ 103-104°, n_D^{20} 1.5369. (From $(EtO)_2PSSeH$ plus CH_2:CHCN and NaOMe catalyst.)[791]

$(EtO)_2P(Se)SCHMeCN$. XXII. (From $(EtO)_2PSSeH$ plus CH_2:CHCN in the absence of catalyst.) Liq. $b_{0.05}$ 105-106°, n_D^{20} 1.5378.[791]

$(PhO)_2P(Se)SBu$. VIII. Liq. $b_{0.001}$ 151-152°, n_D^{20} 1.6050, d^{20} 1.3347.[891]

$(EtO)_2P(Se)SCHMePr$. XXII. (From $(EtO)_2PSSeH$ plus 2-pentene without catalyst.) Liq. $b_{0.2}$ 69°, n_D^{22} 1.5153.[791]

$(EtO)_2P(Se)S$-i-Am. VIII. Liq. $b_{0.015}$ 85-87°, n_D^{20} 1.5200, d^{20} 1.3234.[1433] IR.[1334]

$(PhO)_2P(Se)SPh$. VIII. Liq. $b_{0.002}$ 166-167°, n_D^{20} 1.6490, d^{20} 1.3744.[891]

F.6.9. $(RS)_2P(Se)OR'$

$(BuS)_2P(Se)OEt$. XV. Liq. $b_{0.04}$ 109-110°, n_D^{20} 1.5558, d^{20}

1.2212.[812]

(PrS)$_2$P(Se)OBu. XV. Liq. $b_{0.04}$ 108-109°, n_D^{20} 1.5535, d^{20} 1.2212.[812]

F.6.10. (RS)$_3$P(Se)

(RS)$_3$P(Se). R, ^{31}P: Me, -81.9; Et, -75.8; Pr, -75.7; Bu, -75.1.[715]

(PhS)$_3$P(Se). XV. Solid. m. 95° (from EtOH).[773]

F.6.11. (RO)$_2$P(S)SeR'

(EtO)$_2$P(S)SeCH$_2$OMe. XVII. Liq. $b_{0.01}$ 60°, n_D^{20} 1.5158.[790]

(EtO)$_2$P(S)SeCOOEt. XVII. Liq. $b_{0.03}$ 71-72°, n_D^{20} 1.5032, d^{20} 1.3371.[1434]

(PhO)$_2$P(S)SeCOOEt. XVII. Liq. $b_{0.03}$ 113-114°, n_D^{20} 1.5995, d^{20} 1.3922.[1434]

(EtO)$_2$P(S)SeEt. XVII.[1433] XXVII. (From (EtO)$_2$PSSeNa plus Et_3O^+ PF_4^- in 20% yield, along with 80% P(Se) isomer.) [790] Liq. b_{17} 129°, n_D^{20} 1.4988.[790] IR.[1334] (Ref. 1433 gives: $b_{0.001}$ 67°, n_D^{20} 1.5400, d^{20} 1.3587.)

(EtO)$_2$P(S)SeR. XVII. Liq.[1433]

R	$b_{0.001}$	n_D^{20}	d^{20}
Pr	77-78°	1.5322	1.3587
Bu	80	1.5291	1.3421
i-Am	62	1.5209	1.3236
CH$_2$CH:CH$_2$	104°/0.02	1.6023	1.3627
CH$_2$C≡CH	117°/0.02	1.6208	1.4001

(PhO)$_2$P(S)SeR. XVII. Liq.[1431, 1432]

Et	62-64	1.5942	1.2221
Pr	76-77	1.6018	1.1536
i-Am	77-79°/0.006	1.5415	1.1407

(EtO)$_2$P(S)SeCOR. XVII. Liq.[1434]

R	$b_{0.001}$	n_D^{20}	d^{20}
Me	140°	1.5694	1.3301
Et	59-60	1.5008	1.3356
i-Pr	58°/0.002	1.5132	1.3421

Pr	82-83	1.5050	1.3398
Ph	103-104°/0.07	1.5948	1.3775

(PhO)$_2$P(S)SeCOR. XVII. Liq.[1434]

Me	112-113°/0.002	1.5532	1.3082
Et	72°/0.001	1.5630	1.4026
Ph	88°/0.001	1.5454	1.2908

(PhO)$_2$P(S)SeCH$_2$CH:CH$_2$. XVII. Liq. $b_{0.03}$ 100-110°, n_D^{20} 1.6255, d^{20} 1.3802.[1433]

(PhO)$_2$P(S)SeCH$_2$C≡CH. XVII. Liq. $b_{0.02}$ 135-136°, n_D^{20} 1.6370, d^{20} 1.4165.[1433]

(EtO)$_2$P(S)SeCOCHMe$_2$. XVII. Liq. $b_{0.002}$ 58°, n_D^{20} 1.5132, d^{20} 1.3421.[1434]

(EtO)$_2$P(S)SeC$_6$H$_{13}$. XXIV. Liq. $b_{0.01}$ 68°.[638] IR.[1334]

(RO)$_2$P(S)SePh. VIII, XIX. Liq.[520]

R	b./mm	n_D^{25}	d^{25}
Et	130°/0.5	1.5830	1.3608
i-Pr	104°/0.2	1.5595	1.2750
Pr	110°/0.2	1.5638	1.2842

(EtO)$_2$P(S)SeC$_6$H$_4$NMe$_2$-4. XXIV. Liq. $b_{0.01}$ 132°.[638]

(EtO)$_2$P(S)SeC$_6$H$_4$OMe-2. XXIV.[638] XXVIII.[297, 647] Liq. $b_{0.01}$ 106°,[638] 108°.[297, 647]

(EtO)$_2$P(S)SeCH$_2$Ph. XXVIII. (From (EtO)$_2$P(S)H plus PhCH$_2$-SeSeCH$_2$Ph and NaOMe.) Liq. $b_{0.01}$ 98°.[297, 647]

(EtO)$_2$P(S)SeCH$_2$C$_6$H$_4$Cl-4. XXIV. Liq. $b_{0.01}$ 108°.[638]

F.6.12. Miscellaneous

EtO(BuS)P(O)SePh. XIX. Liq. $b_{0.2}$ 114°, n_D^{25} 1.5434, d_4^{25} 1.3371.[520]

(PhSe)$_3$P(O). VIII. Solid. m. 110° (dec.) (from petroleum ether).[939]

(EtO)$_2$P(S)SeSeP(S)(OEt)$_2$. XXIII. (From (EtO)$_2$PSSeK plus I$_2$.) Liq. $b_{0.003}$ 122°, n_D^{20} 1.4629. Decomp. rapidly in air so no IR available.[1430]

(PhO)$_2$P(S)SeSeP(S)(OPh)$_2$. XXIII. (From (PhO)$_2$PSSeK plus I$_2$.) IR shows both P(S) and P(Se) bands.[1430]

$(EtO)_2P(S)OP(Se)(OEt)_2$. XXXIII. (From $(EtO)_2PSSeH$ plus $(EtO)_2P(O)Cl$.) Liq. $b_{0.65}$ 60-61°, n_D^{20} 1.5281.[790]

F.6.13. Phosphorotelluronates

O,O,O-Triphenylphosphorotelluronate, iron complexes.[153a] (From $Fe_3(CO)_9Te_2$ plus $(PhO)_3P \rightarrow$ (i) $\xrightarrow{-CO}$ (ii) $\xrightarrow{-CO}$ (iii)

(i) $Fe_3(CO)_9(PhO)_3PTe_2$, deep red crystals, m. 128°.

(ii) $Fe_3(CO)_8(PhO)_3PTe_2$, black crystals, m. 99°.

(iii) $Fe_3(CO)_7(PhO)_3PTe_2$, red amorphous solid, m. 130-131°.

(Received June 24, 1974)

REFERENCES

1. Abel, E. W., D. A. Armitage, and R. P. Bush, J. Chem. Soc., Suppl. No. 1, 5584 (1964).; C. A., 63, 1813d (1965).
2. Addor, R. W., S. D. Levy, and R. J. Magee, Belg. 637, 658 (1964); C. A., 62, 6490f (1965).
3. Aichenegg, P. C., U. S. 3,463,836 (1969); C. A., 71, 123551d (1969).
4. Aichenegg, P. C., and L. E. Gillen, U. S. 3,454,679 (1969); C. A., 71, 90816p (1969).
5. Ailman, D. E., U. S. 3,113,957 (1963); C. A., 60, 5336g (1964).
6. Ailman, D. E., J. Org. Chem., 30, 1074 (1965); C. A. 62, 14483f (1965).
7. Ailman, D. E., and R. W. Young, U. S. 3,081,332 (1963); C. A., 59, 3776f (1963).
8. Airapetyan, G. M., P. G. Zherebchenko, V. M. Bystrova, Z. V. Benevolenskaya, N. D. Kuleshova, M. G. Lin'kova, O. V. Kil'disheva, and I. L. Knunyants, Bull. Acad. Sci. USSR, Ser. Chem., 1967, 317; C. A., 61, 21570j (1964).
9. Akerfeldt, S., Acta Chem. Scand., 16, 1897 (1962); C. A., 59, 3760c (1963).
10. Akerfeldt, S., and L. Fagerlind, J. Med. Chem., 10, 115 (1966); C. A., 66, 75619 (1966).
11. Aksenov, V. I., E. A. Chernyshev, E. F. Bugerenko, and A. A. Borisenko, J. Gen. Chem., 41, 478 (1971); C. A. 75, 19838 (1971).
12. Aleksandrov, V. N., V. I. Emel'yanov, J. Gen. Chem., 37, 2383 (1967); C. A., 69 18521 (1968).
13. Alford, D. O., A. Menefee, and C. B. Scott, Chem.

Ind., *1959*, 514; C. A., *53*, 19844i (1959).

14. Alimov, P. I., and I. V. Cheplanova, Izv. Kasansk. Filiala Akad. Nauk SSSR, Ser. Khim. Nauk, *1961*, 54; C. A., *59*, 9858b (1963).
15. Almasi, L., and A. Hantz, Acad. Rep. Populare Romine, Filiala Cluj, Studii Cercetari Chim., *11*, 135 (1960); C. A., *55*, 8335h (1961).
16. Almasi, L., and A. Hantz, Acad. Rep. Populare Romine, Filiala Cluj, Studii Cercetari Chim., *11*, 297 (1960); C. A., *58*, 5556g (1963).
17. Almasi, L., and A. Hantz, Acad. Rep. Populare Romine, Filiala Cluj, Studii Cercetari Chim., *12*, 129 (1961); C. A., *58*, 6731f (1963).
18. Almasi, L., and A. Hantz, Acad. Rep. Populare Romine, Filiala Cluj, Studii Cercetari Chim., *13*, 299 (1962); C. A., *62*, 2729d (1965).
19. Almasi, L., and A. Hantz, Ber., *97*, 661 (1964); C. A., *60*, 13131c (1964).
20. Almasi, L., and A. Hantz, Rev. Roum. Chim., *9*, 155 (1964); C. A., *61*, 14564c (1964).
21. Almasi, L., and A. Hantz, Rev. Roum. Chim., *10*, 287 (1965); C. A., *63*, 11409e (1965).
22. Almasi, L., and A. Hantz, Chem. Ber., *99*, 3288 (1966); C. A., *66*, 10830z (1967).
23. Almasi, L., and A. Hantz, Omagiu Raluca Ripan., *1966*, 59; C. A., *67*, 116856w (1967).
24. Almasi, L., and A. Hantz. Rom. 45,836 (1967); C. A., *68*, 104742p (1968).
25. Almasi, L., and A. Hantz, Rev. Roum. Chim., *13*, 653 (1968); C. A., *70*, 11256j (1969).
26. Almasi, L., and A. Hantz, Chem. Ber., *103*, 718 (1970); C. A., *72*, 110927 (1970).
27. Almasi, L., A. Hantz, and E. Hamburg, Chem. Ber., *103*, 2976 (1970); C. A., *73*, 108914 (1970).
28. Almasi, L., A. Hantz, and L. Paskucz, Acad. Rep. Populare Romine, Filiala Cluj, Studii Cercetari Chim., *12*, 291 (1961); C. A., *59*, 1515h (1963).
29. Almasi, L., A. Hantz, and L. Paskucz, Chem. Ber., *95*, 1582 (1962); C. A., *57*, 16463d (1962).
30. Almasi, L., and L. Paskucz, J. Prakt. Chem., *19*, 138 (1963); C. A., *59*, 7528d (1963).
31. Almasi, L., and L. Paskucz, Ber., *97*, 623 (1964); C. A., *60*, 13130h (1964).
32. Almasi, L., and L. Paskucz, Chem. Ber., *98*, 613 (1965); C. A., *62* 14553d (1965).
33. Almasi, L., and L. Paskucz, Chem. Ber., *98*, 3546 (1965); C. A., *64* 3338c (1966).
34. Almasi, L., L. Paskucz, and L. Fey, Chem. Ber., *103*, 2972 (1970); C. A., *73*, 109419 (1970).
35. American Cyanamid Co., Brit. 644,616 (1950); C. A., *45*, 3862i (1951).
36. American Cyanamid Co., Brit. 646,188 (1950); C. A., *45*, 5713d (1951).

37. American Cyanamid Co., Brit. 948,522 (1964); C. A., 60, 12028a (1964).
38. American Cyanamid Co., Neth. Appl. 6,401,093 (1964); C. A., 62, 2740g (1965).
39. American Cyanamid Co., Neth. Appl. 6,403,092 (1964); C. A., 63, 14767h (1965).
40. American Cyanamid Co., Brit. 1,142,340 (1969); C. A., 70, 87036h (1969).
41. Andrianov, K. A., and I. K. Kuznetsova, Izv. Akad. Nauk SSSR, Otd. Khim. Nauk, 1962, 456; C. A., 67, 15141d (1967).
42. Andrianov, K. A., I. K. Kuznetsova, and I. Pakhomova, Bull, Acad. Sci. USSR, Div. Chem. Sci., 1963, 448; C. A., 59, 2850cd (1963).
43. Anschütz, L., and W. Broeker, Ber., 61, 1265 (1928); C. A. 22, 4113 (1928).
44. Anschütz, L., and W. Marquardt, Chem. Ber., 89, 1119 (1956); C. A., 51, 2661i (1957).
45. Anschütz, L., and H. Walbrecht, J. Prakt. Chem., 133, 65 (1932); C. A., 26, 2437 (1932).
46. Anschütz, L., and F. Wenger, Ann., 482, 25 (1930; C. A., 25, 501 (1931).
47. Anschütz, R., and W. O. Emery, Ann., 253, 105 (1889).
48. Arbisman, Ya.S., N. S. Rylyakova, V. V. Tarasov, Yu.A. Kondrat'ov, and S. Z. Ivin, J. Gen. Chem., 37, 1763 (1967); C. A., 68, 48964 (1968).
49. Arbuzov, A. E., and P. I. Alimov, Izv. Akad. Nauk SSSR, Otd. Khim. Nauk, 1951, 409; C. A., 49, 160b (1955).
50. Arbuzov, A. E., and M. M. Azanovskaya, Izv. Akad. Nauk SSSR, Otd. Khim. Nauk, 1951, 544; C. A., 47, 98f (1953).
51. Arbuzov, A. E., and N. A. Razumova, Bull. Acad. Sci., USSR, Div. Chem. Sci., 1956, 179; C. A., 50, 13735e (1956).
52. Arbuzov, A. E., and V. M. Zoroastrova, Izv. Akad. Nauk SSSR, Otd. Khim. Nauk, 1950, 357; C. A., 45, 1512e (1951).
53. Arbuzov, A. E., V. M. Zoroastrova, and T. N. Myasoedova, Izv. Akad. Nauk SSSR, Otd. Khim. Nauk, 1960, 2127; C. A., 55, 14305b (1961).
54. Arbuzov, A. E., V. M. Zoroastrova, and N. I. Rizpolozhenskii, Bull Acad. Sci. USSR, Classe Sci. Chim., 1948, 208; C. A., 42, 4933f (1948).
55. Arbuzov, B. A., and R. P. Arshinova, Proc. Acad. Sci. USSR, 195, 859 (1970); C. A., 74, 87220w (1961).
56. Arbuzov, B. A., K. V. Nikonorov, and Z. G. Shishova, Bull. Acad. Sci. USSR, Div. Chem. Sci., 1954, 711; C. A., 49, 13891b (1955).
57. Arbuzov, B. A., K. V. Nikonorov, and G. M. Vinokurova, Bull. Acad. Sci. USSR, Div. Chem. Sci., 1955, 597; C. A., 50, 7050b (1956).
58. Arbuzov, B. A., L. Z. Nikonova, O. N. Nuretdinova,

and V. V. Pomazanov, Bull. Acad. Sci. USSR, Ser. Chem., 1970, 1350; C. A., 74, 53355d (1961).
59. Arbuzov, B. A., and O. N. Nuretdinova, Bull. Acad. Sci. USSR, Ser. Chem., 1969, 1212; C. A., 71, 81482e (1969).
60. Arbuzov, B. A., and T. G. Shavsha, Izv. Akad. Nauk SSSR, Otd. Khim. Nauk, 1951, 795; C. A., 46, 3817d (1952).
61. Arbuzov, B. A., and V. S. Vinogradova, Bull. Acad. Sci. USSR, Classe Sci. Chim., 1947, 459; C. A., 42, 3312a (1948).
62. Arbuzov, B. A., and V. S. Vinogradova, Izv. Akad. Nauk SSSR., Otd. Khim. Nauk, 1951, 733; C. A., 46, 7515e (1952).
63. Arbuzov, B. A., and V. S. Vinogradova, Izv. Akad. Nauk SSSR., Otd. Khim. Nauk, 1952, 865; C. A., 47, 10458e (1953).
64. Arbuzov, B. A., and D. Kh. Yarmukhametova, Bull. Acad. Sci. USSR, Div. Chem. Sci., 1957, 307; C. A., 51, 14542c (1957).
65. Arbuzov, B. A., and D. Kh. Yarmukhametova, Izv. Akad. Nauk SSSR., Otd. Khim. Nauk, 1960, 1881; C. A., 55, 16410f (1961).
66. Arbuzov, B. A., and D. Kh. Yarmukhametova, Bull. Acad. Sci. USSR, Div. Chem. Sci., 1962, 1320; C. A., 58, 2468g (1963).
67. Arbuzov, B. A., and V. M. Zoroastrova, Izv. Akad. Nauk SSSR., Otd. Khim. Nauk, 1958, 1331; C. A., 53, 7182h (1959).
68. Arbuzov, B. A., V. M. Zoroastrova, and N. D. Ibragimova, Bull. Acad. Sci. USSR, Ser. Chem., 1964, 611; C. A., 61, 2959c (1964).
69. Aroney, M. J., R. J. W. LeFevre, and J. Saxby, J. Chem. Soc., 1963, 4938. C. A., 59, 12265b (1963).
70. Ashbolt, R. F., and H. Coates, Brit. 656,303 (1951); C. A., 46, 7581g (1952).
71. Asinger, F., M. Thiel, and W. Schaefer, Ann., 637, 146 (1960); C. A., 56, 15330h (1962).
72. Autenrieth, W., and O. Hildebrand, Ber., 31, 1094 (1898).
73. Autenrieth, W., and W. Meyer, Ber., 58, 840 (1925); C. A., 19, 2325 (1925).
74. Babad, H., W. Herbert, and M. C. Goldberg, Anal. Chim. Acta, 41, 259 (1968); C. A., 68, 77181 (1968).
75. Bachinskii, T. P., and N. I. Zemlyanskii, J. Gen. Chem., 38, 2019 (1968); C. A. 70, 11018 (1969).
76. Bakanova, Z. M., Ya. A. Mandel'baum, and N. N. Mel'nikov, J. Gen. Chem., USSR., 26, 2867 (1956); C. A. 51, 1824i (1957).
77. Bakanova, Z. M., Ya. A. Mandel'baum, N. N. Mel'nikov, and E. I. Sventsitskii, J. Gen. Chem., 26, 519 (1956); C. A., 50, 13725g (1956).

77a. Barnas, E. F., and S. B. Richter, Ger. 1,300,113 (1969); C. A., 71, 90869 (1969).
78. Bartlett, J. H., S. B. Lippincott, and L. A. Mikeska, U. S. 3,192,162 (1965); C. A., 63, 14704c (1965).
79. Baum, B. O., U. S. 3,065,197 (1962); C. A., 58, 5845e (1963).
80. Bellamy, L. J., and L. Beecher, J. Chem. Soc., 1952, 475; C. A., 47, 43i (1953).
81. Bel'skii, V. E., N. N. Bezzubova, Z. V. Lustina, V. N. Eliseenkov, and A. N. Pudovik, Zh. Obshch. Khim., 39, 181 (1969); C. A., 70, 95931m. (1969).
82. Beriger, E., Ger. 1,138,977 (1962); C. A., 58, 11224e (1963).
83. Beriger, E., Swiss 378,872 (1964); C. A., 62, 10338f (1965).
84. Beriger, E., Swiss 431,186 (1967); C. A., 68, 59095s (1968).
85. Beriger, E., Ger. Offen. 1,812,399 (1969); C. A., 71, 101526x (1969).
85a. Beriger, E., Ger. Offen. 1,917,922 (1969); C. A., 72, 43710 (1969).
86. Berkelhammer, G., U. S. 3,005,004 (1961); C. A., 56, 5887e (1962).
87. Berkelhammer, G., W. C. Dauterman, and R. D. O'Brien, J. Agr. Food Chem., 11, 307 (1963); C. A., 59, 5009d (1963).
88. Berkelhammer, G., S. DuBreuil, and R. W. Young, J. Org. Chem., 26, 2281 (1961); C. A., 55, 27047c (1961).
89. Berkelhammer, G., and F. A. Wagner, Jr., U. S. 2,992,158 (1961); C. A., 56, 8632i (1962).
90. Berkelhammer G., and F. A. Wagner, Jr., Belg. 645,921 (1964); C. A., 63, 16263c (1965).
91. Berkelhammer G., F. A. Wagner, Jr., and R. J. Magee, U. S. 3,102,885 (1963); C. A., 60, 535a (1964).
92. Bertin, D., and J. Perronnet, Ger. Offen. 1,923,103 (1969); C. A., 72, 31303s (1970).
93. Best, R. J., and N. R. Hersch, J. Agr. Food Chem., 12, 546 (1964).
94. Bianchetti, G., Rend. Ist. Lombardo Sci. Pt. I, 91, 68 (1957); C. A., 52, 11769b (1958).
95. Birum, G. H., U. S. 2,836,534 (1958); C. A., 52, 20056b (1958).
96. Birum, G. H., U. S. 2,931,755 (1960); C. A., 54, 5245a (1960).
97. Birum, G. H., Brit. 871,695 (1961); C. A., 56, 12743e (1962).
98. Birum, G. H., U. S. 3,046,296 (1962); C. A., 58, 482f (1963).
99. Blackman, L. C. F., and M. J. S. Dewar, J. Chem. Soc., 1957, 169; C. A., 51, 9490c (1957).
100. Blagoveshchenskii, V. S., S. N. Kudryavtseva, and Yu.

N. Mitrokhina, Vestn. Mosk. Univ., Khim., 24, 68 (1969); C. A., 71, 112339q (1969)

101. Blagoveshchenskii, V. S., Yu. N. Mitrokhina, and S. N. Kudryavtseva, Vestn. Mosk. Univ., Khim., 24, 69 (1969); C. A., 70, 86944v (1969).

102. Blagoveshchenskii, V. S., and S. N. Vlasova, J. Gen. Chem., 41, 1036 (1971); C. A., 75, 76066 (1971).

103. Blair, E. H., U. S. 2,887,505 (1959); C. A., 54, 2257h (1960).

104. Blair, E. H., and E. C. Britton, U. S. 2,954,394 (1960); C. A., 55, 4430a (1961).

105. Blair, E. H., and B. Fischback, U. S. 3,455,938 (1969); C. A., 71, 91318q (1969).

106. Blair, E. H., K. C. Kauer, and E. E. Kenaga, J. Agr. Food Chem., 11, 237 (1963); C. A., 59, 487a (1963).

107. Blair, J. S., Anal. Chem., 41, 1497 (1969); C. A., 71, 97944 (1969).

108. Bliznyuk, N. K., P. S. Khoskhlov, R. V. Strel'stov, and G. V. Dotsev, J. Gen. Chem., 37, 1064 (1967); C. A., 68, 105297 (1968).

109. Bochwic, B., and A. Frankowski, Tetrahedron, 24, 6653 (1968); C. A., 70, 28775 (1969).

110. Bochwic, B., and J. Kapuscinski, Rocz. Chem., 39, 1251 (1965); C. A., 64, 12656g (1966).

111. Bochwic, B., J. Kapuscinski, and S. Markowicz, Bull. Acad. Pol. Sci., Ser. Sci. Chim., 12, 531 (1964); C. A., 62, 4056h (1965).

112. Bodkin, C., and P. Simpson, Chem. Commun., 1969, 829; C. A., 71 90630 (1969).

113. C. H. Boehringer Sohn, Belg. 614,739 (1962); C. A., 58, 12426c (1963).

114. C. H. Boehringer Sohn, Belg. 619,497 (1962); C. A., 59, 11274d (1963).

115. C. H. Boehringer Sohn, Fr. 1,333,944 (1963); C. A., 60, 2872f (1964).

116. C. H. Boehringer Sohn, Belg. 625,198 (1963); C. A., 60, 13187b (1964).

117. Bolotova, G. L., G. G. Kotova, K. I. Zimina, and V. I. Isagulyants, Izv. Vyssh. Ucheb. Zaved., Neft Gas, 8, 62 (1965); C. A., 63, 6897b (1965).

118. Booth, H. S., D. R. Martin, and F. E. Kendall, J. Am. Chem. Soc., 70, 2523 (1948); C. A., 43, 1313h (1949).

119. Boter, H. L., and H. J. Toet, Rec. Trav. Chim., 84, 1279 (1965); C. A. 64, 3587b (1966).

120. Boter, H. L., and G. R. van den Berg, Rec. Trav. Chim. Pays-Bas, 85, 1099 (1966); C. A., 66, 28846 (1967).

121. Boyer, W. P., U. S. 2,761,806 (1956); C. A., 51, 663g (1957).

122. Brahler, B., J. Reese, and R. Zimmerman, U. S. 2,984,669 (1961); C. A., 55, 22341h (1961).

123. Brannock, K. C., J. Am. Chem. Soc., 73, 4953 (1951); C. A., 46, 11100h (1952); U. S. 2,622,095 (1952); C. A., 47, 9343a (1953).
124. Brevilet, J., P. Appriou, and J. Teste, Bull. Soc. Chim. Fr., 1971, 1344; C. A., 75, 49040a (1971).
125. Bright, N. F. H., J. C. Cuthill, and N. H. Woodbury, J. Sci. Food Agr., 1, 344 (1950; C. A., 45 4868h (1951).
126. Britton, E. C., and E. H. Blair, U. S. 2,922,811 (1960); C. A., 54, 9848h (1960).
126a. Britton, E. C., E. H. Blair, and H. Tolkmith, U. S. 2,891,085 (1959); C. A., 53, 21814 (1959).
127. Broeker, W., J. Prakt. Chem. (2), 118, 287 (1928); C. A. 22, 1964 (1928).
128. Brois, S. J., U. S. 3,476,833 (1969); C. A., 72, 21302b (1970).
129. Brois, S. J., U. S. 3,501,557 (1970); C. A., 72, 132040j (1970).
129a. Brokke, M. E., T. E. Elward, and T. B. Williamson, Fr. 1,506,399 (1967); C. A. 70, 105956 (1959).
130. Brooks, J. W., E. G. Howard, and J. J. Wehrle, J. Am. Chem. Soc., 72, 1289 (1950), C. A., 44, 6411c (1950).
131. Brown, N. C., and D. T. Hollinshead, Brit. 1,007,332 (1965); C. A., 64, 2061d (1966); U. S. 3,294,636 (1966); C. A., 66, 85699 (1967).
132. Brown, N. C., and G. S. Pall, S. African 6,159 (1968); C. A., 70, 81585m (1969).
133. Brown, T. L., J. G. Verkade, and T. S. Piper, J. Phys. Chem., 65, 2051 (1961); C. A., 56, 6759i (1962).
134. Buchner, B., and G. G. Curtis, U. S. 3,286,001 (1966); C. A., 66, 18508u (1967).
135. Buchner, B., and E. Jacoves, U. S. 3,328,494 (1967); C. A., 67, 73440m (1967).
135a. Buckler, S. A., L. Doll, F. K. Lind, and M. Epstein, J. Org. Chem., 27, 794 (1962); C. A., 57, 3336c (1962).
136. Buechel, K. H., H. Roechling, and F. W. A. G. Korte, U. S. 3,130,203 (1964). Ann. Chem., 685, 10 (1965); C. A., 61, 8204c (1964).
137. Bugerenko, E. F., E. A. Chernyshev, and E. M. Popov, Bull. Acad. Sci. USSR, Ser. Chem., 1966, 1334; C. A., 66, 16078q (1967).
138. Bunyan, P. J., and J. I. G. Cadogan, J. Chem. Soc., 1962, 2953; C. A., 57, 9641i (1962).
139. Burn, A. J., and J. I. G. Cadogan, J. Chem. Soc., 1961, 5532; C. A., 56, 12928a (1962).
140. Burn, A. J., J. I. G. Cadogan, and H. N. Moulden, J. Chem. Soc., 1961, 5542; C. A., 56, 12929b (1962).
141. Cadogan, J. I. G., J. Chem. Soc., 1961, 3067; C. A., 56, 3508b (1962).
141a. Cadogan, J. I. G., J. Chem. Soc., 1962, 18; C. A.,

57, 646e (1962).
142. Cadogan, J. I. G., and H. N. Moulden, J. Chem. Soc., 1961, 5524; C. A., 56, 12926d (1962).
143. Cadogan, J. I. G., and L. C. Thomas, J. Chem. Soc., 1960, 2248; C. A., 54, 22319g (1960).
144. Calderbank, A., J. Chem. Soc., C, 1966, 56; C. A., 64, 6653a (1966).
145. Calderbank, A., E. C. Edgar, and J. A. Silk, Ger. 1,118,789 (1961); Ger. 1,132,560 (1962); C. A., 57, 9864g (1962); C. A., 57, 15132a (1962).
145a. Calderbank, A., Ger. 1,132,560 (1962); C. A., 57, 15132a (1962).
146. Calderbank, A., and R. Ghosh, Brit. 832,990 (1960); C. A., 54, 20876c (1960).
147. Calderbank, A., R. Ghosh, and A. Taylor, Brit. 814,264 (1959); C. A., 53, 19974c (1959).
148. Cambi, L, Chim. Ind., 26, 97 (1944); C. A., 40, 3734 (1946).
149. Carter, P. L., A. J. Lambie, and D. W. J. Lane, J. Appl. Chem., 18, 257 (1968); C. A., 69, 95901 (1968).
150. Cassaday, J. T., U. S. 2,578,652 (1951); C. A., 46, 6139c (1952).
151. Cassaday, J. T., E. I. Hoegberg, and B. D. Gleissner, U. S. 2,494,283 (1950); C. A., 44, 3516c (1950).
152. Cassaday, J. T., E. I. Hoegberg, and B. D. Gleissner, U. S. 2,494,284 (1950); C. A., 44, 3516f (1950).
153. Cebrian, G. R., Arch. Inst. Farmacol. Exp., 8, 61 (1956); C. A., 51, 12020d (1957).
153a. Cetini, G., P. L. Stanghellini, R. Rossetti, and O. Gambino, J. Organometal. Chem., 15, 373 (1968); C. A., 70, 53516 (1969).
154. Chabrier, P., and N. T. Thuong, Compt. Rend., 258, 3738 (1964); C. A., 61, 567g (1964).
155. Chabrier, P., and N. T. Thuong, Compt. Rend., 261, 2229 (1965); C. A., 64, 12537h (1966).
156. Chabrier, P., N. T. Thuong, and J. P. Chabrier, C. R. Acad. Sci., Paris, Ser. C, 263, 1556 (1966); C. A., 66, 64763 (1967).
157. Chabrier de Lassauniere, P., N. T. Thuong, and P. R. Savignac, Fr. 1,537,175 (1968); C. A., 71, 30585c (1969).
158. Chang, W. H., J. Org. Chem., 29, 3711 (1964); C. A., 62, 7665f (1965).
159. Chapman, N. B., and B. C. Saunders, J. Chem. Soc., 1948, 1010; C. A., 43, 121f (1949).
159a. Charlton, T. L., and R. G. Cavell, Inorg, Chem., 7, 2195 (1968); C. A., 69, 113068 (1968).
160. Chemische Werke Albert, Brit. 820,358 (1959); C. A., 54, 14154h (1960).
161. Chemische Werke Albert, Brit. 875,828 (1961); C. A., 56, 2453d (1962).
162. Chen, J. Y. T., and R. W. Dority, J. AOAC, 50, 426

(1967); C. A., 67, 72757 (1967).
163. Chen, J. Y., and H. T. Yang, Hua Hsueh Pao, 25, 289 (1969); C. A., 54, 16383g (1960).
164. Chernaya, N. M., and N. I. Zemlyanskii, J. Gen. Chem., 36, 1361 (1966); C. A., 65, 16850a (1966).
165. Chernyshev, E. A., E. F. Bugerenko, and V. I. Aksenov, J. Gen. Chem., 40, 1409 (1970); C. A., 74, 53931 (1971).
166. Cheymol, G., and N. T. Thuong, Med. Pharmacol. Exp., 12, 333 (1965); C. A., 63, 13311h (1965).
167. Cheymol, J., P. Chabrier, N. T. Thuong, and G. Cheymol, Therapie, 21, 1003 (1966); C. A., 65, 11205h (1966).
168. Chittenden, R. A., and L. C. Thomas, Spectrochim. Acta, 20, 1679 (1964); C. A., 61, 14048a (1964).
169. Christman, D. L., U. S. 2,884,354 (1959); C. A., 53, 15465f (1959).
170. Chwalinski, S., and A. Chrzaszczewska, Lodz. Towarz. Nauk, Wydzial III, Acta Chim., 10, 71 (1965); C. A., 65, 13708h (1966).
171. CIBA Ltd., Swiss 308,894 (1955); C. A., 51, 10559h (1957).
172. CIBA Ltd., Swiss 311,463 (1956); C. A., 51, 10560a (1957).
173. CIBA Ltd., Swiss 311,474 (1956); C. A., 51, 10560d (1957).
174. CIBA Ltd., Brit. 892,326 (1962); C. A., 58, 7833e (1963).
175. CIBA Ltd., Brit. 923,702 (1963); C. A., 59, 9794e (1963).
176. CIBA Ltd., Brit. 970,581 (1964); U. S. 3,267,181 (1966); C. A., 62, 3938h (1965).
177. CIBA Ltd., Fr. 1,383,053 (1964); C. A., 62, 9011b (1965).
178. CIBA Ltd., Neth. Appl. 6,515,066 (1966); C. A., 65, 13762a (1966).
179. CIBA Ltd., Neth. Appl. 6,516,265 (1966); C. A., 65, 17003g (1966).
179a. CIBA Ltd., Fr. 1,502,537 (1967); C. A., 70, 19790 (1969).
180. CIBA Ltd., Fr. 1,530,955 (1968); C. A., 71, 80710k (1969).
181. Cierpka, H., Fr. 1,527,848 (1968); C. A., 71, 21863q (1969).
182. Coelln, R., and H. Scheinpflug, Ger. 1,250,429 (1967); C. A., 68, 87064z (1968).
182a. Colinese, D. L., and H. J. Terry, Chem. Ind. (London), 44, 1507 (1968); C. A., 70, 19164 (1969).
183. Cook. A. F., M. J. Holman, and A. L. Nussbaum, J. Am. Chem. Soc., 91, 1522 (1969); C. A., 70, 115,464 (1969).
184. Cooks, R. G., and A. F. Gerrard, J. Chem. Soc. B,

1968, 1327; C. A., 70, 15285 (1969).
185. Cooper, McDougall, and Robertson Ltd., Belg. 610,896 (1962); C. A., 57, 13729g (1962).
186. Cooper, McDougall, and Robertson Ltd., Fr. 1,332,859 (1963); C. A., 60, 1709g (1964).
187. Cooper, McDougall, and Robertson Ltd., Belg. 633,832 (1963); C. A., 61, 13315d (1964).
188. Cooper, McDougall, and Robertson Ltd., Fr. 1,431,701 (1966); U. S. 3,294,630 (1966); C. A., 65, 15337f (1966).
189. Counselman, C. J., U. S. 3,496,270 (1970); C. A., 72, 90102y (1970).
190. Cowen, F. M., U. S. 2,729,675 (1956); C. A., 50, 12103e (1956).
191. Cremlyn, R. J. W., and N. A. Olson, J. Chem. Soc. C, 1969, 2305; C. A., 72, 3631d (1970).
192. Curry, S. Du Breuil, R. W. Young, G. Berkelhammer, and D. E. Ailman, U. S. 3,309,371 (1967); C. A., 68, 87013g (1968).
193. Cyba, H. A., U. S. 3,263,000 (1966); C. A., 65, 8764d (1966).
194. Cyba, H. A., U. S. 3,349,103 (1967); C. A., 68, 87128y (1968).
195. Dale, S. W., and M. E. Hobbs, J. Phys. Chem., 75, 3537 (1971); C. A., 75, 156742 (1971).
196. Damico, J. N., J. AOAC, 49, 1027 (1966); C. A., 66, 1834k (1967).
197. Darlington, W. A., and J. P. Chupp, S. African 6364 (1969); C. A., 74, 111744t (1971).
197a. Delepine, M., Bull. Soc. Chim. (4), 11, 577 (1912).
198. Delventhal, H. J., and W. Kuchen, Z. Naturforsch. B, 26, 190 (1971); C. A., 75, 48357x (1971).
199. Demarcq. M. C., and J. Sleziona, Fr. Addn. 86,531 (1966); C. A., 65, 13763a (1966); Addn. to Fr. 1,413,874; C. A., 64, 4941h (1966).
200. Demosay, D., D. Pillon, and J. Ducret, Ger. Offen. 1,939,418 (1970); C. A., 72, 111304h (1970).
201. Derkach, G. I., and V. P. Belaya, J. Gen. Chem., 36, 1936 (1966); C. A., 66, 65207e (1967); cf. Speziale and Smith, C. A., 58, 5546e (1963).
202. Derkach, G. I., E. S. Gubnitskaya, and L. I. Samarai, J. Gen. Chem., 36, 1942 (1966); C. A., 66 , 65208 (1967).
203. Dewald, C. L., and L. W. Fancher, Fr. 1,327,963 (1963); C. A., 59, 13886c (1963).
204. De Witt, E. G., U. S. 2,866,805 (1958); C. A., 53, 15988c (1959).
205. Diamond, M. J., U. S. 2,957,016 (1960); C. A., 55, 3431g (1961).
206. Dickhaeuser, H., H. Pohlemann, H. Stummeyer, H. Adolphi, and S. Winderl, Belg. 611,303 (1962); C. A., 58, 4425h (1963).

207. Dickhaeuser, H., G. Scheuerer, and H. Adolphi, Ger. 1,217,134 (1966); C. A., 65, 5374d (1966).
208. Dickore, K., K. Sasse, L. Eue, and R. Heiss, Ger. 1,174,757 (1964); C. A., 61, 13244e (1964).
209. Diveley, W. R., U. S. 2,864,741 (1958); C. A., 53, 9057i (1959).
209a. Diveley, W. R., U. S. 2,725,327 (1955); C. A. 50 5973i (1956).
210. Diveley, W. R., U. S. 2,864,826 (1958); C. A., 53, 9058b (1959).
211. Diveley, W. R., and A. D. Lohr, U. S. 2,725,328 (1955); C. A., 50, 5972h (1956). Diveley, W. R., U. S. 2,725,327 (1955).
212. Diveley, W. R., A. H. Haubein, A. D. Lohr, and P. B. Moseley, J. Am. Chem. Soc., 81, 139 (1959); C. A., 53 13152i (1959).
213. Divinskii, A. F., M. I. Kabachnik, and V. V. Sidorenko, Dokl. Akad. Nauk SSSR, 60, 999 (1948); C. A., 43, 560g (1949).
214. Dixon, J. K., S. Du Breuil, and N. L. Boardway, U. S. 2,918,468 (1959); C. A., 54, 9971b (1960).
215. Doerken, A., Fr. 1,374,714 (1964); C. A., 62, 3944a (1965).
216. Doerken, A., and G. Schrader, Ger. 947,367 (1956); C. A., 51, 4424g (1957).
217. Doerken, A., and G. Schrader, U. S. 2,927,123 (1960); C. A., 54, 17267g (1960).
218. Doerken, A., and G. Schrader, Ger. 1,047,776 (1958); C. A., 55, 5348e (1961).
219. Doerken, A., and G. Schrader, Brit. 941,631 (1963); C. A., 60, 6870c (1964).
220. Doerken, A., and G. Schrader, Ger. 1,186,467 (1965); C. A., 62, 16149b (1965).
221. Donninger, C., and D. H. Hutson, Tetrahedron Lett., 47, 4871 (1968); C. A., 70, 11388 (1969).
222. Donninger, C., J. A. Schofield, and P. D. Regan, Ger. Offen. 1,815,399 (1969); C. A., 72, 12091e (1970).
223. Doroshenko, V. V., E. A. Stukalo, and A. V. Kirsanov, J. Gen Chem., 41, 1653 (1971); C. A., 75, 110,380 (1971).
224. Dow Chemical Co., Brit. 807,598 (1959); C. A., 53, 15009a (1959).
225. Drabek, J., and J. Macko, Czech. 122,143 (1967); C. A., 68, 49180b (1968).
226. Drabek, J., I. Pastorek, and S. Gaher, Czech. 126,088 (1968); C. A., 70, 19799j (1969).
227. Drabek, J., I. Pastorek, A. Jaras, S. Gaher, S. Priehradny, S. Misiga, J. Macko, and V. Batora, Czech. 124,138 (1967); C. A., 69, 96227c (1968).
228. Drabek, J., and Z. Vesela, Czech. 129,660 (1968); C. A., 71, 91069j (1969).
229. Drake, L. R., U. S. 2,615,039 (1952); C. A., 47,

9343g (1953).
229a. Drake, L. R., and A. J. Erbel, U. S. 2,599,512 (1952); C. A., 46, 8321i (1952).
230. Dubini, M., Ann. Chim., 53, 1421 (1963); C. A., 60, 6777f (1964).
231. Du Breuil, S., U. S. 2,759,937 (1956); C. A., 51, 2884j (1957).
232. Du Breuil, S., U. S. 2,759,938 (1956); C. A., 51, 2885d (1957).
233. Du Breuil, S., U. S. 2,938,902 (1960); C. A., 54, 21146e (1960).
234. Du Breuil, S., J. Org. Chem., 26, 3382 (1961); C. A., 56, 11591a (1962).
235. Du Breuil, S., R. W. Young, D. E. Ailman, and G. Berkelhammer, Ger. 1,164,408 (1964); C. A., 60, 15735a (1964).
236. Durig, J. R., and J. W. Clark, J. Chem. Phys., 50, 107 (1969); C. A., 70, 62610 (1969).
237. Durig, J. R., and J. W. Clark, J. Cryst. Mol. Struct. 1, 43 (1971); C. A., 75 12757 (1971).
238. Durig, J. R., and J. S. Di Yorio, J. Mol. Struct., 3, 179 (1969); C. A., 70 72369 (1969).
239. Dvornikoff, M. N., and E. J. Young, U. S. 2,633,721 (1953); C. A., 48, 13717i (1954).
240. Dwek, R. A., and R. E. Richards, Chem. Commun., 1966, 581; C. A., 56, 16277d (1966).
241. Dwek, R. A., R. E. Richards, R. Taylor, and R. A. Shaw, J. Chem. Soc. A, 1970, 1173; C. A., 73, 19855 (1970).
242. Dye, W. T., Jr., U. S. 2,730,541 (1956); C. A., 50 11362 (1956).
243. Eckstein, F., J. Am. Chem. Soc., 88, 4292 (1966); C. A., 65, 18674 (1966).
244. Eckstein, F., Tetrahedron Lett., 1967, 1157; C. A., 67, 82357 (1967).
245. Edmundson, R. S., Tetrahedron, 20, 2781 (1964); C. A., 62, 6369b (1965).
246. Edmundson, R. S., Chem. Ind., 1965, 1220; C. A., 63, 7014a (1965).
247. Edmundson, R. S., Tetrahedron, 21, 2379 (1965); C. A., 63, 17875g (1965).
248. Edmundson, R. S., and A. J. Lambie, Chem. Ind., 1959, 1048; C. A., 54 3392e (1960).
249. Edmundson, R. S., and A. J. Lambie, J. Chem. Soc., C, Org., 1966, 1997; C. A., 66, 10512e (1967).
250. Edmundson, R. S., and A. J. Lambie, J. Chem. Soc., C, Org., 1966, 2001; C. A., 66, 10513h (1967).
251. Edmundson, R. S., and E. W. Mitchell, J. Chem. Soc., C, 1968, 2091; C. A., 69, 76438 (1968).
252. Eliseenkov, V. N., J. Gen. Chem., 37, 1946 (1967); C. A., 68, 48962 (1968).
253. Emmett, W. G., and H. O. Jones, J. Chem. Soc., 99, 715 (1911); C. A., 5, 3048 (1911).

254. Enders, E., and G. Ünterstenhoefer, U. S. 3,071,594 (1963); C. A., 58, 11330a (1963).
255. English. J. P., and J. J. Hand, U. S. 3,526,681 (1970); C. A., 73, 98609k (1970).
256. Ephraim, F., Ber., 44, 633 (1911); C. A., 5, 1783 (1911).
257. Epstein, P. F., U. S. 3,341,553 (1967); C. A., 68, 29585j (1968).
258. Epstein, P. F., and M. E. Brokke, U. S. 3,304,226 (1967); C. A., 67, 32598a (1967).
259. Esso Research and Engineering Co., Brit. 1,071,033 (1967); C. A., 67, 63766h (1967).
260. Esso Research and Engineering Co., Fr. 1,567,457 (1969); C. A., 72, 121008d (1970).
261. Estok, G. K., and W. W. Wendlandt, J. Am. Chem. Soc., 77, 4767 (1955); C. A., 50, 3821i (1956).
262. Ethyl Corp., Neth. Appl. 6,615,236 (1967); C. A., 68, 29424f (1968).
263. Eto, M., Y. Kinoshita, T. Kato, and Y. Oshima, Agr. Biol. Chem. (Tokyo), 27, 789 (1963); C. A., 60, 7944c (1964).
264. Eto, M., K. Kobayashi, T. Sasamoto, H. M. Cheng, T. Aikawa, T. Kume, and Y. Oshima, Bochu-Kagaku, 33, 73 (1968); C. A., 70, 28895q (1969).
265. Ettel, V., and M. Zbirovsky, Chem. Listy, 50, 1261 (1956); C. A., 50, 16025f (1956).
266. Ettel, V., and M. Zbirovsky, Chem. Listy, 50, 1265 (1956); C. A., 50, 16025h (1956).
267. Evdakov, V. P., and E. I. Alipova, J. Gen. Chem., 37, 412 (1967); C. A., 67 43880 (1967).
268. Fancher, L. W., U. S. 2,767,194 (1956); C. A., 51, 3915b (1957).
269. Fancher, L. W., U. S. 2,793,224 (1957); C. A., 51, 14196i (1957).
270. Fancher, L. W., U. S. 3,591,600 (1971); C. A., 75, 77030 (1971).
271. Fancher, L. W., and C. L. Dewald, U. S. 3,205,253 (1965); C. A., 64, 2015d (1966).
272. Fancher, L. W., and S. C. Dorman, U. S. 3,128,225 (1965); C. A., 60, 15778c (1964).
273. Fancher, L. W., and R. A. Gray, U. S. 3,504,057 (1970); C. A., 72, 132288w (1970).
274. Fancher, L. W., and R. A. Gray, U. S. 3,520,956 (1970); C. A., 73, 109484u (1970).
275. Fancher, L. W., and J. T. Hallett, Belg. 630,933 (1963); C. A., 60, 15788h (1964).
276. Fancher, L. W., A. M. Imel, and R. C. Maxwell, U. S. 3,076,807 (1963); C. A., 59, 5180h (1963).
277. Fancher, L. W., and G. G. Patchet, Belg. 609,383 (1962); C. A., 57, 16496e (1962).
278. Farbenfabriken Bayer A.G., Brit. 786,013 (1957); C. A., 52, 5450h (1958).
279. Farbenfabriken Bayer A.G., Brit. 797,307 (1958); C.

A., 53, 3058e (1959).
280. Farbenfabriken Bayer A.G., Brit. 803,441 (1958); C. A., 54, 4502a (1960).
281. Farbenfabriken Bayer A.G., Brit. 804,141 (1958); C. A., 53, 11224d (1959).
282. Farbenfabriken Bayer A.G., Brit. 806,148 (1958); C. A., 53, 14937h (1959).
283. Farbenfabriken Bayer A.G., Brit. 808,259 (1959); C. A., 54, 1303b (1960).
284. Farbenfabriken Bayer A.G., Brit. 808,304 (1959); C. A., 53, 10652h (1959).
285. Farbenfabriken Bayer A.G., Brit. 808,879 (1959); C. A., 53, 16069i (1959).
286. Farbenfabriken Bayer A.G., Brit. 812,065 (1959); C. A., 54, 415d (1960).
287. Farbenfabriken Bayer A.G., Brit. 813,966 (1959); C. A., 53, 198786d (1959).
288. Farbenfabriken Bayer A.G., Brit. 814,332 (1959); C. A., 54, 17330b (1960).
289. Farbenfabriken Bayer A.G., Brit. 815,160 (1959); C. A., 54, 1315h (1960).
290. Farbenfabriken Bayer A.G., Brit. 816,286 (1959); C. A., 54, 1299b (1960).
291. Farbenfabriken Bayer A.G., Brit. 819,672 (1959); C. A., 54, 4497f (1960).
292. Farbenfabriken Bayer A.G., Brit. 822,476 (1959); C. A., 54, 7559a (1960).
293. Farbenfabriken Bayer A.G., Brit. 823,732 (1959); C. A., 54, 10860d (1960.
294. Farbenfabriken Bayer A.G., Brit. 825,477 (1959); C. A., 54, 12998b (1960).
295. Farbenfabriken Bayer A.G., Brit. 834,262 (1960); C. A., 55, 380d (1961).
296. Farbenfabriken Bayer A.G., Brit. 866,422 (1961); C. A., 51, 13805a (1957).
297. Farbenfabriken Bayer A.G., Brit. 922,378 (1963); C. A., 59, 10123h (1963).
298. Farbenfabriken Bayer A.G., Brit. 924,034 (1963); C. A., 61, 5566h (1964).
299. Farbenfabriken Bayer A.G., Brit. 930,918 (1963); C. A., 60, 1645f (1964).
300. Farbenfabriken Bayer A.G., Brit. 967,081 (1964); C. A., 62, 11854a (1965).
301. Farbenfabriken Bayer A.G., Neth. Appl. 6,400,465 (1964); C. A., 62, 9139e (1965).
302. Farbenfabriken Bayer A.G., Neth. Appl. 6,400,564 (1964); C. A., 62, 10443f (1965).
303. Farbenfabriken Bayer A.G., Neth. Appl. 6,408,976 (1965); C. A., 62, 16213d (1965).
304. Farbenfabriken Bayer A.G., Neth. Appl. 6,415,326 (1965); C. A., 64, 1962f (1966).
305. Farbenfabriken Bayer A.G., Neth. Appl. 6,605,512 (1966); C. A., 66, 95091e (1967).

306. Farbenfabriken Bayer A.G., Neth. Appl. 6,605,907 (1966): C. A., 67, 22013s (1967).
307. Farbenfabriken Bayer A.G., Neth. Appl. 6,607,054 (1966); C. A., 66, 95085f (1967).
308. Farbenfabriken Bayer A.G., Neth. Appl. 6,607,675 (1966); C. A., 67, 64532 (1967).
309. Farbenfabriken Bayer A.G., Neth. Appl. 6,607,832 (1967): C. A., 67, 11199f (1967).
310. Farbenfabriken Bayer A.G., Neth. Appl. 6,611,860 (1967); C. A., 68, 49297 (1968).
311. Farbenfabriken Bayer A.G., Fr. 1,528,547 (1968); C. A., 71, 3467j (1969).
312. Farbenfabriken Bayer A.G., Fr. 1,570,778 (1969); C. A., 72, 132285t (1970).
313. Farbenfabriken Bayer A.G., Fr. 1,583,320 (1969); C. A., 73, 56231a (1970).
314. Farbenfabriken Bayer, A.G., Fr. 1,600,932 (1970); C. A., 74, 125186z (1971).
315. Farbenfabriken Bayer, A.G., Fr. Demande 2,011,171 (1970); C. A., 73, 109491u (1970).
316. Farbwerke Hoechst A.G., Neth. Appl. 6,607,822 (1966); C. A., 68; 21922w (1968).
317. Fearing, R. B., S. African 2,305 (1968); C. A., 71, 21702m (1969).
318. Fearing, R. B., and M. B. McClellan, U. S. 3,391,230 (1968); C. A., 69, 51579e (1968).
319. Feher, F., and A. Blumcke, Chem. Ber., 90, 1934 (1957); C. A., 54, 19461c (1960).
320. Feher, F., and K. Lippert, Chem. Ber., 94, 2437 (1961); C. A., 56, 3511d (1962).
321. Fest, C., Belg. 636,509 (1964); C. A., 62, 2796c (1965).
322. Fest, C., I. Hammann, W. Stendel, and G. Unterstenhoefer, S. African 69 00,042 (1969); C. A., 72, 100734 (1970).
323. Fest, C., and G. Schrader, Ger. 1,161,275 (1964); C. A., 60, 10718a (1964).
324. Fields, E. K., J. Am. Chem. Soc., 78, 5821 (1956); C. A., 51, 2586e (1957).
325. Fife, T. H., and S. Milstien, J. Org. Chem., 34, 4007 (1969); C. A., 72, 42465 (1970).
326. Finegold, H., Ann. N.Y. Acad. Sci., 70, 875 (1958); C. A., 52, 15238i (1958).
327. Fiszer, B., J. Michalski, and J. Wieczorkowski, Rocz. Chem., 27, 482 (1953); C. A., 49, 3786e (1955); Michalski, J., and J. Wieczorkowski, Bull. Acad. Pol. Sci, Classe III, 5, 917-21 (1957); C. A., 52, 6157g (1958).
328. Fitch, H. M., Brit. 681,102 (1952); C. A., 47, 11221i (1953); U. S. 2,759,961 (1956); C. A., 51 482b (1957).
329. Fitch, H. M., U. S. 2,901,503 (1959); C. A., 54, 1432b (1960).

330. Fitch, H. M., U. S. 2,911,430 (1959); C. A., 54, 4386h (1960).
331. Fletcher, J. H., U. S. 2,520,393 (1950); C. A., 45, 655a (1951).
332. Fletcher, J. H., U. S. 2,664,437 (1953); C. A., 48, 13717h (1954).
333. Fletcher, J. H., J. C. Hamilton, I. Hechenbleikner, E. I. Hoegberg, B. J. Sertl, and J. T. Cassaday, J. Am. Chem. Soc., 70, 3943 (1948); C. A., 43, 1313i (1949).
334. Fletcher, J. H., J. C. Hamilton, I. Hechenbleikner, E. I. Hoegberg, B. J. Sertl, and J. T. Cassaday, J. Am. Chem. Soc., 72, 2461 (1950); C. A., 44, 9933c (1950).
335. Fletcher, J. H., and E. I. Hoegberg, U. S. 2,630,451 (1953); C. A., 48, 7046h (1954).
336. Floyd, A. J., U. S. 3,260,719 (1966); C. A., 65, 15403b (1966).
337. Floyd, A. J., and R. Ghosh, Brit. 923,253 (1963); C. A., 59, 9908c (1963).
338. Floyd, A. J., and R. C. Hinton, Belg. 616,760 (1962); C. A., 59, 14024e (1963).
339. Fluck, E., H. Gross, H. Binder, and J. Gloede, Z. Naturforsch., b 21, 1125 (1966); C. A., 66, 89995m (1967).
340. Ford-Moore, A. H., and G. W. Wood, Brit. 851,590 (1960); C. A., 55 , 10316d (1961).
341. Foss, O., J. Am. Chem. Soc., 69, 2236 (1947); C. A., 42, 142c (1948).
342. Foss, O., Acta Chem. Scand., 1, 8 (1947); C. A., 42, 2537c (1948).
343. Foss, O., Acta Chem. Scand., 1, 307 (1947); C. A., 42 2240f (1948).
344. Friederang, A. W., D. S. Tarbell, and S. Ebine, J. Org. Chem., 34, 3825 (1969); C. A., 72, 43075 (1970).
345. Fukuto, T. R., and R. L. Metcalf, J. Am. Chem. Soc., 76, 5103 (1954); C. A., 49, 12271g (1955).
346. Fukuto, T. R., R. L. Metcalf, M. Frederickson, and M. Y. Winton, J. Agr. Food Chem., 12, 228 (1964); C. A., 61, 1808d (1964).
347. Fukuto, T. R., R. L. Metcalf, R. B. March, and M. Maxon, J. Am. Chem. Soc., 77, 3670 (1955); C. A., 49 13577a (1955).
348. Furdik, M., and V. Sutoris, Acta Fac. Rerum Natur. Univ. Comenianae, Chim., 9, 15 (1965); C. A., 65, 16855e (1966).
349. Fusco, R., G. Losco, and C. A. Peri, U. S. 2,860,085 (1958); C. A., 53, 9560d (1959).
350. Fusco, R., G. Losco, and M. Perini, Ital. 568,602 (1957); C. A., 53, 14940a (1959).
351. Fusco, R., G. Losco, and M. Perini, Ital. 724,419 (1966); C. A., 70, 11366 (1969).

352. Fusco, R., G. Rossi, P. de P. Tonelli, and A. Barontini, Belg. 621,355 (1963); C. A., 59, 11351b (1963).
353. Gaetzi, K., Ger. 1,137,006 (1962); C. A., 59, 9908h (1963).
353a. Gaetzi, K. and P. Mueller, D. B. P. 957,213 (1955).
354. Galashina, M. A., M. V. Sobolevskii, K. A. Andrianov, and T. P. Alekseeva, Plast. Massy, 1962, 16; C. A., 58, 544e (1963).
355. Galashina, M. L., and N. N. Mel'nikov, Zh. Obshch. Khim., 23, 1539 (1953); C. A., 48, 10649i (1954).
356. Galashina, M. L., I. L. Vladimirova, Ya. A. Mandel'baum, and N. N. Mel'nikov, J. Gen Chem. USSR, 23, 441 (1953); C. A., 48, 3887h (1954).
357. Gauss, W., and O. Bayer, Belg. 619,334 (1962); C. A., 59, 11382d (1963).
358. Gauss, W., and O. Bayer, Ger. 1,167,838 (1964); C. A., 61, 1841c (1964).
359. Gauss, W., C. W. Schellhammer, and M. Broemmelhues, Ger. 1,154,113 (1963); C. A., 60, 7968b (1964).
360. Gay, D. C., and N. K. Hamer, J. Chem. Soc. B, 1970, 1123; C. A., 73, 65659y (1970).
361. Gazieva, N. I., L. V. Abramova, A. I. Shchekotikhin, and V. A. Ginsburg, J. Gen, Chem., 38, 1154 (1968); C. A., 69, 86259 (1968).
362. Geigy, A. G., U. S. 2,736,726 (1956).
363. Geigy, A. G., Fr. 1,331,721 (1963); C. A., 60, 1762f (1964).
364. Geigy, A. G., Brit. 978,854 (1964); C. A., 63, 1796d (1965).
365. Ghosh, R., Brit. 738,839 (1955), U. S. 2,863,901 (1958); C. A., 50, 13983h (1956).
366. Ghosh, R., Brit. 763,516 (1956); C. A., 51, 15598d (1957).
367. Gilbert, E. E., and C. J. McGough, U. S. 2,690,450 (1954); C. A., 49, 11682i; (1955); U. S. 2,690,451; C. A., 49, 11683c (1955).
368. Ginak, A. I., K. A. V'yunov, and E. G. Sochilin, J. Gen. Chem., 40, 1410 (1970); C. A., 74, 22752 (1971).
369. Gleim, W. K. T., and R. B. Thompson, U. S. 3,071,548 (1963); C. A., 58, 12474 (1963).
370. Glushkova, L. V., N. I. Zemlyanskii, and O. A. Prib, J. Gen. Chem., 37, 1575 (1967); C. A., 68, 12593n (1968).
371. Godovikov, N. N., and M. I. Kabachnik, J. Gen. Chem., 31, 1516 (1961); C. A., 55, 22200f (1961).
372. Gololobov, Yu. G., T. F. Dmitrieva, and L. Z. Soborovski, Probl. Org. Sinteza, Akad. Nauk SSSR, Otd. Obshch. Tekh. Khim., 1965, 314; C. A., 64, 6684a (1966).
373. Gololobov, Yu. G., V. N. Kulakova, Yu. M. Zinov'ev, and L. Z. Soborovskii, Khim. Org. Soedin. Fosfora, Akad. Nauk SSSR, Otd. Obshch. Tekh. Khim., 1967, 197;

C. A., 68, 114710 (1968).
374. Gore, R. C., Discussions Faraday Soc., 1950, No. 9, 138; C. A., 46, 3408e (1952).
375. Gore, R. C., R. W. Hannah, S. C. Pattacini, and T. J. Porro, J. AOAC, 54, 1040 (1971); C. A., 75, 117202g (1971).
376. Gottlieb, H. B., J. Am. Chem. Soc., 54, 748 (1932); C. A., 26, 1590 (1932).
377. Gozman, I. P., Bull. Acad. Sci. USSR, Ser. Chem., 1968, 2229; C. A., 70, 37090a (1969).
378. Greenbaum, S. B., Belg. 624,225 (1963); C. A., 59, 11429h (1963).
379. Greenbaum, S. D., Fr. 1,347,337 (1963); C. A., 60, 10651a (1964).
380. Greenbaum, S. B., and N. E. Boyer, U. S. 3,146,253 (1964); C. A., 62, 1678a (1965).
381. Griesbaum, K., A. A. Oswald, and D. N. Hall, U. S. 3,591,475 (1971); C. A., 75, 98157c (1971).
382. Grimmer, F., W. Dedek, and E. Leibnitz, Z. Naturforsch., B 23, 10 (1968); C. A., 68, 113716k (1968).
383. Gritsai, N. I., G. G. Vil'danova, G. A. Bokalo, and N. I. Zemlyanskii, J. Gen. Chem., 40, 1961 (1970); C. A., 74, 53209 (1971).
384. Gubler, K., U. Meyer, and H. U. Brechbuehler, U. S. 3,535,325 (1970); C. A. 71, 124424b (1969).
385. Gunar, M. I., T. N. Shumyatskaya, E. B. Mikhalyutina, K. D. Shvetsova-Shilovskaya, and N. N. Mel'nikov, J. Gen. Chem., 38, 2182 (1968); C. A., 70, 28524t (1969).
386. Gupalo, A. P., and N. I. Zemlyanskii, J. Gen. Chem., 37, 2596 (1967); C. A., 69, 66828g (1968).
387. Gupalo, A. P., and N. I. Zemlyanskii, J. Gen. Chem., 38, 2464 (1968); C. A., 70, 68070y (1969).
388. Gupalo, A. P., and N. I. Zemlyanskii, J. Gen. Chem., 38, 2683 (1968); C. A., 70, 77228h (1969).
389. Gupalo, A. P., N. I. Zemlyanskii, and I. V. Murav'ev, Khim. Org. Soedin. Fosfora, Akad. Nauk SSSR, Otd. Obshch. Tekh. Khim., 1967, 227; C. A., 69, 76552b (1968).
390. Gutman, A. D., S. African 3662/68 (1968); C. A., 71, 30236q (1969).
391. Gutman, A. D., and D. J. Broadbent, U. S. 3,299,099 (1967); C. A., 66, 115716t (1967).
392. Gutman, A. D., and J. P. Orr, U. S. 3,351,680 (1967); C. A., 68, 49101b (1968).
393. Gutman, A. D., D. G. Stoffey, and J. T. Hallett, U. S. 3,268,628 (1966); C. A., 65, 16869h (1966).
394. Gysin, H., and A. Margot, U. S. 2,754,243 and U. S. 2,754,302 (1956); C. A., 50, 12122g (1956); Brit. 713,278 (1954); C. A., 50, 1092i (1956).
395. Hackmann, J. T., and J. Wood, Brit. 986,196 (1965); C. A., 62, 16070a (1965).
396. Hackmann, J. T. and J. Wood, Ger. 1,229,776 (1966); C. A., 66, 94804c (1967).

397. Haga, K., M. Kainosho, and M. Yoshikawa, Bull. Chem. Soc. Jap., 44, 460 (1971); C. A., 74, 112364 (1971).
398. Hahn, W., H. G. Schicke, and W. Behrenz, Belg. 663,781 (1965); C. A., 65, 8960f (1966).
399. Hall, S. A., J. Am. Chem. Soc., 72, 2768 (1950); C. A., 44, 10673g (1950).
400. Hamer, N. K., and D. C. Gay, J. Chem. Soc ., D, 1970, 1564; C. A., 74, 99168 (1971).
401. Hanic, F., and J. A. Cakajdova, Acta Cryst., 11, 127 (1958); C. A., 52, 19332d (1958).
402. Harding, D., and G. O. Osborne, Aust. J. Chem., 21, 1093 (1968); C. A., 69, 36084u (1968).
403. Harris, R. K., A. R. Katritsky, S. Musierowicz, and B. Ternai, J. Chem. Soc., A. 1967, 37; C. A., 66, 42143s (1967).
404. Harukawa, C., Japan 13,079 (1963); C. A., 60, 2910e (1964).
405. Harukawa, T., and T. Ishikawa, Japan 10,508 (1960); C. A., 55, 9441f (1961).
406. Harukawa, C., and K. Konishi, Japan 7,976 (1962); C. A., 59, 5141g (1963).
407. Harukawa, C., and K. Konishi, Japan 4,870 (1962); C. A., 59, 1488h (1963).
408. Harukawa, C., and K. Konishi, Japan 18,736 (1962); C. A., 59, 11565a (1963).
409. Harukawa, C., and K. Konishi, Japan 5,672 (1963); C. A., 59, 12766b (1963).
410. Harukawa, C., and K. Konishi, Japan 20,565 (1963); C. A., 60, 2766f (1964).
411. Harukawa, C., and K. Konishi, Japan 20,567 (1963); C. A., 60, 2766g (1964).
412. Harukawa, C., and K. Konishi, Japan 19,294 (1963); C. A., 60, 4160d (1964).
413. Harukawa, C., and K. Konishi, Japan 26,180 (1963); C. A., 60, 5516a (1964).
414. Harukawa, D., and K. Konishi, Japan 3,022 (1964); C. A., 60, 15734h (1964).
415. Harukawa, C., and K. Konishi, Japan 1,917 (1964); C. A., 61, 576c (1964).
416. Harukawa, C., and K. Konishi, Japan 24,420 (1967); C. A., 69, 35702g (1968).
417. Harvey, R. G., H. I. Jacobson, and E. V. Jensen, J. Am. Chem. Soc., 85, 1623 (1963); C. A., 59, 1467h (1963).
418. Harvey, R. G., and E. V. Jensen, J. Org. Chem., 28, 470 (1963); C. A., 58, 9130e (1963).
419. Haubein, A. H., J. Am. Chem. Soc., 81, 144 (1959); C. A., 53, 13154 (1959).
420. Haubein, A. H., U. S. 3,316,145 (1967); C. A., 66, 2230p (1967).
421. Hechenbleikner, I., U. S. 2,482,063 (1949); C. A., 44, 4022a (1950).
422. Hechenbleikner, I., C. W. Pause, and F. C. Lanoue, U.

S. 2,834,798 (1958); C. A., 52 17297g (1958).

423. Heiss, R., K. Mannes, H. Pelster, G. Unterstenhoefer, and W. Behrenz, Belg. 641,528 (1964); C. A., 63, 4208b (1965).

424. Hensel, J., and P. C. Aichenegg, U. S. 3,184,377 (1965); C. A., 63, 4163b (1965).

425. Herail, F., Compt. Rend., 261, 3375 (1965); C. A., 64, 7536b (1966).

426. Herail, F., Compt. Rend., Ser. C, 262, 1493 (1966); C. A., 65, 8195e (1966).

427. Herail, F., Compt. Rend., Ser. C, 262, 1624 (1966); C. A., 65, 11554d (1966).

428. Herail, F., J. Chim. Phys. Physicochim. Biol., 68, 274 (1971); C. A., 74, 148807s (1971).

429. Herail, F., and V. Viossat, Compt. Rend., 259, 4629 (1964); C. A., 62, 11305f (1965).

430. Hercules Powder Co., Neth. Appl. 6,516,125 (1966); C. A., 65, 16910e (1966).

431. Hewitt, R. I., and G. Berkelhammer, U. S. 3,179,560 (1965); C. A., 63, 1738d (1965).

432. Hilgetag, G., and G. Lehmann, J. Prakt. Chem. (4), 9, 3 (1959); C. A., 54, 9811d (1960).

433. Hilgetag, G., and G. Lehmann, Angew. Chem., 69, 506 (1957); C. A., 54, 24487c (1960).

434. Hilgetag, G., G. Lehmann, and W. Feldheim, J. Prakt. Chem., 12, 1 (1960); C. A., 55, 9256g (1961).

435. Hilgetag, G., G. Lehmann, A. Martini, G. Schramm, and H. Teichmann, J. Prakt. Chem., 8, 207 (1959); C. A., 53, 21748b (1959).

436. Hilgetag, G., G. Schramm, and H. Teichmann, J. Prakt. Chem., 8, 73 (1959); C. A., 53, 19847f (1959).

437. Hilgetag, G., and H. Teichmann, J. Prakt. Chem., 8, 90 (1959); C. A., 53, 19848c (1959).

438. Hilgetag, G., and H. Teichmann, J. Prakt. Chem (4), 8, 121 (1959); C. A., 53, 21615f (1959).

439. Hilgetag, G., H. Teichmann, and M. Krueger, Ger. 36,439 (1965); C. A., 63, 13154b (1965).

440. Hilgetag, G., H. Teichmann, and M. Krueger, Ger. 1,443,711 (1970); C. A., 72, 111011k (1970).

441. Hilgetag, G., H. Teichmann, and I. Schwandt, Monatsber, Deut. Akad. Wiss. Berlin, 8, 747 (1966); C. A., 68, 86791x (1968).

442. Hodan, J. J., and H. Tieckelmann, J. Org. Chem., 26, 4429 (1961); C. A., 56, 15507e (1962).

443. Hodnett, E. M., T. E. Moore, and J. E. Lothers, Jr., Proc. Okla. Acad. Sci., 39, 141 (1959); C. A., 54, 3175i (1960).

444. Hoegberg, E. I., U. S. 2,494,126 (1950); C. A., 44, 3515g (1950).

445. Hoegberg, E. I., U. S. 2,632,020 (1953); C. A., 48, 2759h (1954).

446. Hoegberg, E. I., U. S. 2,644,002 (1953); C. A., 48, 5206d (1954).

447. Hoegberg, E. I., and J. T. Cassaday, J. Am. Chem. Soc., 73, 557 (1951); C. A., 45, 5609h (1951).
448. Hoffmann, F. W., J. W. King, and H. O. Michel, J. Am. Chem. Soc., 83, 706 (1961); C. A., 55, 12280h (1961).
449. Hoffmann, F. W., and T. R. Moore, J. Am. Chem. Soc., 80, 1150 (1958); C. A., 52, 13613i (1958).
450. Hoffmann, F. W., T. R. Moore, and B. Kagan, J. Am. Chem. Soc., 78, 6413 (1956); C. A., 51, 4932b (1957).
451. Hoffmann, H., Brit. 1,158,709 (1969); C. A., 72, 43126z (1970).
452. Hoffmann, H., Brit. 1,188,430 (1970); C. A., 72, 132043n (1970).
453. Hoffmann, H., and H. Scheinpflug, S. African 2053 (1968); C. A., 70, 87308y (1969).
454. Hoffmann, H., and H. Scheinpflug, S. African 1040 (1969); C. A., 72, 78648y (1970).
455. Hoffmann, H., H. Scheinpflug, and T. Y. Kume, Ger. Offen. 1,930,264 (1970); C. A., 73, 35035z (1970).
456. Holtschmidt, H., U. S. 3,013,048 (1961); C. A., 57, 11105c (1962).
457. Holtschmidt, H., and H. Wilms, U. S. 3,014,049 (1961); C. A., 57, 4594i (1962).
458. Hook, E. O., and P. H. Moss, U. S. 2,566,288 (1941); C. A., 46, 2563b (1952).
458a. Hook, E. O., and P. H. Moss, U. S. 2,596,076 (1952); C. A., 46, 8322g (1952).
459. Hooker Chemical Corp., Brit. 889,338 (1962); C. A., 58, 8906h (1963).
460. Hooker Chemical Corp., Neth. Appl. 285,931 (1965); C. A., 63, 8212f (1965).
461. Hooker Chemical Corp., Brit. 1,153,741 (1969); C. A., 72, 21626k (1970).
462. Hu, P. F., and W. Y. Chen, Hua Hsueh Hsueh Pao, 22, 215 (1956); C. A., 52, 7186c (1958).
463. Hu, P. F., and W. Y. Chen, Hua Hsueh Hsueh Pao, 24, 112 (1958); C. A., 53, 3120h (1959).
464. Hu, P. F., S. C. Li, and W. Y. Chen, Hua Hsueh Hsueh Pao, 22, 49 (1956); C. A., 52, 6156h (1958).
465. Hubele, A., S. African 4420 (1968); C. A., 71, 123964 (1969).
466. Hurwic, J., S. Smialek-Kazmierowska, and B. Maliszewski, Rocz. Chem., 40, 487 (1966); C. A., 65, 3132d (1966).
466a. Ikeda, S., S. Takahashi, H. Ueda, M. Kondo, and M. Hayashi, Japan 16,779 (1966); C. A., 66, 18510p (1967).
467. Imaev, M. G., J. Gen. Chem., 35, 1857 (1965); C. A., 64, 1991h (1966).
468. Imaev, M. G., and R. A. Faskhutdinova, J. Gen. Chem., 31, 2736 (1961); C. A., 57, 7090c (1962).
469. Imaev, M. G., S. V. Sokolova, and S. D. Felkyaeva, J. Gen. Chem., 35, 741 (1965); C. A., 63, 5550b (1965).
470. ICI Ltd., Belg. 657,705 (1965); C. A., 64, 11250g

(1966).
471. ICI Ltd., Belg. 659,279 (1965); C. A., 64, 3599c (1966).
472. Inoue, T., Y. Kato, N. Kiyama, and T. Kaneki, Japan 3,583 (1969); C. A., 70, 87274j (1969).
473. Ishida, H., and S. Asaka, Japan 6737 (1967); C. A., 67, 73355n (1967).
474. Ishida, H., and S. Asaka, Japan 14,933 (1968); C. A., 70, 57397j (1969).
475. Ishida, H., S. Asaka, and Y. Kawamura, Japan 15364 (1963); C. A., 60, 2766d (1964).
476. Ishida, H., M. Chibu, S. Asaka, and M. Abiko, Japan 21535 (1965); C. A., 64, 2009e (1966).
477. Ishida, H., T. Yoneda, and B. Sekine, Japan 15405 (1968); C. A., 70, 68504t (1969).
478. Ishmaeva, E. A., R. A. Cherkasov, V. V. Ovchinnikov, and A. N. Pudovik, Bull. Acad. Sci. USSR, Ser. Chem., 1971, 1220; C. A., 75, 98043n (1971).
479. Ito, H., Y. Miyazaki, and Y. Uchiyama, Japan 17028/60 (1960); C. A., 55, 18684g (1961).
480. Itskova, A. L., R. S. Soifer, Ya. A. Mandel'baum, and N. N. Mel'nikov, J. Gen. Chem., 38, 2471 (1968); C. A., 70, 57061 (1969).
481. Jacobson, H. I., R. G. Harvey, and E. V. Jensen, J. Am. Chem. Soc., 77, 6064 (1955); C. A., 50, 8442g (1956).
482. Jamison, J. D., U. S. 3,355,353 (1967); C. A., 69, 2725p (1968).
483. Jamison, J. D., U. S. 3,406,179 (1968); C. A., 70, 11700z (1969).
484. Jarás, A., Chem. Zvesti, 10, 617 (1956), C. A., 51, 12017e (1957).
485. Johnson, G. A., U. S. 2,713,018 (1955); C. A., 50, 2113h (1956).
486. Johnston, T. P., J. R. Piper, C. R. Stringfellow, Jr., and R. D. Elliott, J. Med. Chem., 14, 345 (1971); C. A., 74, 125093 (1971).
487. Johnston, T. P., J. R. Piper, and C. R. Stringfellow, Jr., J. Med. Chem., 14, 350 (1971); C. A., 74, 124939 (1971).
488. Jones, R. A. Y., and A. R. Katritsky, Angew. Chem. Int. Edit., 1, 32 (1962); C. A., 56, 6813d (1962).
489. Joshi, K. C., and S. C. Bahel, J. Indian Chem. Soc., 39, 5 (1962); C. A., 57, 3335h (1962).
490. Kabachnik, M. I., E. I. Golubeva, D. M. Paikin, M. P. Shabanova, N. M. Gamper, and L. F. Efimova, Zh. Obshch. Khim., 29, 1671 (1959); C. A., 54, 8594c (1960).
491. Kabachnik, M. I., S. T. Ioffe, and T. A. Mastryukova, J. Gen. Chem. USSR, 25, 653 (1955); C. A., 50, 3850 (1956).
492. Kabachnik, M. I., and T. A. Mastryukova, J. Gen.

Chem. USSR, 25, 1867 (1955); C. A., 50, 8499c (1956).
493. Kabachnik, M. I., T. A. Mastryukova, and V. N. Odnoralova, J. Gen. Chem. USSR, 25, 2241 (1955); C. A., 50, 9281b (1956).
494. Kabachnik, M. I., T. A. Mastryukova, N. P. Rodionova, and E. M. Popov, J. Gen. Chem. USSR, 26, 119 (1956); C. A., 50, 13723f (1956).
495. Kabachnik, M. I., T. A. Mastryukova, and A. E. Shipov, J. Gen. Chem., 33, 315 (1963); C. A., 59, 658d (1963).
496. Kabachnik, M. I., T. A. Mastryukova, A. E. Shipov, and T. A. Melent'eva, Tetrahedron, 9, 10 (1960); C. A., 54, 19555f (1960).
497. Kabachnik, M. I., P. A. Rossiiskaya, M. P. Shabanova, D. M. Paikin, L. F. Efimova, and N. M. Gamper, J. Gen. Chem., 30, 2201 (1960); C. A., 57, 4531i (1962).
498. Kado, M., and T. Maeda, Japan 3,972 (1965); C. A., 62, 16056c (1965).
499. Kado, M., T. Maeda, and E. Yoshinaga, Fr. 1,445,830 (1966); C. A., 66, 115445d (1967).
500. Kainosho, M., J. Phys. Chem., 74, 2853 (1970); C. A., 73, 60993b (1970).
501. Kalabina, A. V., L. D. Asalkhaeva, E. F. Kolmakova, and T. I. Bychkova, Khim. Atsetilena, 1968, 285; C. A., 71, 3086 (1969).
502. Kalabina, A. V., E. F. Kolmakova, M-Y. Liu, and E. P. Matyunich, Izv. Nauch.-Issled. Inst. Nefte-Uglekhim. Sin. Irkutsk. Univ., 8, 111 (1966); C. A., 72, 121117 (1970).
503. Kalabina, A. V., M-Y. Liu, and L. D. Asalkhaeva, J. Gen. Chem., 34, 1104 (1964); C. A., 61, 1783a (1964).
504. Kalabina, A. V., M-Y. Liu, and L. D. Asalkhaeva, J. Gen Chem., 35, 60 (1965); C. A., 62, 13072a (1965).
505. Kalabina, A. V., M-Y. Liu, L. D. Asalkhaeva, and T. I. Bychkova, J. Gen. Chem., 35, 339 (1965); C. A., 62, 13070g (1965).
506. Kalabina, A. V., M-Y. Liu, and N. N. Pugacheva, Izv. Sib. Otd. Akad. Nauk SSSR, Ser. Khim. Nauk, 1963, 141; C. A., 59, 11308 (1963).
507. Kalashnikov. V. P., J. Gen. Chem., 40, 1939 (1970); C. A., 74, 53212 (1971).
508. Kalashnikov. V. P., and N. I. Zemlyanskii, J. Gen. Chem., 39, 570 (1969); C. A., 71, 38869 (1969).
509. Kamai, G., and A. P. Bogdanov, Tr. Kazan. Khim.-Tekhnol. Inst. Im. S. M. Kirova, 1953, 22; C. A., 51, 5721 (1957).
510. Kamai, G., and F. M. Kharrasova, Zh. Obshch. Khim., 27, 953 (1957); C. A., 52, 3666 (1958).
511. Kamai, G., and F. M. Kharrasova, Tr. Kazan. Khim. Tekhnol. Inst. Im. S. M. Kirova, 23, 122 (1957); C. A., 52, 9981 (1958).
512. Kamai, G., and A. S. Khasanov, J. Gen. Chem., 34,

440 (1964); C. A., 60, 13203c (1964).
513. Kamai, G., and E. S. Koshkina, Tr. Kazan. Khim.-Tekhnol. Inst., 1953, 11; C. A., 50, 6346 (1956).
514. Kano, S., T. Nishide, and T. Noguchi, Japan 10,734 (1970); C. A., 73, 34787j (1970).
515. Kaplan, F., and C. O. Schulz, Chem. Commun., 1967, 376; C. A., 67, 59239 (1967).
516. Karabinos, J. V., R. A. Paulson, and W. H. Smith, J. Res. Nat. Bur. Stand., 48, 322 (1952); C. A., 47, 5877g (1953).
517. Kas'yanova, E. F., and S. M. Gurvich, J. Gen. Chem., 39, 342 (1969); C. A., 71, 3242 (1969).
518. Kataev, E. G., and T. G. Mannafov, J. Gen. Chem., 36, 263 (1966); C. A., 64, 15784d (1966).
519. Kataev, E. G., T. G. Mannafov, and G. I. Kostina, J. Gen. Chem., 37, 1953 (1967); C. A., 68, 49246 (1968).
520. Kataev, E. G., T. G. Mannafov, and G. I. Kostina, J. Gen. Chem., 38, 361 (1968); C. A., 69, 51785 (1968).
521. Kauer, K. C., U. S. 2,884,438 (1959); C. A., 53, 14407b (1959).
522. Kauer, K. C., U. S. 3,365,532 (1968); C. A., 68, 77731 (1968).
523. Kauer, K. C., and H. R. Slagh, U. S. 2,885,429 (1959); C. A., 53, 17061c (1959).
524. Keith, L. H., and A. L. Alford, Anal. Chim. Acta, 44, 447 (1969); C. A., 70, 76682q (1969).
525. Keith, L. H., and A. L. Alford, J. AOAC, 53, 157 (1970); C. A., 72, 77842 (1970).
526. Keith, L. H., A. W. Garrison, and A. L. Alford, J. AOAC, 51, 1063 (1968); C. A., 69, 75849 (1968).
527. Ketelaar, J. A. A., and H. R. Gersmann, J. Am. Chem. Soc., 72, 5777 (1950); C. A., 45, 5653a (1951).
528. Ketelaar, J. A. A., and H. R. Gersmann, Rec. Trav. Chim., 78, 190 (1959); C. A., 53, 16697i (1959).
529. Ketelaar, J. A. A., H. R. Gersmann, and F. Hartog, Rec. Trav. Chim., 77, 982 (1958); C. A., 53, 10920d (1959).
530. Kezerian, C., U. S. 3,562,362 (1971); C. A., 75, 76175e (1971).
531. Kharasch, N., U. S. 2,929,820 (1960); C. A., 54, 15318e (1960).
532. Khasanov, A. S., I. N. Azerbaev, R. D. Kovaleko, Z. A. Navrezova, N. Z. Gabdullina, and M. A. Talasbaeva, Tr. Khim.-Met. Inst., Akad. Nauk Kaz. SSR, 5, 71 (1969); C. A., 72, 121113 (1970).
533. Khaskin, B. A., N. N. Mel'nikov, and I. V. Sablina, J. Gen. Chem., 38, 1512 (1968); C. A., 70, 11511 (1969).
534. Khaskin, B. A., N. N. Mel'nikov, and N. A. Torgasheva, J. Gen. Chem., 41, 524 (1971); C. A., 75, 35647 (1971).
535. Kimura, M., I. Saikawa, T. Maeda, and M. Chanoki, Japan 1,379 (1963); C. A., 59, 11253h (1963).

536. Kimura, M., I. Saikawa, T. Maeda, and M. Chanoki, Japan 17,224 (1962); C. A., 59, 11337e (1963).
537. Kimura, M., I. Saikawa, T. Maeda, and M. Chanoki, Japan 2,524 (1963); C. A., 59, 11341f (1963).
538. Kirby, P., and J. Wood, Brit. 1,175,608 (1969); C. A., 72, 66700 (1970).
539. Kishikawa, Z., K. Takehana, and S. Maekawa, Ger. Offen. 1,924,972 (1969); C. A., 72, 90042d (1970).
540. Klement, R., H. D. Hahne, H. Schneider, and A. Wild, Ber., 97, 1716 (1964); C. A., 61, 4199h (1964).
541. Knunyants, I. L., and E. G. Bykhovskaya, Bull. Acad. Sci. USSR, Ser. Chem., 1966, 1514; C. A., 66, 94647d (1967).
542. Kobayashi, K., M. Eto, S. Hirai, and Y. Oshima, Nippon Nogei Kagaku Kaishi, 40, 315 (1966); C. A., 66, 10883 (1967).
543. Kobayashi, K., M. Eto, Y. Oshima, T. Hirano, T. Hosoi, and S. Wakamori, Bochu-Kagaku, 34, 165 (1969); C. A., 72, 100196 (1970).
544. Kobayashi, K., T. Hirana, S. Wakamori, M. Eto, and Y. Oshima, Bochu-Kagaku, 34, 66 (1969); C. A., 71 , 112896 (1969).
545. Kolesnikova, N. A., N. I. Zemlyanskii, and B. P. Kotovich, J. Gen. Chem., 41, 1218 (1971); C. A., 75, 88044 (1971).
546. Kolesnikova, N. A., N. I. Zemlyanskii, and B. P. Kotovich, J. Gen. Chem., 41, 1440 (1971); C. A., 75, 129433 (1971).
546a. Kolka, A. J., U. S. 2,866,809 (1959); C. A., 53 19879h (1959).
547. Kolmakova, E. F., A. V. Kalabina, Yu. K. Maksyutin and L. N. Spiridonova, Zh. Org. Khim., 2, 2048 (1966); C. A., 66, 75472 (1967).
548. Komkov, I. P., and V. M. Levitskaya, Izv. Vyssh. Ucheb. Zaved., Khim. Khim. Tekhnol., 9, 424 (1966); C. A., 66, 18488n (1967).
549. Komkov, I. P., and V. M. Levitskaya, Izv. Vyssh. Ucheb. Zaved., Khim. Khim. Tekhnol., 10, 1014 (1967); C. A., 68, 39241 (1968).
550. Konecny, V., J. Drabek, and I. Pastorek, Czech. 128,778 (1968); C. A., 70, 114818y (1969).
551. Konecny, V., J. Drabek, and I. Pastorek, Czech. 128,784 (1968); C. A., 70, 114819 (1969).
552. Konecny, V., J. Drabek, and I. Pastorek, Czech. 128,779 (1968); C. A., 70, 115327c (1969).
553. Konishi, K., Takeda Kenkyusho Nempo, 24, 221 (1965); C. A., 64, 8076 (1966).
554. Konishi, K., Takeda Kenkyusho Nempo, 24, 229 (1965); C. A., 64, 8076 (1966).
555. Konishi, K., Japan 8,024 (1967); C. A., 67, 73345j (1967).
556. Konishi, K., Agr. Biol. Chem., 34, 926 (1970); C. A., 73, 44837 (1970).

557. Konishi, K., T. Soma, and R. Iwatani, Takeda Kenkyusho Nempo, 25, 101 (1966); C. A., 66, 65005n (1967).
557a. Korak, R. D., and N. I. Zemlyanskii, Zhur. Obshch. Khim., 41, 1211 (1971); C. A., 75 109772 (1971).
558. Koremura, M., Japan 26,160 (1963); C. A., 60, 5336c (1964).
559. Kosolapoff, G. M., U. S. 2,536,647 (1951); C. A., 45, 5185a (1951).
560. Kotovich, B. P., and N. I. Zemlyanskii, J. Gen. Chem., 38, 1718 (1968); C. A., 70, 3420 (1969).
561. Kotovich, B. P., N. I. Zemlyanskii, I. V. Murav'ev, and M. P. Voloshin, J. Gen. Chem., 38, 1235 (1968); C. A., 69, 58785 (1968).
562. Krawiecki, C., J. Michalski, R. A. Y. Jones, and A. R. Katritzky, Rocz. Chem., 43, 869 (1969); C. A., 71, 61484 (1969).
563. Kremura, M., H. Oku, and T. Masano, Japan 6,473 (1963); C. A., 59, 11337g (1963).
564. Krenzer, J., and S. B. Richter, U. S. 3,595,945 (1971); C. A., 75, 98336 (1971).
565. Kruglyak, Yu. L., G. A. Leibovskaya, O. G. Strukov, and I. V. Martynov, J. Gen. Chem., 39, 970 (1969); C. A., 71, 70544 (1969).
566. Kudchadker, M. V., R. A. Zingaro, and K. J. Irgolic, Can. J. Chem., 46, 1415 (1968); C. A., 68, 118883x (1968).
567. Kuehn, G., G. W. Fischer, and Kh. Lohs, Arch. Pharm. (Weinheim), 300, 363 (1967); C. A., 67, 73497k (1967).
568. Kuhlow, G., H. Teichmann, and G. Hilgetag, Z. Chem., 5, 179 (1965); C. A., 63, 5488 (1965).
568a. Kukalenko, S. S., and N. N. Mel'nikov, Zhur. Obshch. Khim., 28, 157 (1958); C. A., 52, 12776a (1958).
569. Kukhta, E. P., and Yu. N. Forostyan, Khim. Prir. Soedin., 6, 383 (1970); C. A., 73, 99078 (1970).
570. Kuliev, A. M., Z. A. Alizade, and R. K. Velieva, Prisadki Smaz. Maslam, 1969, No. 2, 57; C. A., 72, 42656 (1970).
571. Kuliev, A. M., K. I. Sadykhov, and R. K. Mamedova, Prisadki Smaz. Maslam, 1967, 55; C. A., 70, 3181 (1969).
572. Kuliev, A. M., K. I. Sadykhov, R. K. Mamedova, Prisadki Smaz. Maslam, 1967, 59; C. A., 70, 11027 (1969).
573. Kuliev, A. M., K. I. Sadykhov, R. K. Mamedova, and E. Kh. Guseinova, Azerb. Khim. Zh., 1966, 21; C. A., 65, 8745 (1966).
574. Kuramoto, S., K. Fujimoto, Y. Okuno, H. Sakamoto, M. Nakagawa, and T. Mizutani, Ger. 1,212,526 (1966); C. A., 64, 17491g (1966).
575. Kuramoto, S., Y. Nishizawa, K. Fujimoto, H. Sakamoto, M. Nakagawa, and T. Mizutani, Brit. 979,390 (1965); C. A., 62, 16125h (1965).

576. Kuramoto, S., Y. Nishizawa, and T. Mizutani, Japan 5024 (1964); C. A., 61, 6956g (1964).
577. Kuramoto, S., Y. Nishizawa, and T. Mizutani, Japan 5025 (1964); C. A., 61, 6956h (1964).
578. Kuramoto, S., Y. Nishizawa, and M. Nakagawa, Japan 26,196 (1964); C. A., 62, 10376 (1965).
579. Kuramoto, S., Y. Nishizawa, H. Sakamoto, and T. Mizutani, U. S. 3,150,040 (1964); C. A., 64, 2008h (1966).
579a. Kuznetsov, E. V., and B. K. Valetdinov, Trudy Kazan. Khim. Tekhnol. Inst. S. M. Kirova, 1956, No. 21, 167-9; C. A., 51, 11985h (1957).
580. Labarre, M. C., D. Voigt, and F. Gallais, Bull. Soc. Chim. Fr., 1967, 3328; C. A., 68, 44451 (1968).
581. Lane, D. W. J., and P. L. Carter, Brit. 928,303 (1963); C. A., 60, 6756 (1964).
582. Lane, D. W. J., and D. F. Heath, U. S. 2,791,599 (1957); C. A., 51, 13307 (1957).
583. Lanham, W. M., Brit. 759,334 (1956); C. A., 51, 9676i (1957).
583a. Lanham, W. M., Brit. 766,766 (1957).
584. Lanham, W. M., Brit. 770,420 (1957); C. A., 51, 14784 (1957).
585. Lanham, W. M., U. S. 2,827,468 (1958); C. A., 52, 13804 (1958).
586. Lanham, W. M., Brit. 786,601 (1957); C. A., 52, 10211 (1958).
587. Lanham, W. M., Brit. 788,954 (1958); C. A., 52, 14713g (1958).
588. Lanham, W. M., Brit. 807,896 (1969); C. A., 53, 14027a (1959).
589. Lanham, W. M., U. S. 2,876,244 (1959); C. A., 53, 15108h (1959).
590. Lanham, W. M., U. S. 2,876,245 (1959); C. A., 53, 15109c (1959).
591. Lanham, W. M., U. S. 2,894,016 (1959); C. A., 54, 1572h (1960).
592. Lanham, W. M., U. S. 2,894,974 (1959); C. A., 54, 595g (1960).
593. Lanham, W. M., U. S. 3,022,330 (1962); C. A., 57, 7287d (1962).
594. Leber, J. P., Helv. Chim. Acta, 49, 607 (1966); C. A., 64, 7996e (1966).
595. Leber, J. P., and K. Lutz, U. S. 3,041,367 (1962); C. A., 57, 16399c (1962).
596. Lecocq, J., and A. R. Todd, J. Chem. Soc., 1954, 2381; C. A., 49, 12272d (1955).
597. Lee, J. D., and G. W. Goodacre, Acta Crystallogr., Sect. B, 27, 1055 (1971); C. A., 75, 81379 (1971).
598. Lemetre, G., Ital. 624,158 (1961); C. A., 57, 5807h (1962).
599. Lemetre, G., Ital. 639,015 (1962); C. A., 59, 2649e (1963).
600. Lenard-Borecka, B., T. Kapecka, and J. Michalski,

Rocz. Chem., 36, 87 (1962); C. A., 57, 12300i (1962).
601. Lenard-Borecka, B., and J. Michalski, Rocz. Chem., 31, 1167 (1957); C. A., 52, 9945 (1958).
602. Lenard-Borecka, B., J. Michalski, and S. Musierowicz, Rocz. Chem., 32, 1301 (1958); C. A., 53, 17884 (1959).
603. Levy, S. D., R. W. Addor, and R. J. Magee, U. S. 3,317,561 (1967); C. A., 67, 100141y (1967).
604. Lewis, G. L., and C. P. Smyth, J. Am. Chem. Soc., 62, 1529 (1940); C. A., 34, 4951 (1940).
605. Lippman, A. E., J. Org. Chem., 30, 3217 (1965); C. A., 63, 14687f (1965).
606. Lippman. A. E., J. Org. Chem., 31, 471 (1966); C. A., 64, 9580f (1966).
607. Lorenz, W., Ger. 817,753 (1951); C. A., 47, 3879f (1953).
608. Lorenz, W., Ger. 930,446 (1955); C. A., 50, 7865f (1956).
609. Lorenz, W., Ger. 933,627 (1955); C. A., 52, 1729lc (1958).
610. Lorenz, W., U. S. 2,758,115 (1956); C. A., 51, 2888h (1957).
611. Lorenz, W., U. S. 2,843,588 (1958); C. A., 53, 420 (1959).
612. Lorenz, W., Ger. 1,046,062 (1958); C. A. , 55, 4432g (1961).
613. Lorenz, W., Ger. 1,064,072 (1959); C. A., 55, 17664f (1961).
614. Lorenz, W., Ger. 1,112,068 (1961); C. A., 56, 9967h (1962).
615. Lorenz, W., Ger. 1,134,369 (1962); C. A., 58, 3316h (1963).
616. Lorenz, W., Belg. 615,670 (1962); C. A., 58, 12467 (1963).
617. Lorenz, W., Fr. 1,335,759 (1963); C. A., 60, 558c (1964).
618. Lorenz, W., Belg. 625,216 (1963); C. A., 60, 10600e (1964).
619. Lorenz, W., Belg. 627,817 (1963); C. A., 61, 5692f (1964).
620. Lorenz, W., Belg. 633,560 (1963); C. A., 60, 15788c (1964).
621. Lorenz, W., Belg. 635,443 (1964); C. A., 61, 16096e (1964).
622. Lorenz, W., Fr. 1,404,889 (1965); C. A., 63, 16385g (1965).
623. Lorenz, W., R. Coelln, G. Schrader, and G. Unterstenhöfer, U. S. 3,294,631 (1966); C. A., 64, 2114b (1966).
624. Lorenz, W., and C. Fest, Ger. 1,187,622 (1965); C. A., 63, 3000c (1965).
625. Lorenz, W., I. Hammann, and G. Unterstenhoefer, S.

African 4,109 (1968); C. A., 71, 70738m (1969).
626. Lorenz, W., K. Mannes, and G. Schrader, Brit. 928,417 (1963); C. A., 60, 5553b (1964).
627. Lorenz, W., R. Mühlmann, and G. Schrader, Ger. 964,045 (1957); C. A., 53, 11229h (1959).
628. Lorenz, W., R. Mühlmann, G. Schrader, and K. Tettweiler, Ger. 949,229 (1956); C. A., 51, 12957c (1957).
629. Lorenz, W., and H. G. Schicke, Belg. 651,368 (1964); C. A., 64, 15895h (1966).
630. Lorenz, W., H. G. Schicke, and G. Schrader, Ger. 1,116,657 (1961); C. A., 57, 666f (1962).
631. Lorenz, W., H. G. Schicke, and G. Schrader, Belg. 616,096 (1962); C. A., 58, 13995 (1963).
632. Lorenz, W., H. G. Schicke, and G. Schrader, Belg. 616,998 (1962); C. A., 58, 12602b (1963).
633. Lorenz, W., and G. Schrader, Ger. 917,668 (1954); C. A., 49, 12528i (1955).
634. Lorenz, W., and G. Schrader, Ger. 954,960 (1956); C. A., 51, 12958c (1957).
635. Lorenz, W., and G. Schrader, Ger. 1,067,017 (1959); C. A., 56, 1482d (1962).
636. Lorenz, W., and G. Schrader, Ger. 1,071,696 (1959); C. A., 55, 12359h (1961).
637. Lorenz, W., and G. Schrader, Ger. 1,074,034 (1960); C. A., 55, 16485i (1961).
638. Lorenz, W., and G. Schrader, Ger. 1,074,035 (1960); C. A., 55, 12756f (1961).
639. Lorenz, W., and G. Schrader, Ger. 1,083,811 (1960); C. A., 55, 16483f (1961).
640. Lorenz, W., and G. Schrader, Ger. 1,083,827 (1960); C. A., 55, 17669f, (1961).
641. Lorenz, W., and G. Schrader, Ger. 1,087,591 (1960); C. A., 55, 17499h (1961).
642. Lorenz, W., and G. Schrader, Ger. 1,114,478 (1961); C. A., 56, 10045g (1962).
643. Lorenz, W., and G. Schrader, Ger. 1,124,034 (1962); C. A., 57, 3485b (1962).
644. Lorenz, W., and G. Schrader, Belg. 609,155 (1962); C. A., 57, 16402 (1962).
645. Lorenz, W., and G. Schrader, Belg. 609,802 (1962); C. A., 58, 11402g (1963).
646. Lorenz, W., and G. Schrader, Ger. 1,144,265 (1963); C. A., 59, 3778d (1963).
647. Lorenz, W., and G. Schrader, U. S. 3,082,240 (1963); C. A., 59, 5077h (1963).
648. Lorenz, W., and G. Schrader, Belg. 623,551 (1963); C. A., 60, 6748 (1964).
649. Lorenz, W., and G. Schrader, Belg. 631,729 (1963); C. A., 61, 8316a (1964).
650. Lorenz, W., and G. Schrader, U. S. 3,371,095 (1968); C. A., 64, 17554a (1966).
651. Lorenz, W., G. Unterstenhofer, and I. Hammann, Brit.

1,162,088 (1969); C. A., 72, 12876q (1970).
652. Lorenz, W., and R. Wegler, Ger. 962,608 (1957); C. A., 51, 15588a (1957).
653. Losco, G., and C. A. Peri, Ital. 572,321 (1958); C. A., 53, 16065d (1959).
654. Losco, G., and C. A. Peri, U. S. 3,024,162 (1962); C. A., 57, 3296 (1962).
655. Losco, G., and C. A. Peri, U. S. 3,032,579 (1962); C. A., 59, 8601h (1963).
656. Losco, G., G. Rossi, and G. Michieli, U. S. 3,033,744 (1962); C. A., 57, 12333a (1962).
657. Losco, G., G. Rossi, and G. Michieli, Fr. Addn. 81,056 (1963) (Addn. to Fr. 1,153,596); C. A., 60, 1657f (1964).
658. Losco, G., G. Rossi, and G. Michieli, Ital. 707,577 (1966); C. A., 68, 39143c (1968).
659. Loughran, G. A., and E. O. Hook, U. S. 3,135,695 (1964); C. A., 61, 2972b (1964).
660. Lovell, J. B., and R. W. Baer, Belg. 648,531 (1964); C. A., 63, 11433f (1965).
661. Lui, H., H-F. Leng, and P-N. Yang, K'un Ch'ung Hsueh Pao, 14, 339 (1965); C. A., 64, 14896 (1966).
662. Lutz, K., Swiss 318,185 (1957); C. A., 51, 18012d (1957).
663. Lutz, K., Swiss 322,871 (1957); C. A., 52, 2092g (1958).
664. Lutz, K., and O. Jucker, U. S. 2,777,792 (1957); C. A., 51, 12129a (1957).
665. Lutz, K., and M. Schuler, Brit. 900,557 (1962); C. A., 57, 16404g (1962).
666. Lutz, K., M. Schuler, and O. Jucker, Swiss 319,579 (1957); C. A., 51, 17979i (1957).
667. McCombie, H., B. C. Saunders, N. B. Chapman, and R. Heap, U. S. 2,489,917 (1949); C. A., 44, 3005h (1950).
668. McConnell, R. L., and H. W. Coover, J. Org. Chem., 24, 630 (1959); C. A., 54, 20862i (1960).
669. McConnell, R. L., and H. W. Coover, U. S. 2,972,628 (1961); C. A., 55, 12296c (1961).
670. McConnell, R. L., and M. A. McCall, U. S. 2,972,621 (1961); C. A., 56, 14244b (1962).
671. McConnell, R. L., and N. H. Shearer, Jr., U. S. 3,336,421 (1967); C. A., 68, 77972e (1968).
672. McHattie, G. V., Brit. 1,019,227 (1966); C. A., 64, 14197d (1966).
673. McIvor, R. A., G. A. Grant, and C. E. Hubley, Can. J. Chem., 34, 1611 (1956); C. A., 51, 6338h (1957).
674. McIvor, R. A., C. E. Hubley, G. A. Grant, and A. A. Grey, Can. J. Chem., 36, 820 (1958); C. A., 52, 14330f (1958).
675. McIvor, R. A., G. D. McCarthy, and G. A. Grant, Can. J. Chem., 34, 1819 (1956); C. A., 51, 8640b (1957).

676. Macko, J., S. Truchlik, J. Drabek, S. Gaher, and T. Sirota, Czech. 120,424 (1966); C. A., 68, 77848n (1968).
677. Magee, R. J., U. S. 3,493,655 (1970); C. A., 72, 100271 (1970).
678. Magee, R. J., and F. A. Wagner, Jr., U. S. 3,131,186 (1964); C. A., 61, 671d (1964).
679. Maier, L., U. S. 3,179,688 (1965); C. A., 63, 495 (1965).
679a. Maier, L., private communication.
679b. Maier, L., and J. J. Daly, Chimia, 18, 217 (1964); C. A., 61, 9421a (1964).
680. Maier, L., and J. R. Van Wazer, J. Am. Chem. Soc., 84, 3054 (1962); C. A., 57, 12083i (1962).
681. Majoral, J. P., and J. Navech, Bull. Soc. Chim. Fr., 1971, 95; C. A., 74, 99329b (1971).
682. Majoral, J. P., and J. Navech, Bull. Soc. Chim. Fr., 1971, 1331; C. A., 75, 48300y (1971).
683. Malatesta, L., Ital. 458,770 (1950); C. A., 45, 9555g (1951).
684. Malatesta, P., and B. D'Atri, Farmaco (Pavia), Ed. Sci., 8, 398 (1953); C. A., 48, 9312g (1954).
685. Malinovskii, M. S., and Z. F. Solomko, J. Gen. Chem., 30, 673 (1960); C. A., 54, 24339e (1960).
686. Malinovskii, M. S., Z. F. Solomko, and E. I. Evtushenko, J. Gen. Chem., 30, 2574 (1960); C. A., 55, 14287i (1961).
687. Malinovskii, M. S., Z. F. Solomko, and L. M. Yurilina, J. Gen. Chem., 30, 3422 (1960); C. A., 55, 19775f (1961).
688. Malinowski, R., W. Adamczyk, and F. Pastwa, Pol. 48,542 (1964); C. A., 63, 5560e (1965).
689. Malz, H., O. Bayer, H. Freytag, and E. Kuehle, Ger. 1,127,906 (1962); C. A., 57, 8437c (1962).
690. Malz, H., O. Bayer, H. Freytag, and F. Lober, U. S. 2,891,059 (1959); C. A., 54, 4387g (1960).
691. Malz, H., O. Bayer, and W. Neumann, Ger. 1,163,801 (1964); C. A., 60, 11945 (1964).
692. Malz, H., O. Bayer, and R. Wegler, U. S. 2,995,568 (1961); C. A., 57, 11021h (1962).
693. Malz, H., E. Kuehle, and O. Bayer, U. S. 3,053,876 (1962); C. A., 58, 4470d (1963).
694. Malz, H., H. Kuekenthal, W. Behrenz, E. Klauke, and E. Kuehle, Belg. 632,757 (1963); C. A., 61, 9402f (1964).
695. Malz, H., G. Oertel, H. Holtschmidt, and K. Wagner, U. S. 3,234,305 (1966); C. A., 59, 2657a (1963).
696. Mandel'baum, Ya. A., Khim. v Sel'sk. Khoz., 1964, 34; C. A., 61, 11883c (1964).
697. Mandel'baum, Ya. A., Z. M. Bakanova, and N. N. Mel'nikov, J. Gen. Chem., 33, 3757 (1963); C. A., 60, 9135e (1964).

698. Mandel'baum, Ya. A., A. L. Itskova, and N. N. Mel'nikov, Biol. Aktivn. Soedin., Akad. Nauk SSSR, 1965, 252; C. A., 64, 3411 (1966).
699. Mandel'baum, Ya. A., V. I. Lomakina, and N. N. Mel'nikov, Dokl. Akad. Nauk SSSR., 96, 1173 (1954); C. A., 49, 8843 (1955).
700. Mandel'baum, Ya. A., N. N. Mel'nikov and Z. M. Bakanova, Zh. Obshch. Khim., 29, 1149 (1959); C. A., 54, 1377 (1960).
701. Mandel'baum, Ya. A., N. N. Mel'nikov, and Z. M. Bakanova, J. Gen. Chem., 30, 207 (1960).
702. Mandel'baum, Ya. A., N. N. Mel'nikov, Z. M. Bakanova, and P. G. Zaks, J. Gen. Chem., 31, 3682 (1961); C. A., 57, 11072a (1962).
703. Mandel'baum, Ya. A., N. N. Mel'nikov, and V. I. Lomakina, J. Gen. Chem., 26, 2877 (1956); C. A., 51, 1825f (1957).
704. Mandel'baum, Ya. A., N. N. Mel'nikov, and N. I. Petrova, Zh. Obshch. Khim., 28, 479 (1958); C. A., 52, 12776 (1958).
705. Mandel'baum, Ya. A., N. N. Mel'nikov, and P. G. Zaks, Zh. Obschch. Khim., 29, 283 (1959); C. A., 53, 21771 (1959).
706. Mandel'baum, Ya. A., R. S. Soifer, N. N. Mel'nikov, and L. A. Belova, J. Gen. Chem., 37, 2173 (1967); C. A., 68, 86790 (1968).
707. Mandel'baum, Ya. A., I. L. Vladimirova, and N. N. Mel'nikov, J. Gen. Chem., USSR, 23, 437 (1953); C. A., 48, 3887 (1954).
708. Mandel'baum, Ya. A., I. L. Vladimirova, and N. N. Mel'nikov, Dokl. Akad. Nauk SSSR, 100, 77 (1955); C. A., 50, 1650 (1956).
709. Mandel'baum, Ya. A., P. G. Zaks, N. N. Mel'nikov, V. V. Ivanov, Khim. Org. Soedin. Fosfora, Akad. Nauk SSSR, Otd. Obshch. Khim., 1967, 262; C. A., 69, 2457 (1968).
710. Mannes, K., R. Heiss, H. Pelster, and G. Unterstenhoefer, Belg. 642,764 (1964); C. A., 63, 6980a (1965).
711. Mannes, K., G. Schrader, and K. Wedemeyer, Belg. 610,433 (1960); C. A., 58, 10238d (1963).
712. Mannes, K., and G. Unterstenhoefer, Fr. 1,381,661 (1964); C. A., 62, 9072h (1965).
713. Manske, R. H. F., and M. Kulka, U. S. 2,575,225 (1951); C. A., 46, 3566d (1952).
714. March, R. B., T. R. Fukuto, R. L. Metcalf, and M. G. Maxon, J. Econ. Entomol., 49, 185 (1956); C. A., 50, 13360f (1956).
715. Mark, V., C. H. Dungan, M. M. Crutchfield, and J. R. Van Wazer, in Topics in Phosphorus Chemistry, M. Grayson and E. J. Griffith, Eds., Interscience, Vol. V, Chap. 4, 1967.

716. Mark, V., and J. R. Van Wazer, J. Org. Chem., 29, 1006 (1964); C. A., 60, 15723d (1964).
717. Mark, V., and J. R. Van Wazer, J. Org. Chem., 32, 1187 (1967); C. A., 66, 104559 (1967).
718. Markley, F. X., U. S. 2,965,467 (1960); C. A., 55, 9772h (1961).
719. Markley, F. X., and M. L. Larson, U. S. 2,724,710 (1955); C. A., 50, 10122g (1956).
720. Markley, F. X., and M. L. Larson, U. S. 2,866,806 (1958); C. A., 53, 19878a (1959).
721. Markowska, A., and J. Michalski, Rocz. Chem., 38, 1141 (1964); C. A., 61, 15967g (1964).
721a. Markowska, A., Bull. Acad. Pol. Sci., Ser. Sci. Chim., 15, 153 (1967); C. A., 90891b (1967).
722. Martin, G., and A. Besnard, Compt. Rend., 257, 898 (1963); C. A., 59, 13394h (1963).
723. Martynov, I. V., Yu L. Kruglyak, G. A. Leibovskaya, Z. I. Khromova, and O. G. Strukov, J. Gen. Chem., 39, 966 (1969); C. A., 71, 61297 (1969).
724. Martynov, I. V., L. N. Shitov, and E. A. Mordvintseva, J. Gen. Chem., 40, 540 (1970); C. A., 73, 14072 (1970).
725. Mastin, T. W., G. R. Norman, and E. A. Weilmuenster, J. Am. Chem. Soc., 67, 1662 (1945); C. A., 40, 55 (1946).
726. Mastryukova, T. A., E. L. Gefter, Yu. S. Kagan, D. M. Paikin, M. P. Shabanova, N. M. Gamper, L. F. Efimova, and M. I. Kabachnik, J. Gen. Chem., 30, 2790 (1960); C. A., 55, 18574d (1961).
727. Mastryukova, T. A., V. N. Odnoralova, and M. I. Kabachnik, Zh. Obshch. Khim., 28, 1563 (1958); C. A., 53, 1117e (1959).
728. Mastryukova, T. A., E. N. Prilezhaeva, N. I. Uvarova, M. F. Shostakovakii, and M. I. Kabachnik, Izv. Akad. Nauk SSSR, Otd. Khim. Nauk, 1956, 443; C. A., 50, 16662 (1956).
729. Mastyukova, T. A., T. B. Sakharova, and M. I. Kabachnik, J. Gen. Chem., 34, 92 (1964); C. A., 60, 10530 (1964).
730. Mastryukova, T. A., T. B. Sakharova, and M. I. Kabachnik, J. Gen. Chem., 41, 236 (1971); C. A., 74, 141203 (1971).
731. Mastryukova, T. A., A. E. Shipov, V. V. Abalyaeva, E. M. Popov, and M. I. Kabachnik, Proc. Acad. Sci. USSR, 158, 1113 (1964); C. A., 62, 2700h (1965).
732. Matsubara, H., and S. Nabekawa, Japan 4,546 (1969); C. A., 71, 3480h (1969).
733. Matyuk, L. N., K. D. Shvetsova-Shilovskaya, and N. N. Mel'nikov, J. Gen. Chem., 37, 1243 (1967); C. A., 68, 29232 (1968).
734. Mavel, G., and G. Martin, Compt. Rend., 257, 1703 (1963); C. A., 60, 139g (1964).

735. Maxwell, R. C., and D. G. Stoffey, U. S. 3,018,216 (1962); C. A., 56, 9976d (1962).
736. Medved, T. Ya., and M. I. Kabachnik, Izv. Akad. Nauk SSSR, Otd. Khim. Nauk, 1958, 1212; C. A., 53, 4111d (1959).
737. Meinhardt, N. A., S. Z. Cardon, and P. W. Vogel, J. Org. Chem., 25, 1991 (1960); C. A., 55, 12334i (1961).
738. Melles, J. L., Ger. 1,159,687 (1963); C. A., 60, 11987d (1964).
739. Mel'nik, Ya. I., and N. I. Zemlyanski, J. Gen. Chem., 40, 768 (1970); C. A., 73, 41302 (1970).
740. Mel'nik, Ya. I., and N. I. Zemlyanski, J. Gen. Chem., 40, 1001 (1970); C. A., 73, 76604 (1970).
741. Mel'nikov, N. N., Khim. Primen. Fosfororg. Soed., Akad. Nauk SSSR, Tr. 1-oi Konf., 1955, 50-61; C. A., 52, 393h (1958).
742. Mel'nikov, N. N., A. F. Grapov, and K. D. Shvetsova-Shilovskaya, J. Gen. Chem. USSR, 27, 1967 (1957); C. A., 52, 4533c (1958).
743. Mel'nikov, N. N., A. F. Grapov, and K. D. Shvetsova-Shilovskaya, Zhur. Obshch. Khim., 29, 3291 (1959); C. A., 54, 14216c (1960).
744. Mel'nikov, N. N., B. A. Khaskin, and I. V. Sablina, J. Gen. Chem., 38, 1509 (1968); C. A., 70, 3803 (1969).
745. Mel'nikov. N. N., B. A. Khaskin, and K. D. Shvetsova-Shilovskaya, J. Gen. Chem., 33, 2394 (1963); C. A., 59, 15165e (1963).
746. Mel'nikov, N. N., B. A. Khaskin, and K. D. Shvetsova-Shilovskaya, Aspirantsk. Raboty Nauchn.-Issled., 1963 12; C. A., 62, 3924h (1965).
747. Mel'nikov, N. N., and D. N. Khokhlov, Zh. Obshch. Khim., 23, 1357 (1953); C. A., 48, 9903a (1954).
748. Mel'nikov. N. N., and Ya. A. Mandel'baum, Org. Insektofungitsidy Gerbitsidy, 1958, 37; C. A., 54, 7050i (1960).
749. Mel'nikov, N. N., Ya. A. Mandel'baum, Z. M. Bakanova, and P. G. Zaks, Zh. Obshch. Khim., 29, 3286 (1959); C. A., 54, 14215f (1960).
750. Mel'nikov, N. N., Ya. A. Mandel'baum, and V. I. Lomakina, Zh. Obshch. Khim., 28, 476 (1958); C. A., 52, 12776 d (1958).
751. Mel'nikov, N. N., Ya. A. Mandel'baum, V. I. Lomakina, and Z. M. Bakanova, J. Gen. Chem. USSR, 26, 2871 (1956); C. A., 51, 1825a (1957).
752. Mel'nikov, N. N., Ya. A. Mandel'baum, E. I. Sventsitskii, and Z. M. Bakanova, J. Gen. Chem., 27, 1970 (1957); C. A., 52, 4533f (1958).
753. Mel'nikov, N. N., Ya. A. Mandel'baum, and P. G. Zaks, Zh, Oschch. Khim., 29, 522 (1959); C. A., 53, 21771f (1959).

754. Mel'nikov. N. N., O. B. Mikhailova, K. D. Shvetsova-Shilovskaya, and E. B. Mikhalyutina, J. Gen. Chem., 38, 345 (1968); C. A., 69, 59170h (1968).
755. Mel'nikov, N. N., A. F. Prokof'eva, T. P. Krylova, N. A. Popovkina, N. N. Khvorostukhina, and I. L. Vladimirova, Khim. Org, Soedin. Fosfora, 1957, 256 (C. A., 69, 2627h (1968).
756. Mel'nikov, N. N., and K. D. Shvetsova-Shilovskaya, Zh. Obshch. Khim., 23, 1352 (1953); C. A., 48, 9902d (1954).
757. Mel'nikov, N. N., and K. D. Shvetsova-Shilovskaya, Zh. Obshch. Khim., 28, 474 (1958); C. A., 52, 14542g (1958).
758. Mel'nikov, N. N., K. D. Shvetsova-Shilovskaya, and A. F. Grapov, Org. Insektofungitsidy Gerbitsidy, 1958, 108; C. A., 55, 3600 (1961).
759. Mel'nikov, N. N., K. D. Shvetsova-Shilovskaya, and M. Ya. Kagan, J. Gen. Chem., 30, 213 (1960); C. A., 54, 22322c (1960).
760. Mel'nikov, N. N., K. D. Shvetsova-Shilovskaya, and M. Ya. Kagan, J. Gen. Chem., 30, 2300 (1960); C. A., 55, 9320i (1961).
761. Mel'nikov, N. N., K. D. Shvetsova-Shilovskaya, M. Ya. Kagan, and I. M. Mil'shtein, Zh. Obshch. Khim., 29, 1612 (1959); C. A., 54, 8683h (1960).
762. Mel'nikov, N. N., K. D. Shvetsova-Shilovskaya, and I. M. Mil'stein, J. Gen. Chem., 30, 210 (1960); C. A., 54, 22322a (1960).
763. Meltsner, B. R., U. S. 3,451,932 (1969); C. A., 71, 60997f (1969).
764. Menefee, A., D. O. Alford, and C. B. Scott, J. Org. Chem., 22, 792 (1957); C. A., 52, 2740c (1958).
765. Metcalf, R. L., T. R. Fukuto, and R. B. March, J. Econ. Entomol., 50, 338 (1957); C. A., 55, 3912b (161).
766. Metcalf, R. L., and R. B. March, J. Econ. Entomol., 46, 288 (1953); C. A., 48, 4166d (1954).
767. Metivier, J., U. S. 2,752,283 (1956); C. A., 51, 1291i (1957).
768. Metivier, J., U. S. 2,811,476 (1957); C. A., 52, 2930b (1958).
769. Metivier, J., U. S. 2,905,700 (1959); C. A., 54, 2364c (1960).
770. Metivier, J., Fr. 1,277,401 (1962); C. A., 57, 12498 (1962).
771. Metivier, J., Brit. 935,982 (1963); C. A., 60, 1607 (1964).
772. Metivier, J., and M. Sauli, Fr. 1,451,294 (1966); C. A., 66, 115713q (1967); Fr. Addn. 90,257; C. A., 70, 28921 (1969), U. S. 3,432,519 (1969).
773. Michaelis, A., and G. L. Linke, Ber., 40, 3419 (1907); C. A., 2, 84 (1908).

774. Michalski, J., Rocz. Chem., 33, 835 (1959); C. A., 54, 4356h (1960).
775. Michalski, J., B. Borecka, and S. Musierowicz, Bull. Acad. Pol. Sci., Ser. Sci. Chim., Geol. Geograph., 6, 159 (1958); C. A., 52, 18194i (1958).
776. Michalski, J., and C. Krawiecki, Rocz. Chem., 31, 715 (1957); C. A., 52, 5284h (1958).
777. Michalski, J., and B. Lenard, Rocz. Chem., 30, 655 (1956); C. A., 51, 2535h (1957).
778. Michalski, J., and A. Markowska, Proc. Acad. Sci. USSR, 136, 21 (1961); C. A., 55, 17475 (1961).
779. Michalski, J., A. Markowska, and H. Strzelecka, Rocz. Chem., 33, 1251 (1959); C. A., 54, 10827d (1960).
780. Michalski, J., T. Modro, and J. Wieczorkowski, J. Chem. Soc., 1960, 1665; C. A., 54, 19562f (1960).
781. Michalski, J., and S. Musierowicz, Chem. Ind. (London), 1959, 565; C. A., 54, 3182 (1960).
782. Michalski, J., and S. Musierowicz, Rocz. Chem., 36, 1655 (1962); C. A., 59, 8634a (1963).
783. Michalski, J., and B. Pliszka, Chem. Ind. (London), 1962, 1052; C. A., 60, 10532b (1964).
784. Michalski, J., and B. Pliszka-Krawiecka, J. Chem. Soc., C. 1966, 2249; C. A., 66, 28340 (1967).
785. Michalski, J., B. Pliszka-Krawiecka, and A. Skowronska, Rocz. Chem., 37, 1479 (1963); C. A., 60, 9135 (1964).
786. Michalski, J., and A. Ratajczak, Rocz. Chem., 36, 775 (1962); C. A., 58, 6681 (1963).
787. Michalski, J., A. Ratajczak, and Z. Tulimowski, Bull. Acad. Pol. Sci., Ser. Sci. Chim., 11, 237 (1963); C. A., 59, 8579f (1963).
788. Michalski, J., and A, Skowronska, Chem. Ind. (London), 1958, 1199; C. A., 53, 13987d (1959).
789. Michalski, J., and A. Skowronska, J. Chem. Soc., C, 1970, 703; C. A., 72, 111563s (1970).
790. Michalski, J., and Z. Tulimowski, Bull. Acad. Pol. Sci., Ser. Sci. Chim., 14, 217 (1966); C. A., 65, 8952e (1966).
791. Michalski, J., and Z. Tulimowski, Bull. Acad. Pol. Sci., Ser. Sci. Chim., 14, 303 (1966); C. A., 65, 10450d (1966).
792. Michalski, J., and J. Wieczorkowski, Bull. Acad. Pol. Sci., Classe III, 3, 4279 (1956); C. A., 51, 4266 (1957).
793. Michalski, J., and J. Wieczorkowski, Bull. Acad. Pol. Sci. Classe III, 5, 917 (1957); C. A., 52, 6157h (1958).
794. Michalski, J., J. Wieczorkowski, and T. Modro, Rocz. Chem., 32, 1409 (1958); C. A., 53, 15951g (1959).
795. Michalski, J., J. Wieczorkowski, J. Wasiak, and B. Pliszka, Rocz. Chem., 33, 247 (1959); C. A., 53, 17884g (1959).

796. Mikolajczyk, M., Chem. Ber., 99, 2083 (1966); C. A., 65, 10612d (1966).
797. Mikolajczyk, M., Angew. Chem., Int. Ed. Engl., 8, 511 (1969); C. A., 71, 61351 (1969).
798. Mikolajczyk, M., J. Chem. Soc., D, 1969, 1221; C. A., 72, 21208 (1970).
799. Mikolajczyk, M., J. Omelanczuk, and J. Michalski, Bull. Acad. Pol. Sci., Ser. Sci. Chim., 17, 155 (1969); C. A., 71, 60633j (1969).
800. Mikolajczyk, M., and H. M. Schiebel, Angew. Chem., Int. Ed. Engl., 8, 511 (1969); C. A., 71, 61298 (1969).
801. Miller, B., J. Am. Chem. Soc., 82, 3924 (1960); C. A., 55, 11287b (1961).
802. Miller, B., J. Am. Chem. Soc., 82, 6205 (1960); C. A., 55, 11283f (1961).
803. Miller, B., J. Am. Chem. Soc., 84, 403 (1962); C. A. 56, 13590d (1962).
804. Miller, B., U. S. 3,019,159 (1962); C. A., 56, 15367 (1962).
805. Miller, B., U. S. 3,021,352 (1962); C. A., 57, 666i (1962).
806. Miller, B., U. S. 3,127,435 (1964); C. A., 60, 15734 (1964).
807. Miller, B., Tetrahedron, 20, 2069 (1964); C. A., 61, 14564g (1964).
808. Miller, B., U. S. 3,518,279 (1970); C. A., 73, 130990 (1970).
808a. Miller, B., and D. W. Long, U. S. 3,172,888 (1965); C. A., 64, 6669d (1966).
809. Miller, B., and H. Margulies, U. S. 3,480,697 (1969); C. A., 72, 55046m (1970).
810. Miller, L. H., and G. H. Birum, U. S. 3,059,041 (1962); C. A., 58, 5519h (1963).
811. Millward, B. B., Brit. 977,484 (1964); C. A., 62, 9061g (1965).
812. Minich, D., N. I. Rizpolozhenskii, and V. D. Akamsin, Bull. Acad. Sci. USSR, Ser. Chem., 1968, 1694; C. A., 70, 37096g (1969).
813. Minich, D., N. I. Rizpolozhenskii, V. D. Akamsin, and O. A. Raevskii, Bull. Acad. Sci. USSR, Ser. Chem., 1969, 795; C. A., 71, 49447f (1969).
814. Minieri, P. P., Ger. Offen. 1,816,740 (1969); C. A., 71, 124431b (1969).
814a. Mirviss, S. B., and M. M. Schlechter, Fr. 1,506,966 (1967); C. A., 70, 20220 (1969).
815. Mitsurube, N., K. Tomita, T. Yanouchi, and H. Oka, Japan 16,137 (1968); C. A., 70, 57819y (1969).
816. Miyazima, G., K. Takahashi, and T. Yamasaki, Bull. Chem. Soc. Jap., 40, 1540 (1967); C. A., 67, 86356p (1967).
817. Moedritzer, K., L. Maier, and L. C. D. Groenweghe, J. Chem. Eng. Data, 7, 307 (1962); C. A., 57, 308e

(1962).
818. Moen, R. V., and W. H. Mueller, J. Org. Chem., 31, 1971 (1966); C. A., 65, 12090f (1966).
819. Montecatini, Brit. 791,824 (1958); C. A., 52, 18222f (1958).
820. Montecatini, Brit. 808,853 (1959); C. A., 53, 16079h (1959).
821. Montecatini, Brit. 825,397 (1959); C. A., 54, 14127h (1960).
822. Montecatini, Brit. 834,814 (1960); C. A., 55, 457c (1961).
823. Montecatini, Brit. 903,185 (1962); C. A., 58, 482d (1963).
824. Montecatini, Brit. 904,401 (1962); C. A., 58, 3324c (1963).
825. Montecatini, Ital. 595,317 (1959); C. A., 56, 7141e (1962).
826. Montecatini, Ital. 598,913 (1959); C. A., 59, 10057h (1963).
827. Montecatini, Ital. 625,074 (1961); C. A., 57, 8445b (1962).
828. Montecatini, Fr. 1,327,314 (1963); C. A., 59, 11272g (1963).
829. Montecatini, Brit. 924,039 (1963); C. A., 59, 11275e (1963).
830. Montecatini, Brit. 957,829 (1964); C. A., 64, 5004h (1966).
831. Montecatini, Brit. 985,466 (1965); C. A., 62, 16062h (1965).
832. Montecatini, Neth. Appl. 6,514,810 (1966); C. A., 66, 10772 (1967).
832a. Morgan, W. E., W. J. Stec, R. G. Albridge, and J. R. Van Wazer, Inorg. Chem., 10, 926 (1971); C. A., 74, 148462a (1971).
833. Morgun, G. E., M. I. Zemlyans'kii, and M. L. Dovgosheya, Visn. L'vivs'k. Derzh. Univ., Ser. Khim., 1963, 94; C. A., 61, 8245b (1964).
834. Morgun, G. E., N. I. Zemlyanskii, and N. S. Lisovskaya, J. Gen. Chem., 38, 1475 (1968); C. A., 70, 3610e (1969).
835. Morgun, G. E., M. I. Zemlyans'kii, O. D. Voznyak, and M. M. Kishko, Visn. L'vivs'k. Derzh. Univ., Ser. Khim., 1963, 97; C. A., 61, 8245a (1964).
836. Morrison, D. C., J. Am. Chem. Soc., 77, 181 (1966); C. A., 50, 789h (1956).
837. Moyle, C. L., U. S. 2,250,049 (1941); C. A., 35, 7062 (1941).
838. Moyle, C. L., U. S. 2,599,516 (1952); C. A., 46, 8322 (1952).
839. Mueller, W. H., and A. A. Oswald, J. Org. Chem., 31, 1894 (1966). C. A., 65, 3729a (1966).

840. Mueller, W. H., and A. A. Oswald, J. Org. Chem., 32, 1730 (1967); C. A., 67, 21244f (1967).
841. Mueller, W. H., and A. A. Oswald, Brit. 1,198,192 (1970); C. A., 73, 76653 (1970).
842. Mueller, W. H., R. M. Rubin, and P. E. Butler, J. Org. Chem., 31, 3537 (1966); C. A., 66, 1996m (1967).
843. Mühlmann, R., and G. Schrader, Ger. 948,241 (1956); C. A., 51, 4426h (1957).
844. Mühlmann, R., and G. Schrader, Z. Naturforsch., 12, b, 196 (1957); C. A., 51, 9995i (1957).
845. Mühlmann, R., and G. Schrader, Ger. 1,092,464 (1960); C. A., 55, 20957a (1961).
846. Mühlmann, R., and G. Schrader, U. S. 3,082,239 (1963); C. A., 58, 1404a (1963).
847. Mukai, T., S. Inamasu, and T. Yamanaka, Japan 27,973 (1969); C. A., 72, 79067 (1970).
848. Mukaiyama, T., and M. Hashimoto, Tetrahedron Lett., 1971, 2425; C. A., 75, 76560 (1971).
849. Mukmeneva, N. A., P. A. Kirpichnikov, and A. N. Pudovik, J. Gen. Chem., 32, 2159 (1962); C. A., 58, 8943c (1963).
850. Muller, N., P. C. Lauterbur, and J. Goldenson, J. Chem. Soc., 78, 3557 (1956); C. A., 50, 16372c (1956).
851. Müller, P., Ger. 956,503 (1957); C. A., 53, 21664h (1959).
852. Murav'ev, I. V., N. I. Zemlyanskii, and E. P. Panov, J. Gen. Chem., 38, 133 (1968); C. A., 69, 58789e (1968).
853. Murdock, L. L., and T. L. Hopkins, J. Org. Chem., 33, 907 (1968); C. A., 68, 68560 (1968).
854. Mustafa, A., M. M. Sidky, and M. R. Mahran, Ann. Chem., 684, 187 (1965); C. A., 63, 2963e (1965).
855. Myannik, A. O., E. S. Shepeleva, and P. I. Sanin, Neftekhimiya, 4, 899 (1964); C. A., 62, 11675h (1965).
856. Nagasawa, M., Y. Imamiya, and H. Sugiyama, Japan 16,464 (1963); C. A., 60, 2789b (1964).
857. Nakagawa, M., and H. Sakamoto, Japan 4,454 (1967); C. A., 67, 11200z (1967).
858. Nakagawa, M., H. Sakamoto, H. Tsuchiya, K. Hanemichi, T. Fujimoto, and Y. Okuno, Japan 6,731 (1967); C. A., 67, 11206f (1967).
859. Nakanishi, M., and S. Inamasu, Japan 11,858 (1966); C. A., 65, 15229d (1966).
860. Nakanishi, M., and S. Inamasu, Japan 4,330 (1967); C. A., 67, 11207g (1967).
861. Nakanishi, M., and S. Inamasu, Japan 11,894 (1969); C. A., 71, 112948n (1969).
862. Nakanishi, M., and S. Inamasu, Japan 11,895 (1969); C. A., 71, 101870c (1969).

863. Nakanishi, M., S. Inamasu, and T. Yamanaka, Japan 8,508 (1969); C. A., 71, 39003 (1969).
864. Nakanishi, M., T. Mukai, and S. Inamasu, Japan 12,145 (1969); C. A., 71, 91496 (1969).
865. Nakanishi, M., A. Tsuda, and S. Inamasu, U. S. 3,467,655 (1969); C. A., 72, 121556f (1970).
866. Nakaoka, T., H. Matsubara, and M. Takada, Japan 3,381 (1967); C. A., 67, 21709m (1967).
867. Neely, W. B., W. E. Allison, W. B. Crummett, K. Bauer, and W. Reifschneider, J. Agr. Food Chem., 18, 45 (1970); C. A., 72, 54019 (1970).
868. Nesmeyanov, A. N., I. F. Lutsenko, Z. S. Kraits, and A. P. Bokovoi, Dokl. Akad. Nauk SSSR, 124, 1251 (1959); C. A., 53, 16931g (1959).
869. Neuhoeffer, O., and W. Maiwald, Chem. Ber., 95, 108 (1962); C. A., 56, 12721a (1962).
870. Nguyen, H. P., N. T. Thuong, and P. Chabrier, C. R. Acad. Sci., Ser. C, 271, 1465 (1970); C. A., 74, 53928t (1971).
871. Nguyen, H. P., N. T. Thuong, and P. Chabrier, C. R. Acad. Sci., Ser. C, 272, 1145 (1971).
872. Nguyen, H. P., N. T. Thuong, and P. Chabrier, C. R. Acad. Sci., Ser. C, 272, 1588 (1971); C. A., 75, 48178q (1971).
873. Nifant'eva, L. V., K. D. Shvetsova-Shilovskaya, N. N. Mel'nikov, and O. N. Vlasov, J. Gen. Chem., 38, 1403 (1968); C. A., 69, 86495 (1968).
874. Nikonorov, K. V., E. A. Gurylev, and F. F. Mertsalova Bull. Acad. Sci. USSR, Ser. Chem., 1970, 1093; C. A., 73, 65960q (1970).
875. Nikonorov, K. V., E. Z. Gurylev, R. R. Shagidullin, and A. V. Chernova, Bull. Acad. Sci. USSR, Ser. Chem. 1968, 574; C. A., 69, 95861t (1968).
876. Nishimura, N., and J. Nakano, Japan 14,350 (1969); C. A., 72, 3495 (1970).
877. Nishimura, T., S. Nagasawa, H. Shinohara, and M. Kaod Fr. 1,572,192 (1969); C. A., 72, 121006b (1970).
878. Nishimura, T., I. Okuda, S. Nagasawa, and M. Yoshido, Japan 11,289 (1970); C. A., 73, 14444s (1970).
879. Nishimura, T., I. Okuda, and M. Yoshido, Japan 9,583 (1971); C. A., 75, 20204w (1971).
880. Nishiwaki, T., J. Chem. Soc., C, 1967, 2680; C. A., 68, 29186 (1968).
881. Nishizawa, Y., Agr. Biol. Chem. (Tokyo), 25, 61 (1961), 25, 66 (1961); C. A., 55, 10358c (1961).
882. Nishizawa, Y., Agr. Biol. Chem. (Tokyo), 25, 150 (1961); C. A., 55, 14352e (1961).
883. Nishizawa, Y., and T. Mizutani, Japan 2,926 (1960); C. A., 54, 24552c (1960).
884. Nishizawa, Y., and T. Mizutani, Japan 6,375 (1960); C. A., 55, 6381g (1961).
885. Nishizawa, Y., and T. Mizutani, Japan 15,130 (1964); C. A., 61, 16016h (1964).

886. Nishizawa, Y., and M. Nakagawa, Japan 3,367 (1960); C. A., 55, 379b (1961).
887. Nishizawa, Y., and M. Nakagawa, Japan 6,374 (1960); C. A., 55, 6381f (1961).
888. Nishizawa, Y., and M. Nakagawa, Japan 15,167 (1961); C. A., 56, 9974 (1962).
889. Nishizawa, Y., M. Nakagawa, Y. Suzuki, H. Sakamoto, and T. Mizutani, Agr. Biol. Chem., 25, 597 (1961); C. A., 55, 23913a (1961).
890. Norman, G. R., W. M. LeSuer, and T. W. Mastin, J. Am. Chem. Soc., 74, 161 (1952); C. A., 47, 6862a (1953).
891. Nuretdinov, I. A., N. A. Buina, N. P. Grechkin, and E. I. Loginova, Bull. Acad. Sci. USSR, Ser. Chem., 1970, 666; C. A., 73, 14372 (1970).
892. Nuretdinov, I. A., N. A. Buina, N. P. Grechkin, and E. I. Loginova, Bull. Acad. Sci. USSR, Ser. Chem., 1971, 111; C. A., 75, 118066c (1971).
893. Nuretdinov, I. A., N. A. Buina, N. P. Grechkin, and S. G. Salikhov, J. Gen. Chem., 39, 897 (1969); C. A., 71, 50097 (1969).
894. Nuretdinov, I. A., and N. P. Grechkin, Bull. Acad. Sci. USSR, Ser. Chem., 1968, 2685; C. A., 70, 77241g (1969).
895. Nuretdinov, I. A., I. D. Neklesova, M. A. Kudrina, I. S. Iraidova, Bull. Acad. Sci. USSR, Ser. Chem., 1971, 1170; C. A., 75, 76308a (1971).
896. Nuretdinova, O. N., and B. A. Arbuzov, Bull. Acad. Sci. USSR, Ser. Chem., 1970, 134; C. A., 72, 111178 (1970).
897. Nuretdinova, O. N., and B. A. Arbuzov, Bull Acad. Sci. USSR, Ser. Chem., 1971, 287; C. A., 75, 36256a (1971).
898. Nuretdinova, O. N., F. F. Guseva, and B. A. Arbuzov, Bull. Acad. Sci. USSR, Ser. Chem., 1969, 2687; C. A., 72, 78781 (1970).
899. Nussbaum, A. L., and R. Tiberi, J. Am. Chem. Soc., 87, 2513 (1965); C. A., 63, 8473b (1965).
900. Nyquist, R. A., Spectrochim. Acta, Part A, 25, 47 (1969); C. A., 70, 82596f (1969).
901. Nyquist, R. A., Spectrochim. Acta, Part A, 27, 697 (1971); C. A., 75, 27651y (1971).
902. Nyquist, R. A., W. W. Mueider, and M. N. Wass, Spectrochim. Acta, Part A, 26, 769 (1970); C. A., 73, 13691h (1970).
903. Oertel, G., and H. Malz, Ger. 1,163,310 (1964); C. A., 60, 10549b (1964).
904. Oertel, G., H. Malz, and A. Doerken, Belg. 633,306 (1963); C. A., 61, 579d (1964).
905. Okuda, I., K. Takida, and M. Yoshido, Japan 12,586 (1968); C. A., 70, 11350k (1969).
906. Olah, G., and A. Oswald, J. Org. Chem., 25, 603 C. A., 54, 19560h (1960).
907. Olah, G. A., and A. A. Oswald, Can. J. Chem., 40,

1917 (1962); C. A., 58, 5551f (1963).
908. Olah, G., and A. A. Oswald, Ann., 602, 118 (1957); C. A., 51, 11985i (1957).
909. Oláh, G., A. Pávlath, and G. Hosszang, Acta Chim. Acad. Sci. Hung., 8, 41 (1955); C. A., 52, 8139e (1958).
910. Olifirenko, S. P., and N. I. Zemlyanski, J. Gen. Chem., 30, 3455 (1960); C. A., 55, 19757i (1961).
911. Olifirenko, S. P., M. I. Zemlyans'kii, and L. G. Fedosyuk, Visn. L'vivs'k. Derzh. Univ., Ser. Khim., 1964, 59; C. A., 64, 11109h (1966).
912. Olifirenko, S. P., M. I. Zemlyans'kii, and L. A. Vdovichenko, Visn. L'vivs'k. Derzh. Univ., Ser. Khim., 1963, 120; C. A., 61, 8213g (1964).
913. Orlov, N. F., M. A. Belokrinitskii, E. V. Sudakova, and B. L. Kaufman, J. Gen. Chem., 38, 1611 (1968); C. A., 69, 87073 (1968).
914. Orlov, N. F., M. S. Sorokin, and E. E. Shestakov, J. Gen. Chem., 40, 687 (1970); C. A., 73, 14904 (1970).
915. Osborne, G. O., and G. Page, J. Chem. Soc., C, 1967, 1192; C. A., 67, 43785k (1967).
916. Oshima, Y., and M. Eto, Belg. 633,481 (1963); C. A., 61, 9514b (1964).
917. Oshima, Y., K. Kobayashi, and Y. Ishii, Japan 11,301 (1970); C. A., 73, 14,445t (1970).
918. Oswald, A. A., Fr. 1,386,726 (1965); C. A., 63, 547d (1965).
919. Oswald, A. A., Fr. 1,509,248 (1968); C. A., 70, 37172d (1969).
920. Oswald, A. A., K. Griesbaum, and B. E. Hudson, Jr., J. Org. Chem., 28, 1262 (1963); C. A., 58, 12436e (1963).
921. Oswald, A. A., W. H. Mueller, and F. A. Daniher, U. S. 3,574,795 (1971); C. A., 75, 19694 (1971).
922. Oswald, A. A., and P. L. Valint, Jr., Ger. Offen. 1,937,439 (1970); C. A., 72, 100022p (1970).
923. Ottmann, G. F., and H. Hooks, Jr., Angew. Chem., 78, 748 (1966) (Ger.); Angew. Chem. Ed. Int., Eng., 5, 725 (1966); C. A., 66, 28459t (1967).
924. Pannell, J., Chem. Ind. (London), 1962, 1797; C. A., 57, 16020a (1962).
925. Pare, P. J., U. S. 3,073,859 (1963); C. A., 56, 13793h (1962).
926. Pastorek, I., J. Drabek, and S. Truchlik, Chem. Zvesti, 19, 413 (1965); C. A., 64, 9621f (1966).
927. Payne, G. B., W. J. Sullivan, P. R. Van Ess, and P. H. Williams, U. S. 3,062,890 (1962); C. A., 58, 5517e (1963).
928. Peach, M. E., and H. G. Spinney, Can. J. Chem., 49, 644 (1971); C. A., 74, 125036 (1971).
929. Pelchowicz, Z., J. Chem. Soc., 1961, 241; C. A., 55, 16399g (1961).

930. Peri, C. A., Fr. 1,342,565 (1963); C. A., 60, 10556b (1964).
931. Peri, C. A., and G. Rossi, Belg. 617,079 (1962); C. A., 58, 13800 (1963).
932. Perini, M., and G. Speroni, U. S. 3,004,055 (1961); C. A., 56, 15369 (1962).
933. Perini, M., and G. Speroni, U. S. 3,047,459 (1962); C. A., 58, 1349b (1963).
934. Pesin, V. G., and A. M. Khaletskii, J. Gen. Chem., 31, 2337 (1961); C. A., 56, 14045a (1962).
935. Pesin, V. G., and A. M. Khaletskii, J. Gen. Chem., 31, 2347 (1961); C. A., 56, 14046g (1962).
936. Pesin, V. G., and I. G. Vitenberg, J. Gen. Chem., 35, 934 (1965); C. A., 63, 6897d (1965).
937. Pesin, V. G., and I. G. Vitenberg, J. Gen. Chem., 36, 1283 (1966); C. A., 65, 16956h (1966).
938. Pesin. V. G., I. G. Vitenberg, and A. M. Khaletskii, J. Gen. Chem., 34, 2792 (1964); C. A., 61, 14663b (1964).
939. Petragnani, N., V. G. Toscano, and M. De Moura Campos, Chem. Ber., 101, 3070 (1968); C. A., 69, 87108 (1968).
940. Petrov, A. D., V. F. Mironov, and V. G. Glukhovtsev, Zh. Obshch. Khim., 27, 1535 (1957); C. A., 52, 3669b (1958).
941. Petrov, K. A., N. K. Bliznyuk, and I. Yu. Mansurov, J. Gen. Chem., 31, 164 (1961); C. A., 55, 22098a (1961).
942. Petrov, K. A., N. K. Bliznyuk, and V. A. Savostenok, J. Gen. Chem., 31, 1260 (1961); C. A., 55, 23317e (1961).
943. Petrov, K. A., V. P. Evdakov, K. A. Bilevich, V. P. Radchenko, and E. E. Nifant'ev, J. Gen. Chem., 32, 909 (1962); C. A., 58, 2392 (1963).
944. Petrov, K. A., and A. A. Neimysheva, Zh. Obshch. Khim., 29, 3030 (1959); C. A., 54, 11972 (1960).
944a. Petrov, K. A., and A. A. Neimvsheva, Zhur. Obshch. Khim., 29, 3206 (1959); C. A., 54, 11982f (1960).
945. Petrov, K. A., A. A. Neimysheva, G. V. Dotsev, and G. Varich, J. Gen. Chem., 31, 1265 (1961); C. A., 55, 27018f (1961).
946. Petrov, K. A., E. E. Nifant'ev, L. V. Khorkhoyanu, and I. G . Shcherba, J. Gen. Chem., 34, 69 (1934); C. A., 60, 10530h (1964).
947. Petrov, K. A., G. A. Sokol'skii, and B. M. Polees, J. Gen. Chem., 26, 3765 (1956); C. A., 51, 9474a (1957).
948. Pianka, M., Chem. Ind. (London), 1961, 324.
949. Pianka, M., Ger. Offen. 1,925,468 (1970); C. A., 72, 100883v (1970).
950. Pianka, M., and J. D. Edwards, J. Sci. Food Agr., 19, 399 (1968); C. A., 69, 77365 (1968).

951. Pianka, M., and D. J. Polton, Brit. 867,780 (1961); C. A., 55, 25756h (1961).
952. Pianka, M., and D. J. Polton, Brit. 917,924 (1963); C. A., 59, 1487g 1963.
953. Pilgram, K., Tetrahedron, 23, 3007 (1967); C. A., 67, 53813 (1967).
954. Pilgram, K., and F. Korte, Tetrahedron, 20, 177 (1964); C. A., 60, 11928d (1964).
955. Pilgram, K., and F. Korte, Tetrahedron, 21, 1999 (1965); C. A., 63, 12991h (1965).
956. Pilgram, K., and D. D. Phillips, J. Org. Chem., 30, 2388 (1965); C. A., 63, 5464f (1965).
957. Piper, J. R., C. R. Stringfellow, Jr., R. D. Elliott, and T. P. Johnston, J. Med. Chem., 12, 236 (1969); C. A., 70, 96090y (1969).
958. Piper, J. R., C. R. Stringfellow, Jr., and T. P. Johnston, J. Med. Chem., 9, 911 (1966); C. A., 66, 2137c (1967).
959. Pishchimuka, P., Ber., 41, 3854 (1908); C. A., 3, 431 (1909).
960. Pishchimuka, P., J. Prakt. Chem. (2), 84, 746 (1911); C. A., 6, 989 (1912).
961. Pishchimuka, P., J. Russ. Phys. Chem. Soc., 44, 1406 (1912): C. A., 7, 987 (1913).
961a. Plets, V. M., Zhur. Obshch. Khim., 6, 1198 (1936), 8, 1296 (1938); C. A., 33, 4193 (1939).
962. Pohlemann, H., H. Dickhaeuser, G. Scheuerer, H. Adolphi, and H. Stummeyer, Fr. 1,382,442 (1964); C. A., 62, 11694h (1965).
963. Pohlemann, H., H. Schroeder, H. Stummeyer, and H. Adolphi, Ger. 1,029,367 (1958); C. A., 54, 18393d (1960).
964. Pohlemann, H., H. Schroeder, H. Stummeyer, and H. Adolphi, Ger. 1,058,045 (1959); C. A., 56, 11445g (1962).
965. Pohlemann, H., H. Stummeyer, and H. Adolphi, Brit. 815,965 (1959); C. A., 54, 1297h (1960).
966. Pohlemann, H., S. Winderl, H. Stummeyer, and H. Adolphi, Brit. 846,229 (1960); C. A., 55, 7289d (1961).
967. Pohlemann, H., S. Winderl, H. Stummeyer, and H. Adolphi, Brit. 869,399 (1961); C. A., 56, 10104c (1962).
968. Popov, E. M., T. A. Mastryukova, N. P. Rodionova, and M. I. Kabachnik, J. Gen. Chem., 29, 1967 (1959); C. A., 54, 8282d (1960).
969. Porret, D., P. Kohler, and R. Sallmann, Swiss 332,805 (1958); C. A., 53, 19974d (1959).
970. Potenza, J. A., E. H. Poindexter, P. J. Caplan, and R. A. Dwek, J. Am. Chem. Soc., 91, 4356 (1969); C. A., 71, 55340 (1969).
971. Pound, D. W., Brit. 668,536 (1952); C. A., 47, 5438g (1953).

972. Prib, O. A., L. V. Glushkova, and R. V. Sendega, Ukr. Khim. Zh., 34, 497 (1968); C. A., 69, 86506d (1968).
973. Prib, O. A., and M. S. Malinovskii, J. Gen. Chem., 33, 647 (1963); C. A., 59, 2687c (1963).
974. Price G. R., E. N. Walsh, and J. T. Hallett, U. S. 3,094,406 (1963); C. A., 59, 13893a (1963).
975. Price, G. R., E. N. Walsh, and J. T. Hallett, U. S. 3,109,770 (1963); C. A., 60, 2841g (1964).
976. Price, G. R., E. N. Wlash, and J. T. Hallett, U. S. 3,114,761 (1963); C. A., 60, 6790h (1964).
977. Price, G. R., E. N. Walsh, and J. T. Hallett, U. S. 3,258,463 (1966); C. A., 65, 7101d (1966).
978. Prilezhaeva, E. N., L. V. Tsymbal, and M. F. Shostakovskii, Bull. Acad. Sci. USSR, Div. Chem. Sci., 1962, 1593; C. A., 58, 5721e (1963).
979. Pritchard, J. G., Org. Mass Spectrom., 3, 163 (1970); C. A., 72, 115864 (1970).
980. Protsenko, L. D., and N. Ya. Skul'skaya, J. Gen. Chem., 34, 2244 (1964); C. A., 61, 13262h (1964).
981. Pudovik, A. N., and I. M. Aladzheva, J. Gen. Chem., 30, 2599 (1960); C. A., 55, 17474i (1961).
982. Pudovik, A. N., and I. M. Aladzheva, J. Gen. Chem., 32, 1986 (1962); C. A., 58, 4595c (1963).
983. Pudovik, A. N., I. M. Aladzheva, I. A. Sokolova, and G. A. Kozlova, J. Gen. Chem., 33, 95 (1963); C. A., 59, 425f (1963).
984. Pudovik, A. N., I. M. Aladzheva, and L. N. Yakovenko, J. Gen. Chem., 35, 1214 (1965); C. A., 63, 11609g (1965).
985. Pudovik, A. N., and R. A. Cherkasov, J. Gen. Chem., 37, 1252 (1967); C. A., 68, 105303 (1968).
986. Pudovik, A. N., and R. A. Cherkasov, J. Gen. Chem., 38, 2448 (1968); C. A., 70, 57079 (1969).
987. Pudovik, A. N., R. A. Cherkasov, and R. M. Kondrat'eva, J. Gen. Chem., 38, 308 (1968); C. A., 69, 35323 (1968).
988. Pudovik, A. N., R. A. Cherkasov, G. A. Kutyrev, Yu. Yu. Samitov, A. A. Musina, and E. I. Gol'dfarb, J. Gen. Chem., 40, 1970 (1970); C. A., 74, 124443 (1971).
989. Pudovik, A. N., and O. S. Durova, J. Gen. Chem., 36, 1465 (1966); C. A., 66, 11009 (1967).
990. Pudovik, A. N., and E. M. Faizullin, J. Gen. Chem., 34, 876 (1964); C. A. 60, 15903h (1964).
991. Pudovik, A. N., E. M. Faizullin, and V. P. Zhukov, J. Gen. Chem., 36, 319 (1966); C. A., 64, 15916d (1966).
992. Pudovik, A. N., E. M. Faizullin, and G. I. Zhuravlev, Proc. Acad. Sci. USSR, 165, 1132 (1965); C. A., 64, 6481f (1966).
993. Pudovik, A. N., E. M. Faizullin, and G. I. Zhuravlev, Zh. Obshch. Khim., 36, 718 (1966); C. A., 65, 8745e

(1966).
994. Pudovik, A. N., and I. V. Gur'yanova, J. Gen. Chem., 37, 1566 (1967); C. A., 68, 13093 (1968).
995. Pudovik, A. N., I. V. Gur'yanova, M. G. Zimin, and O. E. Raevskaya, J. Gen. Chem., 38, 1488 (1968); C. A., 70, 87915 (1969).
996. Pudovik, A. N., and E. A. Ishmaeva, J. Gen. Chem., 35, 2070 (1965); C. A., 64, 8021b (1966).
997. Pudovik, A. N., and E. J. Kashevarova, Proc. Acad. Sci. USSR, 158, 859 (1964); C. A., 61, 13222e (1964)
998. Pudovik, A. N., E. I. Kashevarova, and G. L. Goloven'kin, J. Gen. Chem., 34, 3283 (1964); C. A., 62, 3922c (1965).
999. Pudovik, A. N., I. V. Konovalova, and L. V. Banderova, J. Gen. Chem., 35, 1210 (1965); C. A., 63 13064h (1965).
1000. Pudovik, A. N., I. V. Konovalova, and L. V. Dedova, J. Gen. Chem., 34, 2935 (1964); C. A., 61, 15968h (1964).
1001. Pudovik, A. N., and V. K. Krupnov, J. Gen. Chem., 38, 306 (1968); C. A., 69, 77369; (1968).
1002. Pudovik, A. N., and V. K. Krupnov, J. Gen. Chem., 38, 1605 (1968); C. A., 70, 86945 (1969).
1003. Pudovik, A. N., and M. K. Sergeeva, J. Gen. Chem. USSR, 25, 1713 (1955); C. A., 50, 7073i (1956).
1004. Raetz, R. F. W., U. S. 3,287,446 (1966); C. A., 66, 65535 (1967).
1005. Raetz, R. F. W., U. S. 3,355,522 (1967); C. A., 68, 39660 (1968).
1006. Raetz, R. F. W., and A. D. Bliss, U. S. 3,340,330 (1967); C. A., 68, 78317 (1968).
1007. Raetz, R. F. W., and A. D. Bliss, U. S. 3,445,468 (1969); C. A., 71, 61430 (1969).
1008. Raetz, R. F. W., and J. F. Cronan, U. S. 3,574,223 (1971); C. A., 75, 20409 (1971).
1009. Raetz, R. F. W., and O. J. Sweeting, J. Org. Chem., 30, 438 (1965); C. A., 62, 8991h (1965).
1010. Ralls, J. W., and A. Cortes, J. Econ. Entomol., 59, 1296 (1966); C. A., 66, 1841 (1967).
1011. Rapun, R., Proc. Int. Conf. Methods Prep. Stor. Label. Compounds, 2nd., 1966, 155; C. A., 71, 49901 (1969).
1012. Razumova, N. A., Zh. L. Evtikhov, and A. A. Petrov, J. Gen. Chem., 38, 1072 (1968); C. A., 69, 9660 (1968).
1013. Reddy, G. S., and R. Schmutzler, Z. Naturforsch., 25, 1199 (1970); C. A., 74, 17830 (1971).
1014. Regel, E., U. S. 3,294,876 (1966); C. A., 66, 45692 (1967).
1015. Regel, E. K., and M. F. Botts, U. S. 3,502,771 (1970); C. A., 73, 24903 (1970).
1016. Reznik, V. S., and N. G. Pashkurov, Dokl. Akad. Nauk SSSR, 177, 604, (1967); C. A., 69, 27368 (1968).

1017. Reznik, V. S., N. G. Pashkurov, R. R. Shagidullin, and R. A. Bulgakova, Khim. Geterotsikl. Soedin., Sb. 1: Azotsoderzhashchie Geterotsikly, 1967, 384; C. A., 70, 87716 (1969).
1018. Rhone-Poulenc, Fr. 1,125,943 (1956); C. A., 53, 17152 (1959).
1019. Rhone-Poulenc, Fr. Addn. 76,457 (1962); C. A., 58, 1357 (1963).
1020. Rhone-Poulenc, Belg. 609,209 (1962); C. A., 58, 11368 (1963).
1021. Rhone-Poulenc S. A., Brit. 920,420 (1963); C. A., 59, 3778h (1963).
1022. Rhone-Poulenc S. A., Neth. Appl. 6,400,757 (1964); C. A., 62, 7729e (1965).
1023. Richert, H., G. Schrader, and H. Jonas, Ger. 1,153,747 (1963); C. A., 60, 558h (1964).
1024. Richter, S. B., U. S. 3,062,704 (1962); C. A., 59, 9797 (1963).
1024a. Richter, S. B., U. S. 2,967,123 (1961); C. A., 55, 6382 (1961).
1025. Richter, S. B., U. S. 3,075,873 (1963); C. A., 59, 9797 (1963).
1026. Richter, S. B., U. S. 3,092,543 (1963); C. A., 59, 11351 (1963).
1027. Richter, S. B., U. S. 3,244,775 (1966); C. A., 64, 19492 (1966).
1028. Richter, S. B., U. S. 3,356,484 (1967); C. A., 69, 10233 (1968).
1029. Richter, S. B., Brit. 1,115,028 (1968); C. A., 69, 76912 (1968).
1030. Richter, S. B., U. S. 3,450,520 (1969); C. A., 71, 60996 (1969).
1031. Rigterink, R. H., Fr. 1,360,901 (1964); C. A., 61, 16052d (1964); U. S. 3,244,586 (1966).
1032. Rigterink, R. H., U. S. 3,228,960 (1966); C. A., 64, 17542 (1966).
1033. Rigterink, R. H., U. S. 3,326,752 (1967); C. A., 67, 72806 (1967).
1034. Rigterink, R. H., U. S. 3,385,859 (1968); C. A., 69, 59105 (1968).
1035. Rigterink, R. H., U. S. 3,502,670 (1970); C. A., 72, 121587 (1970).
1036. Rizpolozhenskii, N. I., L. V. Boiko, and M. A. Zvereva, Proc. Acad. Sci. USSR, 155, 388 (1964); C. A., 61, 1817 (1964).
1037. Rizpolozhenskii, N. I., and F. S. Mukhametov, Bull. Acad. Sci. USSR, Ser. Chem., 1968, 2609; C. A., 70, 77229 (1969).
1038. Rizpolozhenskii, N. I., L. V. Stepashkina, and R. R. Shagidullin, Bull. Acad. Sci. USSR, Ser. Chem., 1967, 1928; C. A., 68, 29790 (1968).
1039. Rizpolozhenskii, N. I., M. A. Zvereva, and L. V. Stepashkina, Khim. Org. Soedin. Fosfora, Acad. Nauk

SSSR, Otd. Obshch. Tekn. Khim., 1967, 202; C. A., 69, 27124 (1968).
1040. Roesky, H. W., Chem. Ber., 101, 636 (1968); C. A., 68, 69087 (1968).
1041. Roesky, H., Chem. Ber., 101, 2977 (1968); C. A., 69, 87088 (1968).
1042. Roesky, H. W., U. S. 3,449,473 (1969); C. A., 71, 49389 (1969).
1043. Roesky, H., Z. Naturforsch., B, 24, 818 (1969); C. A., 71, 112322 (1969).
1044. Roesky, H. W., and H. Beyer, Chem. Ber., 102, 2588 (1969); C. A., 71, 90714 (1969).
1045. Roesky, H. W., and M. Dietl, Z. Naturforsch., B, 25, 316 (1970); C. A., 72, 139167 (1970).
1046. Rosnati, L., Gazz. Chim. Ital., 76, 272 (1946); C. A., 42, 876 (1948).
1047. Ross, R. T., and F. J. Biros, Anal. Chim. Acta, 52, 139 (1970); C. A. 74, 31000 (1971).
1048. Rossi, G., Gazz. Chim. Ital, 89, 1324 (1959); C. A., 54, 22320 (1960).
1049. Roussel-UCLAF, Fr. 1,589,898 (1970); C. A., 74, 41900 (1971).
1050. Rudel, A., and J. M. Boyle, U. S. 2,523,146 (1950); C. A., 45, 1157 (1951).
1051. Ruefenacht, K., Fr. 1,335,755 (1963); C. A., 60, 1764 (1964); U. S. 3,230,230 (1966).
1052. Ruefenacht, K., Ger. 1,184,552 (1964); C. A., 62, 13154 (1965)
1053. Ruefenacht, K., Helv. Chim. Acta, 51, 518 (1968); C. A., 69, 2897 (1968).
1054. Ruf, E., Ger. 1,211,167 (1966); C. A., 64, 17425 (1966).
1055. Saito, T., G. Nishikawa, S. Watanabe, and Y. Kataoka, Japan 21,514 (1964); C. A., 62, 10338d (1965).
1056. Saito, T., G. Nishikawa, S. Watanabe, and Y. Kataoka, Japan 21,831 (1964); C. A., 62, 11688 (1965).
1057. Sakai, M., M. Kato, Y. Sato, K. Konishi, and T. Okutani, Japan 14,625 (1967); C. A., 68, 104740 (1968).
1058. Sakamoto, H., M. Nakagawa, T. Mizutani, S. Kuramoto, T. Fujimoto, and Y. Okuno, Japan 24,806 (1964); C. A., 62, 10439 (1965).
1059. Sakamoto, H., M. Nakagawa, T. Mizutani, S. Kuramoto, T. Fujimoto, and Y. Okuno, Japan 2,345 (1965); C. A., 62, 14570 (1965).
1060. Sakamoto, H., M. Nakagawa, T. Mizutani, H. Tsuchiya, T. Fujimoto, and Y. Okuno, Japan 14,623 (1967); C. A., 68, 95563 (1968).
1061. Sakamoto, H., M. Nakagawa, H. Tsuchiya, T. Fujimoto, and Y. Okuno, Japan 9010 (1966); C. A., 65, 12239d (1966).
1062. Sakamoto, H., H. Tsuchiya, M. Nakagawa, T. Fujimoto,

and Y. Okuno, Japan 5707 (1965); C. A., 63, 1799h (1965).
1063. Sakamoto, H., H. Tsuchiya, M. Nakagawa, T. Fujimoto, Y. Okuno, and K. Hanemichi, Japan 5633 (1965); C. A., 63, 1800a (1965).
1064. Sallmann, R., Swiss 318,815 (1957); C. A., 51, 17988e (1957).
1065. Sallmann, R., Swiss 323,228 (1957); C. A., 52, 14959c (1958).
1066. Sallmann, R., Swiss 324,980 (1957); C. A., 52, 14960a (1958).
1067. Sallmann, R., U. S. 2,959,516 (1960); C. A., 55, 8296b (1961).
1068. Salzberg, P. L., and J. H. Werntz, U. S. 2,063,629 (1936); C. A., 31, 702 (1937).
1069. Sandoz Ltd., Neth. Appl. 6,611,511 (1967); C. A., 67, 73620 (1967).
1070. Sanin, P. I., E. S. Shepeleva, and V. V. Sher, Neftekhimiya, 3, 781 (1963); C. A., 60, 1512f (1964).
1071. Sankyo Co., Ltd., Fr. 1,489,695 (1967); C. A., 69, 36110z (1968).
1072. Sarkar, A. K., Brit. 1,038,611 (1966); C. A., 65, 15275h (1966).
1073. Sauli, M., S. African 4,230 (1968); C. A., 71, 38970 (1969).
1074. Saunders, B. C., G. J. Stacey, F. Wild, and I. G. E. Wilding, J. Chem. Soc., 1948, 699; C. A., 42, 6741c (1948).
1075. Sawa, N., and Y. Tsujino, Japan 12,620 (1965); C. A., 63, 11573a (1965).
1076. Sawa, N., and Y. Tsujino, Japan 13,873 (1965); C. A., 63, 13274h (1965).
1077. Sawa, N., and Y. Tsujino, Japan 17,583 (1965); C. A., 63, 18101c (1965).
1078. Sayre, R., J. Am. Chem. Soc., 80, 5438 (1958); C. A., 53, 9998e (1959).
1079. Schegk, E., and G. Schrader, Brit. 819,689 (1959); C. A., 55, 11751h, (1961); U. S. 3,042,703 (1962).
1080. Schegk, E., and G. Schrader, Ger. 1,063,157 (1959); C. A., 55, 23564 (1961).
1081. Schegk, E., and G. Schrader, Ger. 1,078,117 (1960); C. A., 55, 15821h (1961).
1082. Schegk, E., and G. Schrader, Ger. 1,092,030 (1960); C. A., 55, 22704 (1961).
1083. Schegk, E., and G. Schrader, Ger. 1,099,533 (1961); C. A., 56, 3417 (1962).
1084. Schegk, E., and G. Schrader, Ger. 1,116,656 (1961); C. A., 56, 14170f (1962); U. S. 3,042,703 (1962).
1085. Schegk, E., and G. Schrader, Ger. 1,117,110 (1961); C. A., 56, 14170d (1962).
1086. Scheinpflug, H., H. F. Jung, and G. Schrader, Belg.

666,504 (1966); C. A., 65, 13543d (1966).
1087. Scherer, O., H. Hahn, and G. Stähler, Ger. 1,015,794 (1957); C. A., 53, 19877 (1959).
1088. Scherer, O, and H. Mildenberger, Neth. Appl. 6,602,131 (1966); C. A., 66, 37946 (1967); U. S. 3,496,178 (1970).
1089. Scherer, O. J., and J. Wokulat, Z. Anorg. Allg. Chem., 357, 92 (1968); C. A., 68, 78339 (1968).
1090. Scheuerer, G., and H. Dickhaeuser, Fr. 1,379,964 (1964); C. A., 62, 10381a (1965).
1091. Scheuerer, G., A. Zeidler, H. Dickhaeuser, and H. Adolphi, Ger. 1,276,638 (1968); C. A., 70, 28616 (1969).
1092. Schicke, H. G., Ger. 1,134,382 (1962); C. A., 57, 16624 (1962).
1093. Schicke, H. G., Belg. 631,211 (1963); C. A., 61, 1757 (1964).
1094. Schicke, H. G., Belg. 631,914 (1963); C. A., 60, 13149 (1964).
1095. Schicke, H. G., Ger. 1,183,494 (1964); C. A., 62, 7684g (1965).
1096. Schicke, H. G., Ger. 1,183,506 (1964); C. A., 62, 7798 (1965).
1097. Schicke, H. G., Belg. 641,818 (1964); C. A., 62, 16276 (1965).
1098. Schicke, H. G., Ger. 1,192,202 (1965); C. A., 63, 9990 (1965).
1099. Schicke, H. G., Ger. 1,198,359 (1965); C. A., 63, 18155 (1965).
1100. Schicke, H. G., Belg. 652,189 (1964); C. A., 64, 8238 (1966).
1101. Schicke, H. G., Neth. Appl. 6,516,907 (1966), U. S. 3,402,176 (1968); C. A., 65, 13732 (1966).
1102. Schicke, H. G., and A. Berger, Ger. 1,197,878 (1965) C. A., 63, 13319 (1965).
1103. Schicke, H. G., and W. Lorenz, Ger. 1,144,264 (1963) C. A., 60, 4184 (1964).
1104. Schicke, H. G., and G. Schrader, Ger. 1,111,172 (1961); C. A., 56, 2474 (1962).
1105. Schicke, H. G., and G. Schrader, Ger. 1,122,060 (1962); C. A., 57, 12325 (1962).
1106. Schicke, H. G., and G. Schrader, Ger. 1,129,484 (1962); C. A., 57, 13806 (1962).
1107. Schicke, H., and G. Schrader, Ger. 1,156,070 (1963); C. A., 60, 3014 (1964).
1108. Schicke, H. G., and G. Schrader, Ger. 1,206,900 (1965); C. A., 64, 9641 (1966).
1109. Schicke, H. G., and C. Stoelzer, Belg. 657,941 (1965); C. A., 65, 654 (1966).
1110. Schlichting, H. L., and E. D. Weil, Belg. 622,421 (1962); C. A., 59, 11255 (1963); Brit. 1,011,982 (1965); C. A., 64, 14091d (1966).

1111. Schliebs, R., Ger. 1,210,834 (1966); C. A., 64, 17639a (1966).

1112. Schloer, H., Ger. 1,217,955 (1966); C. A., 66, 2640 (1967).

1113. Schloer, H., B. Homeyer, and I. Hammann, S. African 5,486 (1968); C. A., 70, 12871 (1969).

1114. Schloer, H., and G. Schrader, Ger. 1,083,808 (1960); C. A., 55, 17499 (1961).

1115. Schloer, H., and G. Schrader, Ger. 1,083,809 (1960); C. A., 55, 17500a (1961).

1116. Schloer, H., and G. Schrader, Belg. 616,335 (1962); C. A., 59, 1488 (1963).

1117. Schmidt, K. J., C. Fest, and I. Hammann, S. African 4,505 (1968); C. A., 71, 91644 (1969).

1118. Schmidt, K. J., and I. Hammann, Brit. 1,160,493 (1969); C. A., 71, 113084 (1969).

1119. Schmidt, K. J., I. Hammann, and G. Unterstenhoefer, S. African 4,112 (1968); C. A., 71, 70737 (1969).

1120. Schoenbeck, R., E. Kloimstein, W. Beck, and A. Diskus, Austrian 252,950 (1967); C. A., 67, 11497 (1967).

1121. Schoenbeck, R., E. Kloimstein, W. Beck, and A. Diskus, U. S. 3,310,560 (1967); C. A., 67, 90825h (1967).

1122. Schrader, G., U. S. 2,571,989 (1951); C. A., 46, 3066a (1952).

1123. Schrader, G., U. S. 2,597,534 (1952); C. A., 47, 4357h (1953).

1124. Schrader, G., Ger. 830,262 (1952); C. A., 52, 8447 (1958).

1125. Schrader, G., Brit. 670,030 (1952); C. A., 47, 5438i (1953); U. S. 2,624,745 (1953).

1126. Schrader, G., Ger. 885,176 (1953); C. A., 48, 6644 (1954).

1127. Schrader, G., Ger. 896,643 (1953); C. A., 49, 2482f (1955).

1128. Schrader, G., Brit. 713,142 (1954); C. A., 50, 411 (1956); U. S. 2,748,146 (1956).

1129. Schrader, G., Ger. 927,092 (1955); C. A., 50, 2653 (1956).

1130. Schrader, G., Ger. 935,432 (1955); C. A., 50, 4448 (1956).

1131. Schrader, G., Ger. 942,988 (1956); C. A., 50, 16825h (1956).

1132. Schrader, G., Ger. 946,056 (1956); C. A., 51, 4424 (1957).

1133. Schrader, G., Ger. 951,717 (1956); C. A., 51, 12958g (1957).

1134. Schrader, G., Ger. 954,415 (1956); C. A., 51, 12958 (1957).

1135. Schrader, G., Ger. 961,083 (1957); C. A., 51, 15549g (1957).

1136. Schrader, G., Israeli 8839 (1956); C. A., 52, 11891a (1958); U. S. 2,927,122; C. A., 54, 15245d (1960).
1137. Schrader, G., U. S. 2,829,111 (1958); C. A., 52, 13776e (1958).
1138. Schrader, G., Ger. 932,064 (1955); C. A., 52, 16290e (1958).
1139. Schrader, G., Ger. 963,872 (1957); C. A., 53, 11234g (1959).
1140. Schrader, G., Ger. 1,014,987 (1957); C. A., 53, 19878h (1959).
1141. Schrader, G., U. S. 2,862,019 (1958); C. A., 54, 1313d (1960).
1142. Schrader, G., U. S. 2,870,191 (1959); C. A., 53, 11223f (1959).
1143. Schrader, G., Brit. 811,529 (1959); C. A., 53, 16962g (1959); U. S. 2,923,730 (1960).
1144. Schrader, G., Ger. 1,005,058 (1957); C. A., 53, 16962l (1959); U. S. 2,928,863 (1960).
1145. Schrader, G., Ger. 1,016,260 (1957); C. A., 54, 296 (1960).
1146. Schrader, G., U. S. 2,894,973 (1959); C. A., 54, 11992 (1960).
1147. Schrader, G., U. S. 2,909,557 (1959); C. A., 54, 11992 (1960).
1148. Schrader, G., Ger. 1,014,988 (1957); C. A., 54, 19490e (1960).
1149. Schrader, G., Ger. 1,022,582 (1958); C. A., 54, 4390f (1960).
1150. Schrader, G., Ger. 1,024,509 (1958); C. A., 54, 19596i (1960).
1151. Schrader, G., Ger. 1,039,528 (1958); C. A., 54, 24551h (1960).
1152. Schrader, G., Ger. 1,045,391 (1958); C. A., 55, 2488g (1961).
1153. Schrader, G., Ger. 1,056,121 (1959); C. A., 55, 14383g (1961).
1154. Schrader, G., Ger. 1,058,046 (1959); C. A., 55, 13319i (1961).
1155. Schrader, G., Ger. 1,063,148 (1959); C. A., 55, 5346c (1961).
1156. Schrader, G., Ger. 1,075,109 (1960); C. A., 55, 17540i (1961).
1157. Schrader, G., Ger. 1,100,019 (1961); C. A., 56, 324b (1962).
1158. Schrader, G., Ger. 1,102,138 (1961); C. A., 55, 19787d (1961).
1159. Schrader, G., Ger. 1,108,234 (1961); C. A., 57, 4597b (1962).
1160. Schrader, G., Ger. 1,115,243 (1961); C. A., 56, 10045 (1962).
1161. Schrader, G., Belg. 608,801 (1962); C. A., 57, 5852f (1962).

1162. Schrader, G., Ger. 1,126,382 (1962); C. A., 57, 7108 g (1962).
1163. Schrader, G., Ger. 1,127,892 (1962); C. A., 57, 8504b (1962).
1164. Schrader, G., Belg. 610,970 (1962); C. A., 57, 15155i (1962).
1165. Schrader, G., Ger. 1,132,147 (1962); C. A., 57, 16496a (1962).
1166. Schrader, G., Ger. 1,136,328 (1962); C. A., 58, 1350h (1963).
1167. Schrader, G., Ger. 1,138,041 (1962); C. A., 58, 11276h (1963).
1168. Schrader, G., Belg. 615,668 (1962); C. A., 58, 11403c (1963).
1168a. Schrader, G., "Die Entwicklung neuer insektizider Phosphorsäure-Ester," Verlag Chemie GMBH., Weinheim/ Bergstr., 3d edit., (1963).
1169. Schrader, G., Belg. 620,758 (1963); C. A., 59, 11344 (1963).
1170. Schrader, G., Ger. 1,156,409 (1963); C. A., 60, 4184 (1964).
1171. Schrader, G., Belg. 627,458 (1963); C. A., 60, 10556 (1964).
1172. Schrader, G., Ger. 1,159,935 (1963): C. A., 60, 11945d (1964).
1173. Schrader, G., Ger. 1,161,556 (1964); C. A., 60, 10548h (1964).
1174. Schrader, G., Belg. 636,198 (1964); C. A., 61, 14530h (1964).
1175. Schrader, G., U. S. 2,976,312 (1961); C. A., 62, 446d (1965).
1176. Schrader, G., Ger. 1,178,425 (1964); C. A., 62, 459 (1965).
1177. Schrader, G., Ger. 1,183,495 (1964); C. A., 62, 9174 (1965).
1178. Schrader, G., Ger. 1,187,849 (1965); C. A., 62, 16060g (1965).
1179. Schrader, G., Fr. 1,391,206 (1965); C. A., 62, 16129 (1965).
1180. Schrader, G., U. S. 3,201,444 (1965); C. A., 63, 13154a (1965).
1181. Schrader, G., Belg. 665,504 (1965); C. A., 65, 5400g (1966).
1182. Schrader, G., Ger. 1,215,171 (1966); C. A., 65, 13607f (1966).
1183. Schrader, G., and A. Dörken, Ger. 936,037 (1955); C. A., 50, 4448 (1956).
1184. Schrader, G., and A. Dörken, U. S. 2,852,514 (1958); C. A., 53, 4215 (1959).
1185. Schrader, G., and R. Gönnert, Ger. 949,230 (1956); C. A., 51, 4425c (1957).
1186. Schrader, G., I. Hammann, and W. Behrenz, Ger. 1,262,266 (1968); C. A., 68, 114735c (1968).

1187. Schrader, G., I. Hammann, and W. Stendel, S. African 7,378 (1968); C. A., 72, 3239g (1970).
1188. Schrader, G., I. Hammann, and W. Stendel, S. African 2,317 (1969); C. A., 72, 111017 (1970).
1189. Schrader, G., and W. Lorenz, Ger. 817,057 (1951); C. A., 48, 6643 (1954).
1190. Schrader, G., and W. Lorenz, Brit. 691,267 (1953); C. A., 48, 7047e (1954); Ger. 824,046 (1951).
1191. Schrader, G., and W. Lorenz, Ger. 926,488 (1955); C. A., 50, 2653 (1956).
1192. Schrader, G., and W. Lorenz, Ger. 947,369 (1956); C. A., 51, 4425e (1957).
1193. Schrader, G., and W. Lorenz, Ger. 949,231 (1956); C. A., 51, 4426a (1957).
1194. Schrader, G., and W. Lorenz, U. S. 2,862,017 (1958); C. A., 54, 1437i (1960).
1195. Schrader, G., and W. Lorenz, U. S. 2,929,832 (1960); C. A., 54, 12994g (1960).
1196. Schrader, G., and W. Lorenz, Ger. 1,082,915 (1960); C. A., 55, 25983c (1961).
1197. Schrader, G., and W. Lorenz, Ger. 1,103,324 (1961); C. A., 56, 5888c (1962).
1198. Schrader, G., and W. Lorenz, Neth. Appl. 6,405,586 (1964); C. A., 62, 14728b (1965); U. S. 3,277,215 (1966).
1199. Schrader, G., and R. Mersch, Brit. 793,758 (1958); C. A., 53, 1253c (1959); U. S. 2,870,187 (1959).
1200. Schrader, G., and R. Mühlmann, Ger. 931,106 (1955); C. A., 50, 8714b (1956).
1201. Schrader, G., and R. Mühlmann, Ger. 887,814 (1953); C. A., 52, 13791i (1958).
1202. Schrader, G., and H. Scheinpflug, Ger. 1,300,935 (1969); C. A., 71, 101533x (1969).
1202a. Schrader, G., and H. Scheinpflug, Brit. 1,115,549 (1968); C. A., 69, 58940 (1968).
1203. Schrader, G., and H. G. Schicke, Ger. 1,106,317 (1961); C. A., 56, 4796c (1962).
1204. Schrader, G., and R. Schliebs, Ger. 1,233,390 (1967); C. A., 66, 115324p (1967).
1205. Schroeder, P. H., S. African 7,133 (1968); C. A., 70, 47309t (1969).
1206. Schuler, M., U. S. 3,022,215 (1962); C. A., 57, 4551 (1962).
1207. Schuler, M., Ger. 1,122,935 (1962); C. A., 57, 11101d (1962).
1208. Schulze, H., and A. Mueller, Z. Naturforsch. B, 25, 148 (1970); C. A., 72, 94818j (1970).
1209. Scott, C. B., U. S. 2,769,831 (1956); C. A., 51, 7401a (1957).
1210. Scott, C. B., A. Menefee, and D. O. Alford, J. Org. Chem., 22, 789 (1957); C. A., 52, 2739f (1958).
1211. Seel, F., K. Ballreich, and R. Schmutzler, Chem.

Ber., 95, 199 (1962); C. A., 56, 12932i (1962).
1212. Sehring, R., and W. Buck, S. African 7,026 (1968); C. A., 70, 78144 (1969).
1213. Sehring, R., and K. Zeile, Ger. 1,076,662 (1960); C. A., 55, 12303b (1961).
1214. Sehring, R., and K. Zeile, Ger. 1,146,486 (1963); C. A., 59, 8790h (1963).
1215. Sehring, R., and K. Zeile, Belg. 619,496 (1962); C. A., 60, 2276b (1964); Ger. 1,168,159 (1964).
1216. Sehring, R., and K. Zeile, U. S. 3,227,610 (1966); C. A., 64, 8086a (1966).
1217. Semidetko, V. V., and Yu. G. Gololobov, J. Gen. Chem., 37, 2659 (1967); C. A., 69, 43334 (1968).
1218. Semin, G. K., and T. A. Babushkina, Teor. Eksp. Khim., 4, 835 (1968); C. A., 71, 44253e (1969).
1219. Sen, A. B., and A. K. Sen Gupta, J. Indian Chem. Soc., 32, 120 (1955); C. A., 50, 4135a (1956).
1220. Sen Gupta, A. K., and J. Kumar, J. Indian Chem. Soc., 43, 575 (1966); C. A., 66, 94752j (1967).
1221. Senkbeil, H. O., U. S. 3,041,366 (1962); C. A., 57, 14966i (1962).
1222. Serazutdinova, L. T., and V. I. Isagulyants, Nauch. -Tekh. Konf., Moskov., 1956-7, 215 (Pub. 1958); C. A., 55, 13345e (1961).
1223. Shabanova, M. P., Yu. S. Kagan, E. N. Prilezhaeva, L. V. Tsymbal, and E. Ya. Makhlina, Tr. Vses. Nauchn. -Issled. Inst. Zashchity Rast. No. 21, Pt. 1, 114 (1964); C. A., 65, 12807d (1966).
1223a. Shafik, M. T., D. Bradway, F. J. Biros, and H. F. Enos, J. Agr. Food Chem., 18, 1174 (1970); C. A., 74, 22396 (1971).
1224. Sharma, K. D., and J. B. Lal, Indian J. Chem., 3, 46 (1965); C. A., 62, 14483b (1965).
1225. Sharpe, S. P., and B. K. Snell, S. African 5,808 (1968); C. A., 73, 131022p (1970).
1226. Shell Internationale Research Maatschappij N.W., Fr. 1,383,645 (1964); C. A., 62, 9062f (1965).
1227. Shell Internationale, Neth. Appl. 6,414,234 (1965); C. A., 63, 16268f (1965).
1228. Shell Internationale, Neth. Appl. 6,504,109 (1965); C. A., 64, 14218e (1966).
1229. Shell Internationale, Neth. Appl. 6,510,031 (1965); C. A., 64, 11217 (1966).
1230. Shell Internationale, Neth. Appl. 6,515,923 (1966); C. A., 65, 15390 (1966).
1231. Shell Internationale Research Maatschappij N.V., Neth. Appl. 6,613,894 (1967); C. A., 68, 49435p (1968).
1232. Shell Internationale Research Maatschappij N.V., Neth. Appl. 950 (1969); C. A., 72, 12877 (1970).
1233. Sheppard, W. A., J. Org. Chem., 26, 1460 (1961); C. A., 55, 24627g (1961).

1234. Sherlock, E., Brit. 932,388 (1963); C. A., 60, 2992 (1964).
1235. Shindo, N., K. Ura, and H. Takahashi, Japan 6,740 (1967); C. A., 67, 90933s (1967).
1236. Shitov, L. N., and B. M. Gladshtein, J. Gen. Chem., 38, 2268 (1968); C. A., 70, 29010 (1969).
1237. Shlenkova, E. K., and V. P. Evdakov, J. Gen. Chem., 37, 848 (1967); C. A., 67, 99570 (1967).
1238. Shokol, V. A., V. V. Doroshenko, and G. I. Derkach, J. Gen. Chem., 40, 1678 (1970); C. A., 74, 75967 (1971).
1239. Shostakovskii, M. F., E. N. Prilezhaeva, L. V. Tsymbal, V. A. Azovskaya, and N. G. Starova, Izvest. Akad. Nauk SSSR., Otd. Khim. Nauk, 1959, 2239; C. A., 54, 10847f (1960).
1240. Shvetsova-Shilovskaya, K. D., E. I. Lebedeva, and N. N. Mel'nikov, J. Gen. Chem., 34, 2150 (1964); C. A., 61, 11883e (1964).
1241. Shvetsova-Shilovskaya, K. D., L. N. Matyuk, and N. N. Mel'nikov, J. Gen. Chem., 35, 1498 (1965); C. A., 63, 16236g (1965).
1242. Shvetsova-Shilovskaya, K. D., N. N. Mel'nikov, E. N. Andreeva, L. P. Bocharova, and Yu. N. Sapozhkov, J. Gen. Chem., 31, 776 (1961); C. A., 55, 23554i (1961).
1243. Shvetsova-Shilovskaya, K. D., N. N. Mel'nikov, and V. A. Glushenkov, Zh. Obshch. Khim., 29, 3593 (1959); C. A., 54, 19458h (1960).
1244. Shvetsova-Shilovskaya, K. D., N. N. Mel'nikov, and A. F. Grapov, J. Gen. Chem. USSR, 26, 925 (1956); C. A., 50, 14769g (1956).
1245. Shvetsova-Shilovskaya, K. D., N. N. Mel'nikov, M. Ya Kagan, and V. A. Glushenkov, J. Gen. Chem., 30, 205 (1960); C. A., 54, 22321f (1960).
1246. Shvetsova-Shilovskaya, K. D., N. N. Mel'nikov, and N. I. Martem-yanova, J. Gen. Chem., 26, 523 (1956); C. A., 50, 13726 (1956).
1247. Simone, R. A., and M. E. Brokke, U. S. 3,328,405 (1967); C. A., 68, 49638g (1968).
1248. Simpson, P., and C. L. Bodkin, J. Chem. Soc., B, 1971, 1136; C. A., 75, 62984f (1971).
1249. Singhal, G. H., Ger. Offen. 2,064,474 (1971); C. A., 75, 110318 (1971).
1250. Sirrenberg, W., and R. Coelln, Ger. 1,263,748 (1968); C. A., 68, 114063g (1968).
1251. Sirrenberg, W., and W. Lorenz, Ger. 1,170,401 (1964); C. A., 62, 3977e (1965).
1252. Sirrenberg, W., W. Lorenz, H. Schloer, R. Coelln, G. Schrader, Belg. 618,589 (1962); C. A., 59, 5078e (1963).
1253. Slagh, H. R., U. S. 2,897,227 (1959); C. A., 54, 1428h (1960).

1254. Slagh, H. R., and E. C. Britton, U. S. 2,807,637 (1957); C. A., 52, 2066 (1958).
1255. Sledzinski, B., D. Narkiewicz, and F. Pastwa, Pol. 53,681 (1967); C. A., 71, 12802 (1969).
1256. Smithey, W. R., Jr., U. S. 3,004,054 (1961); C. A., 56, 8636f (1962).
1257. Snell, B. K., Brit. 1,129,797 (1968); C. A., 70, 28936d (1969).
1258. Snell, B. K., J. Chem. Soc., C, 1968, 2358; C. A., 69, 86937 (1968).
1259. Speroni, G., and M. Perini, Ital. 570,506 (1957); C. A., 53, 15464 (1959).
1260. Srivastava, K. C., Aust. J. Chem., 19, 2397 (1966); C. A., 66, 64990t (1967).
1261. Staeubli, S., L. Schiener, and K. Ruefenacht, S. African 4,012 (1967); C. A., 70, 77977b (1969).
1262. Stauffer Chemical Co., Neth. Appl. 6,414,045 (1965); C. A., 63, 17969h (1965).
1263. Stauffer Chemical Co., Neth. Appl. 6,512,363 (1966); C. A., 65, 18619f (1966).
1264. Stauffer Chemical Co., Neth. Appl. 6,614,646 (1967); C. A., 68, 12699b (1968).
1265. Stec. W., and J. Michalski, Z. Naturforsch., B, 25, 554 (1970); C. A., 73, 44831h (1970).
1266. Stepashkina, L. V., and N. I. Rizpolozhenskii, Bull. Acad. Sci. USSR, Ser. Chem., 1967, 584; C. A., 67, 108491h (1967).
1267. Stetter, H., and K. H. Steinacker, Chem. Ber., 85, 451 (1952); C. A., 47, 1583a (1953).
1268. Stiasni, M., and K. Schweikert, J. Label. Compounds, 2, 406 (1966); C. A., 66, 115391h (1967).
1269. Stoelzer, C., and A. Simon, Chem. Ber., 93, 1323 (1960); C. A., 54, 20927h (1960).
1270. Stoelzer, C., and A. Simon, Chem. Ber., 96, 881 (1963); C. A., 58, 13777a (1963).
1271. Stoelzer, C., and A. Simon, Chem. Ber., 96, 1335 (1963); C. A., 59, 2683e (1963).
1272. Stoffey, D. G., J. Org. Chem., 33, 1651 (1968); C. A., 68, 104515 (1968).
1273. Stoffey, D. G., R. C. Maxwell, A. D. Gutman, L. W. Fancher, and J. T. Hallett, U. S. 3,242,498 (1966); C. A., 64, 17434e (1966).
1274. Stothers, J. B., and J. R. Robinson, Can. J. Chem., 42, 967 (1964); C. A., 61, 196g (1964).
1275. Stothers, J. B., and E. Y. Spencer, Can J. Chem., 39, 1389 (1961); C. A., 56, 6815b (1962).
1276. Stoutamire, D. W., Belg. 660,045 (1965); C. A., 63, 17969g (1965).
1277. Strecker, W., and C. Grossmann, Ber., 49, 63 (1916); C. A., 10, 897 (1916).
1278. Sumiki, Y., and S. Tamura, Japan 14,368 (1961); C. A., 56, 10156i (1962).

1279. Sumiki, Y., and S. Tamura, Japan 21,521 (1961); C. A., 57, 14946c (1962).
1280. Sumiki, Y., and S. Tamura, Japan 9,983 (1962); C. A., 59, 3834e (1963).
1281. Sumitomo Chemical Industry Co., Ltd., Belg. 624,058 (1963); C. A., 59, 9814g (1963).
1282. Sumitomo Chemical Co., Ltd., Fr. 1,323,570 (1963); C. A., 59, 11340c (1963).
1283. Sumitomo Chemical Co., Ltd., Belg. 645,888 (1964); C. A., 63, 13157 (1965).
1284. Sumitomo Chemical Co., Ltd., Neth. Appl. 6,615,246 (1967); C. A., 67, 116710u (1967).
1285. Sumitomo Chemical Co., Ltd., Fr. 1,559,485 (1969); C. A., 72, 43118y (1970).
1286. Sumitomo Chemical Co., Ltd., Fr. 1,560,374 (1969); C. A., 72, 90043e (1970).
1287. Sutoris, V., Chem. Zvesti, 19, 379 (1965); C. A., 63, 14721e (1965).
1288. Sutoris, V., Acta Fac. Rerum Nat. Univ. Comenianae, Chimica, 9, 521 (1966); C. A., 66, 18539e (1967).
1289. Sutoris, V., Chem. Zvesti, 23, 356 (1969); C. A., 72, 21239m (1970).
1290. Suzuki, S., J. Miyamoto, K. Fujimoto, H. Sakamoto, and Y. Nishizawa, Agr. Biol. Chem., 34, 1697 (1970); C. A., 74, 42072 (1971).
1291. Szabo, K., and J. G. Brady, U. S. 3,106,510 (1963); C. A., 60, 4010b (1964).
1292. Szabo, K., and J. G. Brady, Fr. 1,360,378 (1964); C. A., 61,11898 (1964).
1293. Szabo, K., and J. G. Brady, U. S. 3,368,002 (1968); C. A., 69, 27519n (1968).
1294. Szabo, K., and L. Gorog, Magyar Kem. Folyoirat, 61, 84 (1955); C. A., 49, 16309g (1955).
1295. Szabo, K., and E. Oswald, Acta Chem. Acad. Sci. Hung., 15, 1 (1958); C. A., 53, 5276h (1959).
1296. Tabor, E. J., U. S. 3,005,005 (1961); C. A., 56, 5836 (1962).
1297. Takahashi, S., H. Ueda, and M. Kondo, Japan 11,860 (1966); C. A., 65, 15231b (1966).
1298. Takahashi, S., H. Ueda and M. Kondo, Japan 16,780 (1966); C. A., 66, 18511q (1967).
1299. Takahashi, K., T. Yamasaki, and G. Miyazima, Bull. Chem. Soc. Japan, 39, 2787 (1966); C. A., 66, 42148x (1967).
1300. Tammelin, L. E., Acta Chem. Scand., 11, 1738 (1957); C. A., 52, 10867i (1958).
1301. Tcherkerzoff, N., Rev. Inst. Fr. Petrole Ann. Combust. Liquides, 18, 438 (1963); C. A., 59, 12800g (1963).
1302. Teichmann, H., and G. Hilgetag, Ber., 96, 1454 (1963); C. A., 59, 2633g (1963).

1303. Teichmann, H., and Phi Trong Lam, Z. Chem., 9, 310 (1969); C. A., 71, 91575w (1969).
1304. Teplov, N. E., N. N. Godovikov, and M. I. Kabachnik, Zh. Organ. Khim., 1, 1658 (1965); C. A., 64, 4976g (1966).
1305. Thomae G.m.b.H., Karl, Brit. 948,039 (1964); C. A., 60, 13146g (1964).
1306. Thompson, A. C., K. Szabo, M. E. Brokke, and J. J. Menn, U. S. 3,313,814 (1967); C. A., 67, 90835m (1967).
1307. Thuong, N. T., Fr. 1,450,400 (1966); C. A., 66, 94695t (1967).
1308. Thuong, N. T., and D. Demozay, Fr. 1,450,559 (1966); C. A., 67, 32327m (1967).
1309. Tichý, V., V. Rattay, J. Janok, and I. Valentinová, Chem. Zvesti, 11, 398 (1957); C. A., 52, 7191i (1958).
1310. Timmler, H., I. Hammann, and R. Wegler, Brit. 1,149,159 (1969); C. A., 72, 132735q (1970).
1311. Timmler, H., I. Hammann, and R. Wegler, Brit. 1,161,381 (1969); C. A., 72, 31806h (1970).
1312. Toepfl, W., Ger. Offen. 1,812,497 (1969); C. A., 71, 124251t (1970).
1313. Tolkmith, H., U. S. 2,693,483 (1954); C. A., 49, 9867g (1955).
1314. Tolkmith, H., J. Org. Chem., 23, 1685 (1958); C. A., 53, 21759d (1959).
1315. Tolkmith, H., E. H. Blair, K. C. Kauer, and E. C. Britton, U. S. 2,887,506 (1959); C. A., 53, 17973f (1959).
1316. Tolkmith, H., H. R. Slagh, and K. C. Kauer, U. S. 3,022,329 (1962); C. A., 57, 9743i (1962).
1317. Tou, J. C., J. Phys. Chem., 75, 1903 (1971); C. A., 75, 133470 (1971).
1318. Toy, A. D. F., J. Am. Chem. Soc., 73, 4670 (1951); C. A., 46, 7697e (1951).
1319. Toy, A. D. F., and G. A. McDonald, U. S. 2,715,136 (1955); C. A., 50, 5724d (1956).
1320. Treub, W., and F. Reisser, Ger. Offen. 1,934,459 (1970); C. A., 72, 90307u (1970).
1321. Trofimov, B. A., A. S. Atavin, A. V. Gusarov, M. F. Shostakovskii, and N. I. Golovanova, Bull. Acad. Sci. USSR, Ser. Chem., 1967, 1529; C. A., 68, 48973a (1968).
1322. Truchlik, S., and J. Masek, Chem. Zvesti, 10, 516 (1956); C. A., 51, 8021h (1957).
1323. Truchlik, S., J. Masek and J. Drabek, Chem. Zvesti, 11, 579 (1957); C. A., 52, 10868e (1958).
1324. Truchlik, S., and V. Tichy, Chem. Zvesti, 11, 119 (1957); C. A., 51, 12829d (1957).
1325. Tseng, C. K., and J. H. H. Chan, Tetrahedron Lett.,

1971, 699; C. A., 74, 140821w (1971).
1326. Tsuchiya, H., A. Kimura, T. Fujimoto, Y. Okuno, T. Ogawa, and Y. Nishizawa, Japan 16,610 (1967); C. A., 68, 68468t (1968).
1327. Tsuchiya, H., K. Mukai, A. Kimura, T. Fujimoto, T. Ozaki, S. Yamamoto, Y. Okuno, T. Koyama, T. Wakatsuki, Y. Nishizawa, Japan 29,847 (1969); C. A., 72, 90039h (1970).
1328. Tsuchiya, H., K. Mukai, A. Kimura, T. Fujimoto, T. Ozaki, S. Yamamoto, Y. Okuno, T. Ogawa, T. Wakatsuki, and Y. Nishizawa, Japan 3,607 (1968); C. A., 70, 11341h (1969).
1329. Tsuchiya, H., Japan 12,587 (1968); C. A., 70, 28672q (1969).
1330. Tsuchiya, H., Japan 15,421 (1968); C. A., 70, 58008 (1969).
1331. Tsuchiya, H., K. Mukai, A. Kimura, H. Taya, K. Fujimoto, T. Ozaki, S. Yamamoto, T. Ogawa, T. Ooishi, and Y. Okuno, S. African 3,116 (1968); C. A., 71, 61004 (1969).
1332. Tsuchiya, H., Japan 31,569 (1969); C. A., 72, 111013n (1970).
1333. Tuppack, H. J., and G. Schrader, Ger. 1,056,132 (1959); C. A., 55, 6499g (1961).
1334. Turkevich, V. V., and I. F. Viblii, Zh. Prikl. Spektrosk. Akad. Nauk Belorussk. SSR, 4, 77 (1966); C. A., 64, 18700f (1966).
1335. Tyrkina, T. S., K. D. Shvetsova-Shilovskaya, and N. N. Mel'nikov, J. Gen. Chem., 38, 167 (1968); C. A., 69, 58893 (1968).
1336. Trykina, T. S., K. D. Shvetsova-Shilovskaya, O. B. Mikhailova, and N. N. Mel'nikov, J. Gen. Chem., 36, 667 (1966); C. A., 65, 8801f (1966).
1337. Uchiyama. Y., and Y. Arima, Japan 3,604 (1968); C. A., 69, 86402s (1968).
1338. Uchiyama, Y., Y. Arima, H. Hasegawa, and T. Noguchi, Japan 29,647 (1969); C. A., 72, 54823a (1970).
1339. Uchiyama, Y., Y. Arima, and K. Taniguchi, Japan 7,891 (1967); C. A., 67, 53664j (1967).
1340. Uchiyama, Y., and S. Kosaka, Japan 12,572 (1968); C. A., 70, 11123p (1969).
1341. Uchiyama, Y., and K. Taniguchi, Japan 14,832 (1966); C. A., 66, 2344 (1967).
1342. Ugi, I., U. Fetzer, G. Unterstenhoefer, and W. Behrenz, Fr. 1,379,916 (1964); C. A., 62, 10370d (1965).
1343. Union Carbide Corp., Ger. 1,129,961 (1962); C. A., 58, 475b (1963).
1344. Usher, D. A., D. I. Richardson, Jr., and F. Eckstein, Nature (London), 228, 663 (1970); C. A., 74, 38643c (1971).
1345. Van Wazer, J. R., C. F. Callis, J. N. Shoolery, and R. C. Jones, J. Am. Chem. Soc., 78, 5715 (1956); C. A., 51, 3286b (1957).

1346. Vasil'ev, A. F., Zh. Prikl. Spektrosk., 5, 524 (1966); C. A., 66, 60452w (1967).
1347. Vasil'ev, A. F., Zh. Prikl. Spektrosk., 6, 485 (1967); C. A., 67, 53491a (1967).
1348. Vasil'ev, A. F., and B. A. Khaskin, J. Gen. Chem., 34, 2333 (1964); C. A., 61, 12794h (1964).
1349. Vasil'ev, A. F., and T. F. Tulyakova, Zh. Prikl. Spektrosk., 8, 102 (1968); C. A., 68, 118162m (1968).
1350. VEB Farbenfabrik Wolfen, Fr. 1,528,125 (1968); C. A., 71, 49255s (1969).
1351. Vegter, G. C., U. S. 3,166,581 (1965); C. A., 63, 2899b (1965).
1352. Vegter, G. C., and A. M. Thrush, Brit. 868,575 (1961); C. A., 55, 24569h (1961).
1353. Velsicol Chemical Corp., Brit. 947,484 (1964); C. A., 60, 15734d (1964).
1354. Verizhnikov, L. V., and P. A. Kirpichnikov, J. Gen. Chem., 37, 1281 (1967); C. A., 68, 12597s (1968).
1355. Verkade, J. G., and R. W. King, Inorg. Chem., 1, 948 (1962); C. A., 58, 2044a (1963).
1356. Verkade, J. G., and L. T. Reynolds, J. Org. Chem., 25, 663 (1960); C. A., 54, 18541h (1960).
1357. Vil'chinskaya, A. R., and V. A. Frinovskaya, J. Gen. Chem., 30, 2565 (1960); C. A., 55, 15534c (1961).
1358. Vitenberg, I. G., and V. G. Pesin, J. Gen. Chem., 39, 1241 (1969); C. A., 71, 70547 (1969).
1359. Vitenberg, I. G., and V. G. Pesin, J. Gen. Chem., 39, 1238 (1969); C. A., 71, 112553 (1969).
1360. Vitenberg, I. G., V. G. Pesin, and R. F. Preser, J. Gen. Chem., 37, 1746 (1967); C. A., 68, 12595 (1968).
1361. Vives, J. P., and F. Mathis, Compt. Rend., 246, 1879 (1958); C. A., 52, 17098d (1958).
1362. Vladimirova, I. L., and N. N. Mel'nikov, J. Gen. Chem. USSR, 26, 2861 (1956); C. A., 51, 1824a (1957).
1363. Vladimirova, I. L., and N. N. Mel'nikov, Tr. Nauchn. Inst. Udobr. Insektofung., 1958, 158, 42; C. A., 58, 4411e (1963).
1364. Vogel, P. W., and N. A. Meinhardt, U. S. 2,900,406 (1959); C. A., 54, 294i (1960).
1365. Voigt, D., M. C. Labarre, and L. Fournes, Compt. Rend., Ser., C, 262, 1113 (1966); C. A., 65, 160h (1966).
1366. Voigt, D., M. C. Labarre, and M. Therasse, Compt. Rend., 260, 2210 (1965); C. A., 62, 14483e (1965).
1367. Volodina, L. N., J. Gen. Chem., 37, 1755 (1967); C. A., 68, 13059 (1968).
1368. Wadsworth, W. S., Jr., and W. D. Emmons, J. Am. Chem. Soc., 84, 610 (1962); C. A., 56, 15342i (1962).
1369. Wagner, F. A., Jr., U. S. 3,737,529 (1973); C. A., 88, 112402v (1973).
1370. Wagner-Jauregg, T., T. Lennartz, and H. Kothny,

Ber., 74, 1513 (1941); C. A., 37, 133 (1943).
1371. Walling, C. T., and R. Rabinowitz, J. Am. Chem. Soc., 79, 5326 (1957); C. A., 52, 2739d (1958).
1372. Walling, C., and R. Rabinowitz, J. Am. Chem. Soc., 81, 1243 (1959); C. A., 53, 16932f (1959).
1373. Walsh, E. N, U. S. 3,354,195 (1967); C. A., 68, 49044k (1968).
1374. Walsh, E. N., U. S. 3,397,269 (1967); C. A., 69, 86353b (1968).
1375. Walsh, E. N., and J. G. Brady, U. S. 3,034,951 (1962); C. A., 57, 12320h (1962).
1376. Walsh, E. N., and J. G. Brady, U. S. 3,274,299 (1966); C. A., 65, 16899e (1966).
1377. Walsh, E. N., and J. T. Hallett, U. S. 3,105,003 (1963); C. A., 60, 6748h (1964).
1378. Walsh, E. N., and J. T. Hallett, Belg. 633,592 (1963); C. A., 61, 3114b (1964).
1379. Ward, L. F., Jr., and D. D. Phillips, U. S. 3,174,990 (1965); C. A., 62, 16123h (1965).
1380. Ward, L. F., Jr., D. D. Phillips, and R. R. Whetstone, U. S. 3,567,803 (1971); C. A., 75, 19913v (1971).
1381. Watanabe, S., and T. Shimamura, Japan 644 (1966); C. A., 64, 11232d (1966).
1382. Weesner, W. E., and J. A. Webster, U. S. 3,075,874 (1963); C. A., 59, 9899e (1963).
1383. Weil, E. D., and S. B. Greenbaum, Belg. 624,224 (1962); C. A., 59, 11429d (1963).
1384. Weil, E. D., and H. L. Schlichting, U. S. 3,336,393 (1967); C. A., 68, 68710 (1968).
1385. Weil, E. D., and K. J. Smith, Fr. 1,412,397 (1965); C. A., 65, 13670c (1966).
1386. Wen-Hsuan Chang, and M. Wismer, U. S. 3,189,633 (1965); C. A., 63, 9971b (1965).
1387. Werbel, L. M., and P. E. Thompson, J. Med. Chem., 10, 32 (1967); C. A., 66, 27574q (1967).
1388. White, W. E., and A. Hood, J. Am. Chem. Soc., 74, 853 (1952); C. A. 47, 9905i (1953).
1389. Whitehead, M. A., and R. M. Hart, J. Chem. Soc., A, 1971, 1738; C. A., 75, 42899 (1971).
1390. Wildgrube, W., M. Born, and D. Meissner, Ger. 34,855 (1965); C. A., 63, 11363f (1965).
1391. Willard, J. R., and J. F. Henahan, U. S. 2,873,228 (1959); C. A., 53, 10650g (1959).
1392. Willard, J. R., and J. F. Henahan, U. S. 2,916,415 (1959); C. A., 54, 7650g (1960).
1393. Willard, J. R., and J. F. Henahan, U. S. 3,014,058 (1961); C. A., 57, 4543h (1962).
1394. Wilson, J. H., U. S. 3,268,393 (1966); C. A., 65, 14362d (1966).
1395. Williamson, M. P., and C. E. Griffin, J. Phys. Chem., 72, 4043 (1968); C. A., 70, 24509 (1969).

1396. Witt, E. R., U. S. 3,157,675 (1964); C. A., 62, 2708f (1965).

1397. VEB Farbenfabrik Wolfen, Brit. 1,070,689 (1967); C. A., 67, 54170g (1967).

1398. Wolf, F., S. Heidenreich, and M. Born, Brit. 1,347,730 (1968); C. A., 70, 57130s (1969).

1399. Wood, J., Ger Offen. 1,915,495 (1969); C. A., 72, 79062q (1970).

1400. Worrel, C. J., U. S. 3,487,131 (1969); C. A., 72, 67144e (1970).

1401. Wu, C-H., Ting-Wei Tong, Tao-Tao Chang, and Yung-Min Sun, Hua Hsueh Hsueh Pao, 30, 233 (1964); C. A., 65, 5386g (1966).

1402. Yagnyukova, Z. I., K. D. Shvetsova-Shilovskaya, and N. N. Mel'nikov, J. Gen. Chem., 41, 80 (1971); C. A., 75, 19872 (1971).

1403 Yamaguchi, T., T. Yamamoto, and K. Imamura, Japan 9,767 (1960); C. A., 55, 8297c (1961).

1404. Yamaguchi, T., T. Yamamoto, and K. Imamura, Japan 5,275 (1961); C. A., 58, 10126e (1963).

1405. Yamasaki, T., Sci. Rep. Res. Inst., Tohoku Univ., Ser. A, 6, 172 (1954); C. A., 49, 6858i (1955).

1406. Yamasaki, T., Bull. Res. Inst. Mineral Dressing Met., Tohoku Univ., 10, 21 (1954); C. A., 49, 13887b (1955).

1407. Yamasaki, T., and T. Sato, Sci. Rep. Res. Inst., Tohoku Univ., Ser. A, 6, 384 (1954); C. A., 50, 314e (1956).

1408. Yamasaki, T., and T. Sato, Sci. Rep. Res. Inst., Tohoku Univ., Ser. A, 8, 45 (1956); C. A., 51, 377i (1957).

1409. Yang, S-H., T-C. Chen, C-M. Li, Y-K. Li, H-S. Tung, S-I. Kao, and S. C. Tung, Hua Hsueh Hsueh Pao, 1962, 187; C. A., 59, 3758h (1963).

1410. Yarmukhametova, D. Kh., and I. V. Cheplanova, Bull. Acad. Sci. USSR, Ser. Chem., 1964, 1900; C. A., 62, 7667h (1965).

1411. Yarmukhameteva, D. Kh., and I. V. Cheplanova, Bull. Acad. Sci. USSR, Ser. Chem., 1966, 459; C. A., 65, 7083b (1966).

1412. Yarmukhametova, D. Kh., and I. V. Cheplanova, Bull. Acad. Sci. USSR, Ser. Chem., 1967, 579; C. A., 67, 99773x (1967).

1413. Yarmukhametova, D. Kh., B. V. Kudryavtsev, and V. S. Buzlama, Bull. Acad. Sci. USSR, Ser. Chem., 1967, 1455; C. A., 68, 49530 (1968).

1414. Yarmukhametova, D. Kh., Z. G. Speranskaya, B. V. Kudryavtsev, and V. D. Ermakova, Bull. Acad. Sci. USSR, Ser. Chem., 1971, 719; C. A., 75, 49003r (1971).

1415. Yashima Optical Industry Co., Ltd., Brit. 915,440 (1963); C. A., 58, 12475b (1963).

1416. Yoshido, M., and T. Maeda, Japan 26,393 (1964); C. A., 62, 10371f (1965).
1417. Yoshido, M., T. Maeda, and H. Sugiyama, Japan 1,541 (1967); C. A., 66, 115455g (1967).
1418. Yoshido, M., K. Takida, T. Maeda, and I. Okuda, Japan 2,345 (1968); C. A., 69, 86607n (1968).
1419. Yoshioka, H., and S. Horie, Japan 24,956 (1964); C. A., 62, 11689e (1965).
1420. Yoshioka, H., and S. Horie, Japan 24,957 (1964); C. A., 62, 11694c (1965).
1421. Yoshioka, H., S. Tanaka, and K. Hamichi, Japan 12,348 (1968); C. A., 70, 19791a (1969).
1422. Young, R. W., U. S. 2,996,531 (1961); C. A., 56, 332g (1962).
1423. Young, R. W., and G. Berkelhammer, U. S. 2,959,610 (1960); C. A., 55, 7288b (1961).
1424. Young, R. W., and E. L. Clark, U. S. 3,210,242 (1965); C. A., 63, 17901h (1965).
1425. Zaks, P. G., Ya. A. Mandel'baum, and N. N. Mel'nikov, J. Gen. Chem., 34, 1218 (1964); C. A., 61, 1762b (1964).
1426. Zaks, P. G., Ya. A. Mandel'baum, N. N. Mel'nikov, and V. V. Ivanov, J. Gen. Chem., 36, 872 (1966); C. A., 65, 12098h (1966).
1427. Zemlyanskii, N. I., Visn. L'vivs'k. Derzh. Univ., Ser. Khim., No. 6, 110 (1963); C. A., 64, 4977g (1966).
1428. Zemlyanskii, N. I., and T. P. Bachinskii, J. Gen. Chem., 37, 2151 (1967); C. A., 68, 86933v (1968).
1429. Zemlyanski, N. I., and T. P. Bachinskii, J. Gen. Chem., 40, 1005 (1970); C. A., 73, 76771e (1970).
1430. Zemlyanskii, N. I., and N. M. Chernaya, J. Gen. Chem., 36, 1710 (1966); C. A., 66, 75775c (1967).
1431. Zemlyanskii, N. I., and N. M. Chernaya, Ukr. Khim. Zh., 33, 182 (1967); C. A., 67, 7571d (1967).
1432. Zemlyanskii, N. I., N. M. Chernaya, and V. V. Turkevich, Proc. Acad. Sci., 163, 808 (1965); C. A., 63, 16240c (1965).
1433. Zemlyanskii, N. I., N. M. Chernaya, V. V. Turkevich, and V. I. Krasnoshchek, J. Gen. Chem., 36, 1255 (1966); C. A., 65, 16848g (1966).
1434. Zemlyanskii, N. I., N. M. Chernaya, and V. V. Turkevich, J. Gen. Chem., 37, 464 (1967); C. A., 67, 53817m (1967).
1435. Zemlyanski, N. I., and B. S. Drach, J. Gen. Chem., 32, 1942 (1962); C. A., 58, 4450d (1963).
1436. Zemlyanskii, N. I., and L. V. Glushkova, J. Gen. Chem., 35, 1482 (1965); C. A., 63, 14689f (1965).
1437. Zemlyanskii, N. I., and L. V. Glushkova, J. Gen. Chem., 36, 2187 (1966); C. A., 66, 75776d (1967).
1438. Zemlyanskii, N. I., and L. V. Glushkova, J. Gen. Chem., 37, 728 (1967); C. A., 68, 49225v (1968).

1439. Zemlyanskii, N. I., and V. P. Kalashnikov, J. Gen. Chem., 37, 1082 (1967); C. A., 68, 78259q (1968).
1440. Zemlyanskii, N. I., and V. P. Kalashnikov, J. Gen. Chem., 39, 584 (1969); C. A., 71, 38924h (1969).
1441. Zemlyanskii, N. I., V. P. Kalashnikov, and A. P. Gupalo, J. Gen. Chem., 38, 819 (1968); C. A., 69, 58783y (1968).
1442. Zemlyanskii, N. I., V. P. Kalashnikov, and V. K. Yarimovich, J. Gen. Chem., 39, 1559 (1969); C. A., 71, 101221u (1969).
1443. Zemlyanskii, N. I., and L. K. Klimovskaya, J. Gen. Chem., 30, 4018 (1961); C. A., 55, 25466a (1961).
1444. Zemlyanskii, N. I., L. K. Klimovskaya, V. I. Galibei, B. S. Drach, I. V. Murav'ev, and V. V. Turkevich, J. Gen. Chem., 32, 3990 (1962); C. A., 58, 13780g (1963).
1445. Zemlyanskii, N. I., and Ya. I. Mel'nik, J. Gen. Chem., 38, 312 (1968); C. A., 69, 77370c (1968).
1446. Zemlyanskii, N. I., and Ya. I. Mel'nik, J. Gen. Chem., 39, 2401 (1969); C. A., 72, 78313k (1970).
1447. Zemlyanskii, N. I., and Ya. I. Mel'nik, J. Gen. Chem., 40, 37 (1970); C. A., 72, 99934d (1970).
1448. Zemlyanskii, N. I., and Ya. I. Mel'nik, J. Gen. Chem., 40, 1701 (1970); C. A., 74, 41810k (1971).
1449. Zemlyanskii, N. I., G. O. Morgun, and B. P. Kotovich, Visn. L'vivs'k Derzh. Univ., Ser. Khim., 1963 104; C. A., 61, 14577b (1964).
1450. Zemlyanskii, N. I., and I. V. Murav'ev, J. Gen. Chem., 34, 96 (1964); C. A., 60, 10577b (1964).
1451. Zemlyanskii, N. I., O. A. Prib, and B. S. Drach, J. Gen. Chem., 31, 811 (1961); C. A., 55, 23314b (1961).
1452. Zemlyanskii, N. I., O. A. Prib, and L. V. Glushkova, J. Gen. Chem., 36, 1131 (1966); C. A., 65, 13586b (1966).
1453. Zemlyanskii, N. I., G. G. Vil'danova, N. I. Gritsai, and V. V. Turkevich, J. Gen. Chem., 40, 1964 (1970); C. A., 74, 53204d (1971).
1454. Zenftman, H., Brit. 713,120 (1954); C. A., 49, 12537g (1955).
1455. Zgadzai, E. A., and A. I. Maklakov, J. Gen. Chem., 34, 1156 (1964); C. A., 61, 2591f (1964).
1456. Zwierzak, A., Tetrahedron, 25, 5177 (1969); C. A., 72, 11985 (1970).
1457. Zwierzak, A., and R. Gramze, Z. Naturforsch., B, 26, 386 (1971); C. A., 75, 48355v (1971).

Errata for Volume 1 Chapter 1

Page

iv. Line 8. Change phosphorous to phosphorus.

vii. Chapter 2. Change Organophosphorous-Metal to Organophosphorus-Metal.

x. Line 1. Change P-C, P-S, to P-O, P-S,.

xi. Line 9 from the bottom. Change Quarternary to Quaternary.

xii. Chapter 18 add: and derivatives.
Reference 5. Change T.L. Becker to Th. Becker.

37. Line 12. Change RP'_2P to RR'_2P.

46. Line 2 from the top. Change $LiAlH^4$ to $LiAlH_4{}^4$.

55. Line 1 after formula (<u>30</u>). Change LiAlH to $LiAlH_4$.

77. formula <u>62</u>. Change 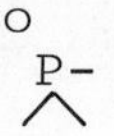 to 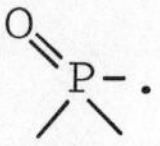.

84. Line 5. Change trimethylphosphite[25] to trimethylphosphite[250].

190. Last formula. Change 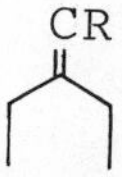 to 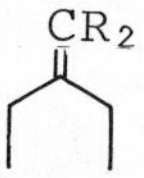.

193. Second formula. Change  to 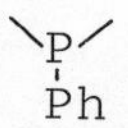.

200. After second formula. Change bimethyllyl to bimethallyl

274. Ref. 1164a. Change Bernheimer to R. Bernheimer.

Chapter 2

328. Second equation. Change 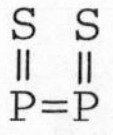 to 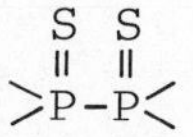.

347. Below equation line 3. Change $(CF_3)_{4,5}$ to $(CF_3P)_{4,5}$.

394. Line 13 from the bottom. Change -40.2 ppm (P) to -40.2 ppm (P_γ).

396. Line 6 from the bottom. Change 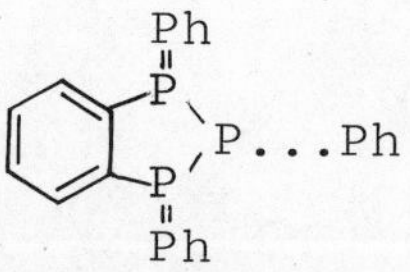 to

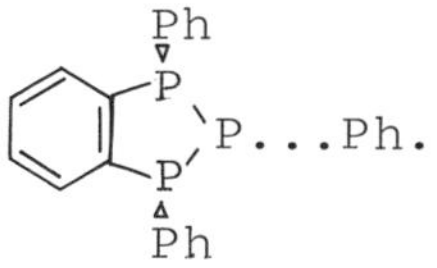

Chapter 3A

Page
459. Line 15. Change ME to Me.

467. Line 7. Change high-192 to high-low.

Errata for Volume 2

Dust jacket. Line 2 from bottom. Change Quarternary to Quaternary.

iv. Line 8. Change phosphorous to phosphorus.

Chapter 4

Page
221. Line 5 from bottom. Should read: $Mo^{III}(SCN)_4.(PBu_3)_2$.

Line 4 from bottom. Should read: $(CH_2{=}CH{-}CH_2)_2PdBr_2$.

Line 3 from bottom. Should read: $(CH_2{=}CH{-}CH_2)_2PdCl_2$.

Line 2 from bottom. Should read: 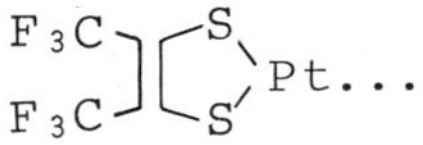

.

Errata for Volume 3

Page
iv. Line 8. Change phosphorous to phosphorus.

Chapter 5B

Page
210. Formula scheme. Change 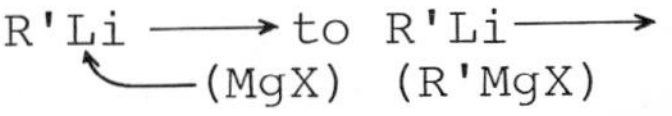

.

296. Last formula at the bottom.

Change 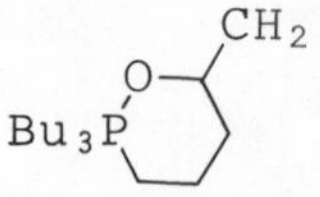 to 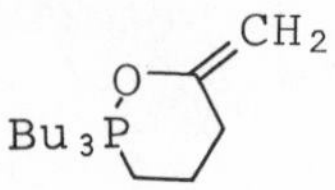

329. Reference 488 should read:
Ramirez,F., Colloques Intern. du C.N.R.S., 182, 67 (1970); Bull.Soc.Chim.France, 1970, 3497.

333. Ref. 572. Change Am. 743, to Ann. 743,.

Errata for Volume 4

Page

Dust jacket. Line 3 from the bottom. } Change W.A. Frank to Arlen W. Frank.

v. Line 21 } Change W.A. Frank to Arlen W. Frank.

iv. Line B. Change phosphorous to phosphorus.

Chapter 7

12. Line 5. Change PbMgBr to PhMgBr.

50. 4th formula, text. I. R = Et etc. close up with text below.

53. Line 19 from the bottom. Change B_{γ} to Br.

55. Before References add: (received July 9, 1970).

Chapter 10

255. Line 3. Change A.W. Frank to Arlen W. Frank.

257. Line 34. In A.5.4.2., change Phosphinic to Phosphonic.

366. Lines 33,34. Delete "t-BuP(OEt)$_2$.....(CH$_2$).[133]". The compound described is actually t-BuP(O)(OEt)$_2$ (Crofts, P.C., and D.M. Parker, J. Chem. Soc., C, 1970, 2342).

263. Lines 28-32. Delete the paragraph "Partial hydrolysis ....HCl(15)". The phosphinic chloride RPH(O)Cl (R = 2,4,6-t-Bu$_3$C$_6$H$_2$) obtained by Cook (Ref. 120) from 1,3,5-tri-t-butylbenzene has been identified as RR"P(O)Cl (R = 3,5-t-Bu$_2$C$_6$H$_3$, R" = t-Bu) (Yoshifuji, M., R.

Okazaki and N. Inamoto, J. Chem. Soc., Perkin Trans. I, 1972, 559). All subject matter relating to this phosphinic chloride and its derivatives should therefore be deleted.

285. Lines 13-19. Delete the paragraph "Only in the case of .....2HCl (2)".

336. Lines 7-10. Delete "$C_{15}H_{23}O_4P$.....IR.[120]".

429. Lines 17-19. Delete "$C_{36}H_{60}O_3P_2$.....IR.[120]".

437. Delete Ref. 120.
523. Before references add: (Received September 4, 1970).

Errata for Volume 5

Dust jacket. Line 6 from the bottom. Change Hypophosphorus, Hypodiphosphorus to Hypophosphorous, Hypodiphosphorous
Line 3 from the bottom. Change Phosphorus Acid and Thiophosphorus Acid to Phosphorous Acid and Thiophosphorous Acid.

Chapter 13

Page
77. Line 3. Change hesea- to hexa- .

101. Fourth product from bottom. Change $(Me_2N)_3\overset{+}{P}C\bar{C}l_3Cl$ to $(Me_2N)_3\overset{+}{P}CCl_3Cl^-$.

122. Line 16. Change $(Et\text{-}3\text{-}oxetanylidene\text{-}CH_2O)_3$ to $(Et\text{-}3\text{-}oxetanylidene\text{-}CH_2O)_3P$.

241. Ref. 39. Change Pakucz to Paskucz.

242. Ref. 53. Change Broecker to Broeker.

Dust jacket. Volume 2 Line 3. Change Quarternary to Quaternary (back outside).

Errata for Volume 6 Chapter 14

Page

3. Line 20 from the bottom. Change Diaryphosphinothioic to Diarylphosphinothioic.

60. Line 7. Change PHOSPHONIC to PHOSPHINIC.

73. Line 17 from the bottom. Change (in 95% EtOH).[503] to (in 95% EtOH).[584].

98. Line 2 from the bottom. Change: "From product from" to "Minor product from".

122. Line 14. Change n^{20} to n_D^{20}.

139. Second formula. Change 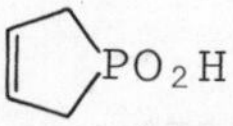 to 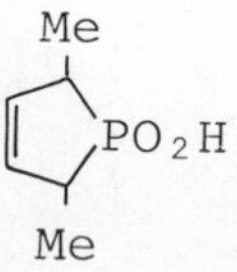 .

160. Line 23. Should be indented with lines 21 and 25.

Chapter 15

Page

234. Line 1. Change patent. to patent.[436].

287. Below first range of formulas. Change (R to R.

356. LIne 6. Change I.l.b.δ - 1 to I.l.b.γ - 1.

363. Section Glucosamine phosphates, Line 12. Change I.l.e -2 to I.l.e.α-2.

Line 19 from the bottom. Change I.l.d. to I.l.d.δ.

364. Line 2. Change I.l.d. to I.l.d.δ.

Line 5. Change I,l.c. -2 to I.l.c.β-2.

365. Section Phosphoglyceric Acids, Line 10. Change -13.27°.[1038] to -13.27°.[1035].

Line 14. Change (PhH O) POCl to $(PhCH_2O)_2POCl$.

365. Line 12 from the bottom. Change (lN HCl).[20x] to (lN HCl).[198a].

Line 11 from the bottom. Change D-Erythronolactone to D-Erythronic acid.

Line 9 from the bottom. Change Cryst. to Di-c-$C_6H_{11}NH_3$ salt, and -20°.[20x] to -20°.[198a].

381. Line 13. Change <u>3. R = Heterocyclic Nucleus, n = 2</u> to <u>3. Heterocyclic Nuclei.</u>

383. Between Lines 12 and 13. Insert 3. R = Heterocyclic Nucleus, n = 2.

Line 15. Change 3. to 4..

386. Line 9. Change $(2\text{-Et-hexylo})_2PO_2H$ to $(2\text{-Et-hexyl-O})_2PO_2H$.

408. Line 27. Change ...(uridine-(3' →5')-)$_2$-uridine... to ...(uridine-(3' →5')-)$_n$-uridine... .

411. Line 14. Change I.l.d.δ,[349,1284] to I.l.d.δ,[349,1484].

474. Line 12. Add Cryst., m. 267-8°.

475. Line 8 from the bottom. Change idene-5'-deoxy=5'... to idene-5'-deoxy-5'... .

477. Lines 24 and 23 from the bottom. Delete.

480 Line 6 from the bottom. Change $MeC(O)CH_2CH_2OHgOC(O)Me$ to $MeC(O)OCH_2CH_2HgO$... .

508. Line 6 from the bottom. Change O-phenylendioxy... to o-phenylenedioxy... .

510. Line 5. Change $(BuO)_2{}^{32}PO$ to $(BuO)_3{}^{32}PO$.

518. Between Refs. 198 and 199. Insert 198a R. Barker and F. Wold, J. Org. Chem., 28, 1847 (1963).

541. Ref. 709. Line 1, change 53, 47 (1953) to 53, 47 (1963), and Line 2, change 13742d to 13741d.

559. Ref. 1099a, Line 3. Change (1956) to (1957).